INTERNATIONAL BIBLIOGRAPHY
OF HISTORICAL SCIENCES

INTERNATIONAL BIBLIOGRAPHY OF HISTORICAL SCIENCES

INTERNATIONALE BIBLIOGRAPHIE DER GESCHICHTSWISSENSCHAFTEN
BIBLIOGRAFIA INTERNACIONAL DE CIENCIAS HISTORICAS
BIBLIOGRAPHIE INTERNATIONALE DES SCIENCES HISTORIQUES
BIBLIOGRAFIA INTERNAZIONALE DELLE SCIENZE STORICHE

VOLUME LXXIV
2005

Edited by Massimo Mastrogregori

with the contribution of a number of scholars,
under the auspices of the
International Committee of Historical Sciences

De Gruyter · Berlin · New York

The IBOHS for the years 1978 to 1992 (Vol. 47 – 61) was edited by
Michel François and Michael Keul for Vol. 47/48 (1978/1979) and
Jean Glénisson and Michael Keul for Vol. 49 – 61 (1980 – 1992)
on behalf of the International Committee of Historical Sciences
and was published by K. G. Saur Munich.

⊚ Gedruckt auf säurefreiem Papier,
das die US-ANSI-Norm über Haltbarkeit erfüllt.

ISBN 978-3-598-20437-1
ISSN 0074-2015

Bibliographic information published by the Deutsche Nationalbibliothek
The Deutsche Nationalbibliothek lists this publication in the Deutsche Nationalbibliografie;
detailed bibliographic data are available in the Internet at http://dnb.d-nb.de.

© 2009 by Walter de Gruyter GmbH & Co. KG, 10785 Berlin,
www.degruyter.com

All rights reserved, including those of translation into foreign languages. No Parts of
this book may be reproduced in any form or by any means, electronic or mechanical,
including photocopy, recording, or any information storage and retrieval system,
without permission in writing from the publisher.

Printed in Germany

Technical partner: Dr. Rainer Ostermann, München
Managing partner and technical support: Ellediemme libri dal mondo, Roma
Printed and Bound by Hubert & Co. GmbH & Co. KG, Göttingen

General editor

Massimo MASTROGREGORI, Università di Roma 'La Sapienza'

Assistant editor

Carlo COLELLA, Rome

Advisory board

Maria Tereza AMADO, Universidad de Evora
Girolamo ARNALDI, Istituto storico italiano per il Medioevo, Rome
† Yuri BESSMERTNY, Institute of General History, Russian Academy of Sciences, Moscow
† Wiesław BIEŃKOWSKI, Polska Akademia Nauk
László BIRÓ, Hungarian Academy of sciences, Budapest
Corinne BONNET, Université de Toulouse-Mirail
Luciano CANFORA, Università di Bari
Alejandro CATTARUZZA, University of Buenos Aires
Anne EIDSFELDT, University of Oslo Library
Ilse FREDERIKSEN VÄHÄKYRÖ, Turku University Library
Jean GLÉNISSON, Comité International des Sciences Historiques, Paris
Kazuhiko KONDO, University of Tokyo
Mario MAZZA, Università di Roma 'La Sapienza'
Vilém PREČAN, Institute of Contemporary History, Prague
Matjaz REBOLJ, Ljubljana
Jacques REVEL, Ecole des Hautes Etudes en Sciences Sociales, Paris
† Ruggiero ROMANO, Ecole des Hautes Etudes en Sciences Sociales, Paris
Gabrielle M. SPIEGEL, Johns Hopkins University, Baltimore
Martina STERCKEN, Universität Zurich
Natasa STERGAR, Ljubliana
Şerban TURCUŞ, Università 'Babeş-Bolyai', Cluj Napoca
Nenad VEKARIĆ, Dubrovnik
Bahaeddin YEDİYILDIZ, Hacettepe Universitesi, Ankara

Contributing editors

Kira E. AGEEVA, Institute of Archaeology of the Russian Academy of Sciences, Moscow (*Russian historiography*)
Maria Tereza AMADO, Universidad de Evora (*Portuguese historiography*)
Vassili N. BABENKO, Russian Academy of Sciences, Moscow (*Russian Historiography*)
Grigory N. BIBIKOV, Moscow State University after M.V. Lomonosov, Moscow (*Russian Historiography*)
† Wiesław BIEŃKOWSKI, Polska Akademia Nauk (*Polish historiography*)
Wolfdieter BIHL, Institut für Geschichte, Universität Wien (*Austrian historiography*)
László BIRÓ, Hungarian Academy of sciences, Budapest (*Hungarian historiography*)
Věra BŘEŇOVÁ, Institute of Contemporary History, Prague (*Czech historiography*)
Alejandro CATTARUZZA, University of Buenos Aires (*Latin American historiography*)

Carlo COLELLA, Rome (*History by countries, History of international relations*)
Darko DAROVEC, Science and Research Centre of Koper, University of Primorska
(*Slovenian historiography*)
Laura DE GIORGI, Università di Venezia (*Chinese historiography*)
Anne EIDSFELDT, University of Oslo Library (*Norwegian historiography*)
Ilse FREDERIKSEN VÄHÄKYRÖ, Turku University Library (*Finnish historiography*)
Ekaterina V. GORBUNOVA, Institute of Universal History of the Russian Academy
of Sciences, Moscow (*Russian historiography*)
Nikolay V. GORDIYCHUK, Russian State University for Humanities
(*Russian historiography*)
Anna GRUCA, Instytut Historii PAN. Pracownia Bibliografii Bieżącej
(*Polish historiography*)
Timophey GUIMON, Institute of Universal History of the Russian Academy of Sciences,
Moscow (*Russian historiography*)
Teymur K. GUSEJNOV, Russian State University for Humanities (*Russian historiography*)
Libby KAHANE, The Jewish National and University Library, Jerusalem
(*Historiography of Israel*)
Kazuhiko KONDO, University of Tokyo (*Japanese historiography*)
Olga V. KOROBEYNIKOVA, Moscow State University after M.V. Lomonosov, Moscow
(*Russian historiography*)
Dmitry S. KOZHEVNIKOV, Institute of the Countries of Asia and Africa of the Moscow
State University (*Russian Historiography*)
Marina KOZLOVA, International University, Moscow (*Russian historiography*)
Anna KRASOVA, Institute of Universal History of the Russian Academy of Sciences,
Moscow (*Russian historiography*)
Mauro LENZI, Società Romana di Storia Patria, Rome (*Palaeography, Diplomatics,
History of the book, Medieval history*)
Jean Marie MAILLEFER, Université Charles-De-Gaulle Lille 3
(*Danish and Swedish historiography*)
Maria M. MARISOVA, State University for Human Sciences (*Russian historiography*)
Massimo MASTROGREGORI, Università di Roma 'La Sapienza'
(*Auxiliary sciences, General works, Modern history*)
Stjepan MATKOVIĆ, Zagreb (*Croatian historiography*)
Anastasija MAYER, Institute of Universal History of the Russian Academy of Sciences,
Moscow (*Russian historiography*)
Matjaz REBOLJ, Ljubljana (*Slovenian historiography*)
Marco SANTUCCI, Università di Roma 'La Sapienza' (*Ancient history*)
Evgeny E. SAVITSKI, Institute of Universal History of the Russian Academy of Sciences,
Moscow (*Russian historiography*)
Alžbeta SEDLIAKOVÁ, Historický ústav Slovenskej akadémie vied, Bratislava
(*Slovak historiography*)
A. SLIVA, Rossijskaja Akademija Nauk, Moskva (*Russian historiography*)
Natale SPINETO, Università di Torino (*History of religions*)
Natasa STERGAR, Ljubliana (*Slovenian historiography*)
Paola STIRPE, Università di Roma 'La Sapienza' (*Ancient history*)

Şerban TURCUŞ, Università 'Babeş-Bolyai', Cluj Napoca (*Romanian historiography*)
Samuel B. VOLFSON, Institute of Universal History of the Russian Academy of Sciences, Moscow (*Russian historiography*)
Galina A. YANKOVSKAYA, Perm State University, Perm (*Russian historiography*)
Bahaeddin YEDİYILDIZ, Hacettepe Universitesi, Ankara (*Historiography of Turkey*)

Consulting editors

Maurice AYMARD, Maison des sciences de l'homme, Paris
Eric BRIAN, Centre Alexandre Koyré, Paris
Louis CHATELLIER, Université de Nancy II
Sten EBBESEN, University of Copenhagen
Carlo FRANCO, Università di Venezia
Olivier GUYOTJEANNIN, Ecole nationale des Chartes, Paris
Michel MORINEAU, Paris
Brian TIERNEY, Cornell University, Ithaca
Giusto TRAINA, Università di Rouen
Pietro VANNICELLI, Università di Urbino
André VAUCHEZ, Paris

Special Assistant editors

Alessio ESPOSITO, Rome
Daniele MASTROGREGORI, Rome

CONTENTS

	Pages
FOREWORD	XI
SCHEME	XIII
GENERAL HISTORICAL BIBLIOGRAPHIES	XIX
BIBLIOGRAPHY	1
INDEX OF NAMES	337
GEOGRAPHICAL INDEX	403

FOREWORD

The International Bibliography of Historical Sciences (I. B. O. H. S.) is a selective and descriptive bibliography, and the works it mentions, both books and articles, are arranged according to a methodical and chronological scheme originally drawn up and established by the Bibliographical Commission of the International Committee of Historical Sciences; the scheme has been revised only in details.

An exposition of the principles which were followed in the choice of works included and of the rules which were observed for their presentation in the present volume is set out below.

A. Manner of Selection.

In agreement with the wish expressed by the Bibliographical Commission of the C. I. S. H., the selection is actuated by the twin concern of preserving for the I. B. O. H. S. its character of a general bibliography comprehending the whole field of historical sciences, and of putting at the disposal of historians as also librarians the essential facts of historical production throughout the world, in one complete volume appearing annually.

In view of the multiplication of specialized bibliographies, it has in fact appeared more than ever necessary to offer to isolated scholars and even scientific establishments unable to obtain all these bibliographies the means of keeping informed, each year, of the advancement of historical science. But it was also desirable that these bibliographies be mentioned, and this has been done in two different ways: firstly, we have listed, outside the systematic inventory and immediately preceding it, the great international or national bibliographies giving the historical production of a country and in which a conspectus of works connected with this country is given; on the other hand, in the systematic inventory, at the head of each division or subdivision are mentioned general bibliographies dealing with one of the historical disciplines or particular bibliographies devoted to one question, one author or one province or state, and which find their logical position in that division or subdivision; in the latter case, the bibliographies are preceded by an asterisk (*).

In order to justify its existence as a working instrument of a high scientific standard and of international application, the I. B. O. H. S. only mentioned books or articles with a wider scope than the narrow field of local preoccupations, and rejects also reviews which are mere presentations or of courtesies. Like-wise re-editings, translations, descriptions of research which do not include new elements of information, exhibition catalogues without commentary, typed or stencilled works and works of popularization and propaganda have been normally eliminated.

On the other hand, the contributing editors have been careful to describe those works which, through slight or of apparently only local interest, make an obvious contribution to general history or to the solution of current problems; this is the case of certain reports on excavations and of articles bearing on controversial subjects touching the history of institutions or civilization; in this case, as whenever the title of an article was too vague, it has been followed whenever possible by a brief remark or by a date in brackets for the reader's orientation. Herein can be found an effort which will not fail to be useful to those who use the I. B. O. H. S., and which can be increased in future volumes without incurring the temptation to transform this essentially *selective* and *descriptive* bibliography into an analytical and critical one, this double character being in fact reserved to specialized bibliographies.

Unlike the greater number of national bibliographies, the I. B. O. H. S. does not limit the works included by any fixed date; that is to say that works connected with the most recent history find a place in it, notably those connected with international relations (P 8); at the same time the selection had to be correspondingly strictly.

By this conception, the I. O. B. H. S. keeps a physiognomy peculiar to itself; it has no tendency towards substituting for any existing bibliography, but while avoiding as far as possible a double role, it allows to be necessary that amount or overlapping which is profitable in the scholarly world.

B. Rules of Presentation.

The volume LXXIV, 2005 mentions the works published with the date: 2005. Within each division or subdivision, the works are presented in alphabetical order of their authors. Slavonic, Greek, Japanese, Hebraic and Arabic names are transcribed into Latin characters and placed according to the order of the Latin alphabet, but characters with diacritics, for instance ć, č, ś, š are considered as if ordinaries c, s. Germanic and Scandinavian names are classed according to the function of the developed value of letters of inflection: ä, ö, ø, ü become ae, oe, ue. Mc and M' are indexed as if Mac.

Anonymous or collective works are classed alphabetically according to the initial of the key word in the title, for instance: "Congress (Fourteenth) of the Learned Societies..." At the same time, in heavy type are the names of scholars, or philosophers, who have been the object of an important biographical or historiographical study (B § 2 b, M § 5 b) and those of Saints (G § 4, I § 13 d); in the first case, the works are indicated in the alphabetical order of the people concerned.

As it has been done for the bibliographies peculiar to a division or subdivision, the publications of texts which had their place in the alphabetical list of each division or subdivision have been extracted and transferred to the head of the alphabetical list and immediately following the list of bibliographies; these publications of texts have been distinguished by being preceded by two asterisks. Thus the reader has immediately before his eyes the bibliographies and editions of the most recent texts bearing on a particular question or period. However, concerning the texts, the procedure of two asterisks (**) has not been adopted in chapter E, F, G and H, each of which already has a division especially devoted to the texts.

When the current year has been marked by the commemoration of an important historical event, the works to which this commemoration has led are grouped separately and under a special title at the end of the subdivision where this event finds its normal place.

When a work which has been in circulation for three or four years has been the object of a review every succeeding year, only the name of the author is cited plus the essential of the title, preceded by a reference to the number of the last volume of the I. B. O. H. S. in which it was quoted; it is thus possible to follow year to year the state of the criticism which the publication of a book has provoked.

Where the "collation" of works is concerned, the unification of references to pages, plates and illustrations, etc. has been sought as far as possible by putting them into either French or English, these being the two languages which have the most words or initial letters of words of an identical meaning in common.The transferrings of works interesting for one part to a section other than their logical one, transferences indicated by "*Cf. n°...*", have been grouped at the end of each section.

In the index of names of authors and persons, the names of Classical authors, Saints and Popes are written in their Latin form.

SCHEME

GENERAL HISTORICAL BIBLIOGRAPHIES
(p. XIX)

A

AUXILIARY SCIENCES
(p. 1-13)

§ 1. Palaeography. 1-7. – § 2. Diplomatics. 8-16. – § 3. History of the book (*a*. Manuscripts; *b*. Printed books). 17-101. – § 4. Chronology. 102-107. – § 5. Genealogy and family history. 108-128. – § 6. Sigillography and heraldry. 129-133. – § 7. Numismatics and metrology. 134-150.– § 8. Linguistics. 151-203. – § 9. Historical geography, travels and discoveries. 204-252. – § 10. Iconography and images. 253-273.

B

MANUALS, GENERAL WORKS AND WORKS ON LARGE PERIODS
(p. 15-43)

§ 1. Archives, libraries and museums (*a*. Archives; *b*. Libraries; *c*. Museums). 274-319. – § 2. History of historiography and memory (*a*. General; *b*. Special studies). 320-490. – § 3. Methodology, philosophy, and teaching of history. 491-543. – § 4. Ethnology, folklore and historical anthropology. 544-588. – § 5. General history. 589-699. – § 6. Theory of the state and of society. 700-723. – § 7. Constitutional and legal history. 724-750. – § 8. Economic and social history. 751-783. – § 9. History of civilization, sciences and education. 784-825. – § 10. History of art. 826-855. – § 11. History of religions (*a*. General; *b*. Special studies). 856-954. – § 12. History of philosophy. 955-969. – § 13. History of literature. 970-1005.

C

PREHISTORY
(p. 45-49)

§ 1. General. 1006-1022. – § 2. Palaeolithic and Mesolithic. 1023-1039. – § 3. Neolithic. 1040-1052. – § 4. Bronze age. 1053-1075. – § 5. Iron age. 1076-1100.

D

THE ANCIENT EAST
(the Hellenistic states included)
(p. 51-60)

§ 1. General. 1101-1117. – § 2. The Near East. 1118-1150. – § 3. Egypt. 1151-1201. – § 4. Mesopotamia. 1202-1251. – § 5. Hittites. 1252-1272. – § 6. Jews and Semitic peoples to the end of the ancient world. 1273-1342. – § 7. Iran. 1343-1368.

E

GREEK HISTORY
(p. 61-74)

§ 1. Classical world in general. 1369-1387. – § 2. Prehellenic epoch. 1388-1401. – § 3. Sources and criticism of sources (*a.* Epigraphical sources; *b.* Literary sources). 1402-1443. – § 4. General and political history. 1444-1478. – § 5. History of law and institutions. 1479-1508. – § 6. Economic and social history. 1509-1528. – § 7. History of literature, philosophy and science (*a.* Literature; *b.* Philosophy and sciences). 1529-1686. – § 8. Religion and mythology. 1687-1723. – § 9. Archaeology and history of art. 1724-1751.

F

HISTORY OF ROME, ANCIENT ITALY AND THE ROMAN EMPIRE
(p. 75-90)

§ 1. The peoples of Italy. 1752-1759. – § 2. The Etruscans. 1760-1777. – § 3. Sources and criticism of sources. (*a.* Epigraphical sources; *b.* Literary sources). 1778-1851. – § 4. General and political history. 1852-1957. – § 5. History of law and institutions. 1958-1992. – § 6. Economic and social history. 1993-2054. – § 7. History of literature, philosophy and science (*a.* Literature; *b.* Philosophy and science). 2055-2126. – § 8. Religion and mythology. 2127-2150. – § 9. Archaeology and history of art. 2151-2193. – § 10. Late antiquity. Transformation of the Roman world. 2194-2209.

G

EARLY HISTORY OF THE CHURCH TO GREGORY THE GREAT
(p. 91-97)

§ 1. Sources. 2210-2272. – § 2. General. 2273-2281. – § 3. Special studies. 2282-2376. – § 4. Hagiography. 2377-2382.

H

BYZANTINE HISTORY
(since Justinian)
(p. 99-104)

§ 1. Sources. 2383-2417. – § 2. General. 2418-2431. – § 3. Special studies. 2432-2518.

I

HISTORY OF THE MIDDLE AGES
(p. 105-146)

§ 1. Sources and criticism of sources (*a*. Non-literary sources; *b*. Literary sources). 2519-2584. – § 2. General works. 2585-2604. – § 3. Political history (*a*. General; *b*. 476–900; *c*. 900–1300; *d*. 1300–1500). 2605-2675. – § 4. Jews. 2676-2692. – § 5. Islam. 2693-2717. – § 6. Vikings. 2718-2731. – § 7. History of law and institutions. 2732-2777. – § 8. Economic and social history. 2778-2875. – § 9. History of civilization, literature, technology and education (*a*. Civilization; *b*. Literature; *c*. Technology; *d*. Education). 2876-3151. – § 10. History of art. 3152-3223. – § 11. History of music. 3224-3245. – § 12. History of philosophy, theology and science. 3246-3316. – § 13. History of the Church and religion (*a*. General; *b*. History of the Popes; *c*. Monastic history; *d*. Hagiography; *e*. Special studies). 3317-3483. – § 14. Settlements. Place names. Town planning. 3484-3514.

K

MODERN HISTORY, GENERAL WORKS
(p. 147-200)

§ 1. General. 3515-3569. – § 2. History by countries. 3570-4807.

L

MODERN RELIGIOUS HISTORY
(p. 201-211)

§ 1. General. 4808-4825. – § 2. Roman Catholicism (*a*. General; *b*. History of the Popes; *c*. Special studies; *d*. Religious orders; *e*. Missions). 4826-4950. – § 3. Orthodox Church. 4951-4956. – § 4. Protestantism. 4957-5013. – § 5. Non-Christian religions and sects. 5014-5048.

M

HISTORY OF MODERN CULTURE
(p. 213-240)

§ 1. General. 5049-5171. – § 2. Academies, universities and intellectual organizations. 5172-5208. – § 3. Education. 5209-5260. – § 4. The Press. 5261-5310. – § 5. Philosophy (*a*. General, *b*. Special studies). 5311-5360. – § 6. Exact, natural, medical sciences and technique. 5361-5473. – § 7. Literature (*a*. General; *b*. Renaissance; *c*. Classicism; *d*. Romanticism and after). 5474-5544. – § 8. Art and industrial art (*a*. General; *b*. Architecture; *c*. Sculpture, painting, etching and drawing; *d*. Decorative, photography, popular and industrial art). 5545-5617. – § 9. Music, theatre, cinema and broadcasting. 5618-5719.

N

MODERN ECONOMIC AND SOCIAL HISTORY
(p. 241-270)

§ 1. General. 5720-5776. – § 2. Political economy. 5777-5786. – § 3. Industry, mining and transportation. 5787-5885. – § 4. Trade. 5886-5926. – § 5. Agriculture and agricultural problems. 5927-5976. – § 6. Money and finance. 5977-6035. – § 7. Demography and urban history. 6036-6124. – § 8. Social history. 6125-6405. – § 9. Working-class movement and socialism. 6406-6469.

O

MODERN LEGAL AND CONSTITUTIONAL HISTORY
(p. 271-277)

§ 1. General. 6470-6501. – § 2. History of constitutional law. 6502-6526. – § 3. Public law and institutions. 6527-6548. – § 4. Civil and penal law. 6549-6613. – § 5. International law. 6614-6618.

P

HISTORY OF INTERNATIONAL RELATIONS
(p. 279-318)

§ 1. General. 6619-6702. – § 2. History of colonization and decolonization (a. General; b. Asia; c. Africa; d. America; e. Oceania). 6703-6867. – § 3. From 1500 to 1789 (a. General; b. 1500–1648; c. 1648–1789). 6868-6914. – § 4. From 1789 to 1815. 6915-6943. – § 5. From 1815 to 1910. 6944-7005. – § 6. From 1910 to 1935. The First World War. 7006-7107. – § 7. From 1935 to 1945. The Second World War (a. General; b. Diplomacy. Economy; c. Military operations; d. Resistance). 7108-7308. – § 8. From 1945 (a. General; b. 1945–1956; c. From 1956). 7309-7561.

R

ASIA
(p. 319-330)

§ 1. General. 7562-7563. – § 2. Western and central Asia. 7564-7581. – § 3. South Asia and Southeast Asia. 7582-7595. – § 4. China. 7596-7801. – § 5. Japan (esp. before 1868). 7802-7849. – § 6. Korea. 7850-7853.

S

AFRICA
(esp. to its colonization)
(p. 331)

N°s 7854-7873.

T

AMERICA
(esp. to its colonization)
(p. 333)

N°s 7874-7888.

U

OCEANIA
(esp. to its colonization)
(p. 335)

N°s 7889-7894.

GENERAL HISTORICAL BIBLIOGRAPHIES

I. [Austria]. Österreichische Historische Bibliographie. Austrian Historical Bibliography 2003. [2002. Cf. Bibl. 2004, n° *II*.] Hrsg. von Johannes GRABMAYER. Bearb. von Martha JAUERNIG. Unter Mitarbeit von Tatjana VALENTINITSCH und Uta HÖDL. Graz, Wolfgang Neugebauer u. Santa Barbara, Clio Press, 2005, 648 p.

II. [Belgium]. Bibliographie de l'histoire de Belgique. Bibliografie van de geschiedenis van België 2003. [2003. Cf. Bibl. 2004, n° *III*.]. Ed. par Romain VAN EENOO [et al.]. *Revue Belge de philologie et d'histoire - Belgisch Tijdschrift voor Filologie en Geschiedenis*, 2005, 83, 2*, [s. p.].

III. [Bulgaria]. POPOVA (Nina). Littérature scientifique historique bulgare en 2004 (suite). *Bulgarian historical review*, 2005, 33, 1-2, p. 250-264.

IV. [Czech Republic]. BŘEŇOVÁ (Věra), ROHLÍKOVÁ (Slavěna). Bibliografie českých/československých dějin 1918–2004. Výběr knih, sborníků a článků vydaných v letech 2000–2004 a doplňky za roky 1996–1999 . (Czech and Czechoslovak History, 1918–2004. A Bibliography of Select Monographs, Collections of Essays, Articles Published from 2000–2004). Tomo 1-3. Praha, Ústav pro soudobé dějiny AV ČR, 2005, 3 vol., 414 p., 466 p., 426 p. (1 CD). – BŘEŇOVÁ (Věra), ROHLÍKOVÁ (Slavěna). Czech and Czechoslovak History, 1918–2004. A Bibliography of Select Monographs, Volumes of Essays, and Articles Published from 2000 to 2004. Prague, Institute of Contemporary History, 2005, 453 p. – HORČÁKOVÁ (Václava), REXOVÁ (Kristina). Select Bibliography on Czech History. Books and Articles 2000–2004. Prague, Institute of History of the Academy of Sciences of the Czech Republic, 2005, 373 p. (1 CD). (Práce Historického ústavu AV ČR. D-Bibliographia, 13).

V. [France]. Bibliographie annuelle de l'histoire de France du Ve siècle à 1958. Vol. 50. Année 2004. [Vol. 49. 2003. Cf. Bibl. 2004, n° *IV*.] Redigée par Christophe CHARLE, Isabelle HAVELANGE, Brigitte KÉRIVEN et Claude GHIATI. Paris, Ed. du C. N. R. S., 2005, 947 p.

VI. [Germany] Historische Bibliographie : Berichtsjahr 2004. [2003. Cf. Bibl. 2004, n° *V*.] Herausgegeben von der Arbeitsgemeinschaft ausseruniversitärer historischer Forschungseinrichtungen in der Bundesrepublik Deutschland. München, Oldenbourg, 2005, 1064 p. (Historische Zeitschrift; Sonderausgabe). – Jahresberichte für Deutsche Geschichte. Neue Folge. 56. Jahrgang 2004. Mit Nachträgen. [55. Jahrgang 2003. Cf. Bibl. 2004, n° *V*.] Hrsg. von der Berlin-Brandeburgischen Akademie der Wissenschaften. Berlin, Akademie Verlag, 2005, XXII-1824 p.

VII. [Great Britain]. Annual bibliography of British and Irish history. On-line edition. [Cf. Bibl. 2004, n° *VI*.] – Annual bulletin of historical literature, 2005, 89, VIII-264 p. [2004. Cf. Bibl. 2004, n° *VI*.]

VIII. International bibliography of historical sciences. Internationale Bibliographie der Geschichtswissenschaften. Bibliografia internacional de ciencias historicas. Bibliographie internationale des sciences historiques. Bibliografia internazionale delle scienze storiche. Vol. LXIX, 2000 [Vol. LXVIII, 1999. Cf. Bibl. 2004, n° *VII*.] Ed. by Massimo MASTROGREGORI with the contribution of a number of scholars. München, K. G. Saur, 2005, XX-439 p.

IX. [Luxembourg]. Bibliographie d'histoire luxembourgeoise pour l'année 2004 avec complément des années précédentes. [2003. Cf. Bibl. 2004 n°, *VIII*.] *Hémecht: Zeitschrift für Luxemburger Geschichte = revue d'histoire luxembourgeoise*, 2005, 57, 2, p. 211-270.

X. [Slovakia] Slovenská historiografia 2003. Výberová bibliografia. [2002. Cf. Bibl. 2004, n° *IX*.] (Historiography in Slovakia. Selected bibliography). Zost. Alžbeta SEDLIAKOVÁ. *Historický časopis*, 2005, 53, 4, p. 768-809.

XI. [Spain]. Indice histórico español. Publicación semestral del centro de estudios históricos internacionales. Ed. por Pere MOLAS RIBALTA y Rosa ORTEGA CANADELL. Vol. 41, n. 117, Año 2003. [Vol. 40, n. 115, 2002. Vol. 41, n. 116, 2003. Cf. Bibl. 2004, n° *XI*.] Barcelona, Publicacions de la Universitat de Barcelona, 2005, 436 p.

A

AUXILIARY SCIENCES

§ 1. Palaeography. 1-7. – § 2. Diplomatics. 8-16. – § 3. History of the book (*a.* Manuscripts; *b.* Printed books). 17-101. – § 4. Chronology. 102-107. – § 5. Genealogy and family history. 108-128. – § 6. Sigillography and heraldry. 129-133. – § 7. Numismatics and metrology. 134-150.– § 8. Linguistics. 151-203. – § 9. Historical geography, travels and discoveries. 204-252. – § 10. Iconography and images. 253-273.

§ 1. Palaeography.

** 1. Chartae Latinae antiquiores: facsimile-edition of the Latin charters. 2nd Series: Ninth century. Ed. by Guglielmo CAVALLO and Giovanna NICOLAJ. Part 66. Italy XXXVIII. Piacenza 3. Publ. by Cristina CARBONETTI VENDITTELLI. Part 67. Italy XXXIX. Piacenza 4. Publ. by Paolo RADICIOTTI. Part 75. Italy XLVII. Lucca 4. Publ. by Francesco MAGISTRALE, Pasquale CORDASCO and Corinna DRAGO. Zürich, Graf, 2005, 3 vol., 136 p., 128 p., 164 p. (facs.).

2. ACHARYA (Subrata Kumar). Palaeography of Orissa. New Delhi, D. K. Printworld (P) Ltd., 2005, XX-340 p. (ill.).

3. BROWN (Virginia). Terra Sancti Benedicti: studies in the paleography, history and liturgy of medieval Southern Italy. Roma, Edizioni di storia e letteratura, 2005, 784 p. (ill.). (Storia e letteratura, 219).

4. CAVALLO (Guglielmo). Il calamo e il papiro: la scrittura Greca dall'età Ellenistica ai primi secoli di Bisanzio. Firenze, Edizioni Gonnelli, 2005, IX-256 p. (pl., facs.). (Papyrologica Florentina, 36).

5. ECKHART (Hans Wilhelm), STÜBER (Gabriele), TRUMPP (Thomas). Paläographie-Aktenkunde-Archivalische Textsorten. Neustadt an der Aisch, Degener, 2005, 276 p. (ill.). (Historische Hilfswissenschaften bei Degener & Co., 1).

6. ORSINI (Pasquale). Manoscritti in maiuscola biblica: materiali per un aggiornamento. Cassino, Università degli Studi di Cassino, 2005, 368 p. (pl., ill., facs.). (Collana scientifica. Studi archeologici, artistici, filologici, letterari e storici, 7).

7. ROBERTS (Jane Annette). Guide to scripts used in English writings up to 1500. London, British Library, 2005, XV-294 p. (ill., pl., facs.).

Cf. n°s 8-16, 32, 2519-2584

§ 2. Diplomatics.

** 8. ODART MORCHESNE. Le formulaire d'Odart Morchesne dans la version du ms BNF fr. 5024. Éd. par Olivier GUYOTJEANNIN et Serge LUSIGNAN. Paris, École des Chartes, 2005, 479 p. (Mémoires et documents de l'École des chartes, 80).

9. FRENZ (Thomas). L'introduzione della scrittura umanistica nei documenti e negli atti della Curia pontificia del secolo XV. Ed. italiana a cura di Marco MAIORINO. Città del Vaticano, Scuola vaticana di paleografia, diplomatica e archivistica, 2005, XLVII-297 p. (pl., facs.). (Littera antiqua, 12. Subsidia studiorum, 4).

10. JANKOVSKAJA (Ninel' B.). Klinopisnaja diplomatika. (The cuneiform diplomatics). *Vspomogatel'nye istoricheskie distsipliny*, 2005, 29, p. 15-39. [English summary]

11. KÖLZER (Theo). Kaiser Ludwig der Fromme (814–840) im Spiegel seiner Urkunden. Paderborn, Ferdinand Schöningh, 2005, 34 p. (pl., ill.). (Geisteswissenschafte. Nordrhein-Westfälische Akademie der Wissenschaften, Vorträge G 401).

12. MORELLI (Serena). Le carte di Léon Cadier alla Bibliothèque nationale de France. Contributo alla ricostruzione della Cancelleria angioina. Rome, École française de Rome e Istituto storico italiano per il Medio Evo, 2005, 354 p. (Sources et documents d'histoire du Moyen Âge, 9. Fonti per la storia dell'Italia medievale, Antiquitates, 20).

13. NOGUEIRA (Bernardo Guimarães Fisher de Sá). Portugaliae tabellionum instrumenta: documentação notarial portuguesa. Lisboa, Centro de História da Universidade de Lisboa, 2005, 335 p.

14. PALL (Francisc). Diplomatica latină din Transilvania medievală. Cluj-Napoca, Argonaut, 2005, 295 p. (Seria Documente, istorie, mărturii).

15. TOCK (Benoît-Michel). Scribes, souscripteurs et témoins dans les actes privés en France (VIIe–début du XIIe siècle). Turnhout, Brepols, 2005, 490 p. (ill.). (ARTEM – Atelier de Recherches sur les Textes Médiévaux, 9).

16. Wege zur Urkunde, Wege der Urkunde, Wege der Forschung. Beiträge zur europäischen Diplomatik des Mittelalters. Hrsg. v. Karel HRUZA und Paul A. HEROLD. Wien, Böhlau, 2005, 305 p. (ill.). (Forschungen zur Kaiser- und Papstgeschichte des Mittelalters, 24).

Cf. n°s 1-7, 2519-2584

§ 3. History of the book.

a. Manuscripts

* 17. CERESA (Massimo). Bibliografia dei fondi manoscritti della Biblioteca vaticana (1991–2000). Città del Vaticano, Biblioteca apostolica Vaticana, 2005, 737 p. (Studi e testi, 426).

** 18. Liber (Der) illuministarum aus Kloster Tegernsee: Edition, Übersetzung und Kommentar der kunsttechnologischen Rezepte. Hrsg. v. Anna BARTL. Stuttgart, Franz Steiner, 2005, 833 p. (ill.). (Veröffentlichung des Instituts für Kunsttechnik und Konservierung im Germanischen Nationalmuseum, 8).

---

19. Beato (El) de San Millán, Códice 33: original conservado en la Biblioteca de la Real Academia de la Historia. Vol. 2. Estudio. Ed. por Manuel C. DÍAZ Y DÍAZ y John WILLIAMS. Madrid, Testimonio Compañía Editorial, 2005, 132 p. (Colección Scriptorium).

20. Beato (El) del abad Banzo del monasterio de San Andrés de Fanio, un "Apocalipsis" aragonés recuperado: facsímil et estudios. Coordinator científico Fernando GALTIER MARTÍ. Zaragoza Caja Inmaculada, 2005, 386 p. (ill.).

21. BRAUN-NIEHR (Beate). Das Brandenburger Evangelistar. Regensburg, Schnell & Steiner, 2005, 104 p. (ill., facs.). (Schriften des Domstifts Brandenburg, 2).

22. BUZI (Paola). Manoscritti latini nell'Egitto tardoantico. Con un censimento dei testi letterari e semiletterari a cura di Simona CIVES. Imola, La mandragora, 2005, 191 p. (Università di Bologna, Dipartimento di Archeologia. Archeologia e storia della civiltà egiziana e del Vicino Oriente antico. Materiali e studi, 9).

23. CAPASSO (Mario). Introduzione alla papirologia: dalla pianta di papiro all'informatica papirologica. Bologna, Il Mulino, 2005, 260 p. (Manuali. Storia).

24. Care and conservation of manuscripts, 8: proceedings of the Eighth International Seminar held at the University of Copenhagen, 16th–17th October 2003. Ed. by Gillian FELLOWS-JENSEN and Peter SPRINGBORG. København, Museum Tusculanum Press a. University of Copenhagen, 2005, 300 p. (ill., map).

25. DÉROCHE (François). Islamic codicology: an introduction to the study of manuscripts in Arabic script. With contributions by Annie BERTHIER [et al.]; edited by Muhammad Isa WALEY. London, Al-Furqān Islamic Heritage Foundation, 2005, [s. p.]. (ill.). (Publications/Al-Furqān Islamic Heritage Foundation, 102).

26. Egbert-Codex (Der). Das Leben Jesu. Ein Höhepunkt der Buchmalerei vor 1000 Jahren: Handschrift 24 der Stadtbibliothek Trier. Hrsg. v. Gunther FRANZ. Darmstadt, Wissenschaftliche Buchgesellschaft, 2005, 248 p. (ill.).

27. Festschrift in honor of Eleazar Birnbaum. Ed. by Virginia H. AKSAN. *Journal of Turkish studies*, 2005, 29, LVI-325 p.

28. Forme e modelli della tradizione manoscritta della Bibbia. A cura di Paolo CHERUBINI. Città del Vaticano, Scuola vaticana di paleografia, diplomaticae archivistica, 2005, XV-562 p. (Littera antiqua, 13).

29. GEDDES (Jane). The St Albans Psalter: a book for Christina of Markyate. London, British Library, 2005, 136 p. (ill.).

30. HOSSAM MUJTĀR AL ABADI. Las artes del libro en al-Andalus y el Magreb (siglos IV h/X dC–VIII h/XV dC). Madrid, Ediciones El Viso, 2005, 219 p. (ill.).

31. Imagining the book. Ed. by Stephen KELLY and John J. THOMPSON. Turnhout, Brepols, 2005, XVIII-253 p. (Medieval texts and cultures of Northern Europe, 7).

32. Iskusstvo zapadnoevropejskoj rukopisnoj knigi V-XVI vv. (The art of 5th–16th-century European manuscripts). Ed. G.V. VILINBAKHOV, L.I. KISELEVA. Authors: V.N. ZAJTSEV, O.G. ZIMINA, L.I. KISELEVA, I.L. TOKAREVA [et al.]. Intr. by Mikhail B. PIOTROVSKIJ. The State Hermitage (Saint-Petersburg). Sankt-Peterburg, Gos. Ermitazh, 2005, 335 p. (ill.; pers. ind. p. 324-327; bibl. p. 330-335).

33. LEMAÎTRE (Jean Loup). Un calendrier retrouvé. Le calendrier des heures de Saint-Pierre-du-Queyroix: Musée du pays d'Ussel, Ms. 6. Ussel, Musée du pays d'Ussel a. Paris, Diffusion de Boccard, 2005, XIV-318 p. (ill., facs.). (Mémoires et documents sur le Bas-Limousin, 25).

34. Liber/libra: il mercato del libro manoscritto nel medioevo italiano. A cura di Caterina TRISTANO e Francesca CENNI. Roma, Jouvence, 2005, 237 p. (Quaderni CISLAB, 1).

35. Libro (Il) d'ore di Lorenzo de' Medici: volume di commento. A cura di Franca ARDUINI. Modena, F.C. Panini, 2005, 295 p. (ill., facs.).

36. Lire le manuscrit médiéval: observer et décrire. Sous la direction de Paul GÉHIN. Paris, Armand Colin, 2005, 283 p. (ill.). (Collection U. Histoire).

37. Livres et bibliothèques au Moyen Âge. Éd par, Antoine-Jean-Victor LE ROUX DE LINCY. Saint-Denis, Pecia, 2005, 118 p. (Le livre médiéval, 1).

38. Manoscritti antichi e moderni. Roma, Biblioteca Nazionale Centrale, 2005, 240 p. (Quaderni della Biblioteca nazionale centrale di Roma, 9).

39. Manuscripts in transition: recycling manuscripts, texts and images. Proceedings of the International Congres held in Brussels (5–9 November 2002). Ed. by Brigitte DEKEYZER and Jan VAN DER STOCK. Leuven, Peeters, 2005, IX-490 p. (ill.). (Corpus of illuminated manuscripts, 15. Low Countries series, 10).

40. Manuscrits (Les) grecs datés des XIIIe et XIVe siècles conservés dans les bibliothèques publiques de France. Vol. 2. Première moitié du XIVe siècle. Paris, Bibliothèque Nationale de France, IRHT et Turnhout, Brepols, 2005, 199 p. (ill.). (Monumenta palaeographica Medii Aevi. Series Graeca).

41. Medieval book fragments in Sweden: an International Seminar in Stockholm, 13–16 November 2003. Ed. by Jan BRUNIUS. Stockholm, Kungl. Vitterhets Historie och Antikvitets Akademien and Almqvist & Wiksell International, 2005, 242 p. (Kungl. Vitterhets, historie och antikvitets akademien. Konferenser, 58).

42. Miniatura (La) in Italia. Vol. 1. Dal tardoantico al Trecento con riferimenti al Medio Oriente e all'Occidente europeo. A cura di Antonella PUTATURO DONATI MURANO e Alessandra PERRICCIOLI SAGGESE. Napoli, Edizioni Scientifiche Italiane e Città del Vaticano, Biblioteca Apostolica Vaticana, 2005, 256 p. (ill.).

43. Mittelalterliche Handschriften der Kölner Dombibliothek: erstes Symposion der Diözesan- und Dombibliothek Köln zu den Dom-Manuskripten (26. bis 27. November 2004). Hrsg. v. Heinz FINGER. Köln, Erzbischöfliche Diözesan- und Dombibliothek, 2005, 338 p. (ill.). (Libelli Rhenani, 12).

44. MURANO (Giovanna). Opere diffuse per exemplar e pecia. Turnhout, Brepols, 2005, 897 p. (Textes et études du moyen âge, 29).

45. Pamplona-Bibel (Die): die Bilderbibel des Königs Sancho el Fuerte (1153–1234) von Navarra: Universitätsbibliothek Augsburg, Sammlung Oettingen-Wallerstein, Cod.I.2.4°15. Kommentarband zum Faksimile. Simbach am Inn, Verlag Müller & Schindler u. Reinbeck, Verlag Coron bei Kindler Verlag, 2005, 336 p. (ill.).

46. POWITZ (Gerhardt). Handschriften und frühe Drucke: ausgewählte Aufsätze zur mittelalterlichen Buch- und Bibliotheksgeschichte. Frankfurt am Main, Klostermann, 2005, 230 p. (Frankfurter Bibliotheksschriften, 12).

47. Scribes and transmission in English manuscripts, 1400–1700. Ed. by Peter BEAL and Anthony Stockwell Garfield EDWARDS. *English manuscript studies 1100–1700*, 2005, 12, VI-266 p.

48. SLAVIN (Michael). The ancient books of Ireland. Montreal a. Ithaca, McGill-Queen's U. P., 2005, X-198 p. (ill., maps).

49. Splendor solis: Handschrift 78 D 3 des Kupferstichkabinetts der Staatlichen Museen zu Berlin, Preußischer Kulturbesitz. Gütersloh, Coron Exclusiv, 2005, 128 p. (pl., ill.).

50. SZKIET (Christine). Reichenauer Codices in Schaffhausen: die frühen Handschriften des Schaffhauser Allerheiligenklosters und ihre Stellung in der südwestdeutschen Buchmalerei des 11. Jahrhunderts. Kiel, Ludwig, 2005, 235 p. (ill.). (Kieler kunsthistorische Studien, 9).

51. VELKOV (Asparukh). Les filigranes dans les documents ottomans: divers types d'images. Sofia, Éditions "Texte – A. Trayanov", 2005, 409 p. (ill.).

52. Vie et miracles de saint Maur = Life and miracles of Saint Maurus. Éd. par Charlotte DENOËL. Paris, Réunion des Musées Nationaux et Troyes, Médiathèque de l'agglomeration troyenne, 2005, 1 CD-ROM (facs.).

53. Villani illustrato (Il). Firenze e l'Italia medievale nelle 253 immagini del ms. Chigiano L VIII 296 della Biblioteca Vaticana. A cura di Chiara FRUGONI. Firenze, Le lettere e Roma, Biblioteca apostolica vaticana, 2005, 263 p.

54. Zisterziensisches Schreiben im Mittelalter: das Skriptorium der Reiner Mönche. Beiträge der internationalen Tagung im Zisterzienserstift Rein, Mai 2003. Hrsg. v. Anton SCHWOB und Karin KRANICH-HOFBAUER. Bern u. New York, P. Lang, 2005, 450 p. (Jahrbuch für internationale Germanistik. Reihe A. Kongressberichte, 71).

b. Printed books

** 55. MAC KENZIE (Donald Francis), BELL (Maureen). A chronology and calendar of documents relating to the London book trade, 1641–1700. Vol. 1. 1641–1670. Vol. 2. 1671–1685. Vol. 3. 1686–1700. Oxford, Oxford U. P., 2005, 3 vol., XVIII-643 p., VII-458 p., VII-468 p.

56. BOORMAN (Stanley). Studies in the printing, publishing and performance of music in the 16th century. Burlington, Ashgate, 2005, [s. p.]. (ill.). (Variorum collected studies series, 815).

57. BUCHANAN-BROWN (John). Early Victorian illustrated books: Britain, France and Germany, 1820–1860. London, British Library a. New Castle, Oak Knoll Press, 2005, 320 p. (ill.).

58. Buchkultur im Mittelalter: Schrift, Bild, Kommunikation. Hrsg. v. Michael STOLZ und Adrian METTAUER. Berlin, De Gruyter, 2005, VIII-374 p. (pl., ill.).

59. BUCHWALD-PELCOWA (Paulina). Historia literatury i historia książki: studia nad książką i literaturą od średniowiecza po wiek XVIII. Kraków: Tow. Autorów i Wydawców Prac Naukowych "Universitas", 2005, 734 p. (ill.).

60. CAYUELA (Anne). Alonso Pérez de Montalbán: un librero en el Madrid de los Austrias. Madrid: Calambur, 2005, 382 p. (Biblioteca litterae, 6).

61. CIARAMELLI (Giancarlo), GUERRA (Cesare). Tipografi, editori e librai mantovani dell'Ottocento. Milano, F. Angeli, 2005, 426 p. (Studi e ricerche di storia dell'editoria, 31).

62. Cinque secoli di carta: produzione, commercio e consumi della carta nella "Regio Insubrica" e in Lombardia dal Medioevo all'età contemporanea: atti del convegno, Varese, 21 aprile 2005. A cura di Renzo P. CORRITORE e Luisa PICCINNO. Varese, Insubria U. P., 2005, VIII-219 p.

63. Cognition and the book: typologies of formal organisation of knowledge in the printed book of the early modern period. Ed. by Karl. A. E. ENENKEL and Wolfgang NEUBER. Leiden, Brill, 2005, XVI-641 p. (Intersections: yearbook for early modern studies, 4).

64. DESGRAVES (Louis). Dictionnaire des imprimeurs, libraires et relieurs de la Dordogne, des Landes, du Lot-et-Garonne et des Pyrénées-Atlantiques: (XVe–XVIIe siècles). Baden-Baden, Éditions Valentin Koerner, 2005, 156 p. (Bibliotheca bibliographica Aureliana, 175).

65. Dictionnaire encyclopédique du livre. Vol. 2. E-M. Sous la dir. de Pascal FOUCHÉ, Daniel PÉCHOIN et Philippe SCHUWER. Paris, Cercle de la librairie, 2005, XI-1074 p. (ill., map).

66. Dintorni (I) del testo: approcci alle periferie del libro: atti del convegno internazionale, Roma, 15–17 novembre 2004, Bologna, 18–19 novembre 2004. A cura di Marco SANTORO e Maria Gioia TAVONI. Roma, Edizioni dell'Ateneo, 2005, 2 vol., XX-IX-789 p. (ill.). (Biblioteca di "paratesto"; 1).

67. FARENGA (Paola). Editori ed edizioni a Roma nel Rinascimento. Roma, Roma nel Rinascimento, 2005, XIII-156 p. (RR inedita, 34).

68. FINKELSTEIN (David), MAC CLEERY (Alistair). An introduction to book history. New York a. London: Routledge, 2005, V-160 p.

69. FINOCCHIARO (Giuseppe). Cesare Baronio e la tipografia dell'Oratorio: impresa e ideologia. Firenze, Leo S. Olschki, 2005, 180 p. (Storia della tipografia e del commercio librario, 6).

70. FLÜGGE (Lars). Die Auswirkungen des Buchdrucks auf die Praxis des Schreibens. Marburg, Tectum, 2005, 142 p. (ill.).

71. Geselligkeit und Bibliothek. Lesekultur im 18. Jahrhundert. Hrsg. v. Wolfgang ADAM und Markus FAUSER; in Zusammenarbeit mit Ute POTT. Göttingen, Wallstein, 2005, 331 p. (Schriften des Gleimhauses Halberstadt, 4).

72. HAUSBERGHER (Mauro). Volendo questo illustrissimo Magistrato Consolato: trecento anni di editoria pubblica a Trento. Trento, Provincia autonoma di Trento, Soprintendenza per i Beni librari e archivistici, 2005, LXXXV-303 p. (Quaderni, 6).

73. HAUSMANN (Albrecht). Überlieferungsvarianz und Medienwechsel. Die deutschen Artes dictandi des 15. Jahrhunderts zwischen Manuskript und Buchdruck. *Revue belge de philologie et d'histoire*, 2005, 83, 1, p. 747-768.

74. History of printing and publishing in the languages and countries of the Middle East. Ed. by Philip C. SADGROVE. Oxford, Oxford University Press on behalf of the University of Manchester, 2005, VI-209 p. (Journal of Semitic studies, 15. Supplement).

75. History of the book in Canada. Vol. 2. 1840–1918. Ed. by Patricia FLEMING, Gilles GALLICHAN and Fiona A. BLACK. Toronto a. London, University of Toronto Press, 2005, XXXIII-659 p. (ill., facs.., maps, ports).

76. Irish (The) book in English, 1550–1800. Ed. by Raymond GILLESPIE and Andrew HADFIELD. Oxford a. New York, Oxford U. P., 2005, XXI-477 p. (Oxford history of the Irish book, 3).

77. KIRON (Arthur). La casa editrice Belforte e l'arte della stampa in ladino: in occasione del bicentenario della casa editrice Salomone Belforte & c. / The Belforte publishing house and the art of Ladino printing: on the occasion of the bicentenary of the Salomone Belforte & c. publishing house: 1805–2005. Livorno, S. Belforte, 2005, 61 p. (Collana di studi ebraici, 2).

78. Lecture et lecteurs en Bourgogne du Moyen Âge à l'époque contemporaine. Ed. par Vincent TABBACH. *Annales de Bourgogne*, 2005, 77, 1-2, 304 p.

79. LEWIS (Mary S.). Antonio Gardano, Venetian music printer, 1538–1569: a descriptive bibliography and historical study. Vol. 3. 1560–1569. New York a. London, Garland, 2005, XI-601 p. (facs.). (Garland reference library of the humanities, 718).

80. MÄKELER (Hendrik). Das Rechnungsbuch des Speyerer Druckherrn Peter Drach d. M. (um 1450– 1504). St. Katharinen, Scripta Mercaturae, 2005, X-382 p. (ill., facs.). (Sachüberlieferung und Geschichte, 38).

81. MERMIER (Franck). Le livre et la ville: Beyrout et l'édition arabe. Essai. Arles: Actes sud-Sindbad, 2005, 244 p. (La Bibliothèque arabe. Collection Hommes et sociétés).

82. MIRANDOLA (Giorgio). Storia del libro botanico. Bergamo, Università di Bergamo, 2005, 171 p. (ill.). (Lezioni di bibliografia).

83. MONTECCHI (Giorgio). Il libro nel Rinascimento. Vol. 2. Scrittura, immagine, testo e contesto. Roma, Viella, 2005, XVII-227 p. (I libri di Viella, 48).

84. Music publishing in Europe 1600–1900: concepts and issues, bibliography. Ed. by Rudolf RASCH. Berlin, BWV, 2005, XVI-314 p. (ill., music). (Musical life in Europe 1600–1900: circulation, institutions, representation. Circulation of music, 1).

85. NEMIROVSKIJ (Evgenij L.). Nachalo knigopechatanija u juzhnykh slavjan. (The beginnings of South Slavonic book-printing). Part 1. Istochniki i istoriografija. Predposylki. (Sources and historiography. Preconditions). Part 2. Izdanija pervoj chernogorskoj tipografii. (The editions of the first print shop of Montenegro). Moskva, Nauka, 2005, 2 vol., 474 p., 542 p. (ill.; portr.; bibl. p. 452-456, 526-529; ind. p. 457-472, 530-540). (Istorija slavjanskogo knigopechatanija XV – nachala XVII veka [A history of Slavonic book-printing of the 15th – the early 17th centuries], 2).

86. NOVA (Giuseppe). Stampatori, librai ed editori bresciani in Italia nel Seicento. Brescia, Fondazione civiltà bresciana, 2005, 275 p. (Strumenti di lavoro, 9).

87. PEARSON (David). English bookbinding styles, 1450–1800: a handbook. London, British Library a. New Castle, Oak Knoll Press, 2005, XII-221 p. (pl., ill.).

88. PERROUSSEAUX (Yves). Histoire de l'écriture typographique de Gutenberg au XVIIe siècle. Méolans-Revel, Atelier Perrousseaux, 2005, 427 p. (ill.).

89. PEŠEK (Jiří). "Litterae et libri prohibiti" in der kommunistisch beherrschten Tschechoslowakei. *In*: Propaganda, (Selbst-)Zensur, Sensation. Grenzen von Presse- und Wissenschaftsfreiheit in Deutschland und Tschechien seit 1871 [Cf. n° 5297], p. 245-252.

90. Prensa, impresos, lectura en el mundo hispánico contemporáneo: homenaje a Jean-François Botrel. Ed. por Jean-Michel DESVOIS. Pessac, PILAR, 2005, IV-584 p. (ill.).

91. Printing and parenting in early modern England. Ed. by Douglas A. BROOKS. Aldershot, Ashgate, 2005, XVIII-436 p. (ill., ports.). (Women and gender in the early modern world).

92. Printing places: locations of book production & distribution since 1500. Ed. by John HINKS, Catherine ARMSTRONG and David HOWNSLOW. New Castle, Oak Knoll Press a. London, British Library, 2005, XII-208 p. (ill.).

93. VASIL'EV (Vladimir I.). Kniga i knizhnaja kul'tura na perelomnykh etapakh istorii Rossii: Teorija. Istorija. Sovremennost'. (Book in the crucial periods of the Russian history: theory, history and present situation). RAN, Otd. ist.-filol. nauk, etc. Moskva, Nauka, 2005, 270 p. (ill.; bibl. p. 237-248; pers. ind. p.249-256).

94. Venezia 1501: Petrucci e la stampa musicale = Venice 1501: Petrucci, music, print and publishing. Atti del Convegno internazionale di studi, Venezia, Palazzo Giustinian Lolin, 10–13 ottobre 2001. A cura di Giulio CATTIN e Patrizia DALLA VECCHIA. Venezia, Fondazione Levi, 2005, XIV-800 p. (music). (Studi musicologici, III.B.6).

b. Addenda 1996–2003

* 95. ORLANDI (Antonella). Studi sulla bibliografia testuale. *Accademie e biblioteche d'Italia*, 96, 64, 1, p. 39-46.

* 96. STUSSI (Alfredo). Bibliografia testuale con Conor Fahy. *Belfagor*, 2000, 55, 3, p. 313-321.

97. Libri a stampa postillati. Atti del colloquio internazionale (Milano, 3–5 maggio 2001). A cura di Edoardo BARBIERI e Giuseppe FRASSO. Milano, CUSL, 2003, VI-410 p. (Humanae litterae, 8).

98. Mondo (Nel) delle postille. Libri a stampa con note manoscritte. Una raccolta di studi. A cura di Edoardo BARBIERI. Milano, CUSL, 2002, XVII-191 p. (Humanae litterae, 6).

99. MONTECCHI (Giorgio), SORELLA (Antonio). I nuovi modi della tradizione: la stampa fra Quattro e Cinquecento. *In*: Storia della letteratura italiana. Diretta da Enrico MALATO. Vol. X. La tradizione dei testi. Coordinato da C. CIOCIOLA. Roma, Salerno Editrice, 2001, p. 633-673.

100. SANTORO (Marco). L'apporto della bibliografia testuale all'indagine filologica. *Esperienze letterarie*, 2000, 25, 1, p. 3-32.

101. TANSELLE (George T.). La storia della stampa e gli studi storici. *La Bibliofilia*, 96, 98, 3, p. 209-231.

Cf. n°s 293-300, 3123, 4898, 5261-5310, 5493, 5542-5544

§ 4. Chronology.

102. Mémoire (La) du temps au Moyen Âge. Études réunies par Agostino PARAVICINI-BAGLIANI. Firenze, SISMEL, Edizioni del Galluzzo, 2005, 448 p. (Micrologus' Library, 12).

103. RUDWICK (Martin J. S.). Bursting the limits of time: the reconstruction of geohistory in the age of revolution. Chicago, University of Chicago Press, 2005, XXIV-708 p.

104. SCHWARZBAUER (Fabian). Geschichtszeit. Über Zeitvorstellungen in den Universalchroniken Frutolfs von Michelsberg, Honorius' Augustodunensis und Ottos von Freising. Berlin, Akademie, 2005, 305 p. (Orbis mediaevalis – Vorstellungswelten des Mittelalters, 6).

105. Time and history: papers of the 28th International Ludwig Wittgenstein Symposium, August 7–13, 2005, Kirchberg am Wechsel. Editors, Friedrich STADLER and Michael STÖLTZNER = Zeit und Geschichte: Beiträge des 28. Internationalen Wittgenstein Symposiums, 7.–13. August 2005, Kirchberg am Wechsel / Herausgeber, Friedrich STADLER und Michael STÖLTZNER. Kirchberg am Wechsel, Austrian Ludwig Wittgenstein Society, 2005, 351 p. (ill.). (Contributions of the Austrian Ludwig Wittgenstein Society = Beiträge der Österreichischen Ludwig Wittgenstein Gesellschaft, 13).

106. *Vacat.*

§ 4. Addendum 2002.

107. ENGLISH (Brigitte). Zeiterfassung und Kalenderprogrammatik in der frühen Karolingerzeit. Das Kalendarium der Hs. Köln DB 83-2 und die Synode von Soissons 744. Stuttgart, Thorbecke, 2002, 182 p. (Instrumenta, 8).

§ 5. Genealogy and family history.

** 108. KARASOY (Yakup). Türklerde şecere geleneği ve anonim Şibanî-nâme. (Tradition de l'arbre généalogique chez les Turcs et Şibanî-nâme anonyme). Konya, Tablet, 2005, 179 p.

———

109. BERNSTEIN (Gail Lee). Isami's house: three centuries of a Japanese family. Berkeley, University of California Press, 2005, XXVIII-283 p.

110. BLAKELY (Ruth Margaret). The Brus family in England and Scotland, 1100–1295. Woodbridge, Boydell Press, 2005, XIII-271 p. (ill., maps).

111. BRUNET (Serge), LEMAÎTRE (Nicole). Clergés, communautés et familles des montagnes d'Europe. Paris, Publications de la Sorbonne, 2005, 421 p.

112. CAINE (Barbara). Bombay to Bloomsbury. A biography of the Strachey family. New York, Oxford U. P., 2005, XVII-488 p.

113. CHIPMAN (Donald E.). Moctezuma's children: Aztec royalty under Spanish rule, 1520–1700. Austin, University of Texas Press, 2005, XXIII-200 p. (ill., maps).

114. DACHER (Michèle). Cent ans au village: chronique familiale gouin (Burkina Faso). Paris, Karthala, 2005, 399 p. (ill., maps). (Hommes et sociétés).

115. DEBRIS (Cyrille). "Tu, felix Austria, nube". La dynastie de Habsbourg et sa politique matrimoniale à la fin du moyen âge, XIIIe–XVIe siècles. Turnhout, Brepols, 2005, 674 p. (Histoires de famille. La parenté au moyen âge).

116. Deutsche Familien. Historische Porträts von Bismarck bis Weizsäcker. Hrsg. v. Volker REINHARDT; unter Mitarbeit von Thomas LAU. München, Beck, 2005, 384 p.

117. DUBERT (Isidro). De la géographie des structures familiales aux stratégies adaptatives des familles en Espagne 1752–1860. *Annales de démographie historique*, 2005, 109, p. 199-226.

118. FARQUHARSON (Geoffrey). Clan Farquharson: a history. Stroud, Tempus, 2005, 319 p. (ill., maps, ports., geneal. Tables).

119. FOYSTER (Elizabeth). Marital violence. An English family history, 1660–1857. New York, Cambridge U. P., 2005, XIII-282 p.

120. FRANZ (Eckhart G.). Das Haus Hessen. Eine europäische Familie. Stuttgart, Kohlhammer, 2005, 253 p.

121. Hochmittelalterliche Adelsfamilien in Altbayern, Franken und Schwaben. Hrsg. v. Ferdinand KRAMER, Wilhelm STÖMER und Elisabeth LUKAS-GÖTZ. München, Kommission für Bayerische Landesgeschichte, 2005, XIV-862 p. (ill., geneal. tab., maps). (Studien zur bayerischen Verfassungs- und Sozialgeschichte. 20).

122. KOCAMEMI (Fazıl Bülent). Bir Türk ailesinin 450 yıllık öyküsü: Koca-Memi (Memi-Can) Paşa ailesinin Tarihçesi. [Récit d'une famille turque de 450 années: histoire de Koca-Memi (Memi-Can) Paşa. İstanbul, Ötüken, 2005, 220 [39] p.

123. LEBRETON (Philippe). La généalogie, source de documentation démographique et socio-historique. *Mémoires de l'Académie des sciences, belles-lettres et arts de Lyon*, 2005, 4, 5, p. 205-222.

124. LOFSTROM M. (William Lee). Tres familias de Charcas: fines del Virreinato, principios de la República. Sucre, Fundación Cultural del Banco Central de Bolivia, Archivo y Biblioteca Nacionales de Bolivia, 2005, 207 p.

125. Médiéviste (Le) et la monographie familiale: sources, méthodes et problématiques. Ed. par Martin AURELL. Turnhout, Brepols, 2005, 310 p. (ill., maps, geneal. Tables). (Histoires de famille. 1 Parenté au Moyen Âge).

126. SAUNT (Claudio). Black, white, and Indian: race and the unmaking of an American family [Grayson family]. New York a. Oxford, Oxford U. P., 2005, IX-300 p. (ill., maps).

127. SCHRAUT (Sylvia). Das Haus Schönborn: eine Familienbiographie: katholischer Reichsadel 1640–1840. Paderborn, Schöningh, 2005, 451 p. (ill., geneal. table). (Veröffentlichungen der Gesellschaft für Fränkische Geschichte, 47).

§ 5. Addendum 2003.

128. NOËL (Erick). Les Beauharnais. Une fortune antillaise, 1756–1796. Préface de Jean CHAGNIOT. Genève, Droz, 2003, 420 p. (Hautes études médiévales et modernes, 83).

Cf. n°s 2040, 5814, 6087

§ 6. Sigillography and heraldry.

———

129. DEMANGE (Jean-François). Glossaire historique et héraldique: l'archéologie des mots: dictionnaire héraldique, symbolique, militaire, nobiliaire, maritime, rural, artisanal et fiscal: ouvrage à l'usage des enseignants, des historiens, des généalogistes, des archéologues et des chercheurs. Anglet, Atlantica, 2005, 477 p.

130. Emblem scholarship: directions and developments: a tribute to Gabriel Hornstein. Ed. by Peter M. DALY. Turnhout, Brepols, 2005, XIII-264 p. (ill., port.). (Imago figurata, 5. Studies).

131. Heraldický register Slovenskej republiky. Zv. 4. (The Heraldry register of the Slovak Republic. Volume 3). Ed. Peter KARTOUS, Ladislav VRTEĽ. Bratislava, Ministerstvo vnútra SR a. Martin, Matica slovenská, 2005, 295 p.

132. HÖLTGEN (Karl Josef). Aspects of the emblem: studies in the English emblem tradition and the European context = Eikoku ni okeru: enburenu no dento: runesansu shikakubunka no ichimen. Tokyo, Keiogijuku Daigaku Shuppankai, 2005, 313 p. (ill.).

133. TURREL (Denise). Le Blanc de France. La construction des signes identitaires pendant les guerres de Religion (1562–1629). Genève, Droz, 2005, 256 p. (Travaux d'humanisme et Renaissance, 396).

§ 7. Numismatics and metrology.

134. AUPIAIS (Grégory). Iconographie monétaire du régime de Vichy. *Hypothèses 2004. Trav. Ecole doct. Hist.*, 2005, p. 23-33.

135. Body counts: medical quantification in historical and sociological perspective = La quantification médicale, perspectives historiques et sociologiques. Ed. by Gérard JORLAND, Annick OPINEL and George WEISZ. Montréal a. Ithaca, Published for Fondation Mérieux by McGill-Queen's U. P., 2005, X-417 p. (ill.).

136. Coinage and identity in the Roman provinces. Ed. by Christopher HOWGEGO, Volker HEUCHERT and Andrew BURNETT. Oxford, Oxford U. P., 2005, XV-228 p. [Cf. n^{os} <choice> 2017, 2019.]

137. Europäische numismatische Literatur im 17. Jahrhundert. Hrsg. v. Christian DEKESEL und Thomas STÄCKER. Wiesbaden, Harrassowitz in Kommission, 2005, 375 p. (ill.). (Wolfenbütteler Arbeiten zur Barockforschung, 42).

138. FROLOVA (Nina A.), ABRAMZOM (Mikhail G.). Monety Ol'vii v sobranii Gosudarstvennogo istoricheskogo muzeja: Katalog. (Coins of Pontic Olbia in the State Historical Museum, Moscow: a catalogue). Moskva, ROSSPEN, 2005, 360 p. (ill.; plates; bibl. p. 12-16).

139. GOODWIN (Tony). Arab-Byzantine coinage. London a. New York, Nour Foundation in association with Azimuth Editions, 2005, 168 p. (ill.). (Studies in the Khalili Collection, 4).

140. HAHN (Wolfgang R. O.). Zur Münzprägung des frühbyzantinischen Reiches: Anastasius I. bis Phocas und Heraclius-Revolte, 491–610. Wien, Money Trend Verlag, 2005, 222 p.

141. Imaginary kings. Royal images in the ancient Near East, Greece and Rome. Ed. by Olivier HEKSTER und Richard FOWLER. Stuttgart, Steiner, 2005, 231 p. (Oriens et occidens, 11).

142. Immaginario (L') del potere: studi di iconografia monetale. A cura di Rossella PERA. Roma, G. Bretschneider, 2005, X-279 p. (ill.). (Serta antiqua et mediaevalia, 8. Scienze documentarie, 1).

143. Juste mesure (La): quantifier, évaluer, mesurer entre Orient et Occident, (VIIIe–XVIIIe siècle). Sous la dir. de Laurence MOULINIER [et al.]. Saint-Denis, Presses universitaires de Vincennes, 2005, 200 p. (ill.). (Temps et espaces).

144. MEL'NIKOVA (Alla S.). Ocherki po istorii russkogo denezhnogo obrashchenija XVI–XVII vekov. (Essays on monetary circulation in Russia in the 16th and the 17th centuries). Gos. istoricheskij muzej. Moskva, Strelets, 2005, 319 p. (ill.; plates; bibl. incl.).

145. MICHAILIDOU (Anna). Weight and value in pre-coinage societies: an introduction. Athens, Diffusion de Boccard, 2005, 172 p. (ill.). (Meletēmata, 42).

146. Moneda (La) al final de la República: entre la tradició i la innovació. IX Curs d'Historia Monetària d'Hispània. Barcelona, Museu Nacional d'Art de Catalunya, Gabinet Numismàtic de Catalunya,, 2005, 147 p. (ill., map).

147. SAPRYKIN (Sergej Ju.). Denezhnoe obrashchenie na khore Hersonesa Tavricheskogo v antichnuju epokhu (istoriko-numizmaticheskoe issledovanie). (The monetary circulation in the chora of Chersonesos Taurica in the Antique time: a historical and numismatic study). RAN, In-t vseobshchej istorii. Moskva, IVI RAN, 2005, 207 p. (bibl. incl.).

148. Simposio Simone Assemani sulla monetazione islamica: Padova, II Congresso internazionale di numismatica e di storia monetale: Padova 17 maggio 2003, Musei civici agli Eremitani, Museo Bottacin (Biblioteca) = Simone Assemani Symposium on Islamic coinage: the 2nd International Congress on Numismatic and Monetary History. Padova, Esedra, 2005, 253 p. (ill., maps). (Numismatica Patavina, 7).

149. TAHBERER (Bekircan), UZEL (İlter). Antik Kilikya Sikkelerinde Asklepios Kültü (1 harita, 79 resim ile birlikte). (The cult of Asklepios as represented in the Ancient coins of Cilicia (with 1 map, 79 illustrations). *Belleten*, 2005, 69, 254, p. 9-58.

150. ZIESMANN (Sonja). Autonomie und Münzprägung in Griechenland und Kleinasien in der Zeit Philipps II. und Alexanders des Großen. Trier, Wissenschaftlicher Verlag Trier, 2005, 286 p. (ill.). (Bochumer altertumswissenschaftliches Colloquium, 67).

Cf. n°s 253-273, 1328

§ 8. Linguistics.

* 151. NIEDEREHE (Hans-Josef). Bibliografía cronológica de la lingüística, la gramática y la lexicografía del español (BICRES III): desde el año 1701 hasta el año 1800. Amsterdam, J. Benjamins, 2005, 474 p. (Amsterdam studies in the theory and history of linguistic science, 108 Series III, Studies in the history of the language sciences).

152. AKAR (Ali). Türk dili tarihi: dönem-eser-bibliyografya. (Histoire de la langue turque: époque – ouvrage – bibliographie). İstanbul, Ötüken, 2005, 339 p.

153. ALBANESE (Umberto). Il latino giuridico. Massime, locuzioni e formule giuridiche latine, traduzione, commento, fonti e riferimenti sistematici alla legislazione italiana. Napoli, Edizioni scientifiche italiane, 2005, 810 p.

154. AVANZA (Martina). La Ligue du Nord: de la défense des dialects à la recherche d'une langue nationale padane. *Mélanges de l'École française de Rome: Italie et mediterranée (MEFRIM)*, 2005, 117, 1, p. 314-330.

155. BAUDOU (Alban). Les noces de philologie et de propagande: l'étymologie dans le corpus annalistique romain. *Revue belge de philologie et d'histoire*, 2005, 83, 1, p. 131-148.

156. BOCHMANN (Klaus). Wie Sprachen gemacht werden: zur Entstehung neuer romanischer Sprachen im 20. Jahrhundert. Leipzig, Verlag der Sächsischen Akademie der Wissenschaften u. Stuttgart, In Kommission bei S. Hirzel, 2005, 24 p. (Sitzungsberichte der Sächsischen Akademie der Wissenschaften zu Leipzig, Philologisch-Historische Klasse, 139, Heft 4).

157. BOURGAIN (Pascale), HUBERT (Marie-Clotilde). Le latin médiéval. Turnhout, Brepols, 2005, 578 p. (Atelier du médiéviste, 10).

158. Britannia Latina: Latin in the culture of Great Britain from the Middle Ages to the twentieth century. Ed. by Charles BURNETT and Nicholas MANN. London, Warburg Institute a. Torino, Nino Aragno Editore, 2005, X- 230 p. (Warburg Institute colloquia, 8).

159. CALBOLI (Gualtiero). Les aphorismes et la langue juridique latine. *Prometheus*, 2005, 31, p. 157-168.

160. CASTRO (Ivo). Introdução à história do português: geografia da língua: português antigo. Lisboa, Edições Colibri, 2005, 159 p. (maps).

161. Deutsche Wortforschung als Kulturgeschichte: Beiträge des Internationalen Symposiums aus Anlass des 90-jährigen Bestandes der Wörterbuchkanzlei der Österreichischen Akademie der Wissenschaften Wien, 25.–27. September 2003. Hrsg. v. Isolde HAUSNER und Peter WIESINGER; unter Mitwirkung von Katharina KORECKY-KRÖLL. Wien, Verlag der Österreichischen Akademie der Wissenschaften, 2005, VI-432 p. (ill., maps). (Sitzungsberichte / Österreichische Akademie der Wissenschaften, Philosophisch-Historische Klasse, 720).

162. Dicționarul toponimic al României. Muntenia (DTRM), Vol. 1. A-B. (Toponymic dictionary of Romania. Wallachia). Sub redacția lui Nicolae SARAMANDU. București, Editura Academiei Române, 2005, 380 p.

163. Dictionary of medieval Latin from British sources. Fasc. 9. P-Pel. Ed. by D. R. HOWLETT and T. CHRISTCHEV. Oxford, Oxford University Press for The British Academy, 2005, p. 2073-2168.

164. Dictionnaire toponymique des communes suisses – DTS = Lexikon der schweizerischen Gemeindenamen – LSG = Dizionario toponomastico dei comuni svizzeri – DTS. Redaction: Florence CATTIN [et al.]; documentation: Dorothee AQUINO-WEBER [et al.]; comite scientifique: Rolf Max KULLY [et al.]; Centre de dialectologie Université de Neuchâtel; direction: Andres KRISTOL. Frauenfeld, Huber et Lausanne, Editions Payot, 2005, 1102 p. (ill.).

165. ERNST (Peter). Deutsche Sprachgeschichte: eine Einführung in die diachrone Sprachwissenschaft des Deutschen. Wien, WUV Facultas, 2005, 254 p. (UTB basics, 2583).

166. Éveil (L') des nationalités et les revendications linguistiques en Europe (1830–1930). Coordonné par Carmen ALÉN GARABATO. Paris, Harmattan, 2005, 287 p. (ill.). (Sociolinguistique).

167. FARGE (Arlette), CHAUMONT (Michel). Les mots pour résister: voyage de notre vocabulaire politique de la Résistance à aujourd'hui. Paris, Bayard, 2005, 209 p.

168. FORSSNER (Thorvald). Englische Familiennamenphilologie: mit einem etymologischen Familiennamenbuch = Philology of English family names: with an etymological dictionary of family names. Hamburg, Baar, 2005, 155 p. (Beiträge zur Lexikographie und Namenforschung, 3).

169. FRAGNITO (Gigliola). Proibito capire: la Chiesa e il volgare nella prima età moderna. Bologna, Il Mulino, 2005, 325 p. (Saggi, 640).

170. FRASER (P. M.), MATTHEWS (E.). A lexicon of Greek personal names. Volume IV: Macedonia, Thrace, Northern regions of the Black Sea. Oxford, Oxford U.P., 2005, XXIX-387 p.

171. GERHARDS (Jürgen). The name game: cultural modernization and first names. New Brunswick a. London, Transaction Publishers, 2005, VIII-148 p. (ill.).

172. Handbook (The) of historical linguistics. Ed. by Brian D. JOSEPH and Richard D. JANDA. Oxford a. Malden, Blackwell Pub., 2005, XVIII-881 p. (ill., maps). (Blackwell handbooks in linguistics).

173. Historical dictionary of Gaelic placenames =: Foclóir stairiúl áitainmneacha na Gaeilge. Fasc. 2. Names in B- = Ainmneacha i B-. Editors/eagarthóirí Pádraig Ó RIAIN, Diarmuid Ó MURCHADHA, Kevin MURRAY. London, Irish Texts Society, 2005, XXXII-258 p.

174. HÜLLEN (Werner). Kleine Geschichte des Fremdsprachenlernens. Berlin, Erich Schmidt, 2005, 184 p.

175. Impérialismes linguistiques hier et aujourd'hui: actes du colloque franco-japonais de Tôkyô (21–23 novembre 1999). Sous la dir. de Louis-Jean CALVET et

Pascal GRIOLET. Aix-en-Provence, Edisud et Paris, INALCO, 2005, 383 p.

176. Langues et identités culturelles dans l'Europe des XVIe et XVIIe siècles: actes du colloque international organisé à Nancy, 13–15 novembre 2003. Sous la dir. de Marie-Sol ORTOLÁ et Marie ROIG MIRANDA. Nancy, Groupe "XVIe et XVIIe siècles en Europe", Université Nancy 2, 2005, 2 vol., [s. p.]. (Europe XVI–XVII, 7-8).

177. LE DU (Jean), LE BERRE (Yves), BRUN-TRIGAUD (Guylaine). Lectures de l'atlas linguistiques de la France de Gilliéron et Edmont: du temps dans l'espace: essai d'interprétation des cartes de l'atlas linguistique de la France de Jules Gilliéron et Edmond Edmont, augmenté de quelques cartes de l'atlas linguistiques de la Basse-Bretagne de Pierre Le Roux. Paris, CTHS, 2005, 363 p. (ill., maps, ports.).

178. Lessicografia (La) a Torino dal Tommaseo al Battaglia: atti del convegno su "La lessicografia a Torino, dal Tommaseo al Battaglia", Torino-Vercelli, 7–9 novembre 2002. A cura di Gian Luigi BECCARIA e Elisabetta SOLETTI. Alessandria, Edizioni dell'Orso, 2005, X-426 p. (In forma di parola, 11).

179. Lexicon mediae et infimae Latinitatis Polonorum. Vol. VIII. Fasc. 3 (65): remunerativus-reticulum. Ed. Christina WEYSSENHOFF BROŻKOWA. Kraków, Polska Akademia Nauk, Instytut Języka Polskiego, 2005, p. 321-480.

180. LIETZ (Gero). Zum Umgang mit dem nationalsozialistischen Ortsnamen-Erbe in der SBZ/DDR. Leipzig, Leipziger Universitätsverlag, 2005, 298 p. (Onomastica Lipsiensia, 4).

181. MARFANY (Joan-Llus). Religion and the survival of 'minority' languages: the Catalan case. *Social history*, 2005, 30, 2, p. 154-174.

182. MARMARA (Rinaldo). Les levantins et la Grécisation des emprunts Turcs-Ottomans: lexique étymologique. İstanbul, Isis, 2005, 276 p.

183. MATARD-BONUCCI (Marie-Anne). Langue, fascisme et race: considération autour d'un dessin totalitaire. *Mélanges de l'École française de Rome: Italie et mediterranée (MEFRIM)*, 2005, 117, 1, p. 299-311. – EADEM. Langue, langages et question nationale en Italie. *Mélanges de l'École française de Rome: Italie et mediterranée (MEFRIM)*, 2005, 117, 1, p. 219-222.

184. MORENO FERNÁNDEZ (Francisco). Historia social de las lenguas de España. Barcelona, Ariel, 2005, 287 p. (map). (Ariel lingüística).

185. MÜCKEL (Wenke). "Trübners Deutsches Wörterbuch" (Band 1-4), ein Wörterbuch aus der Zeit des Nationalsozialismus: eine lexikographische Analyse der ersten vier Bände (erschienen 1939–1943). Tübingen, Niemeyer, 2005, VIII-220 p. (Lexicographica. 125 Series maior).

186. Nom (Le) dans les sociétés occidentales contemporaines. Dirigé par Agnès FINE et Françoise-Romaine OUELLETTE. Toulouse, Presses universitaires du Mirail, 2005, 252 p. (ill.). (Anthropologiques).

187. Nordic languages (The): an international handbook of the history of the North Germanic languages. Vol. 2. Ed. by Oskar BANDLE [et al.]. Berlin, Mouton de Gruyter, 2005, XXIX-[1059]-2208 p. (Handbooks of linguistics and communication science, 22.2).

188. PITTAU (Massimo). Dizionario della lingua etrusca. Sassari, Libreria editrice Dessi, 2005, 525 p.

189. PORCELLI (Bruno). In principio o in fine il nome: studi onomastici su Verga, Pirandello e altro Novecento. Pisa, Giardini, 2005, 217 p. (Bibliotechina di studi, ricerche e testi, 34).

190. RADNER (Karen). Die Macht des Namens: altorientalische Strategien zur Selbsterhaltung. Wiesbaden, Harrassowitz, 2005, VI-341 p. (ill.). (Santag: Arbeiten und Untersuchungen zur Keilschriftkunde, 8).

191. RASCH (Gerhard). Antike geographische Namen nördlich der Alpen. Hrsg. v. Stefan ZIMMER. Berlin, Walter de Gruyter, 2005, XVIII-284 p. (ill., maps). (Ergänzungsbände zum Reallexikon der germanischen Altertumskunde, 47).

192. REANEY (Percy Hide). A dictionary of English surnames. Oxford, Oxford U. P., 2005, LXX-520 p. (maps).

193. ROCHE-PÉZARD (Fanette). L'état d'urgence marinettien: "Tout vite en deux mots". Une langue nouvelle, pour une Italie nouvelle (1912–1915). *Mélanges de l'École française de Rome: Italie et mediterranée (MEFRIM)*, 2005, 117, 1, p. 295-298.

194. RODRÍGUEZ ADRADOS (Francisco). A history of the Greek language. From its origin to the present. Leiden a. Boston, Brill, 2005, XIX-345 p.

195. Role (The) of Latin in early modern Europe: texts and contexts: Akten der Tagung, The role of Latin in early modern Europe: texts and contexts II, Salzburg 2.–4. Mai 2003. Hrsg. v. Gerhard PETERSMANN und Veronika OBERPARLEITER, in Zusammenarbeit mit Geoffrey EATOUGH, Heinz HOFMANN und Keith SIDWELL. Salzburg, Horn, Universität Salzburg u. F. Berger u. Söhne, 2005, IX-179 p. (ill.). (Grazer Beiträge. 9 Supplementband).

196. SARDSHWELADSE (Surab), FÄHNRICH (Heinz). Handbuch der Orientalistik = Handbook of oriental studies. Abt. 8. Bd. 12. Central Asia. Altgeorgisch-deutsches Wörterbuch. Leiden, Brill, 2005, X-1632 p.

197. SCHLOSSER (Horst Dieter). Es wird zwei Deutschlands geben: Zeitgeschichte und Sprache in Nachkriegsdeutschland, 1945–1949. Frankfurt am Main u. Oxford, P. Lang, 2005, 291 p. (Frankfurter Forschungen zur Kultur- und Sprachwissenschaft, 10).

198. SEMERANO (Giovanni). La favola dell'indoeuropeo. A cura di Maria Felicia IAROSSI. Milano, Bruno Mondatori, 2005, IX-117 p. (Testi e pretesti).

199. Sprache der Geschichte. Hrsg. v. Jürgen TRABANT; unter Mitarbeit von Elisabeth MÜLLER-LUCKNER. München, Oldenbourg, 2005, XXII-166 p. (Schriften des historischen Kollegs, Kolloquien, 62).

200. STEPANOVA (Larisa G.). Iz istorii pervykh ital'janskikh grammatik: Neizdannye zametki sovremennika na poljakh traktata P'etro Bembo "Besedy o narodnom jazyke" (1525, kn.III). (To the history of early Italian books of grammar: Unpublished contemporary's notes in the margins of Pietro Bembo's Prose della volgar lingua, 1525, book III: [With a facsimile of the text]). RAN. In-t lingvist. issled. Sankt-Peterburg, Nauka, 2005, 277 p. (bibl. p. 266-276).

201. TESI (Riccardo). Storia dell'italiano: la lingua moderna e contemporanea. Bologna, Zanichelli, 2005, VI-277 p.

202. Tous vos gens à latin: le latin, langue savante, langue mondaine (XIVe–XVIIe siècles). Études réunies et editées par Emmanuel BURY. Genève, Droz, 2005, 463 p. (Travaux d'humanisme et Renaissance, 405).

203. VALE (Malcolm). Language, politics and society: the uses of the vernacular in the later Middle Ages. *English historical review*, 2005, 120, 485, p. 15-34.

Cf. n°s 389, 1460, 2901, 3484-3514, 5097, 5121

§ 9. Historical geography, travels and discoveries.

204. ALAI (Cyrus). Handbook of Oriental studies = Handbuch der Orientalistik. Section 1. Vol. 80. The Near and Middle East. General maps of Persia, 1477–1925. Leiden, Brill, 2005, XV-317 p. (ill., maps).

205. *Vacat.*

206. BAGROV (Lev S.). Istorija russkoj kartografii. (A history of Russian cartography). Moskva, Tsentrpoligraf, 2005, 523 p. (bibl. incl.).

207. Berlin, Paris, Moskau: Reiseliteratur und die Metropolen. Hrsg. v. Walter FÄHNDERS [et al.]. Bielefeld, Aisthesis, 2005, 289 p. (ill.). (Reisen, Texte, Metropolen, 1).

208. Böse Orte: Stätten nationalsozialistischer Selbstdarstellung – heute. Hrsg. v. Stephan POROMBKA und Hilmar SCHMUNDT. Berlin, Claassen, 2005, 222 p. (maps).

209. CALZADILLA (Pedro Enrique). Por los caminos de América en el Siglo de las Luces: la sociedad colonial hispanoamericana del siglo XVIII a través de los viajeros europeos. Caracas, Ministerio de la Cultura, CONAC, Fondo Editorial Trópykos, 2005, 202 p.

210. Chemin (Le), la route, la voie: figures de l'imaginaire occidental à l'époque moderne. Sous la dir. de Marie-Madeleine MARTINET [et al.]. Paris, Presses de l'Université Paris-Sorbonne, 2005, 563 p. (ill., maps). (Recherches actuelles en littérature comparée).

211. Colonização Atlântica (A). Coordenação de Artur Teodoro DE MATOS. Lisboa, Editorial Estampa, 2005, 2 vol., [s. p.]. (ill., maps). (Nova história da expansão portuguesa, 3).

212. Confine (Il) nel tempo: atti del Convegno, Ancarano, 22–24 maggio 2000. A cura di Roberto RICCI e Andrea ANSELMI. L'Aquila, Libreria Colacchi, 2005, 790 p. (ill.).

213. Continental crossroads. Remapping U.S.-Mexico borderlands history. Ed. by Samuel TRUETT and Elliott YOUNG. Durham, Duke U. P., 2005, XIV-344 p.

214. COURTHÈS (Eric). La ínsula paraguaya. Asunción, Centro de Estudios Antropológicos de la Universidad Católica, 2005, 88 p. (Biblioteca paraguaya de antropología, 49).

215. DESPOIX (Philippe). Le monde mesuré: dispositifs de l'exploration à l'âge des Lumières. Genève, Droz, 2005, 271 p. (ill., maps, port.). (Bibliothèque des Lumières, 67).

216. Entre oriente y occidente: ciudades y viajeros en la edad media. Ed. por Juan Pedro MONFERRER SALA y María Dolores RODRÍGUEZ GÓMEZ. Granada, Universidad de Granada, 2005, 359 p. (ill., map). (Monográfica, 1 Biblioteca de humanidades. Estudios arabes).

217. ERMAKOVA (Ljudmila M.). Vesti o Japan-ostrove v starodavnej Rossii i drugoe. (News on the 'Japan-Island' in Old Russia and other). Moskva, Jazyki slavjanskoj kul'tury, 2005, 269 p. (ill.; portr.). (Studia historica, Series minor).

218. Expansionen in der Frühen Neuzeit. Hrsg. v. Renate DÜRR, Gisela ENGEL und Johannes SÜSSMANN. Berlin, Duncker & Humblot, 2005, 392 p. (ill.). (Zeitschrift für historische Forschung, 34 Beiheft).

219. FASCE (Ferdinando). Viaggiatori italiani alla Grande Fiera di New York del 1939–1940. *Storia urbana*, 2005, 28, 109, 4, p. 51-70.

220. GARCÍA DE CORTÁZAR (Fernando). Atlas de historia de España. Barcelona, Editorial Planeta S.A., 2005, XV-544 p. (ill., maps). (Planeta Historia y sociedad).

221. Grand Tour: adeliges Reisen und europäische Kultur vom 14. bis zum 18. Jahrhundert: Akten der internationalen Kolloquien in der Villa Vigoni 1999 und im Deutschen Historischen Institut Paris 2000. Hrsg. v. Rainer BABEL und Werner PARAVICINI. Ostfildern, Thorbecke, 2005, 677 p. (maps, ill.). (Beihefte der Francia, 60).

222. GUYOTJEANNIN (Olivier). Atlas de l'histoire de France: la France médiévale IXe–XVe siècle. Paris, Éditions Autrement, 2005, 1 atlas (103 p.).

223. HENSCHEL (Christine). Italienische und französische Reiseberichte des 16. Jahrhunderts und ihre Übersetzungen: über ein vernachlässigtes Kapitel der europäischen Übersetzungsgeschichte. Darmstadt, Wissenschaftliche Buchgesellschaft, 2005, 315 p. (Beiträge zur Romanistik, 9).

224. Homenaje a Alejandro de Humboldt: Literatura de viajes desde y hacia Latinoamérica, siglos XV al XXI = Homage to Alexander von Humboldt: travel literature to and from Latin America XV through XXI centuries. Oaxaca, Universidad Autónoma Benito Juárez de Oaxaca, Humbolt State University, 2005, 759 p. (ill.).

225. Irish historic towns atlas. No. 15. Derry-Londonderry. By Avril THOMAS; editors: Anngret SIMMS, H.B. CLARKE, Raymond GILLESPIE. Dublin, Royal Irish Academy, 2005, 100 p.

226. Itineraria Posoniensia. Zborník z medzinárodnej konferencie Cestopisy v novoveku, ktorá sa konala v dňoch 3.–5. novembra 2003 v Bratislave. = Itineraria Posoniensia. Akten der Tagung Reisebeschreibungen in der Neuzeit, Bratislava, 3.–5. November 2003. Ed. Eva FRIMMOVÁ, Elisabeth KLECKER. Bratislava, Historický ústav Slovenskej akadémie vied, 2005, 408 p.

227. KENNEDY (Dane). The highly civilized man: Richard Burton and the Victorian world. Cambridge, Harvard U. P., 2005, 354 p.

228. LEVINE (Mark). Overthrowing geography: Jaffa, Tel Aviv, and the struggle for Palestine, 1880–1948. Berkeley a. Los Angeles, University of California Press, 2005, XV-442 p.

229. LOISEAUX (Olivier), DUCLOS (France). L'Afrique au coeur: carnets d'explorateurs français au XIXe siècle. Paris, Éd. du Seuil et Bibliothèque nationale de France avec la collaboration de la Société de géographie, 2005, 188 p. (ill., maps).

230. Mapas (Los) del Quijote. Madrid, Ministerio de Cultura, Biblioteca Nacional, 2005, 168 p. (ill., maps, facsims).

231. Mapping and empire: soldier-engineers on the southwestern frontier. Ed. by Dennis REINHARTZ and Gerald D. SAXON. Austin, University of Texas Press, 2005, XX-204 p.

232. MARTÍN ASUERO (Pablo). Viajeros hispánicos en Estambul: de la cuestión de Oriente al reencuentro con los Sefardíes (1784–1918). İstanbul, Editorial Isis, 2005, 299 p. (Cuadernos del Bósforo, 3).

233. Mobilität, Raum, Kultur: Erfahrungswandel vom Mittelalter bis zur Gegenwart. Hrsg. v. Karl-Siegbert REHBERG, Walter SCHMITZ und Peter STROHSCHNEIDER. Dresden, Thelem, 2005, XI-280 p. (ill.). (Kulturstudien, 1).

234. MÚNERA (Alfonso). Fronteras imaginadas: la construcción de las razas y de la geografía en el siglo XIX colombiano. Bogotá, Planeta, 2005, 225 p.

235. OLAUS MAGNUS. Carta Marina 1539. Ed. par Elena BALZAMO. Paris, José Corti, 2005,190 p. (Collection Merveilleux, 26).

236. Où en est la géographie historique? Économie et culture. Sous la dir. de Philippe BOULANGER et Jean-René TROCHET. Paris, L'Harmattan, 2005, 346 p.

237. PARADELA ALONSO (Nieves). El otro laberinto español: viajeros árabes a España entre el siglo XVII y 1936. Madrid, Siglo XXI de España Editores, 2005, 265 p. (ill.).

238. PEDLEY (Mary Sponberg). The commerce of cartography. Making and marketing maps in eighteenth-century France and England. Chicago, University of Chicago Press, 2005, XV-345 p. (Kenneth Nebenzahl, jr., lectures in the history of cartography).

239. QUANCHI (Max), ROBSON (John). Historical dictionary of the discovery and exploration of the Pacific islands. Lanham, Scarecrow Press, 2005, LXX-297 p. (ill., maps). (Historical dictionaries of discovery and exploration, 2).

240. Raumerfahrung, Raumerfindung: erzählte Welten des Mittelalters zwischen Orient und Okzident. Hrsg. v. Laetitia RIMPAU und Peter IHRING. Berlin, Akademie Verlag, 2005, 325 p.

241. REES (Joachim), SIEBERS (Winfried). Erfahrungsraum Europa: Reisen politischer Funktionsträger des Alten Reichs 1750–1800: ein kommentiertes Verzeichnis handschriftlicher Quellen. Berlin, Berliner Wissenschafts-Verlag, 2005, 480 p. (ill.). (Aufklärung und Europa, 18).

242. RICHARDSON (Brian W.). Longitude and empire: how Captain Cook's voyages changed the world. Vancouver, University of British Columbia Press, 2005, XVI-240 p.

243. ROUSSILLON (Alain). Identité et modernité: les voyageurs égyptiens au Japon, XIXe–XXe siècle. Arles, Actes sud – Sindbad, 2005, 249 p. (La Bibliothèque arabe. série Hommes et sociétés).

244. ROZWADOWSKI (Helen M.). Fathoming the ocean. The discovery and exploration of the deep sea. Cambridge, Belknap Press of Harvard U. P., 2005, XII-276 p.

245. SETTESOLDI (Laura), TARDELLI (Marcello), RAFFAELLI (Mauro). Esploratori italiani nell'Africa Orientale fra il 1870 ed il 1930: missioni scientifiche con raccolte botaniche, rilievi geografici ed etnografici. Firenze, Centro studi erbario tropicale, 2005, 141 p. (ill., maps, ports.). (Pubblicazione, 104).

246. Studi biografici e bibliografici sulla storia della geografia in Italia: pubblicati in occasione del IIIo Congresso geografico internazionale. Staten Island, Maurizio Martino, 2005, 2 vol., [s. p.]. (maps).

247. TALL (Aminatou). Reise und Forschung im westlichen Afrika: deutschsprachige Reiseliteratur im 19. und 20. Jahrhundert. Frankfurt am Main u. Oxford, P. Lang, 2005, 294 p. (maps). (Europäische Hochschulschriften. Reihe 1, Deutsche Sprache und Literatur, 1923 = Publications universitaires Europeennes. Serie 1, Langue et litterature allemandes, 1923 = European university studies. Series 1, German language and literature, 1923).

248. TANCK DE ESTRADA (Dorothy). Atlas ilustrado de los pueblos de indios: Nueva España, 1800. Mapas de Jorge Luis MIRANDA GARCÍA y Dorothy TANCK DE ESTRADA; con la colaboración de Tania Lilia CHÁVEZ SOTO. México, El Colegio de México, 2005, 1 atlas (269 p.) (ill., maps) 1 CD-ROM.

249. Testi cosmografici, geografici e odeporici del medioevo germanico: atti del XXXI Convegno dell'Associazione italiana di filologia germanica, Lecce, 26–28 maggio 2004. A cura di Dagmar GOTTSCHALL. Louvain-la-Neuve, Fédération internationale des instituts d'études médiévales, 2005, XV-275 p. (Textes et études du moyen âge, 33).

250. Welt-Räume: Geschichte, Geographie und Globalisierung seit 1900. Hrsg. v. Iris SCHRÖDER und Sabine HÖHLER. Frankfurt am Main, Campus, 2005, 323 p. (ill., maps). (Campus Historische Studien, 39).

§ 9. Addenda 2001–2004.

251. Imagologie des Nordens. Kulturelle Konstruktionen von Nördlichkeit in interdisziplinärer Perspektive. Hrsg. v. Astrid ARNDT, Andreas BLÖDORN, David FRAESDORFF, Annette WEISNER und Thomas WINKELMANN. Frankfurt am Main, Berlin u. Bern, Lang 2004, 309 p. (Imaginatio Borealis, 7).

252. PAGDEN (Anthony). Peoples and empires: a short history of European migration, exploration, and conquest, from Greece to the present. New York, Modern Library, 2001, XXV-216 p. (Modern library chronicles).

Cf. n°s 364, 7564

§ 10. Iconography and images.

** 253. BOLTANSKI (Christian). 6 Septembres. Milano, Charta, 2005, [s. p.]. (ill.).

254. AMALVI (Christian). Du bon usage des chefs d'œuvre: la reproduction et l'exploitation populaire et scolaire des classiques de la peinture universelle au XXe siècle. *Bibliothèque de l'école des Chartes*, 2005, 163, 1, p. 13-49.

255. BUJOLI (Marina). Louis XVI dans les documents iconographiques et objets produits en Grande-Bretagne: une certaine image de la monarchie, de la France et des français. Lille, ANRT, Atelier national de reproduction des thèses, 2005, 709 p. (ill.). (Thèse à la carte).

256. CENSER (Jack), HUNT (Lynn). Imaging the French Revolution: depictions of the French revolutionary crowd. *American historical review*, 2005, 110, 1, p. 38-45.

257. COSKI (John M.). The Confederate battle flag: America's most embattled emblem. Cambridge a. London, Belknap Press of Harvard U. P., 2005, XI-401 p. (ill.).

258. D'AUTILIA (Gabriele). L'indizio e la prova: la storia nella fotografia. Milano, Bruno Mondadori, 2005, 211 p. (Le scene del tempo).

259. DEKONINCK (Ralph). Ad imaginem: statuts, fonctions et usages de l'image dans la littérature spirituelle jésuite du XVIIe siècle. Genève, Droz, 2005, 423 p. (ill.). (Travaux du Grand Siècle, 26).

260. DIEHL (Paula). Macht, Mythos, Utopie: die Körperbilder der SS-Männer. Berlin, Akademie Verlag, 2005, 286 p. (ill.). (Politische Ideen, 17).

261. DONADIEU-RIGAUT (Dominique). Penser en images les ordres religieux: XIIe–XVe siècles. Paris, Arguments, 2005, II-385 p. (ill.).

262. EDOUARD (Sylvène). L' empire imaginaire de Philippe II: pouvoir des images et discours du pouvoir sous les Habsbourg d'Espagne au XVIe siècle. Paris, H. Champion, 2005, 416 p. (ill., map). (Bibliothèque d'histoire moderne et contemporaine, 17). – IDEM. Un songe pour triompher: la décoration de la galère royale de don Juan d'Autriche à Lépante (1571). *Revue historique*, 2005, 636, p. 821-848.

263. GHERMANI (Naïma). Une difficile représentation? Les portraits de princes calvinistes dans l'Empire allemand à la fin du XVIe siècle. *Revue historique*, 2005, 635, p. 561-592.

264. GROVE (Laurence). Text/image mosaics in French culture: emblems and comic strips. Aldershot a. Burlington, Ashgate, 2005, XIV-187 p. (ill.). (Studies in European cultural transition, 32).

265. Images of power: iconography, culture and state in Latin America. Ed. by Jens ANDERMANN and William ROWE. New York a. Oxford, Berghahn Books, 2005, X-299 p. (ill.). (Remapping cultural history, 2).

266. KRIM (Arthur J.). Route 66: iconography of the American highway. Ed. by Denis WOOD. Santa Fe, Center for American Places, 2005, 288 p. (ill., maps).

267. MANOR (Dalia). Art in Zion: the genesis of modern national art in Jewish Palestine. London, RoutledgeCurzon, 2005, XVII-260 p. (ill.). (Routledge Curzon Jewish studies series).

268. MATARD-BONUCCI (Marie-Anne). La caricature témoin et vecteur d'internationalisation de l'antisémitisme: la figure du "monde-juif". *In*: Antisémythes. L'image des Juifs entre culture et politique (1848–1939) [Cf. n° 5017], p. 439-458.

269. MILOVANOVIC (Nicolas). Du Louvre à Versailles: lecture des grands décors monarchiques du XVIIe siècle. Paris, Les belles Lettres, 2005, 312 p. (ill.).

270. Moc obrazů, obrazy moci. Politický plakát a propaganda. (Power of images, images of power. Political poster and propaganda. Exhibition). Ed. Tomáš BOJAR, Jan TŘESTÍK, Jakub ZELNÍČEK. Praha, Galerie U Křižovníků, 2005, 283 p. (photogr.).

271. POLASCHEGG (Andrea). Der andere Orientalismus: Regeln deutsch-morgenländischer Imagination im 19. Jahrhundert. Berlin, W. De Gruyter, 2005, XI-613 p. (ill.). (Quellen und Forschungen zur Literatur- und Kulturgeschichte, 35).

10. ICONOGRAPHY AND IMAGES

272. RIGAUX (Dominique). Le Christ du dimanche: histoire d'une image médiévale. Paris, L'Harmattan, 2005, 498 p. (ill., maps). (La librairie des humanités).

273. ROEGIERS (Patrick). Magritte and photography. Aldershot, Lund Humphries, 2005, 167 p. (ill., ports.).
Cf. n°s 32, 134-150, 826-855, 2069

B

MANUALS, GENERAL WORKS AND WORKS ON LARGE PERIODS

§ 1. Archives, libraries and museums (*a.* Archives; *b.* Libraries; *c.* Museums). 274-319. – § 2. History of historiography and memory (*a.* General; *b.* Special studies). 320-490. – § 3. Methodology, philosophy, and teaching of history. 491-543. – § 4. Ethnology, folklore and historical anthropology. 544-588. – § 5. General history. 589-699. – § 6. Theory of the state and of society. 700-723. – § 7. Constitutional and legal history. 724-750. – § 8. Economic and social history. 751-783. – § 9. History of civilization, sciences and education. 784-825. – § 10. History of art. 826-855. – § 11. History of religions (*a.* General; *b.* Special studies). 856-954. – § 12. History of philosophy. 955-969. – § 13. History of literature. 970-1005.

§ 1. Archives, libraries and museums.

a. Archives

* 274. KOVALEVA (L.A.) [et al.]. Gosudarstvennyj arkhiv Kostromskoj oblasti: spravochnik. (The State Archive of the Kostroma Region, [the Russian Federation]: A Handbook). Kom. po delam Administratsii Kostromskoj obl., Gos. arkh. Kostromskoj obl. Part 1. Kostroma, Obl. tip., 2005, 327 p. (ind. p. 245-276).

* 275. LASOCHKO (L.S.). Gosudarstvennyj arkhiv Kurskoj oblasti: putevoditel'. (The State Archive of the Kursk Region, [the Russian Federation]: A Guide). Gl. arkh. upr. Kurskoj obl. Kursk, [s. n.], 2005, 869 p.

* 276. POLUKHINA (M.V.). Ob'edinennyj gosudarstvennyj arkhiv Cheljabinskoj oblasti: putevoditel'. (The United State Archive of the Chelyabinsk Region, [the Russian Federation]: A Guide). Ed. Aleksandr P. FINADEEV. Gos. kom. po delam arkh. Cheljabinskoj obl. Vol. 2. Cheljabinsk, Cheljab. dom pechati, 2005, 630 p. (ind. p. 615-625).

277. Archive stories. Facts, fictions, and the writing of history. Ed. by Antoinette BURTON. Durham, Duke U. P., 2005, X-396 p.

278. Catalogo della donazione Licio Gelli all'Archivio di Stato di Pistoia. A cura di Licio GELLI. Roma e Bari, G. Laterza, 2005, 189 p.

279. CHIRONI (Giuseppe). La mitra e il calamo: il sistema documentario della Chiesa senese in età pretridentina, secoli XIV–XVI. Siena, Accademia senese degli Intronati, 2005, 395 p. (Monografie di storia e letteratura senese / Accademia senese degli Intronati, 13).

280. GRAZIOSI (Andrea). Rivoluzione archivistica e storiografia sovietica. *Contemporanea*, 2005, 8, 1, p. 57-85.

281. HAMILTON (Keith). La diplomatie des archives sous la Troisième République. *Revue d'histoire diplomatique*, 2005, 119, 4, p. 307-342.

282. JOLY (Hervé). Les archives des entreprises sous l'occupation: conservation, accessibilité et apport. Lille, Institut Federatif de Recherches sur les Economies et les Societes Indutrielles, 2005, 319 p.

283. MAC CONNELL (Stuart Charles). Historical research in eastern Uganda: local archives. *History in Africa*, 2005, 32, p. 467-478.

284. MAGIDOV (Vladimir M.). Kinofotofonodokumenty v kontekste istoricheskogo znanija. (Film, photo and audio documents in the context of historical knowledge). Ros. gos. gumanit. un-t. Moskva, RGGU, 2005, 394 p. (bibl. p. 317-343; pers. ind. p. 380-394).

285. MANTELLI (Brunello). Im Reich der Unsicherheit? Italienische Archive und die Erforschung des Faschismus. *Vierteljahrshefte für Zeitgeschichte*, 2005, 53, 4, p. 601-614.

286. PARPAROV (Fyodor). The Hitler book: the secret dossier prepared for Stalin from the interrogations of Hitler's personal aides. Ed. by Henrik EBERLE and Matthias UHL. New York, Public Affairs, 2005, XXX-370 p.

287. Registres paroissiaux, actes notariés et bases de données. Informatisation de source de l'histoire moderne. De la démographie historique et de la généalogie. Sous la dir. de Yves LANDRY. Caen, Centre de Recherche d'Histoire quantitative, Université de Caen Basse-Normandie, CNRS, 2005, 431 p.

288. ROMANELLI (Rita). Le carte in villa: l'Archivio Barbolani da Montauto e "La Barbolana" di Anghiari. *Archivio storico italiano*, 2005, 163, 4, p. 717-733.

289. SHI (Qinghua). Ming Qing gongting dang'an. (Palace archives in Ming and Qing). Xi'an, Shanxi Shifan daxue chubanshe, 2005, 2, 4, 336 p.

290. TUCK (Michael W.), ROWE (John Allen). Phoenix from the ashes: rediscovery of the lost Lukiiko archives. *History in Africa*, 2005, 32, p. 403-414.

a. Addenda 2001–2004

* 291. PONCET (Olivier). Frabrique des archives, fabrique de l'histoire du Moyen Age au XIXe siècle: une bibliographie. *Revue de synthèse*, 2004, 5, 125, p. 183-195.

292. STEEDMAN (Carolyn). Dust: the archive and cultural history. New Brunswick, Rutgers U. P., 2001, XI-195 p.

Cf. n°s 5792, 6416

b. Libraries

293. BIESTER (Björn). Tagebuch der Kulturwissenschaftlichen Bibliothek Warburg 1926–1929: Annotiertes Sach-, Begriffs- und Ortsregister. Erlangen, Filos, 2005, 96 p.

294. CAPRISTO (Annalisa). Un caso di "bonifica" libraria antisemita all'Accademia d'Italia. *Quaderni di storia*, 2005, 61, p. 201-220.

295. DANZI (Massimo). La biblioteca del cardinal Pietro Bembo. Genève, Droz, 2005, 470 p. (Travaux d'humanisme et Renaissance, 399).

296. MATTONE (Antonello). Biblioteche ed editoria universitaria nell'Italia medievale. *Studi storici*, 2005, 46, 4, p. 877-922.

297. MÍŠKOVÁ (Alena). "Politische Säuberungen" der Bestände der wissenschaftlichen Bibliotheken der tschechoslowakischen Akademie der Wissenschaften in den 1950er Jahren. *In*: Propaganda, (Selbst-)Zensur, Sensation. Grenzen von Presse- und Wissenschaftsfreiheit in Deutschland und Tschechien seit 1871 [Cf. n° 5297], p. 235-243.

298. SCHÄFER (Hans-Michael). Die Kulturwissenschaftliche Bibliothek Warburg: Geschichte und Persönlichkeiten der Bibliothek Warburg mit Berücksichtigung der Bibliothekslandschaft und der Stadtsituation der Freien und Hansestadt Hamburg zu Beginn des 20. Jahrhunderts. Berlin, Logos, 2005, XIV-413 p. (ill., ports.). (Berliner Arbeiten zur Bibliothekswissenschaft, 11).

299. SCHOCHOW (Werner). Die Berliner Staatsbibliothek und ihr Umfeld: 20 Kapitel preussisch-deutscher Bibliotheksgeschichte. Mit einem Geleitwort von Peter VODOSEK. Frankfurt am Main, V. Klostermann, 2005, 384 p. (ill.). (Zeitschrift für Bibliothekswesen und Bibliographie, 87. Sonderheft).

300. WEIS (Hélène). Les bibliothèques pour enfants entre 1945 et 1975: modèles et modélisation d'une culture pour l'enfance. Paris, Ed. du Cercle de la Librairie, 2005, 267 p.

Cf. n°s 55-101

c. Museums

* 301. STARN (Randolph). A historian's brief guide to new museum studies. *American historical review*, 2005, 110, 1, p. 68-98.

302. ALONSO RUIZ (Begoña), CRUZ DE CARLOS (María), PEREDA (Felipe). Patronos y coleccionistas: los condestables de Castilla y el arte (siglos XV–XVII). Valladolid, Universidad de Valladolid, Secretariado de Publicaciones e Intercambio Editorial, 2005, 328 p. (ill.). (Serie Historia y sociedad / Universidad de Valladolid, 115).

303. BEIER-DE HAAN (Rosmarie). Erinnerte Geschichte, inszenierte Geschichte: Ausstellungen und Museen in der Zweiten Moderne. Frankfurt am Main, Suhrkamp, 2005, 350 p. (Edition zweite Moderne).

304. BOGDANOS (Matthew). The casualties of war: the truth about the Iraq Museum. *American journal of archaeology*, 2005, 109, p. 477-526.

305. DREESBACH (Anne). Gezähmte Wilde: die Zurschaustellung "exotischer" Menschen in Deutschland 1870–1940. Frankfurt am Main, Campus, 2005, 371 p. (ill.).

306. GÉAL (Pierre). La naissance des musées d'art en Espagne: XVIIIe–XIXe siècles. Madrid, Casa de Velázquez, 2005, XII-557 p. (ill.). (Bibliothèque de la Casa de Velázquez, 33).

307. HENARE (Amiria J. M.). Museums, anthropology and imperial exchange. New York, Cambridge U. P., 2005, XIX-323 p.

308. Histoire de l'art et musées: actes du colloque: Ecole du Louvre, Direction des musées de France, 27 et 28 novembre 2001. Publiés sous la direction de Dominique VIÉVILLE. Paris, Ecole du Louvre, 2005, 235 p. (ill.). (Rencontres de l'Ecole du Louvre, 18).

309. Kommunismus (Der) im Museum: Formen der Auseinandersetzung in Deutschland und Ostmitteleuropa. Herausgegeben von Volkhard KNIGGE und Ulrich MÄHLERT im Auftrag der Stiftung Ettersberg und der Stiftung zur Aufarbeitung der SED-Diktatur; Redaktion: Daniela RUGE. Köln, Böhlau, 2005, 311 p. (ill.). (Europäische Diktaturen und ihre Überwindung, 6).

310. LODY (Raul Giovanni da Motta). O negro no museu brasileiro: construindo identidades. Rio de Janeiro, Bertrand Brasil, 2005, 335 p. (ill.).

311. MENDELSON (Jordana). Documenting Spain: artists, exhibition culture, and the modern nation, 1929–1939. University Park, Pennsylvania State U. P., 2005, XXXV-272 p. (Refiguring modernism, 2).

312. NYS (Liesbet). Particular bezit in het museumstijdperk. Bezoek aan privé-verzamelingen in België, circa 1830–1914. *Revue belge de philologie et d'histoire*, 2005, 83, 1, p. 453-478.

313. PHILLIPS (Ruth B.). Re-placing objects: historical practices for the second museum age. *Canadian historical review*, 2005, 86, 1, p. 83-110.

314. PIMENTEL (Cristina). Sistema museológico Português 1833–1991: em direcção a um novo modelo teórico para o seu estudo. Lisboa, Fundação Calouste Gulbenkian, Fundação para a Ciência e a Tecnologia, Ministério da Ciência e do Ensino Superior, 2005, 295 p. (maps). (Textos universitários de ciências sociais e humanas).

315. POULOT (Dominique). Musée et muséographie. Paris, La Découverte, 2005, 122 p. – IDEM. Une histoire des musées de France, XVIIIe–XXe siècle. Paris, La Découverte, 2005, 197 p. (L'espace de l'histoire).

316. SCHIPPER (Friedrich T.). The protection and preservation of Iraq's archaeological heritage, Spring 1991–2003. *American journal of archaeology*, 2005, 109, p. 251-272.

317. TROILO (Simona). La patria e la memoria. Tutela e patrimonio culturale nell'Italia unita. Milano, Electa Mondadori, 2005, 261 p.

318. Villa Borghese: storia e gestione. A cura di Alberta CAMPITELLI. Milano, Skira, 2005, 311 p. (ill., maps).

319. WAHNICH (Sophie). Les musées d'histoire du XXe siècle en Europe. *Etudes*, 2005, 403, 1-2, p. 29-42.

§ 2. History of historiography.

a. General

* 320. Bollettino di storiografia. 2004–2005. [2003–2004. Cf. Bibl. 2004, n° 321.] Dir. Da Massimo MASTROGREGORI. Pisa e Roma, Istituti Editoriali e Poligrafici Internazionali, 2005, 78 p. (Storiografia, supplemento critico e bibliografico, 9).

* 321. CABRERA (Miguel A.). Developments in contemporary Spanish historiography: from social history to the new cultural history. *Journal of modern history*, 2005, 77, 4, p. 988-1023.

* 322. CHERNOBAEV (Anatolij A.). Istoriki Rossii XX veka: Biobibliograficheskij slovar'. (20th-century historians of Russia: A bio-bibliographical dictionary). Saratov. gos. sots.-ekon. un-t. Vol. 1. A-L. Vol. 2. M-Ja. Saratov, [s. n.], 2005, 2 vol., 576 p., 607 p.

* 323. Historische Hilfswissenschaften: Stand und Perspektiven der Forschung. Hrsg. v. Toni DIEDERICH und Joachim OEPEN. Köln, Böhlau, 2005, IX-188 p. (ill.).

* 324. HYMAN (Paula E.). Recent trends in european Jewish historiography. *Journal of modern history*, 2005, 77, 2, p. 345-356.

* 325. LIVI (Massimiliano). Gli "e-journal" storici: una panoramica internazionale. *Contemporanea*, 2005, 8, 4, p. 757-767.

* 326. Warfare and belligerence. Perspectives in First World War studies. Ed. by Pierre PURSEIGLE. Boston, Brill, 2005, XIV-418 p. (History of warfare, 30).

* 327. WINTER (Jay), PROST (Antoine). The Great War in history. Debates and controversies, 1914 to the present. New York, Cambridge U. P., 2005, VIII-250 p. (Studies in the social and cultural history of modern warfare).

---

328. ADAMOVSKY (Ezequiel). Euro-Orientalism and the making of the concept of Eastern Europe in France, 1810–1880. *Journal of modern history*, 2005, 77, 3, p. 591-628.

329. ADAMTHWAITE (Anthony). La recherche française et la réinvention de l'histoire diplomatique. *Revue d'histoire diplomatique*, 2005, 119, 4, p. 343-360.

330. Alia tempora, alii mores. Storici e storia in età postridentina. Atti del Convegno internazionale Torino 24–27 settembre 2003. A cura di Massimo FIRPO. Firenze, Olschki, 2005, 587 p.

331. ANGLEVIEL (Frédéric). La littérature coloniale à vocation historique et la Nouvelle- Calédonie, 1853–1945, ou comment une colonie de peuplement génère des écrits hagiographiques. *In*: Littérature et histoire coloniale. Sous la dir. de Jacques WEBER. Paris, Indes savantes, 2005, p 155-171.

332. Antike Historiographie (Die) und die Anfänge der christlichen Geschichtsschreibung. Hrsg. v. Eve-Marie BECKER. Berlin u. New York, De Gruyter, 2005, XIII-308 p.

333. BENTLEY (Michael). Modernizing England's past: English historiography in the age of modernism, 1870–1970. Cambridge, Cambridge U. P., 2005, VIII-245 p. (Wiles lectures).

334. BERENSON (Edward), GREEN (Nancy L.). Quand l'oncle Sam ausculte l'Hexagone: les historiens américains et l'histoire de France. *XXe siècle*, 2005, 88, p. 121-131.

335. BERGER (Stefan). A return to the national paradigm? National history writing in Germany, Italy, France, and Britain from 1945 to the present. *Journal of modern history*, 2005, 77, 3, p. 629-678.

336. BISPINCK (Henrik). Die Zukunft der DDR-Geschichte. Potentiale und Probleme zeithistorischer Forschung. *Vierteljahrshefte für Zeitgeschichte*, 2005, 53, 4, p. 547-570.

337. BOJARCHENKOV (Vladislav V.). Istoriki-federalisty: kontseptsija mestnoj istorii v russkoj mysli 20–70-kh godov XIX v. (Historians-federalists: The conception of local history in the Russian thought of the 1820s–1870s). Sankt-Peterburg, Dmitrij Bulanin, 2005, 254 p. (bibl. incl.).

338. BOLKHOVITINOV (Nikolaj N.). Russkie uchenye-emigranty (G.V. Vernadskij, M.M. Karpovich, M.T. Florinskij) i stanovlenie rusistiki v SShA. (The Russian emigrant scholars [Georgy Vernadski, Mikhail Karpovich, Mikhail Florinsky] and the emergence of the Russian studies in the USA). Moskva, ROSSPEN, 2005, 141 p. (ill.; bibl. p. 132-140).

339. BONNET (Corinne). Carthage, l'"autre nation" dans l'historiographie ancienne et moderne. *Anabases*, 2005, 1, p. 139-160.

340. BURNS (Kathryn). Notaries, truth, and consequences. *American historical review*, 2005, 110, 2, p. 350-379.

341. CARR (Graham). Rules of engagement: public history and the drama of legitimation. *Canadian historical review*, 2005, 86, 2, p. 317-354.

342. CEAMANOS LLORENS (Roberto). Militancia y universidad. La construcción de la historia obrera en Francia. Valencia, Centro Francisco Tomas y Valiente y Instituto de Historia Social, 2005, 345 p.

343. CHATTERJEE (Kumkum). The king of controversy: history and nation-making in late colonial India. *American historical review*, 2005, 110, 5, p. 1454-1475.

344. CHIRKOV (Sergej V.). Arkheografija v tvorchestve russkikh uchenykh kontsa XIX–nachala XX veka. (Archeography [collecting, describing and publishing of historical sources] in the activity of Russian scholars of the late 19th and the early 20th century). Ed. Sigurd O. SHMIDT [Smidt]. Moskva, Znak, 2005, 317 p. (bibl. incl.; pers. ind. p. 310-317). (Studia historica).

345. Companion to women's historical writing. Ed. by Mary SPONGBERG, Ann CURTHOYS and Barbara CAINE. Basingstoke a. New York, Palgrave Macmillan, 2005, XVII-712 p.

346. Cultura (La) storica nei primi due secoli dell'impero romano (Milano, 3–5 giugno 2004). A cura di Lucio TROIANI e Giuseppe ZECCHINI. Roma, 'L'Erma' di Bretschneider, 2005, 310 p. (Monografie centro ricerche e documentazione sull'antichità classica, 24).

347. Culture post-coloniale, 1961–2006: traces et mémoires coloniales en France. Sous la dir. de Pascal BLANCHARD et Nicolas BANCEL en collaboration avec Sandrine LEMAIRE. Paris, Autrement, 2005, 287 p. (Collection Mémoires, 126).

348. DABDAB TRABULSI (José Antonio). Marxisme et histoire grecque ancienne en France. Flirts, engagements, influences. *Quaderni di storia*, 2005, 62, p. 63-88.

349. DANILOV (Viktor N.). Vlast' i formirovanie istoricheskogo soznanija sovetskogo obshchestva. (The political power and the making of popular notion of history in the Soviet State). INOTsENTR. Saratov, Nauchnaja kniga, 2005, 186 p. (bibl. p. 184-186). (Mezhregional'nye issledovanija v obshchestvennykh naukakh; 6).

350. DAVIS (Eric). Memories of state: politics, history, and collective identity in modern Iraq. Berkeley, University of California Press, 2005, XIII-385 p. (ill.).

351. DAWSON (Graham). Trauma, place and the politics of memory: Bloody Sunday, Derry, 1972–2004. *History workshop*, 2005, 59, p. 151-178.

352. DE GIORGI (Laura), SAMARANI (Guido). La Cina e la storia. Dal tardo Impero a oggi. Roma, Carocci, 2005, 138 p.

353. DE PALMA (Luigi Michele). Chiesa e ricerca storica. Vita e attività del Pontificio Comitato di scienze storiche (1954–1989). Roma, Libreria Editrice Vaticana, 2005, 400 p.

354. DETIENNE (Marcel). Histoire, mythologie, identité nationale. Un exercice comparatiste. *Quaderni di storia*, 2005, 61, p. 5-24.

355. Diktatur – Krieg – Vertreibung: Erinnerungskulturen in Tschechien, der Slowakei und Deutschland seit 1945. Für die Deutsch-Tschechische und die Deutsch-Slowakische Historikerkommission, herausgegeben von Christoph CORNELISSEN, Roman HOLEC und Jiří PEŠEK. Essen, Klartext, 2005, 500 p. (Veröffentlichungen der Deutsch-Tschechischen und Deutsch-Slowakischen Historikerkommission, 13. Veröffentlichungen zur Kultur und Geschichte im östlichen Europa, 26). [Cf. n° <Auswahl> 5271.]

356. Dittature, opposizioni, resistenze. Italia fascista, Germania nazionalsocialista, Spagna franchista: storiografie a confronto. A cura di Lutz KLINKHAMMER, Claudio NATOLI e Leonardo RAPONE. Milano, Unicopli, 2005, 340 p.

357. DOUZOU (Laurent). La Résistance française: une histoire périlleuse: essai d'historiographie. Paris, Éd. du Seuil, 2005, 365 p. (Histoire en débats Points. Histoire, H348).

358. DUBROVSKIJ (Aleksandr M.). Istorik i vlast': Istoricheskaja nauka i SSSR i kontseptsija istorii feodal'noj Rossii v kontekste politiki i ideologii (1930–1950-e gg.). (Historian and political power: Historical science in the USSR and the conception of feudalism in Russia in the context of politics and ideology, the 1930s–1950s). Brjansk, Izd-vo Brjansk. gos. un-ta, 2005, 798 p. (ill.; portr.; pers. ind. p. 789-799).

359. Ecritures de l'histoire (XIVe–XVIe siècle): actes du colloque du Centre Montaigne, Bordeaux, 19–21 septembre 2002. Réunis et édités par Danièle BOHLER et Chatherine MAGNIEN SIMONIN. Genève, Droz, 2005, 565 p. (ill.). (Travaux d'humanisme et Renaissance, 406).

360. ERDMANN (Karl Dietrich). Toward a global community of historians: the International Historical Congresses and the International Committee of Historical Sciences, 1898–2000. Ed. by Jürgen KOCKA and Wolfgang J. MOMMSEN in collaboration with Agnes BLÄNSDORF. New York: Berghahn Books, 2005, XVI-430 p.

361. Erinnerung, Gedächtnis, Wissen: Studien zur kulturwissenschaftlichen Gedächtnisforschung. Hrsg. v. Günter OESTERLE. Göttingen, Vandenhoeck & Ruprecht, 2005, 685 p. (ill.). (Formen der Erinnerung, 26).

362. ERTL (Thomas). Der China-Spiegel. Gedanken zu Chinas Funktionen in der deutschen Mittelalterforschung des 20. Jahrhunderts. *Historische Zeitschrift*, 2005, 280, 2, p. 305-344.

363. ESPAGNE (Michel). La philhellénisme entre philologie et politique: un transfert franco-allemand. *Revue germanique internationale*, 2005, 1-2, p. 61-75.

364. FERES (João). A história do conceito de "Latin America" nos Estados Unidos. Bauru, EDUSC e ANPOCS, 2005, 317 p. (ill.). (Ciências sociais).

365. FERRETTI (Maria). Mémoires divisées. Résistance et guerre aux civils en Italie. *Annales*, 2005, 60, 3, p. 627-651.

366. FERRO (Marc). Les individus face aux crises du XXe siècle: l'histoire anonyme. Paris, Odile Jacob, 2005, 431 p.

367. FEUCHTWANG (Stephan). Mythical moments in national and other family histories. *History workshop*, 2005, 59, p. 179-193.

368. FOCARDI (Filippo). La guerra della memoria: la Resistenza nel dibattito politico italiano dal 1945 a oggi. Roma e Bari, Laterza, 2005, VII-363 p. (Storia e società).

369. FREI (Norbert). 1945 und wir: das Dritte Reich im Bewusstsein der Deutschen. München, C.H. Beck, 2005, 224 p.

370. GARCÍA (Gervasio Luis). La nación antillana: ¿historia o ficción? *Historia social*, 2005, 52, p. 59-72.

371. German scholars and ethnic cleansing, 1919–1945. Ed. by Ingo HAAR and Michael FAHLBUSCH. New York, Berghahn Books, 2005, XXI-298 p.

372. Geschichte für Leser. Populäre Geschichtsschreibung in Deutschland im 20. Jahrhundert. Hrsg. v. Wolfgang HARDTWIG und Erhard SCHÜTZ. Stuttgart, Franz Steiner, 2005, 408 p. (Stiftung Bundespräsident-Theodor-Heuss-Haus, 7).

373. GRANDIN (Greg). The instruction of great catastrophe: truth commissions, national history, and state formation in Argentina, Chile, and Guatemala. *American historical review*, 2005, 110, 1, p. 46-67.

374. GROßE KRACHT (Klaus). Die zankende Zunft. Historische Kontroversen in Deutschland nach 1945. Göttingen, Vandenhoeck & Ruprecht, 2005, 224 p.

375. GUENÉE (Bernard). Ego, je: l'affirmation de soi par les historiens français (XIVe–XVe s.). *Comptes-rendus des séances de l'Académie des Inscriptions et Belles-Lettres*, 2005, 2, p. 597-611.

376. GÜNEŞ (Ahmet). Tarih, Tarihçi ve Meşruiyet. (Histoire, historien et légitimité). *OTAM Ankara Üniversitesi Osmanlı Tarihi Araştırma Ve Uygulama Merkezi Dergisi*, 2005, 17, p. 131- 202.

377. HAUMANN (Heiko). "Wir waren alle ein klein wenig antisemitisch". Ein Versuch über historische Massstäbe zur Beurteilung von Judengegnerschaft an den Beispielen Karl von Rotteck und Jacob Burckhardt. *Schweizerische Zeitschrift für Geschichte*, 2005, 55, 2, p. 196-214.

378. HAZAREESINGH (Sudhir). Napoleonic memory in nineteenth century France. The making of a liberal legend. *Modern language notes*, 2005, 120, 4, p. 747-773.

379. Historia: empiricism and erudition in Early Modern Europe. Ed. by Gianna POMATA and Nancy G. SIRAISI. Cambridge, MIT Press, 2005, VIII-490 p. (Transformations: studies in the history of science and technology).

380. Historical truth, historical criticism, and ideology: Chinese historiography and historical culture from a new comparative perspective. Ed. by Helwig SCHMIDT-GLINTZER, Achim MITTAG and Jörn RÜSEN. Leiden a. Boston, Brill, 2005, XXIII-489 p. (Leiden series in comparative historiography, 1).

381. HOLSINGER (Bruce). The premodern condition. Medievalism and the making of theory. Chicago a. London, University of Chicago Press, 2005, XII-276 p.

382. IGGERS (Georg G.). Historiography in the twentieth century: from scientific objectivity to the postmodern challenge; with a new epilogue by the author. Middletown, Wesleyan U. P., 2005, X-198 p.

383. Intellectual history in a global age. *Journal of the history of ideas*, 2005, 66, 2, p. 143-200. [Contents: SCHNEIDER (Ulrich Johannes). The International Dictionary of Intellectual Historians. – KELLEY (Donald R.). Intellectual history in a global age. – SCHNEEWIND (Jerome B.). Globalization and the history of philosophy. – MEGILL (Allan). Globalization and the history of ideas. – LEVINE (Joseph M.). Intellectual history as history].

384. IONCIOAIA (Florea). Liberalismul sălbatic. Note metodologice asupra istoriografiei liberalismului românesc. (Wild liberalism. Methodological notes on the historiography of the Romanian liberalism). *Xenopoliana. Buletinul Fundaţiei Academice "A.D. Xenopol"*, 2005, 13, 1-4, p. 20-46.

385. KNAPP (Fritz Peter). Historie und Fiktion in der mittelalterlichen Gattungspoetik. Bd. 2. Zehn neue Studien und ein Vorwort. Heidelberg, Universitätsverlag Winter, 2005, 286 p. (Schriften der Philosophisch-Historischen Klasse der Heidelberger Akademie der Wissenschaften, 35).

386. Kolonialismus und Erinnerungskultur: die Kolonialvergangenheit im kollektiven Gedächtnis der deutschen und niederländischen Einwanderungsgesellschaft. Hrsg. v. Helma LUTZ und Kathrin GAWARECKI. Münster, Waxmann, 2005, 205 p. (ill.). (Niederlande-Studien, 40).

387. KOPEČEK (Michal), KUNŠTÁT (Miroslav). Die so gennante sudetendeutsche Frage im tschechischen akademischen Diskurs nach 1989. *In*: Zwischen Konflikt und Annäherung. Ed. Elizabeth REIF [et al.]. Wien, 2005, p. 84-114.

388. LACH (Jiří). The idea of international cooperation in historical sciences since the World War One. *Moderní dějiny*, 2005, 13, p. 5-27.

389. LAPTEVA (Ljudmila P.). Istorija slavjanovedenija v Rossii v XIX veke. (A history of Slavonic studies in Russia in the 19th century). Moskva, Indrik, 2005, 847 p. (bibl. incl.; pers. ind. p. 833-847).

390. LÉVY-DUMOULIN (Olivier). L'enquête collective en lieu d'identité. Les enquêtes du Centre de recherches historiques [Paris]. *Cahiers du Centre de recherches historiques*, 2005, 36, p. 209-222.

391. Lieux (Les) de l'histoire. Sous la direction de Christian AMALVI, Laurent AVEZOU [et al.]. Paris, Armand Colin, 2005, 410 p.

392. Macht und Memoria. Begräbniskultur europäischer Oberschichten in der Frühen Neuzeit. Hrsg. v. Mark HENGERER. Köln, Weimar u. Wien, Böhlau, 2005, IX-525 p.

393. MARÍN GELABERT (Miquel Ángel). Los historiadores españoles en el franquismo, 1948-1975: la historia local al servicio de la patria. Zaragoza, Prensas Universitarias de Zaragoza, Institución Fernando el Católico, Diputación de Zaragoza, 2005, 395 p. (Institución "Fernando el Católico". Publicaciones Institución "Fernando el Católico". Colección Estudios: Historia).

394. MEZZADRA (Sandro). Tempo storico e semantica politica nella critica postcoloniale. *Storica*, 2005, 31, p. 143-162.

395. MOYN (Samuel). A Holocaust controversy. The Treblinka affair in postwar France. Waltham, Brandeis U. P., 2005, XXII-220 p. (Tauber institute for the study of European Jewry series).

396. NAFISSI (Mohammad). Ancient Athens and modern ideology. Value, theory and evidence in historical sciences. Max Weber, Karl Polanyi and Moses Finley. London, Institute of classical studies, School of advanced study, University of London, 2005, XI-325 p.

397. NAGEL (Anne Christine). Im Schatten des Dritten Reichs. Mittelalterforschung in der Bundesrepublik Deutschland 1945–1970. Göttingen, Vandenhoeck & Ruprecht, 2005, 336 p. (Formen der Erinnerung, 24).

398. NEDERMAN (Cary J.). Empire and the historiography of European political thought: Marsiglio of Padua, Nicholas of Cusa, and the Medieval/Modern divide. *Journal of the history of ideas*, 2005, 66, 1, p. 1-15.

399. Neue Politikgeschichte. Perspektiven einer historischen Politikforschung. Hrsg. v. Ute FREVERT und Heinz-Gerhard HAUPT. Frankfurt am Main, Campus, 2005, 315 p. (Historische Politikforschung, 1).

400. NG (On-Cho), WANG (Q. Edward). Mirroring the past. The writing and use of history in imperial China. Honolulu, University of Hawai'i Press, 2005, XXIII-306 p.

401. Novyj obraz istoricheskoj nauki v vek globalizatsii i informatizatsii. (New image of history in the age of globalization and informatization: [Articles]). Ed. Lorina P. REPINA. RAN, In-t vseobshchej istorii. Moskva, IVI RAN, 2005, 287 p. (bibl. incl.).

402. Nuova storiografia (La) in Africa Sub-sahariana dall'indipendenza alla "transizione alla democrazia": il caso del Mozambico. A cura di Anna Maria GENTILI. Intervengono David HEDGES, Teresa CRUZ E SILVA, Arlindo Gonçalo CHILUNDO, João Paulo BORGES COELHO, Manuel Garrido MENDES DE ARAUJO, Isabel CASIMIRO e Joel DAS NEVES TEMBE. *Contemporanea*, 2005, 8, 3, p. 497-521.

403. Orientalism and the Jews. Ed. by Ivan Davidson KALMAR and Derek J. PENSLAR. Waltham, Brandeis U. P., 2005, XL-285 p.

404. PÁNEK (Jaroslav), RAKOVÁ (Svatava), HORČÁKOVÁ (Václava). Scholars of Bohemian, Czech and Czechoslovak history studies. T. 1-3. Praha, Institute of History, 2005, 3 vol., 445 p., 468 p., 390 p.

405. PÁNEK (Jaroslav). Rozmluvy s historiky. Česká historiografie a soudobé dějiny očima zahraničních kolegů. (Conversations with historians. Czech historiography and contemporary history as seen through the eyes of our foreign colleagues). Ed. Jiří KOCIAN. Praha, Sdružení historiků ČR, 2005, 184 p. (photogr.).

406. PAPAILIAS (Penelope). Genres of recollection: archival poetics and modern Greece. New York, Palgrave MacMillan, 2005, XV-301 p. (Anthropology, history, and the critical imagination).

407. PAYEN (Pascal). Réception des historiens anciens et fabrique de l'Histoire. *Anabases*, 2005, 1, p. 221-230.

408. PÓK (Attila). História v transformačnom procese Maďarska. (History as part of the transformation process in Hungary. Geschichte im Transformierungsprozess Ungarns). *Historický časopis*, 2005, 53, 1, p. 111-122.

409. Politique (La) de la mémoire coloniale en Allemagne et au Cameroun. Sous la dir. de Stefanie MICHELS et Albert-Pascal TEMGOUA. Münster, Lit Verlag, 2005, 154 p.

410. POPKIN (Jeremy D.). History, historians and autobiography. Chicago a. London, University of Chicago Press, 2005, 339 p.

411. Prisoners of war, prisoners of peace: captivity, homecoming, and memory in World War II. Ed. by Bob MOORE and Barbara HATELY-BROAD. Introduction by Pieter LAGROU. Oxford, Berg, 2005, XVIII-270 p.

412. RAPHAEL (Lutz). Lieux et idées pour la recherche collective en histoire, 1949–1975. Les premières

décennies du Centre de recherches historiques [Paris] et les pratiques ouest-allemandes. *Cahiers du Centre de recherches historiques*, 2005, 36, p. 107-120.

413. READMAN (Paul). The place of the past in English culture, c. 1890–1914. *Past & Present*, 2005, 186, p. 147-199.

414. Rekishigaku jiten 12: O to kokka. (Kings and States). Ed. Hideo KURODA. Tokyo, Kobundo, 2005, 814 p. (Encyclopedia of historiography. 12).

415. Remapping knowledge. The making of South Asian studies in India, Europe and America (19th–20th centuries). Ed. by Jackie ASSAYAG and Véronique BÉNÉÏ. Gurgaon, Three Essays Collective, 2005, 135 p.

416. RIBÉMONT (Thomas). Les historiens chartistes au cœur de l'affaire Dreyfus. *Raisons politiques*, 2005, 18, p. 97-116.

417. RICUPERATI (Giuseppe). L' età moderna come periodizzazione e le categorie di periodizzazione dell' età moderna. *Rivista storica italiana*, 2005, 117, 2, p. 569-607. – IDEM. Mnemosyne e Anamnesis: discipline della memoria e conoscenza storica fra passato e futuro. *Rivista storica italiana*, 2005, 117, 1, p. 229-282.

418. ROEMER (Nils H.). Jewish scholarship and culture in nineteenth-century Germany: between history and faith. Madison, University of Wisconsin Press, 2005, X-251 p. (Studies in German Jewish cultural history and literature).

419. SARKAR (Sumit). Une ou plusieurs histoires? Formations identitaires au Bengale à la fin de l'époque coloniale. *Annales*, 2005, 60, 2, p. 293-328.

420. SCHMITZ (Michael). Ein deutsches Denkmal. Das Holocaust-Denkmal im Brennpunkt deutscher Erinnerungspolitik. *Archiv für Kulturgeschichte*, 2005, 87, 1, p. 165-193.

421. SFEIR-KHAYAT (Jihane). Historiographie palestinienne. La construction d'une identité nationale. *Annales*, 2005, 60, 1, p. 35-52.

422. SHMIDT (Sigurd O.). "Fenomen Fomenko" v kontekste izuchenija sovremennogo obshchestvennogo istoricheskogo soznanija. ('The phenomenon of Anatoly Fomenko' in the context of present popular notion of history). RAN, Arkheograficheskaja komissija. Moskva, Nauka, 2005, 73 p. (bibl. p. 67-70).

423. SIDOROVICH (Ol'ga V.). Annalisty i antikvary: Rimskaja istoriografija kontsa III–I v. do n.e. (Annalists and antiquarians: The Roman historiography from the late 3rd to the 1st century B.C.). Moskva, RGGU, 2005, 289 p. (bibl. incl.; ind. p. 280-289).

424. SIGNOLES (Aude). Les représentations du passé en Palestine. Municipalités d'hier, municipalités d'aujourd'hui. *Annales*, 2005, 60, 1, p. 109-126.

425. SIRINELLI (Jean-François). Réflexions sur l'histoire et l'historiographie du XXe siècle français. *Revue historique*, 2005, 635, p. 609-626.

426. SMAIL (Dan). In the grip of sacred history. *American historical review*, 2005, 110, 5, p. 1337-1361.

427. SOLL (Jacob). Publishing the prince. History, reading, and the birth of political criticism. Ann Arbor, University of Michigan Press, 2005, XII-2002 p.

428. Sozial- und Wirtschaftsgeschichte, Arbeitsgebiete – Probleme – Perspektiven. Hrsg. V. Günther SCHULZ [et al.]. Stuttgart, Franz Steiner Verlag, 2005, 661 p.

429. STENHOUSE (William). Reading inscriptions and writing ancient history: historical scholarship in the late Renaissance. London, Institute of Classical Studies, School of Advanced Study, University of London, 2005, X-203 p. (Bulletin of the Institute of Classical Studies. Supplement, 86).

430. STORA (Benjamin). Quand une mémoire (de guerre) peut en cacher une autre (coloniale). *In*: Fracture coloniale (La): la société française au prisme de l'héritage colonial [Cf. n° 6717], p. 57-65.

431. Storia Orale. A cura di Alessandro PORTELLI. *Quaderni storici*, 2005, 3, p. 653-752. [Contiene: GINZBURG (Carlo). Memoria e globalizzazione (p. 657-670). – MUKTA (Parita). Il logoramento della memoria: etica, moralità e futuro (p. 671-684). – THOMSON (Alistair). Le storie di vita nello studio dell'emigrazione femminile (p. 685-708). – ALLAN (Diana). Ricordare e dimenticare il 1948. La politica del ricordo fra i rifugiati palestinesi nel campo di Shatila (p. 709-734). – PORTELLI (Alessandro). Memoria e globalizzazione: la lotta contro la chiusura degli acciai speciali a Terni, 2004–2005 (p. 735-752)].

432. Storiografia, cultura storica e circolazione del sapere nell'Italia fascista. A cura di Margherita ANGELINI e Mirco CARRATTIERI. *Storiografia*, 2005, 9, p. 99-266. [Contiene: ANGELINI (Margherita). I corsi universitari di storia tra oralità e scrittura. – MIGANI (Elena). I manuali di storia contemporanea per i licei. – GUIDI (Elisa). Le recensioni della Storia d'Italia e della Storia d'Europa di Benedetto Croce. – PEDIO (Alessia). Le collane editoriali di storia, da "I prefascisti" a "I grandi italiani". – CARRATTIERI (Mirco). Le riviste di divulgazione storica e l'esperienza di "Popoli" (1941–1942)].

433. TEKLAK (Czeslaw). Le ricerche marxiste su Gesù di Nazareth. Roma, Antonianum, 2005, 552 p. (Spicilegium Pontificii athenaei Antoniani, 39).

434. Territoires (Les) du médiéviste. Sous la dir. de Benoît CURSENTE et Mireille MOUSNIER. Rennes, Presses universitaires de Rennes, 2005, 459 p. (Histoire).

435. TOLSTOGUZOV (Aleksandr A.). Japonskaja istoricheskaja nauka: Ocherki istorii: problemy izuchenija srednikh vekov i feodalizma. (Historical science in Japan: Essays on its history. The study of the Middle Ages and the feudalism). Moskva, Vostochnaja literatura, 2005, 566 p. (bibl. p. 542-560).

436. "Tsep' vremen": Problemy istoricheskogo soznanija. (The chain of times: [Articles dedicated to the

memory of Mikhail A. Barg]). Ed. Lorina P. REPINA. RAN, ln-t vseobshchej istorii. Moskva, IVI RAN, 2005, 255 p. (ill.; bibl. incl.).

437. TYRRELL (Ian). Historians in public: the practice of American history, 1890–1970. Chicago, University of Chicago Press, 2005, XII-348 p.

438. VERGINELLA (Marta). La contribución historiográfica a las prácticas de negociación de la frontera italo-eslovena. *Saitabi (Valencia)*, 2005, 55, p. 247-263.

439. Was heisst Kulturgeschichte des Politischen? Hrsg. v. Barbara STOLLBERG-RILINGER. Berlin, Duncker & Humblot, 2005, 376 p. (Zeitschrift für historische Forschung, 35 Beiheft).

440. WÓYCICKA (Zofia). Zur Internationalität der Gedenkkultur. Die Gedenkstätte Auschwitz-Birkenau im Spannungsfeld zwischen Ost und West 1954–1978. *Archiv für Sozialgeschichte*, 2005, 45, p. 269-293.

441. WRIGHT (Donald A.). The professionalization of history in English Canada. Toronto a. London, University of Toronto Press, 2005, X-270 p.

442. ZHANG (Qing). Zhong Xi lishi de "huitong" yu Zhongguo shixue de zhuanxiang. (The introduction of Western historiography and the change of Chinese historiography). *Lishi yanjiu*, 2005, 2, p. 75-95.

a. Addenda 2001–2004

* 443. Historische Sozialforschung: Auswahlbibliographie 1975–2000 / Historical social research: selected bibliography 1975–2000. Bearbeitet von / prepared by Thomas RAHLF, Cornelia BADDACK, Karl PIERAU. *Historical social research / Historische Sozialforschung*, 2004, 29, 360 p. (HSR-Supplement-Heft 16).

* 444. SIARI-TENGOUR (Ouanassa), SOUFI (Fouad). Les Algériens écrivent, enfin, leur guerre. Bibliographie: mémoires, autobiographies, biographies et témoignages, 1962–2004. *Insaniyat. Revue algérienne d'anthropologie et de sciences sociales (Oran)*, 2004, 8, 25-26, p. 267-283.

---

445. Historiografía francesa (La) del siglo XX y su acogida en España, Madrid, Casa de Velázquez, 2002, 480 p.

446. NATTERMANN (Ruth). Deutsch-jüdische Geschichtsschreibung nach der Shoah. Die Gründungs- und Frühgeschichte des Leo Baeck Institute. Essen, Klartext, 2004, 320 p.

447. OSTERHAMMEL (Jürgen). Geschichtswissenschaft jenseits des Nationalstaats: Studien zu Beziehungsgeschichte und Zivilisationsvergleich. Göttingen, Vandenhoeck & Ruprecht, 2001, 384 p.

448. SCHLEIER (Hans). Geschichte der deutschen Kulturgeschichtsschreibung. Band 1. Vom Ende des 18. bis zum Ende des 19. Jahrhunderts. T. 1 u. 2. Waltrop, Spenner, 2003, 2 vol., X-1191 p. (Wissen und Kritik, 24/1).

Cf. n°s 102, 271, 317, 512, 570, 844, 845, 849, 850, 853, 854, 988, 1712, 2066, 2568, 2582, 2792, 2917, 3198, 4959, 5149, 7571

b. Special studies

449. MÜHLE (Eduard). Für Volk und deutschen Osten. Der Historiker Hermann **Aubin** und die deutsche Ostforschung. Düsseldorf, Droste, 2005, X-732 p. (Schriften des Bundesarvchivs, 65). – TRÜPER (Henning). Die Vierteljahresschrift für Sozial- und Wirtschaftsgeschichte und ihr Herausgeber Hermann **Aubin** im Nationalsozialismus. Stuttgart, Steiner, 2005, 167 p. (VSWG, 181).

450. REID (John G.). Viola Florence **Barnes**, 1885–1979. A historian's biography. Buffalo, University of Toronto Press, 2005, XXIV-228 p. (Studies in gender and history, 25).

451. RICUPERATI (Giuseppe). Francesco **Bianchini** e l'idea di storia universale 'figurata'. *Rivista storica italiana*, 2005, 117, 3, p. 872-973.

452. BURGUIÈRE (André). Archéologie du Centre de recherches historiques [Paris]; les enquêtes collectives de Marc **Bloch** et Lucien Febvre et leur postérité. *Cahiers du Centre de recherches historiques*, 2005, 36, p. 61-79. – MORES (Francesco). Marc **Bloch**, il Collège de France e le forme della comparazione storica. *Quaderni storici*, 2005, 2, p. 555-596. – STEELE (Stephen). Lettres d'un ancien combattant: Marc **Bloch** à Gustave Cohen, 1923–1925. *Lendemains*, 2005, 30, 119-120, p. 155-166.

453. LEMOINE (Yves). Fernand **Braudel**: espaces et temps de l'historien. Paris, Punctum, 2005, 165 p.

454. FELICI (Lucia). Alle origini degli Eretici italiani del Cinquecento. Nuovi documenti del carteggio Bainton-**Cantimori** (1932–1940). *Archivio storico italiano*, 2005, 163, 3, p. 531-594. – TEDESCHI (John). Delio **Cantimori**: "Per la storia degli eretici italiani". *Giornale critico della filosofia italiana*, 2005, 2, p. 236-268.

455. BONNET (Corinne). Le grand atelier de la science, Franz **Cumont** et l'Altertumswissenschaft. Héritages et émancipations. Des études universitaires à la fin de la première guerre mondiale, 1888–1923. Vol. XLI-1 et vol. XLII-2. Bruxelles, Belgisch historisch Instituut, 2005, 2 vol., 419 p., 293 p.

456. Bibliografia di don Giuseppe **De Luca**. A cura di Michele PICCHI e Donatella ROTUNDO. Roma, Edizioni di storia e letteratura, 2005, XVIII-407 p. (Sussidi eruditi, 66).

457. GALINIER (Martin). Mythe et images: Georges **Dumézil** au miroir de l'histoire de l'art. *Revue historique*, 2005, 633, p. 3-30.

458. Höfische Gesellschaft und Zivilisationsprozeß. Norbert **Elias**' Werk in kulturwissenschaftlicher Perspektive. Hrsg. v. Claudia OPITZ. Köln, Weimar u. Wien, Böhlau, 2005, 254 p.

459. PASSINI (Michela). Tra passione civile e arte medievale: Henri **Focillon** e la sua "France". *Intersezioni*, 2005, 25, 3, p. 453-478.

460. PARAS (Eric). A new archivist: Michel **Foucault** and the practice of history, 1968-1984. Harvard, [s. n.], 2005, [s. p.]. [Thesis (Ph. D.) – Harvard University]. – REVEL (Judith). Michel **Foucault** expériences de la pensée. Paris, Bordas, 2005, 252 p.

461. PALLARES-BURKE (Maria Lúcia G.). Gilberto **Freyre**: um vitoriano dos trópicos. São Paulo, Editora UNESP, 2005, 481 p. (facsims.).

462. MARCONE (Arnaldo). Il 'Prospetto storico sul mondo romano' di Giuseppe Maria **Galanti**. *Rivista storica italiana*, 2005, 117, 2, p. 529-542.

463. FLASH (Kurt). Eugenio **Garin** tra Medioevo e Rinascimento. *Giornale critico della filosofia italiana*, 2005, 1, p. 27-39.

464. FIRPO (Giulio). L' Italia romana nell'Istoria civile del Regno di Napoli di Pietro **Giannone**. *Rivista storica italiana*, 2005, 117, 2, p. 423-447.

465. SEHLMEYER (Markus), WALTER (Uwe). Unberührt von jedem Umbruch? Der Althistoriker Ernst **Hohl** zwischen Kaiserreich und früher DDR. Frankfurt am Main, Verlag Antike, 2005, 126 p.

466. SEED (John). The spectre of puritanism: forgetting the seventeenth century in David **Hume**'s history of England. *Social history*, 2005, 30, 4, p. 444-462.

467. PIPPIDI (Andrei). Minoritățile naționale ale României în concepția lui Nicolae **Iorga**. (National minorities living in Romania viewed by Nicolae Iorga). *Revista istorică*, 2005, 16, 3-4, p. 145-154.

468. GALASSO (Giuseppe). **Labriola** e la storia generale d'Italia. *Giornale critico della filosofia italiana*, 2005, 1, p. 49-59.

469. BORGHETTI (Maria Novella). L'oeuvre d'Ernest **Labrousse**: genèse d'un modèle d'histoire économique. Paris, Ecole des hautes études en sciences sociales, 2005, 299 p. (Recherches d'histoire et de sciences sociales = Studies in history and the social sciences, 106). – CULLEN (L.). **Labrousse**, the Annales School, and Histoire sans Frontières. *Journal of European economic history*, 2005, 34, 1, p. 309-350.

470. Émile **Mâle** (1862-1954): la construction de l'oeuvre: Rome et l'Italie. Roma, École française de Rome, 2005, 360 p. (ill.). (Collection de l'École française de Rome, 345).

471. Theodor **Mommsen**. Gelehrter, Politiker und Literat. Hrsg. v. Josef WIESEHÖFER; unter Mitarbeit von Henning BÖRM. Stuttgart, Steiner, 2005, 259 p. – Theodor **Mommsen**. Wissenschaft und Politik im 19. Jahrhundert. Hrsg. v. Alexander DEMANDT, Andreas GOLTZ und Heinrich SCHLANGE-SCHÖNINGEN. Berlin, W. de Gruyter, 2005, X-351 p.

472. HOFFMANN (Peter). Gerhard Friedrich **Müller**, 1705-1783. Historiker, Geograph, Archivar im Dienste Rußlands. Frankfurt am Main, Berlin, Bern, Bruxelles, New York, Oxford u. Wien, Lang, 2005, 393 p.

473. CESERANI (Giovanna). Narrative, interpretation, and plagiarism in Mr. **Robertson**'s 1778 History of Ancient Greece. *Journal of the history of ideas*, 2005, 66, 3, p. 413-436.

474. GOTOR (Miguel). Ricordo di Ruggiero **Romano**: lo storico dei meccanismi e degli uomini al plurale. *Storiografia*, 2005, 9, p. 87-96.

475. ECKEL (Jan). Hans **Rothfels**. Eine intellektuelle Biographie im 20. Jahrhundert. Göttingen, Wallstein, 2005, 479 p. (Moderne Zeit, 10).

476. TURI (Gabriele). Luigi **Salvatorelli**, un intellettuale attraverso il fascismo. *Passato e presente*, 2005, 66, p. 89-109.

477. LIERMANN (Christiane). Porträt des Historikers Ernesto **Sestan** (1898-1986). *Archiv für Kulturgeschichte*, 2005, 87, 1, p. 149-164.

478. SCHMAL (Stephan). **Tacitus**. Hildesheim, Olms, 2005, 240 p. (Studienbücher Antike, 14).

479. TOLOCHKO (Aleksej P.). "Istorija Rossijskaja" Vasilija Tatishcheva: Istochniki i izvestija. (The Russian History by Vasily N. **Tatishchev** [1688-1750]: Sources and evidence). Moskva, Novoe literaturnoe obozrenie i Kiev, Kritika, 2005, 543 p. (bibl. incl.; ind. p. 525-541).

480. DE ROMILLY (Jacqueline). L'invention de l'histoire politique chez **Thucydide**. Paris, Editions Rue d'Ulm, Presses de l'École normale supérieure, 2005, 272 p. – ZAGORIN (Perez). **Thucydides**: an introduction for the common reader. Oxford, Princeton U. P., 2005, XIII-190 p.

481. COLDAGELLI (Umberto). Vita di **Tocqueville** (1805-1859). La democrazia tra storia e politica. Roma, Donzelli, 2005, 340 p. – VUILLEUMIER (Marc). La Suisse de 1848: l'analyse de **Tocqueville**. *Schweizerische Zeitschrift für Geschichte*, 2005, 55, 2, p. 149-174.

482. Salvatore Camporeale, Lorenzo **Valla**, humanism, and theology. *Journal of the history of ideas*, 2005, 66, 4, p. 477-556. [Contents: BULLARD (Melissa Meriam). The Renaissance project of knowing: Lorenzo Valla and Salvatore Camporeale's contributions to the querelle between rhetoric and philosophy. – CELENZA (Christopher S.). Lorenzo Valla and the traditions and transmissions of philosophy. – COPENHAVER (Brian P.). Valla our contemporary: philosophy and philology. – REGOLIOSI (Mariangela). Salvatore Camporeale's contribution to theology and the history of the church. – STRUEVER (Nancy S.). Historical priorities].

483. PASTA (Renato). Franco **Venturi** e le antiche repubbliche italiane. *Passato e presente*, 2005, 65, p. 85-107.

484. MORETTI (Mauro). Pasquale **Villari** storico e politico. Con una nota di Fulvio TESSITORE. Napoli, Liguori, 2005, XII-308 p. (La cultura storica: collana di testi e studi, 23).

485. VOLTAIRE (François Marie Arouet). Le siècle de Louis XIV. Ed. établie, présentée et annotée par Jacqueline HELLEGOUARC'H et Sylvain MENANT; avec la collab. de Philippe BONNICHON et Anne-Sophie BARROVECCHIO. Paris, Librairie générale française, 2005, 1213 p.

486. Asketischer Protestantismus und "Geist" des modernen Kapitalismus. Max **Weber** und Ernst Troeltsch. Hrsg. v. Wolfgang SCHLUCHTER und Friedrich Wilhelm GRAF. Tübingen, Mohr Siebeck, 2005, VIII-311 p. – CAPOGROSSI COLOGNESI (Luigi). Max **Weber** und die Wirtschaft der Antike. Göttingen, Vandenhoeck & Ruprecht, 2005, 432 p. (Abhandlungen der Akademie der Wissenschaft in Göttingen, philologisch-historische Klasse 3, 259). – Max **Weber** Gesamtausgabe. 1. Schriften und Reden. Bd. 22. Wirtschaft und Gesellschaft. Die Wirtschaft und die gesellschaftlichen Ordnungen und Mächte. Teilbd. 4. Herrschaft. Hrsg. v. Edith HANKE in Zusammenarbeit mit Thomas KROLL. Tübingen, Mohr Siebeck, 2005, XXX-943 p. – Max **Weber**'s Economy and society: a critical companion. Ed. by Charles CAMIC, Philip S. GORSKI and David M. TRUBEK. Stanford, Stanford U. P., 2005, XII-403 p. – SCHELLER (Benjamin). Das herrschaftsfremde Charisma der Coniuratio und seine Veralltäglichungen. Idealtypische Entwicklungspfade der mittelalterlichen Stadtverfassung in Max **Webers** "Stadt". *Historische Zeitschrift*, 2005, 281, 2, p. 307-335.

487. BIESTER (Björn). Briefe und Postkarten von Ulrich von **Wilamowitz-Moellendorff** an Paul Wendland 1886 bis 1915. *Quaderni di storia*, 2005, 61, p. 137-200.

b. Addenda 1998–2004

488. BETTHAUSEN (Peter). Georg **Dehio**: ein deutscher Kunsthistoriker. München, Deutscher Kunstverlag, 2004, 464 p. (ill.).

489. Relire **Focillon**: cycle de conférences organisé au musée du Louvre par le Service culturel du 27 novembre au 18 décembre 1995 sous la direction de Matthias Waschek. Textes inédits d'Henri Focillon. Paris, Ecole nationale supérieure des beaux-arts, 98, 202 p. (ill.). (Louvre conférences et colloques. Collection Principes et théories de l'histoire de l'art). – Vie (La) des formes: Henri **Focillon** et les arts. Gand, Snoeck-Ducaju & Zoon, 2004, 314 p. (ill., ports.).

490. DE SENARCLENS (Vanessa). **Montesquieu** historien de Rome: un tournant pour la réflexion sur le statut de l'histoire au XVIIIe siècle. Genève, Droz, 2003, 292 p.

Cf. n^{os} *436, 479, 847, 848, 1110, 1539, 1555, 1634, 2520, 3859*

§ 3. Methodology, philosophy, and teaching of history.

491. AGULHON (Maurice). Histoire et politique à gauche: réflexions et témoignages. Paris, Perrin, 2005, 162 p.

492. ANKERSMIT (Frank). Sublime historical experience. Stanford, Stanford U. P., 2005, XVIII-481 p. (Cultural memory in the present).

493. Arkheologija i estestvennonauchnye metody. (Archaeology and scientific methods: [Proceedings of a conference, Moscow, 2004, November 1-3]). Ed. Evgenij N. CHERNYKH, Vladimir I. ZAV'JALOV. Moskva, Jazyki slavjanskoj kul'tury, 2005, 216 p. (ill.; bibl. incl.).

494. ARMSTRONG (Richard H.). A compulsion for antiquity. Freud and the ancient world. Ithaca a. London, Cornell U. P., 2005, XII-303 p.

495. BABEROWSKI (Jörg). Der Sinn der Geschichte: Geschichtstheorien von Hegel bis Foucault. München, Beck, 2005, 249 p. (Beck'sche Reihe; 1623).

496. BARKAN (Elazar). Engaging history: managing conflict and reconciliation. *History workshop*, 2005, 59, p. 229-236.

497. BERMEJO BARRERA (José Carlos). Introduction to the logic of comparison in mythology. *Quaderni di storia*, 2005, 62, p. 89-106. – IDEM. On history considered as epic poetry. *History and theory*, 2005, 44, 2, p. 182-194.

498. BURKE (Peter). Was ist Kulturgeschichte? Frankfurt am Main, Suhrkamp, 2005, 203 p.

499. BUSINO (Giovanni). Cos'è la prova nelle scienze umane? *Rivista storica italiana*, 2005, 117, 3, p. 1050-1064.

500. CADIOU (François), COULOMB (Clarisse), LEMONDE (Anne). Comment se fait l'histoire: pratiques et enjeux. Paris, La Découverte, 2005, 384 p.

501. CARVALHO (Maria Manuela). Poder e ensino: os manuais de história na política do Estado Novo, 1926–1940. Lisboa, Livros Horizonte, 2005, 175 p. (Biblioteca do educador, 153).

502. CROCE (Benedetto). Filosofia e storiografia. A cura di Stefano MASCHIETTI. Napoli, Bibliopolis, 2005, 421 p. (Edizione nazionale delle opere di Benedetto Croce. Saggi filosofici, 12).

503. CROWE (Benjamin D.). Dilthey's philosophy of religion in the "Critique of Historical Reason": 1880–1910. *Journal of the History of Ideas*, 2005, 66, 2, p. 265-283.

504. DE BAETS (Antoon). Argumenten voor en tegen een ethische code voor historici. (Arguments in favour and against a code of ethics for historians). *Tijdschrift voor Geschiedenis*, 2005, 118, 4, p. 564-571. – IDEM. Ethische codes als kompassen: een nawoord. (Codes of ethics as compasses: an afterword). *Tijdschrift voor Geschiedenis*, 2005, 118, 4, p. 581-582.

505. Erinnerung und Gesellschaft = Mémoire et société. Hommage à Maurice Halbwachs, 1877–1945. Ed. par Hermann KRAPOTH et Denis LABORDE. Wiesbaden, Vs Verlag für Sozialwissenschaften, 2005, 287 p. (Jahrbuch für Sozialgeschichte).

506. Esthétique du témoignage: actes du colloque tenu à la Maison de la recherche en sciences humaines de Caen du 18 au 21 mars 2004. Ouvrage coordonné par Carole DORNIER et Renaud DULONG. Paris, Maison des sciences de l'homme, 2005, XIX-388 p. (ill.).

507. FAZLIOĞLU (İhsan). Şehir Tarihi Çalışmalarında Yazma Eserlerden Nasıl İstifade Edilebilir? (Comment interroger les manuscrits pour les recherches sur l'histoire urbaine?). *Türkiye Araştırmaları Literatür Dergisi*, 2005, 3, 6, p. 487-515.

508. GAUTIER (Claude). Hume et les savoirs de l'histoire. Paris, Librairie philosophique Vrin et Éditions de l'EHESS, 2005, 301 p. (Contextes).

509. Generationen: zur Relevanz eines wissenschaftlichen Grundbegriffs. Hrsg. v. Ulrike JUREIT und Michael WILDT. Hamburg, Hamburger Ed., 2005, 354 p.

510. HARTOG (François). Evidence de l'histoire: ce que voient les historiens. Paris, Éd. de l'École des hautes Etudes en Sciences sociales, 2005, 285 p.

511. Histoire (L') entre mémoire et épistémologie: autour de Paul Ricoeur. Publié sous la direction de Bertrand MÜLLER; avec des contributions de Régine ROBIN [et al.]. Lausanne, Payot, 2005, 220 p. (Sciences humaines).

512. Istorija cherez lichnost': istoricheskaja biografija segodnja. (History through personality: Historical biography today: [Articles]). Ed. Lorina P. REPINA. Moskva, Krug', 2005, 719 p. (ill; bibl. incl.). [English summary]

513. KARAVASHKIN (Andrej Ju.), JURGANOV (Andrej L.). Region Doksa. Istochnikovedenie kultury. (The source-criticism of culture). Ros. gos. gumanit. un-t. Moskva, RGGU, 2005, 210 p. (plates; bibl. p. 174-210).

514. KŘEN (Jan). Dokumenty StB jako pramen poznání minulosti. (Secret police records as a source of knowing the past). *Soudobé dějiny*, 2005, 12, 3/4, p. 708-733.

515. LEMERCIER (Claire). Analyse de réseaux et histoire. *Revue d'histoire moderne et contemporaine*, 2005, 52, 2, p. 88-112.

516. MARKUS (Tomislav). Tumačenje (post)historije i ekološka kriza. (Interpreting (Post)history and the ecological crisis). *Časopis za suvremenu povijest*, 2005, 37, 1, p. 7-36.

517. Mémoires et histoire. Des identités personnelles aux politiques de Reconnaissance. Sous la dir. de Johann MICHEL. Rennes, Presses universitaires de Rennes, 2005, 284 p.

518. MOSES (A. Dirk). Hayden White, Traumatic Nationalism, and the Public Role of History. *History & Theory*, 2005, 44, 3, p. 311-332. – IDEM. The public relevance of historical studies: a rejoinder to Hayden White. *History & Theory*, 2005, 44, 3, p. 339-347. – WHITE (Hayden). The Public Relevance of Historical Studies: A Reply to Dirk Moses. *History & Theory*, 2005, 44, 3, p. 333-338.

519. MUSTÈ (Marcello). La storia: teoria e metodi. Roma, Carocci, 2005, 126 p. – IDEM. La teoria della storia in Benedetto Croce. *Giornale critico della filosofia italiana*, 2005, 1, p. 298-327.

520. NINOMIYA (Hiroyuki). Maruku Burokku wo yomu. (Reading Marc Bloch, 1886–1944). Tokyo, Iwanami Shoten, 2005, 273 p.

521. NOUSCHI (André). Initiation aux sciences historiques. Paris, A. Colin, 2005, 320 p.

522. OLIVER OLMO (Pedro). El concepto de control social en la historia social: estructuración del orden y respuestas al desorden. *Historia social*, 2005, 51, p. 73-92.

523. ORRILL (Robert), SHAPIRO (Linn). From bold beginnings to an uncertain future: the discipline of history and history education. [Forum essay]. *American historical review*, 2005, 110, 3, p. 726-751.

524. Practicing history: new directions in historical writing after the linguistic turn. Ed. by Gabrielle M. SPIEGEL. New York a. London, Routledge, 2005, XIV-274 p. (Rewriting histories Re-writing histories).

525. PROCACCI (Giuliano). Carte d'identità: revisionismi, nazionalismi e fondamentalismi nei manuali di storia. Roma, Carocci e Cagliari, AM&D, 2005, 205 p. (Studi storici Carocci, 74. Storia internazionale del 20. secolo, 6). – IDEM. Un Machiavelli per la Delta Force. *Passato e presente*, 2005, 65, p. 109-114.

526. RICUPERATI (Giuseppe). Apologia di un mestiere difficile: problemi, insegnamenti e responsabilità della storia. Roma e Bari, Laterza, 2005, VII-223 p.

527. RÜTH (Axel). Erzählte Geschichte: narrative Strukturen in der französischen Annales-Geschichtsschreibung. Berlin, De Gruyter, 2005, IX-211 p. (Narratologia, 5).

528. SAALER (Sven). Politics, memory and public opinion: the history textbook controversy and Japanese society. München, Iudicium, 2005, 202 p. (Monographien. Philipp-Franz-von-Siebold-Stiftung. Deutsches Institut feur Japanstudien, 39).

529. SALY-GIOCANTI (Frédéric). Utiliser les statistiques en histoire. Paris, A. Colin, 2005, 191 p.

530. SCHINKEL (Anders). Imagination as a category of history: an essay concerning Koselleck's concepts of Erfahrungsraum and Erwartungshorizont. *History and theory*, 2005, 44, 1, p. 42-54.

531. SCHULIN (Ernst). Zeitgemäße Historie um 1870. Zu Nietzsche, Burckhardt und zum "Historismus". *Historische Zeitschrift*, 2005, 281, 1, p. 33-58.

532. SCUCCIMARRA (Luca). Uscire dal moderno. Storia dei concetti e mutamento epocale. *Storica*, 2005, 32, p. 109-134.

533. SEWELL (William H. Jr.). Logics of history: social theory and social transformation. Chicago, Uni-

versity of Chicago Press, 2005, XI-412 p. (Chicago studies in practices of meaning).

534. Shixue xinlun. (New theories in historical science). Edited by the Institute of History and Literature of the University of Henan. Kaifeng, Henan daxue chubanshe, 2005, 2, 3, 692 p.

535. SKAGESTAD (Peter). Collingwood and Berlin: a comparison. *Journal of the history of ideas*, 2005, 66, 1, p. 99-112.

536. SOUTHGATE (Beverley). What is history for? New York, Routledge, 2005, XIII-214 p.

537. TOURNÈS (Ludovic). L'informatique pour les historiens. Graphiques, calculs, internet, bases de données. Paris, Belin, 2005, 175 p.

538. TRAVERS (Emeric). Benjamin Constant, les principes et l'histoire. Paris, Honoré Champion, 2005, 659 p. (Travaux et recherches de l'Institut Benjamin Constant, 7).

539. TSCHOPP (Silvia Serena). Das Unsichtbare begreifen. Die Rekonstruktion historischer Wahrnehmungsmodi als methodische Herausforderung der Kulturgeschichte. *Historische Zeitschrift*, 2005, 280, 1, p. 39-82.

540. WICKHAM (Chris). Problems in doing comparative history. Southampton, Centre for Antiquity and the Middle Ages, University of Southampton, 2005, 35 p. (Reuter lecture, 2004).

541. YORMAZ (Abdullah). Muhalif bir Metin Nasıl Okunur? Osmanlı Medreseleri'nde Hidâyetü'l-Hikme. (Comment lire un texte contraire? Hidâyetü'l-Hikme dans les medrese ottomans). *Dîvân İlmî Araştırmalar*, 2005, 18, p. 175-192.

§ 3. Addenda 2003–2004.

542. BREISACH (Ernst). On the future of history: the postmodernist challenge and its aftermath. Chicago, University of Chicago Press, 2003, IX-243 p.

543. LEHMANN-BRAUNS (Sicco). Weisheit in der Weltgeschichte. Philosophiegeschichte zwischen Barock und Aufklärung. Tübingen, Niemeyer, 2004, XI-429 p. (Frühe Neuzeit, 99).

Cf. n°s 284, 359, 401, 2359, 3500, 5246, 5251

§ 4. Ethnology, folklore and historical anthropology.

* 544. NEUHEUSER (Hanns Peter). Profane Rituale und Ritualität. Tendenzen der fächerübergreifenden Forschung und der kulturhistorischen Ansätze in den Einzeldisziplinen. *Archiv für Kulturgeschichte*, 2005, 87, 2, p. 427-454.

545. AMELANG (James S.). Mourning becomes eclectic: ritual lament and the problem of continuity. *Past and present*, 2005, 187, p. 3-31.

546. Atlasul etnografic român. Vol. II. Ocupațiile. (Romanian ethnographic atlas. Vol II. Professions). Coordonator Ion GHINOIU. București, Editura Academiei Române, 2005, 295 p.

547. BALIVET (Michel). Mélanges Byzantins, Seldjoukides et Ottomans. İstanbul, Isis, 2005, 184 p.

548. BĂRBULESCU (Constantin). Imaginarul corpului uman între cultura țărănească și cultura savantă (secolele XIX–XX). (The imaginary of the human body in folk culture and academic culture, 19th–20th centuries). București, Editura Paideia, 2005, 271 p.

549. BECKER (Thomas). Mann und Weib, schwarz und weiss: die wissenschaftliche Konstruktion von Geschlecht und Rasse 1650–1900. Frankfurt am Main u. New York, Campus, 2005, 415 p. (ill.).

550. Beiträge zur Rezeptions- und Wirkungsgeschichte der Volkserzählung: Berichte und Referate des zwölften und dreizehnten Symposions zur Volkserzählung, Brunnenburg, Südtirol 1998–1999. Hrsg. v. Leander PETZOLDT und Oliver HAID. Frankfurt am Main u. Oxford, P. Lang, 2005, 448 p. (ill.). (Beiträge zur europäischen Ethnologie und Folklore, 9 Reihe B, Tagungsberichte und Materialien).

551. BERGQUIST (Lars-Göran). En enkel till Himlingøje: dödens mode 1: ett virrvarr av varianter: makt och monumentalitet vid gravens rand eller Döden: en berättelse om livet?: praktgravar i Sydskandinavien under 1000 år: ca 150 f.Kr.–ca 1050 e.Kr. (Un aller simple pour Himlingøje. Une multitude de rites funéraires: puissance et monumentalité dans les riches sépultures de Scandinavie méridionale pendant un millénaire, ca. 150 avant JC–ca 1050 après JC). Uppsala, Department of Archaeology and Ancient History, Uppsala University, 2005, 307 p., 1 CD-ROM (Aun, 32). [English summary].

552. BRATOŽ (Rajko). Gli inizi dell'etnogenesi slovena: fatti, tesi e ipotesi relativi al periodo di transizione dall'età antica al medioevo nel territorio situato tra l'Adriatico e il Danubio. *In*: La cristianizzazione degli Slavi nell'arco alpino orientale: (secoli VI–IX). A cura di Andrea TILATTI. Roma, Istituto storico italiano per il Medio Evo e Gorizia, Istituto di storia sociale e religiosa, 2005, p. 145-188. (Nuovi studi storici, 69).

553. CALCANI (Giuliana). Le esequie del "capo". *Quaderni di storia*, 2005, 62, p. 107-122.

554. Confini. Costruzioni, attraversamenti, rappresentazioni. A cura di Silvia SALVATICI. Soveria Mannelli, Rubbettino, 2005, 245 p.

555. DE LUNA (Giovanni). Il corpo del nemico. *Passato e presente*, 2005, 65, p. 7-15.

556. DESCOLA (Philippe). Par-delà nature et culture. Paris, Gallimard, 2005, 623 p. (Bibliothèque des sciences humaines).

557. DETIENNE (Marcel). Les Grecs et nous: une anthropologie comparée de la Grèce ancienne. Paris, Perrin, 2005, 213 p. (ill.). (Pour l'histoire).

558. DUCHET (Michèle). Essais d'anthropologie: espace, langues, histoire. Préface de Claude BLAN-

CKAERT; postface par Georges BENREKASSA. Paris: Presses universitaires de France, 2005, 339 p. (ill.). (Ecriture).

559. ELDEM (Edhem). İstanbul'da ölüm: Osmanlı-islam kültüründe ölüm ve ritüelleri. (Mort à İstanbul: mort et ses rituels dans la culture musulmane ottomane). İstanbul, Osmanlı Bankası Arşiv ve Araştırma Merkezi, 2005, 299 p.

560. Ett folk, ett land: Sápmi i historia och nutid. (Un peuple, un pays. Le Sapmi dans l'histoire et aujourd'hui). Red. Per AXELSSON, Peter SKÖLD. Umeå, Centrum för samisk forskning, Umeå universitet, 2005, 385 p. (ill.). (Skrifter från Centrum för samisk forskning, 3).

561. Geschwister-Eltern-Grosseltern: Beiträge der historischen, anthropologischen und demographischen Forschung = siblings-parents-grandparents: contributions of historical, anthropological, and demographical research. Hrsg. v. Georg FERTIG. Köln, Zentrum für Historische Sozialforschung, 2005, 320 p. (ill.).

562. GODEA (Ioan). Din etnologia cumpătării. Palinca, țuica și vinarsul la români. (From the ethnology of moderation. Palinka, tzuika and brandy in Romania). București, Editura Coresi, 2005, 136 p.

563. GRUNDBERG (Malin). Ceremoniernas makt: maktöverföring och genus i Vasatidens kungliga ceremonier. (Le pouvoir des cérémonies. Transmission du pouvoir et genre dans les cérémonies royales à l'époque de la dynastie des Vasas en Suède). Lund, Nordic Academic Press, 2005, 334 p. (ill.). [English summary].

564. Histoire du corps. 1 De la Renaissance aux Lumières. Volume dirigé par Georges VIGARELLO; avec Daniel ARASSE, Jean-Jacques COURTINE, Jacques GÉLIS [et al.], sous la direction de Alain CORBIN, Jean-Jacques COURTINE et Georges VIGARELLO. Paris, Ed. du Seuil, 2005, 573 p. (ill.).

565. HUMPHREY (Caroline), HÜRELBAATAR (Altanhuu). Regret as a political intervention: an essay in the historical anthropology of the early Mongols. *Past and present*, 2005, 186, p. 3-45.

566. Investitur- und Krönungsrituale. Herrschaftseinsetzungen im kulturellen Vergleich. Hrsg. v. Marion STEINICKE und Stefan WEINFURTER. Köln, Weimar u. Wien, Böhlau, 2005, VII-496 p.

567. IVANCHIK (Askol'd I.). Nakanune kolonizatsii. Severnoe Prichernomor'e i stepnye kochevniki VI-II–VII vv. do n.e. v antichnoj literaturnoj traditsii: Fol'klor, literatura i istorija. (On the eve of the Colonization: The North Black Sea Region and the step nomades of the 8th and the 7th centuries B.C. in the antique literary tradition: Folklore, literature and history). RAN, In-t arkheologii. Moskva – Berlin, [s. n.], 2005, 312 p. (ill.; bibl. p. 247-279; list of sources p. 283-305). (Pontus Septentrionalis; 3).

568. IVANOVA (Ljudmila A.). Kukovskaja kollektsia Peterburgskoj Kunstkamery: problemy istochnikovedenija i atributsii. (The 'Cook collection' of the Kunstkamera, Saint-Petersburg: Problems of critical study and attribution). RAN, In-t etnologii i antropologii im. N.N. Miklukho-Maklajja. Moskva, Nauka, 2005, 308 p. (ill.; bibl. p. 268-274; ind. p. 298-306). [English summary]

569. JOON-HAI LEE (Christopher). The 'Native' undefined: colonial categories, Anglo-African status and the politics of kinship in British Central Africa, 1929–38. *Journal of African history*, 2005, 46, 3, p. 455-478.

570. KERIMOV (Abusaid K.). Istorija dagestanskikh lezgin. (A history of the Lezghins of Dagestan). Moskva, [s. n.], 2005, 559 p. (ill.; portr.; bibl. p. 550-553).

571. KUWAHARA (Makiko). Tattoo: an anthropology. New York, Berg, 2005, XIX-268 p.

572. MAKAROVA (Irina F.). Bolgarskij narod v XV-XVIII vv.: Etnokul'turnoe issledovanie. (The Bulgarian people in the 15th–18th centuries: An ethno-cultural study). RAN, In-t slavjanovedenija. Moskva, URSS – KomKniga, 2005, 190 p. (bibl. p. 176-190).

573. Meje in konfini: Rakitovec, vas kulturnih, družbenih in naravnih prepletanj (Borders and barriers: Rakitovec, a village of cultural, social and natural intertwining). Ed. Vida ROŽAC-DAROVEC. Koper, Univerza na Primorskem, Znanstveno-raziskovalno središče, Založba Annales, Zgodovinsko društvo za južno Primorsko, 2005, [s. p.]. (Knjižnica Annales Majora).

574. MOKSHIN (Nikolaj F.), MOKSHINA (Elena N.). Mordva i vera. (The Mordvins and the religion). Saransk, Mordov. kn. izd-vo, 2005, 532 p. (ill.; portr.; plates).

575. MOKSHINA (Julia N.). Brak i sem'ja v obychnom prave mordvy. (Marriage and family in the customary law of the Mordvins). Saransk, Mordov. kn. izd-vo, 2005, 253 p. (bibl. p. 228-251).

576. Narody Zapadnoj Sibiri: Khanty. Mansi. Sel'kupy. Nentsy. Entsy. Nganasany. Kety. (The peoples of Western Siberia: the Khanty, Mansi, Nenets, Enets, Nganasan, and Ket peoples: Historical and ethnographical study). Ed. Izmail N. GEMUEV, Vjacheslav I. MOLODIN, Zoja P. SOKOLOVA. RAN, In-t etnologii i antropologii im. N.N. Miklukho-Maklajja; RAN, Sib. otd., In-t arkheologii i etnografii. Moskva, Nauka, 2005, 805 p. (ill.; maps; bibl. p. 741-796). (Narody i kul'tury; 2).

577. NITOBURG (Eduard L.). Russkie v SShA: Istorija i sud'by, 1870–1970: Etnoistoricheskij ocherk. (Russians in the USA: History and fates, 1870–1970: An ethno-historical essay). RAN, In-t etnologii i antropologii im. N.N. Miklukho-Maklajja. Moskva, Nauka, 2005, 421 p. (ill.; portr.; maps; bibl. p. 385-407; pers. ind. p. 408-418).

578. OLENDER (Maurice). La chasse aux évidences. Sur quelques formes de racisme entre mythe et histoire. Paris, Galaade Éditions, 2005, 394 p.

579. POGLIANO (Claudio). L' ossessione della razza: antropologia e genetica nel XX secolo. Pisa, Scuola normale superiore, 2005, 582 p. (Studi, 1).

580. REIKAT (Andrea). Wir und die Anderen. Zur Frage nach der Fremdheit in der Ethnologie. *Historische Zeitschrift*, 2005, 281, 2, p. 281-305.

581. ROCA W. (Demetrio). Cultura andina. Lima, Fondo Editorial de la Facultad de Ciencias Sociales, UNMSM y Cusco, Instituto Nacional de Cultura, Dirección Regional de Cultura de Cusco, 2005, 498 p. (ill.).

582. Role (The) of magic in the past. Learned and popular magic, popular beliefs and diversity of attitudes. Ed. Blanka SZEGHYOVÁ. Bratislava, Pro Historia – Historical Institute of Slovak Academy of Sciences, 2005, 255 p. (História, 1).

583. SCHMUHL (Hans-Walter). Grenzüberschreitungen: das Kaiser-Wilhelm-Institut für Anthropologie, menschliche Erblehre und Eugenik 1927–1945. Göttingen, Wallstein, 2005, 597 p. (Geschichte der Kaiser-Wilhelm-Gesellschaft im Nationalsozialismus, 9).

584. Untaming the frontier in anthropology, archaeology, and history. Ed. by Bradley J. PARKER and Lars RODSETH. Tucson, University of Arizona Press, 2005, VI-294 p.

585. UVAROVA (Tat'jana B.). Nerchinskie evenki v XVIII–XX vekakh. (The Nerchinsk Evenks in the 18th–20th centuries). Moskva, INION RAN, 2005, 164 p. (bibl. p. 146-161).

586. WAGNER (Marc-André). Le cheval dans les croyances germaniques: paganisme, christianisme et traditions. Paris, Champion, 2005, 974 p. (ill., map). (Nouvelle bibliothèque du moyen âge, 73).

587. Welt (Die) der Rituale: von der Antike bis heute. Hrsg. v. Claus AMBOS [et al.]. Darmstadt, Wissenschaftliche Buchgesellschaft, 2005, 276 p. (ill.).

§ 4. Addendum 1996.

588. Historians and race: autobiography and the writing of history. Ed. by Paul A. CIMBALA and Robert F. HIMMELBERG. Bloomington, Indiana U. P., 96, XII-154 p.

Cf. n°s 186, 389, 406, 572, 688, 868, 872, 930, 999, 2005, 2205, 3500, 3549, 3873, 4361, 4462, 4474, 4492, 4509

§ 5. General history.

* 589. Antisemitism: an annotated bibliography. Vol. 18. 2002. [Vol. 17. 2001. Cf. Bibl. 2004, n° 709.] Ed. by Susan Sarah COHEN. München, K. G. Saur, 2005, XXXII-406 p.

* 590. HACKER (Barton C.). World military history annotated bibliography: premodern and nonwestern military institutions (works published before 1967). Leiden a. Boston, Brill, 2005, VII-305 p. (History of warfare, 27).

* 591. KEYNES (Simon). Anglo-Saxon England: a bibliographical handbook for students of Anglo-Saxon history. Cambridge, Department of Anglo-Saxon, Norse, and Celtic, University of Cambridge, 2005, 248 p. (maps). (ASNC guides, texts, and studies, 1).

* 592. KRÜPE (Florian), SCHÄFER (Christoph). Digitalisierte Vergangenheit. Datenbanken und Multimedia von der Antike bis zur frühen Neuzeit. Wiesbaden, Harrassowitz, 2005, XI-147 p.

* 593. SANTAMARÍA GARCÍA (Antonio). Historia económica y social de Puerto Rico: 1750–1902: bibliografía, fuentes publicadas (1745–2002) y balance. Madrid, Fundación Mapfre Tavera, 2005, 375 p. (Documentos Tavera, 19).

---

594. ABAZOV (Rafis). Historical dictionary of Turkmenistan. Lanham, Scarecrow Press, 2005, CIV-240 p. (Historical dictionaries of Asia, Oceania, and the Middle East, 53).

595. ALLEN (Philip M.), COVELL (Maureen). Historical dictionary of Madagascar. Lanham, Scarecrow Press, 2005, LXXXIII-420 p. (maps). (Historical dictionaries of Africa, 98).

596. AMIN (Shahid). Un saint guerrier. Sur la conquête de l'Inde du nord par les turcs au XIe siècle. *Annales*, 2005, 60, 2, p. 265-292.

597. ANDREESCU (Ştefan). Izvoare noi cu privire la istoria Mării Negre (New evidence concerning the history of the Black Sea). Bucureşti, Editura Institutului Cultural Român, 2005, 262 p.

598. Atlas-dicţionar al Daciei Romane. (Atlas-dictionary of Roman Dacia). Coordonator Mihai BĂRBULESCU. Cluj-Napoca, Editura Tribuna, 2005, 147 p.

599. AUBOIN (Michel). Histoire et dictionnaire de la police, du Moyen Age à nos jours. Paris, R. Laffont, 2005, XXII-1059 p.

600. BANDIERI (Susana). Historia de la Patagonia. Buenos Aires, Editorial Sudamericana, 2005, 445 p. (maps). (Colección Historia argentina).

601. BARRAL (Pierre). Pouvoir civil et commandement militaire. Du roi-connétable aux leaders du XXe siècle. Paris, Presses de la FNSP, 2005, 260 p.

602. BEARZOT (Cinzia). Manuale di storia greca. Bologna, Il Mulino, 2005, 298 p. (Manuali. Storia).

603. BIELENSTEIN (Hans). Handbook of oriental studies = Handbuch der Orientalistik. Abt. 4. Bd. 18. China. Diplomacy and trade in the Chinese world, 589–1276. Leiden, Brill, 2005, 725 p.

604. Biographisches Handbuch des deutschen Auswärtigen Dienstes: 1871–1945. Herausgeber, Auswärtiges Amt, Historischer Dienst, Maria KEIPERT, Peter GRUPP. Bd. 2. G-K. Bearbeiter, Gerhard Keiper, Martin KRÖGER. Paderborn, F. Schöningh, 2005, XIV-715 p.

605. BIZZARRO (Salvatore). Historical dictionary of Chile. Lanham a. Oxford, Scarecrow Press, 2005, LVIII-937 p. (ill., map). (Historical dictionaries of Latin America, 28).

606. BLUME (Kenneth J.). Historical dictionary of U.S. diplomacy from the Civil War to World War I. Lanham a. Oxford, Scarecrow Press, 2005, XXV-481 p. (Historical dictionaries of U.S. diplomacy, 1).

607. BREGNSBRO (Michael), VILLADS JENSEN (Kurt). Det danske imperium: storhed og fald. (Grandeur et décadence de l'empire danois). København, Aschehoug, 2005, 255 p. (pl.).

608. Brill's New Pauly: encyclopaedia of the ancient world. Ed. by Hubert CANCIK and Helmuth SCHNEIDER. Vol. 6. Hat-Jus. Managing editor, Christine E. SALAZAR. Leiden, Brill, 2005, XVI-1232 p. (ill., maps).

609. BUZHILOVA (Aleksandra P.). Homo sapiens: Istorija bolezni. (Homo sapiens: A history of diseases: [Paleopathology and history]). RAN, In-t arkheologii. Moskva, Jazyki slavjanskoj kul'tury, 2005, 319 p. (bibl. p. 302-309).

610. Cambridge ancient history (The). Vol. 12. Crisis of empire, a.d. 193–337. Ed. by Alan K. BOWMAN, Peter GARNSEY and Averil CAMERON. Cambridge, New York a. Melbourne, Cambridge U. P., 2005, XVIII-965 p.

611. Cambridge companion (The) to the age of Augustus. Ed. by Karl GALINSKY. Cambridge, Cambridge U. P., 2005, XVII-407 p. (Cambridge companion to the classics, 17).

612. Cambridge history (The) of warfare. Ed. by Geoffrey PARKER. Cambridge, New York a. Melbourne, Cambridge U. P., 2005, VIII-515 p.

613. Chaucer: an Oxford guide. Ed. by Steve ELLIS. Oxford, Oxford U. P., 2005, XXIV-644 p. (An Oxford guide).

614. Companion (A) to Chrétien de Troyes. Ed. by Norris J. LACY and Joan T. GRIMBERT. Cambridge, D.S. Brewer, 2005, XIV-242 p. (Arthurian studies).

615. Companion (A) to contemporary Britain, 1939–2000. Ed. by Paul ADDISON and Harriet JONES. Oxford, Blackwell, 2005, XVI-583 p. (Blackwell companions to British history).

616. Companion (A) to Old Norse-Icelandic literature and culture. Ed. by Rory MAC TURK. Malden a. Oxford, Blackwell, 2005, XIII-567 p. (Blackwell companions to literature and culture, 31).

617. Companion (A) to the ancient Near East. Ed. by Daniel C. SNELL. Malden a. Oxford, Blackwell Pub., 2005, XIX-504 p. (ill., maps). (Blackwell companions to the ancient world. Ancient history).

618. Companion (A) to the history of the Middle East. Ed. by Youssef M. CHOUEIRI. Malden, Blackwell Pub., 2005, XVII-602 p. (maps). (Blackwell companions to world history).

619. Companion (A) to the Middle English lyric. Ed. by Thomas Gibson DUNCAN. Cambridge a. Rochester, D.S. Brewer, 2005, XXV-302 p. (Companions).

620. Companion (A) to the Roman Empire. Ed. by David POTTER. Oxford, Blackwell, 2005, 656 p. (ill.). (Blackwell companions to the ancient world. Ancient history).

621. Companion (A) to the works of Hartmann von Aue. Ed. by Francis G. GENTRY. Rochester a. Woodbridge, Camden House, 2005, VI-291 p. (ill.). (Studies in German literature, linguistics, and culture).

622. Companion (The) to Tasmanian history. Ed. by Alison ALEXANDER. Hobart, Centre for Tasmanian Historical Studies, University of Tasmania, 2005, XII-568 p. (ill., maps).

623. Concise companion (A) to the Restoration and the eighteenth century. Ed. by Cynthia WALL. Oxford, Blackwell, 2005, X-284 p. (ill., maps). (Blackwell concise companions to literature and culture).

624. COOK (Chris), STEVENSON (John). The Routledge companion to world history since 1914. London, Routledge, 2005, X-582 p. (maps). (Routledge companions to history).

625. COOK (Chris). The Routledge companion to Britain in the nineteenth century, 1815–1914. London, Routledge, 2005, XI-352 p. (maps). (Routledge companions to history).

626. Costa Rica reader (The): history, culture, politics. Ed. by Steven PALMER and Iván MOLINA. Durham, Duke U. P., 2005, XIV-383 p.

627. DANGL (Vojtech). Bitky a bojiská v našich dejinách. Od Samovej ríše po vznik stálej armády. (Die Schlächte und die Kampfplätze in unsere Geschichte seit dem Reich von Samo bis zur Entstehung der ständige Armee). Bratislava, Perfekt, 2005, 243 p.

628. DEHN-NIELSEN (Henning). Kongelige mord i Danmarkshistorien. (Meurtres de souverains dans l'histoire danoise). København, Aschehoug, 2005, 176 p.

629. DELSALLE (Paul). Vocabulaire historique de la France moderne, XVIe–XVIIIe siècles. Paris, A. Colin, 2005, 127 p.

630. Diccionario histórico argentino. Publicado bajo la dirección de Fermín CHÁVEZ y la participación de Roberto VILCHEZ, Enrique MANSON y Lorenzo GONZÁLEZ. Buenos Aires, Ediciones Fabro, 2005, 571 p.

631. Diccionario histórico biográfico: peruanos ilustres. Ed. por Alfonso CUEVA SEVILLANO. Lima, A.F.A. Editores, 2005, 738 p. (ill., ports.).

632. Dictionnaire de l'antiquité. Sous la dir. de Jean LECLANT. Paris, Presses universitaires de France, 2005, XLVIII-2389 p.

633. Dictionnaire historique de la France moderne. Sous la dir. de Laurent BOURQUIN [et al.]. Paris, Belin, 2005, 441 p. (ill., maps).

634. Dizionario biografico degli italiani. Vol. 64. Latilla-Levi Montalcini. Vol. 65. Levis-Lorenzetti. Roma, Istituto della Enciclopedia Italiana, 2005, 2 vol., XV-813 p., XV-811 p.

635. Encyclopedia (The) of World War I: a political, social, and military history. Ed. by Spencer C. TUCKER and Priscilla Mary ROBERTS; foreword by John S. D. EISENHOWER. Santa Barbara, ABC-CLIO, 2005, 5 vol., [s. p.].

636. Encyclopedia of racism in the United States. Ed. by Pyong Gap MIN. Westport, Greenwood Press, 2005, 3 vol., LXIX-795 p.

637. Encyclopedie de l'Histoire de France. Vol. 8 La tourmente (1918–1945). Vol. 9. En route vers la modernité (1945–2005). Direction éditoriale: Jean-Marc DUBRAY; édition: Renaud DEGOLS et Patrick VICARD. [S. l.], TXT Média Services, 2005, 2 vol., 115 p., 114 p. (ill.).

638. ENGLISH (Edward D.). Encyclopedia of the medieval world. New York, Facts On File, 2005, 2 vol., XX-920 p. (ill., maps). (Facts on File library of world history).

639. ERSKINE (Andrew). A companion to the Hellenistic world. Oxford a. Carlton, Blackwell, 2005, XXVIII-595 p.

640. FERRERA (Carlos). Diccionario de historia de España. Madrid, Alianza Editorial, 2005, 552 p. (El Libro de bolsillo / Alianza Editorial. BT 8127 Biblioteca temática. Biblioteca de consulta).

641. FINDLEY (Carter Vaughn). The Turks in world history. New York, Oxford U. P., 2005, XVI-300 p.

642. FLEMING (Neil C.), O'DAY (Alan). The Longman handbook of modern Irish history since 1800. Harlow, Pearson Education, 2005, IX-808 p. (Longman handbooks to history).

643. GABUCCI (Ada). Roma. Milano, Electa, 2005, 383 p. (I dizionari delle civiltà).

644. GARAVAGLIA (Juan Carlos), MARCHENA (Juan). América Latina de los orígenes a la Independencia. Vol. 1. América precolombina y la consolidación del espacio colonial. Vol. 2. La sociedad colonial ibérica en el siglo XVIII. Barcelona, Crítica, 2005, 2 vol., [s. p.] (ill., maps). (Serie mayor).

645. GIL NOVALES (Alberto). Diccionario biográfico aragonés 1808–1833. Huesca, Instituto de Estudios Altoaragoneses, 2005, 438 p. (ill.). (Colección de estudios altoaragoneses, 52).

646. Handbook (A) of the communist security apparatus in East Central Europe 1944–1989. Ed. by Krzysztof PERSAK and Łukasz KAMIŃSKI. Warsaw, Institute of National Remembrance, 2005, 352 p. (ill.).

647. Handbuch zur Statistik der Parlamente und Parteien in den westlichen Besatzungszonen und in der Bundesrepublik Deutschland. Teilband 2. CDU und CSU: Mitgliedschaft und Sozialstruktur 1945–1990. Bearb. v. Corinna FRANZ und Oliver GNAD. Teilband 3. FDP sowie kleinere bürgerliche und rechte Parteien: Mitgliedschaft und Sozialstruktur 1945–1990. Bearb. v. Oliver GNAD [et al.]. Teilband 4. SPD, KPD und kleinere Parteien des linken Spektrums sowie DIE GRÜNEN: Mitgliedschaft und Sozialstruktur 1945–1990. Bearb. v. Josef BOYER und Till KÖSSLER. Düsseldorf, Droste, 2005, 3 vol., 875 p., 717 p., 1053 p.

648. Histoire des Togolais. Vol. 2. De 1884 à 1960. Sous la direction de Nicoué Lodjou GAYIBOR. Lomé, Presses de l'Université du Bénin, 2005, 2 vol., 629 p., 754 p. (ill., maps, ports.).

649. Historisches Lexikon der Schweiz. Hrsg. von der Stiftung Historisches Lexikon der Schweiz (HLS). Chefred.: Marco JORIO. Band 4. Dudan – Frowin. Band 5. Fruchtbarkeit – Gyssling. Basel, Schwabe, 2005, 2 vol., XXV-856 p., XXV-854 p.

650. Independência. História e historiografia. Ed. István JANCSÓ. Sao Paulo, Editora Hucitec/Fapesp, 2005, 934 p.

651. JACKSON (William Keith), MAC ROBIE (Alan). Historical dictionary of New Zealand. Lanham a. Oxford, Scarecrow Press, 2005, LXXX-451 p. (maps). (Historical dictionaries of Asia, Oceania, and the Middle East, 56).

652. JAIMOUKHA (Amjad M.). The Chechens: a handbook. New York a. London, RoutledgeCurzon, 2005, XII-320 p. (Caucasus world. Peoples of the Caucasus).

653. JOLY (Bertrand). Dictionnaire biographique et géographique du nationalisme français (1880–1900). Paris, Champion, 2005, 687 p. (Champion classiques: références et dictionnaires).

654. JOTISCHKY (Andrew), HULL (Caroline Susan). The Penguin historical atlas of the medieval world. London, Penguin, 2005, 144 p. (ill., maps).

655. KALCK (Pierre). Historical dictionary of the Central African Republic. Lanham a. Oxford, Scarecrow Press, 2005, LXXIV-233 p. (maps). (Historical dictionaries of Africa, 93).

656. KÉRAUTRET (Michel). Histoire de la Prusse. Paris, Ed. du Seuil, 2005, 510 p. (L'Univers historique).

657. KOHNLE (Armin). Kleine Geschichte der Kurpfalz. Leinfelden-Echterdingen, Braun, 2005, 205 p.

658. KOHUT (Zenon E.), NEBESIO (Bohdan Y.). Historical dictionary of Ukraine. Lanham a. Oxford, Scarecrow Press, 2005, LIII-854 p. (maps). (Historical dictionaries of Europe, 45).

659. KULKE (Hermann). Indische Geschichte bis 1750. München, R. Oldenbourg, 2005, XIV-275 p. (Oldenbourg Grundriss der Geschichte, 34).

660. Leviathans: multinational corporations and the new global history. Ed. by Alfred D. CHANDLER Jr. and

Bruce MAZLISH. New York, Cambridge U. P., 2005, XIII-249 p.

661. Liunga. kaupinga: kulturhistoria och arkeologi i Linköpingsbygden. (Histoire et archéologie dans la région de Linköping). Anders KALIFF, Göran TAGESSON. Stockholm, Riksantikvarieämbetet, 2005, 324 p. (ill.). (Riksantikvarieämbetet, Arkeologiska undersökninger, skrifter, 60). [English summary].

662. MANSINGH (Surjit). Historical dictionary of Nepal. New Delhi, Vision Books, 2005, XXXXIV-511 p. (maps).

663. MAZÍN (Oscar). L'Amérique espagnole. París, Les Belles Lettres, 2005, 312 p.

664. Medieval Ireland: an encyclopedia. Ed. by Seán DUFFY; associate editors, Ailbhe MAC SHAMHRÁIN and James MOYNES. New York a. London, Routledge, 2005, XXXI-546 p. (ill., maps).

665. Medieval Islamic civilisation: an encyclopedia. Ed. by Josef W. MERI. London, Routledge, 2005, 2 vol., 1248 p. (ill.). (Encyclopedias of the Middle Ages).

666. MILLETT (Martin). Roman Britain. London, Batsford, 2005, 144 p. (ill., maps). (English Heritage handbook).

667. Nationaal Biografisch Woordenboek (National biographical dictionnary). Volume 17. Ed. Herman VAN DER WEE. Brussels, Royal Academy, 2005, 916 p.

668. New Cambridge medieval history (The). Vol. 1. c. 500–c. 700. Ed. by Paul FOURACRE. Cambridge, New York a. Melbourne, Cambridge U. P., 2005, XVIII-979 p.

669. NEWITT (Malyn). A history of Portuguese overseas expansion, 1400–1668. London, Routledge, 2005, XV-300 p. (ill., maps).

670. NOLTE (Hans-Heinrich). Weltgeschichte. Imperien, Religionen und Systeme 15.–19. Jahrhundert. Wien, Köln u. Weimar, Böhlau, 2005, 392 p.

671. O'TOOLE (Thomas E.), BAKER (Janice E.). Historical dictionary of Guinea. Lanham a. Oxford, Scarecrow Press, 2005, LXVIII-288 p. (maps). (Historical dictionaries of Africa, 94).

672. Odissej: Chelovek v istorii. (Odysseus: Man in history: [Articles]). 2005: Vremja i prostranstvo prazdnika. (Festival: Time and space). Ed. Aron Ja. GUREVICH. RAN, In-t vseobshchej istorii. Moskva, Nauka, 2005, 479 p. (bibl. incl.). [English summaries]

673. OWUSU-ANSAH (David). Historical dictionary of Ghana. Lanham a. Oxford, Scarecrow Press, 2005, LXXIX-332 p. (maps). (Historical dictionaries of Africa, 97).

674. Oxford companion (The) to World War II. General editor, I.C.B. DEAR; consultant editor, M.R.D. FOOT. Oxford, Oxford U. P., 2005, XXIII-1039 p. (ill., maps).

675. POCOCK (John Greville Agard). The discovery of islands. Essays in British history. Cambridge, Cambridge U. P., 2005, XIII-344 p.

676. Prehistoric and early Ireland. Ed. by Dáibhí Ó CRÓINÍN. Oxford a. New York, Oxford U. P., 2005, LXXXII-1219 p. (A new history of Ireland, 1).

677. PRÉVÉLAKIS (Georges). Le processus de purification ethnique à travers le temps. *Guerres mondiales et conflits contemporains*, 2005, 53, 217, p. 47-60.

678. RIASANOVSKY (Nicholas V.). Russian identities: a historical survey. New York, Oxford U. P., 2005, 278 p.

679. ROECK (Bernd). Geschichte Augsburgs. München, Beck, 2005, 221 p.

680. SACKS (David). Encyclopedia of the ancient Greek world. Editorial consultant, Oswyn MURRAY; revised by Lisa R. BRODY. New York, Facts On File, 2005, XX-412 p. (ill., maps). (Facts on File library of world history).

681. SCHMITT (Hatto H.), VOGT (Ernst). Lexicon des Hellenismus. Wiesbaden, Harrassowitz, 2005, XII-1232 p.

682. Sengoshi daijiten = Encyclopedia of postwar Japan 1945–2004. SASAKI Takeshi, hoka, hen. Tokyo, Sanseido, 2005, 136, 1173 p.

683. SIRINELLI (Jean-François). Comprendre le XXe siècle français. Paris, Fayard, 2005, 528 p.

684. Sirīzu minato machi no sekaishi. 1. Minato machi to kaiiki sekai. (Port cities in world history. Vol. 1. Ports and the maritime world). Eds. Rekishigaku kenkyukai/ Shosuke MURAI. Tokyo, Aoki Shoten, 2005, 409 p.

685. SMITH (Harold Eugene), NIEMINEN (May Kyi Win). Historical dictionary of Thailand. Lanham, Scarecrow Press, 2005, XXXIX-374 p. (Historical dictionaries of Asia, Oceania, and the Middle East, 55).

686. SMITH (Philippa Mein). A concise history of New Zealand. Cambridge, New York a. Port Melbourne, Cambridge U. P., 2005, XVIII-302 p.

687. SPINK (Walter M.). Handbook of oriental studies = Handbuch der Orientalistik. Abt. 2. Bd. 18. Vol. 1. Indien. Ajanta: history and development. The end of the golden age. Leiden, Brill, 2005, XII-423 p.

688. TER-SARKISJANTS (Alla E.). Istorija i kul'tura armjanskogo naroda s drevnejshikh vremen do nachala XIX v. (History and culture of the Armenian people since the ancient times until the early 19th century). RAN, In-t etnologii i antropologii im. N.N. Miklukho-Maklajja. Moskva, Vostochnaja literatura, 2005, 686 p. (ill.; bibl. p. 585-635; ind. p. 638-682). (Etnokul'turnye vzaimodejstvija v Evrazii).

689. TRAMONTANA (Salvatore). Capire il Medioevo: le fonti e i temi. Roma, Carocci, 2005, 333 p. (Università, 625. Studi storici).

690. TUCHTENHAGEN (Ralph). Geschichte der baltischen Länder. München, Beck, 2005, 127 p. (Beck'sche Reihe, 2355).

691. VOCELKA (Karl). Österreichische Geschichte. München, Beck, 2005, 128 p. (maps). (Beck'sche Reihe, 2369).

692. WAGNER-PACIFICI (Robin). The art of surrender. Decomposing sovereignty at conflict's end. Chicago, University of Chicago Press, 2005, XII-210 p.

§ 5. Addenda 1999–2003.

693. BEAUREPAIRE (Pierre-Yves). L' Europe des francs-maçons, XVIIIe–XXIe siècles. Paris, Belin, 2002, 319 p. (maps). (Europe & histoire). – IDEM. La république universelle des francs-maçons: de Newton à Metternich. Rennes, Éd. "Ouest-France", 99, 201 p. (De mémoire d'homme: l'histoire).

694. FERGUSON (Niall). Empire: the rise and demise of the British world order and the lessons for global power. London, Allen Lane, 2002, XXVI-351 p.

695. GUHA (Ranajit). History at the limit of world-history. New York, Columbia U. P., 2002, X-116 p.

696. INNES (Matthew). An introduction to early medieval western Europe, 300–900: the sword, the plough and the book. London, Routledge, 2003, XVI-552 p.

697. LUKOWSKI (Jerzy), ZAWADZKI (Hubert). A concise history of Poland. Cambridge, Cambridge U. P., 2001, XVIII-317 p.

698. Nueva historia argentina. Proyecto editorial, Federico POLOTTO; coordinación general de la obra, Juan SURIANO. Vol. 1. Los pueblos originarios y la conquista. Directora de tomo, Myriam Noemí TARRAGÓ. Buenos Aires, Editorial Sudamericana, 2000, 382 p. (ill.).

699. TRAMPUS (Antonio). La massoneria nell'età moderna. Roma e Bari, Laterza, 2001, 151 p. (Biblioteca essenziale Laterza, 44. Storia moderna).

Cf. n^{os} *93, 206, 574, 575, 807, 1014, 1031, 2194, 2420, 2423, 2430, 3515-3569, 6831, 7562, 7865, 7878, 7888*

§ 6. Theory of the state and of society.

700. BARBUTO (Gennaro Maria). Machiavelli e i totalitarismi. Napoli, Guida, 2005, 164 p. (Prima pagina, nuova serie, 6).

701. BERNS (Thomas). Souveraineté, droit et gouvernementalité: lectures du politique moderne à partir de Bodin. Paris, L. Scheer, 2005, 254 p. (Non & non).

702. Eastern Roman empire (The) and the birth of idea of state in Europe. Ed. by Spyridon FLOGAITIS and Antoine PANTELIS. London, Esperia, 2005, 396 p.

703. FISCHER (David Hackett). Liberty and freedom. New York, Oxford U. P., 2005, 851 p. (A cultural history, 3).

704. FORCADE (Olivier), LAURENT (Sébastien). Secrets d'Etat: pouvoirs et renseignement dans le monde contemporain. Paris, Colin, 2005, 238 p. (L'histoire au présent).

705. GONZÁLEZ CUEVAS (Pedro Carlos). El pensamiento político de la derecha española en el siglo XX: de la crisis de la Restauración al Estado de partidos (1898–2000). Madrid, Tecnos, 2005, 285 p. (ill.). (Colección Biblioteca de historia y pensamiento político).

706. Gouvernement mixte (Le): de l'idéal politique au monstre constitutionnel en Europe (XIIIe–XVIIe siècle): [Actes du colloque, 'La constitution mixte, idéal de gouvernement et variations d'un modèle en Europe à la renaissance' qui s'est tenu les 7 et 8 novembre 2003 à l'ENS lettres et sciences humaines de Lyon]. Études réunies et introduites par Marie GAILLE-NIKODIMOV. Saint-Etienne, Publications de l'Université de Saint-Etienne, 2005, 232 p (Institut Claude Longeon, renaissance et âge classique).

707. HÖCHLI (Daniel). Der Florentiner Republikanismus: Verfassungswirklichkeit und Verfassungsdenken zur Zeit der Renaissance. Bern, Haupt, 2005, X-917 p. (ill.). (St. Galler Studien zur Politikwissenschaft, 28).

708. HUI (Victoria Tin-Bor). War and state formation in ancient China and early modern Europe. New York, Cambridge U. P., 2005, XIV-294 p.

709. Institutionelle Macht. Genese – Verstetigung – Verlust. Hrsg. v. André BRODOCZ [et al.]. Köln, Weimar u. Wien, Böhlau, 2005, 500 p.

710. KOCH (Bettina). Zur Dis-/Kontinuität mittelalterlichen politischen Denkens in der neuzeitlichen politischen Theorie: Marsilius von Padua, Johannes Althusius und Thomas Hobbes im Vergleich. Berlin, Duncker & Humblot, 2005, 381 p. (Beiträge zur politischen Wissenschaft, 137).

711. KUNZE (Rolf-Ulrich). Nation und Nationalismus. Darmstadt, Wissenschaftliche Buchgesellschaft, 2005, VIII-126 p.

712. MASTROPAOLO (Alfio). La mucca pazza della democrazia: nuove destre, populismo, antipolitica. Torino, Bollati Boringhieri, 2005, 201 p. (Temi, 148).

713. MAZOWER (Mark). Violencia y Estado en el siglo XX. *Historia social*, 2005, 51, p. 139-160.

714. MONNIER (Raymonde). Républicanisme, patriotisme et Révolution française. Paris, L'Harmattan, 2005, 356 p. (Logiques historiques).

715. MÜNKLER (Herfried). Imperien. Die Logik der Weltherrschaft – vom Alten Rom bis zu den Vereinigten Staaten. Berlin, Rowohlt, 2005, 332 p.

716. Power and the nation in European history. Ed. by Len SCALES and Oliver ZIMMER. Cambridge, New York a. Melbourne, Cambridge U. P., 2005, XII-389 p.

717. Rationalität im Prozess kultureller Evolution. Rationalitätsunterstellungen als eine Bedingung der Möglichkeit substantieller Rationalität des Handelns. Hrsg. v. Hansjörg SIEGENTHALER. Tübingen, Mohr Siebeck, 2005, IX-363 p. (Die Einheit der Gesellschaftswissenschaften, 132).

718. RÜDIGER (Axel). Staatslehre und Staatsbildung. Die Staatswissenschaft an der Universität Halle im 18. Jahrhundert. Tübingen, Niemeyer, 2005, VII-479 p. (Hallesche Beiträge zur europäischen Aufklärung, 15).

719. SÉMELIN (Jacques). Purifier et détruire. Usages politiques des massacres et des génocides. Paris, Ed. du Seuil, 2005, 486 p.

720. SOL (Thierry). Fallait-il tuer César?: L'argumentation politique de Dante à Machiavel: thèse pour le doctorat en science politique de l'Institut d'études politiques de Paris présentée et soutenue publiquement le 16 juin 2003. Paris, Dalloz, 2005, IX-442 p. (Nouvelle bibliothèque de thèses. Science politique, 1626– 1968).

721. Theorizing empire. Theme issue of *History and theory*, 2005, 44, 4, edited by Philip POMPER. Malden, Blackwell, 2005, 136 p. [Contents: POMPER (Philip). The history and theory of empires. – PAGDEN (Anthony). Fellow citizens and imperial subjects: conquest and sovereignty in Europe's overseas empires. – FILLION (Réal). Moving beyond biopower: Hardt and Negri's post-Foucauldian speculative philosophy og history. – ROSENAU (James N.). Illusions of power and empire. – HELLIE (Richard). The structure of Russian imperial history. – DUBOIS (Thomas David). Hegemony, imperialism, and the construction of religion in East and Southeast Asia].

722. THUMFART (Alexander), WASCHKUHN (Arno). Staatstheorien des italienischen Bürgerhumanismus: politische Theorie von Francesco Petrarca bis Donata Giannotti. Baden-Baden, Nomos, 2005, 343 p.

723. YOFFE (Norman). Myths of the archaic state. Evolution of the earliest cities, states, and civilizations. Cambridge, New York a. Port Melbourne, Cambridge U. P., 2005, XIII-277 p.

§ 7. Constitutional and legal history.

** 724. Legislazione suntuaria (La), secoli XIII–XVI: Umbria. A cura di M. Grazia NICO OTTAVIANI. Roma, Ministero per i beni e le attività culturali, Dipartimento per i beni archivistici e librari, Direzione generale per gli archivi, 2005, XXXVIII-1134 p. (Pubblicazioni degli archivi di Stato. Fonti, 43).

725. BENABDELLAH (Saïd). La justice en Algerie, des orgines à nos jours. T. 1. La justice precoloniale et coloniale et son evolution. T. 2. La justice du FLN et son impact sur l'etat algerien. Oran, Éditions dar el gharb, 2005, 2 vol., 380 p., 456 p.

726. Boundaries of the law. Geography, gender and jurisdiction in medieval and early modern Europe. Ed. by Anthony MUSSON. Aldershot, Ashgate, 2005, X-196 p.

727. BREATNACH (Liam). A companion to the Corpus iuris hibernici. Dublin, School of Celtic Studies, Dublin Institute for Advanced Studies, 2005, XV-499 p. (Early Irish law series, 5).

728. CARAVALE (Mario). Alle origini del diritto europeo: ius commune, droit commun, common law: nella dottrina giuridica della prima età moderna. Bologna, Monduzzi Editore, 2005, 254 p. (Archivio per la storia del diritto medioevale e moderno, 9).

729. Carcer II: prison et privation de liberté dans l'empire romain et l'occident médiéval; actes du colloque de Strasbourg (1er et 2 décembre 2000). Édités par Cécile BERTRAND-DAGENBACH [et al.]. Paris, De Boccard, 2005, 292 p (ill.). (Collections de l'Université Marc Bloch – Strasbourg; Études d'archéologie et d'histoire ancienne).

730. CRÉPIN (Marie-Yvonne). La peine de mort en Bretagne: de l'ancien droit au Code pénal de 1791. *Tijdschrift voor rechtsgeschiedenis*, 2005, 73, 3-4, p. 357-369.

731. Diccionario crítico de juristas españoles, portugueses y latinoamericanos (hispánicos, brasileños, quebequenses. 1. A-L. Ed. por Manuel PELÁEZ. Zaragoza, Facultad de derecho, Universidad de Málaga, 2005, 523 p.

732. Droit (Le) de résistance à l'oppression. Sous la dir. de Dominique GROS et Olivier CAMY. *Le Genre humain*, 2005, 44, 288 p.

733. FEENSTRA (Robert). Histoire du droit savant (13e–18e siècle): doctrines et vulgarisation par incunables. Aldershot, Ashgate, 2005, [s. p.]. (Collected studies, 842).

734. GIOLO (Orsetta). Giudici, giustizia e diritto nella tradizione arabo-musulmana. Torino, G. Giappichelli, 2005, XIII-264 p.

735. Handbuch der Orientalistik = Handbook of oriental studies. Ed. by Wilhelm RÖHL. Abt. 5. Bd. 12. Japan. History of law in Japan since 1868. Leiden a. Boston, Brill, 2005, VI-848 p.

736. HARKE (Jan Dirk). Locatio conductio. Kolonat, Pacht, Landpacht. Berlin, Duncker & Humblot, 2005, 108 p. (Schriften zur Europäischen Rechts- und Verfassungsgeschichte, 48).

737. KÜPPER (Herbert). Einführung in die Rechtsgeschichte Osteuropas. Frankfurt am Main u. Oxford, P. Lang, 2005, 709 p. (Studien des Instituts für Ostrecht München, 54).

738. Menschenrechte und europäische Identität. Die antiken Grundlagen. Hrsg. v. Klaus Martin GIRARDET und Ulrich NORTMANN. Stuttgart, Steiner, 2005, 301 p.

739. Rechtspluralismus in der Islamischen Welt: Gewohnheitsrecht zwischen Staat und Gesellschaft.

Hrsg. v. Michael KEMPER und Maurus REINKOWSKI. Berlin, De Gruyter, 2005, VI-378 p. (Studien zur Geschichte und Kultur des islamischen Orients, 16).

740. Recueils (Les) d'arrêts et dictionnaires de jurisprudence: XVIe–XVIIIe siècles. Sous la dir. de Serge DAUCHY et Véronique DEMARS-SION. Paris, Mémoire du Droit, 2005, 468 p. (Collection bibliographie, 2).

741. RÉMOND (René). L'invention de la laïcité: de 1789 à demain. Paris, Bayard, 2005, 172 p.

742. RITTER (Gerhard Albert). Föderalismus und Parlamentarismus in Deutschland in Geschichte und Gegenwart: vorgetragen in der Gesamtsitzung vom 22. Oktober 2004. München, Verlag der Bayerischen Akademie der Wissenschaften in Kommission beim Verlag C.H. Beck, 2005, 66 p. (Sitzungsberichte / Bayerische Akademie der Wissenschaften, philosophisch-historische Klasse, Jahrg. 2005, 4).

743. SCHIAVONE (Aldo). Ius: l'invenzione del diritto in Occidente. Torino, G. Einaudi, 2005, XVI-520 p. (Biblioteca di cultura storica, 254).

744. SCHMOECKEL (Mathias). Auf der Suche nach der verlorenen Ordnung: 2000 Jahre Recht in Europa: ein Überblick. Köln, Böhlau, 2005, XIX-600 p. (ill., maps, ports).

745. SÖLLNER (Alfred). Bona fides – guter Glaube? *Zeitschrift der Savigny-Stiftung für Rechtsgeschichte Romanistische Abteilung*, 2005, 122, p. 1-61.

746. Zeitliche Dimension (Die) des Rechts. Historische Rechtsforschung und geschichtliche Rechtswissenschaft. Hrsg. v. Louis PAHLOW. Paderborn, Schöningh, 2005, 306 p.

747. Zhongguo falü sixiang shi (History of Chinese legal thought). Ed. by YANG Hegao. Beijing, Beijing daxue chubanshe, 2005, 5, 668 p.

§ 7. Addenda 1998–2004.

748. Alternatives to Athens: varieties of political organization and community in ancient Greece. Ed. by Roger BROCK and Stephen HODKINSON. Oxford, Oxford U. P., 2000, XVI-393 p. (ill.).

749. CONSTANTINEAU (Philippe). La doctrine classique de la politique étrangère. Paris, L'Harmattan, 98, 239 p. (Ouverture philosophique).

750. Poleis e politeiai: esperienze politiche, tradizioni letterarie, progetti costituzionali: atti del Convegno Internazionale di Storia Greca: Torino, 29 maggio–31 maggio 2002. A cura di Silvio CATALDI. Alessandria, Edizioni dell'Orso, 2004, VII-549 p. (Fonti e studi di storia antica, 13).

Cf. n°s 153, 159, 566, 575, 1482, 2400, 2401, 2752

§ 8. Economic and social history.

751. Bardejova (Z) do Prešporku. Spoločnosť, súdnictvo a vzdelanosť v mestách v 13.–17. storočí. (Von Bardejov nach Pressburg. Die Gesellschaft, die Justiz und die Ausbildung in die Städten in 13.–17. Jahrhundert). Ed. Enikő CSUKOVITS, Tünde LENGYELOVÁ. Prešov Bratislava, Prešovská univerzita v Prešove, 2005, 308 p. (Acta Facultatis philosophicae Universitatis Prešoviensis).

752. Bodies in contact: rethinking colonial encounters in world history. Ed. by Tony BALLANTYNE and Antoinette BURTON. Durham, Duke U. P., 2005, XII-445 p.

753. DE LA VAISSIÈRE (Étienne). Handbook of Oriental studies = Handbuch der Orientalistik. Abt. 8. Bd. 10. Central Asia. Sogdian traders: a history. Boston, Brill, 2005, 406 p.

754. Diaspora entrepreneurial networks: four centuries of history. Ed. by Ina BAGHDIANTZ MAC CABE, Gelina HARLAFTIS and Ioanna PEPELASIS MINOGLOU. Oxford, Berg, 2005, XXII-440 p.

755. DUNAWAY (Finis). Natural visions: the power of images in American environmental reform. Chicago, University of Chicago Press, 2005, XXI-246 p.

756. EATON (Richard M.). A social history of the Deccan, 1300–1761. Eight Indian lives. Cambridge, New York a. Melbourne, Cambridge U. P., 2005, XIII-221 p. (New Cambridge history of India, 1, 8).

757. EL KENZ (David). Le massacre, objet d'histoire. Paris, Gallimard, 2005, 554 p. (Folio histoire).

758. Fodnoter. Træk af vandringens historie. (Essais sur l'histoire de la marche). Red. Karsten WIND MEYHOFF. København, Information, 2005, 207 p. (ill.).

759. HALL (Marcus). Earth repair. A transatlantic history of environmental restoration. Charlottesville, University of Virginia Press, 2005, XVI-310 p.

760. KENISTON MAC INTOSH (Marjorie). Working women in English society, 1300–1620. Cambridge, Cambridge U. P., 2005, XIV-291 p. (ill.).

761. KINGSLAND (Sharon E.). The evolution of American ecology, 1890–2000. Baltimore, Johns Hopkins U. P., 2005, X-313 p.

762. Living standards in the past. New perspectives on well-being in Asia and Europe. Ed. by Robert C. ALLEN, Tommy BENGTSSON and Martin DRIBE. New York, Oxford U. P., 2005, XXII-472 p. (Oxford scholarship online).

763. LO CASCIO (Elio), MALANIMA (Paolo). Cycles and stability. Italian population before the demographic transition (225 B.C.–A.D. 1900). *Rivista di storia economica*, 2005, 21, 3, p. 197-232.

764. MIEGGE (Mario). Capitalismo e modernità: una lettura protestante. Torino, Claudiana, 2005, 74 p. (Piccola collana moderna, 112).

765. PORTES (Alejandro), GUARNIZO (Luis E.), LANDOLT (Patricia). The study of transnationalism: pitfalls and promise of an emergent research field. *Ethnic and racial studies*, 99, 22, 2, p. 217-237.

766. SAYAR (Ahmed Güner). İktisat metodolojisi ve düşünce tarihi Yazıları. (Ecrits sur méthodologie d'économie et histoire de pensée). İstanbul, Ötüken, 2005, 283 p.

767. SCHLUCHTER (Wolfgang). Handlung, Ordnung und Kultur. Studien zu einem Forschungsprogramm im Anschluss an Max Weber. Tübingen, Mohr (Siebeck), 2005, X-258 p.

768. SCHREPFER (Susan R.). Nature's altars: mountains, gender, and American environmentalism. Lawrence, University Press of Kansas, 2005, XII-316 p.

769. SHELDON (Kathleen E.). Historical dictionary of women in Sub-Saharan Africa. Lanham a. Oxford, Scarecrow Press, 2005, XLI-405 p. (map). (Historical dictionaries of women in the world, 1).

770. Sozialstruktur (Die) und Sozialtopographie vorindustrieller Städte. Beiträge eines Workshop am Institut für Geschichte der Martin-Luther-Universität Halle-Wittenberg am 27.–28. Januar 2000. Hrsg. v. Matthias MEINHARDT und Andreas RANFT. Berlin, Akademie, 2005, 321 p. (Hallische Beiträge zur Geschichte des Mittelalters und der Frühen Neuzeit, 1).

771. Storia della famiglia in Europa. Il Novecento. A cura di Marzio BARBAGLI e David I. KERTZER. Roma e Bari, Laterza, 2005, VI-586 p.

772. STRATHERN (Marilyn). Kinship, law and the unexpected. relatives are always a surprise. Cambridge, Cambridge U. P., 2005, X-229 p.

773. SYLOS LABINI (Paolo). Storia e teoria economica: due casi degni di riflessione. *Rivista di storia economica*, 2005, 21, 2, p. 181-190.

774. ZELIZER (Viviana A.). La signification sociale de l'argent. Paris, Ed. du Seuil, 2005, 348 p.

§ 8. Addenda 2000–2004.

* 775. KIVISTO (Peter). Theorizing transnational immigration: a critical review of current efforts. *Ethnic and racial studies*, 2001, 24, 4, p. 549-577.

776. Agriculture méditerranéenne. Variété des techniques anciennes. Ed. par Marie-Claire AMOURETTI et Georges COMET. Aix-en-Provence, Publications de l'Université de Provence, 2002, 295 p. (Cahier d'histoire des techniques, 5).

777. CHARVOLIN (Florian). L'invention de l'environnement en France: Chroniques anthropologiques d'une institutionnalisation. Paris, La Découverte, 2003, 134 p.

778. DELORT (Robert), WALTER (François). Histoire de l'environnement européen. Paris, Presses Universitaires de France, 2001, 352 p.

779. ELIASSON (Per). Skog, makt och människor. En miljöhistoria om svensk skog 1800–1875. (Wald, Macht und Menschen. Eine Umweltgeschichte des schwedischen Waldes 1800–1875). Stockholm, Königl. Skogsoch Lantbruksakademien, 2002, 455 p. (Skogs- och lantbrukshistoriska meddelanden, 25).

780. FAIST (Thomas). Transnationalization in international migration: implications for the study of citizenship and culture. *Ethnic and racial studies*, 2000, 23, 2, p. 189-222.

781. GABACCIA (Donna R.). Emigranti. Le diaspore degli italiani dal Medioevo a oggi. Torino, Einaudi, 2003, 312 p.

782. LIGHT (Ivan), GOLD (Steven J.). Ethnic economies. San Diego a. London, Academic, 2000, XIII-302 p. (ill., ports.).

783. MAZOWER (Mark). Salonica, city of ghosts: Christians, Muslims and Jews, 1430–1950. London, Harper Collins, 2004, XIV-525 p. (ill., maps).

Cf. n^{os} *108-128, 486, 609, 1014, 1063, 5720-5776, 5943*

§ 9. History of civilization, sciences and education.

784. BRAMBILLA (Elena). Genealogie del sapere: università, professioni giuridiche e nobiltà togata in Italia, 13.–17. secolo: con un saggio sull'arte della memoria. Milano, UNICOPLI, 2005, 384 p. (Early modern, 19).

785. Chiesa e guerra: dalla benedizione delle armi alla Pacem in terris. A cura di Mimmo FRANZINELLI e Riccardo BOTTONI. Bologna, Il Mulino, 2005, 756 p. (Percorsi).

786. CHISICK (Harvey). Historical dictionary of the Enlightenment. Lanham, Scarecrow Press, 2005, XXXIII-512 p. (Historical dictionaries of ancient civilizations and historical eras, 16).

787. Companion (A) to modernist literature and culture. Ed. by David BRADSHAW and Kevin J. H. DETTMAR. Oxford, Blackwell, 2005, XXI-593 p. (Blackwell companions to literature and culture, 39).

788. COSMACINI (Giorgio). Storia della medicina e della sanità in Italia: dalla peste nera ai giorni nostri. Roma e Bari, Laterza, 2005, XIII-630 p. (Robinson. Letture).

789. Dictionnaire de la presse française pendant la Révolution, 1789–1799. La presse départementale. Sous la direction de Gilles FEYEL. Ferney-Voltaire, Centre international d'étude du XVIIIe siècle, 2005, [s. p.] (ill., facsims.). (Publications du Centre international d'étude du XVIIIe siècle, 15).

790. Dictionnaire historique de la laïcité en Belgique. Sous la dir. de Pol DEFOSSE; coordannateur, Jean-Michel DUFAYS; collaboratrice, Martine GOLDBERG. Bruxelles, Fondation rationaliste et les Éditions Luc Pire, 2005, 343 p. (ill.). (Voix de l'histoire).

791. DUMARÇAY (Jacques). Handbuch der Orientalistik = Handbook of oriental studies. Abt. 3. Bd. 15.

South-East Asia. Construction techniques in South and Southeast Asia: a history. Leiden a. Boston, Brill, 2005, [s. p.].

792. ECKART (Wolfgang Uwe). Geschichte der Medizin. Heidelberg, Springer, 2005, IX-324 p. (ill., port.).

793. Evropa a její duchovní tvář. Eseje – komentáře – diskuse. (Europe's spiritual face). Ed. Jiří HANUŠ, Jan VYBÍRAL. Brno, Centrum pro studium demokracie a kultury, 2005, 331 p. (Politika a náboženství, 2).

794. France and the Americas: culture, politics, and history: a multidisciplinary encyclopedia. Ed. by Bill MARSHALL assisted by Cristina JOHNSTON. Santa Barbara, ABC-CLIO, 2005, 3 vol., [s. p.]. (ill., maps). (Transatlantic relations series).

795. HEßLER (Martina). Bilder zwischen Kunst und Wissenschaft. Neue Herausforderungen für die Forschung. *Geschichte und Gesellschaft*, 2005, 31, 2, p. 266-292.

796. IANCU (Carol). Miturile fondatoare ale antisemitismului din Antichitate până în zilele noastre. (The fundamental myths of the anti-Semitism from Antiquity to our days). București, Editura Hasefer, 2005, 304 p.

797. Internet (The): a historical encyclopedia. Ed. by Hilary W. POOLE. V. 1. LAMBERT (Laura). Biographies. V. 2. WOODFORD (Chris). Issues. V. 3. MOSCHOVITIS (Chris) [et al.]. Chronology. Santa Barbara a. Oxford, ABC-CLIO, 2005, 3 vol., [s. p.]. (ill.).

798. JAY (Martin). Songs of experience. Modern American and European variations on a universal theme. Berkeley a. Los Angeles, University of California Press, 2005, X-431 p.

799. JONES (Karen R.), WILLS (John). The invention of the park: recreational landscapes from the Garden of Eden to Disney's Magic Kingdom. Cambridge, Polity, 2005, 216 p.

800. Jüdisches Leben im Rheinland. Vom Mittelalter bis zur Gegenwart. Hrsg. v. Monika GRÜBEL und Georg MOLICH. Köln, Weimar u. Wien, Böhlau, 2005, XIX-315 p.

801. KRAGH (Helge). Dansk naturvidenskabs historie. Bd. 1. Fra Middelalderlærdom til den nye videnskab: 1000–1730. (Histoire des sciences au Danemark. Tome 1. De la science médiévale au nouvel esprit scientifique: 1000–1730). Århus, Aarhus Universitetsforlag, 2005, 480 p.

802. Kulturněhistorická encyklopedie Slezska a severovýchodní Moravy. (Cultural and historical Encyclopedia of the Silesia and north-east Moravia). Tomo 1-2. Ostrava, Ústav pro regionální studia Ostravské univerzity, 2005, 586 p., 469 p. (photogr., ill.).

803. LLOYD (Geoffrey E.R.). The institutions of censure: China, Greece and the modern world. *Quaderni di storia*, 2005, 62, p. 7-52.

804. MAC GUIRE (Brian Patrick). Den levende middelalder: fortællinger om dansk og europæisk identitet. (Le Moyen Age vivant: récits illustrant l'idée danoise et européenne). København, Gyldendal, 2005, 301 p.

805. MAREȘ (Alexandru). Scriere și cultura românească veche. (Old Romanian writing and culture). București, Editura Academiei Române, 2005, 486 p.

806. MATTHEE (Rudi). The pursuit of pleasure. Drugs and stimulants in Iranian history, 1500–1900. Princeton, Princeton U. P., 2005, XV-346 p.

807. Mir i vojna: kul'turnye konteksty social'noj agressii: Vyborgskie chtenija (1), 1–3 sentjabrja 2003 g. (War and peace: Cultural contexts of social agression: Proceedings of the First Vyborg Readings, 1–3 September, 2003). Ed. Igor' O. ERMACHENKO, Lorina P. REPINA. Moskva, IVI RAN, 2005, 265 p. (ill.; bibl. incl.).

808. NESTORESCU-BĂLCEȘTI (Horia). Enciclopedia ilustrată a francmasoneriei din România. Vol. I-III. (The illustrated encyclopaedia of the freemasonry in Romania). București, Editura Phobos, 2005, 3 vol., 462 p., 440 p., 480 p.

809. ONUR (Bekir). Türkiye'de çocukluğun tarihi: çocukluğun sosyo-kültürel tarihine giriş. (Histoire de l'enfance: introduction à l'histoire socio-culturel de l'enfance). Ankara, İmge Kitabevi, 2005, 575 p.

810. Orta Karadeniz Kültürü. (La culture de la région de la Mer Noire). Ed. Bahaeddin YEDİYILDIZ, Hakan KAYNAR, Serhat KÜÇÜK. Ankara, Siyasal Kitabevi, 2005, XIII-722 p.

811. POCOCK (John Greville Agard). Barbarism and religion. Vol. 4. Barbarians, savages and empires. Cambridge, Cambridge U. P., 2005, XII-372 p.

812. Political culture in Central Europe (10th–20th century). 1. Middle Ages and Early Modern era. 2. 19th and 20th centuries). Ed. by Halina MANIKOWSKA, Magdalena HULAS and Jaroslav PÁNEK in collab. with Martin HOLÝ and Roman BARON. Praha, Institute of History, Academy of Sciences of the Czech Republic; Warszawa, Institute of History of the Polish Academy of Sciences, 2005, 2 vol., 389 p., 363 p.

813. Representations of the "other/s" in the Mediterranean world and their impact on the region. Ed. Nudret KURAN-BURÇOĞLU, Susan Gilson MILER. İstanbul, Isis, 2005, 308 p.

814. Säkularisation (Die) im Prozess der Säkularisierung Europas. Hrsg. Peter BLICKLE und Rudolf SCHLÖGL. Epfendorf, Bibliotheca Academica, 2005, 574 p. (Oberschwaben – Geschichte und Kultur, 13).

815. SCHULZ (Andreas). Lebenswelt und Kultur des Bürgertums im 19. und 20. Jahrhundert. München, Oldenbourg, 2005, IX-144 p. (Enzyklopädie deutscher Geschichte, 75).

816. Secolarizzazioni (Le) nel Sacro romano impero e negli antichi stati italiani: premesse, confronti, conseguenze. A cura di/hrsg. von Claudio DONATI, Helmut FLACHENECKER. Bologna, Il Mulino e Berlin,

Duncker & Humblot, 2005, 337 p. (Collezione Annali dell'Istituto storico italo-germanico in Trento. Contributi, 16).

817. SHORTER (Edward). A historical dictionary of psychiatry. Oxford, Oxford U. P., 2005, IX-338 p.

818. STRASSER (Bruno J.), BÜRGI (Michael). L'histoire des sciences, une histoire à part entière? *Schweizerische Zeitschrift für Geschichte*, 2005, 55, 1, p. 3-16.

819. Svenska folkets hälsa i historiskt perspektiv. (La santé des Suédois dans une perspective historique). Red. Jan SUNDIN [et al.]. Stockholm, Statens folkhälsoinstitut, 2005, 474 p. (Statens folkhälsoinstitut, 2005:8).

820. Transmission (La) des savoirs au Moyen Age et à la Renaissance. Sous la dir. de Alfredo PERIFANO. Besançon, Presses universitaires de Franche-Comté, 2005, [s. p.]. (Littéraires).

821. Universités (Les) en Europe du XIIIe à nos jours. Espaces, modèles et fonctions. Actes du colloque international d'Orléans, 16 et 17 octobre 2003. Éd. par Frédéric ATTAL, Jean GARRIGUES, Thierry KOUAMÉ et Jean-Pierre VITTU. Paris, Publications de la Sorbonne, 2005, 294 p. (Homme et société, 31).

822. Zivilisierungsmissionen. Imperiale Weltverbesserung seit dem 18. Jahrhundert. Hsrg. v. Boris BARTH und Jürgen OSTERHAMMEL. Konstanz, Universitätsverlag Konstanz, 2005, 438 p.

§ 9. Addenda 1999–2002.

823. Civiltà (La) dei greci: forme, luoghi, contesti. A cura di Massimo VETTA. Roma, Carocci, 2001, 400 p. (ill.). (Università, 256).

824. POTTS (Annie). The Science/fiction of sex: feminist deconstruction and the vocabularies of heterosex. London, Routledge, 2002, XI-292 p.

825. Sexual cultures in Europe. Ed. by Franz X. EDER, Leslie HALL and Gert HEKMA. Vol. 1. National Histories. Manchester, Manchester U. P., 99, X-270 p.

Cf. n°s 251, 672

§ 10. History of art.

* 826. BHA. Bibliography of the history of Art. Bibliographie de l'histoire de l'art. Vol. 15. 1-4. 2005. [Vol. 14. 1-4, 2004. Cf. Bibl. 2004, n° 942.] Paris, Centre National de la Recherche Scientifique a. Santa Monica, J. Paul Getty Trust, 2005, CD-ROM.

* 827. SCHÄFER (Jörgen). Exquisite Dada: a comprehensive bibliography. Gen. ed. Stephen C. FOSTER. New York, G.K. Hall, 2005, 689 p. (Crisis and the arts: the history of Dada, 10).

* 828. TARUSKIN (Richard). The Oxford history of Western music. Vol. 6. Resources: chronology, bibliography, master index. Oxford, Oxford U. P., 2005, XIX-329 p. (ill.).

829. BARAGLI (Sandra). Il Trecento: [le parole chiave, i luoghi, i protagonisti]. Milano, Electa, 2005, 383 p. (ill.). (I secoli dell'arte).

830. Cambridge companion (The) to Liszt. Ed. by Kenneth HAMILTON. Cambridge, Cambridge U. P., 2005, XIV-282 p. (ill., music). (Cambridge companions to music).

831. Cambridge companion (The) to Raphael. Ed. by Marcia B. HALL. New York, Cambridge U. P., 2005, XIII-415 p. (ill.). (Cambridge companions to the history of art).

832. COSMA (Viorel). Enciclopedia muzicii românești de la origini până în zilele noastre. Vol. 1. A-B. (Encyclopedia of Romanian music since its origins until present. Vol. 1. A-B). În colaborare cu Luminița VARTOLOMEI și Constantin CATRINA. București, Editura Arc 2000, 2005, 540 p.

833. CURL (James Stevens). The Egyptian revival. Ancient Egypt as the inspiration for design motifs in the West. London a. New York, Routledge, 2005, XXXVI-572 p.

834. Early Netherlandish paintings: rediscovery, reception, and research. Ed. by Bernhard RIDDERBOS, Anne VAN BUREN and Henk VAN VEEN. Amsterdam, Amsterdam U. P., 2005, X-481 p. (ill.).

835. EDWARDS (Gwynne). A companion to Luis Buñuel. Woodbridge, Tamesis, 2005, 176 p. (ill.). (Colección Támesis. 210 Serie A, Monografías).

836. Encyclopedia of recorded sound. Ed. by Frank HOFFMANN; technical editor Howard FERSTLER. New York a. London, Routledge, 2005, 2 vol., XII-1289 p. (ill.).

837. GÉNETIOT (Alain). Le classicisme. Paris, Presses universitaires de France, 2005, X-475 p. (Quadrige Manuels).

838. Greenwood encyclopedia (The) of rock history. V. 1. SCRIVANI-TIDD (Lisa). The early years, 1951–1959. – V. 2. MARKOWITZ (Rhonda). The British invasion years, 1960–1966. – V. 3. SMITH (Chris). The rise of album rock, 1967–1973. – V. 4. SMITH (Chris). From arenas to the underground, 1974–1980. – V. 5. JANOSIK (MaryAnn). The video generation, 1981–1990. – V. 6. GULLA (Bob). The grunge and post-grunge years, 1991–2005. Westport, Greenwood Press, 2005, 6 vol., [s. p.]. (ill., ports.).

839. KIRCHNER (Bill). The Oxford companion to jazz. New York a. Oxford, Oxford U. P., 2005, XI-852 p. (ill.).

840. MODORCEA (Grid). Dicționarul cinematografic al artelor românești. (Cinematographic dictionary of Romanian arts). București, Editura Tibo, 2005, 415 p.

841. NAYROLLES (Jean). L'invention de l'art roman à l'époque moderne (XVIIIe–XIXe siècles). Rennes, Presses universitaires de Rennes, 2005, 402 p. (ill.). (Collection Art & société).

842. Originale assente (L'): introduzione allo studio della tradizione classica. A cura di Monica CENTANNI. Milano, Bruno Mondadori, 2005, VIII-439 p. (ill., map). (Sintesi).

843. STOIANOV (Carmen Antoaneta). Istoria muzicii românești. (The history of the Romanian music). București, Editura Fundației România de Mâine, 2005, 152 p.

§ 10. Addenda 1995–2004.

844. BRUSH (Kathryn). The shaping of art history: Wilhelm Vöge, Adolph Goldschmidt, and the study of medieval art. Cambridge, Cambridge U. P., 96, XIII-263 p. (ill.).

845. Écrire l'histoire de l'art: France-Allemagne, 1750–1920. *Revue Germanique internationale*, 2000, 13 285 p.

846. GEBHARDT (Volker). Das Deutsche in der deutschen Kunst. Köln, DuMont, 2004, 511 p. (ill.).

847. GHELARDI (Maurizio) [et al.]. Relire Burckhardt. Textes inédits en français de Jacob Burckhardt. Paris, École nationale supérieure des beaux-arts, 97, 254 p. (ill.). (Louvre conférences et colloques. Collection Principes et théories de l'histoire de l'art).

848. HART (Joan), RECHT (Roland), WARNKE (Martin). Relire Wölfflin: cycle de conférences organisé au musée du Louvre par le Service culturel du 29 novembre au 20 décembre 1993 sous la direction de Matthias Waschek. Introduction de Jacques THUILLIER. Paris, Ecole nationale supérieure des beaux-arts, 95, 181 p. (ill.). (Louvre conférences et colloques. Collection Principes et théories de l'histoire de l'art).

849. Histoire de l'histoire de l'art. *Revue de l'art*, 2004, 146, 4, [s. p.].

850. Kunsthistoriographien (Die) in Ostmitteleuropa und der nationale Diskurs. Hrsg. v. Robert BORN, Alena JANATKOVÁ und Adam S. LABUDA. Berlin, Gebr. Mann, 2004, 479 p. (ill.). (Humboldt-Schriften zur Kunst- und Bildgeschichte, 1).

851. MARESCA (Paola). Giardini incantati, boschi sacri e architetture magiche. Presentazione di Litta Maria MEDRI. Firenze, A. Pontecorboli, 2004, 215 p. (ill.).

852. Renoir (Da) a De Staël: Roberto Longhi e il moderno. A cura di Claudio SPADONI. Milano, Mazzotta, 2003, 415 p. (ill., ports.).

853. ROMANO (Giovanni). Storie dell'arte: Toesca, Longhi, Wittkower, Previtali. Roma, Donzelli, 98, 112 p. (Saggi. Arti e lettere).

854. Toesca, Venturi, Argan: storia dell'arte a Torino 1907–1931. A cura di Michela DI MACCO. *Ricerche di storia dell'arte*, 96, 59, 95 p.

855. Wilhelm Worringer: Schriften. Hrsg. v. Hannes BÖHRINGER, Helga GREBING und Beate SÖNTGEN unter Mitarbeit von Arne ZERBST. München, Wilhelm Fink Verlag, 2004, 2 vol., 1500 p. mit CD-ROM.

Cf. n°s 470, 795

§ 11. History of religions.

a. General

* 856. Chroniques bibliographiques. *Hieros*, 2005, 10, [s. p.]. [2004, 9, Cf. Bibl. 2004, n° 970.]

* 857. Ephemerides theologicae lovanienses. Elenchus bibliographicus. Editae cura J.-M. AUWERS, E. BRITO, L. DE FLEURQUIN, J. FAMERÉE, É. GAZIAUX, J. HAERS, A. HAQUIN, M. LAMBERIGTS, J. LUST, G. VAN BELLE, J. VERHEYDEN. Tomus LXXXI. [Tomus LXXX. Cf. Bibl. n° 971.] Leuven, Peeters, 2005, 848 p.

* 858. *Religious Studies Review*, 31, 1, 2, 3, 4, 2005, [s. p.]. [2004, 30, 1-4, Cf. Bibl. 2004, n° 973.]

* 859. *Science of Religion*, 30, 2005, 1-2, 285 p. 231 p.

860. CALDWELL AMES (Christine). Does Inquisition belong to religious history? *American historical review*, 2005, 110, 1, p. 11-37.

861. Companion (A) to second-century Christian 'Heretics'. Ed. by Antti MARJANEN and Petri LUOMANEN. Leiden, Brill, 2005, XIII-385 p. (Supplements to Vigiliae Christianae, 76).

862. DIEZ DE VELASCO (Francisco). La historia de las religiones. Métodos y perspectivas. Madrid, Akal, 2005, 287 p.

863. Encyclopaedia of Religion. Ed. Lindsay JONES. Detroit, Thomson Gale, 2005, 14 vol., CXLVIII-10017 p.

864. Grandes religions (Les). Éd. Xavier DUFOUR. Paris, Ed. du Cerf, 2005, 206 p.

865. Hedendomen i historiens spegel: bilder av det förkristna Norden. (Le paganisme au miroir de l'histoire: images de la Scandinavie païenne). Red. Catharina RAUDVERE, Anders ANDRÉN, Kristina JENNBERT. Lund, Nordic Academic Press, 2005, 205 p. (pl.). (Vägar till Midgård, 6).

866. How to do comparative religion? Three ways, many goals. Ed. by René GOTHONI. Berlin a. New York, de Gruyter, 2005, 221 p.

867. Introduction to World Religions. Ed. Christopher H. PARTRIDGE. Minneapolis MN, Fortress, 2005, VI-495 p. a. CD-ROM.

868. KOZOLUPENKO (Dar'ja P.). Mif: Na granjakh kul'tury (sistemnyj i mezhdistsiplinarnyj analiz mifa v ego razlichnykh aspektakh: estestvenno-nauchnaja, psikhologicheskaja, kul'turno-poeticheskaja, filosofskaja i sotsial'naja grani mifa kak kompleksnogo javlenija kul'tury). (Myth on the frontiers of culture: A system and interdisciplinary analysis of myth in its various aspects: scientific, cultural and political, philosophical, and social aspects of myth as a complex phenomenon of culture). Moskva, Kanon+, 2005, 212 p.

869. MASUZAWA (Tomoko). The invention of world religions. Or, how European universalism was preserved in the language of pluralism. Chicago, University of Chicago Press, 2005, 384 p.

870. MELTON (J. Gordon). Encyclopedia of Protestantism. New York, Facts On File, 2005, XXVI-628 p. (ill.). (Encyclopedia of world religions).

871. Mythes (Les). Langages et messages. Éd. par Julien RIES. Rodez, Rouergue, 2005, 237 p.

872. NAGOVITSYN (Aleksej E.). Drevnie tsivilizatsii: obshchaja teorija mifa. (Ancient civilizations: a general theory of myth). Moskva, Akademicheskij proekt, 2005, 656 p. (ill). (Kul'tury).

873. Pilgrimage in Graeco-Roman and early Christian Antiquity. Seeing the gods. Ed. by Jas ELSNER and Ian RUTHERFORD. Oxford a. New York, Oxford U. P., 2005, XVII-513 p.

874. Religion in Geschichte und Gegenwart. Hrsg. v. Hans Dieter BETZ, Don S. BROWNING, Berndt JANOWSKI und Eberhard JUNGEL. Band 8. T-Z. Tübingen, Mohr, 2005, LXXXVIII p.- 1966 col.

875. Religion/s between covers. Dilemmas of the world religions textbook. *Religious studies review*, 2005, 31, p. 1-36.

876. Routledge companion (The) to the study of religion. Ed. John R. HINNELLS. London a. New York, Routledge, 2005, 556 p.

877. Stiftungen im Christentum, Judentum und Islam vor der Moderne. Auf der Suche nach ihren Gemeinsamkeiten und Unterschieden in religiösen Grundlagen, praktischen Zwecken und historischen Transformationen. Hrsg. v. Michael BORGOLTE. Berlin, Akademie, 2005, 297 p. (Stiftungsgeschichten, 4).

Cf. n°s 1108, 1326, 1713, 1723, 4808-4825

b. Special studies

* 878. ALTHANN (Robert). Elenchus of Biblica. 2002. [2001. Cf. Bibl. 2004, n° 1002.] Roma, Ed. Istituto Pontificio Biblico, 2005, 846 p.

* 879. Bibliographia internationalis spiritualitatis. Vol. 37. 2002. [Vol. 36. 2001. Cf. Bibl. 2004, n° 1003.] Roma, Ed. del Teresianum, 2005, XXXI-502 p.

* 880. Index Islamicus 2003. [2002. Cf. Bibl. 2004, n° 972.] Ed. by Heather BLEANEY. Leiden, Boston, Brill, 2005, XXXIII-729 p.

* 881. Internationale Zeitschriftenschau für Bibelwissenschaft und Grenzgebiete. International Review of Biblical Studies. Revue internationale des études bibliques. Bd. 50, 2003–2004. [Bd. 49. 2001–2002. Cf. Bibl. 2004, n° 1004.] Düsseldorf, Patmos Verlag, 2005, XII-592 p.

* 882. New Testament Abstracts. Vol. 49, 2005, 699 p. [Vol. 48. 2004. Cf. Bibl. 2004, n° 1005.]

* 883. Old Testament Abstracts. Vol. 28, 2005, 434 p. [Vol. 27. 2004. Cf. Bibl. 2004, n° 1006.]

884. ALBANI (Riccardo). Tempo e storia in Ernesto de Martino. *Studi e materiali di storia delle religioni*, 2005, 71, p. 355-383.

885. Angelo Brelich e la storia delle religioni. Temi, problemi e prospettive. A cura di Maria Grazia LANCELLOTTI e Paolo XELLA. Verona, Essedue edizioni, 2005, 197 p.

886. ANGLEVIEL (Frédéric). Des géants de pierre à l'homme-oiseau. Les manifestations de la religion ancienne de l'île de Pâques. *Religions & sociétés*, 2005, 4, p. 72-81.

887. ANTTONEN (Veikko). Space, body, and the notion of boundary. A category-theoretical approach to religion. *Temenos*, 2005, 41, p. 185-201.

888. Bibliographie zur Symbolik, Iconographie und Mythologie. Internationales Referateorgan. Jahrgang 33. 2000. Baden-Baden, Valentin Koerner, 2005, 115 p.

889. Biographisch-Bibliographisches Kirchenlexicon. Begründet und herausgegeben von Friedrich Wilhelm BAUTZ. Fortgeführt von Traugott BAUTZ. Band 24. Ergänzungen XI-XII. Herzberg, T. Bautz, 2005, XL-1600 col.

890. BLASCO HERRANZ (Inmaculada). Género y religión: de la feminización de la religión a la movilización católica femenina. Una revisión crítica. *Historia social*, 2005, 53, p. 119-136.

891. Bridge or barrier. Religion, violence and visions of peace. Ed. Gerrie ter HAAR and James J. BUSUTTIL. Leiden, Brill, 2005, XII-396 p.

892. BULBILIA (Joseph). Are there any religion? An evolutionary exploration. *Method & theory in the study of religion*, 2005, 17, 2, p. 71-100.

893. CAPOMACCHIA (Anna Maria G.). I contributi di Angelo Brelich in "Studi e materiali di storia delle religioni". *Studi e materiali di storia delle religioni*, 2005, 71, p. 221-235.

894. Catholicisme hier aujourd'hui demain. Encyclopédie publiée sous le patronage de l'Institut Catholique de Lille par G. MATHON et G. H. BAUDRY. Fasc. 76-77. C-D-E. Tables complements et mises à jour. Paris, Letouzey & Ané, 2005, col. 254-763.

895. Christelijke encyclopedie. Ed. George HARINCK [et al.]. Kampen, Kok, 2005, 3 vol., XI-692 p., 693-1373 p., 1376-2026 p.

896. CLARK (Elizabeth A.). Engaging Bruce Lincoln. *Method & theory in the study of religion*, 2005, 17, 1, p. 11-17.

897. COLLINS (Peter). Thirteen ways of looking a "Ritual". *Contemporary religion*, 2005, 20, 3, p. 323-342.

898. CRAIG BRITTAIN (Christopher). The "Secular" as tragic category: on Taal Asad, religion and representation. *Method & theory in the study of religion*, 2005, 17, 2, p. 149-165.

899. Cuisine (La) et l'autel. Les sacrifices en question dans les sociétés de la Méditerranée ancienne. Éd. par Stella GEORGOUDI, Renée KOCH et Francis

SCHMIDT. Turnhout, Brepols, 2005, XVII-455 p. (Bibliothèque de l'École des Hautes Études. Section des sciences religieuses, 124).

900. Encyclopédie de l'Islam. Ed. par P. J. BEARMAN, T. BIANQUIS, C. E. BOSWORTH, E. VAN DONZEL et W. P. HEINRICHS. Tome XX. V-Z. Leiden, Brill, 2005, XVI-621 p.

901. Figure (La) du prêtre dans les grandes traditions religieuses. Actes du colloque organisé en hommage à M. l'abbé Julien Ries à l'occasion de ses 80 ans. Éd. par P. MARCHETTI et André MOTTE. Louvain, Paris et Dudley, Peeters, 2005, VI-231 p.

902. GANDINI (Mario). Raffaele Pettazzoni dall'estate 1946 all'inverno 1947–1948. Materiali per una biografia. *Strada maestra*, 2005, 58, p. 53-250. – IDEM. Raffaele Pettazzoni nel 1948. Materiali per una biografia. *Strada maestra*, 2005, 59, p. 51-207.

903. GELEZ (Philippe). Petit guide pour servir à l'etude de l'islamisation en Bosnie et en Herzegovine: avec le recueil des sources connues, ainsi qu'un commentaire sur l'utilisation qui a été faite de celles-ci. İstanbul, Isis, 2005, 237 p.

904. GELLER (Jay). En jeu: Lincoln Logs or Pick-Up Sticks. *Method & theory in the study of religion*, 2005, 17, 1, p. 18-26.

905. GRIMSHAW (Mike). Notes towards a Loos-ian Theory of religion in modernity. *Method & theory in the study of religion*, 2005, 17, 4, p. 382-392.

906. GROTTANELLI (Cristiano). Fruitful Death: Mircea Eliade and Ernst Jünger on Human sacrifice, 1937–1945. *Numen*, 2005, 52, p. 116-145.

907. HAYDEN (J. Michael), GREENSHIELDS (Malcolm R.). Six hundred years of Reform: bishops and the French Church, 1190–1789. Ithaca, McGill-Queen's U. P., 2005, XX-604 p. (McGill-Queen's studies in the history of religion, series two, 37).

908. Henry Corbin. Philosophies et sagesses des religions du livre. Actes du colloque "Henry Corbin". Sous la direction de Mohammad Alì AMIR-MOEZZI, Christian JAMBET et Pierre LORY. Turnhout, Brepols, 2005, 248 p. (Bibliothèque de l'École des Hautes Études. Section des sciences religieuses, 126).

909. Historicizing "Tradition" in the study of religion. Ed. by Steven ENGLER and Gregory P. GREVE. Berlin a. New York, de Gruyter, 2005, 395 p.

910. Întâlniri cu Mircea Eliade. Ed. Mihaela GLIGOR şi Mac Linscott RICKETTS. Cluj-Napoca, Casa Cărţii de Stinta, 2005, 255 p.

911. Interrompere il quotidiano. La costruzione del tempo nell'esperienza religiosa. A cura di Natale SPINETO. Milano, Jaca Book, 2005, 214 p.

912. Întotdeauna Orientul. Corespondenţă Mircea Eliade-Stig Wikander (1948–1977). Ed. Mihaela TIMUŞ. Iaşi, [s. n.], 2005, [s. p.].

913. Invisibile (L') e lo spazio. L'ordine dell'invisibile. *Religioni e società*, 2005, 52, p. 6-55; 57-98.

914. JACKSON (Peter). Retracing the path. Gesture, memory, and the exegesis of tradition. *History of religions*, 2005, 45, 1, p. 1-28.

915. KARA (Mustafa). Türk tasavvuf tarihi araştırmaları: tarikatlar/tekkeler/şeyhler. (Recherches sur l'histoire de mysticisme islamique turc: ordres/tekke/şeyh). İstanbul, Dergâh, 2005, 615 p.

916. KNOTT (Kim). Spatial theory and method fort the study of religion. *Temenos*, 2005, 41, p. 153-184.

917. KONG (Lily). Religious processions. Urban politics and poetics. *Temenos*, 2005, 41, p. 225-249.

918. KRAFFT (Otfried). Papsturkunde und Heiligsprechung: die päpstlichen Kanonisationen vom Mittelalter bis zur Reformation: ein Handbuch. Köln, Böhlau, 2005, XII-1247 p. (ill.). (Archiv für Diplomatik, 9. Beiheft).

919. LAMOTHE (Kimerer L.). Why dance? Towards a theory of religion as practice and performance. *Method & theory in the study of religion*, 2005, 17, 2, p. 101-133.

920. LE GALL (Dina). A culture of Sufism. Naqhbandis in the Ottoman world, 1450–1700. Albany, State University of New York Press, 2005, XII-285 p. (Medieval Middle East history).

921. LE TOURNEAU (Dominique). Les mots du christianisme: Catholicisme-orthodoxie-protestantisme. Paris, Fayard, 2005, 742 p. (Bibliothèque de culture religieuse).

922. LEUSTEAN (Lucian N.). Towards an integrative theory of religion and politics. *Method & theory in the study of religion*, 2005, 17, 4, p. 364-381.

923. Lexikon neureligiöser Gruppen. Szenen und Weltanschauungen: Orientierungen im religiösen Pluralismus. Hrsg. v. Harald BEER [et al.]. Freiburg-Basel-Wien. Herder, 2005, XII-1474 col.

924. LINCOLN (Bruce). Responsa Miniscula. *Method & theory in the study of religion*, 2005, 17, 1, p. 59-67. – IDEM. Theses on method. *Method & theory in the study of religion*, 2005, 17, 1, p. 8-10.

925. LORY (Pierre). Note sour l'ouvrage "Religion after Religion" – Gershom Scholem, Mircea Eliade and Henry Corbin at Eranos, par Steven Wasserstorm. *Achaeus*, 2005, 9, 1-4, p. 107-113.

926. LOUTH (Andrew). Mysticism. Name and thing. *Achaeus*, 2005, 9, 1-4, p. 9-21..

927. MAC CUTCHEON (Russel T.). Affinities, benefits, and costs: the ABCs of good scholars gone public. *Method & theory in the study of religion*, 2005, 17, 1, p. 27-43.

928. MASSENZIO (Marcello). The Italian school of "History of Religions". *Religion*, 2005, 35, p. 209-222.

929. MAZZOLENI (Gilberto). Pettazzoni. L'eredità e l'oblio. *Studi e materiali di storia delle religioni*, 2005, 71, p. 13-16.

930. Mifologema zhenshchiny-sud'by u drevnikh kel'tov i germantsev. (The mythologem of woman-fortune in ancient Celtic and Germanic beliefs). Ed. Tatiana A. MIKHAJLOVA. Moskva, Indrik, 2005, 336 p. (bibl. incl.).

931. MOLENDIJK (Arie L.). The emergence of the science of religion in the Netherlands. Leiden a. Boston, Brill, 2005, XII-316 p.

932. OLSON (Carl). Politics, power, discourse and representation: a critical look at Said and some of his children. *Method & theory in the study of religion*, 2005, 17, 4, p. 317-336.

933. ORYE (Lieve). Reappropriating "religion"? Constructively reconceptualising (human) science and the study of religion. *Method & theory in the study of religion*, 2005, 17, 4, 337-363.

934. Osmanlı toplumunda tasavvuf ve sufiler: kaynaklar-doktrin-ayin ve erkân-tarikatlar-edebiyat-mimari-güzel sanatlar-modernizm. (Mysticisme islamique et les soufi: sources – doctrine – cérémonie – ordres – litérature – beaux-arts – modernisme). Ed. Ahmet Yaşar OCAK. Ankara, Türk Tarih Kurumu, 2005, XXXII-656 p.

935. PENNINGTON (Brian K.). Introduction: a critical evaluation of the work of Bruce Lincoln. *Method & theory in the study of religion*, 2005, 17, 1, p. 1-7.

936. PYYSIÄINEN (Ilkka). God. A brief history with a cognitive explanation of the concept. *Temenos*, 2005, 41, p. 77-128.

937. REFF (Daniel T.). Plagues, priests, and demons. Sacred narratives and the rise of christianity in the Old World and the New. New York, Cambridge U. P., 2005, XIII-290 p.

938. Religiones andinas. Enciclopedia Iberoamericana de Religiones, 4. Ed. Manuel M. MARZAL. Madrid, Trotta, 2005, 381 p.

939. Religioni, democrazie e diritti. *Daimon*, 2005, p. 5- 136.

940. Religious polemics in context. Papers presented to the Second International Conference of the Leiden Institute fort he Study of Religions (LISOR) held at Leiden, 27–28 April 2000. Ed. by T. L. HETTEMA and A. VAN DER KOOIJ. Leiden, Brill, 2005, 624 p.

941. Religiousness (The) of violence. *Numen*, 2005, 52, p. 1-145.

942. RIES (Julien). Il mito e il suo significato. Milano, Jaca Book, 2005, 287 p.

943. ROUSSEAU (Vanessa). Le goût du sang. Croyances et polémiques dans la chrétienté occidentale. Paris, Armand Colin, 2005, 318 p. (L'Histoire à l'œuvre).

944. SANTIEMMA (Adriano). Il mito, il rito, il pensiero e l'azione. *Studi e materiali di storia delle religioni*, 2005, 71, p. 237-248.

945. SATLOW (Micheal L.). Disappearing categories: using categories in the study of religion. *Method & theory in the study of religion*, 2005, 17, 4, p. 287-298.

946. SCHILBRACK (Kevin). Bruce Lincoln's philosophy. *Method & theory in the study of religion*, 2005, 17, 1, p. 44-58.

947. SHEPHERD (John J.). The Ninian smart archive and bibliography. *Religion*, 2005, 35, p. 167-197.

948. SØRENSEN (Jesper). Religion in mind. A review article of the cognitive science of religion. *Numen*, 2005, 52, p. 465-494.

949. STRINGER (Martin D.). A sociological history of Christian worship. Cambridge, Cambridge U. P., 2005, 268 p.

950. TINAZ (Nuri). A social analysis of religious organisations: the cases of church, sect, denomination, cult and New Religious Movements (NRMs) and their typologies. *İslâm Araştırmaları Dergisi /Turkish journal of Islamic studies*, 2005, 13, p. 63-108.

951. TWEED (Thomas A.). Marking religion's boundaries. Constitutive terms, orienting tropes, and exegetical fussiness. *History of religions*, 2005, 44, 3, p. 252-276.

952. WIEBE (Donald). The politics of wishful thinking? Disentangling the role of the scholar-scientist from that of the public intellectual in the modern academic study of religion. *Temenos*, 2005, 41, p. 7-38.

953. ZHELEZNOVA (Natal'ja A.). Uchenie Kundakundy v filosofsko-religioznoj traditsii dzhajnisma. (The doctrine of Kundakunda in the philosophic and religious tradition of Jainism). RAN, In-t vostokovedeinja, In-t filosofii. Moskva, Vostochnaja literatura, 2005, 343 p. (bibl. p. 323-331; ind. p. 332-338). [English summary]

954. ZINCONE (Sergio). 80 anni dalla fondazione di SMSR. *Studi e materiali di storia delle religioni*, 2005, 71, p. 5-11.

Cf. n° 574

§ 12. History of philosophy.

955. BIÇAK (Ayhan). Türk Düşüncesi araştırmalarında Yöntem Sorunu ve Devlet. (Etat et problème méthodologique dans les recherches sur la pensée turque). *Dîvân İlmî Araştırmalar*, 2005, 18, p. 59-78.

956. Cambridge companion (The) to Arabic philosophy. Ed. by Peter ADAMSON and Richard C. TAYLOR. Cambridge a. New York, Cambridge U. P., 2005, XVIII-448 p. (Cambridge companions to philosophy).

957. Cambridge companion (The) to Berkeley. Ed. by Kenneth P. WINKLER. Cambridge, Cambridge U. P.,

2005, XIV-454 p. (ill., charts). (Cambridge companions to philosophy).

958. CARTWRIGHT (David E.). Historical dictionary of Schopenhauer's philosophy. Lanham a. Oxford, Scarecrow Press, 2005, LXV-239 p. (Historical dictionaries of religions, philosophies, and movements, 55).

959. Companion (A) to Heidegger. Ed. by Hubert L. DREYFUS and Mark A. WRATHALL. Malden a. Oxford, Blackwell Pub., 2005, XVII-540 p. (Blackwell companions to philosophy, 29).

960. CRAIG (Edward). The shorter Routledge Encyclopedia of philosophy. London, Routledge, 2005, XXVI-1077.

961. DEMIR (Remzi). Philosophia Ottomanica: Osmanlı İmparatorluğu döneminde Türk felsefesi. 1. cilt. Eski felsefe. (Philosophia Ottomanica: la philosophie turque au temps de l'Empire Ottoman, 1. La philosophie ancienne). 2. cilt. Eski ile yeni felsefe arasında. (2. Entre la philosophie ancienne et moderne). Ankara, Lotus, 2005, 2 vol., 233 p., 240 p.

962. Encyclopedia of nineteenth-century thought. Ed. by Gregory CLAEYS. London, Routledge, 2005, XVI-549 p.

963. GETHMANN-SIEFERT (Annemarie). Einführung in Hegels Ästhetik. München, W. Fink, 2005, 376 p. (UTB Philosophie, 2646).

964. HOLZHEY (Helmut), MUDROCH (Vilem). Historical dictionary of Kant and Kantianism. Lanham a. Oxford, Scarecrow Press, 2005, XV-374 p. (ill.). (Historical dictionaries of religions, philosophies, and movements, 60).

965. KENNY (Anthony John Patrick). A new history of Western philosophy. Vol. 2. Medieval philosophy. Oxford, Claredon Press, 2005, XIV-252 p. (ill.).

966. KERVÉGAN (Jean-François). Hegel et l'hégélianisme. Paris, Presses universitaires de France, 2005, 127 p. (Que sais-je? 1029).

967. Lexikon zum aufgeklärten Absolutismus in Europa: Herrscher-Denker-Sachbegriffe. Hrsg. v. Helmut REINALTER. Wien, Böhlau, 2005, 663 p. (UTB, 8316).

968. SAYHI-PÉRIGOT (Béatrice). Dialectique et littérature: les avatars de la dispute entre Moyen Age et Renaissance. Paris, H. Champion, 2005, 736 p. (Bibliothèque littéraire de la Renaissance, 58).

§ 12. Addendum 1999.

969. CAZZANIGA (Gian Mario). La religione dei moderni. Pisa, ETS, 99, 348 p. (Filosofia, 23).

Cf. n°s 32, 868, 5311-5360

§ 13. History of literature.

* 970. Bibliographie der deutschen Sprach- und Literaturwissenschaft. Hrsg. v. Wilhelm R. SCHMIDT.

XLIV. 2004. [XLIII. 2003. Cf. Bibl. 2004, n° 1100.] Bearb. v. Doris MAREK und Susanne PRÖGER. Frankfurt am Main, Klostermann, 2005, [s. p.].

* 971. Bibliographie der französischen Literaturwissenschaft. XLII. 2004. [XLI. 2003. Cf. Bibl. 2004, n° 1101.] Begrundet von Otto KLAPP; bearbeitet und herausgegeben von Astrid KLAPP-LEHRMANN. Frankfurt am Main, Klostermann, 2005, [s. p.].

* 972. BIGLI. Bibliografia generale della lingua e della letteratura italiana. Vol. 13. 2003. [Vol. 12. 2002. Cf. Bibl. 2004, n° 1102.] Diretta da Enrico MALATO; condirettori Massimiliano MALAVASI e Debora PISANO. Roma, Salerno, 2005, 2 vol., 2200 p.

* 973. BURNS (Grant). The railroad in American fiction: an annotated bibliography. Jefferson a. London, McFarland, 2005, IX-281 p.

* 974. COULON (Virginia). Bibliographie francophone de littérature africaine: (Afrique subsaharienne). Vanves, EDICEF, 2005, 479 p.

975. Cambridge companion (The) to American modernism. Ed. by Walter KALAIDJIAN. Cambridge a. New York, Cambridge U. P., 2005, XIX-333 p. (ill.). (Cambridge companions to literature).

976. Cambridge companion (The) to Baudelaire. Ed. by Rosemary LLOYD. Cambridge, Cambridge U. P., 2005, XVIII-234 p. (Cambridge companions to literature).

977. Cambridge companion (The) to Montaigne. Ed. by Ullrich LANGER. Cambridge, Cambridge U. P., 2005, XVII-247 p. (Cambridge companions to philosophy).

978. Cambridge companion (The) to Native American literature. Ed. by Joy PORTER and Kenneth M. ROEMER. Cambridge a. New York, Cambridge U. P., 2005, XVIII-343 p. (ill.). (Cambridge Collections Online).

979. Cambridge companion (The) to the Latin American novel. Ed. by Efraín KRISTAL. New York, Cambridge U. P., 2005, XVIII-336 p. (Cambridge companions to literature).

980. Cambridge companion (The) to the literature of the First World War. Ed. by Vincent SHERRY. Cambridge, Cambridge U. P., 2005, XVIII-322 p. (ill.). (Cambridge companions to literature).

981. CHARTIER (Roger). Inscrire et effacer. Culture écrite et littérature, XIe–XVIIIe siècle. Paris, Gallimard et Ed. du Seuil, 2005, 192 p. (Hautes études).

982. Companion (A) to ancient epic. Ed. by John Miles FOLEY. Oxford, Blackwell, 2005, XXIV-664 p. (ill., map). (Blackwell companions to the ancient world. Literature and culture).

983. Companion (A) to magical realism. Ed. by Stephen M. HART and Wen-chin OUYANG. Woodbridge,

Tamesis, 2005, VII-293 p. (Colección Támesis. 220 Serie A, Monografías).

984. Companion (A) to Mark Twain. Ed. by Peter MESSENT and Louis J. BUDD. Malden a. Oxford, Blackwell Pub., 2005, XVIII-568 p. (Blackwell companions to literature and culture, 37. Blackwell reference online).

985. Concise companion (A) to the Victorian novel. Ed. by Francis O'GORMAN. Oxford, Blackwell, 2005, XXIII-277 p. (Blackwell concise companions to literature and culture. Blackwell reference online).

986. COOK (James Wyatt). Encyclopedia of Renaissance literature. New York, Facts On File, 2005, XXVI-598 p.

987. Dictionnaire des pièces de théâtre françaises du XXe siècle. Sous la dir. de Jeanyves GUÉRIN. Paris, Champion, 2005, 729 p (Dictionnaires et références, 13).

988. Drevnejshie gosudarstva Vostochnoj Evropy, 2003 god: Mnimye real'nosti v antichnykh i srednevekovykh tekstakh. (The Earliest States of Eastern Europe, 2003: Imaginary realities in antique and medieval texts: [Articles]). Ed. Tat'jana N. DZHAKSON [JACKSON]. RAN, In-t vseobshchej istorii. Moskva, Vostochnaja literatura, 2005, 440 p. (bibl. incl.).

989. Encyclopedia of post-colonial literatures in English. Ed. by Eugene BENSON and L. W. CONOLLY. London, Routledge, 2005, 3 vol., [s. p.].

990. Gran enciclopedia cervantina. Vol. 1. A buen bocado – Aubigné. Director Carlos ALVAR. Alcalá de Henares, Centro de Estudios Cervantinos; Editorial Castalia, 2005, XXXV-924 p.

991. Greenwood encyclopedia (The) of multiethnic American literature. Ed. by Emmanuel S. NELSON. Westport a. London, Greenwood Press, 2005, 5 vol., [s. p.]. (ill.).

992. Historical companion (A) to postcolonial literature. Ed. by Prem PODDAR and David JOHNSON. Edinburgh, Edinburgh U. P., 2005, XXVII-574 p. (maps).

993. KUT (Günay). Yazmalar arasında Eski Türk edebiyatı araştırmaları. (Recherches sur la littérature ancienne turque parmi les manuscrits). İstanbul, Simurg, 2005, XXX-380 p.

994. Lachgemeinschaften. Kulturelle Inszenierungen und soziale Wirkungen von Gelächter im Mittelalter und in der Frühen Neuzeit. Hrsg. v. Werner RÖCKE und Hans Rudolf VELTEN. Berlin u. New York, de Gruyter, 2005, XXXI-392 p. (Trends in Medieval philology, 4).

995. LASSAVE (Pierre). Bible, la traduction des alliances. Enquête sur un événement littéraire. Paris, L'Harmattan, 2005, 267 p. (Logiques sociales. Littératures et société).

996. MORETTI (Franco). La letteratura vista da lontano. Con un saggio di Alberto PIAZZA. Torino, Einaudi, 2005, X-147 p. (Saggi, 865).

997. OVIEDO (José Miguel). Historia de la literatura hispanoamericana. 4. De Borges al presente. Madrid, Alianza Editorial, 2005, 492 p. (Alianza universidad textos, 170).

998. Pushkin handbook (The). Ed. by David M. BETHEA. Madison, University of Wisconsin Press, 2005, XLII-665 p. (Publications of the Wisconsin Center for Pushkin Studies).

999. SCHÜTTPELZ (Erhard). Die Moderne im Spiegel des Primitiven: Weltliteratur und Ethnologie (1870–1960). München, Wilhelm Fink, 2005, 449 p. (ill.).

1000. STABLEFORD (Brian M.). Historical dictionary of fantasy literature. Lanham a. Oxford, Scarecrow Press, 2005, LXV-499 p. (Historical dictionaries of literature and the arts, 5).

1001. STRUPP (Christoph), ZISCHKE (Birgit). German Americana, 1800–1955: a comprehensive bibliography of German, Austrian, and Swiss books and dissertations on the United States. With the assistance of Kai DREISBACH. Washington, German Historical Institute, 2005, 552 p. (Reference guides of the German Historical Institute, 18).

1002. Thomas-Mann-Handbuch. Hrsg. v. Helmut KOOPMANN. Frankfurt am Main, Fischer Taschenbuch Verlag, 2005, XVIII-1036 p.

1003. UEDA (Atsuko). The production of literature and the effaced realm of the political. *Journal of Japanese studies*, 2005, 31, 1, p. 61-88.

1004. VIALA (Alain). Lettre à Rousseau sur l'intérêt littéraire. Paris, Presses universitaires de France, 2005, 125 p. (Quadrige. Essais, débats).

1005. VIDAL MANZANARES (César). Diccionario del Quijote: la obra para entender uno de los libros esenciales de la cultura universal. Barcelona, Planeta, 2005, 565 p.

Cf. n°s 1513, 1514, 1571, 2059, 2064, 2946

C

PREHISTORY

§ 1. General. 1006-1022. – § 2. Palaeolithic and Mesolithic. 1023-1039. – § 3. Neolithic. 1040-1052. – § 4. Bronze age. 1053-1075. – § 5. Iron age. 1076-1100.

§ 1. General.

* 1006. PROCELLI (Enrico). Bibliografia della preistoria e protostoria della Sicilia. Firenze, Istituto Italiano di Preistoria e Protostoria, 2005, 209 p.

1007. ARCA (Andrea). Archeologia rupestre in Valcamonica: "Dos Cüi", un caso di studio. *Rivista di scienze preistoriche*, 2005, 55, p. 323-384.

1008. BERROCAL (María Cruz), GIL-CARLES ESTEBAN (Josè Manuel), GIL ESTEBAN (Manuel), MARTÍNEZ NAVARRETE (M.ª Isabel). Martin Almagro Basch, Fernando Gil Carles y el Corpus de Arte Rupestre Levantino. *Trabajos de prehistoria*, 2005, 62, 1, p. 27-46.

1009. BURROUGHS (William J.). Climate change in Prehistory. The end of the reign of Chaos. Cambridge, Cambridge U. P., 2005, 368 p.

1010. CERDEÑO (M.ª Luisa), CASTILLO (Alicia), SAGARDOY (Teresa). La evaluación del impacto ambiental y su repercusión sobre el patrimonio arqueológico en España. *Trabajos de prehistoria*, 2005, 62, 2, p. 25-40.

1011. ÇEVIK (Özlem). Arkeolojik kalıntılar ışığında tarihte ilk kentler ve kentleşme süreci: kuramsal bir değerlendirme. (Villes premières d'après les traces archéologiques et processus de l'urbanisation: une étude institutionnel). İstanbul, Arkeoloji ve Sanat, 2005, XI-100 p.

1012. COCCHI GENICK (Daniela). Considerazioni sull'uso del termine "facies" e sulla definizione delle facies archeologiche. *Rivista di scienze preistoriche*, 2005, 55, p. 5-27.

1013. DEVLET (Ekaterina G.), DEVLET (Marianna A.). Mify v kamne. Mir naskal'nogo iskusstva Rossii. (Myths in stone: The world of petroglyphic art of Russia). RAN, In-t arkheologii. Moskva, Aleteja, 2005, 471 p. (ill.; bibl. p. 436-468). – Mir naskal'nogo iskusstva: Sb. dokl. Mezhdunar. konf. (The world of petroglyphs: Proceedings of an international conference). Ed. Ekaterina G. DEVLET. Sibirskaja assotsiatsija issledovatelej pervobytnogo obshchestva; RAN, In-t arkheologii; Ros. gos. gumanit. un-t. Moskva, IA RAN, 2005, 427 p. (ill.; plates; bibl. incl.). [Text partly in English]

1014. DOBROVOL'SKAJA (Maria V.). Chelovek i ego pishcha. (Man and his food). Moskva, Nauchnyj mir, 2005, 367 p. (ill.; plates).

1015. DURU (Refik). Höyücek: 1989–1992 yılları arasında yapılan kazıların sonuçları. (Höyücek: results of the excavations 1989–1992). Ankara, Türk Tarih Kurumu, 2005, 2 vol., XII-242 p., 202 p.

1016. FLEMING (Andrew). Megaliths and post-modernism: the case of Wales. *Antiquity*, 2005, 79, p. 921-932.

1017. FULLER (Dorian Q.). Ceramics, seeds and culinary change in prehistoric India. *Antiquity*, 2005, 79, p. 761-777.

1018. GONZALEZ RUIBAL (Alfredo). Etnoarqueología de la cerámica en el oeste de Etiopía. *Trabajos de prehistoria*, 2005, 62, 2, p. 41-66.

1019. GRAZIOSI (Paolo). Arte rupestre del Fezzan (Missioni Graziosi 1967 e 1968). A cura di A. VIGLIARDI. Firenze, Istituto Italiano di Preistoria e Protostoria, 2005, 199 p.

1020. MANDOLESI (Alessandro), BURANELLI (Francesco), SANNIBALE (Maurizio). Materiale protostorico. Etruria et Latium Vetus. Roma, L'Erma di Bretschneider, 2005, 568 p.

1021. PIANA AGOSTINETTI (Paola), SOMMACAL (Matteo). Il problema della seriazione in archeologia. *Rivista di scienze preistoriche*, 2005, 55, p. 29-69.

1022. SKEATES (Robin). Visual culture and archaeology. Art and social life in Prehistoric South-East Italy. London, Duckworth, 2005, 240 p.

Cf. n°s 493, 609, 7564

§ 2. Palaeolithic and Mesolithic.

1023. ALPERSON-AFIL (Nira), HOVERS (Erella). Differential use of space at the Neandertal site of Amud Cave, Israel. *Eurasian prehistory*, 2005, 3, p. 3-22.

1024. ASTUTI (Paola), DINI (Mario), GRIFONI CREMONESI (Renata), KOZLOWSKI (Stefan), TOZZI (Carlo). L'industria mesolitica di Grotta Marisa (Lecce, Puglia) nel quadro delle industrie litiche dell'Italia meridionale. *Rivista di scienze preistoriche*, 2005, 55, p. 185-208.

1025. COLONESE (André Carlo), MARTINI (Fabio). Molluschi terrestri e disturbi antropici: evidenze epigravettiane a Grotta del Romito (Cosenza). *Bullettino di paletnologia*, 2005, 96, p. 1-15.

1026. DAVIES (Paul), ROBB (John G.), LADBROOK (Dave). Woodland clearance in the Mesolithic: the social aspects. *Antiquity*, 2005, 79, p. 280-288.

1027. FERRARI (Silvia), PERESANI (Marco), PERRONE (Raffaele). Un'industria litica musteriana di superficie nella pianura perieuganea (Colli Euganei, Veneto). *Rivista di scienze preistoriche*, 2005, 55, p. 169-184.

1028. GARCÍA-ARGÜELLES ANDREU (Pilar), NADAL I LORENZO (Jordi), FULLOLA I PERICOT (Josep M.ª). El abrigo del Filador (Margalef de Montsant, Tarragona) y su contextualización cultural y cronológica en el Nordeste peninsular. *Trabajos de prehistoria*, 2005, 62, 1, p. 65-84.

1029. GEBAUER (Anne Birgitte), PRICE (Douglas). Smakkerup Huse. A Late Mesolithic coastal site in Northwest Zealand, Denmark. Aarhus, Aarhus U. P., 2005, 288 p.

1030. GÓMEZ HERNANZ (Juan), MÁRQUEZ MORA (Belén), NICOLÁS CHECA (Elena), PÉREZ-GONZÁLEZ (Alfredo), RUIZ ZAPATA (Blanca). San Isidro (Madrid): 1862–2002. Nuevos hallazgos paleolíticos en la terraza de +30 m del río Manzanares. *Trabajos de prehistoria*, 2005, 62, 1, p. 157-164.

1031. KOL'TSOV (Lev V.). Final'nyj paleolit i mezolit Britanskikh ostrovov. (The Final Paleolithic and Mesolithic of the British Isles). Moskva, Lira, 2005, 300 p.

1032. MAÍLLO FERNÁNDEZ (José Manuel). La producción laminar en el Chatelperroniense de Cueva Morín: modalidades, intenciones y objetivos. *Trabajos de prehistoria*, 2005, 62, 1, p. 47-64.

1033. MARTINI (Fabio), MARTINO (Gabriele), FILIPPI (Omar), SEGID (Amha), KIROS (Asmeret), YOSIEF (Desale), YAMANE (Samuel), LIBSEKAL (Yosief), TEKA (Zelalem). Le industrie paleolitiche del bacino di Buya (Dancalia, Eritrea): prime osservazioni. *Rivista di scienze preistoriche*, 2005, 55, p. 71-137.

1034. O'CONNOR (Sue), VETH (Peter). Early Holocene shell fish hooks from Lene Hara Cave, East Timor establish complex fishing technology was in use in Island South East Asia five thousand years before Austronesian settlement. *Antiquity*, 2005, 79, p. 249-256.

1035. PALMA DI CESNOLA (Arturo), FREGUGLIA (Margherita). L'origine del Musteriano nel Gargano. I-Il premusteriano degli strati 26-28 della grotta Paglicci. *Rivista di scienze preistoriche*, 2005, 55, p. 139-168.

1036. PEARSON (Richard). The social context of early pottery in the Lingnan region of South China. *Antiquity*, 2005, 79, p. 819-828.

1037. Problemy kamennogo veka Russkoj ravniny. (Problems of the Stone Age of the East European Plain: [Articles]). Ed. Elena V. LEONOVA, K.N. GAVRILOV, Khizri A. AMIRKHANOV. RAN, In-t arkheologii. Moskva, Nauchnyj mir, 2004, 331 p. (ill.; bibl. incl.).

1038. VERPOORTE (Alexander). The first modern humans in Europe? A closer look at the dating evidence from the Swabian Jura (Germany). *Antiquity*, 2005, 79, p. 269-279.

1039. WALSH (Kevin). Risk and marginality at high altitudes: new interpretations from fieldwork on the Faravel Plateau, Hautes-Alpes. *Antiquity*, 2005, 79, p. 289-305.

§ 3. Neolithic.

1040. ANGELI (Lucia), FABBRI (Cristina). Analisi archeometriche applicate allo studio della ceramica decorata del villaggio neolitico di Trasano (Matera). *Rivista di scienze preistoriche*, 2005, 55, p. 209-223.

1041. BELFER-COHEN (Anna), HOVERS (Erella). The ground stone assemblages of the Natufian and Neolithic societies in the Levant–Current status. *Journal of the Israel Prehistoric Society*, 2005, 35, p. 299-308.

1042. BENTLEY (R. Alexander), PIETRUSEWSKY (Michael), DOUGLAS (Michele T.), ATKINSON (Tim C.). Matrilocality during the prehistoric transition to agriculture in Thailand? *Antiquity*, 2005, 79, p. 865-881.

1043. BRADLEY (Richard), EDMONDS (Mark). Interpreting the axe trade. Production and exchange in Neolithic Britain. Cambridge, Cambridge U. P., 2005, 250 p.

1044. CALVI REZIA (Gabriella), AGOSTINI (Lucia), ROSINI (Martina), SARTI (Lucia). Il Neolitico di Pienza-Cava Barbieri: la ceramica del saggio VI. *Bullettino di paletnologia*, 2005, 96, p. 37-61.

1045. CRAIG (Oliver E.), CHAPMAN (John), HERON (Carl), WILLIS (Laura H.), BARTOSIEWICZ (László), TAYLOR (Gillian), WHITTLE (Alasdair), COLLINS (Matthew). Did the first farmers of Central and Eastern Europe produce dairy foods? *Antiquity*, 2005, 79, p. 882-894.

1046. GARCÍA ATIÉNZAR (Gabriel). Occupazione e sfruttamento del territorio nel Neolitico: l'alto e il medio bacino del fiume Serpis (Alicante, Spagna). *Bullettino di paletnologia*, 2005, 96, p. 17-36.

1047. GIMBUTAS (Marija). Le Dee viventi. Milano, Medusa Edizioni, 2005, 330 p.

1048. GUILAINE (Jean). Du Proche-Orient à l'Atlantique. Actualité de la recherche sur le Néolithique. *Annales*, 2005, 60, 5, p. 925-952.

1049. NAKAMURA (Shin'ichi). Le riz, le jade et la ville. Évolution des sociétés néolithiques du Yangzi. *Annales*, 2005, 60, 5, p. 1009-1034.

1050. PÉTREQUIN (Pierre), PÉTREQUIN (Anne-Marie), ERRERA (Michel), CASSEN (Serge), CROUTSCH (Christophe), KLASSEN (Lutz), ROSSY (Michel), GARIBALDI (Patrizia), ISETTI (Eugenia), ROSSI (Guido), DELCARO (Dino). Beigua, Monviso e Valais. All'origine delle grandi asce levigate di origine in Europa occidentale durante il V millennio. *Rivista di scienze preistoriche*, 2005, 55, p. 265-322.

1051. ROSINI (Martina), SARTI (Lucia), SILVESTRINI (Mara). La ceramica del sito di Ripabianca di Monterado (Ancona) e le coeve produzioni dell'Italia centro-settentrionale. *Rivista di scienze preistoriche*, 2005, 55, p. 225-263.

1052. TODD (Ian A.). Vasilikos valley project 7: excavations at Kalavasos-Tenta. Volume II. Sävedalen, Astrom Paul, 2005, XVII-396 p.

Cf. n° 1037

§ 4. Bronze age.

1053. ALBA (Elisabetta). La donna nuragica. Roma, Carocci editore, 2005, 222 p.

1054. ARANDA JIMÉNEZ (Gonzalo), MOLINA GONZÁLEZ (Fernando). Intervenciones arqueológicas en el yacimiento de la Edad del Bronce del Cerro de la Encina (Monachil, Granada). *Trabajos de prehistoria*, 2005, 62, 1, p. 165-180.

1055. ARMADA PITA (Xosé-Lois), HUNT ORTIZ (Mark A.), TRESSERRAS (Jordi Juan), MONTERO RUIZ (Ignacio), RAFEL FONTANALS (Núria), RUIZ DE ARBULO (Joaquín). Primeros datos arqueométricos sobre la metalurgia del poblado y necrópolis de Calvari del Molar (Priorat, Tarragona). *Trabajos de prehistoria*, 2005, 62, 1, p. 139-155.

1056. BAGLIONI (Lapo). Aspetti tecnologici della produzione foliata neo-eneolitica: il caso-studio di Pianacci dei Fossi nelle Marche. *Bullettino di paletnologia*, 2005, 96, p. 109-128.

1057. BUENO RAMIREZ (Primitiva), BARROSO BERMEJO (Rosa), DE BALBÍN BEHRMANN (Rodrigo). Ritual campaniforme, ritual colectivo: la necrópolis de cuevas artificiales de Valle de las Higueras (Huecas, Toledo). *Trabajos de prehistoria*, 2005, 62, 2, p. 67-90.

1058. CAZZELLA (Alberto), DE DOMINICIS (Alessandro), RECCHIA (Giulia), RUGGINI (Cristiana). Il sito dell'età del Bronzo recente di Monteroduni-Paradiso (Isernia). *Rivista di scienze preistoriche*, 2005, 55, p. 385-438.

1059. CORK (Edward). Peaceful Harappans? Reviewing the evidence for the absence of warfare in the Indus civilisation of North-West India and Pakistan (c. 2500–1900 BC). *Antiquity*, 2005, 79, p. 411-423.

1060. GARCÍA BORJA (Pablo), DE PEDRO MICHÓ (María Jesús), SÁNCHEZ MOLINA (Ángel). Conjunto de metales procedente del poblado de la Edad del Bronce de L'Arbocer (Font de la Figuera, Valencia). *Trabajos de prehistoria*, 2005, 62, 1, p. 181-192.

1061. GARCÍA SANJUÁN (Leonardo). Las piedras de la memoria. La permanencia del megalitismo en el Suroeste de la Península Ibérica durante el II y I milenios ANE. *Trabajos de prehistoria*, 2005, 62, 1, p. 85-109.

1062. KARASIK (Avshalom), SMILANSKY (Uzy), BEIT-ARIEH (Itzhaq). New typological analyses of Holemouth Jars from the Early Bronze age from Tel Arad and Southern Sinai. *Tel-Aviv*, 2005, 32, p. 20-31.

1063. Kargaly, [a Bronze Age mining centre in the South Urals: Proceedings of the Kargaly Archaeological Expedition]. Vol. 4. Nekropoli na Kargalakh. Naselenie Kargalov: Paleoantropologicheskie issledovanija. (Cemetries of Kargaly. The population of Kargaly: Essays in physical anthropology). Ed. Evgenij N. CHERNYKH. Moskva, Jazyki slavjanskoj kul'tury, 2005, 240 p.

1064. KRISTIANSEN (Kristian). L'Âge du bronze, une période historique. Les relations entre Europe, Méditerranée et Proche-Orient. *Annales*, 2005, 60, 5, p. 975-1007.

1065. KYRIAKIDIS (Evangelos). Ritual in the Bronze Age Aegean. The Minoan Peak sanctuaries. London, Duckworth, 2005, 224 p.

1066. MARIETHOZ (François). Enquête autour d'un tumulus de l'age du bronze. Lausanne, Cahiers d'archéologie romande, 2005, 160 p.

1067. MARTINELLI (Maria Clara). Il Villaggio dell'età del Bronzo medio di Portella a Salina nelle Isole Eolie. Firenze, Istituto Italiano di Preistoria e Protostoria, 2005, 339 p.

1068. MOUSAVI (Ali). Comments on the early Bronze Age in Iran. *Iranica antiqua*, 2005, 40, p. 87-99.

1069. Parker Pearson (Mike), Chamberlain (Andrew), CRAIG (Oliver), MARSHALL (Peter), MULVILLE (Jacqui), SMITH (Helen), CHENERY (Carolyn), COLLINS (Matthew), COOK (Gordon), CRAIG (Geoffrey), EVANS (Jane), HILLER (Jen), MONTGOMERY (Janet), SCHWENNINGER (Jean-Luc), TAYLOR (Gillian), WESS (Timothy). Evidence for mummification in Bronze Age Britain. *Antiquity*, 2005, 79, p. 529–546.

1070. SÁNCHEZ (Alberto), BELLÓN (Juan Pedro), RUEDA (Carmen). Nuevos datos sobre la Zona Arqueológica de Marroquíes Bajos: el quinto Foso. *Trabajos de prehistoria*, 2005, 62, 2, p. 151-164.

1071. SCHUSTER KESWANI (Priscilla). Death, prestige, and copper in Bronze Age Cyprus. *American journal of archaeology*, 2005, 109, p. 341-401.

1072. STEELMAN (K.L.), CARRERA RAMIREZ (F.), FÁBREGAS VALCARCE (R.), GUILDERSON (T.), ROWE (M.W.). Direct radiocarbon dating of megalithic paints from North-West Iberia. *Antiquity*, 2005, 79, p. 379-389.

1073. Territorio nurágico y paisaje antiguo. La meseta de Pranemuru (Cerdaña) en la Edad del Bronce. Ed. por Marisa RUIZ-GÁLVEZ. Madrid, Universidad Complutense de Madrid, 2005, 251 p. (Anejos de Complutum, 10).

1074. TRAVERSO (Antonella). Il santuario prenuragico di Monte d'Accoddi (Sassari): tipologia e cronologia dei materiali ceramici dai saggi di scavo sul monumento (1984–2001). *Bullettino di paletnologia*, 2005, 96, p. 63-108.

1075. ZIFFER (Irit). From Acemhöyük to Megiddo. The banquet scene in the art of the Levant in the second millennium BCE. *Tel Aviv*, 2005, 32, p. 133-167.

§ 5. Iron age.

1076. ABOAL FERNÁNDEZ (Roberto), AYÁN VILA (Xurxo M.), CRIADO BOADO (Felipe), PRIETO MARTÍNEZ (M.ª Pilar), TABARÉS DOMÍNGUEZ (Marta). Yacimientos sin estratigrafía: Devesa do Rei, ¿un sitio cultual de la Prehistoria Reciente y la Protohistoria de Galicia? *Trabajos de prehistoria*, 2005, 62, 2, p. 165-180.

1077. BOUCHARLAT (Rémy), FRANCFORT (Henri-Paul), LECOMTE (Olivier). The citadel of Ulug Depe and the Iron Age archaeological sequence in Southern Central Asia. *Iranica antiqua*, 2005, 40, p. 479-514.

1078. COPLEY (M.S.), BERSTAN (R.), DUDD (S.N.), AILLAUD (S.), MUKHERJEE (A.J.), STRAKER (V.), Payne (S.), Evershed (R.P.). Processing of milk products in pottery vessels through British prehistory. *Antiquity*, 2005, 79, p. 895-908.

1079. CORDOBA (Joaquín María), DEL CERRO (María del Carmen). Archéologie de l'eau dans Al Madam (Sharjah, Emirats Arabes Unis). Puits, Aljah et secheresse pendant l'âge du Fer. *Iranica antiqua*, 2005, 40, p. 515-533.

1080. CURTIS (John). The material culture of Tepe Nush-I Jan and the end of the Iron Age III Period in Western Iran. *Iranica antiqua*, 2005, 40, p. 233-248.

1081. DE POLIGNAC (François). Forms and processes. Some thoughts on the meaning of urbanization in early archaic Greece. *In*: Mediterranean urbanization 800–600 B.C. [Cf. n° 1094], p. 45-69.

1082. DÖNMEZ (Şevket). 1997–1999 Yılları Yüzey Araştırmalarında İncelenen Amasya İli Demir Çağı Yerleşmeleri (42 resim ile birlikte). [The Iron Age settlements in Amasya Province, surveyed results 1997–1999 (with 42 illustrations)]. *Belleten*, 2005, 69, 255, p. 467-498.

1083. FERNÁNDEZ FLORES (Álvaro), RODRÍGUEZ AZOGUE (Araceli). El complejo monumental del Carambolo Alto, Camas (Sevilla). Un santuario orientalizante en la paleodesembocadura del Guadalquivir. *Trabajos de prehistoria*, 2005, 62, 1, p. 111-138.

1084. FORD (L. A.), POLLARD (A. M.), CONINGHAM (R. A. E.), STERN (B.). A geochemical investigation of the origin of Rouletted and other related South Asian fine wares. *Antiquity*, 2005, 79, p. 909-920.

1085. GARCIA (Dominique). Urbanization and spatial organization in Southern France and North-Eastern Spain during the Iron age. *In*: Mediterranean urbanization 800–600 B.C. [Cf. n° 1094], p. 169-186.

1086. HERNÁNDEZ-GASCH (Jordi), ARAMBURU-ZABALA HIGUERA (Javier). Murallas de la Edad del Hierro en la cultura Talayótica. El recinto fortificado del poblado de Ses Païsses (Artà, Mallorca). *Trabajos de prehistoria*, 2005, 62, 2, p. 125-150.

1087. HODOS (Tamar), KNAPPETT (Carl), KILIKOGLOU (Vassilis). Middle and Late Iron Age painted ceramics from Kinet Höyük: macro, micro and elemental analyses. *Anatolian studies*, 2005, 55, p. 61-87.

1088. IACOVOU (Maria). The early Iron age urban forms of Cyprus. *In*: Mediterranean urbanization 800–600 B.C. [Cf. n° 1094], p. 17-43.

1089. JIMÉNEZ ÁVILA (Javier). Cancho Roano: el proceso de privatización de un espacio ideológico. *Trabajos de prehistoria*, 2005, 62, 2, p. 105-124.

1090. KOVALEVSKAJA (Vera B.). Kavkaz – skify, sarmaty, alany. I tys. do n.e. – I tys. n.e. (The Caucasus: the Scythians, the Sarmatians, and the Alans, the 1st century D.C. and the 1st century A.D.). RAN, In-t arkheologii. Moskva, IA RAN, 2005, 395 p. (ill.; maps.; bibl. p. 181-212).

1091. LECOMTE (Olivier). The Iron Age of Northern Hyrcania. *Iranica antiqua*, 2005, 40, p. 461-478.

1092. MAGEE (Peter). The chronology and environmental background of Iron Age settlement in Southeastern Iran and the question of the origin of the Qanat Irrigation System. *Iranica antiqua*, 2005, 40, p. 217-231.

1093. MARINO (Domenico). Kroton prima dei Greci. La prima età del Ferro nelle Calabria centrale ionica. *Rivista di scienze preistoriche*, 2005, 55, p. 439-465.

1094. Mediterranean urbanization 800–600 B.C. Ed. by Robin OSBORNE and Barry CUNLIFFE. Oxford, Oxford U. P., 2005, 279 p. [Cf. n°s <choice> 1081, 1085, 1088, 1099.]

1095. MIZOGUCHI (Koji). Genealogy in the ground: observations of jar burials of the Yayoi period, Northern Kyushu, Japan. *Antiquity*, 2005, 79, p. 316-326.

1096. NIAKAN (Lily). Tappeh Shizar. An Iron Age population center. *Iranica antiqua*, 2005, 40, p. 101-106.

1097. PEREA CAVEDA (Alicia). Mecanismos identitarios y de construcción de poder en la transición Bronce-Hierro. *Trabajos de prehistoria*, 2005, 62, 2, p. 91-104.

1098. PONS (Nina). Une tête de javeline dans une tombe du Fer Ancien en Arménie. Évocation du 'bris rituel' au Moyen Orient. *Iranica antiqua*, 2005, 40, p. 423-435.

1099. RIVA (Corinna). The culture of urbanization in the Mediterranean c. 800-600 B.C. *In*: Mediterranean urbanization 800–600 B.C. [Cf. n° 1094], p. 203-232.

1100. TSETSKHLADZE (Gocha R.). The Caucasus and the Iranian world in the early Iron Age. Two graves from Treli. *Iranica antiqua*, 2005, 40, p. 437-446.

Cf. n° 3500

D

THE ANCIENT EAST

(the Hellenistic states included)

---

§ 1. General. 1101-1117. – § 2. The Near East. 1118-1150. – § 3. Egypt. 1151-1201. – § 4. Mesopotamia. 1202-1251. – § 5. Hittites. 1252-1272. – § 6. Jews and Semitic peoples to the end of the ancient world. 1273-1342. – § 7. Iran. 1343-1368.

§ 1. General.

1101. ABAY (Eşref), ÇEVIK (Özlem). "Interaction and migration". Issues in archaeological theory. *Altorientalische Forschungen*, 2005, 32, p. 62-73.

1102. AILZADEH (Abbas). Excavations of Tall-e Bakun. Seasons of 1932 and '37: the origins of State organisations. Chicago, The Oriental Institute of the University of Chicago, 2005, 300 p.

1103. ARCHI (Alfonso). Remarks on the Early Empire documents. *Altorientalische Forschungen*, 2005, 32, p. 225-229.

1104. BOROVKOVA (Ljudmila A.). Kushanskoe tsarstvo (po drevnim kitajskim istochnikam). (The Kushan Empire according the Ancient Chinese sources). Moskva, [s. n.], 2005, 320 p. (maps; plates; bibl. p. 278-295; ind. p. 296-310).

1105. ÇEVIK (Özlem). The change of settlement patterns in Lake Van basin: ecological constrains caused by Highland landscapes. *Altorientalische Forschungen*, 2005, 32, p. 74-96.

1106. CHANIOTIS (Angelos). War in the Hellenistic world. A social and cultural history. Malden, Oxford a. Carlton, Blackwell, 2005, XIV-308 p.

1107. DEL OLMO LETE (Gregorio). The fundamental problems of comparative linguistics. A forgotten Spanish contribution from the early 20th century. *Aula orientalis*, 2005, 23, p. 233-273.

1108. Drevnee Sredizemnomor'e: Religija, obshchestvo, kul'tura. (La Mediterrannée antique: religion, societe, culture: [Articles]). Ed. Ol'ga P. SMIRNOVA, Aleksandr L. SMYSHLJAEV. RAN, In-t vseobshchej istorii. Moskva, IVI RAN, 2005, 286 p. (bibl. incl.). [Resumé français]

1109. DURNFORD (S.P.B.), AKEROYD (J.R.). Anatolian marashanha and the many uses of fennel. *Anatolian studies*, 2005, 55, p. 1-13.

1110. Edubba vechna i postojanna: Mat-ly konf., posv. 90-letiju so dnja rozhdenija I.M. D'jakonova. (Materials of the conference dedicated to the 90th anniversary of Igor' M. Dyakonov). Ed. Mariam M. DANDAMAEVA, L.E. KOGAN, N.V. KOZLOVA. Gos. Ermitazh. Sankt-Peterburg, Izd-vo Gos. Ermitazha, 2005, 268 p. (ill.; portr.; bibl. incl.). [Text partly in English]

1111. FAIRBAIRN (Andrew), OMURA (Sachihiro). Archaeological identification and significance of ÉSAG (agricultural storage pits) at Kaman-Kalehöyük, central Anatolia. *Anatolian studies*, 2005, 55, p. 15-23.

1112. HEEßEL (Nils P.). Stein, Pflanze und Holz: ein neuer Text zur 'medizinischen Astrologie'. *Orientalia*, 2005, 74, p. 1-22.

1113. HITCHNER (R. Bruce). The Garamantes and the archaeology of Fazzan. *Journal of Roman archaeology*, 2005, 18, p. 717-719.

1114. HUBER (Irene). Rituale der Seuchen- und Schadensabwehr im Vorderen Orient und Griechenland. Formen kollektiver Krisenbewältigung in der Antike. Stuttgart, Steiner, 2005, 287 p.

1115. KNAPPETT (Carl), KILIKOGLOU (Vassilis), STEELE (Val), STERN (Ben). The circulation and consumption of Red Lustrous Wheelmade ware: petrographic, chemical and residue analysis. *Anatolian studies*, 2005, 55, p. 25-59.

1116. LUCCHESI (Enzo). Nouvelles glanures pachômiennes. *Orientalia*, 2005, 74, p. 86-90.

1117. MATTHIAE (Paolo). Prima lezione di archeologia orientale. Roma e Bari, Laterza, 2005, 188 p.

Cf. n°s 872, 953, 7564, 7597

§ 2. The Near East.

1118. BAVANT (B.). Les églises du Massif Calcaire de Syrie du Nord (VIe–VIIe s.). *Journal of Roman archaeology*, 2005, 18, p. 756-770.

1119. BENDA WEBER (Isabella). Lykier und Karer: zwei autochthone Ethnien Kleinasiens zwischen Orient und Okzident. Bonn, Habelt, 2005, XVII-436 p.

1120. BLÖMER (Michael), WINTER (Engelbert). Doliche und das Heiligtum des Iupiter Dolichenus auf dem Dülük Baba Tepesi. 1. Vorbericht (2001–2003). *Istanbuler Mitteilungen*, 2005, 55, p. 191-214.

1121. BOL (Renate). Der ‚Torso von Milet' und das Kultbild des Apollon Termintheus in Myus. *Istanbuler Mitteilungen*, 2005, 55, p. 37-64.

1122. CAHILL (Nicholas), KROLL (John H.). New archaic coin finds at Sardis. *American journal of archaeology*, 2005, 109, p. 589-617.

1123. CARNEY (Elizabeth Donnelly). Women and dunasteia in Caria. *American journal of philology*, 2005, 126, p. 65-91.

1124. CAVALIER (Laurence). Architecture romaine d'Asie Mineure. Les monuments de Xanthos et leur ornementation. Bordeaux, Ausonius éditions et Paris, de Boccard, 2005, 324 p.

1125. CHRISTOF (Eva), ERATH-KOINER (Gabriele). Antike Architekturfragmente aus Tavium. Erste Ergebnisse. *Istanbuler Mitteilungen*, 2005, 55, p. 271-288.

1126. COULTON (J.J.). Pedestals as 'altars' in Roman Asia Minor. *Anatolian studies*, 2005, 55, p. 127-157.

1127. CROWTHER (Charles). Hydisos in Caria. *Anatolian studies*, 2005, 55, p. 99-105.

1128. DMITRIEV (Sviatoslav). City government in Hellenistic and Roman Asia Minor. New York, Oxford U. P., 2005, XVI-428 p. – IDEM. The history and geography of the Province of Asia during its first hundred years and the provincialization of Asia Minor. *Athenaeum*, 2005, 93, p. 71-133.

1129. DURUKAN (Murat). Monumental tomb forms in the Olba region. *Anatolian studies*, 2005, 55, p. 107-126.

1130. DUYRAT (Frédérique). Arados hellénistique. Étude historique et monetaire. Beyrouth, IFPO publications, 2005, XII-433 p.

1131. FREYDANK (Helmut). Zu den Eponymenfolgen des 13. Jahrhunderts v.Chr. in Dur-Katlimmu. *Altorientalische Forschungen*, 2005, 32, p. 45-46.

1132. GABELKO (Oleg L.). Istorija Vifinskogo tsarstva. (A history of the Bithynian Kingdom). Sankt-Peterburg, Gumanitarnaja akademija, 2005, 576 p. (ill.; bibl. p. 532-571).

1133. GLIWITZKY (Christian). Die Kirche im sog. Bischofspalast zu Side. *Istanbuler Mitteilungen*, 2005, 55, p. 337-408.

1134. GOLDHAUSEN (Marco), RICCI (Andrea). Political centralisation in the Syrian Jezira during the 3rd millennium. A case study in settlement hierarchy. *Altorientalische Forschungen*, 2005, 32, p. 132-157.

1135. GYGAX (Marc Domingo), TIETZ (Werner). 'He who of all mankind set up the most numerous trophies to Zeus'. The inscribed pillar of Xanthos reconsidered. *Anatolian studies*, 2005, 55, p. 89-98.

1136. HELD (Winfried). Kult auf dem Dach. Eine Deutung der Tempel mit Treppenhäusern und Giebeltüren als Zeugnis seleukidischer Sakralarchitektur. *Istanbuler Mitteilungen*, 2005, 55, p. 119-160.

1137. KADIOĞLU (Musa). Die Opus Sectile-Wandverkleidung der Latrine in Magnesia a. M. *Istanbuler Mitteilungen*, 2005, 55, p. 309-336.

1138. KOSMETATOU (Elisabeth). Macedonians in Pisidia. *Historia*, 2005, 54, p. 216-221.

1139. MÄGELE (Semra). Ein besonderer Ort für Votive: Anmerkungen zu einem ungewöhnlichen Befund an drei Statuen aus einem Nymphäum von Sagalassos. *Istanbuler Mitteilungen*, 2005, 55, p. 289-307.

1140. PESCHLOW-BINDOKAT (Anneliese). Feldforschungen im Latmos. Die Karische Stadt Latmos. Berlin u. New York, de Gruyter, 2005, XII-62 p.

1141. QUACK (Joachim F.). Ein Unterweltsbuch der solar-sirianischen Einheit? *Die Welt des Orients*, 2005, 35, p. 22-47.

1142. RADT (Wolfgang). Eine antike Wasseruhr im Gymnasion von Pergamon. *Istanbuler Mitteilungen*, 2005, 55, p. 179-190.

1143. RIES (Gerhard). Calumnia und Talion-Einfluß altorientalischen Rechts auf das Syrisch-römische Rechtsbuch. *In*: Antike Rechtsgeschichte [Cf. n° 1479], p. 1-10.

1144. RIZZA (Alfredo). Licia e Lidia prima dell'ellenizzazione. A proposito degli Atti del convegno internazionale, Roma, 11–12 ottobre 1999. *Athenaeum*, 2005, 93, p. 243-251.

1145. ROUAULT (Olivier), MORA (Clelia). Progetto "Terqa e la sua regione": rapporto preliminare 2004. *Athenaeum*, 2005, 93, p. 657-696.

1146. ŞAHIN (Mustafa). Kinderstatuetten aus Ton auf der Rundtempelterrasse in Knidos. *Istanbuler Mitteilungen*, 2005, 55, p. 95-118. – IDEM. Terrakotten aus Knidos: erste Ergebnisse. Die Kulte auf der Rundtempelterrasse. *Istanbuler Mitteilungen*, 2005, 55, p. 65-93.

1147. SEMERANO (Grazia). Per un approccio contestuale alla lettura delle immagni: le ceramiche a rilievo di Hierapolis di Frigia. *Mélanges de l'École Française de Rome-Antiquité*, 2005, 117, p. 83-98.

1148. SÖĞÜT (Bilal). Ein hellenistisches Kapitell aus dem Rauhen Kilikien: das korintische Kapitell von Efrenk. *Istanbuler Mitteilungen*, 2005, 55, p. 161-177.

1149. SOMMER (Michael). The archaeology and history of Roman Syria. *Journal of Roman archaeology*, 2005, 18, p. 726-728.

1150. ZIMMERMANN (Martin). Eine Stadt und ihr kulturelles Erbe. Vorbericht über Feldforschungen im zentrallykischen Phellos 2002–2004. *Istanbuler Mitteilungen*, 2005, 55, p. 215-270.

Cf. n°s 10, 1328, 1852

§ 3. Egypt.

1151. AGER (Sheila L.). Familiarity breeds: incest and the Ptolemaic dynasty. *Journal of Hellenic studies*, 2005, 125, p. 1-34.

1152. AMATO (Eugenio). Influenze egizie nella Descriptio orbis di Dionisio d'Alessandria? *Kernos*, 2005, 18, p. 97-111.

1153. BAIKIE (James). Egyptian antiquities in the Nile Valley. A descriptive handbook. London, Kegan Paul, 2005, 964 p.

1154. BONNET (Charles), VALBELLE (Dominique). Les Pharaons venus d'Afrique. Paris, Citadelles & Mazenod, 2005, 216 p.

1155. BOUD'HORS (Anne), CALAMENT (Florence). Épigraphie fayoumique: addenda et corrigenda. *Journal of Coptic studies*, 2005, 7, p. 131-135.

1156. BOUD'HORS (Anne), NAKANO (Chièmi). Vestiges bibliques en copte fayoumique. *Journal of Coptic studies*, 2005, 7, p. 137-139.

1157. BRESCIANI (Edda), BETRÒ (Marilina). Egypt in India. Egyptian antiquities in Indian Museums. Pisa, PLUS Pisana Libraria Universitatis Studiorum, 2005, 320 p.

1158. BRESCIANI (Edda). La porta dei sogni. Interpreti e sognatori nell'Egitto antico. Torino, Einaudi, 2005, X-190 p.

1159. BUCKLEY (Ian M.), BUCKLEY (Peter), COOKE (Ashley). Fieldwork in Theban tomb KV 39: the 2002 season. *Journal of Egyptian archaeology*, 2005, 91, p. 71-82.

1160. CAPPOZZO (Mario). La collezione egizia del Museo Nazionale Preistorico Etnografico "Luigi Pigorini". *Bullettino di paletnologia*, 2005, 96, p. 131-202.

1161. CHAPPAZ (Jean-Luc). Une stèle de donation de Ramsès III. *Bulletin de la Société d'Égyptologie*, 2005, 27, p. 5-19.

1162. CHOAT (Malcolm). Thomas the 'Wanderer' in a Coptic list of the Apostles. *Orientalia*, 2005, 74, p. 83-85.

1163. COLLOMBERT (Philippe). Renenoutet et Renenet. *Bulletin de la Société d'Égyptologie*, 2005, 27, p. 21-32.

1164. DEMIDCHIK (Arkadij E.). Bezymjannaja piramida. Gosudarstvennaja doktrina drevneegipetskoj Gerakleopol'skoj monarkhii. (The nameless pyramid: The state doctrine of the Ancient Egyptian Heracleopolis Monarchy). Sankt-Peterburg, Aletejja, 2005, 272 p. (ill.; bibl. p. 237-256). [English summary]

1165. DROUX (Xavier). Une représentation de prisonniers décapités en provenance de Hiérakonpolis. *Bulletin de la Société d'Égyptologie*, 2005, 27, p. 33-42.

1166. ELDAMATY (Mamdough). Ein ptolemäisches Priesterdekret aus dem Jahr 186 v. Chr. Eine neue Version von Philensis II in Kairo. München u. Leipzig, Saur, 2005, XVII-92 p.

1167. EL-MASRI (Yahia). An unfinished stela of the earliest Heracleopolitan period. *Bulletin de la Société d'Égyptologie*, 2005, 27, p. 61-73.

1168. EMMEL (Stephen). Ein altes Evangelium der Apostel taucht in Fragmenten aus Ägypten und Nubien auf. *Zeitschrift für antikes Christentum*, 2005, 9, p. 85-99.

1169. ESPINEL (Andrés D.). A newly identified stela from Wadi El-Hudi (Cairo JE 86119). *Journal of Egyptian archaeology*, 2005, 91, p. 55-70.

1170. FILER (Joyce M.). Health in ancient Egypt and Nubia: sources and issues. London, Duckworth, 2005, 176 p.

1171. FROSCHAUER (Harald). Griechische und koptische Schulübungen aus dem byzantinischen und früharabischen Ägypten. *Mitteilungen zur Christlichen Archäologie*, 2005, 11, p. 87-102.

1172. GERMOND (Philippe). De l'Oeil vert d'Horus au Pressoir mystique: réflexions autour de la symbolique du vin dans l'Egypte pharaonique. *Bulletin de la Société d'Égyptologie*, 2005, 27, p. 43-59.

1173. GIANNOBILE (Sergio). Il dio egizio Ptah nella documentazione magica: amuleti e papiri. *Zeitschrift für Papyrologie und Epigraphik*, 2005, 152, p. 161-167.

1174. GILLAM (Robyn). Performance and drama in ancient Egypt. London, Duckworth, 2005, 208 p.

1175. GONIS (Nikolaos). Some curious prescripts (native languages in Greek dress?). *Bulletin of the American Society of Papyrologists*, 2005, 42, p. 41-44.

1176. GOYON (Jean-Claude). De l'Afrique à l'Orient: l'Égypte des pharaons et son rôle historique (1800–300 avant notre ère). Paris, Ellipses Éditions Marketing, 2005, 382 p. – IDEM. La construction pharaonique du Moyen Empire à l'époque romaine. Paris, Picard, 2005, 456 p.

1177. GRAJETZKI (Wolfram). Middle Kingdom of ancient Egypt: history, archaeology and society. London, Duckworth, 2005, 208 p.

1178. GRIMAL (Nicolas), ADLY (Emad). Fouilles et travaux en Égypte et au Soudan, 2003–2004. *Orientalia*, 2005, 74, p. 195-314.

1179. HOMOTH-KUHS (Clemens). Phylakes und Phylakon-Steuer im griechisch-römischen Ägypten. Ein Beitrag zur Geschichte des antiken Sicherheitswesens. München, Saur, 2005, XIV-222 p.

1180. JASNOW (Richard), ZAUZICH (Karl-Theodor). The ancient Egyptian book of Thoth. A demotic discourse on knowledge and pendant to the classical Hermetica. Wiesbaden, Harrassowitz, 2005, 2 vol., XX-581 p.

1181. KLEINER (Diana E.E.). Cleopatra and Rome. Cambridge a. London, The Belknap Press of Harvard U. P., 2005, 340 p.

1182. KORMYSHEVA (Eleonora E.). Drevnij Egipet. (Ancient Egypt). Moskva, Ves' mir, 2005, 192 p. (ill.; maps; bibl. p. 189-191).

1183. KROL (Aleksej A.). Egipet pervykh faraonov. Kheb-Sed i stanovlenie drevneegipetskogo gosudarstva. (Egypt of the first Pharaohs: Heb Sed and the emergence of the Ancient Egyptian state). Moskva, Rudomino, 2005, 224 p. (ill.; maps; bibl. p. 194-211). [English summary]

1184. LEERSON (Benoît). La conception du décor d'un temple au début du règne de Ramsès II: analyse du deuxième registre de la moitié sud du mur ouest de la grande salle hypostyle de Karnak. *Journal of Egyptian archaeology*, 2005, 91, p. 107-124.

1185. MALAISE (Michel). Pour une terminologie et une analyse des cultes isiaques. Bruxelles, Académie royale de Belgique, 2005, 282 p.

1186. OCKINGA (Boyo G.). An eighteenth dynasty h3.ty-ᶜ.w of Heliopolis in Adelaide, South Australia. *Journal of Egyptian archaeology*, 2005, 91, p. 83-94.

1187. PAPACONSTANTINOU (Arietta). Aux marges de l'empire ou au centre du monde? De l'Égypte des Byzantins à celle des historiens. *Journal of juristic papyrology*, 2005, 35, p. 195-236.

1188. PAULET (Aurelie). À propos de la désignation des personnages dans l'histoire d'Apophis et Seqenenrê. *Bulletin de la Société d'Égyptologie*, 2005, 27, p. 75-79.

1189. PINO (Cristina). The market scene in the tomb of Khaemhat (TT 57). *Journal of Egyptian archaeology*, 2005, 91, p. 95-105.

1190. POOLE (Federico). 'All that has been done to the shabtis': some considerations on the decree for the Shabtis of Neskhons and P.BM EA 10800. *Journal of Egyptian archaeology*, 2005, 91, p. 165-170.

1191. QUACK (Joachim F.). Ein neuer Zugang zur Lehre des Ptahhotep? *Die Welt des Orients*, 2005, 35, p. 7- 21.

1192. RAVEN (Maarten J.). Egyptian concepts on the orientation of the human body. *Journal of Egyptian archaeology*, 2005, 91, p. 37-53.

1193. RAY (John D.). An inscribed linen plea from the sacred Animal necropolis, North Saqqara. *Journal of Egyptian archaeology*, 2005, 91, p. 171-179. – IDEM. Demotic papyri and ostraca from Qasr Ibrim. London, The Egypt Exploration Society, 2005, XI-62 p.

1194. REDDÉ (Michel). L'Égypte à l'époque romaine. *Journal of Roman archaeology*, 2005, 18, p. 729-730.

1195. RUPPRECHT (Hans-Albert). Griechen und Ägypter – Vielfalt des Rechtslebens nach den Papyri. *In*: Antike Rechtsgeschichte [Cf. n° 1479], p. 17-25.

1196. SCHENKEL (Wolfgang). Tübinger Einführung in die klassisch-ägyptische Sprache und Schrift. Tübingen, im Selbstverlag, 2005, 413 p.

1197. SHAW (Ian). Sudan's ancient treasures: an exhibition of recent discoveries. *American journal of archaeology*, 2005, 109, p. 81-86.

1198. SMITH (Marc). Papyrus Harkness (MMA 31.9.7). Oxford, Griffith Institute, 2005, XI-366 p.

1199. VEISSE (Anne-Emanuelle). Le discours sur les violences dans l'Égypte hellénistique: le clergé face aux révoltes. *In*: Violence (La) dans les mondes grec et romaine [Cf. n° 1527], p. 213-223.

1200. VOLOKHINE (Youri). Tithoès et Lamarès. *Bulletin de la Société d'Égyptologie*, 2005, 27, p. 81-92.

1201. WILSON (Penelope), JEFFREYS (David), BURNBURY (Judith), NICHOLSON (Paul T.), KEMP (Barry), ROSE (Pamela). Fieldwork, 2004–2005: Sais, Memphis, Saqqara bronzes project, Tell el-Amarna, Tell el-Amarna glass project, Qasr Ibrim. *Journal of Egyptian archaeology*, 2005, 91, p. 1-36.

§ 4. Mesopotamia.

* 1202. CHARPIN (Dominique). Chroniques bibliographiques 5. Économie et société à Sippar et en Babylonie du nord à l'époque paléo-babylonienne. *Revue d'assyriologie et d'archéologie orientale*, 2005, 99, p. 133-176.

1203. AMIET (Pierre). Les sceaux de l'administration princière de Suse à l'époque d'Agadé. *Revue d'assyriologie et d'archéologie orientale*, 2005, 99, p. 1-12.

1204. ARCHI (Alfonso). The head of Kura-the head of Adabal. *Journal of Near Eastern studies*, 2005, 64, p. 81-100.

1205. AR-RAWI (F.), D'AGOSTINO (Franco). Neo-Sumerian administrative texts from Umma kept in the British Museum, Part One. Messina, Dipartimento di Scienze dell'Antichità-Università di Messina, 2005, 264 p.

1206. ATTINGER (Pascal). A propos de AK "faire" (II). *Zeitschrift für Assyriologie und vorderasiatische Archäologie*, 2005, 95, p. 208-275.

1207. AYVAZIAN (Alina). Observations on dynastic continuity in the kingdom of Urartu. *Iranica antiqua*, 2005, 40, p. 197-205.

1208. BEN-DOV (Jonathan), HOROWITZ (Wayne). The Babylonian lunar three in calendrical scrolls from Qumran. *Zeitschrift für Assyriologie und vorderasiatische Archäologie*, 2005, 95, p. 104-120.

1209. BILGE BAŞTÜRK (Mahmut), KONAKÇI (Erim). Settlement patterns in the Malatya Elazig region in the IV. & III. millennium BC. *Altorientalische Forschungen*, 2005, 32, p. 97-114.

1210. BOIY (Tom). The fifth and sixth generation of the Nikarchos=Anu-uballit family. *Revue d'assyriologie et d'archéologie orientale*, 2005, 99, p. 105-110.

1211. BRY (Paul). Des regles administratives et techniques à Mari. Sabadell, Editorial AUSA, 2005, 416 p.

1212. CHARPIN (Dominique). Données nouvelles sur la vie économique et sociale de l'époque paléobabylonienne. *Orientalia*, 2005, 74, p. 409-421.

1213. CLANCIER (Philippe). Les scribes sur parchemindu temple d'Anu. *Revue d'assyriologie et d'archéologie orientale*, 2005, 99, p. 85-104.

1214. COHEN (Eran). The modal system of old Babylonian. Winona Lake, Eisenbrauns, 2005, XIV-225 p.

1215. COHEN (Yoram). Feet of clay at Emar: a happy end? *Orientalia*, 2005, 74, p. 165-170.

1216. D'AGOSTINO (Franco), POMPONIO (Francesco). Due bilanci di entrate e uscite di argento da Umma. *Zeitschrift für Assyriologie und vorderasiatische Archäologie*, 2005, 95, p. 172-207. – IIDEM. Sa3-bi-ta. Texts from Girsu kept in the British Museum. Messina, Dipartimento di Scienze dell'Antichità-Università di Messina, 2005, 205 p.

1217. DALLEY (Stephanie). Old Babylonian texts in the Ashmolean Museum mainly from Larsa, Sippir, Kish and Lagaba. Oxford, Clarendon Press, 2005, X-234 p.

1218. DICK (Michael B.). A Neo-Sumerian ritual tablet in Philadelphia. *Journal of Near Eastern studies*, 2005, 64, p. 271-280.

1219. DURAND (Jean Marie), MARTI (Lionel). Chroniques du Moyen-Euphrate 5. Une attaque de Qatna par le Suhum et la question du "pays de Mari". *Revue d'assyriologie et d'archéologie orientale*, 2005, 99, p. 123-132.

1220. FARAJ (Ali H.), MORIGGI (Marco). Two incantation bowes from the Iraq Museum (Baghdad). *Orientalia*, 2005, 74, p. 71-82.

1221. FARBER (Walter). Neues aus Sippar-Iaārurum *Die Welt des Orients*, 2005, 35, p. 48-55.

1222. GLASSNER (Jean-Jacques). Des dieux, des scribes et des savants. Circulation des idées et transmission des écrits en Mésopotamie. *Annales*, 2005, 60, 3, p. 483-506. – IDEM. L'aruspicine paléo-babylonienne et le témoignage des sources de Mari. *Zeitschrift für Assyriologie und vorderasiatische Archäologie*, 2005, 95, p. 276-300.

1223. GRÄFF (Andreas). Thoughts about the Assyrian presence in Anatolia in the early 2nd millennium. *Altorientalische Forschungen*, 2005, 32, p. 158-167.

1224. GUICHARD (Michael). Archives royales de Mari XXXI. London, ERC, 2005, 569 p.

1225. HAASE (Richard). Darf man den sog. Telipinu-Erlaß eine Verfassung nennen? *Die Welt des Orients*, 2005, 35, p. 56-61.

1226. HAGENS (Graham). The Assyrian king list and chronology. *Orientalia*, 2005, 74, p. 23-41.

1227. HUOT (Jean-Louis). Vers l'apparition de l'état en Mésopotamie. Bilan des recherches récentes. *Annales*, 2005, 60, 5, p. 953-975.

1228. JAHN (Beate). Altbabylonische Wohnhäuser: eine Gegenüberstellung philologischer und archäologischer Quellen. Rahden, Verlag Marie Leidorf, 2005, IX-169 p.

1229. KEETMAN (Jan). Wie hoch war die 6. Stufe von Etemenanki? *Revue d'assyriologie et d'archéologie orientale*, 2005, 99, p. 77-84.

1230. KOUWENBERG (N.J.C.). Reflections on the Gt-Stem in Akkadian. *Zeitschrift für Assyriologie und vorderasiatische Archäologie*, 2005, 95, p. 77-103.

1231. MARTI (Lionel). Chroniques du Moyen-Euphrate 4. Relecture de ARM IX 291. *Revue d'assyriologie et d'archéologie orientale*, 2005, 99, p. 111-122.

1232. MAUL (Stefan M.). Das Gilgamesch-Epos. Neu übersetzt und kommentiert. München, C.H. Beck, 2005, 192 p.

1233. MAYER (Werner R.). Das Gebet des Eingeweideschauers an Ninurta. *Orientalia*, 2005, 74, p. 51-56. – IDEM. Die altbabylonischen Keilschrifttexte in der Sammlung des Päpstlichen Bibelinstituts. *Orientalia*, 2005, 74, p. 317-351. – IDEM. Lexikalische Listen aus Ebla und Uruk. *Orientalia*, 2005, 74, p. 157-164.

1234. MICHALOWSKI (Piotr). Iddin-Dagan and his family. *Zeitschrift für Assyriologie und vorderasiatische Archäologie*, 2005, 95, p. 65-76.

1235. MOLINA (Manuel), SUCH-GUTIÉRREZ (Marcos). Neo-Sumerian administrative texts in the British Museum. Messina, Dipartimento di Scienze dell'Antichità-Università di Messina, 2005, 312 p.

1236. MOORE (Andrew M.T.). A new exhibit on ancient Mesopotamia at the Oriental Institute, University of Chicago. *American journal of archaeology*, 2005, 109, p. 281-285.

1237. NASRABADI (Behzad Mofidi). Eine Steininschrift des Amar-Suena aus Tappeh Bormi (Iran). *Zeitschrift für Assyriologie und vorderasiatische Archäologie*, 2005, 95, p. 161-171.

1238. NEVLING PORTER (Barbara). Ritual and politics in ancient Mesopotamia. New Haven, American Oriental Society, 2005, XI-120 p.

1239. NIEDERREITER (Zoltán). L'insigne de pouvoir et le sceau du grand vizir Sin-ah-usur. *Revue d'assyriologie et d'archéologie orientale*, 2005, 99, p. 57-76.

1240. NOVOTNY (Jamie R.). Assurbanipal inscriptions in the Oriental Institute: prisms E, H and J. *Orientalia*, 2005, 74, p. 352-371.

1241. PAOLETTI (Paolo), SPADA (Gabriella). Testi se-ur5ra da Girsu conservati al British Museum. Messina, Dipartimento di Scienze dell'Antichità-Università di Messina, 2005, 177 p.

1242. PECHA (Lukas). Zum altbabylonischen Abgabensystem: Die "nemettum"-Abgabe. *Revue d'assyriologie et d'archéologie orientale*, 2005, 99, p. 39-56.

1243. POLITI (Janet), VERDERAME (Lorenzo). The Drehem texts in the British Museum. Messina, Dipartimento di Scienze dell'Antichità-Università di Messina, 2005, 335 p.

1244. RICHARDSON (Seth). Axes against Ešnunna. *Orientalia*, 2005, 74, p. 42-50.

1245. ROBSON (Eleanor). A new manuscript of Ninmešara (ETCSL 4.07.02), lines 109-139. *Orientalia*, 2005, 74, p. 382-388. – EADEM. Four old Babylonian school tablets in the Collection of the Catholic University of America. *Orientalia*, 2005, 74, p. 389-398.

1246. SCHOLTEN (Helga). Akkulturationsprozesse in der Euphrat-Region am Beispiel der griechich-makedonischen siedlung Dura-Europos. *Historia*, 2005, 54, p. 18-36.

1247. SCHREIBER (Gebhard J.). Sumerer und Akkader. Geschichte, Gesellschaft, Kultur. München, Beck, 2005, 126 p.

1248. SCURLOCK (Jo Ann), ANDERSEN (Burton R.). Diagnoses in Assyrian and Babylonian medicine: ancient sources, translations and modern medical analyses. Urbana a. Chicago, University of Illinois Press, 2005, XXVI-879 p.

1249. TAYLOR (Jon). The sumerian proverb collections. *Revue d'assyriologie et d'archéologie orientale*, 2005, 99, p. 13-38.

1250. TROPPER (Josef), VITA (Juan Pablo). Der Energikus an Jussiven im Kanaano-Akkadischen der Amarna-Periode. *Orientalia*, 2005, 74, p. 57-64.

1251. WOODS (Christopher). On the Euphrates. *Zeitschrift für Assyriologie und vorderasiatische Archäologie*, 2005, 95, p. 7-45.

§ 5. Hittites.

1252. ALAURA (Silvia). Fleh- und Unterwerfungsgesten in den hethitischen Texten. *Altorientalische Forschungen*, 2005, 32, p. 375-385.

1253. BRYCE (Trevor). The kingdom of the Hittites. Oxford, Oxford U. P., 2005, XIX-554 p.

1254. CARRUBA (Onofrio). Tuthalija 00I. (and Hattusili II.). *Altorientalische Forschungen*, 2005, 32, p. 246-271.

1255. COHEN (Yoram). Change and innovation in the administration and scribal practices of Emar during the Hittite dominion. *Tel Aviv*, 2005, 32, p. 192-203.

1256. DARDANO (Paola). I costrutti perifrastici con il verbo har(k)- dell'ittito: stato della questione e prospettive di metodo. *Orientalia*, 2005, 74, p. 93-113.

1257. DE MARTINO (Stefano). Hittite letters from the time of Tuthaliya I/II, Arnuwanda I and Tuthaliya III. *Altorientalische Forschungen*, 2005, 32, p. 291-321.

1258. FORLANINI (Massimo). Hattušili II. – Geschöpf der Forscher oder vergessener König? Ein Vorschlag zu seiner Stellung in der hethitischen Geschichte. *Altorientalische Forschungen*, 2005, 32, p. 230-245.

1259. GARSTANG (John). The land of the Hittites. An account of explorations and discoveries in Asia Minor. London, Kegan Paul, 2005, 440 p.

1260. GIORGIERI (Mauro). Zu den Treueiden mittelhethitischer Zeit. *Altorientalische Forschungen*, 2005, 32, p. 322-346.

1261. GONZÁLEZ SALAZAR (Juan Manuel). Hethitica. Notas sobre lexicografía hitita. *Cuadernos de filología clásica. Estudios griegos e indoeuropeos*, 2005, 15, p. 5-17.

1262. HAAS (Volkert). Die Erzählungen von den zwei Brüdern, vom Fischer und dem Findelkind sowie vom Jäger Kešše. *Altorientalische Forschungen*, 2005, 32, p. 360-374.

1263. HAASE (Richard). Nekromantie in der hethitischen Rechtssatzung? *Die Welt des Orients*, 2005, 35, p. 62-67.

1264. KLENGEL (Horst). Studien zur hethitischen Wirtschaft. Einleitende Bemerkungen. *Altorientalische Forschungen*, 2005, 32, p. 3-22.

1265. KLINGER (Jörg W.). Das Korpus der Kaskäer-Texte. *Altorientalische Forschungen*, 2005, 32, p. 347-359.

1266. PECCHIOLI DADDI (Franca). Die mittelhethitischen išhiul-Texte. *Altorientalische Forschungen*, 2005, 32, p. 280-290.

1267. POLVANI (Anna Maria). The deity IMIN. IMIN.BI in Hittite texts. *Orientalia*, 2005, 74, p. 181-194.

1268. RICHTER (Thomas). Kleine Beiträge zum hurritischen Wörterbuch. *Altorientalische Forschungen*, 2005, 32, p. 23-44.

1269. SEEHER (Jürgen). Bohren wie die Hethiter: Rekonstruktion von Bohrmaschinen der Spätbronzezeit und Beispiele ihrer Verwendung. *Istanbuler Mitteilungen*, 2005, 55, p. 17-36.

1270. SOYSAL (Oğuz). Beiträge zur althethischen Geschichte (III). Kleine Fragmente historischen Inhalts. *Zeitschrift für Assyriologie und vorderasiatische Archäologie*, 2005, 95, p. 121-144.

1271. TORRI (Giulia). Militärische Feldzüge nach Ostanatolien in der mittelhethitischen Zeit. *Altorientalische Forschungen*, 2005, 32, p. 386-400.

1272. WILHELM (Gernot). Zur Datierung der älteren hethitischen Landschenkungsurkunden. *Altorientalische Forschungen*, 2005, 32, p. 272-279.

Cf. n° 10

§ 6. Jews and Semitic peoples to the end of the ancient world.

1273. AOULAD TAHER (Mohamed). L'hellénisme dans le royaume numide au IIe siècle av. J.-C. *Antiquités Africaines*, 2005, 41, p. 29-42.

1274. ARNAUDIÈS (Alain). Translittération ou transcription? Propositions d'écriture des noms arabes égyptiens en archéologie. *Muséon*, 2005, 118, p. 241-268.

1275. ASHKENAZI (S.), MOTRO (U.), GOREN-INBAR (N.), BITTON (R.), RABINOVICH (R.). New morphometric parameters for assessment of body size and population structure in freshwater fossil crab assemblage from the Pleistocene site of Gesher Benot Ya'aqov (GBY), Israel. *Journal of African Earth sciences*, 2005, 32, p. 675-689.

1276. AVNI (Gideon). The urban limits of Roman and Byzantine Jerusalem: a view from the necropoleis. *Journal of Roman archaeology*, 2005, 18, p. 373-396.

1277. BEN-TOR (Amnon). Hazor and chronology. *Egypt and the Levant*, 2005, 15, p. 45-67. – IDEM. Tel-Hazor 2005. *Israel exploration journal*, 2005, 55, p. 209-216.

1278. BOUDOUHOU (Nouzha). À la recherche d'Assada. *Mélanges de l'École Française de Rome-Antiquité*, 2005, 117, p. 827-838.

1279. BRANKAER (Johanna). Marsanes: un texte séthien platonisant? *Muséon*, 2005, 118, p 21-41.

1280. BRINGMANN (Klaus). Geschichte der Juden im Altertum. Vom babylonischen Exil bis zur arabischen Eroberung. Stuttgart, Klett-Cotta, 2005, 365 p.

1281. BRIQUEL CHATONNET (Françoise). Les cités de la côte phénicienne et leurs sanctuaires de montagne. *Archiv für Religionsgeschichte*, 2005, 7, p. 20-33.

1282. COLLINS (John J.). Anti-Semitism in antiquity? The case of Alexandria. *Archiv für Religionsgeschichte*, 2005, 7, p. 86-101.

1283. CORRIENTE (Federico). The phonemic system of Semitic from the advantage point of Arabic and its dialectology. *Aula orientalis*, 2005, 23, p. 169-173.

1284. DEL OLMO LETE (Gregorio). An etymological and comparative Semitic dictionary. Phonology versus semantics: questions of method. *Aula orientalis*, 2005, 23, p. 185-190. – IDEM. Opening session: the international group on Comparative Semitics. *Aula orientalis*, 2005, 23, p. 5-7.

1285. DI SEGNI (Leah). Monastery, city and village in Byzantine Gaza. *Proche Orient chrétien*, 2005, 55, p. 24-51. – EADEM. The Mosaic inscription in the Nile festival building at Sepphoris: response to G.W. Bowersock. *Journal of Roman archaeology*, 2005, 18, p. 781-784.

1286. DOCHHORN (Jan). Die phönizischen Personennamen in den bei Josephus überlieferten Quellen zur Geschichte von Tyrus. *Die Welt des Orients*, 2005, 35, p. 68-117.

1287. DOWNEY (Susan B.). Petra rediscovered: an exhibition on Petra and Nabataean sanctuaries in Jordan. *American journal of archaeology*, 2005, 109, p. 783-787.

1288. FARAONE (Chris A.), GARNAND (Brien), LÓPEZ-RUIZ (Carolina). Micah's mother (Judg. 17:1-40) and a curse from Carthage (KAI 89): Canaanite precedents for Greek and Latin curses against thieves? *Journal of Near Eastern studies*, 2005, 64, p. 161-186.

1289. FAUST (Avraham). The Israelite village: cultural conservatism and technological innovation. *Tel Aviv*, 2005, 32, p. 204-219.

1290. FINKELSTEIN (Israel). Khirbat en-Nahas, Edom and Biblical history. *Tel Aviv* 32 (2005), pp. 119-125.

1291. FIRPO (Giulio). La "guerra di Quieto" e l'ultima fase della rivolta giudaica del 115–117 d.C. *Rivista storica dell'antichità*, 2005, 35, p. 99-116.

1292. FRONZAROLI (Pelio). Etymologies. *Aula orientalis*, 2005, 23, p. 35-43.

1293. GARR (W. Randall). The comparative method in Semitic linguistics. *Aula orientalis*, 2005, 23, p. 17-21.

1294. GRAGG (Gene). Morphology and root structure: a Beja perspective. *Aula orientalis*, 2005, 23, p. 23-33.

1295. GZELLA (Holger). Erscheinungsformen des historischen Präsens im Aramäischen. *Orientalia*, 2005, 74, p. 399-408.

1296. HEZSER (Catherine). Jewish slavery in antiquity. Oxford, Oxford U. P., 2005, XI-439 p.

1297. JACKSON (Bernard S.). The divorces of the Herodian princesses: Jewish law, Roman law or palace law? *In*: Josephus and Jewish history in Flavian Rome and beyond [Cf. n° 1299], p. 343-368.

1298. JONQUIÈRE (Tessel). Josephus' use of prayers. Between narrative and theology. *In*: Josephus and Jewish history in Flavian Rome and beyond [Cf. n° 1299], p. 229-243.

1299. Josephus and Jewish history in Flavian Rome and beyond. Ed. by Joseph SIEVERS, Gaia LEMBI. Leiden a. Boston, Brill, 2005, XIV-454 p. [Cf. n^{os} <choice> 1297, 1298, 1312, 1319, 1327, 1333, 1335, 1583.]

1300. KALIMI (Isaac). The reshaping of ancient Israelite history in Chronicles. Winona Lake, Eisenbrauns, 2005, XIII-473.

1301. KAYE (Alan S.). Two alleged Arabic etymologies. *Journal of Near Eastern studies*, 2005, 64, p. 109-111.

1302. KOGAN (Leonid). Observations on Proto-Semitic vocalism. *Aula orientalis*, 2005, 23, p. 131-167.

1303. KRUGER (Paul A.). Depression in the Hebrew Bible: an update. *Journal of Near Eastern studies*, 2005, 64, p. 187-192.

1304. LAMBERT (Mayer). Termes massorétiques, prosodie hébraïque et autres études. Appendices à la grammaire hébraïque. Genève, Librairie Droz, 2005, XVI-158 p.

1305. LEVINE (Lee I.). Figurative art in Jewish society. *Ars Judaica*, 2005, 1, p. 9-26.

1306. LEVISON (John R.), EWALD (Owen M.). Josephus and burial of women at public expense. *Athenaeum*, 2005, 93, p. 635-645.

1307. LODS (Adolphe). Israel from its beginnings to the middle of the eighth century. London, Kegan Paul, 2005, 533 p.

1308. LONNET (Antoine). Emprunts intra-sémitiques: l'exemple des emprunts arabes en Sudarabique moderne. *Aula orientalis*, 2005, 23, p. 199-206.

1309. LUCCHESI (Enzo). La 'Vorlage' arabe du Livre du coq éthiopien. *Orientalia*, 2005, 74, p. 91-92.

1310. MAGNESS (Jodi). The date of the Sardis synagogue in light of the numismatic evidence. *American journal of archaeology*, 2005, 109, p. 443-475.

1311. MANSOUR (Wisam). The reality beyond the hyperbolic accentuation of self in al-Shanfarā's poem Lāmiyyatu'l ᶜArab. *Journal of Near Eastern studies*, 2005, 64, p. 257-269.

1312. MENDELS (Doron). The formation of an historical canon of the Greco-Roman period. From the beginnings to Josephus. *In*: Josephus and Jewish history in Flavian Rome and beyond [Cf. n° 1299], p. 3-19.

1313. MIGLIETTA (Massimo). Gesù e il suo processo 'nella prospettiva ebraica'. *Athenaeum*, 2005, 93, p. 497-526.

1314. MILITAREV (Alexander). Root extension and root formation in Semitic and Afrasian. *Aula orientalis*, 2005, 23, p. 83-129.

1315. MOMRAK (Kristoffer). The Phoenicians in the Mediterranean: trade, interaction and cultural transfer. *Altorientalische Forschungen*, 2005, 32, p. 168-181.

1316. MOSCA (Paul G.). The indipendent object pronoun in Punic. *Orientalia*, 2005, 74, p. 65-70.

1317. NEVILLE (Ann). Mountains of silver and rivers of gold: the Phoenicians in Iberia. Oxford, Oxbow books, 2005, 240 p.

1318. PANITZ-COHEN (Nava). A salvage excavation in the New Market in Beer-Sheaba: new light on Iron Age IIB occupation at Beer-Sheaba. *Israel exploration journal*, 2005, 55, p. 143-155.

1319. PARENTE (Fausto). The impotence of Titus, or Josephus' Bellum Judaicum as an example of 'pathetic' historiography. *In*: Josephus and Jewish history in Flavian Rome and beyond [Cf. n° 1299], p. 45-69.

1320. PENNACCHIETTI (Fabrizio A.). Ripercussioni sintattiche in conseguenza dell'introduzione dell'articolo determinativo proclitico in semitico. *Aula orientalis*, 2005, 23, p. 175-184.

1321. RUBIN (Aaron D.). Studies in Semitic gramaticalization. Winona Lake, Eisenbrauns, 2005, XVII-177 p.

1322. RUBIO (Gonzalo). Chasing the Semitic root: the skeleton in the closet. *Aula orientalis*, 2005, 23, p. 45-63.

1323. SAIDEL (Benjamin Adam). On the periphery of an agricultural hinterland in the Negev Highlands: Rekhes Napha 396 in the sixth through the eighth centuries C.E. *Journal of Near Eastern studies*, 2005, 64, p. 241-255.

1324. SANMARTÍN (Joaquín). The semantic potential of bases ('root') and themes ('patterns'): a cognitive approach. *Aula orientalis*, 2005, 23, p. 65-81.

1325. SAVAGE (Stephen H.), ZAMORA (Kurt A.), KELLER (Donald R.). Archaeology in Jordan, 2004 season. *American journal of archaeology*, 2005, 109, p. 527-555.

1326. SCHEDROVITSKIJ (Dmitrij V.). Besedy o knige Iova. (On the Book of Job). Moskva, Oklik, 2005, 240 p. (bibl. p. 231-232; ind. p. 227-230).

1327. SCHIMANOWSKI (Gottfried). Alexandrien als Drehscheibe zwischen Jerusalem und Rom: die Bedeutung der Stadt im Werk des Josephus. *In*: Josephus and Jewish history in Flavian Rome and beyond [Cf. n° 1299], p. 317-330.

1328. SEDOV (Aleksandr V.). Drevnij Khadramaut: Ocherki arkheologii i numizmatiki. (Ancient Hadhramaut: Essays in archaeology and numismatics). RAN, In-t vostokovedeinja; Gos. Ermitazh. Moskva, Vostochnaja literatura, 2005, 528 p. (ill.; maps; bibl. p. 434-466; ind. p. 457-466). [English summary]

1329. STEINER (Richard C.). Påtaḥ and Qåmeṣ: on the etymology and evolution of the names of the Hebrew vowels. *Orientalia*, 2005, 74, p. 372-381.

1330. STIEGLITZ (Robert R.). Tel Tanninim excavations at Krokodeilon Polis, 1996–1999. Boston, American School of Oriental Research publication, 2005, 304 p.

1331. TAKÁCS (Gábor). Recent problems of Semitic-Egyptian and Semito-Cushitic and -Chadic consonant correspondences. *Aula orientalis*, 2005, 23, p. 207-231.

1332. TANTLEVSKIJ (Igor' R.). Istorija Izrailja i Iudei do razrushenija Pervogo Khrama. (A history of Israel and Judea till the destruction of the First Temple). Sankt-Peterburg, Izd-vo Sankt-Peterburgskogo gos. un-ta, 2005, 408 p. (ill.; bibl. p. 372-397).

1333. TROIANI (Lucio). La genèse historique des Antiquités juives. In: Josephus and Jewish history in Flavian Rome and beyond [Cf. n° 1299], p. 21-28.

1334. VAN DER STEEN (Eveline J.). Megiddo in the Early Bronze Age. American journal of archaeology, 2005, 109, p. 1-20.

1335. VAN HENTEN (Jan Willem). Commonplaces in Herod's commander speech in Josephus' A.J. 15. 127-146. In: Josephus and Jewish history in Flavian Rome and beyond [Cf. n° 1299], p. 183-206.

1336. VELLA (Nicholas C.). Phoenician and Punic Malta. Journal of Roman archaeology, 2005, 18, p. 436-450.

1337. WATSON (Wilfred G.E.). Loanwords in Semitic. Aula orientalis, 2005, 23, p. 191-198.

1338. WAZANA (Nili). Natives, immigrants and the Biblical perception of origins in historical times. Tel Aviv, 2005, 32, p. 220-244.

1339. WEISS (Zeev). Sepphoris (Sippori) 2005. Israel exploration journal, 2005, 55, p. 219-227.

1340. WEITZMAN (Steven). Surviving sacrilege. Cultural persistence in Jewish antiquity. Cambridge a. London, Harvard U. P., 2005, VIII-193 p.

1341. YASUR-LANDAU (Assaf). Old wine in New Vessels: intercultural contact, innovation and Aegean, Canaanite and Philistine foodways. Tel Aviv, 2005, 32, p. 168-191.

1342. ZABORSKI (Andrzej). Comparative Semitic studies: status quaestionis. Aula orientalis, 2005, 23, p. 9-15.

§ 7. Iran.

1343. ABAY (Eşref). The expansion of early Transcaucasian culture: cultural interaction or migration? Altorientalische Forschungen, 2005, 32, p. 115-131.

1344. ALDROVANDI (Cibele), HIRATA (Elaine). Buddhism, Pax Kushana and Greco-Roman motifs: pattern and purpose in Gandharan iconography. Antiquity, 2005, 79, p. 306-315.

1345. ALVAREZ-MON (Javier). Aspects of Elamite wall painting. New evidence from Kabnak (Haft Tappeh). Iranica antiqua, 2005, 40, p. 149-164.

1346. BIVAR (A. David H.). Mithraism. A religion for the ancient Medes. Iranica antiqua, 2005, 40, p. 341-358.

1347. FLEMING (S. J.), PIGOTT (V. C.), SWANN (C. P.), NASH (S. K.). Bronze in Luristan. Preliminary analytical evidence from copper/bronze artifacts excvated by the Belgian Mission in Iran. Iranica antiqua, 2005, 40, p. 35-64.

1348. GATES (Charles). Tracking the Achaemenid Persians in Anatolia. American journal of archaeology, 2005, 109, p. 789-792.

1349. GENITO (Bruno). The archaeology of the Median period. An outline and a research perspective. Iranica antiqua, 2005, 40, p. 315-340.

1350. GOPNIK (Hilary). The Shape of Sherds. Function and style at Godin II. Iranica antiqua, 2005, 40, p. 249–269.

1351. GULJAEV (Valerij I.). Skify: Rastsvet i padenie velikogo tsarstva. (The Scythians: The growth and the fall of the great kingdom). Moskva, Aletejja, 2005, 400 p. (ill.; maps; bibl. p. 385; ind. p. 391-399).

1352. HUFF (Dietrich). From Median to Achaemenian palace architecture. Iranica antiqua, 2005, 40, p. 371-395.

1353. IVANTCHIK (Askold). La chronologie des cultures pre-scythe et scythe. Les données proche orientales et caucasiennes. Iranica antiqua, 2005, 40, p. 447-460.

1354. KAWAMI (Trudy S.). Deer in art, life and death in Northwestern Iran. Iranica antiqua, 2005, 40, p. 107-131.

1355. KLEISS (Wolfram). Urartäische Architektur in der Entwicklungsgeschichte der Architektur Irans. Iranica antiqua, 2005, 40, p. 207-215.

1356. KLINKOTT (Hilmar). Der Satrap. Ein achaimenidischer Amtsträger und seine Handlungsspielräume. Frankfurt am Main, Verlag Antike, 2005, 578 p. (Oikumene. Studien zur antiken Weltgeschichte, 1).

1357. KONTANI (Ryoichi). Searching for the origin of the 'Bronze swords with Iron Core' in Northwestern Iran and the Caucasus region. Iranica antiqua, 2005, 40, p. 397-421.

1358. KROLL (Stephan). The Southern Urmia basin in the early Iron Age. Iranica antiqua, 2005, 40, p. 65-85.

1359. LANDSKRON (Alice). Parther und Sasaniden. Das Bild der Orientalen in der römischen Kaiserzeit. Wien, Phoibos Verlag, 2005, 225 p.

1360. MAGEE (Peter), PETRIE (Cameron), KNOX (Robert), KHAN (Farid), THOMAS (Ken). The Achaemenid empire in South Asia and recent excavations at Akra in Northwest Pakistan. American journal of archaeology, 2005, 109, p. 711-741.

1361. MARAS (Sabrina). Notes on seals and seal use in Western Iran from c. 900–600 BCE. Iranica antiqua, 2005, 40, p. 133-147.

1362. NEUJAHR (Matthew). When Darius defeated Alexander: composition and redaction in the dynastic prophecies. Journal of Near Eastern studies, 2005, 64, p. 101-107.

1363. OVERLAET (Bruno). The chronology of the Iron Age in the Pusht-i Kuh, Luristan. Iranica antiqua, 2005, 40, p. 1-33.

1364. POTTS (Daniel T.). Neo-Elamite problems. *Iranica antiqua*, 2005, 40, p. 165-177.

1365. RAZJMOU (Shakrokh). In search of lost Median art. *Iranica antiqua*, 2005, 40, p. 271-314.

1366. RITTER (Nils C.). Had the Sasanians been interested in Roman culture? *Altorientalische Forschungen*, 2005, 32, p. 182-199.

1367. STRONACH (David). The Arjan tomb. Innovation and acculturation in the last days of Elam. *Iranica antiqua*, 2005, 40, p. 179-196.

1368. TOUROVETS (Alexandre). Some reflections about the relation between the architecture of Northwestern Iran and Urartu. The layout of the central temple of Nush-i Djan. *Iranica antiqua*, 2005, 40, p. 359-370.

E

GREEK HISTORY

§ 1. Classical world in general. 1369-1387. – § 2. Prehellenic epoch. 1388-1401. – § 3. Sources and criticism of sources (*a.* Epigraphical sources; *b.* Literary sources). 1402-1443. – § 4. General and political history. 1444-1478. – § 5. History of law and institutions. 1479-1508. – § 6. Economic and social history. 1509-1528. – § 7. History of literature, philosophy and science (*a.* Literature; *b.* Philosophy and sciences). 1529-1686. – § 8. Religion and mythology. 1687-1723. – § 9. Archaeology and history of art. 1724-1751.

§ 1. Classical world in general.

* 1369. Année (L') philologique. Bibliographie critique et analytique de l'antiquité gréco-latine. Publiée par la Société internationale de bibliographie classique. Tome LXXIV: bibliographie de l'année 2003 et compléments d'années anteérieures. [Tome LXXIII. 2002. Cf. Bibl. 2004, n° 1484.] Paris, Les Belles Lettres, 2005, LXV-2019 p.

* 1370. DECKER (Wolfgang), RIEGER (Barbara). Bibliographie zum Sport im Altertum für Jahre 1989 bis 2002. Köln, Verlag Sport und Buch Strauß, 2005, 272 p.

1371. AMIGUES (Suzanne). Glanes naturalistes dans la collection des Universités de France. *Revue des études grecques*, 2005, 118, p. 382-390.

1372. Biographie und Prosopographie. Internationales Kolloquium zum 65. Geburtstag von Anthony R. Birley (28. September 2002, Schloß Mickeln, Düsseldorf). Hrsg. v. Konrad VÖSSING. Stuttgart, Steiner, 2005, 146 p. (Historia Einzelschriften, 178) [Cf. n[os] <Auswahl> 1779, 1972, 2103.]

1373. CARBONE (Gabriella). Tabliope. Ricerche su gioco e letteratura nel mondo greco-romano. Napoli, F. Giannini, 2005, 463 p.

1374. EVSEENKO (Timur P.). Ot obshchiny k slozhnoj gosudarstvennosti v antichnom Sredizemnomor'e. (From common to complicated state in the Antique Mediteranean Region). Assots. "Juridicheskij tsentr". Sankt-Peterburg, Juridicheskij tsentr Press, 2005, 264 p. (bibl. p. 248-261).

1375. Föderalismus in der griechischen und römischen Antike. Hrsg. v. Peter SIEWERT, Luciana AIGNER-FORESTI. Stuttgart, Steiner, 2005, 169 p. [Cf. n[os] <Auswahl> 1449, 1470.]

1376. HANSON (Ann). The state of ancient medicine. *Journal of Roman archaeology*, 2005, 18, p. 495-501.

1377. JOHNSON (Marguerite), RYAN (Terry). Sexuality in Greek and Roman society and literature: a sourcebook. London, Routledge, 2005, XXVI-244 p.

1378. KRON (Geoffrey). Anthropometry, physical anthropometry, and the reconstruction of ancient health, nutrition, and living standards. *Historia*, 2005, 54, p. 68-83.

1379. LENDON (Jon E.). Soldiers and ghosts: a history of battle in classical antiquity. New Haven a. London, Yale U. P., XII-468 p.

1380. LLOYD-JONES (Hugh). Supplementum supplementi Hellenistici. Berlin et New York, de Gruyter, 2005, IX-159 p.

1381. LUND (Allan A.). Hellenentum und Hellenizität: zur Ethnogenese und zur Ethnizität der antiken Hellenen. *Historia*, 2005, 54, p. 1-17.

1382. Nike. Ideologia, iconografia e feste della Vittoria in età antica. A cura di Domenico MUSTI. Roma, L'Erma di Bretschneider, 2005, 358 p. [Cf. n[os] <scelta> 1523, 1525.]

1383. RHODES (P.J.). A history of the classical Greek world, 478–323 BC. Oxford, Blackwell, 2005, XII-407 p.

1384. RUTHERFORD (Richard). Classical literature: a concise history. Malden, Blackwell, 2005, XVIII-350 p.

1385. SCULLION (Scott). Actors and acting in the ancient world. *Journal of Roman archaeology*, 2005, 18, p. 539-544.

1386. SKINNER (Marilyn B.). Sexuality in Greek and Roman culture. Malden, Blackwell, 2005, XXX-343 p.

1387. STOREY (Ian C.), ARLENE (Allen). A guide to ancient Greek drama. Malden, Blackwell, 2005, XVI-311 p.

Cf. n°s 138, 1108

§ 2. Prehellenic epoch.

1388. ARAVANTINOS (Vassilis L.), DEL FREO (Maurizio), GODART (Louis), SACCONI (Anna). Thèbes. Fouilles de la Cadmée. Vol. IV. Les textes de Thèbes (1-433). Translitération et tableaux des scribes. Pisa, Istituti editoriali e poligrafici internazionali, 2005, XII-344 p.

1389. ARDA (Basak), KNAPP (Bernard A.), WEBB (Jennifer M.). The collection of Cypriote antiquities in the Hunterian Museum-University of Glasgow. Sävedalen, Astrom Paul, 2005, 57 p.

1390. BENDALL (Lisa M.). Studies in Mycenaean inscriptions and dialect, 1980–1997. *American journal of archaeology*, 2005, 109, p. 91-94.

1391. BURKE (Brendan). Materialization of Mycenaean ideology and the Ayia Triada sarcophagus. *American journal of archaeology*, 2005, 109, p. 403-422.

1392. DEL FREO (Maurizio). I censimenti di terreni nei testi in lineare B. Pisa e Roma, Istituti editoriali e poligrafici internazionali, 2005, XXVIII-287 p.

1393. DUHOUX (Yves). Les nouvelles tablettes en linéaire B de Thebes et la religion grecque. *L'antiquité classique*, 2005, 74, p. 1-20.

1394. FINKELBERG (Margalit). Greeks and pre-Greeks. Aegean prehistory and Greek heroic tradition. Cambridge, Cambridge U. P., 2005, XV-203 p.

1395. HALSTEAD (Paul). Food, cuisine and society in Prehistoric Greece. Oxford, Oxbow books, 2005, 206 p.

1396. KYRIAKIDIS (Evangelos). Unidentified floating objects on Minoan seals. *American journal of archaeology*, 2005, 109, p. 137-154.

1397. MANGANI (Elisabetta). Materiali micenei, geometrici e orientalizzanti di Rodi. *Bullettino di paletnologia*, 2005, 96, p. 203-310.

1398. MANTZOURANI (Eleni), VAVOURANAKIS (Giorgos), KANELLOPOULOS (Chrysanthos). The Klimataria-Manares building reconsidered. *American journal of archaeology*, 2005, 109, p. 743-776.

1399. PALAIMA (Thomas). The triple invention of writing in Cyprus and written sources for Cypriote history. Nicosia, Anastasios G. Leventis Foundation, 2005, 64 p.

1400. SHAW (Joseph W.), SHAW (Mary C.). Kommos. Vol. V. The monumental Minoan buildings at Kommos. Princeton, Princeton U. P., 2005, 704 p.

1401. WEILHARTNER (Jörg). Mykenische Opfergaben nach Aussage der Linear B-Texte. Wien, Verlag der Österreichischen Akademie der Wissenschaften, 2005, 262 p.

§ 3. Sources and criticism of sources.

a. Epigraphical sources

* 1402. CHANIOTIS (Angelos), MYLONOPOULOS (Jannis). Epigraphic bulletin 2002. *Kernos*, 2005, 18, p. 425-474.

* 1403. GAUTHIER (Philippe), DUBOIS (Laurent). Bulletin épigraphique. *Revue des etudes grecques*, 2005, 118, p. 436-591. [2004, 117. Cf. Bibl. 2004, n° 1530.]

* 1404. Supplementum Epigraphicum Graecum (SEG), 2001, 51. Ed. by Angelos CHANIOTIS, Thomas CORSTEN, Ronald S. STROUD and Rolf A. TYBOUT. Amsterdam, J.C. Gieben, 2005, XXXIV-937 p.

1405. BAKER (Patrick), THÉRIAULT (Gaétan). Les Lyciens, Xanthos et Rome dans la première moitié du Ier s. a.C. Nouvelles inscriptions. *Revue des etudes grecques*, 2005, 118, p. 329-366.

1406. BETTARINI (Luca). Corpus delle defixiones di Selinunte. Alessandria, Edizioni dell'Orso, 2005, VII-178 p.

1407. BRICAULT (Laurent). Recueil des inscriptions concernant les cultes Isiaques. Paris, de Boccard, 2005, 3 vol., XXX-837 p.

1408. DOBIAS-LALOU (Catherine). À propos du nom de femme Timakrátea à Rhodes. *Revue des etudes grecques*, 2005, 118, p. 592-599.

1409. DUBOIS (Laurent). Alphabet, onomastique et dialecte des Îles Lipari. *Revue des etudes grecques*, 2005, 118, p. 214-228.

1410. FOLLET (Simone). Deux inscriptions attiques inédites copiées par l'abbé Michel Fourmont (Parisinus Suppl. gr. 854). *Revue des etudes grecques*, 2005, 118, p. 1-14.

1411. FOURNIER (Julien). Sparte et la justice romaine sous le Haut-Empire. A propos de IG V 1, 21. *Revue des etudes grecques*, 2005, 118, p. 117-137.

1412. Inscriptiones antiquae partis Thraciae quae ad ora maris Aegaei sita est. Prafecturae Xantes, Rhodopes et Hebri. Epigraphes tēs Thrakēs tou Aigaiou metax tōn potamōn nestou kai ebrou: nomoi Xanthēs kai ebrou. Ediderunt et commentariis sermone graeco conscriptis instruxerunt Louisa D. LOUKOPOULOU, Maria Gabriella PARISSAKI, Selene PSOMA [et al.]. Athēnai et Paris, de Boccard, 2005, 688 p.

1413. MANGANARO (Giacomo). La stele in pietra scura (IG XIV 7) con l'espistola di Gerone II ai Siracusani. *Zeitschrift für Papyrologie und Epigraphik*, 2005, 152, p. 141-151.

1414. MARGINESCU (Giovanni). Gortina di Creta. Prospettive epigrafiche per lo studio della forma ur-

3. SOURCES AND CRITICISM OF SOURCES

bana. Atene, Scuola Archeologica Italiana di Atene, 2005, 149 p.

1415. PÉBARTHE (Christophe). La perception des droit de passage à Chalcis (IG I3 40, 446 a.C.). *Historia*, 2005, 54, p. 84-92.

1416. SOSIN (Joshua D.). A common market on Syros: two imperial letters (IG XIII.5 658). *Historia*, 2005, 54, p. 222-226.

b. Literary sources

* 1417. Poiesis. Bibliografia della poesia greca 2003-2004. Direzione Massimo DI MARCO, Bruna M. PALUMBO STRACCA. Pisa e Roma, Istituti editoriali e poligrafici internazionali, 2005, XXIII-671 p.

1418. [Alexander Aphrodisiensis] ACCATTINO (Paolo). Alessandro di Afrodisia, De anima II (Mantissa). Alessandria, Edizioni dell'Orso, 2005, 219 p.

1419. [Andocides; Antiphon; Isaeus; Isocrates; Lysias] JEBB (Richard C.). Selections from the Attic orators Antiphon, Andocides, Lysias, Isocrates, Isaeus. New introduction and bibliography. Bristol, Phoenix Press, 2005, 434 p.

1420. [Apollonius Dyscolus] BRANDENBURG (Philipp). Apollonios Dyskolos. Über das Pronomen. Einführung, Text, Übersetzung und Erläuterung. München u. Leipzig, Saur, 2005, XIV-676 p.

1421. [Aristoteles] RASHED (Marwan). Aristote. De la génération et la corruption. Paris, Les Belles Lettres, 2005, CCLV-195 p.

1422. [Aristoteles] SCHÜTRUMPF (Eckart). Aristoteles. Politik. Buch VII/VIII. Berlin, Akademieverlag, 2005, 685 p.

1423. [Arrianus; Diodorus Siculus; Plutarchus; Quintus Curtius Rufus] ROMM (James). Alexander the Great: selections from Arrian, Diodorus, Plutarch and Quintus Curtius. Indianapolis, Hackett, 2005, XXX-193 p.

1424. [Chariton Aphrodisiensis, Longus, Xenophon Ephesius] BORGOGNO (Alberto). Romanzi greci. Caritone d'Afrodisia, Senofonte Efesio, Longo Sofista. Torino, UTET, 2005, 709 p.

1425. [Demosthenes] YUNIS (Harvey). Demosthenes: speeches 18 and 19. Austin, University of Texas Press, 2005, XXX-243 p.

1426. [Epictetus] SEDDON (Keith). Epictetus' Handbook and the Tablet of Cebes. Guide to Stoic living. London a. New York, Routledge, 2005, XIII-282 p.

1427. [Euripides] PADUANO (Guido). Euripide. Il Ciclope. Milano, BUR, 2005, 133 p.

1428. [Euripides] RIGO (Georges). Euripide. Opera et fragmenta omnia. Index verborum. Listes de fréquence. Liège, C.I.P.L., 2005, XXV-1355 p.

1429. [Favorinus] AMATO (Eugenio), JULIEN (Yvette). Favorine d'Arles. Oeuvres. Tome 1: Introduction générale-Témoignages-Discours aux Corinthiens-Sur la fortune. Paris, Les Belles Lettres, 2005, XIV-607 p.

1430. [Hesychius Alexandrinus] HANSEN (Peter Allan). Hesychii Alexandrini Lexicon. Volumen III. Berlin et New York, de Gruyter, 2005, XXXIII-404 p.

1431. [Lucianus Samosatensis] BALSLEV (Ole), TORTZEN (Chr. Gorm). Lukian. Anacharsis. En entik Rousseau oghans samfundskritik. Inledning, oversættelse og kommentar. København, København Tusculanums Forlag, 2005, 80 p.

1432. [Lucianus Samosatensis] COSTA (Desmond). Lucian. Selected dialogues translated with introduction and notes. Oxford, Oxford U. P., 2005, XVI-278 p.

1433. [Lucianus Samosatensis] PILHOFER (Peter), BAUMBACH (Manuel), GERLACH (Jens), HANSEN (Dirk Uwe). Lukian. Der Tod des Peregrinos. Ein Scharlatan auf dem Scheiteraufen. Darmstadt, Wissenschaftliche Buchgesellschaft, 2005, X-257 p.

1434. [Lycophron] LAMBIN (Gérard). L'Alexandre de Lycophron. Rennes, Presses Universitaires de Rennes, 2005, 303 p.

1435. [Philostratus] JONES (Christopher P.). Philostratus. The life of Apollonios of Tyana. Books I-IV. Books V-VIII. Cambridge, Cambridge a. London, Harvard U. P., 2005, 2 vol., VII-423 p., VII-440 p.

1436. [Pindarus] BURNETT (Anne Pippin). Pindar's songs for young athletes of Aigina. Oxford, Oxford U. P., 2005, X-276 p.

1437. [Pindarus] HENRY (W. Ben). Pindar's Nemeans. A selection. München a. Leipzig, Saur, 2005, XII-133 p.

1438. [Plutarchus] TIRELLI (Aldo). Plutarco. Ad un governante incolto. Introduzione, testo critico, traduzione e commento. Napoli, D'Auria, 2005, 139 p.

1439. [Poetae epici Graeci] BERNABÉ (Albertus). Poetae epici Graeci. Testimonia et fragmenta. Pars II: Orphicorum et Orphicis similium testimonia et fragmenta. Fasciculus z. München et Leipzig, Saur, 2005, XXV-553 p.

1440. [Solon] IRWIN (Elizabeth). Solon and early Greek poetry. The politics of exhortation. Cambridge, Cambridge U. P., 2005, XIII-350 p.

1441. [Sophocles] ROISMAN (Hanna M.). Sophocles. Philoctetes. London, Duckworth, 2005, 159 p.

1442. [Strabo] RADT (Stefan). Strabons Geographika. Band 4. Buch XIV-XVII. Text und Übersetzung. Göttingen, Vandenhoeck & Ruprecht, 2005, 574 p.

1443. [Xenophon Ephesius] O'SULLIVAN (James N.). Xenophon Ephesius. De Anthia et Habrocome Ephesiacorum libri V. München u. Leipzig, Saur, 2005, XXXIV-128 p.

Cf. n°s 1529-1686

§ 4. General and political history.

1444. BOWDEN (Hugh). Classical Athens and the Delphic Oracle. Divination and democracy. Cambridge, Cambridge U. P., 2005, XVIII-188 p.

1445. CAGNAZZI (Silvana). Aspiranti tiranni e tiranni rinunciatari di Atene. *Rivista storica dell'antichità*, 2005, 35, p. 7-22. – EADEM. Il grande Alessandro. *Historia*, 2005, 54, p. 132-143.

1446. CAWKWELL (George). The Greek wars. The failure of Persia. Oxford a. New York, Oxford U. P., 2005, VIII-316 p.

1447. Citoyenneté et participation à la basse époque hellénistique. Actes de la table ronde des 22 et 23 mai 2004, Paris, BNF organisée par le groupe de recherche dirigé par Philippe Gauthier de l'UMR 8585 (Centre Gustave Glotz). Ed. par Pierre FRÖHLICH et Christel MÜLLER. Genève, Droz, 2005, 310 p. (Hautes études du monde gréco-romain, 35).

1448. DE ROMILLY (Jacqueline). L'élan démocratique dans l'Athènes ancienne. Paris, Fallois, 2005, 153 p.

1449. DOUKELLIS (Panagiotis). Föderalismus in hellenistischer und römischer Zeit. Theorien und Praktiken. *In*: Föderalismus in der griechischen und römischen Antike [Cf. n° 1375], p. 43-79.

1450. FORSDYKE (Sara). Exile, ostracism, and democracy: the politics of expulsion in ancient Greece. Oxford, Princeton U. P., 2005, XIV-344 p.

1451. FROST (Frank J.). Politics and the Athenians: essays on Athenian history and historiography. Toronto, Edgar Kent, 2005, 302 p.

1452. GESKE (Norbert). Nikias und das Volk von Athen im Archidamischen Krieg. Stuttgart, Steiner, 2005, 224 p.

1453. Greeks (The) in the East. Ed. by Alexandra VILLING. British Museum Classical Colloquium (21st: 1997: London, England). London, British Museum, 2005, V-123 p. (ill., plates, maps). (Research publication, 157).

1454. HANSEN (Mogens Herman). Der athenske demokrati-og vores. København, Museum Tusculanums Forlag, 2005, 203 p.

1455. HOLT (Frank L.). Into the land of bones. Alexander the Great in Afghanistan. Berkeley, Los Angeles a. London, University of California Press, 2005, XIII-241 p.

1456. Imaginary polis (The). Symposium, Januara 7–10 2004. Ed. by Mogens Herman HANSEN. København, Det Kongelige Danske videnskabernes selskab, 2005, 444 p. (Acts of the Copenhagen polis centre 7. Historisk-filosofiske meddelelser, 91).

1457. KIENAST (Dietmar). Die Zahl der Demen in der Kleisthenischen Staatsordnung. *Historia*, 2005, 54, p. 495-498.

1458. LAVELLE (B. M.). Fame, money, and power: the rise of Peisistratos and "Democratic" tyranny at Athens. Ann Arbor, University of Michigan Press, 2005, XIV-370 p.

1459. LEHMLER (Caroline). Syrakus unter Agathokles und Hieron II. Die Verbindung von Kultur und Macht in einer hellenistischen Metropole. Frankfurt am Main, Verlag Antike, 2005, 254 p.

1460. LOW (Polly). Looking for the language of Athenian imperialism. *Journal of Hellenic studies*, 2005, 125, p. 93-111.

1461. MAC KECHNIE (Paul). Beau monde and demimonde in Alexandria, 323–116 BC. *L'antiquité classique*, 2005, 74, p. 69-82.

1462. MARCHENKO (Konstantin K.), VAKHTINA (Marina Ju.), VINOGRADOV (Jurij A.), ROGOV (Evgenij Ja.), ANDREEV (Jurij V.). Greki i varvary Severnogo Prichernomor'ja v skifskuju epokhu. (Greeks and Barbarians of the North Black Sea Region in the Scythian Age). Ed. Konstantin K. MARCHENKO. Sankt-Peterburg, Aletejja, 2005, 464 p. (ill.; maps; bibl. p. 418-461).

1463. MOSSÉ (Claude). Périclès: l'inventeur de la démocratie. Paris, Payot & Rivages, 2005, 280 p. (ill.). (Biographie Payot).

1464. NESSELRATH (Hans-Günter). 'Where the lord of the sea grants passage to sailors through the deep-blue mere no more'. The Greeks and the Western seas. *Greece and Rome*, 2005, 52, p. 153-171.

1465. PÉBARTHE (Christophe). Clisthène a-t-il été archonte en 525/4? Mémoire, oubli et histoire des Athéniens à l'époque classique. *Revue belge de philologie et d'histoire*, 2005, 83, 1, p. 25-53.

1466. RATHMANN (Michael). Perdikkas zwischen 323 und 320. Nachlassverwalter des Alexanderreiches oder Autokrat? Wien, Verlag der Österreichischen Akademie der Wissenschaften, 2005, 100 p. (Österreichische Akademie der Wissenschaften, philos.-histor. Klasse, Sitzungsberichte, 724).

1467. ROSCALLA (Fabio). Biaios didaskalos. Rappresentazioni della crisi di Atene della fine del V secolo. Pisa, Edizioni ETS, 2005, 134 p.

1468. SARTRE (Maurice). Histoires grecques. Paris, Ed. du Seuil, 2005, 462 p. (L'Univers historique).

1469. SCHEPENS (Guido). À la recherche d'Agésilas. Le roi de Sparte dans le jugement des historiens du IVe siècle av. J.-C. *Revue des etudes grecques*, 2005, 118, p. 31-78.

1470. SIEWERT (Peter). Föderalismus in der griechischen Welt bis 338 v.Chr. *In*: Föderalismus in der griechischen und römischen Antike [Cf. n° 1375], p. 17-41.

1471. SPIVEY (Nigel). The ancient olympics. New York, Oxford a. Auckland, Oxford U. P., 2005, XXI-273 p.

1472. SURIKOV (Igor' E.). Antichnaja Gretsija. Politiki v kontekste epokhi: Arkhaika i rannjaja klassika.

(Ancient Greece: Politicians in the context of their epokh: The Archaic and the Early Classical periods). RAN, In-t vseobshchej istorii. Moskva, Nauka, 2005, 352 p. (bibl. p. 326-349).

1473. WALLINGA (H. T.). Xerxes' Greek adventure. The naval perspective. Leiden a. Boston, Brill, 2005, XIII-174 p. (Mnemosyne, 264).

1474. WIEMER (Hans-Ulrich). Alexander der Große. München, Beck, 2005, 243 p.

§ 4. Addenda 1994–2002.

1475. Even more studies in the ancient Greek polis. Ed. by Thomas HEINE NIELSEN. Stuttgart, Steiner, 2002, 294 p. (ill.). (Historia. Einzelschriften, 162. Papers from the Copenhagen Polis Centre, 6).

1476. Further studies in the ancient Greek polis. Ed. by Pernille FLENSTED-JENSEN. Stuttgart, F. Steiner, 2000, 262 p. (ill.). (Historia. Einzelschriften, 138. Papers from the Copenhagen Polis Centre, 5).

1477. Introduction to an inventory of Poleis: Acts of the Copenhagen Polis Centre vol. 3, Symposium August, 23–26 1995. Ed. by Mogens Herman HANSEN. København, Det Kongelige Danske Videnskabernes Selskab, 96, 411 p. (ill.). (Det Kongelige Danske Videnskabernes Selskab. Historisk-filosofiske Meddelelser, 3).

1478. Political architecture (From) to Stephanus Byzantius: sources for the ancient Greek polis. Ed. by David WHITEHEAD. Stuttgart, Steiner, 94, 124 p. (ill.). (Historia. Einzelschriften, 1).

Cf. n°s 138, 750, 1132, 1351

§ 5. History of law and institutions.

1479. Antike Rechtsgeschichte. Einheit und Vielfalt. Hrsg. v. Gerhard THÜR. Wien, Verlag der Österreichischen Akademie der Wissenschaften, 2005, VIII-95 p. [Cf. n[os] <Auswahl> 1143, 1195, 1500.]

1480. AVRAMOVIČ (Sima). The rhetra of Epitadeus and testament in Spartan law. In: Vorträge zur griechischen und hellenistischen Rechtsgeschichte [Cf. n° 1507], p. 175-186.

1481. BREITBACH (Michael). Der Prozess des Sokrates-Verteidigung der oder Anschlag auf die Athenische Demokratie? Ein Beitrag aus rechtswissenschaftlicher Perspektive. Gymnasium, 2005, 112, p. 321-343.

1482. Cambridge companion (The) to ancient Greek law. Ed. by Michael GAGARIN and David COHEN. Cambridge, Cambridge U. P., 2005, XIII-480 p. [Cf. n[os] <choice> 1483, 1485, 1487, 1490, 1498, 1500, 1502, 1504.]

1483. CANTARELLA (Eva). Gender, sexuality and law. In: Cambridge companion (The) to ancient Greek law [Cf. n° 1482], p. 236-253. – EADEM. Violence privée et procès. In: Violence (La) dans les mondes grec et romaine [Cf. n° 1527], p. 339-347.

1484. CHANIOTIS (Angelos). Victory's verdict: the violent occupation of territory in Hellenistic interstate relations. In: Violence (La) dans les mondes grec et romaine [Cf. n° 1527], p. 455-464.

1485. COHEN (David). Crime, punishment and the rule of law in classical Athens. In: Cambridge companion (The) to ancient Greek law [Cf. n° 1482], p. 211-235. – IDEM. Theories of punishment. In: Cambridge companion (The) to ancient Greek law [Cf. n° 1482], p. 170-190. – IDEM. Women in public: gender, citizenship and social status in classical Athens. In: Vorträge zur griechischen und hellenistischen Rechtsgeschichte [Cf. n° 1507], p. 33-45.

1486. CUDJOE (Richard V.). The purpose of the "epidikasia" for an "epikleros" in classical Athens. Dike, 2005, 8, p. 55-88.

1487. DAVIES (John). The Gortyn laws. In: Cambridge companion (The) to ancient Greek law [Cf. n° 1482], p. 305-327.

1488. ENGEN (Darel Tai). "Ancient Greenbacks": Athenian owls, the law of Nikophon, and the Greek economy. Historia, 2005, 54, p. 359-381.

1489. FARAGUNA (Michele). La figura dell'aisymnetes tra realtà storica e teoria politica. In: Vorträge zur griechischen und hellenistischen Rechtsgeschichte [Cf. n° 1507], p. 321-338. – IDEM. Terra pubblica e vendite di immobili confiscati a Chio nel V secolo a.C. Dike, 2005, 8, p. 89-99.

1490. GAGARIN (Michael). Early Greek law. In: Cambridge companion (The) to ancient Greek law [Cf. n° 1482], p. 82-94.

1491. GHEZZI (Vania). I Locresi e la legge del laccio. Dike, 2005, 8, p. 101-113.

1492. GLAZEBROOK (Allison). Prostituting female kin (Plut. Sol. 23.1-2). Dike, 2005, 8, p. 33-53.

1493. HAMMER (Dean). Plebiscitary politics in archaic Greece. Historia, 2005, 54, p. 107-131.

1494. HARRIS (Edward M.). Feuding or the rule of law? The nature of litigation in classical Athens. An essay in legal sociology. In: Vorträge zur griechischen und hellenistischen Rechtsgeschichte [Cf. n° 1507], p. 125-141.

1495. HAßKAMP (Dorothee). Oligarchische Willkür – demokratische Ordnung. Zur athenischen Verfassung im 4. Jahrhundert v. Chr. Darmstadt, Wissenschaftliche Buchgesellschaft, 2005, 192 p.

1496. KIMMERLE (Ralph). Völkerrechtliche Beziehungen Spartas in spätarchaischer und frühklassischer Zeit. Münster, Lit, 2005, 165 p. (Antike Kultur und Geschichte, 6).

1497. KOIV (Mait). The origins, development, and reliability of the ancient tradition about the formation of Spartan constitution. Historia, 2005, 54, p. 233-264.

1498. LANNI (Adriaan). Relevance in Athenian court. *In*: Cambridge companion (The) to ancient Greek law [Cf. n° 1482], p. 112-128.

1499. LEÃO (Delfim F.). Sólon e a legislação em matéria de direito familiar. *Dike*, 2005, 8, p. 5-31.

1500. MAFFI (Alberto). Family and property law. *In*: Cambridge companion (The) to ancient Greek law [Cf. n° 1482], p. 254-266. – IDEM. Klassisches griechisches und hellenistisches Recht. *In*: Antike Rechtsgeschichte [Cf. n° 1479], p. 11-16.

1501. MIGEOTTE (Léopold). Les pouvoirs des agronomes dans les cités grecques. *In*: Vorträge zur griechischen und hellenistischen Rechtsgeschichte [Cf. n° 1507], p. 287-301.

1502. PARKER (Robert). Law and religion. *In*: Cambridge companion (The) to ancient Greek law [Cf. n° 1482], p. 61-81.

1503. RUBINSTEIN (Lene). Main litigants and witnesses in Athenian courts: procedural variations. *In*: Vorträge zur griechischen und hellenistischen Rechtsgeschichte [Cf. n° 1507], p. 99-120.

1504. RUPPRECHT (Hans-Albert). Greek law in foreign surroundings: continuity and development. *In*: Cambridge companion (The) to ancient Greek law [Cf. n° 1482], p. 328-342.

1505. RUSCHENBUSCH (Eberhard). Kleine Schriften zur griechischen Rechtsgeschichte. Wiesbaden, Harrassowitz, 2005, 248 p.

1506. SCAFURO (Adele C.). Parent abusers, military shirkers and accused killers: the authenticity of the second law inserted at Dem. 24. 105. *In*: Vorträge zur griechischen und hellenistischen Rechtsgeschichte [Cf. n° 1507], p. 51-69.

1507. Vorträge zur griechischen und hellenistischen Rechtsgeschichte (Evanston, Illinois, 5–8 September 2001). Hrsg. v. Robert W. WALLACE und Michael GAGARIN. Wien, Verlag der Österreichischen Akademie der Wissenschaften, 2005, X-369 p. [Cf. n^{os} <Auswahl> 1480, 1485, 1489, 1494, 1501, 1503, 1506.]

1508. ZELNICK-ABRAMOVITZ (Rachel). Not wholly free. The concept of manumission and the status of manumitted slaves in the ancient Greek world. Boston, Brill, 2005, VI-385 p. (Mnemosyne supplementa, 266).

Cf. n° 1374

§ 6. Economic and social history.

1509. ANDÒ (Valeria). L'ape che tesse. Saperi femminili nella Grecia antica. Roma, Carocci, 2005, 292 p.

1510. BOEDEKER (Deborah), RAAFLAUB (Kurt). Tragedy and city. *In*: Companion (A) to tragedy [Cf. n° 1514], p. 109-127.

1511. CARTER (Michael). Neokoria and imperial cult in the Greek East. *Journal of Roman archaeology*, 2005, 18, p. 635-637.

1512. COARELLI (Filippo). Selling people: five papers on Roman slave-traders and the buildings they used. L"Agora des Italiens': lo statarion di Delo? *Journal of Roman archaeology*, 2005, 18, p. 196-212.

1513. Companion (A) to Greek tragedy. Ed. by Justina GREGORY. Oxford, Blackwell, 2005, XVIII-552 p. [Cf. n^{os} <choice> 1557, 1577, 1580, 1622, 1624, 1625, 1629, 1640, 1714.]

1514. Companion (A) to tragedy. Ed. by Rebecca BUSHNELL. Oxford a. Malden, Blackwell, 2005, XII-556 p. [Cf. n^{os} <choice> 1510, 1528, 1619, 1624, 1653, 1715, 1717, 1718.]

1515. COUVENHES (Jean-Christophe). De disciplina Graecorum: les relations de violence entres chefs militaires grecs et leur soldats. *In*: Violence (La) dans les mondes grec et romaine [Cf. n° 1527], p. 431-454.

1516. DE CALLATAY (François). The Graeco-Roman economy in the super long run: lead, copper, and shipwrecks. *Journal of Roman archaeology*, 2005, 18, p. 361-372.

1517. DUPLOUY (Alain). Pouvoir ou prestige? Apports et limites de l'histoire politique à la définition des élites grecques. *Revue belge de philologie et d'histoire*, 2005, 83, 1, p. 5-23.

1518. FORSDYKE (Sara). Revelry and riot in archaic Megara: democratic disorder or ritual reversal? *Journal of Hellenic studies*, 2005, 125, p. 73-92.

1519. HUGONIOT (Christophe). Les noms d'aristocrates et de notables gravés sur les gradins de l'amphithéâtre de Carthage au Bas-Empire. *Antiquités Africaines*, 2005, 41, p. 205-258.

1520. INGLESE (Alessandra). Note sull'onomastica di Thera arcaica: astynomos e gli antroponimi a primo elemento asty-. *Rivista di filologia e di istruzione classica*, 2005, 133, p. 129-155.

1521. MORRIS (Sarah P.), PAPADOPOULOS (John K.). Greek towers and slaves: an archaeology of exploitation. *American journal of archaeology*, 2005, 109, p. 155-225.

1522. ROLLER (Duane W.). Seleukos of Seleukeia. *L'antiquité classique*, 2005, 74, p. 111-118.

1523. SANTUCCI (Marco). Tempi del sacro, tempi della politica. Festeggiare, giurare, 'contare' dia trieteridos e pentaeteridos. *In*: Nike [Cf. n° 1382], p. 173-225.

1524. STAFFORD (Emma J.). Nemesis, hybris and violence. *In*: Violence (La) dans les mondes grec et romaine [Cf. n° 1527], p. 195-212.

1525. STIRPE (Paola). Concomitanze di feste greche e romane con grandi feste panelleniche tra l'età ellenistica e la prima età imperiale. *In*: Nike [Cf. n° 1382], p. 227-280.

1526. THOMAS (Carol G.). Finding people in early Greece. Columbia, University of Missouri Press, 2005, XIV-154 p.

1527. Violence (La) dans les mondes grec et romaine. Actes du colloque international (Paris, 2–4 mai 2002). Ed. par Jean-Marie BERTRAND. Paris, Publications de la Sorbonne, 2005, 467 p. [Cf. n^{os} <sélection> 1199, 1483, 1484, 1515, 1524, 1698, 1722, 1727.]

1528. WOHL (Victoria). Tragedy and feminism. *In*: Companion (A) to tragedy [Cf. n° 1514], p. 145-160.

Cf. n° 147

§ 7. History of literature, philosophy and science.

a. Literature

1529. AHLRICHS (Bernhard). 'Prüfstein der Gemüter'. Untersuchungen zu den ethischen Vorstellungen in den Parallelbiographien Plutarchs am Beispiel des 'Coriolan'. Hildesheim, Zürich u. New York, Olms, 2005, 549 p.

1530. AKUJÄRVI (Johanna). Researcher, traveller, narrator. Studies in Pausanias' Periegesis. Lund, Almqvist & Wiksell, 2005, XXVIII-314 p.

1531. ANTIPENKO (Anton L.). Put' predkov. Traditsionnye motivy v "Argonavtike" Apollonija Rodosskogo. (The way of ancestors: Traditional motives in the Argonautica of Apollonius of Rhodes). Moskva, Ladomir, 2005, 318 p.

1532. ASCHERI (Paola). Un elenco di grammatici greci nel palimps. Lipsiensis Gr. 2: problemi di identificazione. *Rivista di filologia e di istruzione classica*, 2005, 133, p. 413-442.

1533. ASIRVATHAM (Sulochana Ruth). Classicism and Romanitas in Plutarch's De Alexandri fortuna aut virtute. *American journal of philology*, 2005, 126, p. 107-125.

1534. BANNERT (H.). Drei Textstellen in den Choephoren des Aischylos. *Wiener Studien*, 2005, 118, p. 21-30.

1535. BARCHIESI (Alessandro). The search for the perfect book: a PS to the new Posidippus. *In*: New Posidippus (The) [Cf. n° 1606], p. 320-342.

1536. BARTLEY (Adam). Techniques of composition in Lucian's minor dialogues. *Hermes*, 2005, 133, p. 358-367.

1537. BENEDETTI (Francesco). Studi su Oppiano. Amsterdam, Hakkert, 2005, 190 p.

1538. BERTOLINI (Francesco). Dialetti, ripartizione dialettale, lingue letterarie. *In*: Dialetti e lingue letterarie nella Grecia arcaica [Cf. n° 1564], p. 89-106.

1539. BIAGINI (Lorenzo). I momenti della storia greca in Thuc. II 36, 1-3. *Eikasmos*, 2005, 16, p. 93-104.

1540. BITTLESTONE (Robert), DIGGLE (James), UNDERHILL (John). Odysseus unbound. The search for Homer's Ithaca. Cambridge, Cambridge U. P., 2005, XX-598 p.

1541. BRANDENBURG (Philipp). The second stasimon in Sophocles' Oedipus Tyrannus. *L'antiquité classique*, 2005, 74, p. 29-40.

1542. BROCCIA (Giuseppe). Archiloco 1 W. [= 1 D.3, 1 T.]. Un tentativo di messa a punto. *Rivista di filologia e di istruzione classica*, 2005, 133, p. 385-391.

1543. BURZACCHINI (Gabriele). Fenomenologia innodica nella poesia di Saffo. *Eikasmos*, 2005, 16, p. 11-39.

1544. CAIRNS (Douglas L.). Myth and the "polis" in Bacchylides' eleventh Ode. *Journal of Hellenic studies*, 2005, 125, p. 35-50.

1545. CALAME (Claude). Masques d'autorité: fiction et pragmatique dans la poétique grecque antique. Paris, Les Belles Lettres, 2005, 335 p.

1546. CALCANTE (Cesare Marco). Da Cratilo a Dionigi d'Alicarnasso: eufonia e iconismo. *Athenaeum*, 2005, 93, p. 5-50.

1547. CAMEROTTO (Alberto). Il grido di Diomedes. Epiteti eroici e composizione tematica. *In*: Dialetti e lingue letterarie nella Grecia arcaica [Cf. n° 1564], p. 107-129.

1548. CASPERS (Christiaan). Artemis in de derde Hymne van Callimachus. *Lampas*, 2005, 38, p. 262-279.

1549. CASSIO (Albio Cesare). I dialetti eolici e la lingua della lirica corale. *In*: Dialetti e lingue letterarie nella Grecia arcaica [Cf. n° 1564], p. 13-44.

1550. CHIASSON (Charles C.). Myth, ritual, and authorial control in Herodotus' story of Cleobis and Biton (Hist. 1. 31). *American journal of philology*, 2005, 126, p. 41-64.

1551. CINGANO (Ettore). A catalogue within a catalogue: Helen's suitors in the Hesiodic Catalogue of women (frr. 196-204). *In*: Hesiodic catalogue (The) of women [Cf. n° 1582], p. 118-152.

1552. COLOMO (Daniela). Recenti contributi alla storia del testo di Isocrate (vd. Aa. Vv., Studi sulla tradizione del testo di Isocrate). *Rivista di filologia e di istruzione classica*, 2005, 133, p. 356-379.

1553. CONSANI (Carlo). Dialettalità genuina e dialettalità riflessa nel quadro delle più antiche attestazioni dei dialetti greci. *In*: Dialetti e lingue letterarie nella Grecia arcaica [Cf. n° 1564], p. 71-88.

1554. COPANI (Fabio). La Nemea IX di Pindaro e lo scontro tra Geloi e Siracusani all'Eloro. *Mélanges de l'École Française de Rome-Antiquité*, 2005, 117, p. 651-676.

1555. CORCELLA (Aldo). Note ai libri settimo, ottavo e nono di Erodoto. *Rivista di filologia e di istruzione classica*, 2005, 133, p. 5-22.

1556. CORTÉS GABAUDAN (Francisco). La oratoria griega como género literario. *In*: Géneros grecolatinos en prosa [Cf. n° 1574], p. 205-232.

1557. CROPP (Martin). Lost tragedies: a survey. *In*: Companion (A) to Greek tragedy [Cf. n° 1513], p. 271-292.

1558. CURRIE (Bruno). Pindar and the cult of heroes. Oxford, Oxford U. P., 2005, XV-487 p.

1559. D'ALESSIO (Giovan Battista). Ordered from the Catalogue: Pindar, Bacchylides and the Hesiodic genealogical poetry. *In*: Hesiodic catalogue (The) of women [Cf. n° 1582], p. 217-238. – IDEM. The Megalai Ehoiai: a survey of the fragments. *In*: Hesiodic catalogue (The) of women [Cf. n° 1582], p. 176-216.

1560. DANIELEWICZ (Jerzy). Further Hellenistic acrostics: Aratus and others. *Mnemosyne*, 2005, 58, p. 321-334.

1561. DHUGA (Umit Singh). Choral identity in Sophocles' Oedipus Coloneus. *American journal of philology*, 2005, 126, p. 333-362.

1562. DI BENEDETTO (Vincenzo). La nuova Saffo e dintorni. *Zeitschrift für Papyrologie und Epigraphik*, 2005, 153, p. 7-20.

1563. DI DONATO (Riccardo). Problemi di lingua poetica arcaica: le espressioni epiche della parentela. *In*: Dialetti e lingue letterarie nella Grecia arcaica [Cf. n° 1564], p. 143-158.

1564. Dialetti e lingue letterarie nella Grecia arcaica. Atti della IV Giornata ghisleriana di Filologia classica (Pavia, 1–2 aprile 2004). A cura di Francesco BERTOLINI e Fabio GASTI. Como e Pavia, Ibis edizioni, 2005, 158 p. [Cf. n^os^ <scelta> 1538, 1547, 1549, 1553, 1563, 1570, 1670.]

1565. DIETZ (Günter), KICK (Hermes Andreas). Grenzsituationen und neues Ethos. Von Homers Weltsicht zum modernen Menschbild. Heidelberg, Winter, 2005, 112 p.

1566. DURBEC (Yannick). "Kyon, Kyon": lectures métapoétiques d'une apostrophe (Callimaque, Aitia, fr. 75, 4 Pfeiffer et "Hymne à Déméter" 63). *Revue des etudes grecques*, 2005, 118, p. 600-604.

1567. ENGLISH (Mary). Aristophanes' Frogs: Brekkek-kek-kek! on Broadway. *American journal of philology*, 2005, 126, p. 127-133.

1568. FANTUZZI (Marco). Posidippus at court: the contribution of the Hippika of P. Mil. Vogl. VIII 309 of the ideology of Ptolemaic kingship. *In*: New Posidippus (The) [Cf. n° 1606], p. 249-268.

1569. FENNO (Jonathan). "A great wave against the stream": water imagery in Iliadic battle scenes. *American journal of philology*, 2005, 126, p. 475-504.

1570. FENOGLIO (Silvia). Epiteti e formularità della notte nei poemi omerici. *In*: Dialetti e lingue letterarie nella Grecia arcaica [Cf. n° 1564], p. 131-141.

1571. FOLI (John Miles). A companion to ancient epic. Oxford a. Carlton, Blackwell, 2005, XXIV-664 p.

1572. GALLÉ CEJUDO (R. J.). Reflexiones sobre la epistolografía griega. *In*: Géneros grecolatinos en prosa [Cf. n° 1574], p. 263-299.

1573. GARZYA (Antonio). Sul problema della rappresentazione della individualità nella tragedia. *Cuadernos de filología clásica. Estudios griegos e indoeuropeos*, 2005, 15, p. 35-47.

1574. Géneros grecolatinos en prosa. Ed. por Dulce ESTEFANÍA, Mª Teresa AMADO, Cecilia CRIADO, Mª Teresa MIÑAMBRES, Álvaro PÉREZ VILARIÑO y Carmen RIOBÓ. Santiago de Compostela, Universidad de Alcalá de Henares, 2005, 360 p. [Cf. n^os^ <selección> 1556, 1572, 1621, 1650, 1712.]

1575. GERMANY (Robert). The figure of Echo in the Homeric Hymn to Pan. *American journal of philology*, 2005, 126, p. 187-208.

1576. GIUROVICH (Sara). Considerazioni sul procedere etnografico posidoniano: "ethos" vs. "historia". L'esempio dei Galli Scordisci e dei Galli Tectosagi. *Rivista storica dell'antichità*, 2005, 35, p. 23-52.

1577. GREGORY (Justina). Euripidean tragedy. *In*: Companion (A) to Greek tragedy [Cf. n° 1513], p. 251-270.

1578. GRETHLEIN (Jonas). Eine Anthropologie des Essens: der Essensstreit in der 'Ilias' und die Erntemetapher in Il. 19, 221-224. *Hermes*, 2005, 133, p. 257-279.

1579. GUTZWILLER (Kathryn). The literariness of the Milan Papyrus or 'what difference a book?'. *In*: New Posidippus (The) [Cf. n° 1606], p. 287-319.

1580. HALLIWELL (Stephen). Learning from suffering: ancient responses to tragedy. *In*: Companion (A) to Greek tragedy [Cf. n° 1513], p. 394-412.

1581. HEITSCH (Ernst). Ilias und Aithiopis. *Gymnasium*, 2005, 112, p. 431-441.

1582. Hesiodic catalogue (The) of women. Constructions and reconstructions. Ed. by Richard HUNTER. Cambridge, Cambridge U. P., 2005, X-349 p. [Cf. n^os^ <choice> 1551, 1559, 1584, 1609, 1632, 1697.]

1583. HOWELL CHAPMAN (Honora). 'By the waters of Babylon': Josephus and Greek poetry. *In*: Josephus and Jewish history in Flavian Rome and beyond [Cf. n° 1299], p. 121-146.

1584. HUNTER (Richard). The Hesiodic Catalogue and Hellenistic poetry. *In*: Hesiodic catalogue (The) of women [Cf. n° 1582], p. 239-265.

1585. HUTTON (William). Describing Greece. Landscape and literature in the Periegesis of Pausanias. Cambridge, Cambridge U. P., 2005, XIII-371 p.

1586. INGLESE (Lionello). Il logos e la trophe (o le tryphai) in Plutarco, es. carn. 993 a-b. *Rivista di filologia e di istruzione classica*, 2005, 133, p. 168-176.

1587. JOUANNO (Corinne). La "Vie d'Ésope": une biographie comique. *Revue des etudes grecques*, 2005, 118, p. 391-425.

1588. KANTZIOS (Ippokratis). The trajectory of archaic Greek trimeters. Leiden a. Boston, Brill, 2005, IX-208 p.

1589. KOWERSKI (Lawrence M.). Simonides on the Persian Wars. A study of the Elegiac Verses of the "New Simonides". London a. New York, Routledge, 2005, XVI-243 p.

1590. KRÜCK (M.-P.). Homère au prisme de ses relecteurs. Le motif de la Nekuia. *Mouseion*, 2005, 49, p. 59-73.

1591. LADA-RICHARDS (Ismene). 'In the mirror of the dance': a Lucianic metaphor in its performative and ethical contexts. *Mnemosyne*, 2005, 58, p. 335-357.

1592. LAPINI (Walter). Due note su POxy. 4030 col. V, 21-31 (Eschine, Contro Timarco 51-52) e alcuni appunti su kai ge ed ede ge nei testi greci. *Zeitschrift für Papyrologie und Epigraphik*, 2005, 152, p. 31-41.

1593. LATEINER (Donald). Proxemic and chronemic in Homeric epic: time and space in heroic social interaction. *Classical world*, 2005, 98, p. 413-421.

1594. LAWRENCE (Stuart). Ancient ethics, the heroic code, and the morality of Sophocles' Ajax. *Greece and Rome*, 2005, 52, p. 18-33.

1595. LUCARINI (Carlo Martino). Libri di scuola della tarda antichità (vd. Flammini G. [ed.], Hermeneumata Pseudodositheana Leidensia). *Rivista di filologia e di istruzione classica*, 2005, 133, p. 484-500.

1596. LURAGHI (Nino). Pausania e i Messenii. Interpretazioni minime. *Rivista di filologia e di istruzione classica*, 2005, 133, p. 177-201.

1597. LUZ (Christine), D'ANGOUR (Armand). Some Sapphic and Anacreontic verses. *Greece and Rome*, 2005, 52, p. 45.

1598. MAC NEIL (Lynda). Bridal cloths, cover-ups, and Kharis: the 'carpet scene' in Aeschylus' Agamemnon. *Greece and Rome*, 2005, 52, p. 1-17.

1599. MALTOMINI (Francesca). Due testimonianze trascurate dell'epigramma di Posidippo sul Kairos (Plan. 275; Posidippus 142 A.-B.; XIX G.-P.). *Rivista di filologia e di istruzione classica*, 2005, 133, p. 283-306.

1600. MANUWALD (Bernd). Jasons dynastische Pläne und Medeas Rachekalkül. Zur Konzeption der Rachehandlung in der ‚Medea' des Euripides. *Gymnasium*, 2005, 112, p. 515-530.

1601. MARKS (J.). The ongoing Neikos: Thersites, Odysseus, and Achilleus. *American journal of philology*, 2005, 126, p. 1-31.

1602. MAURACH (Gregor). Kleine Geschichte der antiken Komödie. Darmstadt, Wissenschaftliche Buchgesellschaft, 2005, 160 p.

1603. MEDDA (Enrico). Il coro straniato. Considerazioni sulla voce corale nelle Fenicie di Euripide. *Prometheus*, 2005, 31, p. 119-131.

1604. MOLES (John). The thirteenth oration of Dio Chrysostom: complexity and simplicity, rhetoric and moralism, literature and life. *Journal of Hellenic studies*, 2005, 125, p. 112-138.

1605. MORI (Anatole). Jason's reconciliation with Telamon: a moral exemplar in Apollonius' Argonautica (1. 1286-1344). *American journal of philology*, 2005, 126, p. 209-236.

1606. New Posidippus (The). A Hellenistic poetry book. Ed. by Kathryn GUTZWILLER. Oxford, Oxford U. P., 2005, XVI-394 p. [Cf. n[os] <choice> 1535, 1568, 1579, 1607, 1608, 1626, 1627, 1631.]

1607. NISETICH (Frank). The poems of Posidippus translated. *In*: New Posidippus (The) [Cf. n° 1606], p. 17-64.

1608. OBBINK (Dirk). New old Posidippus and old new Posidippus: from occasion to edition in the epigrams. *In*: New Posidippus (The) [Cf. n° 1606], p. 97-115.

1609. OSBORNE (Robin). Ordering women in Hesiod's Catalogue. *In*: Hesiodic catalogue (The) of women [Cf. n° 1582], p. 5-24.

1610. OTRANTO (Rosa). La più antica edizione superstite delle Elleniche di Senofonte. *Quaderni di storia*, 2005, 62, p. 167-191.

1611. PADUANO (Guido). Il teatro antico. Guida alle opere. Bari, Laterza, 2005, VIII-358 p.

1612. PALLANTZA (Elena). Der Troische Krieg in der nachhomerischen Literatur bis zum 5. Jahrhundert v. Chr. Stuttgart, Steiner, 2005, VII-358 p.

1613. PONTANI (Filippo Maria). Il mito, la lingua, la morale: tre piccole introduzioni a Omero (1. Antehomerica in Omero – 2. Le figure grammaticali in Omero – 3. Omero etico – 4. Appendice: quattro hypotheseis metriche bizantine). *Rivista di filologia e di istruzione classica*, 2005, 133, p. 23-74. – IDEM. Questioni omeriche sulle allegorie di Omero in merito agli dei. Pisa, Edizioni ETS, 2005, 237 p.

1614. PORTER (David H.). Aeschylus' Eumenides: some contrapuntal lines. *American journal of philology*, 2005, 126, p. 301-331.

1615. PRIVITERA (G. Aurelio). Il ritorno del guerriero. Lettura dell'Odissea. Torino, Einaudi, 2005, IX-297 p.

1616. PUIGGALI (Jacques). "Daimôn" et les mots de la même famille dans les Antiquités romaines de Denys d'Halicarnasse. *Latomus*, 2005, 64, p. 626-630.

1617. RICHTER (Daniel). Lives and afterlives of Lucian of Samosata. *Arion*, 2005, 13, p. 75-100.

1618. ROISMAN (Joseph). The rhetoric of manhood. Masculinity in the Attic orators. Berkeley, Los Angeles a. London, University of California Press, 2005, XIV-283 p.

1619. ROSEN (Ralph M.). Aristophanes, Old comedy and Greek tragedy. *In*: Companion (A) to tragedy [Cf. n° 1514], p. 251-268.

1620. RUBINCAM (Catherine). A tale of two "Magni": Justin/Trogus on Alexander and Pompey. *Historia*, 2005, 54, p. 265-274.

1621. RUIZ MONTERO (Consuelo). La novela griega. Panorama general. *In*: Géneros grecolatinos en prosa [Cf. n° 1574], p. 313-342.

1622. SAÚD (S.). Aeschylean tragedy. *In*: Companion (A) to Greek tragedy [Cf. n° 1513], p. 215-232.

1623. SCIARRA (Elisabetta). La tradizione degli scholia iliadici in terra d'Otranto. *Bollettino dei classici*, 2005, 23, p. 5-302.

1624. SCODEL (Ruth). Sophoclean tragedy. *In*: Companion (A) to Greek tragedy [Cf. n° 1513], p. 233-250. – EADEM. Tragedy and epic. *In*: Companion (A) to tragedy [Cf. n° 1514], p. 181-197.

1625. SEIDENSTICKER (Bernd). Dithyramb, comedy and satyr-play. *In*: Companion (A) to Greek tragedy [Cf. n° 1513], p. 38-54.

1626. SENS (Alexander). The art of poetry and the poetry of art: the unity and poetics of Posidippus' statue poems. *In*: New Posidippus (The) [Cf. n° 1606], p. 206-225.

1627. SIDER (David). Posidippus on weather signs and the tradition of didactic poetry. *In*: New Posidippus (The) [Cf. n° 1606], p. 164-182.

1628. SIMMS (R. Clinton). The missing bones of Thersites: a note on Iliad 2. 212-19. *American journal of philology*, 2005, 126, p. 33-40.

1629. SMALL (Jocelyn Penny). Pictures of tragedy. *In*: Companion (A) to Greek tragedy [Cf. n° 1513], p. 103-118.

1630. SOMVILLE (Pierre). Deux joyaux poétiques. *L'antiquité classique*, 2005, 74, p. 21-28.

1631. STEWART (Andrew). Posidippus and the truth in sculpture. *In*: New Posidippus (The) [Cf. n° 1606], p. 183-205.

1632. STRAUSS CLAY (Jenny). The beginning and the end of the Catalogue of women and its relation to Hesiod. *In*: Hesiodic catalogue (The) of women [Cf. n° 1582], p. 25-34.

1633. TURKELTAUB (Daniel). The syntax and semantics of Homeric glowing eyes: Iliad 1, 200. *American journal of philology*, 2005, 126, p. 157-186.

1634. VÖSSING (Konrad). Objektivität oder subjektivität, sinn oder überlegung? Zu Thukidides' "gnóme" im Methodenkapitel (I, 22). *Historia*, 2005, 54, p. 210-215.

1635. WHITEHORN (John). O City of Kranaos! Athenian identity in Aristophanes' Acharnians. *Greece and Rome*, 2005, 52, p. 34-44.

1636. WHITMARSH (Tim). The Greek novel: titles and genre. *American journal of philology*, 2005, 126, p. 587-611. – IDEM. The lexicon of love: Longus and Philetas "grammatikos". *Journal of Hellenic studies*, 2005, 125, p. 145-148. – IDEM. The Second Sophistic. Oxford, Oxford U. P., 2005, 106 p.

1637. WOERTHER (Frédérique). Aux origines de la Notion Rhétorique d'èthos. *Revue des etudes grecques*, 2005, 118, p. 79-116.

1638. WRIGHT (Matthew). Euripides' escape-tragedies: a study of Helen, Andromeda, and Iphigenia among the Taurians. Oxford, Oxford U. P., 2005, XIII-433 p.

1639. ZORODDU (Donatella). Posidippo miniatore. *Athenaeum*, 2005, 93, p. 577-596.

Cf. n°s 1417-1443

b. Philosophy and sciences

1640. ALLEN (William). Tragedy and early Greek philosophical tradition. *In*: Companion (A) to Greek tragedy [Cf. n° 1513], p. 71-82.

1641. BOTTER (Barbara). Dio e divino in Aristotele. Sankt Augustin, Academia Verlag, 2005, 305 p.

1642. BOUDON-MILLOT (Véronique), PIETROBELLI (Antoine). Galien ressuscité: Édition "Princeps" du texte grec du "De Propriis Placitis". *Revue des etudes grecques*, 2005, 118, p. 168-213.

1643. BRICKHOUSE (Thomas C.), SMITH (Nicolas D.). Socrates' daimonion and rationality. *In*: Socrates' divine sign [Cf. n° 1678], p. 43-62.

1644. BRISSON (Luc). Socrates and the divine signal according to Plato's testimony: philosophical practice as rooted in religious tradition. *In*: Socrates' divine sign [Cf. n° 1678], p. 1-12.

1645. CAPRIGLIONE (Jolanda). L'acqua è fredda e umida. *Cuadernos de filología clásica. Estudios griegos e indoeuropeos*, 2005, 15, p. 107-123.

1646. COLLOUD-STREIT (Marlis). Fünf platonische Mythen im Verhältnis zu ihren Textumfeldern. Fribourg, Academie Press, 2005, 256 p.

1647. CORRADI (Michele). Protagora facchino e l'invenzione del cercine. *Rivista di filologia e di istruzione classica*, 2005, 133, p. 392-412.

1648. DENNINGMAN (Susanne). Die astrologische Lehre der Doryphorie. Eine soziomorphe Metapher in der antiken Planetenastrologie. München u. Leipzig, Saur, 2005, XIV-548 p.

1649. DESTRÉE (Pierre). The daimonion and the philosophical mission: should the divine sign remain unique to Socrates? *In*: Socrates' divine sign [Cf. n° 1678], p. 63-79.

1650. DÍAZ DE CERIO (Mercedes). Filosofía griega antigua. *In*: Géneros grecolatinos en prosa [Cf. n° 1574], p. 9-84.

1651. DORION (Louis-André). The daimonion and the megalegoria of Socrates in Xenophon's Apology. *In*: Socrates' divine sign [Cf. n° 1678], p. 127-142.

1652. DROZDEK (Adam). Protagoras and instrumentality of religion. *L'antiquité classique*, 2005, 74, p. 41-50.

1653. EDEN (Kathy). Aristotle's Poetics: a defense of tragic fiction. *In*: Companion (A) to tragedy [Cf. n° 1514], p. 41-50.

1654. FERRARI (Franco). L'officina epica di Parmenide. *Seminari romani di cultura greca*, 2005, 8, p. 113-129. – IDEM. Parmenide, il Parmenide di Platone e la teoria delle idee. *Athenaeum*, 2005, 93, p. 367-396.

1655. GERSON (Lloyd P.). Aristotle and the other Platonists. Ithaca, Cornell U. P., 2005, XII-335 p.

1656. HATZISTAVROU (Antony). Socrates' deliberative authoritarianism. *Oxford studies in ancient philosophy*, 2005, 29, p. 75-113.

1657. JOURDAN (Fabienne). Porphyre, lecteur et citateur du traité de Plutarque "Manger de la viande". *Revue des etudes grecques*, 2005, 118, p. 426-435.

1658. KING (Karen L.). What is gnosticism? Cambridge, Harvard U. P., 2005, XIV-343 p.

1659. LAURAND (Valéry). La politique stoïcienne. Paris, Presses universitaires de France, 2005, 153 p.

1660. LEE (Mi-Kyoung). Epistemology after Protagoras: responses to relativism in Plato, Aristotle and Democritus. Oxford, Oxford U. P., 2005, XII-291 p.

1661. LEFKA (Aikaterini). Religion publique et croyances personnelles: Platon contre Socrate? *Kernos*, 2005, 18, p. 85-95.

1662. LORUSSO (Vito). Nuovi frammenti di Galeno (in Hp. Epid. VI Comm. VII; In Plat. Tim. Comm.). *Zeitschrift für Papyrologie und Epigraphik*, 2005, 152, p. 43-56.

1663. MAC PHERRAN (Mark L.). Introducing a new god: Socrates and his daimonion. *In*: Socrates' divine sign [Cf. n° 1678], p. 13-30.

1664. MANSFELD (Jaap). 'Illuminating what is thought': a middle Platonist placitum on 'voice' in context. *Mnemosyne*, 2005, 58, p. 358-407.

1665. MANUWALD (Bernd). The unity of virtue in Plato's Protagoras. *Oxford studies in ancient philosophy*, 2005, 29, p. 115-135.

1666. MAZZARA (Giuseppe). La rhétorique éléatico-gorgienne d'Alcidamas chez Diogéne Laërce (IX, 54). *L'antiquité classique*, 2005, 74, p. 51-68.

1667. MEGINO RODRÍGUEZ (Carlos). Orfeo y el orfismo en la poesia de Empédocles: influencias y paralelismos. Madrid, Universidad Autónoma de Madrid, 2005, 104 p.

1668. MOURACADE (John). Virtue and pleasure in Plato's Laws. *Apeiron*, 2005, 38, p. 73-85.

1669. O'BRIEN (Denis). "Einai" copulatif et existentiel dans le Parménide de Platon. *Revue des etudes grecques*, 2005, 118, p. 229-245.

1670. PASSA (Enzo). Dialetti parlati, tradizioni letterarie, livelli di ricezione antica nel testo dei Presocratici: gli Eleati e la cultura attica. *In*: Dialetti e lingue letterarie nella Grecia arcaica [Cf. n° 1564], p. 45-69.

1671. PETIT (Caroline). Un nouveau témoin du "De Constitutione artis medicae" de Galien: le "Mutinensis" gr. 213 (ff. 149v-154r). *Revue des etudes grecques*, 2005, 118, p. 266-270.

1672. PETZL (Georg). Furchterregende Götter? Eine Notiz zu Diogenes von Oinoanda NF 126. *Zeitschrift für Papyrologie und Epigraphik*, 2005, 153, p. 103-107.

1673. QUARANTOTTO (Diana). Causa finale sostanza essenza in Aristotele. Saggio sulla struttura dei processi teleologici naturali e sulla funzione del telos. Roma, Istituto per il Lessico Intellettuale Europeo e Storia delle Idee-C.N.R., 2005, 372 p.

1674. ROSKAM (Geert). On the path to virtue. The Stoic doctrine of moral progress and its reception in (middle-)Platonism. Leuven, Leuven U. P., 2005, VIII-507 p.

1675. ROSLER (Andrés). Political authority and obligation in Aristotle. Oxford, Clarendon Press, 2005, XIV-298 p.

1676. SCHMIDT (Arno). Das Elend des Logos-antike Philosophie nach Aristoteles-(360 v. Chr.–500 n. Chr.). Berlin, Logos Verlag, 2005, 257 p.

1677. SCHUBERT (Charlotte). Der hippokratische Eid. Medizin und Ethik von der Antike bis heute. Darmstadt, Wissenschaftliche Buchgesellschaft, 2005, 128 p.

1678. Socrates' divine sign: religion, practice and value in Socratic philosophy. Ed. by Pierre DESTRÉE and Nicolas D. SMITH. Kelowna, Academic printing and publishing, 2005, XII-180 p. [Cf. n[os] <choice> 1643, 1644, 1649, 1651, 1663, 1679, 1681.]

1679. VAN RIEL (Gerd). Socrates' daemon: internalisation of the divine and knowledge of the self. *In*: Socrates' divine sign [Cf. n° 1678], p. 31-42.

1680. VILLERS (Jürgen). Das Paradigma des Alphabets. Platon und die Schriftbedingtheit der Philosophie. Würzburg, Königshausen & Neumann, 2005, 497 p.

1681. WEISS (Roslyn). For whom the daimonion tolls. *In*: Socrates' divine sign [Cf. n° 1678], p. 81-96.

1682. WOLFSDORF (David). Eutyphro 10a2-11b1: a study in Platonic metaphysics and its reception since 1960. *Apeiron*, 2005, 38, p. 1-71.

1683. YUNIS (Harvey). Eros in Plato's Phaedrus and the shape of Greek rhetoric. *Arion*, 2005, 13, p. 101-125.

1684. ZAGDOUN (Mary-Anne). Dion de Pruse et la philosophie stoïcienne de l'art. *Revue des etudes grecques*, 2005, 118, p. 605-612.

1685. ZUCKER (Arnaud). Aristote et les classifications zoologiques. Louvain la Neuve, Paris et Dudley, Peeters, 2005, 368 p.

1686. ZUNTZ (Günther). Griechische philosophische Hymnen. Tübingen, Mohr Siebeck, 2005, XXV-227 p.

Cf. n°ˢ 1417-1443

§ 8. Religion and mythology.

1687. BRUIT-ZAIDMAN (Louise). Les Grecs et leurs dieux. Pratiques et représentations religieuses dans la cité à l'époque classique. Paris, Armand Colin, 2005, X-198 p. (U).

1688. BRULÉ (Pierre). Dans le nom, tout n'est-il pas déjà dit? Histoire et géographie dans les récits généalogiques. *Kernos*, 2005, 18, p. 241-268.

1689. CAMASSA (Giorgio). La Sibilla giudaica di Alessandria: ricerche di storia delle religioni. Firenze, Le Monnier, 2005, 237 p.

1690. DANEK (Georg). Antenor und die Bittgesandtschaft. Ilias, Bakchylides 15 und der Astarita-Krater. *Wiener Studien*, 2005, 118, p. 5-20.

1691. DARTHOU (Sonia). Retour à la terre: la fin de la Geste d'Érechthée. *Kernos*, 2005, 18, p. 69-83.

1692. DEBORD (Pierre). La déesse Ma et les hirondelles blanches. *Revue des etudes grecques*, 2005, 118, p. 15-30.

1693. DES BOUVRIES (Synnove). Myth as a mobilizing force in Attic warrior society. *Kernos*, 2005, 18, p. 185-201.

1694. DILLERY (John). Greek sacred history. *American journal of philology*, 2005, 126, p. 505-526.

1695. EBBINGHAUS (Susanne). Protector of the city, or the art of storage in early Greece. *Journal of Hellenic studies*, 2005, 125, p. 51-72.

1696. Initiation in ancient Greek rituals and narratives. New critical perspectives. Ed. by David B. DODD and Christopher A. FARAONE. London a. New York, Routledge, 2005, XIX-294 p.

1697. IRWIN (Elizabeth). Gods among men? The social and politic dynamics of the Hesiodic Catalogue of women. *In*: Hesiodic catalogue (The) of women [Cf. n° 1582], p. 35-84.

1698. JACQUEMIN (Anne). Images de violence et offrandes de victoire en Grèce ancienne. *In*: Violence (La) dans les mondes grec et romaine [Cf. n° 1527], p. 121-135.

1699. JOST (Madeleine). Deux mythes de métamorphose en animal et leurs interprétations: Lykaon et Kallisto. *Kernos*, 2005, 18, p. 347-370.

1700. KARILA-COHEN (Karine). Apollon, Athènes et la Pythaïde: mise en scène "mythique" de la cité au IIe siècle av. notre ère. *Kernos*, 2005, 18, p. 219-239.

1701. LAFOND (Yves). Le mythe, référence identitaire pour les cités grecques d'époque impériale. L'exemple du Péloponnèse. *Kernos*, 2005, 18, p. 329-346.

1702. LEDUC (Claudine). 'Le pseudo-sacrifice d'Hermès'. Hymne homérique à Hermès I, vers 112-142. Poésie rituelle, théologie et histoire. *Kernos*, 2005, 18, p. 141-165.

1703. LINKE (Bernhard). Religion und Herrschaft im archaischen Griechenland. *Historische Zeitschrift*, 2005, 280, 1, p. 1-37.

1704. PARKER (Robert). Polytheism and society at Athens. Oxford, Oxford U. P., 2005, XXXII-544 p.

1705. PEDLEY (John Griffiths). Sanctuaries and the sacred in the ancient Greek world. Cambridge, Cambridge U. P., 2005, XVIII-272 p. (ill., maps).

1706. PEDUCCI (Giulia). I santuari rupestri metroaci fra Sicilia e Anatolia. *Rivista storica dell'antichità*, 2005, 35, p. 165-180.

1707. PIOLOT (Laurent). Nom d'une Artémis! À propos de l'Artémis Phôsphoros de Messène (Pausanias IV 31, 10). *Kernos*, 2005, 18, p. 113-140.

1708. PIRONTI (Gabriella). Aphrodite dans le domaine d'Arès. Éléments pour un dialogue entre mythe et culte. *Kernos*, 2005, 18, p. 167-184.

1709. PRÊTEUX (Franck). Priapos Bébrykès dans la Propontide et les Détroits: succès d'un mythe local. *Revue des etudes grecques*, 2005, 118, p. 246-265.

1710. ROCCHI (Maria). Culti sui monti della Grecia. Osservazioni da una lettura di Pausania. *Archiv für Religionsgeschichte*, 2005, 7, p. 56-61.

1711. ROSSIGNOLI (Benedetta). Ulisse, Laerte e Germanico. *Athenaeum*, 2005, 93, p. 305-308.

1712. RUIZ PÉREZ (Angel). La historiografía griega y el mito. De la genealogía a la mitología. *In*: Géneros grecolatinos en prosa [Cf. n° 1574], p. 109-130.

1713. RUSJAEVA (Anna S.). Religija pontijskikh ellinov v antichnuju epokhu: Mify. Svjatilishcha. Kul'ty olimpijskikh bogov i geroev. (The religion of the Greeks of the Black Sea Region in the Antique age: Myths, temples, cults of Olympic gods and heroes). NAN Ukrainy, In-t arkheologii. Kiev, Stilos, 2005, 560 p. (ill.; bibl. p. 499-534; ind. p. 538-557).

1714. SCULLION (Scott). Tragedy and religion: the problem of origins. *In*: Companion (A) to Greek tragedy [Cf. n° 1513], p. 23-37.

1715. SEAFORD (Richard). Tragedy and Dionysus. *In*: Companion (A) to tragedy [Cf. n° 1514], p. 25-38.

1716. SEBILLOTTE CUCHET (Violaine). La terre-mère: une lecture par le genre et la rhétorique patriotique. *Kernos*, 2005, 18, p. 203-218.

1717. SOMMERSTEIN (Alan H.). Tragedy and myth. *In*: Companion (A) to tragedy [Cf. n° 1514], p. 163-180.

1718. SOURVINOU-INWOOD (Christiane). Greek tragedy and ritual. *In*: Companion (A) to tragedy [Cf. n° 1514], p. 7-24.

9. ARCHAEOLOGY AND HISTORY OF ART

1719. STAFFORD (Emma J.). Héraklès: encore et toujours le probléme du heros theos. *Kernos*, 2005, 18, p. 391-406.

1720. TURCAN (Robert). Un bige de centaures dionysiaques. *Latomus*, 2005, 64, p. 125-131.

1721. VALDÉS GUÍA (Miriam), MARTÍNEZ NIETO (Roxana). Los pequeños misterios de Agras: unos misterios órficos en época de Pisístrato. *Kernos*, 2005, 18, p. 43-68.

1722. VILLANUEVA (Marie-Christine). Des ménades et de la violence dans la céramique attique. *In*: Violence (La) dans les mondes grec et romaine [Cf. n° 1527], p. 225-243.

1723. ZUBAR' (Vitalij M.). Bogi i geroi antichnogo Khersonesa. (Gods and heroes of Antique Chersonesos Taurica). Kiev, Stilos, 2005, 192 p. (ill.; maps; bibl. p. 177-178). – ZUBAR' (Vitalij M.), SOROCHAN (Sergej B.). U istokov khristianstva v Jugo-Zapadnoj Tavrike: Epokha i vera. Poslednjaja chetvert' III–VI v. (The background of Christianity in South-West Taurica: the epoch and the belief, the late 3rd–the 6th centuries A.D.). Kiev, Stilos, 2005, 182 p. (ill.; maps; portr.; bibl. p. 177-178).

Cf. n°s 868, 872, 1351, 1462

§ 9. Archaeology and history of art.

* 1724. Activités archéologiques de l'École Française de Rome: année 2004. *Mélanges de l'École Française de Rome-Antiquité*, 2005, 117, p. 269-427.

1725. AMORE (Maria Grazia), CEROVA (Ylli), BEJKO (Lorenc). Via Egnatia (Albania) Project: results of fieldwork 2002. *Journal of Roman archaeology*, 2005, 18, p. 336-360.

1726. BERGEMANN (Johannes). Luigi Maria Ugolini und das Theater von Butrint: neue Forschungen im Land der Skipetaren. *Journal of Roman archaeology*, 2005, 18, p. 720-722.

1727. CASSIMATIS (Hélène). La violence dans les figurations des scènes théâtrales portées par la céramique italiote. *In*: Violence (La) dans les mondes grec et romaine [Cf. n° 1527], p. 39-65.

1728. ERICKSON (Brice). Archaeology of empire: Athens and Crete in the fifth century B.C. *American journal of archaeology*, 2005, 109, p. 619-663.

1729. FERRARA (Enzo). Earth science, soil chemistry, and archaeology. *American journal of archaeology*, 2005, 109, p. 87-90.

1730. GEORGOULA (Electra). Greek treasures from the Benaki museum in Athens. Sydney, Powerhouse publishing a. Athens, Benaki Museum, 2005, 264 p.

1731. HASELBERGER (Lothar). Bending the truth: curvature and other refinements of the Parthenon. *In*: Parthenon (The) [Cf. n° 1742], p. 101-157.

1732. HURST (Henry R.). Rescuing Durres (Dyrrachium). *Journal of Roman archaeology*, 2005, 18, p. 723-725.

1733. KOUSSER (Rachel). Creating the past: the Vénus de Milo and the Hellenistic reception of classical Greece. *American journal of archaeology*, 2005, 109, p. 227-250.

1734. KÜNZE-GÖTTE (Erika). Corpus vasorum antiquorum. Deutschland, 78: München, Antiken-sammlungen ehemals Museum Antiker Kleinkunst. Band 14. Attisch-schwarz-figurige Halsamphoren. München, Beck, 2005, 94 p.

1735. LAPATIN (Kenneth D.S.). The statue of Athena and other treasures in the Parthenon. *In*: Parthenon (The) [Cf. n° 1742], p. 261-291.

1736. MALFITANA (Daniele). Fatiche erculee nella ceramica corinzia di età romana: coppe abbinate per un ciclo figurativo incompiuto. *Mélanges de l'École Française de Rome-Antiquité*, 2005, 117, p. 17-53.

1737. NEILS (Jenifer). 'With noblest images on all sides': the Ionic frieze of the Parthenon. *In*: Parthenon (The) [Cf. n° 1742], p. 199-223.

1738. NYS (Karin), ASTROEM (Paul). Cypriote antiquities in public collections in Sweden: Malmoe, Lund and Goeteborg. Sävedalen, Astrom Paul, 2005, 54 p.

1739. OAKLEY (John H.), REITZAMMER (Laurialan). A Hellenistic terracotta and the gardens of Adonis. *Journal of Hellenic studies*, 2005, 125, p. 142-144.

1740. OHNESORG (Aenne). Ionische Altäre. Formen und Varianten einer Architekturgattung aus Insel- und Ostionien. Berlin, Mann, 2005, XIV-259 p.

1741. PALAGIA (Olga). Fire from heaven: pediments and akroteria of the Parthenon. *In*: Parthenon (The) [Cf. n° 1742], p. 225-259.

1742. Parthenon (The). From antiquity to the present. Ed. by Jenifer NEILS. Cambridge, Cambridge U. P., 2005, XVI-430 p. [Cf. n^{os} <choice> 1731, 1735, 1737, 1741, 1748.]

1743. PRICE (Simon), NIXON (Lucia). Ancient agricultural terraces: evidence from texts and archaeological survey. *American journal of archaeology*, 2005, 109, p. 665-694.

1744. PSOMA (Selene), TOURATSOGLOU (Yannis). Sylloge nummorum Graecorum. Greece. 4. Numismatic Museum, Athens. The Petros Z. Saroglos Collection. Volume I. Macedonia. Athens, Academy of Athens, 2005, 144 p.

1745. RADICI COLACE (Paola), MONDIO (Alessandra). Lexicon vasorum Graecorum. Vol. 5. Pisa, Scuola Normale Superiore, 2005, 209 p.

1746. ROSIVACH (Vincent J.). How safe was travel abroad? Some evidence from Athenian vases. *Historia*, 2005, 54, p. 343.

1747. SARAGUSTI (Idit), KARASIK (Avshalom), SHARON (Ilan), SMILANSKY (Uzy). Quantitative analysis of shape attributes based on contours and section profiles in archaeological research. *Journal of archaeological science*, 2005, 32, p. 841-853.

1748. SCHWAB (Katherine A.). Celebrations of victory: the metopes of the Parthenon. *In*: Parthenon (The) [Cf. n° 1742], p. 159-197.

1749. SHAYA (Josephine). The Greek temple as museum: the case of the legendary treasure of Athena from Lindos. *American journal of archaeology*, 2005, 109, p. 423-442.

1750. TASSIGNON (Isabelle). Réflexions sur les fragments statuaires du "Maître des Lions" d'Amathonte (Chypre). *Revue des etudes grecques*, 2005, 118, p. 367-381.

1751. TORTORELLA (Stefano). Il repertorio iconografico della ceramica africana a rilievo del IV–V secolo d.C. *Mélanges de l'École Française de Rome-Antiquité*, 2005, 117, p. 173-198.

Cf. n°s 138, 147

F

HISTORY OF ROME, ANCIENT ITALY AND THE ROMAN EMPIRE

§ 1. The peoples of Italy. 1752-1759. – § 2. The Etruscans. 1760-1777. – § 3. Sources and criticism of sources. (*a*. Epigraphical sources; *b*. Literary sources). 1778-1851. – § 4. General and political history. 1852-1957. – § 5. History of law and institutions. 1958-1992. – § 6. Economic and social history. 1993-2054. – § 7. History of literature, philosophy and science (*a*. Literature; *b*. Philosophy and science). 2055-2126. – § 8. Religion and mythology. 2127-2150. – § 9. Archaeology and history of art. 2151-2193. – § 10. Late antiquity. Transformation of the Roman world. 2194-2209.

§ 1. The peoples of Italy.

1752. D'ERCOLE (Maria Cecilia). Identités, mobilités et frontières dans la Méditerranée antique. L'Italie adriatique, VIIIe–Ve siècle avant J.C. *Annales*, 2005, 60, 1, p. 165-181.

1753. D'ORIANO (Rubens). I Serdaioi da Olbia. *La parola del passato*, 2005, 340, p. 58-74.

1754. DE HOZ (Javier). Los monumentos de la lengua mesápica y los problemas de la edición de inscripciones en lenguas fragmentarias atestiguadas. *Cuadernos de filología clásica*, 2005, 15, p. 225-236.

1755. GULLÌ (Domenica). Caratteri di un centro indigeno nella valle del Platani. Nuove ricerche. *Sicilia antiqua*, 2005, 2, p. 9-62.

1756. Italia (L') antica. Culture e forme del popolamento nel I millennio a.C. A cura di Fabrizio PESANDO, Roma, Carocci, 2005, 326 p.

1757. Mastino (Attilio), Spanu (Pier Giorgio), ZUCCA (Raimondo). Mare Sardum. Merci, mercati e scambi marittimi della Sardegna antica. Roma, Carocci, 2005, 254 p. (Collana del dipartimento di storia dell'Università degli Studi di Sassari. Nuova serie, 26).

1758. Storia della Sardegna antica. A cura di Attilio MASTINO. Nuoro, Il maestrale, 2005, 583 p. (La Sardegna e la sua storia, 2).

1759. TATARANNI (Francesca). Il toro, la lupa e il guerriero: l'immagine marziale dei Sanniti nella monetazione degli insorti italici durante la guerra sociale (90–88 a.C.). *Athenaeum*, 2005, 93, p. 291-304.

§ 2. The Etruscans.

1760. CHIESA (Federica). Tarquinia. Archeologia e prosopografia tra ellenismo e romanizzazione. Roma, 'L'Erma' di Bretschneider, 2005, 406 p. (Bibliotheca archaeologica, 39).

1761. D'AGOSTINO (Bruno). La città. *In*: Dinamiche di sviluppo delle città nell'Etruria meridionale [Cf. n° 1765], p. 21-25.

1762. DELLA FINA (Giuseppe M.). Etruschi. La vita quotidiana. Roma, 'L'Erma' di Bretschneider, 2005, 59 p.

1763. DI FAZIO (Massimiliano). Uno, nessuno e centomila Mezenzio. *Athenaeum*, 2005, 93, p. 51-69.

1764. DI GIUSEPPE (Helga). Un confronto tra l'Etruria settentrionale e meridionale dal punto di vista della ceramica a vernice nera. *Papers of the British school at Rome*, 2005, 73, p. 31-84.

1765. Dinamiche di sviluppo delle città nell'Etruria meridionale. Veio, Caere, Tarquinia, Vulci. Atti del XXIII convegno di studi etruschi ed italici. A cura di Orazio PAOLETTI. Pisa e Roma, Istituti editoriali e poligrafici internazionali, 2005, 745 p. [Cf. n^{os} <scelta> 1761, 1769, 1770.]

1766. HAYNES (Sybille). Kulturgeschichte der Etrusker. Mainz am Rhein, Von Zabern, 2005, 490 p. (Kulturgeschichte der antiken Welt, 108).

1767. JANNOT (Jean-René). Banqueteurs aux mains vides. Absences, immatérialités, gestes vides. A propos de quelques images du banquet funéraire étrusque. *Revue des études anciennes*, 2005, 107, p. 527-541.

1768. MAC INTOSH TURFA (Jean). Catalogue of the Etruscan gallery of the University of Pennsylvania museum of archaeology and anthropology. Philadelphia, University of Pennsylvania museum of archaeology and anthropology, 2005, XVI-329 p.

1769. MASSA-PAIRAULT (Françoise-Hélène). Athènes-Etrurie: brèves considérations à partir de Caere. *In*: Dinamiche di sviluppo delle città nell'Etruria meridionale [Cf. n° 1765], p. 247-255.

1770. MUSTI (Domenico). Temi etici e politici nella decorazione pittorica della Tomba François. *In*: Dinamiche di sviluppo delle città nell'Etruria meridionale [Cf. n° 1765], p. 485-508.

1771. Orvieto, l'Etruria meridionale interna e l'agro falisco. Atti del XII convegno internazionale di studi sulla storia e l'archeologia dell'etruria. A cura di Giuseppe M. DELLA FINA. Roma, Quasar, 2005, 482 p.

1772. PATURZO (Franco). Etruschi. L'enigma delle origini. Arezzo, Letizia, 2005, 287 p.

1773. PEREGO (Lucio G.). Il territorio tarquiniese. Ricerche di topografia storica. Milano, LED, 2005, 278 p.

1774. RODRÍGUEZ ADRADOS (Francisco). El etrusco como indoeuropaeo anatolio: viejos y nuevos argumentos. *Emérita*, 2005, 75, p. 45-56.

1775. WARDEN (P. Gregory), THOMAS (Michael L.), STEINER (Ann), MEYERS (Gretchen). Poggio Colla: an Etruscan settlement of the $7^{th}-2^{nd}$ c. B. C. (1998–2004 excavations). *Journal of Roman archaeology*, 2005, 18, p. 252-266.

1776. WERNER (Ingrid). Dionysos in Etruria. The ivy leaf group. Stockholm a. Sävedalen, Åströms förl., 2005, 84 p.

1777. WINTER (Nancy A.). Gods walking on the roof: the evolution of terracotta statuary in archaic Etruscan architecture in the light of the kings of Rome. *Journal of Roman archaeology*, 2005, 18, p. 241-251.

§ 3. Sources and criticism of sources.

a. Epigraphical sources

* 1778. Année (L') épigraphique 2002. [2001. Cf. Bibl. 2004, n° 1916.] Éd. par Mireille CORBIER. Paris, Presses universitaires de France, 2005, 891 p.

1779. ALFÖLDY (Géza). Inschriften und Biographie in der römischen Welt. *In*: Biographie und Prosopographie [Cf. n° 1372], p. 29-52.

1780. BEN ABDALLAH (Z. Benzina), CARANDE HERRERO (Rocio), FERNÁNDEZ MARTÍNEZ (Concepción), GÓMEZ PALLARÈS (Joan), JORBA (Norma). Carmina Latina Epigraphica Inedita Ammaedarae. *Zeitschrift für Papyrologie und Epigraphik*, 2005, 152, p. 89-113.

1781. FEHÉR (Bence). Lexicon epigraphicum Pannonicum supplementum 1994–2003. *Acta archaeologica academiae scientiarum Hungaricae*, 2005, 56, p. 213-244.

1782. FERONE (Claudio). Subigit omne Loucanam: a proposito dell'elegio di Scipione Barbato (CIL I 6, 7 = ILLRP 309). Klio, 2005, 77, p. 116-122.

1783. FERRARY (Jean-Louis). L'épigraphie juridique romaine: historiographie, bilan et perspectives. *In*: Monde romain (Le) à travers l'épigraphie [Cf. n° 1789], p. 35-70.

1784. HASEGAWA (Kinuko). The familia urbana during the Early Empire. A study of Columbaria inscriptions. Oxford, Archaeopress, 2005, IV-115 p. (British archaeological reports. International series, 1440).

1785. Inscriptions latines de l'Ain (ILAin). Éd. par François BERTRANDY. Chambéry, Université de Savoie, Laboratoire langages, littératures, sociétés, 2005, 299 p.

1786. KOLB (Anne), WALSER (Gerold), WINKLER (Gerhard), SCHMIDT (Manfred G.), JANSEN (Ulrike). Corpus inscriptionum Latinarum consilio et auctoritate Academiae scientiarum Berolinensis et Brandenburgensis editum. Vol. XVII. Miliaria imperii Romani cura Anne KOLB. Pars IV. Illyricum et provinciae Europae Graecae. Fasc. I. Miliaria provinciarum Raetiae et Norici. Ediderunt Anne KOLB, Gerold WALSER, Gerhard WINKLER. Edenda curaverunt Manfred G. SCHMIDT, Ulrike JANSEN. Berlin et New York, De Gruyter, 2005, XXX-122 p.

1787. KRÓLCZYK (Krzysztof). Tituli veteranorum. Veteraneninschriften aus den Donauprovinzen des Römischen Reiches (1.–3. Jh. n. Chr.). Poznan, Wydawn, Contact, 2005, 217 p.

1788. LIEBERG (Godo). Tiburtini versus Pompeiani. CIL IV 4966-4973. *Museum Helveticum*, 2005, 62, p. 56-64.

1789. Monde romain (Le) à travers l'épigraphie. Méthodes et pratiques; actes du XXIVe colloque international de Lille (8–10 novembre 2001). Éd. par Janine DESMULLIEZ. Villeneuve d'Ascq, CeGes, 2005, 411 p. (Collection UL3: Travaux et recherches). [Cf. n° <sélection> 1783.]

1790. NEMETI (Sorin). Magische Inschriften aus Dakien. *Latomus*, 2005, 64, p. 397-403.

1791. RÉMY (Bernard). Antonin le Pieux et les siens dans les inscriptions des provinces romaines d'Afrique. *Revue des études anciennes*, 2005, 107, p. 745-800.

1792. SEGENNI (Simonetta). Frontino, gli archivi della cura aquarum e l'acquedotto tardo repubblicano di Amiternum (CIL, I² 1853 = ILLRP, 487). *Athenaeum*, 2005, 93, p. 603-618.

1793. WALLACE (Rex). An introduction to wall inscriptions from Pompeii and Herculaneum. Introduction, inscriptions with notes, historical commentary, vocabulary. Wauconda, Bolchazy-Carducci, 2005, XLVI-136 p.

1794. WASE (Dick). Erotiska inskrifter från Pompeji. I urval av Dick Wase. Enskede, Oeisspeis, 2005, 33 p.

b. Literary sources

1795. [Ammianus Marcellinus] DEN BOEFT (Jan), DRIJVERS (Jan Willem), DEN HENGST (Daniel), TEITLER (Hans Carel). Philological and historical commentary on Ammianus Marcellinus. Leiden a. Boston, Brill, 2005, XXV-415 p.

1796. [Ammianus Marcellinus] KELLY (Gavin). Constantius II, Julian, and the example of Marcus Aurelius (Ammianus Marcellinus XXI, 16, 11-12). *Latomus*, 2005, 64, p. 409-416.

1797. [Anthologia Latina] ZURLI (Loriano). Unius poetae sylloge. Verso un'edizione di Anthologia Latina, cc. 90-197 Riese² = 78-188 Shackleton Bailey. Hildesheim, Olms, 2005, XXIV-132 p. (Spudasmata, 105).

1798. [Appendix Vergiliana] Appendix (Die) Vergiliana. Pseudepigraphen im literarischen Kontext. Hrsg. v. Niklas HOLZBERG. Tübingen, Narr, 2005, XIX-294 p. (Classica monacensia, 30).

1799. [Appianus] ŠAŠEL KOS (Marjeta). Appian and Illyricum. Ljubljana, Narodni muzej Slovenije, 2005, 671 p. (Situla, 43).

1800. [Apuleius] Apuleius. Metamorphosi. Volume I. Libri I-II. A cura di Alessandro BARCHIESI con un saggio introduttivo di Charles SEGAL. Milano, Mondadori, 2005, CXC-310 p. (Fondazione Lorenzo Valla. Scrittori greci e latini).

1801. [Apuleius] LEE (Benjamin Todd). Apuleius. Florida. A commentary. Berlin a. New York, De Gruyter, 2005, XI-215 p. (Texte und Kommentare, 25).

1802. [Cassiodorus] Variae. Cassiodoro Senatore. Introduzione, traduzione e note. A cura di Lorenzo VISCIDO. Cosenza, Pellegrini, 2005, 269 p.

1803. [Catullus] Catullus' poem on Attis. Text and contexts. Ed. by Ruurd R. NAUTA and Annette HARDER. Leiden a. Boston, Brill, 2005, 157 p. [Cf. n° <choice> 2128.]

1804. [Cicero] GARCEA (Alessandro). Cicerone in esilio. L'epistolario e le passioni. Hildesheim, Olms, 2005, VIII-323 p. (Spudasmata, 103).

1805. [Cicero] M. Tullius Cicero. Scripta quae manserunt omnia. Fasc. 43. De finibus bonoroum et malorum. Hrsg. v. Claudio MORESCHINI. München u. Leipzig, K. G. Saur, 2005, XVIII-215 p.

1806. [Cicero] MONTELEONE (Ciro). Prassi assembleare e retorica libertaria. La quarta Filippica di Cicerone. Bari, Palomar, 2005, 226 p.

1807. [Cicero] PLATSCHEK (Johannes). Studien zu Ciceros Rede für P. Quinctius. München, Beck, 2005, IX-300 p.

1808. [Cicero] TEDESCHI (Antonella). Lezione di buon governo per un dittatore. Cicerone, Pro Marcello: saggio di commento. Bari, Edipuglia, 2005, 181 p.

1809. [Claudius Quadrigarius] LACONI (Sonia). Q. Claudii Quadrigarii Annalium reliquiae. Introduzione, testo critico e commento filologico. Roma, Herder, 2005, 253 p.

1810. [Fabulae] Favole di Fedro e Aviano. A cura di Giannina SOLIMANO. Torino, UTET, 2005, 429 p.

1811. [Florus] EMBERGER (Peter). Catilina und Caesar. Ein historisch-philologischer Kommentar zu Florus (epit. 2, 12-13). Hamburg, Kovac, 2005, 670 p. (Studien zur Geschichtsforschung des Altertums, 12).

1812. [Florus] Florus. Römische Geschichte. Lateinisch und deutsch. Eingel., übers. und komment. Hrsg. v. Günter LASER. Darmstadt, Wiss. Buchgesellschaft, 2005, XVI-330 p. (Edition Antike).

1813. [Grillius] JAKOBI (Rainer). Grillius. Überlieferung und Kommentar. Berlin, De Gruyter, 2005, X-297 p. (Untersuchungen zur antiken Literatur und Geschichte, 77).

1814. [Historia Augusta] Historiae Augustae Colloquium Barcinonense. Centro Interuniverstiario per gli Studi sulla Historia Augusta, Università di Macerata e di Perugia. A cura di Giorgio BONAMENTE e Marc MAYER. Bari, Edipuglia, 2005, 323 p. (Munera, 22).

1815. [Historia Augusta] MADER (Gottfried). History as carnival, or method and madness in the Vita Heliogabali. *Classical antiquity*, 2005, 24, p. 131-169.

1816. [Historia Augusta] POTTIER (Bruno). Un pamphlet contre Stilichon dans l'Histoire Auguste. La vie de Maximin le Thrace. *Mélanges de l'école française de Rome. Antiquité*, 2005, 117, p. 223-267.

1817. Historia de ortu atque iuventute Constantini Magni eiusque matre Helena. Incerti auctoris. Historie über Herkunft und Jugend Constantins des Grossen und seine Mutter Helena / von einem unbekannten Vervasser. Hrsg. v. Paul DRÄGER. Trier, Kliomedia, 2005, 238 p.

1818. [Iosephus] LABOW (Dagmar). Flavius Josephus: Contra Apionem, Buch I. Einleitung, Text, textkritischer Apparat, Übersetzung und Kommentar. Stuttgart, Kohlhammer, 2005, LXXXIV-395 p.

1819. [Iustinus] RUBINCAM (Catherine). A Tale of two 'magni': JustinTrogus on Alexander and Pompey. *Historia*, 2005, 54, p. 265-274.

1820. [Livius] OAKLEY (Stephen P.). A commentary on Livy, books VI-X. Vol. III. Book IX. Oxford, Clarendon press, 2005, XV-758 p.

1821. [Martialis] Marco Valerio Marcial. Epigramas. Vol. II (libros 8-14). Ed. por Rosario MORENO SOLDEVILA, Juan FERNÁNDEZ VALVERDE y Enrique MONTERO CARTELLE. Madrid, Consejo superior de investigaciones cientificas, 2005, XI-330 p.

1822. [Ovidius] GAERTNER (Jan Felix). Ovid, Epistulae ex Ponto. Book I. Introd., transl., and commentary. Oxford, Oxford U. P., 2005, XV-606 p. (Oxford classical monographs, XV).

1823. [Ovidius] GREEN (Peter). Ovid. The poems of exile. Tristia and the Black Sea letters. Translated with an introduction, notes, and glossary. Berkeley, California U. P., 2005, 451 p.

1824. [Ovidius] LIVELEY (Genevieve). Ovid: Love songs. London, Bristol classical press, 2005, 141 p.

1825. [Phaedrus] CASCÓN DORADO (Antonio). Fábulas. Fedro. Introducciones, trad. y notas. Madrid, Gredos, 2005, 403 p. (Biblioteca clásica Gredos, 343).

1826. [Plautus] ÁLVAREZ HERNÁNDEZ (Arturo). Pséudolo. Tito Maccio Plauto. Introducción, traducción y notas. Buenos Aires, Losada, 2005, 228 p.

1827. [Plautus] Tito Maccio Plauto. Stico. Testo latino a fronte. A cura di Cesare QUESTA, Elena ROSSI e Mario SCÀNDOLA. Milano, Rizzoli, 2005, 177 p.

1828. [Plinius maior] BEAGON (Mary). The elder Pliny on the human animal. Natural history, book 7. Translated with introduction and historical commentary by Mary BEAGON. Oxford a. New York, Clarendon a. Oxford U. P., 2005, XX-515 p. (Clarendon ancient history series).

1829. [Plinius maior] COTTA RAMOSINO (Luisa). Mamilio Sura o Emilio Sura? Alcune considerazioni sulla teoria della successione degli imperi nella Naturalis Historia di Plinio il Vecchio. *Latomus*, 2005, 64, p. 945-958.

1830. [Plinius maior] GIEBEL (Marion). Plinius der Ältere, Naturalis historia. Naturgeschichte. Lateinisch/ Deutsch. Stuttgart, Reclam, 2005, 165 p.

1831. [Probus] M. Valeri Probi Beryti fragmenta. Ed. por Javier VELAZA. Barcelona, Univ. de Barcelona, 2005, 152 p.

1832. [Propertius] FEDELI (Paolo). Properzio, Elegie. Libro II. Introduzione, testo e commento. Cambridge, Cairns, 2005, IX-1070 p.

1833. [Propertius] GIARDINA (Giancarlo). Properzio, Elegie. Edizione critica e traduzione. Roma, Edizioni dell'Ateneo, 2005, 418 p. (Testi e commenti, 19).

1834. [Propertius] VIARRE (Simone). Élégies. Properce. Texte établi, traduit et commenté. Paris, Les belles lettres, 2005, LXVII-254 p. (Collection des universités de France. Série latine, 382).

1835. [Pseudo-Cassiodorus] Sergius (PS.-Cassiodorus), commentarium de oratione et de octo partibus orationis artis secundae donati. Überlieferung, Text und Kommentar. Hrsg. v. Christian STOCK. München, K. G. Saur, 2005, 439 p.

1836. [Pseudo-Ulpianus] AVENARIUS (Martin). Der pseudo-ulpianische liber singularis regularum. Entstehung, Eigenart und Überlieferung einer hochklassischen Juristenschrift. Analyse, Neuedition und deutsche Übersetzung. Göttingen, Wallstein Verlag, 2005, 640 p.

1837. [Pseudo-Ulpianus] Pseudo-ulpianische (Die) Einzelschrift der Rechtsregeln (liber singularis regularum). Hrsg. v. Martin AVENARIUS. Göttingen, Wallstein Verlag, 2005, 71 p.

1838. [Quintilianus] Quintilianus. Die Bienen des armen Mannes (Größere Deklamationen, 13). Text und Übersetzung. Hrsg. v. Gernot KRAPINGER. Cassino, Edizioni dell'Università degli studi di Cassino, 2005, 185 p.

1839. [Rutilius Namatianus] SQUILLANTE (Marisa). Il viaggio, la memoria, il ritorno. Rutilio Namaziano e le trasformazioni del tema odeporico. Napoli, D'Auria, 2005, 262 p.

1840. [Seneca] CHAUMARTIN (François-Régis). Sénèque, De la clémence. Texte établi et traduit. Paris, Les belles lettres, 2005, XCII-124 p. (Collection des universités de France. Série latine, 379).

1841. [Seneca] MARINO (Rosanna). Lucio Anneo Seneca. Ad Lucilium epistula 85. Palermo, Palumbo, 2005, VIII-164 p.

1842. [Statius] STEINIGER (Judith). P. Papinius Statius, Thebais. Kommentar zu Buch 4, 1-344. Stuttgart, Steiner, 2005, 181 p. (Altertumswissenschaftliches Kolloquium, 14).

1843. [Tacitus] HELLER (Erich). Tacitus. Annalen. Tacitus. Düsseldorf, Artemis & Winkler, 2005, 589 p.

1844. [Tacitus] ÖNNERFORS (Alf). Cornelius Tacitus, Germania. Översättning fran latinet, med inledning och kommentarer. Stockholm, Wahlström & Widstrand, 2005, 139 p. (Wahlström & Widstrands Klassikerserie, 11).

1845. [Valerius Flaccus] KLEYWEGT (Adrianus Jan). Valerius Flaccus, Argonautica, book I. A. commentary. Leiden a. Boston, Brill, 2005, XIX-506 p. (Mnemosyne. Suppl. 262).

1846. [Valerius Flaccus] WACHT (Manfred). Gai Valerii Flacci Argonauticon concordantia. Hildesheim, Olms-Weidmann, 2005, 680 p.

1847. [Valerius Maximus] WARDLE (David). Valerius Maximus and the end of the first Punic war. *Latomus*, 2005, 64, p. 377-384.

1848. [Velleius Paterculus] KRAMER (Emil A.). Book one of Velleius' History: scope, levels of treatment, and non-Roman elements. *Historia*, 2005, 54, p. 144-161.

1849. [Vergilius] COBBOLD (G. B.). Vergil's Aeneid. Hero, war, humanity (G. B. COBBOLD, translator). Wauconda, Bolchazy-Carducci, 2005, XVIII-366 p.

1850. [Vergilius] FINK (Gerhard). Aeneis. Lateinisch-deutsch. P. Vergilius Maro. Düsseldorf, Artemis & Winkler, 2005, 739 p.

1851. [Vergilius] LEMBKE (Janet). Virgil's Georgics. A new verse translation. New Haven, Yale U. P., 2005, XXIV-114 p.

Cf. n°s 155, 2055-2126

§ 4. **General and political history.**

1852. ABRAMZON (Mikhail G.). Rimskoe vladychestvo na Vostoke: Rim i Kilikija (II v. do n.e. – 74 g. n.e.). (The Roman rule in the East: Rome and Cilicia, the 2nd century B.C.–74 A.D.). Sankt-Peterburg, Gumanitarnaja akademija, 2005, 252 p. (ill., maps; bibl. p. 199-207).

1853. Afrique (L') romaine. Ier siècle avant J.-C.–début Ve siècle après J.-C. Actes du colloque de la SOPHAU (Poitiers, 1–3 avril 2005). Éd. par Hélène GUIRAUD. Toulouse, Presses universitaires du Mirail, 2005, 360 p.

1854. AMHERDT (David). Le vie de Cicéron, d'après l'humaniste Juan Luis Vivès (1492–1540): un portrait idéal au service d'une conception de l'action politique. *Latomus*, 2005, 64, p. 995-1007.

1855. Ancient Rome: from the early Republic to the assassination of Julius Caesar. Ed. by Matthew DILLON and Lynda GARLAND. London, Routledge, 2005, IX-784 p.

1856. Aquitaine (L') et l'Hispanie à l'époque julio-claudienne. Organisation et exploitation des espaces provinciaux. Colloque Aquitania (Saintes, 11–13 septembre 2003). Éd. par Pierre SILLIÈRES. Bordeaux, Aquitania, 2005, 536 p.

1857. ARENA (Gaetano). Città di Panfilia e Pisidia sotto il dominio romano. Continuità strutturali e cambiamenti funzionali. Catania, Edizioni del Prisma, 2005, 476 p. (Testi e studi di storia antica, 16).

1858. ASMIS (Elizabeth). A new kind of model: Cicero's Roman constitution in De re publica. *Classical philology*, 2005, 100, p. 377-416.

1859. BADEL (Christophe). La noblesse de l'empire romain. Les masques et la vertu. Seyssel, Champ Vallon, 2005, 507 p.

1860. BARCELÓ (Pedro A.). Kleine römische Geschichte. Darmstadt, Primus Verlag, 2005, 160 p.

1861. BARRETT (Anthony A.). Vespasian's wife. *Latomus*, 2005, 64, p. 385-396.

1862. BENESS (J. Lea). Scipio Aemilianus and the crisis of 129 B.C. *Historia*, 2005, 54, p. 37-48.

1863. BENOIST (Stéphane). Rome, le prince et la Cité. Pouvoir impérial et cérémonies publiques (Ier siècle av. J.-C.–début du IVe siècle apr. J.-C.). Paris, Presses universitaires de France, 2005, XI-397 p. (Le Nœud gordien).

1864. BERRENDONNER (Clara). Les interventions du peuple dans les cités d'Étrurie et d'Ombrie à l'époque impériale. *Mélanges de l'école française de Rome. Antiquité*, 2005, 117, p. 517-539.

1865. BERTRANDY (François). Le proconsulat de Salluste en Africa noua: ombres et lumières. *Latomus*, 2005, 64, p. 33-48.

1866. BIRLEY (Anthony Richard). The Roman government of Britain. Oxford, Oxford U. P., 2005, XIII-532 p. (Oxford scholarship online).

1867. BOCCHIOLA (Massimo), SARTORI (Marco). Teutoburgo. La grande disfatta delle legioni di Augusto. Milano, Rizzoli, 2005, 331 p.

1868. BRACKMANN (Stephan). Die militärische Selbstdarstellung des Caligula. Das Zeugnis der Münzen im Widerspruch zur antiken Geschichtsschreibung. *Gymnasium*, 2005, 112, p. 375-383.

1869. CABOURET (Bernadette), ARNAUD-LINDET (Marie-Pierre). L'Afrique romaine de 69 à 439. Romanisation et christianisation. Nantes, Éditions du Temps, 2005, 345 p.

1870. CADOUX (T. J.). Catiline and the Vestal virgins. *Historia*, 2005, 54, p. 162-179.

1871. CAMERON (Hamish), PARKER (Victor). A mobile people? Sallust's presentation of the Numidians and their manner of fighting. *La parola del passato*, 2005, 60, p. 33-57.

1872. CHEKANOVA (Nina V.). Rimskaja diktatura poslednego veka Respubliki. (The dictatorship in Rome in the last century of the Republic). Sankt-Peterburg, Gumanitarnaja akademija, 2005, 480 p. (bibl. p. 462-476). (Studia classica).

1873. CHRIST (Karl). Kaiserideal und Geschichtsbild bei Sextus Aurelius Victor. *Klio*, 2005, 87, p. 177-200.

1874. COARELLI (Filippo). P. Faianius Plebeius, Forum Novum and Tacitus. *Papers of the British school at Rome*, 2005, 73, p. 85-98.

1875. COSME (Pierre). Auguste. Paris, Perrin, 2005, 345 p.

1876. CURRAN (John). 'The long hesitation': some reflections on the Romans in Judaea. *Greece and Rome*, 2005, 52, p. 70-98.

1877. DAHL (Gunnar). Tiberius. Kejsare mot sin vilja. Lund, Historiska Media, 2005, 335 p.

1878. DAHLHEIM (Werner). Julius Caesar. Die Ehre des Kriegers und die Not des Staates. Paderborn, München u. Wien, Schöningh, 2005, 321 p.

1879. DANZIGER (Danny), PURCELL (Nicholas). Hadrian's empire. When Rome ruled the world. London, Hodder & Stoughton, 2005, XV-302 p.

1880. DE GALBERT (Geoffroy). Hannibal en Gaule. Nouvelle hypothèse basée sur des découvertes géographiques et archéologiques. Grenoble, Editions de Belledonne, 2005, 194 p.

1881. DENCH (Emma). Romulus' asylum. Roman identities from the age of Alexander to the age of Hadrian. Oxford, Oxford U. P., 2005, XI-441 p.

1882. *Vacat.*

1883. DOCTER (Roald Fritjof). Een spectaculair einde: Carthago 246–146 voor Chr. *Lampas*, 2005, 38, p. 313-329.

1884. EDMONDSON (Jonathan C.). Flavius Josephus and Flavian Rome. Oxford, Oxford U. P., 2005, XVI-400 p.

1885. EICH (Peter). Zur Metamorphose des politischen Systems in der römischen Kaiserzeit. Die Entstehung einer "personalen Bürokratie" im langen dritten

Jahrhundert. Berlin, Akademie verlag, 2005, 467 p. (Klio, Beih. n. F., 9).

1886. FARNUM (Jerome H.). The positioning of the Roman imperial legions. Oxford, Archaeopress, 2005, 121 p. (British archaeological reports. International series, 1458).

1887. FELD (Karl). Barbarische Bürger. Die Isaurier und das Römische Reich. Berlin u. New York, de Gruyter, 2005, XII-411 p. (Millennium-Studien / Millenium Studies, 8).

1888. FERDIÈRE (Alain). Les Gaules. Provinces des Gaules et Germanies, provinces alpines (IIe siècle av.–Ve siècle ap. J.-C.). Paris, A. Colin, 2005, 446 p.

1889. FISCHER (Svante). Roman imperialism and runic literacy. The westernization of Northern Europe (150–800 AD). Uppsala, Department of archaeology and ancient history, Uppsala University, 2005, 262 p.

1890. FORSYTHE (Gary). A critical history of early Rome. From prehistory to the first Punic War. Berkeley, University of California Press, 2005, XVI-400 p.

1891. FRASCHETTI (Augusto). Giulio Cesare. Roma e Bari, Laterza, 2005, 158 p. (Biblioteca essenziale Laterza, 68. Storia antica). – IDEM. Roma e il principe. Roma e Bari, Laterza, 2005, XV-362 p.

1892. FRASER (James Earle). The Roman conquest of Scotland. The battle of Mons Graupius AD 84. Stroud, Tempus, 2005, 159 p.

1893. GELZER (Matthias). Pompeius. Lebensbild eines Römers. Stuttgart, Steiner, 2005, 244 p.

1894. GILLIVER (Catherine Mary), GOLDSWORTHY (Adrian Keith), WHITBY (Michael). Rome at war. Oxford, Osprey, 2005, 288 p.

1895. GOWING (Alain Michael). Empire and memory. The representation of the Roman Republic in imperial culture. Cambridge, Cambridge U. P., 2005, XIV-178 p.

1896. GRAINGE (Gerald). The Roman invasions of Britain. Stroud, Tempus, 2005, 192 p.

1897. Guerre et diplomatie romaines (IVe–IIIe siècles). Pour un réexamen des sources. Éd. par Emmanuèle CAIRE et Sylvie PITTIA. Aix-en-Provence, Publications de l'Université de Provence, 2005, 322 p.

1898. HALEY (Evan W.). Hadrian as Romulus or the self-representation of a Roman emperor. *Latomus*, 2005, 64, p. 969-980.

1899. HEALY (Mark). Cannae 216 BC. Hannibal smashes Rome's army. Westport, Praeger, 2005, 96 p.

1900. HEINZE (Traudel). Konstantin der Große und das konstantinische Zeitalter in den Urteilen und Wegen der deutsch-italienischen Forschungsdiskussion. München, Utz, 2005, 377 p.

1901. HERRMANN (Horst). Nero. Eine Biographie. Berlin, Aufbau-Taschenbuch-Verlag, 2005, 456 p.

1902. Hispania (La) de Antoninos (98–180), Actas del II Congreso Internacional de Historia Antigua (Valladolid, 10, 11 y 12 de noviembre de 2004). Ed por Liborio HERNÁNDEZ GUERRA. Valladolid, Univ. de Valladolid, 2005, 627 p.

1903. Histoire de la civilisation romaine. Sous la direction de Hervé INGLEBERT; avec la collaboration de Pierre GROS et Gilles SAURON. Paris, Presses universitaires de France, 2005, XCII-512 p. (Nouvelle Clio).

1904. HUMM (Michel). Appius Claudius Caecus. La République accomplie. Rome, École française de Rome, 2005, X-779 p. (Bibliothèque des Écoles françaises d'Athènes et de Rome, 322).

1905. ITGENSHORST (Tanja). Tota illa pompa. Der Triumph in der römischen Republik. Mit einer CD-ROM: Katalog der Triumphe von 340 bis 19 vor Christus. Göttingen, Vandenhoeck & Ruprecht, 2005, 301 p. (Hypomnemata, 161).

1906. JACKOB (Nikolaus). Öffentliche Kommunikation bei Cicero. Publizistik und Rhetorik in der späten römischen Republik. Baden-Baden, Nomos, 2005, 353 p.

1907. Julio César y Corduba. Tiempo y espacio en la campaña de Munda (49–45 a.C.). Actas del simposio organizado por la Facultad de Filosofía y Letras de la Universidad de Córdoba (Córdoba, 21–25 de abril de 2003). Ed por Enrique MELCHOR GIL. Córdoba, Servicio de publ. Univ. de Córdoba, 2005, 500 p.

1908. KADRA (Haouaria). Jugurtha. Un Berbère contre Rome. Paris, Arléa, 2005, 223 p.

1909. KEAVENEY (Arthur). Sulla. The last republican. London, Routledge, 2005, X-233 p. – IDEM. The terminal date of Sulla's dictatorship. *Athenaeum*, 2005, 93, p. 423-439.

1910. LAEBEN-ROSÉN (Viktoria). Age of rust. Court and power in the Severan age (188–238 AD). Uppsala, Uppsala University, 2005, 263 p.

1911. LE BOHEC (Yann). Histoire de l'Afrique romaine (146 avant J.-C.–439 après J.-C.). Paris, Picard, 2005, XLIII-282 p. – IDEM. Le clergé celtique et la guerre des Gaules. Historiographie et politique. *Latomus*, 2005, 64, p. 871-881.

1912. Limes XIX. Proceedings of the XIXth international congress of Roman frontier studies held in Pécs, Hungary, September 2003. Ed. by Zsolt VISY. Pécs, University of Pécs, 2005, 1004 p.

1913. LINKE (Bernhard). Die römische Republik von den Gracchen bis Sulla. Darmstadt, Wiss. Buchgesellschaft, 2005, 150 p.

1914. LUIK (Martin). Der schwierige Weg zur Weltmacht. Roms Eroberung der Iberischen Halbinsel 218–19 v. Chr. Mainz, von Zabern, 2005, 117 p.

1915. MALITZ (Jürgen). Nero. Oxford, Blackwell, 2005, 174 p.

1916. Marcus Porcius Cato, über den Ackerbau. Hrsg. v. Dieter FLACH. Stuttgart, Steiner, 2005, 204 p.

1917. MARKS (Raymond David). From republic to empire. Scipio Africanus in the Punica of Silius Italicus. Frankfurt am Main, Lang, 2005, 308 p. (Studien zur klassischen Philologie, 152).

1918. MORGAN (M. Gwyn). The opening stages in the battle for Cremona, or the devil in the details (Tacitus, Histories 3, 15-18). *Historia*, 2005, 54, p. 189-209.

1919. MOSIG-WALBURG (Karin). Hanniballianus rex. *Millennium*, 2005, 2, p. 229-254.

1920. NEMETH (Eduard), RUSTOIU (Aurel), POP (Horea). Limes Dacicus Occidentalis. Die Befestigungen im Westen Dakiens vor und nach der römischen Eroberung. Cluj-Napoca, Mega, 2005, 180 p.

1921. NEMETH (Eduard). Armata în sud-vestul Daciei Romane. Die Armee im Südwesten des Römischen Dakien. (The army in the South-West of the Roman Dacia). Timişoara, Editura Mirton, 2005, 228 p.

1922. PAPANGELES (Theodoros D.). E Rome kai o kosmos tes. Thessalonike, Instituto Neoellenikon Spudon, 2005, 235 p.

1923. PARADISO (Annalisa). Sesto Pompeo tra storia ed "elegia triste" ovidiana. *Aufidus*, 2005, 57, p. 147-175.

1924. PFEILSCHIFTER (Rene). Titus Quinctius Flamininus. Untersuchungen zur römischen Griechenlandpolitik. Göttingen, Vandenhoeck & Ruprecht, 2005, 442 p. (Hypomnemata, 162).

1925. PISO (Ioan). An der Nordgrenze des Römischen Reiches. Ausgewählte Studien, 1972–2003. Stuttgart, Steiner, 2005, 527 p. (Heidelberger althistorische Beiträge und epigraphische Studien, 41).

1926. Politica e cultura in Roma antica. Atti dell'incontro di studio in ricordo di Italo Lana (Torino 16–17 ottobre 2003). A cura di Federica BESSONE. Bologna, Pàtron, 2005, 171 p. (Pubblicazioni del Dipartimento di filologia, linguistica e tradizione classica, 22).

1927. Politica e partecipazione nelle città dell'impero romano. A cura di Francesco AMARELLI. Roma, 'L'Erma' di Bretschneider, 2005, IX-201 p. (Saggi di storia antica, 25).

1928. RÉMY (Bernard). Antonin le Pieux. 138–161; le siècle d'or de Rome. Paris, Fayard, 2005, 452 p.

1929. RENUCCI (Pierre). Tibère. L'empereur malgré lui. Paris, Mare & Martin, 2005, 491 p.

1930. Romanisierung-Romanisation. Theoretische Modelle und praktische Fallbeispiele. Hrsg. v. Günther SCHÖRNER. Oxford, Archaeopress, 2005, XVI-264 p. (British archaeological reports. International series, 1427).

1931. Römische Geschichte und Geschichtsschreibung. Hrsg. v. Ursula GÄRTNER. Potsdam, Universitätsverlag Potsdam, 2005, 89 p. (Potsdamer Lateintage, 1).

1932. Römische Werte als Gegenstand der Altertumswissenschaft. Hrsg. v. Andreas HALTENHOFF, Andreas HEIL und Fritz-Heiner MUTSCHLER. München u. Leipzig, Saur, 2005, XIII-286 p. (Beiträge zur Altertumskunde, 227).

1933. Roms auswärtige Freunde in der späten Republik und im frühen Prinzipat. Hrsg. v. Altay COSKUN. Göttingen, Duehrkohp & Radicke, 2005, IX-300 p. (Göttinger Forum für Altertumswissenschaften. Beihefte, 19, Geschichte).

1934. ROSE (Charles Brian). The Parthians in Augustan Rome. *American journal of archaeology*, 2005, 109, p. 21-76.

1935. ROUSSOT (Thomas). Marc-Aurèle et l'Empire romain. Paris, Harmattan, 2005, 115 p.

1936. RUSSO (Flavio). 89 a. C. assedio a Pompei. La dinamica e le tecnologie belliche della conquista sillana di Pompei. Pompei, Flavius, 2005, 96 p.

1937. SAINT-HILAIRE (Janine Cels). La République romaine 133–44 av. J.-C. Paris, Armand Colin, 2005, 239 p. (Cursus).

1938. SCHLANGE-SCHÖNINGEN (Heinrich). Augustus. Darmstadt, Wiss. Buchgesellschaft, 2005, 157 p.

1939. SCHMIDT (Joël). Jules César. Paris, Gallimard, 2005, 359 p.

1940. SCHMIDT (Michael). Der römische Limes im Hohen Odenwald, am Mittleren Neckar und in der Schwäbischen Alb. Frankfurt am Main, Artaunon-Verlag, 2005, III-432 p.

1941. SCHMITZER (Ulrich). Der Tod auf offener Szene. Tacitus über Nero und die Ermordung des Britannicus. *Hermes*, 2005, 133, p. 337-357.

1942. SEAGER (Robin). Tiberius. Malden, Blackwell, 2005, XVIII-310 p.

1943. Senatores populi Romani. Realität und mediale Präsentation einer Führungsschicht. Kolloquium der Prosopographia Imperii Romani vom 11.–13. Juni 2004. Hrsg. v. Werner ECK und Matthäus HEIL. Stuttgart, Steiner, 2005, VIII-329 p. (Heidelberger althistorische Beiträge und epigraphische Studien, 40). [Cf. n[os] <Auswahl> 1976, 2038.]

1944. SHELDON (Rose Mary). Intelligence activities in ancient Rome. Trust in the gods, but verify. London, Cass., 2005, XXVII-317 p.

1945. SHEPPARD (Simon). Pharsalus, 48 BC. Caesar and Pompey – clash of the titans. Oxford, Osprey Publ., 2005, 96 p.

1946. SIDEBOTTOM (Harry). Roman imperialism: the changed outward trajectory of the Roman empire. *Historia*, 2005, 54, p. 315-330.

1947. SIMPSON (Christopher J.). Rome's 'official imperial seal'? The rings of Augustus and his first century successors. *Historia*, 2005, 54, p. 180-188.

1948. SPEIDEL (Michael Alexander). Early Roman rule in Commagene. *Scripta classica Israelica*, 2005, 24, p. 85-100.

1949. STEFAN (Alexandre Simon). Les guerres daciques de Domitien et de Trajan. Architecture militaire, topographie, images et histoire. Rome, École Française de Rome, 2005, XIII-811 p. (Collection de l'École française de Rome, 353).

1950. SUMI (Geoffrey S.). Ceremony and power: performing politics in Rome between Republic and Empire. Ann Arbor, University of Michigan Press, 2005, XII-360 p.

1951. TEDESCHI (A.). Giuliano e il koinòs katheghemón. *Quaderni di storia*, 2005, 62, p. 123-129.

1952. WHITE (John F.). Restorer of the world. The Roman emperor Aurelian. Staplehurst, Spellmount, 2005, XXVI-198 p.

1953. WILKINSON (Sam). Caligula. London a. New York, Routledge, 2005, VIII-110 p.

1954. WITTCHOW (Frank). Vater und Onkel: Julius Caesar und das Finale der Aeneis. *Gymnasium*, 2005, 112, p. 45-70.

1955. WÜRFEL (Walter). Die Schlachtfelder der Varus-Armee. Studie zur römisch-germanischen Geschichte. Frankfurt am Main, R. G. Fischer, 2005, 56 p.

1956. YASUI (Moyuru). Kyōwasei rōma no katōseijitaisei: nobiritasushihai no kenkyū. (Oligarchical regime in the Roman republic: Studies in the Nobilitätsherrschaft). Kyoto, Minerva Shobo, 2005, 400 p.

1957. ZIMMERMANN (Klaus). Rom und Karthago. Darmstadt, Wissenschaftliche Buchgesellschaft, 2005, VII-152 p.

Cf. n°s 138, 1108, 1132

§ 5. History of law and institutions.

** 1958. Corpus der römischen Rechtsquellen zur antiken Sklaverei: (CRRS). Hrsg. v. Johanna FILIP-FRÖSCHL [et al.]. Teil 10. Juristisch speziell definierte Sklavengruppen. 6. Servus fugitivus. Bearb. v. Georg KLINGENBERG. Stuttgart, Steiner, 2005, XVI-233 p. (Forschungen zur antiken Sklaverei, Beiheft; 3, 10, 6).

1959. BANDELLI (Gino), CHIABÀ (Monica). Le amministrazioni locali nella transpadana orientale. Dalla provincia repubblicana della Gallia Cisalpina alla provincia tardoantica della Venetia et Histria. *Mélanges de l'école française de Rome. Amtiquité*, 2005, 117, p. 439-463.

1960. BAR (Doron). Roman legislation as reflected in the settlement history of late antique Palestine. *Scripta classica Israelica*, 2005, 24, p. 195-206.

1961. BARRETT (Anthony A.). Aulus Caecina Severus and the military woman. *Historia*, 2005, 54, p. 301-314.

1962. BECK (Hans). Karriere und Hierarchie. Die römische Aristokratie und die Anfänge des cursus honorum in der mittleren Republik. Berlin, Akad.-Verlag, 2005, 452 p. (Klio Beihefte. Neue Folge, 10).

1963. BOUDEWIJN SIRKS (Adriaan Johan). Der Zweck des Senatus Consultum Claudianum von 52 n.Chr.. *Zeitschrift der Savigny-Stiftung für Rechtsgeschichte Romanistische Abteilung*, 2005, 122, p. 138-149.

1964. BRÉLAZ (Cédric). La sécurité publique en Asie Mineure sous le Principat (Ier–IIIe s. ap. J.-C.). Institutions municipales et institutions impériales dans l'Orient romain. Basel, Schwabe, 2005, XI-530 p. (Schweizerische Beiträge zur Altertumswissenschaft, 32).

1965. CASINOS MORA (Francisco Javier). Lexicografía y derecho romano. Valor y uso de los instrumenta studiorum iuris romani. *Estudios clásicos*, 2005, 128, p. 69-89.

1966. CHEVREAU (Emmanuelle). La lex Rhodia de iactu: un exemple de la réception d'une institution étrangère dans le droit romain. *Tijdschrift voor rechtsgeschiedenis*, 2005, 73, 1-2, p. 67-80.

1967. CHRISTOL (Michel). À propos d'hommages publics en Gaule Narbonnaise. *Mélanges de l'école française de Rome. Antiquité*, 2005, 117, p. 555-566.

1968. DALLA MASSARA (Tommaso). La domanda parziale nel processo civile romano. Padova, CEDAM, 2005, X-151 p.

1969. DANDO-COLLINS (Stephen). Nero's killing machine. The true story of Rome's remarkable fourteenth legion. Hoboken, Wiley, 2005, XIV-322 p.

1970. DE CARLO (Antonella). I cavalieri e l'amministrazione cittadina nelle città dell'Italia meridionale. La Campania e le Regiones II e III. *Mélanges de l'école française de Rome. Antiquité*, 2005, 117, p. 491-506.

1971. DOVERE (Elio). De iure. L'esordio delle epitomi di Ermogeniano. Prefazione di Franco CASAVOLA. Napoli, Jovene, 2005, XXXI-217 p. (Collezione di opere giuridiche e storiche scelte, 6).

1972. ECK (Werner). Auf der Suche nach Personen und Persönlichkeiten. Cursus honorum und Biographie. *In*: Biographie und Prosopographie [Cf. n° 1372], p. 53-72.

1973. GRELLE (Francesco). Diritto e società nel mondo romano. A cura di Lucia FANIZZA. Roma, 'L'Erma' di Bretschneider, 2005, XVI-560 p. (Saggi di storia antica, 26).

1974. HARKE (Jan Dirk). Societas als Geschäftsführung und das römische Obligationensystem. *Tijdschrift voor rechtsgeschiedenis*, 2005, 73, 1-2, p. 43-66.

1975. HEMELRIJK (Emily A.). Octavian and the introduction of public statues for women in Rome. *Athenaeum*, 2005, 93, p. 309-317.

1976. HESBERG (Henner von). Die Häuser der Senatoren in Rom: gesellschaftliche und politische Funktion. *In*: Senatores populi Romani [Cf. n° 1943], p. 19-52.

1977. KNOCH (Stefan). Sklavenfürsorge im Römischen Reich. Formen und Motive. Hildesheim, Zürich

u. New York, Olms, 2005, VII-338 p. (Sklaverei – Knechtschaft – Zwangsarbeit, 2).

1978. KOPTEV (Alexandr). 'Three brothers' at the head of archaic Rome: the king and his 'Consuls'. *Historia*, 2005, 54, p. 382-423.

1979. MIRKOVIC (Miroslava). Child labour and taxes in the agriculture of Roman Egypt. *Scripta classica Israelica*, 2005, 24, p. 139-150.

1980. NASTI (Fara). M. Cn. Licinius Rufinus e i suoi Regularum libri. *Index*, 2005, 33, p. 263-292.

1981. PANI (Mario), TODISCO (Elisabetta). Società e istituzioni di Roma antica. Roma, Carocci, 2005, 247 p.

1982. PESARESI (Roberto). Studi sul processo penale in età repubblicana. Dai tribunali rivoluzionari alla difesa della legalità democratica. Napoli, Jovene, 2005, IX-233 p.

1983. RAMPAZZO (Natale). Il bellum iustum e le sue cause. *Index*, 2005, 33, p. 235-261.

1984. RAMSEY (John T.). Mark Antony's judiciary reform and its revival under the triumvirs. *Journal of Roman studies*, 2005, 95, p. 20-37.

1985. ROUGÉ (Jean). Le lois religieuses des empereurs romains de Constantin à Théodose II (312–438). I. Code Théodosien XVI (Texte latin Theodor MOMMSEN, traduction Jean ROUGÉ, introduction et notes Roland DELMAIRE, collab. François RICHARD). Paris, Du Cerf, 2005, 524 p.

1986. RUNDEL (Tobias). Mandatum zwischen utilitas und amicitia. Perspektiven zur Mandatarhaftung im klassischen römischen Recht. Münster, Lit Verlag, 2005, XXX-221 p.

1987. SACCHI (Osvaldo). La nozione di ager publicus populi romani nella lex agraria del 111 a.C. come espressione dell'ideologia del suo tempo. *Tijdschrift voor rechtsgeschiedenis*, 2005, 73, 1-2, p. 19-42.

1988. SIMKINS (Michael), EMBLETON (Ron). Die römische Armee von Caesar bis Constantin (44 v. Chr.–333 n. Chr.). Sankt Augustin, Siegler, 2005, 95 p.

1989. TUCCILLO (Fabiana). Le prescrizione acquisitiva delle servitù. *Index*, 2005, 33, p. 433-447.

1990. TUCK (Steven L.). The origins of Roman imperial hunting imagery: Domitian and the redefinition of Virtus under the principate. *Greece and Rome*, 2005, 52, p. 221-245.

1991. VITALI (Camilla). ... manumissus liber(um)ve iussus erit...; sul capitolo 28 della lex Irnitana. *Index*, 2005, 33, p. 389-431.

1992. WILLIAMSON (Callie). The laws of the Roman people. Public law in the expansion and decline of the Roman Republic. Ann Arbor, University of Michigan Press, 2005, XXVIII-506 p.

Cf. n°s 1374, 2089, 2752

§ 6. Economic and social history.

1993. ALLASON-JONES (Lindsay). Women in Roman Britain. York, Council for British Archaeology, 2005, XII-209 p.

1994. Artisanat et économie romaine. Italie et provinces occidentales de l'empire. Actes du 3e Colloque International d'Erpeldange (Luxembourg) sur l'artisanat romain (14–16 octobre 2004). Éd. par Michel POLFER. Montagnac, Mergoil, 2005, 194 p.

1995. BOTERMANN (Helga). Wie aus Galliern Römer wurden. Leben im Römischen Reich. Stuttgart, Klett-Cotta, 2005, 474 p.

1996. BOUNEGRU (Octavian). Trafiquants et navigateurs sur le Bas Danube et dans le Pont Gauche à l'époque romaine. Wiesbaden, Harrassowitz, 2005, 197 p.

1997. BOURDIN (Stéphane). Ardée et les Rutules. Réflexions sur l'émergence et le maintien des identités ethniques des populations du Latium préromain. *Mélanges de l'école française de Rome. Antiquité*, 2005, 117, p. 585-631.

1998. BRADLEY (Keith). The Roman child in sickness and in health. *In*: Roman family (The) in the empire [Cf. n° 2040], p. 67-92.

1999. BRUN (Jean-Pierre). Archéologie du vin et de l'huile en Gaule romaine. Paris, Errance, 2005, 268 p.

2000. BURNAND (Yves). Primores Galliarum. Sénateurs et chevaliers romains originaires de Gaule de la fin de la République au IIIe siècle. Bruxelles, Latomus, 2005, 450 p.

2001. CHAMBERT (Régine). Rome: le mouvement et l'ancrage. Morale et philosophie du voyage au début du Principat. Bruxelles, Éditions Latomus, 2005, 411 p. (Collection Latomus, 288).

2002. Comerţ şi civilizaţie. Transilvania în contextul schimburilor comerciale şi culturale în antichitate. (Trade and civilization. Transylvania in the context of commercial and cultural exchanges in ancient time). Editori Călin COSMA şi Aurel RUSTOIU. Cluj-Napoca, Editura Mega, 2005, 296 p.

2003. Concepts, pratiques et enjeux environnementaux dans l'Empire romain. Ed. by Robert BEDON. Limoges, Presses universitaires de Limoges, 2005, 400 p. (Caesarodunum, 39).

2004. CONNOLLY (Peter). Colosseum. Arena der Gladiatoren. Stuttgart, Reclam, 2005, 224 p.

2005. CORDIER (Pierre). Gymnase et nudité à Rome. *Métis*, 2005, 3, p. 253-269. – IDEM. Nudités romaines. Un problème d'histoire et d'anthropologie. Paris, Les belles lettres, 2005, 428 p. (ill.). (Collection d'études anciennes. Série latine, 63).

2006. COSTANZA (Salvatore). Aspetti e problemi della fiscalità nel tardo impero romano. Normativa imperiale, fonti pagane e fonti cristiane a confronto. Caltanissetta, Centro studi Cammarata e Lussografica, 2005, 127 p.

2007. CRISTOFOLI (Roberto). L'alimentazione nell'antica Roma. Aspetti storici, scientifici e sociali. Roma, Il Calamo, 2005, 135 p. (Episteme dell'antichità e oltre, 11).

2008. DES BOSCS-PLATEAUX (Françoise). Un parti hispanique à Rome? Ascension des élites hispaniques et pouvoir politique d'Auguste à Hadrien (27 av. J.-C.– 138 ap. J.-C.). Madrid, Casa de Velázquez, 2005, XVII-763 p. (Bibliothèque de la Casa de Velázquez, 32).

2009. DOMINGUEZ PÉREZ (Juan Carlos). El potencial económico de Saiganthé como 'casus belli' en el estallido de la segunda guerra Punica. *Latomus*, 2005, 64, p. 590-600.

2010. DUNCAN-JONES (Richard P.). Implications of Roman coinage: debates and differences. *Klio*, 2005, 87, p. 459-487.

2011. ERDKAMP (Paul P. M.). The grain market in the Roman empire. A social, political and economic study. Cambridge, Cambridge U. P., 2005, VIII-364 p.

2012. FERTL (Evelyn). Von Musen, Miminnen und leichten Mädchen. Die Schauspielerin in der römischen Antike. Wien, Braumülle, 2005, VIII-228 p. (Blickpunkte, 9).

2013. FODOREAN (Florin). Drumurile la romani. (Roman roads). Prefață de Mihai BĂRBULESCU. Cluj-Napoca, Editura Napoca Star, 2005, 157 p.

2014. GEORGE (Michele). Family imagery and family values in Roman Italy. *In*: Roman family (The) in the empire [Cf. n° 2040], p. 37-66.

2015. HABINEK (Thomas Noel). The world of Roman song. From ritualized speech to social order. Baltimore, Johns Hopkins U. P., 2005, VI-329 p.

2016. HAECK (Tom). The quinquennales in Italy: social status of a Roman municipal magistrate. *Latomus*, 2005, 64, p. 601-618.

2017. HEUCHERT (Volker). The chronological development of Roman provincial coin iconography. *In*: Coinage and identity in the Roman provinces [Cf. n° 136], p. 29-56.

2018. HINGLEY (Richard). Globalizing Roman culture. Unity, diversity and empire. London, Routledge, 2005, XIII-208 p.

2019. HOWGEGO (Christopher). Coinage and identity in the Roman provinces. *In*: Coinage and identity in the Roman provinces [Cf. n° 136], p. 1-18.

2020. IONESCU (Mihai), PAPUC (Gheorghe). Sistemul de apărare a litoralului Dobrogei Romane (sec. I–VII p. Chr.). (The defense system of the shores of Roman Dobruja. 1st–7th century AD). Prefață de Alexandru SUCEVEANU. Constanța, Editura Ex Ponto, 2005, 205 p.

2021. JACOB (Christian). "La table et le cercle". Sociabilités savantes sous l'Empire Romain. *Annales*, 2005, 60, 3, p. 507-530.

2022. KASTER (Robert Andrew). Emotion, restraint, and community in ancient Rome. London, Oxford U. P., 2005, 245 p. (Classical culture and society, IX).

2023. KRIECKHAUS (Andreas). Senatorische Familien und ihre patriae, 1./2. Jahrhundert n. Chr. Hamburg, Kovac, 2005, XI-247 p. (Studien zur Geschichtsforschung des Altertums; 14).

2024. KRON (Geoffrey). The Augustan census figures and the population of Italy. *Athenaeum*, 2005, 93, p. 441-495.

2025. KUNST (Christiane). Römische Adoption. Zur Strategie einer Familienorganisation. Clauss, Hennef, 2005, 351 p. (Frankfurter althistorische Beiträge, 10).

2026. LOVATT (Helen). Statius and epic games. Sport, politics, and poetics in the Thebaid. New York, Cambridge U. P., 2005, XII-336 p. (Cambridge classical studies).

2027. MAYER (Jochen Werner). Imus ad villam. Studien zur Villeggiatur im stadtrömischen Suburbium in der späten Republik und frühen Kaiserzeit. Stuttgart, Steiner, 2005, 266 p. (Geographica historica, 20).

2028. MILNOR (Kristina). Gender, domesticity and the age of Augustus. Inventing private life. Oxford, Oxford U. P., 2005, X-360 p.

2029. MINAUD (Gérard). La comptabilité à Rome. Essai d'histoire économique sur la pensée comptable commerciale et privée dans le monde antique romain. Lausanne, Presses polytechniques et universitaires romandes, 2005, 383 p.

2030. MOHLE (Ingvar Brandvik). Masse og elite i den romerske republikk. (Masses et élites dans la république romaine). Olso, Unipub Forl., 2005, 297 p.

2031. MOURITSEN (Henrik). Freedmen and decurions: epitaphs and social history in imperial Italy. *Journal of Roman studies*, 2005, 95, p. 38-63.

2032. NACO DEL HOYO (Toni). Vectigal incertum: guerra y fiscalidad republicana en el siglo II a. C. *Klio*, 2005, 87, p. 366-395.

2033. NDIAYE (Émilia). L'étranger "barbare" à Rome: essai d'analyse sémique. *L'antiquité classique*, 2005, 74, p. 119-135.

2034. NEWBY (Zahra). Greek athletics in the Roman world: victory and virtue. Oxford a. New York, Oxford U. P., 2005, XIV-314 p. (ill., maps, plans). (Oxford studies in ancient culture and representation).

2035. PERETZ (Daniel). Military burial and the identification of the Roman fallen soldiers. *Klio*, 2005, 87, p. 123-138.

2036. PILLON (Michel). Les Daces, Trajan et les origines du peuple roumain: aspects et étapes d'une controverse européenne. *Anabases*, 2005, 1, p. 75-104.

2037. Prosopographia Imperii Romani (PIR). Elektronische Ressource. ein 'Who is Who' des römischen Kaiserreichs. Berlin, Berlin-Brandenburgische Akademie der Wissenschaften, 2005, 1 CD-ROM.

2038. RAEPSAET-CHARLIER (Marie-Thérèse). Les activités publiques des femmes sénatoriales et équestres sous le Haut-Empire romain. *In*: Senatores populi Romani [Cf. n° 1943], 169-212.

2039. RICCI (Cecilia). Orbis in urbe: fenomeni migratori nella Roma imperiale. Roma, Quasar, 2005, 105 p. (ill.). (Vita e costumi nel mondo romano antico, 26).

2040. Roman family (The) in the empire. Rome, Italy and beyond. Ed. by Michele GEORGE. Oxford, Oxford U. P., 2005, XX-358 p. [Cf. nos <choice> 1998, 2014, 2054.]

2041. Roman working lives and urban living. Ed. by Ardle MAC MAHON. Oxford, Oxbow, 2005, VIII-224 p.

2042. ROSILLO LÓPEZ (Cristina). La corruption à la fin de la république romaine (IIe–Ier s. av. J.-C.). Aspects politiques et financiers. Neuchâtel, Univ. diss., 2005, 381 p.

2043. ROTH (Ulrike). Food, status, and the peculium of agricultural slaves. *Journal of Roman archaeology*, 2005, 18, p. 278-292.

2044. ROTHENHÖFER (Peter). Die Wirtschaftsstrukturen im südlichen Niedergermanien: Untersuchungen zur Entwicklung eines Wirtschaftsraumes an der Peripherie des Imperium Romanum. Rahden, Leidorf, 2005, 320 p. (ill.). (Kölner Studien zur Archäologie der römischen Provinzen, 7).

2045. SCHEIDEL (Walter). Human mobility in Roman Italy, II. The slave population. *Journal of Roman studies*, 2005, 95, p. 64-79.

2046. SILVESTRINI (Marina). Le città della Puglia romana. Un profilo sociale. Bari, Edipuglia, 2005, 253 p. (Scavi e ricerche, 15).

2047. Social struggles in archaic Rome. New perspectives on the conflict of the orders. Ed. by Kurt A. RAAFLAUB. Malden, Oxford a. Carlton, Blackwell, 2005, XXIX-418 p.

2048. STEIN-HÖLKESKAMP (Elke). Das römische Gastmahl. Eine Kulturgeschichte. München, Beck, 2005, 364 p.

2049. SZAIVERT (Wolfgang), WOLTERS (Reinhard). Löhne, Preise, Werte. Quellen zur römischen Geldwirtschaft. Darmstadt, Wissenschaftliche Buchgesellschaft, 2005, X-376 p.

2050. TEYSSIER (Éric), LOPEZ (Brice). Gladiateurs. Des sources de l'expérimentation. Paris, Errance, 2005, 154 p.

2051. UDOH (Fabian E.). To Caesar what is Caesar's. Tribute, taxes and imperial administration in early Roman Palestine (63 B. C. E.–70 C. E.). Providence, Brown Judaic studies, 2005, XIII-350 p.

2052. VALLEJO RUIZ (José María). Antroponimia indígena de la Lusitania romana. Vitoria Gasteiz, UPV, Servicio ed., 2005, 788 p.

2053. WATSON (Patricia Anne). Non tristis torus et tamen pudicus. The sexuality of the matrona in Martial. *Mnemosyne*, 2005, 58, p. 62-87.

2054. WOOLF (Greg D.). Family history in the Roman North-West. *In*: Roman family (The) in the empire [Cf. n° 2040], p. 231-254.

Cf. nos 136, 147, 1943

§ 7. History of literature, philosophy and science.

a. Literature

2055. ARENA (Antonella). Tibullo II, 5: la celebrazione di Messalino? *Latomus*, 2005, 64, p. 362-376.

2056. BALDINI MOSCADI (Loretta). Magica musa. La magia dei poeti latini; figure e funzioni. Bologna, Pàtron, 2005, 305 p. (Testi e manuali per l'insegnamento universitario del latino, 87).

2057. BROWN (Sarah Annes). Ovid. Myth and metamorphosis. London, Bristol classical press a. Duckworth, 2005, 159 p.

2058. BURGESS (Richard W.). A common source for Jerome, Eutropius, Festus, Ammianus and the Epitome de Caesaribus between 358 and 378, along with further thoughts on the date and nature of the Kaisergeschichte. *Classical philology*, 2005, 100, p. 166-192.

2059. Cambridge companion (The) to Roman satire. Ed. by Kirk FREUDENBURG. Cambridge, Cambridge U. P., 2005, 352 p. (Cambridge companions to literature, XVI).

2060. CARLUCCI (Nadia). Presenza delle Bucoliche nel XII libro dell'Eneide. Lexis, 2005, 23, p. 255-269.

2061. CASAMENTO (Alfredo). La parola e la guerra. Rappresentazioni letterarie del Bellum civile in Lucano. Bologna, Pàtron, 2005, 245 p.

2062. CHAMPLIN (Edward). Phaedrus the fabulous. *Journal of Roman studies*, 2005, 95, p. 97-123.

2063. CIRILLO (Olga). Sulla interlocuzione della puella nella poesia elegiaca. Napoli, Loffredo, 2005, 155 p.

2064. Companion (A) to Latin literature. Ed. by Stephen J. HARRISON. Oxford a. Malden, Blackwell publishing Ltd, 2005, 450 p. (Blackwell companions to the ancient world. Ancient history).

2065. COPPOLINO (Nina Carmel). The death of Lausus: Lucretian intertext as propaganda foil in Aeneid 10, 801-832. *New England Classical Journal*, 2005, 32, p. 5-18.

2066. CORDIER (Pierre). Rome n'est plus dans Rome, ou pourquoi l'histoire en grec? L'exemple de Dion Cassius. *Métis*, 2005, 3, p. 337-348.

2067. COUTELLE (Éric). Poétique et métapoésie chez Properce. De l'Ars amandi à l'Ars scribendi. Louvain, Paris et Dudley, Peeters, 2005, 668 p.

2068. DEE (James H.). Repertorium Vergilianae poesis hexametricum. A repertory of the hexameter patterns in Vergil, Bucolica, Georgica, Aeneis. Hildesheim, Olms-Weidmann, 2005, XVIII-458 p.

2069. Demonstrare: Voir et faire voir: forme de la démonstration à Rome. Actes du colloque international de Toulouse, 18–20 novembre 2004. Réunis par Mireille ARMISEN-MARCHETTI. Toulouse, Presses universitaires du Mirail, 2005, 445 p. (ill.). (Pallas: revue d'études antiques, 69).

2070. DESY (Ph.). Les vraies et les fausses angoisses du choeur dans la Médée de Sénèque: une nouvelle interprétation. *Latomus*, 2005, 64, p. 926-944.

2071. DRÄGER (Paul). Incerti auctoris Historia de ortu iuventute Constantini Magni eiusque matre Helena. Historie über Herkunft und Jugend Constantins des Großen und seiner Mutter Helena. Von einem unbekannten Verfasser. Ein Beitrag zum Constantin-Jahr 2006. Trier, Kliomedia, 2005, 238 p.

2072. DUGAN (John). Making a new man. Ciceronian self-fashioning in the rhetorical works. New York, Oxford a. Auckland, Oxford U. P., 2005, X-388 p.

2073. Eloquenza e astuzie della persuasione in Cicerone. Atti del V Symposium Ciceronianum Arpinas (Arpino 7 maggio 2004). A cura di Emanuele NARDUCCI. Firenze, Le Monnier, 2005, XI-72 p.

2074. ESTÈVES (Aline). Color épique et color tragique dans la Thébaïde de Stace: récits de nefas et stratégies narratives (VIII, 751-765 et XI, 524-579). *Latomus*, 2005, 64, p. 96-120.

2075. Ethopoiia. La représentation de caractères entre fiction scolaire et réalité vivante à l' époque impériale et tardive. A cura di Eugenio AMATO. Salerno, Helios, 2005, 231 p.

2076. FRASCHETTI (Augusto). Poesia anonima latina. Roma, 'L'Erma'di Bretschneider, 2005, XX-366 p. (Saggi di storia antica, 23).

2077. FRINGS (Irene). Das Spiel mit eigenen Texten. Wiederholung und Selbstzitat bei Ovid. München, Beck, 200, 302 p. (Zetemata, 124).

2078. FULKERSON (Laurel). The Ovidian heroine as author. Reading, writing, and community in the Heroides. Cambridge, Cambridge U. P., 2005, XI-187 p.

2079. GÄRTNER (Thomas). Zur Bedeutung der mythologischen Erzählung über Daedalus und Icarus am Anfang des Zweiten Buchs von Ovids Ars amatoria. *Latomus*, 2005, 64, p. 649-660.

2080. GENTILI (Bruno), CERRI (Giovanni). La letteratura di Roma arcaica e l'ellenismo. Torino, Aragno, 2005, XII-352 p.

2081. GOLDBERG (Sander Michael). Constructing literature in the Roman republic. Poetry and its reception. Cambridge, Cambridge U. P., 2005, XI-249 p.

2082. GUILLAUMIN (Jean-Yves). Les arpenteurs romains. Tome I. Hygin le gromatique. Frontin. Teste établie et traduit. Paris, Les belles lettres, 2005, 263 p.

2083. HILLS (Philip). Horace. London, Bristol classical press a. Duckworth, 2005, 160 p.

2084. HOLFORD-STREVENS (Leofranc). Aulus Gellius. An Antonine scholar and his achievement. Oxford, Oxford U. P., 2005, XXIII-440 p.

2085. JOHNSON (Timothy Scott). A symposion of praise. Horace returns to lyric in Odes IV. Madison, Wisconsin U. P., 2005, XXI-320 p.

2086. JONES (Prudence J.). Reading rivers in Roman literature and culture. Lanham, Lexington books, 2005, XIV-123 p.

2087. KARAKASIS (Evangelos). Terence and the language of Roman comedy. Cambridge, Cambridge U. P., 2005, 309 p. (Cambridge classical studies, XIV).

2088. KÖNIG (Jason). Athletics and literature in the Roman empire. Cambridge, Cambridge U. P., 2005, XIX-398 p.

2089. LEITNER (Philipp). Nasonis Relegatio. Zu den Hintergründen der Verbannung Ovids. *Zeitschrift der Savigny-Stiftung für Rechtsgeschichte Romanistische Abteilung*, 2005, 122, p. 150-165.

2090. LÜTKEMEYER (Sabine). Ovids Exildichtung im Spannungsfeld von Ekloge und Elegie. Eine poetologische Deutung der Tristia und Epistulae ex Ponto. Frankfurt am Main, Lang, 2005, 177 p. (Studien zur klassischen Philologie, 150).

2091. LYNE (R. O. A. M.). Structure and allusion in Horace's book of Epodes. *Journal of Roman studies*, 2005, 95, p. 1-19.

2092. MARTINDALE (Charles Anthony). Latin poetry and the judgment of taste. An essay in aesthetics. Oxford, Oxford U. P., 2005, 265 p.

2093. MAURACH (Gregor). Die charmanten Spiele des Catull. *Gymnasium*, 2005, 112, p. 211-228.

2094. MAURACH (Gregor). Seneca. Leben und Werk. Darmstadt, Wiss. Buchgesellschaft, 2005, X-236 p.

2095. Modelli letterari e ideologia nell'età flavia. Atti della III giornata ghisleriana di filologia classica (Pavia, 30–31 ottobre 2003). A cura di Fabio GASTI. Como, Ibis, 2005, 203 p.

2096. MURGATROYD (Paul). Mythical and legendary narrative in Ovid's Fasti. Leiden a. Boston, Brill, 2005, XIII-299 p.

2097. NAPPA (Christopher). Reading after Actium. Vergil's Georgics, Octavian, and Rome. Ann Arbor, Michigan U. P., 2005, XII-293 p.

2098. NARDUCCI (Emanuele). Introduzione a Cicerone. Roma e Bari, Laterza, 2005, VII-243 p. (Universale Laterza, 851).

2099. NEWMAN (John Kevin), NEWMAN (Frances Stickney). Troy's children. Lost generations in Virgil's Aeneid. Hildesheim, Olms, 2005, XII-387 p. (Spudasmata, 101).

2100. O'HARA (James J.). Trying not to cheat: responses to inconsistencies in Roman epic. *Transactions and proceedings of the American Philological Association*, 2005, 135, p. 15-33.

2101. PAPAIOANNOU (Sophia). Epic succession and dissension. Ovid, Metamorphoses 13. 623 – 14. 582, and the reinvention of the Aeneid. Berlin a. New York, De Gruyter, 2005, XII-218 p. (Untersuchungen zur antiken Literatur und Geschichte, 73).

2102. PÂRVULESCU (Adrian). The golden bough, Aeneas' piety, and the suppliant branch. *Latomus*, 2005, 64, p. 882-909.

2103. PASCHOUD (François). Biographie und Panegyricus: Wie spricht man vom lebenden Kaiser? *In*: Biographie und Prosopographie [Cf. n° 1372], p. 103-118.

2104. PEROTTI (Pier Angelo). L'eroismo 'privato' di Eurialo e Niso. *Latomus*, 2005, 64, p. 56-69.

2105. Représentation (La) du temps dans la poésie augustéenne. Zur Poetik der Zeit in augusteischer Dichtung. Hrsg. v. Jürgen Paul SCHWINDT. Heidelberg, Universitätsverlag Winter, 2005, X-230 p. p.

2106. RIVOLTELLA (Massimo). Le forme del morire. La gestualità nelle scene di morte dell'Eneide. Milano, Vita e Pensiero, 2005, 129 p.

2107. SALEMME (Carmelo). Marziale e la poesia delle cose. Napoli, Loffredo, 2005, 111 p. (Studi latini, 58).

2108. SEITA (Mario). La vita è sogno? Lettura della Rudens di Plauto. Alessandria, Edizioni dell'Orso, 2005, VI-172 p.

2109. SPENCER (Diana). Lucan's follies: memory and ruin in a civil-war landscape. *Greece and Rome*, 2005, 52, p. 46-69.

2110. STÄRK (Ekkehard). Kleine Schriften zur römischen Literatur. Hrsg. v. Ursula GÄRTNER, Eckard LEFÈVRE und Kurt SIER. Tübingen, Gunter Narr Verlag, 2005, 326 p. (Leipziger Studien zur Klassischen Philologie, 2).

2111. STEEL (Catherine E. W.). Reading Cicero. London, Duckworth, 2005, 176 p.

2112. Studies in Latin literature and Roman history. XII. Ed. by Carl DEROUX. Bruxelles, Latomus, 2005, 496 p.

2113. Studies of Roman literature. Ed. by Jerzy STYKA. Kraków, Kiegarnia Akad., 2005, 167 p. (Classica Cracoviensia, 9).

2114. SYED (Yasmin). Vergil's Aeneid and the Roman self. Subject and nation in literary discourse. Ann Arbor, Michigan U. P., 2005, 277 p.

2115. TAISNE (Anne-Maria). L'art de Stace au chant I de la Thébaïde. *Latomus*, 2005, 64, p. 661-677.

2116. YPSILANTI (Maria). Literary loves as cycles: from Meleager to Ovid. *L'antiquité classique*, 2005, 74, p. 83-110.

Cf. n°s 346, 1795-1851

b. Philosophy and sciences

2117. ANDRÉ (Jean-Marie). Du serment hippocratique à la déontologie de la médecine romaine. *Revue des études latines*, 2005, 83, p. 140-153.

2118. BENFERHAT (Yasmina). Cives Epicurei. Les épicuriens et l'idée de monarchie à Rome et en Italie de Sylla à Octave. Bruxelles, Éditions Latomus, 2005, 369 p. (Collection Latomus, 292).

2119. DUTSCH (Dorota). Roman pharmacology: Plautus' blanda venena. *Greece and Rome*, 2005, 52, p. 205-220.

2120. FLACH (Dieter). Cato. Über den Ackerbau. Stuttgart, Steiner, 2005, 204 p.

2121. INWOOD (Brad). Reading Seneca. Stoic philosophy at Rome. Oxford, Clarendon press, 2005, XVI-376 p.

2122. KASULKE (Christoph Tobias). Fronto, Marc Aurel und kein Konflikt zwischen Rhetorik und Philosophie im 2. Jh. n. Chr. München u. Leipzig, Saur, 2005, 456 p. (Beiträge zur Altertumskunde, 218).

2123. LA ROCCA (Adolfo). Il filosofo e la città. Commento storico ai Florida di Apuleio. Roma, 'L'Erma' di Bretschneider, 2005, 301 p. (Saggi di storia antica, 24).

2124. PIAZZI (Lisa). Lucrezio e i presocratici. Un commento a De rerum natura 1, 635-920. Pisa, Edizioni della Normale, 2005, 322 p. (Testi e commenti, 1).

2125. REYDAMS-SCHILS (Gretchen J.). The Roman stoics. Self, responsibility, and affection. Chicago, Chicago U. P., 2005, X-210 p.

2126. SPAHLINGER (Lothar). Tulliana simplicitas. Zu Form und Funktion des Zitats in den philosophischen Dialogen Ciceros. Göttingen, Vandenhoeck & Ruprecht, 2005, 360 p. (Hypomnemata, 159).

Cf. n°s 1795-1851

§ 8. Religion and mythology.

2127. AMATO (Eugenio). Ovidio e l'aurea aetas: continuità di miti, continuazione di storie (a proposito di Met. XV, 104). *Latomus*, 2005, 64, p. 910-918.

2128. BREMMER (Jan N.). Attis: a Greek god in Anatolian Pessinous and Catullan Rome. *In*: Catullus' poem on Attis [Cf. n° 1803], p. 26-64.

2129. CIBU (Simina). Mercure dans les provinces romaines des Alpes occidentales. *Mélanges de l'école française de Rome. Antiquité*, 2005, 117, p. 747-776.

2130. DE CAZANOVE (Olivier). Mont et citadelle, temple et templum. Quelques réflexions sur l'usage religieux des hauteurs dans l'Italie républicaine. *Archiv für Religionsgeschichte*, 2005, 7, p. 183-197.

2131. Depositi votivi e culti dell'Italia antica dall'età arcaica a quella tardo-repubblicana. Atti del Convegno di Studi (Perugia, 1–4 giugno 2000). A cura di Annamaria COMELLA. Bari, Edipuglia, 2005, 750 p. (Bibliotheca archaeologica, 16).

2132. Fasti sacerdotum. Die Mitglieder der Priesterschaften und das sakrale Funktionspersonal römischer, griechischer, orientalischer und jüdisch-christlicher Kulte in der Stadt Rom um 300 v. Chr. bis 499 n. Chr. T. 1. Jahres- und Kollegienlisten. Von Jörg RÜPKE. T. 2. Biographien. Von Jörg RÜPKE und Anne GLOCK; unter Mitarbeit von Christa FRATEANTONIO [et al.]. T. 3. Beiträge zur Quellenkunde und Organisationsgeschichte. Bibliographie. Register. Von Jörg RÜPKE. Stuttgart, Steiner, 2005, 3 vol., 1860 p. (Potsdamer Altertumswissenschaftl. Beitr. 12/1-3).

2133. FISHWICK (Duncan). The imperial cult in the Latin west. Studies in the ruler cult of western provinces of the Roman empire. Vol. III. Provincial cult. Part 1. Institution and evolution. Part 2. The provincial priesthood. Part 3. The provincial centre; Provincial cult. Part 4. Bibliography, indices, addenda. Leiden a. Boston, Brill, 2005, 4 vol., XVI-259, XVI-324, XXIII-397, VII-256 p.

2134. FORMICOLA (Crescenzo). L'Eneide di Giunone. Una divinità in progress. Napoli, Loffredo, 2005, 167 p.

2135. FRATANTUONO (Lee). Diana in the Aeneid. *New England classical journal*, 2005, 32, p. 101-115.

2136. GUALERZI (Saverio). Né uomo, né donna, né dio, né dea. Ruolo sessuale e ruolo religioso dell'imperatore Elagabalo. Bologna, Pàtron, 2005, 115 p.

2137. HAUDRY (Jean). Le préhistoire de Janus. *Revue des études latines*, 2005, 83, p. 36-54.

2138. HEMELRIJK (Emily A.). Priestesses of the imperial cult in the Latin west: titles and function. *L'antiquité classique*, 2005, 74, p. 137-170.

2139. HOFENEDER (Andreas). Die Religion der Kelten in den antiken literarischen Zeugnissen. Band I: Von den Anfängen bis Caesar. Wien, Verlag der Österreichischen Akademie der Wissenschaften, 2005, 349 p.

2140. Keltische Götter im Römischen Reich. Akten des 4. Internationalen Workshops „Fontes Epigraphici Religionis Celticae Antiquae" (F.E.R.C.A.N.) vom 4.–6. 10. 2002 an der Universität Osnabrück. Hrsg. v. Wolfgang SPICKERMANN. Möhnesee, Bibliopolis, 2005, XI-188 p.

2141. MASTROCINQUE (Attilio). Cosmologia e impero in Giuliano l'Apostata. *Klio*, 2005, 87, p. 154-176.

2142. MONACA (Mariangela). La Sibilla a Roma. I Libri sibillini fra religione e politica. Cosenza, Giordano, 2005, 324 p. (Hierá, 8).

2143. NEMETI (Sorin). Sincretismul religios în Dacia Romană. (Religious syncretism in Roman Dacia). Cluj-Napoca, Editura Presa Universitară Clujeană, 2005, 423 p.

2144. RIDLEY (Ronald T.). The absent Pontifex Maximus. *Historia*, 2005, 54, p. 275-300.

2145. SANTANGELO (Federico). The religious tradition of the Gracchi. *Archiv für Religionsgeschichte*, 2005, 7, p. 198-214.

2146. SCHEID (John). Quand faire, c'est croire. Les rites sacrificiels des Romains. Paris, Aubier, 2005, 348 p. (Collection historique). – IDEM. Un elément original de l'identité romaine: les cultes selon le rite grec. *Métis*, 2005, 3, p. 25-34.

2147. SCHMID (Alfred). Augustus und die Macht der Sterne. Antike Astrologie und die Etablierung der Monarchie in Rom. Köln, Weimar u. Wien, Böhlau, 2005, IX-469 p.

2148. VAN HAEPEREN (Françoise). Représentations chrétiennes du pontificat païen. *Latomus*, 2005, 64, p. 678-703.

2149. VERHULST (Gilliane). Répertoire mythologique dans les métamorphoses d'Ovide. Paris, Ellipses, 2005, 125 p.

2150. WELLINGTON (I. J.). Gifts to the Gods? Votive deposition in north-eastern France from 250 BC to the age of Augustus, a numismatic perspective. Durham, Ph. Diss., 2005, [s. p.].

Cf. n°s 868, 872, 1713, 1723

§ 9. Archaeology and history of art.

2151. Arredi di lusso di età romana. Da Roma alla Cisalpina. A cura di Fabrizio SLAVAZZI. Borgo S. Lorenzo, All'insegna del giglio, 2005, 209 p.

2152. BARTMAN (Elizabeth). The mock face of battle. *Journal of Roman archaeology*, 2005, 18, p. 99-119.

2153. BECKMANN (Martin). The border of the frieze of the Column of Marcus Aurelius and its implications. *Journal of Roman archaeology*, 2005, 18, p. 302-312.

2154. BOURGEOIS (Ariane). Céramiques romaines en Gaule (productions – exportations – importations) (années 2003–2004). *Revue des études anciennes*, 2005, 107, p. 131-148.

2155. CHIŞ (Silvius Ovidiu). Monumentele funerare cu imaginea Sfinxului din Dacia romană. (Funerary monuments imaging the Sphinx from Roman Dacia). *Revista Bistriţei. Complexul Muzeal Bistriţa-Năsăud*, 2005, 19, p. 87-94.

2156. COUHADE-BEYNEIX (Cynthia). L'évocation de la trahison dans l'iconographie monétaire à la fin de la république romaine et au début du règne d'Auguste. *Revue des études anciennes*, 2005, 107, p. 675-696.

2157. CROISILLE (Jean-Michel). La peinture romaine. Paris, Picard, 2005, 375 p.

2158. DAGUET-GAGEY (Anne). L'arc des argentiers à Rome. *Revue historique*, 2005, 307, p. 499-520.

2159. DARK (Ken). The archaeological implications of fourth- and fifth-century descriptions of villas in the northwest provinces of the Roman empire. *Historia*, 2005, 54, p. 331-342.

2160. DE LUCIA BROLLI (Maria Anna), MICHETTI (Laura Maria). La ceramica a rilievo di produzione falisca. *Mélanges de l'école française de Rome. Antiquité*, 2005, 117, p. 137-171.

2161. DICKMANN (Jens-Arne). Pompeji. Archäologie und Geschichte. München, Beck, 2005, 128 p.

2162. ECKARDT (Hella). The social distribution of Roman artefacts: the case of nailcleaners and brooches in Britain. *Journal of Roman archaeology*, 2005, 18, p. 139-160.

2163. ELSNER (Jas). Sacrifice and narrative on the arch of the Argentarii at Rome. *Journal of Roman archaeology*, 2005, 18, p. 83-98.

2164. FÄHNDRICH (Sabine). Bogenmonumente in der römischen Kunst. Ausstattung, Funktion und Bedeutung antiker Bogen- und Torbauten. Rahden, Leidorf, 2005, XII-262 p.

2165. GEYER (Angelika). Moneta Augusti. Römische Münzen der Kaiserzeit und Spätantike im Akademischen Münzkabinett der Friedrich-Schiller-Universität Jena; die Sammlung Schmidt der Stiftung Weimarer Klassik und Kunstsammlungen und Eigenbestände; Katalog zur Ausstellung im Stadtmuseum Göhre in Jena, vom 10. Februar bis 1. Mai 2005. Jena, Glaux-Verlag, 2005, 358 p. (Jenaer Hefte zur klassischen Archäologie, 6).

2166. GIAVARINI (Carlo). La Basilica di Massenzio. Il monumento, i materiali, le strutture, la stabilità. Roma, 'L'Erma' di Bretschneider, 2005, 263 p.

2167. GREENE (Kevin T.). Roman pottery: models, proxies and economic interpretation. *Journal of Roman archaeology*, 2005, 18, p. 34-56.

2168. HALLET (Christopher H.). The Roman nude. Heroic portrait statuary 200 BC–AD 300. Oxford, Oxford U. P., 2005, XXI-391 p.

2169. HESBERG (Henner von). Römische Baukunst. München, Beck, 2005, 294 p.

2170. HØJTE (Jakob Munk). Roman imperial statue bases from Augustus to Commodus. Aarhus, Aarhus U. P., 2005, 658 p.

2171. HÖPKEN (Constanze). Die römische Keramikproduktion in Köln. Mainz am Rhein, von Zabern, 2005, 181 p.

2172. HOPKINS (Keith), BEARD (Mary). The Colosseum. Cambridge, Harvard U. P., 2005, X-214 p. (Wonders of the world).

2173. Interpretare i bolli laterizi di Roma e della valle del Tevere: produzione, storia economica e topografia. Atti del convegno all'École française de Rome e all'Institutum Romanum Finlandiae (31 marzo e 1 aprile 2000). A cura di Christer BRUUN e François CHAUSSON. Roma, Institutum Romanum Finlandiae, 2005, X-323 p. (Acta Instituti Romani Finlandiae. 32.).

2174. JONES (Rick), ROBINSON (Damian). Water, wealth, and social status at Pompeii: the house of the Vestals in the first century A. D. *American journal of archaeology*, 2005, 109, p. 695-710.

2175. LANCASTER (Lynne C.). The process of building the Colosseum: the site, materials and construction techniques. *Journal of Roman archaeology*, 2005, 18, p. 57-82.

2176. LETZNER (Wolfram). Das römische Pula. Bilder einer Stadt in Istrien. Mainz, von Zabern, 2005, 107 p.

2177. LING (Roger), LING (Lesley). The insula of the Menander at Pompei. Vol. II. The decorations. Oxford, Oxford U. P., 2005, 541 p.

2178. MATTUSCH (Carol C.), LIE (Henry). The villa dei papiri at Herculaneum. Life and afterlife of a sculpture collection. Los Angeles, J. Paul Getty Museum, 2005, XXIII-390 p.

2179. Nuove ricerche archeologiche a Pompei ed Ercolano. Atti del Convegno Internazionale, Roma 28–30 novembre 2002. A cura di Pietro Giovanni GUZZO e Maria Paola GUIDOBALDI. Napoli, Electa, 395 p. (Studi della Soprintendenza archeologica di Pompei, 10).

2180. OLOVSDOTTER (Cecilia). The consular image. An iconological study of the consular diptychs. Oxford, Archaeopress, 2005, IV-238 p.

2181. Pompei. Rinvenimenti monetali nella Regio IX. A cura di Marina TALIERCIO MENSITIERI. Roma, Istituto Italiano di Numismatica, 2005, 345 p.

2182. Portus. An archaeological survey of the port of imperial Rome. Ed. by Simon J. KEAY. London, British School at Rome, 2005, XVIII-360 p.

2183. REIS PILAR (Maria). Las termas y balnea romanos de Lusitania. Mérida, Libreria Museo Nacional de Arte Romano, José Ramón Mélida, 2005, 205 p.

2184. SCHOLLMEYER (Patrick). Römische Plastik. Eine Einführung. Darmstadt, Wissenschaftliche Buchgesellschaft, 2005, 160 p.

2185. SCHWEIZER (Jürg). Baukörper und Raum in tetrarchischer und konstantinischer Zeit. Der Außenaspekt der weströmischen Architektur im 4. Jahrhundert. Bern, Lang, 2005, 276 p.

2186. SIMPSON (Christopher J.). Where is the Parthian? The Prima Porta statue of Augustus revisited. *Latomus*, 2005, 64, p. 82-90.

2187. Théorie et pratique d'architecture romaine. Études offertes à Pierre Gros. Éd. par Xavier LAFON et Gilles SAURON. Aix-en-Provence, Publications de l'Université de Provence, 2005, 342 p.

2188. TIMOC (Călin). Templul palmyrienilor din castrul de la Tibiscum – Jupa. (Palmyrian temple in the Roman castrum at Tibiscum – Jupa). *Patrimonium Banaticum. Anuar de Arheologie, Istorie, Istoria Artei, Istoria Culturii, Arhitectură – Timişoara*, 2005, 4, p. 115-122.

2189. TUCCI (Pier Luigi). 'Where high Moneta leads her steps sublime': the 'Tabularium' and the temple of Juno Moneta. *Journal of Roman archaeology*, 2005, 18, p. 6-33.

2190. VOUT (Caroline). Antinous, archaeology and history. *Journal of Roman studies*, 2005, 95, p. 80-96.

2191. WEBSTER (Jane). Archaeologies of slavery and servitude: bringing 'New World' perspectives to Roman Britain. *Journal of Roman archaeology*, 2005, 18, p. 161-179.

2192. WILKES (J. J.). The Roman Danube: an archaeological survey. *Journal of Roman studies*, 2005, 95, p. 124-225.

2193. WITCHER (Rob). The extended metropolis: urbs, suburbium and population. *Journal of Roman archaeology*, 2005, 18, p. 120-138.

Cf. n°s 138, 147

§ 10. Late antiquity.
Transformation of the Roman world.

2194. Cambridge companion (The) to the age of Constantine. Ed. by Noel LENSKI. Cambridge, Cambridge U. P., 2005, XVIII-469 p.

2195. DI PAOLA (Lucietta). Per la storia degli 'occhi del re'. I servizi ispettivi nella tarda antichità. Messina, Dipartimento di scienze dell'antichità dell'Università degli Studi di Messina, 2005, 159 p.

2196. HEATHER (Peter J.). The fall of the Roman empire. A new history of Rome and the barbarians. London, Macmillan, 2005, XVI-572 p.

2197. Hispania in late antiquity. Current perspectives. Edited and translated by Kim BOWES and Michael KULIKOWSKI. Boston, Brill, 2005, XII-645 p. (Medieval and Early Modern Iberian world, 24).

2198. KAKRIDI (Christina). Cassiodors Variae: Literatur und Politik im ostgotischen Italien. München, Saur, 2005, 419 p. (Beiträge zur Altertumskunde, 223).

2199. KORELIN (Mikhail S.). Padenie antichnogo mirosozertsanija. Kul'turnyj krizis v Rimskoj imperii. (The fall the antique notion of the world: The cultural crisis in the Roman Empire). Sankt-Peterburg, Kolo, 2005, 192 p. (ill.; portr.; bibl. p. 188-190; pers. ind. p. 181-187). (Aleksandrijskaja biblioteka. Antichnost').

2200. LOTZ (Almuth). Der Magienkonflikt in der Spätantike. Bonn, Dr. Rudolf Habelt GmbH, 2005, II-284 p.

2201. LUHTALA (Anneli). Grammar and philosophy in late antiquity. A study of Priscian's sources. Amsterdam, Benjamins, 2005, X-171 p. (Amsterdam studies in the theory and history of linguistic science, ser. III: Studies in the history of linguistics, 107).

2202. MAIER (Gideon). Amtsträger und Herrscher in der Romania Gothica. Vergleichende Untersuchungen zu den Institutionen der ostgermanischen Völkerwanderungsreiche. Stuttgart, Steiner, 2005, 363 p. (Historia, 181).

2203. MAZZA (Mario). Cultura, guerra e diplomazia nella tarda antichità. Tre studi. Catania, Prisma, 2005, VII-251 p.

2204. NOBLE (Thomas F. X.). From Roman provinces to Medieval kingdoms. London, Routledge, 2005, XXV-402 p. (maps). (Rewriting histories).

2205. POHL (Walter). Aux origines d'une Europe ethnique. Transformations d'identités entre Antiquité et Moyen Âge. *Annales*, 2005, 60, 1, p. 183-208.

2206. RICHARDOT (Philippe). La fin de l'armée romaine (284-476). Paris, Économica, 2005, XII-408 p.

2207. SAVINO (Eliodoro). Campania tardoantica (284-604 d.C.). Bari, Edipuglia, 2005, 398 p. (Munera, 20).

2208. STROUMSA (Guy G.). La fin du sacrifice. Les mutations religieuses de l'Antiquité tardive. Paris, Jacob, 2005, 213 p.

2209. WARD-PERKINS (Bryan). The fall of Rome and the end of civilization. Oxford, Oxford U. P., 2005, VI-239 p.

Cf. n°s 2210-2382

G

EARLY HISTORY OF THE CHURCH TO GREGORY THE GREAT

§ 1. Sources. 2210-2272. – § 2. General. 2273-2281. – § 3. Special studies. 2282-2376. – § 4. Hagiography. 2377-2382.

§ 1. Sources.

* 2210. LÉVY (Antoine). Bulletin d'histoire et de théologie comparées. Occident latin et Orient byzantin. *Revue des sciences philosophiques et théologiques*, 2005, 89, p. 337-365.

2211. [Ambrosius] RASCHLE (Christian R.). Ambrosius' Predigt gegen Magnus Maximus. Eine historische Interpretation der explanatio in psalmum 61 (62). *Historia*, 2005, 54, p. 49-67.

2212. [Apocrypha] KLAUCK (Hans-Josef). Apokryphe Apostelakten. Eine Einführung. Stuttgart, Verl. Kath. Bibelwerk, 2005, 291 p.

2213. [Athanasius Alexandrinus] SCHNEIDER (Carolyn). The intimate connection between Christ and Christians in Athanasius. *Scottish Journal of Theology*, 2005, 58, p. 1-12.

2214. [Augustinus] GORI (Franco). Augustinus. Enarrationes in Psalmos 141-150. Wien, Verlag der Österreichischen Akademie der Wissenschaften, 2005, 304 p.

2215. [Basilius Caesariensis] SILVAS (Anna M.). The Asketikon of St Basil the Great. Oxford, Oxford U. P., 2005, 517 p.

2216. [Biblia] Biblische Aufklärung – die Entdeckung einer Tradition. Sechstes Sankt Georgener Symposion getragen von der Stiftung Hochschule Sankt Georgen in Verbindung mit dem Fachbereich Katholische Theologie an der Johann Wolfgang Goethe-Universität in Frankfurt am Main10. / 11. Oktober 2003. Hrsg. v. Martin FRÜHAUF und Werner LÖSER. Frankfurt am Main, Knecht, 2005, 128 p.

2217. [Biblia] BORMANN (Lukas). Bibelkunde. Altes und Neues Testament. Göttingen, Vandenhoeck & Ruprecht, 2005, 293 p.

2218. [Biblia] Forme e modelli della tradizione manoscritta della Bibbia. A cura di Paolo CHERUBINI. Città del Vaticano, Scuola Vaticana di paleografia, 2005, XV-562 p.

2219. [Biblia] Ideales Königtum. Studien zu David und Salomo. Hrsg. v. Rüdiger LUX. Leipzig, Evang. Verl.-Anstalt, 2005, 177 p. (Arbeiten zur Bibel und ihrer Geschichte, 16).

2220. [Biblia] Vetus Latina. Die Reste der altlateinischen Bibel. Nach Petrus Sabatier neu gesammelt und hrsg. von der Erzabtei Beuron 4/5: Ruth. Einleitung und Text. Hrsg. v. Bonifatia GESCHE. Freiburg, Herder, 2005, 83 p.

2221. [Boethius] Boethius. De consolatione philosophiae. Opuscula theologica. A cura di Claudio MORESCHINI. München, K. G. Saur, 2005, 262 p.

2222. [Cyrillus Alexandrinus] SCHURIG (Sebastian). Theologie des Kreuzes beim frühen Cyrill von Alexandria. Dargestellt an seiner Schrift „De adoratione et cultu in spiritu et veritate". Tübingen, Mohr Siebeck, 2005, VIII-361 p.

2223. [Diodorus Tarsensis] Diodorus of Tarsus. Commentary on Psalms 1-51. Translated with an introduction and notes. Ed. by Robert C. HILL. Atlanta, Society of biblical literature, 2005, XXXVII-181 p.

2224. [Ennodius] DI RIENZO (Daniele). Gli epigrammi di Magno Felice Ennodio. Napoli, Pubblicazioni del Dipartimento di Filologia Classica "Francesco Arnaldi" dell'Università degli Studi di Napoli Federico II, 2005, 263 p.

2225. [Eugippius] DEGÓRSKI (Bazyli). La regola. Eugippio. Introduzione, traduzione e note. Roma, Città nuova, 2005, 222 p. (Collana di testi patristici, 183).

2226. [Eusebius Caesariensis] Eusebius von Caesarea. Das Onomastikon der biblischen Ortsnamen. Hrsg. v. Stefan TIMM. Berlin, De Gruyter, 2005, VIII-253 p. (Texte und Untersuchungen zur Geschichte der altchristlichen Literatur, 152).

2227. [Eusebius Caesariensis] MORGAN (Teresa). Eusebius of Caesarea and christian historiography. *Athenaeum*, 2005, 93, p. 193-208.

2228. [Eusebius Caesariensis] NOTLEY (R. Steven), SAFRAI (Zeev). Eusebius, Onomasticon. The place names of divine scripture. Boston a. Leiden, Brill, 2005, XXXVII-212 p.

2229. [Eusebius Caesariensis] ZAMAGNI (Claudio). Eusebii Caesarensis quaestionum concordantia. Textus iuxta Vat. Pal. Gr. 220. Hildesheim, Olms, 2005, 426 p.

2230. [Evagrius Ponticus] DYSINGER (Luke). Psalmody and prayer in the writings of Evagrius Ponticus. New York a. Oxford, Oxford U. P., 2005, VIII-245 p.

2231. [Gregorius Magnus] FLORYSZCZAK (Silke). Die Regula Pastoralis Gregors des Großen. Studien zu Text, kirchenpolitischer Bedeutung und Rezeption in der Karolingerzeit. Tübingen, Mohr Siebeck, 2005, X-444 p.

2232. [Gregorius Nazianzenus] Sancti Gregorii Nazianzeni opera. Versio arabica antiqua III. Oratio XL (arab. 4). Ed. by Jacques GRAND'HENRY. Turnhout, Brepols a. Leuven, Leuven U. P., 2005, XXXVII-373 p.

2233. [Gregorius Nyssenus] TURCESCU (Lucian). Gregory of Nyssa and the concept of divine persons. New York, Oxford U. P., 2005, XI-171 p.

2234. [Hilarius Pictaviensis] DOIGNON (Jean). Hilaire de Poitiers. Disciple et témoin de la vérité (356-367). Texte revu par Marc MILHAU. Paris, Institut d'études augustiniennes, 2005, 243 p.

2235. [Iohannes Antiochenus] Ioannis Antiocheni Fragmenta ex Historia Chronica. Hrsg. v. Umberto ROBERTO. Berlin, De Gruyter, 2005, CC-656 p. (Texte und Untersuchungen zur Geschichte der altchristlichen Literatur, 154).

2236. [Iohannes Chrysostomus] BARONE (Francesca Prometea). Per un'edizione critica delle omelie De Davide et Saule di Giovanni Crisostomo. *Augustinianum*, 2005, 45, p. 231-258.

2237. [Itinerarium Egeriae] SPEVAK (Olga). Itinerarium Egeriae: l'ordre des constituants obligatoires. *Mnemosyne*, 2005, 58, p. 235-261.

2238. [Iustinus] Iustini martyris apologiae pro christianis. Ed. by Miroslav MARCOVICH. Berlin a. New York, De Gruyter, 2005, XII-339 p. (Patristische Texte und Studien, 38).

2239. [Lactantius] SPINELLI (Mario). Come muoiono i persecutori. Lattanzio. Introduzione, traduzione e note. Roma, Città nuova, 2005, 149 p. (Collana di testi patristici, 180).

2240. [Marius Mercator] KONOPPA (Claudia). Die Werke des Marius Mercator. Übersetzung und Kommentierung seiner Schriften. Frankfurt am Main, Lang, 2005, 479 p.

2241. [Maximus Homologetes] CHARPIN-PLOIX (Marie-Lucie). La mystagogie. Maxime le Confesseur. Introduction, traduction, notes, glossaires et index. Paris, Migne, 2005, 201 p. (Collection "Les Pères dans la foi", 92).

2242. [Origenes] Origene. Commentario al Cantico dei cantici. Testi in lingua greca. Introduzione, testo, traduzione e commento. A cura di Maria Antonietta BARBÀRA. Bologna, Centro editoriale dehoniano, 2005, 617 p.

2243. [Patres Ecclesiae] GUY (Jean-Claude). Les apophthegmes des pères. Colléction systématique. Chapitres XVII-XXI. Paris, Ed. du Cerf, 2005, 470 p.

2244. [Paulinus Nolanus] CIENFUEGOS GARCÍA (Juan José). Poemas. Paulino de Nola. Introd., trad. y notas. Madrid, Gredos, 2005, 475 p. (Biblioteca clásica Gredos, 335).

2245. [Paulinus Nolanus] SURMANN (Beate). Licht-Blick: Paulinus Nolanus, carm. 23. Edition, Übersetzung, Kommentar. Trier, Wissenschaftlicher Verlag Trier, 2005, 426 p.

2246. [Porphyrius] BERCHMAN (Robert M.). Porphyry against the Christians. Leiden, Brill, 2005, XVI-242 p.

2247. [Prudentius] CANALI (Luca), PELLEGRINI (Maria). Le corone. Con testo a fronte. Aurelio Prudenzio Clemente. Firenze, Le lettere, 2005, 330 p.

2248. [Prudentius] GARUTI (Giovanni). Apotheosis. Prudentius. Testo critico, traduzione, commento e indici. Modena, Mucchi, 2005, 178 p. (Collana di studi Accademia Nazionale di Scienze, Lettere e Arti, Modena, 26).

2249. [Pseudo-Basilius] BURGSMÜLLER (Anne). Die Askeseschrift des Pseudo-Basilius. Untersuchungen zum Brief "Über die wahre Reinheit in der Jungsfräulichkeit". Tübingen, Mohr Siebeck, 2005, 477 p.

2250. [Sozomen] Sozomène. Histoire ecclésiastique. Livres V-VI. Éd. par André Jean FESTUGIÈRE, Bernard GRILLET et Guy SABBAH. Paris, Ed. du Cerf, 2005, 491 p.

2251. [Synesius Cyrenaicus] LUCHNER (K.). Gott' und Selbstpräsentation in den Briefen des Synesios von Kyrene. *Millennium*, 2005, 2, p. 33-62.

2252. [Tertullianus] Tertullian. De Pallio. A commentary. Ed. by Vincent HUNINK. Amsterdam, Gieben, 2005, 332 p.

2253. [Testamentum novum] BERGER (Klaus). Formen und Gattungen im Neuen Testament. Tübingen, Francke, 2005, XII-483 p.

2254. [Testamentum novum] DELGADO JARA (Inmaculada). Evidencias en el Nuevo Testamento de la evolución del attic a la koine. *Helmantica*, 2005, 56, p. 105-124.

2255. [Testamentum novum] DORMEYER (Detlev). Das Markusevangelium. Darmstadt, Wiss. Buchgesellschaft, 2005, 239 p.

2256. [Testamentum novum] FRANKEMÖLLE (Hubert). Studien zum jüdischen Kontext neutestamentlicher Theologien. Stuttgart, Verl. Kath. Bibelwerk, 2005, X-316 p. (Stuttgarter biblische Aufsatzbände, 37: Neues Testament).

2257. [Testamentum novum] KELHOFFER (James A.). The diet of John the Baptist. "Locusts and wild honey" in synoptic and patristic interpretation. Tübingen, Mohr Siebeck, 2005, XXIII-256 p. (Wissenschaftliche Untersuchungen zum Neuen Testament, 176).

2258. [Testamentum novum] Luke and his readers. Festschrift A. Denaux. Ed. by Reimund BIERINGER, Gilbert VAN BELLE and Jozef VERHEYDEN. Leuven, Leuven U. P., 2005, XXVIII-470 p. (Bibliotheca ephemeridum theologicarum Lovaniensium, 182).

2259. [Testamentum novum] MEYER (Annegret). Kommt und seht. Mystagogie im Johannesevangelium ausgehend von Joh 1, 35-51. Würzburg, Echter, 2005, VIII-395 p. (Forschung zur Bibel, 103).

2260. [Testamentum novum] MIN (Kyµong-sik). Die früheste Überlieferung des Matthäusevangeliums (bis zum 3./4. Jh.). Edition und Untersuchung. Berlin, De Gruyter, 2005, X-357 p. (Arbeiten zur Neutestamentlichen Textforschung, 34).

2261. [Testamentum novum] Picturing the New Testament. Ed. by Annette WEISSENRIEDER, Friederike WENDT and Petra von GEMÜNDEN. Tübingen, Mohr Siebeck, 2005, XVII-445 p.

2262. [Testamentum novum] Prägende (Die) Kraft der Texte. Hermeneutik und Wirkungsgeschichte des Neuen Testaments (ein Symposium zu Ehren von Ulrich Luz). Hrsg. v. Moisés MAYORDOMO. Stuttgart, Verl. Kath. Bibelwerk, 2005, 177 p. (Stuttgarter Bibelstudien, 199).

2263. [Testamentum novum] SCHENKE (Ludger). Das Markusevangelium. Literarische Eigenart – Text und Kommentierung. Stuttgart, Kohlhammer, 2005, 357 p.

2264. [Testamentum novum] Textkritik des Markusevangeliums. Hrsg. v. Heinrich GREEVEN und Eberhard GÜTING. Münster, Lit Verlag, 2005, 768 p.

2265. [Testamentum novum] THYEN (Hartwig). Das Johannesevangelium. Tübingen, Mohr (Paul Siebeck), 2005, XII-796 p. (Handbuch zum Neuen Testament, 6).

2266. [Testamentum novum] WENGST (Klaus). Der Brief an Philemon. Stuttgart, Kohlhammer, 2005, 120 p. (Theologischer Kommentar zum Neuen Testament, 16).

2267. [Testamentum vetus] Book (The) of Psalms. Composition and reception. Ed. by Peter W. FLINT. Leiden, Brill, 2005, XX-680 p. (Supplements to Vetus Testamentum, 99).

2268. [Testamentum vetus] DIM (Emmanuel Uchenna). The eschatological implications of Isa. 65 and 66 as the conclusion of the book of Isaiah. New York, Lang, 2005, XVIII-409 p. (Bible in history, 3).

2269. [Testamentum vetus] MARX (Alfred). Les systèmes sacrificiels de l'Ancien Testament. Formes et fonctions du culte sacrificiel à Yhwh. Leiden, Brill, 2005, VI-263 p.

2270. [Testamentum vetus] VAN KEULEN (Percy S. F.). Two versions of the Solomon narrative. An inquiry into the relationship between MT 1 Kgs. 2-11 and LXX 3 Reg. 2-11. Leiden, Brill, 2005, VI-338 p. (Supplements to Vetus Testamentum, 104).

2271. [Theodorus Mopsuestenus] MAC LEOD (Frederick G.). The roles of Christ's humanity in salvation: insights from Theodore of Mopsuestia. Washington, Catholic University of America press, 2005, 296 p.

2272. [Vigilius Thapsensis] GUIDI (Patrizia). Contro gli ariani. Vigilio di Tapso. Introduzione, traduzione e note. Roma, Città Nuova, 2005, 216 p. (Testi patristici, 184).

Cf. n° 433

§ 2. General.

* 2273. Historical (The) Jesus in recent research. Ed by James G. DUNN. Winona Lake, Eisenbrauns, 2005, XVI-618 p. (Sources for biblical and theological study, 10).

2274. Biblisch-historisches Handwörterbuch (BHH). Landeskunde, Geschichte, Religion, Kultur, Literatur. Hrsg. v. Bo REICKE und Leonhard ROST. Göttingen, Vandenhoeck & Ruprecht, 2005, 1 CD-ROM.

2275. CUMMINGS (Owen F.). Eucharistic doctors. A theological history. New York a. Mahwah, Paulist press, 2005, VIII-274 p.

2276. Dictionary (A) of Jewish-Christian relations. Ed. by Edward KESSLER. Cambridge, Cambridge U. P., 2005, XXIX-507 p.

2277. Formation (The) of the early church. June 14–18, 2003, Seventh nordic New Testament conference at Solborg Folkehøgskole in Stavanger, Norway. Ed. by Jostein ÅDNA. Tübingen, Mohr Siebeck, 2005, XII-451 p. (Wissenschaftliche Untersuchungen zum Neuen Testament, 183).

2278. HEINZ (Werner). Der Aufstieg des Christentums. Geschichte und Archäologie einer Weltreligion. Darmstadt, Wiss. Buchgesellschaft, 2005, 126 p.

2279. PIEPENBRINK (Karen). Christliche Identität und Assimilation in der Spätantike. Probleme des Christseins in der Reflexion der Zeitgenossen. Frankfurt am Main, Verlag Antike, 2005, 432 p. (Studien zur alten Geschichte, 3).

2280. RAPP (Claudia). Holy bishops in Late Antiquity. The nature of Christian leadership in an age of transition. Berkeley a. Los Angeles, University of California Press, 2005, XII-346 p. (Transformation of the classical heritage, 37).

2281. Spread (The) of Christianity in the first four centuries: essays in explanation. Ed. by William V.

HARRIS. Leiden, Brill, 2005, XIV-176 p. (Columbia studies in the classical tradition, 27).

Cf. n° 1723

§ 3. Special studies.

2282. ALAND (Barbara). Frühe direkte Auseinandersetzung zwischen Christen, Heiden und Häretikern. Berlin, de Gruyter, 2005, XI-48 p. (Hans-Lietzmann-Vorlesungen, 8).

2283. AMARELLI (Francesco). In tema di processo di designazione dei vescovi. L'autodesignazione costantiniana. *Studia et documenta historiae et iuris*, 2005, 71, p. 517-524.

2284. AMERISE (Marilena). Il battesimo di Costantino il Grande. Storia di una scomoda eredità. Stuttgart, Steiner, 2005, 177 p.

2285. AMHERDT (David). Le locus amoenus de Paulin de Nole: la rhétorique au service du Christianisme. *Mouseion*, 2005, 49, p. 143-158.

2286. AMICI (Angela). Cassiodoro a Costantinopoli. Da magister officiorum a religiosus vir. *Vetera christianorum*, 2005, 42, p. 215-231.

2287. ANDRIST (Patrick). Les protagonistes égyptiens du débat apollinariste: le "Dialogue d'Athanase et Zachée" et les dialogues pseudoathanasiens. Intertextualité et polémique religieuse en Égypte vers la fin du IVe siècle. *Recherches augustiniennes et patristiques*, 2005, 34, p. 63-141.

2288. Apologétique (L') chrétienne gréco-latine à l'époque prénicénienne. Sept exposés suivis de discussions. Éd. par Antoine WLOSOK. Genève, Fondation Hardt, 2005, VI-316 p.

2289. Augustine and the disciplines. From Cassiciacum to Confessions. Ed. by Karla POLLMANN and Mark VESSEY. Oxford, Oxford U. P., 2005, XI-258 p.

2290. BALDWIN (Matthew C.). Whose acts of Peter? Text and historical context of the Actus Vercellenses. Tübingen, Mohr Siebeck, 2005, XVI-339 p.

2291. BANFI (Antonio). "Habent illi iudices suos". Studi sull'esclusività della giurisdizione ecclesiastica e sulle origini del privilegium fori in diritto romano e bizantino. Milano, Giuffrè, 2005, XII-401 p.

2292. BAR (Doron). Rural monasticism as a key element in the christianization of Byzantine Palestine. *Harvard theological review*, 2005, 98, p. 49-65.

2293. BEDUHN (Jason D.). Augustine, Manichaeism and the logic of persecution. *Archiv für Religionsgeschichte*, 2005, 7, p. 153-166.

2294. BENNEMA (Cornelis). The sword of the Messiah and the concept of liberation in the fourth gospel. *Biblica*, 2005, 86, p. 174-191.

2295. Bibel (Die) im Dialog der Schriften. Konzepte intertextueller Bibellektüre. Hrsg. v. Stefan ALKIER. Tübingen, Francke, 2005, VIII-281 p. (Neutestamentliche Entwürfe zur Theologie, 10).

2296. BIERNATH (Andrea). Mißverstandene Gleichheit. Die Frau in der frühen Kirche zwischen Charisma und Amt. Stuttgart, Steiner, 2005, 179 p.

2297. BINDER (Timon). Semen est sanguis Christianorum. Literarische Inszenierungen von Macht und Herrschaft in frühchristlicher Passionsliteratur. Berlin, Logos-Verlag, 2005, 244 p.

2298. BITTON-ASHKELONY (Brouria). Encountering the sacred: the debate on Christian pilgrimage in Late Antiquity. Berkeley a. Los Angeles, University of California Press, 2005, XV-250 p. (Transformation of the classical heritage, 38).

2299. BROTTIER (Laurence). L'appel des "demi-chrétiens" à la vie angélique: Jean Chrysostome prédicateur entre idéal monastique et réalité mondaine. Paris, Ed. du Cerf, 2005, 421 p.

2300. BRYAN (Christopher). Render to Caesar. Jesus, the early church, and the Roman superpower. Oxford, Oxford U. P., 2005, XII-185 p.

2301. CASTELLI (Elizabeth). Persecution and spectacle. Cultural appropriation in the christian commemoration of martyrdom. *Archiv für Religionsgeschichte*, 2005, 7, p. 102-136.

2302. CHANCEY (Mark A.). Greco-roman culture and the Galilee of Jesus. Cambridge, Cambridge U. P., 2005, XVII-285 p. (Monograph series. Society for New Testament Studies, 134).

2303. CHASE (Michael). Porphyre et Augustin: des trois sortes de "visions" au corps de résurrection. *Revue d'études augustiniennes et patristiques*, 2005, 51, p. 233-256.

2304. Christliche Wandmalereien in Syrien. Qara und das Kloster Mar Yakub. Hrsg. v. Andrea SCHMIDT und Stephan WESTPHALEN. Wiesbaden, Reichert, 2005, 240 p.

2305. COCO (Virginia Giuseppina). Il matrimonio in Sant'Agostino. Firenze, Firenze Atheneum, 2005, 79 p.

2306. Coliseo (Del) al Vaticano. Claves del cristianismo primitivo. Ed. por Elena MUÑIZ GRIJALVO y Rafael URÍAS MARTÍNEZ. Sevilla, Fundación José Manuel Lara, 2005, 313 p.

2307. COLOT (Blandine). Historiographie chrétienne et romanesque: le De mortibus persecutorum de Lactance (250–325 ap. J.-C.). *Vigiliae christianae*, 2005, 59, p. 135-151.

2308. DALMON (Laurence). Les lettres échangées entre l'Afrique et Rome à l'occasion de la controverse pélagienne. Genèse et fortunes d'un dossier de chancellerie ecclésiastique. *Mélanges de l'école française de Rome. Antiquité*, 2005, 117, p. 791-826.

2309. DE ROTEN (Philippe). Baptême et mystagogie: enquête sur l'initiation chrétienne selon S. Jean Chrysostome. Münster, Aschendorff, 2005, 498 p.

2310. Deutungen des Todes Jesu im Neuen Testament. Hrsg v. Jörg FREY. Tübingen, Mohr Siebeck, 2005, IX-707 p. (Wissenschaftliche Untersuchungen zum Neuen Testament, 181).

2311. DULAEY (Martine). L'apprentissage de l'exégèse biblique par Augustin (3). Années 393–394. *Revue d'études augustiniennes et patristiques*, 2005, 51, p. 21-65.

2312. DUNN (Geoffrey B.). The date of Innocent I's Epistula 12 and the second exile of John Chrysostom. *Greek, Roman and Byzantine studies*, 2005, 45, p. 155-170.

2313. DUVAL (Yves-Marie). La décrétale Ad Gallos episcopos. Son texte et son auteur; texte critique, traduction française et commentaire. Leiden, Brill, 2005, IX-177 p. (Supplements to vigiliae christianae, 73).

2314. FELLE (Antonio Enrico). Epigrafia cristiana e pagana in Sicilia: consonanze e peculiarità. *Vetera christianorum*, 2005, 42, p. 233-250.

2315. FIEDLER (Peter). Studien zur biblischen Grundlegung des christlich-jüdischen Verhältnisses. Stuttgart, Verl. Kath. Bibelwerk, 2005, X-291 p. (Stuttgarter biblische Aufsatzbände, 35).

2316. GADDIS (Michael). There is no crime for those who have Christ. Religious violence in the Christian Roman empire. Berkeley, University of California Press, 2005, 396 p.

2317. GELJON (Albert-Kees). Divine infinity in Gregory of Nyssa and Philo of Alexandria. *Vigiliae christianae*, 2005, 59, p. 152-177.

2318. GERBER (Christine). Paulus und seine "Kinder". Studien zur Beziehungsmetaphorik der paulinischen Briefe. Berlin, De Gruyter, 2005, XVII-576 p. (Beihefte zur Zeitschrift für die neutestamentliche Wissenschaft und die Kunde der älteren Kirche, 136).

2319. Giovanni Crisostomo. Oriente e Occidente tra IV e V secolo. XXIII incontro di studiosi dell'antichità cristiana. Augustinianum 6–8 maggio 2004. Roma, Institutum patristicum Augustinianum, 2005, 1049 p.

2320. GRADL (Hans-Georg). Zwischen Arm und Reich. Das lukanische Doppelwerk in leserorientierter und textpragmatischer Perspektive. Würzburg, Echter-Verlag, 2005, 500 p. (Forschung zur Bibel, 107).

2321. Griechische Mythologie und frühes Christentum. Hrsg. v. Raban von HAEHLING. Darmstadt, Wissenschaftliche Buchgesellschaft, 2005, 399 p.

2322. GRINDHEIM (Sigurd). The crux of election. Paul's critique of the Jewish confidence in the election of Israel. Tübingen, Mohr Siebeck, 2005, XI-282 p.

2323. GUINOT (Jean-Noël). Doit-on glorifier le Christ ou le Fils Monogène? La défense par Théodoret de Cyr d'une doxologie incriminée (ep. 147). *Revue d'études augustiniennes et patristiques*, 2005, 51, p. 327-356.

2324. HARRIS (William V.). The spread of Christianity in the first four centuries. Essays in explanation. Leiden, Brill, 2005, XIV-176 p. (Columbia studies in the classical tradition, 27).

2325. HEVELONE-HARPER (Jennifer L.). Disciples of the desert. Monks, laity, and spiritual authority in sixth-century Gaza. Baltimore, Johns Hopkins U. P., 2005, XII-211 p.

2326. HILL (Robert C.). Reading the Old Testament in Antioch. Leiden a. Boston, Brill, 2005, XIII-220 p.

2327. HIRSCHMANN (Vera-Elisabeth). Horrenda secta: Untersuchungen zum frühchristlichen Montanismus und seinen Verbindungen zur paganen Religion Phrygiens. Stuttgart, Steiner, 2005, 168 p. (Historia: Einzelschriften, 179).

2328. HOFFMANN (Matthias Reinhard). The destroyer and the lamb. The relationship between angelomorphic and lamb christology in the book of relevation. Tübingen, Mohr Siebeck, 2005, XVI-311 p.

2329. HOSAKA (Takaya). Tabunka kūkan no naka no kodai kyōkai. (Ikyō sekai to kirisutokyō 2). (Ancient church in the multi-cultural space. Christianity face to the heretical world 2). Tokyo, Kyobunkan, 2005, 320 p.

2330. HÜBNER (Sabine). Der Klerus in der Gesellschaft des spätantiken Kleinasiens. Stuttgart, Steiner, 2005, 318 p.

2331. KATOS (Demetrios). Origenists in the desert. Palladius of Helenopolis and the Alexandrian theological tradition. *American benedictine review*, 2005, 56, p. 167-193.

2332. Kontinuität und Unterbrechung. Gottesdienst und Gebet in Judentum und Christentum. Hrsg. v. Albert GERHARDS und Stephan WAHLE. Paderborn u. München, Schöningh, 2005, 285 p. (Studien zu Judentum und Christentum).

2333. KRANNICH (Torsten). Von Leporius bis zu Leo dem Großen. Studien zur lateinisch-sprachigen Christologie im fünften Jahrhundert nach Christus. Tübingen, Mohr Siebeck, 2005, XII-295 p. (Studien und Texte zu Antike und Christentum, 32).

2334. LALONDE (Gerald V.). Pagan cult to Christian ritual. The case of Agia Marina Theseiou. *Greek, Roman and Byzantine studies*, 2005, 45, p. 91-125.

2335. LEE (Aquila H. I.). From Messiah to preexistent son. Jesus' self-consciousness and early Christian exegesis of Messianic psalms. Tübingen, Mohr Siebeck, 2005, XII-375 p.

2336. LÖHR (Winrich Alfried). Arius reconsidered (Part 1). *Zeitschrift für antikes Christentum*, 2005, 9, p. 524-560.

2337. MADIGAN (Kevin), OSIEK (Carolyn). Ordained women in the early church. A documentary history. Baltimore, Johns Hopkins U. P., 2005, XIII-220 p.

2338. MARASCO (Gabriele). Filostorgio. Cultura, fede e politica in uno storico ecclesiastico del V secolo.

Roma, Institutum Patristicum Augustinianum, 2005, 288 p.

2339. MAUGANS DRIVER (Lisa D.). The cult of martyrs in Asterius of Amaseia's vision of the Christian city. *Church history*, 2005, 74, p. 236-254.

2340. MEYENDORFF (John). Lo scisma tra Roma e Costantinopoli. Magnano, Qiqajon, 2005, 150 p.

2341. MICAELLI (Claudio). La cristianizzazione dell'ellenismo. Brescia, Morcelliana, 2005, 236 p. (Letteratura cristiana antica. Nuova serie, 9: Studi).

2342. NIKOLAOU (Theodor). Theologische Konstanten in der patristischen Tradition und die Einheit der Kirche. *Orthodoxes forum*, 2005, 19, p. 153-167.

2343. OTRANTO (Giorgio). Il vescovo siculo Evagrio (IV secolo) tra filologia e storia. *Vetera christianorum*, 2005, 42, p. 5-14.

2344. Pères (Les) de l'Église face à la science médicale de leur temps. Actes du troisieme colloque d'études patristiques (Paris, 9–11 septembre 2004). Sous la direction de Véronique BOUDON-MILLOT et Bernard POUDERON. Paris, Beauchesne, 2005, XIX-582 p. (Théologie historique, 117).

2345. PITRE (Brant). Jesus, the tribulation, and the end of the exile. Restoration eschatology and the origin of the atonement. Tübingen, Mohr Siebeck a. Grand Rapids, Baker Acad., 2005, XIII-586 p.

2346. POPKES (Enno Edzard). Die Theologie der Liebe Gottes in den johanneischen Schriften. Studien zu Semantik der Liebe und zum Motivkreis des Dualismus. Tübingen, Mohr Siebeck, 2005, XX-466 p.

2347. PORCARO (M.). Psicologie e caratteri in Agostino d'Ippona: l'amante. *Auctores nostri*, 2005, 2, p. 177-184.

2348. RAMBAUX (Claude). L'accès à la vérité chez Tertullien. Bruxelles, Éditions Latomus, 2005, 264 p. (Collection Latomus, 293).

2349. RAUNIG (Walter). Das christliche Äthiopien. Geschichte – Architektur – Kunst. Regensburg, Schnell & Steiner, 2005, 319 p.

2350. Reception (The) of the New Testament in the Apostolic fathers. Ed. by Andrew GREGORY. Oxford, Oxford U. P., 2005, XIII-375 p. (The New Testament and the Apostolic fathers, 1).

2351. REED (Annette Yoshiko). Fallen angels and the history of Judaism and Christianity. The reception of Enochic literature. Cambridge, Cambridge U. P., 2005, XII-318 p.

2352. Religiöses Lernen in der biblischen, frühjüdischen und frühchristlichen Überlieferung. Hrsg. v. Beate EGO und Helmut MERKEL. Tübingen, Mohr Siebeck, 2005, XI-336 p. (Wissenschaftliche Untersuchungen zum Neuen Testament, 180).

2353. RICHTER (Gerhard). Oikonomia. Der Gebrauch des Wortes Oikonomia im Neuen Testament, bei den Kirchenvätern und in der theologischen Literatur bis ins 20. Jahrhundert. Berlin, De Gruyter, 2005, IX-753 p.

2354. RIEDO-EMMENEGGER (Christoph). Prophetisch-messianische Provokateure der Pax Romana. Jesus von Nazaret und andere Störenfriede im Konflikt mit dem Römischen Reich. Fribourg, Acad. Press, 2005, XXI-381 p. (Novum Testamentum et Orbis Antiquus. Studien zur Umwelt des Neuen Testaments, 56).

2355. RIZZO (Francesco Paolo). Sicilia christiana. Dal I al V secolo. Volume I. Roma, Giorgio Bretschneider, 2005, 265 p.

2356. Saint Augustin, la Numidie et la société de son temps. Actes du colloque SEMPAM-Ausonius (Bordeaux, 10–11 octobre 2003). Éd. par Serge LANCEL. Paris, De Boccard, 2005, 182 p.

2357. SCHLUEP (Christoph). Der Ort des Christus. Soteriologische Metaphern bei Paulus als Lebensregeln. Zürich, Theologischer Verlag Zürich, 2005, 453 p.

2358. SCHULZE (Christian). Medizin und Christentum in Spätantike und frühem Mittelalter. Christliche Ärzte und ihr Wirken. Tübingen, Mohr Siebeck, 2005, 280 p.

2359. SHAUF (Scott). Theology as history, history as theology. Paul in Ephesus in Acts 19. Berlin, De Gruyter, 2005, X-377 p. (Beihefte zur Zeitschrift für die neutestamentliche Wissenschaft und die Kunde der älteren Kirche, 133).

2360. SICK (David H.). Apuleius, Christianity, and virgin birth. *Wiener Studien*, 2005, 62, p. 864-874.

2361. SIMONS (Roswitha). Dracontius und der Mythos. Christliche Weltsicht und pagane Kultur in der ausgehenden Spätantike. München, K. G. Saur, 2005, 430 p. (Beiträge zur Altertumskunde, 186).

2362. SOLER (Joëlle). Religion et récit de voyage. Le Peristephanon de Prudence et le De reditu suo de Rutilius Namatianus. *Revue d'études augustiniennes et patristiques*, 2005, 51, p. 297-326.

2363. SOTINEL (Claire). Identité civique et christianisme. Aquilée du IIIe au VIe siècle. Rome, École française de Rome, 2005, 458 p. (Bibliothèque des Écoles françaises d'Athènes et de Rome, 324).

2364. SPIELVOGEL (Jörg). Arianische Vandalen, katholische Römer: die reichspolitische und kulturelle Dimension des christlichen Glaubenskonflikts im spätantiken Nordafrica. *Klio*, 2005, 87, p. 201-222.

2365. STELLADORO (Maria). La tanto discussa origine apostolica della Chiesa di Catania. *Bollettino della badia greca di Grottaferrata*, 2005, 2, p. 143-172.

2366. STETTBERGER (Herbert). Nichts haben – alles geben? Eine kognitiv-linguistisch orientierte Studie zur Besitzethik im lukanischen Doppelwerk. Freiburg, Herder, 2005, 568 p.

2367. STEYMANS (Hans Ulrich). Psalm 89 und der Davidbund. Eine strukturale und redaktionsgeschichtliche

Untersuchung. Frankfurt am Main, Lang, 2005, 492 p. (Österreichische Biblische Studien, 27).

2368. THIEL (Andreas). Die Johanneskirche in Ephesos. Wiesbaden, Reichert, 2005, 240 p.

2369. TLOKA (Jutta). Griechische Christen – Christliche Griechen. Plausibilisierungsstrategien des Antiken Christentums bei Origenes und Johannes Chrysostomos. Tübingen, Mohr Siebeck, 2005, XII-295 p. (Antike und Christentum, 30).

2370. TRAMPEDACH (Kai). Reichsmönchtum? Das politische Selbstverständnis der Mönche Palästinas im 6. Jahrhundert und die historische Methode des Kyrill von Skythopolis. *Millennium*, 2005, 2, p. 271-296.

2371. UTHEMANN (Karl-Heinz). Christus, Kosmos, Diatribe. Themen der frühen Kirche als Beiträge zu einer historischen Theologie. Berlin u. New York, De Gruyter, 2005, 665 p.

2372. VAN DER WATT (Jan Gabriël). Salvation in the New Testament. Perspectives on soteriology. Leiden, Brill, 2005, XIII-529 p. (Supplements to Novum Testamentum, 121).

2373. WARE (James P.). The mission of the church in Paul's letter to the Philippians in the context of ancient Judaism. Leiden, Brill, 2005, XV-380 p. (Supplements to Novum Testamentum, 120).

2374. WOYKE (Johannes). Götter, "Götzen", Götterbilder. Aspekte einer paulinischen "Theologie der Religionen". Berlin, De Gruyter, 2005, XVI-570 p. (Beihefte zur Zeitschrift für die neutestamentliche Wissenschaft und die Kunde der älteren Kirche, 132).

2375. Written (The) Gospel. Ed. by Markus BOCKMUEHL and Donald A. HAGNER. Cambridge, Cambridge U. P., 2005, xx p.

2376. YEE (Tet-Lim N.). Jews, gentiles and ethnic reconciliation. Paul's Jewish identity and Ephesians. Cambridge, Cambridge U. P., 2005, XXI-302 p. (Monograph series. Society for New Testament studies, 130).

Cf. n° 332

§ 4. Hagiography.

2377. Biographie und Persönlichkeit des **Paulus**. Hrsg. v. Eve-Marie BECKER und Peter PILHOFER. Tübingen, Mohr Siebeck, 2005, VIII-392 p. (Wissenschaftliche Untersuchungen zum Neuen Testament, 187).

2378. Consonantia salutis. Studi su **Ireneo** di Lione. Proceedings, Naples, 2003. A cura di Enrico CATTANEO. Trapani, Pozzo di Giacobbe, 2005, 267 p. (Oi christianoi, 1).

2379. **Mariam**, the Magdalen, and the Mother. Ed. by Deirdre GOOD. Bloomington, Indiana U. P., 2005, XVII-240 p.

2380. ROBINSON (Joseph Armitage). The passion of S. **Perpetua**. Newly edited from the mss. with an introduction and notes; together with an appendix containing the original Latin text of the Scillitan martyrdom. Piscataway, Gorgias press, 2005, VIII-103 p.

2381. VAN UYTFANGHE (Marc). La biographie classique et l'hagiographie chrétienne antique tardive. *Hagiographica*, 2005, 12, p. 223-248.

2382. Zwischen Historiographie und Hagiographie. Hrsg. v. Jürgen DUMMER und Meinolf VIELBERG. Stuttgart, Steiner, 2005, 107 p.

H

BYZANTINE HISTORY

(since Justinian)

---

§ 1. Sources. 2383-2417. – § 2. General. 2418-2431. – § 3. Special studies. 2432-2518.

§ 1. Sources.

* 2383. Byzantinische Zeitschrift. Band 98. [Band 97. Cf. Bibl. 2004, n° 2519.] Hrsg. v. Albrecht BERGER. München u. Leipzig, 2005, 884 p.

* 2384. Byzantinische Zeitschrift. Bibliographie 2. Begründet v. Karl KRUMBACHER, hrsg. v. Peter SCHREINER, programmiert und bearbeitet v. Reinhard HIß. München u. Leipzig, Saur, 2005, CD-ROM-Edition. [Cf. Bibl. 2004, n° 2520.]

2385. [Critobulus of Imbros] REINSCH (Diether Roderich), KOLOBU (Photeine). Kritoboúlou toû Imbríou Historía. Eisagoghé – Metáphrase – Schólia. Athena, Kanáke, 2005, 678 p.

2386. Discours annuels en l'honneur du patriarche Georges Xiphilin. Éd. par Marina LOUKAKI et Corinne JOUANNO. Paris, Association des amis du centre d'histoire et civilisation de Byzance, 2005, 235 p.

2387. [Eustratius Constantinopolitanus] VAN DEUN (Peter). Le De anima et angelis attribué à Eustrate de Constantinople (CPG 7523): un texte fantôme? *Sacris erudiri*, 2005, 44, p. 219-226.

2388. Fontes minores. 11. Hrsg. v. Ludwig BURGMANN. Frankfurt am Main, Löwenklau-Ges., 2005, XI-483 p. (ill.). (Forschungen zur byzantinischen Rechtsgeschichte, 26).

2389. [Georgius Gemistus Plethon] BALOGLOU (Christos P.), KARAYIANNIS (Anastassios D.). The economic thought of Georgios Gemistos-Plethon and Niccolò Machiavelli. Some comparative parallels and links. *Archeion oikonomikés historías*, 2005, 17, p. 5-19.

2390. [Georgius Gemistus Plethon] Georgios Gemistos Plethon (1355–1452). Reformpolitiker, Philosoph, Verehrer der alten Götter. Hrsg. v. Wilhelm. BLUM und Walter SEITTER. Zürich u. Berlin, Diaphanes, 2005, 144 p.

2391. [Georgius Gemistus Plethon] TAMBRUN-KRASKER (Brigitte). Métaphysique et politique chez Pléthon. *Athenâ*, 2005, 83, p. 141-165.

2392. [Georgius Pachymeres] Georgios Pachymeres, Philosophia. Biblion hendekaton. Ta Etika, etoi ta Nikomacheia. Editio princeps. Prolegomena, keimeno, eureteria. Ed. Konstantinou OIKONOMAKU. Athena, Akademia Athenon, 2005, 151 p.

2393. [Gregorius Antiochus] SIDERAS (Alexander). Der unedierte Brief des Gregorios Antiochos an Eustathios von Thessalonike. *Byzantinoslavica*, 2005, 63, p. 153-186.

2394. GRÜNBART (Michael). Formen der Anrede im byzantinischen Brief vom 6. bis zum 12. Jahrhundert. Wien, Verlag der Österreichischen Akademie der Wissenschaften, 2005, 403 p. (Wiener byzantinistische Studien, 25).

2395. [Halosis tes Konstantinoupoleos] MATZUKES (C.). He Halosis tes Konstantinoupóleos. Tétarte Staurophoría. Athena, Hellenikés ekdóseis, 2005, 357 p.

2396. [Iohannes Lydus] KALDELLIS (Anthony). Republican theory and political dissidence in Ioannes Lydos. *Byzantine and modern Greek studies*, 2005, 29, p. 1-16.

2397. [Iohannes Mauropus] CORTASSA (Guido). "Signore e padrone della terra e del mare". Poesia e ideologia del potere imperiale in Giovanni Mauropode. *Néa Rhóme*, 2005, 2, p. 205-226.

2398. [Iohannes Philoponus] SCHOLTEN (Clemens). Unbeachtete Zitate und doxographische Nachrichten in der Schrift De aeternitate mundi des Johannes Philoponos. *Rheinisches Museum für Philologie*, 2005, 148, p. 202-219.

2399. [Iohannes Philoponus] SHARE (Michael John). Philoponus. Against Proclus' on the eternity of the world 1-5. Ithaca, Cornell U. P., 2005, IX-154 p.

2400. [Iustinianus] Iustiniani Augusti Digesta seu Pandectae. Digesti o Pandette dell'imperatore Giustiniano. Testo e traduzione. A cura di Sandro SCHIPANI. Milano, Giuffrè, 2005, XXXII-384 p.

2401. [Iustinianus] SCOTTI (Francesca). Antologia del Digesto di Giustiniano. Testi tradotti e annotati ad uso degli studenti. Milano, Università Cattolica, 2005, 229 p.

2402. [Leo Diaconus] TALBOT (Alice-Mary). The History of Leo the Deacon. Byzantine military expansion in the tenth century. Introduction, translation, and annotation. Washington, Dumbarton Oaks research library and collection, 2005, XIX-264 p.

2403. [Leontius Scholasticus] SCHULTE (Hendrich). Die Epigramme des Leontios Scholastikos Minotauros. Text, Übersetzung, Kommentar. Trier, WVT, 2005, 88 p.

2404. [Marinus Falierus] CARPINATO (Caterina). Una lettura del threnos ... del Cretese Marinos Falieros. *Byzantinische Zeitschrift*, 2005, 98, p. 403-421.

2405. [Maximus Homologetes] MUELLER-JOURDAN (Pascal). Typologie spatio-temporelle de l'Ecclesia byzantine: la Mystagogie de Maxime le Confesseur dans la culture philosophique de l'antiquité tardive. Leiden, Brill, 2005, 215 p.

2406. MEL'NIKOVA (Elena A.). Vizantija v svete skandinavskikh runicheskikh nadpisej. (Byzantium in the Old Norse Runic inscriptions). *Vizantijskij vremennik = BYZANTINA XRONIKA*, 2005, 64 (89), p. 160-180.

2407. [Nonnus Panopolitanus] Nonnos de Panopolis. Les dionysiaques. Tome XI. Chants XXXII-XXXIV. Texte établi et traduit. Éd. par Bernard GERLAUD. Paris, Les belles lettres, 2005, XIII-267 p.

2408. [Notitia dignitatum] Notitia Dignitatum (La). Nueva edición crítica y comentario histórico. Ed. por Concepción NEIRA FALEIRO. Madrid, Consejo superior de investigaciones científicas, 2005, 697 p.

2409. [Paulus Silentiarius] FOBELLI (Maria Luigia). Un tempio per Giustiniano. Santa Sofia di Costantinopoli e la descrizione di Paolo Silenziario. Testo greco e traduzione italiana a fronte. Roma, Viella, 2005, 240 p.

2410. [Proclus] DAY (Juliette). Proclus on baptism in Constantinople. Norwich, SCM-Canterbury press, 2005, 48 p.

2411. [Procopius Gazaeus] MATINO (Giuseppina). Procopio di Gaza. Panegirico per l'imperatore Anastasio. Introduzione, testo critico, traduzione e commentario. *Quaderni dell'accademia pontaniana*, 2005, 41, 135 p.

2412. [Psellus] PIETSCH (Efthymia). Die Chronographia des Michael Psellos. Kaisergeschichte, Autobiographie und Apologie. Wiesbaden, Reichert, 2005, 151 p.

2413. [Severus Alexandrinus] AMATO (Eugenio). Prolegomeni all'edizione critica dei Progimnasmi di Severo Alessandrino. *Medioevo greco*, 2005, 5, p. 31-72.

2414. SHUVALOV (Petr V.). Urbikij i "Strategikon" psevdo-Mavrikija. (Urbicius and the *Strategicon* of Pseudo-Mauricius). *Vizantijskij vremennik = BYZANTINA XRONIKA*, 2005, 64 (89), p. 34-60.

2415. [Symeon Magister] PRATSCH (Thomas). Zum Briefcorpus des Symeon Magistros: Edition, Ordnung, Datierung. *Jahrbuch der Österreichischen Byzantinistik*, 2005, 55, p. 71-86.

2416. [Urbicius] GREATREX (Geoffrey), Elton (Hugh), BURGESS (Richard). Urbicius' Epitedeuma: an edition, translation and commentary. *Byzantinische Zeitschrift*, 2005, 98, p. 35-74.

2417. VASSIS (Ioannis). Initia carminum Byzantinorum. Berlin u. New York, De Gruyter, 2005, LIV-932 p. (Supplementa Byzantina. Texte und Untersuchungen, 8).

§ 2. General.

2418. Ägäis und Europa. Hrsg. v. Evangelos KONSTANTINOU. Frankfurt am Main, Lang, 2005, 471 p.

2419. Byzantino-Nordica 2004. Papers presented at the International symposium of Byzantine studies held on 7 May 2004 in Tartu, Estonia. Ed. by Ivo VOLT and Janika PÄLL. Tartu, Tartu U. P., 2005, 200 p.

2420. Cambridge companion (The) to the age of Justinian. Ed. by Michael Robert MAAS. Cambridge, Cambridge U. P., 2005, XXVII-626 p.

2421. CHEYNET (Jean-Claude). Histoire de Byzance. Paris, Presses universitaires de France, 2005, 128 p. (Que sais-je ?).

2422. GREGORY (Timothy E.). A history of Byzantium. Oxford, Blackwell, 2005, XIV-382 p. (Blackwell history of the ancient world, 49).

2423. HALDON (John F.). The Palgrave atlas of Byzantine history. Basingstoke, Palgrave Macmillan, 2005, 187 p.

2424. KHVOSTOVA (Ksenija V.). Osobennosti vizantijskoj tsivilizatsii. (The features of the Byzantine Civilization). RAN, In-t vseobshchej istorii. Moskva, Nauka, 2005, 197 p.

2425. LOUNGHIS (T.), BLYSIDU (B.), LAMPAKES (S.). Regesten der Kaiserurkunden des oströmischen Reiches von 476 bis 565. Nicosia, Zyprisches Forschungszentrum, 2005, 358 p.

2426. Palgrave advances in Byzantine history. Ed. by Jonathan HARRIS. Basingstoke, Palgrave Macmillan, 2005, XIII-252 p.

2427. PATEL (Ismail Adam). Madina to Jerusalem. Encounters with the Byzantine empire. Leicester, Islamic foundation, 2005, VIII-160 p.

2428. PATLAGEAN (Évelyne). Byzance dans le millénaire médiéval. *Annales*, 2005, 60, 4, p. 721-732.

2429. Society, culture and politics in Byzantium. Ed. by Nikolas OIKONOMIDES and Elizabeth ZACHARIADOU. Aldershot, Ashgate/Variorum, 2005, XII-356 p.

2430. VENNING (Timothy), HARRIS (Jonathan). Chronology of the Byzantine empire. London, Palgrave, 2005, 1000 p.

2431. Zwischen Polis, Provinz und Peripherie. Beiträge zur byzantinischen Geschichte und Kultur. Hrsg. v. Lars M. HOFFMANN und Anuscha MONCHIZADEH. Wiesbaden, Harrassowitz, 2005, XIX-968 p.

§ 3. Special studies.

2432. AMATO (Eugenio). Sei epistole mutuae inedite di Procopio di Gaza ed il retore Megezio. *Byzantinische Zeitschrift*, 2005, 98, p. 367-382.

2433. AMENGUAL I BATLE (Josep). Ubi pars Graecorum est. Medio milenio de historia relegada de las Baleares y las Pitiusas. *Pyrenae*, 2005, 36, p. 87-103.

2434. ANCA (Alexandru S.). Ehrerweisung durch Geschenke in der Komnenenzeit: Gewohnheiten und Regeln des herrscherlichen Schenkens. *Mitteilungen zur Spätantiken Archäologie und Byzantinischen Kunstgeschichte*, 2005, 4, p. 185-194.

2435. ANDREOPOULOS (Andreas). Metamorphosis: the transfiguration in Byzantine theology and iconography. Crestwood, St. Vladimir's seminary press, 2005, 286 p.

2436. AOUN (Marc). Jésubokt, métropolitain et juriste de l'Église d'Orient (Nestorienne), Auteur au VIIIe siècle du premier traité systématique de droit séculier. *Tijdschrift voor rechtsgeschiedenis*, 2005, 73, 1-2, p. 81-92.

2437. Arab-Byzantine relations in early Islamic times. Ed. by Michael BONNER. Ashgate, Variorum, 2005, LV-465 p.

2438. ARGYRIOU (Astérios). Perception de l'Islam et traductions du Coran dans le monde byzantin grec. *Byzantion*, 2005, 75, p. 25-69.

2439. AVENARIUS (Alexander).The Byzantine struggle over the icon. On the problem of Eastern European symbolism. Ed. Elena MANNOVÁ. Bratislava, Historický ústav Slovenskej akadémie vied, 2005, 210 p. (Studia historica Slovaca, 23).

2440. BALL (Jennifer L.). Byzantine dress. Representations of secular dress in eighth to twelfth century painting. New York, Palgrave MacMillan, 2005, VII-176 p.

2441. BIANCONI (Daniele). Tessalonica nell'età dei Paleologi. Le pratiche intellettuali nel riflesso della cultura scritta. Paris, Centre d'études byzantines, néohelléniques et sud-est européennes, 2005, 341 p.

2442. BRACCINI (Tommaso). L'imperatore Giovanni VIII Paleologo a Pistoia. *Byzantinische Zeitschrift*, 2005, 98, p. 383-397.

2443. BRUBAKER (Leslie). Byzantium in the iconoclast era (ca 680–850): a history. Cambridge, Cambridge U. P., 2005, 300 p.

2444. Byzance et le monde extérieur. Contacts, relations, échanges. Actes de trois séances du XXe Congrès international des études byzantines (Paris, 19–25 août 2001). Éd. par Michel BALARD, Élisabeth MALAMUT et Jean-Michel SPIESER. Paris, Publications de la Sorbonne, 2005, 288 p.

2445. CAÑIZAR PALACIOS (J. L.). About iureconsulti and emperors in Justinian's legislative labour. *Byzantion*, 2005, 75, p. 104-116.

2446. CASTILLO (R.). La dinastia de los Comneno. *Porphyra*, 2005, 5, p. 57-71.

2447. CESARETTI (Paolo). Ravenna. Gli splendori di un Impero (testi di Paolo CESARETTI e Gianni GUADALUPI, fotografie di Alfredo DAGLI ORTI). Villanova di Castenaso, FMR, 2005, 238 p.

2448. CONGOURDEAU (Marie-Hélène). L'empereur et le patriarche dans l'empire byzantin. *Istina*, 2005, 50, p. 8-21.

2449. CONSTANTINOU (Stavroula). Female corporeal performances: reading the body in Byzantine passions and lives of holy women. Uppsala, Uppsala University, 2005, 225 p. (Acta Universitatis Upsaliensis, 9).

2450. COSTANZA (Salvatore). Una syntaxis mantica pitagorica. *Byzantinische Zeitschrift*, 2005, 98, p. 5-21.

2451. CREAZZO (Tiziana). Coinvolgimenti politici e sociali nell'affaire di Leone metropolita di Calcedonia. *Orpheus*, 2005, 26, p. 66-85.

2452. CROKE (Brian). Leo I and the palace guard. *Byzantion*, 2005, 75, p. 117-151.

2453. CUOMO (Valentina). Athos Dionysiou 180 + Paris, Suppl. Grec. 495: un nuovo manoscritto di Teodosio Principe. *Byzantinische Zeitschrift*, 2005, 98, p. 23-34.

2454. DOMARADZKI (Kamil). Kariera polityczna Michała Psellosa na dworze bizantyńskim w XI wieku. (Political career of Michael Psellos at the Byzantine court in the 11th century). *Acta Universitatis Lodziensis. Folia historica*, 2005, 80, p. 117-138.

2455. EVANS (James A.). The emperor Justinian and the Byzantine empire. Westport, Greenwood press, 2005, XXXVI-178 p.

2456. FATTI (Federico). Il seme del diavolo. La parabola della zizzania e i conflitti politico-dottrinali a Bisanzio (IV–V secolo). *Cristianesimo nella storia*, 2005, 26, p. 123-172.

2457. Feast, fast or famine: food and drink in Byzantium. Ed. by Wendy MAYER and Silke TRZCIONKA. Brisbane, Australian association for Byzantine studies, 2005, IX-215 p. (Byzantina Australiensia, 15).

2458. FRANSES (Henri). Portraits of patrons in Byzantine religious manuscripts. Ann Arbor, UMI dissertation services, 2005, 105 p.

2459. GOLDSTEIN (Ivo). Funkcija Jadrana u ratu Bizantskog carstva protiv Ostrogota 535–555. godine. (How did the Byzantines use the Adriatic Sea in their war against the Ostrogoths, 535–555). *Radovi*, 2005, 37, p. 23-34.

2460. GRÜNBART (Michael). 'Tis love that has warm'd us'. Reconstructing networks in 12th century Byzantium. *Revue belge de philologie et d'histoire*, 2005, 83, 1, p. 301-314.

2461. HADERMANN-MISGUICH (Lydie). Les temps des Anges. Recueil d'études sur la peinture byzantine du XIIe siècle, ses antécédents, son rayonnement. Bruxelles, Le Livre Timpermann, 2005, 270 p.

2462. HEERS (Jacques). Chute et mort de Constantinople (1204–1453). Paris, Perrin, 2005, 345 p.

2463. HOLMES (Catherine). Basil II and the governance of empire (976–1025). Oxford, Oxford U. P., 2005, XIV-625 p.

2464. IERODIAKONOU (Katerina). The Byzantine reception of Aristotle's categories. *Synthesis philosophica*, 2005, 39, p. 7-31.

2465. JACOBY (David). Commercial exchange across the Mediterranean: Byzantium, the crusader Levant, Egypt, and Italy. Aldershot, Ashgate, 2005, [s. p.].

2466. KAPRIEV (Georgi). Philosophie in Byzanz. Würzburg, Königshausen & Neumann, 2005, 383 p.

2467. KARAHAN (Anne). Byzantine holy images and the issue of transcendence and immanence: the theological background of the Late Byzantine Palaiologan iconography and aesthetics of the Chora church, Istanbul. Diss. Stockholm, Konstvetenskapliga institutionen, Stockholm universitet, 2005, 351 p.

2468. KELLY (Christopher). John Lydus and the eastern praetorian prefecture in the sixth century AD. *Byzantinische Zeitschrift*, 2005, 98, p. 431-458.

2469. KHALILIEH (Hassan S.). Human jettison, contribution for lives, and life salvage in Byzantine and early Islamic maritime laws in the Mediterranean. *Byzantion*, 2005, 75, p. 225-235.

2470. KOTALA (Tomasz). Sprawa śmierci cesarza bizantyńskiego Romana III Argyrosa (1028–1034). (The question of the death of the Byzantine emperor Romanos III Argyros). *Acta Universitatis Lodziensis. Folia historica*, 2005, 80, p. 101-116.

2471. KUHOFF (W.). Leo I., (ost-)römischer Kaiser. *Biographisch-bibliographisches Kirchenlexikon*, 2005, 25, p. 810-829.

2472. LAIOU (Angeliki). The Byzantine village (5th–14th century). *In*: Villages (Les) dans l'empire byzantin [Cf. n° 2514], p. 31-53.

2473. Mallorca y Byzancio. Ed. by R. DURÁN TAPIA. Palma de Mallorca, Asociación amigos del Castillo de San Carlos, "Aula General Weyler", 2005, 199 p.

2474. MĂRCULEȚ (Vasile). Imperiul Bizantin la Dunărea de Jos în secolele X–XII. Spațiul carpato-balcanic în politica vest-pontică a Imperiului Bizantin. (L'empire byzantin au bas-Danube aux siècles X–XII). Târgu Mureș, [s. n.], 2005, 205 p.

2475. MARIN (Șerban). Giustiniano Partecipazio and the representation of the first Venetian embassy to Constantinople in the Chronicles of the Serenessima. *Historical yearbook*, 2005, 2, p. 75-92.

2476. MARTIN (Jean-Marie), NOYÉ (Ghislaine). Les villages de l'Italie méridionale byzantine. *In*: Villages (Les) dans l'empire byzantin [Cf. n° 2514], p. 149-164.

2477. MARTIN (Jean-Marie). L'empreinte de Byzance dans l'Italie normande. Occupation du sol et institutions. *Annales*, 2005, 60, 4, p. 733-765.

2478. MASTOROPOULOS (Geōrgios S.). Naxos, to allo kallos: periēgēseis se Vyzantina mnēmeia (Naxos: Byzantine monuments). Athēna, Hellēnikes homoiographikes ekd., 2005, 253 p.

2479. MÉTIVIER (Sophie). La Cappadoce (IVe–VIe siècle). Une histoire provinciale de l'empire romain d'Orient. Paris, Publications de la Sorbonne, 2005, 496 p. (Publications de la Sorbonne. Série Byzantina Sorbonensia, 22).

2480. MINTSĒS (Geōrgios I.). Hē "erōtikē" Anna Komnēnē: mia meletē koinōnikēs tautotētas tou phylou ("Erotic" Anna Comnena: a study on gender). Thessalonikē, Ekdotikos oikos ant. stamoulē, 2005, 126 p.

2481. MOROZOV (Maksim A.). Monastyri srednevekovoj Vizantii: Khozjajstvo, sotsial'nyj i pravovoj statusy. (The monasteries of Byzantium: Economy, social and legal position). Sankt-Peterburgskij gos. un-t. Sankt-Peterburg, Izd-vo Sankt-Peterburgskogo un-ta, 2005, 173 p. (bibl. p. 150-159).

2482. MUELLER-JOURDAN (Pascal). Typologie spatio-temporelle de l'Ecclesia byzantine: la mystagogie de Maxime le Confesseur dans la culture philosophique de l'antiquité tardive. Leiden et Boston, Brill, 2005, IX-215 p. (Supplements to Vigiliae christianae, 74).

2483. NIKOLAOU (Katerina). Ē gynaika stē mesē vyzantinē epochē: koinōnika protypa kai kathēmerinos vios sta hagiologika keimena. (Woman in the middle byzantine period: social models and everyday life in the hagiographical texts). Athēna, Ethniko hidryma ereunōn. Instituto Vyzantinōn ereunōn, 2005, 376 p. (Monographs, 6).

2484. NILSSON (Ingela). Narrating images in Byzantine literature. The ekphraseis of Konstantinos Manasses. *Jahrbuch der Österreichischen Byzantinistik*, 2005, 55, p. 121-146.

2485. NIYOGI (Ruma). Gender, politics and rhetoric in Byzantium. Chicago, [s. n.], 2005, [s. p.]. [Thesis (Ph. D.) – University of Chicago].

2486. ORSINI (Pasquale). Pratiche collettive di scrittura a Bisanzio nei secoli IX e X. *Segno e testo*, 2005, 3, p. 265-342.

2487. OTSUKI (Yasuhiro). Teikoku to Jizen: Bizantsu. (Byzantine empire and charity). Tokyo, Sobunsha, 2005, 467 p.

2488. OUSTERHOUT (Robert G.). A Byzantine settlement in Cappadocia. Washington, Dumbarton oaks research library and collection, 2005, XX-474 p. (Dumbarton oaks studies, 42).

2489. PIERI (Dominique). Le commerce du vin oriental à l'époque byzantine, V^e–VIIe siècles: le témoignage des amphores en Gaule. Beyrouth, Institut français du Proche-Orient, 2005, VI-329 p. (Bibliothèque archéologique et historique, 174).

2490. PILLON (Michel). Armée et défense de l'Illyricum byzantin de Justinien à Héraclius (527–641). De la réorganisation justinienne à l'émergence des "armées de cité". *Erytheia*, 2005, 26, p. 7-85.

2491. PILTZ (Elisabeth). Byzantium in the mirror: the message of Skylitzes Matritensis and Hagia Sophia in Constantinople. Oxford, Archaeopress, 2005, II-92 p. (BAR international series, 1334).

2492. POPOVA (Olga). The ascetic trend in Byzantine art of the second quarter of the eleventh century and its subsequent fate. *Néa Rhóme*, 2005, 2, p. 243-257.

2493. PRATSCH (Thomas). Der hagiographische Topos. Griechische Heiligenviten in mittelbyzantinischer Zeit. Berlin u. New York, De Gruyter, 2005, XVI-454 p.

2494. REINSCH (Diether Roderich). Die Kultur des Schenkens in den Texten der Historiker der Komnenzeit. *Mitteilungen zur Spätantiken Archäologie und Byzantinischen Kunst-geschichte*, 2005, 4, p. 173-183.

2495. RUGGIERI (Vincenzo). La Caria bizantina: topografia, archeologia ed arte. Soveria Mannelli, Rubbettino, 2005, 267 p.

2496. SÄNGER (P.). Die Eirenarchen im römischen und byzantinischen Ägypten. *Tyche*, 2005, 20, p. 143-204.

2497. SANSARIDOU-HENDRICKX (Thekla). The study of the Weltanschauung of anonymous late Byzantine chroniclers in the framework of metahistory. *Acta Patristica et Byzantina*, 2005, 16, p. 255-273.

2498. SCHAMP (Jacques). Les "petits-fils" de Jean le Lydien ou le parfum du scorpion. *L'antiquité classique*, 2005, 74, p. 171-187.

2499. SERIN (Ufuk). Some observations on the middle Byzantine church outside the east gate at Iasos. *La parola del passato*, 2005, 60, p. 156-178.

2500. SHOJU (Keitaro). Bizantsu teikoku no seiji seido. (Political institution of the Byzantine empire). Hatano, Tokai U. P., 2005, 192 p.

2501. SIBILIO (Vito). La chiesa bizantina nell'età dei Comneni. *Porphyra*, 2005, 5, p. 96-109.

2502. SIDERAS (Alexander). Die unedierte Trostrede des Gregorios Antiochos an den Logothetes Michael Hagiotheodoritos. *Jahrbuch der Österreichischen Byzantinistik*, 2005, 55, p. 147-190.

2503. SOPHOULIS (Pananos). A study of Byzantine-Bulgar relations, 775–816. Oxford, [s. n.], 2005, XVI-390 p. [Thesis (D. Phil) – University of Oxford].

2504. STONE (Andrew F.). Nerses IV "the Gracious", Manuel I Komnenos, The Patriarch Michael III Anchialos and negotiations for church union between Byzantium and the Armenian Church. *Jahrbuch der Österreichischen Byzantinistik*, 2005, 55, p. 191-208.

2505. STOURAITIS (Ioannis). Neue Aspekte des Machtkampfes im Zeitraum 976–986 in Byzanz. *Jahrbuch der Österreichischen Byzantinistik*, 2005, 55, p. 50-69.

2506. STRANO (Gioacchino). Il patriarca Fozio e le epistole agli Armeni: disputa religiosa e finalità politiche. *Jahrbuch der Österreichischen Byzantinistik*, 2005, 55, p. 43-58.

2507. STUDER (Walter). Byzanz in Disentis. Fragmente frühbyzantinischer Monumentalmalerei. Chur, Rätisches Museum, 2005, 75 p.

2508. THEIS (Lioba). Flankenräume in mittelbyzantinischen Kirchenbau: zur Befundsicherung, Rekonstruktion und Bedeutung einer verschwundenen architektonischen Form in Konstantinopel. Wiesbaden, Reichert, 2005, 216 p.

2509. TINNEFELD (F.). Kirche und Stadt im byzantinischen Reich. *Ostkirchliche Studien*, 2005, 54, p. 56-78.

2510. TOCCI (Raimondo). Zu Genese und Kompositionsvorgang der Synopsis chroniké des Theodoros Skutariotes. *Byzantinische Zeitschrift*, 2005, 98, p. 551-568.

2511. Urbs capta: the fourth crusade and its consequences = la IVe croisade et ses conséquences. Ed. by Angeliki LAIOU. Paris, Lethielleux, 2005, 371 p. (Réalités byzantines, 10).

2512. VANDERHEYDE (Catherine). La sculpture architecturale byzantine dans le thème de Nikopolis: du Xe au début du XIIIe siècle (Épire, Étolie-Acarnanie et Sud de l'Albanie). Athena, École française d'Athènes, 2005, XIII-183 p. (Bulletin de correspondance hellénique, 45. Supplément).

2513. Venezia e Bisanzio: aspetti della cultura artistica bizantina da Ravenna a Venezia (V–XIV secolo). A cura di Clementina RIZZARDI. Venezia, Istituto veneto di scienze, lettere ed arti, 2005, VI-658 p.

2514. Villages (Les) dans l'empire byzantin: IVe–XVe siècles. Éd. par Jacques LEFORT, Cécile MORRISSON et Jean-Pierre SODINI. Paris, Lethellieux, 2005, 591 p. [Cf. n^{os} <sélection> 2472, 2476.]

2515. VITIELLO (Massimiliano). "Cui Iustinus Imperator venienti ita occurrit ac si beato Petro". Das Ritual Beim ersten Papst-Kaiser-Treffen in Konstantinopel: eine römische Auslegung? *Byzantinische Zeitschrift*, 2005, 98, p. 81-96.

2516. VOLAN (Angela M.). Last judgements and last emperors. Illustrated apocalyptic history in late and post-Byzantine art. Chicago, [s. n.], 2005, [s. p.]. [Thesis (Ph. D.) – University of Chicago].

2517. WORTLEY (John). The Marian relics at Constantinople. *Greek, Roman and Byzantine studies*, 2005, 45, p. 171-187.

2518. ZUCKERMAN (Constantine). Learning from the enemy and more. Studies in 'Dark Centuries' Byzantium. *Millennium*, 2005, 2, p. 79-135.

Cf. n°s 139, 140, 2406, 2592, 2598, 2715, 2752, 3222

I

HISTORY OF THE MIDDLE AGES

§ 1. Sources and criticism of sources (*a*. Non-literary sources; *b*. Literary sources). 2519-2584. – § 2. General works. 2585-2604. – § 3. Political history (*a*. General; *b*. 476–900; *c*. 900–1300; *d*. 1300–1500). 2605-2675. – § 4. Jews. 2676-2692. – § 5. Islam. 2693-2717. – § 6. Vikings. 2718-2731. – § 7. History of law and institutions. 2732-2777. – § 8. Economic and social history. 2778-2875. – § 9. History of civilization, literature, technology and education (*a*. Civilization; *b*. Literature; *c*. Technology; *d*. Education). 2876-3151. – § 10. History of art. 3152-3223. – § 11. History of music. 3224-3245. – § 12. History of philosophy, theology and science. 3246-3316. – § 13. History of the Church and religion (*a*. General; *b*. History of the Popes; *c*. Monastic history; *d*. Hagiography; *e*. Special studies). 3317-3483. – § 14. Settlements. Place names. Town planning. 3484-3514.

§ 1. Sources and criticism of sources.

a. Non-literary sources

2519. Acta vectigalia Regni Navarrae: documentos financieros para el estudio de la Hacienda Real de Navarra. Serie I. Comptos reales. Registros. Vol. 10. Registros de la Casa de Francia. Luis I el Hutín, Felipe II el Largo: 1315–1318. Ed. por Juan CARRASCO, Marcelino BEROIZ y Iñigo MUGUETA. Pamplona, Gobierno de Navarra, Departamento de Economía y Hacienda, 2005, 820 p.

2520. Ad fontem – U istochnika: Sb. st. v chest' Sergeja Mikhajlovicha Kashtanova. (Source-critical studies presented to Sergey M. Kashtanov). Ed. Sigurd O. SHMIDT, Ljubov' V. STOLJAROVA [et al.]. RAN, In-t vseobshchej istorii. Moskva, Nauka, 2005, 495 p. (bibl. incl.; ill.; biographical essay by Ljubov' V. Stoljarova p. 7-77).

2521. Bulario aragonés de Benedicto XIII. Vol. 2. La curia itinerante (1404–1411). Ed. por Ovidio CUELLA ESTEBAN. Zaragoza, Institución Fernando el Católico, 2005, 686 p. (Colección Fuentes históricas aragonesas, 36).

2522. CÁRCEL ORTI (María Milagros). Un formulari i un registre del bisbe de València jaume d'Aragó (segle XIV). Valencia, Universitat de Valencia, 2005, 2 vol, 439 p., 59 p. (Fonts Històriques Valencianes, 17).

2523. Cartulaire (Le) de Bigorre (XIe–XIIIe siècle). Ed. par Xavier RAVIER en collaboration avec Benoît CURSENTE. Paris, CTHS, 2005, 317 p. (Collection de documents inédits sur l'histoire de France, section histoire et philologie des civilisations médiévales, série in-8°, 36).

2524. Catàleg de pergamins del fons de l'Ajuntament de Girona (1144–1862). Ed. por Joan VILLAR I TORRENT, Núria SURIÁ I VENTURA, Joan BOADAS I RASET i Lluís-Esteve CASELLAS I SERRA. Barcelona, Fundació Noguera i Lleida, Pagès Editors, 2005, 3 vol., 2012 p. (Col·lecció Diplomataris, 32-34).

2525. Censos de población del territorio de Barcelona en la década de 1360. Ed. por Esperança PIQUER FERRER. Tübingen, Miemeyer, 2005, 236 p. (Patronymica Romanica, 22).

2526. Charters of Malmesbury Abbey. Ed. by Susan E. KELLY. Oxford, Oxford U. P. a. British Academy, 2005, XXII-328 p. (map). (Anglo-Saxon charters, 11).

2527. Chartes (Les) de Gérard Ier, Liébert et Gérard II, évêques de Cambrai et d'Arras, comtes du Cambrésis (1012–1092/93). Éd. par Erik VAN MINGROOT. Leuven, Leuven U. P., 2005, XXXVI-382 p. (Mediaevalia Lovaniensia. Studia, 8).

2528. Chartes (Les) de l'Abbaye d'Anchin (1079–1201). Éd. par Jean-Pierre GERZAGUET. Turnhout, Brepols, 2005, 511 p. (Atelier de recherches sur les textes médiévaux, 6).

2529. Colección diplomática do mosteiro de Santiago de Mens: edición e estudo. Ed. por María Pilar ZAPICO BARBEITO. Noia, Toxosoutos, 2005, 356 p. (Trivium).

2530. Corrispondenza dell'ambasciatore Giovanni Lanfredini. Serie seconda, Vol. 1. 13 aprile 1484–9 maggio 1485). A cura di Elisabetta SCARTON. Salerno, Carlone e Istituto italiano per gli studi filosofici, 2005, LIX-721 p. (Fonti per la storia di Napoli aragonese).

2531. Digital (The) Domesday Book. Hartley Wintney, Alecto Historical Editions, 2005, 4 CD-ROM.

2532. Diocesi (Una) di confine tra Regno di Napoli e Stato pontificio: documenti e regesti del fondo pergamenaceo della Curia vescovile dei Marsi, secc. XIII–XVI. A cura di Maria Rita BERARDI. L'Aquila, Libreria Colacchi, 2005, XXXIX-397 p. (Documenti per la storia d'Abruzzo, 18).

2533. Diplomatari de la Catedral de Vic. Segle XI, fascicle 3. Ed. por Ramon ORDEIG I MATA. Vic, Arxiu i Biblioteca Episcopals, Patronat d'Estudis Osonencs, 2005, 263 p.

2534. Diplomatari del monestir de Santa Maria de Santes Creus (975–1225). Ed. por Joan PAPELL I TARDIU. Lleida, Pagès Editors, 2005, 2 vol., 973 p. (Diplomataris, 35-36).

2535. Documentación medieval de Estella: siglos XII–XVI. Ed. por Merche OSÉS URRICELQUI. Pamplona, Gobierno de Navarra, Departamento de Cultura y Turismo, Institución Príncipe de Viana, 2005, 839 p. (Corpus documental para la historia del Reino de Navarra. Serie II. Documentación municipal, buenas villas).

2536. Documentación medieval de la cuadrilla de Salvatierra: Barrundia, Elburgo-Burgelu e Iruraiz-Gauna. Ed. por Felipe POZUELO RODRÍGUEZ. Donostia, Eusko Ikaskuntza, 2005, II-623-LXXIX p. (Fuentes documentales medievales del País Vasco, 125).

2537. Early Northampton wills preserved in Northampton Record Office. Northampton, Northamptonshire Record Society, 2005, 289 p. (ill., maps). (Publication of the Northamptonshire Record Society, 42).

2538. EDWARDS (Nancy). A corpus of early medieval inscribed stones and stone sculpture in Wales. Vol. 2. South-West Wales. Cardiff, University of Wales Press, 2005, XVIII-568 p.

2539. English episcopal acta. Vol. 30. Carlisle, 1133–1292. Ed. by David M. SMITH. Vol. 31. Ely 1109–1197. Ed. by Nicholas KARN. Oxford, Oxford U. P. for the British Academy, 2005, 2 vol., LXI-253 p., CXLIX-288 p.

2540. Extenta dominii de Longdendale anno xxxiiij° Edwardi tercij = Extent of the lordship of Longdendale 1360. Ed. by Paul HARROP, Paul BOOTH and Sylvia A. HARROP. [S. l.], Record Society of Lancashire and Cheshire, 2005, LII-116 p. (Record Society of Lancashire and Cheshire, 140).

2541. Formulari (Un) i un registre del bisbe de València En Jaume d'Aragó (segle XIV). Ed. a cura de M. Milagros CARCEL ORTÍ. Valencia, Universitat de València, 2005, 439-59 p. (Fonts històriques valencianes, 17).

2542. Germania pontificia sive repertorium privilegiorum et litterarum a Romanis pontificibus ante annum MCLXXXXVIII Germaniae ecclesiis monasteriis civitatibus singulisque personis concessorum. Vol. 5/2. Pars 6. Provincia Maguntinensis. Dioeceses Hildesheimensis et Halberstadensis, Appendix Saxonia. Congessit Hermannus JAKOBS usus Heinrici BÜTTNER schedis. Gottingae, Vandenhoeck et Ruprecht, 2005, 530 p.

2543. Grand (Le) cartulaire de Conches et sa copie: transcription et analyse. Éd. par Claire DE HAAS. [S. l.], Firmin-Didot, 2005, 727 p.

2544. Grand cartulaire de Fontevraud: pancarta et cartularium abbatissae et ordinis Fontis Ebraudi. Éd. par Jean-Marc BIENVENU, Robert FAVREAU et Georges PON. Poitiers, Société des Antiquaires de l'ouest, 2005, 480 p.

2545. Great (The) roll of the pipe for the reign of King Henry III: Michaelmas 1224 (Pipe roll 68), now first printed from the originals in the National Archives. London, Pipe Roll Society, 2005, XIV-384 p. (Publications of the Pipe Roll Society, 92 = new ser., 54).

2546. KEHR (Paul Fridolin). Ausgewählte Schriften. Hrsg. v. Rudolf HIESTAND. Göttingen, Vandenhoeck & Ruprecht, 2005, 2 vol., XXXIII-1418 p. (ill.). (Abhandlungen der Akademie der Wissenschaften in Göttingen. Philologisch-Historische Klasse, 3. Folge, 250).

2547. L'OCCASO (Stefano). Fonti archivistiche per le arti a Mantova tra medioevo e rinascimento (1382–1459). Mantova, G. Arcari, 2005, XXI-438 p. (Strumenti e fonti, 9).

2548. Libro das posesións do Cabido Catedral de Ourense (1453): edición, transcrición e indices. Ed. por María Beatriz VAQUERO DÍAZ. Vigo, Universidade de Vigo, Servicio de Publicacións, 2005, 223 p. (Monografías da Universidade de Vigo. Humanidades e ciencias xurídico-sociais, 65).

2549. Llibre de deutes, trameses i rebudes de Jaume de Mitjavila i Compnayi 1345–1370: edició studi comptable i econòmic. Ed. por Víctor HURTADO OVIEDO. Barcelona, Consejo Superior de Investigaciones Científicas, 2005, 653 p. (Anuario de Estudios Medievales, 60 Annex).

2550. Manual (El) de Consells de Gandìa a la fi del segle XV. Ed. por Vicent OLASO. València, Universitat de València, 2005, 399 p. (Fonts històriques valencianes, 20).

2551. Monasterio (El) de San Isidro de Dueñas en la Edad Media: un priorato cluniacense hispano (911–1478), estudio y colección documental. Ed. por Carlos Manuel REGLERO DE LA FUENTE. León, Centro de Estudios e Investigación "San Isidoro", Caja España de Inversiones y Archivo Histórico Diocesano, 2005, 630 p. (Fuentes y estudios de historia leonesa, 106).

2552. OBERMAIR (Hannes). Bozen Süd, Bolzano Nord: Schriftlichkeit und urkundliche Überlieferung der Stadt Bozen bis 1500 / Scritturalità e documentazione archivistica della città di Bolzano fino al 1500. Bd. 1. Regesten der kommunalen Bestände 1210–1400 / Vol. 1. Regesti dei fondi comunali 1210–1400. Bolzano, Città di Bolzano / Stadt Bozen, 2005, 472 p. (ill.).

2553. OSTOS SALCEDO (Pilar). Notariado, documentos notariales y Pedro González de Hoces, veinticuatro de Córdoba. Sevilla, Universidad de Sevilla y Córdoba,

Universidad de Córdoba, 2005, 494 p. (Historia y geografía / Universidad de Sevilla, 104).

2554. *Vacat.*

2555. Paston letters and papers of the fifteenth century. Part III. Ed. by Norman DAVIS. Oxford, Early English Text Society by the Oxford U. P., 2005, CIII-319 p. (facs., maps, pl.). (Early English Text Society, 20-22).

2556. Pergamene (Le) dell'Ospedale di S. Maria della Misericordia di Perugia dalle origini al 1400: regesti. A cura di Alberto Maria SARTORE. Roma, Ministero per i beni e le attività culturali, Dipartimento per i beni archivistici e librari, Direzione generale per gli archivi, 2005, LXIII-862 p. (ill., facs.). (Pubblicazioni degli archivi di Stato. Strumenti, 169).

2557. Pergamene (Le) e i libri dei conti del secolo XIII del Monastero di S. Radegonda di Milano conservati presso l'Archivio di Stato di Milano. A cura di Maria Franca BARONI. Milano: Università degli studi di Milano, 2005, XI-175 p. (Pergamene milanesi dei secoli XII-XIII, 18).

2558. Poll (The) taxes of 1377, 1379 and 1381. Vol. 3. Wiltshire-Yorkshire: unidentified documents and additional data. Ed. by Carolyn C. FENWICK. Oxford, Oxford U. P. a. Academy, X-794 p. (Records of social and economic history, 37).

2559. Recueil des actes de Philippe Auguste roi de France. Éd. par Jean FAVIER et Michel NORTIER. Tome 6. Lettres mises sous le nom de Philippe Auguste dans le recueils de formulaires d'école, ou pouvant être considérées, quoique anonymes, comme lui ayant été attribuées. Paris, Académie des inscriptions et belles-lettres et Diffusion Brocard, 2005, 214 p. (Chartes et diplômes relatifs à l'histoire de France).

2560. Recueil des rouleaux des morts (VIIIe siècle-vers 1536). Sous la direction de Jean FAVIER. Éd. par Jean DUFOUR. Vol. 1. VIIIe siècle–1180. Paris, Diffusion de Boccard, 2005, XLVIII-725 p. (Recueil des historiens de la France. Obituaires. Série in-4°, 8).

2561. Regestes (Les) des comtes de Habsbourg en Alsace avant 1273. Éd. par Philippe NUSS. Altkirch, Société d'histoire du Sundgau, 2005, VIII-513 p.

2562. Ser ciabattus: imbreviature lucchesi del Duecento. Regesti. Vol. 1. Anni 1222–1232. A cura di Andreas MEYER. Lucca, Istituto storico lucchese, 2005, 699 p. (ill., map, geneal. tav.). (Strumenti per la ricerca, 7).

2563. Statuti (Gli) del contado di Imola (1341–1347). A cura di Corrado BENATTI. Imola, La mandragora, 2005, 628 p. (ill.). (Fonti per la storia e l'arte di Imola, 6).

2564. Villes d'Italie: textes et documents des XIIe, XIIIe, XIVe siècles. Éd. par Jean-Louis GAULIN, Armand JAMME et Véronique ROUCHON-MOUILLERON. Lyon, Presses universitaires de Lyon, 2005, 329 p. (Collection d'histoire et d'archéologie médiévales).

2565. Westfälisches Urkundenbuch. Bd. 11. Die Urkunden des Kölnischen Westfalen 1301–1325. Lieferung 3. 1321–1325. Hrsg. v. Manfred WOLF. Münster, Aschendorff, 2005, 670 p. (Veröffentlichungen der Historischen Kommission für Westfalen, I.11).

2566. Wheathampstead manor, Hertfordshire: Account roll 1405–1406. Ed. by V. S. WHITE. Wheathampstead, V. S. White, 2005, 345 p. (facs).

2567. WOLF (Manfred). Westfälisches Urkundenbuch. 11. Die Urkunden des kölnischen Westfalen, 1301–1325. Lief. 3. 1321–1325, Indices. Münster, Aschendorff, 2005, VIII-667 p. (Veröffentlichungen der Historischen Kommission für Westfalen, 1).

Cf. n°s 1-16

b. Literary sources

** 2568. KONJAVSKAJA (Elena L.). Novgorodskaja letopis' XVI v. iz sobranija T.F. Bol'shakova. (A 16th-century Novgorodian chronicle from the collection of T.F. Bolshakov: [Study and text]). *Novgorodskij istoricheskij sbornik*, 2005, 10 (20), p. 322-383.

** 2569. RICCIARDI (Alberto). L'epistolario di Lupo Di Ferrières: intellettuali, relazioni culturali e politica nell'età di Carlo il Calvo. Spoleto, Fondazione Centro italiano di studi sull'alto Medioevo, 2005, XX-396 p. (ill.). (Istituzioni e società, 7).

2570. AELRED OF RIEVAULX. The historical works. Ed. by Jane Patricia FREELAND and Marsha L. DUTTON. Kalamazoo, Cistercian Publications, 2005, XII-306 p. (Cistercian Fathers series, 56).

2571. Brenhinoedd y Saeson = The kings of the English, A.D. 682–954: texts P, R, S in parallel. Ed. by David N. DUMVILLE. Aberdeen, Dept. of History, University of Aberdeen, 2005, XVI-56 p. (Basic texts for mediaeval British history, 1).

2572. EINAR SKÚLASON. Einarr Skúlason's Geisli: a critical edition. Ed. by Martin CHASE. Toronto a. Buffalo, University of Toronto Press, 2005, VIII-249 p. (Toronto Old Norse and Icelandic studies, 1).

2573. Erikskrönika: première chronique rimée suédoise (première moitié du XIVe siècle). Introduction, traduction et commentaires de Corinne PÉNEAU. Paris, Publications de la Sorbonne, 2005, 258 p. (Textes et documents d'histoire médiévale, 5).

2574. GRAY (Thomas). Scalacronica: 1272–1363. Ed. by Andy KING. Woodbridge, Boydell Press a. Surtees Society, 2005, LXIV-288 p. (Publications of the Surtees Society, 209).

2575. Harrogate great chronicle, 1332–1841. Ed. by Malcolm George NEESAM. Lancaster, Carnegie, 2005, XXXIII-414 p. (ill).

2576. "History" (The) of Leo the Deacon. Byzantine military expansion in the tenth century. Introduction, translation, and annotations by Alice-Mary TALBOT and Denis F. SULLIVAN with the assistance of George T. DENNIS and Stamatina MAC GRATH. Washington, Dumbarton Oaks research library and collection, 2005, XIX-264 p.

2577. JULIAN, ARCHBISHOP OF TOLEDO. The story of Wamba: Julian of Toledo's Historia Wambae Regis. Ed. by Joaquín MARTÍNEZ PIZARRO. Washington, Catholic University of America Press, 2005, XVII-262 p. (ill., map).

2578. MORI (Nobuyoshi). Sukarudo sijin no saga: korumaku no saga/haruhurezu no saga. (Saga of the skald: Kormaks saga/ Halfrreoar saga). Hatano, University of Tokai Press, 2005, 202 p.

2579. POTTHAST (August). Repertorium fontium historiae medii aevi primum ab Augusto Potthast digestum, nunc cura collegii historicum e pluribus nationibus emendatum et auctum. Vol. X/3. Fontes. Sa-Si. Fontes. Sj-Sz. Romae, Istituto Storico Italiano per il Medio Evo, 2005, 2 vol., p. 249-390, p. 392-554.

2580. Segunda leyenda de la muy noble, leal y antigua ciudad de Avila. Ed. por Angel BARRIOS GARCÍA. Avila, Ediciones de la Institución "Gran Duque de Alba" de la Excma, Diputación Provincial de Avila y Obra Cultural de la Caja de Ahorros de Avila, 2005, 239 p. (Fuentes historicas abulenses, 63).

2581. Story (The) of Wamba: Julian of Toledo's Historia Wambae regis. Translated with an introduction and notes by Joaquín MARTÍNEZ PIZARRO. Washington, Catholic University of America Press, 2005, XVIII-262 p. (ill., map).

2582. VILKUL (Tat'jana L.). O proiskhozhdenii obshchego teksta Ipat'evskoj i Lavrent'evskoj letopisi za XII v. (predvaritel'nye zametki). (On the origin of the common text of the Laurentian and the Hypatian [Old Rus'] chronicles for the 12th century: Preliminary notes). *Palaeoslavica*, 2005, 13, p. 21-80.

2583. VILLANI (Giovanni). Il Villani illustrato: Firenze e l'Italia medievale nelle 253 immagini del ms. Chigiano L VIII 296 della Biblioteca vaticana. A cura di Chiara FRUGONI. Firenze, Le Lettere e Città del Vaticano, Biblioteca apostolica vaticana, 2005, 263 p. (ill.).

2584. ZANGARO (Pierantonio). La fortuna di due false cronache medievali bresciane. *Archivio storico italiano*, 2005, 163, 2, p. 283-312.

Cf. n°s 1-16, 2947-3123, 3198

§ 2. General works.

2585. ABRAMSON (Meri L.). Chelovek ital'janskogo Vozrozhdenija. Chastnaja zhizn' i kul'tura. (Man of the Italian Renaissance: Private life and culture). Ros. gos. gumanit. un-t. Moskva, RGGU, 2005, 428 p. (bibl. incl.).

2586. BAGGE (Sverre). Christianization and state formation in early Medieval Norway. *Scandinavian journal of history*, 2005, 30, 2, p. 107-134.

2587. CONTAMINE (Philippe). Pages d'histoire militaire médiévale, XIVe–XVe siècles. Paris, Diffusion De Boccard, 2005, XIII-342 p. (Mémoires de l'Academie des inscriptions et belles-lettres, 32).

2588. East Central and Eastern Europe in the early Middle Ages. Ed. by Florin CURTA. Ann Arbor, University of Michigan Press, 2005, VIII-391 p. (ill., maps).

2589. FONT (Márta). Ugarsko kraljevstvo i Hrvatska u srednjem vijeku. (Hungarian Kingdom and Croatia in the Middle Ages). *Povijesni prilozi*, 2005, 28, p. 7-22.

2590. FRANCE (John). The Crusades and the expansion of Catholic Christendom, 1000–1714. London, Routledge, 2005, XI-380 p. (ill., maps).

2591. GARCÍA DE CORTÁZAR Y RUIZ DE AGUIRRE (José Angel). Investigaciones sobre historia medieval del País Vasco, 1965–2005: 20 artículos y una entrevista. Ed. por José Ramón DÍAZ DE DURANA. Bilbao, Universidad del País Vasco, Servicio Editorial, 2005, 678 p. (ill., maps). (Historia medieval y moderna).

2592. IORGULESCU (Vasile). Le sud-est européen entre Byzance et Occident aux Xe–XIVe siècles. Le cas des roumains. Iași, Editura Trinitas, 2005, 463 p.

2593. Jornadas (2^a) de Historia Medieval de Extremadura: ponencias y comunicaciones. Ed. por Julián CLEMENTE RAMOS y Juan Luis de la MONTAÑA CONCHIÑA. Mérida, Editora Regional de Extremadura, 2005, 214 p. (ill., maps). (Documentos/actas).

2594. Medieval East Anglia. Ed. by Christopher HARPER-BILL. Woodbridge, Boydell Press, 2005, XIV-341 p. (ill.).

2595. MULLETT (Margaret). Power, relations and networks in Medieval Europe. *Revue belge de philologie et d'histoire*, 2005, 83, 1, p. 255-260.

2596. MUSIN (Aleksandr E.). Milites Christi Drevnej Rusi: Voinskaja kul'tura russkogo Srednevekov'ja v kontekste religioznogo mentaliteta. (Warrior culture of Old Rus' in the context of religious mentality). Sankt-Peterburg, Peterburgskoe vostokovedenie, 2005, 368 p. (Militaria antiqua; 8). [English summary]

2597. Omnia disce: medieval studies in memory of Leonard Boyle, O.P. Ed. by Anne J. DUGGAN, Joan GREATREX and Brenda BOLTON. Aldershot, Ashgate, 2005, XVI-322 p. (Church, faith and culture in the Medieval West).

2598. Prichernomor'e v Srednie Veka. (The Black Sea Region in the Middle Ages). Ed. Sergej P. KARPOV. Vol. 6. Moskva – Sankt-Peterburg, Aletejja, 2005, 248 p. (facs.; plates; bibl. incl.). (Trudy Istoricheskogo fakul'teta MGU). [English summaries; text partly in Italian]

2599. Rus' v IX-XIV vekakh. Vzaimodejstvie Severa i Juga. (Rus' in the 9th–14th centuries: The interrelations of the North and the South: [Essays in archaeology and history]). Ed. Vladimir Ju. KOVAL', Inna N. KUZINA, Nikolaj A. MAKAROV, Aleksej V. CHERNETSOV. RAN,

In-t arkheologii. Moskva, Nauka, 2005, 326 p. (ill.; portr.; maps; bibl. incl.). [English summaries]

2600. SATO (Shoichi), IKEGAMI (Shunichi), TAKAYAMA (Hiroshi). Seiyō chūseishi kenkyū nyūmon, zohokaiteiban. (Introduction to european medieval history). Nagoya, University of Nagoya Press, 2005, 400 p.

2601. Seiyō chūseigaku nyūmon. (Introduction to medieval studies in Europe). Ed. by Hiroshi TAKAYAMA and Shunichi IKEGAMI. Tokyo, University of Tokyo Press, 2005, 395 p.

2602. WEINFURTER (Stefan). Gelebte Ordnung, gedachte Ordnung: ausgewählte Beiträge zu König, Kirche und Reich: aus Anlass des 60. Geburtstages. Hrsg. v. Helmuth KLUGER, Hubertus SEIBERT und Werner BOMM. Ostfildern, Thorbecke, 2005, XIV-405 p. (ill., map, pl.).

2603. WICKHAM (Chris). Framing the early Middle Ages: Europe and the Mediterranean, 400–800. Oxford, Oxford U. P., 2005, XXVIII-990 p. (maps).

2604. WINKS (Robin W.), RUIZ (Teofilo F.). Medieval Europe and the world: from late antiquity to modernity, 400–1500. Oxford a. New York, Oxford U. P., 2005, XVII-302 p. (ill., maps).

Cf. n°s 358, 436, 672, 1090, 3222, 3500

§ 3. Political history.

a. General

2605. JÄCKEL (Dirk). Der Herrscher als Löwe: Ursprung und Gebrauch eines politischen Symbols im Früh- und Hochmittelalter. Köln, Böhlau, 2005, 377 p. (ill., pl.). (Beihefte zum Archiv für Kulturgeschichte, 60).

2606. LECUPPRE (Gilles). L'imposture politique au Moyen Âge: la seconde vie des rois. Paris, Presses Universitaires de France, 2005, 405 p. (Le Nœud gordien).

2607. MARTIN (Jean-Marie). Guerre, accords et frontières en Italie méridionale pendant le haut Moyen Âge: Pacta de liburia, Divisio principatus beneventani et autres actes. Rome, École française de Rome, 2005, 257 p. (Sources et documents d'histoire du Moyen Âge, 7).

2608. MARTÍNEZ DÍEZ (Gonzalo). El Condado de Castilla (711-1038). La historia frente a la leyenda. Valladolid-Madrid, Junta de Castilla y León. Consejeria de Cultura y Turismo-Marcial Pons, 2005, 2 vol., 819 p.

2609. PETTI BALBI (Giovanna). Negoziare fuori patria. Nazioni e genovesi in età medievale. Bologna, Clueb, 2005, XII-305 p.

2610. SAUL (Nigel). The three Richards: Richard I, Richard II and Richard III. London, Hambledon Continuum, 2005, VIII-287 p.

2611. ŠTIH (Peter). Istra na začetku frankovske oblasti in v kontekstu razmer na širšem prostoru med severnim Jadranom in srednjo Donavo. (Istria at the beginning of Frankish rule and in the context of the state of affairs in the wider region between the northern Adriatic and the central Danubian area). Acta Histriae, 2005, 13, 1, p. 1-20.

b. 476–900

2612. Charlemagne: empire and society. Ed. by Joanna STORY. Manchester, Manchester U. P., 2005, XVI-330 p. (ill., maps).

2613. HUMMER (Hans J.). Politics and power in early medieval Europe: Alsace and the Frankish Realm, 600–1000. Cambridge, Cambridge U. P., 2005, XIV-299 p. (Cambridge studies in medieval life and thought, 65).

2614. KOCH (Armin). Kaiserin Judith: eine politische Biographie. Husum, Matthiesen, 2005, 245 p. (Historische Studien, 486).

2615. PESIRI (Giovanni). Per una definizione dei confini del ducato di Gaeta secondo il preceptum di papa Giovanni VIII. *Bullettino dell'istituto storico italiano per il medio evo*, 2005, 107, p. 169-183 (map).

2616. RICCIARDI (Alberto). L'epistolario di Lupo di Ferrières: intellettuali, relazioni culturali e politica nell'età di Carlo il Calvo. Spoleto, Centro italiano di studi sull'alto Medioevo, 2005, XX-396 p. (ill.). (Istituzioni e società, 7).

2617. SCHMIDT (Michael). Germanien, Skandinavien, die West- und Südslawen unter den Karolingern (von 746 bis 911/987). Frankfurt am Main, Artaunon-Verlag, 2005, VI-596 p. (maps).

2618. Tassilo III von Bayern. Großmacht und Ohnmacht im 8. Jahrhundert. Hrsg. v. Lothar KOLMER und Christian ROHR. Regensburg, Pustet, 2005, 256 p.

2619. VITIELLO (Massimiliano). Momenti di Roma ostrogota: aduentus, feste, politica. Stuttgart, F. Steiner, 2005, 162 p. (Historia. Einzelschriften, 188).

Cf. n° 2685

c. 900–1300

** 2620. Recueil des actes de Philippe Auguste, roi de France. Publié sous la direction de Jean FAVIER par Michel NORTIER. Tome 6. Lettres mises sous le nom de Philippe Auguste dans les recueils de formulaires d'école, ou pouvant être considérées, quoique anonymes, comme lui ayant été attribuées. Paris, Boccard, 2005, 214 p. (Chartes et diplômes relatifs à l'histoire de France).

2621. ANDERSEN (Per). Rex imperator in regno suo. Dansk kongemagt og rigslovgivning i 1200-tallets Europa. (Le pouvoir royal et la législation au Danemark dans l'Europe du 13e siècle). Odense, Syddansk Universitetsforlag, 2005, 206 p. (ill, carte).

2622. BERTIN (Emiliano). La pace di Castelnuovo Magra (6 ottobre 1306). Otto argomenti per la paternità dantesca. *Italia medioevale e umanistica*, 2005, 56, p. 1-34.

2623. BURT (Caroline). The demise of the general Eyre in the reign of Edward I. *English historical review*, 2005, 120, 485, p. 1-14.

2624. Cyprus. Society and culture, 1191–1374. Ed. by Angel NICOLAOU-KONNARI and Chris SCHABEL Boston, Brill, 2005, XVI-403 p. (Medieval mediterranean, 58).

2625. DALEWSKI (Zbigniew). Rytuał i polityka: opowieść Galla Anonima o konflikcie Bolesława Krzywoustego ze Zbigniewem. (Rituale e politica: la storia di Gallus Anonymus sul conflitto di Bolesław Krzywousty con Zbigniew). Warszawa, Instytut Historii PAN, 2005, 259 p.

2626. DUNLOP (Eileen). Queen Margaret of Scotland. Edinburgh, National Museums of Scotland, 2005, XI-109 p. (pl., ill., geneal. tab., maps).

2627. Experience (The) of power in medieval Europe: 950–1350. Ed. by Robert F. BERKHOFER, Alan COOPER and Adam J. KOSTO. Aldershot, Ashgate, 2005, VII-292 p. (ill.).

2628. FRIEDL (Christian). Studien zur Beamtenschaft Kaiser Friedrichs II. im Königreich Sizilien (1220–1250). Wien, Verlag der Österreichischen Akademie der Wissenschaften, 2005, IX-633 p. (ill.). (Denkschriften / Österreichische Akademie der Wissenschaften. Philosophisch-Historische Klasse, 337).

2629. García Sánchez III "el de Nájera", un rey un reino en la Europa del siglo XI. XV Semana de Estudios Medievales (Nájera, Tricio y San Millán de la Cogolla 2004). Actas. Ed. por José Ignacio DE LA IGLESIA DUARTE. Logroño, Gobierno de la Rioja, Instituto de Estudios Riojanos, 2005, 539 p.

2630. Grafen, Herzöge, Könige. Der Aufstieg der frühen Staufer und das Reich 1079–1152. Hrsg. v. Hubertus SEIBERT und Jürgen DENDORFER. Ostfildern, Thorbecke, 2005, VIII-440 p. (Mittelalter-Forschungen, 18).

2631. HOLLÝ (Karol). Kňažná Salomea a uhorsko-poľské vzťahy v rokoch 1214–1241. (Die Fürstin Salomea und die ungarisch-polnischen Beziehungen in den Jahren 1214–1241). *Historický časopis*, 2005, 53, 1, p. 3-27. [Deutsche Zsfassung].

2632. HUSCROFT (Richard). Ruling England, 1042–1217. Harlow, Longman, 2005, 232 p.

2633. KINTZINGER (Martin). Die Erben Karls des Großen: Frankreich und Deutschland im Mittelalter. Ostfildern, Thorbecke, 2005, 208 p. (ill., maps).

2634. LOWER (Michael). The Barons' Crusade: a call to arms and its consequences. Philadelphia, University of Pennsylvania Press, 2005, VIII-256 p. (maps). (Middle Ages series).

2635. MATTHEW (Donald). Britain and the continent, 1000–1300. The impact of the Norman conquest. London, Hodder Arnold, 2005, X-326 p. (Britain and Europe).

2636. MENANT (François). L'Italie des communes: 1100–1350. Paris, Belin, 2005, 398 p. (Histoire).

2637. MILANI (Giuliano). I comuni italiani: secoli XII–XIV. Roma. Laterza, 2005, 200 p. (Quadrante, 126).

2638. MOORE (David). The Welsh wars of independence, c. 410–c. 1415. Stroud, Tempus, 2005, 287 p. (pl., ill., maps).

2639. MORRIS (Marc). The Bigod earls of Norfolk in the thirteenth century. Woodbridge, Boydell Press, 2005, XVI-261 p. (ill., map, geneal. tab.).

2640. Muslims, Mongols and crusaders: an anthology of articles published in the Bulletin of the School of Oriental and African Studies. Ed. by Gerald R. HAWTING. London, Routledge Curzon, 2005, XXVI-282 p.

2641. NIEUS (Jean-François). Un pouvoir comtal entre Flandre et France. Saint-Pol, 1000–1300. Bruxelles, De Boeck, 2005, 512 p. (Bibliothèque du Moyen Age, 23).

2642. ORAM (Richard D.). The reign of Alexander II, 1214–1249. Leiden, Brill, 2005, XII-343 p. (ill., pl., maps). (Northern world, 16).

2643. PINCELLI (Agata). Le liste dei ghibellini banditi e confinati da Firenze nel 1268–1269. Premessa all'edizione critica. *Bullettino dell'istituto storico italiano per il medio evo*, 2005, 107, p. 283-340 (app).

2644. PRESTWICH (Michael). Plantagenet England 1225–1360. Oxford, Clarendon Press, 2005, XXII-638 p. (New Oxford history of England).

2645. RUSSO (Luigi). Il viaggio di Boemondo d'Altavilla in Francia (1106): un riesame. *Archivio storico italiano*, 2005, 163, 1, p. 3-42.

2646. SCHEIN (Sylvia). Gateway to the heavenly city: crusader Jerusalem and the Catholic West (1099–1187). Burlington, Ashgate Publishing Company, 2005, XVI-239 p. (Church, faith and culture in the Medieval west).

2647. WEBBER (Nick). The evolution of Norman identity, 911–1154. Rochester, Boydell Press, 2005, XI-195 p.

Cf. n° 2685

d. 1300–1500

** 2648. Ordonnances de Philippe le Bon pour les Duchés de Brabant et de Limbourg et les pays d'Outre-Meuse 1430–1467. Ed. par Philippe GODDING. Bruxelles, Service public fédéral justice, 2005, 671 p. (Recueildes ordonnances des Pays-Bas. sér 1381–1506 2., 2).

2649. *Vacat.*

3. POLITICAL HISTORY

2650. Battle (The) of Crécy, 1346. Ed. by Andrew AYTON and Philip PRESTON BART. Woodbridge: Boydell Press, 2005, XI-390 p. (ill., maps). (Warfare in history).

2651. BRENNER (Bernhard). Ludwig der Bayer, ein Motor für die Urbanisierung Ostschwabens? Zu den Auswirkungen herrscherlicher Städtepolitik auf die Entwicklung der schwäbischen Städtelandschaft im ausgehenden Mittelalter. Augsburg, Wissner, 2005, VI-II-184 p. (ill., maps). (Materialien zur Geschichte des Bayerischen Schwaben, 27).

2652. BROWN (Michael). The black Douglases: war and lordship in late medieval Scotland, 1300–1455. Edinburgh, Birlinn, 2005, VI-358 p.

2653. Charles the Bold and Italy (1467–1477): politics and personnel. Ed. by Richard J. WALSH, Werner PARAVICINI and Cecil H. CLOUGH. Liverpool, Liverpool U. P., 2005, XXXIV-478 p. (Liverpool historical studies, 19).

2654. CHEVALIER (Bernard). Guillaume Briçonnet (v. 1445–1514). Un cardinal-ministre au début de la Renaissance: marchand, financier, homme d'État et prince de l'Église. Rennes, Presses universitaires de Rennes, 2005, 444 p.

2655. CHITTOLINI (Giorgio). La formazione dello Stato regionale e le istituzioni del contado: secoli XIV e XV. Milano, UNICOPLI, 2005, X-284 p. (Early modern, 18).

2656. Coups d'état à la fin du Moyen Âge? Aux fondements du pouvoir politique en Europe occidentale. Colloque international, 25–27 novembre 2002 [Casa de Velazquez et à l'Université Complutense]. Sous la dir. de François FORONDA, Jean-Philippe GENET et José Manuel NIETO SORIA. Madrid, Casa de Velásquez, 2005, 644 p.

2657. DEL VAL VALDIVIELSO (María Isabel). Isabel la Católica y su tiempo. Granada, Universidad de Granada, 2005, 448 p.

2658. Diplomatische Forschungen in Mitteldeutschland. Hrsg. v. Tom GRABER. Leipzig, Leipziger Universitätsverlag, 2005, 390 p. (ill., map). (Schriften zur sächsischen Geschichte und Volkskunde, 12).

2659. EVANGELISTI (Paolo). Politica e credibilità personale. Un diplomatico francescano tra Tabriz e la Borgogna (1450 circa–1479). *Quaderni storici*, 2005, 1, p. 3-40.

2660. FERRER MALLOL (María Teresa). Entre la paz y la guerra. La Corona catalano-aragonesa y Castilla en la Baja Edad Media. Barcelona, Consejo Superior de Investigaciones Científicas, Institución Milá y Fontanals, Departamento de Estudios Medievales, 2005, 662 p. (Anuario de estudios medievales, 39).

2661. FRANZINI (Antoine). La Corse du XVe siècle. Politique et société, 1433–1483. Ajaccio, Éditions Alain Piazzola, 2005, 750 p.

2662. Guerra y Diplomacia en la Europa Occidental, 1280–1480. XXXI Semana de Estudios Medievales. Estella 19–23 de julio de 2004. Pamplona, Gobierno de Navarra, 2005, 467 p.

2663. HARRISS (Gerald L.). Shaping the nation: England 1360–1461. Oxford, Clarendon Press, 2005, XXI-705 p. (pl., ill., maps, ports.). (New Oxford history of England).

2664. Hundred Years War. A wider focus. Ed. by L. J. Andrew VILLANON and Donald J. KAGAY. Leiden and Boston, Brill, 2005, 520 p. (ill., tab., maps).

2665. KEKEWICH (Margaret Lucille), ROSE (Susan). Britain, France, and the empire, 1350–1500. Basingstoke a. New York, Palgrave Macmillan, 2005, XXIII-308 p.

2666. KOLLER (Heinrich). Kaiser Friedrich III. Darmstadt, Wissenschaftliche Buchgesellschaft, 2005, 311 p. (Gestalten des Mittelalters und der Renaissance).

2667. KREY (Hans-Josef). Herrschaftskrisen und Landeseinheit: die Straubinger und Münchner Landstände unter Herzog Albrecht IV. von Bayern-München. Aachen, Shaker, 2005, XL-288 p. (Berichte aus der Geschichtswissenschaft).

2668. LEGUAI (André). Les ducs de Bourbon, le Bourbonnais et le royaume de France à la fin du Moyen Âge. Avant-propos d'Olivier MATTÉONI. Yzeure, Société bourbonnaise des études locales, 2005, 224 p.

2669. LIDDY (Christian Drummond). War, politics and finance in late medieval English towns: Bristol, York and the Crown, 1350–1400. Woodbridge, Royal Historical Society a. The Boydell Press, 2005, X-264 p. (ill., tab.). (Royal Historical Society studies in history).

2670. MONIÉ NORDIN (Jonas). När makten blev synlig: senmedeltid i södra Dalarna. (Quand le pouvoir devint visible: la province de Dalécarlie à la fin du Moyen Age). Stockholm, Stockholms universitet, 2005, 238 p. (ill., cartes). (Stockholm studies in archaeology, 36). [English summary].

2671. MORTIMER (Ian). The death of Edward II in Berkeley Castle. *English historical review*, 2005, 120, 489, p. 1175-1214.

2672. SAVY (Pierre). Gli stati italiani del XV secolo: una proposta sulle tipologie. *Archivio storico italiano*, 2005, 163, 4, p. 735-760.

2673. TANG (Frank). The 'rex fidelissimus': Spanish-French rivalry in the reign of Alfonso XI of Castile (1312–1350). *Tijdschrift voor rechtsgeschiedenis*, 2005, 73, 1-2, p. 93-110.

2674. TITONE (Fabrizio). Il tumulto popularis del 1450. Conflitto politico e società urbana a Palermo. *Archivio storico italiano*, 2005, 163, 1, p. 43-86.

2675. WALSH (Richard J.). Charles the Bold and Italy 1467–1477. Politics and personnel. With a postscript and bibliographical supplement by Werner PARAVICINI and an editorial preface by Cecil H. CLOUGH. Liverpool, Liverpool U. P., 2005, XXIV-478 p. (Liverpool historical studies, 19).

Cf. n°s 572, 2627

§ 4. Jews.

** 2676. BEN YOM TOV (David). Kelal qātan: original Hebrew text, medieval Latin translation, modern English translation. Ed. by Gerrit BOS, Charles BURNETT and Y. Tzvi LANGERMANN. Philadelphia, American Philosophical Society, 2005, IX-121 p. (Transactions of the American Philosophical Society, 95).

** 2677. HAVERKAMP (Eva). Hebräische Berichte über die Judenverfolgungen während des Ersten Kreuzzugs. Hannover, Hahnsche Buchhandlung, 2005, L-626 p. (Monumenta Germaniae Historica. Hebräische Texte aus dem mittelalterlichen Deutschland, 1).

** 2678. HOLLENDER (Elisabeth). Clavis commentariorum of Hebrew liturgical poetry in manuscript. Leiden a. Boston: Brill, 2005, XI-979 p. (Clavis commentariorum. Antiquitatis et Medii Aevi).

** 2679. IBN DAUD (Abraham ben David, Halevi). Se fer ha-Qabbalah: the Book of tradition. Ed. by Gerson David COHEN. London, Routledge & K. Paul, 2005, LXIII-348-75-XXIII p. (The Littman library of Jewish civilization).

2680. BRENER (Ann). Judah Halevi and his circle of Hebrew poets in Granada. Leiden a. Boston: Brill, 2005, X-155 p. (Hebrew language and literature series, 6).

2681. Cambridge companion (The) to Maimonides. Ed. by Kenneth SEESKIN. Cambridge a. New York, Cambridge U. P., 2005, XV-406 p. (Cambridge companions to philosophy).

2682. COHEN (Mark R.). Poverty and charity in the Jewish community of medieval Egypt. Princeton a. Oxford, Princeton U. P., 2005, XI-287 p. (ill.). (Jews, Christians, and Muslims from the ancient to the modern world).

2683. DAVIDSON (Herbert Alan). Moses Maimonides: the man and his works. New York a. Oxford, Oxford U. P., 2005, X-567 p.

2684. HECKER (Joel). Mystical bodies, mystical meals: eating and embodiment in medieval Kabbalah. Detroit, Wayne State U. P., 2005, X-282 p. (Raphael Patai series in Jewish folklore and anthropology).

2685. Khazary. (Khazars: [Articles]). Ed. Vladimir Ja. PETRUKHIN, Vol'f A. MOSKOVICH, Dan SHAPIRA. Evrejskij un-t v Ierusalime, Tsentr slavjan. jaz. i lit., Tsentr. Chejza po razvitiju iudaiki na rus. jaz., etc. Moskva, Mosty kul'tury – Jerusalem, Gesharim, 2005, 567 p. (ill.; bibl. incl.). (Evrei i slavjane; 16. Bibliotheca judaica).

2686. KRUGER (Steven F.). The spectral Jew: conversion and embodiment in medieval Europe. Minneapolis a. London, University of Minnesota Press, 2005, XXX-320 p. (Medieval cultures, 40).

2687. NIESNER (Manuela). Wer mit juden well disputiren: deutschsprachige Adversus-Judaeos-Literatur des 14. Jahrhunderts. Tübingen, M. Niemeyer, 2005, X-650 p. (Münchener Texte und Untersuchungen zur deutschen Literatur des Mittelalters, 128).

2688. PÉREZ PONS (E.). Fonts per l'estudi de la comunitat jueva de Mallorca: regesta i bibliografia. Barcelona, PPU, 2005, 333 p. (Catalonia hebraica; 6).

2689. ROTH (Norman). Daily life of the Jews in the Middle Ages. Westport, Greenwood Press, 2005, XXII-231 p. (ill.). (Daily life through history).

2690. SCHWARTZ (Dov). Central problems of medieval Jewish philosophy. Leiden a. Boston, Brill, 2005, VII-282 p. (The Brill reference library of Judaism, 26).

2691. SEESKIN (Kenneth). Maimonides on the origin of the world. Cambridge, Cambridge U. P., 2005, VIII-215 p.

2692. Trias (The) of Maimonides: Jewish, Arabic, and ancient culture of knowledge = Die Trias des Maimonides: jüdische, arabische und antike Wissenskultur. Ed. by / Hrsg. v. Georges TAMER. Berlin a. New York, De Gruyter, 2005, VII-455 p. (Studia Judaica, 30).

§ 5. Islam.

** 2693. Muslim-Christian polemic during the Crusades: the letter from the people of Cyprus and Ibn Abī Tālib al-Dimashqī's response. Ed. by Rifaat. Y. EBIED and David THOMAS. Leiden a. Boston, Brill, 2005, VII-516 p. (The history of Christian-Muslim relations, 2).

2694. Arabia Vitalis. Arabskij Vostok, islam, drevnjaja Aravija: Sb. st., posvjashch. 60-letiju V. V. Naumkina. (Arabian East, Islam, and Ancient Arabia: Articles dedicated to the 60th anniversary of Vitaly V. Naumkin). Ed. Aleksandr V. SEDOV, Irina M. SMILJANSKAJA. Moskva, [s. n.], 2005, 411 p. (portr., ill.; bibl. p. 40-63).

2695. Banquete (El) de las palabras: la alimentación de los textos árabes. Ed. por Manuela MARÍN y Cristina DE LA PUENTE. Madrid, Consejo Superior de Investigaciones Científicas, 2005, 274 p. (Estudios árabes e islámicos. Monografías, 10).

2696. DAFTARY (Farhad). Ismailis in medieval Muslim societies: collected studies. London, I. B. Tauris, 2005, XII-296 p. (Ismaili heritage series).

2697. DE PRÉMARE (Alfred-Louis). Wahb b. Munabbin, une figure singulière du premier Islam. *Annales*, 2005, 60, 3, p. 531-549.

2698. DIETZ (Gunther). Muslim women in southern Spain: stepdaughters of Al-Andalus. La Jolla, Center for Comparative Immigration Studies a. UCSD, 2005, 169 p. (Monograph series, 4).

2699. FRANK (Richard M.). Texts and studies on the development and history of kalām. Vol. 1. Philosophy, theology and mysticism in medieval Islam. Ed. by Dimitri GUTAS. Aldershot, Ashgate Variorum, 2005, [s. p.]. (Variorum collected studies series, 833).

2700. GLICK (Thomas F.). Islam and Christian Spain. Boston, Brill, 2005, XXII-402 p. (The medieval and early modern Iberian world, 27).

2701. GORDON (Matthew). The rise of Islam. Westport, Greenwood Press, 2005, XXXV-180 p. (ill., maps). (Greenwood guides to historic events of the medieval world).

2702. GRABAR (Oleg). Constructing the study of Islamic art. Vol. 1. Early Islamic art, 650–1100. Aldershot, Ashgate, 2005, 326 p. (Collected studies, 809).

2703. GRUBE (Ernst J.). The painted ceilings of the Cappella Palatina. Genova, Bruschettini Foundation for Islamic and Asian Art, 2005, X-518 p. (ill.). (Islamic art. Supplement, 1).

2704. Image and meaning in Islamic art. Ed. by Robert HILLENBRAND and Ernst J. GRUBE. London, Altajir Trust, 2005, 320 p. (ill.).

2705. JANIN (Hunt). The pursuit of learning in the Islamic world, 610–2003. Jefferson a. London, McFarland, 2005, V-229 p. (ill., maps, ports.).

2706. KAHYA (Esin). İbn Sina'nın Mineroloji Çalışmaları. (Avicenna's mineralogy studies). *Belleten*, 2005, 69, 256, p. 801-824.

2707. KUMAR (Sunil). La communauté musulmane et les relations hindous-musulmans dans l'Inde du nord au début du XIIIe siècle: une réévaluation politique. *Annales*, 2005, 60, 2, p. 239-264.

2708. LEV (Yaacov). Charity, endowments, and charitable institutions in medieval Islam. Gainesville, University Press of Florida, 2005, X-214 p.

2709. MELHAOUI (Mohammed). Peste, contagion et martyre: histoire du fléau en occident musulman médiéval. Paris, Publisud, 2005, 217 p. (L'Europe au fil des siècles).

2710. MOOSA (Ebrahim). Ghazālī and the poetics of imagination. Chapel Hill a. London, University of North Carolina Press, 2005, XI-349 p. (Islamic civilization and Muslim networks).

2711. MUHAMMAD IBN 'ABDĀLLAH AL-BĀZYĀR. Das Falken- und Hundebuch des Kalifen al-Mutawakkil. Ein arabischer Traktat aus dem 9. Jahrhundert. Herausgegeben, übersetzt und eingeleitet von Anna AKASOY und Stefan GEORGES. Berlin, Akademie, 2005, 198 p. (Wissenskultur und gesellschaftlicher Wandel, 11).

2712. On fiction and adab in medieval Arabic literature. Ed. by Philip F. KENNEDY. Wiesbaden, Harrassowitz, 2005, XXII-326 p. (ill.). (Studies in Arabic language and literature).

2713. Patronate and patronage in early and classical Islam. Leiden, Brill, 2005, XIV-511 p. (ill., maps). (Islamic history and civilization. Studies and texts).

2714. RAPOPORT (Yossef). Marriage, money and divorce in medieval islamic society. Cambridge, Cambridge U. P., 2005, 137 p.

2715. ROTMAN (Youval). Byzance face à l'Islam arabe, VIIe-Xe siècle. D'un droit territorial à l'identité par la foi. *Annales*, 2005, 60, 4, p. 767-788.

2716. SUHRAWARDY (Shahid). The art of the Mussulmans in Spain. Karachi, Oxford U. P., 2005, XI-190 p. (ill.).

2717. VAN GELDER (Geert Jan H.). Close relationships: incest and inbreeding in classical Arabic literature. London a. New York, I.B. Tauris, 2005, VIII-278 p. (Library of Middle East history, 9).

§ 6. Vikings.

** 2718. EINAR SKÚLASON. Einarr Skúlason's Geisli: a critical edition. Ed. by Martin CHASE. Toronto a. Buffalo, University of Toronto Press, 2005, VIII-249 p. (Toronto Old Norse-Icelandic series, 1).

** 2719. Leyes (Las) del Gulathing. Ed. por Maria Pilar FERNÁNDEZ ALVAREZ y Teodoro MANRIQUE ANTÓN. Salamanca, Ediciones Universidad de Salamanca, 2005, 230 p. (Acta Salmanticensia. Estudios filológicos, 306).

———

2720. ASHMAN ROWE (Elizabeth). The development of Flateyjarbók: Iceland and the Norwegian dynastic crisis of 1389. Odense, University Press of Southern Denmark, 2005, 486 p. (ill.). (Viking collection, 15).

2721. BOULHOSA (Patricia Pires). Icelanders and the kings of Norway: mediaeval sagas and legal texts. Leiden a. Boston, Brill, 2005, XV-256 p. (The Northern world, 17).

2722. Fondations (Les) scandinaves en Occident et les débuts du duché de Normandie. Colloque de Cerisy-la-Salle, 25–29 septembre 2002. Éd. par Pierre BAUDUIN. Caen, Publications du Centre de recherches archéologiques et historiques médiévales, 2005, 271 p. (Colloque de Cerisy-la-Salle, 11).

2723. FORTE (Angelo D. M.), ORAM (Richard D.), PEDERSEN (Frederik). Viking empires. Cambridge a. New York, Cambridge U. P., 2005, XIV-447 p. (ill., maps).

2724. Från Bysans till Norden: östliga kyrkoinfluenser under vikingatid och tidig medeltid. (De Byzance à la Scandinavie. Influences orientales à l'époque viking et au Moyen Age). Red. Henrik JANSON. Skellefteå, Artos, 2005, 224 p. (pl.).

2725. HUDSON (Benjamin T.). Viking pirates and Christian princes: dynasty, religion, and empire in the North Atlantic. Oxford, Oxford U. P., 2005, XIV-278 p. (ill., maps).

2726. O'BRIEN (Harriet). Queen Emma and the Vikings: a history of power, love and greed in eleventh-century England. London, Bloomsbury, 2005, XXI-264 p. (ill., pl., geneal. tab., maps).

2727. SINDBÆK (Søren Michael). Ruter og rutinisering: vikingetidens fjernhandel i Nordeuropa. (Les routes du grand commerce en Europe du nord à l'époque viking). København, Multivers, 2005, 356 p.

2728. TROW (M. J.). Cnut emperor of the north. Stroud, Sutton, 2005, 260 p. (ill., maps, pl.).

2729. Viking and Norse in the North Atlantic: select papers from the fourteenth Viking Congress. Ed. by Andras MORTENSEN and Símun V. ARGE. Torshavn, Foroya, 2005, 445 p. (ill.). (Annales Societatis Scientiarum Faeroensis, 44. Supplementum).

2730. Vikings (Les), premiers européens: VIIIe–XIe siècle. Les nouvelles découvertes de l'archéologie. Éd. par Régis BOYER. Paris, Autrement, 2005, 284 p.

2731. VINDING (Niels). The Viking discovery of America, 985 to 1008: the Greenland Norse and their voyages to Newfoundland. Ed. by Birgitte MOYER-VINDING. Lewiston a. Lampeter, Edwin Mellen Press, 2005, X-129 p. (ill., pl., maps).

Cf. n° 2406

§ 7. History of law and institutions.

** 2732. Archivo Foral de Bizkaia: Sección Judicial. Documentación Medieval (1284–1520). Ed. por Javier ENRÍQUEZ FERNÁNDEZ, Concepción HIDALGO DE CISNEROS AMESTOY y Adela MARTÍNEZ LAHIDALGA. Donostia, Eusko Ikaskuntza, 2005, III-454 p.-XLIV (Fuentes documentales medievales del País Vasco, 126).

** 2733. Belluno. Statuti del 1392. Testi scelti, tradotti e annotati da Enrico BACCHETTI; presentazione di Gherardo ORTALLI. Roma, Viella, 2005, 262 p. (Quaderni del Corpus statutario delle Venezie, 1).

** 2734. Capbreu (El) de Banifallet de 1373. Ed. por Ventura CASTELLVELL. Barcelona, Societat Catalana de Llengua i Literatura. Filial de l'Institut d'Estudis Catalans, 2005, 108 p.

** 2735. Cavarzere. Statuti del 1401–1402. Testi scelti, tradotti e annotati da Enrico BACCHETTI; presentazione di Gherardo ORTALLI. Roma, Viella, 2005, 80 p. (Quaderni del Corpus statutario delle Venezie, 2).

** 2736. Clavis Canonum: selected canon law collections before 1140. Access with data processing. Ed. by Linda FOWLER-MAGERL. Hannover: Hahnsche, 2005, 282 p. a. 1 CD-ROM (ill.). (Monumenta Germaniae historica. Hilfsmittel, 21).

** 2737. Curia Regis rolls of the reign of Henry III preserved in the Public Record Office. Vol. 19. 33 to 34 Henry III (1249–1250). Ed. by David CROOK. Woodbridge, Boydell Press, 2005, XVI-521 p. (Curia Regis rolls 135-140, 229).

** 2738. Decretum civitatis danceke: gdański kodeks prawa lubeckiego z 1263 roku. Hrsg. v. Tadeusz DOMAGAŁA. Gdańsk, Muzeum Historyczne Miasta Gdańska, 2005, 143 p.

** 2739. DEUS (Valdemar de). Cartas de foral de Torre de Moncorvo: Junqueira, Santa Cruz da Vilariça, e Torre de Moncorvo (sua génese e cronologia medieval). Torre de Moncorvo, Câmara Municipal de Torre de Moncorvo, 2005, 84 p.

** 2740. Earliest (The) English law reports. Ed. by Paul BRAND. London, Selden Society, 2005, CXLIX-322 p. (The publications of the Selden Society, 122).

** 2741. Kolberger (Das) Rechtsbuch: der Kolberger Kodex des lübischen Rechts von 1297. Faksimiledruck der verschollenen Handschrift mit hochdeutscher Übersetzung und Glossar. Hamburg, P. Jancke, 2005, 247 p. (Beiträge zur Geschichte der Stadt Kolberg und des Kreises Kolberg-Körlin, 32).

** 2742. Liber (Il) iurium del comune di Monselice (secoli XII–XIV). A cura di Sante BORTOLAMI e Luigi CABERLIN. Roma, Viella, 2005, LXX-838 p. (ill.). (Fonti per la storia della Terraferma veneta, 21).

** 2743. Llibre (El) Vermell de Falset: privilegis i ordinacions de la vila als segles XIII–XIV. Ed. por Ezequiel GORT JUANPERE. Reus, Ediciones del Migdia CB, 2005, 69 p. (ill.).

** 2744. Odluke dubrovačkih vijeća 1390–1392. Ed. Nella LONZA, Z. ŠUNDRICA. Zagreb, HAZU, 2005, 454 p. (Monumenta historica Ragusina, 6).

** 2745. Ordonnances de Philippe le Bon pour les duchés de Brabant et de Limbourg et les Pays d'Outre-Meuse, 1430–1467 = Verordeningen van Filips de Goede voor de hertogdommen Brabant en Limburg en de landen van Overmaas, 1430–1467. Éd. par Philippe GODDING. Bruxelles, Service public fédéral Justice, 2005, 671 p. (Recueil des ordonnances des Pays-Bas. Sect. 2, t. 2, 1 série, 1381-1506).

** 2746. Parliament rolls (The) of Medieval England, 1275–1504. General editor: Chris GIVEN-WILSON. Vol. 1. Edward I, 1275–1294. Ed. by Paul BRAND with the assistance of Shelagh SNEDDON. Vol. 2. Edward I, 1294–1307. Ed. by Paul BRAND with the assistance of Shelagh SNEDDON. Vol. 3. Edward II, 1307–1327. Ed. by Seymour PHILLIPS. Vol. 4. Edward III, 1327–1348. Ed. by Seymour PHILLIPS and Mark ORMROD. Vol. 5. Edward III, 1351–1377. Ed. by Mark ORMROD. Vol. 6. Richard II, 1377–1384. Ed. by Geoffrey MARTIN and Chris GIVEN-WILSON. Vol. 7. Richard II, 1385–1397. Ed. by Chris GIVEN-WILSON. Vol. 8. Henry IV, 1399–1413. Ed. by Chris GIVEN-WILSON. Vol. 9. Henry V, 1413–1422. Ed. by Chris GIVEN-WILSON. Vol. 10. Henry VI, 1422–1431. Ed. by Anne CURRY. Vol. 11. Henry VI, 1432–1445. Ed. by Anne CURRY. Vol. 12. Henry VI, 1447–1460. Ed. by Anne CURRY and Rosemary HORROX. Vol. 13. Edward IV, 1461–1470. Ed. by Rosemary HORROX. Vol. 14. Edward IV, 1472–1483. Ed. by Rosemary HORROX. Vol. 15.

Richard III, 1484–1485, Henry VII, 1485–1487. Ed. by Rosemary HORROX. Vol. 16. Henry VII, 1489–1504. Ed. by Rosemary HORROX. Woodbridge, Boydell & Brewer, 2005, 16 vol., IV-688 p., II-688 p., II-468 p., IV-462 p., II-428 p., II-427 p., II-428 p., II-558 p., II-329 p., II-482 p., II-508 p., II-545 p., I-394 p., I-469 p., I-395 p., I-435 p.

** 2747. Regesten der Reichsstadt Aachen. Einschließliche des Aachener Reiches und der Reichsabtei Burtscheid. 5. 1381–1395. Bearb. v. Thomas R. KRAUS. Düsseldorf, Droste, 2005. XXXIX-527 p. (Publikationen der Gesellschaft für rheinische Geschichtskunde, 47).

** 2748. Statuten (Die) der Reichsstadt Mühlhausen in Thüringen. Hrsg. v. Wolfgang WEBER und Gerhard LINGELBACH. Köln, Böhlau, 2005, XXXVI-121 p. (ill., pl.).

** 2749. Statuti (Gli) di Dronero, 1478. A cura di Giuseppe GULLINO. Cuneo, Società per gli studi storici archeologici ed artistici della provincia di Cuneo, 2005, 247 p. (Marchionatus Saluciarum monumenta. Fonti, 4).

** 2750. Statuti di Cologna Veneta del 1432, con le aggiunte quattro-cinquecentesche e la ristampa anastatica dell'edizione del 1593. A cura di Bruno CHIAPPA e Gian Maria VARANINI. Roma, Viella, 2005, 466 p. (Corpus statutario delle Venezie, 19).

** 2751. Weimarer (Die) Stadtbücher des späten Mittelalters: Edition und Kommentar. Hrsg. v. Henning STEINFÜHRER. Köln, Böhlau, 2005, XXXVI-266 p. (facs., ill., pl.). (Veröffentlichungen der Historischen Kommission für Thüringen, 11. Grosse Reihe).

2752. BELJAKOVA (Elena V.), SHCHAPOV (Jaroslav N.). Novelly imperatora Justiniana v russkoj pis'mennoj traditsii. K istorii retseptsii rimskogo prava v Rossii. (Novellae of Emperor Justinian in Rus' written tradition: To the history of the reception of the Roman law in Russia). RAN, In-t rossijskoj istorii. Moskva, IRI RAN, 2005, 59 p.

2753. BIANCALANA (Joseph). Monetary penalty clauses in thirteenth-century England. *Tijdschrift voor rechtsgeschiedenis*, 2005, 73, 3-4, p. 231-265.

2754. Centenario (VI) della morte di Baldo degli Ubaldi, 1400–2000. A cura di Carla FROVA, Maria Grazia NICO OTTAVIANI e Stefania ZUCCHINI. Perugia, Università degli studi, 2005, 562 p. (pl., ill., map).

2755. D'AVRAY (David L.). Authentication of marital status: a thirteenth-century English Royal annulment process and late Medieval cases from the Papal penitentiary. *English historical review*, 2005, 120, 488, p. 987-1013.

2756. GALLEGOS VÁZQUEZ (Federico). Estatuto jurídico de los peregrinos en la España medieval. Santiago de Compostela, Xunta de Galicia, 2005, [s. p.]. (ill.). (Colección científica).

2757. GIULIODORI (Serena). De rebus uxoris. Dote e successione negli statuti bolognesi (1250–1454). *Archivio storico italiano*, 2005, 163, 4, p. 651-686.

2758. GROSSI (Paolo). Il sistema giuridico medievale e la civiltà comunale. *Rivista di storia del diritto italiano*, 2005, 78, p. 31-52.

2759. Hofordnungen (Die) der Herzöge von Burgund. 1. Herzog Philipp der Gute 1407–1467. Hrsg. v. Holger KRUSE und Werner PARAVICINI. Ostfildern, Thorbecke, 2005, 507 p. (Instrumenta, 15).

2760. Justice temporelle (La) dans les territoires angevin aux XIIIe et XIVe siècles. théories et pratiques. Actes du colloque d'Aix-en-Provence, 21–23 février 2002. Sous la dir. de Jean-Paul BOYER, Anne MAILLOUX et Laure VERDON. Roma, École française de Rome, 2005, 470 p. (Collection de l'École française de Rome, 354).

2761. KAZBEKOVA (Elena V.). Priemy ritmizirovannoj prozy (cursus) v svodakh dekretal'nogo prava i ispol'zovanie cursus v novellakh Innokentija IV (1243–1254). (The method of rhythmic prose (cursus) in the collections of decretal law and the use of cursus in Novellae of Pope Innocent IV, 1243–1254). *Srednie veka*, 2005, 66, p. 218-264.

2762. Land, lords and peasants. Peasants' right to control land in the Middle Ages and the Early modern period: Norway, Scandinavia and the Alpine region. Report from a seminar in Trondheim, November 2004. Ed. by Tore IVERSEN and John Ragnar MYKING. Trondheim, Norwegian University of Science and Technology, Department of History and Classical Studies, 2005, 309 p. (Trondheim studies in history).

2763. LOGAN (Francis Donald). The Medieval Court of Arches. Woodbridge, Boydell Press, 2005, XLVIII-240 p. (Canterbury and York society, 95).

2764. MAIHOLD (Harald). Strafe für fremde Schuld? Die Systematisierung des Strafbegriffs in der spanischen Spätscholastik und Naturrechtslehre. Köln, Böhlau, 2005, XVI-393 p. (ill.). (Konflikt, Verbrechen und Sanktion in der Gesellschaft Alteuropas. Symposien und Synthesen, 9).

2765. MAYADE-CLAUSTRE (Julie). Le corps lié de l'ouvrier. Le travail et la dette à Paris au XVe siècle. *Annales*, 2005, 60, 2, p. 383-408.

2766. MUSSON (Anthony). Boundaries of the law: geography, gender, and jurisdiction in medieval and early modern Europe. Aldershot, Ashgate, 2005, X-196 p. (map).

2767. O'Sullivan (Carolin). Die Ahndung von Rechtsbrüchen der Seeleute im mittelalterlichen hamburgischen und hansischen Seerecht, 1301–1482. Frankfurt am Main, Lang, 2005, LVIII-458 p. (Rechtshistorische Reihe, 305).

2768. RAO (Riccardo). Risorse collettive e tensioni giurisdizionali nella pianura vercellese e novarese (XII-XIII secolo). *Quaderni storici*, 2005, 3, p. 753-776.

2769. RIBALTA I HARO (Jaume). Dret urbanístic medieval de la Mediterrània. Barcelona, Universitat Pompeu Fabra, Institut d'Estudis Catalans, 2005, 309 p. (Memóries de la Secció Històrico-Arqueològica, 66. Estudis d'història del dret, 1).

2770. SEGURA URRA (Félix). Fazer justicia: fuero, poder público y delito en Navarra (siglos XIII–XIV). Pamplona, Gobierno de Navarra, Departamento de Cultura y Turismo, Institución Príncipe de Viana, 2005, 502 p. (ill., maps). (Serie Historia, 115).

2771. Stagnation oder Fortbildung? Aspekte des allgemeinen Kirchenrechts im 14. und 15. Jahrhundert. Hrsg. v. Martin BERTRAM. Tübingen, Niemeyer, 2005, XV-425 p. (Bibliothek des Deutschen Historischen Instituts in Rom, 108).

2772. SZUROMI (Anzelm Sz.). 'Work in progress' – The transition from cathedral teaching to university instruction of canon law in the 11th and 12th centuries – Some notes on Anselm's Collection compared with Ivo's works. *Zeitschrift der Savigny-Stiftung für Rechtsgeschichte Kanonistische Abteilung*, 2005, 91, p. 758-766.

2773. TAMM (Ditlev), VOGT (Helle). How Nordic are the Nordic medieval laws? København, University of Copenhagen Press, 2005, 237 p. (Medieval legal history, 1).

2774. TAN (Elaine S.). Origins and evolution of the Medieval church's usury laws: economic self interest or systematic theology? *Journal of European economic history*, 2005, 34, 1, p. 263-281.

2775. TELLIEZ (Romain). "Per potentiam officii". Les officiers devant la justice dans le royaume de France au XIVe siècle. Paris, Honoré Champion, 2005, 704 p. (Études d'histoire médiévale, 8).

2776. THEISEN (Frank). Die Bedeutung des SC Velleianum in der Rechtspraxis des Hochmittelalters. *Zeitschrift der Savigny-Stiftung für Rechtsgeschichte Romanistische Abteilung*, 2005, 122, p. 103-137.

2777. VALLERANI (Massimo). La giustizia pubblica medievale. Bologna, Il Mulino, 2005, 304 p.

Cf. n° 6618

§ 8. Economic and social history.

* 2778. INGRAM (Martin). Men and women in late Medieval and Early Modern times. *English historical review*, 2005, 120, 487, p. 732-758.

2779. ABERTH (John). The Black Death. The great mortality of 1348-1350: a brief history with documents. Boston, Bedford/St. Martin's, 2005, XV-199 p. (ill., map, port.). (Bedford series in history and culture).

2780. ANDENMATTEN (Bernard). La Maison de Savoie et la noblesse vaudoise (XIIIe–XIVe s.). Supériorité féodale et autorité princière. Lausanne, Société d'histoire de la Suisse romande, 2005, XIII-722 p. (Mémoires et documents publiés par la Société d'histoire de la Suisse romande. Quatrième série, 8).

2781. ARAI (Yukio). Jentori kara mita chūsei kōki igirisu shakai. (Gentry and society in fifteenth century England). Tokyo, Tosui Shobo, 2005, 430 p.

2782. ARSLAN (Ermanno A.). Repertorio dei ritrovamenti di moneta altomedievale in Italia (489–1002). Spoleto, Centro italiano di studi sull'alto Medioevo, 2005, X-182 p. (Testi, studi, strumenti, 18).

2783. ASTARITA (Carlos). Del feudalismo al capitalismo: cambio social y político en Castilla y Europa Occidental, 1250–1520. Valencia, Universitat de València y Granada, Editorial Universidad de Granada, 2005, 264 p. (Història).

2784. At the margins: minority groups in premodern Italy. Ed. by Stephen J. MILNER. Minneapolis, University of Minnesota Press, 2005, X-283 p. (ill.). (Medieval cultures, 39).

2785. AURELL (Martin), BOYER (Jean-Paul), COULET (Noël). La Provence au Moyen Âge. Aix-en-Provence, Publications de l'Université de Provence, 2005, 360 p.

2786. BLANCHARD (Ian). Mining, metallurgy and minting in the Middle Ages. 3. Continuing Afro-European supremacy, 1250–1450. Stuttgart, Franz Steiner Verlag, 2005, 781 p.

2787. BLOMQUIST (Thomas W.). Merchant families, banking and money in medieval Lucca. Aldershot, Ashgate, 2005, [s. p.]. (Variorum collected studies series).

2788. BOJOVIC (Bosko). Entre Venise et l'Empire Ottoman, les métaux précieux des Balkans (XVe–XVIe siècle). *Annales*, 2005, 60, 6, p. 1277-1297.

2789. BRIGGS (Chris). Taxation, warfare, and the early fourteenth century 'crisis' in the north: Cumberland lay subsidies, 1332–1348. *Economic history review*, 2005, 58, 4, p. 639-672.

2790. Campagne medievali: strutture materiali, economia e società nell'insediamento rurale dell'Italia settentrionale (VIII–X secolo). Atti del convegno, Nonantola (MO), San Giovanni in Persiceto (BO), 14–15 marzo 2003. A cura di Sauro GELICHI. Mantova, SAP, 2005, 290 p. (ill., maps). (Università degli studi di Venezia. Dipartimento di scienze dell'antichità e del Vicino Oriente. Documenti di archeologia, 37).

2791. CAMPBELL (Bruce M. S.). The Agrarian problem in the early fourteenth century. *Past and present*, 2005, 188, p. 3-70.

2792. Carte di famiglia: strategie, rappresentazione e memoria del gruppo familiare di Totone di Campione (721–877). A cura di Stefano GASPARRI e Cristina LA ROCCA. Roma, Viella, 2005, 390 p. (pl., ill., maps). (Altomedioevo, 5).

2793. Città e contado nel Mezzogiorno tra Medioevo ed età moderna. A cura di Giovanni VITOLO. Salerno, Laveglia, 2005, 350 p. (facs., maps). (Centro interuniversitario per la storia delle città campane nel Medioevo. Quaderni, 1).

2794. Comercio (El) en la Edad Media: XVI Semana de Estudios Medievales (Nájera y Tricio, agosto 2005). Ed. por José Ignacio DE LA IGLESIA DUARTE. Logroño, Gobierno de la Rioja, Instituto de Estudios Riojanos, 2005, 647 p.

2795. Contabilità (La) delle Case dell'Ordine Teutonico in Puglia e in Sicilia nel Quattrocento. A cura di Kristjan TOOMASPEG; presentazione di Hubert HOUBEN. Galatina, Congedo, 2005, 415 p. (Acta Theutonica, 2).

2796. Credito e usura fra teologia, diritto e amministrazione: linguaggi a confronto, sec. XII-XVI. A cura di Diego QUAGLIONI, Giacomo TODESCHINI e Gian Maria VARANINI. Rome, École française de Rome, 2005, 308 p. (Collection de l'École française de Rome, 346).

2797. CROUCH (David). The birth of nobility: social change in England and France, 900–1300. Harlow, Pearson a. Longman, 2005, XIII-361 p. (ill., ports.).

2798. D'ADAMI (Luisa Maria). Alimentazione e malattie infantili nel pieno e nel tardo Medioevo. Firenze, Atheneum, 2005, 91 p. (Mercator, 75).

2799. DE LA RONCIÈRE (Charles Marie). Firenze e le sue campagne nel trecento. Mercanti, produzione, traffici. Firenze, Olschki, 2005, XVI-440 p.

2800. DIRLMEIER (Ulf), FUHRMANN (Bernd). Räumliche Aspekte sozialer Ungleichheit in der spätmittelalterlichen Stadt. *Vierteljahresschrift für Sozial- und Wirtschaftsgeschichte*, 2005, 92, 4, p. 424-439.

2801. DUMOLYN (Jan), HAEMERS (Jelle). Patterns of urban rebellion in medieval Flanders. *Journal of Medieval history*, 2005, 31, 4, p. 369-393.

2802. DYER (Christopher). An age of transition? Economy and society in England in the later Middle Ages. Oxford, Clarendon Press, 2005, VI-293 p. (ill., maps). (The Ford lectures 2001).

2803. EMANUELSSON (Anders). Kyrkojorden och dess ursprung: Oslo biskopsdöme perioden ca 1000-ca 1400. (Origine du domaine foncier de l'évêché d'Oslo entre l'an 1000 et le début du 15e siècle). Göteborg, Historiska institutionen, Göteborgs universitet, 2005, 339 p. (ill., diagr., cartes, tab.). (Avhandlingar från Historiska institutionen, Göteborgs universitet, 44). [English summary].

2804. ERDEM (İlhan). İlk Dönem Selçuklu-Moğol İlişkilerinin İktisadi Boyutu (1243–1258). [The economic dimension of the Saljukid-Mongols relationships in the first period (1243–1258)]. *Tarih Araştırmaları Dergisi (Ankara Üniversitesi Dil Ve Tarih Coğrafya Fakültesi)*, 2005, 24, 38, p. 1-10.

2805. FEHSE (Monika). Dortmund um 1400: Hausbesitz, Wohnverhältnisse und Arbeitsstätten in der spätmittelalterlichen Stadt. Bielefeld, Verlag für Regionalgeschichte, 2005, 397 p. (Dortmunder Mittelalter-Forschungen, 4).

2806. FERNÁNDEZ TRABAL (Josep). Política, societat i economia una vila Catalana medieval: Molins de Rei, 1190–1512. Barcelona, Ajuntament de Molins de Rei, 2005, 248 p. (ill.).

2807. FIETZE (Katharina). Im Gefolge Dianas: Frauen und höfische Jagd im Mittelalter (1200–1500). Köln, Böhlau, 2005, X-176 p. (ill., pl.). (Beihefte zum Archiv für Kulturgeschichte, 59).

2808. FLIGHT (Colin). The survey of the whole of England: studies of the documentation resulting from the survey conducted in 1086. Oxford, Archaeopress, 2005, X-157 p. (charts). (BAR British series, 405).

2809. Forms of servitude in northern und central Europe. Decline, resistance, and expansion. Ed. by Paul FREEDMAN and Monique BOURIN. Turnhout, Brepols, 2005, X-449 p. (Medieval texts and cultures of northern europe; 9).

2810. Funktions- und Strukturwandel spätmittelalterlicher Hospitäer im europäischen Vergleich. Hrsg. v. Michael MATHEUS. Stuttgart, Steiner, 2005, XII-260 p. (ill, maps, plans). (Geschichtliche Landeskunde).

2811. GARCÍA FERNÁNDEZ (Ernesto). Bilbao, Vitoria y San Sebastián: espacios para mercaderes, clérigos y gobernantes en el Medievo y la modernidad. Bilbao, Universidad del País Vasco, Servicio Editorial = Euskal Herriko Unibertsitatea, Argitalpen Zerbitzua, 2005, 542 p. (Historia medieval y moderna).

2812. GAUDE-FERRAGU (Murielle). D'or et de cendres. La mort et les funérailles des princes dans le royaume de France au bas Moyen Âge. Villeneuve-d'Ascq, Presses universitaires du Septentrion, 2005, 397 p.

2813. GIAGNACOVO (Maria). Mercanti toscani a Genova: traffici, merci e prezzi nel XIV secolo. Napoli, Edizioni scientifiche italiane, 2005, 285 p. (ill.). (Università degli studi del Molise. Dipartimento di scienze economiche, gestionali e sociali, 7).

2814. GILLI (Patrick). Villes et sociétés urbaines en Italie: milieu XIIe–milieu XIVe siècle. Paris, Sedes, 2005, 302 p. (ill., maps). (Regards sur l'histoire. Histoire médiévale).

2815. GRAHAM-LEIGH (Elaine). The southern French nobility and the Albigensian Crusade. Rochester, Boydell Press, 2005, 187 p.

2816. GULLBEKK (Svein H.). Natural and money economy in medieval Norway. *Scandinavian journal of history*, 2005, 30, 1, p. 3-19.

2817. HAMMOND (P. W.). Food and Feast in Medieval England. Stroud, Sutton, 2005, VIII-200 p. (ill.).

2818. HANSSON (Anders). Agrarkris och ödegårdar i Jämtland. (Crise agraire et abandon de fermes au bas Moyen Age). Östersund, Jamtli, 2005, 177 p. (ill.). [English summary].

2819. HINTON (David Alban). Gold and gilt, pots and pins: possessions and people in medieval Britain. Oxford, Oxford U. P., 2005, XI-439 p. (ill., pl.). (Medieval history and archaeology).

2820. Honra de hidalgos, yugo de labradores: nuevos textos para el estudio de la sociedad rural alavesa (1332–1521). Ed. por Francisco Javier GOICOLEA JULIÁN. Bilbao, Universidad del País Vasco, Servicio Editorial = Euskal Herriko Unibertsitatea, Argitalpen Zerbitzua, 2005, 231 p. (Historia medieval y moderna).

2821. HURST (J. D.). Sheep in the Cotswolds: the medieval wool trade. Stroud, Tempus, 2005, 224 p. (ill., maps). (Revealing history).

2822. HUTTON (Ronald). Seasonal festivity in late Medieval England: some further reflections. *English historical review*, 2005, 120, 485, p. 66-79.

2823. Impôt (L') dans les villes de l'Occident méditerranéen, XIIIe–XVe siècle: colloque tenu à Bercy les 3, 4 et 5 octobre 2001. Éd. par Denis MENJOT, Albert RIGAUDIÈRE et Manuel SÁNCHEZ MARTÍNEZ. Paris, Ministère de l'économie, des finances et de l'industrie, Comité pour l'histoire économique et financière de la France, 2005, 609 p. (Histoire économique et financiére de la France. Animation de la recherche).

2824. Individu (L') au Moyen Âge. Individuation et individualisation avant la modernité. Sous la dir. de Brigitte Miriam BEDOS-REZAK et Dominique IOGNA-PRAT. Paris, Aubier, 2005, 380 p.

2825. ISRAEL (Uwe). Fremde aus dem Norden: transalpine Zuwanderer im spätmittelalterlichen Italien. Tübingen, M. Niemeyer, 2005, VIII-380 p. (Bibliothek des Deutschen Historischen Instituts in Rom, 111).

2826. IVANIČ (Peter). Komunikácie na Pohroní v období stredoveku. (Die Kommunikationen im Pohronie [Mittelslowakei] im Mittelalter). *Historický časopis*, 2005, 53, 4, p. 617-632. [Deutsche Zsfassung].

2827. JOMINI (Marie-Noëlle), MOSER (Marie Hélène), ROD (Yann). Les Hôpitaux vaudois au Moyen Age: Lausanne, Lutry, Yverdon. Lausanne, Université de Lausanne, 2005, XI-432 p. (Cahiers lausannois d'histoire médiévale, 37).

2828. JOSEFSON (Kristina). Sökandet efter en tidsanda: i spåren efter Absalon. (A la recherche de l'air d'une époque: dans les traces de l'archevêque Absalon [1128–1201]). Stockholm, Almqvist & Wiksell International, 2005, 227 p. (ill.). (Lund studies in medieval archaeology, 38). [English summary].

2829. KANJASCHIN (Jurij). Parzelle als Flurbestandteil des mittelalterlichen Burgund (nach den Quellen des Mâconbezirkes im X.–XI. Jahrhundert). *Vierteljahresschrift für Sozial- und Wirtschaftsgeschichte*, 2005, 92, 4, p. 440-452.

2830. KNAPP (Keith Nathaniel). Selfless offspring. Filial children and social order in Medieval China. Honolulu, University of Hawai'i Press, 2005, X-300 p.

2831. LLUCH BRAMON (Rosa). Els Remences. La senyoria de l'Almoina de Girona als segles XIV i XV. Girona, Documenta universitaria, Associació d'història rural de les comarques girondine i Centre de Recerca d'història rural dela Universitat de Girona (ILCC, secció Vicens Vives), 2005, 420 p.

2832. LÓPEZ RODRÍGUEZ (Carlos). Nobleza y poder político en el Reino de Valencia: 1416–1446. Valencia, Universidad de Valencia, 2005, 438 p. (Col·lecció Oberta. Sèrie Història).

2833. MAC INTOSH (Marjorie Keniston). Working women in English society, 1300–1620. Cambridge a. New York, Cambridge U. P., 2005, XIV-291 p. (ill., map.).

2834. Marché (Le) de la terre au Moyen Âge. Ed. par Laurent FELLER et Chris WICKHAM. Roma, École française de Rome, 2005, XII-670 p. (Coll. de l'École française de Rome, 350).

2835. Marco Tangheroni (Per): studi su Pisa e sul Mediterraneo medievale offerti dai suoi ultimi allievi. A cura di Cecilia IANNELLA, Michele CAMPOPIANO e Marco TANGHERONI. Pisa, ETS, 2005, 233 p.

2836. MAZEL (Florian). Amitié et rupture de l'amitié: moines et grands laïcs provençaux au temps de la crise grégorienne (milieu XIe–milieu XIIe siécle). *Revue historique*, 2005, 633, p. 53-96.

2837. MILITZER (Klaus). Die Geschichte des deutschen Ordens. Stuttgart, Kohlhammer, 2005, 225 p.

2838. MIRA JODAR (Antonio José). Entre la renta y el impuesto. Fiscalidad, finanzas y crecimiento económico en las villas reales del sur valenciano (siglos XIV–XVI). València, Publicacions de Universitat València, 2005, 268 p.

2839. MORIMOTO (Yoshiki). Seiō chūsei keiseiki no nōson to toshi. (Villages and towns during the formation of medieval western Europe). Tokyo, Iwanamishoten, 2005, 488 p.

2840. MOURÓN FIGUEROA (Cristina). El ciclo de York: sociedad y cultura en la Inglaterra bajomedieval. Santiago de Compostela, Universidade de Santiago de Compostela, 2005, 310 p. (Monografias de la Universidad de Santiago de Compostela, 215).

2841. MÜLLER (Miriam). Social control and the hue and cry in two fourteenth-century villages. *Journal of Medieval history*, 2005, 31, 1, p. 29-53 (tab.).

2842. MUNRO (John). Spanish merino wools and the nouvelles draperies: an industrial transformation in the late medieval Low Countries. *Economic history review*, 2005, 58, 3, p. 431-484.

2843. MURRAY (James M.). Bruges, cradle of capitalism, 1280–1390. Cambridge, Cambridge U. P., 2005, XII-409 p. (ill., maps).

2844. Negociar en la Edad Media; actas del coloquio celebrado en Barcelona los días 14, 15 y 16 de octubre de 2004 = Négocier au Moyen Âge: actes du colloque tenu à Barcelone du 14 au 16 octobre 2004. Ed. por María Teresa FERRER MALLOL [et al.]. Barcelona, Consejo Superior de Investigaciones Científicas, Institución Milá y Fontanals, Depto. de Estudios Medievales y Madrid, Casa de Velázquez y Val-de-Marne, Université de Paris-XII, 2005, 593 p. (Anuario de estudios medievales, 61. Anejo).

2845. NEVILLE (Cynthia J.). Native lordship in medieval Scotland: the earldoms of Strathearn and Lennox, c. 1140–1365. Dublin, Four Courts Press, 2005, XV-255 p. (maps).

2846. NIGHTINGALE (Pamela). Some new evidence of crises and trends of mortality in late Medieval England. *Past and present*, 2005, 187, p. 33-68. – EADEM. The economic, political, and social influences on levels of credit in late medieval England. Swindon, Economic and Social Research Council, 2005, [s. p.]. (End of award report).

2847. North-east England in the later Middle Ages. Ed. by Christian Drummond LIDDY and R. H. BRITNELL. Woodbridge a. Rochester, Boydell Press, 2005, XIII-250 p. (ill., maps). (Regions and regionalism in history, 3).

2848. ORMROD (W. M.). The royal nursery: a household for the younger children of Edward III. *English historical review*, 2005, 120, 486, p. 398-415.

2849. ORUÇ (Hatice). 15. Yüzyılda Bosna Sancağı ve İdari Dağılımı. (Sandjak de Bosnie au 15e siècle et sa structure administrative). *OTAM Ankara Üniversitesi Osmanlı Tarihi Araştırma Ve Uygulama Merkezi Dergisi*, 2005, 18, p. 249-272.

2850. PAOLILLI (Antonio Luigi). Development and crisis in the late Middle Ages: the role of trade. *Journal of European economic history*, 2005, 34, 3, p. 687-720.

2851. PAPAJÍK (David). Stavy a majetek na Moravě ve 14.–16. století. (Estates and property in Moravia in the 14th–16th centuries). *Historický časopis*, 2005, 53, 3, p. 545-558.

2852. PARKS (Tim). Medici money: banking, metaphysics, and art in fifteenth-century Florence. New York, W.W. Norton & Company, 2005, XII-273 p. (Enterprise).

2853. PÉREZ GONZÁLEZ (Silvia María). La mujer en la Sevilla de finales de la Edad Media: solteras, casadas y vírgenes consagradas. Sevilla, Secretariado de Publicaciones de la Universidad de Sevilla y Ateneo de Sevilla, 2005, 199 p. (ill.). (Premios Historia "Ateneo de Sevilla", 4).

2854. PIRILLO (Paolo). Forme e strutture del popolamento nel contado fiorentino. Vol. 1. Gli insediamenti nell'organizzazione dei populi (prima metà del XIV secolo), Firenze, Olschki, 2005, 2 vol., X-727 p. (Cultura e Memoria, 27).

2855. Poteri signorili e feudali nelle campagne dell'Italia settentrionale fra Tre e Quattrocento: fondamenti di legittimità e forme di esercizio. Atti del Convegno di studi, Milano, 11-12 aprile 2003. A cura di Federica CENGARLE, Giorgio CHITTOLINI e Gian Maria VARANINI. Firenze, Firenze U. P., 2005, 263 p. (maps). (Quaderni di Reti medievali rivista, 1).

2856. REITEMEIER (Arnd). Pfarrkirchen in der Stadt des späten Mittelalters. Politik, Wirtschaft und Verwaltung. Wiesbaden, Steiner, 2005, 722 p. (Vierteljahrschrift für Sozial- und Wirtschaftsgeschichte, Beihefte, 177).

2857. RICHARDSON (Gary). The prudent village: risk pooling institutions in Medieval English agriculture. *Journal of economic history*, 2005, 65, 2, p. 386-413.

2858. RIGBY (Stephen Henry). The overseas trade of Boston in the reign of Richard II. Woodbridge a. Rochester, Boydell Press, 2005, XXXVIII-302 p. (The publications of the Lincoln Record Society, 93).

2859. SCORDIA (Lydwine). "Le roi doit vivre du sien": la théorie de l'impôt en France (XIIIe–XVe siècles). Paris, Institute des Études Augustiniennes, 2005, 539 p. (Collection des Études Augustiniennes; Série Moyen Âge et Temps Modernes, 40).

2860. SEGEŠ (Vladimír). Prešporský pitaval. Zločin a trest v stredovekej Bratislave. (Der Pitaval von Pressburg. Das Verbrechen und die Strafe in mittelalterliche Bratislava). Bratislava, Perfekt, 2005, 220 p.

2861. SHAW (David Gary). Necessary conjunctions: the social self in medieval England. New York a. Basingstoke, Palgrave Macmillan, 2005, XII-292 p. (The new Middle Ages).

2862. SŁOŃ (Marek). "Kein stat findt man der gelich". Zur vergleichenden mittelalterlichen Spital- und Stadtgeschichte: Polen – Rhein-Maas-Raum. *Vierteljahresschrift für Sozial- und Wirtschaftsgeschichte*, 2005, 92, 4, p. 453-473.

2863. SOLCAN (Şarolta). Femeile din Moldova, Transilvania și Țara Românească în evul mediu. (Women in Moldavia, Transylvania and Wallachia in the Middle Ages). București, Editura Universității, 2005, 292 p.

2864. SUTTON (Anne F.). The mercery of London: trade, goods and people, 1130–1578. Aldershot, Ashgate, 2005, XVII-670 p.

2865. SZABÓ (Péter). Woodland and forests in medieval Hungary. Oxford, Archaeopress, 2005, 187 p. (ill., maps). (British Archaeological reports international series, 1348).

2866. THOMAS (Richard). Animals, economy and status: integrating zooarchaeological and historical data in the study of Dudley Castle, West Midlands (c. 1100–1750). Oxford, Archaeopress, 2005, VIII- 232 p. a. 1 CD-ROM (British Archaeological reports, 392).

2867. TOGNETTI (Sergio). Il ruolo della Sardegna nel commercio mediterraneo del Quattrocento. Alcune considerazioni sulla base di fonti toscane. *Archivio storico italiano*, 2005, 163, 1, p. 87-132.

2868. Town and country in the Middle Ages: contrasts, contact and interconnections, 1100–1500. Ed. by Katherine GILES and Christopher DYER. Leeds, Maney, 2005, VI-330 p. (ill.). (Society for Medieval Archaeology monograph, 22).

2869. Villes et campagnes en Austrasie du IVe–Xe siècle: sociétés, economies, territoires, christianisation. XXVIe Journées Internationales d'Archéologie mérovingienne: Nancy, 22–25 septembre 2005. Paris, Association française d'archéologie mérovingienne, 2005, 128 p. (Bulletin de liaison, 29).

2870. VINYOLES I VIDAL (Teresa-Maria). Història de les dones a la Catalunya medieval. Lleida, Pagès editors i Vic, Eumo Editorial, 2005, 264 p. (Biblioteca d'historia de Catalunya, 6).

2871. VOGT (Helle). Slægtens funktion i nordisk højmiddelalderret: kanonisk retsideologi og fredsskabende lovgivning. (La fonction du lignage dans le haut-Moyen Age scandinave: idéologie du droit ecclésiastique et législation de paix). København, Jurist- og økonomforbundet, 2005, 344 p.

2872. Voies d'eau, commerce, et artisanat en Gaule mérovingienne. Éd. par Jean PLUMIER et Maude REGNARD. Namur, Ministère de la Région wallonne, 2005, 445 p. (ill.). (Études et documents. Archéologie, 10).

2873. Water management in medieval rural economy = les usages de l'eau en milieu rural au moyen âge. Ruralia V, 27e septembre–2e octobre 2003, Lyon Villard-Sallet Région Rhône-Alpes France. Éd. par Jan KLÁPSTE. Prague, Institute of Archaeology, 2005, 269 p. (ill., maps, plans). (Památky archeologické, 17 Supplementum).

2874. ZILMER (Kristel). "He drowned in Holmr's Sea, his cargo-ship drifted to the sea-bottom, only three came out alive": records and representations of Baltic traffic in the Viking age and the early Middle Ages in early Nordic sources. Tartu, Tartu U. P., 2005, 403 p. (ill.). (Dissertationes philologiae scandinavicae Universitatis Tartuensis. Nordistica Tartuensia).

§ 8. Addendum 2003.

2875. TAKAHASHI (Kiyonori). Kokka to mibunsei gikai: Furansu kokuseishi kenkyū. (State and estates: studies in the states institution of France in the old regime). Tokyo, Toyo Shorin, 2003, 448 p.

Cf. n°s 121, 144, 2205, 3500

§ 9. History of civilization, literature, technology and education.

a. Civilization

** 2876. BROZYNA (Martha A.). Gender and sexuality in the Middle Ages: a medieval source documents reader. Jefferson a. London, McFarland, 2005, XII-316 p.

2877. ATZBACH (Rainer). Leder und Pelz am Ende des Mittelalters und zu Beginn der Neuzeit: die Funde aus den Gebäudehohlräumen des Mühlberg-Ensembles in Kempten (Allgäu). Bonn, R. Habelt, 2005, 282 p. (ill., pl., maps). (Bamberger Schriften zur Archäologie des Mittelalters und der Neuzeit, 2. Mühlbergforschungen Kempten, 1).

2878. BAUMGARTNER (Emmanuéle), HARF-LANCNER (Laurence). Dire et penser le temps au Moyen Âge: frontières de l'histoire et du roman. Paris, Presses Sorbonne nouvelle, 2005, 264 p. (ill., pl.).

2879. *Vacat.*

2880. Bestiaires médiévaux: nouvelles perspectives sur les manuscrits et les traditions textuelles: communications présentées au XVe Colloque de la Société Internationale Renardienne (Louvain-la-Neuve, 19-22.8. 2003). Éd. par Baudouin VAN DEN ABEELE. Louvain-la-Neuve, Institut d'Études Médiévales de l'Université Catholique de Louvain, 2005, IX-318 p. (ill., pl.). (Publications de l'Institut d'études médiévales. Textes, études, congrès, 21).

2881. BEZNER (Frank). Vela Veritatis: Hermeneutik, Wissen und Sprache in der Intellectual History des 12. Jahrhunderts. Leiden u. Boston, Brill, 2005, XIV-695 p. (Studien und Texte zur Geistesgeschichte des Mittelalters).

2882. BOQUET (Damien). L'ordre de l'affect au Moyen Âge. Autour de l'anthropologie affective d'Aelred de Rievaulx. Caen, Publications du CRAHM, 2005, 382 p.

2883. Borders, barriers, and ethnogenesis: frontiers in late Antiquity and the Middle Ages. Ed. by Florin CURTA. Turnhout, Brepols, 2005, 265 p. (Studies in the early Middle Ages, 12).

2884. BRAUN (Manuel), HERBERICHS (Cornelia). Gewalt im Mittelalter: Realitäten, Imaginationen. München, Fink, 2005, 436 p. (ill.).

2885. CALKIN (Siobhain Bly). Saracens and the making of English identity: the Auchinleck manuscript. New York a. London, Routledge, 2005, XII-299 p. (Studies in medieval history and culture).

2886. Care for the here and the hereafter: memoria, art and ritual in the Middle Ages. Ed. by Truus VAN BUEREN and Andrea VAN LEERDAM. Turnhout, Brepols, 2005, 332 p. (ill., plan).

2887. CAVALLARI (Cinzia). Oggetti di ornamento personale dall'Emilia Romagna bizantina: i contesti di rinvenimento. Bologna, Ante quem, 2005, 226 p. (Università di Bologna, Dipartimento di archeologia. Studi e scavi, 13).

2888. CHEWNING (Susannah Mary). Intersections of sexuality and the divine in medieval culture: the word made flesh. Aldershot, Ashgate, 2005, XII-213 p.

2889. Comunicare e significare nell'alto medioevo. Spoleto, 15–20 aprile 2004. Spoleto, Centro Italiano di Studi sull'Alto Medioevo, 2005, 2 vol., XVI-1188 p.

(Settimane di studio del Centro italiano di studi sull'alto Medioevo, 52).

2890. Cultural (The) heritage of medieval rituals: genre and ritual. Ed. by Eyolf ØSTREM. København, Museum Tusculanum Press, University of Copenhagen, 2005, 336 p. (ill.).

2891. D'AVRAY (David L.). Medieval marriage. Symbolism and society. Oxford a. New York, Oxford U. P., 2005, XII-322 p. (Oxford scholarship online).

2892. DANGLER (Jean). Making difference in medieval and early modern Iberia. Notre Dame, University of Notre Dame Press, 2005, VIII-218 p. (maps).

2893. DENERY (Dallas George). Seeing and being seen in the later medieval world: optics, theology and religious life. Cambridge, Cambridge U. P., 2005, X-202 p. (Cambridge studies in medieval life and thought, 63).

2894. DI VENOSA (Elena). Die deutschen Steinbücher des Mittelalters: magische und medizinische Einblicke in die Welt der Steine. Göppingen, Kümmerle, 2005, 138 p. (Göppinger Arbeiten zur Germanistik, 714).

2895. DUSHIN (Oleg E.). Ispoved' i sovest' v zapadnoevropejskoj kul'ture XIII–XVI vv. (Confession and conscience in the West European culture of the 13th–16th centuries). Sankt-Peterburg, Izd-vo Sankt-Peterburg. un-ta, 2005, 152 p.

2896. Emotions in the heart of the city (14th–16th century). Ed. by Elodie LECUPPRE-DESJARDIN and Anne-Laure VAN BRUAENE. Turnhout, Brepols, 2005, VIII-298 p. (ill.). (Studies in European urban history, 5).

2897. Gender and sexuality in the Middle Ages: a medieval source documents reader. Ed. by Martha A. BROZYNA. Jefferson a. London, McFarland, 2005, XII-316 p.

2898. Gentry culture in late medieval England. Ed. by Raluca RADULESCU and Alison TRUELOVE. Manchester, Manchester U. P., 2005, XII-220 p. (Manchester medieval studies).

2899. GRAGNOLATI (Manuele). Experiencing the afterlife: soul and body in Dante and medieval culture. Notre Dame, University of Notre Dame Press, 2005, XVII-279 p. (ill.). (The William and Katherine Devers series in Dante studies).

2900. Grèce (La) antique sous le regard du Moyen Âge occidental: colloque. Actes. Éd. par Jean LECLANT et Michel ZINK. Paris, Académie des inscriptions et belles-lettres et Boccard, 2005, XII-220 p. (Cahiers de la villa "Kérylos", 16).

2901. HAYE (Thomas). Lateinische Oralität: gelehrte Sprache in der mündlichen Kommunikation des hohen und späten Mittelalters. Berlin, W. De Gruyter, 2005, VI-176 p.

2902. Hofkultur in Frankreich und Europa im Spätmittelalter = La culture de cour en France et en Europe a la fin du Moyen-Age. Hrsg. v. Christian FREIGANG, Jean Claude SCHMITT, Chrystele BLONDEAU und Werner PARAVICINI. Berlin, Akademie Verlag, 2005, 451 p. (Passagen / Deutsches Forum für Kunstgeschichte, 11 = Passages / Centre Allemand d'Histoire de l'Art, 11).

2903. Homo risibilis: capacità di ridere e pratica del riso nelle civiltà medievali, Siena, 2–4 ottobre 2002: Atti delle I Giornate internazionali interdisciplinari di studio sul Medioevo. A cura di Francesco MOSETTI CASARETTO. Alessandria, Edizioni dell'Orso, 2005, XIV-403 p.

2904. Household, women, and Christianities in late antiquity and the Middle Ages. Ed. by Anneke B. MULDER-BAKKER and Jocelyn WOGAN-BROWNE. Turnhout, Brepols, 2005, 260 p. (Medieval women: texts and contexts, 14).

2905. HÜLSEN-ESCH (Andrea von). Inszenierung und Ritual in Mittelalter und Renaissance. Düsseldorf, Droste, 2005, 322 p. (Studia humaniora, 40).

2906. KARRAS (Ruth Mazo). Sexuality in medieval Europe: doing unto others. New York a. London, Routledge, 2005, VIII-200 p.

2907. KLEINSCHMIDT (Harald). Perception and action in Medieval Europe. Rochester, Boydell, 2005, VIII-198 p.

2908. Komik und Sakralität: Aspekte einer ästhetischen Paradoxie in Mittelalter und früher Neuzeit. Hrsg. v. Anja GREBE und Nikolaus STAUBACH. Frankfurt am Main u. Oxford, P. Lang, 2005, 249 p. (Tradition, Reform, Innovation).

2909. Kommunikation im Spätmittelalter: Spielarten, Wahrnehmungen, Deutungen. Hrsg. v. Romy GÜNTHART und Michael JUCKER. Zürich, Chronos, 2005, 159 p. (ill.).

2910. LAURIOUX (Bruno). Une histoire culinaire du Moyen Âge. Paris, Champion, 2005, 476 p. (ill., maps). (Sciences, techniques et civilisations du Moyen Âge à l'aube des Lumiéres, 8).

2911. LAUWERS (Michel). Naissance du cimetière. Lieux sacrés et terre des morts dans l'Occident médiéval. Paris, Aubier, 2005, 393 p.

2912. LOCHRIE (Karma). Heterosyncrasies: female sexuality when normal wasn't. Minneapolis a. London, University of Minnesota Press, 2005, XXVIII-178 p.

2913. LOWE (Jeremy). Desiring truth: the process of judgment in fourteenth-century art and literature. New York a. London, Routledge, 2005, IX-259 p. (Studies in medieval history and culture, 30).

2914. LOWNEY (Chris). A vanished world: medieval Spain's golden age of enlightenment. New York a. London, Free Press, 2005, 320 p. (ill., map).

2915. MÄND (Anu). Urban carnival: festive culture in the Hanseatic cities of the eastern Baltic, 1350–1550. Turnhout, Brepols, 2005, XXV-374 p. (ill.). (Medieval texts and cultures of Northern Europe, 8).

2916. Mediation (The) of symbol in late medieval and early modern times. Ed. by Rudolf SUNTRUP, Jan R. VEENSTRA and Anne M. BOLLMANN. Frankfurt am Main a. New York, P. Lang, 2005, XVIII-312 p. (ill.). (Medieval to early modern culture = Kultureller Wandel vom Mittelalter zur frühen Neuzeit, 5).

2917. Memoria: ricordare e dimenticare nella cultura del Medioevo = Memoria: erinnern und vergessen in der Kultur des Mittelalters. A cura di / Hrsg. v. Michael BORGOLTE, Cosimo Damiano FONSECA, Hubert HOUBEN. Bologna, Il Mulino, 2005, 405 p. (ill.). (Annali dell'Istituto storico italo-germanico in Trento. Contributi = Jahrbuch des italienisch-deutschen historischen Instituts in Trient. Beiträge, 15).

2918. Métier (Du) des armes à la vie de cour, de la forteresse au château de séjour: familles et demeures aux XIVe–XVIe siècles. Éd. par Jean-Marie CAUCHIES et Jacqueline GUISSET. Turnhout, Brepols, 2005, X-258 p. (ill.).

2919. Mice (Of) and men: image, belief and regulation in late medieval England. Ed. by Linda CLARK. Woodbridge, Boydell Press, 2005, 181 p. (ill., map, plan). (Fifteenth century, 5).

2920. MILIS (Ludovicus). Religion, culture, and mentalities in the medieval Low Countries: selected essays. Ed. by Jeroen DEPLOIGE. Turnhout, Brepols, 2005, 338 p. (ill., maps).

2921. MILLS (Robert). Suspended animation: pain, pleasure and punishment in medieval culture. London, Reaktion, 2005, 248 p. (ill., pl.).

2922. Mort (La) écrite: rites et réthoriques du trépas au Moyen Âge. Éd. par Estelle DOUDET. Paris, Presses de l'Université Paris-Sorbonne, 2005, 186 p. (Cultures et civilisations médiévales, 30).

2923. Multicultural Europe and cultural exchange in the Middle Ages and Renaissance: 9th Arizona Center for Medieval and Renaissance Studies conference. Ed. by James Peter HELFERS. Turnhout, Brepols, 2005, 182 p. (ill.). (Arizona studies in the Middle Ages and the Renaissance, 12).

2924. NEJIME (Kenichi). Firenze kyōwakoku no Hyūmanisuto: itaria runesansu kenkyū. (Humanists of the Republic of Florence: studies in the Italian renaissance). Tokyo, Sobunsha, 2005, 364 p.

2925. NEWHAUSER (Richard). In the garden of evil: the vices and culture in the Middle Ages. Toronto, Pontifical Institute of Mediaeval Studies, 2005, XXIII-568 p. (Papers in mediaeval studies, 18).

2926. NOLAN (Maura). John Lydgate and the making of public culture. Cambridge a. New York, Cambridge U. P., 2005, IX-276 p. (Cambridge studies in medieval literature, 58).

2927. ÖZYETGIN (A. Melek). Orta zaman Türk dili ve kültürü üzerine incelemeler. (Recherches sur la langue et la culture turques du moyen âge). İstanbul, Ötüken, 2005, 272 p.

2928. Prêcher la paix et discipliner la société: Italie, France, Angleterre, XIIIe–XVe siècle. Éd. par Rosa Maria DESSÌ. Turnhout, Brepols, 2005, 462 p. (ill.). (Collection d'études médiévales de Nice, 5).

2929. Reading images and texts: medieval images and texts as forms of communication: papers from the Third Utrecht Symposium on Medieval Literacy, Utrecht, 7–9 December 2000. Ed. by Mariëlle HAGEMAN and Marco MOSTERT. Turnhout, Brepols, 2005, X-563 p. (ill., pl.). (Utrecht studies in medieval literac, 8).

2930. ROSENWEIN (Barbara H.). Y avait-il un "moi" au haut Moyen Âge? *Revue historique*, 2005, 633, p. 31-52.

2931. SCHLOTHEUBER (Eva). Die Autobiographie Karls IV. und die mittelalterlichen Vorstellungen vom Menschen am Scheideweg. *Historische Zeitschrift*, 2005, 281, 3, p. 561-591.

2932. Scientia in margine: études sur les marginalia dans les manuscrits scientifiques du Moyen Age à la Renaissance. Éd. par Danielle JACQUART et Charles S. F. BURNETT. Genève, Droz, 2005, XII-402 p. (ill.). (Hautes études médiévales et modernes, 88).

2933. SÈRE (Bénédicte). De la vérité en amitié: une phénoménologie médiévale du sentiment dans les commentaires de "l'Éthique à Nicomaque" (XIIIe–XVe siècle). *Revue historique*, 2005, 636, p. 793-820.

2934. Sexuality and culture in medieval and renaissance Europe. Ed. by Philip M. SOERGEL. New York, AMS Press, 2005, XV-285 p. (ill.). (Studies in medieval and renaissance history, 2).

2935. SLOCUM (Kay Brainerd). Medieval civilization. London, Laurence King Pub., 2005, 448 p. (ill.).

2936. SMITH (Julia M. H.). Europe after Rome: a new cultural history 500–1000. Oxford, Oxford U. P., 2005, XIII-384 p. (ill., maps).

2937. STUARD (Susan Mosher). Gilding the market: luxury and fashion in fourteenth century Italy. Philadelphia, University of Pennsylvania Press a. Bristol, University Presses Marketing, 2005, VIII-322 p. (ill.). (Middle Ages series).

2938. Sutton Hoo: a seventh-century princely burial ground and its context. Ed. by M. O. H. CARVER, with contr. by Angela Care EVANS [et al.]. London, British Museum Press, 2005, XL-536 p. (Reports of the Research Committee of the Society of Antiquaries of London, 69).

2939. Thirteenth century England: proceedings of the Durham conference 2003. Ed. by Michael PRESTWICH, R. H. BRITNELL and Robin FRAME. Woodbridge, Boydell & Brewer, 2005, XII-226 p. (ill., pl.). (Thirteenth Century Conference, 10).

2940. Trasmissione (La) dei saperi nel Medioevo, secoli XII-XV: diciannovesimo Convegno internazionale di studi, Pistoia, 16–19 maggio 2003. A cura di Anna BENVENUTI PAPI. Pistoia, Centro italiano di studi di storia e d'arte, 2005, X-422 p. (ill.).

2941. Virtuelle Räume: Raumwahrnehmung und Raumvorstellung im Mittelalter. Akten des 10. Symposiums des Mediävistenverbandes, Krems, 24.–26. März 2003. Hrsg. v. Elisabeth VAVRA. Berlin, Akademie Verlag, 2005, X- 386 p. (ill., maps).

2942. Visual culture and the German Middle Ages. Ed. by Kathryn STARKEY and Horst WENZEL. New York, Palgrave Macmillan, 2005, XII-290 p. (ill.).

2943. Voci (Le) del Medioevo: testi, immagini, tradizioni. Atti del VII Convegno internazionale (Rocca Grimalda, 21–22 settembre 2002). A cura di Nicolò PASERO e Sonia Maura BARILLARI. Alessandria, Edizioni dell'Orso, 2005, 270 p. (ill.). (L'immagine riflessa. Quaderni. Serie miscellanea, 8).

2944. YOUNG (Simon). A.D. 500: a journey through the dark isles of Britain and Ireland. London, Weidenfeld & Nicolson, 2005, 260 p. (maps).

2945. ZABBIA (Marino). Dalla propaganda alla periodizzazione. L'invenzione del "buon tempo antico". *Bullettino dell'istituto storico italiano per il medio evo*, 2005, 107, p. 247-82.

2946. Zhanry i formy v pis'mennoj kul'ture Srednevekov'ja. (Genres and forms in the written culture of the Middle Ages: [Articles]). Ed. Julija IVANOVA. RAN, In-t mirovoj literatury im. A.M. Gor'kogo. Moskva, IMLI RAN, 2005, 271 p. (bibl. incl.).

b. Literature

* 2947. BISANTI (Armando). Un ventennio di studi su Rosvita di Gandersheim. Spoleto, Centro italiano di studi sull'alto Medioevo, 2005, IX-208 p. (Studi, 12).

* 2948. SCHALLER (Dieter). Initia carminum Latinorum saeculo undecimo antiquiorum: bibliographisches Repertorium für die lateinische Dichtung der Antike und des früheren Mittelalters. Supplementband. Hrsg. v. Ewald KÖNSGEN, John TAGLIABUE und Thomas KLEIN. Göttingen, Vandenhoeck & Ruprecht, 2005, XLVIII-492 p.

** 2949. ANDREA DA BARBERINO. Il guerrin meschino. A cura di Mauro CURSIETTI. Roma, Antenore, 2005, XLVII-704 p. (Medioevo e umanesimo, 109).

** 2950. Anglo-Saxon manuscripts in microfiche facsimile. Interim index (volumes 1-10). Ed. by Alger Nicolaus DOANE and Matthew T. HUSSEY. Tempe, Arizona Center for Medieval and Renaissance Studies, 2005, 186 p. (Medieval & Renaissance Texts & Studies, 309).

** 2951. BOIARDO (Matteo Maria). Pastorali. A cura di Marina RICCUCCI. Milano, Fondazione Pietro Bembo e Parma, U. Guanda, 2005, LIII-277 p. (Biblioteca di scrittori italiani).

** 2952. Breudwyt Maxen Wledic. Ed. by Brynley F. ROBERTS. Dublin, School of Celtic Studies, Dublin Institute for Advanced Studies, 2005, XCVII-80 p. (Mediaeval and modern Welsh series, 11).

** 2953. Chansonnier (The) of Oxford Bodleian MS Douce 308: essays and complete edition of texts. Ed. by Mary ATCHISON. Aldershot, Ashgate, 2005, 580 p.

** 2954. CHAUCER (Geoffrey). Troilus and Criseyde with facing page Il Filostrato: context, criticism. Ed. by Stephen A. BARNEY. New York, W.W. Norton, 2005, XXVII-628 p. (Norton critical edition).

** 2955. Chaucerian (The) apocrypha: a selection. Ed. by Kathleen FORNI. Kalamazoo, Medieval Institute Publications, 2005, VII-169 p. (Middle English texts series).

** 2956. Concordance of Medieval Occitan: the troubadours narrative verse = Concordance de l'Occitan Médiéval: les troubadours les textes narratifs en vers. Turnhout, Brepols, 2005, 1 CD-ROM 1 users' guide.

** 2957. Corpus of Hispanic chivalric romances: texts and concordances. Vol. 1. Ed. by Pablo ANCOS-GARCÍA and Ivy A. CORFIS. New York, Hispanic Seminary of Medieval Studies, 2005, 1 CD-ROM (Spanish series, 134).

** 2958. DANTE ALIGHIERI. Rime su CD-ROM. A cura di Domenico DE ROBERTIS, Paolo MASTANDREA e Luigi TESSAROLO. Firenze, Edizioni del Galluzzo per la Fondazione Ezio Franceschini, 2005, LI-628 p. e 1 CD-ROM (Archivio romanzo, 7).

** 2959. ELEONORE VON ÖSTERREICH. „Pontus und Sidonia" in der Eleonore von Österreich zugeschriebenen Fassung (A): nach der Gothaer Handschrift Chart. A 590. Hrsg. v. Reinhard HAHN. Göppingen, Kümmerle Verlag, 2005, XXVII-183 p. (Göppinger Arbeiten zur Germanistik, 726).

** 2960. EUSTACHE LE MOINE. Le roman d'Eustache le Moine: nouvelle édition, traduction, présentation et notes. Éd. par Anthony J. HOLDEN et Jacques MONFRIN. Louvain, Peeters, 2005, V-171 p. (Ktémata, 18).

** 2961. FLAVIO BIONDO. Italy illuminated. Ed. by Jeffrey A. WHITE. Cambridge, Harvard U. P., 2005, XXVII-489 p. (The I Tatti Renaissance library, 20).

** 2962. Fragment (Un) de la Genèse en vers: fin XIIIe-début XIVe siècle: édition du Ms. Brit. Libr. Harley 3775. Éd. par Julia C. SZIRMAI. Genève, Librairie Droz, 2005, 284 p. (Textes littéraires français, 574).

** 2963. GRUFFUDD AP MAREDUDD AP DAFYDD. 2. Cerddi crefyddol. Golygwyd gan Barry J. LEWIS. Aberystwyth, Canolfan Uwchefrydiau Cymreig a Cheltaidd Prifysgol Cymru, 2005, XIX-227 p. (Cyfres Beirdd yr Uchelwyr).

** 2964. HAY (Gilbert). The prose works of Sir Gilbert Hay. Vol. 2. The buke of the law of armys. Ed. by Jonathan A. GLENN. Edinburgh, Scottish Text Society, XII-401 p. (Scottish Text Society, 3).

** 2965. HEINRICH VON DEM TÜRLIN. Die Krone (Verse 12282-30042): nach der Handschrift Cod. Pal. germ. 374 der Universitätsbibliothek Heidelberg. Hrsg. v. Fritz Peter KNAPP, Klaus ZATLOUKAL, Alfred EBEN-

BAUER und Florian KRAGL. Tübingen, M. Niemeyer, 2005, XXXI-514 p. (Altdeutsche Textbibliothek, Nr. 118).

** 2966. Katalog der althochdeutschen und altsächsischen Glossenhandschriften. Hrsg. v. Stefanie STRICKER, Yvonne GOLDAMMER und Claudia WICH-REIF. Berlin, Walter de Gruyter, 2005, 6 vol., XIV-3016 p. (ill.).

** 2967. KÖBLER (Gerhard). Altdeutsch: Katalog aller allgemein bekannten Altdeutschhandschriften: Althochdeutsch, Altsächsisch, Altniederfränkisch. Giessen-Lahn, Arbeiten zur Rechts- und Sprachwissenschaft Verlag, 2005, XLV-1019 p. (Arbeiten zur Rechts- und Sprachwissenschaft, 60).

** 2968. MARBODE. Lapidario = Liber lapidum. Éd. par María Esther HERRERA. Paris, Belles Lettres, 2005, CXVIII-226 p. (Auteurs latins du Moyen Âge, 15).

** 2969. Millstätter (Der) Physiologus: Text, Übersetzung, Kommentar. Hrsg. v. Christian SCHRÖDER. Würzburg, Königshausen & Neumann, 2005, 385 p. (Würzburger Beiträge zur deutschen Philologie, 24).

** 2970. Nibelungenlied (Das): nach der St. Galler Handschrift. Hrsg. v. Hermann REICHERT. Berlin: W. de Gruyter, 2005, VII-549 p. (De Gruyter Texte).

** 2971. Rime due e trecentesche tratte dall'Archivio di Stato di Bologna. A cura di Sandro ORLANDO. Bologna, Commissione per i testi di lingua, 2005, LXXX-327 p. (Collezione di opere inedite o rare, 161).

** 2972. SACCHETTI (Franco). Il pataffio. A cura di Federico DELLA CORTE. Bologna, Commissione per i testi di lingua, 2005, CXVI-170 p. (Collezione di opere inedite o rare, 160).

** 2973. SACCHETTI (Giannozzo). Rime. A cura di Tiziana ARVIGO. Bologna, Commissione per i testi di lingua, 2005, CCIV-64 p. (Scelta di curiosità letterarie inedite o rare dal secolo XIII al XIX, 296).

** 2974. Studente (Uno) alla scuola del Pontano a Napoli: le Recollecte del ms. 1368 (T. 5.5) della Biblioteca angelica di Roma. Edizione critica con introduzione e commento. A cura di Antonietta IACONO. Napoli, Loffredo editore, 2005, 184 p. (Nova itinera humanitatis latinae, 4).

** 2975. Testi veronesi dell'età scaligera: edizione, commento linguistico e glossario. A cura di Nello BERTOLETTI. Padova, Esedra, 2005, 576 p. (ill.). (Vocabolario storico dei dialetti veneti, 6).

** 2976. Welsh poetry and English pilgrimage: Gruffudd ap Maredudd and the Rood of Chester. Ed. by Barry James LEWIS. Aberystwyth, University of Wales Centre for Advanced Welsh and Celtic Studies, 2005, 51 p. (Research papers, 23).

** 2977. WIRNT VON GRAFENBERG. Wigalois. Hrsg. v. Johannes Marie Neele KAPTEYN, Sabine HEIMANN-SEELBACH und Ulrich SEELBACH. Berlin u. New York, De Gruyter, 2005, 329 p. (De Gruyter Texte).

2978. ALLEN (Elizabeth). False fables and exemplary truth in later Middle English literature. New York, Palgrave Macmillan, 2005, VIII-225 p. (The new Middle Ages).

2979. ANDERSON (John Julian). Language and imagination in the Gawain-poems. Manchester, Manchester U. P., 2005, 247 p. (Manchester medieval literature).

2980. ARCHER (Robert). The problem of woman in medieval Hispanic literature. Woodbridge, Tamesis, 2005, 227 p. (Colección Támesis, Serie A, Monografías, 214).

2981. ARNULFUS AURELIANENSIS. Arnulfi Aurelianensis Glosule Ovidii Fastorum. A cura di Jörg Rudolf RIEKER. Firenze, SISMEL edizioni del Galluzzo, 2005, LXXX-307 p. (Millennio medievale, 54. Testi, 14).

2982. Articulation (L') langue-littérature dans les textes médiévaux anglais. Actes des journées d'étude de juin 2000 et 2001 à l'Université de Nancy II. Éd. par Colette STEVANOVITCH. Nancy, AMAES, 2005, XI-474 p. (Publications de l'Association des médiévistes anglicistes de l'enseignement supérieur. Collection GRENDEL, 5).

2983. AVIANUS. Il Novus Avianus di Vienna. A cura di Emanuela SALVADORI. Genova, Università di Genova, Facoltà di lettere, 2005, 218 p. (Pubblicazioni del Dipartimento di archeologia, filologia classica e loro tradizioni, 221. Favolisti latini medievali, 12).

2984. BARTH (Ferdinand). Dante zwischen Franziskus und Reformation. Darmstadt, Bogen, 2005, 122 p. (ill.).

2985. BERNDT (Rainer). Schrift, Schreiber, Schenker. Studien zur Abtei Sankt Viktor in Paris und den Viktorinern. Berlin, Akademie, 2005, 394 p. (Corpus victorinum, instrumenta, 1).

2986. BERTAU (Karl). Schrift, Macht, Heiligkeit in den Literaturen des jüdisch-christlich-muslimischen Mittelalters. Hrsg. v. Sonja GLAUCH. Berlin, De Gruyter, 2005, XXXIII-678 p. (ill.).

2987. BETHLEHEM (Ulrike). Guinevere, a medieval puzzle: images of Arthur's queen in the medieval literature of England and France. Heidelberg, Winter, 2005, VI-441 p. (ill.). (Anglistische Forschungen, 345).

2988. BETTELLA (Patrizia). The ugly woman: transgressive aesthetic models in Italian poetry from the Middle Ages to the Baroque. Toronto, University of Toronto Press, 2005, VIII-259 p. (Toronto Italian studies).

2989. BIGONGIARI (Dino). Backgrounds of the Divine comedy: a series of lectures. Ed. by Henry PAOLUCCI and Anne PAOLUCCI. Dover, Griffon House, 2005, 322 p.

2990. BRUSCAGLI (Riccardo). Il Quattrocento e il Cinquecento. Bologna, Il Mulino, 2005, 180 p. (Storia della letteratura italiana, 2. Itinerari. Critica letteraria).

2991. BURROWS (Daron Lee). The stereotype of the priest in the Old French fabliaux: anticlerical satire and lay identity. Oxford a. New York, P. Lang, 2005, 265 p.

2992. CAILLE (Jacqueline). Medieval Narbonne: a city at the heart of the troubadour world. Ed. by Kathryn REYERSON. Aldershot a. Burlington, Ashgate, 2005, [s. p.]. (ill., maps). (Variorum Collected studies series, 792).

2993. CÁTEDRA (Pedro M.). Liturgia, poesía y teatro en la edad media: estudios sobre prácticas culturales y literarias. Madrid: Gredos, 2005, 688 p. (ill.). (Biblioteca románica hispánica, 444).

2994. CAZELLES (Brigitte). Soundscape in early French literature. Tempe, Arizona Center for Medieval and Renaissance Texts Studies, 2005, 186 p. (Arizona studies in the Middle Ages and the Renaissance, 17. Medieval & Renaissance texts & studies, 295).

2995. CHAUCER (Geoffrey). Troilus and Criseyde, with facing page Il Filostrato: context, criticism. Ed. by Stephen A. BARNEY. New York a. London, W.W. Norton, 2005, XXVII-628 p.

2996. CORRY (Jennifer M.). Perceptions of magic in medieval Spanish literature. Bethlehem, Lehigh U. P., 2005, 258 p.

2997. CORTI (Maria). La lingua poetica avanti lo stilnovo: studi sul lessico e sulla sintassi. Tavarnuzze, Edizioni del Galluzzo per la fondazione Ezio Franceschini, 2005, XIV-226 p. Archivio romanzo, 5).

2998. COX (Catherine S.). The Judaic other in Dante, the Gawain poet, and Chaucer. Gainesville, University Press of Florida, 2005, X-239 p.

2999. DAUB (Susanne). Von der Bibel zum Epos: poetische Strategien des Laurentius am geistlichen Hof von Durham. Köln, Böhlau, 2005, 283 p.

3000. DAVIDSON (Clifford). The dramatic tradition of the Middle Ages. New York, AMS Press, 2005, XVI-317 p. (AMS studies in the Middle Ages, 26).

3001. Deutsche Texte des Mittelalters zwischen Handschriftennähe und Rekonstruktion: Berliner Fachtagung 1.–3. April 2004. Hrsg. v. Martin J. SCHUBERT. Tübingen, M. Niemeyer, 2005, VI-330 p. (Beihefte zu Editio, 23).

3002. DIECKMANN (Sandra). Variation und Wiederholung: Untersuchungen zur Formelsprache und Laissentechnik in der altfranzözischen Heldenepik. Frankfurt am Main a. Oxford, P. Lang, 2005, 414 p. (Europäische Hochschulschriften. Reihe 13, Französische Sprache und Literatur. 279).

3003. DOBOZY (Maria). Re-membering the present: the medieval German poet-minstrel in cultural context. Turnhout, Brepols, 2005, XII-353 p. (ill.). (Disputatio, 6).

3004. Entra mayo y sale abril: medieval Spanish literary and folklore studies in memory of Harriet Goldberg. Ed. by Manuel da Costa FONTES and Joseph Thomas SNOW. Newark, Juan de la Cuesta, 2005, 422 p.

3005. Épique (L') médiéval et le mélange des genres. Éd. par Caroline CAZANAVE. Besançon, Presses universitaires de Franche-Comté, 2005, 324 p. (Collection littéraire).

3006. Favolisti latini medievali e umanistici. Vol. 13. Genova, Università di Genova, Facoltà di lettere, Dipartimento di archeologia, filologia classica e loro tradizioni, 2005, 230 p. (ill.). (Pubblicazioni del Dipartimento di archeologia, filologia classica e loro tradizioni, 223).

3007. FILIOS (Denise K.). Performing women: sex, gender and the medieval Iberian lyric. New York, Palgrave Macmillan, 2005, 261 p. (New Middle Ages series).

3008. FIORILLA (Maurizio). Marginalia figurata nei codici di Petrarca. Firenze, L.S. Olschki, 2005, 96 p. (ill.). (Biblioteca di "Lettere italiane", 65).

3009. Formas narrativas breves en la Edad Media: actas del IV Congreso, Santiago de Compostela, 8–10 de julio de 2004. Ed. por Elvira FIDALGO. Santiago de Compostela, Universidade de Santiago de Compostela, 2005, 378 p. (Cursos e congresos da Universidade de Santiago de Compostela, 164).

3010. Fortunes (The) of King Arthur. Ed. by Norris J. LACY. Cambridge, D.S. Brewer, 2005, XVI-231 p. (ill.). (Arthurian studies, 64).

3011. GADE (Dietlind). Wissen, Glaube, Dichtung: Kosmologie und Astronomie in der meisterlichen Lieddichtung des vierzehnten und fünfzehnten Jahrhunderts. Tübingen, M. Niemeyer, 2005, VIII-360 p. (ill.). (Münchener Texte und Untersuchungen zur deutschen Literatur des Mittelalters, 130).

3012. GENTILI (Sonia). L'uomo aristotelico alle origini della letteratura italiana. Roma, Carocci e Università degli studi di Roma La Sapienza, 2005, 278 p. (Ricerca letteraria, 3).

3013. GIOVINI (Marco). I viaggi a Costantinopoli di Liutprando da Cremona fra professione storiografica e spunti terenziani. Studi Medievali, 2005, 46, 1, p. 753-81.

3014. GIUNTA (Claudio). Codici: saggi sulla poesia del Medioevo. Bologna, Il Mulino, 2005, 367 p. (ill.). (Ricerca).

3015. GRADY (Frank). Representing righteous heathens in late medieval England. New York a. Basingstoke, Palgrave Macmillan, 2005, 214 p. (New Middle Ages).

3016. GRIGSBY (John). Beowulf and Grendel: the truth behind England's oldest myth. London, Watkins, 2005, VIII-246 p. (ill., pl.).

3017. Grosse Texte des Mittelalters: Erlanger Ringvorlesung 200. Hrsg. v. Sonja GLAUCH. Erlangen, Palm & Enke, 2005, 305 p. (ill.). (Erlanger Studien).

3018. HANNA (Ralph). London literature, 1300–1380. Cambridge, Cambridge U. P., 2005, XXI-359 p. (Cambridge studies in medieval literature, 57).

3019. HARDYMENT (Christina). Malory: the life and times of King Arthur's chronicler. London, Harper-Collins, 2005, XVI-634 p. (ill., pl., maps).

3020. HARVEY (Anthony), POWER (Jane). The non-Classical lexicon of Celtic latinity. Vol. I. Letters A-H. Turnhout, Brepols, 2005, XLVIII-370 p. (Corpus Christianorum. Continuatio Mediaevalis. Royal Irish Academy dictionary of medieval Latin from Celtic sources, 1).

3021. HEMMES-HOOGSTADT (Annette C.). Sies mijn vlien, mijn jaghen: over vorm en inhoud van een corpus Middelnederlandse spreukachtige hoofse lyriek: Lund, UB, Mh 55 en Brussel, KB, Ms.IV 209/II = 'Sies mijn vlien, mijn jaghen' (She is what I flee, what I pursue): on form and content of a corpus of Middle Dutch pseudo-proverbial courtly love lyrics: Lund, UB, Mh 55 and Brussels, KB, Ms.IV 209/II (with a summary in English) = 'Sies mijn vlien, mijn jaghen': Zu form und Inhalt eines Korpus mittelniederländischer spruchhafter höfischer Lyrik: Lund, UB, Mh 55 und Brussel, KB, Ms.IV 209/II (mit einer Zusammenfassung in deutscher Sprache). Hilversum, Verloren, 2005, 314 p. (ill., facs.). (Middeleeuwse studies en bronnen, 86).

3022. Historicist essays on Hispano-Medieval narrative: in memory of Roger M. Walker. Ed. by Barry TAYLOR and Geoffrey WEST. London, Maney Pub. for the Modern Humanities Research Association, 2005, XI-418 p. (Publications of the Modern Humanities Research Association, 16).

3023. HODGES (Laura Fulkerson). Chaucer and clothing: clerical and academic costume in the general prologue to the Canterbury tales. Woodbridge a. Rochester, D.S. Brewer, 2005, XIV-316 p. (ill., pl.). (Chaucer studies, 34).

3024. HOLDENRIED (Anke). The Sibyl and her scribes: manuscripts and interpretation of the Latin Sibylla Tiburtina, c. 1050–1500. Aldershot, Ashgate, 2005, XXVI-254 p. (facs.). (Church, faith, and culture in the Medieval West).

3025. Images of matter: essays on British literature of the Middle Ages and Renaissance. Ed. Yvonne BRUCE. Newark, University of Delaware Press, 2005, 283 p. (ill.).

3026. Index scriptorum novus mediae latinitatis. Supplementum (1973–2005). Ed. Bruno BON et Anne Marie BAUTIER. Genève, Droz, 2005, XI-291 p.

3027. JIMURA (Akiyuki). Studies in Chaucer's words and his narratives. Hiroshima, Keisuisha, 2005, VII-263 p.

3028. JOHNSTON (Dafydd). Llén yr uchelwyr: hanes beirniadol llenyddiaeth Gymraeg 1300–1525. Caerdydd, Gwasg prifysgol Cymru, 2005, XVII-491 p.

3029. KANE (George). Will's visions of Piers Plowman, Do-Well, Do-Better and Do-Best: a glossary of the English vocabulary of the A, B, and C versions as presented in the Athlone editions. London, Continuum, 2005, XIII-240 p. (Piers Plowman: the three versions).

3030. KAPS (Gabriele). Zweisprachigkeit im paraliturgischen Text des Mittelalters. Frankfurt am Main, P. Lang, 2005, 274 p. (Studia Romanica et linguistica).

3031. KLUNDER (Nolanda). Lucidarius: de Middelnederlandse Lucidarius-teksten en hun relatie tot de Europese traditie. Amsterdam, Prometheus, 2005, 558 p. (Nederlandse literatuur en cultuur in de middeleeuwen).

3032. KÖLBL (Angelika). Der Blick auf die Frau: Frauendidaxe in den Reden Heinrichs des Teichners. Wien, Praesens, 2005, 302 p. (ill.).

3033. KOMPATSCHER GUFLER (Gabriela). Herbert von Clairvaux und sein Liber miraculorum: die Kurzversion eines anonymen bayerischen Redaktors. Bern, Lang, 2005, 372 p. (Lateinische Sprache und Literatur des Mittelalters).

3034. KRAGL (Florian). Die Weisheit des Fremden: Studien zur mittelalterlichen Alexandertradition. Mit einem allgemeinen Teil zur Fremdheitswahrnehmung. Bern a. Oxford, Lang, 2005, 500 p. (Wiener Arbeiten zur germanischen Altertumskunde und Philologie, 39).

3035. KUHNS (Richard Francis). Decameron and the philosophy of storytelling: author as midwife and pimp. New York, Columbia U. P., 2005, XXIII-177 p. (ill.).

3036. Langland (William). The Piers Plowman electronic archive. Vol. 5. London, British Library MS additional 35287 (M). Ed. by Eric ELIASON, Thorlac TURVILLE-PETRE and Hoyt N. DUGGAN. Woodbridge, Medieval Academy of America, Society for Early English and Norse Electronic Texts and Boydell & Brewer, 2005, 1 CD-ROM. (Society for Early English and Norse Electronic Texts, 7).

3037. LIO-ITS: repertorio della lirica italiana delle origini: incipitario dei testi a stampa (secoli XIII–XVI) su CD-ROM. A cura di Lino LEONARDI e Giuseppe MARRANI. Firenze, Edizioni del Galluzzo per la Fondazione Ezio Franceschini, 2005, XI-120 p. e 1 CD-ROM (Archivio romanzo, 10).

3038. Literacy in medieval and early modern Scandinavian culture. Ed. by Pernille HERMANN. Odense, University Press of Southern Denmark, 2005, 355 p. (Viking collection, 16).

3039. Literatur und Wandmalerei, II: Konventionalität und Konversation. Burgdorfer Colloquium 2001. Hrsg. v. Eckart Conrad LUTZ, Johanna THALI und René WETZEL. Tübingen, M. Niemeyer, 2005, X-593 p.

3040. LOMBARDI (Chiara). Troilo e Criseida nella letteratura occidentale. Roma, Edizioni di storia e letteratura, 2005, XIII-330 p. (ill., pl.). (Temi e testi, 62).

3041. MAC CULLY (Chris), HILLES (Sharon). The earliest English: an introduction to old English language. Harlow, Pearson Longman, 2005, XV-307 p. (Learning about language).

3042. MAC TURK (Rory). Chaucer and the Norse and Celtic worlds. Aldershot, Ashgate, 2005, IX-218 p.

3043. MALLETTE (Karla). The Kingdom of Sicily, 1100–1250: a literary history. Philadelphia, University of Pennsylvania Press, 2005, 214 p. (ill.). (The Middle Ages series).

3044. Manual (A) of the writings in Middle English, 1050-1500. Ed. by Peter G. BEIDLER. Vol. 11. Sermons and homilies. Ed. by Thomas J. HEFFERNAN and Patrick J. HORNER. Lyrics of the MS Harley 2253 ed. by Susanna FEIN. New Haven, Connecticut Academy of Arts and Sciences, 2005, IX-p. 3970-4373 (port.).

3045. MARI (Paolo). L'armario del filologo. Roma, Istituto storico italiano per il Medio Evo, 2005, XXV-308 p. (Fonti per la storia dell'Italia medievale. Subsidia, 8).

3046. MASI (Michael). Chaucer and gender. New York a. Oxford, P. Lang, 2005, X-165 p.

3047. Medieval Celtic literature and society. Ed. by Helen FULTON. Dublin, Portland a. Four Courts Press, 2005, 304 p.

3048. MIKLAUTSCH (Lydia). Montierte Texte, hybride Helden: zur Poetik der Wolfdietrich-Dichtungen. Berlin, W. De Gruyter, 2005, IX-271 p. (Quellen und Forschungen zur Literatur- und Kulturgeschichte, 36).

3049. Monsters, marvels and miracles: imaginary journeys and landscapes in the Middle Ages. Ed. by Leif SØNDERGAARD and Rasmus Thorning HANSEN. Odense, University Press of Southern Denmark, 2005, 210 p. (ill., pl.).

3050. MONTEIRO DE CASTRO (Bernardo). As cantigas de Santa Maria: um estilo gótico na lírica ibérica medieval. Newark, Juan de la Cuesta, 2005, 234 p.

3051. MOOS (Peter von). Entre histoire et littérature: communication et culture au Moyen Âge. Tavarnuzze, SISMEL Edizioni del Galluzzo, 2005, XVIII-712 p. (Millennio medievale, 58. Strumenti e studi, 2).

3052. MORGAN (Gerald). The tragic argument of Troilus and Criseyde. Lewiston, Edwin Mellen Press, 2005, 2 vol., 700 p. (ill.).

3053. MORTIMER (Nigel). John Lydgate's Fall of princes: narrative tragedy in its literary and political contexts. Oxford a. New York, Clarendon Press a. Oxford U. P., 2005, XV-360 p. (Oxford English monographs).

3054. Motif-index of German secular narratives from the beginning to 1400. Ed. by Helmut BIRKHAN. Vol. 1. Matière de Bretagne: Albrecht, Jüngerer Titurel-Lancelot 2. Ed. by Karin LICHTBLAU and Christa TUCZAY. Vol. 2. Matière de Bretagne: Lancelot 3 – Wolfram von Eschenbach, Titurel. Ed. by Karin LICHTBLAU and Christa TUCZAY. Berlin, W. de Gruyter, 2005, 2 vol., LII-429 p., 407 p.

3055. Mouvances et jointures: du manuscrit au texte médiéval. Éd. par Milena MIKHAILOVA. Orléans, Paradigme, 2005, 334 p. (Medievalia, 55).

3056. New (A) index of Middle English verse. Ed. by Julia BOFFEY, Anthony Stockwell Garfield EDWARDS and Rossell Hope ROBBINS. London, British Library, 2005, XVI-344 p.

3057. New directions in oral theory: essays on ancient and medieval literatures. Ed. by Mark AMODIO. Tempe, Arizona Center for Medieval and Renaissance Studies, 2005, X-341 p. (Medieval and renaissance texts and studies, 287).

3058. NICHOLSON (Peter). Love and ethics in Gower's Confessio amantis. Ann Arbor, University of Michigan Press, 2005, VIII-461 p.

3059. NISSE (Ruth). Defining acts: drama and the politics of interpretation in late medieval England. Notre Dame, University of Notre Dame Press, 2005, X-226 p.

3060. Ó CARRAGÁIN (Éamonn). Ritual and the rood: liturgical images and the Old English poems of the Dream of the rood tradition. London, British Library, 2005, XXXII-427 p. (ill, facs, plates). (The British Library studies in medieval culture).

3061. O'SULLIVAN (Daniel E.). Marian devotion in thirteenth-century French lyric. Toronto a. London, University of Toronto Press, 2005, X-263 p. (music).

3062. Orality and literacy in the Middle Ages: essays on a conjunction and its consequences in honour of D.H. Green. Mark CHINCA, Christopher YOUNG and Dennis Howard GREEN. Turnhout, Brepols, 2005, X-259 p. (Utrecht studies in medieval literacy, 12).

3063. OVER (Kristen Lee). Kingship, conquest, and patria: literary and cultural identities in medieval French and Welsh Arthurian romance. New York a. London, Routledge, 2005, XI-231 p. (Studies in medieval history and culture).

3064. PAKKALA-WECKSTRÖM (Mari). The dialogue of love, marriage and maistrie in Chaucer's Canterbury tales. Helsinki, Société néophilologique, 2005, 265 p. (Mémoires de la Société néophilologique de Helsinki, 67).

3065. PARKER (Will). The four branches of the Mabinogi. Oregon House, Bardic Press, 2005, XIV-694 p. (maps).

3066. Parrhasiana III. "Tocchi da huomini dotti": codici e stampati con postille di umanisti. Atti del III seminario di studi, Roma 27–28, 2002. A cura di Giancarlo ABBAMONTE, Lucia GUALDO ROSA e Luigi MUNZI. Pisa, Istituti editoriali e poligrafici internazionali, 2005, 265 p. (facs). (Annali dell'Università degli studi di Napoli "L'Orientale". Dipartimento di studi del mondo classico e del Mediterraneo antico, Sezione filologico-letteraria, 27).

3067. PERFETTI (Lisa Renée). The representation of women's emotions in medieval and early modern culture. Gainesville, University Press of Florida, 2005, 222 p. (ill.).

3068. Performing medieval narrative. Ed. by Evelyn Birge VITZ, Nancy Freeman REGALADO and Marilyn

LAWRENCE. Cambridge, D.S. Brewer, 2005, XVI-261 p. (ill.).

3069. PETOLETTI (Marco). Il Marziale autografo di Giovanni Boccaccio. *Italia medioevale e umanistica*, 2005, 46, p. 35-55 (tav).

3070. Poesía latina medieval (siglos V–XV): actas del IV Congreso del "Internationales Mittellateinerkomitee", Santiago de Compostela, 12–15 de septiembre de 2002. Ed. por Manuel C. DÍAZ Y DÍAZ y J. M. DÍAZ DE BUSTAMANTE. Tavarnuzze, SISMEL – Edizioni del Galluzzo, 2005, X-1253 p. (facs.). (Millennio medievale, 55. Atti di convegni, 17).

3071. Poesía medieval: historia literaria y transmisión de textos. Ed. por Vitalino VALCÁRCEL MARTINEZ y Carlos PÉREZ GONZÁLEZ. Burgos, Fundación Instituto Castellano y Leonés de la Lengua, 2005, 483 p. (Colección Beltenebros, 12).

3072. Poésie (La) politique dans l'Italie médiévale. Éd. par Anna FONTES-BARATTO, Marina MARIETTI, Claude PERRUS. Paris, Presses Sorbonne nouvelle, 2005, 381 p. (Centre d'études et de recherches sur la littérature italienne médiévale de l'Université Paris 3-Sorbonne nouvelle. Arzanà: cahiers de littérature médiévale italienne, 11).

3073. POULAIN-GAUTRET (Emmanuelle). La tradition littéraire d'Ogier le Danois après le XIIIe siècle: permanence et renouvellement du genre épique médiéval. Paris, H. Champion, 2005, 412 p. (Nouvelle bibliothèque du Moyen Âge, 72).

3074. QUAST (Bruno). Vom Kult zur Kunst: Öffnungen des rituellen Textes in Mittelalter und Früher Neuzeit. Tubingen, A. Francke, 2005, VIII-237 p. (Bibliotheca Germanica).

3075. RAJA (Maria Elisa). Il dolce inmaginar: miti e figure della poesia trecentesca. Piacenza, Vicolo del Pavone, 2005, 137 p.

3076. Raumerfahrung, Raumerfindung: erzählte Welten des Mittelalters zwischen Orient und Okzident. Hrsg. v. Laetitia RIMPAU, Peter IHRING und Friedrich WOLFZETTEL. Berlin, Akademie Verlag, 2005, 325 p. (ill.).

3077. Regesten deutscher Minnesänger des 12. und 13. Jahrhunderts. Berlin, De Gruyter, 2005, CXXXVII-1075 p. u. 1 CD-ROM (ill., facs.).

3078. Rethinking Middle English: linguistic and literary approaches. Ed. by Nikolaus RITT and Herbert SCHENDL. Frankfurt am Main, P. Lang, 2005, XI-339 p. (Studies in English medieval language and literature, 10).

3079. Rhetoric, royalty, and reality: essays on the literary culture of medieval and early modern Scotland. Ed. by Alasdair A. MAC DONALD and Kees DEKKER. Paris a. Dudley, Peeters, 2005, X-224 p. (Mediaevalia Groningana, 7).

3080. RICCI (Alessio). Mercanti scriventi: sintassi e testualità di alcuni libri di famiglia fiorentini fra Tre e Quattrocento. Roma, Aracne, 2005, 281 p. (Scienze dell'antichità, filologico-letterarie e storico-artistiche, 149).

3081. ROBERTS (Anna). Queer love in the Middle Ages. New York, Palgrave Macmillan, 2005, 195 p. (The new Middle Ages).

3082. ROBINSON (Cynthia). Medieval Andalusian courtly culture in the Mediterranean: three ladies and a lover. London, RoutledgeCurzon, 2005, X-225 p. (Routledge Studies in Middle Eastern literatures, 10).

3083. ROSS (Margaret Clunies). A history of old Norse poetry and poetics. Cambridge, D.S. Brewer, 2005, X-283 p.

3084. ROSSI BELLOTTO (Carla). Il manoscritto perduto del Voyage de Charlemagne: il Codice Royal 16 E VIII della British Library. Roma, Salerno, 2005, 136 p. (Piccoli saggi, 25).

3085. ROUSE (Robert Allen). The idea of Anglo-Saxon England in Middle English romance. Cambridge, D.S. Brewer, 2005, VIII-180 p. (Studies in medieval romance, 3).

3086. Royautés imaginaires (XIIe–XVIe siècles): actes du colloque organisé par le Centre de recherche d'histoire sociale et culturelle (CHSCO) de l'Université de Paris X-Nanterre sous la direction de Colette BEAUNE et Henri BRESC (26 et 27 septembre 2003). Turnhout, Brepols, 2005, 227 p.

3087. Sagalands: the Icelandic sagas and oral tradition in the Nordic world. Ed. by David COOPER, Rögnvaldur GUDMUNDSSON and Tom MUIR. Reykjavik, Destination Viking Sagalands, 2005, 192 p. (ill., maps). (Destination Viking).

3088. SCHEUBLE (Robert). Mannes manheit, vrouwen meister: männliche Sozialisation und Formen der Gewalt gegen Frauen im Nibelungenlied und in Wolframs von Eschenbach Parzival. Frankfurt am Main, P. Lang, 2005, 381 p. (Kultur, Wissenschaft, Literatur, 6).

3089. Scots (The) and medieval Arthurian legend. Ed. by Rhiannon PURDIE and Nicola ROYAN. Cambridge, D.S. Brewer, 2005, VIII-156 p. (Arthurian studies, 61).

3090. Sens rassis (De): essays in honor of Rupert T. Pickens. Ed. by Keith BUSBY, Bernard GUIDOT and Logan E. WHALEN. Amsterdam, Rodopi, 2005, XXVIII-753 p. (Faux titre, 259).

3091. Seyd in forme and reverence: essays on Chaucer and Chaucerians in memory of Emerson Brown, Jr. Ed. by T. L. BURTON and John F. PLUMMER. Provo, Chaucer Studio Press, 2005, XIX-249 p.

3092. Sources and analogues of the Canterbury tales. Ed. by Robert M. CORREALE and Mary HAMEL. Vol. 2. Cambridge a. Rochester, D.S. Brewer, 2005, XVI-824 p. (Chaucer studies, 35).

3093. STALEY (Lynn). Languages of power in the age of Richard II. University Park, Pennsylvania State U. P., 2005, XIV-394 p. (ill.).

3094. STRINATI (Maria Gabriella). La Vera historia di Luciano: un volgarizzamento dal greco del secondo Quattrocento. Amsterdam, Adolf M. Hakkert, 2005, 143 p. (Supplementi di Lexis, 32).

3095. Studi su volgarizzamenti italiani due-trecenteschi. A cura di Paolo RINOLDI e Gabriella RONCHI. Roma, Viella, 2005, 213 p. (ill.).

3096. SUITNER (Franco). Dante, Petrarca e altra poesia antica. Fiesole, Cadmo, 2005, 251 p. (I saggi di "Letteratura italiana antica", 11).

3097. SULLIVAN (Karen). Truth and the heretic: crises of knowledge in medieval French literature. Chicago, University of Chicago Press, 2005, XII-281 p.

3098. SURDICH (Luigi). Il Duecento e il Trecento. Bologna, Il Mulino, 2005, 182 p. (Storia della letteratura italiana, 1).

3099. SVERRIR (Jakobsson). Við veröldin: heimsmynd Íslendinga, 1100–1400. Reykjavík, Háskólaútgáfan, 2005, 424 p.

3100. Text and language in medieval English prose: a festschrift for Tadao Kubouchi. Ed. by Akio OIZUMI, Jacek FISIAK and John SCAHILL. Frankfurt am Main a. Oxford, Peter Lang, 2005, XVII-319 p. (Studies in English medieval language and literature, 12).

3101. Textos medievales: recursos, pensamiento e influencia. Trabajos de las IX Jornadas Medievales. Ed. por Concepción COMPANY COMPANY, Aurelio GONZÁLEZ y Lillian von der WALDE MOHENO. México, Colegio de México, Universidad Autónoma Metropolitana y Universidad Nacional Autónoma de México, 2005, 421 p. (Publicaciones de Medievalia, 32).

3102. Texttyp und Textproduktion in der deutschen Literatur des Mittelalters. Hrsg. v. Elizabeth ANDERSEN, Manfred EIKELMANN und Anne SIMON. Berlin, W. de Gruyter, 2005, XXV-498 p. (ill.). (Trends in medieval philology).

3103. TOFTGAARD (Anders). Novellegenrens fødsel: fra Il novellino til Decameron. København, Museum Tusculanum, 2005, 167 p. (ill., pl.). (Renæssance studier, 13).

3104. Translating desire in medieval and early modern literature. Ed. by Craig BERRY and Heather HAYTON. Tempe, Arizona Center for Medieval and Renaissance Studies, 2005, XV-254 p. (Medieval and Renaissance texts and studies, 294).

3105. Trasmissione (La) dei testi latini del Medioevo = Mediaeval Latin texts and their transmission. A cura di Paolo CHIESA e Lucia CASTALDI. Tavarnuzze, SISMEL-Edizioni del Galluzzo, 2005, XI-609 p. (Millennio medievale, 57).

3106. TREMBLAY (Florent A.). A medieval Latin-English dictionary. Lewiston, Edwin Mellen Press, 2005, 471 p. (Mediaeval studies, 25).

3107. Übertragungen: Formen und Konzepte von Reproduktion in Mittelalter und Früher Neuzeit. Berlin, W. de Gruyter, 2005, XX-478 p. (ill., music). (Trends in medieval philology, 5).

3108. Under the influence: questioning the comparative in Medieval Castile. Ed. by Cynthia ROBINSON and Leyla ROUHI. Leiden, Brill, 2005, XIII-332 p. (ill.). (Medieval and early modern Iberian world, 22).

3109. VAN DYKE (Carolynn). Chaucer's agents: cause and representation in Chaucerian narrative. Madison, Fairleigh Dickinson U. P., 2005, 371 p.

3110. VERNER (Lisa). The epistemology of the monstrous in the Middle Ages. New York a. London, Routledge, 2005, XI-173 p. (Studies in medieval history and culture, 33).

3111. VILLORESI (Marco). La fabbrica dei cavalieri: cantari, poemi, romanzi in prosa fra Medioevo e Rinascimento. Roma, Salerno, 2005, 402 p. (Studi e saggi, 35).

3112. Voices in dialogue: reading women in the Middle Ages. Ed. by Linda OLSON and Kathryn KERBY-FULTON. Notre Dame, University of Notre Dame Press, 2005, XVII-508 p. (ill.).

3113. Vom vielfachen Schriftsinn im Mittelalter: Festschrift für Dietrich Schmidtke. Hrsg. v. Freimut LÖSER und Ralf G. PÄSLER. Hamburg, Kovač, 2005, XVIII-647 p. (ill., facs.). (Schriften zur Mediävistik).

3114. Vrai (Le) et le faux au moyen âge: actes du colloque du centre d'études médiévales et dialectales de Lille 3, Université Charles-de-Gaulle-Lille 3, 18, 19, 20 septembre 2003. Éd. par Élisabeth GAUCHER. Villeneuve d'Ascq, Université Charles-de-Gaulle – Lille 3, 2005, 368 p.

3115. Vrai humain entendement (De): Etudes sur la littérature française de la fin du moyen âge offertes en hommage à Jacqueline Cerquiglini-Toulet le 24 janvier 2003. Éd. par Yasmina FOEHR-JANSSENS. Genève, Droz, 2005, 161 p. (Recherches et rencontres, 21).

3116. WATT (Mary Alexandra). The cross that Dante bears: pilgrimage, crusade, and the cruciform church in the Divine comedy. Gainesville, University Press of Florida, 2005, XII-227 p. (ill., maps).

3117. WEGMANN (Milene). Naturwahrnehmung im Mittelalter im Spiegel der lateinischen Historiographie des 12. und 13. Jahrhunderts Bern a. Oxford, P. Lang, 2005, XI-240 p. (Lateinische Sprache und Literatur des Mittelalters).

3118. WENZEL (Horst). Höfische Repräsentation: symbolische Kommunikation und Literatur im Mittelalter. Darmstadt, Wissenschaftliche Buchgesellschaft, 2005, 308 p. (ill.).

3119. WHITE (Paul Andrew). Non-native sources for the Scandinavian kings' sagas. New York a. London, Routledge, 2005, XV-172 p. (Studies in medieval history and culture, 34).

3120. World (The) of Eleanor of Aquitaine: literature and society in southern France between the eleventh and thirteenth centuries. Ed. by Marcus Graham BULL and Catherine LÉGLU. Woodbridge, Boydell Press, 2005, 189 p.

3121. WUNDERLI (Peter). Die franko-italienische Literatur: literarische 'memoria' und sozio-kultureller Kontext. Paderborn, Ferdinand Schöningh, 2005, 64 p. (Geisteswissenschaften / Nordrhein-Westfälische Akademie der Wissenschaften. Vorträge, G 399).

3122. ZOTZ (Nicola). Intégration courtoise: zur Rezeption okzitanischer und französischer Lyrik im klassischen deutschen Minnesang. Heidelberg, Winter, 2005, 270 p. (Germanisch-romanische Monatsschrift. Beiheft, 19).

b. Addendum 2001

3123. LANDINO (Cristoforo). Comento sopra la Comedia. A cura di Paolo PROCACCIOLI. Roma, Salerno Editrice, 2001, 4 vol., 2131 p. (Edizione nazionale dei commenti danteschi, 28).

Cf. n°s 2568-2584

c. Technology

* 3124. DE VRIES (Kelly). A cumulative bibliography of medieval military history and technology. Update 2004. Leiden a. Boston, Brill, 2005, 325 p. (History of Warfare, 26).

―――――

3125. BORK (Robert Odell), MONTGOMERY (Scott Bradford). De re metallica: the uses of metal in the Middle Ages. Aldershot a. Burlington, Ashgate, 2005, XXII-420 p. (ill., maps). (AVISTA studies in the history of medieval technology, science and art, 4).

3126. FRATI (Marco). De bonis lapidibus conciis. La costruzione di Firenze ai tempi di Arnolfo di Cambio: strumenti, tecniche e maestranze nei cantieri fra XIII e XIV secolo. Firenze, Firenze U. P., 2005, 372 p. (ill.). (Monografie scienze tecnologiche, 13).

3127. JOCKENHÖVEL (Albrecht), WILLMS (Christoph). Das Dietzhözetal-Projekt: archämetallurgische Untersuchungen zur Geschichte und Struktur der mittelalterlichen Eisengewinnung im Lahn-Dill-Gebiet (Hessen). Rahden/Westf., VML u. Verlag Marie Leidorf, 2005, XIX-615 p. (ill., maps). (Münstersche Beiträge zur ur- und frühgeschichtlichen Archäologie).

3128. Medieval clothing and textiles. Vol. 1. Ed. by Robin NETHERTON and Gale R. OWEN-CROCKER. Woodbridge, Boydell & Brewer, 2005, XIV-185 p. (ill.).

3129. Ports maritimes et ports fluviaux au Moyen Âge. Actes du XXXVe Congrès de la SHMES, La Rochelle, 5 et 6 juin 2004. Paris, Publications de la Sorbonne, Société des historiens médiévistes de l'Enseignement supérieur public, 2005, 284 p. (Histoire ancienne et médiévale, 81).

3130. SERDON (Valérie). Armes du diable. Arcs et arbalètes au Moyen Âge. Préf. de Philippe CONTAMINE. Rennes, Presses universitaires de Rennes, 2005, 335 p. (Archéologie et culture).

3131. *Vacat.*

3132. SMITH (Robert Douglas), DE VRIES (Kelly). The artillery of the Dukes of Burgundy, 1363–1477. Woodbridge, Boydell Press, 2005, VIII-377 p. (Armour and weapons, 1). (ill.).

3133. STONE (David). Decision-making in medieval agriculture. Oxford a. New York, Oxford U. P., 2005, XVII-303 p. (ill., maps).

3134. Tradition, Innovation, Invention: Fortschrittsverweigerung und Fortschrittsbewusstsein im Mittelalter. Hrsg. v. Hans-Joachim SCHMIDT. Berlin, Walter de Gruyter, 2005, 467 p. (ill.). (Scrinium Friburgense, 18).

3135. TRILLO SAN JOSÉ (Carmen). A social analysis of irrigation in Al-Andalus: Nazari Granada (13th–15th centuries). *Journal of Medieval history*, 2005, 31, 2, p. 163-183 (map., tab.).

d. Education

3136. ARLEGUI SUESCUN (José). Escuela de Grámatica, Universitat Sertoriana de Huesca (siglos XIV– XVII). Instituto de Estudios Altoaragoneses, 2005, 410 p.

3137. BAKKE (Odd Magne). When children became people: the birth of childhood in early Christianity. Minneapolis, Fortress Press, 2005, IX-348 p.

3138. Childhood in the Middle Ages and the Renaissance: the results of a paradigm shift in the history of mentality. Ed. by Albrecht CLASSEN. Berlin, Walter de Gruyter, 2005, VII-444 p. (ill.).

3139. DOYLE (Matthew A.). Bernard of Clairvaux and the schools: the formation of an intellectual milieu in the first half of the twelfth century. Spoleto, Centro italiano di studi sull'alto Medioevo, 2005, X-106 p. (Studi, 11).

3140. FABRIS (Cécile). Étudier et vivre à Paris au Moyen Âge. Le collège de Laon (XIVe–XVe siècles). Paris, École des Chartes, 2005, V-504 p. (Mémoires et documents de l'École des Chartes, 81).

3141. Formation intellectuelle et culture du clergé dans les territoires angevins (milieu du XIIIe–fin du XVe siècle). Sous la dir. de Marie-Madeleine DE CEVINS et Jean-Michel MATZ. Rome, École française de Rome, 2005, 382 p. (Collection de l'École française de Rome, 349).

3142. Hoping for continuity: childhood, education and death in antiquity and the Middle Ages. Ed. by Katariina MUSTAKALLIO. Rome, Institutum Romanum Finlandiae, 2005, XI-252 p. (ill.). (Acta Instituti Romani Finlandiae, 33).

3143. KOUAMÉ (Thierry). Le College de Dormans-Beauvais a la fin du Moyen Age: strategies politiques et parcours individuels a l'Universite de Paris (1370–1458). Leiden a. Boston, Brill, 2005, XXVI-720 p. (ill., maps). (Education and society in the Middle Ages and Renaissance) .

3144. LEFEBVRE-TEILLARD (Anne). Magister B., Étude sur les maîtres parisiens du début du XIIIe siècle. *Tijdschrift voor rechtsgeschiedenis*, 2005, 73, 1-2, p. 1-18.

3145. Medieval education. Ed. by Ronald B. BEGLEY and Joseph W. KOTERSKI. New York, Fordham U. P., 2005, XVII-215 p. (Fordham series in medieval studies, 4).

3146. MURPHY (James Jerome). Latin rhetoric and education in the Middle Ages and Renaissance. Aldershot, Ashgate, 2005, [s. p.]. (Variorum collected studies series, 827).

3147. PINI (Antonio Ivan). Studio, università e città nel medioevo bolognese. Bologna, CLUEB, 2005, 351 p. (Studi / Centro interuniversitario per la storia delle università italianei, 5).

3148. Schullandschaften in Altbayern, Franken und Schwaben: Untersuchungen zur Ausbreitung und Typologie des Bildungswesens in Spätmittelalter und Früher Neuzeit. Hrsg. v. Helmut FLACHENECKER und Rolf KIESSLING. München, C.H. Beck, 2005, VIII-351 p. (ill., maps). (Zeitschrift für bayerische Landesgeschichte. Reihe B. Beiheft., 26).

3149. Stiftsschulen in der Region: Wissenstransfer zwischen Kirche und Territorium. Dritte wissenschaftliche Fachtagung zum Stiftskirchenprojekt des Instituts für Geschichtliche Landeskunde und Historische Hilfswissenschaften der Universität Tübingen (15.–17. März 2002, Weingarten). Ostfildern, Thorbecke, 2005, VIII-251 p. (Schriften zur südwestdeutschen Landeskunde, 50).

3150. WRIEDT (Klaus). Schule und Universität: Bildungsverhältnisse in norddeutschen Städten des Spätmittelalters. Gesammelte Aufsätze. Leiden u. Brill, 2005, IX-267 p. (ill.). (Education and society in the Middle Ages and Renaissance, 23).

3151. ZDANEK (Maciej). Szkoły i studia dominikanów krakowskich w średniowieczu. (Dominican School of Krakow and studies in the Middle Ages). Warszawa, Wydawnictwo Neriton, Instytut Historii PAN, 2005, 231 p.

§ 10. History of art.

** 3152. ALBERTI (Leon Battista). Il testamento di Leon Battista Alberti: il manoscritto Statuti Mss. 87 della Biblioteca del Senato della Repubblica "Giovanni Spadolini": i tempi, i luoghi, i protagonisti. A cura di Enzo BENTIVOGLIO, Giuliana CREVATIN e Marcello CICCUTO. Roma, Gangemi, 2005, 103 p. (Biblioteca di Giano, 5).

** 3153. Diocesi (La) di Sabina. A cura di Fabio BETTI, Giorgio BAZZUCCHI e G. Giacomo PANI. Spoleto, Centro italiano di studi sull'alto Medioevo, 2005, X-360 p. (pl., ill., map). (Corpus della scultura altomedievale, 17).

** 3154. Documents de la pintura valenciana medieval i moderna. Vol. 1. 1238–1400. Ed. a cura de Ximo COMPANY I CLIMENT, Joan ALIAGA i Lluïsa TOLOSA I MAITE FRAMIS. València, Universitat de València, 2005, 511 p.

3155. Abruzzo (L') in età angioina: arte di frontiera tra Medioevo e Rinascimento. Atti del convegno internazionale di studi, Chieti, Campus universitario, 1–2 aprile 2004. A cura di Daniele BENATI, Alessandro TOMEI. Cinisello Balsamo, Silvana, 2005, 319 p. (ill.). (Biblioteca d'arte, 8).

3156. ALTAVISTA (Clara). Lucca e Paolo Guinigi (1400–1430): la costruzione di una corte rinascimentale. Città, architettura, arte. Pisa, Edizioni ETS e Firenze, Distribuzione PDE, 2005, 255 p. (ill., maps). (Accademia lucchese di scienze, lettere ed arti. Saggi e ricerche, 9).

3157. Art and architecture of late medieval pilgrimage in Northern Europe and the British Isles. Text. Ed. by Sarah BLICK and Rita TEKIPPE. Leiden, Brill, 2005, XXXII-876 p. (Studies in medieval and Reformation traditions, 104).

3158. Art and form in Norman Sicily. Proceedings of an international conference, Rome, 6–7 December 2002. Ed. by David KNIPP. München, Hirmer Verlag, 2005, 207 p. (ill., plans). (Römisches Jahrbuch der Bibliotheca Hertziana, 35).

3159. BARAGLI (Sandra). L'art au XIVe siècle. Paris, Hazan, 2005, 383 p. (Guide des arts).

3160. BECK (Nora). Giotto's harmony: music and art in Padua at the crossroads of the Renaissance. Florence, European Press Academic Pub., 2005, 251 p. (ill.).

3161. Between the picture and the word: manuscript studies from the Index of Christian Art. Ed. by Colum HOURIHANE and John PLUMMER. University Park, Index of Christian Art, Dept. of Art and Archaeology, Princeton University, In association with Penn State U. P., 2005, XXVII-216 p. (ill., pl.). (Index of Christian Art. Occasional papers, 8).

3162. BLŒDÉ (James). Paolo Uccello et la représentation du mouvement: regards sur la Bataille de San Romano. Paris, Ecole nationale supérieure des beaux-arts, 2005, 110 p. (ill.). (D'art en questions).

3163. BÖKER (Hans Josef). Architektur der Gotik: Bestandskatalog der weltgrössten Sammlung an gotischen Baurissen (Legat Franz Jäger) im Kupferstichkabinett der Akademie der bildenden Künste Wien: mit einem Anhang über die mittelalterlichen Bauzeichnungen im Wien Museum Karlsplatz = Gothic architecture: catalogue of the world-largest collection of Gothic architectural drawings (bequest Franz Jäger) in the collection of prints and drawings of the Academy of Fine Arts Vienna: with an appendix of the medieval construction drawings in the Wien Museum Karlsplatz. Salzburg, A. Pustet, 2005, 464 p. (ill.).

3164. BÖKER (Hans). Per Graecos operarios: the reception of a Byzantine building type in Western romanesque architecture. The ancient world, 2005, 36, p. 229-246.

3165. BOLVIG (Axel). Kunsten i kalkmaleriet. København, Gyldendal, 2005, 208 p. (ill., maps).

3166. BOUCHERON (Patrick). "Tournez les yeux pour admirer, vous qui exercez le pouvoir, celle qui est peinte ici". La fresque du Bon Gouvernement d'Ambrogio Lorenzetti. Annales, 2005, 60, 6, p. 1137-1199.

3167. BRAUN (Lucien). L'image de la philosophie: méconnaissance et reconnaissance. Strasbourg, Presses universitaires de Strasbourg, 2005, 111 p. (ill.).

3168. BREDOW-KLAUS (Isabel von). Heilsrahmen: spirituelle Wallfahrt und Augentrug in der flämischen Buchmalerei des Spätmittelalters und der frühen Neuzeit. München, H. Utz, 2005, 478 p. (ill.). (Kunstgeschichte, 81. Tuduv Studien, 81).

3169. BROOKER (Robert E.). The impact of manuscript illumination on the evolution of artistic style from the Franco-Gothic to the Italo-Gothic in Castile during the XIVth century. Providence, Brown University, 2005, XII-280 p. (ill.).

3170. BRUCKER (Gene A.). Living on the edge in Leonardo's Florence. Berkeley, University of California Press, 2005, XXVI-211 p.

3171. BRUNEL (Ghislain). Images du pouvoir royal: les chartes décorées des Archives nationales, XIIIe-XVe siècle. Paris, Somogy, 2005, 255 p. (ill.).

3172. CARRERO SANTAMARÍA (Eduardo). Las catedrales de Galicia durante la Edad Media: claustros y entorno urbano. La Coruña, Fundación Pedro Barriè de la Maza, 2005, 435 p. (ill.). (Publicaciones breves de la catalogación arqueológica y artística de Galicia).

3173. CASSAGNES-BROUQUET (Sophie). L'art en famille: les milieux artistiques à Londres à la fin du Moyen Âge. Turnhout, Brepols, 2005, 312 p. (Histoires de famille. La parenté au Moyen Âge, 3).

3174. COOK (William Robert). The Art of the Franciscan Order in Italy. Leiden a. Boston, Brill, 2005, XXII-297 p. (ill., pl., plans). (Medieval Franciscans).

3175. COWEN (Painton). The rose window: splendour and symbol. London, Thames & Hudson, 2005, 276 p. (ill.).

3176. CRISTIANI TESTI (Maria Laura). Arte medievale a Pisa: tra Oriente e Occidente. Roma, Consiglio nazionale delle ricerche, 2005, XVII-620 p. (ill.).

3177. CRIVELLO (Fabrizio). Le Omelie sui Vangeli di Gregorio Magno a Vercelli: le miniature del ms. CXLVIII/ 8 della Biblioteca Capitolare. Tavarnuzze, SISMEL Edizioni del Galluzzo, 2005, XIV-151 p. (Archivum Gregorianum, 6).

3178. EPKING (Simone). Die Entwicklung des Altarstipes in Florenz vom 12. bis 15. Jahrhundert. Weimar, Verlag und Datenbank für Geisteswissenschaften, 2005, 375 p. (ill., pl.).

3179. ESCH (Arnold). Wiederverwendung von Antike im Mittelalter: die Sicht des Archäologen und die Sicht des Historikers. Berlin und New York, De Gruyter, 2005, IX-88 p. (ill.). (Hans-Lietzmann-Vorlesungen, 7).

3180. FOLDA (Jaroslav). Crusader art in the Holy Land: from the Third Crusade to the fall of Acre, 1187-1291. New York, Cambridge U. P., 2005, LXVII-714 p. a. 1 CD-ROM (ill., pl., maps).

3181. GILBERT (Creighton). Lex amoris: la legge dell'amore nell'interpretazione di Fra Angelico. Firenze, Le Lettere, 2005, 117 p. (ill., pl.). (Annali dell'Università di Ferrara. Sezione Storia. Saggi, 1).

3182. GIVENS (Jean Ann). Observation and image-making in Gothic art. Cambridge a. New York, Cambridge U. P., 2005, XIV-231 p. (ill., pl.).

3183. GRANT (Lindy). Architecture and society in Normandy, 1120 to 1270. New Haven a. London, Yale U. P., 2005, VII-274 p. (ill.).

3184. GRUENINGER (Donat). "Deambulatorium Angelorum" oder irdischer Machtanspruch? Der Chorumgang mit Kapellenkranz von der Entstehung, Diffusion und Bedeutung einer architektonischen Form. Wiesbaden, L. Reichert, 2005, 372 p., (ill., pl., maps).

3185. HELTEN (Leonhard). Mittelalterliches Masswerk: Entstehung – Syntax – Topologie. Berlin, Reimer, 2005, 283 p. (ill.).

3186. Hochaltarretabel (Das) der St. Jacobi-Kirche in Göttingen. Hrsg. v. Bernd CARQUÉ und Hedwig RÖCKELEIN. Göttingen, Vandenhoeck & Ruprecht, 2005, 563 p. (ill., map). (Veröffentlichungen des Max-Planck-Instituts für Geschichte, 213. Studien zur Germania Sacra, 27).

3187. Höfe und Residenzen im spätmittelalterlichen Reich. Bilder und Begriffe. Teilbd. 1. Begriffe. Teilbd. 2. Bilder. Hrsg. v. Werner PARAVICINI. Ostfildern, Thorbecke, 2005, XVI-563 p. (Residenzforschung 15, 2).

3188. HOWE (Eunice D.). Art and culture at the Sistine Court: Platina's "Life of Sixtus IV" and the frescoes of the Hospital of Santo Spirito. Città del Vaticano, Biblioteca Apostolica Vaticana, 2005, 268 p. (ill.). (Studi e testi, 422).

3189. ISHINABE (Masumi). Piero Della Francesca. Tokyo, Heibonsha, 2005, 550 p.

3190. KRAUS (Jeremia). Worauf gründet unser Glaube? Jesus von Nazaret im Spiegel des Hitda-Evangeliars. Freiburg im Breisgau, Herder, 2005, 431 p. (ill.). (Freiburger theologische Studien, 168).

3191. KU (Su-Hun). Vision und Realität: die Darstellung des Kirchenraums in der niederländischen Malerei des 15. Jahrhunderts. Marburg, Tectum, 2005, 115 p. (ill.).

3192. Kunst & Region: Architektur und Kunst im Mittelalter, Beiträge einer Forschungsgruppe = Art &

region: architecture and art in the Middle Ages, contributions of a research group. Hrsg. v. Uta Maria BRÄUER, Emanuel S. KLINKENBERG und Jeroen WESTERMAN. Utrecht, Clavis, 2005, 260 p.

3193. LUXFORD (Julian M.). The art and architecture of English Benedictine monasteries, 1300-1540: a patronage history. Woodbridge, Boydell Press, 2005, XXII-281 p. (Studies in the history of medieval religion, 25).

3194. MAC CLENDON (Charles B.). The origins of medieval architecture: building in Europe, A.D. 600-900. New Haven a. London, Yale U. P., 2005, 264 p. (ill., maps, plans).

3195. MARCHANT (Laurence Gérard). Orli, nastri e righe, passamanerie e tessitura nelle vesti fiorentine del Trecento. *Archivio storico italiano*, 2005, 163, 1, p. 133-158.

3196. MARCHENA HIDALGO (Rosario). Pedro de Palma, miniaturista del siglo XVI. Sevilla, Universidad de Sevilla, 2005, 144 p. (Serie Arte, 21).

3197. MARROW (James H.). Pictorial invention in the Netherlandish manuscript illumination of the late Middle Ages: the play of illusion and meaning. Ed. by Brigitte DEKEYZER and Jan VAN DER STOCK. Leuven, Peeters, 2005, 54 p. (Corpus of illuminated manuscripts = Corpus van verluchte handschriften, 16. Low Countries series, 11).

3198. MEIER (Claudia Annette). Chronicon pictum: von den Anfängen der Chronikenillustration zu den narrativen Bilderzyklen in den Weltchroniken des hohen Mittelalters. Mainz, Chorus, 2005, 323 p. (ill).

3199. MICHAEL (Angelika). Das Apsismosaik von S. Apollinare in Classe: seine Deutung im Kontext der Liturgie. Frankfurt am Main, P. Lang, 2005, 270 p. (Europäische Hochschulschriften. Reihe 23, Theologie, 799 = Publications universitaire Europeennes. Serie 23, Theologie, 799 = European university studies. Series 23, Theology, 799).

3200. MONCIATTI (Alessio). Il Palazzo Vaticano nel Medioevo. Firenze, Olschki, 2005, XIX-454 p. (ill., pl.). (Fondazione Carlo Marchi. Studi, 19).

3201. MONTORSI (William). Neobizantino e romanico in Puglia: la Basilica di San Nicola nell'età lanfranchiana. Modena, Aedes Muratoriana, 2005, VIII-301 p. (ill.). (Deputazione di storia patria per le antiche provincie modenesi. Biblioteca. Serie speciale, 27).

3202. Patrimoine (Le) médiéval de Wallonie. Éd. par Julien MAQUET, Guy FOCANT et Fabrice DOR. Namur, Institut du patrimoine wallon, 2005, 632 p. (ill.).

3203. PEARSON (Andrea G.). Envisioning gender in Burgundian devotional art, 1350–1530: experience, authority, resistance. Aldershot, Ashgate, 2005, XIX-236 p. (ill., pl., ports). (Women and gender in the early modern world).

3204. PINI (Raffaella). Il mondo dei pittori a Bologna, 1348–1430. Bologna, CLUEB, 2005, 220 p. (ill.). (Lexis. III, Biblioteca delle arti, 11).

3205. Realität und Projektion: wirklichkeitsnahe Darstellung in Antike und Mittelalter. Hrsg. v. Martin BÜCHSEL. Berlin, Mann, 2005, 239 p. (ill.). (Neue Frankfurter Forschungen zur Kunst, 1).

3206. Reliquiare im Mittelalter. Hrsg. v. Bruno REUDENBACH und Gia TOUSSAINT. Berlin, Akademie Verlag, 2005, IX-221 p. (ill.). (Hamburger Forschungen zur Kunstgeschichte. Studien, Theorien, Quellen, 5).

3207. ROMANINI (Angiola Maria). Arte medievale: interpretazioni storiografiche. A cura di Adriano PERONI e Marina RIGHETTI. Spoleto, Centro italiano di studi sull'alto Medioevo, 2005, XII-303 p. (ill.). (Lezioni spoletine, 1).

3208. ROTHSTEIN (Bret Louis). Sight and spirituality in early Netherlandish painting. Cambridge a. New York, Cambridge U. P., 2005, XII-262 p. (ill.).

3209. ROUX (Guy). Opicinus de Canistris (1296–1352?): prêtre, pape et Christ réssuscité. Paris, Léopard d'or, 2005, 483 p. (ill.).

3210. RUSSO (Daniel). Peintures murales médiévales, XIIe–XVIe siècles: regards comparés. Dijon, Éditions universitaires de Dijon, 2005, 216 p. (ill.). (Collection art & patrimoine).

3211. SCHLICHT (Markus). Un chantier majeur de la fin du Moyen Âge. La cathédrale de Rouen vers 1300. Portail des Libraires, portail de la Calende, chapelle de la Vierge. Caen, Société des Antiquaires de Normandie, 2005, 425 p. (Mémoires de la Société des Antiquaires de Normandie, 41).

3212. SCHMIDT (Gerhard). Malerei der Gotik: Fixpunkte und Ausblicke. Vol. 1. Malerei der Gothik in Mitteleuropa. Vol. 2. Malerei der Gothik in Süd- und Westeuropa: Studien zum Herrscherporträt Hrsg. v. Martin ROLAND. Graz, Akademische Druck- u. Verlagsanstalt, 2005, 2 vol., XII-468 p., VI-406 p. (ill.).

3213. SCHMIDT (Victor Michael). Painted piety: panel paintings for personal devotion in Tuscany, 1250–1400. Firenze, Centro Di, 2005, 350 p. (ill.). (Italia e i Paesi Bassi, 8).

3214. Seeing the invisible in late antiquity and the early Middle Ages: papers from "Verbal and pictorial imaging: representing and accessing experience of the invisible, 400–1000" (Utrecht, 11–13 December 2003). Ed. by Giselle DE NIE, Karl Frederick MORRISON and Marco MOSTERT. Turnhout, Brepols, 2005, X-545 p. (ill.). (Utrecht studies in medieval literacy, 14).

3215. SELTMANN (Ingeborg). Handwerker / Henker / Heilige: Bilder erzählen vom Leben im Mittelalter. Ostfildern, J. Thorbecke, 2005, 192 p. (ill.).

3216. SIGNORI (Gabriela). Räume, Gesten, Andachtsformen: Geschlecht, Konflikt und religiöse Kultur im europäischen Mittelalter. Ostfildern, J. Thorbecke, 2005, 180 p. (ill., pl.).

3217. St George's Chapel, Windsor, in the fourteenth century. Ed. by Nigel SAUL. Woodbridge, Boydell Press, 2005, XVII-241 p. (ill., maps).

3218. STAHULJAK (Zrinka). Bloodless genealogies of the French Middle Ages: translation, kinship, and metaphor. Gainesville, University Press of Florida, 2005, 242 p. (ill.).

3219. STOPFORD (Jennie). Medieval floor tiles of northern England: pattern and purpose. Production between the 13th and 16th centuries. Oxford, Oxbow, 2005, XVII-393 p. (ill., maps). (English Heritage).

3220. TÄNGEBERG (Peter). Retabel und Altarschreine des 14. Jahrhunderts: schwedische Altarausstattungen in ihrem europäischen Kontext. Stockholm, Kungl. Vitterhets, historie och antikvitets akademien, Distributeur Almqvist & Wiksell, 2005, 295 p. (ill.).

3221. Texte et archéologie monumentale: approches de l'architecture médiévale. Actes du colloque, Centre international de congrés, Palais des papes, Avignon, 30 novembre, 1er et 2 décembre 2000. Éd. par Philippe BERNARDI, Andreas HARTMANN-VIRNICH et Dominique VINGTAIN. Montagnac, Monique Mergoil, 2005, 156 p. (ill.). (Europe médiévale, 6).

3222. Tserkov' Spasa na Nereditse: Ot Vizantii k Rusi: K 800-letiju pamjatnika. (The Church of Our Saviour on Nereditsa, [Novgorod]: From Byzantium to Russia: To the 800th anniversary: [Articles]). Ed. Ol'ga E. ETINGOF. RAN, In-t vostokovedenija, etc. Moskva, Indrik, 2005, 327 p. (ill.; bibl. incl.).

3223. WRIGHT (Alison). The Pollaiuolo brothers: the arts of Florence and Rome. New Haven, Yale U. P., 2005, VII-575 p. (ill.).

Cf. n°s 32, 2585

§ 11. History of music.

** 3224. EGAN-BUFFET (Máire). Manuscript sources of French music theory: Paris, Bibliothèque nationale, Arsenal, MS 3042: a facsimile edition with translations, introduction and commentary. Ottawa, Canada, Institute of Mediaeval Music, 2005, V-211 p. (Wissenschaftliche Abhandlungen, 82 = Musicological studies, 82).

3225. BERGER BUSSE (Anna Maria). Medieval music and the art of memory. Berkeley a. London, University of California Press, 2005, XVI-288 p. (ill, music).

3226. Cantus coronatus: 7 cantigas. Ed. por Manuel Pedro FERREIRA. Kassel, Edition Reichenberger, 2005, X-306 p. (DeMusica, 10).

3227. Citation and authority in medieval and renaissance musical culture: learning from the learned. Ed. by Suzannah CLARK and Elizabeth Eva LEACH. Woodbridge, Boydell Press, 2005, XXXII-250 p. (Studies in Medieval and Renaissance music, 4).

3228. COLIN MUSET. Les chansons de Colin Muset: textes et mélodies. Ed. by Christopher J. CALLAHAN and Samuel N. ROSENBERG. Paris, Honoré Champion, 2005, 240 p. (music). (Les classiques français du Moyen Âge, 149).

3229. Collectie middelnederlandse en latijnse geestelijke liederen, ca. 1500 = Collection of Middle Dutch and Latin Sacred Songs, ca. 1500. Ed. Bruno BOUCKAERT. Leuven, Alamire, 2005, XLV-132 p. 1 choir book (facs.). (Monumenta flandriae musica, 7).

3230. Fleury (The) playbook. Vol. 3. Plays of conversion and rebirth. Ed. by Wyndham THOMAS. Moretonhampstead, Antico Edition, 2005, XVIII-36 p. of music (ill.). (Medieval church music).

3231. Gattungen und Formen des europäischen Liedes vom 14. bis zum 16. Jahrhundert: internationale Tagung vom 9. bis 12. Dezember 2001 in Münster. Hrsg. v. Michael ZYWIETZ, Volker HONEMANN und Christian BETTELS. Münster, Waxmann, 2005, 307 p. (ill.). (Studien und Texte zum Mittelalter und zur frühen Neuzeit).

3232. HAAS (Max). Musikalisches Denken im Mittelalter: eine Einführung. Bern a. Oxford., P. Lang, 2005, XIII-687 p. (charts, music).

3233. HUCK (Oliver). Die Musik des frühen Trecento. Hildesheim, Olms, 2005, XII-363 p. (Musica mensurabilis, 1).

3234. HUGLO (Michel). Chant grégorien et musique médiévale. Aldershot, Ashgate, 2005, [s. p.] (ill., facs, maps, music). (Variorum collected studies series, 814). – IDEM. La théorie de la musique antique et médiévale. Aldershot, Ashgate, 2005, [s. p.]. (ill.). (Variorum collected studies series, 822). – IDEM. Les anciens répertoires de plain-chant. Aldershot, Ashgate, 2005, [s. p.]. (ill., facs., music). (Variorum collected studies series, 804).

3235. ILNITCHI (Gabriela). The play of meanings: Aribo's De musica and the hermeneutics of musical thought. Lanham a. Oxford, Scarecrow Press, 2005, XIII-265 p. (ill.).

3236. KANDLER (Johannes). "Gedoene ân wort daz ist ein tôter galm": Studien zur Wechselwirkung von Wort und Ton in einstimmigen Gesängen des hohen und späten Mittelalters. Wiesbaden, Reichert, 2005, XII-324 p. (ill., music). (Elementa musicae, 5).

3237. LÓPEZ ELUM (Pedro). Interpretando la música medieval del siglo XIII: las cantigas de Santa María. Valencia, Universitat de València, 2005, 343 p. (ill.). (Oberta. Historia, 107).

3238. Manuscrito (El) de Munébrega 1: un testimonio aragonés de la cultura litúrgico-musical de los siglos XIII-XIV en el contexto europeo. Ed. por Luis PRENSA. Zaragoza, Institución "Fernando el Católico", 2005, 508 p. (Monumenta monodica aragonesia, 1).

3239. Mittelalter und Mittelalterrezeption: Festschrift für Wolf Frobenius. Hrsg. v. Herbert SCHNEIDER. Hildesheim, Olms, 2005, VIII-440 p. (ill.). (Musikwissenschaftliche Publikationen).

3240. PFAU (Marianne Richert), MORENT (Stefan). Hildegard von Bingen: der Klang des Himmels. Köln,

Böhlau, 2005, 401 p. u. 1 sound CD (ill., music). (Europäische Komponistinnen, 1).

3241. Représentations (Les) de la musique au moyen âge: actes du colloque des 2 et 3 avril 2004. Éd. par Martine CLOUZOT et Christine LALOUE. Paris, Cité de la musique, 2005, 128 p. (ill., music). (Les cahiers du Musée de la musique, 6).

3242. STONE (Anne), TONIOLO (Federica). The manuscript Modena, Biblioteca estense, α.M.5.24: commentary. Lucca, Libreria musicale italiana, 2005, 171 p. (Ars nova, 1).

3243. Tracce di una tradizione sommersa: i primi testi lirici italiani tra poesia e musica. Atti del seminario di studi, Cremona, 19 e 20 febbraio 2004. A cura di Maria Sofia LANNUTTI e Massimiliano LOCANTO. Tavarnuzze, SISMEL – Edizioni del Galluzzo e Firenze, Fondazione Ezio Franceschini, 2005, VIII-280 p. e 1 sound CD (ill.). (Facoltà di musicologia dell'Università di Pavia, Fondazione WalterStauffer. Studi e testi, 3. La tradizione musicale, 9).

3244. VIRET (Jacques). Musique médiévale. Grez-sur-Loing, Pardès, 2005, 127 p.

3245. WAECHTER (Hans). Die geistlichen Lieder des Mönchs von Salzburg: Untersuchungen unter besonderer Berücksichtigung der Melodien. Göppingen, Kümmerle, 2005, V-277 p. (music). (Göppinger Arbeiten zur Germanistik, 724).

§ 12. History of philosophy, theology and science.

** 3246. ARISTOTELES. The Arabic version of the Nicomachean ethics. Ed. by Anna AKASOY, Alexander FIDORA and D. M. DUNLOP. Leiden a. Boston, Brill, 2005, XV-619 p. (Aristoteles Semitico-latinus, 17).

** 3247. EUGENIUS TOLETANUS. Eugenii Toletani Opera omnia. Ed. Alberto Paulo FARMHOUSE. Turnhout, Brepols, 2005, 480 p.

** 3248. GOY (Rudolf). Die handschriftliche Überlieferung der Werke Richards von St. Viktor im Mittelalter. Turnhout, Brepols, 2005, 455 p. (Bibliotheca Victorina, 18).

** 3249. GUILLELMUS A SANCTO THEODORICO. Opera omnia. Pars 4. Meditationes devotissimae. Éd. par Paul VERDEYEN et Stanislaus CEGLAR. Turnhout, Brepols, 2005, XIX-139 p. (Corpus Christianorum, Continuatio Mediaevalis, 89).

** 3250. HENRICUS DE GANDAVO. Henrici de Gandavo Opera omnia. Vol. 21. Summa (Quaestiones ordinariae), art. I-V. Ed. Gordon Anthony WILSON. Leuven, Leuven U. P., 2005, XCIX-402 p. (Ancient and medieval philosophy. Series 2).

** 3251. LEGENDRE (Olivier). Collectaneum exemplorum et visionum Clarevallense e codice Trecensi 946. Turnhout, Brepols, 2005, CXIV-468 p. (Corpus Christianorum, Continuatio Mediaevalis, 208. Exempla medii aevi, 2).

** 3252. LLULL (Ramon). Doctrina pueril. Ed. por Joan SANTANACH I SUÑOL. Palma, Patronat Ramon Llull, 2005, CXII-305 p. (Nova edició de les obres de Ramon Llull, 7).

** 3253. ODONIS (Giraldus). Giraldus Odonis O.F.M. Opera philosophica. Vol II. De intentionibus. Edited by Lambertus Marie DE RIJK. Leiden a. Boston, Brill, 2005, 894 p. (Studien und Texte zur Geistesgeschichte des Mittelalters, 60).

** 3254. RODLER (Klaus). Die Prologe der Reportata Parisiensia des Johannes Duns Scotus: Untersuchungen zur Textüberlieferung und kritische Edition. Innsbruck, Institut für Christliche Philosophie, Abteilung für die Quellenkunde der Philosophie und Theologie des Mittelalters, Katholisch-Theologische Fakultät der Universität Innsbruck, 2005, 134-239 p. (Mediaevalia oenipontana, 2).

** 3255. ROSETH (Roger). Lectura super Sententias: quaestiones 3, 4 & 5. Ed. by Olli HALLAMAA. Helsinki, Luther-Agricola-Seura, 2005, 309 p. (ill.). (Helsingin yliopiston systimaattisen teologian laitoksen julkaisuja / Reports from the Department of Systematic Theology, University of Helsinki, 18).

** 3256. SIRAT (Colette), GEOFFROY (Marc). L'original arabe du Grand commentaire d'Averroès au De anima d'Aristote: prémices de l'édition. Paris, J. Vrin, 2005, 120 p. (ill., pl.). (Sic et non).

3257. ACAR (Rahim). Talking about God and talking about creation: Avicenna's and Thomas Aquinas' positions. Leiden a. Boston, Brill, 2005, X-250 p. (Islamic philosophy, theology and science: texts and studies, 58).

3258. Actes de la II Trobada Internacional d'Estudis sobre Arnau de Vilanova. Barcelona, Institut d'Estudis Catalans, 2005, 873 p. (facs.). (Arxiu de textos Catalans antics 23/24).

3259. Alain de Lille, le docteur universel: philosophie, théologie et littérature au XIIe siécle. Actes du XIe Colloque international de la Société internationale pour l'étude de la philosophie médiévale. Éd. par Jean-Luc SOLÈRE, Anca VASILIU et Alan GALONNIER. Turnhout, Brepols, 2005, XIV-495 p. (Rencontres de philosophie médiévale, 12).

3260. Albertus Magnus und die Anfänge der Aristoteles-Rezeption im lateinischen Mittelalter: von Richardus Rufus bis zu Franciscus de Mayronis = Albertus Magnus and the beginnings of the medieval reception of Aristotle in the Latin West: from Richardus Rufus to Franciscus de Mayronis. Hrsg. v. Ludger HONNEFELDER. Münster, Aschendorff, 2005, 862 p. (Subsidia Albertina, 1).

3261. ALTMEYER (Claudia). Grund und Erkennen in deutschen Predigten von Meister Eckhart. Würzburg, Königshausen & Neumann, 2005, 304 p.

3262. AMERINI (Fabrizio). La logica di Francesco da Prato: con l'edizione critica della Loyca e del Tractatus de voce univoca. Tavarnuzze, SISMEL – Edizioni del Galluzzo, 2005, V-646 p. (Testi e studi per il "Corpus philosophorum Medii Aevi", 19).

3263. Aquinas on scripture: an introduction to his biblical commentaries. Ed. by Thomas G. WEINANDY, Daniel A. KEATING and John YOCUM. London, T & T Clark International, 2005, XII-257 p.

3264. Autour de Guillaume d'Auvergne (1249). Éd par Franco MORENZONI et Jean-Yves TILLIETTE. Turnhout, Brepols, 2005, 424 p. (Bibliothèque d'histoire culturelle du Moyen Age, 2).

3265. Averroès et l'averroïsme, XIIe–XVe siècle. Un itinéraire historique du Haut Atlas à Paris et Padoue. Actes du colloque international organisé à Lyon, les 4 et 5 octobre 1999. Sous la dir. de André BAZZANA, Nicole BÉRIOU et Pierre GUICHARD. Lyon, Presses universitaires de Lyon, 2005, 352 p.

3266. Bartholomaeus Anglicus, De proprietatibus rerum: texte latin et réception vernaculaire. Actes du colloque international, Münster, 9.-11.10.2003 = Lateinischer Text und volkssprachige Rezeption: Akten des Internationalen Kolloquiums, Münster, 9.-11.10.2003. Hrsg. v. Baudouin VAN DEN ABEELE und Heinz MEYER. Turnhout, Brepols, 2005, XII-327 p. (De diversis artibus).

3267. BIARD (Joël), CELEYRETTE (Jean). De la théologie aux mathématiques: l'infini au XIVe siècle. Paris, Belles lettres, 2005, 318 p. (Sagesses médiévales, 3).

3268. Book of nature in antiquity & the Middle Ages. Ed. by Arie Johan VANDERJAGT and Klaas VAN BERKEL. Leuven, Peeters, 2005, XI-188 p.

3269. BOTTIN (Francesco). Filosofia medievale della mente. Padova, Il poligrafo, 2005, 249 p. (Subsidia mediaevalia patavina, 7).

3270. BÜCHNER (Christine). Gottes Kreatur – "ein reines Nichts": Einheit Gottes als Ermöglichung von Geschöpflichkeit und Personalität im Werk Meister Eckharts. Innsbruck u. Wien, Tyrolia-Verlag, 2005, 597 p. (Innsbrucker theologische Studien, 71).

3271. CROSS (Richard). Duns Scotus on God. Aldershot, Ashgate, 2005, XI-289 p. (Ashgate studies in the history of philosophical theology).

3272. EBNETER (Thomas). Exsistere: zur Persondefinition in der Trinitätslehre des Richard von St. Viktor (d. 1173). Fribourg, Academic Press, 2005, 108 p.

3273. ECHAVARRÍA (Martín F.). La Praxis de la psicología y sus niveles epistemologicos según Santo Tomás de Aquino. Girona, Documenta Universitaria, 2005, 858 p. (Documenta Universitaria. Tesis, 2).

3274. FALILEEV (Aleksandr I.). The Leiden leechbook: a study of the earliest neo-Brittonic medical compilation. Innsbruck, Institut für Sprachen und Literaturen der Universität Innsbruck, 2005, XI-115 p. (Innsbrucker Beiträge zur Kulturwissenschaft. Sonderheft, 122).

3275. Felicità (Le) nel Medioevo: atti del convegno della Società italiana per lo studio del pensiero medievale (S.I.S.P.M.), Milano, 12–13 settembre 2003. A cura di Maria BETTETINI e Francesco D. PAPARELLA. Louvain-la-Neuve, Fédération internationale des instituts d'études médiévales, 2005, XV-462 p. (Textes et études du moyen âge, 31).

3276. FEUERLE (Mark). Blide – Mange – Trebuchet. Technik, Entwicklung und Wirkung des Wurfgeschützes im Mittelalter. Eine Studie zur mittelalterlichen Innovationsgeschichte. Diepholz, Stuttgart u. Berlin, GNT-Verlag, 2005, 193 p. (Veröffentl. des 1. Zentrums für experimentelles Mittelalter, Vechta, 1).

3277. GAMBERO (Luigi). Mary in the Middle Ages: the Blessed Virgin Mary in the thought of medieval Latin theologians. San Francisco, Ignatius Press, 2005, 339 p.

3278. GEORGES (Tobias). Quam nos divinitatem nominare consuevimus: die theologische Ethik des Peter Abaelard. Leipzig, Evangelische Verlagsanstalt, 2005, 334 p. (Arbeiten zur Kirchen- und Theologiegeschichte, 16).

3279. GLÄSER (Pascal). Zurechnung bei Thomas von Aquin: eine historisch-systematische Untersuchung mit Bezug auf das aktuelle deutsche Strafrecht. Freiburg, Alber, 2005, 221 p. (ill.). (Symposion, 124).

3280. GRESCHAT (Katharina). Die Moralia n Job Gregors des Grossen: ein christologisch-ekklesiologischer Kommentar. Tübingen, Mohr Siebeck, 2005, IX-298 p. (Studien und Texte zu Antike und Christentum = Studies and texts in antiquity and Christianity, 31).

3281. GUGLIELMETTI (Rossana). La tradizione manoscritta del Policraticus di Giovanni di Salisbury: primo secolo di diffusione. Tavarnuzze, SISMEL – Edizioni del Galluzzo, 2005, XIII-255 p. (Millennio medievale, 60. Strumenti e studi, 13).

3282. Healing the body politic: the political thought of Christine de Pizan. Ed. by Karen GREEN and C. J. MEWS. Turnhout, Brepols, 2005, XXI-264 p. (Disputatio, 7).

3283. HONOLD (Marianne). Studie zur Funktionsgeschichte der spätmittelalterlichen deutschsprachigen Kochrezepthandschriften. Würzburg, Königshausen & Neumann, 2005, 452 p. (Würzburger medizinhistorische Forschungen, 87).

3284. HOROWSKI (Aleksander). La Visio Dei come forma della conoscenza umana in Alessandro di Hales: una lettura della Glossa in quatuor libros sententiarum e delle Quaestiones disputatae. Roma, Istituto storico dei Cappuccini, 2005, 376 p. (Bibliotheca seraphico-capuccina, 73).

3285. LEINKAUF (Thomas). Platons Timaios als Grundtext der Kosmologie in Spätantike, Mittelalter und Renaissance = Plato's Timaeus and the foundations of cosmology in later antiquity, the Middle Ages and Renaissance. Leuven, Leuven U. P., 2005, XXVI-492 p. (Ancient and medieval philosophy. Series 1, 34).

3286. Logik und Theologie: das Organon im arabischen und im lateinischen Mittelalter. Hrsg. v. Dominik PERLER und Ulrich RUDOLPH. Leiden, Brill, 2005, VI-511 p. (Studien und Texte zur Geistesgeschichte des Mittelalters, 84).

3287. MARENBON (John). Le temps, l'éternité et la prescience de Boèce à Thomas d'Aquin. Éd. par Irène ROSIER-CATACH. Paris, Vrin, 2005, 188 p. (Conférences Pierre Abélard).

3288. MARTELLO (Concetto). La dottrina dei teologi: ragione e dialettica nei secoli XI–XII. Catania, CUECM, 2005, 367 p. (Symbolon: studi e testi di filosofia antica e medievale, 29).

3289. Mathematikverständnis (Das) des Nikolaus von Kues: mathematische, naturwissenschaftliche und philosophisch-theologische Dimensionen. Akten der Tagung im Schwäbischen Tagungs- und Bildungszentrum Kloster Irsee vom 8.–10. Dezember 2003. Hrsg. v. Friedrich PUKELSHEIM und Harald SCHWAETZER. Trier, Paulinus-Verlag, 2005, X-342 p. (Mitteilungen und Forschungsbeiträge der Cusanus-Gesellschaft).

3290. Medieval science, technology, and medicine: an encyclopedia. Ed. by Thomas F. GLICK, Steven John LIVESEY and Faith WALLIS. New York a. London, Routledge, 2005, XXV-598 p. (ill.). (Routledge encyclopedias of the Middle Ages, 11).

3291. Mediterraneo medievale (Nel): la medicina. Atti della giornata di studio, Fisciano, Università degli studi di Salerno, 14 maggio 2004. A cura di Alfonso LEONE e Gerardo SANGERMANO. Salerno, Laveglia, 2005, 87 p. (Centro studi scientifico culturale salernitano Trotula de Ruggiero. Quaderni, 2).

3292. Meister Eckhart in Erfurt. Hrsg. v. Andreas SPEER und Lydia WEGENER Berlin, Walter de Gruyter, 2005, XI-612 p. (ill.). (Miscellanea mediaevalia).

3293. MEWS (C. J.). Abelard and Heloise. Oxford: Oxford U. P., 2005, XVIII-308 p. (Great medieval thinkers).

3294. Middle English medical texts. Ed. by Irma TAAVITSAINEN, Päivi PAHTA, Martti MÄKINEN and Raymond HICKEY. Amsterdam, John Benjamins Pub. Co., 2005, 1 CD-ROM.

3295. MONSON (Don Alfred). Andreas Capellanus, scholasticism, and the courtly tradition. Washington, Catholic University of America Press, 2005, IX- 383 p.

3296. NIEDERER (Monica). Der St. Galler Botanicus: ein frühmittelalterliches Herbar. Kritische Edition, Übersetzung und Kommentar. Bern a. Oxford, Peter Lang, 2005, 459 p. (Lateinische Sprache und Literatur des Mittelalters, 38).

3297. NORTH (John David). God's clockmaker: Richard of Wallingford and the invention of time. London, Hambledon and London, 2005, XVII-441 p. (ill.).

3298. OSBORNE (Thomas Michael). Love of self and love of God in thirteenth-century ethics. Notre Dame, University of Notre Dame Press, 2005, IX-325 p.

3299. PANZIG (Erik A.). Geläzenheit und abegescheidenheit: eine Einführung in das theologische Denken des Meister Eckhart. Leipzig, Evangelische Verlagsanstalt, 2005, 298 p.

3300. PARISOLI (Luca). La contraddizione vera: Giovanni Duns Scoto tra le necessità della metafisica e il discorso della filosofia pratica. Roma, Istituto storico dei Cappuccini, 2005, 222 p. (Bibliotheca seraphico-capuccina, 72).

3301. POULIOT (François). La doctrine du miracle chez Thomas d'Aquin: Deus in omnibus intime operatur. Paris, J. Vrin, 2005, 162 p. (Bibliothèque thomiste, 56).

3302. Reading John with St. Thomas Aquinas: theological exegesis and speculative theology. Ed. by Michael DAUPHINAIS and Matthew LEVERING. Washington, Catholic University of America Press, 2005, XXV-371 p.

3303. RIEGER (Reinhold). Contradictio: Theorien und Bewertungen des Widerspruchs in der Theologie des Mittelalters. Tübingen, Mohr Siebeck, 2005, XII-570 p. (Beiträge zur historischen Theologie).

3304. ROREM (Paul). Eriugena's commentary on the Dionysian Celestial hierarchy. Toronto, Pontifical Institute of Mediaeval Studies, 2005, XIV-242 p. (Pontifical Institute of Medieval Studies. Studies and texts, 150).

3305. RZIHACEK-BEDÖ (Andrea). Medizinische Wissenschaftspflege im Benediktinerkloster Admont bis 1500. Wien u. München, Oldenbourg, 2005, 289 p. (Österreichische Geschichtsforschung, 46).

3306. Schrift, Schreiber, Schenker: Studien zur Abtei Sankt Viktor in Paris und den Viktorinern. Hrsg. v. Rainer BERNDT. Berlin, Akademie Verlag, 2005, 394 p. (Corpus Victorinum. Instrumenta, 1).

3307. Selbstbewusstsein und Person im Mittelalter: Symposium des Philosophischen Seminars der Universität Hannover vom 24. bis 26. Februar 2004. Hrsg. v. Günther MENSCHING, Eckhard HOMANN und Anneke MEYER. Würzburg, Königshausen & Neumann, 2005, 268 p. (Contradictio, 6).

3308. SPYRA (Ulrike). Das Buch der Natur Konrads von Megenberg: die illustrierten Handschriften und Inkunabelm. Köln, Böhlau, 2005, vi, 488 p. (pl., ill.). (Pictura et poesis, 19).

3309. Text and controversy from Wyclif to Bale: essays in honor of Anne Hudson. Ed. by Helen BARR and Ann M. HUTCHISON. Turnhout, Brepols, 2005, XXII-448 p. (ill.). (Medieval church studies, 4).

3310. Textual healing: essays on medieval and early modern medicine. Ed. by Elizabeth Lane FURDELL. Leiden a. Boston, Brill, 2005, XVII-287 p. (ill.). (Studies in medieval and Reformation traditions).

3311. Theology (The) of Thomas Aquinas. Ed. by Rik VAN NIEUWENHOVE and Joseph Peter WAWRYKOW. Notre Dame, University of Notre Dame Press, 2005, XX-472 p. (ill.).

3312. TRONCARELLI (Fabio). Cogitatio mentis: l'eredità di Boezio nell'alto Medioevo. Napoli, M. D'Auria, 2005, 372 p. (Storie e testi, 16).

3313. VERLAGUET (Waltraud). L'"éloignance": la théologie de Mechthild de Magdebourg (XIIIe siécle). Bern, Peter Lang, 2005, XVIII-427 p. (ill.).

3314. WANNENMACHER (Julia Eva). Hermeneutik der Heilsgeschichte: De septem sigillis und die sieben Siegel im Werk Joachims von Fiore. Leiden a. Boston, Brill, 2005, X-393 p. (Studies in the history of Christian traditions, 118).

3315. WICKI (Nikolaus). Die Philosophie Philipps des Kanzlers: ein philosophierender Theologe des frühen 13. Jahrhunderts. Fribourg, Academic Press, 2005, VIII-198 p. (Dokimion, 29).

3316. ZAMUNER (Ilaria). La tradizione romanza del Secretum secretorum pseudo-aristotelico, *Studi Medievali*, 2005, 46, 1, p. 31-116 (tab.).

Cf. n° 965

§ 13. History of the Church and religion.

a. General

3317. BERMAN (Constance H.). Medieval religion: new approaches. New York a. London, Routledge, 2005, XXIII-422 p. (ill., pl.). (Re-writing histories).

3318. BLAIR (John). The church in Anglo-Saxon society. Oxford, Oxford U. P., 2005, XIX-604 p. (ill., map).

3319. BLOMKVIST (Nils). The discovery of the Baltic: the reception of a Catholic world-system in the European North (a.d. 1075–1225). Boston, Brill, 2005, VIII-774 p. (Northern world, 15).

3320. CHADWICK (Henry). East and west: the making of a rift in the church: from apostolic times until the Council of Florence. Oxford, Oxford U P, 2005, 306 p. (Oxford history of the Christian Church).

3321. Chiesa, vita religiosa, società nel Medioevo italiano: studi offerti a Giuseppina De Sandre Gasparini. A cura di Mariaclara ROSSI e Gian Maria VARANINI. Roma, Herder, 2005, XXVI-759 p. (ill., pl.). (Italia sacra, 80).

3322. Church, state, vellum, and stone: essays on medieval Spain in honor of John Williams. Ed. by Therese MARTIN and Julie A. HARRIS. Leiden a. Boston, Brill, 2005, VII-571 p. (ill.). (Medieval and early modern Iberian world, 26).

3323. DEMURA (Akira). Chūsei Kirisutokyō no rekishi. (A history of medieval christianity). Tokyo, nihon-kirisutokyodan-shuppankyoku, 2005, 407 p.

3324. DUMÉZIL (Bruno). Les racines chrétiennes de l'Europe: conversion et liberté dans les royaumes barbares, V^e–VIIIe siècle. Paris, Fayard, 2005, 804 p. (geneal. tabl.).

3325. HÄRDELIN (Alf). Världen som yta och fönster. Spiritualitet i medeltidens Sverige. (Le monde comme surface et fenêtre. La spiritualité dans la Suède médiévale). Stockholm, Sällskapet Runica et Mediævalia, 2005, 592 p. (ill.). (Runica et Mediævalia, Scriptora minora, 13).

3326. In principio erat verbum: mélanges offerts en hommage à Paul Tombeur par des anciens étudiants à l'occasion de son éméritat, Éd. par Benoît-Michel TOCK. Turnhout, Brepols, 2005, XXXVIII-485 p. (Textes et études du moyen âge, 25).

3327. JOHRENDT (Jochen). La protezione apostolica alla luce dei documenti pontifici (896–1046). *Bullettino dell'istituto storico italiano per il medio evo*, 2005, 107, p. 135-168.

3328. JORDAN (William Chester). Unceasing strife, unending fear. Jacques de Thérines and the freedom of the Church in the age of the last Capetians. Princeton, Princeton U. P., 2005, XI-154 p.

3329. KEJŘ (Jiří). Die mittelalterlichen Synoden in Böhmen und Mähren. *Zeitschrift der Savigny-Stiftung für Rechtsgeschichte Kanonistische Abteilung*, 2005, 91, p. 767-770.

3330. LEJON (Kjell O.). Diocesis Lincopensis. 1. historik över Linköpings stift. 2, Medeltida internationella influenser: några uttryck för en framväxande östgötsk delaktighet i den västeuropeiska kultursfären. (Le diocèse de Linköping. Tome 1. Histoire. Tome 2. Influences internationales au Moyen Age: expressions de la participation grandissante du diocèse à la sphère culturelle occidentale. Skellefteå, Norma, 2005, 552 p., 224 p. (pl.). (Linköpings stiftshistoriska sällskaps skriftserie, 1-2).

3331. MAC GUIRE (Brian Patrick). Jean Gerson and the last Medieval Reformation. University Park, Pennsylvania State U. P., 2005, XVI-441 p. (ill., maps).

3332. MILES (Margaret Ruth). The Word made flesh: a history of Christian thought. Malden a. Oxford, Blackwell, 2005, XV-435 p. a. 1 CD-ROM (ill., maps).

3333. Oboedientia. Zu Formen und Grenzen von Macht und Unterordnung im mittelalterlichen Religiosentum. Hrsg. v. Sébastien BARRET und Gert MELVILLE. Münster, Lit, 2005, XII-459 p. (Vita regularis, Abhandlungen, 27).

3334. Reforming the Church before Modernity. Patterns, Problems and Approaches. Ed. by Christopher M. BELLITTO et Louis I. HAMILTON. Aldershot, Ashgate, 2005, XXIV--224 p. (Church, Faith and Culture in the Medieval West).

3335. SIEBEN (Hermann Josef). Studien zur Gestalt und Überlieferung der Konzilien. Paderborn, Ferdinand Schöningh, 2005, 428 p. (Konziliengeschichte. Reihe B, Untersuchungen).

3336. SMOLINSKY (Heribert). Im Zeichen von Kirchenreform und Reformation: gesammelte Studien zur

Kirchengeschichte in Spätmittelalter und Früher Neuzeit. Hrsg. v. Karl-Heinz BRAUN, Barbara HENZE und Bernhard SCHNEIDER. Münster, Aschendorff, 2005, VI-469 p. (Reformationsgeschichtliche Studien und Texte. Supplementband, 5).

3337. SUDMANN (Stefan). Das Basler Konzil: synodale Praxis zwischen Routine und Revolution. Frankfurt am Main, P. Lang, 2005, 508 p. (Tradition, Reform, Innovation).

3338. THELIANDER (Claes). Västergötlands kristnande: religionsskifte och gravskickets förändring 700–1200. (La christianisation du Västergötland. Conversion et changement de mode de sépulture entre 700 et 1200). Göteborg, Institutionen för arkeologi, Göteborgs universitet, 2005, 399 p. (GOTARC. Series B, Gothenburg archaeological theses, 41).

3339. THÜMMEL (Hans Georg). Die Konzilien zur Bilderfrage im 8. und 9. Jahrhundert: das 7. Ökumenische Konzil in Nikaia 787. Paderborn, Ferdinand Schöningh, 2005, XXIV-319 p. (Konziliengeschichte. Reihe A, Darstellungen).

3340. TISCHLER (Matthias M.). Die Christus- und Engelweihe im Mittelalter. Texte, Bilder und Studien zu einem ekklesiologischen Erzählmotiv. Berlin, Akademie, 2005, 244 p. (Erudiri sapientia, 5).

3341. TURCUŞ (Şerban). A canonization Process in eleventh century Transylvania and its political role in Hungary and the Venetian Republic. *Nouvelles études d'histoire*, 2005, 11, p. 29-49.

3342. VAN WIJNENDAELE (Jacques). Silences et mensonges autour d'un concile. Le concile de Sutri (1046) en son temps. *Revue belge de philologie et d'histoire*, 2005, 83, 1, p. 315-353.

3343. Vom Schisma zu den Kreuzzügen: 1054–1204. Hrsg. v. Peter BRUNS und Georg GRESSER. Paderborn, Schöningh, 2005, 271 p.

Cf. n°s 272, 4843

b. History of the Popes

** 3344. Catalogus episcoporum Ultrajectinorum =: Lijst van de Utrechtse bisschoppen, 695–1378: uitgave en vertaling. Ed. Jo T. J. JAMAR, C. A. VAN KALVEEN. Utrecht, Utrechts Archief, 2005, 188 p. (ill.). (Toegangen van Het Utrechts Archief).

** 3345. English episcopal acta. Vol. 29. Durham, 1241–1283. Ed. by Philippa M. HOSKIN. Vol. 30. Carlisle, 1133–1292. Ed. by David Michael SMITH. Vol. 31. Ely, 1109–1197. Ed. by Nicholas KARN. Oxford a. New York, British Academy a. Oxford U. P., 2005, 3 vol., LXXXII- 293 p., LXI-253 p., CXLIX-288 p. (ill., pl.).

** 3346. Fasti ecclesiae Gallicanae. Vol. 9. Diocèse de Sées, 1200–1547. Éd. par Pierre DESPORTES. Turnhout, Brepols, 2005, X-193 p.

** 3347. MANCURTI (Francesco Maria). Memorie della chiesa cattedrale d'Imola incomminciando dal quarto secolo sino alla metà del secolo diciottesimo, e più oltre ancora, descritte, e distribuite in sette libri dal canonico Francesco Maria Mancurti, col catalogo in fine de canonici, e de' mansionari della medesima, colla indicazione inoltre di tutti i benefici ecclesiastici in essa eretti, e colle iscrizioni sepolcrali, ed altri monumenti che vi si veggono. A cura di Andrea FERRI. Imola, Diocesi di Imola, 2005, 796 p. (ill., facs, ports.). (Pubblicazioni dell'Archivio diocesano di Imola. Documenti e studi).

** 3348. Medieval (The) Court of Arches. Ed. by Francis Donald LOGAN. Woodbridge, Boydell Press, 2005, XLVIII-240 p. (Canterbury and York society, 95).

** 3349. Records of Convocation. Ed. Gerald Lewis BRAY. Vol. 1. Sodor and Man, 1229–1877. Vol. 3. Canterbury, 1313–1377. Vol. 4. Canterbury, 1377–1414. Vol. 5. Canterbury, 1414–1443. Vol. 6. Canterbury, 1444–1509. Woodbridge, Boydell Press in association with the Church of England Record Society, 2005, 5 vol., 508 p., 449 p., 457 p., 456 p., 511 p.

3350. ARNOLD (Dorothee). Johannes VIII: Päpstliche Herrschaft in den karolingischen Teilreichen am Ende des 9. Jahrhunderts. Frankfurt, Lang, 2005, 267 p. (Europäische Hochschulschriften. Reihe 23, Theologie, 797).

3351. Aux origines de la paroisse rurale en Gaule méridionale, IVe–IXe siècles: Actes du colloque international, 21–23 mars 2003, Salle Tolosa (Toulouse). Éd. par Christine DELAPLACE. Paris, Errance, 2005, 255 p. (ill.).

3352. BARRIO GOZALO (Maximiliano). Iglesia y sociedad en Segovia: siglos XVI–XIX. Valladolid, Universidad de Valladolid, 2005, 307 p. (Historia y sociedad, 113).

3353. BELLENGER (Dominic Aidan), FLETCHER (Stella). The Mitre and the Crown: a history of the archbishops of Canterbury. Stroud, Sutton, 2005, XX-236 p.

3354. BERTRAM (Jerome). The Chrodegang rules: the rules for the common life of the secular clergy from the eighth and ninth centuries. Aldershot, Ashgate, 2005, 293 p. (ill.). (Church, faith, and culture in the Medieval West).

3355. CANTARELLA (Glauco Maria). Il sole e la luna: la rivoluzione di Gregorio VII papa, 1073–1085. Roma, Laterza, 2005, VIII-354 p. (Storia e società).

3356. CAPITANI (Ovidio). Bonifacio VIII. *Bullettino dell'istituto storico italiano per il medio evo*, 2005, 107, p. 229-45.

3357. CUSHING (Kathleen G.). Reform and Papacy in the eleventh century: spirituality and social change. Manchester, Manchester U. P., 2005, XV-172 p. (Manchester medieval studies).

3358. Eredità (L') spirituale di Gregorio Magno tra Occidente e Oriente: atti del Simposio internazionale "Gregorio Magno 604–2004", Roma 10–12 marzo 2004.

A cura di Innocenzo GARGANO. Negarine di S. Pietro in Cariano, Il segno dei Gabrielli, 2005, 386 p.

3359. FLORYSZCZAK (Silke). Die Regula Pastoralis Gregors des Großen: Studien zu Text, kirchenpolitischer Bedeutung und Rezeption in der Karolingerzeit. Tübingen, Mohr Siebeck, 2005, X-444 p. (Studien und Texte zu Antike und Christentum = Studies and texts in antiquity and Christianity, 26).

3360. HIRTE (Markus). Papst Innozenz III., das IV. Lateranum und die Strafverfahren gegen Kleriker: eine registergestützte Untersuchung zur Entwicklung der Verfahrensarten zwischen 1198 und 1216. Tübingen, Edition Diskord, 2005, 349 p. Rothenburger Gespräche zur Strafrechtsgeschichte, 5).

3361. JARITZ (Gerhard), JØRGENSEN (Torsten Bo.), SALONEN (Kirsi). The long arm of papal authority: late medieval Christian peripheries and their communication with the Holy See. New York, Central European U. P., 2005, VI-192 p. (CEU medievalia, 8).

3362. KEHNEL (Annette). Päpstliche Kurie und menschlicher Körper. Zur historischen Kontextualisierung der Schrift "De miseria humane conditionis" des Lothar von Segni (1194). *Archiv für Kulturgeschichte*, 2005, 87, 1, p. 27-52.

3363. KUHN (Lambrecht). Bistum Lebus: das kirchliche Leben im Bistum Lebus in den letzten zwei Jahrhunderten (1385-1555) seines Bestehens unter besonderer Berücksichtigung des Johanniterordens. Leipzig, Evangelische Verlagsanstalt, 2005, 403 p. (ill., maps). (Herbergen der Christenheit. Sonderband, 8).

3364. Kulturarbeit und Kirche: Festschrift Msgr. Dr. Paul Mai zum 70. Geburtstag. Hrsg. v. Werner Johann CHROBAK und Karl HAUSBERGER. Regensburg, Verlag des Vereins für Regensburger Bistumsgeschichte, 2005, 817 p. (Beiträge zur Geschichte des Bistums Regensburg).

3365. Lebensbilder aus dem Bistum Augsburg: vom Mittelalter bis in die neueste Zeit. Hrsg. v. Walter MIXA, Manfred WEITLAUFF und Walter ANSBACHER. Augsburg, Verlag des Vereins für Augsburger Bistumsgeschichte, 2005, XI-680 p. (ill.). (Jahrbuch des Vereins für Augsburger Bistumsgeschichte).

3366. Pastoral care in late Anglo-Saxon England. Ed. by Francesca TINTI. Woodbridge, Boydell Press, 2005, VIII-152 p. (ill.). (Anglo-Saxon studies).

3367. PFLEFKA (Sven). Das Bistum Bamberg, Franken und das Reich in der Stauferzeit: der Bamberger Bischof im Elitengefüge des Reiches, 1138-1245. Schweinfurt: Gesellschaft für fränkische Geschichte, 2005, XII-442 p. (Veröffentlichungen der Gesellschaft für Fränkische Geschichte. Reihe IX, Darstellungen aus der fränkischen Geschichte, 49).

3368. PLÖGER (Karsten). England and the Avignon Popes. The practice of diplomacy in late Medieval Europe. London, Modern Humanities Research Association and Maney Publishing, 2005, XIV-304 p.

3369. Scrittura e storia: per una lettura delle opere di Gregorio Magno. A cura di Lucia CASTALDI. Tavarnuzze, SISMEL-Edizioni del Galluzzo, 2005, VIII-435 p. (Archivum Gregorianum, 7).

Cf. n° 2761

c. Monastic history

* 3370. Livre (Le) de la confrérie de Saint-Jacques d'Overmolen à Bruxelles (1357-1419). Éd. par C. DICKSTEIN-BERNARD. Brussels, Archives de la ville de Bruxelles, 2005, XXXIX-186 p. (Fontes Bruxellae, 1).

** 3371. Ancrene wisse: a four-manuscript parallel text, parts 5-8 with wordlists. Ed. by Tadao KUBOUCHI, Keiko IKEGAMI and John SCAHILL. Frankfurt am Main, Peter Lang, 2005, 360 p. (Studies in English medieval language and literature, 11).

** 3372. Libros (Los) de visita de la Orden de Santiago: Calzadilla de los Barros. Ed. por Manuel LEYGUARDA DOMÍNGUEZ. Badajoz, Junta de Extremadura, Consejería de Cultura, Archiv Histórico Provincial de Badajoz, 2005, 384 p.

3373. BOURGEOIS (Nicolas). Les Cisterciens et la croisade de Livonie. *Revue historique*, 2005, 635, p. 521-560.

3374. BOYNTON (Susan), COCHELIN (Isabelle). From dead of night to end of day: the medieval customs of Cluny = Du cœur de la nuit à la fin du jour: les coutumes clunisiennes au Moyen Âge. Turnhout, Brepols, 2005, 398 p. (ill.). (Disciplina monastica, 3).

3375. CARRAZ (Damien), DEMURGER (Alain). L'Ordre du Temple dans la basse vallée du Rhône: 1124-1312. Ordres militaires, croisades et sociétés méridionales. Lyon, Presses universitaires de Lyon, 2005, 662 p. (ill., maps). (Collection d'histoire et d'archéologie médiévales).

3376. Charisma und religiöse Gemeinschaften im Mittelalter: Akten des 3. Internationalen Kongresses des "Italienisch-deutschen Zentrums für Vergleichende Ordensgeschichte" in Verbindung mit Projekt C "Institutionelle Strukturen religioser Orden im Mittelalter" und Projekt W "Stadtkultur und Klosterkultur in der mittelalterlichen Lombardei. Institutionelle Wechselwirkung zweier politischer und sozialer Felder" des Sonderforschungsbereichs 537 "Institutionalität und Geschichtlichkeit" (Dresden, 10.-12. Juni 2004). Hrsg. v. Giancarlo ANDENNA, Mirko BREITENSTEIN und Gert MELVILLE. Münster, Lit, 2005, XX-495 p. (Vita regularis. Abhandlungen, 26).

3377. DALARUN (Jacques), ZINELLI (Fabio). Le manuscrit des soeurs de Santa Lucia de Foligno, *Studi Medievali*, 2005, 46, 1, p. 117-67 (ill.).

3378. DIETZ (Maribel). Wandering monks, virgins, and pilgrims: ascetic travel in the Mediterranean world, A.D. 300/800. University Park, Pennsylvania State U. P., 2005, IX-270 p.

3379. Domenico di Caleruega e la nascita dell'ordine dei frati predicatori: atti del XLI Convegno storico internazionale, Todi, 10–12 ottobre 2004. Spoleto, Fondazione Centro italiano di studi sull'alto Medioevo, 2005, XI-638 p. (ill., pl.). (Atti dei convegni del Centro italiano di studi sul basso Medioevo – Accademia tudertina e del Centro di studi sulla spiritualità medievale, 18).

3380. Frühformen von Stiftskirchen in Europa: Funktion und Wandel religiöser Gemeinschaften vom 6. bis zum Ende des 11. Jahrhunderts. Festgabe für Dieter Mertens zum 65. Geburtstag. Hrsg. v. Sönke LORENZ und Thomas L. ZOTZ. Leinfelden-Echterdingen, DRW-Verlag, 2005, VIII-424 p. (Schriften zur südwestdeutschen Landeskunde, 54).

3381. GILCHRIST (Roberta), SLOANE (Barney). Requiem: the medieval monastic cemetery in Britain. London, Museum of London Archaeology Service, 2005, XVIII-273 p. (ill., map).

3382. JAMROZIAK (Emilia). Rievaulx Abbey and its social context, 1132–1300: memeory, locality, and networks. Turnhout, Brepols, 2005, XII-252 p. (maps). (Medieval church studies, 8).

3383. KLUETING (Edeltraud). Monasteria semper reformanda: Kloster- und Ordensreformen im Mittelalter. Münster, Lit, 2005, XII-140 p. (Historia profana et ecclesiastica, 12).

3384. Libri, biblioteche e letture dei frati mendicanti (secoli XIII–XIV): atti del XXXII Convegno internazionale, Assisi, 7–9 ottobre 2004. Spoleto, Fondazione Centro italiano di studi sull'alto Medioevo, 2005, X-501 p. (Atti dei convegni della Società internazionale di studi francescani e del Centro interuniversitario di studi francescani, 15).

3385. LLOYD (Joan Barclay). SS. Vincenzo e Anastasio at Tre Fontane near Rome: history and architecture of a medieval Cistercian abbey. Kalamazoo, Cistercian Publications, 2005, XXIII-324 p. (ill., pl.). (Cistercian studies series, 198).

3386. LUTTER (Christina). Geschlecht & Wissen, Norm & Praxis, Lesen & Schreiben: monastische Reformgemeinschaften im 12. Jahrhundert. Wien, Oldenbourg, 2005, X-338 p. (ill., pl., facs.). (Veröffentlichungen des Instituts für Österreichische Geschichtsforschung, 43).

3387. PARENTI (Stefano). Il monastero di Grottaferrata nel Medioevo, 1004–1062: segni e percorsi di una identità. Roma, Pontificio Istituto orientale, 2005, 570 p. (ill., pl.). (Orientalia Christiana analecta).

3388. RAPETTI (Anna Maria). Monachesimo medievale: uomini, donne e istituzioni. Venezia, Marsilio, 2005, 172 p. (Saggi).

3389. Regulae – consuetudines – statuta: studi sulle fonti normative degli ordini religiosi nei secoli centrali del Medioevo. Atti del I e del II Seminario internazionale di studio del Centro italo-tedesco di storia comparata degli ordini religiosi (Bari/Noci/Lecce, 26–27 ottobre 2002 / Castiglione delle Stiviere, 23–24 maggio 2003). A cura di Cristina ANDENNA e Gert MELVILLE. Münster, Lit, 2005, XII-709 p. (Vita regularis. Abhandlungen, 25).

3390. RODRÍGUEZ CASTILLO (Héctor). Los monasterios dúplices en Galicia en la Alta Edad Media: un trabajo sobre modelos sociales. Noia, Editorial Toxosoutos, 2005, 192 p. (Trivium, 16).

3391. ROUET (Dominique). Le cartulaire de l'abbaye bénédictine de Saint-Pierre-de-Préaux (1034–1227). Paris, Éd. du CTHS, 2005, 586 p.

3392. SCHÜRER (Markus). Das Exemplum, oder, Die erzählte Institution: Studien zum Beispielgebrauch bei den Dominikanern und Franziskanern des 13. Jahrhunderts. Münster, Lit, 2005, 365 p. (Vita regularis. Abhandlungen, 23).

3393. SEKIGUCHI (Takehiko). Kuryunī shūdōsei no kenkyū. (Studies in the organizational aspects of the Cluny congregation). Tokyo, Nansosha, 2005, 648 p.

3394. Sepulturae cistercienses: sépulture, mémoire et patronage dans les monastères cisterciens au Moyen Âge = burial, memorial and patronage in medieval Cistercian monasteries = Grablegen, Memoria und Patronatswesen in mittelalterlichen Zisterzienserklöstern. Éd. par Jackie HALL et Christine KRATZKE. Forges-Chimay, Cîteaux, 2005, 418 p. (ill., maps, pl.). (Cîteaux: commentarii Cistercienses, 56/1-4).

3395. SPEAR (Valerie G.). Leadership in Medieval English nunneries. Rochester, Boydell Press, 2005, XIX-244 p. (Studies in the history of Medieval religion, 24).

3396. St. Wulfsige and Sherborne: essays to celebrate the millennium of the Benedictine Abbey, 998-1998. Ed by Katherine BARKER, David Alban HINTON and Alan HUNT. Oxford, Oxbow Books, 2005, XXII-248 p. (Occasionale paper, 8).

3397. St. Wulfstan and his world. Ed. by Julia S. BARROW. Aldershot, Ashgate, 2005, XIX-242 p. (ill.). (Studies in early medieval Britain, 4).

3398. SUGIZAKI (Taiichiro). 12 seiki no Shūdōin to Shakai, zohokaiteiban. (Monasteries and society in twelfth century Europe). Tokyo, Hara Shobo, 2005, 330 p.

3399. UGÉ (Karine). Creating the monastic past in Medieval Flanders. York, York Medieval Press, 2005, XV-196 p.

3400. UNGER (Helga). Die Beginen: eine Geschichte von Aufbruch und Unterdrückung der Frauen. Freiburg im Breisgau, Herder, 2005, 220 p. (Herder Spektrum, 5643).

3401. WARNATSCH-GLEICH (Friederike). Herrschaft und Frömmigkeit: Zisterzienserinnen im Hochmittelalter. Berlin, Lukas, 2005, 268 p. (Studien zur Geschichte, Kunst und Kultur der Zistezienser, 21).

3402. WEIGEL (Petra). Ordensreform und Konziliarismus: der Franziskanerprovinzial Matthias Döring (1427–1461). Frankfurt am Main u. Oxford, Peter Lang, 2005, 540 p. (Jenaer Beiträge zur Geschichte).

3403. WINSTON-ALLEN (Anne). Convent chronicles: women writing about women and reform in the late Middle Ages. University Park, Pennsylvania State U. P., 2005, XVII-345 p. (ill.).

Cf. n°s 261, 2481

d. Hagiography

* 3404. SCAHILL (John), ROGERSON (Margaret). Middle English saints' legends. Woodbridge a. Rochester, D.S. Brewer, 2005, 209 p. (Annotated bibliographies of Old and Middle English literature).

** 3405. DEGL'INNOCENTI (Antonella), FRIOLI (Donatella), GATTI (Paolo). Manoscritti agiografici latini di Trento e Rovereto. Tavarnuzze, SISMEL – Edizioni del Galluzzo, 2005, LXX-317 p. (ill., pl., facs.). (Quaderni di "Hagiographica", 3).

3406. **Chiara** di Assisi: una vita prende forma. Iter storico. Padova, Messaggero, 2005, 221 p. (Secundum perfectionem Sancti Evangelii, 2).

3407. **Christina** of Markyate: a twelfth-century holy woman. Ed. by Samuel FANOUS and Henrietta LEYSER. London a. New York, Routledge, 2005, XV-266 p. (ill.).

3408. DOLBEAU (François). Sanctorum societas: récits latins de sainteté, IIIe–XIIe siècles. Bruxelles, Société des Bollandistes, 2005, 2 vol, XII-1001 p. (ill.). (Subsidia hagiographica, 85).

3409. GOULLET (Monique). Ecriture et réécriture hagiographiques: essai sur les réécritures de vies de saints dans l'occident latin médiéval (VIIe–XIIIe s.). Turnhout, Brepols, 2005, 318 p. (Hagiologia, 4).

3410. GRETSCH (Mechthild). Aelfric and the cult of saints in late Anglo-Saxon England. Cambridge, Cambridge U. P., 2005, XI-263 p. (Cambridge studies in Anglo-Saxon England, 34).

3411. GUIDI (Remo L.). Note sull'agiografia nel Quattrocento. *Archivio storico italiano*, 2005, 163, 2, p. 219-228.

3412. Heiligenleben zur deutsch-slawischen Geschichte. Adalbert von Prag und Otto von Bamberg. Unter Mitarbeit von Jerzy STRZELCZYK herausgegeben von Lorenz WEINRICH. Darmstadt, Wissenschaftliche Buchgesellschaft, 2005, VII-496 p. (Deutsche Geschichte des Mittelalters 23, Freiherr-vom Stein-Gedächtnisausgabe).

3413. KLEINBERG (Aviad M.). Histoires de saints: leur rôle dans la formation de l'Occident. Paris, Gallimard, 2005, 360 p. (Bibliothèque des histoires).

3414. LICCIARDELLO (Pierluigi). Agiografia aretina altomedievale: testi agiografici e contesti socio-culturali ad Arezzo tra VI e IX secolo. Tavarnuzze, SISMEL – Edizioni del Galluzzo, 2005, VI-764 p. (ill., pl.). (Millennio medievale, 56. Strumenti e studi, 9).

3415. PARISH (Helen L.). Monks, miracles, and magic: reformation representations of the medieval church. London a. New York, Routledge, 2005, VIII-224 p.

3416. Saint **Mary** of Egypt: three medieval lives in verse. Ed. by Ronald E. PEPIN and Hugh FEISS. Kalamazoo, Cistercian Publications, 2005, X-159 p. (Cistercian studies series, 209).

3417. Signs, wonders, miracles: representations of divine power in the life of the church. Papers read at the 2003 Summer Meeting and the 2004 Winter Meeting of the Ecclesiastical History Society. Ed. by Catherine Fales COOPER and Jeremy GREGORY. Woodbridge, Ecclesiastical History Society a. Boydell Press, 2005, XIX-475 p. (Studies in church history, 41).

3418. Tempo (Il) dei santi tra Oriente e Occidente: liturgia e agiografia dal tardo antico al concilio di Trento. A cura di Anna BENVENUTI PAPI e Marcello GARZANITI. Roma, Viella, 2005, VIII-501 p. (ill.).

3419. VELÁZQUEZ SORIANO (Isabel). Hagiografía y culto a los santos en la Hispania visigoda: aproximación a sus manifestaciones literarias. Mérida, Museo Nacional de Arte Romano, 2005, 271 p. (ill.). (Cuadernos emeritenses, 32).

e. Special studies

** 3420. Beati Iordanis de Saxonia sermons. Ed. Paul-Bernard HODEL. Romae, Institutum historicum Ordinis fratrum praedicatorum, 2005, 260 p. (Monumenta Ordinis Fratrum Praedicatorum historica, 29).

** 3421. Collectaneum exemplorum et visionum Clarevallense e codice Trecensi 946. Turnhout, Brepols Publishers, 2005, CXVI-468 p. (Corpus Christianorum. Continuatio Mediaevalis, 208).

** 3422. JOHANNES DE RUPESCISSA. Liber Ostensor quod adesse festinant tempora. Éd. par André VAUCHEZ, Clémence THÉVANEZ, Christine MOREROD-FATTEBERT, Marie-Henriette JULLIEN DE POMMEROL et Jeanne BIGNAMI ODIER. Rome, École française de Rome, 2005, XIII-1041 p. (geneal. tab.). (Sources et documents d'histoire du Moyen Âge, 8).

** 3423. JULIAN OF NORWICH. Showings: authoritative text, contexts, criticism Ed. by Denise NOWAKOWSKI BAKER. New York a. London, Norton, 2005, XXVII-213 p. (A Norton critical edition).

** 3424. KRAFFT (Otfried). Papsturkunde und Heiligsprechung: die päpstlichen Kanonisationen vom Mittelalter bis zur Reformation, ein Handbuch. Köln, Böhlau, 2005, XII-1247 p. (ill.). (Archiv für Diplomatik Schriftgeschichte, Siegel-und Wappenkunde. Beiheft, 9)..

** 3425. Liber sacramentorum Paduensis (Padova, Biblioteca capitolare, cod. D 47). A cura di Alceste CATELLA, F. DELL'ORO, Aldo MARTINI e Fabrizio CRI-

VELLO. Roma, CLV-Edizioni liturgiche, 2005, 595 p. (Bibliotheca "Ephemerides liturgicae". Subsidia, 131. Monumenta Italiae liturgica, 3).

** 3426. LOVE (Nicholas). The Mirror of the blessed life of Jesus Christ: a full critical edition. Ed. by Michael G. SARGENT. Exeter, University of Exeter Press, 2005, XLVII-280 p. (Exeter medieval texts and studies).

** 3427. WENZEL (Siegfried). Latin sermon collections from later Medieval England: orthodox preaching in the age of Wyclif. Cambridge a. New York, Cambridge U. P., 2005, XXIV-713 p. (Cambridge studies in medieval literature, 53).

3428. Approaching medieval English anchoritic and mystical texts. Ed. by Dee DYAS, Valerie EDDEN and Roger ELLIS. Cambridge, D.S. Brewer, 2005, XVI-213 p. (Christianity and culture).

3429. ARNOLD (John). Belief and unbelief in medieval Europe. London, Hodder Arnold, 2005, VIII-320 p.

3430. AUFFARTH (Christoph). Die Ketzer: Katharer, Waldenser und andere religiöse Bewegungen. München, Beck, 2005, 128 p. (ill.). (Beck'sche Reihe, 2383).

3431. BACCI (Michele). Lo spazio dell'anima: vita di una chiesa medievale. Roma e Bari, Laterza, 2005, XVII-286 p. (ill., pl.). (Storia e società).

3432. BARNWELL (P. S.), CROSS (M. Claire), RYCRAFT (Ann). Mass and parish in late medieval England: the use of York. Reading, Spire Books, 2005, 224 p. (ill, maps, music, plans).

3433. Bède le Vénérable entre tradition et postérité = The Venerable Bede: tradition and posterity. Colloque organisé à Villeneuve d'Ascq et Amiens par le CRHEN-O (Université de Lille 3) et Textes, Images et Spiritualité (Université de Picardie – Jules Verne) du 3 au 6 juillet 2002. Éd. par Stéphane LEBECQ, Michel PERRIN et Olivier SZERWINIACK. Villeneuve d'Ascq, CEGES, Université Charles-de-Gaulle, Lille III, 2005, 338 p. (ill.). (Histoire de l'Europe du Nord-Ouest, 34).

3434. Biblical studies in the early Middle Ages: proceedings of the Conference on biblical studies in the early Middle Ages. Ed. by Claudio LEONARDI e Giovanni ORLANDI. Tavarnuzze, SISMEL – Edizioni del Galluzzo, 2005, XI-352 p. (Millennio medievale, 52. Atti di convegni, 16).

3435. BIRK (Bonnie A.). Christine de Pizan and biblical wisdom: a feminist-theological point of view. Milwaukee, Marquette U. P., 2005, 202 p. (Marquette studies in theology, 47).

3436. BRONSTEIN (Judith). The Hospitallers and the Holy Land: financing the Latin East, 1187–1274. Woodbridge, Boydell Press, 2005, X-190 p. (maps).

3437. CARATINI (Roger). Les Cathares: de la gloire à la tragédie, 1209–1244. Paris, L'Archipel, 2005, 329 p. (map, geneal. tab.).

3438. Clérigos e religiosos na sociedade medieval. Ed. por Ana Maria C. M. JORGE e Hermínia Vasconcelos VILAR. Lisboa, Centro de Estudos de História Religiosa, Universidade Católica Portuguesa, 2005, 636 p.

3439. DAMERON (George Williamson). Florence and its church in the age of Dante. Philadelphia, University of Pennsylvania Press, 2005, 374 p. (ill., maps). (The Middle Ages series).

3440. Defining the holy: sacred space in medieval and early modern Europe. Ed. by Sarah HAMILTON and Andrew SPICER. Aldershot, Ashgate, 2005, XVIII-345 p. (ill., maps).

3441. DREYER (Elizabeth). Passionate spirituality: Hildegard of Bingen and Hadewijch of Brabant. Mahwah, Paulist Press, 2005, XVII-180 p.

3442. EDWARDS (John). Torquemada & the Inquisitors. Stroud, Tempus, 2005, 255 p. (ill.).

3443. FERNÁNDEZ CONDE (Francisco Javier). La religiosidad medieval en España. Vol. 2. Plena Edad Media (ss. XI–XIII). Gijón, Ediciones Trea, 2005, 640 p. (Estudios históricos La Olmeda. Piedras angulares).

3444. FERNÁNDEZ LLAMAZARES (José). Historia de las cuatro órdenes militares de Santiago, Calatrava, Alcántara y Montesa. Valencina de la Concepción, Espuela de Plata, 2005, 422 p. (Biblioteca de Historia).

3445. FERREIRO (Alberto). Simon Magus in patristic, medieval, and early modern traditions. Leiden a. Boston, Brill, 2005, XI-371 p. (ill.). (Studies in the history of Christian traditions, 125).

3446. FORREST (Ian). The detection of heresy in late medieval England. Oxford, Clarendon Press, 2005, XI-277 p. (Oxford historical monographs).

3447. Frühmittelalterliche Königtum (Das): ideelle und religiöse Grundlagen. Hrsg. v. Franz-Reiner ERKENS. Berlin, W. De Gruyter, 2005, VIII-462 p. (ill.). (Ergänzungsbände zum Reallexikon der germanischen Altertumskunde, 49).

3448. GAGLIARDI (Isabella). Li trofei della croce: l'esperienza gesuata e la società lucchese tra Medioevo ed età moderna. Roma, Edizioni di storia e letteratura, 2005, XV-352 p. (Centro alti studi in scienze religiose, 3).

3449. GRATHOFF (Stefan). Mainzer Erzbischofsburgen. Erwerb und Funktion von Burgherrschaft am Beispiel der Mainzer Erzbischöfe im Hoch- und Spätmittelalter. Stuttgart, Steiner, 2005, XIII-590 p. (Geschichtliche Landeskunde, 58).

3450. HERBERT MAC AVOY (Liz), HUGHES-EDWARDS (Mari). Anchorites, wombs and tombs: intersections of gender and enclosure in the Middle Ages. Cardiff, University of Wales Press, 2005, 240 p. (Religion & culture in the Middle Ages).

3451. Heresy in transition: transforming ideas of heresy in medieval and early modern Europe. Ed. by John

Christian LAURSEN, Cary J. NEDERMAN and Ian HUNTER. Aldershot, Ashgate, 2005, X-205 p. (Catholic Christendom).

3452. JOHNSON (Richard Freeman). Saint Michael the Archangel in medieval English legend. Woodbridge, Boydell Press, 2005, XII-174 p.

3453. LAUWERS (Michel). La naissance du cimetière: lieux sacrés et terre des morts dans l'Occident médiéval. Paris, Aubier, 2005, 393 p. (Collection historique).

3454. LEHANE (Brendan). Early Celtic Christianity. London, Continuum, 2005, 240 p.

3455. Liturgy (The) of the late Anglo-Saxon church. Ed. by Helen GITTOS and M. Bradford BEDINGFIELD. London, Boydell Press, 2005, XI-331 p. (ill.). (Subsidia, 5).

3456. MAC GINN (Bernard). The presence of God: a history of Western Christian mysticism. Vol. 4. The harvest of mysticism in medieval Germany (1300–1500). New York, Crossroad Pub. Co., 2005, XX-738 p. (The presence of God: a history of Western Christian mysticism, 4).

3457. MAKOWSKI (Elizabeth M.). A pernicious sort of woman: quasi-religious women and canon lawyers in the later Middle Ages. Washington, Catholic University of America Press, 2005, XXXIII-170 p. (Studies in medieval and early modern canon law, 6).

3458. MARIN (Olivier). L'archevêque, le maître et le dévot. Genèses du mouvement réformateur pragois. Années 1360–1419. Paris, Honoré Champion, 2005, 605 p. (Études d'histoire médiévale, 9).

3459. Medieval history writing and crusading ideology. Ed. by Tuomas M. S. LEHTONEN and Kurt Villads JENSEN. Helsinki, Finnish Literature Society, 2005, 320 p. (ill., maps). (Studia Fennica. Historica, 9).

3460. MORRIS (Colin). The sepulchre of Christ and the medieval West: from the beginning to 1600. Oxford a. New York, Oxford U. P., 95, XXII-427 p. (ill., maps).

3461. MULDER-BAKKER (Anneke B.). Lives of the anchoresses: the rise of the urban recluse in medieval Europe. Philadelphia, University of Pennsylvania Press, 2005, 300 p. (Middle Ages series).

3462. MUZZARELLI (Maria Giuseppina). Pescatori di uomini: predicatori e piazze alla fine del Medioevo. Bologna, Il Mulino, 2005, 315 p. (ill., pl.). (Biblioteca storica).

3463. Pace (La) fra realtà e utopia. Caselle di Sommacampagna, Cierre, 2005, 310 p. (Quaderni di storia religiosa).

3464. PASCOE (Louis B.). Church and reform. Bishops, theologians, and canon lawyers in the thought of Pierre d'Ailly, 1351–1420. Leiden a. Boston, Brill, 2005, XII-326 p. (Studies in medieval and reformation traditions, 105).

3465. ROACH (Andrew). The Devil's world: heresy and society, 1100–1300. Harlow, Pearson/Longman, 2005, XVIII-262 p. (ill., pl., maps). (The medieval world).

3466. ROCHE (Julien). Une église cathare. L'évêché du Carcassès: Carcassonne, Béziers, Narbonne. Cahors, Hydre, 2005, 558 p. (ill.). (Domaine historique).

3467. ROMAN (Christopher). Domestic mysticism in Margery Kempe and Dame Julian of Norwich: the transformation of Christian spirituality in the late Middle Ages. Lewiston a. Lampeter, E. Mellen Press, 2005, III-236 p. (Mediaeval studies, 24).

3468. Sacramentary (The) of Ratoldus (Paris, Bibliothèque Nationale de France, Lat. 12052). Ed. by Nicholas ORCHARD. London, Henry Bradshaw Society a. Boydell Press, 2005, CCVI-601 p. (Henry Bradshaw Society, 116).

3469. SCHWERHOFF (Gerd). Zungen wie Schwerter: Blasphemie in alteuropäischen Gesellschaften 1200–1650. Konstanz, UVK, 2005, 361 p. (ill.).

3470. Scripture and pluralism: reading the Bible in the religiously plural worlds of the Middle Ages and Renaissance. Ed. by Thomas J. HEFFERNAN and Thomas E. BURMAN. Leiden a. Brill, 2005, VII-246 p. (Studies in the history of Christian traditions, 123).

3471. Sermones (Die) des Nikolaus von Kues: Merkmale und ihre Stellung innerhalb der mittelalterlichen Predigtkultur. Akten des Symposions in Trier vom 21. bis 23. Oktober 2004. Hrsg. v. Klaus KREMER und Klaus REINHARDT. Trier, Paulinus Verlag, 2005, XLI-267 p. (Mitteilungen und Forschungsbeiträge der Cusanus-Gesellschaft, 30).

3472. SOULA (René). Les cathares, entre légende et histoire. Puylaurens, Institut d'estudis occitans, 2005, 592 p. (ill., maps). (Textes & documents).

3473. STÖLTING (Ulrike). Christliche Frauenmystik im Mittelalter: historisch-theologische Analyse. Mainz, Matthias-Grünewald-Verlag, 2005, 551 p.

3474. TACCONI (Marica). Cathedral and civic ritual in late Medieval and Renaissance Florence: the service books of Santa Maria del Fiore. Cambridge, Cambridge U. P., 2005, XX-357 p. (ill., maps, music, plans). (Cambridge studies in palaeography and codicology, 12).

3475. TAYLOR (Claire). Heresy in medieval France: dualism in Aquitaine and the Agenais, 1000–1249. Woodbridge a. Rochester, Royal Historical Society a. Boydell Press, 2005, XI-311 p. (ill., maps). (Royal Historical Society studies in history).

3476. Thesaurus Ratherii Veronensis, necnon Leodiensis: sermones inediti. Ed. François DOLBEAU, Paul TOMBEUR, Hubert MARAITE. Brepols, Turnhout, 2005, XLII-333 p. (Corpus Christianorum. Thesaurus patrum Latinorum. Series A-B, Formae et lemmata).

3477. THOMPSON (Augustine). Cities of God: the religion of the Italian communes, 1125–1325. University Park, Pennsylvania State U. P., 2005, XII-502 p. (ill.).

3478. TILATTI (Andrea). La cristianizzazione degli Slavi nell'arco alpino orientale, secoli VI–IX. Roma, Istituto storico italiano per il medio Evo e Gorizia, Istituto di storia sociale e religiosa, 2005, 201 p. (ill., pl., maps). (Nuovi studi storici, 69).

3479. WARREN (Nancy Bradley). Women of God and arms: female spirituality and political conflict, 1380–1600. Philadelphia, University of Pennsylvania Press, 2005, 264 p. (ill.).

3480. WEBER (Elka). Traveling through text: message and method in late medieval pilgrimage accounts. New York a. London, Routledge, 2005, XIII-204 p. (Studies in medieval history and culture).

3481. WENZEL (Siegfried). Latin sermon collections from later medieval England. Orthodox preaching in the age of Wyclif. Cambridge, Cambridge U. P., 2005, XXIV-713 p. (Cambridge studies in medieval literature, 53).

3482. Women's space: patronage, place, and gender in the medieval church. Ed. by Virginia CHIEFFO RAGUIN and Sarah STANBURY. Albany, State University of New York Press, 2005, X-261 p. (ill.). (SUNY series in medieval studies).

e. Addendum 2004

3483. SATO (Yoshiaki). Kirisutokyō ni okeru junkyō kenkyū. (Martyrdom in the christianity). Tokyo, Sobunsha, 2004, 436 p.

Cf. n°s 43, 2596

§ 14. Settlements. Place names. Town planning.

** 3484. RONCHESE (Gino). Paolino: pianta cronologica di Venezia: VIII–XIV sec. Venezia, Supernova, 2005, 85 p. (ill.).

** 3485. Villes d'Italie: textes et documents des XIIe, XIIIe, XIVe siècles. Éd. par Jean-Louis GAULIN, Armand JAMME et Véronique ROUCHON-MOUILLERON. Lyon, Presses universitaires de Lyon, 2005, 329 p. (Collection d'histoire et d'archéologie médiévales).

3486. Alban's buried towns: an assessment of St. Albans' archaeology up to AD 1600. Ed. by Rosalind NIBLETT and Isobel THOMPSON. Oxford, Oxbow Books a. English Heritage, 2005, XV-413 p. (ill., maps).

3487. Castelli del Veneto tra archeologia e fonti scritte. Atti del convegno, Vittorio Veneto, Ceneda, settembre 2003. A cura di Gian Pietro BROGIOLO e Elisa POSSENTI. Mantova, SAP, 2005, 237 p. (ill., maps). (Documenti di archeologia, 38).

3488. CHERVONOV (P.D.). Ispanskij srednevekovyj gorod. (Medieval Spanish town). Ed. Oleg V. AUROV, E.I. SHCHERBAKOVA; intr. by Oleg V. AUROV. Moskva, IVI RAN, 2005, 300 p.

3489. Città (Le) campane fra tarda antichità e alto Medioevo. A cura di Giovanni VITOLO. Salerno, Laveglia, 2005, 450 p. (ill., maps, plans). (Centro interuniversitario per la storia delle città campane nel Medioevo. Quaderni, 2).

3490. Città di mare del Mediterraneo medievale: tipologie. Atti del convegno di studi in memoria di Robert P. Bergman, Amalfi, 1–3 giugno 2001. Amalfi, Centro di cultura e storia amalfitana, 2005, 457 p. (ill., maps). (Atti, 9).

3491. Ciudades y villas portuarias del Atlántico en la Edad Media. Nàjera, encuentros Internacionales del Medievo 2004. Actas. Ed. por Beatriz ARIZAGA BOLUMBURU y Jesús Ángel SOLÓRZANO TELECHEA. Logroño, Gobierno de la Rioja, Instituto de Estudios Riojanos, 2005, 294 p.

3492. CREIGHTON (Oliver Hamilton), HIGHAM (Robert). Medieval town walls: an archaeological and social history of urban defence. Stroud, Tempus, 2005, 320 p. (ill., pl.).

3493. Domus (Dalle) alla corte regia. S. Giulia di Brescia: gli scavi dal 1980 al 1992. A cura di Gian Pietro BROGIOLO, Francesca MORANDINI e Filli ROSSI. Firenze, All'insegna del giglio, 2005, 440 p.

3494. Dopo la fine delle ville: le campagne dal VI al IX secolo. XI Seminario sul tardo antico e l'alto Medioevo, Gavi, 8–10 maggio 2004. A cura di Gian Pietro BROGIOLO, Alexandra ARNAU CHAVARRÍA e Marco VALENTI. Mantova, SAP, 2005, 358 p. (ill.). (Documenti di archeologia, 40).

3495. IAMBOR (Petru). Aşezări fortificate din Transilvania (secolele IX–XIII). (Fortified settlements in Transylvania (9th–13th centuries). Ediție îngrijită de Tudor SĂLĂGEAN, cuvânt înainte de Nicolae EDROIU. Cluj-Napoca, Editura Argonaut, 2005, 400 p.

3496. ICHIHARA (Koichi). Chūsei zenki hokusei surabujin no teijū to shakai. (Settlement and society of the north-western slavics in the early Middle Ages). Fukuoka, Kyushu U. P., 2005, 244 p.

3497. ISTRIA (Daniel). Pouvoirs et fortifications dans le nord de la Corse: XIe–XIVe siècle. Ajaccio, A. Piazzola, 2005, 517 p.

3498. KOS (Dušan). Vitez in grad: vloga gradov v življenju plemstva na Kranjskem, slovenskem Štajerskem in slovenskem Koroškem do začetka 15. stoletja. (The knight and the castle: the role of castles in the life of nobility in Carniola, Slovene Styria and Carnthia since the early 15th century). Ljubljana, Založba ZRC, ZRC SAZU, 2005. 431 p.

3499. KREUTZ (Bernhard). Städtebünde und Städtenetz am Mittelrhein im 13. und 14. Jahrhundert. Trier, Kliomedia, 2005, 538 p. (maps). (Trierer historische Forschungen).

3500. Kul'tura srednevekovoj Moskvy: Istoricheskie landshafty. (Culture of medieval Moscow: Historical landscapes). Ed. Sergej Z. CHERNOV. RAN, Nauch. sovet "Istorija mirovoj kul'tury; RAN, In-t arkheologii. Vol. 2. Domen moskovskikh knjazej v gorodskikh stanakh, 1271–1505 gody. (The demesne of the princes of Moscow in the neighbourhood of the city, 1271–1505). Vol. 3. Mental'nyj landshaft. Moskovskie sela i slobody. (Mental landscape: Villages and slobodas around Moscow). Moskva, Nauka, 2005, 2 vol., 671 p., 571 p. (ill.; maps; plates; ind. p. 531-626, 519-553). [English summaries]

3501. LEIVERKUS (Yvonne). Köln. Bilder einer spätmittelalterlichen Stadt. Köln, Weimar u. Wien, Böhlau, 2005, 408 p.

3502. LILLIE (Amanda). Florentine villas in the fifteenth century: an architectural and social history. Cambridge a. New York, Cambridge U. P., 2005, 353 p. (ill., maps, plans, geneal. tables).

3503. NICOLAS (Nathalie). La guerre et les fortifications du Haut-Dauphiné: étude archéologique des travaux des châteaux et des villes à la fin du Moyen Âge. Aix-en-Provence, Publications de l'Université de Provence, 2005, 377 p. (ill., maps).

3504. OPLL (Ferdinand). Das Werden der mittelalterlichen Stadt. *Historische Zeitschrift*, 2005, 280, 3, p. 561-589.

3505. Paesaggi e insediamenti rurali in Italia meridionale fra Tardoantico e Altomedioevo: atti del primo Seminario sul Tardoantico e l'Altomedioevo in Italia meridionale (Foggia 12–14 febbraio 2004). A cura di Giuliano VOLPE e Maria TURCHIANO. Bari, Edipuglia, 2005, 718 p. (ill., maps). (Insulae Diomedeae, 4. STAIM, 1).

3506. POUNDS (Norman John Greville). The medieval city. Westport a. London, Greenwood Press, 2005, XXIX-233 p. (ill., maps). (Greenwood guides to historic events of the medieval world).

3507. PROFUMO (Maria Cecilia). Schede di archeologia altomedievale in Italia: Marche. *Studi medievali*, 2005, 46, 1, p. 843-902.

3508. RÁBIK (Vladimír). Rusíni a valašské obyvateľstvo na východnom Slovensku v stredoveku. (Ruthenians and Wallachian inhabitants in the East Slovakia during the Middle Ages). *Historický časopis*, 2005, 53, 2, p. 217-242. [Deutsche Zsfassung].

3509. SCHMIEDER (Felicitas). Die mittelalterliche Stadt. Darmstadt, Wissenschaftliche Buchgesellschaft, 2005, VIII-152 p. (Geschichte Kompakt).

3510. SODEN (Iain). Coventry: the hidden history. Stroud, Tempus, 2005, 256 p. (ill., maps, plans).

3511. TARQUINI (Stefania). Pellegrinaggio e assetto urbano di Roma. *Bullettino dell'Istituto storico italiano per il medio evo*, 2005, 107, p. 1-133 (tab.).

3512. Urbanen Zentren (Die) des hohen und späteren Mittelalters. Vergleichende Untersuchungen zu Städten und Städtelandschaften im Westen des Reiches und in Ostfrankreich. 1. Thematischer Teil. 2. Ortsartikel. 3. Karten, Verzeichnisse, Register. Hrsg. v. Monika ESCHER und Frank G. HIRSCHMANN. Trier, Kliomedia, 2005, 3 vol., 555 p., 704 p., 349 p. (Trierer historische Forschungen, 50, 1-3).

3513. Walls and Memory. The Abbey of San Sebastiano at Alatri (Lazio): from late Roman monastery to Renaissance villa and beyond. Ed. by Elizabeth FENTRESS. Turnhout, Brepols, 2005, 445 p. (ill., pl., map). (Disciplina monastica, 2).

3514. WIDDER (Ellen). Waiblingen. Eine Stadt im Spätmittelalter. Waiblingen, Heimatverein Waiblingen, 2005, X-221 p. (Waiblingen in Vergangenheit und Gegenwart, 16).

Cf. n°s 151-203

K

MODERN HISTORY, GENERAL WORKS

§ 1. General. 3515-3569. – § 2. History by countries. 3570-4807.

§ 1. General.

** 3515. MASARYK (Tomáš Garrigue). Světová revoluce. Za války a ve válce 1914–1918. (World Revolution, 1914–1918). Praha, Masarykův ústav AV ČR, 2005, 639 p. (Spisy T. G. Masaryka, 15). – IDEM. Válka a revoluce. (War and revolution). Tomo 1. Články, memoranda, přednášky, rozhovory 1914–1916. Praha, Masarykův ústav AV ČR, 2005, 346 p. (Spisy T. G. Masaryka, 30).

3516. ANDERSON (Benedict). Under three flags: anarchism and the anti-colonial imagination. London, Verso, 2005, 255 p.

3517. ARTOLA GALLEGO (Miguel), PÉREZ LEDESMA (Manuel). Contemporánea. Una historia del mundo desde 1776. Madrid, Alianza Editorial, 2005, 560 p.

3518. Augsburger Religionsfriede 1555 (Der). Ein Epochenereignis und seine regionale Verankerung. Hrsg. v. Wolfgang WÜST, Georg KREUZER und Nicola SCHÜMANN. Augsburg, Wißner, 2005, 416 p. (Zeitschrift des historischen Vereins für Schwaben, 98).

3519. BAILYN (Bernard). Atlantic history: concept and contours. Cambridge, Harvard U. P., 2005, IX-149 p.

3520. BEHRENDS (Jan C.), KIND (Friederike). Vom Untergrund in den Westen. Samizdat, Tamizdat und die Neuerfindung Mitteleuropas in den achtziger Jahren. *Archiv für Sozialgeschichte*, 2005, 45, p. 427-448.

3521. BELL (Duncan S. A.). Dissolving distance: technology, space, and empire in British political thought, 1770–1900. *Journal of modern history*, 2005, 77, 3, p. 523-562.

3522. BERNAL (Antonio-Miguel). España, proyecto inacabado: los costes-beneficios del imperio. Madrid, Fundación Carolina, Centro de Estudios Hispánicos e Iberoamericanos y Marcial Pons, 2005, 612 p. (Marcial Pons Historia).

3523. Bilanz Balkan. Hrsg. v. Michael DAXNER [et al.]. Wien, Verlag für Geschichte und Politik u. München, Oldenbourg, 2005, 301 p. (Schriftenreihe des österreichischen Ost- und Südosteuropa-Instituts, 30).

3524. BRENDEL (Thomas). Zukunft Europa? Das Europabild und die Idee der internationalen Solidarität bei den deutschen Liberalen im Vormärz 1815–1845. Bochum, Winkler, 2005, 521 p. (Herausforderungen. Historisch-Politischeanalysen, 17).

3525. BREUER (Stefan). Nationalismus und Faschismus. Frankreich, Italien und Deutschland im Vergleich. Darmstadt, Wissenschaftliche Buchgesellschaft, 2005, 202 p.

3526. Buenos (Los), los malos y los feos: poder y resistencia en América Latina. Ed. por Nicolaus BÖTTCHER, Isabel GALAOR y Bernd HAUSBERGER. Frankfurt am Main, Vervuert y Madrid, Ibero-Americana, 2005, 512 p.

3527. ČAPKOVÁ (Kateřina). Češi, Němci, Židé. Národní identita Židů v Čechách 1918–1938. (Tschechen, Deutsche, Juden. Die nationale Identität der Juden in Böhmen, 1918–1938). Praha, Paseka, 2005, 355 p. (photogr.).

3528. CAPOCCIA (Giovanni). Defending democracy: reactions to extremism in interwar Europe. Baltimore, John Hopkins U. P., 2005, VIII-335 p. (ill.).

3529. Chelovek XVII stoletija. (Man of the 17th century: [Articles]). Ed. Ada A. SVANIDZE, Vladimir A. VEDJUSHKIN. RAN, In-t vseobshchej istorii. Parts 1-2. Moskva, IVI RAN, 2005, 202, 155 p. (bibl. incl.).

3530. CONZE (Vanessa). Das Europa der Deutschen: Ideen von Europa in Deutschland zwischen Reichstradition und Westorientierung (1920–1970). München, Oldenbourg, 2005, VII-453 p. (Studien zur Zeitgeschichte, 69).

3531. Democrazia radicale (La) nell'Ottocento europeo. Forme della politica, modelli culturali, riforme sociali. A cura di Maurizio RIDOLFI. Milano, Feltrinelli, 2005, 424 p. (Annali della Fondazione Giangiacomo Feltrinelli).

3532. Empire and nation: the American Revolution in the Atlantic World. Ed. by Eliga H. GOULD and Peter S. ONUF. Baltimore, Johns Hopkins U. P., 2005, VIII-381 p. (Anglo-America in the transatlantic world).

3533. GELVIN (James L.). The modern Middle East: a history. New York a. Oxford, Oxford U. P., 2005, IX-357 p. (ill., maps).

3534. JOANNÈS (Sidonie). Les méridionalistes et l'Europe: l'influence des projets italiens pour le Mezzogiorno sur la construction européenne. *Mélanges de l'École française de Rome: Italie et mediterranée (MEFRIM)*, 2005, 117, 1, p. 423-445.

3535. JUDT (Tony). Postwar: a history of Europe since 1945. New York, Penguin, 2005, XV-933 p.

3536. Kaiser Ferdinand I. ein mitteleuropäischer Herrscher. Hrsg. v. Martina FUCHS, Teréz OBORNI und Gábor UJVÁRY. Münster, Aschendorff, 2005, VI-357 p. (Geschichte in der Epoche Karls V, 5).

3537. KAMRAVA (Mehran). The modern Middle East: a political history since the First World War. Berkeley a. London, University of California Press, 2005, XI-497 p. (ill.).

3538. KŘEN (Jan). Dvě století střední Evropy. (Two centuries of the Central Europe). Praha, Argo, 2005, 1109 p. (graf., mp.). (Dějiny Evropy, 8).

3539. LAWSON (George). Negotiated revolutions: the Czech Republic, South Africa and Chile. Aldershot, Ashgate, 2005, XII-272 p.

3540. LEE VAN COTT (Donna). From movements to parties in Latin America: the evolution of ethnic politics. New York a. Cambridge, Cambridge U. P., 2005, XXII-276 p.

3541. LORY (Bernard). Les Balkans: de la transition post-Ottomane a la transition post-communiste. İstanbul, Isis, 2005, 435 p.

3542. ŁOSSOWSKI (Piotr). Kraje bałtyckie w latach przełomu 1934–1944. (I paesi baltici negli anni 1934–1944). Przy współudziale Broniusa MAKAUSKASA; pod redakcją naukową Andrzeja KORYNA. Warszawa, Instytut Historii PAN, 2005, 184 p. (maps).

3543. MANN (Michael). The dark side of democracy. Explaining ethnic cleansing. New York, Cambridge U. P., 2005, X-580 p.

3544. Memorias del mestizaje: cultura política en Centroamérica de 1920 al presente. Ed. por Dario A. EURAQUE, Jeffery L. GOULD y Charles R. HALE. Ciudad de Guatemala, Centro de Investigaciones Regionales de Mesoamérica, 2005, 625 p.

3545. MIDLARSKY (Manus I.). The killing trap. Genocide in the twentieth century. New York, Cambridge U. P., 2005, XV-463 p.

3546. MOADDEL (Mansoor). Islamic modernism, nationalism and fundamentalism: episode and discourse. Chicago, University of Chicago Press, 2005, 448 p.

3547. MONTRONI (Giovanni). Scenari del mondo contemporaneo dal 1815 ad oggi. Roma e Bari, Laterza, 2005, XXVIII-277 p.

3548. MOUTAFTCHIEVA (Vera). L'anarchie dans les Balkans à la fin du XVIIIe siècle. İstanbul, Isisi, 2005, 360 p.

3549. Natsional'naja ideja na evropejskom prostranstve XX v.: Sb. st. (National idea in Europe in the 20th century: Articles). Ed. Elena Ju. POLJAKOVA. RAN, In-t vseobshchej istorii. Books 1-2. Moskva, IVI RAN, 2005, 250, 212 p. (bibl. incl.).

3550. O'NEILL (Kathleen). Decentralizing the state: elections, parties and local power in the Andes. Cambridge, Cambridge U. P., 2005, 286 p.

3551. Policing interwar Europe: 1918–1940. Ed. by Gerald BLANEY. Basingstoke, Palgrave Macmillan, 2005, XI-240 p. [Cf. n^{os} <choice> 3682, 3783, 4338, 4401, 4418.]

3552. Political cultures in the Andes, 1750–1950. Ed. by Nils JACOBSEN and Cristóbal ALJOVÍN DE LOSADA. Durham, Duke U. P., 2005, XII-386 p. (Latin America otherwise: languages, empires, nations).

3553. Politicheskaja kul'tura XIX v.: Rossija i Evropa: Sb. st. (The political culture of the 19th century: Russia and Europe: Articles). Ed. Marina S. BOBKOVA. RAN, In-t vseobshchej istorii. Moskva, IVI RAN, 2005, 202 p.

3554. POSKONINA (Ol'ga I.). Istorija Latinskoj Ameriki: Do XX veka. (A history of Latin America up to 1900). Moskva, Ves' mir, 2005, 247 p. (bibl. p. 245-247).

3555. Problemy istorii Russkogo zarubezh'ja: Matla i issledovanija. (Problems of history of the Russian diaspora). Ed. Natal'ja T. ENEEVA. Vol. 1. Moskva, Nauka, 2005, 453 p. (ill.; bibl. incl.).

3556. RESÉDENZ (Andrés). Changing national identities at the frontier. Texas and New Mexico, 1800–1850. New York, Cambridge U. P., 2005, XIII-309 p.

3557. RICHTER (Karel), CÍLEK (Roman), BÍLEK (Jiří). Cesty k moci: 1948, 1917, 1933. (The road to power: 1948, 1917, 1933). Praha, Themis, 2005, 398 p. (photogr.).

3558. RYDELL (Robert W.), KROES (Rob). Buffalo Bill in Bologna. the Americanization of the world, 1869–1922. Chicago, University of Chicago Press, 2005, XII-209 p.

3559. SPÄTI (Christina). Kontinuität und Wandel des Antisemitismus und dessen Beurteilung in der Schweiz nach 1945. *Schweizerische Zeitschrift für Geschichte*, 2005, 55, 4, p. 419-440.

3560. Storia della Shoah. La crisi dell'Europa, lo sterminio degli ebrei e la memoria del XX secolo. Vol. 1. La crisi dell'Europa e lo sterminio degli ebrei. A cura di Marina CATTARUZZA, Marcello FLORES, Simon

LEVIS SULLAM e Enzo TRAVERSO. Torino, Utet, 2005, 1188 p.

3561. TORGAL (Luís Reis). De l'Empire atlantique à la Communauté des pays de langue portugaise. Realité, mythe et utopie. *Matériaux pour l'histoire de notre temps*, 2005, 77, p. 61-67.

3562. Trajectories of the Left. Social democratic and (Ex-)Communist parties in Contemporary Europe. Between past and future. Ed. by Lubomír KOPEČEK. Brno, Democracy and Culture Studies Centre, 2005, 179 p. (Comparative Political Science, 3).

3563. VIVARELLI (Roberto). I caratteri dell'età contemporanea. Bologna, Il Mulino, 2005, 296 p.

3564. Weight of history in Central European societies of the 20th century. Central European studies in social sciences. Meeting. Ed. Zora HLAVIČKOVÁ, Nicolas MASLOWSKI. Praha, Central European Seminar, 2005, 187 p. (Central European Studies, 2).

3565. WILDER (Gary). The French imperial nation-state: Negritude and colonial humanism between the two world wars. Chicago, University of Chicago Press, 2005, XI-404 p.

3566. WINKS (Robin W.), NEUBERGER (Joan). Europe and the making of modernity, 1815-1914. New York, Oxford U. P., 2005, XVII-396 p.

3567. WINKS (Robin W.), TALBOTT (John E.). Europe 1945 to the present. New York, Oxford U. P., 2005, XV-176 p. (ill., maps).

3568. Zones of fracture in modern Europe. The Baltic countries, the Balkans, and northern Italy = Zone di frattura in epoca moderna. il Baltico, i Balcani e l'Italia settentrionale. Ed. by Almut BUES. Wiesbaden, Harrassowitz, 2005, 292 p. (Deutsches historisches Institut Warschau, Quellen und Studien, 16).

§ 1. Addendum 2004.

3569. JUNZ (Helen B.), RATHKOLB (Oliver), VENUS (Theodor) [et al.]. Das Vermögen der jüdischen Bevölkerung Österreichs. NS-Raub und Restitution nach 1945. Unt. Mitarb. v. Dieter HOPPENKOTHEN, Edith LEISCH-PROST, Ludmilla LUDOVA [et al.]. Wien u. München, Oldenbourg, 2004, 233 p. (Veröffentlichungen der Österreichischen Historikerkommission. Vermögensentzug während der NSZeit sowie Rückstellungen und Entschädigungen seit 1945 in Österreich, 9).

Cf. n°s 85, 228, 262, 389, 589-699, 711, 2895, 4379, 4402, 4458, 4550, 4630, 4812, 4953, 5034, 5106, 5123, 5130, 5133, 5156, 5503, 6619-6702, 6797, 6824, 6831, 6998, 7018, 7126, 7562, 7585

§ 2. History by countries.

Albania

** 3570. HEATON-ARMSTRONG (Duncan). The six month kingdom: Albania 1914. Ed. by Gervase BELFIELD and Bejtullah DESTANI; with an introduction by Gervase BELFIELD. London a. New York, I.B. Tauris, in association with the Centre for Albanian Studies, 2005, XXXVII-191 p. (ill., map).

3571. Kongresi i Dibrës 1909-2004: referata dhe kumtesa. Redaktorë Gazmend SHPUZA dhe Mahmud HYSA = The Congress of Diber 1909-2004: reports and papers. Editors Gazmend SHPUZA and Mahmud HYSA. Dibër, Shoqata Qytetare "Votras Dibrane"-Dibër, 2005, 238 p. (ill.).

3572. PEARSON (Owen). Albania in occupation and war: from fascism to communism, 1940-1945. London, Centre for Albanian Studies in association with I. B. Tauris, 2005, XV-570 p. (maps). (Albania in the twentieth century: a history, 2).

Algeria

3573. AGGOUN (Lounis), RIVOIRE (Jran-Baptiste). Françalgérie, crimes et mensonges d'états: histoire secrète, de la guerre d'indépendance à la "troisième guerre" d'Algérie. Paris, La Découverte/Poche, 2005, 683 p. (maps). (Découverte/Poche, 215).

3574. SIMON (Jacques). Le PPA: le parti du peuple algérien, 1937-1947. Paris, L'Harmattan, 2005, 270 p. (maps). (Collection CREAC-Histoire).

Cf. n° 6793

Angola

3575. BRINKMAN (Inge). A war for people: civilians, mobility, and legitimacy in South-East Angola during MPLA's war for independence. Köln, Köppe, 2005, 256 p. (ill., maps). (History, cultural traditions, and innovations in Southern Africa, 23).

3576. MABEKO-TALI (Jean-Michel). Barbares et citoyens, l'identité nationale à l'épreuve des transitions africaines: Congo-Brazzaville, Angola. Paris, L'Harmattan, 2005, 334 p. (Etudes africaines).

3577. PIMENTA (Fernando Tavares). Brancos de Angola: autonomismo e nacionalismo (1900-1961). Coimbra, MinervaCoimbra, 2005, 224 p. (Minerva-História, 24).

Argentina

** 3578. Correspondencia de Juan Manuel de Rosas. Estudio preliminar y selección por Marcela TERNAVASIO. Buenos Aires, Eudeba, 2005, 233 p. (Historia argentina).

** 3579. PARRA GARZÓN (Gabriela). El Cabildo de Cordoba del Tucumán a través de sus documentos (1573-1600): estudio diplomático. Cordoba, Centro de Estudios Historicos Prof. Carlos Segreti, 2005, 221 p.

3580. Años setenta (Los): memoria y militancia. Intr. de Juan SURIANO. *Entrepasados*, 2005, 28, [s. p.]. [Contiene: CARNOVALE (Vera). Jugarse al cristo: mandatos y construcción identitaria en el PRT-ERP. – FRANCO (Marina). Derechos humanos, política y fútbol (o de las pasiones argentinas y francesas). – LEVÍN

(Florencia). Arqueología de la memoria. Algunas reflexiones a propósito de Los vecinos del horror. Los otros testigos. – LORENZ (Federico). "Recuerden argentinos". Por una revisión de la Vulgata Procesista. – GONZÁLEZ (M. Paula). La historia argentina reciente en la escuela media: un inventario de preguntas].

3581. Argentina en el siglo XIX. Coord. Pablo YANKELEVICH, Celina BONINI, Jorge CERNADAS, Damián LÓPEZ MARTÍN y Roberto VILLARRUEL. México D.F., Instituto de Investigaciones Dr. José Luis Mora, 2005, 423 p.

3582. ARGERI (María F.). De guerreros a delincuentes: la desarticulación de las jefaturas indígenas y el poder judicial: Norpatagonia, 1880–1930. Madrid, Consejo superior de investigaciones científicas, Instituto de historia, Departamento de historia de América, 2005, 331 p. (ill., maps). (Colección Tierra nueva e cielo nuevo, 51).

3583. BÉJAR (María Dolores). El régimen fraudulento: la política en la provincia de Buenos Aires, 1930–1943. Buenos Aires, Siglo Veintiuno Editores Argentina, 2005, 294 p. (Historia y cultura).

3584. BISSO (Andrés). Acción Argentina: un antifascismo nacional en tiempos de guerra mundial: Acción Argentina y las estrategias de movilización del antifascismo liberal-socialista en torno a la Segunda Guerra Mundial, 1940–1946. Buenos Aires, Prometeo Libros, 2005, 394 p. (maps).

3585. BLEJMAN (Saul). Hegemonías, crisis y corrupción en la política argentina, 1890–2003. Mendoza, EDIUNC, 2005, 532 p. (Universidad Nacional de Cuyo. Editorial. Serie estudios, 41).

3586. CAMOGLI (Pablo), DE PRIVITELLO (Luciano). Batallas por la libertad: todos los combates de la guerra de la independencia. Prólogo de Miguel Angel DE MARCO. Buenos Aires, Aguilar, 2005, 389 p. (ill., maps).

3587. CANTÓN (Darío), JORRAT (Jorge Raúl). Elecciones en la ciudad 1864–2003. Vol. 1. 1864–1910. Buenos Aires, Instituto Histórico de la Ciudad de Buenos Aires, 2005, 505 p.

3588. CERUTI (Leonidas). Historia de la U.C.R. de Rojas, 1890–1943. Rosario, Ediciones del Castillo, 2005, x, 376 p. ill.

3589. CHAVES (Liliana). Sufragio y representación política bajo el régimen oligárquico en Córdoba, 1890–1912: las élites y el debate sobre las instituciones de la igualdad y el pluralismo político. Córdoba, Ferreyra Editor, 2005, 240 p. (map).

3590. DEL CAMPO (Hugo). Sindicalismo y peronismo: los comienzos de un vínculo perdurable. Buenos Aires, Siglo Veintiuno Editores Argentina, 2005, 389 p. (ill.). (Historia y cultura / Siglo Veintiuno Editores Argentina, 14).

3591. ELENA (Eduardo). What the people want: state planning and political participation in Peronist Argentina, 1946–1955. *Journal of Latin American studies*, 2005, 37, 1, p. 81-108.

3592. GARCÍA SEBASTIANI (Marcela). Los antiperonistas en la Argentina peronista: radicales y socialistas en la política Argentina entre 1943 y 1951. Buenos Aires, Prometeo Libros, 2005, 296 p.

3593. GELMAN (Jorge). Rosas, estanciero: gobierno y expansión ganadera. Buenos Aires, Capital Intelectual, 2005, 95 p. (Claves para todos).

3594. Gobierno (El) de Domingo A. Mercante en Buenos Aires, 1946–1952: un caso de peronismo provincial. T. 1. Compilador Claudio PANELLA. La Plata, Asociación Amigos del Archivo Histórico de la Provincia de Buenos Aires, 2005, 379 p. (ill., maps). (Publicaciones del Archivo Histórico de la Provincia de Buenos Aires. Estudios sobre la historia y la geografía histórica de la Provincia de Buenos Aires, 14).

3595. HOLLENSTEINER (Stephan). Aufstieg und Randlage: Linksintellektuelle, demokratische Wende und Politik in Argentinien und Brasilien. Frankfurt am Main, Vervuert Verlag, 2005, 462 p. (Bibliotheca Ibero-Americana, 104).

3596. MORENO (Hugo). Le désastre argentin: péronisme, politique et violence sociale, (1930–2001). Paris, Syllepse, 2005, 222 p. (ill., map). (Histoire, enjeux et débats).

3597. MOYA (Angel Omar). El radicalismo en Catamarca: trayectoria entre 1890 y 1930. San Fernando del Valle de Catamarca, Editorial de la Universidad Nacional de Catamarca, 2005, 298 p. (ill.).

3598. Nueva historia argentina. Proyecto editorial, Federico POLOTTO; coordinación general de la obra, Juan SURIANO. Vol. 10. Dictadura y democracia (1976–2001). Director de tomo Juan SURIANO. Buenos Aires, Editorial Sudamericana, 2005, 557 p. (ill.).

3599. Partido Socialista (El) en Argentina: sociedad, política e ideas a través de un siglo. Ed. por Hernán CAMARERO y Carlos Miguel HERRERA. Buenos Aires, Prometeo Libros, 2005, 413 p. (Colección de historia argentina).

3600. Question libérale (La) en Argentine au XIXe siècle (le libéralisme argentin en héritage). Numéro coordonné par Darío ROLDÁN; ouvrage édité sous la direction d'Isabel SANTI. Paris, ALHIM, 2005, 255 p. (Les cahiers ALHIM, 11).

3601. Representaciones, sociedad y poder: Tucumán en la primera mitad del siglo XIX. Compiladoras Irene GARCÍA DE SALTOR y Cristina del C. LÓPEZ. Tucumán, Instituto de Historia y Pensamiento Argentinos, Facultad de Filosofía y Letras, Universidad Nacional de Tucumán, 2005, 341 p. (ill.).

3602. Representaciones, sociedad y política en los pueblos de la república: primera mitad del siglo XIX. Compiladoras Irene GARCÍA DE SALTOR y Cristina LÓPEZ. Tucumán, Instituto de Historia y Pensamiento

Argentinos, Facultad de Filosofía y Letras, Universidad Nacional de Tucumán, 2005, 143 p.

3603. República oligárquica (De la) a la república democrática: estudio sobre la reforma política de Roque Sáenz Peña. Compilador Mario Justo LÓPEZ. Buenos Aires, Lumière, 2005, 639 p. (ill.). (Nuevas miradas a la Argentina del siglo XX).

3604. ROBBEN (Antonius C. G. M.). Political violence and trauma in Argentina. Philadelphia, University of Pennsylvania Press, 2005, XII-467 p.

3605. SCHNEIDER (Alejandro). Los compañeros: trabajadores, izquierda y peronismo, 1955–1973. Buenos Aires, Imago Mundi, 2005, 430 p. (Colección Bitácora argentina).

3606. SPINELLI (María Estela). Los vencedores vencidos: el antiperonismo y la "revolución libertadora". Buenos Aires, Editorial Biblos, 2005, 345 p. (ill.). (Colección Argentina contemporánea).

3607. Sujetos sociales y políticas: historia reciente de la Norpatagonia argentina. Coordinadora Orietta FAVARO. Buenos Aires, La Colmena, 2005, 305 p. (ill., maps).

Addendum 1998

3608. Nueva historia argentina. Proyecto editorial, Federico POLOTTO; coordinación general de la obra, Juan SURIANO. Vol. 3. Revolución, república, confederación (1806–1852). Directora de tomo Noemí GOLDMAN. Buenos Aires, Editorial Sudamericana, 98, 445 p. (ill.).

Cf. n°s 6045, 6595, 6633

Armenia

Cf. n° 688

Australia

3609. COTTLE (Drew), KEYS (Angela). Douglas Evelyn Darby, MP: anti-Communist internationalist in the Antipodes. *In*: 'Extreme Right' (The) in twentieth century Australia [Cf. n° 3610], p. 87-100.

3610. 'Extreme Right' (The) in twentieth century Australia. Ed. by Andrew MOORE. *Labour history*, 2005, 89, p. 1-124. [Cf. n°s <selection> 3609, 3611, 3613-3616, 6462.]

3611. FISCHER (Nick). The Australian Right, the American Right and the threat of the Left, 1917–1935. *In*: 'Extreme Right' (The) in twentieth century Australia [Cf. n° 3610], p. 17-36.

3612. GOLDER (Hilary). Politics, patronage, and public works: 1842–1900. Sydney, UNSW Press, 2005, X-268 p. (ill., maps). (Administration of New South Wales, 1).

3613. GOOT (Murray). Pauline Hanson's one nation: extreme Right, centre party or extreme Left? *In*: 'Extreme Right' (The) in twentieth century Australia [Cf. n° 3610], p. 101-120.

3614. HENDERSON (Peter). Frank Browne and the Neo-Nazis. *In*: 'Extreme Right' (The) in twentieth century Australia [Cf. n° 3610], p. 73-86.

3615. LA ROOIJ (Marinus). Arthur Nelson Field: Kiwi theoretician of the Australian radical Right. *In*: 'Extreme Right' (The) in twentieth century Australia [Cf. n° 3610], p. 37-54.

3616. MOORE (Andrew). The New Guard and the Labour Movement, 1931–1935. *In*: 'Extreme Right' (The) in twentieth century Australia [Cf. n° 3610], p. 55-72.

3617. PATMORE (Greg), COATES (David). Labour parties and the state in Australia and the UK. *Labour history*, 2005, 88, p. 121-142.

3618. WINTER (Barbara). The Australia-First Movement and the Publicist, 1936–1942. Brisbane, Glass House Books, 2005, XII-292 p. (ill.).

Cf. n° 6504

Austria

** 3619. Österreich! und Front Heil! Aus den Akten des Generalsekretariats der Vaterländischen Front; Innenansichten eines Regimes. Hrsg. v. Robert KRIECHBAUMER. Wien, Böhlau, 2005, 436 p. (Schriftenreihe des Forschungsinstitutes für politisch-historische Studien der Dr.-Wilfried-Haslauer-Bibliothek, Salzburg, 23).

3620. 1703, Der "Bayerische Rummel" in Tirol: Akten des Symposiums des Tiroler Landesarchivs, Innsbruck, 28.–29. November 2003. Hrsg. v. Martin P. SCHENNACH und Richard SCHOBER. Innsbruck, Universitätsverlag Wagner, 2005, 110 p. (ill.). (Veröffentlichungen des Tiroler Landesarchivs, 10).

3621. BAHLCKE (Joachim). Ungarischer Episkopat und österreichische Monarchie: von einer Partnerschaft zur Konfrontation (1686–1790). Stuttgart, Franz Steiner, 2005, 516 p. (ill., maps, ports.). (Forschungen zur Geschichte und Kultur des östlichen Mitteleuropa, 23).

3622. Befreien – besetzen – bestehen: das Burgenland von 1945–1955: Tagungsband des Symposions des Burgenländischen Landesarchivs vom 7./8. April 2005. Herausgeber und Verleger, Amt der Burgenländischen Landesregierung; Hauptreferatsleiter, Roland WIDDER; Redaktion, Felix TOBLER und Michael HESS. Eisenstadt, Amt der Burgenländischen Landesregierung, 2005, 288 p. (ill., maps). (Burgenländische Forschungen, 90).

3623. BINDER (Dieter A.), BRUCKMÜLLER (Ernst). Essay über Österreich: Grundfragen von Identität und Geschichte, 1918–2000. Wien, Verlag für Geschichte und Politik, 2005, 129 p. (ill., maps). (Schriftenreihe des Instituts für Österreichkunde).

3624. BINDER (Harald). Galizien in Wien: Parteien, Wahlen, Fraktionen und Abgeordnete im Übergang zur Massenpolitik. Wien, Verlag der Österreichischen Akademie der Wissenschaften, 2005, 741 p. (Studien zur Geschichte der Österreichisch-Ungarischen Monarchie, 29).

3625. CVIRN (Janez), GAŠPARIČ (Jure). 'Neizbežnost' razpada Habsburške monarhije – slovenski pogled. (The 'Imminence' of the break-down of the Habsburg monarchy – the Slovenian view). *Studia Historica Slovenica*, 2005, 5, 1/3, p. 443-456.

3626. Entnazifizierung zwischen politischem Anspruch. Parteienkonkurrenz und Kaltem Krieg. Das Beispiel der SPÖ [Sozialistische Partei Österreichs]. Hrsg. v. Maria MESNER. Wien u. München, Oldenbourg, 2005, XIII-362 p.

3627. KULENKAMPFF (Angela). Österreich und das Alte Reich: die Reichspolitik des Staatskanzlers Kaunitz unter Maria Theresia und Joseph II. Köln, Böhlau, 2005, 213 p. (ill., map).

3628. NADERER (Otto). Der bewaffnete Aufstand: der Republikanische Schutzbund der österreichischen Sozialdemokratie und die militärische Vorbereitung auf den Bürgerkrieg (1923–1934). Graz, Ares, 2005, 384 p. (ill., maps). (Reihe Hochschulschriften).

3629. NEUGEBAUER (Wolfgang), SCHWARZ (Peter). Der Wille zum aufrechten Gang: Offenlegung der Rolle des BSA [Bund Sozialdemokratischer AkademikerInnen, Intellektueller und KünsterlerInnen] bei der gesellschaftlichen Reintegration ehemaliger Nationalsozialisten. Herausgegeben vom Bund sozialdemokratischer AkademikerInnen, Intellektueller und KünstlerInnen (BSA). Wien, Czernin, 2005, 335 p.

3630. RAKHSHMIR (Pavel Ju.). Knjaz' Metternikh: Chelovek i politik. (Prince von Metternich [1773–1859]: Man and politician). Perm', Kommersant', 2005, 407 p. (bibl. incl.).

3631. SCHAUSBERGER (Franz). Alle an den Galgen! Der politische "Takeoff" der "Hitlerbewegung" bei den Salzburger Gemeindewahlen 1931. Wien, Böhlau, 2005, 278 p. (ill.). (Schriftenreihe des Forschungsinstitutes für politisch-historische Studien der Dr.-Wilfried-Haslauer-Bibliothek, Salzburg, 26).

3632. SCHWANNINGER (Florian). Im Heimatkreis des Führers: Nationalsozialismus, Widerstand und Verfolgung im Bezirk Braunau 1938 bis 1945. Grünbach, Steinmassl, 2005, 360 p. (ill., facsims.). (Edition Geschichte der Heimat).

3633. STOURZH (Gerald). 1945 und 1955: Schlüsseljahre der Zweiten Republik: gab es die Stunde Null?: wie kam es zu Staatsvertrag und Neutralität? Innsbruck, StudienVerlag, 2005, 111 p. (ill.). (Österreich, Zweite Republik, 1).

3634. TIDL (Georg). Streuzettel: illegale Propaganda in Österreich 1933–1938. Wien, Löcker, 2005, 195 p. (ill.).

3635. UNOWSKY (Daniel L.). The pomp and politics of patriotism: imperial celebrations in Habsburg Austria, 1848–1916. West Lafayette, Purdue U. P., 2005, XIII-263 p. (ill.). (Central European studies).

3636. Wittelsbach (Von) zu Habsburg: Maximilian I. und der Übergang der Gerichte Kufstein, Rattenberg und Kitzbühel von Bayern an Tirol 1504–2004: Akten des Symposiums des Tiroler Landesarchivs Innsbruck, 15.–16. Oktober 2004. Hrsg. v. Christoph HAIDACHER und Richard SCHOBER. Innsbruck, Universitätsverlag Wagner, 2005, 174 p. (ill.). (Veröffentlichungen des Tiroler Landesarchivs, 12).

3637. WLADIKA (Michael). Hitlers Vätergeneration: Die Ursprünge des Nationalsozialismus in der k. u. k. Monarchie. Vienna: Böhlau Verlag, 2005, XI-675 p.

Cf. n°s 4124, 4550, 7334, 7417

Barbados

3638. LAMBERT (David). White creole culture. Politics and identity during the age of abolition. Cambridge, Cambridge U. P., 2005, 245 p.

Belgium

3639. Comment (se) sortir de la Grande guerre: regards sur quelques pays vainqueurs, la Belgique, la France et la Grande-Bretagne. Études réunies par Stéphanie CLAISSE et Thierry LEMOINE. Paris, L'Harmattan, 2005, 159 p. (ill.). (Structures et pouvoirs des imaginaires).

3640. DE SCHAEPDRIJVER (Sophie). La Belgique et la Première Guerre mondiale. Bruxelles, Archives et Musée de la littérature, 2005, 336 p. (Documents pour l'histoire des francophonies/Europe, 4).

3641. DI MURO (Giovanni F.). Léon Degrelle et l'aventure rexiste (1927–1940). Bruxelles, L. Pire, 2005, 205 p. (ill., ports.). (Voix de l'histoire).

3642. DUBOIS (Sébastien). L'invention de la Belgique. Genèse d'un État-nation (1648–1830). Bruxelles, Éd. Racine, 2005, 446 p.

3643. DUMONT (Georges Henri). Le miracle belge de 1848. Bruxelles, Le Cri édition, 2005, 226 p. (ill., ports.). (Histoire).

3644. GODDEERIS (Idesbald). De Poolse migratie in België, 1945–1950: politieke mobilisatie en sociale differentiatie. Amsterdam, Aksant, 2005, 281 p. (ill., maps, ports.).

3645. Guerre totale (Une)? La Belgique dans la Première Guerre mondiale. Nouvelles tendances de la recherche historique, Actes du colloque international organisé à l'ULB du 15 au 17 janvier 2003. Sous la dir. de Serge JAUMAIN, Michaël AMARA, Benoît MAJERUS et Antoon VRINTS. Bruxelles, Archives générales du Royaume, 2005, 666 p. (Études sur la Première Guerre mondiale, 11).

3646. Intégration ou représentation? Les exilés polonais en Belgique et la construction européenne = Integration or representation? Polish exiles in Belgium and the European construction. Ed. par Michel DUMOULIN et Idesbald GODDEERIS. Louvain-la-Neuve, Academia-Bruylant, 2005, 240 p.

3647. Nationale Bewegungen in Belgien: ein historischer Überblick. Hrsg. v. Johannes KOLL. Münster,

Waxmann, 2005, 196 p. (ill., maps). (Niederlande-Studien, 37).

3648. Nouvelle histoire de Belgique. Vol. 1. 1830–1905. Sous la direction de Michel DUMOULIN [et al.]. Bruxelles, Editions Complexe, 2005, XV-222 p. (Questions à l'histoire).

Cf. n° 7026

Birmania

3649. MONG (Sai Kam). Kokang and Kachin in the Shan State, 1945–1960. Bangkok, Institute of Asian Studies, Chulalongkorn University, 2005, 193 p. (ill., maps).

Bolivia

3650. BARRAGÁN (Rossana), ROCA (José Luis). Regiones y poder constituyente en Bolivia: una historia de pactos y disputas. Prólogo de James DUNKERLEY. La Paz, PNUD, 2005, 458 p. (ill.). (Cuaderno de futuro, 21).

3651. Vacat.

3652. CHOQUE CANQUI (Roberto). Historia de una lucha desigual: los contenidos ideológicos y políticos de las rebeliones indígenas de la pre-revolución nacional. Con colaboración de Cristina QUISBERT. La Paz, Unidad de Investigaciones Históricas UNIH-PAKAXA, 2005, 180 p. (ill.).

3653. CORTÉS (Jorge). Caciques y hechiceros: huellas en la historia de Mojos. La Paz, Plural Editores y Universidad de la Cordillera, 2005, XV-113 p. (ill., maps).

3654. DELFOUR (Christine). L'invention nationaliste en Bolivie: une culture politique complexe. Avant-propos de François DELPRAT. Paris, L'Harmattan, 2005, 329 p. (maps). (Recherches Amériques latines).

3655. SHCHELCHKOV (Andrej A.). Konservativnaja sotsial'naja utopija v Bolivii: Pravlenie M.I. Belsu. (The conservative social utopia in Bolivia under the rulership of Manuel Isidoro Belzu, 1848–1855). Ed. Nikolaj P. KALMYKOV. RAN, In-t vseobshchej istorii. Moskva, IVI RAN, 2005, 267 p. (ill.; bibl. p. 256-265).

3656. SOLANO CHUQUIMIA (Franz). La revolución nacional y el restablecimiento de la democracia. La Paz, Colegio Nacional de Historiadores de Bolivia y Producciones CIMA, 2005, 230 p. (ill., ports.). (Colección Bolivia, estudios en ciencias sociales, 2).

3657. WIENER S. (Guillermo). La década olvidada de Bolivia: (los años 40). La Paz, Colegio Nacional de Historiadores de Bolivia y Producciones CIMA, 2005, 77 p. (ill., ports.). (Historias de vida, 2).

Cf. n° 3819

Bosnia-Herzegovina

Cf. n°s 4520, 4522, 4523

Brazil

3658. BAIOCCHI (Gianpaolo). Militants and citizens: the politics of participatory democracy in Porto Alegre. Stanford, Stanford U. P., 2005, XVII-224 p. (ill., map, tabs.).

3659. BOITO JÚNIOR (Armando). O sindicalismo na política brasileira. Campinas, Universidade Estadual de Campinas, Instituto de Filosofia e Ciências, 2005, 309 p. (Coleção Trajetória, 8).

3660. BORGES FILHO (Nilson), FILGUERAS (Fernando). Estado autoritário e violência no Brasil. *Revista portuguesa de história*, 2005, 37, p. 105-130.

3661. BOTELHO (André). O Brasil e os dias: estado-nação, modernismo e rotina intelectual. Bauru, EDUSC, 2005, 255 p. (História).

3662. Brasilien im amerikanischen Kontext: vom Kaiserreich zur Republik: Kultur, Gesellschaft und Politik. Hrsg. v. Horst NITSCHACK. Frankfurt am Main, Teo Ferrer de Mesquita, 2005, 303 p. (ill., maps). (Biblioteca luso-brasileira, 23).

3663. CÁNEPA (Mercedes Maria Loguércio). Partidos e representação política: a articulação dos níveis estadual e nacional no Rio Grande do Sul (1945–1965). Porto Alegre, Eitora da UFRGS, 2005, 431 p.

3664. DE ALMEIDA BALDISSERA (Marli). Onde estão os Grupos de Onze? Os comandos nacionalistas na região Alto Uruguai – RS. Passo Fundo, Universidade de Passo Fundo, Editora, 2005, 192 p.

3665. DE SOUZA (Ismara Izepe). Solidariedade internacional: a comunidade espanhola do Estado de São Paulo e a polícia política diante da Guerra civil de Espanha, 1936–1946. São Paulo, Associação Editorial Humanitas, FAPESP, LEI-USP, 2005, 278 p. (ill.). (Histórias da intolerância, 5).

3666. Ditadura (A) em debate: Estado e sociedade nos anos do autoritarismo. Organizadores, Adriano DE FREIXO e Oswaldo MUNTEAL FILHO. Rio de Janeiro, LPPE, Laboratório de Pesquisa e Práticas de Ensino, Novos Mundos, Núcleo de Estudos Americanos, Contraponto, 2005, 203 p.

3667. DO RÊGO (André Heráclio). Famille et pouvoir régional au Brésil: le coronelismo dans le Nordeste, 1850–2000. Préface de Katia M. DE QUEIRÓS MATTOSO. Paris, L'Harmattan, 2005, 319 p. (ill., maps). (Recherches Amériques latines).

3668. DOLHNIKOFF (Miriam). O pacto imperial: origens do federalismo no Brasil. São Paulo, Editora Globo, 2005, 330 p.

3669. FERREIRA (Jorge). O imaginário trabalhista: getulismo, PTB [Partido Trabalhista Brasileiro] e cultura política popular, 1945–1964. Rio de Janeiro, Editoria Civilização Brasileira, 2005, 390 p. (ill.).

3670. KITTLESON (Roger Alan). The practice of politics in postcolonial Brazil: Porto Alegre, 1845–1895.

Pittsburgh, University of Pittsburgh Press, 2005, XI-266 p. (Pitt Latin American series).

3671. LOGUÉRCIO CÁNEPA (Mercedes Maria). Partidos e representação política: a articulação dos níveis estadual e nacional no Rio Grande do Sul (1945–1965). Porto Alegre, Eitora da UFRGS, 2005, 431 p.

3672. Minas e os fundamentos do Brasil moderno. Organizadora Angela DE CASTRO GOMES. Belo Horizonte, Editora UFMG, 2005, 348 p. (ill., ports.). (Humanitas).

3673. MOREL (Marco). As transformações dos espaços públicos: imprensa, atores políticos e sociabilidades na cidade imperial, 1820–1840. São Paulo, Editora Hucitec, 2005, 326 p. (ill.). (Estudos históricos, 57).

3674. PEDROSO (Regina Célia). Estado autoritário e ideologia policial. Sao Paulo, FAPESP, 2005, 209 p.

3675. PINTO DA FONSECA (Francisco César). O consenso forjado: a grande imprensa e a formação da agenda ultraliberal no Brasil. São Paulo, Editora Hucitec, 2005, 461 p. (Estudos brasileiros, 38).

3676. ROSE (R. S.). The unpast: elite violence and social control in Brazil, 1954–2000. Athens, Ohio U. P., 2005, XIX-437 p.

3677. Votos (Os) de Deus: evangélicos, política e eleições no Brasil. Organizadores Joanildo A. BURITY e Maria das Dores C. MACHADO. Recife, Fundação Joaquim Nabuco, Editora Massangana, 2005, 236 p.

3678. WOODARD (James P.). History, sociology and the political conflicts of the 1920s in São Paulo, Brazil. *Journal of Latin American studies*, 2005, 37, 2, p. 333-349.

Cf. n°s 650, 3595, 6595, 6849

Bulgaria

3679. Almanach na bălgarskite nacionalni dviženija sled 1878 g. (Almanac of the Bulgarian national movements after 1878). Bălgarska Akademija na Naukite, Institut po Istorija. Săst.: Georgi MARKOV. Sofija, Akad. Izdat. Marin Drinov, 2005, 623 p. (ill.).

3680. *Vacat.*

3681. DAYIOĞLU (Ali). Toplama kampından Meclis'e: Bulgaristan'da Türk ve Müslüman azınlığı. (Du camp de concentration à l'Assemblée: la minorité turco-musulmane en Bulgarie). İstanbul, İletişim, 2005, 512 p.

3682. DIMOV (Dimcho N.). Dr. Jekyll and Mr. Hyde revisited: policing interwar Bulgaria. *In*: Policing interwar Europe: 1918–1940 [Cf. n° 3551], p. 172-191.

3683. GALUNOV (T.). Konservativna ili liberalna isbiratelna sistema (Bălgarskijat opit ot 1882 g.). (Système de vôte conservateur ou libéral. L'expérience bulgare de 1882). *Istoričeski pregled*, 2005, 5-6, p. 64-88.

3684. KEREN (Zvi). Kehilat Yehude Rusts'uk: mi-"yarkete Tugarmah" le-virat mehoz ha-Danubah: tsemihatah shel kehilah ba-Imperyah ha-Otmanit u ferihatah, 1788–1878. (Jewish community of Ruscuk: from the periphery of the Ottoman Empire to capital of the Tuna Vilayet-i, 1788–1878). Yerushalayim, Mekhon Ben-Tsevi le-heker kehilot Yiśra el ba-mizrah, Yad Yitshak Ben-Tsevi, ha-Universitah ha- Ivrit bi-Yerushalayim, 2005, 358-VIIII p. (ill., facsims., maps, port., tables).

3685. KHRISTOVA (Natalija). Specifika na bălgarskoto "disidentstvo": vlast i inteligencija 1956–1989 g. (Spécificité de la "dissidence" bulgare. Pouvoir et Intelligentsia 1956–1989). Sofija, Letera, 2005, 383 p.

3686. TODEV (Ilija T.). Bălgarskoto nacionalno revoljucionno dviženie: (1853–1878); netradicionni variacii na tradicionna tema. (Le mouvement National révolutionnaire bulgare, 1853–1878). Sofija, Kama, 2005, 127 p.

Cf. n°s 4281, 6620, 7009, 7048

Cambodia

3687. LAVOIX (Hélène). "Nationalism" and "genocide": the case of Cambodia (1861–1979). London, University of London, 2005, 473 p. [Thesis (Ph.D.) – University of London, 2005].

3688. TULLY (John Andrew). A short history of Cambodia: from Empire to survival. Crows Nest, Allen & Unwin, 2005, XIII-268 p. (ill.). (Short history of Asia series).

Cameroon

3689. NKWENGUE (Pierre). L'Union nationale des étudiants du Kamerun: ou la contribution des étudiants africains à l'émancipation de l'Afrique. Paris, L'Harmattan, 2005, 257 p. (Etudes africaines).

3690. TERRETTA (Meredith). 'God of independence, God of peace': village politics and nationalism in the maquis of Cameroon, 1957–1971. *Journal of African history*, 2005, 46, 1, p. 75-101.

Canada

3691. ELLIS (Faron). The limits of participation: members and leaders in Canada's Reform Party. Calgary, University of Calgary Press, 2005, XXII-225 p.

3692. MAC FARLANE (John). Agents of control or chaos? A strike at Arvida helps clarify Canadian policy on using troops against workers during the Second World War. *Canadian historical review*, 2005, 86, 4, p. 619-640.

3693. MALONEY (Sean M.). The roots of soft power: the Trudeau government, de-NATOization, and denuclearization, 1967–1970. Kingston, Centre for International Relations, Queen's University, 2005, X-132 p. (Martello papers, 27).

3694. MARTIN (Pierre). Dynamiques partisanes et réalignements électoraux au Canada: 1867–2004. Avec l'aide d'André BLAIS. Paris, L'Harmattan, 2005, 301 p. (ill.). (Logiques politiques).

3695. MESSAMORE (Barbara J.). 'The line over which he must not pass': defining the office of Governor General, 1878. *Canadian historical review*, 2005, 86, 3, p. 453-484.

3696. MILLS (Sean). When Democratic Socialists discovered democracy: the League for social reconstruction confronts the Quebec problem. *Canadian historical review*, 2005, 86, 1, p. 53-82.

3697. MORTON (Desmond). Billet pour le Front. Histoire sociale des volontaires canadiens (1914–1919). Balma, Athéna Éditions, 2005, 344 p.

Cf. n° 6835

Cape Verde

3698. EVORA (José Silva). Santo Antão no limiar do século XIX: da tensão social às insurreições populares, 1886–1894: uma perspectiva histórica. Praia, Instituto do Arquivo Histórico Nacional, 2005, 163 p. (Colecção Monografias).

Chad

3699. BANGOURA (Mohamed Tétémadi). Violence politique et conflits en Afrique: le cas du Tchad. Paris, Harmattan, 2005, 487 p. (ill., maps). (Etudes africaines).

Chile

3700. ARANCIBIA CLAVEL (Patricia), JARA HINOJOSA (Isabel), NOVOA MACKENNA (Andrea). La Marina en la historia de Chile. T. 1. Siglo XIX. Santiago, Sudamericana, 2005, 589 p.

3701. BLANCPAIN (Jean-Pierre). Immigration et nationalisme au Chili, 1810–1925: un pays à l'écoute de l'Europe. Paris, L'Harmattan, 2005, 319 p. (ill., maps, ports.). (Recherches Amériques latines).

3702. Camino a La moneda: las elecciones presidenciales en la historia de Chile 1920–2000. Ed. por Alejandro SAN FRANCISCO y Ángel SOTO. Santiago, Instituto de Historia, Pontificia Universidad Católica de Chile y Centro de Estudios Bicentenario, 2005, XVII-520 p.

3703. Chile-Perú, Perú-Chile en siglo XIX: la formación del Estado, la economía y la sociedad. Comp. Eduardo CAVIERES FIGUEROA y Cristóbal ALJOVÍN DE LOSADA. Valparaíso, Ediciones Universitarias de Valparaíso, Pontificia Universidad Católica de Valparaíso, 2005, 335 p.

3704. CORREA (Sofía). Con las riendas del poder: la derecha chilena en el siglo XX. Santiago, Editorial Sudamericana, 2005, 313 p. (Biblioteca Todo es historia).

3705. Democracy in Chile: the legacy of September 11, 1973. Ed. by Silvia NAGY-ZEKMI and Fernando LEIVA. Brighton, Sussex Academic Press, 2005, XI-226 p.

3706. GARCÉS (Mario), LEIVAM (Sebastián). El golpe en La Legua: los caminos de la historia y la memoria. Santiago, LOM Ediciones, 2005, 128 p. (ill.). (Historia).

3707. GOICOVIC DONOSO (Igor). Entre el dolor y la ira: la venganza de Antonio Ramón Ramón: Chile, 1914. Osorno, Editorial Universidad de Los Lagos, 2005, 188 p. (ill.). (Colección Monográficos).

3708. LEÓN (Leonardo). La Araucanía: la violencia mestiza y el mito de la "Pacificación", 1880–1900. Santiago, Universidad ARCIS, Escuela de Historia y Ciencias Sociales, 2005, 292 p.

3709. MALLON (Florencia E.). Courage tastes of blood: the Mapuche community of Nicolás Ailío and the Chilean state, 1906–2001. Durham, Duke U. P., 2005, XVII-319 p. (Radical perspectives).

3710. SALAZAR VERGARA (Gabriel). Construcción de estado en Chile (1760–1860): democracia de "los pueblos", militarismo ciudadano, golpismo oligárquico. Santiago, Editorial Sudamericana, 2005, 550 p. (Biblioteca Todo es historia).

3711. TIMMERMANN (Freddy). El factor Pinochet: dispositivos de poder, legetimación, élites, Chile, 1973–1980. Santiago, Ediciones Universidad Católica Silva Henríquez, 2005, 478 p. (Historia).

3712. VAYSSIÈRE (Pierre). Le Chili d'Allende et de Pinochet dans la presse française: passions politiques, informations et désinformation, 1970–2005. Paris, L'Harmattan, 2005, 301 p. (ill.). (Recherches Amériques latines).

Cf. n°s 3539, 3582, 4791, 6122, 6595, 6952

China

3713. FRIEDMAN (Edward), PICKOWICZ (Paul G.), SELDEN (Mark). Revolution, resistance, and reform in village China. New Haven, Yale U. P., 2005, X-340 p. (ill., maps). (Yale agrarian studies).

3714. LI (Huaiyin). Village governance in North China, 1875–1936. Stanford, Stanford U. P., 2005, XII-325 p. (maps).

3715. NEWBY (L. J.). The Empire and the Khanate: a political history of Qing relations with Khoqand c. 1760–1860. Leiden, Brill, 2005, XIV-297 p. (ill.). (Brill's Inner Asian library, 16).

3716. PALTEMAA (Lauri). In the vanguard of history: the Beijing Democracy Wall Movement 1978–1981 and social mobilisation of former Red Guard dissent. Turku, Turun Yliopisto, 2005, 390 p. (Turun yliopiston julkaisuja, 288. Humaniora; Sarja B).

3717. WASMER (Caterina). Qi junzi, Protagonisten einer demokratischen Bewegung: die Entwicklung der Jiuguohui von 1935–1949. Heidelberg, Edition Forum, 2005, 220-XXVI p. (ill.). (Würzburger sinologische Schriften).

Cf. n°s 5948, 7018, 7047, 7372, 7571, 7596-7801

Colombia

** 3718. LLERAS RESTREPO (Carlos), SANTOS (Eduardo). Cartas del exilio. Edición y prólogo de Carlos

LLERAS DE LA FUENTE. Bogotá, Planeta, 2005, 297 p. (ill.).

3719. DELGADO BARÓN (Mariana). El discurso político partidista en Boyacá, 1930–1940. Bogotá, Universidad de los Andes, Facultad de Ciencias Sociales, Departamento de Ciencia Política, 2005, 141 p. (Colección Prometeo).

3720. GONZÁLEZ TRAVIESO (David). El control del orden público y la criminalidad en el cantón Petare de la República de Colombia, 1822–1830. *Anuario de estudios bolivarianos*, 2005, 12, p. 159-187.

3721. LLANO ISAZA (Rodrigo). Los draconianos: origen popular del liberalismo colombiano. Bogotá, Planeta, 2005, 229 p.

3722. MONDRAGÓN CASTAÑEDA (Julio). Las ideas políticas de los radicales boyacenses, 1850–1886. Tunja, Academia Boyacense de Historia, 2005, 209 p. (ill., ports.). (Biblioteca de la Academia Boyacense de Historia, 5. Colección Centenario).

3723. SAETHER (Steinar A.). Independence and the redefinition of Indianness around Santa Marta, Colombia, 1750–1850. *Journal of Latin American studies*, 2005, 37, 1, p. 55-80.

3724. TAUSSIG (Michael). Law in a lawless land: diary of a limpieza in Colombia. Chicago, University of Chicago Press, 2005, XIII-208 p.

3725. WILLIFORD (Thomas J.). Laureano Gómez y los masones, 1936–1942. Bogotá, Planeta, 2005, 249 p.

Cf. n° 6846

Congo

3726. DE VOS (Luc), GERARD (Emmanuel), GÉRARD-LIBOIS (Jules), RAXHON (Philippe). Les secrets de l'affaire Lumumba. Bruxelles, Racine, 2005, 668 p. (ill., maps).

3727. HASKIN (Jeanne M.). The tragic state of the Congo: from decolonization to dictatorship. New York, Algora Pub., 2005, IX-228 p.

3728. TSHONDA OMASOMBO (Jean), VERHAEGEN (Benoît). Patrice Lumumba: acteur politique: de la prison aux portes du pouvoir, juillet 1956–février 1960. Paris, L'Harmattan et Tervuren, Musée royal de l'Afrique centrale, 2005, 406 p. (ill.). (Cahiers africains = Afrika studies, 68/70).

Cf. n° 3576

Costa Rica

3729. ALVARENGA VENUTOLO (Ana Patricia). De vecinos a ciudadanos: movimientos comunales y luchas cívicas en la historia contemporánea de Costa Rica. San José, Editorial de la Universidad de Costa Rica, Editorial de la Universidad Nacional, 2005, XXXIV-320 p.

3730. BOTEY SOBRADO (Ana María). Costa Rica entre guerras: 1914–1940. San José, Editorial de la Universidad de Costa Rica, 2005, 124 p. (ill.). (Serie Cuadernos de historia de las instituciones de Costa Rica, 6).

3731. CRUZ (Consuelo). Political culture and institutional development in Costa Rica and Nicaragua: world making in the tropics. New York, Cambridge U. P., 2005, XVII-281 p.

3732. DÍAZ ARIAS (David). Construcción de un Estado moderno: política, Estado e identidad nacional en Costa Rica, 1821–1914. San José, Editorial Universidad de Costa Rica, 2005, 86 p. (ill.). (Serie Cuadernos de historia de las instituciones de Costa Rica, 18).

3733. DOBLES (Ignacio), LEANDRO ZUÑIGA (Vilma). Militantes: la vivencia de lo político en la segunda ola del marxismo en Costa Rica. San José, Editorial de la Universidad de Costa Rica, 2005, XIX-392 p.

3734. MOLINA JIMÉNEZ (Iván). Demoperfectocracia: la democracia pre-reformada en Costa Rica (1885–1948). Heredia, EUNA, 2005, 484 p. (ill.).

3735. OCONTRILLO (Eduardo). Cien años de política costarricense, 1902–2002: de Ascención Esquivel a Abel Pacheco. San José, Editorial Universidad Estatal a Distancia (EUNED), 2005, 353 p.

Cf. n°⁵ 626, 6323, 6980

Côte d'Ivoire

Cf. n° 6769

Croatia

** 3736. Zapisnici Politbiroa Centralnoga Komiteta Komunističke Partije Hrvatske: 1945–1952. (Verbali del Ufficio politico del Comitato Centrale del Partito comunista croato). Sv. 1. 1945–1948. Priredila Branislava VOJNOVIĆ. Zagreb, Hrvatski državni arhiv, 2005, 656 p.

3737. BARIĆ (Nikica). Srpska pobuna u Hrvatskoj: 1990.–1995. (Serb rebellion in Croatia, 1990–1995). Zagreb, Golden marketing – Tehnička knjiga, 2005, 614 p.

3738. MATIJEVIĆ (Zlatko), MATKOVIĆ (Stjepan), ŠOKČEVIĆ (Dinko). 1903 in Croatia. *Časopis za suvremenu povijest*, 2005, 37, 3, p. 801-806.

3739. MATIJEVIĆ (Zlatko). U sjeni dvaju orlova: prilozi crkveno-nacionalnoj povijesti Hrvata u prvim desetljećima 20. stoljeća. (In the shadow of two eagles: contributions to religious and national history of the Croats in the first decades of 20th century). Zagreb, Golden marketing, Tehnička knjiga, 2005, 359 p.

3740. OGNYANOVA (I.). Croatian nationalism and the breakup of Yugoslavia. *Balkan Studies*, 2005, 1, p. 3-24.

3741. RUMENJAK (Nives). Politička i društvena elita Srba u Hrvatskoj potkraj 19. stoljeća: uspon i pad Srpskoga kluba. (Political and social elite of the Serbs in Croatia at the end of the 19th century: rise and decline of the Serb Club). Zagreb, Hrvatski institut za povijest, 2005, 603 p. (ill.). (Biblioteka Hrvatska povjesnica. Monografije i studije, III/30).

Cf. n°s 4520, 4522, 4523

Cuba

** 3742. Cuba before Castro: British annual reports, 1898–1958. Vol. 1. 1898–1911. Vol. 2. 1912–1931. Vol. 3. 1932–1938. Vol. 4. 1939–1958. With an introductory essay by Hugh THOMAS; research editor, Robert L. JARMAN. London, Archival Publications International, 2005, 4 vol., XII-491 p., VII-520 p., VI-562 p., VII-553 p. (ill., facsims., maps). (British Archives on Cuba).

3743. GUERRA (Lillian). The myth of José Martí: conflicting nationalisms in early twentieth-century Cuba. Chapel Hill a. London, University of North Carolina Press, 2005, XII-310 p. (ill., map). (Envisioning Cuba).

3744. LÉTRILLIART (Philippe). Cuba, l'Eglise et la Révolution: approche d'une concurrence conflictuelle. Paris, L'Harmattan, 2005, 473 p. (Recherches Amériques latines).

3745. MESA-LAGO (Carmelo), PÉREZ-LÓPEZ (Jorge F.). Cuba's aborted reform: socioeconomic effects, international comparisons and transition policies. Gainesville, University Press of Florida, 2005, XX-223 p.

3746. PIQUERAS ARENAS (José Antonio). Sociedad civil y poder en Cuba: colonia y poscolonia. Madrid, Siglo XXI de España Editores, 2005, XV-393 p. (ill., maps).

3747. PRADOS-TORREIRA (Teresa). Mambisas: rebel women in nineteenth-century Cuba. Gainesville, University Press of Florida, 2005, XII-185 p.

3748. SÁENZ ROVNER (Eduardo). La conexión Cubana: narcotráfico, contrabando y juego en Cuba entre los años 20 y comienzos de la Revolución. Bogotá, Universidad Nacional de Colombia, Facultad de Ciencias Humanas, 2005, 265 p.

3749. SHAFFER (Kirwin R.). Anarchism and countercultural politics in early twentieth-century Cuba. Gainesville, University Press of Florida, 2005, XI-279 p.

Cyprus

3750. DEMIRYÜREK (Mehmet). İngiliz devrinde Kıbrıs'ta eşkıyalar ve devlet (1878–1896). [Les brigants et l'Etat au Chypre à l'époque anglaise (1878–1896)]. İstanbul, Deniz Plaza Yayınevi, 2005, 176 p.

3751. HATZIVASSILIOU (Evanthis). Cyprus at the Crossroads, 1959–1963. *European history quarterly*, 2005, 35, 4, p. 523-540.

3752. MALLINSON (William). Cyprus: a modern history. London a. New York, I. B. Tauris, 2005, XX-244 p.

Czech Republic

** 3753. Dějiny české politiky v dokumentech. (The history of the Czech politics in documents). Ed. Zdeněk VESELÝ. Praha, Professional Publ., 2005, 764 p. (photogr.).

** 3754. Vademecum soudobých dějin. Česká republika. Průvodce po archivech, badatelských institucích, knihovnách, sdruženích, muzeích a památnících. (Vademecum contemporary history. Czech Republic. A guide to archives, research institutions, libraries, associations, museums and places of memorial). Ed. Oldřich TŮMA, Jitka SVOBODOVÁ, Ulrich MÄHLERT. Praha, Ústav pro soudobé dějiny AV ČR, 2005, 109 p.

3755. ABRAMS (Bradley F.). The struggle for the soul of the nation: Czech culture and the rise of communism. Lanham, Rowman & Littlefield, 2005, VIII-363 p. (Harvard Cold War studies book series).

3756. Beneš-Dekrete (Die): Nachkriegsordnung oder ethnische Säuberung: kann Europa eine Antwort geben? Hrsg. v. Heiner TIMMERMANN, Emil VORÁČEK und Rüdiger KIPKE. Münster, LIT, 2005, 667 p. (Dokumente und Schriften der Europäischen Akademie Otzenhausen, 108).

3757. BÍLEK (Jiří). Československo a jeho armáda v roce 1945. Deset tezí k 60. výročí konce druhé světové války. (Czechoslovakia and its Army in 1945). *Historie a vojenství*, 2005, 54, 2, p. 37-49.

3758. BLAIVE (Muriel). Une déstalinisation manquée: Tchécoslovaquie 1956. Préface de Krzysztof POMIAN. Bruxelles, Editions Complexe, IHTP, CNRS, 2005, 281 p. (Histoire du temps présent).

3759. BLAŽEK (Petr), ŽÁČEK (Pavel). Czechoslovakia. *In*: A Handbook of the Communist Security Apparatus in East Central Europe 1944–1989. Ed. Krzysztof PERSAK, Łukasz KAMIŃSKI. Warsaw, 2005, p. 87-161.

3760. BLAŽEK (Petr). Politická represe v komunistickém Československu 1948–1989. (The violence typology in the years 1948–1989). *In*: Moc verzus občan. Úloha represie a politického násilia v komunizme. Ed. Pavel ŽÁČEK. Bratislava, 2005, p. 8-22.

3761. Bolševismus, komunismus a radikální socialismus v Československu. (Bolshevism, Communism, and Radical Socialism in Czechoslovakia). Tomo 4-5. Ed. Zdeněk KÁRNÍK and Michal KOPEČEK. Praha, Dokořán a Ústav pro soudobé dějiny AV ČR, 2005, 2 vol., p. 279, 383 p. (KSČ a radikální socialismus v Československu 1918–1989).

3762. BORÁK (Mečislav), JANÁK (Dušan). Die tschechoslowakische Retributionsgerichtsbarkeit und die deutsche Problematik. Die aussordentlichen Volksgerichte in Moravská Ostrava, Opava und Nový Jičín. *In*: Sozialgeschichtliche Kommunismusforschung. Tschechoslowakei, Polen, Ungarn und DDR 1948–1968. Ed. Christiane BRENNER. München, 2005, p. 365-422.

3763. CABADA (Ladislav). Komunismus, levicová kultura a česká politika 1890–1938. (Communism, Left culture and Czech politics, 1890–1938). Plzeň, Vyd. a naklad. A. Čeněk, 2005, 198 p.

3764. Československá armáda jako oběť i nástroj politické perzekuce. Sborník vybraných dokumentů z let 1948–1953. (The Czechoslovak armed forces as victim and instrument of the political persecution, 1948–1953). Ed. Jiří BÍLEK. Praha, Ministerstvo obrany ČR, 2005, 111 p.

3765. Československo na rozhraní dvou epoch nesvobody. Sborník z konference k 60. výročí konce druhé světové války. (Czechoslovakia between totalitarian epochs). Ed. Zdeňka KOKOŠKOVÁ, Jiří KOCIAN, Stanislav KOKOŠKA. Praha, Národní archiv a Ústav pro soudobé dějiny AV ČR, 2005, 419 p. (photogr.).

3766. Chekhija i Slovakija v XX veke: Ocherki istorii. (Czech Lands and Slovakia in the 20th century: Essays on history). RAN, In-t slavjanovedenija. Books 1-2. Moskva, Nauka, 2005, 2 vol., 453 p., 558 p. (ill.; bibl. incl.; pers. ind. p. 439-450, 542-554). (XX vek v dokumentakh i issledovanijakh).

3767. DOSKOČIL (Zdeněk). Vot vož partiji! Hle, vůdce strany! Okolnosti volby Gustáva Husáka prvním tajemníkem ÚV KSČ v dubnu 1969. (Vot vož partiji! Look, the leader of the Party!) Český časopis historický, 2005, 103, 3, p. 572-620. – IDEM. Zahraniční a domácí ohlasy na volbu Gustáva Husáka prvním tajemníkem ÚV KSČ v dubnu 1969. (Foreign and domestic reactions to the election of Gustáv Husák to the position of First Secretary of the Communist Party of Czechoslovakia in April 1969). Moderní dějiny, 2005, 13, p. 245-291.

3768. DVOŘÁK (Jiří). Počátek velkého díla. Z dějin československého regionalismu. (Anfang des grossen Werkes. Aus der Geschichte des tschechoslowakischen Regionalismus). Moderní dějiny, 2005, 13, p. 181-233.

3769. FROMMER (Benjamin). National cleansing. Retribution against nazi collaborators in postwar Czechoslovakia. New York, Cambridge U. P., 2005, XV-387 p. (Studies in the social and cultural history of modern warfare, 20).

3770. GABZDILOVÁ (Soňa), OLEJNÍK (Milan). Od pozitivizmu k negativizmu – Karpatonemecká strana v období predmníchovskej ČSR. (Vom Pozitivizmus zum Negativizmus. Der Anteil der Karpatendeutschen Partei im politischen Leben der deutschen Minorität in der Slowakei in der Zeit der Vormünchener Tschechoslowakischen Republik). Historický časopis, 2005, 53, 3, p. 467-485. [Deutsche Zsfassung].

3771. GLASSHEIM (Eagle). Noble nationalists: the transformation of the Bohemian aristocracy. Cambridge a. London, Harvard U. P., 2005, XIII-299 p. (ill., map).

3772. Jan Masaryk. Úvahy o jeho smrti. (Jan Masaryk. Reflections on his death). Ed. Ladislava KREMLIČKOVÁ. Praha, Úřad dokumentace a vyšetřování zločinů komunismu, 2005, 495 p. (Svědectví).

3773. KABELE (Jiří). Z kapitalismu do socialismu a zpět. Teoretické vyšetřování přerodů Československa a České republiky. (From Capitalism to Socialism and back). Praha, Karolinum, 2005, 582 p.

3774. KAPLAN (Karel). Kronika komunistického Československa. Doba tání 1953–1956. (A chronicle of the communist Czechoslovakia, 1953–1956). Brno, Barrister & Principal, 2005, 768 p.

3775. KOCIAN (Jiří). Die Dekrete des Präsidenten der Republik in der Tschechoslowakei in den Jahren 1948–1989. In: Die Beneš-Dekrete. Nachkriegsordnung oder ethnische Säuberung. Kann Europa eine Antwort geben? Ed. Heiner TIMMERMANN. Münster, 2005, p. 359-372.

3776. MACHOVEC (Milan). Česká státnost. (Czech statehood). Praha, Vyšehrad, 2005, 70 p. (photogr.).

3777. MAREŠ (Miroslav). Terorismus v ČR. (Terrorism in the Czech Republic). Brno, Centrum strategických studií, 2005, 476 p.

3778. MEZNÍK (Jaroslav). Můj život za vlády komunistů 1948–1989. (My life under the communist government, 1948–1989). Brno, Matice moravská, 2005, 329 p. (photogr.). (Prameny dějin moravských, 10).

3779. MICHÁLEK (Slavomír). Československo verzus William Nathan Oatis. (Tschechoslowakei versus William Nathan Oatis). Historický časopis, 2005, 53, 4, p. 697-717. [Deutsche Zsfassung].

3780. Opozice a odpor proti komunistickému režimu v Československu 1968–1989. (Opposition and resistance, 1968–89). Ed. Petr BLAŽEK. Praha, Dokořán a Ústav českých dějin FF UK, 2005, 355 p.

3781. PERNES (Jiří). Die Verfolgung der Teilnehmer an den Arbeiterdemonstrationen in Brünn im Jahre 1951. In: Sozialgeschichtliche Kommunismusforschung. Tschechoslowakei, Polen, Ungarn und DDR 1948–1968. Ed. Christiane BRENNER. München, 2005, p. 355-364.

3782. Politické procesy v Československu po roce 1945 a "Případ Slánský". Sborník příspěvků. (Political Trials in Czechoslovakia after 1945 and the Fall Slánský. Proceedingss of a conference). Ed. Jiří PERNES and Jan FOITZIK. Brno, Prius a Ústav pro soudobé dějiny AV ČR, 2005, 391 p.

3783. RONSIN (Samuel). Police, republic and nation: the Czechoslovak state police and the building of a multinational democracy, 1918–1925. In: Policing interwar Europe: 1918–1940 [Cf. n° 3551], p. 136-158.

3784. SPIRITOVÁ (Markéta). Formen der Repression in der Tschechoslowakei nach 1968. Das Alltagsleben tschechischer Intellektueller zur Zeit der "Normalisierung'. In: Sozialgeschichtliche Kommunismusforschung. Tschechoslowakei, Polen, Ungarn und DDR 1948–1968. Ed. Christiane BRENNER. München, 2005, p. 337-353.

3785. STANĚK (Tomáš). Poválečné "excesy" v českých zemích v roce 1945 a jejich vyšetřování. (Post-War "excesses" in the Bohemian lands from 1945 and their investigation). Praha, Ústav pro soudobé dějiny AV ČR, 2005, 366 p. (Sešity Ústavu pro soudobé dějiny AV ČR, 41).

2. HISTORY BY COUNTRIES

3786. STEHLÍK (Eduard). Československé opevňovací programy 1936–1938. (Czechoslovak fortifications programmes, 1936–1938). Tomo 1-2. *Historie a vojenství*, 2005, 54, 1-2, p. 4-29, p. 50-59.

3787. SZÁRAZ (Peter). Poskytovanie azylu španielskym utečencom na československom vyslanectve v Madride a v Československu v rokoch 1936–1937. (Der Asylangebot den spanischen Flüchtlingen auf der tschechoslowakischen Botschaft in Madrid und in der Tschechoslowakei in den Jahren 1936–1937). *Historický časopis*, 2005, 53, 2, p. 255-281. [Deutsche Zsfassung].

3788. URBÁNEK (Miroslav). Správa sledování Ministerstva vnitra v letech 1948–1989. Stručný nástin organizačního vývoje. (The Ministry of Interior Surveillance Bureau from 1948 to 1989. A brief outline of its organizational development). *Sborník Archivu ministerstva vnitra*, 2005, 3, p. 173-232.

3789. VANÍČKOVÁ (Vladimíra) [et al.]. Informace o archivních materiálech k tématice poválečné československé sociální demokracie, uložených v Archivu Ministerstva vnitra. (Information about the Ministry of Interior Archives materials on the postwar Czechoslovak Social Democratic Party). *Sborník Archivu ministerstva vnitra*, 2005, 3, p. 343-376.

3790. Vítězové? Poražení? Životopisná interview. (Victors? Vanquished?) Tomo 1. Disent v období tzv. Normalizace. Tomo 2. Politické elity v období tzv. normalizace. Ed. Miroslav VANĚK, Pavel URBÁŠEK. Praha, Ústav pro soudobé dějiny AV ČR, 2005, 2 vol., 1121 p., 827 p. (Obzor, 60-61).

3791. VLČEK (Radomír). Slovanství, panslavismus a rusofilství při formování moderního českého národa. (Slavism, Pan-Slavism and Russophilia during the formation of the Modern Czech Nation). *Slovanské historické studie*, 2005, 30, p. 59-109.

3792. ŽÁČEK (Pavel). Nástroj triedneho štátu. Organizácia ministerstiev vnútra a bezpečnostných zborov 1953–1990. (The instrument of the class state. The organization of the Interior Ministries and the Czechoslovak Security Forces, 1953–1990). Bratislava, Ústav pamäti národa, 2005, XXII-389 p. (Dokumenty). – IDEM. Třídní boj po 17. listopadu 1989 v dokumentech politického aparátu ČSLA. (Class war after November 1989 from the viewpoint of the political apparat of the ČSLA). *Historie a vojenství*, 2005, 54, 1, p. 105-119.

Addendum 2004

3793. FAUTH (Tim). Deutsche Kulturpolitik im Protektorat Böhmen und Mähren 1939 bis 1941. Göttingen, V&R unipress, 2004, 98 p. (Berichte und Studien, 45).

Cf. n^{os} *3539, 4127, 4398, 5271, 6086, 7238*

Czechoslovakia

Cf. n^{os} *3753-3793*

Denmark

** 3794. HARDIS (Arne). Forræderens dagbog. En dansk nazist 1941–1945. (Le journal du traître. Un nazi danois 1941–1945). København, Lindhardt og Ringhof, 2005, 304 p. (ill.).

3795. FRIISBERG (Claus). En nation dannes. Danmark og danskerne til 1920. (Naissance d'une nation. Le Danemark et les Danois jusqu'en 1920). Vande, Vestjysk Kulturforlag, 2005, 495 p. (ill.).

3796. Fronten i Danmark, den ekstreme højrefløj fra 1995 til 2005. (L'extrême-droite au Danemark de 1995 à 2005). København, Demos, 2005, 130 p. (ill.).

3797. GAUL (Volker). Möglichkeiten und Grenzen absolutistischer Herrschaft: landesherrliche Kommunikationsstrategien und städtische Interessen während der Pest in den Herzogtümern Schleswig-Holstein-Gottorf, 1709–1713. Tönning, Der Andere Verlag, 2005, 256 p. (ill.).

3798. HANSEN (Peer Henrik). "Firmaets" største bedrift. Den hemmelige krig mod de danske kommunister. (La principale activité de la "Firme". La guerre secrète contre les communistes danois). København, Ascheoug, 2005, 398 p.

3799. HERMANSEN (Karsten). Kirken, kongen og enevælden: en undersøgelse af det danske bispeembede 1660–1746. (L'Eglise, le roi et l'absolutisme: recherches sur la fonction épiscopale au Danemark 1660–1746). Odense, Syddansk Universitetsforlag, 2005, 422 p. (University of Southern Denmark studies in history and social sciences, 298).

3800. HOFFMEYER (Erik). Den politiske historie om den økonomiske og monetære union. (Histoire politique de l'union économique et monétaire). Odense, Syddansk Universitetsforlag, 2005, 143 p. (University of Southern Denmark studies in history and social sciences, 294).

3801. HOVBAKKE SØRENSEN (Lars). Nordenforestillinger i dansk politik 1945–1968. (La politique danoise vis-à-vis de la Scandinavie 1945–1968). Århus, Historia, 2005, 252 p. [English summary].

3802. Hvar er så danskhed? (Où est la danité?) Red. Torben FLEDELIUS KNAP, Morten NIELSEN, Jacob PINDSTRUP. Højbjerg, Hovedland, 2005, 198 p.

3803. PROBST (Niels). Niels Juel: vor største flådefører. (Niels Juel [1626–1697]: le plus grand amiral danois). Köpenhamn, Statens Forsvarshistoriske Museum, 2005, 132 p. (Forsvarshistoriske skrifter, 2).

3804. RÜNITZ (Lone). Af hensyn til konsekvenserne: Danmark og flygtningespørgsmålet 1933–1940. (Tenir compte des conséquences: la question des réfugiés au Danemark 1933–1940). Odense, Syddansk Universitetsforlag, 2005, 577 p. (University of Southern Denmark studies in history and social sciences, 303).

3805. SKOU (Kaare R.). Dansk politik A-Å leksikon. (La politique danoise. Dictionnaire de A à Z). København, Ascheoug, 2005, 783 p. (ill.).

3806. Slesvigske hertuger (De). (Les ducs du Schleswig). Red. Carsten PORSKROG RASMUSSEN, Inge ADRIANSEN, Lennart S. MADSEN. Aabenraa, Historisk Samfund for Sønderjylland, 2005, 383 p. (Skrifter udg. af Historisk Samfund for Sønderjylland, 92).

3807. Veje til danskheden. Bidrag til den moderne nationale selvforståelse. (Les voies de la danité. Contribution à une évidence nationale moderne). Red. Palle Ove CHRISTIANSEN. København, Reitzel, 2005, 163 p. (ill.).

3808. VENBORG PEDERSEN (Mikkel). Hertuger: at synes og at være i Augustenborg, 1700–1850. (Etre et être vu: les ducs d'Augustenborg 1700–1850). København, Museum Tusculanums forlag, 2005, 297 p. (planches). (Etnologiske studier, 12).

3809. VILLAUME (Poul). Lavvækst og frontdannelser: 1970–1985. Gyldendal og Politikens Danmarkshistorie. Bd. 15. (Faible croissance et formation de fronts 1970–1985. Histoire du Danemark. Tome 15). Red. Olaf OLSEN. København, Gyldendalske boghandel, 2005, 401 p.

Cf. n° 7327

Dominican Republic

3810. BONILLA (Walter R.). La revolución dominicana de 1965 y la participación de Puerto Rico. Hato Rey, Publicaciones Puertorriqueñas, 2005, IX-206 p. (ill., ports.).

3811. DE LA ROSA (Jesús). La Revolución de abril de 1965: siete días de guerra civil. Santo Domingo, Secretaría de Estado de Cultura, Editora Nacional, 2005, 188 p. (ill.). (Colección Premios nacionales).

3812. FAXAS (Laura). République dominicaine: système politique et mouvement populaire, 1961–1990. Toulouse, Presses universitaires du Mirail, 2005, 502 p. (maps). (Hespérides. Amérique latine).

3813. LÓPEZ-CALVO (Ignacio). God and Trujillo: literary and cultural representations of the Dominican dictator. Gainesville, University Press of Florida, 2005, XVIII-196 p.

3814. MARTÍNEZ-VERNE (Teresita). Nation and citizen in the Dominican Republic, 1880–1916. Chapel Hill, University of North Carolina Press, 2005, XVIII-235 p.

3815. PEGUERO (Valentina). The militarization of culture in the Dominican Republic, from the Captains General to General Trujillo. Lincoln a. London, University of Nebraska Press, 2005, X-263 p.

East Timor

3816. MENDES (Nuno Canas). A "multidimensionalidade" da construção identitária em Timor-Leste: nacionalismo, Estado e identidade nacional. Lisboa, Instituto Superior de Ciências Sociais e Políticas, 2005, 545 p.

Ecuador

3817. Camino y significación del Partido Social Cristiano. Ed. por Marco LARA GUZMÁN. Quito, Corporación Editora Nacional, Universidad Andina Simón Bolivar, 2005, 215 p. (ill.). (Libros de bolsillo, 22).

3818. LIND (Amy). Gendered paradoxes: women's movements, state restructuring, and global development in Ecuador. University Park, Pennsylvania State U. P., 2005, XVI-182 p.

3819. Mirada esquiva (La): Reflexiones históricas sobre la interacción del estado y la ciudadanía en los Andes (Bolivia, Ecuador y Perú), siglo XIX. Ed. por Marta IRUROZQUI VICTORIANO. Madrid, Consejo Superior de Investigaciones Científicas, 2005, 385 p. (Biblioteca de Historia de América, 35).

3820. QUINTERO (Rafael). Electores contra partidos en un sistema político de mandos. Quito, Abya-Yala, 2005, 268 p. (ill.).

Ecuatorial Guinea

3821. LINIGER-GOUMAZ (Max). La Guinée Equatoriale, opprimée et convoitée: aide-mémoire d'une démocrature, 1968–2005. Préface de Djongele BOKOKO BOKO. Paris, L'Harmattan, 2005, 509 p. (ill.).

Egypt

3822. BARON (Beth). Egypt as a woman. Nationalism, gender, and politics. Berkeley a. Los Angeles, University of California Press, 2005, XV-287 p.

3823. BEN NÉFISSA (Sarah), ARAFAT (Alâ Al-dîn). Vote et démocratie dans l'Egypte contemporaine. Paris, Karthala et Institut de recherche pour le développement, 2005, 279 p. (Hommes et sociétés).

3824. CAMPANINI (Massimo). Storia dell'Egitto contemporaneo: dalla rinascita ottocentesca a Mubarak. Roma, Lavoro, 2005, 295 (maps). (Islam, 9).

3825. DUNN (John P.). Khedive Ismail's army. London, Routledge, 2005, XVII-240 p. (ill.). (Cass military studies series).

3826. HAMAMSY (Chafika Soliman). Zamalek: the changing life of a Cairo elite, 1850–1945. Cairo, American University in Cairo Press, 2005, XVIII-366 p. (ill.).

3827. Re-envisioning Egypt, 1919–1952. Ed. by Arthur GOLDSCHMIDT, Amy J. JOHNSON and Barak A. SALMONI. New York, American University in Cairo Press, 2005, XVIII-510 p.

El Salvador

3828. SPRENKELS (Ralph). The price of peace: the Human Rights Movement in Postwar El Salvador. Amsterdam, Centrum voor Studie en Documentatie van Latijns Amerika, 2005, V-120 p. (Cuadernos, 19).

3829. TILLEY (Virginia Q.). Seeing Indians. A study of race, nation, and power in El Salvador. Albuquerque, University of New Mexico Press, 2005, XVIII-297 p.

Estonia

3830. Eestimaa Kommunistliku Partei kohalikud organisatsioonid, 1940–1991. (Il Partito comunista estone, le organizzazioni locali, 1940–1991). Autorid Olev LIIVIK ja Raili NUGIN. Tallinn, Kistler-Ritso Eesti Sihastus, 2005, 403 p. (ill.).

3831. Hitler-Stalin-Pakt (Vom) bis zu Stalins Tod: [Estland 1939–1953]. Hrsg. v. Olaf MERTELSMANN. Hamburg, Bibliotheca Baltica, 2005, 301 p. (map). (Bibliotheca Baltica).

Ethiopia

3832. CARMICHAEL (Tim). Approaching Ethiopian history: Addis Abäba and local governance in Harär, c. 1900 to 1950. Ann Arbor, UMI Dissertation Services, 2005, XI-292 p.

3833. KĀHSĀY (Barha). Ethiopia, democratization and unity: the role of the Tigray People's Liberation Front. Münster, Monsenstein und Vannerdat, 2005, 331 p.

Finland

3834. SILTALA (Juha). Nation as mother figure for reformers in Finland, 1840–1910. *Scandinavian journal of history*, 2005, 30, 2, p. 135-158.

Cf. n° 7240

France

* 3835. DE BENOIST (Alain). Bibliographie générale des droites françaises. Vol. 3. Vol. 4. Paris, Dualpha, 2005, 2 vol., 648 p., 736 p. (Collection "Patrimoine des lettres").

* 3836. Histoire de la maréchaussée et de la gendarmerie. Guide de recherche. Sous la dir. de Jean-Noël LUC. Maisons-Alfort, Service historique de la gendarmerie nationale, 2005, 1105 p.

* 3837. JENNINGS (Eric T.). Visions and representations of French Empire. *Journal of modern history*, 2005, 77, 3, p. 701-721.

* 3838. LALOUETTE (Jacqueline). Laïcité et séparation des Églises et de l'État: esquisse d'un bilan historiographique (2003–2005). *Revue historique*, 2005, 636, p. 849-870.

3839. Années Giscard (Les): Valéry Giscard d'Estaing et l'Europe, 1974–1981. Actes de la journée d'études organisée par le centre d'histoire de sciences po et l'institut pour la démocratie en Europe le 26 janvier 2004. Centre d'histoire de sciences politique; sous la direction de Serge BERSTEIN et Jean-François SIRINELLI. Paris, A. Colin, 2005, 272 p.

3840. ASBACH (Olaf). Staat und Politik zwischen Absolutismus und Aufklärung: der Abbé de Saint-Pierre und die Herausbildung der französischen Aufklärung bis zur Mitte des 18. Jahrhunderts. Hildesheim, G. Olms, 2005, 332 p. (Europaea memoria: Studien und Texte zur Geschichte der europäischen Ideen. Reihe I, Studien, 37).

3841. AUDIGIER (François). Génération gaulliste: L'union des jeunes pour le progrès, une école de formation politique, 1965–1975. Nancy, Presses universitaires de Nancy, 2005, 479 p. (ill.). (Collection Histoire contemporaine).

3842. BECKER (Jean Jacques). La France de 1914 à 1940: les difficultés de la République. Paris, Presses universitaires de France, 2005, 127 p. (maps). (Que sais-je? 3750).

3843. BEIDERBECK (Friedrich). Zwischen Religionskrieg, Reichskrise und europäischem Hegemoniekampf: Heinrich IV. von Frankreich und die protestantischen Reichsstände. Berlin, BWV, Berliner Wissenschafts-Verlag, 2005, 499 p. (Innovationen, 8).

3844. BEIK (William). The absolutism of Louis XIV as social collaboration. *Past and present*, 2005, 188, p. 195-224.

3845. BONINCHI (Marc). Vichy et l'ordre moral. Paris, Presses universitaires de France, 2005, 319 p.

3846. BONNER (Elizabeth). The Earl of Huntly and the King of France, 1548: man for rent. *English historical review*, 2005, 120, 485, p. 80-103.

3847. BURSTIN (Haim). Une Révolution à l'oeuvre, le faubourg Saint-Marcel (1789–1794). Seyssel, Champ Vallon, 2005, 924 p.

3848. CHAMBON (Pascal). La Loire et l'Aigle: les Foréziens face à l'État napoléonien. Saint-Etienne, Publications de l'université de Saint-Etienne, 2005, 574 p. (ill., maps).

3849. CLÉMENT (Jean Louis). La hiérarchie catholique et les principes de la Révolution nationale. *Guerres mondiales et conflits contemporains*, 2005, 53, 218, p. 27-36. – IDEM. La démocratie chrétienne en France. Un pari à haut risque, de 1900 à nos jours. Paris, François de Guibert, 2005, 199 p.

3850. COINTET (Michèle). De Gaulle et Giraud: l'affrontement, 1942–1944. Paris, Perrin, 2005, 549 p. (ill., maps, ports.).

3851. CRÉPIN (Annie). Défendre la France. Les Français, la guerre et le service militaire, de la guerre de Sept Ans à Verdun. Rennes, Presses universitaires de Rennes, 2005, 424 p.

3852. DANTHIEUX (Dominique). Le département rouge: république, socialisme et communisme en Haute-Vienne, 1895–1940. Limoges, Presses universitaires de Limoges, 2005, 339 p. (ill., maps). (Histoire).

3853. DARD (Olivier). Voyage au cœur de l'OAS. Paris, Perrin, 2005, 425 p.

3854. DAUBRESSE (Sylvie). Le Parlement de Paris, ou, La voix de la raison: (1559–1589). Préf. de Denis CROUZET. Genève, Droz, 2005, XV-558 p. (Travaux d'humanisme et Renaissance, 398).

3855. DRAKE (David). French intellectuals and politics from the Dreyfus Affair to the Occupation. Basingstoke, Palgrave Macmillan, 2005, XII-214 p. (French politics, society, and culture series).

3856. DRÉVILLON (Hervé). L'impôt du sang. Le métier des armes sous Louis XIV. Paris, Tallandier, 2005, 526 p.

3857. FORTESCUE (William). France and 1848: the end of monarchy. London, Routledge, 2005, 215 p.

3858. France in the era of fascism: essays on the French authoritarian right. Ed. by Brian JENKINS. New York, Berghahn Books, 2005, 232 p.

3859. Frantsuzskij ezhegodnik, 2005: Absoljutizm vo Frantsii: K 100-letiju B.F. Porshneva (1905–1972). (The French Yearbook, 2005: Absolutism in France: Dedicated to the 100th anniversary of Boris Porshnev, 1905–1972). Ed. Aleksandr V. CHUDINOV. RAN, In-t vseobshchej istorii. Moskva, KomKniga, 2005, 304 p.

3860. GRÉVY (Jérôme). Le cléricalisme? Voilà l'ennemi: une guerre de religion en France. Préf. de Serge BERSTEIN. Paris, A. Colin, 2005, 247 p. (Enjeux de l'histoire).

3861. HAMMERSLEY (Rachel). French revolutionaries and English republicans. The Cordeliers Club, 1790–1794. Rochester, Boydell Press, 2005, XIII-192 p. (Royal historical society studies in history, n.s., 43).

3862. HANLEY (Wayne). The genesis of Napoleonic propaganda, 1796 to 1799. New York a. Chichester, Columbia U. P., 2005, XIV-234 p.

3863. HATZFELD (Hélène). Faire de la politique autrement. Les expériences inachevées des années 1970. Rennes, Presses Universitaires de Rennes et Paris, ADELS, 2005, 328 p.

3864. Histoire des gauches en France. Vol. 1. L'héritage du XIXe siècle. Vol. 2. XXe siècle: à l'épreuve de l'histoire. Sous la dir. de Jean-Jacques BECKER et Gilles CANDAR. Paris, La Découverte, 2005, 2 vol., 1360 p. (Découverte/Poche, 216-217. Sciences humaines et sociales).

3865. IMLAY (Talbot). Mind the gap: the perception and reality of communist sabotage of French war production during the Phoney War 1939–1940. *Past and present*, 2005, 189, p. 179-224.

3866. JOHANSEN (Anja). Soldiers as police: the French and Prussian armies and the policing of popular protest, 1889–1914. Aldershot, Ashgate, 2005, IX-329 p.

3867. KARILA-COHEN (Pierre). Les fonds secrets ou la méfiance légitime: l'invention paradoxale d'une "tradition républicaine" sous la Restauration et la monarchie de Juillet. *Revue historique*, 2005, 636, p. 731-766.

3868. KOZHOKIN (Evgenij M.). Istorija bednogo kapitalizma: Frantsija XVIII – pervoj poloviny XIX veka. (A history of 'poor capitalism': France in the 18th and the first half of the 19th century). Moskva, ROSSPEN, 2005, 367 p. (ill.; portr.; bibl. p. 333-353; pers. ind. p. 354-365).

3869. LALOUETTE (Jacqueline). La séparation des Églises et de l'État: genèse et développement d'une idée, 1789–1905. Paris, Ed. du Seuil, 2005, 449 p. (L'univers historique).

3870. LAMBAUER (Barbara). Opportunistischer Antisemitismus. Der deutsche Botschafter Otto Abetz und die Judenverfolgung in Frankreich (1940–1942). *Vierteljahrshefte für Zeitgeschichte*, 2005, 53, 2, p. 241-273.

3871. LAURENT (Sébastien). Aux origenes de la "guerre des polices": militaires et policiers du reseignement dans la République (1870–1914). *Revue historique*, 2005, 636, p. 767-792.

3872. LE MOINE (Frederic). Les évêques français de Verdun à Vatican II: une génération en mal d'héroïsme. Rennes, Presses Universitaires de Rennes, 2005, 373 p. (tav., ill.). (Histoire).

3873. LJUBART (Margarita K.). Sem'ja vo frantsuzskom obshchestve: XVIII – nachalo XX veka. (Family in the French society, the 18th–the early 20th century). RAN, In-t etnologii i antropologii im. N.N. Miklukho-Maklaja. Moskva, Nauka, 2005, 296 p. (ill.; bibl. p. 285-293). [Resumé français]

3874. MARTIN (Jean-Philippe). Histoire de la nouvelle gauche paysanne: des contestations des années 1960 à la Confédération paysanne. Paris, Editions La Découverte, 2005, 311 p. (Cahiers libres).

3875. MATONTI (Frédérique). Intellectuels communistes: essai sur l'obéissance politique. La Nouvelle Critique (1967–1980). Paris, La Découverte, 2005, 414 p. (L'espace de l'histoire).

3876. MAYEUR (Jean Marie). La séparation des églises et de l'État. Paris, Les éditions de l'Atelier-Editions Ouvrières, 2005, 255 p. (ill.).

3877. Michel Debré, Premier ministre (1959–1962): actes du colloque organisé les 14, 15 et 16 mars 2002. Sous la direction de Serge BERNSTEIN, Pierre MILZA et Jean-François SIRINELLI. Paris, Presses universitaires de France, 2005, 680 p.

3878. MIDDELL (Matthias). Die Geburt der Konterrevolution in Frankreich 1788–1792. Leipzig, Leipziger Universitätsverlag, 2005, 234 p. (Beiträge zur Universalgeschichte und vergleichenden Gesellschaftsforschung, 15).

3879. MOULINET (Daniel). Genèse de la laïcité: à travers les textes fondamentaux de 1801 à 1959. Paris, Éd. du Cerf, 2005, 289 p. (Droit civil ecclesiastique).

3880. NERI-ULTSCH (Daniela). Sozialisten und Radicaux – eine schwierige Allianz: Linksbündnisse in der Dritten Französischen Republik, 1919–1938. München, Oldenbourg, 2005, X-528 p. (ill., map). (Quellen und Darstellungen zur Zeitgeschichte, 63).

3881. NOVIKOVA (Ol'ga E.). Politika i etika v epoku religioznykh vojn: Just Lipsij (1547–1606). (Politics and ethics during the French Wars of Religion: Justus Lipsius, 1547–1606). Moskva, RKhTU, 2005, 399 p. (ill.; bibl. p. 277-297). [English summary]

3882. NOVOSELOV (Vasilij R.). Poslednij dovod chesti: Duel' vo Frantsii v XVI – nachale XVII stoletija. (The last argument of honour: Duels in France, the 16th to the early 17th century). Sankt-Peterburg, Atlant, 2005, 285 p. (ill.; portr.; bibl. p. 282-284). (Oruzhejnaja akademija).

3883. Origins (The) of the French Revolution. Ed. by Peter R. CAMPBELL. Basingstoke a. New York, Palgrave Macmillan, 2005, 384 p.

3884. OZOUF (Mona). Varennes: la mort de la royauté (21 juin 1791). Paris, Gallimard, 2005, 433 p. (Les journées qui ont fait la France).

3885. PLÉ (Bernhard). Die sakralen Grundlagen der laizistischen Republik Frankreichs. Zur Liturgie der aufgeklärten Bürgerschaft in der Dritten Republik. *Archiv für Kulturgeschichte*, 2005, 87, 2, p. 373-394.

3886. PLESHKOVA (Sof'ja L.). Frantsija XVI – nachala XVII veka: Korolevskij gallikanizm: Tserkovnaja politika monarkhii i formirovanie ofitsial'noj ideologii. (France in the 16th and the early 17th centuries: the royal gallicanism: The religious policy of the monarchy and the formation of the official ideology). Moskva, Izd-vo Moskovskogo un-ta, 2005, 463 p. (ill.; bibl. p. 449-461). (Trudy Istoricheskogo f-ta MGU; 30. Ser. 2: Istoricheskie issledovanija; 10).

3887. RAYSKI (Adam). The choice of the Jews under Vichy. Between submission and resistance. Notre Dame, University of Notre Dame Press, United States Holocaust Memorial Museum, 2005, XVI-388 p.

3888. RÉMOND (René). Les droites aujourd'hui. Paris, Audibert, 2005, 271 p.

3889. SANTAMARIA (Yves). Le pacifisme, une passion française. Paris, Armand Colin, 2005, 351 p. (L'histoire au présent).

3890. SCHÜLE (Klaus). Paris: die politische Geschichte seit der Französischen Revolution; vom Erfinden und Schwinden der Demokratie in der Metropole. Tübingen, Narr, 2005, 315 p. (ill.).

3891. SECONDY (Philippe). Pier Leroy-Beaulieu: un importateur des méthodes électorales américaines en France. *Revue historique*, 2005, 634, p. 309-342.

3892. SERNA (Pierre). La République des girouettes, 1789–1815 et au-delà. Une anomalie politique, la France de l'extrême centre. Seyssel, Champ Vallon, 2005, 570 p.

3893. SHAFER (David A.). The Paris Commune: French politics, culture, and society at the crossroads of the revolutionary tradition and revolutionary socialism. New York, Palgrave Macmillan, 2005, XI-226 p. (ill.). (European history in perspective).

3894. SMITH (Jay M.). Nobility reimagined. The patriotic nation in eighteenth-century France. Ithaca, Cornell U. P., 2005, XIII-307 p.

3895. SMITH (Paul). A history of the French Senate. Vol. 1. The Third Republic, 1870–1940. Lewiston a. Lampeter, E. Mellen Press, 2005, 528 p. (maps).

3896. TELLIER (Thibault). Paul Reynaud, 1878–1966. Un indépendant en politique. Préf. de Jean-François SIRINELLI. Paris, Fayard, 2005, 886 p. (Pour une histoire du XXe siècle).

3897. Vichy, resistance, liberation: new perspectives on wartime France. Ed. by Hanna DIAMOND and Simon KITSON. Oxford, Berg, 2005, XII-207 p.

3898. VOILLIOT (Christophe). La candidature officielle: une pratique d'Etat de la Restauration à la Troisième République. Rennes, Presses Universitaires de Rennes, 2005, 298 p. (Collection Carnot).

3899. Voix (La) et le geste: une approche culturelle de la violence socio-politique: actes du colloque tenu à Clermont-Ferrand en septembre 2003. Sous la direction de Philippe BOURDIN, Mathias BERNARD et Jean-Claude CARON. Clermont-Ferrand, Presses Universitaires Blaise-Pascal, 2005, 381 p. (ill.). (Collection "Histoires croiseés").

3900. VONAU (Jean-Laurent). L'épuration en Alsace: la face méconnue de la Libération, 1944–1953. Préf. d'Alphonse IRJUD. Strasbourg, Rhin, 2005, 221 p. (ill., map).

3901. WEIS (Cédric). Jeanne Alexandre, une pacifiste intégrale. Angers, Presses de l'université d'Angers, 2005, 293 p.

3902. WILLMS (Johannes). Napoleon: eine Biographie. München, Beck, 2005, 839 p. (ill., maps).

3903. YAGIL (Limore). Chrétiens et Juifs sous Vichy (1940–1944). Sauvetage et désobéissance civile. Paris, Ed. du Cerf, 2005, 765 p. (Histoire).

Addenda 2000–2003

** 3904. BASCH (Victor). Le deuxième procès Dreyfus. Rennes dans la tourmente, correspondance. Édition établie par Françoise BASCH et André HÉLARD. Paris, Berg International, 2003, 212 p.

** 3905. Lettres d'Henri III roi de France. T. V. 8 avril 1580–31 décembre 1582. Recueillies par Pierre CHAMPION et Michel FRANÇOIS, publiées par Jacqueline BOUCHER avec la collaboration d'Henri ZUBER. Paris, Champion, Société de l'Histoire de France, 2000, 374 p.

Cf. n°s 3639, 4084, 5168, 6505, 6757, 6920, 6926, 7053, 7150, 7345, 7436, 7446

Gabon

3906. AUGÉ (Axel Eric). Le recrutement des élites politiques en Afrique subsaharienne: une sociologie du pouvoir au Gabon. Préf. de François GAULME. Paris, L'Harmattan, 2005, 299 p. (ill.). (Etudes africaines).

Cf. n° 6801

Georgia

3907. JONES (Stephen F.). Socialism in Georgian colors. The European road to social democracy 1883–1917. Cambridge, Harvard U. P., 2005, XIV-384 p.

Germany

** 3908. Akten der Reichskanzlei. Die Regierung Hitler. Band IV. 1937. Bearb. v. Friedrich HARTMANNSGRUBER. München, Oldenbourg, 2005, 895 p.

** 3909. Dokumente zur Deutschlandpolitik. Reihe 6. 1. Oktober 1969 bis 30. September, 1982. Band. 3. 1. Januar 1973 bis 31. Dezember 1974. Bearb. v. Monika KAISER, Daniel HOFMANN und Hans-Heinrich JANSEN. München, Oldenbourg, 2005, LXVIII-970 p.

** 3910. "Gemeingefährlichen Bestrebungen (Die) der Sozialdemokratie". Teil 1. Die Berichte der Regierungspräsidenten über die sozialdemokratische Bewegung in den Regierungsbezirken Frankfurt/Oder und Potsdam während des Sozialistengesetzes 1878–1890. Bearbeitet und eingeleitet von Beatrice FALK und Ingo MATERNA. Berlin, Berliner Wissenschafts-Verlag, 2005, 325 p. (ill.). (Schriftenreihe des Landesarchivs Berlin, 8. Veröffentlichungen des Brandenburgischen Landeshauptarchivs, 49).

** 3911. Helmuth von Moltke, 1848–1916. Dokumente zu seinem Leben und Wirken. 1. Briefe Helmuth von Moltkes an seine Frau, 1877–1915 mit Schilderungen von Reisen mit dem älteren Moltke und von Aufenthalten am Zarenhof – Briefe und Dokumente zu Kriesausbruch und Kriegschuldfrage. Hrsg. v. Andreas BRACHER und Thomas MEYER. Basel, Perseus, 2005, VI-666 p.

** 3912. Kabinettsprotokolle (Die) der Bundesregierung. Band. 15. 1962. Bearbeitet von Uta RÖSSEL und Christoph SEEMANN, unter Mitwirkung von Ralf BEHRENDT [et al.]. Herausgegeben für das Bundesarchiv von Hartmut WEBER. München, Oldenbourg, 2005, 806 p.

** 3913. Kiesinger. "Wir leben in einer veränderten Welt". Die Protokolle des CDU-Bundesvorstands 1965–1969. Bearbeitet von Günter BUCHSTAB; unter Mitarbeit von Denise LINDSAY. Düsseldorf, Droste, 2005, XV-1580 p. (Forschungen und Quellen zur Zeitgeschichte, 50).

** 3914. "Sehnlich erwarte ich die morgende Post". Amalie und Theodor von Schöns Briefwechsel aus dem Befreiungskrieg, 1813. Hrsg. v. Gustava Alice KLAUSA. Köln, Weimar u. Wien, Böhlau, 2005, VIII-295 p.

** 3915. Tagebücher (Die) von Joseph Goebbels. Im Auftr. des Instituts für Zeitgeschichte und mit Unterstützung des Staatlichen Archivdienstes Rußlands hrsg. von Elke FRÖHLICH. Teil 1. Aufzeichnungen 1923–1941. Band. 1. 2. Dezember 1925–Mai 1928. Bearb. von Elke FRÖHLICH. Bd. 2. 1. Dezember 1929–Mai 1931. Bearb. von Anne MUNDING. Bd. 2. 3. Oktober 1932–März 1934. Bearb. von Angela HERMANN. Band 3. 1. April 1934–Februar 1936. Bearb. von Angela HERMANN [et al.]. München, Saur, 2005, 4 vol., 401 p., 447 p., 429 p., 423 p.

3916. 20. Juli 1944 (Der) und das Erbe des deutschen Widerstands. Hrsg. v. Günter BRAKELMANN und Manfred KELLER. Münster, Lit, 2005, 260 p. (ill.). (Zeitansage, 1).

3917. AASLESTAD (Katherine). Place and politics: local identity, civic culture, and German nationalism in North Germany during the revolutionary era. Leiden a. Boston, Brill, 2005, XIII-384 p. (ill.). (Studies in Central European histories, 36).

3918. ADAM (Thomas). Die Gondelsheimer Rebellion von 1730. Ein Bauernaufstand und seine Folgen. Heidelberg, Verlag Regionalkultur, 2005, 166 p. (Die Gondelsheimer Geschichte, 5).

3919. ALBRECHT (Joachim). Die Avantgarde des "Dritten Reiches": die Coburger NSDAP während der Weimarer Republik 1922–1933. Frankfurt am Main u. Oxford, P. Lang, 2005, 243 p. (ill.). (Europäische Hochschulschriften. Reihe III, Geschichte und ihre Hilfswissenschaften = Publications universitaires européennes. Série III, Histoire, sciences auxiliaires de l'histoire = European University studies. Series III, History and allied studies, 1008).

3920. ALY (Götz). Hitlers Volksstaat: Raub, Rassenkrieg und nationaler Sozialismus. Frankfurt am Main, S. Fischer, 2005, 444 p. (ill.).

3921. BALD (Detlef). Die Bundeswehr: eine kritische Geschichte 1955–2005. München, Beck, 2005, 231 p. (Beck'sche Reihe, 1622).

3922. BAVAJ (Riccardo). Von links gegen Weimar. Linkes antiparlamentarisches Denken in der Weimarer Republik. Bonn, Dietz, 2005, 535 p. (Politik und Gesellschaftsgeschichte, 67).

3923. BEACHY (Robert). The soul of commerce: credit, property, and politics in Leipzig, 1750–1840. Boston, Brill, 2005, X-248 p. (Studies in Central European histories, 34).

3924. BECK (Ralf). Der traurige Patriot. Sebastian Haffner und die Deutsche Frage. Berlin, Be.Bra, 2005, 367 p.

3925. BECQUET LAVOINNE (Claude). Itinéraire du général Walther von Seydlitz-Kurzbach (1878–1976): un officier allemand face aux totalitarismes. *Guerres mondiales et conflits contemporains*, 2005, 53, 218, p. 53-66.

3926. Berlin, Rosenstraße 2-4. Protest in der NS-Diktatur. Neue Forschungen zum Frauenprotest in der Rosenstraße 1943. Hrsg. v. Antonia LEUGERS. Annweiler, Plöger, 2005, 263 p.

3927. BERNHARD (Michael H.). Institutions and the fate of democracy: Germany and Poland in the twentieth century. Pittsburgh, University of Pittsburgh Press, 2005, XV-310 p. (ill.). (Pitt series in Russian and east European studies).

3928. BERNHARD (Patrick). Zivildienst zwischen Reform und Revolte: eine bundesdeutsche Institution im gesellschaftlichen Wandel 1961–1982. München, Oldenbourg, 2005, X-462 p. (Quellen und Darstellungen zur Zeitgeschichte, 64).

3929. BLASIUS (Dirk). Weimars Ende: Bürgerkrieg und Politik; 1930–1933. Göttingen, Vandenhoeck & Ruprecht, 2005, 188 p.

3930. BLED (Jean-Paul). Bismarck. De la Prusse à l'Allemagne. Paris, Alvick, 2005, 332 p.

3931. BREMM (Klaus-Jürgen). Von der Chaussee zur Schiene: Militärstrategie und Eisenbahnen in Preußen von 1833 bis zum Feldzug von 1866. München, Oldenbourg, 2005, XII-295 p. (ill., maps). (Militärgeschichtliche Studien, 40).

3932. Changing legacy (The) of 1945 in Germany: a round-table discussion between Doris BERGEN, Volker BERGHAHN, Robert MOELLER, Dirk MOSES, and Dorothee WIERLING. *German history*, 2005, 23, 4, p. 519-546.

3933. CONZE (Eckart). Sicherheit als Kultur. Überlegungen zu einer "modernen Politikgeschichte" der Bundesrepublik Deutschland. *Vierteljahrshefte für Zeitgeschichte*, 2005, 53, 3, p. 357-380.

3934. Coping with the Nazi past: West German debates on Nazism and generational conflict, 1955–1975. Ed. by Alan S. STEINWEIS and Philipp GASSERT. New York a. Oxford, Berghahn, 2005, VIII-339 p. (Studies in German history, 2)).

3935. DALE (Gareth). Popular protest in East Germany, 1945–1989. London a. New York, Routledge, 2005, X-246 p. (Routledge advances in European politics, 27).

3936. DANKER (Uwe), SCHWABE (Astrid). Schleswig-Holstein und der Nationalsozialismus. Neumünster, Wachholtz, 2005, 224 p. (ill.).

3937. Demokratiewunder: transatlantische Mittler und die kulturelle Öffnung Westdeutschlands 1945–1970. [Konferenz, die am 12. und 13. Dezember 2003 in der American Academy, Berlin, stattfand]. Hrsg. v. Arnd BAUERKÄMPER, Konrad H. JARAUSCH und Marcus M. PAYK. Göttingen, Vandenhoeck & Ruprecht, 2005, 335 p.

3938. Deutsche Marinen im Wandel. Vom Symbol nationaler Einheit zum Instrument nationaler Sicherheit. Hrsg. v. Werner RAHN. München, Oldenbourg, 2005, XIII-738 p. (Militärgeschichte, 63).

3939. Deutsche Reich (Das) und der Zweite Weltkrieg. Band 9. Die Deutsche Kriegsgesellschaft: 1939 bis 1945. Halbband 2. Ausbeutung, deutungen, ausgrenzung. Hrsg. v. Jörg ECHTERNKAMP. München, Deutsche Verlags-Anstalt, 2005, XIII-1110 p.

3940. Deutschland-ein Land ohne revolutionäre Traditionen? Revolutionen im Deutschland des 19. und 20. Jahrhunderts im Lichte neuerer geistes- und kulturgeschichtlicher Erkenntnisse. Hrsg. v. Riccardo BAVAJ und Florentine FRITZEN. Frankfurt am Main u. Oxford, Peter Lang, 2005, 195 p.

3941. ELO (Kimmo). Die Systemkrise eines totalitären Herrschaftssystems und ihre Folgen: eine aktualisierte Totalitarismustheorie am Beispiel der Systemkrise in der DDR 1953. Münster, Lit, 2005, 246 p. (ill.). (Diktatur und Widerstand, 10).

3942. ENGEHAUSEN (Frank). Kleine Geschichte des Großherzogtums Baden 1806–1918. Karlsruhe, Braun, 2005, 208 p.

3943. EVANS (Richard J.). The Third Reich in power: 1933–1939. New York: Penguin, 2005, XVII-941 p.

3944. FINGS (Karola). Krieg, Gesellschaft und KZ. Himmlers SS-Baubrigaden. Paderborn, München u. Wien, Schöningh, 2005, 412 p. (Sammlung Schöningh zur Geschichte und Gegenwart).

3945. FÖLLMER (Moritz). The problem of national solidarity in interwar Germany. *German history*, 2005, 23, 2, p. 202-231.

3946. FRACKOWIAK (Johannes). Soziale Demokratie als Ideal. Die Verfassungsdiskussionen in Sachsen nach 1918 und 1945. Köln, Weimar u. Wien, Böhlau, 2005, 367 p.

3947. FRIEDEBURG (Robert von). The making of patriots: love of fatherland and negotiating monarchy in seventeenth-century Germany. *Journal of modern history*, 2005, 77, 4, p. 881-916.

3948. GEBAUER (Annekatrin). Der Richtungsstreit in der SPD [Sozialdemokratische Partei Deutschlands]: Seeheimer Kreis und Neue Linke im innerparteilichen Machtkampf. Mit einem Geleitwort von Helmut SCHMIDT. Wiesbaden, VS Verlag für Sozialwissenschaften, 2005, 286 p. (Forschung Politik).

3949. GEISEL (Christof). Auf der Suche nach einem dritten Weg: das politische Selbstverständnis der DDR-Opposition in den achtziger Jahren. Berlin, Links, 2005, 330 p. (ill.). (Forschungen zur DDR-Gesellschaft).

3950. Germany's two unifications. Anticipations, experiences, responses. Ed. by John BREUILLY and Ronald SPEIRS. Basingstoke, Palgrave Macmillan, 2005, 360 p. (New perspectives in German studies).

3951. GERWARTH (Robert). The Bismarck myth: Weimar Germany and the legacy of the Iron Chancellor. Oxford, Clarendon Press, 2005, IX-216 p.

3952. GNISS (Daniela). Der Politiker Eugen Gerstenmaier, 1906–1986. Eine Biographie. Düsseldorf, Droste, 2005, 514 p. (Beiträge zur Geschichte des Parlamentarismus und der politischen Parteien, 144).

3953. GOELDEL (Denis). Le tournant occidental de l'Allemagne après 1945: contribution à l'histoire politique et culturelle de la RFA. Strasbourg, Presses universitaires de Strasbourg, 2005, 375 p. (Collection "Les mondes germaniques", 11).

3954. GOSCHLER (Constantin). Schuld und Schulden: die Politik der Wiedergutmachung für NS-Verfolgte seit 1945. Göttingen, Wallstein, 2005, 543 p. (Beiträge zur Geschichte des 20. Jahrhunderts, 3).

3955. GRAU (Andreas). Gegen den Strom. Die Reaktion der CDU[Christlich-Demokratische Union Deutschlands]/CSU[Christlich-Soziale Union]-Opposition auf die Ost- und Deutschlandpolitik der sozial-liberalen Koalition 1969-1973. Düsseldorf, Droste, 2005, 556 p. (Forschungen und quellen zur Zeitgeschichte; 47).

3956. GRUNER (Wolf). Widerstand in der Rosenstraße. Die Fabrik-Action und die Verfolgung der "Mischehen" 1943. Frankfurt am Main, Fischer, 2005, 224 p. (Die Zeit des Nationalsozialismus).

3957. HAARMANN (Lutz). Die deutsche Einheit kommt bestimmt! Zum Spannungsverhältnis von Deutscher Frage, Geschichtspolitik und westdeutscher Dissidenz in den 1980er Jahren. Mit einem Geleitwort von Wolfgang SEIFFERT. Berlin, BWV, Berliner Wissenschafts-Verlag, 2005, XVI-123 p.

3958. HAAS (Stefan). Die Kultur der Verwaltung. Die Umsetzung der preußischen Reformen, 1800-1848. Frankfurt am Main u. New York, Campus, 2005, 479 p.

3959. Hamburg im "Dritten Reich". Herausgegeben von der Forschungsstelle für Zeitgeschichte in Hamburg; Redaktion, Josef SCHMID. Göttingen, Wallstein, 2005, 792 p. (ill., maps, ports.).

3960. HEFTY (Julia). Die Parlamentarischen Staatssekretäre im Bund. Eine Entwicklungsgeschichte seit 1967. Düsseldorf, Droste, 2005, 320 p. (Beiträge zur Geschichte des Parlamentarismus und der politischen Parteien, 145).

3961. HESSE (Hans). Konstruktionen der Unschuld. Die Entnazifizierung am Beispiel von Bremen und Bremerhaven, 1945-1953. Bremen, Staatsarchiv Bremen, 2005, 518 p. (Veröffentlichungen aus dem Staatsarchiv der Freien Hansestadt Bremen, 67).

3962. HIRSCHINGER (Frank). "Gestapoagenten, Trotzkisten, Verräter": kommunistische Parteisäuberungen in Sachsen-Anhalt 1918-1953. Göttingen, Vandenhoeck & Ruprecht, 2005, 412 p. (Schriften des Hannah-Arendt-Instituts für Totalitarismusforschung, 27).

3963. HÖMIG (Herbert). Brüning. Politiker ohne Auftrag. Zwischen Weimarer und Bonner Republik. Paderborn, München u. Wien, Schöningh, 2005, 848 p.

3964. How green were the nazis? Nature, environment, and nation in the Third Reich. Ed. by Franz-Josef BRÜGGEMEIER, Mark CIOC and Thomas ZELLER. Athens, Ohio U. P., 2005, VIII-283 p.

3965. HULL (Isabel V.). Absolute destruction. Military culture and the practices of war in imperial Germany. Ithaca, Cornell U. P., 2005, XI-384 p.

3966. INACHIN (Kyra T.). Nationalstaat und regionale Selbstbehauptung: die preussische Provinz Pommern 1815-1945. Bremen, Edition Temmen, 2005, 400 p. (Quellen und Studien aus den Landesarchiven Mecklenburg-Vorpommerns, 7).

3967. JARDIN (Pierre). Aux racines du mal: 1918, le déni de défaite. Paris, Tallandier, 2005, 639 p.

3968. KARLSCH (Rainer). Hitlers Bombe. Die geheime Geschichte der deutschen Kernwaffenversuche. München, DVA, 2005, 415 p.

3969. KELLERHOFF (Sven Felix). Hitlers Berlin. Geschichte einer Haßliebe. Berlin, Be.Bra, 2005, 223 p.

3970. KELLOGG (Michael). The Russian roots of nazism. White émigrés and the making of National Socialism, 1917-1945. Cambridge, New York a. Port Melbourne, Cambridge U. P., 2005, XIII-328 p.

3971. KIISKINEN (Elina). Die Deutschnationale Volkspartei in Bayern (Bayerische Mittelpartei) in der Regierungspolitik des Freistaats während der Weimarer Zeit. München, Beck, 2005, XXVII-623 p. (ill., maps). (Schriftenreihe zur bayerischen Landesgeschichte, 145).

3972. KIßENER (Michael). Das Dritte Reich. Darmstadt, Wissenschaftliche Buchgesellschaft, 2005, VIII-136 p.

3973. KLAUSCH (Hans-Peter). Tätergeschichten: die SS-Kommandanten in früheren Konzertartionslagern im Emsland. Bremen, Ed. Temmen, 2005, 319 p. (ill.). (DIZ-Schriften, 13).

3974. KOHLRAUSCH (Martin). Der Monarch im Skandal. Die Logik der Massenmedien und die Transformation der wilhelminischen Monarchie. Berlin, Akademie, 2005, 536 p. (Elitenwandel in der Moderne, 7).

3975. Konzentrationslager im Rheinland und in Westfalen 1933-1945: zentrale Steuerung und regionale Initiative. Hrsg. v. Jan Erik SCHULTE für den Arbeitskreis der NS-Gedenkstätten in NRW e.V. Paderborn, F. Schöningh, 2005, XLI-333 p. (ill., maps).

3976. KÖSSLER (Till). Abschied von der Revolution. Kommunisten und Gesellschaft in Westdeutschland, 1945-1968. Düsseldorf, Droste, 2005, 499 p. (Beiträge zur Geschichte des Parlamentarismus und der politischen Parteien, 143).

3977. KRAUSHAAR (Wolfgang), WIELAND (Karin), REEMTSMA (Jan Philipp). Rudi Dutschke, Andreas Baader und die RAF [Rote Armee Fraktion]. Hamburg, Hamburger Ed., 2005, 142 p.

3978. "Krise" (Die) der Weimarer Republik: zur Kritik eines Deutungsmusters. Hrsg. v. Moritz FÖLLMER und Rüdiger GRAF. Frankfurt am Main, Campus Verl., 2005, 367 p.

3979. KUNZ (Andreas). Wehrmacht und Niederlage. Die bewaffnete Macht in der Endphase der nationalsozialistischen Herrschaft 1944 bis 1945. München, Oldenbourg, 2005, IX-390 p. (Militärgeschichte, 64).

3980. Kupierte Alternative (Die): Konservatismus in Deutschland nach 1945. Hrsg. v. Frank-Lothar KROLL. Berlin, Duncker & Humblot, 2005, VIII-347 p. (Studien und Texte zur Erforschung des Konservatismus, 6).

3981. Kurt Georg Kiesinger 1904-1988. Von Ebingen ins Kanzleramt. Hrsg. v. Günter BUCHSTAB, Philipp GASSERT und Peter Thaddäus LANG; herausgegeben im Auftrag der Konrad-Adenauer-Stiftung e. V. Freiburg im Breisgau, Basel u. Wien, Herder, 2005, 568 p.

3982. LAUGKAU-ALEX (Ursula). Deutsche Volksfront 1932-1939. Zwischen Berlin, Paris, Prag und Moskau.

Band 3. Dokumente zur Geschichte des Ausschusses zur Vorbereitung einer deutschen Volksfront, Chronik und Verzeichnisse. Berlin, Akademie, 2005, XVI-544 p.

3983. LEMAY (Benoît). La guerre des généraux de la Wehrmacht: Hitler au service des ambitions de ses élites militaires? *Guerres mondiales et conflits contemporains*, 2005, 53, 220, p. 85-96.

3984. LINK (Stephan). Politischer Katholizismus, Liberalismus, Sozialdemokratie: das politische Bamberg im 19. Jahrhundert. Bamberg, Selbstverlag des Historischen Vereins Bamberg, 2005, 459 p. (ill.). (Schriftenreihe / Historischer Verein Bamberg, 38).

3985. LÖFFLER (Ursula). Dörfliche Amtsträger im Staatswerdungsprozess der Frühen Neuzeit. Die Vermittlung von Herrschaft auf dem Lande im Herzogtum Magdeburg, 17. und 18. Jahrhundert. Münster, Lit, 2005, 266 p. (Herrschaft und Soziale Systeme in der Frühen Neuzeit, 8).

3986. Membra unius capitis. Studien zu Herrschaftsauffassungen und Regierungspraxis in Kurbrandenburg, 1640-1688. Hrsg. v. Michael KAISER und Michael ROHRSCHNEIDER. Berlin, Duncker & Humblot, 2005, 245 p. (Forschungen zur brandenburgische und preußische Geschichte, 7).

3987. METZLER (Gabriele). Konzeptionen politischen Handelns von Adenauer bis Brandt. Politische Planung in der pluralistischen Gesellschaft. Paderborn, Schöningh, 2005, 478 p.

3988. MICHEL (Marco). Die Bundestagswahlkämpfe der FDP [Freie Demokratische Partei] 1949-2002. Wiesbaden, VS Verlag für Sozialwissenschaften, 2005, 309 p.

3989. MIERZEJEWSKI (Alfred C.). Ludwig Erhard. Der Wegbereiter der Sozialen Marktwirtschaft. Biografie. München, Siedler, 2005, 398 p.

3990. MINAGAWA (Taku). Tozokusei-kokka kara kokka-rengo e: kinsei Doitsu kokka no sekkeizu "Shuvâben domei". (From imperial estate to confederation: the Swabian League, a blueprint for the early modern German state). Tokyo, Sobunsha, 2005, 376 p.

3991. MÖLLER (Horst). La République de Weimar. Paris, Tallandier Éditions, 2005, 367 p.

3992. MÜLLER (Anett). Modernisierung in der Stadtverwaltung. Das Beispiel Leipzig im späten 19. Jahrhundert. Köln, Weimar u. Wien, Böhlau, 2005, 486 p. (Geschichte und Politik in Sachsen, 24).

3993. MÜLLER (Jürgen). Deutscher Bund und deutsche Nation 1848-1866. Göttingen, Vandenhoeck & Ruprecht, 2005, 637 p. (Schriftenreihe der Historischen Kommission bei der Bayerischen Akademie der Wissenschaften, 71).

3994. MÜLLER (Markus). Gemeinden und Staat in der Reichsgrafschaft Sayn-Hachenburg 1652-1799. Wiesbaden, Historische Kommission für Nassau, 2005, XIV-561 p. (Beiträge zur Geschichte Nassaus und des Landes Hessen, 3).

3995. MULLIGAN (William). The creation of the modern German Army. General Walther Reinhardt and the Weimar Republic, 1914-1930. New York a. Oxford, Berghahn, 2005, 247 p. (Monographs in German history, 12).

3996. Nazism, war and genocide. Essays in honour of Jeremy Noakes. Ed. by Neil GREGOR. Exeter, University of Exeter Press, 2005, 226 p.

3997. Networks of Nazi persecution: bureaucracy, business, and the organization of the Holocaust. Ed. by Gerald D. FELDMAN and Wolfgang SEIBEL. New York, Berghahn Books, 2005, XII-376 p. (Studies on war and genocide, 6).

3998. NEUBERT (Ehrhart), AUERBACH (Thomas). "Es kann anders werden". Opposition und Widerstand in Thüringen, 1945-1989. Köln, Weimar u. Wien, Böhlau, 2005, 293 p.

3999. NOLZEN (Armin). Charismatic legitimation and bureaucratic rule: the NSDAP in the Third Reich, 1933-1945. *German history*, 2005, 23, 4, p. 494-518.

4000. OLICK (Jeffrey K.). In the house of the hangman: the agonies of German defeat, 1943-1949. Chicago, University of Chicago Press, 2005, XI-380 p.

4001. Ort (Der) des Terrors. Geschichte der nationalsozialistischen Konzentrationslager. Hrsg. v. Wolfgang BENZ und Barbara DISTEL; Redaktion, Angelika KÖNIGSEDER. Band 2. Frühe Lager, Dachau, Emslandlager. München, Beck, 2005, 607 p.

4002. PFLÜGER (Christine). Kommissare und Korrespondenzen: politische Kommunikation im Alten Reich (1552-1558). Köln, Böhlau, 2005, 365 p. (Norm und Struktur, 24).

4003. PIPER (Ernst). Alfred Rosenberg: Hitlers Chefideologe. München, Blessing, 2005, 830 p. (ill.).

4004. Politische Köpfe aus Südwestdeutschland. Hrsg. v. Reinhold WEBER und Ines MAYER. Stuttgart, Kohlhammer, 2005, XI-336 p. (Schriften zur politischen Landeskunde Baden-Württembergs, 33).

4005. PRITCHARD (Gareth). Schwarzenberg 1945: antifascists and the 'third way' in German politics. *European history quarterly*, 2005, 35, 4, p. 499-522.

4006. RATHGEB (Eberhard). Die engagierte Nation: deutsche Debatten 1945-2005. München, Hanser, 2005, 447 p. (ill.).

4007. REMEKE (Stefan). Gewerkschaften und Sozialgesetzgebung: DGB [Deutscher Gewerkschaftsbund] und Arbeitnehmerschutz in der Reformphase der sozialliberalen Koalition. Essen, Klartext, 2005, 519 p. (Veröffentlichungen des Instituts für Soziale Bewegungen, 33. Darstellungen. Schriftenreihe A).

4008. SALEWSKI (Michael). Deutschland und der Zweite Weltkrieg. Paderborn, München, Wien u. Zürich, Schöningh, 2005, 442 p.

4009. SCHWEGEL (Andreas). Der Polizeibegriff im NS-Staat: Polizeirecht, juristische Publizistik und Judikative 1931–1944. Tübingen, Mohr Siebeck, 2005, XI-II-419 p. (Beiträge zur Rechtsgeschichte des 20. Jahrhunderts, 48).

4010. SEARLE (Alaric). Tolsdorff Trials in Traunstein: public and judicial attitudes to the Wehrmacht in the Federal Republic, 1954–1960. *German history*, 2005, 23, 1, p. 50-78.

4011. SEGGERN (Jessica von). Alte und neue Demokraten in Schleswig-Holstein. Demokratisierung und Neubildung einer politischen Elite auf Kreis- und Landesebene 1945 bis 1950. Stuttgart, Steiner, 2005, 243 p. (Historische Mitteilungen, 61).

4012. SERESSE (Volker). Politische Normen in Kleve-Mark während des 17. Jahrhunderts. Argumentationsgeschichtliche und herrschaftstheoretische Zugänge zur politischen Kultur der frühen Neuzeit. Eppendorf, Bibliotheca Academica, 2005, 456 p. (Frühneuzeit-Forschungen, 12).

4013. Staatsgründung auf Raten? Zu den Auswirkungen des Volksaufstandes 1953 und des Mauerbaus 1961 auf Staat, Militär und Gesellschaft in der DDR. Im Auftrag des Militärgeschichtlichen Forschungsamtes und der Bundesbeauftragten für die Unterlagen des Staatssicherheitsdienstes der ehemaligen DDR herausgegeben von Torsten DIEDRICH und Ilko-Sascha KOWALCZUK. Berlin, Links, 2005, XII-435 p. (Militärgeschichte der DDR, 11).

4014. STRASSER (Christian). Der Aufstand im bayerischen Oberland 1705 – Majestätsverbrechen oder Heldentat? Eine Untersuchung der Strafprozesse gegen die Anführer der in der "Mordweihnacht von Sendling" gescheiterten Erhebung. Münster, Lit, 2005, XX-330 p. (Augsburger Schriften zur Rechtsgeschichte, 3).

4015. STRÖTZ (Jürgen). Der Katholizismus im deutschen Kaiserreich 1871 bis 1918. Strukturen eines problematischen Verhältnisses zwischen Widerstand und Integration. 1. Reichsgründung und Kulturkampf, 1871–1890. 2. Wilhelminische Epoche und Erster Weltkrieg 1890–1918. Hamburg, Kovac, 2005, 628 p., 333 p. (Religionspädagogik und Pastoralgeschichte, 6).

4016. TRABA (Robert). "Wschodniopruskość". Tożsamość regionalna i narodowa w kulturze politycznej Niemiec. ("Prusse-orientalisme". Une identité régionale et nationale dans la culture politique allemande). Poznań, Wydawnictwo PTPN, 2005, 471 p.

4017. Überlebenden (Die) des deutschen Widerstandes und ihre Bedeutung für Nachkriegsdeutschland. Hrsg. v. Joachim SCHOLTYSECK und Stephen SCHRÖDER. Münster, Lit, 2005, II-153 p. (Schriftenreihe der Forschungsgemeinschaft 20. Juli, 6).

4018. Unmasking Hitler. Cultural representations of Hitler from the Weimar Republic to the present. Ed. by Klaus L. BERGHAHN and Jost HERMAND. Oxford, Peter Lang, 2005, 264 p. (German life and civilization, 44).

4019. VATLIN (Aleksandr Ju.). Germanija v XX veke. (Germany in the 20th century). Moskva, ROSSPEN, 2005, 332 p. (bibl. p. 309-313; pers. ind. p. 328-333).

4020. VOSSLER (Frank). Propaganda in die eigene Truppe: die Truppenbetreuung in der Wehrmacht 1939–1945. Paderborn, Ferdinand Schöningh, 2005, 430 p. (Krieg in der Geschichte, volume 21).

4021. WEGMANN (Bodo). Die Militäraufklärung der NVA: die zentrale Organisation der militärischen Aufklärung der Streitkräfte der Deutschen Demokratischen Republik. Berlin, Köster, 2005, VIII-714 p. (Beiträge zur Friedensforschung und Sicherheitspolitik, 22).

4022. "Wider die Kriegsmaschinerie": Kriegserfahrungen und Motive des Widerstandes der "Weissen Rose". Hrsg. v. Detlef BALD; mit einem Vorwort von Eugen BISER. Essen, Klartext, 2005, 211 p. (ill., maps).

4023. Wirtschaft, Politik und Freiheit. Freiburger Wirtschaftswissenschaftler und der Widerstand. Hrsg. v. Nils GOLDSCHMIDT. Tübingen, Mohr Siebeck, 2005, XVII-510 p. (Untersuchungen zur Ordnungstheorie und Ordnungspolitik, 48).

4024. WREDE (Martin). Der Kaiser, das Reich, die deutsche Nation – und ihre "Feinde". Natiogenese, Reichsidee und der "Durchbruch des Politischen" im Jahrhundert nach dem Westfälischen Frieden. *Historische Zeitschrift*, 2005, 280, 1, p. 83-116.

Cf. n°s 647, 3797, 3806, 3866, 4127, 4215, 4429, 4490, 4550, 4642, 5732, 6313, 6443, 6466, 6691, 6797, 6993, 7056, 7140, 7363, 7492

Ghana

** 4025. Danish sources for the history of Ghana 1657–1754. Vol. 1. 1657–1735. Vol. 2. 1735–1754. Ed. by Ole JUSTESEN. København, Royal Danish Academy of Sciences and Letters, 2005, 2 vol., XXXVIII-1058 p. (UAI Fontes Historiae Africanae. Series Varia, 8).

4026. BOTWE-ASAMOAH (Kwame). Kwame Nkrumah's politico-cultural thought and policies: an African-centered paradigm for the second phase of the African revolution. New York a. London, Routledge, 2005, XVII-242 p. (African studies: history, politics, economics, and culture).

4027. GOCKING (Roger S.). The history of Ghana. Westport a. London, Greenwood Press, 2005, XXXIII-331 p. (ill., maps). (The Greenwood histories of the modern nations).

4028. OWUSU (Robert Yaw). Kwame Nkrumah's liberation thought: a paradigm for religious advocacy in contemporary Ghana. Trenton, Africa World Press, 2005, [s. p.].

Great Britain

** 4029. Calendar of state papers. Preserved in the Public Record Office. Vol. 3. Domestic series of the reign of Anne. 1704–1705. Compiled by Alexander R. RUMBLE, C. DIMMER [et al.]; edited for the press by C. S. KNIGHTON. Woodbridge, Boydell Press, 2005, XL-671 p.

** 4030. Clarke papers (The). Further selections from the papers of William Clarke. Cambridge, Cambridge U. P., 2005, 400 p.

** 4031. COLVILLE (John Rupert, Sir.). The fringes of power: Downing Street diaries, 1939–1955. London, Phoenix, 2005, XIII-738 p. (ill.).

** 4032. DONOUGHUE (Bernard). Downing Street diary: with Harold Wilson in No. 10. London, Jonathan Cape, 2005, XII-785 p.

** 4033. Duff Cooper diaries 1915–1951 (The). Ed. by John Julius NORWICH. London, Weidenfeld & Nicolson, 2005, XIII-512 p. (ill., ports.).

** 4034. Horatio Nelson, the new letters. Ed. by Colin WHITE. Woodbridge a. Rochester, Boydell Press, National Maritime Museum, Royal Naval Museum, 2005, XXIX-525 p.

** 4035. Neville Chamberlain diary letters (The). Vol. 4. The Downing Street years, 1934–1940. Ed. by Robert SELF. Aldershot, Ashgate, 2005, XI-588 p.

** 4036. Newsletters from the Caroline court, 1631–1638. Catholicism and the politics of the personal rule. Ed. by Michael C. QUESTIER. Cambridge, New York a. Melbourne, Cambridge U. P., 2005, XVI-358 p. (Camden fifth series, 26).

** 4037. Proceedings in the opening session of the Long Parliament: House of Commons. Vol. 5. 7 June–17 July 1641. Vol. 6. 19 July–9 September 1641. Ed. by Maija JANSSON; assisted by Alisa PLANT. Rochester a. Woodbridge, University of Rochester Press, 2005, 2 vol., XXIV-689 p., XII-721 p.

4038. ADDISON (Paul). Churchill, the unexpected hero. Oxford, New York a. Auckland, Oxford U. P., 2005, 308 p.

4039. Anglo-Scottish relations from 1603 to 1900. Ed. by T. Christopher SMOUT. Oxford, Oxford U. P., 2005, X-281 p. (ports.). (Proceedings of the British Academy, 127).

4040. Anglo-Scottish relations from 1900 to devolution and beyond. Ed. by William L. MILLER. Oxford, Published for the British Academy by Oxford U. P., 2005, XI-272 p. (ill.). (Proceedings of the British Academy, 128).

4041. BARR (Niall). The lion and the poppy: British veterans, politics, and society, 1921–1939. Westport a. London, Praeger, 2005, 228 p.

4042. BERGER (Stefan), LAPORTE (Norman). Britische Parlamentarierkontakte nach Osteuropa 1945–1989. Zwischen fellow travelling und ostpolitischer Erneuerung. *Archiv für Sozialgeschichte*, 2005, 45, p. 3-42.

4043. BEST (Geoffrey). Churchill and war. London, Hambledon, 2005, XIII-353 p.

4044. BOR (Michael). The Socialist League in the 1930s. London, Athena Press, 2005, 420 p. (ill.).

4045. British fascism, the labour movement, and the state. Ed. by Nigel COPSEY and David RENTON. Houndmills, Palgrave Macmillan, 2005, VII-209 p.

4046. CHALUS (Elaine). Elite women in English political life, c. 1754–1790. Oxford, Clarendon Press, 2005, VII-278 p. (Oxford historical monographs).

4047. COLCLOUGH (David). Freedom of speech in early Stuart England. Cambridge, Cambridge U. P., 2005, 293 p.

4048. CURRAN (James), GABER (Ivor), PETLEY (Julian). Culture wars: the media and the British left. Edinburgh, Edinburgh U. P., 2005, 316 p.

4049. CUST (Richard). Charles I: a political life. New York, Longman, 2005, XII-488 p.

4050. DE KREY (Gary S.). London and the restoration, 1659–1683. New York, Cambridge U. P., 2005, XIX-472 p. (Cambridge studies in Early Modern British history).

4051. DOUGLAS (Roy). Liberals: a history of the Liberal and Liberal Democratic parties. London, Hambledon and London, 2005, XII-395 p. (ill.).

4052. EDGERTON (David). Warfare state. Britain, 1920–1970. Cambridge, Cambridge U. P., 2005, XV-364 p.

4053. FAIRBURN (Miles), HASLETT (Stephen). Voter behaviour and the decline of the Liberals in Britain and New Zealand, 1911–1929: some comparisons. *Social history*, 2005, 30, 2, p. 195-217.

4054. FEDOROV (Sergej E.). Rannestjuartovskaja aristokratija (1603–1629). (The aristocracy of early Stuarts, 1603–1629). Sankt-Peterburg, Aletejja, 2005, 525 p. (ill.; plates; bibl. p. 500-521). (Biblioteka Srednikh vekov).

4055. FIELDHOUSE (Roger). Anti-apartheid: a history of the movement in Britain, 1959–1994; a study in pressure groups politics. London, Merlin Press, 2005, XIV-546 p.

4056. FRENCH (David). Military identities: the regimental system, the British Army, and the British people, c. 1870–2000. Oxford, Oxford U. P., 2005, X-404 p.

4057. FRY (Geoffrey Kingdon). The politics of decline: an interpretation of British politics from the 1940s to the 1970s. Basingstoke, Palgrave Macmillan, 2005, X-307 p.

4058. GONZÁLEZ ADÁNEZ (Noelia). Crisis de los imperios: monarquía y representación política en Inglaterra y España: 1763–1812. Madrid, Centro de Estudios Políticos y Constitucionales, 2005, 308 p. (Historia de la sociedad política).

4059. GOODRICH (Amanda). Debating England's aristocracy in the 1790s: pamphlets, polemics and political ideas. Rochester, Boydell Press for the Royal Historical Society, 2005, X-213 p. (Royal historical society studies in history new series).

4060. GROVE (Eric). The Royal Navy since 1815: a new short history. Basingstoke, Palgrave Macmillan, 2005, X-300 p.

4061. HARRIS (Tim). Restoration. Charles II and his kingdoms, 1660–1685. London, Allan Lane, Penguin, 2005, XXI-506 p.

4062. HAY (William Anthony). The Whig revival, 1808–1830. New York, Palgrave Macmillan, 2005, XVI-235 p. (Studies in modern history).

4063. JARVIS (Mark). Conservative governments, morality and social change in affluent Britain, 1957–1964. Manchester, Manchester U. P., 2005, X-187 p.

4064. KISHLANSKY (Mark). Charles I: a case of mistaken identity. *Past and present*, 2005, 189, p. 41-80.

4065. KNIGHTS (Mark). Representation and misrepresentation in later Stuart Britain: partisanship and political culture. Oxford: Oxford U. P., 2005, XVI-431 p.

4066. Kūkan no igirisushi. (History of space in Britain). Ed. By Minoru KAWAKITA and Takao FUJIKAWA. Tokyo, Yamakawa Shuppansha, 2005, 295 p.

4067. Labour's grass roots. Essays on the activities of local Labour parties and members, 1918–1945. Ed. by Matthew WORLEY. Aldershot, Ashgate, 2005, IX-267 p. (Studies in Labour history).

4068. LEE (Stephen J.). Gladstone and Disraeli. London, Routledge, 2005, XIV-194 p. (ill.). (Questions and analysis in history).

4069. MAC CARTNEY (Helen B.). Citizen soldiers: the Liverpool territorials in the First World War. New York, Cambridge U. P., 2005, XV-275 p. (Studies in the social and cultural history of modern warfare, 22).

4070. MAC LEAN (Iain), MAC MILLAN (Alistair). State of the union. Unionism and the alternatives in the United Kingdom since 1707. Oxford, Oxford U. P., 2005, 283 p.

4071. MAC PHILLIPS (Kevin). Joseph Burgess (1853–1934) and the founding of the Independent Labour Party. Lewiston a. Lampeter, Edwin Mellen Press, 2005, XII-201 p. (ill.). (Studies in British history, 78).

4072. MACINNES (Allan I.). The British revolution, 1629–1660. Basingstoke, Palgrave Macmillan, 2005, XI-337 p. (British studies series).

4073. MEARS (Natalie). Queenship and political discourse in the Elizabethan realms. Cambridge, New York a. Melbourne, Cambridge U. P., 2005, XIV-311 p.

4074. MERGEL (Thomas). Großbritannien seit 1945. Göttingen, Vandenhoeck & Ruprecht, 2005, 229 p. (Europäische Zeitgeschichte, 1).

4075. MITCHELL (Leslie George). The Whig world: 1760–1837. London, Hambledon Continuum, 2005, XI-211 p. (ill.).

4076. NEWTON (Diana). The making of the Jacobean regime. James VI and I and the government of England, 1603–1605. Rochester, Boydell Press for the Royal Historical Society, 2005, X-164 p. (Studies in history new series).

4077. Parliament and dissent. Ed. by Stephen TAYLOR and David L. WYKES. Edinburgh, Edinburgh University Press for the Parliamentary History Yearbook Trust, 2005, VIII-156 p. (Parliamentary history, 24, pt. 1).

4078. Parliament and politics in Scotland, 1567–1707. Ed. by Keith M. BROWN and Alastair J. MANN. Edinburgh, Edinburgh U. P., 2005, XI-303 p. (ill.). (History of the Scottish Parliament, 2).

4079. Partisan politics, principle and reform in parliament and the constituencies, 1689–1880. Essays in memory of John Phillips. Ed. by Clyve JONES, Philip SALMON and Richard W. DAVIS. Edinburgh, Edinburgh U. P., 2005, XXX-213 p. (Parliamentary history, 24).

4080. Political thought (The) of the Conservative Party since 1945. Ed. by Kevin HICKSON. Houndmills a. New York, Palgrave Macmillan, 2005, IX-232 p.

4081. PUGH (Martin). "Hurrah for the blackshirts!" Fascists and fascism in Britain between the wars. London, Jonathan Cape, 2005, XII-387 p. (ill.).

4082. RODGER (N. A. M.). The command of the ocean: a naval history of Britain, 1649–1815. New York, W. W. Norton, 2005, LXV-907 p.

4083. RUSSELL (Andrew T.), FIELDHOUSE (Edward). Neither left nor right? The Liberal Democrats and the electorate. Manchester a. New York, Manchester U. P., 2005, VII-272 p. (ill.).

4084. SALIBA (Fabrice). Les politiques de recrutement militaire britannique et française (1920–1939): chronique d'un désastre annoncé. Préf. par William J. PHILPOTT et Frédéric ROUSSEAU. Paris, L'Harmattan, 2005, 283 p. (Histoire de la défense).

4085. SANKEY (Margaret). Jacobite prisoners of the 1715 rebellion. Preventing and punishing insurrection in early Hanoverian Britain. Aldershot, Ashgate, 2005, XIX-196 p.

4086. SAUNDERS (Robert). Lord John Russell and Parliamentary Reform, 1848–1867. *English historical review*, 2005, 120, 489, p. 1289-1315.

4087. Scotland in the age of the French Revolution. Ed. by Bob HARRIS. Edinburgh, John Donald, 2005, X-278 p.

4088. SMITH (Angela K.). Suffrage discourse in Britain during the First World War. Aldershot, Ashgate, 2005, 153 p.

4089. STOYLE (Mark). Soldiers and strangers: an ethnic history of the English Civil War. New Haven, Yale U. P., 2005, X-297 p.

4090. SYKES (Alan). The Radical Right in Britain: Social Imperialism to the BNP [British National Party]. Palgrave Macmillan, Basingstoke 2005, XII-184 p. (British history in perspective).

4091. TAYLOR (Andrew). The NUM [National Union of Mineworkers] and British politics. Vol. 2. 1969–1995. Aldershot a. Burlington, Ashgate, 2005, XIV-357 p. (Studies in labour history).

4092. THOMAS (James). Popular newspapers, the Labour Party and British politics. London, Routledge, 2005, 222 p. (British politics and society).

4093. THOMAS (Matthew). Anarchist ideas and counter-cultures in Britain, 1880–1914: revolutions in everyday life. Aldershot, Ashgate, 2005, VIII-264 p. (ill.).

4094. TROOST (Wout). William III, the stadholder-king. A political biography. Aldershot a. Burlington, Ashgate, 2005, XVII-361 p.

4095. VALLANCE (Edward). Revolutionary England and the national covenant. State oaths, Protestantism, and the political nation, 1553–1682. Rochester, Boydell Press, 2005, VIII-263 p.

4096. WARD (Paul). Unionism in the United Kingdom, 1918–1974. New York, Palgrave Macmillan, 2005, XI-243 p.

4097. WEST (Nigel). MASK: MI5's penetration of the Communist Party of Great Britain. London, Routledge, 2005, XII-324 p.

4098. WITHINGTON (Phil). The politics of Commonwealth: citizens and freemen in Early Modern England. New York, Cambridge U. P., 2005, XIV-298 p. (Cambridge social and cultural histories, 4).

4099. WORLEY (Matthew). Labour inside the gate: a history of the British Labour Party between the wars. London a. New York, I.B. Tauris, 2005, 278 p. (International library of political studies, 4).

Cf. n°s 3617, 3639, 3861, 4171, 4175, 6932, 6939, 7074, 7102

Greece

4100. AVDELA (Efi). Engendering "Greekness": women's emancipation and irredentist politics in nineteenth-century Greece. *Mediterranean historical review*, 2005, 20, 1, p. 67-81.

4101. Citizenship and the nation-state in Greece and Turkey. Ed. by Faruk BIRTEK and Thalia DRAGONAS. London, Routledge, 2005, XV-195 p. (Social and historical studies on Greece and Turkey).

4102. GEORGAKIS (Angelo). Ottoman Salonika and Greek Nationalism before 1908. *Balkan Studies*, 2005, 1, p. 111-138.

4103. LAGOS (Katerina). The Metaxas dictatorship and Greek Jewry, 1936–1941. Oxford, [s. n.], 2005, X-311 p. [Thesis (D. Phil.) – University of Oxford, 2005].

4104. PETROV (Bisser). The problem of collaboration in Post-war Greece 1944–1946. *Balkan Studies*, 2005, 3, p. 15-36.

4105. ULUNJAN (Arutjun A.). Posle rezhima. Grecija: ot voenno-politicheskoj sistemy k parlamentskoj demokratii (1974–1981 gg.). (Greece: From military political system to parliamentary democracy, 1974–1981). Moskva, IVI RAN, 2005, 241 p. (ill.; maps; bibl. p. 229-239). (Evropejskie sravnitel'no istoricheskie issledovanija Intstituta vseobshchej istorii RAN).

Guatemala

4106. GARCÍA (Prudencio). El genocidio de Guatemala: a la luz de la sociología militar. Madrid, Sepha Edición y Diseño, 2005, 514 p. (Libros abiertos, 1).

4107. MURGA ARMAS (Jorge). Iglesia católica, movimiento indígena y lucha revolucionaria: Santiago Atitlán, Guatemala. Guatemala, Instituto de Investigaciones Económicas y Sociales, IIES, Facultad de Ciencias Económicas, Universidad de San Carlos de Guatemala, 2005, 219 p.

Cf. n° 6820

Guinea

4108. CAMARA (Mohamed Saliou). His master's voice: mass communication and single-party politics in Guinea under Sékou Touré. Trenton, Africa World Press, 2005, XVI-217 p.

4109. SCHMIDT (Elizabeth). Mobilizing the masses: gender, ethnicity, and class in the nationalist movement in Guinea, 1939–1958. Portsmouth, Heinemann, 2005, XV-293 p. (Social history of Africa series). – EADEM. Top down or bottom up? Nationalist mobilization reconsidered, with special reference to Guinea (French West Africa). *American historical review*, 2005, 110, 4, p. 975-1014.

Haiti

4110. FARRAUDIÈRE (Yvette). La naissance d'Haïti à la croisée de trois voies révolutionnaires. Paris, L'Harmattan, 2005, 248 p. (ill., maps).

4111. FLEURIMOND (Wiener Kerns). Haïti, 1804–2004: le bicentenaire d'une révolution oubliée. Préf. de Antoine Fritz PIERRE. Paris, L'Harmattan, 2005, 204 p. (map). (Recherches Amériques latines).

4112. GIRARD (Philippe R.). Paradise lost. Haiti's tumultuous journey from pearl of the Caribbean to third world hotspot. Basingstoke, Palgrave Macmillan, 2005, X-230 p.

Cf. n° 6930

Honduras

4113. BARAHONA (Marvin). Honduras en el siglo XX: una síntesis histórica. Tegucigalpa, Editorial Guaymuras, 2005, 376 p. (ill.). (Colección Códices. Ciencias sociales).

4114. DODD (Thomas J.). Tiburcio Carías: portrait of a Honduran political leader. With a foreword by Douglas BRINKLEY. Baton Rouge, Louisiana State U. P.,

2005, XVII-268 p. (ill., map). (Eisenhower Center studies on war and peace).

Hungary

** 4115. Revolution (Von der) zur Reaktion. Quellen zur Militärgeschichte der ungarischen Revolution, 1848-1849. Hrsg. Christoph TEPPERBERG und Jolán SZIJJ. Budapest u. Wien, Argumentum, 2005, CX-878 p.

4116. BÉRENGER (Jean), KECSKEMÉTI (Charles). Parlement et vie parlementaire en Hongrie: 1608-1918. Paris, H. Champion, 2005, 570 p. (maps). (Bibliothèque d'histoire moderne et contemporaine, 15).

4117. BIHARI (Mihály). Magyar politika 1944- 2004: Politikai és hatalmi viszonyok. (Hungarian politics 1944-2004: Political and power structures). Budapest, Osiris, 2005, 485 p.

4118. DÁVID (Géza). Pasák és bégek uralma alatt: demográfiai és közigazgatás-történeti tanulmányok. (Under the rule of pashas and officials: demographical and public administration studies). Budapest, Akadémiai Kiadó, Magyar-Török Baráti Társaság, 2005, 379 p. (ill., maps).

4119. HODOS (George H.). Tettesek és áldozatok: koncepciós perek Magyarországon és Közép-Kelet Európában. (Delinquents and victims: Concept causes in Hungary and in Central-Eastern Europe). Budapest, Noran, 2005, 379 p. (ill.).

4120. KELÉNYI (György). A királyi udvar építkezései Pest-Budán a XVIII. században: hatalom és reprezentáció; a hivatalos építészet formaváltozásai. (The constructions of the royal court in Pest-Buda in the 18th century: authority and representation: changes in the style of the official architecture). Budapest, Akadémiai Kiadó, 2005, 98 p. (Művészettörténeti füzetek, 28).

4121. KOZÁRI (Monika). A dualista rendszer, 1867-1918. Budapest, Pannonica Kiadó, 2005, 319 p. (map). (Modern magyar politikai rendszerek).

4122. MARK (James). Society, resistance and revolution: the Budapest middle class and the Hungarian Communist state 1948-1956. *English historical review*, 2005, 120, 488, p. 963-986.

4123. MEVIUS (Martin). Agents of Moscow. The Hungarian Communist Party and the origins of socialist patriotism 1941-1953. Clarendon Press of Oxford U. P., 2005. XV-296 p. (Oxford historical monographs).

4124. MITEV (J.). Istorija na Avstro-Yngarija 1867-1918. (Histoire de l'Autriche-Hongrie 1867-1918). Veliko Tărnovo, [s. n.], 2005, 312 p.

4125. Ráday Pál és a Rákóczi-szabadságharc. (Pál Ráday and the Rákóczi-civil war). Szerkesztette és az előszót írta RÁDAY-PESTHY Pál Frigyes; a tanulmányok szerzői CZIGÁNY István [et al.]. Budapest, Akadémiai Kiadó, 2005, 105 p.

4126. Regimes and transformations: Hungary in the twentieth century. Ed. by István FEITL and Balázs SIPOS. Budapest, Napvilág Kiadó, Institute of Political History, 2005, 464 p.

4127. SAXONBERG (Steven). The fall: a comparative study of the end of communism in Czechoslovakia, East Germany, Hungary, and Poland. London a. New York, Routledge, 2005, XVII-434 p. (International studies in global change, 11).

4128. SIMÁNDI (Irén). Magyarország a Szabad Európa Rádió hullámhosszán, 1951-1956. (Hungary in the Free Europe Radio Channel, 1951-1956). Budapest, Országos Széchényi Könyvtár, Gondolat Kiadó, 2005, 346 p. (ill.). (Nemzeti téka).

4129. SZÁNTAY (Antal). Regionalpolitik im alten Europa: die Verwaltungsreformen Josephs II. in Ungarn, in der Lombardei und in den österreichischen Niederlanden, 1785-1790. Budapest, Akadémiai Kiadó, 2005, 490 p. (maps).

4130. SZIJÁRTÓ (István M.). A diéta. A magyar rendek és az országgyűlés, 1708-1792. (The Diet. The Estates and the Parliament of Hungary, 1708-1792). Budapest, Osiris Kiadó, 2005, 614 p.

Cf. n°s 3625, 6100

Iceland

4131. HERMANNSSON (Birgir). Understanding nationalism: studies in Icelandic nationalism, 1800-2000. Stockholm, Department of Political Science, Stockholm University, 2005, 375 p. (Stockholm studies in politics, 110).

India

** 4132. Punjab politics, 1940-1943: strains of war: governors' fortnightly reports and other key documents. Compiled and edited by Lionel CARTER. New Delhi, Manohar, 2005, 427 p. (map).

4133. AHUJA (M. L.). General elections in India: electoral politics, electoral reforms, and political parties. New Delhi, Icon Publishers, 2005, XVI-660 p. (ill., maps).

4134. ALAM (Abu Yusuf). Muslims and Bengal politics (1912-1924). Kolkata, Raktakarabee, 2005, VIII-440 p.

4135. ANNOUSSAMY (David). L' intermède français en Inde: secousses politiques et mutations juridiques. Pondichéry, Institut français de Pondichéry et Paris, L'Harmattan, 2005, X-412 p. (maps). (Collection Droits et cultures, 11).

4136. BATABYAL (Rakesh). Communalism in Bengal: from famine to Noakhali, 1943-1947. New Delhi a. London, Sage, 2005, 418 p. (Sage series in modern Indian history, 6).

4137. CHISHTI (S. M. A. W.). Political development in Manipur, 1919-1949. Delhi, Kalpaz Publications, 2005, 340 p.

4138. Coalition politics and Hindu nationalism. Ed. by Katharine ADENEY and Lawrence SAEZ. London,

Routledge, 2005, XVI-294 p. (Routledge advances in South Asian studies, 2).

4139. COPLAND (Ian). State, community and neighbourhood in princely North India, c. 1900–1950. Houndmills, Palgrave Macmillan, 2005, XII-259 p. (maps).

4140. DHILLON (Kirpal S.). Police and politics in India: colonial concepts, democratic compulsions, India police 1947–2002. Foreword by Richard H. WARD. New Delhi, Manohar, 2005, 619 p.

4141. JOYKUMAR SINGH (N.). Revolutionary movements in Manipur. New Delhi, Akansha Pub. House, 2005, XVI-189 p.

4142. KRISHAN (Shri). Political mobilization and identity in western India, 1934–1947. New Delhi a. London, Sage, 2005, 279 p. (map). (Sage series in modern Indian history, 7).

4143. KUMARI (Saroj). Role of women in the freedom movement in Bihar, 1912–1947. Patna, Janaki Prakashan, 2005, 195 p.

4144. MANI (Braj Ranjan). Debrahmanising history: dominance and resistance in Indian society. New Delhi, Manohar, 2005, 456 p.

4145. MEHTA (Jaswant Lal). Advanced study in the history of modern India, 1707–1813. Slough a. Elgin, New Dawn Press, 2005, XX-739 p. (maps).

4146. NOORANI (Abdul Gafoor Abdul Majeed). Indian political trials, 1775–1947. New Delhi a. New York, Oxford U. P., 2005, XIV-316 p. (ill.). (Oxford India paperbacks).

4147. ROTHERMUND (Dietmar). Gandhi und Nehru: Kontrastierende Visionen Indiens. *Geschichte und Gesellschaft*, 2005, 31, 3, p. 354-372.

4148. SARILA (Narendra Singh). The shadow of the great game: the untold story of India's partition. New Delhi, HarperCollins Publishers India a joint venture with India Today Group, 2005, 436 p. (ill., maps).

4149. SHAHID (Kamran). Gandhi and the partition of India: a new perspective. Lahore, Ferozsons, 2005, 124 p.

4150. SHARMA (Shalini). The radical response to colonialism: the organised Left in Punjab 1920–1947. London, School of Oriental and African Studies, 2005, 264 p.

4151. TANEJA (Anup). Gandhi, women, and the national movement, 1920–1947. New Delhi, Har-Anand Publications, 2005, 244 p.

4152. TANWAR (Shyam Singh). State administration in Rajasthan (19th century): (with special reference to Jodhpur State). Jodhpur, Maharaja Mansingh Pustak Prakash Research Centre Fort, 2005, 182 p.

4153. VIDYASAGAR (K.). Communist politics in India: struggle for survival. Delhi, Academic Excellence, 2005, X-205 p.

Cf. n°s 6412, 6464, 6742, 7549

Indonesia

4154. ASPINALL (Edward). Opposing Suharto: compromise, resistance, and regime change in Indonesia. Stanford, Stanford U. P., 2005, XIII-328 p. (Contemporary issues in Asia and the Pacific).

4155. GIEBELS (Lambert J.). De stille genocide: de fatale gebeurtenissen rond de val van de Indonesische president Soekarno. Amsterdam, B. Bakker, 2005, 303 p. (ill., maps, ports.).

4156. MASUHARA (Ayako). Indoneshia-Suharuto taisei shoki no daitōryō to zantei kokumin kyōgikai. (MPRS [Madjelis Permusjawaratan Rakjat Sementara] and the presidents in the early years of the Suharto Regime). *Ajia Keizai*, 2005, 45, 10, p. 2-23.

4157. VICKERS (Adrian). A history of modern Indonesia. Cambridge, Cambridge U. P., 2005, XIV-291 p. (ill., maps).

Cf. n° 6739

Iran

4158. AFARY (Janet), ANDERSON (Kevin B.). Foucault and the Iranian Revolution: gender and the seductions of Islamism. Chicago, University of Chicago Press, 2005, 358 p.

4159. CLAWSON (Patrick), RUBIN (Michael). Eternal Iran: continuity and chaos. Basingstoke, Palgrave Macmillan, 2005, 203 p. (maps). (Middle East in focus series).

4160. DAVARI (Mahmood T.). The political thought of Ayatullah Murtaża Muaḥhari: an Iranian theoretician of the Islamic state. New York, Routledge Curzon, 2005, 212 p. (RoutledgeCurzon/BIPS Persian studies series, 5).

4161. MARTIN (Vanessa). The Qajar pact. Bargaining, protest and the state in nineteenth-century Persia. London a. New York, I. B. Tauris, 2005, X-214 p. (International library of Iranian studies, 4).

4162. MOALLEM (Minoo). Between warrior brother and veiled sister: Islamic fundamentalism and the politics of patriarchy in Iran. Berkeley, University of California Press, 2005, IX-269 p.

4163. NEWMAN (Andrew J.). Safavid Iran: Persia between the medieval and the modern. London, I.B. Tauris, 2005, 272 p. (ill.). (Library of Middle East history, 5).

4164. POULSON (Stephen C.). Social movements in twentieth century Iran. Culture, ideology, and mobilizing frameworks. Lanham, Lexington Books, 2005, 349 p.

4165. RIDGEON (Lloyd). Religion and politics in modern Iran. London, I. B. Tauris, 2005, XIV-279 p.

4166. SIEBERTZ (Roman). Die Briefmarken Irans als Mittel der Politischen Propaganda. Vienna, Verlag der Österreichischen Akademie der Wissenschaften, 2005,

274 p. (Österreichische Akademie der Wissenschaften, Philosophisch-Historische Klasse, Sitzungsberichte 722. Veröffentlichungen zur Iranistik 32).

4167. TAQAVĪ (Muhammad'Alī). The flourishing of Islamic reformism in Iran: political Islamic groups in Iran (1941–1961). London, RoutledgeCurzon, 2005, XI-177 p. (RoutledgeCurzon studies in political Islam, 1).

Cf. n° 4169

Iraq

4168. Iraq: defence intelligence 1920–1973. Vol. 1. 1920–1925. Vol. 2. 1926–1932. Vol. 3. 1933–1941. Vol. 4. 1942–1957. Vol. 5. 1958–1973. Vol. 6. Maps. Ed. by Anita L. P. BURDETT. Slough, Archive Editions, 2005, 6 vol., XXXVII-836 p., XXXVIII-739 p., XXXVII-886 p., XXVI-702 p., XXXVI-716 p., 11 maps

4169. NATALI (Denise). The Kurds and the state: evolving national identity in Iraq, Turkey, and Iran. Syracuse, Syracuse U. P., 2005, 268 p. (Modern intellectual and political history of the Middle East).

4170. VISSER (Reidar). Basra, the failed gulf state. Separatism and nationalism in Southern Iraq. Münster, Lit., 2005, X-238 p. (Politik. Forschung und Wissenschaft).

Cf. n° 350

Ireland

4171. ARMSTRONG (Robert). Protestant war. The "British" of Ireland and the wars of the three kingdoms. Manchester a. New York, Manchester U. P., 2005, VIII-261 p.

4172. British interventions in early modern Ireland. Ed. by Ciaran BRADY and Jane OHLMEYER. Cambridge, New York a. Port Melbourne, Cambridge U. P., 2005, XX-371 p.

4173. CAMPBELL (Fergus). Land and revolution: nationalist politics in the west of Ireland, 1891–1921. New York, Oxford U. P., 2005, XVI-356 p.

4174. CRAWFORD (Jon G.). A Star Chamber Court in Ireland. The Court of Castle Chamber, 1571–1641. Dublin, Four Courts Press, 2005, XVI-655 p. (The Irish Legal History Society series, 15).

4175. FLEMING (Neil C.). The Marquess of Londonderry: aristocracy, power and politics in Britain and Ireland. London, Tauris Academic Studies, 2005, 281 p. (International library of twentieth century history, 5).

4176. GARVIN (Tom). Nationalist revolutionaries in Ireland, 1858–1928. Dublin, Gill & Macmillan, 2005, VI-200 p.

4177. HOLTHUSEN (Christoph). Der Nordirlandkonflikt: Geschichte, zentrale Aspekte und Lösungsmodelle unter völkerrechtlicher Betrachtung. Frankfurt am Main u. Oxford, P. Lang, 2005, 378 p. (ill., maps). (Schriften zum Staats- und Völkerrecht, 116).

4178. KISSANE (Bill). The politics of the Irish Civil War. Oxford, Claredon Press, 2005, 264 p.

4179. Lemass era (The): politics and society in the Ireland of Seán Lemass. Ed. by Brian GIRVIN and Gary MURPHY. Dublin, University College Dublin Press, 2005, XX-265 p. (ill.).

4180. LYNCH (David). Radical politics in modern Ireland: the Irish Socialist Republican Party, 1896–1904. Dublin, Irish Academic Press, 2005, X-182 p. (ill.).

4181. MAC GEE (Owen). The IRB: the Irish Republican Brotherhood, from the Land League to Sinn Féin. Dublin, Four Courts Press, 2005, 384 p. (ill.).

4182. MANSERGH (Danny). Grattan's failure: parliamentary opposition and the people in Ireland, 1779–1800. Dublin, Irish Academic Press, 2005, VIII-337 p.

4183. Nineteenth-century Ireland: a guide to recent research. Ed. by Laurence M. GEARY and Margaret KELLEHER. Dublin, University College Dublin Press, 2005, XII-340 p. (ill., maps).

4184. Politics and the Irish working class, 1830–1945. Ed. by Fintan LANE and Donal Ó DRISCEOIL. Houndmills, Palgrave Macmillan, 2005, XI-295 p. (map).

4185. POWELL (Martyn J.). The politics of consumption in eighteenth-century Ireland. Basingstoke a. New York, Palgrave Macmillan, 2005, VIII-293 p.

4186. Ulster crisis (The), 1885–1921. Ed. by D. George BOYCE and Alan O'DAY. London, Palgrave Macmillan, 2005, 368 p.

4187. Was Ireland a colony? Economics, politics, and culture in nineteenth-century Ireland. Ed. by Terrence MAC DONOUGH. Dublin a. Portland, Irish Academic Press, 2005, XIV-356 p.

4188. WHEATLEY (Michael). Nationalism and the Irish party. Provincial Ireland, 1910–1916. Oxford, Oxford U. P., 2005, 295 p.

Israel

** 4189. Yitshak Rabin: rosh memshelet Yiśra'el 1974–1977, 1992–1995: mivhar te'udot mi-pirke ha-yav. (Yitzhak Rabin: prime minister of Israel 1974–1977, 1992–1995). Kerekh 1. 1922–1967. 'Arkhah ve-khatvah mevo'ot, Yemimah ROZENTAL. Yerushalayim, Ganzakh ha-medinah, 2005, 618 p. (ill., maps). (Sidrah le-hantsahhat zikhram shel neśi'e Yiśra'el ve-rashe memshelotcha).

4190. BAR-GIL (Shlomo). Be-reshit hayah halom: bogre tenu'ot ha-no'ar ha-halutsiyot me-Amerikah ha-Latinit ba-tenu'ah ha-kibutsit, 1946–1967. (We started with a dream: graduates of Latin American youth movements in the kibbutz movement 1946–1967). [S. l.], Mekhon Ben Guryon, Universitat Ben-Guryon ba-Negev, Yad Tabenkin, 2005, 160 p. (ill.).

4191. DERORI (Ze'ev). Israel's reprisal policy, 1953–1956: the dynamics of military retaliation. London, Frank Cass, 2005, VIII-232 p. (Cass military studies).

4192. KRAKAU (Constanze). Die Rolle der palästinensischen Minderheit im politischen Leben Israels 1976–1996. Münster, Lit, 2005, VIII-185 p. (Studien zur Zeitgeschichte des Nahen Ostens und Nordafrikas, 14).

4193. RABINOWITZ (Dan), ABU-BAKER (Khawla). Coffins on our shoulders [electronic resource]: the experience of the Palestinian citizens of Israel. Berkeley, University of California Press, 2005, XI-221 p.

4194. SHALOM (Zakai). Ben-Gurion's political struggles: a lion in winter. London a. New York, Routledge, 2005, 142 p. (Israeli history, politics, and society, 44).

4195. YIZHAR (Uri). Ben hazon le-shilton: mifleget Ahdut ha-'avodah-Po'ale Tsiyon bi-tekufat ha-yishuv veha-medinah. (Between vision and power: the history of Ahdut-Ha'avoda-Poalei-Zion Party). Ef'al, Yad Tabenkin, Merkaz mehkari, ra'yoni ve-ti'udi shel ha-tenu'ah ha-kibutsit, 2005, 791 p.

Cf. n° 7464

Italy

* 4196. Bibliografia dell'età del Risorgimento, 1970-2001. Vol. 4. Indici. [Vol. 1-3. Cf. Bibl. 2003, n° 4400.] A cura di Emma MOSCATI. Firenze, Olschki, 2005, 223 p. (Biblioteca di bibliografia italiana, 176).

** 4197. Italia (L') del Novecento. Le fotografie e la storia. Vol. 1. 1. Il potere da Giolitti a Mussolini (1900–1945). Vol. 1. 2. Il potere da De Gasperi a Berlusconi (1945–2000). A cura di Giovanni DE LUNA, Gabriele D'AUTILIA e Luca CRISCENTI. Torino, Einaudi, 2005, 2 vol., 358 p., 568 p.

4198. Alcide De Gasperi: un percorso europeo. A cura di Eckart CONZE, Gustavo CORNI e Paolo POMBENI. Bologna, Il Mulino, 2005, 306 p.

4199. ARFÉ (Gaetano). Scritti di storia e politica. A cura di Giuseppe ARAGNO. Napoli, La città del Sole, 2005, 425 p.

4200. ARMANI (Barbara). Italia anni settanta. Movimenti, violenza politica e lotta armata tra memoria e rappresentazione storiografica. *Storica*, 2005, 32, p. 41-82.

4201. ASTUTO (Giuseppe). "Io sono Crispi": Adua, 1 marzo 1896: governo forte: fallimento di un progetto. Bologna, Il Mulino, 2005, 210 p. (Le grandi date della storia costituzionale, 5).

4202. BANTI (Alberto Mario). L'onore della nazione. Identità sessuali e violenza nel nazionalismo europeo dal XVIII secolo alla Grande Guerra. Torino, Einaudi, 2005, XII-389 p.

4203. BOSWORTH (R. J. B.). Mussolini's Italy: life under the dictatorship 1915–1945. London, Allen Lane, 2005, XXVI-692 p. (ill.). (Allen Lane history).

4204. BROERS (Michael). The Napoleonic empire in Italy, 1796–1814: cultural imperialism in a European context? Basingstoke, Palgrave Macmillan, 2005, XVII-368 p. (maps).

4205. BRUCH (Anne). Italien auf dem Weg zum Nationalstaat: Giuseppe Ferraris Vorstellungen einer föderal-demokratischen Ordnung. Hamburg, Krämer, 2005, 203 p. (ill.). (Beiträge zur deutschen und europäischen Geschichte, 33).

4206. CARRÀ (Ettore). L'ordine pubblico nel periodo napoleonico: Piacenza, 1806–1814. Piacenza, Tip.Le.Co, 2005, XIX-247 p. (ill.). (Saggi. Storia, 3).

4207. CASELLA (Mario). Stato e Chiesa in Italia dalla Conciliazione alla riconciliazione, 1929–1931: aspetti e problemi nella documentazione dell'Archivio storico diplomatico del Ministero degli affari esteri. Galatina, Congedo stampa, 2005, 468 p. (Dipartimento di studi storici dal Medioevo all'età contemporanea, 66 e 73).

4208. CASTIGLIONE (Caroline). Patrons and adversaries. Nobles and villagers in Italian politics, 1640–1760. New York, Oxford U. P., 2005, XII-254 p.

4209. COLARIZI (Simona), GERVASONI (Marco). La cruna dell'ago. Craxi, il Partito socialista e la crisi della Repubblica. Roma e Bari, Laterza, 2005, 292 p.

4210. CORDOVA (Ferdinando). Verso lo Stato totalitario. Sindacati, società e fascismo. Soveria Mannelli, Rubbettino, 2005, XIV-320 p.

4211. Crisi (La) del sistema politico italiano e il Sessantotto. A cura di Giovanni ORSINA e Gaetano QUAGLIARIELLO. Soveria Mannelli, Rubbettino, 2005, XXXVIII-560 p.

4212. D'ANGELO (Augusto). Moro, i vescovi e l'apertura a sinistra. Roma, Studium, 2005, 158 p.

4213. DEGL'INNOCENTI (Maurizio). Identità nazionale e poteri locali in Italia tra '800 e '900. Manduria, P. Lacaita, 2005, 333 p. (Strumenti e fonti, 42). – IDEM. Il mito di Stalin. Comunisti e socialisti nell'Italia del dopoguerra. Manduria, Bari e Roma, Lacaita, 2005, 187 p.

4214. EMICH (Birgit). Territoriale Integration in der Frühen Neuzeit: Ferrara und der Kirchenstaat. Köln, Böhlau, 2005, XI-1178 p. (maps).

4215. Faschismus in Italien und Deutschland: Studien zu Transfer und Vergleich. Hrsg. v. Sven REICHARDT und Armin NOLZEN. Göttingen, Wallstein, 2005, 283 p. (ill.). (Beiträge zur Geschichte des Nationalsozialismus, 21).

4216. GIOVAGNOLI (Agostino). Il caso Moro. Una tragedia repubblicana. Bologna, Il Mulino, 2005, 382 p.

4217. Jews in Italy under fascist and nazi rule, 1922–1945. Ed. by Joshua D. ZIMMERMANN. New York, Cambridge U. P., 2005, XIX-374 p.

4218. KIRK (Thomas Allison). Genoa and the sea: policy and power in an Early Modern Maritime Republic, 1559–1684. Baltimore, Johns Hopkins U. P., 2005, XIII-276 p. (Johns Hopkins University studies in historical and political science, 3).

4219. LAUDANI (Simona). "Quegli strani accadimenti": la rivolta palermitana del 1773. Roma, Viella, 2005, 230 p. (Libri di Viella, 47).

4220. LECHNER (Stefan). "Die Eroberung der Fremdstämmigen". Provinzfaschismus in Südtirol, 1919–1926. Innsbruck, Univ.-Verlag Wagner, 2005, 523 p.

4221. Marine italienne (La) de l'unité à nos jours. Sous la dir. de Michel OSTENC. Paris, Economica-ISC-CFHM, 2005, 317 p.

4222. Movimento (Il) di unità proletaria (1943–1945). Con due contributi su Lelio Basso e il PSI nel dopoguerra. A cura di Giancarlo MONINA. Roma, Carocci, 2005, XIX-248 p.

4223. Nazione (La) in rosso: socialismo, comunismo e "questione nazionale": 1889–1953. A cura di Marina CATTARUZZA. Soneria Mannelli, Rubbettino, 2005, 334 p. (ill.). (Le ragioni degli storici, 9).

4224. O'BRIEN (Paul). Mussolini in the First World War: the journalist, the soldier, the fascist. Oxford a. New York, Berg, 2005, 212 p. (maps).

4225. PARISELLA (Antonio). Cultura cattolica e Resistenza nell'Italia repubblicana. Roma, AVE, 2005, 204 p. (Il seme e l'aratro, 9).

4226. PATRIARCA (Silvana). Indolence and regeneration: tropes and tensions of Risorgimento patriotism. *American historical review*, 2005, 110, 2, p. 380-408.

4227. PICK (Daniel). Rome or death. The obsessions of General Garibaldi. London, Jonathan Cape, 2005, 160 p.

4228. PIRJEVEC (Jože). Politika Italije do slovenske manjšine. (Italian policy towards the Slovene minority). *Javna uprava*, 2005, 41, 2/3, p. 237-256.

4229. SCHLEMMER (Thomas). Der italienische Faschismus und die Juden 1922 bis 1945. *Vierteljahrshefte für Zeitgeschichte*, 2005, 53, 2, p. 164-201.

4230. Toscana (La) in età moderna, secoli XVI–XVIII: politica, istituzioni, società: studi recenti e prospettive di ricerca: atti del convegno, Arezzo, 12–13 ottobre 2000. A cura di Mario ASCHERI e Alessandra CONTINI. Firenze, L.S. Olschki, 2005, X-349 p. (Biblioteca storica toscana / Deputazione di storia patria per la Toscana, 51).

4231. VANDIVER NICASSIO (Susan). Imperial city: Rome, Romans, and Napoleon, 1796–1815. Welwyn Garden City, Linden Publishing/Ravenhall, 2005, 255 p.

4232. VENTRONE (Angelo). Il nemico interno. Immagini, parole e simboli della lotta politica nell'Italia del Novecento. Roma, Donzelli, 2005, X-339 p.

Addenda 2002–2003

4233. GIULIETTI (Fabrizio). Il movimento anarchico italiano nella lotta contro il fascismo, 1927–1945. Manduria, Piero Lacaita Editore, 2002, 449 p. (Società e cultura, 32).

4234. RAGUSA (Andrea). I comunisti e la società italiana: Innovazione e crisi di una cultura politica, 1956–1973. Manduria, Piero Lacaita Editore, 2003, 245 p. (Società e cultura, 29).

Cf. n°s 200, 2585, 4129, 4535, 4828, 5102, 6547, 6975, 7160, 7363, 7392

Japan

4235. BOTSMAN (Daniel V.). Punishment and power in the making of modern Japan. Princeton, Princeton U. P., 2005, XIV-319 p.

4236. KIM (YoungSook). Mansyū-jihen go no kokusai jōsei to nisso hukasin jōyaku. (International affairs after the Mukden incident and the Nonaggression Pact between Japan and Soviet Russia). *Nihon Rekishi*, 2005, 681, p. 52-66.

4237. KOMIYA (Hitoshi). Yoshida Shigeru no seijisidō to tō-soshiki. (Shigeru Yoshida's political leadership and party organization). *Nihon Seiji Kenkyu*, 2005, 2, 1, p. 111-160 [Eng. Summary].

4238. Left (The) in Japanese politics. Ed. by Rikki KERSTEN and David WILLIAMS. New York a. London, RoutledgeCurzon, 2005, 240 p.

4239. MAXEY (Trent Elliott). "The greatest problem": religion in the politics and diplomacy of early Meiji Japan, 1868–1884. Ann Arbor, UMI, 2005, VII-363 p.

4240. NAM (Sang Ho). The formation of the Nakajinm Chikuhei group in the political party 'Seiyukai'. *Journal of Japanese history*, 2005, 22, p. 91-116.

4241. NISHIKAWA (Makoto). 'Hogohirohi: Sasaki Takayuki nikki' no chuki to seiritsu. (Notes and formation of 'Hogohirohi: journal of Sasaki Takayuki'). *Nihon Rekishi*, 2005, 685, p. 85-87.

4242. OKADO (Masakatsu). Mō hitori no nōson josei kenkyū-sha, Yamagishi Masako: Sengo no tōhoku wo kyoten ni shite. (Yamagishi Masako, as a researcher on rural women in postwar Japan). *Josei shigaku*, 2005, 15, p. 15-36.

4243. ONITSUKA (Hiroshi). Nichiro-sensō to chīkishakai no soshikika: nagano-ken kamisato-mura wo jirei ni. (The Russo-Japanese War and organizing of local society: the case of Kamisato village, Nagano). *Nenpō*, 2005, 3, p. 55-7.

4244. SHIMIZU (Yuichiro). Seitō naikaku no seiritsu to sei-kan kankei no henyō: wai-han naikaku~daiyoji itō naikaku. (The formation of the first party cabinet, 1898–1901: changing relationship between the politico-bureaucracy and the cabinet). *Shigaku Zasshi*, 2005, 114, 2, p. 59-83 [Eng. Summary].

4245. SUZUKI (Tamon). Tōjō naikaku sōjishoku no keii ni tsuiteno saikentō: Showa tennō to jūshin. (Reconsideration of the resignation of Tojo Ministry, 1944). *Nihon Rekishi*, 2005, 685, p. 69-84.

4246. TANAKA (Ryuichi). 'Mansyūkoku' ni okeru tōchi kikō no keisei to 'kokumin' no sōsyutsu: zai-man tyōsenjin mondai wo tyūshin ni. (Making of the government machinery and creation of 'the Nation' in 'Manchukuo'). *Journal of Japanese history*, 2005, 511, p. 61-78.

4247. YOKOYAMA (Yuriko). Meiji ishin to kinsei mibunsei no kaitai. (Meiji revolution and the dismantling of early modern Japanese rank system). Tokyo, Yamakawa Shuppansha, 2005, 342 p.

Cf. n°s 217, 6483, 7353, 7802-7849

Jordan

* 4248. AKHRAS (Mahmūd). Al- Bibliyūghrāfiyā al-Urdunīyah al-Filastīnīyah fī al-qarn al-'ishrīn, 1900–1978. Amman, Wizārat al-Thaqāfah, 2005, 2 vol., [s. p.]. (Kitāb al-Shahr, 59).

4249. ABU NOWAR (Ma'an). The development of Transjordan 1929–1939: a history of the Hashemite kingdom of Jordan. Reading: Ithaca, 2005, XI-392 p.

4250. ANDERSON (Betty Signe). Nationalist voices in Jordan: the street and the state. Austin, University of Texas Press, 2005, X-288 p.

4251. KATZ (Kimberly). Jordanian Jerusalem: Holy places and national spaces. Gainesville, University Press of Florida. 2005, XVI-214 p.

4252. TALL (Ahmad Yūsuf). King Hussein's legacy: the components of King Hussein's policy and its impact on the economic development of Jordan, 1953–1990. Amman, Ahmad Yousef Al-Tal, 2005, 167 p. (ill., map).

Kenya

4253. ANDERSON (David). Histories of the hanged. The dirty war in Kenya and the end of empire. New York, W. W. Norton, 2005, VIII-406 p.

4254. COHEN (David William), ODHIAMBO (E. S. Atieno). The risks of knowledge: investigations into the death of the Hon. Minister John Robert Ouko in Kenya, 1990. Athens, Ohio U. P., 2005, XV-344 p.

4255. GATHERU (R. Mugo). Kenya: from colonization to independence, 1888–1970. Jefferson, McFarland & Co., 2005, VIII-236 p. (ill.).

4256. KIHORO (Wanyiri). The price of freedom: the story of political resistance in Kenya. Nairobi, Mvule Africa Publishers, 2005, XIV-234 p. (ill.).

4257. SMITH (David Lovatt). Kenya, the Kikuyu and Mau Mau. East Sussex, Mawenzi Books, 2005, 359 p.

Korea

4258. CHOI (Suk Wan). Japanese Government's proposal for reforming the Korean government. *Journal of Japanese history*, 2005, 21, p. 79-110.

4259. LAN'KOV (Andrei Nikolaevich). Crisis in North Korea: the failure of de-Stalinization, 1956. Honolulu, University of Hawai'i Press and Center for Korean Studies, University of Hawai'i, 2005, XV-274 p. (ill.). (Hawai'i studies on Korea).

4260. LANKOV (Andrei). Crisis in North Korea. The failure of de-stalinization, 1956. Honolulu, University of Hawai'is Press, 2005, XV-274 p. (Hawai'i studies on Korea).

4261. OLSEN (Edward A.). Korea, the divided nation. Westport, Praeger, 2005, 191 p. (Praeger security international).

4262. POPOV (I.M.), LAVRENOV (Sergej Ja.), BOGDANOV (V.N.). Koreja v ogne vojny: K 55-letiju nachala vojny v Koree, 1950–1953 gg. (Korea in the fire of war: To the 55th anniversary of the beginning of the war in Korea, 1950–1953). Moskva, Kuchkovo pole, 2005, 543 p. (ind.). (Istorija XX veka: Sud'by, sobytija, dokumenty, versii).

Cf. n°s 7413, 7850-7853

Kuwait

4263. SLOT (B. J.). Mubarak Al-Sabah: founder of modern Kuwait 1896–1915. London, Arabian Publishing, 2005, 461 p.

Kyrgyzstan

4264. ZHUMANALIEV (Asanbek). Politicheskaja istorija Kyrgyzstana: stanovlenie politicheskoj sistemy kyrgyzskogo obshchestva v 1920–1930-e gody. (Political history of Kyrgyzstan: formation of political system in Kyrgyz society, 1920–1930). Bishkek, Natsional'naja akademija nauk Kyrgyzskoj Respubliki, 2005, 397 p.

Latvia

4265. BLEIERE (Daina), [et al.]. Istorija Latvii: XX vek. Riga, Jumava, 2005, 473 p. (ill.).

4266. Hidden (The) and forbidden history of Latvia under Soviet and Nazi occupations 1940–1991: selected research of the Commission of the Historians of Latvia. Ed. by Valters NOLLENDORFS and Erwin OBERLÄNDER. Rīga, Institute of the History of Latvia, 2005, 383 p. (ill., maps.). (Latvijas Vēsturnieku komisijas raksti, 14. sējums).

4267. KRĒSLIŅŠ (Uldis). Aktīvais nacionālisms Latvijā: 1922–1934. (Active nationalism in Latvia, 1922–1934). Rīga, Latvijas vēstures institūta apgāds, 2005, 319 p. (ill.).

4268. ŠTEIMANIS (Josifs). Iz istorii gosudarstva i prava Latvii. (History of state and law of Latvia). Redaktor A. KUZNETSOV. Riga, Baltijskij russkij institut, 2005, 219 p.

4269. Totalitārie režīmi Baltijā = Totalitarian regimes in the Baltic; research findings and issues: materials of an international conference 3–4 June 2004, Riga. redakcijas kolēģija, Andris CAUNE [et al.]; sastādītājs Dzintars ĒRGLIS. Rīga, Latvijas vēstures institūta apgāds, 2005, 283 p. (Latvijas Vēsturnieku komisijas raksti = Symposium of the Commission of the Historians of Latvia, 15).

Lebanon

4270. BESHARA (Adel). Lebanon: the politics of frustration. The failed coup of 1961. London a. New York, RoutledgeCurzon, 2005, IX-228 p. (History and society in the Islamic world).

4271. KANAAN (Claude Boueiz). Lebanon 1860–1960: a century of myth and politics. London, Saqi, 2005, 320 p. (ill., maps).

4272. SHANAHAN (Rodger). The Shi'a of Lebanon. Clans, parties and clerics. London a. New York, I. B. Tauris, 2005, XI-208 p. (Library of modern Middle East studies, 49).

Libya

4273. AHMIDA (Ali Abdullatif). Forgotten voices: power and agency in colonial and postcolonial Libya. New York a. London, Routledge, 2005, XVI-108 p. (ill., map).

Liechtenstein

4274. JUD (Ursina). Liechtenstein und die Flüchtlinge zur Zeit des Nationalsozialismus: Studie im Auftrag der Unabhängigen Historikerkommission Liechtenstein Zweiter Weltkrieg. Vaduz, Historischer Verein für das Fürstentum Liechtenstein u. Zürich, Chronos, 2005, 310 p. (Veröffentlichungen der Unabhängigen Historikerkommission Liechtenstein Zweiter Weltkrieg, 1).

Lithuania

4275. Aparat represji a opór społeczeństwa wobec systemu komunistycznego w Polsce i na Litwie w latach 1944–1956. Pod redakcją Piotra NIWIŃSKIEGO. Warszawa, Instytut Pamięci Narodowej, 2005, 121 p. (Konferencje IPN, 24).

4276. JANUŽYTĖ (Audronė). Historians as nation state-builders: the formation of Lithuanian University, 1904–1922. Tampere, University of Tampere, 2005, 367 p. (ill., maps). (Studies in European societies and politics. Acta electronica Universitatis Tamperensis, 441).

4277. Lietuva 1940–1990: okupuotos Lietuvos istorija. (Storia della Lituania occupata). Vyriausias redaktorius, Arvydas ANUŠAUSKAS [et al.]. Vilnius, Lietuvos Gyventojų Genocido ir Rezistencijos Tyrimo Centras, 2005, 710 p. (ill., maps).

4278. MAČIULIS (Dangiras). Valstybės kultūros politika Lietuvoje 1927–1940 metais. (State cultural policy in Lithuania 1927–1940). Vilnius, Lietuvos istorijos institutas, 2005, 300 p.

Luxemburg

4279. HACKER (Peter). Die Anfänge eines eigenen Nationalbewußtseins? Eine politische Geschichte Luxemburgs von 1815 bis 1865. Trier, Kliomedia, 2005, 458 p. (maps).

Macedonia

4280. OPFER-KLINGER (Björn). Im Schatten des Krieges: Besatzung oder Anschluss – Befreiung oder Unterdrückung? Eine komparative Untersuchung über die bulgarische Herrschaft in Vardar-Makedonien 1915–1918 und 1941–1944. Münster, Lit, 2005, 373 p. (Studien zur Geschichte, Kultur und Gesellschaft Südosteuropas, 3).

4281. PSILOS (Christopher). From cooperation to alienation: an insight into relations between the Serres Group and the Young Turks during the years 1906–1909. *European history quarterly*, 2005, 35, 4, p. 541-557.

Cf. n°s 4520, 4522, 4523

Madagascar

4282. CAMPBELL (Gwyn). An economic history of Imperial Madagascar 1750–1895: the rise and fall of an island empire. Cambridge, Cambridge U. P., 2005, XVII-413 p. (ill., maps). (African studies series, 106).

Malaysia

Cf. n° 7537

Mali

4283. SOARES (Benjamin F.). Islam and the prayer economy: history and authority in a Malian town. Edinburgh, Edinburgh University Press for the International African Institute, London, 2005, XII-306 p. (ill., maps). (International African library, 32).

Cf. n° 6774

Malta

4284. FENECH (Dominic). Responsibility and power in inter-war Malta: book one: endemic democracy (1919–1930). San Gwann, Publishers Enterprises Group, 2005, XII-458 p. (ill.).

Mexico

** 4285. Correspondencia Juárez-Vidaurri, 1855–1864. Director de este número Artemio BENAVIDES HINOJOSA. Monterrey, AGENL, 2005, 334 p. (ill.). (Historia del noreste mexicano, 3).

** 4286. Revolución mexicana (La): crónicas, documentos, planes y testimonios. Estudio introductorio, selección y notas Javier GARCIADIEGO DANTAN. México D.F., Universidad Nacional Autónoma de México, Coordinación de Humanidades, 2005, XCII-408 p. (maps). (Biblioteca del estudiante universitario, 138).

4287. Actores, espacios y debates en la historia de la esfera pública en la ciudad de México. Coordinadores Cristina SACRISTÁN y Pablo PICCATO. Seminario Internacional La Experiencia Institucional en la Ciudad de México: Esfera Pública y Elites Intelectuales (2002: Mexico D. F.). México D.F., Instituto de Investigaciones Históricas-UNAM, Instituto Mora, 2005, 283 p. (map). (Historia política. Instituto de Investigaciones Dr. José María Luis Mora).

4288. ALANIS ENCISO (Fernando). De factores de inestabilidad nacional a elementos de consolidación del

Estado posrevolucionario: Los exiliados mexicanos en Estados Unidos, 1929–1933. *Historia mexicana*, 2005, 54, 4, p. 1155-1205.

4289. BALLESTEROS GARCÍA (Víctor M.). Síntesis de la Guerra de Independencia en el estado de Hidalgo. Pachuca, Universidad Autónoma del Estado de Hidalgo, 2005, 124 p. (Pasado y presente hidalguense, 1).

4290. BARAJAS (Rafael). El País de "El Ahuizote": la caricatura mexicana de oposición durante el gobierno de Sebastián Lerdo de Tejada (1872–1876). México D.F., Fondo de Cultura Económica, 2005, 324 p. (ill.). (Tezontle).

4291. BARRACCA (Steven). Devolution and the deepening of democracy: explaining outcomes of municipal reform in Mexico. *Journal of Latin American studies*, 2005, 37, 1, p. 1-28.

4292. BREWSTER (Claire). Responding to crisis in contemporary Mexico: the political writings of Paz, Fuentes, Monsiváis, and Poniatowska. Tucson, University of Arizona Press, 2005, VIII-265 p.

4293. BRISEÑO SENOSIAIN (Lillian). La moral en acción: teoría y práctica durante el porfiriato. *Historia mexicana*, 2005, 55, 2, p. 419-460.

4294. Caciquismo in twentieth-century Mexico. Ed. by Alan KNIGHT and Wil G. PANSTERS. London, Institute for the Study of the Americas, 2005, X-409 p.

4295. CADENA INOSTROZA (Cecilia). Administración pública y procesos políticos en México. Zinacantepec, Colegio Mexiquense, 2005, 318 p.

4296. Chiapas: de la independencia a la revolución. Coordinadores Mercedes OLIVERA y María Dolores PALOMO. México D. F., Centro de Investigaciones y Estudios Superiores en Antropología Social (CIESAS) y Tuxtla Gutiérrez, Consejo de Ciencia y Tecnología del Estado de Chiapas, 2005, 484 p. (Publicaciones de la Casa Chata. Historias. Centro de Investigaciones y Estudios Superiores en Antropología Social).

4297. Culturas de pobreza y resistencia: estudios de marginados, proscritos y descontentos. México, 1804–1910. Coord. Romana FALCÓN. México, Colegio de México y Universidad Autónoma de Querétaro, 2005, 358 p.

4298. DE VEGA (Mercedes). Los dilemas de la organización autónoma: Zacatecas 1808–1832. México D.F., El Colegio de México, 2005, 378 p. (map).

4299. DEL LLANO IBÁÑEZ (Ramón). El Partido Católico y el primer gobernador de la revolución en Querétaro, Carlos M. Loyola. Querétaro, Universidad Autónoma de Querétaro, Facultad de Ciencias Políticas y Sociales, 2005, 94 p. (Serie Humanidades / Universidad Autónoma de Querétaro).

4300. Divine charter (The): constitutionalism and liberalism in nineteenth-century Mexico. Ed. by Jaime E. RODRÍGUEZ O. Lanham a. Oxford, Rowman & Littlefield Publishers, 2005, XII-402 p.

4301. Encrucijadas de la ciudadanía y la democracia: Yucatán, 1812–2004. Coordinador Sergio QUEZADA. Mérida, Universidad Autónoma de Yucatán, H. Congreso del Estado de Yucatán, LVII Legislatura, 2005, 245 p. (ill.).

4302. FALCÓN (Romana). El Estado liberal ante las rebeliones populares. México, 1867–1876. *Historia mexicana*, 2005, 54, 4, p. 973-1048.

4303. GALINDO CÁRDENAS (Benjamín). El provincialismo nuevoleonés en la época de Parás Ballesteros, 1822–1850. Monterrey, Universidad Autónoma de Nuevo León, 2005, 255 p. (ill., map). (Ancla de tiempo).

4304. GILLINGHAM (Paul). The Emperor of Ixcateopan: fraud, nationalism and memory in modern Mexico. *Journal of Latin American studies*, 2005, 37, 3, p. 561-584.

4305. GUARDINO (Peter F.). The time of liberty: popular political culture in Oaxaca, 1750–1850. Durham a. London, Duke U. P., 2005, IX-405 p. (maps). (Latin America otherwise).

4306. GÜÉMEZ PINEDA (Arturo). El poder de los cabildos mayas y la venta de propiedades privadas a través del Tribunal de Indios. Yucatán (1750–1821). *Historia mexicana*, 2005, 54, 3, p. 697-760.

4307. HART (Paul). Bitter harvest. The social transformation of Morelos, Mexico, and the origins of the Zapatista revolution, 1840–1910. Albuquerque, University of New Mexico Press, 2005, XI-291 p.

4308. HORNE (Gerald). Black and brown: African Americans and the Mexican Revolution, 1910–1920. New York, New York U. P., 2005, X-274 p. (American history and culture).

4309. LANDAVAZO (Marco Antonio). De la razón moral a la razón de estado: violencia y poder en la insurgencia mexicana. *Historia mexicana*, 2005, 54, 3, p. 833-865.

4310. LEVINSON (Irving W.). Wars within war: Mexican guerrillas, domestic elites, and the United States of America, 1846–1848. Fort Worth, Texas Christian University Press a. London, Europspan, 2005, XVIII-173 p. (ill., maps).

4311. LEWIS (Stephen E.). The ambivalent revolution: forging state and nation in Chiapas, 1910–1945. Albuquerque, University of New Mexico Press, 2005, XXII-283 p. (ill.).

4312. MARTIN (JoAnn). Tepoztlán and the transformation of the Mexican state: the politics of loose connections. Tucson, University of Arizona Press, 2005, XIV-276 p.

4313. OLCOTT (Jocelyn). Revolutionary women in postrevolutionary Mexico. Durham, Duke U. P., 2005, 337 p. (Next wave: new directions in women's studies).

4314. Raíces del federalismo mexicano. Coordinadores Manuel MIÑO GRIJALVA [et al.]. Zacatecas, Universidad Autónoma de Zacatecas y Guadalupe, Secretaría de Educación y Cultura del Gobierno del Estado de Zacatecas, 2005, 199 p.

4315. RATH (Thomas). 'Que el cielo un soldado en cada hijo te dio...': conscription, recalcitrance and resistance in Mexico in the 1940s. *Journal of Latin American studies*, 2005, 37, 3, p. 507-531.

4316. REDINGER (Matthew). American Catholics and the Mexican Revolution, 1924–1936. Notre Dame, University of Notre Dame Press, 2005, XII-260 p.

4317. RÍOS ZÚÑIGA (Rosalinda). Formar ciudadanos: sociedad civil y movilización popular en Zacatecas, 1821–1853. México D.F., ESU, Universidad Nacional Autónoma de México, Plaza y Valdés, 2005, 302 p. (Historia).

4318. Rostros (Los) del conservadurismo mexicano. Compiladores Renée DE LA TORRE, Marta Eugenia GARCÍA UGARTE y Juan Manuel RAMÍREZ SÁIZ. México D.F., Centro de Investigaciones y Estudios Superiores en Antropología Social, 2005, 473 p. (Publicaciones de la Casa Chata).

4319. ROUX (Rhina). El príncipe mexicano: subalternidad, historia y Estado. Prólogo de Adolfo GILLY. México D.F., Ediciones Era, 2005, 258 p. (Biblioteca Era).

4320. Sociedad, milicia y política en Nuevo León, siglos XVIII y XIX. Coordinador Artemio BENAVIDES HINOJOSA. Monterrey, Archivo General del Estado de Nuevo León, 2005, 260 p. (Colección Cuadernos del noreste, 2).

4321. VELÁZQUEZ MORALES (Catalina). Diferencias políticas entre los inmigrantes chinos del noroeste de México (1920–1930): el caso de Francisco L. Yuen. *Historia mexicana*, 2005, 55, 2, p. 461-512.

4322. WEEKS (Charles A.). The Juárez myth in Mexico. Tuscaloosa, University of Alabama Press, 2005, 224 p. (ill.).

4323. ZEITLIN (Judith Francis). Cultural politics in colonial Tehuantepec: community and state among the Isthmus Zapotec, 1500–1750. Stanford, Stanford U. P., 2005, XIX-323 p. (ill., maps, geneal. Tables).

Cf. n°s 4383, 5301, 6023, 6519, 6820

Moldova

4324. CIOBANU (Vitalie). Anatomia unui faliment geopolitic: Republica Moldova. (The anatomy of a geopolitical bankrupcy: Republic of Moldavia). Iaşi, Editura Polirom, 2005, 408 p.

Montenegro

Cf. n°s 4520, 4522, 4523

Morocco

4325. ZEGHAL (Malika). Les islamistes marocains: le défi à la monarchie. Paris, La Découverte, 2005, 332 p. (Cahiers libres).

Mozambique

4326. MAGODE (José). Pouvoir et réseaux sociaux au Mozambique: appartenances, interactivité du social et du politique (1933–1994). Paris, Éditions Connaissances et Savoirs, 2005, 648 p. (ill., maps, ports.).

4327. WEST (Harry G.). Kupilikula: governance and the invisible realm in Mozambique. Based on research conducted in collaboration with Marcos Agostinho MANDUMBINE and with assistance from Eusébio TISSA KAIRO and Felista Elias MKAIMA. Chicago, University of Chicago Press, 2005, XXVIII-362 p.

Namibia

4328. NAMHILA (Ellen Ndeshi). Kaxumba kaNdola: man and myth. The biography of a barefoot soldier. Basel, Basler Afrika Bibliographien, 2005, XI-157 p.

Nepal

4329. DĘBNICKI (Krzysztof). Współczesna historia Królestwa Nepalu. (The modern history of the Kingdom of Nepal). Warszawa, Dialog, 2005, 272 p. (Dzieje Orientu).

4330. WHELPTON (John). A history of Nepal. Cambridge, Cambridge U. P., 2005, XXIII-296 p. (ill., maps).

Netherlands

4331. BARNOW (David), VAN DER STROOM (Gerrold). Wer verriet Anne Frank? Münster, Agenda, 2005, 112 p.

4332. DE CONINCK (Pieter). Een les uit Pruisen: Nederland en de Kulturkampf, 1870–1880. Hilversum, Verloren, 2005, 431 p. (ill.).

4333. DLUGAICZYK (Martina). Der Waffenstillstand (1609–1621) als Medienereignis: politische Bildpropaganda in den Niederlanden. Münster, Waxmann, 2005, 378 p. (ill.). (Niederlande-Studien, 39).

4334. FÜHNER (Harald). Nachspiel: die niederländische Politik und die Verfolgung von Kollaborateuren und NS-Verbrechern, 1945–1989. Münster, Waxmann, 2005, 471 p. (Niederlande-Studien, 35).

4335. HAGEMAN (Maarten). Het kwade exempel van Gelre: de stad Nijmegen, de Beeldenstorm en de Raad van Beroerten, 1566–1568. Nijmegen, Vantilt, 2005, 453 p. (ill.). (Werken uitgegeven door Gelre, Vereeniging tot Beoefening van Geldersche Geschiedenis, Oudheidkunde en Recht, 59).

4336. HOEKSTRA (Hanneke). Het hart van de natie: morele verontwaardiging en politieke verandering in Nederland, 1870–1919. Amsterdam, Wereldbibliotheek, 2005, 237 p. (De natiestaat: politiek in Nederland sinds 1815).

4337. ROSENDAAL (J. G. M. M.). De Nederlandse Revolutie: vrijheid, volk en vaderland, 1783–1799. Nijmegen, Vantilt, 2005, 256 p. (ill.).

4338. SMEETS (Jos). Turbulent times: the Dutch police between the two World Wars. *In*: Policing interwar Europe: 1918–1940 [Cf. n° 3551], p. 192-214.

4339. TRACY (James D.). The Low Countries in the sixteenth century: Erasmus, religion and politics, trade and finance. Aldershot, Ashgate/Variorum, 2005, [s. p.]. (ill.). (Collected studies series, CS808).

4340. VAN DEURSEN (Arie Theodorus). De last van veel geluk: de geschiedenis van Nederland, 1555–1702. Amsterdam, B. Bakker, 2005, 372 p. (ill., facsims., ports.). (De geschiedenis van Nederland, 4).

4341. VAN SCHIE (Patricius Gerardus Cornelis). Vrijheidsstreven in verdrukking: liberale partijpolitiek in Nederland 1901–1940. Amsterdam, Boom, 2005, 503 p. (ill.).

Cf. n°s 4129, 6141

New Zealand

4342. BUTTERWORTH (Susan). More than law and order: policing a changing society, 1945–1992. Dunedin, University of Otago Press, 2005, 348 p. (ill., maps). (History of policing in New Zealand, 5).

4343. Class, gender and the vote: historical perspectives from New Zealand. Ed. by Miles FAIRBURN and Erik OLSSEN. Dunedin, University of Otago Press, 2005, 288 p. (ill.). (Otago history series).

4344. FAIRBURN (Miles), HASLETT (Stephen). Cleavage within the working class? The working-class vote for the Labour Party in New Zealand, 1911–1951. *Labour history*, 2005, 88, p. 183-214.

Cf. n° 4053

Nicaragua

4345. ANDERSON (Leslie E.), DODD (Lawrence C.). Learning democracy: citizen engagement and electoral choice in Nicaragua, 1990–2001. Chicago, University of Chicago Press, 2005, XVI-370 p.

4346. BARACCO (Luciano). Nicaragua: imagining the nation: from nineteenth-century liberals to twentieth-century Sandinistas. New York, Algora Pub., 2005, X-177 p.

4347. BARBOSA (Francisco J.). July 23, 1959: student protest and state violence as myth and memory in León, Nicaragua. *Hispanic American historical review*, 2005, 85, 2, p. 187-222.

4348. BATAILLON (Gilles). De Sandino aux "contras". Formes et pratiques de la guerre au Nicaragua. *Annales*, 2005, 60, 3, p. 653-688.

4349. KINLOCH TIJERINO (Frances). Historia de Nicaragua. Managua, Universidad Centroamericana, Instituto de Historia de Nicaragua y Centroamérica, 2005, 409 p. (ill., maps, ports.).

4350. LUNDQUIST (Jennifer H.), MASSEY (Douglas S.). Politics or economics? International migration during the Nicaraguan Contra war. *Journal of Latin American studies*, 2005, 37, 1, p. 29-53.

Cf. n°s 3731, 6655, 6980, 7539

Nigeria

4351. ACHEBE (Nwando). Farmers, traders, warriors, and kings: female power and authority in northern Igboland, 1900–1960. Portsmouth, Heinemann, 2005, XII-274 p. (Social history of Africa series).

4352. Nigerian history, politics and affairs: the collected essays of Adiele Afigbo. Ed. by Toyin FALOLA. Trenton a. Asmara, Africa World Press, 2005, X-722 p.

4353. NOUHOU (Alhadji Bouba). Islam et politique au Nigeria: genèse et évolution de la chari'a. Paris, Karthala, 2005, 280 p. (maps). (Collection Tropiques).

4354. Religion, history, and politics in Nigeria: essays in honor of Ogbu U. Kalu. Ed. by Chima J. KORIEH and G. Ugo NWOKEJI; foreword by Obioma NNAEMEKA. Lanham, University Press of America, 2005, VIII-282 p.

Norway

** 4355. FURE (Eli). Alt til Norge: den svensknorske arkivsaken 1895–1952. Oslo, Universitetsforlaget, 2005, 119 p. (ill., ports.). (Riksarkivaren skriftserie, 24).

4356. LINDER (Jan). Norge 1940: drama i 9 akter. (1940 en Norvège. Un drame en neuf actes). Stockholm, Infomanager, 2005, 196 p. (ill.).

4357. Paritiernas århundrade: fempartimodellens uppgång och fall i Norge och Sverige. (Le siècle des partis: ascension et décadence du modèle à cinq partis en Norvège et en Suède). Red. Marie DEMKER et Lars SVÅSAND. Stockholm, Santérus, 2005, 474 p. (ill.).

Pakistan

4358. ANSARI (Sarah). Life after partition: migration, community and strife in Sindh, 1947–1962. Karachi, Oxford U. P., 2005, XV-240 p. (ill., map).

4359. FROTSCHER (Ann). Banden- und Bürgerkrieg in Karachi: die Ethnisierung von Politik am Beispiel der Mohajir. Baden-Baden, Nomos, 2005, 288 p. (ill., map). (Studien zu Ethnizität, Religion und Demokratie, 6).

4360. HAQQANI (Husain). Pakistan: between mosque and military. Washington, Carnegie Endowment for International Peace, 2005, XI-397 p. (ill., map).

4361. PANICHKIN (Jurij N.). Obrazovanie Pakistana i pushtunskij vopros. (The creation of Pakistan and the Pashto question). RAN, In-t vostokovedenija. Moskva, Nauchnaja kniga, 2005, 205 p. (bibl. p. 195-202). [English summary]

4362. ZAMARAEVA (Natal'ja A.). Studencheskoe dvizhenie v Pakistane. (Student movement in Pakistan, [the late 1940s–the 1990s]). RAN, In-t vostokovedenija. Moskva, IV RAN, 2005, 211 p. (bibl. p. 203-208). [English summary]

Palestine

4363. JAMAL (Amal). The Palestinian national movement: politics of contention, 1967–2005. Bloomington, Indiana U. P., 2005, 247 p.

4364. TUTEN (Eric Engel). Between capital and land: the Jewish National Fund's Finance and Zionist National Land Purchase priorities in mandatory Palestine, 1939–1945. London, Routledge Curzon, 2005, XIV-248 p. (maps). (Israeli history, politics and society).

Cf. n°s 424, 4248

Paraguay

4365. PANGRAZIO (Miguel Angel). Los fraudes electorales en el Paraguay. Asunción, Paraguay: Intercontinental Editora, 2005, 247 p.

Cf. n° 6838

Peru

4366. AGUIRRE (Carlos). The criminals of Lima and their worlds: the prison experience, 1850–1935. Durham a. London, Duke U. P., 2005, XI-310 p.

4367. ARCE (Moisés). Market reform in society: post-crisis politics and economic change in authoritarian Peru. University Park, Pennsylvania State U. P., 2005, XIV-169 p.

4368. ARROYO REYES (Carlos). Nuestros años diez: la Asociación Pro-Indígena, el levantamiento de Rumi Maqui y el incaísmo modernista. Buenos Aires, LibrosEnRed, 2005, 329 p. (Insumisos Latinoamericanos).

4369. CONAGHAN (Catherine M.). Fujimori's Peru: deception in the public sphere. Pittsburgh, University of Pittsburgh Press, 2005, XII-311 p. (ill.). (Pitt Latin American series).

4370. GARCÍA (María Elena). The making of indigenous citizens: identity, development, and multicultural activism in Peru. Stanford, Stanford U. P., 2005, XI-213 p.

4371. GARRETT (David T.). Shadows of empire: the Indian nobility of Cusco, 1750–1825. Cambridge a. New York, Cambridge U. P., 2005, XVIII-300 p. (geneal. tables, maps). (Cambridge Latin American studies, 90).

4372. GONZÁLES (Osmar). Los orígenes del populismo en el Perú: el gobierno de Guillermo E. Billinghurst, 1912–1914. Lima, Nuevo Mundo, 2005, 331 p.

4373. Historia de las elecciones en el Perú: estudios sobre el gobierno representativo. Ed. por Cristóbal ALJOVÍN DE LOSADA y Sinesio LÓPEZ. Lima, Instituto de Estudios Peruanos, 2005, 568 p. (ill., tabs.). (Serie Estudios históricos, 41).

4374. Luces y reformas en el Perú: siglo XVIII. IV Jornadas de Historia; director José María SESÉ ALEGRE; compiladora Ruth Magali ROSAS. Piura, Universidad de Piura, Facultad de Ciencias y Humanidades, Departamento de Humanidades, 2005, XIV-278 p. (Cuadernos de humanidades, 10).

4375. Más allá de la dominación y la resistencia: estudios de historia peruana, siglos XVI–XX. Ed. por Paulo DRINOT y Leo GAROFALO. Lima, Instituto de Estudios Peruanos, 2005, 379 p. (ill). (Serie Estudios históricos, 40).

4376. MÉNDEZ (Cecilia). The plebeian republic: the Huanta rebellion and the making of the Peruvian state, 1820–1850. Durham, Duke U. P., 2005, XVI-343 p.

4377. Miedo (El) en el Perú: Siglos XVI al XX. Ed. por Claudia ROSAS LAURO. Lima, Fondo Editorial de la Pontificia Universidad Católica del Perú, Seminario Interdisciplinario de Estudios Andinos, 2005, 285 p.

4378. MONTOYA (Rodrigo). De la utopía andina al socialismo mágico: antropología, historia y política en el Perú. Cusco, Instituto Nacional de Cultura, 2005, 252 p.

4379. ORREGO PENAGOS (Juan Luis). La ilusión del progreso: los caminos hacia el estado-nación en el Perú y América Latina (1820–1860). Lima, Pontificia Universidad Católica del Péru, Fondo Editorial, 2005, 266 p.

4380. QUIROZ-PÉREZ (Lissel). Les magistrats péruviens au XIXe siècle. Des hommes de pouvoir au coeur de la transition politique (1808–1825). *Cahiers des Amériques Latines*, 2005, 50, p. 107-126.

4381. SANTA-CRUZ (Arturo). Monitoring elections, redefining sovereignty: the 2000 Peruvian electoral process as an international event. *Journal of Latin American studies*, 2005, 37, 4, p. 739-767.

4382. VELÁZQUEZ CASTRO (Marcel). Las máscaras de la representación: el sujeto esclavista y las rutas del racismo en el Perú (1775–1895). Lima, Banco Central de Reserva del Perú y Universidad Nacional Mayor de San Marcos, 2005, 288 p. (ill.).

Cf. n°s 3703, 3819

Philippines

4383. Cacicazgo (El) en Nueva España y Filipinas. Coordinadores Margarita MENEGUS BORNEMANN y Rodolfo AGUIRRE SALVADOR. México D.F., Plaza y Valdés, Universidad Nacional Autónoma de México, Centro de Estudios Sobre la Universidad (CESU), 2005, 406 p. (ill., maps). (Historia).

4384. SIMBULAN (Dante C.). The modern principalia: the historical evolution of the Philippine ruling oligarchy. Quezon City, University of the Philippines Press, 2005, 350 p. (ill.).

Poland

** 4385. Aparat bezpieczeństwa w Polsce: kadra kierownicza. T. 1. 1944–1956. (The security apparatus in Poland: managers. V. 1. 1944–1956). Redakcja naukowa Krzysztof SZWAGRZYK. Warszawa, Instytut Pamięci Narodowiej. Komisja Ścigania Zbrodni przeciwko Narodowi Polskiemu, 2005, 605 p. (ill.).

** 4386. Marzec 1968 w Krakowie: w dokumentach. (March 1968 in Krakow: documents). Wstęp i opracowanie, Julian KWIEK; wprowadzenie, Tomasz GĄSOWSKI.

Kraków, Księgarnia Akademicka, 2005, 433 p. (ill.). (Biblioteka Centrum Dokumentacji Czynu Niepodległościowego, 30).

** 4387. Walka instytucji państwowych z białoruską działalnością dywersyjną 1920–1925. (Fight of Polish state institutions with Belarusian sabotage groups in north-eastern territories of the Second Republic of Poland, 1920–1925). Wybór i opracowanie Wojciech ŚLESZYŃSKI. Białystok, Polskie Tow. Historyczne, Oddz. w Białymstoku, Wydawn. Prymat, 2005, 182 p. (Dokumenty do dziejów kresów północno-wschodnich II Rzeczypospolitej).

** 4388. Ziemie wschodnie: raporty Biura Wschodniego Delegatury Rządu na Kraj 1943-1944. (Lands east: report of Government for Poland, 1943–1944). Wstęp, wybór i opracowanie, Mieczysław ADAMCZYK, Janusz GMITRUK, Adam KOSESKI. Warszawa, Oficyna Wydawnicza "Aspra-JR", Muzeum Historii Polskiego Ruchu Ludowego i Pułtusk, Wyższa Szkoła Humanistyczna im. Aleksandra Gieysztora, 2005, 242 p. (ill.).

4389. ALVIS (Robert E.). Religion and the rise of nationalism: a profile of an East-Central European city. Syracuse, Syracuse U. P., 2005, XXVI-227 p. (Religion and politics).

4390. BÉGIN (Natalie). Kontakte zwischen Gewerkschaften in Ost und West. Die Gründung von Solidarność und ihre Auswirkungen in Deutschland und Frankreich. Ein Vergleich. *Archiv für Sozialgeschichte*, 2005, 45, p. 293-324.

4391. BINGEN (Dieter). Ostpolitik und demokratischer Wandel in Mittel- und Osteuropa. Der Testfall Polen. *Archiv für Sozialgeschichte*, 2005, 45, p. 117-140.

4392. BUTTERWICK (Richard). Political discourses of the Polish Revolution, 1788–1792. *English historical review*, 2005, 120, 487, p. 695-731.

4393. CURP (T. David). 'Roman Dmowski Understood': ethnic cleansing as permanent revolution. *European history quarterly*, 2005, 35, 3, p. 405-427.

4394. DAJNOWICZ (Małgorzata). Orientacje polityczne ludności polskiej północno-wschodniej części Królestwa Polskiego na przełomie XIX i XX wieku. (Political orientations of the Polish population in the north-eastern part of the Kingdom of Poland at the turn of the nineteenth and twentieth century). Białystok, Wydawn. Uniwersytetu w Białymstoku, 2005, 354 p. (ill., col. maps).

4395. DŁUGAJCZYK (Edward). Polska konspiracja wojskowa na Śląsku Cieszyńskim w latach 1919–1920. (Poland military conspiracy in Cieszyn Silesia in the years 1919–1920). Katowice, Wydawnictwo Uniwersytetu Śląskiego, 2005, 284 p.

4396. GIPPERT (Wolfgang). Kindheit und Jugend in Danzig 1920 bis 1945: Identitätsbildung im sozialistischen und im konservativen Milieu. Essen, Klartext, 2005, 552 p. (maps).

4397. JANECZEK (Zdzisław). Polityczna rola marszałka litewskiego Ignacego Potockiego w okresie Sejmu Wielkiego 1788–1792. (Il ruolo politico di Ignacy Potocki nel periodo della Grande Dieta, 1788–1792). Katowice, Wydawnictwo Akademii Ekonomicznej im. Karola Adamieckiego, 2005, 305 p. (facsims., ports.). (Prace naukowe Akademii Ekonomicznej im. Karola Adamieckiego w Katowicach).

4398. JIRÁSEK (Zdeněk), MALKIEWICZ (Andrzej). Polska i Czechoslowacja w dobie stalinizmu 1948–1956. Studium porównawcze. (Poland and Czechoslovakia during the Stalinist Period, 1948–1956. A Comparative Study). Warszawa, Instytut Studiów Politycznych PAN, 2005, 414 p.

4399. KALLAS (Marian). Historia ustroju Polski. (History of Polish political system). Warszawa, Wydawn. Naukowe PWN, 2005, 463 p. (ill., maps).

4400. KISLUK (Eugene J.). Brothers from the North. The Polish democratic society and the European revolutions of 1848–1849. Boulder, East European Monographs, 2005, VI-272 p. (East European monographs, 665).

4401. MISIUK (Andrzej). Administracja spraw wewnętrznych w Polsce: od połowy XVIII wieku do współczesności: zarys dziejów. (Administration of the Interior in Poland since the mid-eighteenth century to the present: an outline history). Olsztyn, Wydawn. Uniwersytetu Warmińsko-Mazurskiego w Olsztynie, 2005, 316 p. (ill). – IDEM. Police and policing under the Second Polish Republic, 1918–1939. *In*: Policing interwar Europe: 1918–1940 [Cf. n° 3551], p. 159-171.

4402. OST (David). The defeat of Solidarity. Anger and politics in postcommunist Europe. Ithaca, Cornell U. P., 2005, IX-238 p.

4403. PARADOWSKI (Przemysław). W obliczu "nagłych potrzeb" Rzeczypospolitej: Sejmy ekstraordynaryjne za panowania Władysława IV Wazy. (In view of the "urgent needs" Republic: extraordinary Sejm during the reign of Władysław IV Vasa). Toruń, Wydawn. Adam Marszałek, 2005, 282 p.

4404. PENN (Shana). Solidarity's secret: the women who defeated Communism in Poland. Ann Arbor, University of Michigan Press, 2005, XVII-372 p. (ill.).

4405. SIELEZIN (Jan Ryszard). Płaszczyzna konfrontacji politycznej między "Solidarnością" a władzą w latach 1980–1981. (Piano di confronto politico tra Solidarność e le autorità negli anni 1980–1981). Wrocław, Wydawn. Uniwersytetu Wrocławskiego, 2005, 472 p. (Acta Universitatis Wratislaviensis, 2825).

4406. Stan wojenny w Małopolsce: relacje i dokumenty. (Martial law in Małopolsce: relations and documents). Opracowanie Zbigniew SOLAK, Jarosław SZAREK; współpraca Henryk GŁĘBOCKI, Jolanta NOWAK, Adam ROLIŃSKI. Kraków, Instytut Pamięci Narodowej, Oddział w Krakowie, Fundacja Centrum Dokumentacji Czynu Niepodległościowego, Zarząd Regionu NSZZ "Solidarność" Małopolska, 2005, 499 p.

4407. STASIAK (Arkadiusz Michał). Patriotyzm w myśli konfederatów barskich. (Patriotism according to the Confederates of Bar). Lublin, Towarzystwo Naukowe Katolickiego Uniwersytetu Lubelskiego, 2005, 189 p. (Źródła i Monografie / Towarzystwo Naukowe Katolickiego Uniwersytetu Lubelskiego, 287).

4408. WIERZBICKI (Leszek Andrzej). O zgodę w Rzeczypospolitej: zjazd warszawski i sejm pacyfikacyjny 1673 roku. (The agreement of the Republic). Lublin, Wydawnictwo Uniwersytetu Marii Curie-Skłodowskiej, 2005, 328 p.

4409. Wieś w Polsce Ludowej. (Les villages dans la Pologne Populaire). Pod redakcją Grzegorza MIERNIKA. Kielce, Wydawnictwo Akademii świętokrzyskiej im. Jana Kochanowskiego, 2005, 181 p. (tables).

4410. Władze komunistyczne wobec Kościoła katolickiego w Łódzkiem 1945–1967. (The Communist authorities towards the Catholic Church in Łódź 1945–1967). Pod redakcją Janusza WRÓBLA i Leszka PRÓCHNIAKA. Warszawa, Instytut Pamięci Narodowej, 2005, 187 p. (ill.). (Konferencje IPN, 28).

4411. ZDRADA (Jerzy). Historia Polski: 1795–1914. Warszawa, PWN, 2005, 859 p. (maps).

4412. "Zwyczajny" resort: studia o aparacie bezpieczeństwa 1944–1956. ("Normal" resort: studies of the security apparatus 1944–1956). Pod redakcją Kazimierza KRAJEWSKIEGO i Tomasza ŁABUSZEWSKIEGO. Warszawa, Instytut Pamięci Narodowej, 2005, 600 p. (ill., map, facsims.). (Monografie IPN, 14).

Addendum 2000

4413. JAROSZ (Dariusz). Polacy a stalinizm, 1948–1956. Warsaw, Instytut Historii PAN, 2000, 244 p.

Cf. n°s 3927, 3966, 4127, 4275, 4506, 4711, 7057, 7240, 7354, 7510, 7520

Portugal

** 4414. BREYNER (Thomaz de Mello, Conde de Mafra). Diário de um monárquico: 1902–1904. Transcrição, selecção, anotações e nota prévia de Gustavo de Mello Breyner ANDRESEN. Porto, Fundação Eng. António de Almeida, 2005, 327 p. (ill., facsims., ports.).

4415. AFRICANO (António de Freitas). Primores políticos e regalias do nosso rei (1641). Estudo introdutório José Adelino MALTEZ. Lisboa, Universidade de Lisboa, Instituto de História do Direito e do Pensamento Político e São João do Estoril, Principia, 2005, 96 p. (ill.). (Clássicos do pensamento político em Portugal, 1).

4416. BEBIANO (Rui). Contestaçao do regime e tentaçao da luta armada sob o marcelismo. *Revista portuguesa de história*, 2005, 37, p. 65-104.

4417. FERNANDES (Filipe S.), SANTOS (Hermínio). Os excomungados de Abril: os empresários na Revolução. Lisboa, Dom Quixote, 2005, 145 p. (Cadernos DQ. 07. Reportagem).

4418. LLOYD-JONES (Stewart), PALACIOS CEREZALES (Diego). Guardians of the Republic? Portugal's Guarda Nacional Republicana and the Politicians during the 'New Old Republic', 1919–1922. *In*: Policing interwar Europe: 1918–1940 [Cf. n° 3551], p. 91-111.

4419. MALTEZ (José Adelino). Tradição e revolução: uma biografia do Portugal político do século XIX ao XXI. Vol. II. 1910–2005. Lisboa, Tribuna da Historia, 2005, 772 p. (ill.).

4420. MARQUES (Tiago Pires). Crime e castigo: no liberalismo em Portugal. Prefácio de Pedro TAVARES DE ALMEIDA. Lisboa, Livros Horizonte, 2005, 166 p. (ill.).

4421. MARTINS (Susana). Socialistas na oposição ao estado novo: um estudo sobre o movimento socialista português de 1926 a 1974. Cruz Quebrada, Casa das Letras, 2005, 249 p. (ill.). (Biblioteca da história).

4422. MAURÍCIO (Maria José). Mulheres e cidadania: alguns perfis e acção política 1949–1973. Lisboa, Caminho, 2005, 229 p. (ill., ports.). (Colecção universitária).

4423. MITEV (J.). Salazar i raždaneto na "Novata dăržava" (1926–1936). (Salazar et la naissance du "Nouvel Etat", 1926–1936). Veliko Tărnovo, [s. n.], 2005, 159 p.

4424. MONTOITO (Eugénio). Henrique Galvão: ou a dissidência de um cadete do 28 de maio (1927–1952). Lisboa, Centro de História da Universidade de Lisboa, 2005, 211 p. (ill.).

4425. REIS (Maria de Fátima). Santarém no tempo de D. João V: administração, sociedade e cultura. Lisboa, Edições Colibri, 2005, 788 p. (ill., map). (Colibri história, 40).

4426. TAVARES (João Moreira). Indústria militar portuguesa no tempo da guerra, 1961–1974. Casal de Cambra, Caleidoscópio, 2005, 235 p. (ill.).

4427. TORGAL (Luís Reis). António José de Almeida e a República: discurso de uma vida ou vida de um discurso. Selecção de imagens de Alexandre RAMIRES. Porto, Temas e Debates, 2005, 255 p. (ill., map, ports.). – IDEM. História (Contemporânea) de Portugal. *In*: Dicionário Temático da Lusofonia. Dir. e coord. Fernando CRISTÓVÃO. Lisboa, Texto Editores, p. 503-509.

Qatar

4428. RAHMAN (Habibur). The emergence of Qatar: the turbulent years, 1627–1916. London, K. Paul, 2005, XXX-282 p. (ill., maps).

Romania

** 4429. Akten um die Deutsche Volksgruppe in Rumänien 1937–1945: eine Auswahl. Hrsg. v. Klaus POPA. Frankfurt u. Oxford, P. Lang, 2005, 600 p.

** 4430. Poziția politică a Mișcării Legionare în vederile Tribunalului Internațional de la Nürnberg. Documente. (The political position of the Iron Guard Movement at the Nürnberg International Court. Documents). București, Editura Necoz Plus, 2005, 63 p.

** 4431. TRONCOTĂ (Cristian), PINTILIE (Florin), SPÂNU (Alin). Documente SSI despre poziţia şi activităţile politice din România în perioada regimului autoritar. 6 septembrie 1940–23 august 1944. Vol. I. (Secret Service documents about the political activities and stand points in Romania during the autoritarian regime. September 6, 1940–August 23, 1944). Bucureşti, Editura Institutului Naţional pentru Studiul Totalitarismului, 2005, 335 p.

4432. ANCEL (Jean). Preludiu la asasinat. Pogromul de la Iaşi, 29 iunie 1941. Bucureşti, Editura Polirom, 2005, 491 p. (Historia).

4433. BOIA (Lucian). Două secole de mitologie naţională. (Two centuries of national mythology). Bucureşti, Editura Humanitas, 2005, 135 p. – IDEM. Mitologia ştiinţifică a comunismului. (The scientific mythology of communism). Bucureşti, Editura Humanitas, 2005, 234 p.

4434. BUCHET (Constantin). Social-democraţia încarcerata. Evoluţii politice interne şi reacţii internaţionale. 1946–1969. (Imprisoned social-democracy. Internal political evolutions and international reactions. 1946–1969). Bucureşti, Editura Institutului Naţional pentru Studiul Totalitarismului, 2005, 362 p.

4435. Cei care au spus nu: oponenţi şi disidenţi în anii '70 şi '80. (Those who said "no": oponents and dissidents in the 70's and 80's). Editor Romulus RUSAN. Bucureşti, Editura Fundaţiei Academia Civică, 2005, 232 p.

4436. Comandanţi fără armată. Exilul militar românesc (1939–1972). (The Romanian military exile 1939–1972). Editori Dumitru DOBRE, Veronica NANU, Mihaela TOADER. Bucureşti, Editura Pro Historia, 2005, 344 p.

4437. COSMA (Ela). Revoluţia de la 1848. Un catalog de documente şi regeste (Fondul Institutului de Istorie din Cluj). (The revolution of 1848. A catalog of documents and registers. The collections of the history Institute of Cluj). Vol. I-II. Cluj-Napoca, Editura Argonaut, 2005, 279, 318 p.

4438. COSTANTINI (Emanuela). Nae Ionescu, Mircea Eliade, Emil Cioran. Antiliberalismo nazionalista alla periferia d'Europa. Perugia, Morlacchi, 2005, XVI-II-200 p. (Storia).

4439. DENIZE (Eugen), MÂŢĂ (Cezar). România comunistă. Statul şi propaganda (1948–1953). (Communist Romania. The State and the propaganda, 1948–1953). Târgovişte, Editura Cetatea de Scaun, 2005, 422 p.

4440. DENIZE (Eugen). Structura de organizare a propagandei comuniste în România. 1948–1953. (The structure of the communist propaganda in Romania 1948–1953). *Studii şi materiale de istorie contemporană. Academia Română*, 2005, 4, p. 89-102.

4441. ENACHE (George). Ortodoxie şi putere politică în România contemporană. Studii şi eseuri. (Orthodoxy and political power in contemporary Romania. Studies and essays). Bucureşti, Editura Nemira, 2005, 590 p.

4442. GALLAGHER (Tom). Modern Romania: the end of communism, the failure of democratic reform, and the theft of a nation. New York, New York U. P., 2005, XXIII-428 p. (maps). – IDEM. Romanian tyranny seen from above and below. *European history quarterly*, 2005, 35, 4, p. 559-568. – IDEM. Theft of a nation: Romania since Commnunism. London, Hurst, 2005, XXIII-428 p. (maps).

4443. GUIDA (Francesco). Romania. Milano, Unicopli, 2005, 350 p.

4444. GUŢAN (Manuel). Istoria administraţiei publice locale în statul român modern. (The history of the public local administration in the modern Romanian state). Bucureşti, Editura ALL Beck, 2005, 422 p.

4445. HONCIUC BELDIMAN (Dana). Statul naţional legionar. Septembrie 1940–Ianuarie 1941. Cadru legislativ. (National legionary state. September 1940–January 1941. Legislative framework). Bucureşti, Editura Institutul Naţional pentru Studiul Totalitarismului, 2005, 295 p.

4446. IONESCU-GURĂ (Nicoleta). Stalinizarea României. Republica Populară Română 1948–1950: transformări instituţionale. (Stalinizing Romania. Romanian National Republic 1948–1950: institutional changes). Bucureşti, Editura All, 2005, 567 p.

4447. LĂCUSTĂ (Ioan). 1948-1952. Republica Populară şi România. (The People's Republic of Romania). Bucureşti, Editura Curtea Veche, 2005, 265 p.

4448. Mişcarea Legionară în ţară şi în exil. Puncte de reper (1919–1980). (Iron Guard in the country and abroad. Landmarks, 1919–1980). Cronologie documentară de Dinu ZAMFIRESCU. Bucureşti, Editura Pro Historia, 2005, 215 p.

4449. NEAGOE (Stelian). Cazul social-democraţilor români. (The case of the Romanian social-democrats). Bucureşti, Editura Institutului de Ştiinţe Politice şi Relaţii Internaţionale, 2005, 371 p.

4450. ORGHIDAN (I.). Locul şi rolul locotenenţei domneşti în epoca modernă a României. (Place and role of princely lieutenancy in the modern age of Romania). *Studii şi articole de istorie*, 2005, 70, p. 225-235.

4451. PLOSCARU (Cristian). "Cărvunarii" şi primele manifestări politice ale cugetării liberale în Moldova. ("Carbonari" and the first political elements of the liberal thinking in Moldavia). *Anuarul Institutului de Istorie "A.D. Xenopol". Academia Română*, 2005, 42, p. 83-113.

4452. POPOVICI (Vlad). Elite şi strategii politice în mişcarea naţională românească din Transilvania (1869–1894). Studiu de caz: disputa activism-pasivism. (Elites and political strategies in the Romanian national movement in Transylvania, 1869–1894. Case study: the activism-passivism dispute). *Anuarul Institutului de Istorie "George Bariţiu". Academia Română*, 2005, 44, p. 193-212.

4453. RADU (Sorin). Modernizarea sistemului electoral din România (1866–1937). (The modernization of the elective system in Romania (1866–1937). Iaşi, Editura Institutul European, 2005, 284 p.

4454. Sfârşitul perioadei liberale a regimului Ceauşescu: Minirevoluţia culturală din 1971. (The end of the liberal period of the Ceauşescu regime: The cultural minirevolution of 1971). Ediţie îngrijită şi cuvânt înainte de Ana-Maria CĂTĂNUŞ. Bucureşti, Editura Institutului Naţional pentru Studiul Totalitarismului, 2005, 143 p.

4455. SIANI-DAVIES (Peter). The Romanian revolution of December 1989. Ithaca, Cornell U. P., 2005, XI-315 p. (ill., map).

4456. TISMĂNEANU (Vladimir). Stalinism pentru eternitate. O istorie politică a comunismului românesc. (Stalinism for eternity. A political history of the Romanian communism). Iaşi, Editura Polirom, 2005, 415 p.

4457. ZAMFIRESCU (Dinu). Mişcarea legionară în ţară şi în exil. Puncte de vedere (1919–1980). O cronologie documentară. (The Iron Guard Movement at home and in exile. Points of view, 1919–1980. Documentary chronology). Bucureşti, Editura Pro Historia, 2005, 216 p.

Cf. n°s 4524, 5222, 5968, 5970, 7087, 7116, 7462

Russia

* 4458. ANOKHINA (Tat'jana G.), ATABEKOVA (A.G.), KALITKINA (N.L.). Rossija i rossijskaja emigratsija v vospominanijakh i dnevnikakh: Annotirovannyj ukazatel' knig, zhurnal'nykh i gazetnykh publikatsij, izdannykh za rubezhom v 1917–1991 gg. (Russia and the Russian diaspora in memoirs and diaries: An annotated handlist of books and articles in journals and newspapers, issued abroad in 1917–1991). Ed. Andrej G. TARTAKOVSKIJ, Terence EMMONS, Oleg V. BUDNITSKIJ. Gos. publ. ist. bibl. Rossii; University of Stanford. In 4 vols. Vol. 4, part 1. Moskva, ROSSPEN, 2005, 463 p. (bibl. p. 456-459). [English summary]

** 4459. ACTON (Edward), STABLEFORD (Tom). The Soviet Union: a documentary history. Volume 1. 1917–1940. Exeter, University of Exeter Press, 2005, XXVIII-468 p. (ill.). (Exeter studies in history).

** 4460. [BOJKO (Nadezhda V.) et al.]. MAJSKIJ [Maysky] (Ivan M.). Izbrannaja perepiska s rossijskimi korrespondentami. (Selected correspondence with correspondents in Russia). Ed. Vladimir S. MJASNIKOV. Intr. by Nadezhda V. BOJKO. RAN, Arkhiv; RAN, In-t vseobshchej istorii.; MID RF. Book 1: 1900–1934. Book 2: 1935-1975. Moskva, Nauka, 2005, 2 vol., 580 p., 644 p. (ill.; portr.; ind. p. 532-579, 603-643). (Nauchnoe nasledstvo, 31).

** 4461. [BORISOVA (L.V.) et al.] Sovetskaja derevnja glazami VChK – OGPU – NKVD, 1918–1939: dokumenty i materialy. (Les campagnes soviétiques vues par la TCHEKA – O.G.P.U. – N.K.V.D., 1918–1939: documents et matériaux). Ed. André BERELOVICH, Viktor P. DANILOV. RAN, In-t rossijskoj istorii; Maison des sciences de l'homme, etc. Vol. 3. 1930–1934. Book 2. Moskva, ROSSPEN, 2005, 838 p. (ind. p. 726-813). [Resumé français]

** 4462. [GATAGOVA (Ljudmila S.), KOSHELEVA (L.P.), ROGOVAJA (L.A.).]. TsK RKP(b) – VKP(b) i natsional'nyj vopros, 1918–1933. (The Central Comitee of the R.C.P.(b)–C.P.S.U.(b.) and the question of nationalities, 1918–1933). Mosovskij gos. un-t im. M.V. Lomonosova, F-t gos. upravlenija; Ros. gos. arkhiv sots.-polit. ist.; RAN, In-t rossijskoj istorii. Book 1: 1918–1933. Moskva, ROSSPEN, 2005, 826 p. (bibl. p. 727-729, ind. p. 730-754). (Dokumenty sovetskoj istorii).

** 4463. KUMANEV (G.A.). Govorjat stalinskie narkomy: Vstrechi, besedy, interv'ju, dokumenty. (Stalin's commissars speak: Meetings, talks, interviews, documents). Smolensk, Rusich, 2005, 632 p.

** 4464. Obshchestvo i vlast'. Rossijskaja provintsija, 1917–1985: Dokumenty i materialy (Permskaja, Sverdlovskaja, Cheljabinskaja oblasti). (Society and political power: The Russian province, 1917–1985: Documents and materials: The Perm, Sverdlovsk and Chelyabinsk Regions). Ed. Veniamin V. ALEKSEEV. RAN, Ural. otd., In-t istorii i arkheologii, etc. Vol. 1. 1917–1945, Cheljabinskaja oblast'. Vol. 1. 1917–1945, Sverdlovskaja oblast'. Cheljabinsk – Ekaterinburg – Perm', Kniga, 2005, 2 vol., 638 p., 786 p. – ** [KULAKOV (Arkadij A.), KOLODNIKOVA (Ljudmila P.), SMIRNOV (Valerij V.).] Obshchestvo i vlast'. Rossijskaja provintsija, 1917 – 1980-e gody (po materialam nizhegorodskikh arkhivov): Dokumenty. (Society and political power: The Russian province, 1917–1980s: The documents from the archives of Nizhny Novgorod). RAN, In-t rossijskoj istorii, etc. Vol. 2. 1930 g.–ijun' 1941 g. (1930–June 1941). Vol. 3. Ijun' 1941 g.–1953 g. (June 1941–1953). RAN, In-t rossijskoj istorii. Moskva – Nizhnij Novgorod, IRI RAN, 2005, 2 vol., 1149 p., 1079 p. (plates; portr.; ind. p. 1109-1138, 1039-1070).

4465. 50 let bez Stalina: nasledie stalinizma i ego vlijanie na istoriju vtoroj poloviny XX veka. Materialy "kruglogo stola" 4 marta 2003 g. (50 years without Stalin: The heritage of stalinism and its influence on the history of the second half of the 20th century). Ed. Aleksandr S. SENJAVSKIJ. RAN, In-t rossijskoj istorii. Moskva, IRI RAN, 2005, 184 p. (bibl. incl.).

4466. BESPJATYKH (Jurij N.). Aleksandr Danilovich Menshikov: mify i real'nost'. (Aleksander Danilovich Menshikov: mythes and reality). Sankt-Peterburg, Istoricheskaja illjustracija, 2005, 235 p. (ill.; portr.; facs.; bibl. p. 194-222; pers. ind. p. 223-232).

4467. BOGDANOV (Andrej P.). Stikh i obraz izmenjajushchejsja Rossii: Poslednjaja chetvert' XVII – nachalo XVIII v. (Verse and image of changing Russia, the late 17th and the early 18th centuries). RAN, In-t rossijskoj istorii. Moskva, IRI RAN, 2005, 503 p. (pers. ind. p. 474-498).

4468. CIGLIANO (Giovanna). La Russia contemporanea: un profilo storico, 1855–2005. Roma, Carocci, 2005, 258 p. (Studi superiori, 501. Studi storici).

4469. DE MADARIAGA (Isabel). Ivan the Terrible: first Tsar of Russia. New Haven, Yale U. P., 2005, XXII-485 p.

4470. Dilemmas (The) of de-Stalinisation: negotiating cultural and social change in the Khrushchev era. Ed. by Polly JONES. London, Routledge, 2005, XIV-279 p. (BASEES/Routledge series on Russian and East European studies, 23).

4471. ESAKOV (Vladimir D.), LEVINA (Elena S.). Stalinskie "sudy chesti": Delo "KR". (Stalin's 'courts of honor': The Case of the 'counter-revolutionaries': [Microbiologist N.G. Klueva and immunologist G.I. Roskin, June 1947]). RAN, Otd. ist.-filol. nauk. Moskva, Nauka, 2005, 423 p. (ill.; portr.; bibl. p. 391-412; pers. ind. p. 413-421).

4472. GOLDMAN (Wendy). Stalinist terror and democracy: the 1937 union campaign. *American historical review*, 2005, 110, 5, p. 1427-1453.

4473. HAIMSON (Leopold H.). Russia's revolutionary experience, 1905–1917: two essays. With an introduction by David MAC DONALD. New York, Columbia U. P., 2005, XXXII-265 p. (Studies of the Harriman Institute of Columbia University).

4474. HIRSCH (Francine). Empire of nations: ethnographic knowledge and the making of the Soviet Union. Ithaca a. London, Cornell U. P., 2005, 367 p.

4475. HORVATH (Robert). The legacy of Soviet dissent: dissidents, democratisation and radical nationalism in Russia. London a. New York, RoutledgeCurzon, 2005, X-293 p. (BASEES/RoutledgeCurzon series on Russian and East European studies, 17).

4476. IVANOV (Jurij M.). V stalinskom "raju". (In Stalin's 'paradise'). Moskva, [s. n.], 2005, 115 p.

4477. IVANTSOV (Igor' G.). Mesto i rol' vnutripartijnogo kontrolja RKP(b)-VKP(b) v partijnom stroitel'stve na Kubani, 1920–1930 gg. (The place and the role of the party controle of C.P.R.(b.) and C.P.S.U.(b) in the party building in the Kuban Region, the 1920s and the 1930s). Krasnodar, [s. n.], 2005, 208 p. (plates; bibl. p. 165-173).

4478. JENSEN (Bent). Stalin: en biografi. (Biographie de Staline). København, Gyldendal, 2005, 349 p.

4479. KELLY (Catriona). Comrade Pavlik. The rise and fall of a Soviet boy hero. London, Granta Books, 2005, XXXII-352 p.

4480. KEMPER (Michael). Herrschaft, Recht und Islam in Daghestan: von den Khanaten und Gemeindebünden zum ğihād-Staat. Wiesbaden, Reichert, 2005, XI-462 p. (maps). (Kaukasienstudien, 8).

4481. KISELEV (Nikolaj P.). Iz istorii russkogo rozenkrejcerstva. (From the history of Russian Rosicrucianism). Intr. by Andrej I. SERKOV. Sankt-Peterburg, Izd-vo im. N.I. Novikova, 2005, 422 p. (ill.; ind. p. 392-416). (Russkoe masonstvo: Mat-ly i issled.; 5).

4482. KONDRATENKO (D. P.). Samoderzhavie, liberaly i natsional'nyi vopros v Rossii v kontse xix–nachale xx veka. (Autocracy, liberals, and the national question in Russia at the end of the nineteenth century and the beginning of the twentieth century). Kirov, Viatskii Gosudarstvennyi Gumanitarnyi Universitet, 2005, 238 p.

4483. KUHR-KOROLEV (Corinna). "Gezähmte Helden". Die Formierung der Sowjetjugend, 1917–1932. Essen, Klartext, 2005, 365 p.

4484. KUROMIYA (Hiroaki). Stalin. New York, Longman, 2005, XVII-227 p. (Profiles in power).

4485. KUZNECOVA (Tamara E.). "...I primknuvshij k nim..." – neprimknuvshij: (k 100 letiju so dnja rozhdenija D.T. Shepilova). ('And Shepilov joined them...' – joined nobody: To the 100th annuversary of Dmitry Shepilov). RAN, In-t ekonomiki. Moskva, [s. n.], 2005, 165 p. (bibl. p. 92-165).

4486. LARUELLE (Marlène). Mythe aryen et rêve impérial dans la Russie du XIXe siècle. Préf. de Pierre-André TAGUIEFF. Paris, CNRS, 2005, 223 p. (Mondes russes. Etats, sociétés, nations).

4487. Late imperial Russia. Problems and prospects. Ed. by Ian THATCHER. Manchester, Manchester U. P., 2005, 208 p.

4488. LEȘCU (Anatol). Românii în armata imperială rusă. Secolul al XVIII-lea – prima jumătate a secolului al XIX-lea. (Romanians in Russian imperial army. 18th century – first half of the 19th century). București, Editura Militară, 2005, 232 p.

4489. LEWIN (Moshe). The Soviet century. Ed. by Gregory ELLIOTT. London a. New York, Verso, 2005, IX-416 p.

4490. LIPHARDT (Elizaveta). Aporien der Gerechtigkeit: politische Rede der extremen Linken in Deutschland und Russland zwischen 1914 und 1919. Tübingen, Niemeyer, 2005, XI-255 p. (Reihe Germanistische Linguistik, 261).

4491. LITVIN (Alter), KEEP (John). Stalinism. Russian and Western views at the turn of the millenium. London a. New York, Routledge, 2005, XIV-248 p.

4492. MAL'KOVSKAJA (Tat'jana N.). Sem'ja i vlast' v Rossii XVII–XVIII stoletij. (Family and political power in Russia in the 17th and 18th centuries). Moskva, CheRo, 2005, 199 p.

4493. MURPHY (Brian). Rostov in the Russian Civil War, 1917–1920: the key to victory. London a. New York, Routledge, 2005, XIII-196 p. (ill., maps). (Cass military studies series).

4494. Pervaja revoljutsija v Rossii. Vzgljad cherez stoletie. (The First Revolution in Russia: a look from after a century). Ed. Avenir P. KORELIN, Stanislav V. TJUTJUKIN. RAN, In-t rossijskoj istorii. Moskva, Pamjatniki istoricheskoj mysli, 2005, 602 p. (bibl. incl.; pers. ind. p. 589-600).

4495. PIPES (Richard). Russian conservatism and its critics: a study in political culture. New Haven, Yale U. P., 2005, XV-216 p.

4496. Politicheskie partii v rossijskikh revoljutsijakh v nachale XX veka. (Political parties in the Russian revolutions of the early 20th century). Ed. Grigorij N. SEVOST'JANOV, Salavat M. ISKHAKOV. RAN, Nauch. sovet po istorii sots. reform, dvizhenij i revoljutsij. Moskva, Nauka, 2005, 533 p. (bibl. incl.).

4497. POLUNOV (Alexander). Russia in the nineteenth century. Autocracy, reform and social change, 1814–1914. Armonk, M. E. Sharpe, 2005, XVI-286 p. (New Russian history).

4498. READ (Christopher). Lenin: a revolutionary life. London a. New York, Routledge, 2005, XII-311 p. (ill., ports.). (Routledge historical biographies).

4499. REESE (Roger R.). Red commanders: a social history of the Soviet Army officer corps, 1918–1991. Lawrence, University Press of Kansas, 2005, XIII-315 p. (ill.). (Modern war studies).

4500. ROSSMAN (Jeffrey J.). Worker resistance under Stalin: class and revolution on the shop floor. Cambridge, Harvard U. P., 2005, 314 p. (ill., maps). (Russian Research Center studies, 96).

4501. RUZHITSKAJA (Irina V.). Zakonodatel'naja dejatel'nost' v tsarstvovanie imperatora Nikolaja I. (The legislative activity in Russia under Emperor Nicholas I, [1825–1855]). RAN, In-t rossijskoj istorii. Moskva, IRI RAN, 2005, 315 p. (bibl. p. 243-295; pers. ind. p. 309-313).

4502. SERVICE (Robert). Stalin: a biography. Cambridge, Harvard U. P. a. Belknap, 2005, XVIII-715 p.

4503. SHUBIN (Aleksandr V.). Anarkhija – mat' porjadka. Mezhdu krasnymi i belymi. Nestor Makhno kak zerkalo Rossijskoj revoljutsii. (Between the Red and the White: Nestor Makhno as a mirror of the Russian Revolution). Moskva, Eksmo–Jauza, 2005, 412 p. (Vnutrennjaja vojna).

4504. SMITH (Jeremy). The fall of Soviet Communism. Basingstoke, Palgrave Macmillan, 2005, 144 p. (Studies in European history).

4505. Stalin. A new history. Ed. by Sarah DAVIES and James HARRIS. New York, Cambridge U. P., 2005, XIII-295 p.

4506. TJUMENTSEV (Igor' O.), MIRSKIJ (Stanislav V.), RYBALKO (Natal'ja V.) [et al.]. Russkij arkhiv Jana Sapegi 1608–1711 godov: Opyt rekonstruktsii i istochnikovedcheskogo analiza. (The Russian archive of Jan Sapieha, 1608–1611: an attempt in reconstruction and source-critical study). Ed. Oleg V. INSHAKOV. Volgogradskij gos. un-t. Volgograd, Izd-vo Volgogr. gos. un-ta,

2005, 338 p. (ill.; bibl. p. 293-334). (Jubilejnaja serija "Trudy uchenykh VolGU").

4507. TORSTENDAHL (Rolf), SELUNSKAIA (Natal'ia). Zarozhdenie demokraticheskoi kul'tury: Rossiia v nachale XX veka. (The Birth of a democratic culture: Russia at the beginning of the twentieth century). Moskva, Rosspen, 2005, 335 p.

4508. UL'JANOVA (Galina N.). Blagotvoritel'nost' v Rossijskoj imperii, XIX – nachalo XX veka. (Philanthropy in the Russian Empire, the 19th and the early 20th century). RAN, In-t rossijskoj istorii. Moskva, Nauka, 2005, 403 p. (ill.; portr.; pers. ind. p. 390-400). [English summary]

4509. USMANOVA (Diljara M.). Musul'manskie predstaviteli v rossijskom parlamente, 1906–1916. (The Moslem representatives in the Russian parliament, 1906–1916). Kazan', Fen AN RT, 2005, 584 p. (ill.; bibl.).

4510. WADA (Haruki). Teroru to kaikaku: Alekusandoru nisei ansatsu zengo. (Terror and reform: the assassination of Alexander II, 1881). Tokyo, Yamakawa Shuppansha, 2005, 388 p.

4511. WETTIG (Gerhard). Stalins Aufrüstungsbeschluß. Die moskauer Beratungen mit den Parteichefs und Verteidigungsministern der "Volksdemokratien" vom 9. bis 12. Januar 1951. *Vierteljahrshefte für Zeitgeschichte*, 2005, 53, 4, p. 635-650.

4512. Zemskoe samoupravlenie v Rossii, 1864–1918. (Zemstvo and local self-governement in Russia, 1864–1918). Ed. Nadezhda G. KOROLEVA. RAN, In-t rossijskoj istorii. Part 1. 1864–1904. Part 2. 1905–1918. Moskva, Nauka, 2005, 2 vol., 428 p., 384 p. (bibl. incl.; pers. ind. p. 424-426, 379-382).

Addendum 2004

4513. TANIUCHI (Yuzuru). Ue kara no kakumei: Sutârin-shugi no genryu. (Revolution from above: the origin of Stalinism). Tokyo, Iwanami Shoten, 2004, 592 p.

Cf. n°s 93, 144, 206, 217, 274, 275, 276, 284, 322, 337, 338, 349, 358, 389, 422, 479, 568, 572, 574, 575, 576, 577, 585, 3553, 3555, 3907, 4953, 5016, 5020, 5022, 5102, 5115, 5503, 5557, 5837, 5841, 5951, 5969, 6099, 6437, 6455, 6821, 6880, 6898, 6905, 6920, 6932, 6967, 6975, 6989, 6998, 7004, 7018, 7057, 7084, 7161, 7240, 7260, 7354, 7372, 7385, 7520

Saudi Arabia

4514. CORDESMAN (Anthony H.), OBAID (Nawaf). National security in Saudi Arabia: threats, responses, and challenges. Westport, Praeger Security International, 2005, 439 p.

Senegal

4515. SECK (Assane). Sénégal, émergence d'une démocratie moderne, (1945–2005): un itinéraire politique. Préf. de Djibril SAMB. Paris, Karthala, 2005, 360 p. (ill., map). (Hommes et sociétés).

4516. SECK (Mamadou). Les scandales politiques sous la présidence de Abdoulaye Wade: vers un nouveau domaine d'étude en Afrique, la "scandalogie". Paris, L'Harmattan, 2005, 216 p. (Etudes africaines).

Serbia

4517. BARRATT BROWN (Michael). From Tito to Milosevic: Yugoslavia, the lost country. London, Merlin, 2005, IX-198 p. (ill., map).

4518. BEŠLIN (Branko). Evropski uticaji na srpski liberalizam u XIX veku. (European influences on the Serbian liberalism in the nineteenth century). Sremski Karlovci, Izd. knjižarnica Zorana Stojadinovića, 2005, 943 p. (Istorijska biblioteka).

4519. BIEBER (Florian). Nationalismus in Serbien vom Tode Titos bis zum Ende der Ära Milošević. Wien, Lit, 2005, IV-541 p. (Wiener Osteuropa-Studien, 18).

4520. EVANS (James R.). The creation of Yugoslavia: British attitudes to questions of South Slav nationality, 1900–1921. Oxford, University of Oxford, 2005, VII-291 p.

4521. Evolucija populacione politike u Srbiji 1945–2004. (Evolution of population politics in Serbia 1945–2004). Urednici Miloš MACURA, Ana GAVRILOVIĆ. Beograd, Srpska akdemija nauka i umetnosti, 2005, 426 p. (ill.). (Demografski zbornik, 7).

4522. IVETIC (Egidio). Lo jugoslavismo nell' esperienza delle due Jugoslavia. *Rivista storica italiana*, 2005, 117, 3, p. 780-824.

4523. LUTARD-TAVARD (Catherine). La Yougoslavie de Tito écartelée: 1945–1991. Paris, L'Harmattan, 2005, 566 p. (maps).

4524. MÜLLER (Dietmar). Staatsbürger aus Widerruf: Juden und Muslime als Alteritätspartner im rumänischen und serbischen Nationscode: ethnonationale Staatsbürgerschaftskonzepte 1878–1941. Wiesbaden, Harrassowitz, 2005, 537 p. (Balkanologische Veröffentlichungen, 41).

4525. STEFANOVIC (Djordje). Seeing the Albanians through Serbian Eyes: The Inventors of the Tradition of Intolerance and Their Critics, 1804–1939. *European history quarterly*, 2005, 35, 3, p. 465-492.

Slovakia

** 4526. MAČUHA (Maroš). Listy Matúša Ivanku, prefekta turčianskych majetkov rodu Révai, z rokov 1601–1625. (Letters of Matúš Ivanka, the Prefect of the Révay Estate in the Turiec region from 1601–1625). *Historický časopis*, 2005, 53, 1, p. 93-110.

** 4527. PETRUF (Pavol), SEGEŠ (Dušan). Memorandum Karola Sidora a slovenská otázka z júna 1943. (Karol Sidor's memorandum and standing of Slovakia in June 1943). *Historický časopis*, 2005, 53, 1, p. 123-150.

4528. BYSTRICKÝ (Jozef), ČAPLOVIČ (Miloslav), PURDEK (Imrich) [et. al.]. Ozbrojené sily Slovenskej republiky. História a súčasnosť. 1918–2005. = The Armed Forces of the Slovak Republic. The Past and Presence. 1918–2005). Bratislava, Magnet Press Slovakia s. r. o., 2005, 208 p.

4529. DULOVIČ (Erik). Realizácia štátnej politiky županov Košickej župy v rokoch 1923–1928. (Die Staatspolitik aus der Sicht der Gauleiter des Kaschauer Gaus, 1923–1928). *Historický časopis*, 2005, 53, 4, p. 659-695. [Deutsche Zsfassung].

4530. MIHÁLIKOVÁ (Silvia). Sviatky na Slovensku ako súčasť politických rituálov. (Die Feste in der Slowakei als Bestandteil der politischen Rituale). *Historický časopis*, 2005, 53, 2, p. 339-353. [Deutsche Zsfassung].

4531. RYBÁŘOVÁ (Petra). Prejavy politického antisemitizmu v Nitrianskej župe v osemdesiatych rokoch 19. storočia. (Die Anzeichen des politischen Antisemitismus im Nitraer Gau in den 80er Jahren des 19. Jahrhunderts). *Historický časopis*, 2005, 53, 3, p. 443-465. [Deutsche Zsfassung].

4532. ŠTILLA (Miloš). Martin Čulen a slovenská politika v šesťdesiatych a sedemdesiatych rokoch 19. storočia. (Martin Čulen und die slowakische Politik in den 60er un 70er Jahren des 19. Jahrhunderts). *Historický časopis*, 2005, 53, 1, p. 29-43. [Deutsche Zsfassung].

Cf. n°s 3756-3761, 3764-3767, 3769, 3772, 3774, 3775, 3778-3780, 3782-3784, 3786, 3788, 3790, 3792

Slovenia

4533. BLUMENWITZ (Dieter). Okkupation und Revolution in Slowenien (1941–1946): eine völkerrechtliche Untersuchung. Wien, Böhlau, 2005, 162 p. (Studien zu Politik und Verwaltung, 81).

4534. CORSELLIS (John), FERRAR (Marcus). Slovenia 1945: memories of death and survival after World War II. London a. New York, I.B. Tauris, 2005, XI-276 p. (ill., maps).

4535. MARUŠIČ (Branko). Pregled politične zgodovine Slovencev na Goriškem: 1848–1899. (A survey of political history of Slovenes in the Gorica Region: 1848–1899). Nova Gorica, Goriški muzej, 2005, 368 p.

4536. Med politiko in zgodovino: življenje in delo dr. Dušana Kermavnerja (1903–1975). (Among the policy and history: life and work of dr. Dusan Kermavner). Uredila Aleksander ŽIŽEK in Jurij PEROVŠEK. Ljubljana, Slovenska akademija znanosti in umetnosti, Zveza zgodovinskih društev Slovenije, 2005, 208 p.

4537. NEĆAK (Dušan), REPE (Božo). Prelom: 1914–1918: svet in Slovenci v 1. svetovni vojni (The 1914–1918 milestone: the world and Slovenes during WWI). Ljubljana, Sophia, 2005, VI-297 p. (Zbirka Spekter, 2005, 7).

4538. PEROVŠEK (Jurij). Na poti v moderno: poglavja iz zgodovine evropskega in slovenskega liberalizma 19. in 20. stoletja. (Towards modernity: chapters from the history of European and Slovene liberalism in the 19th and 20th centuries). Ljubljana, Inštitut za novejšo zgodovino, 2005, 285 p. (Zbirka Razpoznavanja, 1).

4539. Slovenska novejša zgodovina : od programa Zedinjena Slovenija do mednarodnega priznanja Republike Slovenije : 1848–1992. (Contemporary Slovene history : from the programme "United Slovenia" to the international recognition of the Republic of Slovenia : 1848–1992). Ed. Jasna FISCHER [et al.]. Ljubljana, Mladinska knjiga, Inštitut za novejšo zgodovino, 2005, 2 vol., [s. p.].

4540. VODOPIVEC (Peter). La Slovénie entre la Yougoslavie et l'Union européenne. *In*: La Slovénie et l'Europe: contributions à la connaissance de la Slovénie actuelle. Paris, L'Harmattan, 2005, p. 79-90. (Aujourd'hui l'Europe).

Cf. n°s 4520, 4522, 4523

South Africa

4541. BHANA (Surendra), VAHED (Goolam). The making of a political reformer: Gandhi in South Africa, 1893–1914. New Delhi, Manohar Pub., 2005, 181 p. (ill., facsim., ports.).

4542. BRECKENRIDGE (Keith). Verwoerd's Bureau of Proof: total information in the making of Apartheid. *History workshop*, 2005, 59, p. 83-108.

4543. GUELKE (Adrian). Rethinking the rise and fall of apartheid: South Africa and world politics. Basingstoke a. New York, Palgrave Macmillan, 2005, XVII-248 p. (Rethinking world politics).

4544. KRIKLER (Jeremy). White rising. The 1922 insurrection and racial killing in South Africa. New York, Manchester U. P., 2005, XIV-405 p.

4545. KYNOCH (Gary). We are fighting the world. A history of the Marashea gangs in South Africa, 1947–1999. Athens, Ohio U. P., 2005, XV-200 p. (New African histories series).

4546. NGANDO (B. Alfred). Le prince Mandela: essai d'introduction politique à la renaissance africaine. Paris, Maisonneuve et Larose, 2005, 104 p.

4547. ROOS (Neil). Ordinary springboks. White servicemen and social justice in South Africa, 1939–1961. Burlington, Ashgate, 2005, XVI-233 p.

Cf. n° 3539

Spain

* 4548. MOLINA APARICIO (Fernando). Modernidad e identidad nacional: el nacionalismo español del siglo XIX y su historiografía. *Historia social*, 2005, 52, p. 147-172.

** 4549. Cartas (Las) de Franco: la correspondencia desconocida que marcó el destino de España. Ed. por Jesús PALACIOS; prólogo de Stanley G. PAYNE. Madrid, Esfera de los Libros, 2005, 590 p. (ill.).

** 4550. Deutsche Reichstagsakten. Jüngere Reihe. Hrsg. v. der Historischen Kommission bei der Bayerischen Akademie der Wissenschaften durch Eike WOLGAST. Band 17. Deutsche Reichstagsakten unter Kaiser Karl V. Der Reichstag zu Regensburg 1546. Bearb. V. Rosemarie AULINGER. Band 19. Deutsche Reichstagsakten unter Kaiser Karl V. Der Reichtstag zu Augsburg 1550/51. Bearb v. Erwein ELTZ. München, Oldenbourg, 2005, 3 vol., 596 p., 1681 p.

** 4551. FÉLIX BALLESTA (Mª. Angeles). Relaciones Iglesia-Estado en la España de 1919 a 1923, según el Archivo Secreto Vaticano. Madrid, Dykinson, 2005, 601 p.

4552. ABELLÁN (José Luis). El "problema de España" y la cuestión militar: historia y conciencia de una anomalía. Madrid, Dykinson, 2005, 127 p. (Colección Comentarios y monografías, 5).

4553. ALBAREDA I SALVADÓ (Joaquim). El " cas dels catalans": la conducta dels aliats arran de la Guerra de Successió (1705–1742). Lleida, Pagès Editors i Barcelona, Fundació Noguera, 2005, 437 p. (ill.). (Estudis, 32).

4554. ALCALÁ (César). La represión política en Cataluña (1936–1939). Madrid, Grafite Ediciones, 2005, 332 p. (Biblioteca de historia).

4555. ANGOSTO VÉLEZ (Pedro Luis). Alfonso XIII, un rey contra el pueblo: raíces de la Guerra Civil: una mirada a través de "El Socialista" (1917–1923). Prólogo de Eduardo HARO TECGLEN. Sevilla, Renacimiento, 2005, 267 p. (ill.).

4556. BARRAGÁN MORIANA (Antonio). Crisis del franquismo y transición democrática en la provincia de Córdoba. Córdoba, Servicio de Publicaciones, Universidad de Córdoba, Ayuntamiento de Córdoba, Concejalía de Cultura, 2005, 612 p.

4557. BLESA I DUET (Isaïes). Un nuevo municipio para una nueva monarquía: oligarquías y poder local: Xàtiva, 1707–1808. Valencia, Universitat de València, 2005, 471 p. (Oberta. Historia, 119).

4558. CASALI (Luciano). Franchismo. Sui caratteri del fascismo spagnolo. Bologna, Clueb, 2005, 350 p.

4559. CASALS (Xavier). Franco y los Borbones: la corona de España y sus pretendientes. Barcelona, Planeta, 2005, 667 p. (ill.). (EspañaEscrita, EE:03).

4560. CHRIST (Michel). Le POUM: histoire d'un parti révolutionnaire espagnol, 1935–1952. Préface de Denis BERGER. Paris, L'Harmattan, 2005, 131 p.

4561. COBO ROMERO (Francisco), ORTEGA LÓPEZ (Teresa María). No sólo Franco: la heterogeneidad de los apoyos sociales al régimen franquista y la composición de los poderes locales. Andalucía, 1936–1948. *Historia social*, 2005, 51, p. 49-72.

4562. CORTÉS PEÑA (Antonio Luis). Alojamientos de soldados y levas: dos factores de conflictividad en la Andalucía de los Austrias. *Historia social*, 2005, 52, p. 19-34.

2. HISTORY BY COUNTRIES

4563. DUCH I PLANA (Montserrat). Dones públiques: política i gènere a l'Espanya del segle XX. Tarragona, Arola, 2005, 234 p. (ill.). (Col·lecció Atenea, 1).

4564. EALHAM (Chris). Class, culture, and conflict in Barcelona, 1898–1937. London a. New York, Routledge, 2005, XVI-264 p. (ill. map). (Routledge/Cañada Blanch studies on contemporary Spain, 7).

4565. Entre Clío y Casandra: poder y sociedad en la monarquía hispánica durante la edad moderna. Ed. por F. Javier GUILLAMÓN ÁLVAREZ, Julio D. MUÑOZ RODRÍGUEZ y Domingo CENTENERO DE ARCE. Murcia, Seminario "Floridablanca," Universidad de Murcia, 2005, 301 p. (ill.). (Cuadernos del Seminario "Floridablanca", 6).

4566. ESENWEIN (George Richard). The Spanish Civil War. New York a. London, Taylor & Francis, 2005, XII-302 p. (maps). (Routledge sources in history).

4567. Exilios (Los) en España (Siglos XIX y XX). III Congreso sobre el Republicanismo. Vol. I. Ponencias. Vol. II. Comunicaciones. Coord. José Luis CASAS SÁNCHEZ y Francisco DURÁN ALCALÁ. Córdoba, Priego de Córdoba 2005, 2 vol., 292 p., 713 p.

4568. Felipe IV: el hombre y el reinado. Coordinado por José ALCALÁ-ZAMORA Y QUEIPO DE LLANO. Madrid, Centro de Estudios Europa Hispánica, Real Academia de la Historia, 2005, 330 p. (ill.).

4569. *Vacat.*

4570. FONTANA I LÀZARO (Josep). Aturar el temps: la segona restauració espanyola, 1823–1834. Barcelona, Crítica, 2005, 298 p. (Fontana i Làzaro, Josep. Works. 2002, 2).

4571. FRANCO LANAO (Elena). Denuncias y represión en años de posguerra: el Tribunal de Responsabilidades Políticas en Huesca. Huesca, Instituto de Estudios Altoaragoneses, Diputación de Huesca, 2005, 201 p. (ill.).

4572. Franquisme (El) a Catalunya: 1939–1977. Coordinació Sonsoles MIGUEL. Vol. 1. La dictadura totalitària (1939–1945). Barcelona, Edicions 62, 2005, 270 p. (ill., maps, ports.).

4573. GANDOULPHE (Pascal). Au service du roi: institutions de gouvernement et officiers dans le royaume de Valence, (1556–1624). Montpellier, ETILAL, Université de Montpellier III, 2005, 338 p. (Collection Espagne médiévale et moderne, 7).

4574. GARCÍA FERNÁNDEZ (Hugo). Historia de un mito político: el "peligro comunista" en el discurso de las derechas españolas (1918–1936). *Historia social*, 2005, 51, p. 3-20.

4575. GARCÍA-SANZ MARCOTEGUI (Angel) [et al.]. Los liberales navarros durante el sexenio democrático. Pamplona, Universidad Pública de Navarra, 2005, 388 p.

4576. GLAZER (Peter). Radical nostalgia: Spanish Civil War commemoration in America. Rochester, University of Rochester, 2005, XVII-330 p. (ill.).

4577. GONZÁLEZ CALLEJA (Eduardo). La España de Primo de Rivera: la modernización autoritaria, 1923-1930. Madrid, Alianza, 2005, 463 p. (ill.).

4578. GRAHAM (Helen). The Spanish Civil War. Oxford, Oxford U. P., 2005, 175 p. (ill.). (A very short introduction).

4579. GUERRERO ARJONA (Melchor). Lorca: de ciudad de frontera a ciudad moderna: transformaciones políticas, sociales y económicas (1550–1598). Murcia, Academia Alfonso X el Sabio, 2005, 414 p. (ill.).

4580. HOLGUÍN (Sandie). "National Spain Invites You": battlefield tourism during the Spanish Civil War. *American historical review*, 2005, 110, 5, p. 1399-1426.

4581. JUNQUERAS (Oriol). Guerra, economia i política a la Catalunya de l'alta Edat Moderna. Sant Vicenç de Castellet, Farell, 2005, 173 p. (Col·lecció Nostra història, 8).

4582. Letrados, juristas y burócratas en la España moderna. Coordinador, Francisco José ARANDA PÉREZ. Cuenca, Ediciones de la Universidad de Castilla-La Mancha, 2005, 580 p. (ill.). (Ediciones institucionales, 47).

4583. LUQUE BALLESTEROS (Antonio). Política y fomento en la Andalucía liberal: Agustín Álvarez de Sotomayor Domínguez (Málaga, 1793–Puente Genil, 1855). Córdoba, Servicio de Publicaciones Universidad de Córdoba, 2005, 286 p.

4584. MARÍN (Dolors). Ministros anarquistas: la CNT [Confederación Nacional del Trabajo] en el gobierno de la II república (1936–1939). Barcelona, Random House Mondadori, 2005, 310 p. (Historia, 129).

4585. MARTÍN DE SANTA OLALLA SALUDES (Pablo). La Iglesia que se enfrentó a Franco: Pablo VI, la Conferencia episcopal y el Concordato de 1953. Madrid, Dilex, 2005, 495 p.

4586. MARTÍN RUBIO (Ángel David). Los mitos de la represión en la guerra civil. Madrid, Grafite, 2005, 283 p.

4587. MARTÍNEZ DE SAS (María Teresa). Socialistas, anarquistas e insurreccionalismo republicano (1876–1878). *Historia social*, 2005, 52, p. 59-72.

4588. MARTÍNEZ LEAL (Juan). Los socialistas en acción: la II República en Elche (1931–1936). Alicante, Publicaciones de la Universidad de Alicante, 2005, 166 p. (ill.). (Monografías / Universidad de Alicante).

4589. MARZAL RODRÍGUEZ (Pascual). Magistratura y República: el Tribunal Supremo (1931–1939). Sedaví, Editorial Práctica de Derecho, 2005, 276 p.

4590. MOA (Pío). 1936: el asalto final a la República. Coordinación general, Javier RUIZ PORTELLA. Barcelona, Áltera, 2005, 338 p. (ill., ports.). – IDEM. Franco – un balance histórico. Barcelona, Planeta, 2005, 180 p.

4591. MOLINERO (Carme). La captación de las masas: política social y propaganda en el régimen franquista. Madrid, Cátedra, 2005, 223 p. (ill.). (Historia.. Serie menor).

4592. MUNIESA (Bernat). Dictadura y transición: la España lampedusiana. Volume 1. La dictadura franquista, 1939–1975. Barcelona, Publicacions i Edicions, Universitat de Barcelona, 2005, 277 p. (Historia-Perspectiva, 1).

4593. MUÑOZ ALTABERT (M. Lluïsa). Les corts valencianes de Felip III. Valencia, Universitat de València, 2005, 276 p. (Oberta. Història, 112).

4594. NICOLÁS (Encarna). La libertad encadenada: España en la dictadura franquista, 1939–1975. Madrid, Alianza editorial, 2005, 455 p.

4595. NÚÑEZ SEIXAS (Xosé Manuel). ¿Un nazismo colaboracionista español? Martín de Arrizubieta, Wilhelm Faupel y los últimos de Berlín (1944–1945). *Historia social*, 2005, 51, p. 21-48.

4596. Pobreza, marginación, delicuencia y políticas sociales bajo el franquismo. Ed. por Conxita MIR, Carme AGUSTÍ y Josep GELONCH. Lleida, Edicions de la Universitat de Lleida, 2005, 229 p. (Espai/temps, 45).

4597. PRADA RODRÍGUEZ (Julio). A dereita política ourensá: monárquicos, católicos e fascistas (1934–1937). Vigo, Universidade de Vigo, Servizo de Publicacións, 2005, 364 p. (ill.). (Monografías da Universidade de Vigo. 66. Humanidades e ciencias xurídico-sociais).

4598. Premsa cultural i intervenció política dels intel·lectuals a la Catalunya contemporània (1814–1975). Coordinador Jordi CASASSAS. Barcelona, Publicacions i Edicions Universitat de Barcelona, 2005, 141 p. (Els textos de cercles, 1).

4599. República y republicanas en España. Ed. por Ma. Dolores RAMOS. Madrid, Asociación de Historia Contemporánea y Marcial Pons Ediciones de Historia, 2005, 335 p. (Ayer, 60/2005, 4).

4600. ROMERO SAMPER (Miligrosa). La oposición durante el franquismo. Vol. 3. El exilio republicano. Madrid, Encuentro Ediciones, 2005, 339 p (Ensayos, 207. Historia).

4601. RUIZ (Julius). Franco's justice: repression in Madrid after the Spanish Civil War. Oxford, Clarendon Press, 2005, X-257 p. (ill., maps). (Oxford historical monographs).

4602. SÁNCHEZ GARCÍA (Raquel). Alcalá Galiano y el liberalismo español. Prólogo de Jesús A. MARTÍNEZ MARTÍN. Madrid, Centro de Estudios Políticos y Constitucionales, 2005, 516 p. (Historia de la sociedad política).

4603. SANFELIÚ (Luz). Republicanas: identidades de género en el blasquismo (1895–1910). Valencia, Universitat de València, 2005, 344 p. (Oberta. Historia, 118).

4604. SOTO CARMONA (Alvaro). ¿Atado y bien atado? Institucionalización y crisis del franquismo. Prólogo de Javier TUSELL. Madrid, Biblioteca Nueva, 2005, 316 p. (Colección Historia Biblioteca Nueva). – IDEM. Transición y cambio en España, 1975–1996. Madrid, Alianza Editorial, 2005, 478 p. (ill., ports.).

4605. Spanish Civil War (The). Ed. by Kenneth W. ESTES and Daniel KOWALSKY. Detroit, Thomson/Gale, 2005, XXIX-440 p. (ill., map). (History in dispute, 18).

4606. Transició democràtica (La) als Països Catalans: història i memòria. Dir. Pelai PAGÈS I BLANCH. València, Universitat de València, 2005, 423 p. (Història i memòria del franquisme).

4607. TUSELL (Javier). Dictadura franquista y democracia, 1939–2004. Barcelona, Crítica, 2005, 481 p. (Historia de España, 14).

4608. VÁZQUEZ OSUNA (Federico). La rebellió dels tribunals: l'administració de justícia a Catalunya (1931–1953), la judicatura i el ministeri fiscal. Barcelona, Afers, 2005, 317 p. (Recerca i pensament, 21).

4609. VEGA SOMBRÍA (Santiago). De la esperanza a la persecución: la represión franquista en la provincia de Segovia. Prólogo de Julio ARÓSTEGUI. Barcelona, Crítica, 2005, XXII-543 p. (ill., maps). (Crítica contrastes).

4610. VILANOVA (Francesc). La Barcelona franquista i l'Europa totalitària (1939–1946): lectures polítiques de la segona guerra mundial. Barcelona, Empúries, 2005, 429 p. (Biblioteca universal Empúries, 201).

Addenda 1993–2004

4611. BAHAMONDE MAGRO (Angel), CERVERA GIL (Javier). Así terminó la guerra de España. Madrid, Marcial Pons, Ediciones de Historia, 99, 529 p. (maps). (Estudios).

4612. BOSCH (Aurora). The Spanish republic and the Civil War: rural conflict and collectivization. *Bulletin of Spanish studies*, 98, 75, p. 117-132.

4613. BURDIEL (Isabel), CRUZ ROMEO (M.). The making of the Spanish liberal devolution, 1808–1844. *Bulletin of Hispanic studies*, 98, 75, 5, p. 105-120.

4614. CASANOVA (Julián). De la calle al frente: el anarcosindicalismo en España (1931–1939). Barcelona, Crítica, 97, 265 p. (Libros de historia).

4615. CENARRO LAGUNAS (Angela). Cruzados y camisas azules: los orígenes del franquismo en Aragón, 1936–1945. Zaragoza, Prensas Universitarias de Zaragoza, 97, 499 p. (ill.). (Ciencias sociales, 30).

4616. CHAVES PALACIOS (Julián). La represión en la Provincia de Cáceres durante la Guerra Civil, 1936–1939. Cáceres, Universidad de Extremadura, 95, 324 p. (maps).

4617. GONZÁLEZ CUEVAS (Pedro Carlos). Acción española: teología política y nacionalismo autoritario en España, 1913–1936. Madrid, Tecnos, 98, 411 p. (Serie de historia).

4618. GONZÁLEZ MARTÍNEZ (Carmen). Guerra Civil en Murcia: un análisis sobre el poder y los comportamientos colectivos. Murcia, Universidad de Murcia, 99, 333 p. (ill.).

4619. HERRERÍN LÓPEZ (Angel). La CNT [Confederación Nacional del Trabajo] durante el franquismo: clandestinidad y exilio (1939–1975). Madrid, Siglo XXI de España, 2004, XII-468 p. (ill.). (Historia).

4620. MATEOS (Abdón). El PSOE [Partido Socialista Obrero Español] contra Franco: continuidad y renovación del socialismo español, 1953–1974. Madrid, Editorial P. Iglesias, 93, 500 p. (ill., maps).

4621. Régimen (El) de Franco (1936–1975): política y relaciones exteriores. Ed. por Javier TUSELL [et al.]. Madrid, Universidad Nacional de Educación a Distancia, 93, 2 vol., 592 p., 640 p.

4622. ROMEO MATEO (María Cruz). Entre el orden y la revolución: la formación de la burguesía liberal en la crisis de la monarquía absoluta (1814–1833). Alicante, Instituto de Cultura Juan Gil-Albert, 93, 255 p. (Ensayo e investigación, 47).

4623. RUIZ CARNICER (Miguel Angel). El Sindicato Español Universitario (SEU), 1939–1965: la socialización política de la juventud universitaria en el franquismo. México D.F., Siglo Veintiuno Editores, 96, XVIII-531 p. (ill.). (Historia).

4624. RÚJULA LÓPEZ (Pedro). Contrarrevolución: realismo y carlismo en Aragón y el Maestrazgo, 1820–1840. Zaragoza, Prensas Universidad de Zaragoza, 98, XII-516 p. (maps). (Ciencias sociales, 34).

4625. SÁNCHEZ RECIO (Glicerio). Los cuadros políticos intermedios del régimen franquista, 1936–1959: diversidad de origen e identidad de intereses. Valencia, Generalitat Valenciana, Conselleria de Cultura, Educació i Ciència y Alicante, Instituto de Cultura Juan Gil-Albert, Diputació Provincial de Alicante, 96, 220 p. (ill.). (Textos universitaris).

4626. SEBASTIÀ (Enric). La revolución burguesa: la transición de la cuestión señorial a la cuestión social en el País Valenciano. Prefacio de Javier PANIAGUA; estudio preliminar de José A. PIQUERAS. Valencia, Centro Francisco Tomás y Valiente UNED Alzira-Valencia, Fundación Instituto de Historia Social, 2001, 2 vol., 228 p., 212 p.

4627. SEVILLANO CALERO (Francisco). Propaganda y medios de comunicación en el franquismo (1936–1951). Alicante, Universidad de Alicante, 98, 150 p. (Universidad de Alicante España Publicaciones).

4628. Víctimas de la Guerra Civil. Coordinador Santos JULIÁ. Madrid, Temas de Hoy, 99, 431 p. (Colección Historia).

4629. Violencia política en la España del siglo XX. Dir. por Santos JULIÁ. Madrid, Taurus, 2000, 422 p. (Pensamiento).

4630. YUN (Bartolomé). Marte contra Minerva: El precio del Imperio español, c. 1450–1600. Barcelona, Crítica, 2004, XXVI-623 p. (Serie mayor).

Cf. n°s 262, 3522, 4058, 5261, 6468, 6469, 6652, 6875, 7440, 7521

Sudan

4631. COLLINS (Robert O.). Civil wars and revolution in the Sudan: essays on the Sudan, Southern Sudan and Darfur, 1962–2004. Hollywood, Tsehai, 2005, VI-II-408 p. (maps).

4632. IDRIS (Amir H.). Conflict and politics of identity in Sudan. New York, Palgrave Macmillan, 2005, XIV-143 p. (map).

4633. Land, ethnicity and political legitimacy in eastern Sudan, Kassala and Gedaref states. Ed. By Catherine MILLER. Le Caire, CEDEJ a. Khartoum, DSRC, 2005, XVIII-502 p.

4634. POLJAKOV (Konstantin I.). Istorija Sudana: XX vek. (A history of Sudan in the 20th century). RAN, In-t vostokovedenija. Moskva, IV RAN, 2005, 509 p. (bibl. p. 478-493; ind. p. 494-504). [English summary]

4635. PRUNIER (Gérard). Darfur. The ambiguous genocide. Ithaca, Cornell U. P., 2005, XXIII-212 p. (Crises in world politics).

4636. ROLANDSEN (Øystein). Guerilla government: political changes in the southern Sudan during the 1990s. Uppsala, Nordic Africa Institute, 2005, 201 p.

4637. UDAL (John O.). The Nile in darkness, a flawed unity, 1863–1899. Norwich, Michael Russell, 2005, XVII-685 p.

Sweden

** 4638. Krigsarkivet 200 år. (Deux siècles d'archives militaires suédoises). Red. Kerstin ABUKHAN-FUSA. Stockholm, Riksarkivet, 2005, 368 p. (ill.). (Årsbok för Riksarkivet och landsarkiven, 2005). (Meddelanden från Krigsarkivet, 24).

4639. ALMGREN (Birgitta). Drömmen om Norden: nazistisk infiltration i Sverige 1933–1945. (Le rêve du Nord. Infiltration nazie en Suède 1933–1945). Stockholm, Carlsson, 2005, 434 p. (ill.).

4640. BERLING ÅSELIUS (Ebba). Rösträtt med förhinder: rösträttsstrecken i svensk politik 1900–1920. (Obstacles au droit de vote: les restrictions au droit de vote dans la politique suédoise 1900–1920). Stockholm, Acta Universitatis Stockholmiensis, Almqvist & Wiksell International, 2005, 246 p. (ill.). (Stockholm studies in history, 82). [English summary].

4641. EINONEN (Piia). Poliittiset areenat ja toimintatavat: Tukholman porvaristo vallan käyttäjä ja vallankäytön kohteena n. 1592–1644. (Political arenas and modes of action: the burghers of Stockholm as subjects and objects in exercise of power, ca. 1592–1644). Helsinki, Suomalaisen Kirjallisuuden Seura, 2005, 316 p. (ill., port.). (Bibliotheca historica, 94).

4642. ETZEMÜLLER (Thomas). 1968, ein Riss in der Geschichte? Gesellschaftlicher Umbruch und 68er-Bewegungen in Westdeutschland und Schweden. Konstanz, UVK, 2005, 269 p. (ill.).

4643. FORSSBERG (Anna Marie). Att hålla folket på gott humör: informationsspridning, krigspropaganda och mobilisering i Sverige 1655–1680. (Maintenir le peuple de bonne humeur. Diffusion de l'information, propagande de guerre et mobilisation en Suède, 1665–1680). Stockholm, Acta Universitatis Stockholmiensis, Almqvist & Wiksell International, 2005, 330 p. (Stockholm studies in history, 80). [English summary].

4644. FROM (Peter). Karl XII:s död: gåtans lösning. (La mort de Charles XII. solution de l'énigme). Lund, Historiska media, 2005, 400 p. (pl.).

4645. HADENIUS (Stig). Gustaf V: en biografi. (Biographie du roi Gustave V de Suède). Lund, Historiska media, 2005, 314 p. (ill.).

4646. HÄGG (Göran). Välfärdsåren: svensk historia 1945–1986. (Les années d'Etat-providence: histoire de la Suède de 1945 à 1986). Stockholm, Wahlström & Widstrand, 2005, 475 p.

4647. IHSE (Cecilia). Präst, stånd och stat: kung och kyrka i förhandling 1642–1686. (Clergé, ordre et Etat. Les négociations entre le souverain et l'Eglise suédoise, 1642–1686). Stockholm, Acta Universitatis Stockholmiensis, Almqvist & Wiksell International, 2005, 219 p. (Stockholm studies in history, 78)[English summary].

4648. IRIE (Koji). Suwêden zettai osei kenkyu: zaisei, gunji, Barutokai teikoku. (A study of Swedish absolute monarchy: finance, war, and the Baltic empire). Tokyo, Chisen Shokan, 2005, 286 p.

4649. JONSSON (Alexander). De norrländska landshövdingarna och statsbildningen 1634–1769. (Les gouverneurs de la province du Norrland et la construction de l'Etat suédois, 1634–1769). Umeå, Institutionen för historiska studier, Umeå universitet, 2005, 317 p. (Skrifter från institutionen för historiska studier, 10). [English summary].

4650. LARSSON (Lars-Olof). Arvet efter Gustav Vasa: berättelsen om fyra kungar och ett rike. (L'héritage de Gustave Vasa: histoire de quatre rois et d'un royaume). Stockholm, Prisma, 2005, 497 p. (pl.).

4651. RYDGREN (Jens). Från skattemissnöje till etnisk nationalism: högerpopulism och parlamentarisk högerextremism i Sverige. (Du mécontentement fiscal au nationalisme ethnique. Populisme de droite et extrême-droite parlementaire en Suède). Lund, Studentlitteratur, 2005, 149 p. (Samtidshistoriska studier).

4652. Stormakten som sjömakt: marina bilder från karolinsk tid. (La marine suédoise à l'époque de la Grandeur). Red. Björn ASKER. Lund, Historiska media, 2005, 144 p. (ill.). (Karolinska förbundets årsbok, 2004. Meddelanden från Krigsarkivet, 25. Forum navales skriftserie, 11).

4653. SUNDBERG (Ulf). Stockholms blodbad. (Le bain de sang de Stockholm). Lund, Historiska media, 2005, 111 p.

Addendum 2001

4654. ANDERSSON (Irene). Kvinnor mot krig. Aktioner och nätverk för fred 1914–1940. Lund, Lund Universitetsforlag, 2001, 357 p. (Studia Historica Lundensia, 1).

Cf. n°s 563, 4355, 4357, 6880, 6898

Switzerland

4655. FÉRAL (Thierry). Suisse et nazisme. Postf. du Hanania Alain AMAR. Paris, L'Harmattan, 2005, 196 p. (Allemagne d'hier et d'aujourd'hui).

4656. KAESTLI (Tobias). Selbstbezogenheit und Offenheit: die Schweiz in der Welt des 20. Jahrhunderts: zur politischen Geschichte eines neutralen Kleinstaats. Zürich, Verlag Neue Zürcher Zeitung, 2005, 579 p. (ill.).

4657. KRUMMENACHER-SCHÖLL (Jörg). Flüchtiges Glück: die Flüchtlinge im Grenzkanton St. Gallen zur Zeit des Nationalsozialismus. Vorwort von Kathrin HILBER. Zürich, Limmat, 2005, 415 p. (ill., maps).

4658. MÄCHLER (Stefan). Hilfe und Ohnmacht: der Schweizerische Israelitische Gemeindebund und die nationalsozialistische Verfolgung 1933–1945. Zürich, Chronos, 2005, 569 p. (ill.). (Beiträge zur Geschichte und Kultur der Juden in der Schweiz, 10).

4659. Pouvoirs et société à Fribourg sous la Médiation (1803–1814): actes du colloque de Fribourg (journée du 11 octobre 2003) = Staat und Gesellschaft in Freiburg zur Mediationszeit (1803–1814). Ed. par Francis PYTHON. Fribourg, Academic Press, 2005, XI-463 p. (ill., map). (Religion – Politik – Gesellschaft in der Schweiz, 37).

4660. Suisse de 1848 (La): réalités et représentations. Sous la dir. de Marie-Jeanne HEGER-ETIENVRE. Strasbourg, Presses universitaires de Strasbourg, 2005, 160 p. (Series: Helvetica, 6).

4661. WIPF (Matthias). Bedrohte Grenzregion: die schweizerische Evakuationspolitik 1938–1945 am Beispiel von Schaffhausen. Zürich, Chronos, 2005, 280 p. (ill., maps, ports.). (Schaffhauser Beiträge zur Geschichte, 79).

Syria

4662. GINAT (Rami). Syria and the doctrine of Arab neutralism: from independence to dependence. Brighton a. Portland, Sussex Academic Press, 2005, XIX-310 p.

4663. PROVENCE (Michael). The great Syrian revolt and the rise of Arab nationalism. Austin, University of Texas Press, 2005, XII-209 p. (Modern Middle East series, 22).

4664. SORBY (Karol). Sýria v centre blízkovýchodného mocenského zápasu 1955–1957). (Syria in the Centre of the Middle Easter power struggle). *Historický časopis*, 2005, 53, 3, p. 523-544.

2. HISTORY BY COUNTRIES

Taiwan

4665. TSAI (Shih-Shan Henry). Lee Teng-hui and Taiwan's quest for identity. New York, Palgrave Macmillan, 2005, XVIII-271 p.

Tanzania

4666. GWASSA (Gilbert Clement Kamana). The outbreak and development of the Maji Maji War, 1905–1907. Ed. by Wolfgang APELT; with a supplementary bibliography up to 2005 by Wilhelm J.G. MÖHLIG. Köln, Rüdiger Köppe, 2005, 330 p. (ill., maps).

4667. MBOGONI (Lawrence Ezekiel Yona). The cross vs the crescent: religion and politics in Tanzania from 1890s to the 1990s. Dar es Salaam, Mukuki na Nyota Publishers, 2005, XI-230 p.

4668. Search (In) of a nation: histories of authority and dissidence in Tanzania. Ed. by Gregory H. MADDOX and James L. GIBLIN. Oxford, James Currey, 2005, XIV-337 p. (Eastern African Studies).

Thailand

4669. BAKER (Chris), PHONGPAICHIT (Pasuk). A history of Thailand. Cambridge, New York a. Melbourne, Cambridge U. P., 2005, XVIII-301 p.

4670. TERWIEL (B. J.). Thailand's political history: from the fall of Ayutthaya in 1767 to recent times. Bangkok, River, 2005, 328 p. (ill.).

Togo

Cf. n° 648

Tunisia

4671. ABBASSI (Driss). Entre Bourguiba et Hannibal: identité tunisienne et histoire depuis l'indépendance. Préf. de Robert ILBERT. Paris, Karthala et Aix-en-Provence, IREMAM, 2005, 265 p. (ill., maps). (Hommes et sociétés).

4672. BOUZIDI (Abdelmadjid). Pouvoir et esclavage dans la régence de Tunis: les serviteurs des beys Husseinites, XVIIIe–début XIXe siècles. Tunis, Centre de publication universitaire, 2005, 165 p. (ill.).

4673. HUNGER (Bettina). Wer sind wir? Gruppenidentitäten und nationale Einheit im kolonialen und postkolonialen Tunesien. Bern u. Oxford, P. Lang, 2005, 373 p. (ill.).

4674. SHA'BĀN (al-Sādiq). Processus de gouvernement des societes en transition: essai d'une approche par stratégie. L'exemple de la Tunisie, avant et après 1969. Tunis, Centre de Publication Universitaire, 2005, 339 p. (ill.).

Turkey

** 4675. Fransız diplomatik belgelerinde Ermeni olayları (1914–1918). [Les évenements Armeniens dans les documents diplomatiques Français (1914–1918)]. Ed. Hasan DILAN. Ankara, Türk Tarih Kurumu, 2005, 6 vol., [s. p.]. (Türk Tarih Kurumu Yayınları 106. 16 dizi).

** 4676. Völkermord (Der) an den Armeniern 1915/16: Dokumente aus dem Politischen Archiv des deutschen Auswärtigen Amts. Hrsg. v. Wolfgang GUST. Springe, Zu Klampen, 2005, 674 p.

4677. ÁGOSTON (Gábor). Guns for the sultan: military power and the weapons industry in the Ottoman Empire. Cambridge a. New York, Cambridge U. P., 2005, XVII-277 p. (ill., maps). (Cambridge studies in Islamic civilization).

4678. AKMEŞE (Handan Nezir). The birth of modern Turkey: the Ottoman military and the march to World War I. London a. New York, I.B. Tauris, 2005, XI-224 p. (International library of twentieth century history, 4).

4679. AKÖZ (Alaaddin). Karamanoğlu II. İbrahim Beyin Osmanlı Sultanı II. Murad'a Vermiş Olduğu Ahidnâme. (The Ahidnâme which was given by Karamanoğlu İbrahim Bey II to the Ottoman Sultan Murad II). *Tarih Araştırmaları Dergisi (Ankara Üniversitesi Dil Ve Tarih Coğrafya Fakültesi)*, 2005, 24, 38, p. 71-92.

4680. ANDUZE (Eric). La Franc-maçonnerie de la Turquie Ottomane: 1908–1924. Paris, L'Harmattan, 2005, 178 p. (Inter-national).

4681. BILGE (M. Sadık). Osmanlı Devleti ve Kafkasya: Osmanlı varlığı döneminde Kafkasya'nın siyasî-askerî tarihi ve idarî taksimatı (1454–1829). [Etat Ottoman et Caucase: Histoire politique et militaire de Caucase et sa structure administrative à l'époque ottoman 1454–1829]. İstanbul, Eren, 2005, 295 p.

4682. BLOXHAM (Donald). The great game of genocide. Imperialism, nationalism and destruction of the Ottoman Armenians. Oxford, Oxford U. P., 2005, XIV-329 p.

4683. ÇETINSAYA (Gökhan). The Caliph and Mujtahids: Ottoman policy towards the Shiite community of Iraq in the late nineteenth century. *Middle Eastern studies*, 2005, 41, 4, p. 561-574.

4684. ÇIÇEK (Kemal). Ermenilerin zorunlu göçü 1915–1917. (Migration obligatoire des Arméniens 1915–1917). Ankara, Türk Tarih Kurumu, 2005, XI-368 p.

4685. EROĞLU (Cengiz). Osmanlı vilayet salnamelerinde Musul. (Mossoul d'après les annales provinciales ottomanes). Ankara, Global Strateji Enstitüsü, 2005, 461 p.

4686. FARKAS (Paul). Palace revolution and counterrevolution in Turkey: March–April 1909. İstanbul, Isis, 2005, 77 p.

4687. GEORGELIN (Hervé). La fin de Smyrne: du cosmopolitisme aux nationalismes. Paris, CNRS, 2005, 254 p. (ill., maps). (CNRS histoire).

4688. GÜLALP (Haldun). Enlightenment by Fiat: secularization and democracy in Turkey. *Middle Eastern studies*, 2005, 41, 3, p. 351-372.

4689. KAYLAN (Muammer). The Kemalist: Islamic revival and the fate of secular Turkey. Amherst, Prometheus Books, 2005, 482 p. (ill., maps, ports.).

4690. KECHRIOTIS (Vangelis). Greek-Orthodox, Ottoman Greeks or just Greeks? Theories of coexistence in the aftermath of the Young Turk Revolution. *Balkan Studies*, 2005, 1, p. 51-72.

4691. KIESER (Hans-Lukas). Vorkämpfer der "Neuen Türkei": revolutionäre Bildungseliten am Genfersee (1870–1939). Zürich, Chronos, 2005, 197 p. (ill.).

4692. KUSHNER (David). To be governor of Jerusalem : the city and district during the time of Ali Ekrem Bey, 1906–1908. İstanbul, Isis, 245 p.

4693. Legitimizing the order: the Ottoman rhetoric of state power. Ed. by Hakan T. KARATEKE and Maurus REINKOWSKI. Leiden a. Boston, Brill, 2005, VIII-259 p. (ill.). (Ottoman Empire and its heritage, 34).

4694. LEWY (Guenter). The Armenian massacres in Ottoman Turkey: a disputed genocide. Salt Lake City, University of Utah Press, 2005, 383 p.

4695. MASSICARD (Elise). L'autre Turquie: le mouvement aléviste et ses territoires. Paris, Presses universitaires de France, 2005, 361 p. (map). (Proche Orient).

4696. MILLER (Ruth Austin). Legislating authority: sin and crime in the Ottoman Empire and Turkey. New York a. London, Routledge, 2005, VII-188 p. (Middle East studies: history, politics, and law. Middle East studies).

4697. Ottoman reform and Muslim regeneration. Ed. by Itzchak WEISMANN and Fruma ZACHS. London a. New York, I.B. Tauris, 2005, 233 p. (ill.). (Library of Ottoman studies, 8).

4698. ÖZDEN (Mehmet). Atatürk Döneminde Kemalist Metinler: A'râfda Bir Kemalizm: Tekin Alp ve Kemalizm (1936). [Textes Kémalistes au temps d'Atatürk: un Kémalisme dans l'A'raf (1936)]. *Bilig Türk Dünyası Sosyal Bilimler Dergisi / Journal of Social Sciences of the Turkish World*, 2005, 34, p. 45-81.

4699. POPE (Hugh). Sons of the conquerors: the rise of the Turkic world. New York, Overlook Press, 2005, 432 p.

4700. REINKOWSKI (Maurus). Die Dinge der Ordnung. Eine vergleichende Untersuchung über die osmanische Reformpolitik im 19. Jahrhundert. München, Oldenbourg, 2005, 365 p. (Südosteuropäische Arbeiten, 124).

4701. TÜRKDOĞAN (Orhan). Türk ulus-devlet kimliği. (Identité de nation-état turque). İstanbul, IQ Kültür Sanat Yayıncılık, 2005, 525 p.

4702. ÜNSALDI (Levent). Le militaire et la politique en Turquie. Préf. de Deniz AKAGÜL. Paris, L'Harmattan, 2005, 353 p. (ill., maps).

4703. URSINUS (Michael). Grievance administration (şikayet) in an Ottoman province: the Kaymakam of Rumelia's 'Record book of complaints' of 1781–1783. London, RoutledgeCurzon, 2005, XI-190 p. (Royal Asiatic Society books. Ibrahim Pasha Fund of Egypt series).

4704. VANDERLIPPE (John M.). The politics of Turkish democracy: Ismet Inönü and the formation of the multi-party system, 1938–1950. Albany, State University of New York Press, 2005, 280 p. (State University of New York series in the social and economic history of the Middle East).

4705. YAKUT (Esra). Şeyhülislamlık: yenileşme döneminde devlet ve din. (Shayhulislamat: Etat et religion à l'époque de la modernisation). İstanbul, Kitap, 2005, 268 p.

4706. ZAIM (Sabahattin).Türkiye'nin yirminci yüzyılı: toplum/iktisat/siyaset. (Le vingtième siècle de la Turquie: société / économie / politique). İstanbul, İşaret, 2005, 3 vol., [s. p.].

Cf. n^{os} *4101, 4102, 4169, 5520, 7004, 7017, 7022, 7054, 7418*

Turkmenistan

4707. FÉNOT (Anne), GINTRAC (Cécile). Achgabat, une capitale ostentatoire: autocratie et urbanisme au Turkménistan. Paris, L'Harmattan, 2005, 228 p. (ill., maps).

Uganda

4708. OCITTI (Jim). Press politics and public policy in Uganda: the role of journalism in democratization. Lewiston a. Lampeter, Edwin Mellen Press, 2005, XIV-152 p.

Ukraine

4709. BOECKH (Katrin). Jüdisches Leben in der Ukraine nach dem Zweiten Weltkrieg. Zur Verfolgung einer Religionsgemeinschaft im Spätstalinismus (1945– 1953). *Vierteljahrshefte für Zeitgeschichte*, 2005, 53, 3, p. 421-448.

4710. GRELKA (Frank). Die ukrainische Nationalbewegung unter deutscher Besatzungsherrschaft 1918 und 1941/42. Wiesbaden, Harrassowitz, 2005, 507 p. (Studien der Forschungsstelle Ostmitteleuropa an der Universität Dortmund, 38).

4711. STRUVE (Kai). Bauern und Nation in Galizien: über Zugehörigkeit und soziale Emanzipation im 19. Jahrhundert. Göttingen, Vandenhoeck & Ruprecht, 2005, 485 p. (ill.). (Schriften des Simon-Dubnow-Instituts Leipzig, 4).

4712. VARVARCEV (Mikola Mikolajovic). Dzuzeppe Madzini madzinizm i Ukraina. (Giuseppe Mazzini il mazzinianesimo e l'Ucraina). Kiiv, Pulsary, 2005, 302 p. (ill.).

4713. VULPIUS (Ricarda). Nationalisierung der Religion: Russifizierungspolitik und ukrainische Nationsbildung 1860–1920. Wiesbaden, Harrassowitz, 2005, 475 p. (ill., maps). (Forschungen zur osteuropäischen Geschichte, 64).

Cf. n^{os} *4503, 5102, 6975*

United Arab Emirates

4714. DAVIDSON (Christopher). The United Arab Emirates. A study in survival. Boulder, Lynne Rienner, 2005, 343 p. (Middle East in the international system).

United States

** 4715. Documentary history of the John F. Kennedy presidency. General editor, Lewis GOULD. Vol. 1. The 1960 election and the religion question. V. 2. The 1960 presidential debates. V. 3. Creation of the Presidential Task Force on Vietnam and a drafting of a "Program of Action" on Vietnam, April–May 1961. Bethesda, LexisNexis, 2005, 3 vol., LIV-835 p., XXXIV-644 p., XXXVIII-670 p.

** 4716. Papers (The) of James Madison. Vol. 7. Secretary of state series. 2 April-31 August 1804. Ed. by David B. MATTERN [et al.]. Charlottesville: University Press of Virginia, 2005, XXXV-714 p.

4717. ACKERMAN (Bruce). The failure of the founding fathers: Jefferson, Marshall, and the rise of presidential democracy. Cambridge, Belknap Press of Harvard U. P., 2005, 384 p.

4718. ALENT'EVA (Tat'jana V.). Obshchestvennoe mnenie i nazrevanie "neotvratimogo konflikta" v SShA v otrazhenii "New York Tribune" (1841–1861 gg.). (The public opinion and the brewing of the "inevitable conflict" in the USA: As reflected by New York Tribune, 1841–1861). Regional'nyj otkrytyj sots. in-t. Kursk, ROSI, 2005. 255 p. (bibl. p. 218-250). [English summary]

4719. ALTERMAN (Eric). When presidents lie. A history of official deception and its consequences. New York, Viking Press, 2005, IX-447 p.

4720. American Civil War (The). Ed. by Ethan RAFUSE. Aldershot, Ashgate, 2005, XXXIII-645 p. (ill.). (International library of essays in military history).

4721. ANDERSON (Fred), CAYTON (Andrew). The dominion of war. Empire and liberty in North America, 1500–2000. New York, Viking, 2005, XXIV-520 p.

4722. ANDERSON (Gary Clayton). Conquest of Texas: ethnic cleansing in the promised land, 1820–1875. Norman, University of Oklahoma Press, 2005, X-494 p.

4723. BARRETT (David M.). The CIA and Congress: the untold story from Truman to Kennedy. Lawrence, University Press of Kansas, 2005, X-542 p.

4724. BÉLAND (Daniel). Social security: history and politics from the New Deal to the privatization debate. Lawrence, University Press of Kansas, 2005, XII-252 p. (Studies in government and public policy).

4725. BERG (Manfred). "The Ticket to Freedom": the NAACP [National Association for the Advancement of Colored People] and the struggle for Black political integration. Foreword by John David SMITH. Gainesville, University Press of Florida, 2005, XX-352 p. (New perspectives on the history of the South).

4726. BLUE (Frederick J.). No taint of compromise. Crusaders in antislavery politics. Baton Rouge, Louisiana State U. P., 2005, XIV-301 p. (Antislavery, abolition, and the atlantic world).

4727. BLUM (Edward J.). Reforging the White Republic: race, religion, and American nationalism, 1865–1898. Baton Rouge, Louisiana State U. P., 2005, X-356 p. (Conflicting worlds: new dimensions of the American Civil War).

4728. BOWLES (Nigel). Nixon's business: authority and power in presidential politics. College Station, Texas A&M U. P., 2005, X-305 p.

4729. BRUNDAGE (William Fitzhugh). The Southern past: a clash of race and memory. Cambridge, Belknap Press of Harvard U. P., 2005, XIII-418 p.

4730. BUEL (Richard Jr.). America on the brink. How the political struggle over the war of 1812 almost destroyed the young republic. New York, Palgrave Macmillan, 2005, 302 p.

4731. CARMICHAEL (Peter S.). The last generation. Young Virginians in peace, war, and reunion. Chapel Hill, University of North Carolina Press, 2005, XIV-343 p. (Civil War America).

4732. CREIGHTON (Margaret S.). The colors of courage. Gettysburg's forgotten history; immigrants, women and African Americans in the Civil War's defining battle. New York, Basic Books, 2005, XXVII-321 p.

4733. DUBBER (Markus Dirk). The police power. Patriarchy and the foundations of American government. New York, Columbia U. P., 2005, XVI-268 p.

4734. ELLIS (Joseph J.). Seine Exzellenz George Washington. München, Beck, 2005, 386 p.

4735. FINZSCH (Norbert). Konsolidierung und Dissens: Nordamerika von 1800 bis 1865. Münster, LIT, 2005, VIII-926 p. (ill., maps). (Geschichte Nordamerikas in atlantischer Perspektive von den Anfängen bis zur Gegenwart, 5).

4736. FISHER (Louis). Military tribunals and presidential power. American revolution to the war on terrorism. Lawrence, University Press of Kansas, 2005, XIV-279 p.

4737. GOULD (Lewis L.). The most exclusive club: a history of the modern United States Senate. New York, Basic Books, 2005, XIV-402 p.

4738. GRAUBARD (Stephen). Command of office. How war, secrecy, and deception transformed the presidency from Theodore Roosevelt to George W. Bush. New York, Basic Books, 2005, XIII-722 p.

4739. GRENIER (John). The first way of war. American war making on the frontier, 1607–1814. New York, Cambridge a. Melbourne, Cambridge U. P., 2005, XIV-232 p.

4740. HOLM (Tom). The great confusion in Indian affairs. Native Americans and whites in the progressive era. Austin, University of Texas Press, 2005, XX-244 p.

4741. HORTON (Carol A.). Race and the making of American liberalism. New York, Oxford U. P., 2005, IX-300 p.

4742. KANN (Mark E.). Punishment, prisons, and patriarchy. liberty and power in the early American republic. New York, New York U. P., 2005, IX-337 p.

4743. KRUSE (Kevin M.). White flight. Atlanta and the making of modern conservatism. Princeton, Princeton U. P., 2005, XIV-325 p.

4744. LAFEBER (Walter). The deadly bet: LBJ [Lyndon Baines Johnson], Vietnam, and the 1968 election. Lanham, Rowman & Littlefield Publishers, 2005, X-215 p. (Vietnam-America in the war years).

4745. LAUSE (Mark A.). Young America: land, labor, and the republican community. Urbana a. Chicago, University of Illinois Press, 2005, VIII-240 p.

4746. LEPORE (Jill). New York burning: liberty, slavery, and conspiracy in eighteenth-century Manhattan. New York, Alfred A. Knopf, 2005, XX-323 p.

4747. LEUCHTENBURG (William E.). The White House looks South: Franklin D. Roosevelt, Harry S. Truman, Lyndon B. Johnson. Baton Rouge, Louisiana State U. P., 2005, XI-668 p. (Walter Lynwood Fleming lectures in Southern history).

4748. LEVINE (Bruce). Confederate emancipation: Southern plans to free and arm slaves during the Civil War. New York, Oxford U. P., 2005, 272 p.

4749. Looking back at LBJ [Lyndon B. Johnson]: White House politics in a new light. Ed. by Mitchell B. LERNER. Lawrence, University Press of Kansas, 2005, VIII-303 p.

4750. MAC FARLAND (Keith D.), ROLL (David L.). Louis Johnson and the arming of America. The Roosevelt and Truman years. Bloomington, Indiana U. P., 2005, X-452 p.

4751. MANCKE (Elizabeth). The fault lines of empire. Political differentiation in Massachusetts and Nova Scotia, ca. 1760–1830. New York, Routledge, 2005, XI-214 p. (New World in the Atlantic world).

4752. MAY (Gary). The informant. The FBI, the Ku Klux Klan, and the murder of Viola Liuzzo. New Haven, Yale U. P., 2005, XV-431 p.

4753. METTLER (Suzanne). Soldiers to citizens: the G.I. bill and the making of the greatest generation. New York, Oxford U. P., 2005, XVI-252 p.

4754. MIXON (Gregory). The Atlanta riot. Race, class, and violence in a new South city. Gainesville, University Press of Florida, 2005, XV-197 p. (Southern dissent).

4755. MORGAN (Francesca). Women and patriotism in Jim Crow America. Chapel Hill, University of North Carolina Press, 2005, XIII-293 p. (Gender and American culture).

4756. MORRISON (Jeffry H.). John Witherspoon and the founding of the American republic. Notre Dame, University of Notre Dame Press, 2005, XV-220 p.

4757. MOSER (John E.). Right turn: John T. Flynn and the transformation of American liberalism. New York, New York U. P., 2005, IX-277 p.

4758. NADASEN (Premilla). Welfare warriors: the welfare rights movement in the United States. New York, Routledge, 2005, XIX-310 p.

4759. NASH (Gary B.). The unknown American Revolution. The unruly birth of democracy and the struggle to create America. New York, Viking, 2005, XXIX-512 p.

4760. NEFF (John R.). Honoring the Civil War dead. Commemoration and the problem of reconciliiation. Lawrence, University Press of Kansas, 2005, XIV-328 p. (Modern war studies).

4761. OAKLEY (Christopher Arris). Keeping the circle: American Indian identity in Eastern North Carolina, 1885–2004. Lincoln, University of Nebraska Press, 2005, XII-191 p. (Indians of the Southeast).

4762. OBERLY (James W.). A nation of statesmen. The political culture of the Stockbridge-Munsee Mohicans, 1815–1972. Norman, University of Oklahoma Press, 2005, XV-336 p. (Civilization of the American Indian series, 252).

4763. OGBAR (Jeffrey O. G.). Black power. Radical politics and African American identity. Baltimore, Johns Hopkins U. P., 2005, X-258 p. (Reconfiguring American political history).

4764. OROPEZA (Lorena). "Raza si" "guerra no". Chicano protest and patriotism during the Viet Nam war era. Berkeley a. Los Angeles, University of California Press, 2005, XVIII-278 p.

4765. PFAU (Michael William). The political style of conspiracy. Chase, Sumner, and Lincoln. East Lansing, Michigan State U. P., 2005, VII-248 p. (Rhetoric and public affairs series).

4766. PORTNOY (Alisse). Their right to speak: women's activism in the Indian and slave debates. Cambridge, Harvard U. P., 2005, XII-288 p.

4767. PRUSHANKIN (Jeffery S.). A crisis in Confederate command: Edmund Kirby Smith, Richard Taylor, and the Army of the Trans-Mississippi. Baton Rouge, Louisiana State U. P., 2005, XX-308 p.

4768. RAUS (Edmund J. Jr.). Banners South. A northern commmunity at war. Kent, Kent State U. P., 2005, XIV-333 p. (Civil War in the north).

4769. ROMANOV (Vladimir V.). V poiskakh novogo miroporjadka: Vneshnepoliticheskaja mysl' SShA (1913–1921 gg.). (Searching a new world order: The discussions on foreign policy in the USA in 1913–1921). RAN, In-t vseobshchej istorii; Tambov. gos. un-t im. G.R. Derzhavina. Tambov, Izd-vo TGU im. G.R. Derzhavina, 2005, 515 p. (ill.; portr.; bibl. p. 468-494; pers. ind. p. 496-504). [English summary]

4770. RUBIN (Anne Sarah). A shattered nation. The rise and fall of the Confederacy, 1861–1868. Chapel Hill, University of North Carolina Press, 2005, X-319 p.

4771. Ruling America. A history of wealth and power in America. Ed. by Steve FRASER and Gary GERSTLE. Cambridge, Harvard U. P., 2005, 368 p.

4772. SANDERS (Charles W. Jr.). While in the hands of the enemy. Military prisons of the Civil War. Baton Rouge, Louisiana State U. P., 2005, X-390 p. (Conflicting worlds. New dimensions of the American Civil War).

4773. SEVERANCE (Ben H.). Tennessee's radical army. The state guard and its role in reconstruction, 1867–1869. Knoxville, University of Tennessee Press, 2005, XVIII-327 p.

4774. SHENK (Gerald E.). "Work or Fight!" race, gender, and the draft in World War One. New York, Palgrave Macmillan, 2005, X-194 p.

4775. SILBER (Nina). Daughters of the union. Northern women fight the Civil War. Cambridge, Harvard U. P., 2005, 332 p.

4776. SLOTKIN (Richard). Lost battalions. The Great War and the crisis of American nationality. New York, Henry Holt, 2005, XII-639 p.

4777. SMALL (Melvin). At the water's edge: American politics and the Vietnam War. Chicago, Ivan R. Dee, 2005, XI-241 p. (American ways series).

4778. STALOFF (Darren). Hamilton, Adams, Jefferson. The politics of enlightenment and the American founding. New York, Hill & Wang, 2005, VIII-419 p.

4779. THOMPSON FRIEND (Craig). Along the Maysville Road: the early American republic in the Trans-Appalachian West. Knoxville, University of Tennessee Press, 2005, XVII-378 p.

4780. TREFOUSSE (Hans L.). "First Among Equals": Abraham Lincoln's reputation during his administration. New York, Fordham U. P., 2005, XIV-199 p.

4781. US government (The), citizen groups and the Cold War: the state-private network. Ed. by Helen LAVILLE and Hugh WILFORD. Abingdon: Routledge, 2005, XVIII-240 p. (Cass series-studies in intelligence).

4782. WALDREP (Christopher). Vicksburg's long shadow. The Civil War legacy of race and remembrance. Lanham, Rowman and Littlefield, 2005, XIX-344 p. (The American crisis series).

4783. WARREN (Stephen). The shawnees and their neighbors, 1795–1870. Urbana a. Chicago, University of Illinois Press, 2005, IX-217 p.

4784. WEITZ (Mark A.). More damning than slaughter. Desertion in the Confederate Army. Lincoln, University of Nebraska Press, 2005, XIX-346 p.

4785. WELLS (Cheryl A.). Civil War time. Temporality and identity in America, 1861–1865. Athens, University of Georgia Press, 2005, XII-195 p.

4786. WETHERINGTON (Mark V.). Plain folk's fight. The Civil War and reconstruction in Piney Woods Georgia. Chapel Hill, University of North Carolina Press, 2005, 383 p. (Civil War America).

4787. WHITEHEAD (John S.). Completing the union: Alaska, Hawaii, and the battle for statehood. Albuquerque, University of New Mexico Press, 2005, XVII-438 p. (Histories of the American frontier).

4788. WILENTZ (Sean). The rise of American democracy. Jefferson to Lincoln. New York, W. W. Norton, 2005, XXIII-1044 p.

4789. WILKINSON (Charles). Blood struggle. The rise of modern Indian nations. New York, W. W. Norton, 2005, XVI-543 p.

4790. WORK (Clemens P.). Darkest before dawn. Sedition and free speech in the American West. Albuquerque, University of New Mexico Press, 2005, X-318 p.

Cf. n°s 338, 577, 3611, 6514, 6821, 6945, 7360, 7372, 7415

Uruguay

4791. CASTIGLIONI (Rossana). The politics of social policy change in Chile and Uruguay: retrenchment versus maintenance, 1973–1998. New York a. London, Routledge, 2005, XVI-150 p. (ill.). (Latin American studies/social sciences and law).

4792. MAIZTEGUI CASAS (Lincoln R.). Orientales: una historia política del Uruguay. 1. De los orígenes a 1865. T. 2. De 1865 a 1938. Montevideo, Planeta, 2005, 2 vol., 330 p., 415 p.

4793. MARKARIAN (Vania). Left in transformation: Uruguayan exiles and the Latin American human rights networks, 1967–1984. New York a. London, Routledge, 2005, XI-263 p. (Latin American studies: social sciences and law).

4794. REY TRISTÁN (Eduardo). La izquierda revolucionaria Uruguaya: 1955–1973. Madrid, Consejo Superior de Investigaciones Científicas y Sevilla, Universidad de Sevilla, Diputación de Sevilla, 2005, 472 p. (Catálogo de Publicaciones, Universidad de Sevilla. Serie: Historia y geografía, 66. Catálogo del Consejo Superior de Investigaciones Científicas, Escuela de Estudios Hispano-Americanos, 414. Catálogo Diputación de Sevilla, Servicio de Archivo y Publicaciones. Serie: Nuestra América, 10).

Cf. n°s 5745, 5758

Venezuela

4795. BERRÍOS BERRÍOS (Alexi). 1914: una encrucijada política para Venezuela. Caracas, Universidad Nacional Experimental Simón Rodríguez, Fondo Editorial Tropykos, 2005, 133 p.

4796. CARRERA DAMAS (Germán). El bolivarianismo-militarismo: una ideología de reemplazo. Caracas, Ala de Cuervo, 2005, 216 p. (Colección Ensayo).

4797. Militares y poder en Venezuela: ensayos históricos vinculados con las relaciones civiles y militares venezolanas. Coordinadores, Domingo IRWIN G. y Frédérique LANGUE. Caracas, Universidad Católica Andrés Bello, Universidad Pedagógica Experimental "Libertador", Vicerrectorado de Investigación y Postgrado, Centro de Investigaciones Históricas "Mario Briceño Iragorry", 2005, 375 p. (ill.).

4798. STRAKA (Tomás). Las alas de Ícaro: indagación sobre ética y ciudadanía en Venezuela (1800–1830). Caracas, Universidad Católica Andrés Bello, Fundación Konrad Adenauer Stiftung, 2005, 269 p.

4799. TARVER (Hollis Micheal), FREDERICK (Julia C.). The history of Venezuela. Westport, Greenwood Press, 2005, XXIII-189 p. (map). (The Greenwood histories of the modern nations).

4800. TRINKUNAS (Harold A.). Crafting civilian control of the military in Venezuela: a comparative perspective. Chapel Hill, University of North Carolina Press, 2005, XIV-297 p. (ill.).

Vietnam

** 4801. 75 years of the Communist Party of Viêt Nam (1930–2005): a selection of documents from nine party congresses. Hà Nôi, Thê Giới Publishers, 2005, VII-1295 p. (ill., ports.).

4802. KERKVLIET (Benedict J.). The power of everyday politics: how Vietnamese peasants transformed national policy. Ithaca, Cornell U. P., 2005, XII-305 p.

4803. TRAN (My-Van). A Vietnamese royal exile in Japan: Prince Cu'ò'ng Đê (1882–1951). London, Routledge, 2005, XVI-266 p. (ill., map). (Routledge studies in the modern history of Asia, 29.

Yemen

4804. KÜHN (Thomas). Shaping Ottoman Rule in Yemen, 1872–1919. New York, New York University, Department of Middle Eastern and Islamic Studies, Department of History, 2005, 370 p.

Yugoslavia

Cf. n°s 4517-4525, 7419

Zambia

4805. PHIRI (Bizeck Jube). A political history of Zambia: from the colonial period to the 3rd Republic. Trenton, Africa World Press a. London, Turnaround, 2005, 284 p.

Zimbabwe

4806. BULL-CHRISTIANSEN (Lene). Tales of the nation: feminist nationalism of patriotic history? Defining national history and identity in Zimbabwe. Uppsala, Nordic Africa Institute, 2005, 118 p. (Research report, 132).

4807. SIBANDA (Eliakim M.). The Zimbabwe African People's Union, 1961–1987: a political history of insurgency in Southern Rhodesia. Trenton, Africa World Press, 2005, XI-321 p. (ill., maps).

Cf. n° 6816

L

MODERN RELIGIOUS HISTORY

§ 1. General. 4808-4825. – § 2. Roman Catholicism (*a*. General; *b*. History of the Popes; *c*. Special studies; *d*. Religious orders; *e*. Missions). 4826-4950. – § 3. Orthodox Church. 4951-4956. – § 4. Protestantism. 4957-5013. – § 5. Non-Christian religions and sects. 5014-5048.

§ 1. General.

* 4808. ZIMMER (Petra). Helvetia Sacra. Arbeitsbericht 2004. *Schweizerische Zeitschrift für Geschichte*, 2005, 55, 2, p. 215-222.

** 4809. ERASME DE ROTTERDAM. Exhortation à la lecture de l'evangile. Sous la dir. de Alexandre VANAUTGAERDEN. 1. Le texte Latin. Édition, annotation et introduction, Jean-François COTTIER; traduction, Jean-François COTTIER et Alexandre VANAUTGAERDEN. 2. Les traductions françaises de 1543 et 1563. Introduction et édition des textes français, Guy BEDOUELLE; catalogue des éditions anciennes, Alexandre VANAUTGAERDEN. Bruxelles, Musée de la Maison d'Erasme, 2005, 2 vol., [s. p.]. (Notulae erasmianae, 5).

4810. BASTIEN (Paul). La parole du confesseur auprès des suppliciés (Paris XVIIe–XVIIIe siècles). *Revue historique*, 2005, 634, p. 283-308.

4811. Bistümer (Die) der deutschsprachigen Länder von der Säkularisation bis zur Gegenwart. Hrsg. v. Erwin GATZ; unter Mitwirkung von Clemens BRODKORB und Rudolf ZINNHOBLER. Freiburg im Breisgau, Basel u. Wien, Herder, 2005, 791 p.

4812. Charles Quint face aux réformes: colloque international organisé par le Centre d'histoire des réformes et du protestantisme (11e colloque Jean Boisset), Montpellier, 8–9 juin 2001, Université Paul Valéry – Montpellier III. Textes recueillis par Guy LE THIEC et Alain TALLON. Paris, H. Champion, 2005, 216 p. (ill.). (Colloques, congrès et conférences sur la Renaissance, 49).

4813. GRIFFIN (Clive). Journeymen-printers. Heresy and the Inquisition in sixteenth-century Spain. Oxford, Oxford U. P., 2005, 318 p.

4814. KRAFT (John). Ledung och sockenbildning ("Ledung") et formation des paroisses). Kungsängen, Västerås, Upplands-Bro kulturhistoriska forskningsinstitut, 2005, 256 p. (cartes, tab.).

4815. LURIA (Keith P.). Sacred boundaries. Religious coexistence and conflict in early-modern France. Washington, Catholic University of America Press, 2005, XXXVIII-357 p.

4816. MAC SHEFFREY (Shannon). Heresy, orthodoxy and English vernacular religion 1480–1525. *Past and present*, 2005, 186, p. 47-80.

4817. MAYEUR-JAOUEN (Catherine). Pèlerinages d'Égypte. Histoire de la piété copte et musulmane XVe–XXe siècles. Paris, Éditions de l'EHESS, 2005, 445 p.

4818. Origins (The) of sectarianism in Early Modern Ireland. Ed. by Alam FORD and John MAC CAFERTY. Cambridge, New York a. Melbourne, Cambridge U. P., 2005, X-249 p.

4819. PROSPERI (Adriano). Dare l'anima: storia di un infanticidio. Torino, G. Einaudi, 2005, X-373 p., (tav., ill.). (Einaudi storia, 4).

4820. Religion et enfermements (XVIIe–XXe siècles). Sous la direction de Bernard DELPAL et Olivier FAURE. Rennes, Presses universitaires de Rennes, 2005, 240 p. (Histoire).

4821. RÉMOND (Réne). Le nouvel antichristianisme. Entretiens avec Marc LEBOUCHER. Paris, Desclee de Brouwer, 2005, 149 p.

4822. SALOMÓN CHÉLIZ (María del Pilar). Las mujeres en la cultura política republicana: religión y anticlericalismo. *Historia social*, 2005, 53, p. 103-118.

4823. SCHLOESSER (Stephen). Jazz age Catholicism: mystic modernism in postwar Paris, 1919–1933. Toronto, University of Toronto press, 2005, XI-449 p.

4824. WEISS (Dieter J.). Katholische Reform und Gegenreformation. Darmstadt, Wissenschaftliche Buchgesellschaft, 2005, 216 p.

4825. WHELAN (Irene). The Bible war in Ireland: the "Second Reformation" and the polarization of Protestant-Catholic relations, 1800–1840. Madison, University of Wisconsin Press, 2005, XX-347 p. (History of Ireland and the Irish diaspora).

Cf. n°s *814, 816, 856-877, 3518, 4354, 4481, 4898, 5228, 6405, 7443*

§ 2. Roman Catholicism.

a. General

4826. ALBERIGO (Giuseppe). Breve storia del Concilio Vaticano II. Bologna, Il Mulino, 2005, 201 p.

4827. BRECHENMACHER (Thomas). Der Vatikan und die Juden. Geschichte einer unheiligen Beziehung vom 16. Jahrhundert bis zur Gegenwart. München, Beck, 2005, 326 p.

4828. Chiesa e azione cattolica alle origini della Costituzione repubblicana. A cura di Francesco MALGERI e Ernesto PREZIOSI, Roma, AVE, 2005, 495 p. (Ricerche e documenti, 18).

4829. CONGAR (Yves). Diario del Concilio. Cinisello Balsamo, San Paolo, 2005, 2 vol. (Storia della Chiesa).

4830. COSTIGAN (Richard F.). The consensus of the Church and papal infallibility: a study in the background of Vatican I. Washington, Catholic University of America press, 2005, XI-218 p.

4831. GRAN (John W.). Aufbruch und Erneuerung: Rückblick auf das Zweite Vatikanische Konzil, 1962–1965. Hrsg. v. Norbert HAUNSCHILD. [S. l.], [s. n.], 2005, 114 p.

4832. GUTTERMAN (David S.). Prophetic politics. Christian social movements and American democracy. Ithaca, Cornell U. P., 2005, XII-222 p.

4833. GWENAEL (Murphy). Les religieuses dans la Révolution française. Paris, Bayard, 2005, 328 p.

4834. HANUŠ (Jiří). Malý slovník osobností českého katolicismu 20. století. S antologií textů . (The little encyclopedia of the Czech Catholicism in the twentieth century). Brno, Centrum pro studium demokracie a kultury, 2005, 307 p. (Dějiny a kultura, 10).

4835. Katholische Kirche in SBZ und DDR. Hrsg. v. Christoph KÖSTERS und Wolfgang TISCHNER. Paderborn, F. Schoningh, 414 p. (tav., ill.).

4836. KEITT (Andrew W.). Inventing the sacred. Imposture, inquisition, and the boundaries of the supernatural in golden age Spain. Boston, Brill, 2005, VIII-229 p. (Medieval and Early Modern Iberian world).

4837. Kontinuität und Innovation um 1803. Säkularisation als Transformationsprozeß. Kirche – Theologie – Kultur – Staat. Hrsg. v. Rolf DECOT. Mainz, von Zabern, 2005, IX-324 p. (Veröffentlichungen des Instituts für Europäische Religionsgeschichte, 65).

4838. LAMBERTS BENDROTH (Margaret). Fundamentalists in the city: conflict and division in Boston's churches, 1885–1950. New York, Oxford U. P., 2005, X-250 p. (Religion in America series).

4839. LANDI (Fiorenzo). Storia economica del clero in Europa: secoli XV–XIX. Roma, Carocci, 2005, 209 p. (Studi superiori, 505).

4840. MARCHETTO (Agostino). Il Concilio ecumenico Vaticano II. Contrappunto per la sua storia. Città del Vaticano, Libreria editrice vaticana, 2005, 407 p. (Storia e attualità, 17).

4841. MAREK (Pavel). Církevní krize na počátku první Československé republiky 1918–1924. (The Church crisis in the early years of the Czechoslovak First Republic, 1918–1924). Brno, Nakl. L. Marek, 2005, 335 p. (Pontes pragenses, 36).

4842. MAROTTI (Arthur F.). Religious ideology and cultural fantasy. Catholic and anti-Catholic discourses in early modern England. Notre Dame, University of Notre Dame Press, 2005, 307 p.

4843. Offices et papauté (XIVe–XVIIe siècles). Charges, hommes destins. Sous la dir. d'Armand JAMME et Olivier PONCET. Roma, École française de Rome, 2005, 1049 p. (Coll. de l'École française de Rome, 334).

4844. PESCH (Otto Hermann). Il Concilio vaticano secondo: preistoria, svolgimento, risultati, storia postconciliare. Brescia, Queriniana, 2005, 446 p. (Biblioteca di teologia contemporanea, 131).

4845. Politics and religion in the white South. Ed. by Glenn FELDMAN. Lexington, University Press of Kentucky, 2005, XIII-386 p. (Religion in the South).

4846. Politische Religion und Religionspolitik: zwischen Totalitarismus und Bürgerfreiheit. Hrsg. v. Gerhard BESIER und Hermann LÜBBE. Göttingen, Vandenhoeck und Ruprecht, 2005, 415 p. (Schriften des Hannah-Arendt-Instituts für Totalitarismusforschung, 28).

4847. POP (Ovidiu H.). La Chiesa rumena unita, 1830–1853. Roma, Pontificia università gregoriana, 2005, 704 p. (Tesi Gregoriana. Ser. Storia ecclesiastica, 7).

4848. POULAT (Émile). La question religieuse et ses turbulences au 20e siècle: trois generations de catholiques en France. Paris, Berg International, 2005, 328 p.

4849. RICCARDI (Andrea). Santuari italiani e Vaticano II: tradizioni, crisi, ripresa. *Mélanges de l'École française de Rome: Italie et mediterranée (MEFRIM)*, 2005, 117, 2, p. 519-536.

Cf. n°s *2895, 3872, 3876, 3886, 4551, 6831*

b. History of the Popes

4850. CAFFIERO (Marina). La repubblica nella città del papa: Roma 1798. Roma, Donzelli, 2005, VI-184 p. (ill.). (Saggi. Storia e scienze sociali).

4851. GALAVOTTI (Enrico). Processo a papa Giovanni: la causa di canonizzazione di A. G. Roncalli, 1965–

2000. Bologna, Il mulino, 2005, 529 p. (Testi e ricerche di scienze religiose, 35).

4852. MELLONI (Alberto). Il conclave: storia dell'elezione del papa. Bologna, Il Mulino, 2005, 207 p. (Saggi, 543).

4853. Pontificate (The) of Clement VII: history, politics, culture. Ed. by Kenneth GOUWENS and Sheryl E. REISS. Aldershot, Ashgate, 2005, XXVI-437 p. (tav., ill.). (Catholic Christendom, 1300–1700).

4854. RONCALLI (Marco). Sotto il monte Giovanni XXIII: la memoria delle radici. *Mélanges de l'École française de Rome: Italie et mediterranée (MEFRIM)*, 2005, 117, 2, p. 727-751.

4855. Secolo fa (Un): il pontificato di Leone XIII nel confronto con il potere, la cultura, la storia. A cura di Lorenzo CAPPELLETTI e Alessandro RECCHIA, contributi di Gianpaolo ROMANATO. Reggio Emilia, Edizioni San Lorenzo e Anagni, Istituto Teologico Leoniano, 2005, 109 p.

4856. WENZEL (Knut). Kleine Geschichte des Zweiten Vatikanischen Konzils. Freiburg, Basel u. Wien, Herder, 2005, 256 p.

4857. WOLF (Hubert). Pius XI, und die "Zeitirrtümer". Die Initiativen der römischen Inquisition gegen Rassismus und Nationalismus. *Vierteljahrshefte für Zeitgeschichte*, 2005, 53, 1, p. 1-42.

Cf. n° 4585

c. Special studies

** 4858. Katholische Reform im Niederstift Münster. Die Akten der Generalvikare Johannes Hartmann und Petrus Nicolartius über ihre Visitationen im Niederstift Münster in den Jahren 1613 bis 1631/32. Herausgegeben und eingeleitet von Heinrich LACKMANN. Münster, Aschendorff, 2005, 437 p. (Westfalia sacra, 14).

** 4859. NETZHAMMER (Raymund). Episcop în România într-o epocă a conflictelor naționale și religioase (1905–1924). Jurnal. Vol. I-II. (Bishop in Romania in an epoch of national and religious conflicts, 1905–1924. Journal). Editori Nikolaus NETZHAMMER și Krista ZACH. București, Editura Academiei Române, 2005, 1737 p.

4860. BARNAY (Sylvie). La Vierge des sanctuaires de France dans la séconde moitié du XXe siècle: une histoire d'incorporation. *Mélanges de l'École française de Rome: Italie et mediterranée (MEFRIM)*, 2005, 117, 2, p. 661-675.

4861. BAUTISTA GARCÍA (Cecilia Adriana). Hacia la romanización de la iglesia mexicana a fines del siglo XIX. *Historia mexicana*, 2005, 55, 1, p. 99-144.

4862. BEOZZO (Jose Oscar). A Igreja do Brasil no Concilio Vaticano II, 1959–1965. Sao Paulo, Paulinas, 2005, 611 p. (ill.). (Presenca de Alceu).

4863. Bischof (Ein) vor Gericht: der Prozeß gegen den Danziger Bischof Carl Maria Splett 1946. Hrsg. v. Ulrich BRÄUEL und Stefan SAMERSKI. Osnabrück, Fibre, 2005, 313 p.

4864. BROTÁNKOVÁ (Helena). Změna religiozity v České republice mezi lety 1921 a 2001. (Changes in the Religiousness in the Czech Republic between 1921 and 2001). *Historická geografie*, 2005, 33, p. 301-321.

4865. BURKARD (Dominik). Häresie und Mythus des 20. Jahrhunderts. Rosenbergs nationalsozialistische Weltanschauung vor dem Tribunal der Römischen Inquisition. Paderborn, München, Wien u. Zürich, Schöningh, 2005, 416 p. (Römische Inquisition und Indexkongresgation, 5).

4866. BYRNES (Joseph F.). Catholic and French forever: religious and national identity in Modern France. University Park, Pennsylvania State U. P., 2005, XXIII-278 p.

4867. CALIÒ (Tommaso). Santuari, reti sociali e sacralizzazione nella Roma del dopoguerra. *Mélanges de l'École française de Rome: Italie et mediterranée (MEFRIM)*, 2005, 117, 2, p. 635-660.

4868. CANTA (Caterina Chiara). Identità culturale e religiosa nei pellegrinaggi della Sicilia contemporanea. *Mélanges de l'École française de Rome: Italie et mediterranée (MEFRIM)*, 2005, 117, 2, p. 537-564.

4869. CHOWNING (Margaret). Convent reform, Catholic reform, and Bourbon reform in eighteenth-century New Spain: the view from the nunnery. *Hispanic American historical review*, 2005, 85, 1, p. 1-38.

4870. CONFESSORE (Ornella). Padre Pio e il santuario di San Giovanni Rotondo. *Mélanges de l'École française de Rome: Italie et mediterranée (MEFRIM)*, 2005, 117, 2, p. 713-726.

4871. CONNELL (William J.), CONSTABLE (Giles). Sacrilege and redemption in Renaissance Florence: the case of Antonio Rinaldeschi. Toronto, Centre for Reformation and Renaissance Studies, 2005, 125 p. (ill.). (Essays and studies, 8).

4872. COȘA (Anton). Catolicii din Moldova în Arhiva Congregației de Propaganda Fide. Documente inedite. (Catholics from Moldavia in the archive of the Sacred Congregation of Propaganda Fide. New documents). *Carpica. Muzeul Județean de Istorie Bacău*, 2005, 34, p. 151-173.

4873. CUSSEN (Celia L.). The search for idols and saints in colonial Peru: linking extirpation and beatification. *Hispanic American historical review*, 2005, 85, 3, p. 417-448.

4874. DE DAINVILLE BARBICHE (Ségolène). Devenir curé a Paris: institutions et carrières écclesiastiques (1695–1789). Paris, Presses universitaires de France, 2005, 550 p. (Le nœud gordien).

4875. DEFFAYET (Laurence). "Amici Israel": les raisons d'un échec: des éléments nouveaux apportés par l'ouverture des archives du Saint-Office. *Mélanges de l'École française de Rome: Italie et mediterranée (MEFRIM)*, 2005, 117, 2, p. 831-851.

4876. DESCIMON (Robert), RUIZ IBÁÑEZ (José Javier). Les ligueurs de l'exil: le refuge catholique français après 1594. Seyssel, Champ Vallon, 2005, 317 p. (Epoques).

4877. DUNCAN (Jason K.). Citizens or papists? The politics of anti-Catholicism in New York, 1685–1821. New York, Fordham U. P., 2005, XVIII-253 p. (Hudson Valley heritage series, 3).

4878. FABER (Martin). Scipione Borghese als Kardinalprotektor: Studien zur römischen Mikropolitik in der frühen Neuzeit. Mainz, Philipp von Zabern, 2005, X-544 p. (Veröffentlichungen des Instituts für Europäische Geschichte Mainz. Abteilung für Abendländische Religionsgeschichte, 204).

4879. FATTORINI (Emma). Santuari mariani in Italia tra Otto e Novecento. *Mélanges de l'École française de Rome: Italie et mediterranée (MEFRIM)*, 2005, 117, 2, p. 457-486.

4880. FIRPO (Massimo). Inquisizione romana e Controriforma: studi sul cardinal Giovanni Morone (1509–1580) e il suo processo d'eresia. Brescia, Morcelliana, 2005, 620 p. (Storia, 9). – IDEM. Le ambiguità della porpora e i "diavoli" del Sant' Ufficio: identità e storia nei ritratti di Giovanni Grimani. *Rivista storica italiana*, 2005, 117, 3, p. 825-871.

4881. GAUVREAU (Michael). The Catholic origins of Quebec's Quiet Revolution, 1931–1970. Ithaca: McGill-Queen's U. P., 2005, XIV-501 p. (McGill-Queen's studies in the history of religion, series two, 41).

4882. HARRIS (Rita). Lourdes: les femmes et la spiritualité. *Mélanges de l'École française de Rome: Italie et mediterranée (MEFRIM)*, 2005, 117, 2, p. 622-634.

4883. HEBEISEN (Erika). Leidenschaftlich Fromm. Die pietische Bewegung in Basel 1750–1830. Köln, Weimar u. Wien, Böhlau, 2005, X-334 p.

4884. HEITZ (Claudius). Volksmission und badischer Katholizismus im 19. Jahrhundert. Freiburg u. München, Alber, 2005, 456 p. (Forschungen zur oberrheinischen Landesgeschichte, 50).

4885. Hielp Maria. En bok om biskop Brynolf Gerlaksson. (Etudes sur l'évêque suédois Brynolf Gerlaksson). Red. Johnny HAGBERG. Skara, Skara stiftshistoriska sällskap, 2005, 288 p. (Skara stiftshistoriska sällskaps skriftserie, 21).

4886. HUMMEL (Karl Joseph). Der Heilige Stuhl, deutsche und polnische Katholiken 1945–1978. *Archiv für Sozialgeschichte*, 2005, 45, p. 165-214.

4887. ISENMANN (Moritz). Die Verwaltung der päpstlichen Staatsschuld in der Frühen Neuzeit. Sekretariat, Computisterie und Depositerie vom 16. bis zum ausgehenden 18. Jahrhundert. Stuttgart, Steiner, 2005, 182 p. (Vierteljahrschrift für Sozial- und Wirtschaftsgeschichte, Beih., 179).

4888. KRÖGER (Bernward). Der französische Exilklerus im Fürstbistum Münster, 1794–1802. Mainz, von Zabern, 2005, IX-299 p. (Abendländische Religionsgeschichte, 203).

4889. LAGERLF NILSSON (Ulrika). The bishops in the church of Sweden. An elite in society during the first half of the 20th century. *Scandinavian journal of history*, 2005, 30, 3-4, p. 308-319.

4890. LANGLOIS (Claude). Une romanisation des pèlerinages? Le couronnement des statues de la Vierge en France dans la seconde moitié du XIXe siècle. *Mélanges de l'École française de Rome: Italie et mediterranée (MEFRIM)*, 2005, 117, 2, p. 601-620.

4891. LEONARD (Amy). Nails in the wall: Catholic nuns in Reformation Germany. Chicago a. London, University of Chicago Press, 2005, XIII-218 p. (Women in culture and society).

4892. LIDA (Miranda). Iglesia y sociedad porteñas. El proceso de parroquialización de la arquidiócesis de Buenos Aires. *Entrepasados*, 2005, 28, [s. p.].

4893. LUPI (Regina). Gli studia del papa: nuova cultura e tentativi di riforma tra Sei e Settecento. Firenze, Centro editoriale toscano, 2005, 299 p. (Politica e storia, 58).

4894. MAC NAMARA (Patrick). A catholic cold war: Edmund A. Walsh, S.J., and the politics of American anticommunism. New York, Fordham University press, 2005, XIX-280 p.

4895. MATOVINA (Timothy). Guadalupe and her faithful: Latino Catholics in San Antonio, from colonial origins to the present. Baltimore, Johns Hopkins U. P., 2005, XV-232 p. (Lived religions).

4896. MORENO SECO (Mónica). Cristianas por el feminismo y la democracia: catolicismo femenino y movilización en los años setenta. *Historia social*, 2005, 53, p. 137-154.

4897. PERRY (Mary Elizabeth). The handless maiden. Moriscos and the politics of religion in Early Modern Spain. Princeton, Princeton U. P., 2005, XVI-202 p. (Jews, Christians, and Muslims from the ancient to the modern world).

4898. Römische Inquisition und Indexkongregation: Grundlagenforschung 1814–1917. Hrsg. v. Hubert WOLF. Band 1. Romische Bucherverbote: Edition der Bandi von Inquisition und Indexkongregation, 1814–1917. Auf der Basis von Vorarbeiten von Herman H. SCHWEDT bearbeitet von Judith SCHEPERS und Dominik BURKARD. Band 2. Systematisches Repertorium zur Buchzensur: 1814–1917. T. 1. Indexkongregation. T. 2. Inquisition. Bearb. v. Sabine SCHRATZ, Jan Dirk BUSEMANN und Andreas PIETSCH. Band 3. Prosopographie von romischer Inquisition und Indexkongregation: 1814–1917. T. 1. A-K. T. 2. L-Z. Von Herman H. SCHWEDT; unter Mitarbeit von Tobias LAGATZ. Paderborn, Schöningh, 2005, 5 vol., XVII-604 p., XX-1087 p., XXIII-1636 p.

4899. RUFF (Mark Edward). Catholic youth in postwar West Germany, 1945–1965. Chapel Hill, University of North Carolina Press, 2005, XVI-284 p.

4900. SCHIERSNER (Dietmar). Politik, Konfession und Kommunikation: Studien zur katholischen Konfessionalisierung der Markgrafschaft Burgau 1550–1650. Berlin, Akademie Verlag, 2005, 523 p. (Colloquia Augustana, 19).

4901. SCHOLZ (Stephan). Der deutsche Katholizismus und Polen, 1830–1849. Identitätsausbildung zwischen konfessioneller Solidarität und antirevolutionärer Abgrenzung. Osnabrück, Fibre, 2005, 430 p. (Einzelveröffentlichungen des Deutschen Historischen Instituts Warschau, 13).

4902. SCIUTI RUSSI (Vittorio). L' abbé Grégoire e l'inquisizione di Spagna: la Lettre del 1798 e la reazione del partito inquisitoriale. *Rivista storica italiana*, 2005, 117, 2, p. 494-528.

4903. SENSI (Mario). Assisi da città-santuario a città dei santuari. *Mélanges de l'École française de Rome: Italie et mediterranée (MEFRIM)*, 2005, 117, 2, p. 753-786.

4904. SÉVENET (Jacques). Les paroisses parisiennes devant la separation des Églises et de l'Etat: 1901-1908. Préface de Valentine ZUBER. Paris, Letouzey & Ane, 2005, 316 p. (Mémoire chrétienne au présent, 2).

4905. SORREL (Christian). Les sanctuaires contemporains entre ancrage régional et enjeux nationaux: les mutations du pèlerinage dans les diocèses de Savoie. *Mélanges de l'École française de Rome: Italie et mediterranée (MEFRIM)*, 2005, 117, 2, p. 565-600.

4906. VLADIMIR (Angelo). Les curés de Paris au 16e siècle. Paris, Ed. du Cerf, 2005, V-893 p. (Histoire religieuse de la France, 26).

4907. YANNOU (Hervé). Les assemblées plénières de l'épiscopat français (1906-1907): travaux, organisation et signification. *Mélanges de l'École française de Rome: Italie et mediterranée (MEFRIM)*, 2005, 117, 2, p. 787-829.

4908. ZÖBERLEIN (Renate). Religionsfreiheit und Staatskirchenrecht im Polen der Dritten Republik. *Zeitschrift der Savigny-Stiftung für Rechtsgeschichte Kanonistische Abteilung*, 2005, 91, p. 650-684.

4909. ZUREK (Robert). Die Rolle der Katholischen Kirche Polens bei der deutsch-polnischen Aussöhnung 1966–1972. *Archiv für Sozialgeschichte*, 2005, 45, p. 141-164.

Cf. n°s 3621, 3744, 3799, 4015, 4107, 4410

d. Religious orders

* 4910. MOTTA (Franco). La compagine sacra: elementi di un mito delle origini nella storiografia sulla Compagnia di Gesù. *Rivista storica italiana*, 2005, 117, 1, p. 5-25.

4911. AVON (Dominique). Les Frères prêcheurs en Orient. Les dominicains du Caire (années 1910–années 1960). Paris, Éd. du Cerf, 2005, 1029 p.

4912. BUFFON (Giuseppe). Les franciscains en Terre Sainte (1869–1889): religion et politique: une recherche institutionelle. Paris, Ed. du Cerf, 2005, 604 p. (ill.). (Histoire).

4913. Cappuccini (I) nell'Umbria tra Sei e Settecento: Convegno internazionale di studi: Todi, 24–26 giugno 2004. A cura di Gabriele INGEGNERI. Roma, Istituto storico dei Cappuccini, 2005, 300 p. (ill.). (Biblioteca Seraphico-Capuccina, 74).

4914. CARAVALE (Giorgio). Ambrogio Catarino Politi e i primi gesuiti. *Rivista storica italiana*, 2005, 117, 1, p. 80-109.

4915. DINET-LECOMTE (Marie-Claude). Les sœurs hospitalières en France aux XVIIe et XVIIIe siècles. La charité en action. Paris, Honoré Champion, 2005, 595 p.

4916. Grand (Le) exil des congregations religieuses francaises, 1901–1914: Colloque international de Lyon: Universite Jean-Moulin-Lyon-3., 12–13 juin 2003. Sous la dir. de Patrick CABANEL et Jean-Dominique DURAND. Paris, Ed. du Cerf, 2005, 489 p. (Histoire).

4917. Jésuites (Les) à Lyon, XVIe–XXe siècle. Sous la dir. Étienne FOUILLOUX et Bernard HOURS. Lyon, ENS Éd., 2005, 274 p.

4918. LANGENBACHER (Ferdy), JIMÉNEZ (O.F.M.). Origen, desarrollo e influjo de los colegios De propaganda fide en la Iglesia y sociedad de la recién fundada Republica boliviana (1834–1877). Intr. de Francisco de BORJA MEDINA. Grottaferrata, Frati editori di Quaracchi, 2005, LIV-498 p. (Analecta francescana sive cronica aliaque varia documenta ad historiam fratrum minorum spectantia, 15. Analecta francescana. Nova series, documenta et studia, 3).

4919. LAZAR (Lance Gabriel). Working in the vineyard of the Lord: Jesuit confraternities in early modern Italy. Toronto, University of Toronto Press, 2005, 377 p.

4920. LOZANO NAVARRO (Julián J.). La Compañía de Jesús y el poder en la España de los Austrias. Madrid, Cátedra, 2005, 430 p. (Historia).

4921. LUX-STERRITT (Laurence). Redifining female religious life. French ursulines and English ladies in seventeenth-century Catholicism. Aldershot, Ashgate, 2005, 244 p.

4922. MALLINCKRODT (Rebekka Von). Struktur und kollektiver Eigensinn. Kölner Laienbruderschaften im Zeitalter der Konfessionalierung. Göttingen, Vandenhoeck & Ruprecht, 2005, 513 p. (Geschichte, 209).

4923. MONGINI (Guido). Per un profilo dell'eresia gesuitica: la Compagnia di Gesù sotto processo. *Rivista storica italiana*, 2005, 117, 1, p. 26-63.

4924. NELSON (Eric). The Jesuits and the monarchy: Catholic reform and political authority in France (1590–

1615). Aldershot, Ashgate a. Roma, Institutum historicum Societatis Iesu, 2005, XIV-275 p. (ill.). (Bibliotheca Istituti historici Societatis Iesu, 58).

4925. Orden und Klöster im Zeitalter von Reformation und katholischer Reform 1500–1700. Hsg. Friedhelm JÜRGENSMEIER und Regina Elisabeth SCHWERDTFEGER. Münster, Aschendorff, 2005, 254 p. (Katholisches Leben und Kirchenreform im Zeitalter der Glaubensspaltung, 65).

4926. Ordini regolari. A cura di Simona FECI e Angelo TORRE. *Quaderni storici*, 2005, 2, p. 319-554. [Contiene: FECI (Simona). Frati di S. Agostino: conflitti di comunità e poteri a Genova alla metà del Cinquecento (p. 333-368). – ARMSTRONG (Megan). Predicare la politica. L'evangelizzazione francescana nella Francia della Riforma (p. 369-388). – FABER (Martin). Meglio la tirannide o l'indifferenza? I cardinali protettori degli Olivetani (1591–1633) (p. 389-412). – GIANA (Luca). Complicazioni giurisdizionali. Un convento domenicano e la Repubblica di Genova nel XVII secolo (p. 413-440). – ZUCCARELLO (Ugo). Una periferia modello. La "Istoria" di Astino del Mazzoleni e la riforma vallombrosana (p. 441-460). – MODICA (Marilena). Direzione spirituale, misticismo e quietismo alla fine del Seicento. Il caso degli Agostiniani scalzi di Palermo (p. 461-484). – BARZAZI (Antonella). Una cultura per gli ordini religiosi: l'erudizione (p. 485-518). – MEYER (Fédéric). Religiosi fuorilegge: i regolari di fronte alla giustizia in Savoia nel secolo XVIII (p. 519-554)].

4927. PASTORE (Stefania). I primi gesuiti e la Spagna: strategie, compromessi, ambiguità. *Rivista storica italiana*, 2005, 117, 1, p. 158-178.

4928. PAVONE (Sabina). Preti riformati e riforma della Chiesa: i gesuiti al concilio di Trento. *Rivista storica italiana*, 2005, 117, 1, p. 110-134.

4929. SCARAMELLA (Pierroberto). I primi gesuiti e l'Inquisizione romana (1547–1562). *Rivista storica italiana*, 2005, 117, 1, p. 135-157.

4930. STRÖBELE (Ute). Zwischen Kloster und Welt: die Aufhebung südwestdeutscher Frauenklöster unter Kaiser Joseph II. Köln, Böhlau, 2005, X-347 p. (ill.). (Stuttgarter historische Forschungen, 1).

4931. Ženské řehole za komunismu (1948–1989). Sborník příspěvků z konference pořádané v kostele sv. Voršily v Praze. (Women's religious orders during the communist régime, 1948–1989). Ed. Vojtěch VLČEK. Olomouc, Matice cyrilometodějská, 2005, 447 p. (ill.).

e. Missions

4932. Canadian missionaries, indigenous peoples: representing religion at home and abroad. Ed. by Alvyn AUSTIN and Jamie S. SCOTT. Buffalo, University of Toronto Press, 2005, VIII-326 p.

4933. Chiesa cattolica e mondo cinese: tra colonialismo ed evangelizzazione, 1840–1911. A cura di Agostino GIOVAGNOLI ed Elisa GIUNIPERO, prefazione di Crescenzio card. SEPE. Città del Vaticano, Urbaniana university press, 2005, 319 p.

4934. Europa (L') e l'evangelizzazione delle Indie Orientali. A cura di Luciano VACCARO. Milano, Centro ambrosiano, 2005, 538 p. (Europa ricerche, 10).

4935. FERRO (Teresa). I missionari cattolici in Moldavia. Studi storici e linguistici. Cluj-Napoca, Editura Clusium, 2005, 192 p.

4936. GALGANO (Robert C.). Feast of souls: Indians and Spaniards in the seventeenth-century missions of Florida and New Mexico. Albuquerque, University of New Mexico Press, 2005, XII-212 p.

4937. HACKEL (Steven W.). Children of coyote, missionaries of Saint Francis: Indian-Spanish relations in colonial California, 1769–1850. Chapel Hill, University of North Carolina Press for the Omohundro Institute of Early American History and Culture, 2005, XX-476 p.

4938. HODGSON (Dorothy L.). The church of women. Gendered encounters between Maasai and missionaries. Bloomington, Indiana U. P., 2005, XVII-307 p.

4939. HVIID JENSEN (Anne). I lys og skygge. Dansk mission i Kina. (La mission religieuse danoise en Chine). Frederiksborg, Unitas, 2005, 168 p. (ill.). (Mission på vej, 3).

4940. JACKSON (Robert H.). Missions and the frontiers of Spanish America. A comparative study of the impact of environmental, economic, political, and socio-cultural variations on the missions in the Rio de la Plata region and on the northern frontier of New Spain. Scottsdale, Pentacle Press, 2005, XXII-568 p.

4941. Missions and empire. Ed. by Norman ETHERINGTON. Oxford a. New York, Oxford U. P., 2005, XIII-332 p. (Oxford history of the British empire, companion ser.).

4942. MUTLU (Şamil). Osmanlı Devleti'nde misyoner Okulları. (Ecoles de missionnaire dans l'Empire Ottoman). İstanbul, Gökkubbe, 2005, 472 p.

4943. OKAFOR (Eddie E.). Francophone Catholic achievements in Igboland, 1883–1905. *History in Africa*, 2005, 32, p. 307-319.

4944. PARENTE (Ulderico). Note sull'attività missionaria di Nicolás Bobadilla nel Mezzogiorno d'Italia prima del concilio di Trento (1540–1541). *Rivista storica italiana*, 2005, 117, 1, p. 64-79.

4945. PIZZORUSSO (Giovanni), SANFILIPPO (Matteo). Dagli indiani agli emigranti: l'attenzione della Chiesa romana al Nuovo mondo, 1492–1908. Viterbo, Sette città, 2005, 246 p. (Archivio storico dell'emigrazione italiana. Quaderni, 1).

4946. SCHULTZE (Andrea). "In Gottes Namen Hütten bauen". Kirchlicher Landbesitz in Südafrika. Die Berliner Mission und die Evangelisch-Lutherische Kirche Südafrikas zwischen 1834 und 2005. Stuttgart, Steiner, 2005, 619 p. (Missionsgeschichtliches Archiv, 9).

4947. SIEBER (Dominik). Jesuitische Missionierung, priesterliche Liebe, sakramentale Magie. Volkskulturen in Luzern, 1563–1614. Basel, Schwabe, 2005, 298 p. (Luzerner Historische Veröffentlichungen, 40).

4948. SIVASUNDARAM (Sujit). Nature and the godly empire: science and evangelical mission in the Pacific, 1795–1850. New York, Cambridge U. P., 2005, XI-244 p. (Cambridge social and cultural histories).

4949. VAN DER HEYDEN (Ulrich), STOECKER (Holger). Mission und Macht im Wandel politischer Orientierungen. Europäische Missionsgesellschaften in politischen Spannungsfeldern in Afrika und Asien zwischen 1800 und 1945. Stuttgart, Steiner, 2005, 700 p. (Missionsgeschichtliches Archiv, 10).

4950. ŽUPANOV (Ines G.). Missionary tropics: the Catholic frontier in India (sixteenth-seventeenth centuries). Ann Arbor, University of Michigan Press, 2005, XII-374 p. (History, languages, and cultures of the Spanish and Portuguese worlds).

§ 3. Orthodox Church.

4951. GARUTI (Adriano). Libertà religiosa ed ecumenismo: la questione del territorio canonico in Russia. Siena, Cantagalli, 2005, 212 p. (Cristianesimo e cultura, 6).

4952. HAMAMOTO (Mami). 17 seiki roshia ni okeru hi-roshiaseikyōto erīto seisaku. (Policies for non-Russian elites in 17th century Russia). *Slavic studies*, 2005, 52, p. 63-96 [Eng. Summary].

4953. POPOV (Andrej V.). Rossijskoe pravoslavnoe zarubezh'e: Istorija i istochniki. S prilozheniem sistematicheskoj bibliografii. (The Russian Orthodox diaspora: History and sources. With a systematic bibliography). Moskva, Institut politicheskogo i voennogo analiza, 2005, 620 p. (bibl. p. 340-590; ind. p. 595-616). (Materialy k istorii russkoj politicheskoj emigratsii; 10).

4954. SCHULZ (Günther), SCHRÖDER (Gisela A.), RICHTER (Timm). Bolschewistische Herrschaft und Orthodoxe Kirche in Russland: das Landeskonzil 1917/1918: Quellen und Analysen. Munster, LIT, 2005, V-802 p. (Theologie, 4).

4955. SOROȘTINEANU (Valeria). Manifestări ale sentimentului religios la românii ortodocși din Transilvania: 1899–1916. (Manifesting the religious feeling at the Orthodox Romanians from Transylvania: 1899–1916). Sibiu, Editura Universității "Lucian Blaga", 2005, 182 p.

4956. VASILE (Cristian). Biserica Ortodoxă Română în primul deceniu comunist. (The Romanian Orthodox Church in the first communist decade). București, Editura Curtea Veche, 2005, 296 p.

Cf. n°s 574, 4441

§ 4. Protestantism.

* 4957. Martin Bucer (1491–1551): Bibliographie. Hrsg. v. Holger PILS, Stephan RUDERER und Petra SCHAFFRODT unter Mitarbeit von Zita FARAGÓ-GÜNTHER; mit Unterstützung der Heidelberger Akademie der Wissenschaften hrsg. von Gottfried Seebass. Gütersloh, Gütersloher Verlagshaus, 2005, 751 p.

* 4958. VAN DER GRIJP (Klaus). Bibliografía de la historia del protestantismo español. Salamanca, Centro de estudios orientales y ecuménicos Juan 23, 2005, 306 p. (Biblioteca oecumenica Salmanticensis, 32).

* 4959. VIDMAR (John C.). English Catholic historians and the English Reformation, 1585–1954. Brighton, Sussex academic press, 2005, VII-184 p.

** 4960. BULLINGER (Heinrich). Briefe des Jahres 1541. Hrsg. v. Henrich RAINER. Zürich, Theologischer Verlag, 2005, 385 p. (Abt. 2: Briefwechsel, 11).

4961. BERNARD (G. W.). The kings reformation: Henry VIII. and the remaking of the English Church. New Haven, London, Yale U. P., 2005, X-736 p. (ill.).

4962. BERTI (Silvia). Bernard Picart e Jean Frédéric Bernard dalla religione riformata al deismo: un incontro con il mondo ebraico nell'Amsterdam del primo Settecento. *Rivista storica italiana*, 2005, 117, 3, p. 974-1001.

4963. BROADHEAD (P. J.). Public worship, liturgy and the introduction of the Lutheran Reformation in the territorial lands of Nuremberg. *English historical review*, 2005, 120, 486, p. 277-302.

4964. BROHED (Ingmar). Sveriges Kyrkohistoria. Bd 8. Religionsfrihetens och ekumenikens tid. (Histoire de l'Eglise de Suède. Tome 8. Le temps de la liberté de religion et de l'œcuménisme). Stockholm, Verbum, 2005, 447 p.

4965. BROWN (Christopher Boyd). Singing the gospel. Luthran hymns and the success of the Reformation. Cambridge, Harvard U. P., 2005, 298 p.

4966. BRUENING (Michael W.). Calvinism's first battleground. Conflict and reform in the Pays de Vaud, 1528–1559. Heidelberg, Springer Verlag, 2005, XVI-286 p.

4967. Catholics and the Protestant nation: religious politics and identity in early modern England. Ed. by Ethan SHAGAN. Manchester a. New York, Manchester U. P., 2005, VIII-213 p. (Politics, culture and society in early modern Britain).

4968. CHRISTIN (Olivier). Mort et mémoire: les portraits de réformateurs protestants au XVIe siècle. *Schweizerische Zeitschrift für Geschichte*, 2005, 55, 4, p. 383-400.

4969. COLEMAN (Heather J.). Russian Baptists and spiritual revolution, 1905–1929. Bloomington, Indiana U. P., 2005, XI-304 p. (Indiana-Michigan series in Russian and East European studies).

4970. DÜRR (Renate). Prophetie und Wunderglauben – zu den kulturellen Folgen der Reformation. *Historische Zeitschrift*, 2005, 281, 1, p. 3-32.

4971. ENCREVÉ (André). Témoignages: des Protestants à l'aube du XXe siècle. *Revue historique*, 2005, 635, p. 593-608.

4972. ESTES (James M.). Peace, order and the glory of God: secular authority and the Church in the thought of Luther and Melanchthon: 1518–1559. Leiden, Brill, 2005, XVIII-234 p. (Studies in medieval and reformation traditions, 11).

4973. FLÜGEL (Wolfgang). Konfession und Jubiläum. Zur Institutionalisierung der lutherischen Gedenkkultur in Sachsen 1617–1830. Leipzig, Leipziger Universitätsverlag, 2005, 335 p. (Schriften zur sächsischen Geschichte Undvolkskunde, 14).

4974. GILMONT (Jean François). Le livre réformé au XVIe siècle. Paris, Bibliothèque Nationale de France, 2005, 151 p. (ill.). (Conférences Léopold Delisle, 16).

4975. Glaube und Macht. Theologie, Politik und Kunst im Jahrhundert der Reformation. Hrsg. v. Enno BÜNZ, Stefan RHEIN und Günther WARTENBERG. Leipzig, Evangelische Verlagsanstalt, 2005, 288 p. (Schriften der Stiftung Luthergedenkstätten in Sachsen-Anhalt, 5).

4976. GRESCHAT (Martin). Protestantismus in Europa. Geschichte – Gegenwart – Zukunft. Darmstadt, Wissenschaftliche Buchgesellschaft, 2005, 175 p.

4977. HARISMENDY (Patrick). Le Parlement des Huguenots: organisations et synodes réformés français au XIXe siècle. Rennes, Presses Universitaires de Rennes, 2005, 435 p. (Collection Carnot).

4978. HARVEY (Paul). Freedom's coming: religious culture and the shaping of the South from the Civil War through the civil rights era. Chapel Hill, University of North Carolina Press, 2005, XVI-338 p.

4979. HEAL (Felicity). What can King Lucius do for you? The Reformation and the early British Church. *English historical review*, 2005, 120, 487, p. 593-614.

4980. HEMPTON (David). Methodism. Empire of the spirit. New Haven, Yale U. P., 2005, XIII-278 p.

4981. HEYEN (Elk Volkmar). Pastorale Beamtenethik 1650–1700: Amtstugenden in lutherischen Regentenpredigten. *Historische Zeitschrift*, 2005, 280, 2, p. 345-380.

4982. HINDMARSH (D. Bruce). The evangelical conversion narrative. Spiritual autobiography in early modern England. New York, Oxford U. P., 2005, X-384 p.

4983. HÖLSCHER (Lucian). Geschichte der protestantischen Frömmigkeit in Deutschland. München, C. H. Beck, 2005, 466 p.

4984. KAMIL (Neil). Fortress of the soul: violence, metaphysics, and material life in the Huguenots' New World, 1517–1751. Baltimore, Johns Hopkins U. P., 2005, XXIV-1058 p. (Early America: history, context, culture).

4985. KEMPFER (Jacqueline). Die Entwicklung der Grundordnung in der Evangelischen Kirche in Berlin-Brandenburg 1959–1990 – Ein Beitrag zur Diskussion um die Einheit der EKiBB. *Zeitschrift der Savigny-Stiftung für Rechtsgeschichte Kanonistische Abteilung*, 2005, 91, p. 785-796.

4986. "Kirchengeschichte in Lebensbildern". Lebenszeugnisse aus den evangelischen Kirchen im östlichen Europa des 20. Jahrhunderts. Hrsg. v. Peter MASER und Christian-Erdmann SCHOTT. Münster, Verein für Ostdeutsche Kirchengeschichte in Verbindung mit dem Ostkirchen-Institut Münster, 2005, V-279 p. (Ostdeutsche Kirchengeschichte, 7).

4987. LEPP (Claudia). Tabu der Einheit? Die Ost-West-Gemeinschaft der evangelischen Christen und die deutsche Teilung (1945–1969). Göttingen, Vandenhoeck & Ruprecht, 2005, 1028 p. (Arbeiten zur kirchlichen Zeitgeschichte: Reihe B, Darstellungen, 42).

4988. Melanchthon und der Calvinismus. Hrsg. v. Günter FRANK und Herman J. SELDERHUIS; unter Mitarbeit von Sebastian LALLA. Stuttgart-Bad Cannstatt, Frommann-Holzboog, 2005, 375 p. (ill.). (Melanchthon-Schriften der Stadt Bretten, 9).

4989. MILLER (John). 'A Suffering People': English Quakers and their neighbours c.1650–c.1700. *Past and present*, 2005, 188, p. 71-103.

4990. MOOS (Thorsten). Staatszweck und Staatsaufgaben in den protestantischen Ethiken des 19. Jahrhunderts. Münster, Lit, 2005, 300 p. (Bochumer Forum zur Geschichte des sozialen Protestantismus, 5).

4991. MÖRKE (Olaf). Die Reformation. Voraussetzungen und Durchsetzung. München, Oldenbourg, 2005, X-174 p. (Enzyklopädie deutscher Geschichte, 74).

4992. Nationalprotestantische Mentalitäten. Konturen, Entwicklungslinien und Umbrüche eines Weltbildes. Hrsg. v. Manfred GAILUS und Hartmut LEHMANN. Göttingen, Vandenhoeck & Ruprecht, 2005, 472 p. (Veröffentl. des Max-Planck-Instituts für Geschichte, 214).

4993. NEWMAN (Mark). Divine agitators. The delta ministry and civil rights in Mississippi. Athens, University of Georgia Press, 2005, XVII-352 p.

4994. NURSER (John S.). For all peoples and all nations: the ecumenical church and human rights. Forew. by David LITTLE. Washington, Georgetown U. P., 2005, 240 p.

4995. OPP (James). The lord for the body: religion, medicine, and protestant faith healing in Canada, 1880–1930. Ithaca, McGill-Queen's U. P., 2005, X-274 p. (McGill-Queen's studies in the history of religion).

4996. PETTEGREE (Andrew). Reformation and the culture of persuasion. New York, Cambridge U. P., 2005, XI-237 p.

4997. PLATH (Christian). Konfessionskampf und fremde Besatzung. Stadt und Hochstift Hildesheim im Zeitalter der Gegenreformation und des Dreißigjährigen Krieges, ca. 1580–1660. Münster, Aschendorff, 2005, XIII-732 p. (Reformationsgeschichtliche Studien und Texte, 147).

4998. PRIOR (Charles W. A.). Defining the Jacobean church. The politics of religious controversy, 1603–1625. Cambridge, New York a. Melbourne, Cambridge U. P., 2005, XIV-294 p.

4999. Protestants, protestantisme et pensée clandestine. Dossier thématique établi par Geneviève ARTIGAS-MENANT et Antony MAC KENNA; avec la collaboration de Maria Susana SEGUIN. Paris, Presses de l'université Paris-Sorbonne, 2005, 485 p. (Lettre clandestine; 13. (La lettre clandestine. 13).

5000. REYNOLDS (Matthew). Godly reformers and their opponents in Early Modern England: religion in Norwich c. 1560–1643. Rochester, Boydell Press, 2005, XVI-310 p. (Studies in modern British religious history, 10).

5001. ROBERSON (Houston Bryan). Fighting the good fight. The story of the Dexter Avenue King Memorial Baptist Church, 1865–1977. New York, Routledge, 2005, XXI-248 p.

5002. RUBLACK (Ulinka). Reformation Europe. Cambridge a. New York, Cambridge U. P., 2005, XIV-208 p. (ill.). (New approaches to European history).

5003. RYRIE (Alec). The gospel and Henry VIII. Evangelicals in the early English Reformation. New York, Cambridge U. P., 2005, XIX-306 p. (Cambridge studies in early modern British history).

5004. SALVATORE (Nick). Singing in a strange land. C. L. Franklin, the Black Church, and the transformation of America. New York, Little Brown, 2005, XII-419 p.

5005. SCHLÜTER (Theodor C.). Flug- und Streitschriften zur "Kölner Reformation". Die Publizistik um den Reformationsversuch des Kölner Erzbischofs und Kurfürsten Hermann von Wied, 1515-1547. Wiesbaden, Harrassowitz, 2005, XVI-461 p. (Buchwissenschaftliche Beiträge aus dem deutschen Bucharchiv München, 73).

5006. SCHWARZ (Karl W.). Ius circa sacra und ius in sacra im Spiegel der Protestantenpolitik der Habsburger im 19. Jahrhundert. *Zeitschrift der Savigny-Stiftung für Rechtsgeschichte Kanonistische Abteilung*, 2005, 91, p. 578-624.

5007. SERVET (Miguel). Servet frente a Calvino, a Roma y al luteranismo. Ed. par Angel ALCALÁ. Zaragoza, Prensas Universitarias de Zaragoza, 2005, C-444 p. (La rumbe clasicos aragoneses. Historia y pensamento, 40).

5008. SKINNER (S. A.). Tractarians and the "Condition of England". The social and political thought of the Oxford movement. Oxford, Clarendon Press, 2005, 330 p.

5009. Socinianism and arminianism: antitrinitarians, calvinists and cultural exchange in seventeenth-century Europe. Ed. by Martin MULSOW and Jan ROHLS. Leiden, Boston, Brill, 2005, IX-306 p. (Brills studies in intellectual history, 134).

5010. SPIERLING (Karen E.). Infant Baptism in Reformation Geneva. The shaping of a community, 1536–1564. Aldershot, Ashgate, 2005, 253 p.

5011. THOMPSON (Nicholas). Eucharistic sacrifice and patristic tradition in the theology of Martin Bucer, 1534–1546. Boston, Brill, 2005, XV-315 p. (Studies in the history of Christian traditions, 119).

5012. WILLIAMS (Michael E. Sr.). Isaac Taylor Tichenor: the creation of the Baptist new South. Tuscaloosa, University of Alabama Press, 2005, 240 p. (Religion and American culture).

5013. WILLIS (Alan Scot). All according to god's plan: Southern Baptist Missions and race, 1945–1970. Lexington, University Press of Kentucky, 2005, XIII-260 p. (Religion in the South).

Cf. n°s 3677, 3886, 4095, 4339, 5842

§ 5. Non-Christian religions and sects.

* 5014. VAN RAHDEN (Till). Jews and the ambivalences of civil society in Germany, 1800–1933: assessment and reassessment. *Journal of modern history*, 2005, 77, 4, p. 1024-1047.

** 5015. Jüdische Quellen zur Reform und Akkulturation der Juden in Westfalen. Bearb. v. Arno HERZIG. Münster, Aschendorff, 2005, 232 p. (Veröffentlichungen der Historischen Kommission für Westfalen, 1. Quellen und Forschungen zur jüdischen Geschichte in Westfalen, XLV).

** 5016. [KOSTYRCHENKO (Gennadij V.).]. Gosudarstvennyj antisemitizm v SSSR. Ot nachala do kul'minatsii, 1938–1953. (The state anti-semitism in the USSR: From the beginning to the culmination, 1918–1953). Ed. Aleksandr N. JAKOVLEV. Mezhdunar. fond "Demokratija". Moskva, MFD – Materik, 2005, 592 p. (plates; pers. ind. p. 555-572). (Rossija. XX vek. Dokumenty). (Rossija. XX vek. Dokumenty). [English summary]

5017. Antisémythes. L'image des Juifs entre culture et politique (1848–1939). Sous la dir. de Marie-Anne MATARD-BONUCCI. Paris, Nouveau Monde Éditions, 2005, 463 p. [Cf. n° <sélection> 268.]

5018. BARTAL (Israel). The Jews of Eastern Europe, 1772–1881. Philadelphia, University of Pennsylvania Press, 2005, (Jewish culture and contexts).

5019. BIRNBAUM (Pierre). Prier pour l'Etat: les Juifs, l'alliance royale et la démocratie. Paris, Calmann-Lévy, 2005, 222 p.

5020. BUDNITSKIJ (Oleg V.). Rossijskie evrei mezhdu krasnymi i belymi (1917–1920). (The Russian Jews between the Red and the White, 1917–1920). Moskva, ROSSPEN, 2005, 551 p. (ill.; bibl. p. 503-521; pers. ind. p. 525-548). [English summary]

5021. COHEN (Shaye J. D.). Why aren't Jewish women circumcised? Gender and covenant in Judaism. Berkeley a. Los Angeles, University of California Press, 2005, XVII-317 p.

5022. Evrei v Sibiri i na Dal'nem Vostoke: istorija i sovremennost': Materialy VI region. nauch.-prakt. konf. (22–23 avg. 2005 g.). (Jews in Siberia and the Far East: History and modernity: Proceedings of the 6th conference, Barnaul, 22–23 August 2005). Ed. Jakov M. KOFMAN. Krasnojarsk – Barnaul, Klaretianum, 2005, 249 p. (plates; bibl. incl.).

5023. FERZIGER (Adam S.). Exclusion and hierarchy. Orthodoxy, nonobservance, and the emergence of modern Jewish identity. Philadelphia, University of Pennsylvania Press, 2005, X-303 p. (Jewish culture and contexts).

5024. GELLER (Jay Howard). Jews in post-Holocaust Germany, 1945–1953. Cambridge a. New York, Cambridge U. P., 2005, XIII-330 p. (ill.).

5025. GUROCK (Jeffrey S.). Judaism's encounter with American sports. Bloomington, Indiana U. P., 2005, X-234 p. (The Modern Jewish experience).

5026. HALLAQ (Wael B.). The origins and evolution of Islamic law. New York, Cambridge U. P., 2005, IX-234 p. (Themes in Islamic law, 1).

5027. HARVEY (L. P.). Muslims in Spain, 1500 to 1614. Chicago, University of Chicago Press, 2005, XII-448 p.

5028. JENSEN (Uffa). Gebildete Doppelgänger: Bürgerliche Juden und Protestanten im 19. Jahrhundert. Göttingen, Vandenhoeck & Ruprecht, 2005, 383 p. (Kritische Studien zur Geschichtswissenschaft, 167).

5029. Jewish daily life in Germany, 1618–1945. Ed. by Marion A. KAPLAN. Oxford, Oxford U. P., 2005, XIV-529 p.

5030. Jüdische Welten: Juden in Deutschland vom 18. Jahrhundert bis in die Gegenwart. Hrsg. v. Marion KAPLAN und Beate MEYER. Göttingen, Wallstein-Verl., 2005, 489 p. (Hamburger Beiträge zur Geschichte der deutschen Juden, 27).

5031. KLAPPER (Melissa R.). Jewish girls coming of age in America, 1860–1920. New York, New York U. P., 2005, X-310 p.

5032. LAGROU (Pieter). Return to a vanished world. European societies and the remnants and their Jewish communities, 1945–1947. *In*: The Jews are coming back. The return of the Jews in their countries of origin after World War II. Ed. by David BANKIER. New York a. Oxford, Berghahn Books a. Jerusalem, Yad Vashem, 2005, p. 1-24.

5033. MEYER (Ahlrich). Täter im Verhör. Die "Endlösung der Judenfrage" in Frankreich 1940–1944. Darmstadt, Wissenschaftliche Buchgesellschaft, 2005, 471 p.

5034. Mirovoj krizis 1914–1920 godov i sud'ba vostochnoevropejskogo evrejstva. (The world crisis of 1914–1920 and the fate of the East European Jewry: [Articles]). Ed. Oleg V. BUDNITSKIJ. Mezhdunar. issled. tsentr ros. i vostochnoevr. evrejstva. M.: ROSSPEN, 2005, 447 p. (bibl. incl.; pers. ind. p. 437-445). (Isotrija i kul'tura rossijskogo i vostochnoevropejskogo evrejstva: Novye istochniki i podkhody). [English summary]

5035. MUCHNIK (Natalia). Une vie de marrane. Les pérégrinations de Juan de Prado dans l'Europe du XVIIe siècle. Paris, Honoré Champion, 2005, 597 p. (Bibliothèque d'études juives).

5036. MUSULLAM (Adnan A.). From secularism to Jihad. Sayyid Qutb and the foundation of radical Islamism. Westport, Praeger, 2005, XIII-261 p.

5037. Musulmans de France et d'Europe. Sous la dir. Rémy LEVEAU et Khadija MOHSEN-FINAN. Paris, CNRS Éditions, 2005, 187 p.

5038. PALAU (Josep). El pensamiento judío en el exilio ante la Inquisición. *Historia social*, 2005, 52, p. 3-18.

5039. Percorsi di storia ebraica: fonti per la storia degli ebrei in Italia nell'età moderna e contemporanea, 8. centenario della morte di Maimonide: atti del 18. Convegno internazionale, Cividale del Friuli – Gorizia, 7/9 settembre 2004. A cura di Pier Cesare IOLY ZORATTINI. Udine, Forum, 2005, 464 p. Testi e studi, 15).

5040. PREUSS (Monika). ...aber die Krone des guten Namens überragt sie: jüdische Ehrvorstellungen im 18. Jahrhundert im Kraichgau. Stuttgart, Kohlhammer, 2005, XVIII-149 p. (ill., map). (Veröffentlichungen der Kommission für Geschichtliche Landeskunde in Baden-Württemberg, 160. Forschungen; Reihe B).

5041. SHILO (Margalit). Princess or prisoner? Jewish women in Jerusalem, 1840–1914. Waltham, Brandeis U. P., 2005, XXVIII-330 p. (Brandeis series on Jewish women and the Tauber Institute for the Study of European Jewry Series).

5042. THING (Morten). Jiddishland i København: den jødiske invandring 1905–1914. (Yiddischland à Copenhague: l'immigration juive 1905–1914). København, Nemos Bibliotek, 2005, 192 p.

5043. Til et folk af alle høre. Den jødiske minoritet i Danmark. (La minorité juive au Danemark). Red. Henri GOLDSTEIN. København, Gyldendal, 2005, 178 p.

5044. VILHJALMSSON (Viljamur Örn). Medaljens bagside: jødiske flygtningeskæbner i Danmark 1933–1945. (Le revers de la médaille. Destins de réfugiés juifs au Danemark 1933–1945). København, Vandkunsten, 2005, 471 p.

5045. Židé a Morava. Sborník z konference konané v muzeu Kroměřížska. (Jews and Moravia. Proceedings of a conference). Ed. Petr PÁLKA. Kroměříž, Muzeum Kroměřížska, 2005, 195 p. (photogr.).

§ 5. Addenda 2002–2004.

5046. FATHI (Habiba). Femmes d'autorité dans l'Asie centrale contemporaine. Quête des ancêtres et recompositions identitaires dans l'Islam postsoviétique. Paris, Maisonneuve & Larose et Institut français d'études sur l'Asie centrale, 2004, 348 p.

5047. NATHANS (Benjamin). Beyond the pale: the Jewish encounter with late imperial Russia. Berkeley, University of California Press, 2002, XVII-424 p.

5048. SINKOFF (Nancy). Out of the shtetl: making Jews modern in the Polish borderlands. Hanover, Brown U. P., 2004, XII-320 p. (Brown Judaic studies, 336).

Cf. n°s 403, 574, 3560, 3887, 4134, 4165, 4217, 4229, 4283, 4353, 4509, 4709, 5937

M

HISTORY OF MODERN CULTURE

§ 1. General. 5049-5171. – § 2. Academies, universities and intellectual organizations. 5172-5208. – § 3. Education. 5209-5260. – § 4. The Press. 5261-5310. – § 5. Philosophy (*a*. General, *b*. Special studies). 5311-5360. – § 6. Exact, natural, medical sciences and technique. 5361-5473. – § 7. Literature (*a*. General; *b*. Renaissance; *c*. Classicism; *d*. Romanticism and after). 5474-5544. – § 8. Art and industrial art (*a*. General; *b*. Architecture; *c*. Sculpture, painting, etching and drawing; *d*. Decorative, photography, popular and industrial art). 5545-5617. – § 9. Music, theatre, cinema and broadcasting. 5618-5719.

§ 1. General.

* 5049. Bibliographie du dix-neuvième siècle.: lettres, arts, sciences, histoire. Année 2003. Société des études romantiques et dix-neuviémistes; rédigé par Claude DUCHET, Dominique PETY et Philippe RÉGNIER. Paris, Presses Sorbonne nouvelle, 2005, 332 p.

* 5050. Bibliographie zum Fortwirken der Antike in den deutschsprachigen Literaturen des 19. und 20. Jahrhunderts. Hrsg. v. Michael von ALBRECHT, Walter KIßEL und Werner SCHUBERT. Frankfurt am Main u. Oxford, P. Lang, 2005, XV-277 p. (Studien zur klassischen Philologie, 149).

* 5051. CONLON (Pierre M.). Le siècle des lumières: bibliographie chronologique. T. 23. 1788. Genève, Librairie Droz, 2005, XXVII-456 p. (Histoire des idées et critique littéraire, 419).

** 5052. BAYLE (Pierre). 4 Janvier 1684–juillet 1684: lettres 242-308. Ed. par Elisabeth LABROUSSE, avec la collaboration de Eric-Olivier LOCHARD et Dominique TAURISSON. Oxford, Voltaire foundation, 2005, XXI-287 p. (ill.). Correspondance de Pierre Bayle, 4).

** 5053. Correspondance de Théodore de Bèze. Recueillie par Hippolyte AUBERT. Tome 27. 1586. Publiée par Alain DUFOUR, Béatrice NICOLLIER et Hervé GENTON. Genève, Droz, 2005, XXVI-319 p. (ill.). (Travaux d'humanisme et Renaissance, 401).

** 5054. FLAVIO BIONDO. Rome restaurée = Roma instaurata. T. 1. Livre 1 = Liber 1. Édition, traduction, présentation et notes par Anne RAFFARIN-DUPUIS. Paris, Belles lettres, 2005, CLXXIX-195 p. (Les classiques de l'humanisme, 24).

5055. ALBRECHT (Andrea). Kosmopolitismus: Weltbürgerdiskurse in Literatur, Philosophie und Publizistik um 1800. Berlin, W. de Gruyter, 2005, IX-442 p. (Spectrum Literaturwissenschaft = Spectrum Literature, 1).

5056. Antiamerikanismus im 20. Jahrhundert: Studien zu Ost- und Westeuropa. Hrsg. v. Jan C. BEHRENDS, Árpád von KLIMÓ und Patrice G. POUTRUS. Bonn, Dietz, 2005, 365 p. (ill.). (Reihe: Politik- und Gesellschaftsgeschichte, 68).

5057. Antiquae lectiones: el legado clásico desde la Antigüedad hasta la Revolución Francesa. Ed. por Juan SIGNES CODOÑER [et al.]. Madrid, Cátedra, 2005, 610 p. (Crítica y estudios literarios).

5058. ANTOINE (Jacques). Histoire des sondages. Paris, O. Jacob, 2005, 282 p.

5059. ARNDT (Richard T.). The first resort of kings. American cultural diplomacy in the twentieth century. Dulles, Potomac Books, 2005, XXI-602 p.

5060. AYMES (Jean René). Ilustración y revolución francesa en España. Prólogo de Alberto GIL NOVALES. Lleida, Editorial Milenio, 2005, 336 p. (Coleccion Hispania, 18).

5061. BABEROWSKI (Jörg). Zivilisation der Gewalt. Die kulturellen Ursprünge des Stalinismus. *Historische Zeitschrift*, 2005, 281, 1, p. 59-102.

5062. BAR-YOSEF (Eitan). The Holy Land in English culture, 1799–1917: Palestine and the question of Orientalism. Oxford, Clarendon press, 2005, XIII-319 p. (ill.). (Oxford English monographs).

5063. BELARDELLI (Giovanni). Il ventennio degli intellettuali: cultura, politica, ideologia nell'Italia fascista. Roma e Bari, Laterza, 2005, X-310 p. (Storia e società).

5064. BELGE (Murat). Osmanlı'da kurumlar ve kültür. (Institutions et culture dans l'Empire Ottoman). İstanbul, İstanbul Bilgi Üniversitesi, 2005, XV-563 p.

5065. BLACK (Jeremy). A subject for taste. Culture in eigtheenth-century England. London, Hambledon, 2005, XIX-272 p.

5066. BOSCANI LEONI (Simona). Riflessioni sulla corrispondenza erudita tra Sei e Settecento. *Schweizerische Zeitschrift für Geschichte*, 2005, 55, 4, p. 441-447.

5067. BROUILLET (Eugénie). La négation de la nation: l'identité culturelle québécoise et le fédéralisme canadien. Sillery, Septentrion, 2005, 478 p. (Cahiers des Amériques, 12).

5068. CANFORA (Davide). Prima di Machiavelli: politica e cultura in età umanistica. Roma e Bari, Laterza, 2005, XI-177 p. (Biblioteca universale Laterza, 580).

5069. CESEREANU (Ruxandra). Made in Romania. Subculturi urbane la sfârşit de secol XX şi început de secol XXI. (Made in Romania. Urban subcultures at the end of the 20th century and beginning of the 21st). Cluj-Napoca, Editura. Limes, 2005, 186 p.

5070. Cinque dita (Le) del sultano. Turchi armeni arabi greci ed ebrei nel continente mediterraneo del '900. A cura di Stefano TRINCHESE. Mare nostrum. L'Aquila, Textus, 2005, 166 p.

5071. CLARK (Michael D.). The American discovery of tradition, 1865–1942. Baton Rouge, Louisiana State U. P., 2005, IX-268 p.

5072. DE LA FLOR (Fernando R.). Pasiones frías: secreto y disimulación en el Barroco hispano. Madrid, Marcial Pons Historia, 2005, 330 p. (ill.). (Historia / Marcial Pons. Estudios).

5073. DE VENUTO (Liliana). Il dibattito sulla felicità a metà del Settecento in Italia. *Quaderni di storia*, 2005, 62, p. 131-166.

5074. DELTOMBE (Thomas). L' islam imaginaire: la construction médiatique de l'islamophobie en France, 1975–2005. Paris, La Découverte, 2005, 382 p. (Cahiers libres).

5075. Deutschland und Italien, 1860–1960: politische und kulturelle Aspekte im Vergleich. Hrsg. v. Christof DIPPER unter Mitarbeit von Elisabeth MÜLLER-LUCKNER. München, Oldenbourg, 2005, VIII-284 p. (Schriften des Historischen Kollegs. 52. Kolloquien).

5076. DUGAC (Željko). Protiv bolesti i neznanja: Rockfellerova fondacija u međuratnoj Jugoslaviji. (Against illness and ignorance: Rockefeller Foundation in the interwar Yugoslavia). Zagreb, Srednja Europa, 2005, 196 p.

5077. EISENBERG (Christiane). Medienfußball. Entstehung und Entwicklung einer transnationalen Kultur. *Geschichte und Gesellschaft*, 2005, 31, 4, p. 586-609.

5078. ENGBERG (Jens). Magten og kulturen: dansk kulturpolitik 1750–1900. Bd. 1. Under enevælden. Bd. 2. Mellem enevælde og grundlovsstyre. Bd. 3. Under grundlovsstyret. (Pouvoir et culture: la politique culturelle danoise 1750–1900. Tome 1. Sous l'absolutisme. Tome 2. Entre l'absolutisme et la monarchie constitutionnelle. Tome 3. Sous la monarchie constitutionnelle). København, Gads Forlag, 2005, 3 vol., 515 p., 601 p., 621 p.

5079. Französische Kultur im Berlin der Weimarer Republik: kultureller Austausch und diplomatische Beziehungen. Hrsg. v. Hans Manfred BOCK. Tübingen, Narr, 2005, 334 p. (Edition lendemains, 1).

5080. Frühaufklärung. Hrsg. v. Karl EIBL. Hamburg, Felix Meiner, 2005, 255 p. (Aufklärung; Jahrg. 2005, 17).

5081. Germanistik und Kunstwissenschaften im "Dritten Reich": Marburger Entwicklungen, 1920–1950. Hrsg. v. Kai KÖHLER, Burghard DEDNER und Waltraud STRICKHAUSEN. München, Saur, 2005, 490 p. (ill.). (Academia Marburgensis, 10).

5082. Germany's nature. Cultural landscapes and environmental history. Ed. by Thomas LEKAN and Thomas ZELLER. New Brunswick a. London, Rutgers U. P., 2005, VIII-266 p.

5083. GRANGE (Daniel J.). La società "Dante Alighieri" et la défense de" l'italianità". *Mélanges de l'École française de Rome: Italie et mediterranée (MEFRIM)*, 2005, 117, 1, p. 261-267.

5084. GRDINA (Igor). Preroki, doktrinarji, epigoni : idejni boji na Slovenskem v prvi polovici 20. stoletja. (Prophets, doctrinists, epigones : the struggle of minds on the Slovene territory in the first half of the 20th century). Ljubljana, Inštitut za civilizacijo in kulturo – ICK, 2005, 214 p. (Zbirka Zbiralnik, 11).

5085. GREGOR (A. James). Mussolini's intellectuals. Fascist social and political thought. Princeton, Princeton U. P., 2005, X-282 p.

5086. GUDIS (Catherine). Buyways: billboards, automobiles, and the American landscape. New York, Routledge, 2005, VIII-333 p. (Cultural Spaces).

5087. HAAS (Mark L.). The ideological origins of great power politics, 1789–1989. Ithaca a. London, Cornell U. P., 2005, X-232 p. (ill.). (Cornell studies in security affairs).

5088. HARDTWIG (Wolfgang). Hochkultur des bürgerlichen Zeitalters. Göttingen, Vandenhoeck & Rurecht, 2005, 387 p. (Kritische Studien zur Geschichtswissenschaft, 169).

5089. Heimat abroad (The): the boundaries of Germanness. Ed. by Krista O'DONNELL, Renate BRIDENTHAL and Nancy REAGIN. Ann Arbor, University of Michigan Press, 2005, X-326 p. (Social history, popular culture, and politics in Germany).

5090. HEINICH (Nathalie). L' élite artiste: excellence et singularité en régime démocratique. Paris, Gallimard, 2005, 370 p. (Bibliothèque des sciences humaines).

5091. HIRSCHI (Caspar). Wettkampf der Nationen. Konstruktionen einer deutschen Ehrgemeinschaft an der Wende vom Mittelalter zur Neuzeit. Göttingen, Wallstein, 2005, 555 p.

5092. Histoire culturelle de la France. Sous la direction de Jean-Pierre RIOUX et Jean-François SIRINELLI. 2. CROIX (Alain), QUÉNIART (Jean). De la Renaissance a l'aube des Lumieres. Paris, Ed. du Seuil, 2005, 490 p. (Points, H349).

5093. HRADSKÁ (Katarína). Slobodomurárske lóže v Bratislave. (Freimaurerlogen in Bratislava.) Bratislava, Albert Marenčin vydavateľstvo PT, 2005, 158 p. (Bratislava – Pressburg).

5094. Interdisciplinary century (The): tensions and convergences in eighteenth-century art, history and literature. Ed. by Julia V. DOUTHWAITE and Mary VIDAL. Oxford, Voltaire Foundation, 2005, XXXIV-312 p. (ill., facsims., ports.). (Studies on Voltaire and the eighteenth century, 2005:04).

5095. KAISER (Wolfram). Cultural transfer of free trade at the world exhibitions, 1851–1862. *Journal of modern history*, 2005, 77, 3, p. 563-590.

5096. KALIFA (Dominique). Crime et culture au XIXe siècle. Paris, Perrin, 2005, 331 p.

5097. KATHE (Steffen R.). Kulturpolitik um jeden Preis: die Geschichte des Goethe-Instituts von 1951 bis 1990. München, M. Meidenbauer, 2005, 525 p. (ill.).

5098. KLIMASMITH (Betsy). At home in the city: urban domesticity in American literature and culture, 1850–1930. Hanover, University of New Hampshire Press, 2005, XII-293 p. (Becoming modern: new nineteenth-century studies).

5099. KNAPÍK (Jiří). Arbeiter versus Künstler. Gewerkschaft und neue Elemente in der tschechoslowakischen Kulturpolitik im Jahr 1948. *In*: Sozialgeschichtliche Kommunismusforschung. Tschechoslowakei, Polen, Ungarn und DDR 1948–1968. Ed. Christiane BRENNER. München, 2005, p. 243-262.

5100. KNAPP (Marion). Österreichische Kulturpolitik und das Bild der Kulturnation: Kontinuität und Diskontinuität in der Kulturpolitik des Bundes seit 1945. Frankfurt am Main, Lang, 2005, 398 p. (ill.). (Politik und Demokratie, 4).

5101. KOCIAN (Jiří). Political Culture in Czechoslovakia 1945–1968. *In*: Political culture in Central Europe (10th–20th century). Tomo 2. 19th and 20th centuries. Ed. Magdalena HULAS, Jaroslav PÁNEK. Praha a. Warsaw, 2005, p. 271-281.

5102. KOMOLOVA (Nelli P.). Italija v russkoj kul'ture Serebrjanogo veka. (Italy in the Russian culture of the 'Silver Age', [the early 20th century]). RAN, In-t vseobshchej istorii. Moskva, Nauka, 2005, 470 p. (ill.; portr.; bibl. incl.). – EADEM. Koktebel' v russkoj kul'ture. (Koktebel [the Crimea] in the Russian culture). RAN, In-t vseobshchej istorii. Moskva, IVI RAN, 2005, 306 p.

5103. KOPEČEK (Michal). The ups and downs of Central Europe. Chapters from Czech symbolic. *In*: The weight of history in Central European societies of the 20th century. Ed. Zora HLAVIČKOVÁ. Praha, 2005, p. 41-59.

5104. KREJČÍ (Oskar). Geopolitics of the Central European region. The view from Prague and Bratislava. Bratislava, Veda, 2005, 493 p.

5105. Kultur und Kommunikation. Die europäische Gelehrtenrepublik im Zeitalter von Leibniz und Lessing. Hrsg. v. Ulrich Johannes SCHNEIDER. Wiesbaden, Harrassowitz, 2005, 364 p. (Wolfenbütteler Forschungen, 109).

5106. LABUTINA (Tat'jana L.). Kul'tura i vlast' v epokhu Prosveshchenija. (Culture and political power in the Age of Enlightenment). RAN, In-t vseobshchej istorii. Moskva, Nauka, 2005, 458 p. (ill.; portr.; maps; bibl. p. 443-450; pers. ind. p. 453-458).

5107. LANDON (William J.). Politics, patriotism and language: Niccolò Machiavelli's "secular patria" and the creation of an Italian national identity. New York a. Oxford, Peter Lang, 2005, XIV-300 p. (Studies in modern European history, 57).

5108. LATZIN (Ellen). Lernen von Amerika? Das US-Kulturaustauschprogramm für Bayern und seine Absolventen. Stuttgart, Steiner, 2005, 496 p. (Transatlantische historische Studien, 23).

5109. LOYER (Emmanuelle). Paris à New York. Intellectuels et artistes français en exil, 1940–1947. Paris, Grasset, 2005, 497 p.

5110. LUZZI (Serena). Il processo a Carlo Antonio Pilati (1768–1769), ovvero della censura di stato nell'Austria di Maria Teresa. *Rivista storica italiana*, 2005, 117, 3, p. 687-740.

5111. MACHAČOVÁ (Jana), MATĚJČEK (Jiří). K vývoji obecné kultury v českých zemích v období 1781–1989. (Zur Entwicklung der allgemeinen Kultur in den böhmischen Ländern 1781–1989). *Slezský sborník*, 2005, 103, 2, p. 81-117.

5112. MADDOX (Lucy). Citizen Indians. Native American intellectuals, race, and reform. Ithaca, Cornell U. P., 2005, 205 p.

5113. MALIS (Christian). Raymond Aron et le débat stratégique français: 1930–1966. Paris, Institut de stratégie comparée et Commission française d'histoire militaire et Economica, 2005, 821 p. (Bibliothèque stratégique).

5114. MALTARICH (Bill). Samurai and supermen. National socialist views of Japan. Oxford, Bern a. Berlin, Lang, 2005, 406 p. (German life and civilization, 42).

5115. MALYSHEVA (Svetlana Ju.). Sovetskaja prazdnichnaja kul'tura v provintsii: Prostranstvo, simvoly, istoricheskie mify (1917–1927). (The Soviet provincial festival culture: Space, symbols, historical myths, 1917–1927). Kazan', Ruten, 2004, 399 p. (ill.; bibl. p. 386-399).

5116. MARCONE (Arnaldo). La crisi dell'impero romano come paradigma di quella europea: Ortega y Gasset. *Anabases*, 2005, 2, p. 101-114.

5117. Mare nostrum: percezione ottomana e mito mediterraneo in Italia all'alba del '900. A cura di Stefano TRINCHESE; prefazione di Andrea RICCARDI. Milano, Guerini studio, 2005, 314 p. (Contemporanea; 12).

5118. MARTINI (Mauro). L'utopia spodestata: le trasformazioni culturali della Russia dopo il crollo dell'Urss. Torino, Einaudi, 2005, 181 p. (Piccola biblioteca Einaudi. Nuova serie. Saggistica letteraria e linguistica, 295).

5119. MEHLMAN (Jeffrey). Les intellectuels français à Manhattan 1940–1944. Paris, Albin Michel, 2005, 253 p.

5120. MÉNAGER (Daniel). La Renaissance et la nuit. Genève, Droz, 2005, 270 p. (ill.). (Les seuils de la modernité, 10).

5121. MICHELS (Eckard). Von der Deutschen Akademie zum Goethe-Institut. Sprach- und auswärtige Kulturpolitik, 1923–1960. München, Oldenbourg, 2005, VI-266 p. (Zeitgeschichte, 70).

5122. MUNTEAN (Ovidiu). Imaginea românilor în Franța la mijlocul secolului al XIX-lea. (The image of the Romanians in France at the middle of the 19th century). Cluj-Napoca, Editura Napoca Star, 2005, 283 p.

5123. Mýty naše slovenské. (Mythen unseren slowakischen). Ed. Eduard KREKOVIČ, Elena MANNOVÁ, Eva KREKOVIČOVÁ. Bratislava, Historický ústav Slovenskej akadémie vied, 2005, 246 p.

5124. Naples, Rome, Florence: une histoire comparée des milieux intellectuels italiens, XVIIe–XVIIIe siècles. Sous la dir. de Jean BOUTIER, Brigitte MARIN et Antonella ROMANO. Roma, Ecole française de Rome, 2005, 815 p. (Collection de l'École française de Rome, 355).

5125. NICCOLI (Ottavia). Rinascimento anticlericale: infamia, propaganda e satira in Italia tra Quattro e Cinquecento. Roma e Bari, Laterza, 2005, IX-218 p. (tav., ill.). (Storia e società).

5126. NIZ (Faruk). Osmanlı-Türkiye Batılılaşmasının Kararsızlığı: 'Köhne' Geçmiş, 'Şanlı' Gelecek. (Indécision de l'occidentalisation turco-ottomane: Passé 'moyenâgeux', Avenir 'glorieux'). *Dîvân İlmî Araştırmalar*, 2005, 19, p. 61-115.

5127. NOLAN (Michael E.). The inverted mirror. Mythologizing the enemy in France and Germany, 1898–1914. New York a. Oxford, Berghahn, 2005, IX-141 p.

5128. Norsk telekommunikasjonshistorie. Bind 1. RINDE (Harald). Et telesystem tar form, 1855–1920. Bind 2. ESPELI (Harald). Det statsdominerte teleregimet 1920–1970. Oslo, Gyldendal Fakta, 2005, 2 vol., 480 p., 592 p.

5129. NÜTZENADEL (Alexander). Stunde der Ökonomen. Wissenschaft, Politik und Expertenkultur in der Bundesrepublik, 1949–1974. Göttingen, Vandenhoeck & Ruprecht, 2005, 427 p. (Geschichtswissenschaft, 166).

5130. *Vacat*.

5131. Philologie humaniste (La) et ses représentations dans la théorie et dans la fiction. Sous la dir. de Perrine GALAND-HALLYN, Fernand HALLYN, Gilbert TOURNOY. Genève, Librairie Droz, 2005, 2 vol., VII-654 p. (Romanica Gandensia, 32).

5132. *Vacat*.

5133. Politische Kulturgeschichte der Zwischenkriegszeit 1918–1939. Hrsg. v. Wolfgang HARDTWIG. Göttingen, Vandenhoeck & Ruprecht, 2005, 376 p. (Geschichte und Gesellschaft, Sonderh., 21).

5134. PORTER (Martin). Windows of the soul. The art of physiognomy in European culture 1470–1780. Oxford, Clarendon Press, 2005, 386. p.

5135. POSTLER (Vicky). De Gaumont Italia à Arte: la politique culturelle française en Europe. Paris, Connaissances et savoirs, 2005, 314 p.

5136. RATAJ (Jan). Das Deutschlandbild im Protektorat und im tschechoslowakischen Exil 1939–1945. *Zeitschrift für Geschichtswissenschaft*, 2005, 53, 5, p. 434-454.

5137. Redes intelectuales y formación de naciones en España y América Latina (1890–1940). Ed. por Marta Elena CASAÚS ARZÚ y Manuel PÉREZ LEDESMA. Madrid, Ediciones Universidad Autónoma de Madrid, 2005, 450 p. (Colección de estudios, 101).

5138. RÉV (István). Retroactive justice: prehistory of post-communism. Stanford, Stanford U. P., 2005, 340 p. (Cultural memory in the present).

5139. RICUPERATI (Giuseppe). La lettera dedicatoria e i suoi problemi nel tempo e nello spazio. *Rivista storica italiana*, 2005, 117, 2, p. 552-568.

5140. Rinascimento italiano (Il) di fronte alla Riforma, letteratura e arte = Sixteenth-century Italian art and literature and the Reformation: atti del colloquio internazionale, London, The Warburg Institute, 30–31 gennaio 2004. A cura di Chrysa DAMIANAKI, Paolo PROCACCIOLI e Angelo ROMANO. Manziana, Vecchiarelli, 2005, XII-343 p. (ill., facsims., ports.). (Cinquecento. 12 Studi).

5141. Rinascimento, mito e concetto. A cura di Renzo RAGGHIANTI e Alessandro SAVORELLI. Pisa, Edizioni della Normale, 2005, XXVIII-301 p. (Seminari e convegni, 2).

5142. ROGER (Philippe). The American enemy: a story of French anti-Americanism. Chicago, University of Chicago Press, 2005, XVIII – 518 p.

5143. ROSA (Mario). Cultura toscana e cultura europea nel Settecento: intorno ad Angelo Maria Bandini. *Archivio storico italiano*, 2005, 163, 2, p. 259-282.

5144. ROSSBACH (Sabine). Moderner Manierismus: Literatur – Film – Bildende Kunst. Frankfurt am Main, Peter Lang, 2005, 363 p. (ill.).

5145. RUBERG (Willemijn). Letter writing and elite identity in the Netherlands, 1770–1850. *Scandinavian journal of history*, 2005, 30, 3-4, p. 249-258.

5146. Ruralismus, jeho kořeny a dědictví. Osobnosti – díla – ideje. Sborník referátů z vědecké konference. (Ruralism, its roots and inheritance. Personalities – works – ideas). Semily, Státní okresní archiv, 2005, 342 p. (ill.). (Z Českého ráje a Podkrkonoší. Suppl., 10).

5147. SANTORO (Stefano). L'Italia e l'Europa orientale. Diplomazia culturale e propaganda 1918–1943. Milano, Franco Angeli, 2005, 422 p.

5148. SCHMIDT (Alexander). Kultur in Nürnberg 1918–1933: die Weimarer Moderne in der Provinz. Nürnberg, Sandberg, 2005, 403 p. (ill.).

5149. SCHULZE (Thies). Dante Alighieri als nationales Symbol Italiens, 1793–1915. Tübingen, Niemeyer, 2005, VIII-275 p. (Bibliothek des Deutschen Historischen Instituts in Rom, 109).

5150. SEGURA (Mauricio). La faucille et le condor: le discours français sur l'Amérique latine, 1950–1985. Montréal, Presses de l'Université de Montréal, 2005, 247 p. (Socius (Montréal, Québec).

5151. Shiryo de yomu Amerika bunkashi 1. Shokuminchi jidai: 15 seikimatsu kara 1770 nendai. (United States cultural history: reading historical sources. Vol. 1. The colonial period: late 15th century to the 1770s). Ed. by Yasuo ENDO. 2. Dokuritsu kara Nanbokusenso made. 1770 nendai kara 1850 nendai. (Vol. 2. From the Revolution to the Civil War: the 1770s to the 1850s). Ed. by Konomi ARA. Tokyo, University of Tokyo Press, 2005, 2 vol., 380 p., 408 p.

5152. Signums svenska kulturhistoria. Renässansen. (Histoire culturelle de la Suède. La Renaissance). Stormaktstiden. (La période de Grandeur). Red. Jakob CHRISTENSSON. Lund, Signum, 2005, 2 vol., 508 p., 639 p. (ill.).

5153. Slovanství a věda v 19. a 20. století. (Slav national feeling and science in the 19th and 20th centuries). Ed. Marek ĎURČANSKÝ, Hana BARVÍKOVÁ. Praha, Archiv AV ČR, 2005, 287 p. (photogr.). (Práce z Archivu Akademie věd. Řada A, 8).

5154. SOAVE (Sergio). Senza tradirsi, senza tradire. Silone e Tasca dal comunismo al socialismo cristiano (1900–1940). Torino, Aragno, 2005, 660 p.

5155. Splintering (The) of Spain: cultural history and the Spanish Civil War, 1936–1939. Ed. by Chris EALHAM and Michael RICHARDS. New York, Cambridge U. P., 2005, XXIII-282 p.

5156. Staatsbildung als kultureller Prozess. Strukturwandel und Legitimation von Herrschaft in der Frühen Neuzeit. Hrsg. v. Ronald G. ASCH und Dagmar FREIST. Köln, Weimar u. Wien, Böhlau, 2005, 442 p.

5157. STOLL (Ulrike). Kulturpolitik als Beruf: Dieter Sattler (1906–1968) in München, Bonn und Rom. Paderborn, Schöningh, 2005, 594 p. (ill.). (Veröffentlichungen der Kommission für Zeitgeschichte, 98 Reihe B, Forschungen).

5158. THOMAS (Martin). Albert Sarraut, French colonial development, and the communist threat, 1919–1930. *Journal of modern history*, 2005, 77, 4, p. 917-955.

5159. VAJNSHTEJN (Ol'ga B.). Dendi: Moda, literatura, stil' zhizni. (Dandy: Mode, literature and the style of life). Moskva, Novoe literaturnoe obozrenie, 2005, 638 p. (ill.; bibl. p. 610-623; pers. ind. p. 624-639). (Kul'tura povsednevnosti).

5160. VAN DAMME (Stéphane). Le temple de la sagesse: savoirs, écriture et sociabilité urbaine, Lyon, XVIIe–XVIIIe siècle. Paris, Editions de l'Ecole des hautes études en sciences sociales, 2005, 514 p. (map). (Civilisations et sociétés, 119).

5161. Vertreibung (Die) der Deutschen aus dem Osten in der Erinnerungskultur. Hrsg. v. Jörg-Dieter GAUGER und Manfred KITTEL. Sankt Augustin, München u. Berlin, Konrad-Adenauer- Stiftung, Institut für Zeitgeschichte, 2005, 146 p.

5162. WALGENBACH (Katharina). "Die weiße Frau als Trägerin deutscher Kultur": koloniale Diskurse über Geschlecht, "Rasse" und Klasse im Kaiserreich. Frankfurt am Main u. New York, Campus, 2005, 297 p. (Campus Forschung, 891).

5163. WALSH (Margaret). The American West. Visions and revisions. Cambridge, Cambridge U. P., 2005, X-161 p.

5164. WARREN (Louis S.). Buffalo Bill's America: William Cody and the Wild West Show. New York, Alfred A. Knopf, 2005, XVI-652 p.

5165. WATENPAUGH (Keith David). Cleansing the cosmopolitan city: historicism, journalism and the Arab nation in the post-Ottoman eastern Mediterranean. *Social history*, 2005, 30, 1, p. 1-24.

5166. WEINBERGER (Jerry). Benjamin Franklin unmasked. On the unity of his moral, religious, and political thought. Lawrence, University Press of Kansas, 2005, XVI-336 p. (American political thought).

5167. WERNER (Yvonne Maria). Kvinnligt klosterliv i Sverige och Norden: en motkultur i det moderna samhället. (Moniales en Scandinavie après la Réforme: une contre-culture dans la société moderne). Uppsala, Ängelholm, Katolsk historisk förening, Catholica, 2005, 282 p. (ill.). (Katolsk historisk förening i Sverige, 3).

5168. YARDENI (Myriam). Enquêtes sur l'identité de la "nation France": de la Renaissance aux Lumières. Seyssel, Champ Vallon, 2005, 374 p. (Epoques).

5169. Zukunftsvoraussagen in der Renaissance. Hrsg. v. Klaus BERGDOLT und Walther LUDWIG; unter Mitwirkung von Daniel SCHÄFER. Wiesbaden, Harrassowitz, 2005, 444 p. (Wolfenbütteler Abhandlungen zur Renaissanceforschung, 23).

§ 1. Addenda 2004.

5170. BERGHAHN (Volker). Transatlantische Kulturkriege. Shepard Stone, die Ford-Stiftung und der europäische Antiamerikanismus. Stuttgart, Steiner, 2004, 392 p. (Transatlantische Historische Studien, 21).

5171. SÜNDERHAUF (Eshter Sophia). Griechensehnsucht und Kulturkritik. Die deutsche Rezeption von Winckelmanns Antikenideal 1840–1945. Berlin, Akademie, 2004, XXX-413 p.

Cf. n°s 85, 93, 133, 200, 206, 215, 271, 322, 338, 349, 358, 359, 389, 401, 422, 479, 558, 572, 811, 814, 815, 816, 2895, 3500, 3802, 3807, 3813, 3886, 4433, 4467, 4481, 4492, 4701, 4865, 4893, 5262, 6183, 6831, 7318, 7390

§ 2. Academies, universities and intellectual organizations.

* 5172. MAC CLELLAND (Charles). Modern German universities and their historians since the fall of the Wall. *Journal of modern history*, 2005, 77, 1, p. 138-159.

** 5173. Archivio storico (L') dell'Università degli studi di Milano: inventario. A cura di Stefano TWARDZIK. Milano, Cisalpino, 2005, XIX-279 p (Quaderni di Acme / Università degli studi di Milano, Facoltà di lettere e filosofia, 69).

5174. Berliner Universität (Die) in der NS-Zeit. Hrsg. v. Rüdiger vom BRUCH und Christoph JAHR im Auftrag der Senatskommission "Die Berliner Universität und die NS-Zeit. Erinnerung, Verantwortung, Gedenken"; unter Mitarbeit von Rebecca SCHAARSCHMIDT. Band 1. Strukturen und Personen. Hrsg. v. Christoph JAHR; unter Mitarbeit von Rebecca SCHAARSCHMIDT. Band 2. Fachbereiche und Fakultäten. Hrsg. v. Rüdiger vom BRUCH; unter Mitarbeit von Rebecca SCHAARSCHMIDT. Stuttgart, Steiner, 2005, 2 vol., 257 p., 308 p.

5175. Collèges jésuites (Les) de Bruxelles: histoire et pédagogie, 1604, 1835, 1905, 2005. Sous la direction de Bernard STENUIT. Bruxelles, Lessius, Collège St-Michel, 2005, 656 p. (ill., ports., plans).

5176. DESLANDES (Paul R.). Oxbridge men: British masculinity and the undergraduate experience, 1850–1920. Bloomington, Indiana U. P., 2005, XVIII-319 p. (ill.).

5177. "Doktorgrad entzogen!" Aberkennungen akademischer Titel an der Universität Köln 1933 bis 1945. Verfasst und herausgegeben von Margit SZÖLLÖSI-JANZE und Andreas FREITÄGER, und den TeilnehmerInnen des Hauptseminars "Die Universität Köln im Nationalsozialismus", Wintersemester 2003/2004. Nümbrecht, Kirsch, 2005, 132 p. (ill., ports.).

5178. FERNÁNDEZ LUZÓN (Antonio). La universidad de Barcelona en el siglo XVI. Barcelona, Publicacions i Edicions, Universitat de Barcelona, 2005, 342 p. (Universitat de Barcelona, 14).

5179. FRIEDMAN (Rebecca). Masculinity, autocracy, and the Russian university, 1804–1863. New York, Palgrave Macmillan, 2005, X-195 p.

5180. Gendered Academia: Wissenschaft und Geschlechterdifferenz, 1890–1945. Hrsg. v. Miriam KAUKO, Sylvia MIESZKOWSKI und Alexandra TISCHEL. Göttingen, Wallstein, 2005, 368 p. (ill.). (Münchener komparatistische Studien, 6).

5181. GÖMÖRI (George). Magyarországi diákok angol és skót egyetemeken 1526-1789 = Hungarian students in England and Scotland, 1526–1789. Budapest, Eötvös Loránd Tudományegyetem Levéltára, 2005, 116 p. (Magyarországi diákok egyetemjárása az újkorban, 14).

5182. HEHL (Ulrich von). Sachsens Landesuniversität in Monarchie, Republik und Diktatur. Beiträge zur Geschichte der Universität Leipzig vom Kaiserreich bis zur Auflösung des Landes Sachsen 1952. Leipzig, Evangelische Verlagsanstalt, 2005, 587 p.

5183. HOPKINS (Clare). Trinity: 450 years of an Oxford college community. Oxford a. New York, Oxford U. P., 2005, XXII-500 p. (ill.).

5184. ISRAËL (Stéphane). Les études et la guerre: les normaliens dans la tourmente (1939–1945). Préf. de Jean-François SIRINELLI. Paris, Rue d'Ulm, 2005, 334 p. (ill.).

5185. KARABEL (Jerome). The chosen: the hidden history of admission and exclusion at Harvard, Yale, and Princeton. Boston, Houghton Mifflin, 2005, VIII-711 p. (ill.).

5186. "Klassische Universität" und "akademische Provinz". Studien zur Universität Jena von der Mitte des 19. bis in die dreißiger Jahre des 20. Jahrhunderts. Hrsg. v. Matthias STEINBACH und Stefan GERBER. Jena, Bussert & Stadeler, 2005, 572 p.

5187. KLEINEN (Karin). Ringen um Demokratie. Studieren in der Nachkriegszeit. Die akademische Jugend Kölns, 1945–1950. Köln, Weimar u. Wien, Böhlau, 2005, IX-453 p. (Studien zur Geschichte der Universität zu Köln, 17).

5188. LARSSON (Esbjörn). Från adlig uppfostran till borgerlig utbildning: Kungl. krigsakademien mellan åren 1792 och 1866. (De l'éducation noble à la formation bourgeoise: l'Académie royale militaire suédoise de 1792 à 1866. Uppsala, Acta Universitatis Upsaliensis, 2005, 411 p. (Studia historica Upsaliensia, 220).

5189. LENGWILER (Martin). Im Schatten Humboldts: Angewandte Forschung im Wissenschaftssystem Westdeutschlands (1945–1970). *Schweizerische Zeitschrift für Geschichte*, 2005, 55, 1, p. 46-59.

5190. LESLIE (William Bruce). Gentlemen and scholars: colleges and community in the "age of the university". New Brunswick a. London, Transaction Publishers, 2005, 284 p. (Foundations of higher education).

5191. LIN (Xiaoqing Diana). Peking University. Chinese scholarship and intellectuals, 1898–1937. Albany, State University of New York Press, 2005, XI-233 p. (Chinese philosophy and culture).

5192. Physics in Oxford, 1839–1939: laboratories, learning, and college life. Ed. by Robert FOX, Graeme GOODAY. Oxford, Oxford U. P., 2005, XIX-363 p. (ill.).

5193. Reichsuniversitäten (Les) de Strasbourg et de Poznan et les résistances universitaires, 1941–1944. Textes réunis par Christian BAECHLER, François IGERSHEIM, Pierre RACINE; avec le concours de Simone HERRY. Strasbourg, Presses universitaires de Strasbourg, 2005, 283 p. (Collection "Les mondes germaniques", 12).

5194. RODRÍGUEZ CRUZ (Agueda María). La Universidad de Salamanca en Hispanoamérica. Salamanca, Ediciones Universidad de Salamanca, 2005, 108 p. (Historia de la Universidad, 76. Acta Salmanticensia).

5195. ROSSO (Paolo). "Rotulus legere debentium": professori e cattedre all'Università di Torino nel Quattrocento. Torino, Deputazione subalpina di storia patria, 2005, 256 p. (Studi e fonti per la storia della Università di Torino, 14).

5196. SCHÜBL (Almar). Der Universitätsbau in der Zweiten Republik. Ein Beitrag zur Entwicklung der universitären Landschaft in Österreich. Wien, Berger, 2005, XV-495 p.

5197. Science (La) sous influence: l'université de Strasbourg enjeu des conflits franco-allemands, 1872–1945. Sous la direction d'Elisabeth CRAWFORD et de Josiane OLFF-NATHAN. Strasbourg, Nuée Bleue, 2005, 319 p. (ill., ports.).

5198. SCOT (Marie). La London School of Economics et le welfare state: science et politique, 1940–1979. Paris, L'Harmattan, 2005, 288 p. (Collection "Inter-national." Série Centre d'histoire de Sciences po).

5199. SHAPIRO (Harold T.). A larger sense of purpose: higher education and society. Princeton, Princeton U. P., 2005, XVI-183 p.

5200. TERVOORT (Ad). The "iter italicum" and the northern Netherlands. Dutch students at Italian universities and their role in the Netherlands' society, 1426–1575. Boston, Brill, 2005, XXII-438 p. (Education and society in the Middle Ages and Renaissance, 21).

5201. Universität der Gelehrten – Universität der Experten. Adaptionen deutscher Wissenschaft in den USA des neunzehnten Jahrhunderts. Hrsg. v. Philipp LÖSER und Christoph STRUPP. Stuttgart, Steiner, 2005, 171 p. (Transatlantische historische Studien; 24).

5202. Universities under dictatorship. Ed. by John CONNELLY and Michael GRÜTTNER. University Park, Pennsylvania State U. P., 2005, X-305 p.

5203. VERNEUIL (Yves). Les agrégés: histoire d'une exception française. Paris, Belin, 2005, 367 p. (Histoire de l'éducation).

5204. Vysoká škola uměleckoprůmyslová v Praze 1885–2005. (Academy of arts, architecture and design in Prague). Ed. Martina PACHMANOVÁ, Markéta PRAŽANOVÁ. Praha, Vysoká škola uměleckoprůmyslová, 2005, 339 p. (ill.).

5205. WRIGHT (Conrad Edick). Revolutionary generation. Harvard men and the consequences of independence. Amherst, University of Massachusetts Press, 2005, XI-298 p.

§ 2. Addenda 1994–2001.

5206. PALETSCHEK (Sylvia). Die permanente Erfindung einer Tradition: die Universität Tübingen im Kaiserreich und in der Weimarer Republik. Stuttgart, Steiner, 2001, XIV-608 p. (Contubernium, 53).

5207. SCHMEISER (Martin). Akademischer Hasard. Das Berufsschiksal der deutschen Universität, 1870–1920: Eine verstehend soziologische Untersuchung. Stuttgart, Klett-Cotta, 1994, 436 p.

5208. Universität Jena (Die): Zwischen Tradition und Innovation um 1800. Hrsg. V. Gerhard MÜLLER, Klaus RIES und Paul ZICHE. Stuttgart, Steiner, 2001, 237 p. (Pallas Athene, 2).

Cf. n°s 4276, 4362, 5209-5260, 5437, 5841

§ 3. Education.

** 5209. CHARMASSON (Thérèse). Archives et sources pour l'histoire de l'enseignement. Paris, Comité des travaux historiques et scientifiques, 2005, 391 p. (maps). (Collection Orientations et méthodes, 5).

5210. BALDZUHN (Michael). "Cato" bei Hofe. Transformationen eines Schultextes in den Händen adeliger Laien. *Archiv für Kulturgeschichte*, 2005, 87, 2, p. 315-349.

5211. BASDEVANT-GAUDEMET (Brigitte), SICARD (Germain). Les communes françaises: l'enseignement et les cultes de la fin de l'ancien régime à nos jours. Paris, H. Champion, 2005, 414 p. (Bibliothèque de l'Ecole des hautes études. Sciences historiques et philologiques. Hautes études d'histoire contemporaine, 4).

5212. BHATTACHARYA (Tithi). The sentinels of culture: class, education, and the colonial intellectual in Bengal (1848–1885). New York, Oxford U. P., 2005, XIII-272 p.

5213. BOLTON (Charles C.). The hardest deal of all: the battle over school integration in Mississippi, 1870–1980. Jackson: University Press of Mississippi, 2005, XXII-278 p.

5214. CHIRHART (Ann Short). Torches of light: Georgia teachers and the coming of the modern South. Athens, University of Georgia Press, 2005, XV-334 p.

5215. Crise (La) de la culture scolaire: origines, interprétations, perspectives. Sous la direction de François JACQUET-FRANCILLON et Denis KAMBOUCHNER. Paris, Presses universitaires de France, 2005, VI-504 p.

5216. DEER (Cécile). L' Empire britannique et l'instruction en Inde (1780–1854). Paris, L'Harmattan, 2005, 165 p. (Educations et sociétés).

5217. DEMİR (Hüseyin). Die osmanischen Medresen: das Bildungswesen und seine historischen Wurzeln im Osmanischen Reich von 1331–1600. Frankfurt am Main, Lang, 2005, 127 p. (Leipziger Beiträge zur Orientforschung, 17).

5218. DOUGHERTY (Jack). More than one struggle: the evolution of Black school reform in Milwaukee. Chapel Hill, University of North Carolina Press, 2005, XIII-253 p.

5219. DYHOUSE (Carol). Students: a gendered history. Abingdon, Routledge, 2005, XIV-273 p. (Women's and gender history).

5220. EGGERT (Heinz-Ulrich). Schul-Zeit 1938 bis 1949. Zur Vorgeschichte des Wilhelm-Hittorf-Gymnasiums Münster im NS-Staat und in der Nachkriegszeit. Münster, Aschendorff, 2005, 525 p. (Geschichte der Stadt Münster, 22).

5221. Elementarbildung und Berufsausbildung 1450–1750. Hrsg. v. Alwin HANSCHMIDT und Hans-Ulrich MUSOLFF. Köln u. Weimar, Böhlau, 2005, 348 p. (Beiträge zur historischen Bildungsforschung, 31).

5222. ENE (Sorin). Pregătirea profesională și instruirea intelectuală a membrilor C. C. ai P. C. R. între anii 1945–1989. (Professional training and intellectual education of the members of the Central Committee of the Romanian Communist Party between 1945–1989). *Studia Universitatis Cibisensis Historia*, 2005, 2, p. 239-299.

5223. FAZLIOĞLU (Şükran). Ta'lim ile İrşad Arasında: Erzurumlu İbrahim Hakkı'nın Medrese Ders Müfredatı. (Entre l'enseignement et l'action d'éclairer: Programme d'enseignement d'İbrahim Hakkı d'Erzurum dans le medrese). *Dîvân İlmî Araştırmalar*, 2005, 18, p. 115-173.

5224. GASS-BOLM (Torsten). Das Gymnasium 1945–1980: Bildungsreform und gesellschaftlicher Wandel in Westdeutschland. Göttingen, Wallstein, 2005, 490 p. (ill.). (Moderne Zeit, 7).

5225. GRAHAM (Patricia Albjerg). Schooling America: how the public schools meet the nation's changing needs. Oxford a. New York, Oxford U. P., 2005, XI-273 p. (ill.).

5226. Handbuch der deutschen Bildungsgeschichte.: vom späten 17 Jahrhundert bis zur Neuordnung Deutschlands um 1800. Bd. 2. 18 Jahrhundert. Hrsg. v. Notker HAMMERSTEIN und Ulrich G. HERRMANN. München, Beck, 2005, XVIII-583 p.

5227. JAGEMANN (Norbert). "Der Studienführer": zur Wissenschaftspolitik der SS. Hamburg, Dr. Kovač, 2005, 195 p. (Studien zur Zeitgeschichte, 47).

5228. JUSTICE (Benjamin). The war that wasn't. Religious conflict and compromise in the common schools of New York state, 1865–1900. Albany, State University of New York Press, 2005, XIV-285 p.

5229. KIRMIZI (Abdulhamit). "Usûl-i Tedrîs Hâlâ Tarz-ı Kadîm Üzre": Konya Valisi Ferid Paşa'nın Eğitimi Islah Çalışmaları. (Méthodologie d'enseignement est déjà à la manière antique: travaux de Ferid Pacha, gouverneur de Konya, sur le réforme de l'enseignement). *Dîvân İlmî Araştırmalar*, 2005, 19, p. 195-229.

5230. LASSONDE (Stephen). Learning to forget: schooling and family life in New Haven's working class, 1870–1940. New Haven, Yale U. P., 2005, XVI-301 p.

5231. LÉON (Antoine), ROCHE (Pierre). Histoire de l'enseignement en France. Paris, Presses universitaires de France, 2005, 127 p. (Que sais-je? 393).

5232. LEUȘTEAN (Lucian). Evreii ardeleni și conflictul dintre români și maghiari pe teme educaționale din primii ani postbelici. (The Jews from Transylvania and the conflict between Romanians and Magyars for educational reasons, during the first post war years). *Studia et Acta Historiae Iudeorum Romaniae*, 2005, 9, p. 188-214.

5233. LEYDER (Dirk). L'état de la classe, l'état dans la classe. Une tentative de contrôle sur l'enseignement moyen dans les Pays-Bas autrichiens (1777–1794). *Revue belge de philologie et d'histoire*, 2005, 83, 1, p. 1155-1174.

5234. Lycées, lycéens, lycéennes. Deux siècles d'histoire. Sous la dir. de Pierre CASPARD, Jean-Noël LUC et Philippe SAVOIE. Lyon, INRP, 2005, 501 p.

5235. MARCÍLIO (Maria Luiza). História da escola em São Paulo e no Brasil. São Paulo, Instituto Braudel e Imprensa Oficial, 2005, XVII-485 p. (ill.).

5236. MÂRZA (Iacob). Ecole et nation: les écoles de Blaj à l'époque de la renaissance nationale. Cluj-Napoca, Institut culturel roumain, Centre d'études transylvaines, 2005, 259 p. (Bibliotheca rerum Transsilvaniæ, 33).

5237. NEGRUZZO (Simona). L'armonia contesa: identità ed educazione nell'Alsazia moderna. Bologna, Il Mulino, 2005, 396 p. (map). (Percorsi).

5238. NELSON (Adam R.). The elusive ideal: equal educational opportunity and the Federal role in Boston's public schools, 1950–1985. Chicago, University of Chicago Press, 2005, XVII-332 p. (Historical studies of urban America).

5239. NIKRMAJER (Leoš). Problémy českého a německého školství v jihočeském regionu v období Protektorátu Čechy a Morava. (Problems of Czech and German schools in southern Bohemia during the Bohemian and Moravian Protectorate). *Sborník Archivu ministerstva vnitra*, 2005, 3, p. 111-139.

5240. Praxis (Die) der Reformpädagogik: Dokumente und Kommentare zur Reform der öffentlichen Schulen in der Weimarer Republik. Hrsg. v. Inge HANSEN-SCHABERG. Bad Heilbrunn/Obb., Klinkhardt, 2005, 285 p. (ill.).

5241. PRAZ (Anne-Françoise). De l'enfant utile à l'enfant précieux. Lausanne, Antipodes, 2005, 652 p. (Histoire).

5242. RAICICH (Marino). Storie di scuola da un'Italia lontana. A cura di Simonetta SOLDANI. Roma, Archivio Guido Izzi, 2005, 324 p.

5243. RIEDI (Eliza). Teaching empire: British and Dominions women teachers in the South African War concentration camps. *English historical review*, 2005, 120, 489, p. 1316-1347.

5244. RODRÍGUEZ ROSALES (Isolda). Historia de la educación en Nicaragua: restauración conservadora (1910–1930). Managua, Hispamer, 2005, 294 p. (ill.).

5245. ROGERS (Rebecca). From the salon to the schoolroom. Educating bourgeois girls in nineteenth-century France. University Park, Pennsylvania State U. P., 2005, XV-335 p.

5246. RUCHNIEWICZ (Krzysztof). Der Entstehungsprozess der Gemeinsamen deutsch-polnischen Schulbuchkommission 1937/38–1972. *Archiv für Sozialgeschichte*, 2005, 45, p. 237-252.

5247. SALAÜN (Marie). L'école indigène. Nouvelle-Calédonie, 1885–1945. Rennes, Presses universitaires de Rennes, 2005, 279 p.

5248. SAMUELS (Albert L.). Is separate unequal? Black colleges and the challenge to desegregation. Lawrence, University Press of Kansas, 2005, X-246 p.

5249. SLAWSON (Douglas J.). The Department of Education battle, 1918–1932: public schools, Catholic schools, and the social order. Notre Dame, University of Notre Dame Press, 2005, XVI-332 p. (Notre Dame studies in American Catholicism).

5250. SPECHT (Minna). Gesinnungswandel: Beiträge zur Pädagogik im Exil und zur Erneuerung von Erziehung und Bildung im Nachkriegsdeutschland. Hrsg. und eingeleitet von Inge HANSEN-SCHABERG unter Mitarb. von Sigrid RATHGENS. Frankfurt am Main, Berlin, Bern, Bruxelles, New York, Oxford u. Wien, Lang, 2005, XXVIII-255 p. (ill.). (Schriften des Exils zur Bildungsgeschichte und Bildungspolitik, 2).

5251. STROBEL (Thomas). Die Gemeinsame deutsch-polnische Schulbuchkommission. Ein spezifischer Beitrag zur Ost-West-Verständigung 1972–1989. *Archiv für Sozialgeschichte*, 2005, 45, p. 253-269.

5252. TIMMERMANS (Linda). L' accès des femmes à la culture sous l'ancien régime. Paris, H. Champion, 2005, 967 p. (Champion classiques. 1 Série "Essais").

5253. TROGER (Vincent), RUANO-BORBALAN (Jean-Claude). Histoire du système éducatif. Paris, Presses universitaires de France, 2005, 126 p. (Que sais-je? 3729).

5254. UNGERN-STERNBERG (Jürgen von). Wilhelm von Humboldts Bildungsideen. Von der freien Entfaltung des Individuums zum Schulmodell. *Archiv für Kulturgeschichte*, 2005, 87, 1, p. 127-148.

5255. VINOVSKIS (Maris A.). The birth of head start: preschool education politics in the Kennedy and Johnson Administrations. Chicago, University of Chicago Press, 2005, XII-205 p.

5256. VOURI (Sophia). Oikotrofeia kai Ypotrofies sti Makedonia 1903–1913. Tekmiria historias, (Internate schools and scolarships in Macedonia 1903–1913. Historical sources). Athina, Gutenberg 2005, 500 p.

5257. WILLIAMS (Heather Andrea). Self-taught. African American education in slavery and freedom. Chapel Hill, University of North Carolina Press, 2005, XIII-304 p. (John Hope Franklin series in African American history and culture).

§ 3. Addenda 2001–2004.

5258. Personal effects: the social character of scholarly writing. Ed. by Deborah HOLDSTEIN and David BLEICH. Logan, Utah State U. P., 2001, 392 p.

5259. SCHLEIMER (Ute). Die Opera Nazionale Balilla bzw. Gioventù Italiana del Littorio und die Hitlerjugend. Eine vergleichende Darstellung. Münster, New York u. München, Waxmann, 2004, 301 p. (Internationale Hochschulschriften, 435).

5260. TURNER (George). Hochschule zwischen Vorstellung und Wirklichkeit: zur Geschichte der Hochschulreform im letzten Drittel des 20. Jahrhunderts. Berlin, Duncker & Humblot, 2001, 294 p. (Abhandlungen zu Bildungsforschung und Bildungsrecht, 7).

Cf. n°s 5172-5208, 7869

§ 4. The Press.

** 5261. República, periodismo y literatura: la cuestión política en el periodismo literario durante la Segunda República española: antología (1931–1936). Edición a cargo de Javier GUTIÉRREZ PALACIO. Madrid, Tecnos, Asociación de la Prensa de Madrid y Villanueva, Centro universitario Villanueva, 2005, 991 p. (ill.).

5262. BIANCHIN (Lucia). Dove non arriva la legge: dottrine della censura nella prima età moderna. Bologna, Il Mulino, 2005, 389 p. (Annali dell'Istituto storico italo-germanico in Trento; Monografie, 41).

5263. BLATMAN (Daniel). En direct du ghetto: la presse clandestine juive dans le ghetto de Varsovie (1940–1943). Paris, Ed. du Cerf, 2005, 541 p. (ill.). (Histoires-Judaïsmes).

5264. BOSMA (Ulbe), SENS (Angelie), TERMORSHUIZEN (Gerard). Journalistiek in de tropen. De Indischen Indonesisch-Nederlandse pers, 1850–1958. Amsterdam, Aksant/Het Persmuseum, 2005, 96 p.

5265. CHAPMAN (Jane). Comparative media history. An introduction: 1789 to the present. Cambridge, Polity, 2005, XII-302 p. (ill.).

5266. CHIABURU (Elena). Carte şi tipar în Ţara Moldovei pînă la 1829. (Books and printing in Moldavia until 1829). Iaşi, Editura Universităţii "Alexandru Ioan Cuza", 2005, 402 p.

5267. COLOMBANI (Marie-Françoise), FITOUSSI (Michèle). Elle 1945-2005: une histoire des femmes. Paris, Filipacchi, 2005, 267 p. (ill.).

5268. Constantinescu (Jean Leontin). România între Seceră şi Ciocan. Publicistica Exilului. (Romania between the hammer and the sickle. The journalism of the exile). Ediţie îngrijita şi prefaţă de Nicolae FLORESCU. Bucureşti, Editura Jurnalul literar, 2005, 204 p.

5269. CROUSAZ (Karine). Érasme et le pouvoir de l'imprimerie. Lausanne, Editions Antipodes, 2005, 197 p.

5270. DAUM (Werner). Oszillationen des Gemeingeistes: Öffentlichkeit, Buchhandel und Kommunikation in der Revolution des Königreichs beider Sizilien, 1820–1821. Köln, SH-Verlag, 2005, 569 p. (Italien in der Moderne, 12). – IDEM. Zeit der Drucker und Buchhändler: die Produktion und Rezeption von Publizistik in der Verfassungsrevolution Neapel-Siziliens 1820–1821. Frankfurt am Main u. Oxford, P. Lang, 2005, 255 p. (ill.). (Italien in Geschichte und Gegenwart, 21).

5271. DRÁPALA (Milan). Freiheitsgrenzen der tschechoslowakischen Presse in der Zeit der Nationalen Front (1945–1948). In: Propaganda, (Selbst-)Zensur, Sensation. Grenzen von Presse- und Wissenschaftsfreiheit in Deutschland und Tschechien seit 1871 [Cf. n° 5297], p. 211-222. – IDEM. Glosse am Rande eines großen Themas. Die Aussiedlung der Deutschen und Deutschland in der tschechischen nichtsozialistischen Publizistik 1945–1948. In: Diktatur – Krieg – Vertreibung. Erinnerungskulturen in Tschechien, der Slowakei und Deutschland seit 1945–1948 [Cf. n° 355], p. 355-368.

5272. Editori ed edizioni a Roma nel Rinascimento. A cura di Paola FARENGA. Roma, Roma nel Rinascimento, 2005, XIII-156 p. (RR inedita, 34 saggi).

5273. FORNO (Mauro). La stampa del ventennio: strutture e trasformazioni nello Stato totalitario. Soveria Mannelli, Rubbettino, 2005, XVI-304 p.

5274. GABRIEL MELÉNDEZ (A.). Spanish-language newspapers in New Mexico, 1834–1958. Tucson, University of Arizona Press, 2005, XIII-268 p.

5275. GALFRÉ (Monica). Il regime degli editori. libri, scuola e fascismo. Roma, Laterza, 2005, XV-255 p. (Quadrante Laterza, 130).

5276. GAVRILOVA (Stella). Die Darstellung der UdSSR und Russlands in der Bild-Zeitung 1985–1999: eine Untersuchung zu Kontinuität und Wandel deutscher Russlandbilder unter Berücksichtigung der Zeitungen Die Welt, Süddeutsche Zeitung und Frankfurter Rundschau. Frankfurt am Main u. Oxford, P. Lang, 2005, 261 p. (ill.). (Europäische Hochschulschriften. Reihe III, Geschichte und ihre Hilfswissenschaften, 1019 = Publications universitaires européennes. Série III, Histoire, sciences auxiliares de l'histoire, 1019 = European university studies. Series III, History and allied studies, 1019).

5277. HALLEWELL (Laurence). O livro no Brasil: sua história. São Paulo, Editora da Universidade de São Paulo, 2005, 809 p. (ill.).

5278. HERZSTEIN (Robert E.). Henry R. Luce, Time, and the American crusade in Asia. New York, Cambridge U. P., 2005, XV-346 p.

5279. Histoire du livre et de l'imprimé au Canada. Vol. 2. 1840–1918. Directeurs: Patricia LOCKHART FLEMING, Gilles GALLICHAN, Yvan LAMONDE. Montréal, Presses de l'Université de Montréal, 2005, XXXIII-659 p. (ill., maps, ports.).

5280. Historische Presse und ihre Leser: Studien zu Zeitungen, Zeitschriften, Intelligenzblättern und Kalendern in Nordwestdeutschland. Hrsg. v. Peter ALBRECHT und Holger BÖNING. Bremen, Edition Lumiere, 2005, 362 p. (ill.). (Presse und Geschichte, 14).

5281. KOLOĞLU (Orhan). 1908 basın patlaması. (Explosion de presse à 1908). İstanbul, Bas-Haş, 2005, 190 p.

5282. KRÜGER (Gunnar). "Wir sind doch kein exklusiver Club!": die Bundespressekonferenz in der Ära Adenauer. Mit einem Geleitwort von Reinhard APPEL. Münster, Lit, 2005, II-237 p. (ill., ports.).

5283. LEÓN LIRA (Matías). El periodismo que no calló: historia de la revista Análisis (1977–1993). Santiago, La Nación, 2005, 261 p.

5284. MAC CUSKER (John J.). The demise of distance: the business press and the origins of the Information Revolution in the early Modern Atlantic world. *American historical review*, 2005, 110, 2, p. 295-321.

5285. MAISSEN (Thomas). Die Geschichte der NZZ [Neue Zürcher Zeitung], 1780–2005. Mit einem Anhang von Konrad STAMM über die Auslandsberichterstattung. Zürich, Neue Zürcher Zeitung, 2005, 385 p. (ill.).

5286. MALÝ (Karel). Presserecht und Zensur in der Tschechoslowakei in den Jahren 1945–1990. In: Propaganda, (Selbst-)Zensur, Sensation. Grenzen von Presse- und Wissenschaftsfreiheit in Deutschland und Tschechien seit 1871 [Cf. n° 5297], p. 223-233.

5287. MARTIN (Laurent). La presse écrite en France au XXe siècle. Paris, Librairie générale française, 2005, 256 p. (ill., facsims., ports.). (La France contemporaine. Le Livre de poche, 610 Références).

5288. MARTIN (Marc). Les grands reporters. Les débuts du journalisme moderne. Paris, Audibert, 2005, 399 p.

5289. Media and political culture in the eighteenth century. Ed. by Marie-Christine SKUNCKE. Stockholm, Kungl. Vitterhets Historie och Antikvitets Akademien, 2005, 130 p. (ill., map, ports., facsims.).

5290. MICHÁLEK (Slavomír). Prípad Oatis. Československý komunistický režim verzus dopisovateľ Associated Press. (Der Fall Oatis [Oatis Wiliam Nathan 1914–1997]. Das tschechoslowakische kommunistische Regime versus der Associated Press Korrespondent). Bratislava, Ústav pamäti národa, 2005, 293 p. (Ed. Monografie). [Eng. Summary].

5291. Ministri e giornalisti: la guerra e il Minculpop, 1939–1943. Introduzione e cura di Nicola TRANFAGLIA; note al testo di Bruno MAIDA. Torino, Einaudi, 2005, XXXIII-331 p. (Biblioteca Einaudi; 203).

5292. MÜLLER (Philipp). Auf der Suche nach dem Täter: die öffentliche Dramatisierung von Verbrechen im Berlin des Kaiserreichs. Frankfurt am Main, Campus, 2005, 423 p.

5293. NUOVO (Angela), COPPENS (Chris). I Giolito e la stampa: nell'Italia del XVI secolo. Genève, Droz, 2005, 634 p. (ill.). (Travaux d'humanisme et Renaissance, 402).

5294. PETERS (Kate). Print culture and the early Quakers. New York, Cambridge U. P., 2005, XIII-273 p. (Cambridge studies in early modern British history).

5295. PETTAS (William A.). A history & bibliography of the Giunti (Junta) printing family in Spain 1526–1628. New Castle, Oak Knoll Press, 2005, XII-1047 p (ill.).

5296. PILLININI (Giovanni). Il giornalismo politico a Venezia nel 1848–1849. Roma, Archivio Guido Izzi, 2005, 119 p. (Biblioteca scientifica. vol. 52 Serie II, Memorie).

5297. Propaganda, (Selbst-)Zensur, Sensation: Grenzen von Presse- und Wissenschaftsfreiheit in Deutschland und Tschechien seit 1871. Hrsg. v. Michal ANDĚL [et al.]. Essen, Klartext, 2005, 309 p. (ill.). (Veröffentlichungen zur Kultur und Geschichte im östlichen Europa, 27). [Cf. n^os <Auswahl> 89, 297, 5271, 5286.]

5298. Rayonnement (Le) d'une maison d'édition dans l'Europe des Lumières: la Société typographique de Neuchâtel, 1769–1789: actes du colloque organisé par la Bibliothèque publique et universitaire de Neuchâtel et la Faculté des lettres de l'Université de Neuchâtel, Neuchâtel, 31 octobre–2 novembre 2002. Textes publiés par Robert DARNTON et Michel SCHLUP; avec la collaboration de Jacques RYCHNER; index établi par Marie-Christine HAUSER avec le concours de Marcel GUERDAT et Michael SCHMIDT. Neuchâtel, Hauterive, Bibliothèque publique et universitaire, G. Attinger, 2005, 620 p. (ill.).

5299. RUEDA RAMÍREZ (Pedro J.). Negocio e intercambio cultural: el comercio de libros con América en la Carrera de Indias (siglo XVII). Sevilla, Universidad de Sevilla, Diputación de Sevilla y Madrid, Consejo Superior de Investigaciones Científicas, Escuela de Estudios Hispano-Americanos, 2005, 524 p. (ill.). (Catálogo del Consejo Superior de Investigaciones Científicas, 440. Serie Historia y geografía, 99. Nuestra América, 16).

5300. SCHNEIDER (Ute). Der unsichtbare Zweite. Die Berufsgeschichte des Lektors im literarischen Verlag. Göttingen, Wallstein, 2005, 399 p.

5301. SEVILLA SOLER (María Rosario). La Revolución Mexicana y la opinión pública española: la prensa sevillana frente al proceso de insurrección. Madrid, Consejo Superior de Investigaciones Científicas, 2005, 249 p. (Colección Biblioteca de historia de América, 34).

5302. Stampa (La) del regime, 1932–1943: le veline del Minculpop per orientare l'informazione. A cura di Nicola TRANFAGLIA; con la collaborazione di Bruno MAIDA. Milano, Bompiani, 2005, 456 p. (Saggi Bompiani).

5303. STÖBER (Rudolf). Deutsche Pressegeschichte: von den Anfängen bis zur Gegenwart. Konstanz, UVK Verlagsgesellschaft, 2005, 395 p. (ill., charts, facsims., map). (Uni-Taschenbücher; 2716).

5304. THIEMEYER (Guido). "Wandel durch Annäherung": Westdeutsche Journalisten in Osteuropa 1956–1977. *Archiv für Sozialgeschichte*, 2005, 45, p. 101-116.

5305. TUSAN (Michelle Elizabeth). Women making news: gender and journalism in modern Britain. Urbana a. Chicago, University of Illinois Press, 2005, X-306 p. (History of communication).

5306. VAN DEN DUNGEN (Pierre). Milieux de presse et journalistes en Belgique (1828–1914). Bruxelles, Académie royale de Belgique, 2005, 562 p. (ill.). (Mémoires de la Classe des lettres. 3e sér., 38 Collection in-8o).

5307. VITCU (Dumitru). Lumea românească şi Balcanii în reportajele corespondenţilor americani de război (1877–1878). (Romanian world and the Balkans in the coverage of the American war correspondents, 1877–1878). Iaşi, Editura Junimea, 2005, 441 p.

5308. WALLSTEN (Scott). Returning to Victorian competition, ownership, and regulation: an empirical study of European telecommunications at the turn of the twentieth century. *Journal of economic history*, 2005, 65, 3, p. 693-722.

5309. WITTEK (Thomas). Auf ewig Feind? Das Deutschlandbild in den britischen Massenmedien nach dem Ersten Weltkrieg. München, R. Oldenbourg, 2005, 437 p. (Veröffentlichungen des Deutschen Historischen Instituts in London, 59).

5310. WOLFE (Thomas C.). Governing Soviet journalism: the press and the Socialist person after Stalin. Bloomington, Indiana U. P., 2005, XXI-240 p.

Cf. n^os *55-101, 3123, 3712, 3875, 4048, 4627, 4708, 4718, 4898, 5493, 5542, 5543, 5544, 5621, 7125, 7221*

§ 5. Philosophy.

a. General

5311. CILIBERTO (Michele). Pensare per contrari: disincanto e utopia nel Rinascimento. Roma, Edizioni di storia e letteratura, 2005, X-509 p. (Storia e letteratura, 226).

5312. CLÉMENT (Michèle). Le cynisme à la Renaissance d'Erasme à Montaigne. Genève, Droz, 2005, 284 p. (Les seuils de la modernité, 9. Cahiers d'humanisme et Renaissance, 72).

5313. Einfluss (Der) des Hellenismus auf die Philosophie der Frühen Neuzeit. Hrsg. v. Gábor BOROS. Wiesbaden, Harrassowitz in Kommission, 2005, 198 p. (Wolfenbütteler Forschungen, 108).

5314. GÜNAY (Mustafa). Cumhuriyet döneminde felsefe tarihçiliği. (Recherches sur l'histoire de la philosophie dans la République de la Turquie). Ankara, Kültür ve Turizm Bakanlığı, 2005, 185 p.

5315. HARTWICH (Wolf-Daniel). Romantischer Antisemitismus: von Klopstock bis Richard Wagner. Göttingen, Vandenhoeck & Ruprecht, 2005, 277 p.

5316. LOMONACO (Fabrizio). Tolleranza: momenti e percorsi della modernità fino a Voltaire. Napoli, Guida, 2005, 143 p. (Parole chiave della filosofia, 10).

5317. ROBERTSON (John). The case for Enlightenment. Scotland and Naples, 1680–1760. Cambridge, New York a. Melbourne, Cambridge U. P., 2005, XII-455 p. (Ideas in context, 73).

5318. ROSSO (Maxime). La renaissance des institutions de Sparte dans la pensée française: XVIe–XVIIIe siècle. Aix-en-Provence, Presses universitaires d'Aix-Marseille, 2005, 587 p. (Collection d'histoire du droit, 8 Série "Thèses et travaux").

5319. SEIGEL (Jerrold). The idea of the self: thought and experience in Western Europe since the seventeenth century. New York, Cambridge U. P., 2005, VIII-724 p.

5320. VAN DAMME (Stéphane). Paris, capitale philosophique: de la Fronde à la Révolution. Paris, Odile Jacob, 2005, 311 p.

5321. WILLIAMS (David). Defending Japan's Pacific war. The Kyoto school philosophers and post-white power. London, Routledge Curzon, 2005, XXVI-238 p.

5322. ZOUHAR (Jan), PAVLINCOVÁ (Helena), GABRIEL (Jiří). Demokracie je diskuse. Česká filosofie 1918–1938. (Demokracy is discussion. Czech Philosophy, 1918–1938). Olomouc, Nakl. Olomouc, 2005, 197 p.

Cf. n°s 700-723, 955-969

b. Special studies

5323. ADLER (Laure). Dans les pas de Hannah **Arendt**. Paris, Gallimard, 2005, VI-645 p. – AHRENS (Stefan). Die Gründung der Freiheit: Hannah **Arendts** politisches Denken über die Legitimität demokratischer Ordnungen. Frankfurt am Main u. New York, Lang, 2005, 297 p. (Hannah Arendt-Studien, 2).

5324. **BAYLE** (Pierre). Correspondance de Pierre Bayle. Tome 4 janvier 1684–juillet 1684, Lettres 242-308. Publiée et annotée par Elisabeth LABROUSSE. Oxford, Voltaire Foundation, 2005, XXI-287 p. (ill., ports.).

5325. CATANA (Leo). The concept of contraction in Giordano **Bruno**'s philosophy. Aldershot, Ashgate, 2005, VII-209 p.

5326. **CASSIRER** (Ernst). Descartes: Lehre, Persönlichkeit, Wirkung. Text und Anmerkugen bearbeitet von Tobias BERBEN. Hamburg, Felix Meiner, 2005, 228 p. (Gesammelte Werke. Hamburger Ausgabe / Ernst Cassirer, 20). – **CASSIRER** (Ernst). Vorlesungen und Studien zur philosophischen Anthropologie. Hrsg. v. Gerald HARTUNG und Herbert KOPP-OBERSTEBRINK; unter Mitwirkung von Jutta FAEHNDRICH. Hamburg, Felix Meiner, 2005, XII-774 p. (Selections. 1995, 6. Nachgelassene Manuskripte und Texte / Ernst Cassirer, 6).

5327. Carteggio **Croce**-Laurini. A cura di Gianluca GENOVESE; con un'appendice di scritti di Gerardo LAURINI. Napoli, Bibliopolis, 2005, XXXII-163 p. (ill.). – CONTE (Domenico). Storia universale e patologia dello spirito: saggio su **Croce**. Bologna, Il Mulino, 2005, XI-241 p. – **CROCE** (Benedetto), LATERZA (Giovanni). Carteggio. 2. 1911–1920. A cura di Antonella POMPILIO. Roma e Bari, Laterza, 2005, 967 p. (ill.). – DESSÌ SCHMID (Sarah). Ernst Cassirer und Benedetto **Croce**: die Wiederentdeckung des Geistes: ein Vergleich ihrer Sprachtheorien. Mit einem Vorwort von Jürgen TRABANT. Tübingen, Francke, 2005, 275 p. (ill.). – MARTINA (Rossella). **Croce** giornalista: dal "biennio rosso" all'antifascismo. Napoli, Editoriale scientifica, 2005, XVIII-390 p. (Crociana; 5).

5328. **DILTHEY** (Wilhelm). Das Erlebnis und die Dichtung: Lessing, Goethe, Novalis, Hölderlin. Hrsg. v. Gabriele MALSCH. Göttingen, Vandenhoeck & Rurecht, 2005, V-535 p. (Gesammelte Schriften / Wilhelm Dilthey, 26).

5329. Victor **Ehrenberg** und Georg Jellinek Briefwechsel, 1872–1911. Hrsg. v. Christian KELLER. Frankfurt am Main, Klostermann, 2005, VIII-478 p.

5330. BONACINA (Giovanni). Due prestiti da **Gibbon** per la descrizione hegeliana dell'Impero romano d'Oriente. Quaderni di storia, 2005, 61, p. 25-58.

5331. LIGUORI (Guido), META (Chiara). **Gramsci**: guida alla lettura. Milano, UNICOPLI, 2005, 111 p. (Prospettive/strumenti; 11). – SANTUCCI (Antonio A.). Antonio **Gramsci**, 1891–1937. Palermo, Sellerio, 2005, 192 p. (Tutto e subito, 3). – Togliatti editore di **Gramsci**. A cura di Chiara DANIELE; introduzione di Giuseppe VACCA. Roma, Carocci, 2005, 293 p. (ill.). (Fondazione Istituto Gramsci. Annali, 13).

5332. MAUTNER (Thomas). **Grotius** and the skeptics. Journal of the history of ideas, 2005, 66, 4, p. 577-601.

5333. **HALBWACHS** (Maurice), SAUVY (Alfred). Le point de vue du nombre: 1936. Précédé de l'avant-propos au tome VII de l'encyclopédie française de Lucien Febvre et suivi de trois articles de Maurice Halbwachs. Paris, Institut national d'études démographiques, 2005, VI-469 p. (ill., maps). (Classiques de l'économie et de la population. Éditions critiques).

5334. Dix ans de critique: notes inédites de Flaubert sur l'Esthétique de **Hegel**. Textes réunis et présentés par Gisèle SÉGINGER. Paris et Caen, Lettres modernes Minard, 2005, 363 p. (Revue des lettres modernes. 5 Gustave Flaubert). – **HEGEL** (Georg Wilhelm Friedrich). Grundlinien der Philosophie des Rechts. Berlin, Akademie-Verlag, 2005, XI-313 p. (Klassiker auslegen, 9). – **HEGEL** (Georg Wilhelm Friedrich). Philosophie der Kunst: Vorlesung von 1826. Hrsg. v. Annemarie GETHMANN-SIEFERT, Jeong-Im KWON und Karsten BERR. Frankfurt am Main, Suhrkamp, 2005, 296 p. (Suhrkamp Taschenbuch Wissenschaft; 1722). – VEGETTI (Matteo). **Hegel** e i confini dell'Occidente: la Fenomenologia nelle interpretazioni di Heidegger, Marcuse, Löwith, Kojève, Schmitt. Napoli, Bibliopolis, 2005, 357 p. (Serie studi / Istituto italiano per gli studi filosofici, 29).

5335. FAYE (Emmanuel). **Heidegger**, l'introduction du nazisme dans la philosophie: autour des séminaires inedits de 1933–1935. Paris, Albin Michel, 2005, 567 p. (Biblioteque Albin Michel. Idees). – KLEINBERG (Ethan). Generation existential. **Heidegger**'s philosophy in France 1927–1961. Ithaca, Cornell U. P., 2005, X-294 p.

5336. Johann Gottfried **Herder**: Aspekte seines Lebenswerkes. Hrsg. v. Martin KESSLER und Volker LEPPIN. Berlin, De Gruyter, 2005, X-437 p. (Arbeiten zur Kirchengeschichte, 92).

5337. COLLINS (Jeffrey R.). The allegiance of Thomas **Hobbes**. Oxford, Oxford U. P., 2005, XII-313 p.

5338. ENGELHARD (Kristina). Das Einfache und die Materie: Untersuchungen zu **Kants** Antinomie der Teilung. Berlin, W. de Gruyter, 2005, X-459 p. (Kantstudien. 148 Ergänzungshefte). – **KANT** (Immanuel). Notes and fragments. Ed. by Paul GUYER, Curtis BOWMAN and Frederick RAUSCHER. Cambridge, Cambridge U. P., 2005, XXX-663 p (Cambridge edition of the works of Immanuel Kant). – **Kant** et la France = Kant und Frankreich. Hildesheim, G. Olms, 2005, 420 p. (Europaea memoria. Bd. 46 Reihe 1, Studien). – Kant im Streit der Fakultäten. Hrsg. v. Volker GERHARDT. Berlin u. New York, De Gruyter, 2005, XVI-306 p. – KLIMMEK (Nikolai F.). **Kants** System der transzendentalen Ideen. Berlin, W. de Gruyter, 2005, IX-235 p. (Kantstudien. 147 Ergänzungshefte). – LA VOPA (Anthony J.). Thinking about marriage: **Kant**'s liberalism and the peculiar morality of conjugal union. *Journal of modern history*, 2005, 77, 1, p. 1-34. – MASCHIETTI (Stefano). L' interpretazione heideggeriana di **Kant**: sulla disarmonia di verità e differenza. Bologna, Il Mulino, 2005, XXIII-238 p. (Istituto italiano per gli studi storici in Napoli, 51). – PONSO (Marzia). Cosmopoliti e patrioti : trasformazioni dell'ideologia nazionale tedesca tra **Kant** e Hegel, 1795–1815. Milano, F. Angeli, 2005, 451 p. (Collana Gioele Solari, 46).

5339. PIROTTE (Dominique). Alexandre **Kojève**: un système anthropologique. Paris, Presses universitaires de France, 2005, 236 p. (Philosophie d'aujourd'hui).

5340. Übersetzung von Alexandre **Koyré**, Descartes und die Scholastik. Übersetzt von Edith STEIN mit Hedwig CONRAD-MARTIUS; Einführung, Bearbeitung und Anmerkungen von Hanna-Barbara GERL-FALKOVITZ. Freiburg im Breisgau, Herder, 2005, XXIX-223 p. (Works. 2000; 25. (Edith Stein Gesamtausgabe; 25).

5341. Antonio **Labriola** nella storia e nella cultura della nuova Italia. A cura di Alberto BURGIO. Macerata, Quodlibet, 2005, 381 p. (ill.). (Fonti e commenti, 6). – GUARAGNELLA (Pasquale). Il pensatore e l'artista: prosa del moderno in Antonio **Labriola** e Luigi Pirandello. Roma, Bulzoni, 2005, 269 p. (Strumenti di ricerca; 77-78). – MICCOLIS (Stefano). Antonio **Labriola**. *Belfagor*, 2005, 60, 355, p. 55-71.

5342. BERKOWITZ (Roger). The gift of science: **Leibniz** and the modern legal tradition. Cambridge, Harvard U. P., 2005, XVIII-214 p. – **LEIBNIZ** (Gottfried Wilhelm, Freiherr von). Essais scientifiques et philosophiques: les articles publiés dans les journaux savants. Recueillis par Antonio LAMARRA et Roberto PALAIA; préface de Heinrich SCHEPERS. Hildesheim, Georg Olms, 2005, 3 vol., XXVII-1354 p. (ill.).

5343. MOYN (Samuel). Origins of the other: Emmanuel **Levinas** between revelation and ethics. Ithaca, Cornell U. P., 2005, XI-268 p.

5344. ANGLO (Sydney). **Machiavelli** – the first century: studies in enthusiasm, hostility, and irrelevance. Oxford, Oxford U. P., 2005, X-765 p. (Oxford-Warburg studies). – BAUSI (Francesco). **Machiavelli**. Roma, Salerno, 2005, 407 p. (Sestante, 9). – **MACHIAVELLI** (Niccolò). Legazioni. Commissarie. Scritti di governo. Tomo 3. 1503–1504. A cura di Jean-Jacques MARCHAND e Matteo MELERA-MORETTINI. Roma, Salerno, 2005, 591 p. (Edizione nazionale delle opere di Niccolò Machiavelli. 3 Sezione 5, Legazioni. Comissarie. Scritti di governo. – ROSA (Mario). Dispotismo e libertà nel Settecento: interpretazioni repubblicane di **Machiavelli**. Pisa, Edizioni della Normale, 2005, X-82 p.

5345. JOURNET (Charles), **MARITAIN** (Jacques). Correspondance. 4. 1950–1957. Saint-Maurice, Saint-Augustin, 2005, 952 p.

5346. **MARX** (Karl), ENGELS (Friedrich). Briefwechsel Juni 1860 bis Dezember 1861. Bearb. v. Rolf DLUBEK und Vera MOROZOVA; unter Mitwirkung von Galina GOLOVINA und Elena VAŠČENKO. Berlin, Akademie Verlag, 2005, 2 vol., XXI-1467 p. (facsims.). (Karl Marx, Friedrich Engels Gesamtausgabe, 11 3 Abt. Briefwechsel). – **MARX** (Karl), ENGELS (Friedrich). Das Kapital: Kritik der politischen Ökonomie, Zweites Buch Redaktionsmanuskript von Friedrich Engels, 1884/1885. Bearb. v. Izumi OMURA [et al.], unter Mitwirkung von Ljudmila VASINA, Kenji ITIHARA und Kenji MORI. Berlin, Akademie Verlag, 2005, 2 vol., [s. p.]. (Karl Marx Friedrich Engels Gesamtausgabe, 12. 2 Abt., 'Das Kapital' und Vorarbeiten).

5347. FELICE (Domenico). Per una scienza universale dei sistemi politico-sociali : dispotismo, autonomia della giustizia e carattere delle nazioni nell'Esprit des lois di **Montesquieu**. Firenze, L. S. Olschki, 2005, IX-210 p. (Pansophia, 6). – **MONTESQUIEU** (Charles de Secondat, baron de). Collectio juris. Textes établis, présentés et annotés par Iris COX et Andrew LEWIS; coordination éditoriale Caroline VERDIER. Oxford, Voltaire Foundation et Napoli, Istituto italiano per gli studi filosofici, 2005, 2 vol., LXIII-1069 p. (facsims.). (Oeuvres complètes de Montesquieu; 11-12). – **Montesquieu**-Traditionen in Deutschland. Beiträge zur Erforschung eines Klassikers. Hrsg. v. Edgar MASS und Paul-Ludwig WEINACHT. Berlin, Duncker & Humblot, 2005, 289 p. (Beiträge zur politischen Wissenschaft, 139). – PLATANIA (Marco). Robert Shackleton e gli studi su **Montesquieu**: scenari interpretativi tra Otto e Novecento. *Rivista storica italiana*, 2005, 117, 1, p. 283-308.

5348. BENNE (Christian). **Nietzsche** und die historisch-kritische Philologie. Berlin, W. de Gruyter, 2005, XII-428 p. (Monographien und Texte zur Nietzsche-Forschung, 49). – **NIETZSCHE** (Friedrich Wilhelm). Arbeitsheft WI 8. Bearb. von Marie-Luise HAASE [et al.]. Berlin, Walter de Gruyter, 2005, X-290 p. (facsims.). (Nietzsche Werke: kritische Gesamtausgabe, 5 Neunte Abteilung, Der handschriftliche Nachlass ab Frühjahr 1885).

5349. WETTERSTEN (John). New insights on young **Popper**. *Journal of the history of ideas*, 2005, 66, 4, p. 603-631.

5350. KUSTER (Friederike). **Rousseau** – die Konstitution des Privaten: zur Genese der bürgerlichen Familie. Berlin, Akademie Verlag, 2005, 232 p. (Deutsche Zeitschrift für Philosophie. 11 Sonderband). – WILLIAMS (David Lay). Justice and the general will: affirming **Rousseau**'s ancient orientation. *Journal of the history of ideas*, 2005, 66, 3, p. 383-411.

5351. PROCHASSON (Christophe). **Saint-Simon**, ou, L'anti-Marx: figures du saint-simonisme français, XIXe-XXe siècles. Paris, Perrin, 2005, 344 p.

5352. BEISER (Frederick). **Schiller** as philosopher. A re-examination. Oxford, Clarendon Press, 2005, 283 p.

5353. BLASIUS (Dirk). Carl **Schmitt** und der "Heereskonflikt" des Dritten Reiches 1934. *Historische Zeitschrift*, 2005, 281, 3, p. 659-682. – CALDWELL (Peter C.). Controversies over Carl **Schmitt**: a review of recent literature. *Journal of modern history*, 2005, 77, 2, p. 357-387. – **SCHMITT** (Carl). Die Militärzeit 1915–1919: Tagebuch Februar bis Dezember 1915: Aufsätze und Materialen. Hrsg. v. Ernst HÜSMERT und Gerd GIESLER. Berlin, Akademie Verlag, 2005, X-587 p. (ill.). – **SCHMITT** (Carl). Frieden oder Pazifismus? Arbeiten zum Völkerrecht und zur internationalen Politik, 1924–1978. Herausgegeben, mit einem Vorwort und mit Anmerkungen versehen von Günter MASCHKE. Berlin, Duncker & Humblot, 2005, XXX-1010 p. – ZARKA (Yves Charles). Un détail nazi dans la pensée de Carl **Schmitt**: la justification des lois de Nuremberg du 15 septembre 1935. Paris, Presses universitaires de France, 2005, 95 p. (ill.). (Intervention philosophique).

5354. Raison dévoilée (La): études **schopenhaue**-riennes. Sous la direction de Christian BONNET et Jean SALEM. Paris, Vrin, 2005, 256 p. (Bibliothèque d'histoire de la philosophie).

5355. JAQUET (Chantal). Les expressions de la puissance d'agir chez **Spinoza**. Paris, Publications de la Sorbonne, 2005, 304 p. (Série Philosophie, 14). – SÉVÉRAC (Pascal). Le devenir actif chez **Spinoza**. Paris, Champion, 2005, 476 p. (Travaux de philosophie, 7). – **Spinoza** to the letter: studies in words, texts and books. Ed. by Fokke AKKERMAN and Piet STEENBAKKERS. Leiden, Brill, 2005, XIII-344 p. (ill.). (Brill's studies in intellectual history, 137). – VINCIGUERRA (Lorenzo). **Spinoza** et le signe: la genèse de l'imagination. Paris, Vrin, 2005, 334 p. (Bibliothèque d'histoire de la philosophie. Nouvelle série. Age classique).

5356. KOSLOWSKI (Stefan). Zur Philosophie von Wirtschaft und Recht. Lorenz von **Stein** im Spannungsfeld zwischen Idealismus, Historismus und Positivismus. Berlin, Duncker & Humblot, 2005, 479 p. (Philosophische Schriften, 60).

5357. ALLEN (Barbara). **Tocqueville**, covenant, and the democratic revolution. Harmonizing earth with heaven. Lanham, Lexington books, 2005, XIX-393 p.

5358. Complete works (The) of **Voltaire**: Les Oeuvres complètes de Voltaire. 68. Histoire du parlement de Paris. Édition critique par John RENWICK. Oxford, Voltaire Foundation, 2005, XXIV-604 p.

5359. GERNET (Jacques). La raison des choses. Essai sur la philosophie de **Wang Fuzhi** (1619–1692). Paris, Gallimard, 2005, 436 p. (Bibliothèque de philosophie).

5360. PUCHNER (Martin). Doing logic with a hammer: **Wittgenstein**'s Tractatus and the polemics of logical positivism. *Journal of the history of ideas*, 2005, 66, 2, p. 285-300. – **WITTGENSTEIN** (Ludwig). The Big Typescript, TS. 213. Ed. and translated by C. Grant LUCKHARDT and Maximilian AUE. Malden, Blackwell Pub., 2005, XVIII-516 p.

Cf. n°s 502, 519

§ 6. Exact, natural, medical sciences and technique.

** 5361. Hallers Netz: ein europäischer Gelehrtenbriefwechsel zur Zeit der Aufklärung. Hrsg. von Martin STUBER, Stefan HÄCHLER und Luc LIENHARD. Basel, Schwabe, 2005, X-592 p. (ill., maps, ports.). (Studia Halleriana, IX).

5362. AGRAWAL (Arun). Environmentality. Technologies of government and the making of subjects. Durham, Duke U. P., 2005, XVI-325 p. (New ecologies for the twenty-first century).

5363. ANDERSON (Katharine). Predicting the weather: victorians and the science of meteorology. Chicago, University of Chicago Press, 2005, X-331 p.

5364. BAGUS (Anita). Volkskultur in der bildungsbürgerlichen Welt: zum Institutionalisierungsprozess wissenschaftlicher Volkskunde im wilhelminischen Kaiserreich am Beispiel der Hessischen Vereinigung für Volkskunde. Giessen, Universitätsbibliothek, 2005, 446 p. (Berichte und Arbeiten aus der Universitätsbibliothek und dem Universitätsarchiv Giessen, 54).

5365. BERAN (Jiří). Vytváření členské základny ČSAV v roce 1952. (The Formation of the Membership Base of the Czechoslovak Academy of Sciences in 1952). *Soudobé dějiny*, 2005, 12, 1, p. 102-139.

5366. BIEBER (Hans-Joachim). Zur Frühgeschichte der indischen Nuklearpolitik. *Geschichte und Gesellschaft*, 2005, 31, 3, p. 373-414.

5367. BIGG (Charlotte). L'optique de preécision et la Première Guerre mondiale. *Schweizerische Zeitschrift für Geschichte*, 2005, 55, 1, p. 34-45.

5368. BÖHME-KAßLER (Katrin). Gemeinschaftsunternehmen Naturforschung. Modifikation und Tradition in der Gesellschaft Naturforschender Freunde zu Berlin, 1773–1906. Stuttgart, Steiner, 2005, 218 p. (Pallas Athene, Universitäts-und Wissenschaftsgeschichte, 15).

5369. BRIGGS (Asa). A history of Royal College of Physicians of London. Vol. 4. 1948–1983. Oxford, Oxford U. P., 2005, p. 1250-1735.

5370. BROSSEDER (Claudia). The writing in the Wittenberg sky: astrology in sixteenth-century Germany. *Journal of the history of ideas*, 2005, 66, 4, p. 557-576.

5371. BÜCHEL (Jochen). Psychologie der Materie: Vorstellungen und Bildmuster von der Assimilation von Nahrung im 17. und 18. Jahrhundert unter besonderer Berücksichtigung des Paracelsismus. Würzburg, Königshausen & Neumann, 2005, 284 p. (ill., port.). (Epistemata, 375 Reihe Philosophie).

5372. BURKHARDT (Richard W., Jr.). Patterns of behavior: Konrad Lorenz, Niko Tinbergen, and the founding of ethology. Chicago, University of Chicago Press, 2005, XII-636 p.

5373. BYERLY (Carol R.). The fever of war. The influenza epidemic in the U.S. Army during World War I. New York, New York U. P., 2005, XV-250 p.

5374. CADEDDU (Antonio). Les vérités de la science: pratique, récit, histoire; le cas Pasteur. Firenze, L. S. Olschki, 2005, XVII-279 p. (Biblioteca di Nuncius, 57 Studi e testi).

5375. CARRILLO (Ana María). ¿Estado de peste o estado de sitio?: Sinaloa y Baja California, 1902–1903. *Historia mexicana*, 2005, 54, 4, p. 1049-1103.

5376. CARVALHO DA SILVA (Paulo José). La médecine de l'âme: trois cas de convergence entre psychologie aristotélicienne et savoirs médicaux dans l'ancienne Compagnie de Jésus (Europe et Nouveau Monde). *Mélanges de l'École française de Rome: Italie et mediterranée (MEFRIM)*, 2005, 117, 1, p. 351-369.

5377. CASTRO-GÓMEZ (Santiago). La hybris del punto cero: ciencia, raza e ilustración en la Nueva Granada (1750–1816). Bogotá, Editorial Pontificia Universidad Javeriana, 2005, 345 p. (ill., maps). (Pensar).

5378. CHEN-MORRIS (Raz). Shadows of instruction: optics and classical authorities in Kepler's Somnium. *Journal of the history of ideas*, 2005, 66, 2, p. 223-243.

5379. Church (The) and Galileo. Ed. by Ernam MAC MULLIN. Notre Dame, University of Notre Dame Press, 2005, XII-391 p. (Studies in science and the humanities from the Reilly Center for Science, Technology, and Values).

5380. DAMOUSI (Joy). Freud in the antipodes. A cultural history of psychoanalysis in Australia. Sydney, University of New South Wales Press, 2005, X-374 p.

5381. DANIEL (Pete). Toxic drift: pesticides and health in the post-World War II South. Baton rouge, Louisiana State University Press in association with Smithsonian National Museum of American History, Washington, D.C, 2005, XII-209 p. (Walter Lynwood Fleming lectures in Southern history).

5382. DANTO (Elizabeth Ann). Freud's free clinics: psychoanalysis and social justice, 1918–1938. New York, Columbia U. P., 2005, XI-348 p.

5383. DRAY (Philip). Stealing God's thunder. Benjamin Franklin's lightning rod and the invention of America. New York, Random House, 2005, XVIII-279 p.

5384. DUBOURG GLATIGNY (Pascal), ROMANO (Antonella). La Trinité-des-Monts dans la "République romaine des sciences et des arts". *Mélanges de l'École française de Rome: Italie et mediterranée (MEFRIM)*, 2005, 117, 1, p. 7-43.

5385. EISENSTAEDT (Jean). Avant Einstein: relativité, lumière, gravitation. Paris, Ed. du Seuil, 2005, 349 p. (Science ouverte).

5386. ELMAN (Benjamin A.). On their own terms. Science in China, 1550–1900. Cambridge, Harvard U. P., 2005, XXXVIII-567 p.

5387. ESPAHANGIZI (Kijan Malte). Experimentalsysteme, Erinnerungskulturen und die transatlantische Quantenrevolution: die "Entdeckung der Materiewellen" und die Bell Telephone Laboratories (1925–1927). Münster, LIT, 2005, 122 p. (ill.). (Studien zu Geschichte, Politik und Gesellschaft Nordamerikas, 22).

5388. Fascismo e scienza. Le celebrazioni voltiane e il Congresso internazionale dei fisici del 1927. A cura Aldo GAMBA e Pierangelo SCHIERA. Bologna, Il Mulino, 2005, 243 p.

5389. FAVINO (Federica). Minimi in "Sapienza": François Jacquier, Thomas Le Seur e il rinnovamento dell'insegnamento scientifico allo "Studium Urbis". *Mélanges de l'École française de Rome: Italie et mediterranée (MEFRIM)*, 2005, 117, 1, p. 160-187.

5390. Figurationen des Experten: Ambivalenzen der wissenschaftlichen Expertise im ausgehenden 18. und frühen 19. Jahrhundert. Hrsg. v. Eric J. ENGSTROM, Volker HESS und Ulrike THOMS. Frankfurt am Main u. Oxford, Lang, 2005, 229 p. (ill.). (Berliner Beiträge zur Wissenschaftsgeschichte, 7).

5391. Francesco Bianchini (1662–1729) und die europäische gelehrte Welt um 1700. Hrsg. v. Valentin KOCKEL und Brigitte SÖLCH; Redaktion, Martin HEISE und Eva-Maria LANDWEHR. Berlin, Akademie Verlag, 2005, 273 p. (ill.). (Colloquia Augustana, 21).

5392. FUNIGIELLO (Philip J.). Chronic politics: health care security from FDR to George W. Bush. Lawrence, University Press of Kansas, 2005, XII-395 p.

5393. GARDNER NAKAMURA (Ellen). Practical pursuits: Takano Chei, Takahashi Keisaku, and Western medicine in nineteenth-century Japan. Cambridge, Harvard U. P., 2005, XIV-236 p. (Harvard East Asian monographs).

5394. GAUSEMEIER (Bernd). Natürliche Ordnungen und politische Allianzen: biologische und biochemische Forschung an Kaiser-Wilhelm-Instituten, 1933–1945. Göttingen, Wallstein, 2005, 352 p. (Geschichte der Kaiser-Wilhelm-Gesellschaft im Nationalsozialismus, 12).

5395. GONZÁLEZ LEANDRI (Ricardo). Madurez y poder. Médicos e Instituciones sanitarias en Argentina a fines del siglo XIX. *Entrepasados*, 2005, 27, [s. p.].

5396. GRADMANN (Christoph). Krankheit im Labor. Robert Koch und die medizinische Bakteriologie. Göttingen, Wallstein, 2005, 376 p.

5397. GUERRAGGIO (Angelo), NASTASI (Piero). Matematica in camicia nera: il regime e gli scienziati. Milano, B. Mondatori, 2005, 279 p. (Matematica e dintorni).

5398. GÜNERGUN (Feza). Salih Zeki ve astronomi: Rasathane-i Amire Müdürlüğü'nden 1914 tam güneş tutulmasına. (Salih Zeki and astronomy: From meteorological observatory to the total solar eclipse of 1914). *Osmanlı Bilimi Araştırmaları / Studies ın Ottoman science*, 2005, 7, 1, p. 97-120.

5399. HÅARD (Mikael), JAMISON (Andrew). Hubris and Hybrids: a cultural history of technology and science. New York, Routledge, 2005, XV-335 p.

5400. HAHN (Roger). Pierre Simon Laplace 1749–1827. A determined scientist. Cambridge, Harvard U. P., 2005, X-310 p.

5401. HAMBLIN (Jacob Darwin). Oceanographers and the Cold War: Disciples of marine science. Seattle, University of Washington Press, 2005, XXIX-346 p.

5402. HELLYER (Marcus). Catholic physics. Jesuit natural philosophy in Early Modern Germany. Notre Dame, University of Notre Dame Press, 2005, XII-336 p.

5403. HENTSCHEL (Klaus). Die Mentalität deutscher Physiker in der frühen Nachkriegszeit (1945–1949). Heidelberg, Synchron, 2005, 191 p. (Studien zur Wissenschafts- und Universitätsgeschichte, 11).

5404. HERMANN (Tomáš), KLEISNER (Karel). Biolog Michail M. Novikov (1846–1965). Kapitola z dějin ruské vědy v emigraci. (Michail M. Novikov (1876–1965), Biologist. A chapter from the history of Russian science in exile). *Český časopis historický*, 2005, 103, 2, p. 313-353.

5405. Histoire et médicament aux XIXe et XXe siècles. Sous la dir. de Christian BONAH et Anne RASMUSSEN. Paris, Biotem, Éditions glyphe, 2005, 273 p.

5406. HOQUET (Thierry). Buffon: histoire naturelle et philosophie. Paris, Honoré Champion, 2005, 809 p. (Dix-huitièmes siècles 92).

5407. HUNNER (Jon). Inventing Los Alamos: the growth of an atomic community. Norman, University of Oklahoma Press, 2005, XI-288 p.

5408. JEATER (Diana). Imagining Africans: scholarship, fantasy, and science in colonial administration, 1920s southern Rhodesia. *International journal of African historical studies*, 2005, 38, 2, p. 1-26.

5409. KAPILA (Shruti). Masculinity and madness: princely personhood and colonial sciences of the mind in western India 1871–1940. *Past and present*, 2005, 187, p. 121-156.

5410. KASSELL (Lauren). Medicine and magic in Elizabethan London: Simon Forman; astrologer, alchemist, and physician. New York, Clarendon Press of Oxford U. P., 2005, XVIII-281 p. (Oxford historical monographs).

5411. KRAGH (Helge). Dansk naturvidenskabs historie. Bd. 2. Natur, nytte og ånd: 1730–1850. (Histoire des sciences au Danemark. Tome 2. Nature, utilitarisme, esprit: 1730–1850). Århus, Aarhus Universitetsforlag, 2005, 486 p.

5412. LAWRENCE (Christopher). Rockefeller money, the laboratory, and medicine in Edinburgh, 1919–1930: new science in an old country. Rochester, University of Rochester Press, 2005, VIII-373 p. (Rochester studies in medical history).

5413. LEISER (Gary). The dawn of aviation in the Middle East: The first flying machines over İstanbul (16 şekil ile birlikte). *Belleten*, 2005, 69, 256, p. 937-1014.

5414. LEMOV (Rebecca). World as laboratory: experiments with mice, mazes and men. New York, Hill and Wang, 2005, 291 p.

5415. LEVIN (Miriam R.). Defining women's scientific enterprise. Mount Holyoke faculty and the rise of American science. Hanover, University Press of New England, 2005, XIII-209 p.

5416. LICHTENBERG (Georg Christoph). Vorlesungen zur Naturlehre: Lichtenbergs annotiertes Handexemplar der vierten Auflage von Johann Christian Polykarp Erxleben: "Anfangsgründe der Naturlehre". Herausgegeben von der Akademie der Wissenschaften zu Göttingen. Göttingen, Wallstein, 2005, XXX-1103 p. (ill.). (Gesammelte Schriften / Georg Christoph Lichtenberg, 1).

5417. LINTON (Derek S.). Emil von Behring: infectious disease, immunology, serum therapy. Philadelphia, American Philosophical Society, 2005, XI-580 p. (Memoirs of the American philosophical society, 255).

5418. LUCAS (A. A.). Bombe atomique et croix gammée. Bruxelles, Académie royale de Belgique, Classe des sciences, 2005, 119 p. (ports.). (Mémoire de la Classe des sciences. Collection in-8o, 3e sér., 22).

5419. MAC DONALD (Helen). Human remains: dissection and its histories. New Haven, Yale U. P., 2005, XIV-220 p.

5420. MARSDEN (Ben), SMITH (Crosbie). Engineering empires. A cultural history of technology in nineteenth-century Britain. Basingstoke, Palgrave Macmillan, 2005, XI-351 p.

5421. MAYAUD (Pierre-Noël). Le conflict entre l'Astronomie Nouvelle et l'Écriture Sainte aux XVIe et XVIIe siècles. Un moment de l'histoire des idées autour de l'affaire Galilée. Vol. 1. Présentation des dossiers restituant le conflit et arrière-plan historique. Vol. 2. Dossier A. Extraits d'ouvrages exégétiques. Vol. 3. Dossier B. Extraits d'ouvrages astronomiques, philosophiques et autres. V. 4-5. Notes concernant les extraits des dossiers A et B. V. 6. Une analyse du conflit. Paris, Honoré Champion, 2005, 6 vol., 3416 p. (Bibliothéque littéraire de la renaissance; 55).

5422. Medizin im Nationalsozialismus und das System der Konzentrationslager: Beiträge eines interdisziplinären Symposiums. Hrsg. v. Judith HAHN, Silvija KAVČIČ und Christoph KOPKE; mit einem Geleitwort von Gerhard BAADER. Frankfurt am Main, Mabuse-Verlag, 2005, 213 p. (ill.). (Mabuse-Verlag Wissenschaft; 82).

5423. Medizin, Geschichte und Geschlecht: körperhistorische Rekonstruktionen von Identitäten und Differenzen. Hrsg. v. Frank STAHNISCH und Florian STEGER. Stuttgart, Franz Steiner, 2005, 297 p. (ill.). (Geschichte und Philosophie der Medizin, 1 = History and philosophy of medicine, 1).

5424. Men, women, and the birthing of modern science. Ed. by Judith P. ZINSSER. Dekalb, Northern Illinois U. P., 2005, VIII-215 p.

5425. MINNA STERN (Alexandra). Eugenic nation: faults and frontiers of better breeding in modern America. Berkeley a. Los Angeles, University of California Press, 2005, XIV-347 p. (American crossroads, 17).

5426. MOHR (James C.). Plague and fire: battling Black Death and the 1900 burning of Honolulu's Chinatown. New York, Oxford U. P., 2005, XI-235 p.

5427. MOM (Gijs). The electric vehicle. Technology and expectations in the automobile age. Baltimore, Johns Hopkins U. P., 2005, XIII-423 p.

5428. MONTI (Maria Teresa). Spallanzani e le rigenerazioni animali: l'inchiesta, la comunicazione, la rete. Florence, L.S. Olschki, 2005, IX-424 p. (ill.). (Bibliothèque d'histoire des sciences, 7).

5429. MORAN (Bruce T.). Distilling knowledge: alchemy, chemistry, and the scientific revolution. Cambridge, Harvard U. P., 2005, 210 p. (New histories of science, technology, and medicine).

5430. NADIS (Fred). Wonder shows. Performing science, magic and religion in America. New Brunswick, Rutgers U. P., 2005, XIV-318 p.

5431. NIEKERK (Carl). Zwischen Naturgeschichte und Anthropologie: Lichtenberg im Kontext der Spätaufklärung. Tübingen, Niemeyer, 2005, VII-395 p. (Studien zur deutschen Literatur, 176).

5432. Niels Bohr: collected works. Vol. 11. The political arena (1934–1961). Ed. by Finn ASERUD. Amsterdam a. London, Elsevier, 2005, XXIII-753 p. (ill, ports.).

5433. Nouvelles pratiques (Les) de santé. Acteurs, objets, logiques sociales, XVIIIe–XXe siècles. Sous la dir. de Patrice BOURDELAIS et Olivier FAURE. Paris, Belin, 2005, 383 p.

5434. O'MARA (Margaret Pugh). Cities of knowledge. Cold War science and the search for the next Silicon Valley. Princeton, Princeton U. P., 2005, XIII-298 p. (Politics and society in twentieth-century America).

5435. Organisation (The) of knowledge in Victorian Britain. Ed. by Martin J. DAUNTON. Oxford, Oxford U. P., 2005, VIII-424 p. (British Academy centenary monographs).

5436. OSHINSKY (David M.). Polio. An American story. New York, Oxford U. P., 2005, VIII-342 p.

5437. PEPE (Luigi). Istituti nazionali, accademie e società scientifiche nell'Europa di Napoleone. Firenze, L.S. Olschki, 2005, XXX-521 p. (Biblioteca di Nuncius. 59 Studi e testi / Istituto e museo di storia della scienza, Firenze, 59).

5438. PIHLAJA (Pivi Maria). Sweden and l'Académie des sciences. Scientific elites in 18th-century Europe. *Scandinavian journal of history*, 2005, 30, 3-4, p. 271-285.

5439. PIZZOLATO (Nicola). "Una situazione sado-masochistica ad incastro". Il dibattito scientifico sull'immigrazione meridionale (1950–1970). *Quaderni storici*, 2005, 1, p. 97-120.

5440. Practice (The) of reform in health, medicine and science, 1500–2000. Essays for Charles Webster. Ed. by Margaret PELLING and Scott MANDELBROTE. Aldershot, Ashgate, 2005, XV-376 p.

5441. Primi Lincei (I) e il Sant'Uffizio: questioni di scienza e di fede: Roma, 12–13 giugno 2003. Roma, Bardi, 2005, 524 p. (Atti dei convegni lincei; 215).

5442. QUENET (Grégory). Les tremblements de terre aux XVIIe et XVIIIe siècles. La naissance d'un risque. Seyssel, Champ Vallon, 2005, 587 p. (Époques).

5443. RIEGER (Bernhard). Technology and the culture of modernity in Britain and Germany, 1890–1945. Cambridge, New York a. Melbourne, Cambridge U. P., 2005, X-319 p.

5444. RINALDI (Massimo). La cultura delle accademie: immaginario urbano e scienze della natura tra Cinquecento e Seicento. Milano, Unicopli, 2005, 213 p. (Mappe dell'immaginario, 10).

5445. ROTHSCHILD (Joan). The dream of the perfect child. Bloomington, Indiana U. P., 2005, X-343 p. (Bioethics and the humanities).

5446. RUFFINI (Marco). Le imprese del drago: politica, emblematica e scienze naturali alla corte di Gregorio XIII, 1572–1585. Roma, Bulzoni, 2005, 165 p. (ill.). (Biblioteca del Cinquecento, 118).

5447. RUIZ SALGUERO (Magda). Anticoncepción y salud reproductiva en España: crónica de una (r)evolución. Madrid, Consejo Superior de Investigaciones Científicas, Instituto de Economía y Geografía, 2005, 302 p. (ill.). (Colección de estudios ambientales y socioeconómicos; 6).

5448. SAMUELI (Jean-Jacques), BOUDENOT (Jean-Claude). H.A. Lorentz (1853–1928): la naissance de la physique moderne. Paris, Ellipses, 2005, 348 p. (ill., ports.).

5449. SCHAFFER (Simon). L'inventaire de l'astronome. Le commerce d'instruments scientifiques au XVIIIe siècle (Angleterre-Chine-Pacifique). *Annales*, 2005, 60, 4, p. 791-815.

5450. SCHEIDELER (Britta). Albert Einstein in der Weimarer Republik. Demokratisches und elitäres Denken im Widerspruch. *Vierteljahrshefte für Zeitgeschichte*, 2005, 53, 3, p. 381-419.

5451. SCHULZ (Matthias). Integration durch eine europäische Atomstreitmacht? Nuklearambitionen und die deutsche Europa-Initiative vom Herbst 1964. *Vierteljahrshefte für Zeitgeschichte*, 2005, 53, 2, p. 275-313.

5452. Science across the European empires, 1800–1950. Ed. by Benedikt STUCHTEY. Oxford, Oxford U. P., 2005, 376 p.

5453. SHELL (Marc). Polio and its aftermath. The paralysis of culture. Cambridge, Harvard U. P., 2005, 324 p.

5454. SIMON (Jonathan). Chemistry, pharmacy and Revolution in France, 1777–1809. Burlington, Ashgate Publishing Company, 2005, VI-189 p. (Science, technology and culture, 1700–1945).

5455. ŠIMŮNEK (Michal). Between "Eugenics" and "Racial Hygiene". plans for the regulation of human heredity in the Czech Lands, 1900–1925. *In*: "Blood and Homeland". Eugenics and racial nationalism in Central and Southeast Europe, 1900–1940. Budapest, 2005, p. 168-194.

5456. Sins of the flesh. Responding to sexual disease in early modern Europe. Ed. by Kevin SIENA. Toronto, Centre for reformation and Renaissance studies, 2005, 292 p. (Essays and studies, 7).

5457. SMIL (Vaclav). Creating the twentieth century: technical innovations of 1867–1914 and their lasting impact. New York, Oxford U. P., 2005, IX-350 p.

5458. STENZEL (Oliver). Medikale Differenzierung. Der Konflikt zwischen akademischer Medizin und Laienheilkunde im 18. Jahrhundert. Heidelberg, Carl Auer, 2005, 188 p.

5459. STRASSER (Bruno J.), JOYE (Frédéric). Une science "neutre" dans la Guerre froide? La Suisse et la coopération scientifique européenne (1951–1969). *Schweizerische Zeitschrift für Geschichte*, 2005, 55, 1, p. 95-112.

5460. SZRETER (Simon). Health and wealth: studies in history and policy. Rochester, University of Rochester Press, 2005, XIV-506 p. (Rochester studies in medical history).

5461. TUCKER (Jennifer). Nature exposed: photography as eyewitness in Victorian science. Baltimore, Johns Hopkins U. P., 2005, IX-294 p.

5462. TWOHIG (Peter L.). Labour in the laboratory. Medical laboratory workers in the maritimes, 1900–1950. Ithaca, McGill-Queen's U. P., 2005, XVI-241 p. (Studies in the history of medicine, health, and society, 23).

5463. VIGNAUD (Laurent-Henri). Des mathématiques à la botanique: la conversion scientifique du P. Charles Plumier durant son séjour à Rome (1676–1681). *Mélanges de l'École française de Rome: Italie et mediterranée (MEFRIM)*, 2005, 117, 1, p. 131-157.

5464. VILENSKY (Joel A.). Dew of death. The story of Lewisite, America's World War I weapon of mass destruction. Bloomington, Indiana U. P., 2005, XXIII-213 p.

5465. VOELKE (Jean-Daniel). Renaissance de la géométrie non euclidienne entre 1860 et 1900. Bern, P. Lang, 2005, XII-459 p. (ill.).

5466. WATZKA (Carlos). Vom Hospital zum Krankenhaus. Zum Ungang mit psychisch und somatisch Kranken im frühneuzeitlichen Europa. Köln, Weimar u. Wien, Böhlau, 2005, XIII-385 p. (Beihefte zum Saeculum, 1).

5467. WEBER (Thomas P.). Darwin und die neuen Biowissenschaften: ein Einführung. Köln, DuMont, 2005, 270 p.

5468. WEISZ (George). Divide and conquer. A comparative history of medical specialization. Oxford, Oxford U. P., 2005, 329 p.

5469. WEITEKAMP (Margaret A.). Right stuff, wrong sex. America's first women in space program. Baltimore, Johns Hopkins U. P., 2005, XI-232 p. (Gender relations in the American experience).

5470. WILDI (Tobias). Die Reaktor AG: Atomtechnologie zwischen Industrie, Hochschule und Staat. *Schweizerische Zeitschrift für Geschichte*, 2005, 55, 1, p. 70-83.

5471. WILSON (Daniel J.). Living with polio. The epidemic and its survivors. Chicago, University of Chicago Press, 2005, XII-300 p.

5472. WOHL (Robert). The spectacle of flight: aviation and the Western imagination, 1920–1950. New Haven, Yale U. P., 2005, X-364 p.

§ 6. Addendum 2003.

5473. Médecins et ingénieurs ottomans à l'âge des nationalismes. Sous la dir. de Méropi ANASTASSIADOU-DUMONT. Paris, Maisonneuve & Larose et İstanbul, Institut français d'études anatoliennes, 2003, 387 p.

Cf. n°s 401, 422, 579, 4471, 7263

§ 7. Literature.

a. General

5474. GREENE (Jody). The trouble with ownership. Literary property and authorial liability in England, 1660–1730. Philadelphia, University of Pennsylvania Press, 2005, 272 p. (Material texts).

5475. Kulturgeschichte der englischen Literatur: von der Renaissance bis zur Gegenwart. Hrsg. v. Vera NÜNNING. Tübingen, A. Francke, 2005, IX-346 p. (ill.). (Uni-Taschenbücher, 2663).

5476. Principe philosophique (D'un) à un genre littéraire: les "secrets": actes du colloque de la Newberry Library de Chicago, 11–14 septembre 2002. Publiés par Dominique DE COURCELLES. Paris, H. Champion, 2005, 501 p. (ill.). (Colloques, congrès et conférences sur la Renaissance, 45).

5477. PUJOL (Stéphane). Le dialogue d'idées au dix-huitième siècle. Oxford, Voltaire Foundation, 2005, XVII-336 p. (Studies on Voltaire and the eighteenth century, 2005:06).

5478. SEEGER (Tatyana). Literarischer Detailismus: dargestellt an deutschen und russischen Beispielen erzählender Prosa. Frankfurt am Main u. Oxford, P. Lang, 2005, 233 p. (ill.). (Heidelberger Beiträge zur deutschen Literatur, 16).

5479. STAUBER (Reinhard). Kultur – Raum – Politik. Italiens Bild von sich selbst in der Renaissance. *Archiv für Kulturgeschichte*, 2005, 87, 2, p. 285-314.

5480. Traum-Diskurse der Romantik. Hrsg. v. Peter-André ALT und Christiane LEITERITZ. Berlin, W. de Gruyter, 2005, VI-391 p. (Spectrum Literaturwissenschaft, 4 = Spectrum literature, 4).

5481. WARREN (Joyce W.). Women, money, and the law. Nineteenth-century fiction, gender, and the courts. Iowa City, University of Iowa Press, 2005, VIII-373 p.

5482. YOUNGMAN (Paul A.). Black devil and iron angel : the railway in nineteenth-century German realism. Washington, Catholic University of America Press, 2005, XIII-173 p.

Cf. n°s 2946, 5568

b. Renaissance

* 5483. PARADA (Alejandro E.). Bibliografía cervantina editada en la Argentina: una primera aproximación. Presentación de Pedro Luis BARCIA. Buenos Aires, Academia Argentina de Letras, 2005, 256 p. (ill.). (Biblioteca Academia Argentina de Letras. Serie Prácticas y representaciones bibliográficas, 1).

5484. BASTOS DA SILVA (Jorge). Shakespeare no romantismo português: factos, problemas, interpretações. Porto, Campo das Letras, 2005, 331 p. (Colecção Campo da literatura. 125 Ensaio).

5485. CERVANTES (Fernando). Cervantes in Italy: Christian humanism and the visual impact of Renaissance Rome. *Journal of the history of ideas*, 2005, 66, 3, p. 325-350.

5486. COLON (Germà). Las primeras traducciones europeas del Quijote. Bellaterra, Universitat Autònoma de Barcelona, Servei de Publicacions, 2005, 137 p. (Cuadernos de filología; 6 (Cuadernos de filología (Bellaterra, Spain); 6).

5487. Don Quijote húngaro (El). Editors Ádám ANDERLE. Szeged, Universidad de Szeged, Departamento de Estudios Hispánicos, 2005, 127 p. (Acta Universitatis Szegediensis. Acta hispanica, 10).

5488. Donoso (Del) y grande escrutinio del cervantismo en Cuba. A cargo de José Antonio BAUJÍN [et al.]. La Habana, Letras Cubanas, Academia Cubana de la Lengua, Universidad de La Habana, Facultad de Artes y Letras, 2005, XLV-1411 p. (ill.).

5489. FORSYTH (Elliott Christopher). La justice de Dieu: Les Tragiques d'Agrippa d'Aubigné et la Réforme protestante en France au XVIe siècle. Paris, H. Champion, 2005, 567 p. (Etudes et essais sur la Renaissance, 57).

5490. FRAZIER (Alison Knowles). Possible lives. Authors and saints in Renaissance Italy. New York, Columbia U. P., 2005, XX-527 p.

5491. GARCÉS (María Antonia). Cervantes en Argel: historia de un cautivo. Madrid, Editorial Gredos, 2005, 457 p. (ill.). (Biblioteca románica hispánica. II, Estudios y ensayos; 446).

5492. GONZÁLEZ GADEA (Diego N.). Cervantes en el Uruguay: historia y bibliografía. Montevideo, Ediciones El Galeón, 2005, 104 p. (ill.). (Ensayos bibliográficos; 9).

5493. HARRIS (Neil). Sopravvivenze e scomparse delle testimonianze del Morgante di Luigi Pulci. *Rinascimento*, 2005, 45, p. 179-245.

5494. MANFERLOTTI (Stefano). Amleto in parodia. Roma, Bulzoni, 2005, 139 p. (ill.). (Piccola biblioteca shakespeariana; 36).

5495. Miguel de Cervantes' Don Quijote: explizite und implizite Diskurse im Don Quijote. Hrsg. v. Christoph STROSETZKI; mit Beiträgen von Mariano DELGADO [et al.]. Berlin, E. Schmidt, 2005, 401 p. (ill.). (Studienreihe Romania, 22).

5496. UECKER (Heiko). Der nordische Hamlet. Frankfurt am Main, Lang, 2005, XXXVII-264 p. (Texte und Untersuchungen zur Germanistik und Skandinavistik, 56).

5497. WALKER (Greg). Writing under tyranny. English literature and the Henrician reformation. Oxford, Oxford U. P., 2005, 556 p.

Cf. n° 200

c. Classicism

5498. Deutsche Klassik: Epoche, Autoren, Werke. Hrsg. v. Rolf SELBMANN. Darmstadt, Wissenschaftliche Buchgesellschaft, 2005, 235 p.

5499. Poésie (La) à l'âge baroque, 1598–1660. Édition établie et présentée par Alain NIDERST. Paris, Laffont, 2005, XLVIII-877 p. (Bouquins).

5500. ROULIN (Jean-Marie). L' épopée de Voltaire à Chateaubriand: poésie, histoire et politique. Oxford, Voltaire Foundation, 2005, XX-277 p (Studies on Voltaire and the eighteenth century, 2005:03).

5501. WILLIAMS (Abigail). Poetry and the creation of a Whig literary culture, 1681–1714. Oxford, Oxford U. P., 2005, 303 p.

c. Addendum 2002

5502. CAZZANIGA (Gian Mario), TURCHI (Roberta), TOCCHINI (Gerardo). Le Muse in loggia: massoneria e letteratura nel Settecento. Milano, UNICOPLI, 2002, 104 p. (A tre voci, 3).

d. Romanticism and after

* 5503. BABICHEVA (Majja E.). Pisateli vtoroj volny russkoj emigracii: Biobibliograficheskie ocherki. (The writers of the Second Wave of the Russian Emigration: Bio-bibliographical essays). Ros. gos. bibl., NIO bibliografii. Moskva, Pashkov dom, 2005. 447 p. (bibl. incl.; ind. p. 441-445).

* 5504. Bibliografia su Renato Serra (1909–2005). A cura di Dino PIERI, saggio introduttivo di Marino BIONDI. Roma, Edizioni di storia e letteratura, 2005, LXXXI-328 p. (Sussidi eruditi, 68).

** 5505. Fondo (Il) "Renato Serra" della biblioteca malatestiana di Cesena. A cura di Manuela RICCI: premessa di Renzo CREMANTE. Roma, Edizioni di storia e letteratura, 2005, XII-173 p. (Sussidi eruditi, 69).

5506. Atelier (L') de Baudelaire: les fleurs du mal. Éd. par Claude PICHOIS et Jacques DUPONT. Paris, H. Champion, 2005, 4 vol., [s. p.]. (ill., facsims.). (Textes de littérature moderne et contemporaine, 83).

5507. BODEI (Remo) [et al.]. Pier Paolo Pasolini: palabra de corsario. [S. l.], Caja Duero y Madrid, Círculo de Bellas Artes de Madrid, 2005, 349 p. (ill., facsims., ports.).

5508. BOTTACIN (Annalisa). Stendhal e Firenze: 1811–1841. Moncalieri, Centro interuniversitario di ricerche sul viaggio in Italia, 2005, 269 p. (Bibliothèque Stendhal. Etudes = Biblioteca Stendhal. Studi, 6).

5509. CANFORA (Davide). Una fonte de "I promessi sposi". *Quaderni di storia*, 2005, 61, p. 225-244.

5510. COMPAGNON (Antoine). Les antimodernes: de Joseph de Maistre à Roland Barthes. Paris, Gallimard, 2005, 464 p. (Bibliothèque des idées).

5511. D'ELIA (Gianni). L' eresia di Pasolini: l'avanguardia della tradizione dopo Leopardi. Milano, Effigie, 2005, 167 p. (ill.). (Stellefilanti; 12. Luoghi letterari).

5512. DEBENEDETTI (Giacomo). Proust. Progetto editoriale e saggio introduttivo di Mario LAVAGETTO; testi e note a cura di Vanessa PIETRANTONIO. Torino, Bollati Boringhieri, 2005, XLI-446 p. (Nuova cultura, 111).

5513. Figlio prigioniero (Il): carteggio tra Luigi e Stefano Pirandello durante la guerra 1915–1918. A cura di Andrea PIRANDELLO. Milano, Mondadori, 2005, 371 p. (ill.).

5514. German literature of the nineteenth century, 1832–1899. Ed. by Clayton KOELB and Eric DOWNING. Rochester, Camden House, 2005, VI-348 p. (ill.). (Camden House history of German literature, 9).

5515. GIDE (André), ALLÉGRET (Marc). Correspondance, 1917–1949. Éd. établie, présentée et annotée par Jean CLAUDE et Pierre MASSON. Paris, Gallimard, 2005, 886 p. (Cahiers André Gide, 19. Cahiers de la NRF).

5516. GIDE (André), SCHIFFRIN (Jacques). Correspondance: 1922–1950. Avant-propos d'André SCHIFFRIN; édition établie par Alban CERISIER. Paris, Gallimard, 2005, 364 p. (ill.). (Cahiers de la NRF).

5517. HEIDEMANN (Gudrun). Das schreibende Ich in der Fremde: Il'ja Erenburgs und Vladimir Nabokovs Berliner Prosa der 1920er Jahre. Bielefeld, Aisthesis, 2005, 415 p. (ill.). (Schrift und Bild in Bewegung, 11).

5518. Hugo et l'histoire. Textes édités par Léon-François HOFFMANN et Suzanne NASH. Fasano, Schena y Paris, Presses de l'Université de Paris-Sorbonne, 2005, 244 p. (ill.). (Biblioteca della ricerca, 2. Transatlantique).

5519. KOÇ (Murat). Türk romanında İttihat ve Terakki: 1908–2004. (Union et Progrès dans le roman turc). İstanbul, Temel, 2005, 635 p.

5520. KÖSOĞLU (Nevzat). Türk milliyetçiliğinin doğuşu ve Ziya Gökalp. (Naissance du nationalisme turc et Ziya Gökalp). İstanbul, Ötüken, 2005, 212 p.

5521. LARKIN (Edward). Thomas Paine and the literature of Revolution. Cambridge, Cambridge U. P., 2005, 205 p.

5522. Leopardi en los poetas españoles. Edición de Pedro Luis LADRÓN DE GUEVARA MELLADO. Madrid,

Huerga & Fierro Editores, 2005, 214 p. (Fenice textos; 16).

5523. MANZONI (Alessandro). Discorso sopra alcuni punti della storia longobardica in Italia. A cura di Isabella BECHERUCCI; premessa di Dario MANTOVANI. Milano, Centro nazionale studi manzoniani, 2005, CXI-462 p. (ill.). (Edizione nazionale ed europea delle opere di Alessandro Manzoni, 5).

5524. MANZONI (Alessandro). Postille al Vocabolario della Crusca nell'edizione veronese. A cura di Dante ISELLA. Milano, Centro nazionale studi manzoniani, 2005, XXXVII-616 p. (ill.). (Edizione nazionale ed europea delle opere di Alessandro Manzoni, 24).

5525. MARZEL (Shoshana-Rose). L'ésprit du chiffon: le vêtement dans le roman français du XIXe siècle. Bern et Oxford, Peter Lang, 2005, X-384 p. (ill.).

5526. MEYNARD (Cécile). Stendhal et la province. Paris, H. Champion, 2005, 682 p. (maps). (Romantisme et modernités, 87).

5527. MOZET (Nicole). Balzac et le temps: littérature, histoire et psychanalyse. Saint-Cyr-sur-Loire, C. Pirot, 2005, 260 p. (Collection Balzac / Groupe international de recherches balzaciennes).

5528. PETERSON (Brent O.). History, fiction, and Germany: writing the nineteenth-century nation. Detroit, Wayne State U. P., 2005, VII-360 p. (Kritik: German literary theory and cultural studies).

5529. Poésie populaire (La) en France au XIXe siècle. Théories, pratiques et réception. Sous la dir. de Hélène MILLOT, Nathalie VINCENT-MUNNIA et Marie-Claude SCHAPIRA. Tusson, Editions du Lérot, 2005, 767 p. (Idéographies).

5530. PROVIDENTI (Elio). Colloqui con Pirandello. Firenze, Polistampa, 2005, 207 p. (ill.).

5531. RABONI (Giovanni). La poesia che si fa: cronaca e storia del Novecento poetico italiano, 1959–2004. A cura di Andrea CORTELLESSA. Milano, Garzanti, 2005, VII-415 p. (Saggi).

5532. ROUSSIN (Philippe). Misère de la littérature, terreur de l'histoire: Céline et la littérature contemporaine. Paris, Gallimard, 2005, 754 p. (NRF essais).

5533. SAPELLI (Giulio). Modernizzazione senza sviluppo: il capitalismo secondo Pasolini. A cura di Veronica RONCHI. Milano, B. Mondadori, 2005, XIV-248 p. (Testi e pretesti).

5534. Schatten Wagners (Im): Thomas Mann über Richard Wagner: Texte und Zeugnisse 1895–1955. Ausgewählt, kommentiert und mit einem Essay von Hans Rudolf VAGET. Frankfurt am Main, Fischer Taschenbuch Verlag, 2005, 367 p.

5535. SCHMIDT-BURKHARDT (Astrit). Stammbäume der Kunst: zur Genealogie der Avantgarde. Berlin, Akademie Verlag, 2005, IX-473 p. (ill.).

5536. TIMMS (Edward). Karl Kraus, apocalyptic satirist: the post-war crisis and the rise of the Swastika. New Haven, Yale U. P., 2005, XXI-639 p.

5537. TORRANCE (Richard). Literacy and literature in Osaka, 1890–1940. *Journal of Japanese studies*, 2005, 31, 1, p. 27-60.

5538. TRICOMI (Antonio). Sull'opera mancata di Pasolini: un autore irrisolto e il suo laboratorio. Roma, Carocci, 2005, 447 p. (Lingue e letterature Carocci, 52).

5539. VAILLANT (Alain). La crise de la littérature: Romantisme et modernité. Grenoble, ELLUG, 2005, 395 p. (ill.). (Bibliothèque stendhalienne et romantique).

5540. WAISMAN (Sergio Gabriel). Borges and translation: the irreverence of the periphery. Lewisburg, Bucknell U. P., 2005, 267 p. (Bucknell studies in Latin American literature and theory).

5541. WALKER (Barbara). Maximilian Voloshin and the Russian literary circle: culture and survival in revolutionary times. Bloomington, Indiana U. P., 2005, XIV-235 p.

d. Addenda 1998–2004

5542. FAHY (Conor). La carta dell'esemplare veronese dell'"Orlando Furioso" 1532. *La Bibliofilia*, 98, 100, p. 283-300.

5543. HARRIS (Neil). Filologia e bibliologia a confronto nell'"Orlando Furioso" del 1532. *In*: Libri, tipografi, biblioteche: ricerche storiche dedicate a Luigi Balsamo [Cf. Bibl. 97, n° 402], p. 105-122. – IDEM. Per una filologia del titolo corrente: il caso dell'Orlando Furioso del 1532. *In*: Bibliografia testuale o filologia dei testi a stampa? Definizioni metodologiche e prospettive future. Convegno di studi in onore di Conor Fahy (Udine, 24–26 febbraio 1997). A cura di Neil HARRIS. Udine, Forum, 99, p. 139-204.

5544. PIERAZZO (Elena). Nel laboratorio del censore: Girolamo Giovannini da Capuagnano editore della "Zucca" del Doni. *In*: Storia della lingua e filologia. Per Alfredo Stussi nel suo sessantacinquesimo compleanno. A cura di M. ZACCARELLO e L. TOMASIN. Firenze, Edizioni del Galluzzo per la Fondazione Ezio Franceschini, 2004, p. 291-311.

Cf. n°s 4292, 6762

§ 8. Art and industrial art.

a. General

5545. *Vacat.*

5546. Barock als Aufgabe. Hrsg. v. Andreas KREUL. Wiesbaden, Harrassowitz in Kommission, 2005, 287 p. (ill.). (Wolfenbütteler Arbeiten zur Barockforschung, 40).

5547. Barock im Vatikan: Kunst und Kultur im Rom der Päpste II, 1572–1676. Koordination, Jutta FRINGS. Leipzig, E.A. Seemann, 2005, 535 p. (ill., plans).

5548. BATTISTI (Eugenio). L'antirinascimento. Torino, Nino Aragno, 2005, 2 vol., XLVIII-1006 p. (ill.). (Biblioteca Aragno).

5549. BEYME (Klaus von). Das Zeitalter der Avantgarden: Kunst und Gesellschaft 1905–1955. München, C.H. Beck, 2005, 995 p. (ill., ports.).

5550. BODON (Giulio). Veneranda antiquitas: studi sull'eredità dell'antico nella Rinascenza veneta. Bern; Oxford, P. Lang, 2005, 359 p. (ill., facsims.). (Studies in early modern European culture, 1 = Studi sulla cultura europea della prima età moderna, 1).

5551. BRISON (Jeffrey D.). Rockefeller, Carnegie, and Canada: American philanthrophy and the arts and letters in Canada. Ithaca: McGill-Queen's U. P., 2005, XIII-281 p.

5552. DALTON (Margaret Stieg). Catholicism, popular culture, and the arts in Germany, 1880–1933. Notre Dame, University of Notre Dame Press, 2005, XI-378 p.

5553. DE MAMBRO SANTOS (Ricardo). Le virtù romane: temi e motivi dello stoicismo nell'arte nordica del Cinquecento. Roma, Edilazio, 2005, 269 p. (ill.). (Percorsi d'arte, 2).

5554. Dějiny českého výtvarného umění. (History of Czech fine arts). Tomo 5. 1939–1958. Ed. Rostislav ŠVÁCHA, Marie PLATOVSKÁ. Praha, Academia, 2005, 525 p. (ill.).

5555. Donatello among the Blackshirts: history and modernity in the visual culture of Fascist Italy. Ed. by Claudia LAZZARO and Roger J. CRUM. Ithaca, Cornell U. P., 2005, VI-293 p. (ill., maps).

5556. DUPONT (Christine A.). Modèles italiens et traditions nationales. Les artistes belges en Italie, 1830–1914. Bruxelles et Roma, Institut historique belge de Rome, 2005, LXXVII-682 p. (Bibliothèque de l'IHBR, 54).

5557. EKSHTUT (Semen A.). Shajka peredvizhnikov: Istorija odnogo tvorcheskogo sojuza. (The band of Peredvizhniki: A history of an artistic union). Moskva, Belyj gorod, 2005, 464 p. (ill.).

5558. IRBOUH (Hamid). Art in the service of colonialism: French art education in Morocco, 1912–1956. London a. New York, Tauris Academic Studies, 2005, 280 p. (ill., maps). (International library of colonial history, 2).

5559. JENKS (Andrew L.). Russia in a box: art and identity in an age of Revolution. DeKalb, Northern Illinois U. P., 2005, IX-264 p.

5560. KESSLER (Hans-Ulrich). Pietro Bernini (1562–1629). München, Hirmer, 2005, 494 p. (ill.). (Römische Studien der Bibliotheca Hertziana, 16. Veröffentlichungen der Bibliotheca Hertziana, Max-Planck-Institut für Kunstgeschichte in Rom).

5561. KRENN (Michael L.). Fall-out shelters for the human spirit: American art and the Cold War. Chapel Hill, University of North Carolina Press, 2005, XII-300 p.

5562. LYONS (Maura). William Dunlap and the construction of an American art history. Amherst, University of Massachusetts Press, 2005, XII-182 p.

5563. Manierismo y transición al Barroco: memoria del III Encuentro Internacional sobre Barroco (La Paz, Bolivia). La Paz, Unión Latina, 2005, 384 p. (ill.).

5564. Nation och union: konst och konstnärer i Sverige, Norge och Värmland under unionstiden. (Nation et Union. Art et artistes en Suède, en Norvège et dans le Värmland, 1814–1905). Red. Hans-Olof BOSTRÖM. Arvika, Ideella föreningen Rackstadmuseet, 2005, 185 p. (ill.).

5565. ONUK (Taciser). Osmanlı çadır sanatı: XVII–XIX. Yüzyıl. (Ottoman tent art: XVII–XIX. Centuries). Ankara, Atatürk Kültür Merkezi, 2005, XIII-297 p.

5566. RIEDL (Peter Philipp). Epochenbilder, Künstlertypologien: Beiträge zu Traditionsentwürfen in Literatur und Wissenschaft 1860 bis 1930. Frankfurt am Main, Vittorio Klostermann, 2005, 803 p. (Das Abendland: Forschungen zur Geschichte europäische Geisteslebens, 33).

5567. STITES (Richard). Serfdom, society, and the arts in imperial Russia. The pleasure and the power. New Haven, Yale U. P., 2005, XIII-586 p.

5568. STRÄTLING (Susanne). Allegorien der Imagination: Lesbarkeit und Sichtbarkeit im russischen Barock. München, W. Fink, 2005, 452 p. (ill.). (Theorie und Geschichte der Literatur und der schönen Künste, 111).

5569. TELESKO (Werner). Einführung in die Ikonographie der barocken Kunst. Wien, Böhlau, 2005, 286 p. (ill.). (UTB, 8301: Kunstwissenschaft Kunstgeschichte).

5570. TÍO BELLÍDO (Ramón). L' art et les expositions en Espagne pendant le Franquisme. Paris, Isthme, 2005, 317 p. (ill.).

5571. ZANOBONI (M. Paola). Rinascimento sforzesco: innovazioni tecniche, arte e società nella Milano del secondo Quattrocento. Milano, CUEM, 2005, 322 p. (ill.). (CUEM storia).

5572. ZUGASTI (Miguel). La alegoría de América en el barroco hispánico: del arte efímero al teatro. Valencia, Pre-Textos, Fundación Amado Alonso, 2005, 193 p. (ill.). (Pre-textos, 783).

a. Addenda 1997–2004

5573. CLAYSON (Hollis). Paris in despair: art and everyday life under siege, 1870–1871. Chicago, University of Chicago Press, 2002, XXXI-485 p.

5574. ERBEN (Dietrich). Paris und Rom. Die staatlich gelenkten Kunstbeziehungen unter Ludwig XIV. Berlin, Akademie, 2004, XX-409 p. (Studien aus dem Warburg-Haus, 9).

5575. FALASCA-ZAMPONI (Simonetta). Fascist spectacle: the aesthetics of power in Mussolini's Italy. Berkeley, University of California Press, 97, XI-303 p.

Cf. n°s 5479, 5482

b. Architecture

5576. ABRAMSON (Daniel M.). Building the bank of England. Money, architecture, society, 1694-1942. New Haven a. London, Yale U. P., 2005, 282 p.

5577. BATUR (Afife). Rasathane-i Amire binası için 1895 projeleri. [Architectural projects for a new observatory (Rasathane-i Amire) in Istanbul (abstract)]. *Osmanlı Bilimi Araştırmaları / Studies In Ottoman Science*, 2005, 6, 2, p. 125-138.

5578. ÇAKIRHAN (Nail). Geleneksel mimarinin şiiri: yapı sanatında yarım yüz yıl. (The poetry of traditional architecture: half a century in the art of building). İstanbul, Ege, 2005, 295 p.

5579. COMBA (Michele), OLMO (Carlo). In presenza del Lingotto. Costruzione e ricostruzione di un architettura nella Torino del Novecento. *Quaderni storici*, 2005, 1, p. 121-168.

5580. DUBOURG GLATIGNY (Pascal), LE BLANC (Marianne). Architecture et expertise mathématique: la contribution des Minimes Jacquier et Le Seur aux polémiques de 1742 sur la coupole de Saint-Pierre de Rome. *Mélanges de l'École française de Rome: Italie et mediterranée (MEFRIM)*, 2005, 117, 1, p. 189-218.

5581. ENGELBERG (Meinrad von). Renovatio ecclesiae: die "Barockisierung" mittelalterlicher Kirchen. Petersberg, Imhof, 2005, 671 p. (ill., maps). (Studien zur internationalen Architektur- und Kunstgeschichte, 23).

5582. HERMAN (Bernard L.). Town house: architecture and material life in the early American city, 1780-1830. Chapel Hill, University of North Carolina Press, 2005, XVIII-295 p.

5583. HOFER (Sigrid). Reformarchitektur, 1900-1918: deutsche Baukünstler auf der Suche nach dem nationalen Stil. Stuttgart u. London, Axel Menges, 2005, 175 p. (ill., map, plans).

5584. KROPP (Alexander). Die politische Bedeutung der NS-Repräsentationsarchitektur: die Neugestaltungspläne Albert Speers für den Umbau Berlins zur "Welthauptstadt Germania" 1936-1942/43. Neuried, Ars Una, 2005, 193 p. (ill.). (Deutsche Hochschuledition, 135).

5585. LEMERLE (Frédérique). La Renaissance et les antiquités de la Gaule: l'architecture gallo-romaine vue par les architectes, antiquaires et voyageurs des guerres d'Italie à la Fronde. Turnhout, Brepols, 2005, 290 p. (ill., map). (Études renaissantes / Centre d'études supérieures de la Renaissance, 1).

5586. LENIAUD (Jean-Michel). Les basiliques de pèlerinage en France et leur architecture (XIXe-début XXe siècle). *Mélanges de l'École française de Rome: Italie et mediterranée (MEFRIM)*, 2005, 117, 2, p. 487-496.

5587. MENNEKES (Ralf). Die Renaissance der deutschen Renaissance. Petersberg, M. Imhof, 2005, 622 p. (ill.). (Studien zur internationalen Architektur- und Kunstgeschichte, 27).

5588. PUPPI (Lionello). Palladio: introduzione alle architetture e al pensiero teorico. Fotografie di Piero CODATO e Massimo VENCHIERUTTI. Venezia, Arsenale, 2005, 463 p. (ill.).

5589. RITTER (Markus). Moscheen und Madrasabauten in Iran, 1785-1848: Architektur zwischen Rückgriff und Neuerung. Leiden, Brill, 2005, XVIII-1001 p. (ill., maps, plans). (Islamic history and civilization. Studies and texts, 62).

5590. ROBERTO (Sebastiano). San Luigi dei Francesi: la fabbrica di una chiesa nazionale nella Roma del '500. Roma: Gangemi, 2005, XVIII-297 p. (ill., plans). (Roma, storia, cultura, immagine, 14).

5591. ROSER (Hannes). St. Peter in Rom im 15. Jahrhundert: Studien zu Architektur und skulpturaler Ausstattung. München, Hirmer Verlag, 2005, 295 p. (ill., plans). (Römische Studien der Bibliotheca Hertziana, 19).

Cf. n° 4120

c. Sculpture, painting, etching and drawing

5592. ALPERS (Svetlana). Les vexations de l'art. Velázquez et les autres. Paris, Gallimard, 2005, 291 p.

5593. ARLT (Thomas). Andrea Mantegna, Triumph Caesars: ein Meisterwerk der Renaissance in neuem Licht. Wien, Böhlau, 2005, 153 p. (ill.). (Ars viva, 9).

5594. BIÈRE-CHAUVEL (Delphine). Le réseau artistique de Robert Delaunay: échanges, diffusion et création au sein des avant-gardes entre 1909 et 1939. Aix-en-Provence, Publications de l'Université de Provence, 2005, 309 p. (ill.). (Arts. Histoire des arts).

5595. BURCKHARDT (Jacob). Italian Renaissance painting according to genres. Introduction by Maurizio GHELARDI. Los Angeles, Getty Research Institute, 2005, IX-235 p. (ill.). (Texts & documents).

5596. CÀNDITO (Cristina). Corrispondenze otticoprospettiche tra le opere di Maignan e di Borromini a palazzo Spada. *Mélanges de l'École française de Rome: Italie et mediterranée (MEFRIM)*, 2005, 117, 1, p. 73-89.

5597. CORNEJO (Francisco J.). Pintura y teatro en la Sevilla del Siglo de Oro: la "sacra monarquía". Sevilla, Fundación El Monte, 2005, 390 p. (ill.).

5598. COŞKUN (Rıdvan). Resimde zaman kavramı. (Concepte de temps dans la peinture). Eskişehir, Anadolu Üniversitesi Güzel Sanatlar Fakültesi, 2005, 103 p.

5599. DA COSTA KAUFMANN (Thomas). Painterly enlightenment. The art of Franz Anton Maulbertsch, 1724-1796. Chapel Hill, University of North Carolina Press, 2005, XI-162 p. (Bettie Allison Rand lectures in art history).

5600. EHRESMANN (Nina). Paint misbehavin': Neoexpressionismus und die Rezeption und Produktion figurativer, expressiver Malerei in New York zwischen 1977 und 1984. Frankfurt am Main u. Oxford, Peter

Lang, 2005, 278 p. (ill.). (Europäische Hochschulschriften. Reihe XXVIII, Kunstgeschichte, 408 = Publications universitaires européennes. Série XXVIII, Histoire de l'art, 408 = European university studies. Series XXVIII, History of art, 408).

5601. GIANFREDA (Sandra). Caravaggio, Guercino, Mattia Preti: das halbfigurige Historienbild und die Sammler des Seicento. Emsdetten, Edition Imorde, 2005, 247 p. (ill.). (Zephir, 4).

5602. HENNING (Andreas). Raffaels Transfiguration und der Wettstreit um die Farbe: koloritgeschichtliche Untersuchung zur römischen Hochrenaissance. München, Deutscher Kunstverlag, 2005, 295 p. (ill.). (Kunstwissenschaftliche Studien, 125).

5603. JUNTUNEN (Eveliina). Peter Paul Rubens' bildimplizite Kunsttheorie in ausgewählten mythologischen Historien (1611–1618). Petersberg, M. Imhof, 2005, 187 p. (ill.). (Studien zur internationalen Architektur- und Kunstgeschichte, 39).

5604. KUBERSKY-PIREDDA (Susanne). Kunstwerke-Kunstwerte: die Florentiner Maler der Renaissance und der Kunstmarkt ihrer Zeit. Norderstedt, Books on Demand, 2005, 577 p. (ill.).

5605. LEUSCHNER (Eckhard). Antonio Tempesta: ein Bahnbrecher des römischen Barock und seine europäische Wirkung. Petersberg, Michael Imhof, 2005, 640 p. (ill., ports.). (Studien zur internationalen Architektur- und Kunstgeschichte, 26).

5606. LO RE (Salvatore). Da Pontormo a Tiziano: incidenti di percorso in alcuni studi recenti. *Rivista storica italiana*, 2005, 117, 1, p. 201-228.

5607. LOVELL (Margaretta M.). Art in a season of revolution. Painters, artisans, and patrons in early America. Philadelphia, University of Pennsylvania Press, 2005, X-341 p. (Early American studies).

5608. OPINEL (Annick). Le peintre et le mal. France, XIXe siècle. Paris, Presses universitaires de France, 2005, 363 p. (Science, histoire et société).

5609. OY-MARRA (Elisabeth). Profane Repräsentationskunst in Rom von Clemens VIII. Aldobrandini bis Alexander VII. Chigi: Studien zur Funktion und Semantik römischer Deckenfresken im höfischen Kontext. München, Deutscher Kunstverlag, 2005, 385 p. (ill.). (Italienische Forschungen des Kunsthistorischen Institutes in Florenz, Max-Planck-Institut, 4. Folge, 5).

5610. VALE (Teresa Leonor M.). Escultura barroca italiana em Portugal: obras dos séculos XVII e XVIII em colecções públicas e particulares. Lisboa, Livros Horizonte, 2005, 159 p. (ill.).

5611. VERGINE (Lea). L'altra metà dell'avanguardia, 1910–1940: pittrici e scultrici nei movimenti delle avanguardie storiche. Milano, Il saggiatore, 2005, 404 p. (ill.). (Opere e libri).

5612. WELLMANN (Marc). Die Entdeckung der Unschärfe in Optik und Malerei: zum Verhältnis von Kunst und Wissenschaft zwischen dem 15. und dem 19. Jahrhundert. Frankfurt am Main u. Oxford, P. Lang, 2005, 261 p. (ill.).

d. Decorative, popular and industrial art

5613. BARIŞTA (Örçün). Türkiye Cumhuriyeti dönemi halk plastik sanatları. (Arts plastiques populaires à l'époque de la République de Turquie). Ankara, Kültür ve Turizm Bakanlığı, 2005, 415 p.

5614. BERTRAND (Pascal-François). Les tapisseries des Barberini et la décoration d'intérieur dans la Rome baroque. Turnhout, Brepols, 2005, 343 p. (ill.). (Studies in western tapestry).

5615. KARLEKAR (Malavika). Re-visioning the past: early photography in Bengal 1875–1915. New York, Oxford U. P., 2005, XII-197 p.

5616. LÓPEZ MONDÉJAR (Publio). Historia de la fotografía en España: fotografía y sociedad, desde sus orígenes hasta el siglo XXI. Barcelona, Lunwerg Editores, 2005, 689 p. (ill.).

5617. MEIKLE (Jeffrey L.). Design in the USA. New York, Oxford U. P., 2005, 252 p.

Cf. n°s 32, 85

§ 9. Music, theatre, cinema and broadcasting.

* 5618. GONZÁLEZ SUBÍAS (José Luis). Catálogo de estudios sobre el teatro romántico español y sus autores: fuentes bibliográficas. Madrid, Fundación Universitaria Española, 2005, 351 p. (Publicaciones de la Fundación Universitaria Española. 9 Colección de investigaciones bibliográficas sobre autores españoles).

5619. ACATRINEI (Filaret). Radiodifuziunea Română de la înfiintare la "etatizare" (1925–1948). (The Romanian Radio Broadcast from foundation to "nationalization", 1925–1948). Bucureşti, Editura Tritonic, 2005, 354 p.

5620. AGGIO (Regina). Cinema novo – ein kulturpolitisches Projekt in Brasilien: Ursprünge des neuen brasilianischen Films im Kontext der Entwicklungspolitik zwischen 1956 und 1964. Remscheid, Gardez!, 2005, 281 p. (ill.). (Filmstudien, 31).

5621. ANDERSEN (Ole Ejnar). Nye kurver i medialandskabet. En analyse i mediaudbud og -forbrug i Danmark 1994–2004. (Nouvelles tendances dans le paysage médiatique. Analyse de l'offre et de l'usage des médias au Danemark 1994–2004). Frederiksberg, Markedsføring, 2005, 240 p. (ill.).

5622. APPLEGATE (Celia). Bach in Berlin: nation and culture in Mendelssohn's revival of the St. Matthew Passion. Ithaca, Cornell U. P., 2005, XII-288 p.

5623. Arte (L') del risparmio: stile e tecnologia: il cinema a basso costo in Italia negli anni Sessanta. A cura di Giacomo MANZOLI e Guglielmo PESCATORE. Roma, Carocci, 2005, 195 p. (ill.). (Cinema/tecnologia, 3).

5624. AUFENANGER (Stephan). Die Oper während der Französischen Revolution: Studien zur Gattungs- und Sozialgeschichte der französischen Oper. Tutzing, Hans Schneider, 2005, 538 p. (Frankfurter Beiträge zur Musikwissenschaft, 31).

5625. BAKHLE (Janaki). Two men and music. Nationalism in the making of an Indian classical tradition. New York, Oxford U. P., 2005, XVI-338 p.

5626. BAKKER (Gerben). The decline and fall of the European film industry: sunk costs, market size, and market structure, 1890–1927. *Economic history review*, 2005, 58, 2, p. 310-351.

5627. BÄR (Gerald). Das Motiv des Doppelgängers als Spaltungsphantasie in der Literatur und im deutschen Stummfilm. Amsterdam, Rodopi, 2005, 718 p. (ill.). (Internationale Forschungen zur allgemeinen und vergleichenden Literaturwissenschaft, 84).

5628. BARTOV (Omer). The "Jew" in cinema. From "The golem" to "Don't touch my Holocaust". Bloomington, Indiana U. P., 2005, XV-374 p.

5629. BIESEN (Sheri Chenin). Blackout: World War II and the origins of Film Noir. Baltimore, Johns Hopkins U. P., 2005, XII-243 p.

5630. BOSCHI (Alberto) [et al.]. I greci al cinema: dal peplum d'autore alla grafica computerizzata. Bologna, D. U. Press, 2005, 108 p. (ill.).

5631. BRAMANI (Lidia). Mozart massone e rivoluzionario. Milano, Bruno Mondadori, 2005, IX-465 p. (Sintesi).

5632. BRENNAN (James R.). Democratizing cinema and censorship in Tanzania, 1920–1980. *International journal of African historical studies*, 2005, 38, 3, p. 481-512.

5633. BROWN (Gregory S.). A field of honor: writers, court culture, and public theater in French Literary life from Racine to the Revolution. New York, Columbia U. P., 2005, 512 p.

5634. CABEZA SAN DEOGRACIAS (José). El descanso del guerrero: cine en Madrid durante la Guerra Civil española (1936–1939). Madrid, Ediciones Rialp, 2005, 239 p. (ill.). (Libros de cine).

5635. CANNING (Charlotte M.). The most American thing in America. Circuit Chautauqua as performance. Iowa City, University of Iowa Press, 2005, XI-268 p. (Studies in theatre history and culture).

5636. CAROTTI (Carlo). America on the road: tre film, tre registi, tre sguardi. *Storia urbana*, 2005, 28, 109, 4, p. 71-90.

5637. CHARNOW (Sally Debra). Theatre, politics, and markets in fin-de-siècle Paris. Staging modernity. New York, Palgrave Macmillan, 2005, XI-268 p. (Palgrave studies in theatre and performance history).

5638. Cinema e televisão durante a ditadura militar: depoimentos e reflexões. Org. Anita SIMIS. São Paulo, Cultura Acadêmica Editora, FCL, Laboratório Editorial, UNESP, 2005, 134 p. (Série temas em sociologia. Série do Programa de Pós-Graduação em Sociologia, 4, 2005).

5639. CLASSEN (Steven D.). Watching Jim Crow. The struggles over Mississippi tv, 1955–1969. Durham, Duke U. P., 2005, X-275 p.

5640. CORKIN (Stanley). Cowboys as cold warriors: the Western and U.S. history. Philadelphia, Temple U. P., 2005, VII-273 p. (Culture and the moving image).

5641. CYRINO (Monica Silveira). Big screen Rome. Oxford, Blackwell, 2005, 274 p. (ill).

5642. DAMMANN (Clas). Stimme aus dem Äther, Fenster zur Welt: die Anfänge von Radio und Fernsehen in Deutschland. Köln, Böhlau, 2005, 283 p.

5643. DAWSON (Mark S.). Gentility and the comic theatre of late Stuart London. New York, Cambridge U. P., 2005, XVI-300 p. (Cambridge social and cultural histories, 5).

5644. DECHERNEY (Peter). Hollywood and the cultural elite: how the movies became American. New York, Columbia U. P., 2005, X-269 p. (Film and culture).

5645. DITTMANN (Gudrun). Oper zwischen Anpassung und Integrität: zu den Uraufführungen zeitgenössischer deutscher Opern am Leipziger Neuen Theater im NS-Staat. Essen, Die Blaue Eule, 2005, 333 p. (ill., music). (Musikwissenschaft/Musikpädagogik in der Blauen Eule, 70).

5646. DOERKSEN (Clifford J.). American Babel: rogue radio broadcasters of the jazz age. Philadelphia, University of Pennsylvania Press, 2005, XI-157 p.

5647. DREMEL (Erik). Pastorale Träume: die Idealisierung von Natur in der englischen Musik, 1900–1950. Köln, Böhlau, 2005, VII-355 p. (ill., map, music, ports.).

5648. ELGEMYR (Göran). Får jag be om en kommentar? Yttrandefriheten i svensk radio 1925–1960. (Puis-je faire un commentaire? La liberté d'expression à la radio suédoise entre 1925 et 1960). Stockholm, Prisma, 2005, 534 p. (Skrifter utgivna av Stiftelsen Etermedierna i Sverige, 20).

5649. FALKENBERG (Karin). Radiohören: zu einer Bewusstseinsgeschichte 1933 bis 1950. Nürnberg, Institut für Alltagskultur, 2005, 368 p. (ill., ports.).

5650. FAUSER (Annegret). Musical encounters at the 1889 Paris World's Fair. Rochester, University of Rochester Press, 2005, XIX-391 p. (Eastman studies in music, 32).

5651. FEIGELSON (Kristian). Caméra politique: cinéma et stalinisme. Paris, Presses Sorbonne nouvelle, 2005, 317 p. (ill.). (Théorème, 8).

5652. FELBER (Andreas). Die Wiener Free-Jazz-Avantgarde: Revolution im Hinterzimmer. Wien, Böhlau, 2005, 512 p. (ill., ports.).

5653. FELICI (Candida). Musica italiana nella Germania del Seicento: i ricercari dell'intavolatura d'organo tedesca di Torino. Firenze, L.S. Olschki, 2005, 260 p. (music). (Historiae Musicae Cultores, 107).

5654. FENERICK (José Adriano). Nem do morro, nem da cidade: as transformações do samba e a indústria cultural (1920–1945). São Paulo, Fapesp, Annablume, 2005, 281 p.

5655. FERNÁNDEZ SANDE (Manuel). Los orígenes de la radio en España. Madrid, Editorial Fragua, 2005, 433 p. (ill.). (Fragua comunicación, 22).

5656. Filmat în România. Repertoriul filmelor de acțiune 1911–2004. (Shot in Romania. Romanian fiction movies 1911–2004). Vol. I. 1911–1969 cercetare filmografică de B. T. RÎPEANU. București, Editura Fundației Pro, 2005, 302 p.

5657. FORSTER (Ralf). Ufa und Nordmark: zwei Firmengeschichten und der deutsche Werbefilm 1919–1945. Trier, Wissenschaftlicher Verlag Trier, 2005, 387 p. (ill.).

5658. FRANCFORT (Didier). Le crépuscule des héros: opéra et nation après Verdi. *Mélanges de l'École française de Rome: Italie et mediterranée (MEFRIM)*, 2005, 117, 1, p. 269-293.

5659. FUBINI (Enrico). Il pensiero musicale del Romanticismo. Torino, EDT, 2005, XIII-221 p. (Biblioteca di cultura musicale. 4 Risonanze).

5660. FUHRIMANN (Daniel). Herzohren für die Tonkunst: Opern- und Konzertpublikum in der deutschen Literatur des langen 19. Jahrhunderts. Freiburg, Rombach, 2005, 405 p. (ill.). (Rombach Wissenschaften. Reihe Litterae, 134).

5661. FULCHER (Jane F.). The composer as intellectual. Music and ideology in France 1914–1940. New York, Oxford U. P., 2005, XIV-473 p.

5662. GALLO (Klaus). Una sociedad volteriana? Política, religión y teatro en Buenos Aires 1821–1827. *Entrepasados*, 2005, 27, [s. p.].

5663. GAUDELUS (Sébastien). Les offices de Ténèbres en France, 1650–1790. Paris, CNRS éditions, 2005, 300 p. (ill.). (Collection Sciences de la musique. Série Études).

5664. Geschichte (Die) des Nordwestdeutschen Rundfunks. Hrsg. v. Peter von RÜDEN und Hans-Ulrich WAGNER. Hamburg, Hoffmann und Campe, 2005, 463 p. (ill.).

5665. Geschichte der Musik im 20. Jahrhundert, 1945–1975. Hrsg. v. Hanns-Werner HEISTER. Laaber, Laaber-Verlag, 2005, 381 p. (ill., music). (Handbuch der Musik im 20. Jahrhundert, 3).

5666. GILBERT (Shirli). Music in the Holocaust. Confronting life in the nazi ghettos and camps. New York, Clarendon Press of Oxford U. P., 2005, XII-243 p. (Oxford historical monographs).

5667. GOOSMAN (Stuart L.). Group harmony. The black urban roots of rhythm and blues. Philadelphia, University of Pennsylvania Press, 2005, XI-291 p.

5668. GORDON (Robert). Not quite cricket: "Civilization on trial in South Africa": a note on the first "Protest Film" made in Southern Africa. *History in Africa*, 2005, 32, p. 457-466.

5669. Govori London. Predavanija na Bi Bi Si za Bălgarija prez Vtorata svetovna vojna. (Parle Londres. Emission du BBC pour la Bulgarie pendant la Deuxième guerre mondiale). Săstavitel Borislav DIČEV. Sofija, Knigoizdatelska kušta "Trud", 2005, 712 p.

5670. HAHN (Hans-Joachim). Repräsentationen des Holocaust: zur westdeutschen Erinnerungskultur seit 1979. Heidelberg, Universitätsverlag Winter, 2005, 310 p. (Probleme der Dichtung, 33).

5671. HEFFERNAN (Kevin). Ghouls, gimmicks, and gold. Horror films and the American movie business, 1953–1968. Durham, Duke U. P., 2005, VIII-323 p.

5672. HERRERA NAVARRO (Javier). El cine en su historia: manual de recursos bibliográficos e Internet. Madrid, Arco Libros, 2005, 463 p. (Instrumenta bibliologica).

5673. JACKSON (Jerma A.). Singing in my soul. Black gospel music in a secular age. Chapel Hill, University of North Carolina Press, 2005, XII-193 p.

5674. JIRGENS (Eckhard). Der deutsche Rundfunk der 1. Tschechoslowakischen Republik: eine Bestandsaufnahme. Frankfurt am Main u. Oxford, P. Lang, 2005, 2 vol., [s. p.]. (ill.).

5675. KANNAPIN (Detlef). Dialektik der Bilder: der Nationalsozialismus im deutschen Film: ein Ost-West-Vergleich. Berlin, Karl Dietz, 2005, 289 p. (Manuskripte, 58).

5676. KENNEY (William Howland). Jazz on the river. Chicago, University of Chicago Press, 2005, X-229 p.

5677. Kinematografie a město. Studie z dějin lokální kultury. (The cinematography and the town). Ed. Pavel SKOPAL. Brno, Masarykova univerzita, 2005, 248 p. (ill., photogr.). (Sborník prací filozofické fak. Brněnské univerzity. Řada filmologická, Otázky filmu a audiovizuální kultury, 2).

5678. KNAPP (Raymond). The American musical and the formation of national identity. Princeton, Princeton U. P., 2005, XXI-361 p.

5679. KOCH (Hans Jürgen), GLASER (Hermann). Ganz Ohr. Eine Kulturgeschichte des Radios in Deutschland. Köln, Böhlau, 2005, 376 p.

5680. LIPPE (Marcus Chr). Rossinis opere serie: zur musikalisch-dramatischen Konzeption. Stuttgart, F. Steiner, 2005, 369 p. (music). (Beihefte zum Archiv für Musikwissenschaft, 55).

5681. LOSKANT (Alexander). Der neue europäische Großfilm: Ökonomie und Ästhetik europäischer Kinogroßproduktionen der 90er Jahre. Frankfurt am Main u. Oxford, P. Lang, 2005, XIII-352 p. (Europäische Hochschulschriften. Reihe XXX, Theater-, Film- und Fernsehwissenschaften, 88 = Publications universitaires européennes. Série XXX, Théâtre, chinéma, television, 88 = European university studies. Series XXX, Theatre, film and television, 88).

5682. Louis Grénon. Un musicien d'église au XVIIIe siècle. Sous la dir. de Bernard DOMPNIER. Clermont-Ferrand, Presses universitaires Blaise-Pascal, 2005, 206 p. (Études sur le Massif central).

5683. LUNEAU (Aurélie). Radio Londres: les voix de la liberté, 1940–1944. Paris, Perrin, 2005, 349 p. (ill.).

5684. MAC KAY (George). Circular breathing: the cultural politics of Jazz in Britain. Durham, Duke U. P., 2005, XIV-357 p.

5685. MADESANI (Angela). Storia della fotografia. Milano, Bruno Mondatori, 2005, X-406 p. (Campus).

5686. MARANGHELLO (César). Breve historia del cine argentino. Barcelona, Laertes, 2005, 364 p. (ill., ports.).

5687. MARCIAK (Dorothée). La place du prince: perspective et pouvoir dans le théâtre de cour des Médicis, Florence, 1539–1600. Paris, Champion, 2005, 381 p. (Etudes et essais sur la Renaissance, 50).

5688. MARCUS (Kenneth H.). Musical metropolis: Los Angeles and the creation of a music culture. New York, Palgrave Macmillan, 2005, XIII-274 p.

5689. MARIE (Laurent). Le cinéma est à nous: le PCF [Parti communiste français] et le cinéma français de la Libération à nos jours. Paris, L'Harmattan, 2005, 363 p. (Champs visuels).

5690. MARINO (Natalia), MARINO (Emanuele Valerio). L'Ovra a Cinecittà: polizia politica e spie in camicia nera. Torino, Bollati Boringhieri, 2005, XIII-332 p.

5691. MASLAN (Susan). Revolutionary acts: theater, democracy, and the French Revolution. Baltimore, Johns Hopkins U. P., 2005, XII-275 p. (Parallax: Revisions of culture and society).

5692. MATTHEWS (Jill Julius). Dance hall and picture palace. Sydney's romance with modernity. Sydney, Currency Press, 2005, X-342 p.

5693. MIQUEL (Angel). Disolvencias: Literatura, cine y radio en México (1900–1950). México D.F., Fondo de Cultura Económica, 2005, 207 p. (Colección popular, 448).

5694. MONOD (David). Settling scores. German music, denazification, and the Americans, 1945–1953. Chapel Hill, University of North Carolina Press, 2005, XIV-325 p.

5695. MOON (Krystyn R.). Yellowface. Creating the Chinese in American popular music and performance, 1850s–1920s. New Brunswick, Rutgers U. P., 2005, XI-220 p.

5696. Musical education in Europe (1770–1914): compositional, institutional, and political challenges. Ed. by Michael FEND and Michel NOIRAY. Berlin, Berliner Wissenschafts-Verlag, 2005, 2 vol., XVIII-727 p. (Musical life in Europe, 1600–1900: circulation, institutions, representation).

5697. NADEL (Alan). Television in Black and White America: race and national identity. Lawrence, University Press of Kansas, 2005, IX-224 p. (Culture America).

5698. NYSTRÖM (Esbjörn). Libretto im Progress: Brechts und Weills Aufstieg und Fall der Stadt Mahagonny aus textgeschichtlicher Sicht. Bern, Lang, 2005, 709 p. (ill.). (Arbeiten zur Editionswissenschaft, 6).

5699. PISANI (Michael V.). Imagining native America in music. New Haven, Yale U. P., 2005, XIV-422 p.

5700. RABER (Christine). Der Filmkomponist Wolfgang Zeller: propagandistische Funktionen seiner Filmmusik im Dritten Reich. Laaber, Laaber-Verlag, 2005, IX-259 p. (ill.).

5701. RIVERO (Yeidi M.). Tuning out blackness. Race and nation in the history of Puerto Rican television. Durham, Duke U. P., 2005, XII-264 p. (Consoling passions, television and cultural power).

5702. ROFFAT (Sébastien). Animation et propagande: les dessins animés pendant la seconde guerre mondiale. Paris, L'Harmattan, 2005, 325 p. (Champs visuels).

5703. Royal chapel (The) in the time of the Habsburgs: music and ceremony in the early modern European court. Ed. by Juan José CARRERAS and Bernardo GARCÍA GARCÍA. Woodbridge, Suffolk, Boydell Press, 2005, VIII-402 p. (ill.). (Studies in Medieval and Renaissance music, 3).

5704. SALMI (Hannu). Wagner and Wagnerism in nineteenth-century Sweden, Finland, and the Baltic Provinces. Reception, enthusiasm, cult. Rochester, University of Rochester Press, 2005, XI-310 p.

5705. SAMMOND (Nicholas). Babes in tomorrowland. Walt Disney and the making of the American child, 1930–1960. Durham, Duke U. P., 2005, X-472 p.

5706. SCOTT (Allen J.). On Hollywood: the place, the industry. Princeton, Princeton U. P., 2005, XIII-200 p.

5707. SEDGWICK (John), POKORNY (Michael). The film business in the United States and Britain during the 1930s. *Economic history review*, 2005, 58, 1, p. 79-112.

5708. SORIA (Claudia). Los cuerpos de Eva: anatomía del deseo femenino. Rosario, Beatriz Viterbo, 2005, 223 p. (Estudios culturales, 15).

5709. SPRINGER (Christian). Verdi-Studien: Verdi in Wien; Hanslick versus Verdi; Verdi und Wagner; zur Interpretation der Werke Verdis; Re Lear – Shakespeare bei Verdi. Wien, Edition Praesens, 2005, 435 p. (ill., music).

5710. STEINEGGER (Catherine). La musique à la Comédie-Française de 1921 à 1964: aspects de l'évolution d'un genre. Sprimont, Mardaga, 2005, 232 p. (ill., ports.). (Collection Musique, musicologie).

5711. STEINER (Kilian J. L.). Ortsempfänger, Volksfernseher und Optaphon: die Entwicklung der deutschen Radiound Fernsehindustrie und das Unternehmen Loewe, 1923–1962. Essen, Klartext Verlag, 2005, 381 p.

5712. VAN ORDEN (Kate). Music, discipline, and arms in Early Modern France. Chicago, University of Chicago Press, 2005, XIV-322 p.

5713. VASSALLO (Aude). La télévision sous de Gaulle: le contrôle gouvernemental de l'information, (1958–1969). Bruxelles, Bry-sur-Marne, De Boeck, INA, 2005, 310 p. (ill.). (Collection Médias-recherches).

5714. WASSON (Haidee). Museum movies. The museum of modern art and the birth of art cinema. Berkeley a. Los Angeles, University of California Press, 2005, XIII-314 p.

5715. WATERMAN (David). Hollywood's road to riches. Cambridge, Harvard U. P., 2005, XVI-393 p.

5716. WIERMANN (Barbara). Die Entwicklung vokal-instrumentalen Komponierens im protestantischen Deutschland bis zur Mitte des 17. Jahrhunderts. Göttingen, Vandenhoeck & Ruprecht, 2005, IX-650 p. (music). (Abhandlungen zur Musikgeschichte, 14).

5717. WOLBOLD (Matthias). Reden über Deutschland: die Rundfunkreden Thomas Manns, Paul Tillichs und Sir Robert Vansittarts aus dem Zweiten Weltkrieg. Münster, Lit, 2005, XIV-380 p. (ill.). (Tillich-Studien, 17).

5718. ZHEN (Zhang). An amorous history of the silver screen: Shanghai cinema, 1896–1937. Chicago, University of Chicago Press, 2005, XXXIII-488 p. (Cinema and modernity).

§ 9. Addendum 1998.

5719. TOCCHINI (Gerardo). I fratelli d'Orfeo: Gluck e il teatro musicale massonico tra Vienna e Parigi. Firenze, L. S. Olschki, 98, XVI-367 p. (ill.). (Studi, 174).

Cf. n° 4128

N

MODERN ECONOMIC AND SOCIAL HISTORY

§ 1. General. 5720-5776. – § 2. Political economy. 5777-5786. – § 3. Industry, mining and transportation. 5787-5885. – § 4. Trade. 5886-5926. – § 5. Agriculture and agricultural problems. 5927-5976. – § 6. Money and finance. 5977-6035. – § 7. Demography and urban history. 6036-6124. – § 8. Social history. 6125-6405. – § 9. Working-class movement and socialism. 6406-6469.

§ 1. General.

* 5720. STELLNER (František), SOBĚHART (Radek), VÁŇA (Daniel). Bibliografie hospodářských a sociálních dějin v České republice v devadesátých letech 20. století. (The Bibliography of the Economic and Social History in the Czech Republic in the 1990s). *Acta Oeconomica Pragensia*, 2005, 13, 3, p. 268-316.

5721. ADAMS (Julia). The familial state: ruling families and merchant capitalism in Early Modern Europe. Ithaca, Cornell U. P., 2005, XI-235 p. (Wilder House series in politics, history, and culture).

5722. AHONEN (Kalevi). From sugar triangle to cotton triangle: trade and shipping between America and Baltic Russia, 1783–1860. Jyväskylä, University of Jyväskylä, 2005, 572 p. (ill.). (Jyväskylä studies in humanities, 38).

5723. AKYILDIZ (Ali). Anka'nın sonbaharı: Osmanlı'da iktisadî modernleşme ve uluslarararası sermaye. (Automne d'Anka: modernisation économique et capital international dans l'Empire Ottoman). İstanbul, İletişim, 2005, 240 p.

5724. AMITH (Jonathan D.). The Möbius strip: a spatial history of colonial society in Guerrero, Mexico. Stanford, Stanford U. P., 2005, XVII-661 p. (ill., maps, geneal. tables).

5725. Atlantic economy (The) during the seventeenth and eighteenth centuries. Organization, operation, practice, and personnel. Ed. by Peter A. COCLANIS. Columbia, University of South Carolina Press, 2005, XIX-377 p. (Carolina lowcountry and the Atlantic World).

5726. BASKICI (Mehmet Murat). 1800–1914 Yıllarında Anadolu'da iktisadi değişim. (Changement économique en Anatolie dans les années de 1800–1914). Ankara, Turhan Kitabevi, 2005, XII-272 p.

5727. BURHOP (Carsten), WOLFF (Guntram B.). A compromise estimate of German net national product, 1851–1913, and its implications for growth and business cycles. *Journal of economic history*, 2005, 65, 3, p. 613-657.

5728. CASELLI (Gian Paolo), THOMA (Grid). The Albanian economy during World War II and the first attempt at planning. *Journal of European economic history*, 2005, 34, 1, p. 93-120.

5729. CHUKU (Gloria). Igbo women and economic transformation in southeastern Nigeria, 1900–1960. New York, Routledge, 2005, XIII-320 p. (African studies).

5730. CUFF (Timothy). The hidden cost of economic development. The biological standard of living in antebellum Pennsylvania. Burlington, Ashgate, 2005, XVII-277 p. (Modern economic and social history series).

5731. DI GIACINTO (Valter), NUZZO (Giorgio). I fattori dello sviluppo economico abruzzese: un'analisi storica. *Rivista di storia economica*, 2005, 21, 1, p. 31-62.

5732. DIEHL (Markus Albert). Von der Marktwirtschaft zur nationalsozialistischen Kriegswirtschaft: die Transformation der deutschen Wirtschaftsordnung 1933–1945. Stuttgart, Steiner, 2005, 195 p. (ill.). (Beiträge zur Wirtschafts- und Sozialgeschichte, 104).

5733. Dokumentation: Wirtschaftsgeschichte in der Bundesrepublik und in Österreich 1951 – ein zeitgenössischer Bericht. *Vierteljahresschrift für Sozial- und Wirtschaftsgeschichte*, 2005, 92, 1, p. 31-46.

5734. Economics (The) of World War I. Ed. by Stephen BROADBERRY and Mark HARRISON. New York, Cambridge U. P., 2005, XVI-345 p.

5735. ELMQUIST JØRGENSEN (Kasper). Studier i samspillet mellem stat og erhvervsliv under 1. verdenskrig. (Etudes sur l'interaction entre Etat et vie économique pendant la première Guerre mondiale). København, Handelshøjskolen i København, 2005, 286 p.

5736. Entrepreneurship in theory and history. Ed. by Youssef CASSIS and Ioanna PEPELASIS MINOGLOU. Basingstoke, Palgrave Macmillan, 2005, XIII-211 p.

5737. FAGERFJÄLL (Ronald). De gjorde Sverige rikt: 1900-talets entreprenörer, företagsledare och riskkapitalister. (Ils ont fait de la Suède un pays riche: entrepreneurs, capitaines d'industrie et capitalistes au 20e siècle). Stockholm, Kalla kulor, 2005, 612 p.

5738. FEINSTEIN (Charles H.). An economic history of South Africa. Conquest, discrimination and development. Cambridge, Cambridge U. P., 2005, XVIII-302 p.

5739. FELICE (Emanuele). Il reddito delle regioni italiane nel 1938 e nel 1951. Una stima basata sul costo del lavoro. *Rivista di storia economica*, 2005, 21, 1, p. 3-30. – IDEM. Il valore aggiunto regionale. Una stima per il 1891 e per il 1911 e alcune elaborazioni di lungo periodo (1891–1971). *Rivista di storia economica*, 2005, 21, 3, p. 273-314.

5740. FENOALTEA (Stefano). La crescita economica dell'Italia postunitaria: le nuove serie storiche. *Rivista di storia economica*, 2005, 21, 2, p. 91-122.

5741. GATRELL (Peter). Russia's First World War. A social and economic history. London, Pearson education, 2005, XX-318 p.

5742. GERLACH (Christian). Die Welternährungskrise 1972–1975. *Geschichte und Gesellschaft*, 2005, 31, 4, p. 546-585.

5743. HATTON (Timothy J.), WILLIAMSON (Jeffrey G.). Global migration and the world economy: two centuries of policy and performance. Cambridge, MIT Press, 2005, XI-471 p.

5744. História económica de Portugal, 1700–2000. V. 1. O século XVIII. V. 2. O século XIX. V. 3. O século XX. Organização Pedro LAINS e Alvaro FERREIRA DA SILVA. Lisboa, Imprensa de Ciências Sociais, 2005, 3 vol., 425 p., 491 p., 502 p. [Cf. n° <seleção> 5838.]

5745. Historia económica del Uruguay. T. 3. BERTINO (Magdalena) [et al.]. La economía del primer batllismo y los años veinte: auge y crisis del modelo agroexportador (1911–1930). Con la colaboración de Paola AZAR y Henry WILLEBALD. Montevideo, Fin de Siglo, 2005, 435 p. (ill.).

5746. JACOBS (Meg). Pocketbook politics. Economic citizenship in twentieth-century America. Princeton, Princeton U. P., 2005, XII-349 p. (Politics and society in twentieth-century America).

5747. JAKUBEC (Ivan). Československo-západoněmecké obchodněpolitické a dopravněpolitické vztahy v období 1949–1967. (The Czechoslovak-Westgerman relations of trade, transport and politics in the period 1949–1967). *Acta Oeconomica Pragensia*, 2005, 13, 3, p. 190-209.

5748. Japan, China, and the growth of the Asian international economy, 1850–1949. Ed. by Kaoru SUGIHARA. New York, Oxford U. P., 2005, XII-295 p.

5749. KAZGAN (Gülten). Türkiye ekonomisinde krizler (1929–2001): "ekonomi politik" açısından bir İrdeleme. [Crises économiques en Turquie (1929–2001): une étude du point de vue de l'"économie politique"]. İstanbul, İstanbul Bilgi Üniversitesi, 2005, XXII-347 p.

5750. KOSTA (Jiří). Česká/československá ekonomika ve světle měnících se systémů. (Czech/Czechoslovak economy in the economic changes of the systems). Ostrava, Vysoká škola báňská, 2005, 217 p. – IDEM. K historii a koncepci československé ekonomické reformy v letech 1965–1969. (History and concept of the Czechoslovak economic reform, 1965–1969). *Acta Oeconomica Pragensia*, 2005, 13, 3, p. 27-47.

5751. KOSTELENOS (G.), VASILIOU (D.). The evolution of tertiary production in pre World War II Greece: 1833–1939. *Journal of European economic history*, 2005, 34, 1, p. 187-214.

5752. LIPSEY (Richard G.), CARLAW (Kenneth I.), BEKAR (C. T.). Economic transformations. General purpose technologies and long-term economic growth. Oxford, Oxford U. P., 2005, XXI-595 p.

5753. MAGARIÑOS (Gustavo). Integración económica latinoamericana: Proceso ALALC/ALADI, 1950–2000. Montevideo, Asociación Latinoamericano de Integración, 2005, 2 vol., II-569 p., IV-479 p.

5754. MILWARD (Alan S.). Politics and economics in the history of the European Union. London a. New York, Routledge, 2005, X-127 p.

5755. MITCHENER (Kris James), WEIDENMIER (Marc). Empire, public goods, and the Roosevelt corollary. *Journal of economic history*, 2005, 65, 3, p. 658-692.

5756. MOKYR (Joel). The intellectual origins of modern economic growth. *Journal of economic history*, 2005, 65, 2, p. 285-351.

5757. MORRIS (R. J.). Men, women and property in England, 1780–1870. A social and economic history of family strategies amongst the Leeds middle classes. Cambridge, New York a. Port Melbourne, Cambridge U. P., 2005, XIII-445 p.

5758. NAHUM (Benjamín). Nacionalización de empresas británicas de servicios públicos, 1947–1949. Montevideo, Departamento de Publicaciones, Universidad de la República, 2005, 397 p. (Serie Escritos de historia económica, 6).

5759. OUDIN-BASTIDE (Caroline). Travail, capitalisme et société esclavagiste. Guadeloupe, Martinique (XVIIe–XIXe siècles). Paris, La Découverte, 2005, 345 p. (Textes à l'appui. Histoire contemporaine).

5760. ÖZKUL (Ali Efdal). Kıbrıs'ın sosyo-ekonomik tarihi (1726–1750). [Histoire socio-économique de Chypre (1726–1750)]. İstanbul, İletişim, 2005, 452 p.

5761. RINALDI (Alberto). The Emilian model revisited: twenty years after. *Business history*, 2005, 47, 2, p. 244-266.

5762. ROSEN (Elliott A.). Roosevelt, the Great Depression, and the economics of recovery. Charlottesville, University of Virginia Press, 2005, X-308 p.

5763. SALVUCCI (Richard). Algunas consideraciones económicas (1836): Análisis mexicano de la depresión a principios del siglo XIX. *Historia mexicana*, 2005, 55, 1, p. 67-97.

5764. SCHLARP (Karl-Heinz). Die ökonomische Untermauerung der Entspannungspolitik. Visionen und Realitäten einer deutsch-sowjetischen Wirtschaftskooperation im Zeichen der Neuen Ostpolitik. *Archiv für Sozialgeschichte*, 2005, 45, p. 77-100.

5765. Service public (Le), l'économie, la République (1780–1960). Dossier coordonné par Michel MARGAIRAZ et Olivier DARD. *Revue d'histoire moderne et contemporaine*, 2005, 52, 3, 206 p.

5766. SLUYTERMAN (Keetie E.). Dutch enterprise in the twentieth century: business strategies in a small open economy. New York, Routledge, 2005, XIII-319 p.

5767. SMITH (Frederick H.). Caribbean Rum: a social and economic history. Gainesville, University of Florida Press, 2005, XIII-339 p.

5768. SPOERER (Mark). Demontage eines Mythos? Zu der Kontroverse über das nationalsozialistische "Wirtschaftswunder". *Geschichte und Gesellschaft*, 2005, 31, 3, p. 415-438.

5769. TORP (Cornelius). Die Herausforderung der Globalisierung: Wirtschaft und Politik in Deutschland 1860–1914. Göttingen, Vandenhoeck und Ruprecht, 2005, 430 p. (Kritische Studien zur Geschichtswissenschaft, 168).

5770. TUDORANCEA (Radu). Relațiile economice româno-elene în perioada interbelică. (The Romanian-Hellenic economic relations during interwar). *Studii și materiale de istorie contemporană*. Academia Română, 2005, 4, p. 5-24.

5771. VAN DER WEE (Herman). Das 20. Jahrhundert: eine wirtschaftliche Retrospektive. Wissenschaftszentrum Berlin für Sozialforschung. Berlin, WZB, 30 p. (Beim Präsidenten: Discussion Paper 2005.003).

5772. WILLS (Jocelyn). Boosters, hustlers, and speculators. Entrepreneurial culture and the rise of Minneapolis and St. Paul, 1849–1883. St. Paul, Minnesota historical society press, 2005, XI-290 p.

5773. Wirtschaftskontrolle und Recht in der nationalsozialistischen Diktatur. Hrsg. v. Dieter GOSEWINKEL. Frankfurt am Main, Klostermann, 2005, LIX-427 p. (Das Europa der Diktatur 4, Studien zur europäischen Rechtsgeschichte, 180).

5774. WU (Yongping). A political explanation of economic growth: state survival, bureaucratic politics, and private enterprises in the making of Taiwan's economy, 1950–1985. Cambridge, Harvard U. P., 2005, X-410 p.

5775. ZANINI (Andrea). Strategie politiche ed economia feudale ai confini della Repubblica di Genova (secoli XVI–XVIII). Un buon negotio con qualche contrarietà. Genoa, Società Ligure di Storia Patria, 2005, 269 p. (Atti della Società Ligure di Storia Patria, XLV/3).

5776. Zhongguo jingji de xiatian (L'été de l'économie chinoise). Ed. par YU Yongding et HE Fan. Beijing, Zhongguo qingnian chubanshe, 2005, 259 p.

Cf. n° 4282

§ 2. Political economy.

5777. Action publique (L') et ses dispositifs. Institutions, économie, politique. Sous la dir. Élisabeth CHATEL, Thierry KIRAT et Robert SALAIS. Paris, L'Harmattan, 2005, 459 p.

5778. DAVIS (Timothy). Ricardo's macroeconomics: money, trade cycles, and growth. New York, Cambridge U. P., 2005, 328 p.

5779. Economists in Parliament in the liberal age 1848–1920. Ed. by Massimo M. AUGELLO and Marco E.L. GUIDI. Aldershot, Ashgate, 2005, XVIII-375 p.

5780. FINCH (Henry). La economía política del Uruguay contemporáneo, 1870–2000. Montevideo, Ediciones de la Banda oriental, 2005, 351 p.

5781. GUARINO (Giuseppe). Reflections on economic theory and the theory of institutions. *Rivista di storia economica*, 2005, 21, 1, p. 63-84.

5782. HOLL (Richard E.). From the boardroom to the war room: America's corporate liberals and FDR's preparedness program. Rochester, N.Y.: University of Rochester Press, 2005, X-191 p.

5783. LARSEN (Erling Røed). The top three questions in economics? When theory is confronted with history. *Scandinavian economic history review*, 2005, 53, 3, p. 34-57.

5784. MAAS (Harro). William Stanley Jevons and the making of modern economics. New York, Cambridge U. P., 2005, XXII-330 p. (Historical perspectives on modern economics).

5785. SAVONA (Paolo). On the definition of political economy and on the method of investigation proper to it: reflections on the bicentennial of the birth of John Stuart Mill. *Journal of European economic history*, 2005, 34, 3, p. 573-600.

5786. SCHABAS (Margaret). The natural origins of economics. Chicago, University of Chicago Press, 2005, XI-231 p.

Cf. n°ˢ 3868, 4417, 4508, 5822

§ 3. Industry, mining and transportation.

5787. AIKEN (Katherine G.). Idaho's Bunker Hill. The rise and fall of a great mining company, 1885–

1981. Norman, University of Oklahoma Press, 2005, XIX-284 p.

5788. ALAJOUTSIJÄRVI (Kimmo), HOLMA (Heikki), NYBERG (Kjell), TIKKANEN (Henrikki). Cyclicality in the Finnish and Swedish sawmill industry, 1970–2000. *Scandinavian economic history review*, 2005, 53, 1, p. 66-90.

5789. ÁLVAREZ DE LA BORDA (Joel). Los orígenes de la industria petrolera en México, 1900–1925. México D.F., Petróleos Mexicanos, 2005, 308 p.

5790. ANASTAKIS (Dimitry). Auto pact: creating a borderless North American auto industry. Toronto, University of Toronto Press, 2005, XIV-285 p.

5791. ANDERSEN (Steen). De gjorde Danmark større. De multinationale danske entreprenørfirmær i krise og krig 1919–1947. (Ils ont grandi le Danemark. Les entreprises multinationales danoises dans la crise et la guerre 1919–1947). København, Lindhardt og Ringhof, 2005, 573 p. (ill.).

5792. Archives (Les) des entreprises sous l'Occupation: conservation, accessibilité et apport. Sous la dir. de Hervé JOLY. Lille, Institut fédératif de recherche sur les économies et sociétés industrielles, 2005, 319 p.

5793. ATACK (Jeremy), BATEMAN (Fred), MARGO (Robert A.). Capital deepening and the rise of the factory: the American experience during the nineteenth century. *Economic history review*, 2005, 58, 3, p. 586-595.

5794. BATEN (Joerg), SCHULZ (Rainer). Making profits in wartime: corporate profits, inequality, and GDP in Germany during the First World War. *Economic history review*, 2005, 58, 1, p. 34-56.

5795. BÖSCH (Frank). Krupps "Kornwalzer". Formen und Wahrnehmungen der Korruption im Kaiserreich. *Historische Zeitschrift*, 2005, 281, 2, p. 337-379.

5796. BREDE (Christina). Das Instrument der Sauberkeit. Die Entwicklung der Massenproduktion von Feinseifen in Deutschland, 1850–2000. Münster, Waxmann, 2005, 365 p. (Cottbuser Studien zur Geschichte der Technik, Arbeit und Umwelt, 26).

5797. BURIDANT (Jérôme). Espaces forestiers et industrie verrière, XVIIe–XIXe siècle. Préface d'Yves-Marie BERCÉ, avant-propos d'Andrée CORVOL. Paris, L'Harmattan, 2005, 416 p.

5798. CAIN (Louis P.), HADDOCK (David D.). Similar economic histories, different industrial structures: transatlantic contrasts in the evolution of professional sports leagues. *Journal of economic history*, 2005, 65, 4, p. 1116-1147.

5799. CARON (François). Histoire des chemins de fer en France, 1883–1937. Paris, Fayard, 2005, 1029 p. – IDEM. Les grandes compagnies de chemin de fer en France, 1823–1937. Genève, Droz, 2005, 411 p.

5800. CHANDLER (Alfred D. Jr.). Shaping the industrial century. The remarkable story of the evolution of the modern chemical and pharmaceutical industries. Cambridge, Harvard U. P., 2005, VIII-366 p. (Harvard studiesin business history, 46).

5801. CHILDS (William R.). The Texas Railroad Commission: understanding regulation in America to the mid-twentieth century. College Station, Texas A&M U. P., 2005, X-323 p. (Kenneth E. Montague series in oil and business history, 17).

5802. CONDRAU (Flurin). Die Industrialisierung in Deutschland. Darmstadt, Wissenschaftliche Buchgesellschaft, 2005, VII-139 p.

5803. CREPAS (Nicola). Le premesse dell'industrializzazione. *In*: Storia d'Italia. Annali 15. L'industria. A cura di Franco AMATORI [et al.]. Torino, Einaudi, 99, p. 85-179.

5804. DAVIDS (Mila). The privatisation and liberalisation of Dutch telecommunications in the 1980s. *Business history*, 2005, 47, 2, p. 219-243.

5805. DESPORTES (Marc). Paysages en mouvement. Transports et perception de l'êspace, XVIIIe–XXe siècle. Paris, Gallimard, 2005, 414 p. (Bibliothèque illustrée des histoires).

5806. DUBLIN (Thomas), LICHT (Walter). The face of decline. The Pennsylvania anthracite region in the twentieth century. Ithaca, Cornell U. P., 2005, 277 p.

5807. DUFEK (Pavel). Vliv právního prostředí na soukromé podnikání v Československu v 50. a 60. letech 20. století. (The Impact of the Legal Milieu on the Private Enterprise in Czechoslovakia in the 1950s and the 1960s). *Acta Oeconomica Pragensia*, 2005, 13, 3, p. 163-189.

5808. DVOŘÁK (Tomáš). Těžba uranu versus "očista" pohraničí. Německé pracovní síly v Jáchymovských dolech na přelomu čtyřicátých a padesátých let 20. století. (Uranium mining versus the "Purging" of the borderlands. German Labour in the Jáchymov mines in the late 1940s and early 1950s). *Soudobé dějiny*, 2005, 12, 3/4, p. 626-671.

5809. EDIGER (Volkan Ş.). Enerji ekonomi-politiği perspektifinden Osmanlı'da neft ve petrol. (Naphte et pétrole dans l'Empire Ottoman du point de vue de l'économie politique de l'énergie). Ankara, ODTÜ Yayıncılık, 2005, XXIII-472 p.

5810. EMINGER (Stefan). Das Gewerbe in Österreich, 1930–1938. Organisationsformen, Interessenpolitik und politische Mobilität. Innsbruck, Wien u. Bozen, Studien Verlag, 2005, 324 p.

5811. ENOKI (Kazue). Nihon seishigyō no tajōki dōnyū ni kansuru ichi-kosatsu. (A study of the change-over to the multi-ends reeling machine in the silk-reeling industry of Japan). *Shakai Keizai Shigaku*, 2005, 71, 2, p. 3-24 [Eng. Summary].

5812. ERKER (Paul). Vom nationalen zum globalen Wettbewerb: die deutsche und die amerikanische Reifenindustrie im 19. und 20. Jahrhundert. Paderborn, Ferdinand Schöningh, 2005, 710 p.

3. INDUSTRY, MINING AND TRANSPORTATION

5813. FEAR (Jeffrey R.). Organizing control: August Thyssen and the construction of German corporate management. Cambridge, Harvard U. P., 2005, XV-956 p. (Harvard studies in business history, 45).

5814. FERRER I ALÓS (Llorenç). Familia e industrialización en Catalunya: la trayectoria de los Pons y Enrich de Manresa. *Historia social*, 2005, 53, p. 3-30.

5815. FINE (Lisa M.). The story of Reo Joe. Work, kin and community in Autotown, U.S.A. Philadelphia, Temple U. P., 2005, XII-239 p. (Critical perspectives on the past).

5816. FRANK (Alison Fleig). Oil empire. Visions of prosperity in Austrian Galicia. Cambridge, Harvard U. P., 2005, XX-343 p.

5817. FRANZ (Kathleen). Tinkering: consumers reinvent the early automobile. Philadelphia, University of Pennsylvania Press, 2005, 224 p.

5818. GAVIRA MÁRQUEZ (María Concepción). Historia de una crisis: la minería en Oruro a fines del periodo colonial. La Paz, Instituto de Estudios Bolivianos-IEB y Lima, Instituto Francés de Estudios Andinos-IFEA, 2005, 333 p. (ill., maps). (Travaux de l'Institut français d'études andines, 213. Colección Cuarto centenario de la fundación de Oruro).

5819. GERBER (Larry G.). The irony of state intervention: American industrial relations policy in comparative perspective, 1914-1939. DeKalb, Northern Illinois U. P., 2005, VII-212 p.

5820. GERICKE (Hans Otto). Braunes Gold in Anhalt. Zur Geschichte der Braunkohlengewinnung und Verarbeitung im Gebiet von Anhalt von etwa 1700 bis zur Mitte des 20. Jahrhunderts. Dessau, Funk, 2005, 104 p.

5821. Global perspectives on industrial transformation in the American South. Ed. by Susanna DELFINO and Michele GILLESPIE. Columbia, University of Missouri Press, 2005, X-240 p. (New currents in the history of Southern economy and society).

5822. GREAVES (Julian). Industrial reorganization and government policy in interwar Britain. Aldershot, Ashgate, 2005, XIX-296 p.

5823. GUPTA (Bishnupriya). Why did collusion fail? The Indian jute industry in the inter-war years. *Business history*, 2005, 47, 3, p. 532-552.

5824. HARDY-HÉMERY (Odette). Eternit et l'amiante, 1922-2000. Aux sources du profit, une industrie du risque. Villeneuve-d'Ascq, Presses universitaires du Septentrion, 2005, 272 p.

5825. HATANO (Isamu). Kindai-Nihon no gun-sangaku fukugotai. (Formation of the military-industrial-academic complex in science and technology in modern Japan: the interrelation of Japanese navy, heavy industry, and university). Tokyo, Sobunsha, 2005, 246 p.

5826. History (A) of corporate governance around the world. Family business groups to professional managers. Ed. by Randall K. MORCK. Chicago a. London, University of Chicago Press for National Bureau of economic research, 2005, XII-687 p.

5827. HOCHSTETTER (Dorothee). Motorisierung und "Volksgemeinschaft": das Nationalsozialistische Kraftfahrkorps (NSKK), 1931-1945. München, R. Oldenbourg, 2005, VIII-537 p. (ill.). (Studien zur Zeitgeschichte, 68).

5828. HYLDTOFT (Ole), JOHANSEN (Hans Christian). Dansk industris historie efter 1870. (Histoire de l'industrie danoise après 1870). Odense, Odense Universitets Forlag, 2005, 440 p.

5829. Industrie (L') du gaz en Europe aux XIXe et XXe siècles. L'innovation entre marchés privés et collectivités publiques. Sous la dir. de Serge PAQUIER et Jean-Pierre WILLIOT. Brxelles, P. Lang, 2005, 600 p. (Euroclio. Etudes et documents).

5830. ISENBERG (Andrew C.). Mining California. An ecological history. New York, Hill & Wang, 2005, 242 p.

5831. JACOBY (Sanford M.). The embedded corporation: corporate governance and employment relations in Japan and the United States. Princeton a. Oxford, Princeton U. P., 2005, IX-216 p.

5832. JOHNMAN (Lewis), MURPHY (Hugh). Scott Lithgow: Déjà Vu all over again! The rise and fall of a shipbuilding company. St. John's, International Maritime Economic History Association, 2005, XI-364 p.

5833. JONES (Geoffrey), MISKELL (Peter). European integration and corporate restructuring: the strategy of Unilever, c.1957-c.1990. *Economic history review*, 2005, 58, 1, p. 113-139.

5834. KOCKEL (Titus). Deutsche Ölpolitik 1928-1938. Berlin, Anadeuwe Verlag, 2005, 393 p. (Jahrbuch für Wirtschaftsgeschichte, 7).

5835. KRIGER (Colleen). Mapping the history of cotton textile production in precolonial West Africa. *African economic history*, 2005, 33, p. 87-116.

5836. KUČERA (Radek). Ke vzniku Veřejné bezpečnosti na železnici. (Establishing public security on the railways). *Sborník Archivu ministerstva vnitra*, 2005, 3, p. 9-40.

5837. KURLAEV (Evgenij A.), MAN'KOVA (Irina L.). Osvoenie rudnykh mestorozhdenij Urala i Sibiri v XVII veke: U istokov rossijskoj promyshlennoj politiki. (The development of ore deposits of Urals and Siberia in the 17th century: The beginnings of the Russian industrial policy). RAN, Ural'skoe otd., In-t istorii i arkheologii. Moskva, Drevlekhranilishche, 2005, 323 p. (ill.; bibl. incl.; ind. p. 300-319). [English summary]

5838. LAINS (Pedro). A indústria. *In*: História económica de Portugal, 1700-2000 [Cf. n° 5744], [s. p.].

5839. LINDNER (Stephan H.). Ein I. G. Farben Werk im Dritten Reich. München, Beck, 2005, XVIII-460 p.

5840. LUU (Lien Bich). Immigrants and the industries of London, 1500–1700. Aldershot, Ashgate, 2005, XIII-366 p.

5841. MARKEVICH (Andrej M.), SOKOLOV (Andrej K.). "Magnitka bliz Sadovogo kol'tsa": Stimuly k rabote na Moskovskom zavode "Serp i molot", 1883-2001 gg. (Incentives for labor in the Moscow plant 'Hammer and sickle', 1883-2001). RAN, In-t rossijskoj istorii, Mezhdunar. in-t sots. istorii (Amsterdam). Moskva, ROSSPEN, 2005, 367 p. (ill.; plates; bibl. incl.). (Sotsial'naja istorija Rossii XX veka: Motivatsija truda).

5842. Vacat.

5843. MARTY (Nicolas). Perrier c'est nous ! Histoire de la source Perrier et de son personnel. Paris, Les Éditions de l'Atelier, 2005, 254 p.

5844. MELANDER (Anders). Industry-wide belief structures and strategic behaviour. *Scandinavian economic history review*, 2005, 53, 1, p. 91-118.

5845. MÉNDEZ BELTRÁN (Luz María). La exportación minera en Chile, 1800–1840: un estudio de historia económica y social en la transición de la Colonia a la República. Santiago de Chile, Editorial Universitaria, 2005, 216 p.

5846. MILLWARD (Robert). Private and public enterprise in Europe: energy, telecommunications and transport, 1830–1990. Cambridge, Cambridge U. P., 2005, XIX-351 p.

5847. MORGAN (Chad). Planters' progress: modernizing confederate Georgia. Gainesville, University Press of Florida, 2005, XI-163 p. (New perspectives on the history of the South).

5848. NAVICKAS (Katrina). The search for 'General Ludd': the mythology of Luddism. *Social history*, 2005, 30, 3, p. 281-295.

5849. NORBERG (Arthur L.). Computers and commerce: a study of technology and management at Eckert-Mauchly Computer Company, Engineering Research Associates, and Remington Rand, 1946–1957. Cambridge, MIT Press, 2005, X-347 p.

5850. OPRIȘ (Petre). Industria românească de apărare înainte de înființarea organizației Tratatului de la Varșovia. (Romanian Defence Industry before the foundation of the organization of the Warsaw Treaty). *Anuarul Institutului de Istorie "George Barițiu"*. Academia Română, 2005, 44, p. 463-482.

5851. ORSI (Richard J.). Sunset limited. The Southern Pacific Railroad and the development of the American West, 1850–1930. Berkeley a. Los Angeles, University of California Press, 2005, XXII-615 p.

5852. Österreichische Industriegeschichte. Herausgegeben von Österreichische Industriegeschichte GmbH, Linz. Band 3. 1955 bis 2005: die ergriffene Chance. Unter Mitarbeit von Peter MOOSLECHNER und Christoph NEUMAYER. Wien, Ueberreuter, 2005, 343 p.

5853. OWEN (Thomas C.). Dilemmas of Russian capitalism: Fedor Chizhov and corporate enterprise in the railroad age. Cambridge, Harvard U. P., 2005, XIV-275 p.

5854. PFISTERER (Stephan). Maschinenbau im Ruhrgebiet. Stuttgart, Steiner, 2005, 372 p. (Beiträge zur Unternehmensgeschichte, 21).

5855. PINTO (Anthony), FONQUERNIE (Laurent), PRACA (Edwige), TOSTI (Jean), CABANAS (Nathalie), QUINTA (Françoise), QUINTA (Henri), VITOU (Luc). La fibre catalane. Industrie et textile en Roussillon au fil du temps. Actes du Colloque organisé le 15 novembre 2005 par l'APHPO en partenariat avec le CRHiSM. Perpignan, Trabucaire, 2005, 152 p. (História).

5856. PORTE (Rémy). La mobilisation industrielle: "premier front" de la Grande Guerre? Avant-propos de Jean-Jacques BECKER. Saint-Cloud, 14–18 éditions, 2005, 365 p.

5857. RAWE (Kai). "... wir werden sie schon zur Arbeit bringen!" Ausländerbeschäftigung und Zwangsarbeit im Ruhrkohlenbergbau während des Ersten Weltkrieges. Essen, Klartext, 2005, 284 p. (Veröffentlichungen des Instituts für Soziale Bewegungen R. C., 3).

5858. RICH (Jeremy). Forging permits and failing hopes: African participation in the Gabonese timber industry, ca. 1920–1940. *African economic history*, 2005, 33, p. 149-173.

5859. RICHTER (Amy G.). Home on the rails. Women, the railroad, and the rise of public domesticity. Chapel Hill, University of North Carolina Press, 2005, XIII-272 p. (Gender and American culture).

5860. ROCCHI (Fernando). Chimneys in the desert: industrialization in Argentina during the export boom years, 1870–1930. Stanford, Stanford U. P., 2005, XVIII-394 p.

5861. ROTH (Ralf). Das Jahrhundert der Eisenbahn. Die Herrschaft über Raum und Zeit, 1800–1914. Ostfildern, Thorbecke, 2005, 288 p.

5862. RÜBNER (Hartmut). Konzentration und Krise in der deutschen Schifffahrt. Maritime Wirtschaft und Politik im Kaiserreich, in der Weimarer Republik und im Nationalsozialismus. Bremen, Hauschild, 2005, 524 p. (Deutschemaritime Studien, 1).

5863. SABIN (Paul). Crude politics: the California oil market, 1900–1940. Berkeley a. Los Angeles, University of California Press, 2005, XX-307 p.

5864. SABOL (Miroslav). Elektrifikácia Slovenska v rokoch 1938–1943 a vznik celoštátneho elektrárenského podniku. (Electrification of Slovakia in the Years 1938–1943 and the origin of state electricity enterprise). *Historický časopis*, 2005, 53, 2, p. 327-338.

5865. SÁNCHEZ ROMÁN (José Antonio). La dulce crisis: estado, empresarios e industria azucarera en Tucumán, Argentina (1853–1914). Sevilla, Diputación de Sevilla, Universidad de Sevilla y Madrid, Consejo Superior de Investigaciones Científicas, Escuela de Estudios Hispano-Americanos, 2005, 383 p. (ill., maps). (Colección Americana, 23. Nuestra América, 18).

5866. SCHAAL (Dirk). Rübenzuckerindustrie und regionale Industrialisierung. Der Industrialisierungsprozess im mitteldeutschen Raum, 1799–1930. Münster, Lit, 2005, 283 p. (Forschungen zur neuesten Geschichte, 4).

5867. SCHNEIDER (Michael C.). Unternehmensstrategien zwischen Weltwirtschaftskrise und Kriegswirtschaft: Chemnitzer Maschinenbauindustrie in der NS-Zeit 1933–1945. Essen, Klartext, 2005, 543 p. (ill.). (Bochumer Schriften zur Unternehmens- und Industriegeschichte, 14).

5868. SEGAL (Howard P.). Recasting the machine age: Henry Ford's village industries. Amherst, University of Massachusetts Press, 2005, XV-244 p.

5869. ŠIMONČIĆ-BOBETKO (Zdenka). Industrija Hrvatske: 1918. do 1941. godine. (Industry of Croatia, 1918–1941). Ed. Mira KOLAR-DIMITRIJEVIĆ. Zagreb, AGM, 2005, 648 p.

5870. SPADAVECCHIA (Anna). Financing industrial districts in Italy, 1971–1991: a private venture? *Business history*, 2005, 47, 3, p. 569-593.

5871. STOIDE (Constantin A.). Manufacturile din Țara Bârsei între 1750 și 1850. (The manufactures from Țara Bârsei 1750–1850). Ediție, cuvânt înainte și biobibliografie de Ioan CAPROȘU. Cluj-Napoca, Editura Centrului de Studii Transilvane, 2005, 189 p.

5872. Storia della Pininfarina, 1930–2005: un'industria italiana nel mondo. A cura di Valerio CASTRONOVO. Roma e Bari, Laterza, 2005, X-636 p.

5873. STRAUMANN (Lukas). Nützliche Schädlinge: angewandte Entomologie, chemische Industrie und Landwirtschaftspolitik in der Schweiz 1874–1952. Zürich, Chronos, 2005, 392 p. (ill.). (Interferenzen, 9).

5874. THOMSON (J. K. J.). Explaining the 'take-off' of the Catalan cotton industry. *Economic history review*, 2005, 58, 4, p. 701-735.

5875. TOOZE (J. Adam). No room for miracles. German industrial output in World War II reassessed. *Geschichte und Gesellschaft*, 2005, 31, 3, p. 439-464.

5876. TREMBLAY (Victor J.), HORTON TREMBLAY (Carol). The U.S. brewing industry: data and analysis. Cambridge, MIT Press, 2005, XV-379 p.

5877. TRUJILLO BOLIO (Mario). El Golfo de México en la centuria decimonónica: entornos geográficos, formación portuaria y configuración marítima. México D.F., Centro de Investigaciones y Estudios Superiores en Antropología Social y Miguel Ángel Porrúa y Cámara de Diputados LIX Legislatura, 2005, 196 p. (Sociedades, historias, lenguajes).

5878. TURNER (Henry Ashby Jr.). General Motors and the Nazis. The struggle for control of Opel, Europe's biggest carmaker. New Haven, Yale U. P., 2005, VIII-200 p.

5879. Überlieferung (Die) der preußischen Bergverwaltung. Erfahrungen und Perspektiven zur Bearbeitung des sachthematischen Inventars der preußischen Berg-, Hütten und Salinenverwaltung, 1783–1865. Hrsg. v. Mechthild BLACK-VELDTRUP, Michael FARRENKOPF und Wilfried REININGHAUS. Bochum u. Münster, Selbstverlag des Deutschen Bergbau-Museums Bochum, 2005, 148 p. (Veröffentlichungen aus dem Deutschen Bergbau-Museum Bochum, 131. Schriften des Bergbau-Archivs, 17. Veröffentlichungen des Landesarchivs Nordrhein-Westfalen, 1).

5880. Understanding industrial and corporate change. Ed. by Giovanni DOSI, David J. TEECE and Josef CHYTRY. Oxford, Oxford U. P., 2005, XX-419 p.

5881. WEINBERG (Carl R.). Labor, loyalty, and rebellion: Southwestern Illinois coal miners and World War I. Carbondale, Southern Illinois U. P., 2005, XIII-246 p.

5882. ZELIN (Madeleine). The merchants of Zigong. Industrial entrepreneurship in early modern China. New York, Columbia U. P., 2005, XXIV-406 p. (Studies of the Weatherhead East Asian Institute).

5883. ZIEGLER (Dieter). Die Industrielle Revolution. Darmstadt, Wissenschaftliche Buchgesellschaft, 2005, VII-152 p.

§ 3. Addenda 1996–1998.

5884. "... das einzige Land in Europa, das eine grosse Zukunft vor sich hat": Deutsche Unternehmen und Unternehmer im Russischen Reich im 19. und frühen 20. Jahrhundert. Hrsg. v. Dittmar DAHLMANN und Carmen SCHEIDE. Essen, Klartext-Verlag, 98, [s. p.]. (Veröffentlichungen des Instituts für Kultur und Geschichte der Deutschen im östlichen Europa, 8).

5885. OSWALD (Anne von). Die deutsche Industrie auf dem italienischen Markt 1882 bis 1945: außenwirtschaftliche Strategien am Beispiel Mailands und Umgebung. Frankfurt am Main, P. Lang, 96, 340 p. (Italien in Geschichte und Gegenwart, 4).

Cf. n°s 4426, 7018, 7200

§ 4. Trade.

* 5886. HIGGINS (David M.). British business history: a review of the periodical literature for 2003. *Business history*, 2005, 47, 2, p. 159-173.

5887. ABILDGREN (Kim). Real effective exchange rates and purchasing-power-parity convergence: empirical evidence for Denmark, 1875–2002. *Scandinavian economic history review*, 2005, 53, 3, p. 58-70.

5888. AHUJA (Ravi). Die "Lenksamkeit" des "Lacars". Regulierungsszenarien eines transterritorialen Arbeitsmarktes in der ersten Hälfte des 20. Jahrhunderts. *Geschichte und Gesellschaft*, 2005, 31, 3, p. 323-353.

5889. AYGÜN (Necmettin). Onsekizinci yüzyılda Trabzon'da ticaret. (Commerce à Trébizonde au 17e siècle). Trabzon, Serander, 2005, XV-461 p.

5890. BATTISTINI (Francesco). La produzione, il commercio e i prezzi della seta grezza nello Stato di Firenze 1489–1859. *Rivista di storia economica*, 2005, 21, 3, p. 233-372.

5891. BUȘĂ (Daniela). Consecvență și concesii în politica comercială a României (sfârșitul secolului al XIX-lea–începutul secolului al XX-lea). (Consistency and concessions in Romania's trade policy, end of the 19th–beginning of the 20th century). *Studii și Materiale de Istorie Modernă. Academia Română*, 2005, 18, p. 5-24.

5892. CHAUVARD (Jean-François). La circulation des biens à Venise: Stratégies patrimoniales et marché immobilier (1600–1750). Roma, École française de Rome, 2005, X-629 p. (Bibliothèque des Écoles françaises d'Athènes et de Rome, 323).

5893. COLE (Shawn). Capitalism and freedom: manumissions and the slave market in Louisiana, 1725– 1820. *Journal of economic history*, 2005, 65, 4, p. 1008-1027.

5894. COOPEY (Richard), O'CONNELL (Sean), PORTER (Dilwyn). Mail order retailing in Britain. A business and social history. Oxford, Oxford U. P., 2005, X-248 p.

5895. CRUZ BARNEY (Oscar). El comercio exterior de México, 1821–1928: sistemas arancelarios y disposiciones aduanales. México D.F., Universidad Nacional Autónoma de México, Instituto de Investigaciones Jurídicas, 2005, XXII-205 p. (Serie Doctrina jurídica, 246).

5896. Cultural change and the market revolution in America, 1789–1860. Ed. by Scott C. MARTIN. Lanham, Rowman and Littlefield, 2005, VI-298 p.

5897. DAUDIN (Guillaume). Commerce et prospérité: la France au XVIIe siècle. Paris, Presses de l'Université Paris-Sorbonne, 2005, 611 p. (Collection Roland Mousnier, 19).

5898. DUPLESSIS (Robert S.). Market makers and market takers: a history of natural fibres textiles in the Central Apennine region (the Marche and Umbria). Bari, Mario Adda Editore, 2005, 172 p.

5899. EITRHEIM (Øyvind), ERLANDSEN (Solveig K.). House prices in Norway, 1819–1989. *Scandinavian economic history review*, 2005, 53, 3, p. 7-33.

5900. FAGOAGA HERNÁNDEZ (Ricardo A.), ESCOBAR OHMSTEDE (Antonio). Indígenas y comercio en la Huasteca (México): siglo XVIII. *Historia mexicana*, 2005, 55, 2, p. 333-417.

5901. FITZGERALD (Robert). Products, firms and consumption: Cadbury and the development of marketing, 1900–1939. *Business history*, 2005, 47, 3, p. 511-531.

5902. FRENCH (Michael). Commercials, careers, and culture: travelling salesmen in Britain, 1890s–1930s. *Economic history review*, 2005, 58, 2, p. 352-377.

5903. HONT (Istvan). Jealousy of trade: international competition and the nation-state in historical perspective. Cambridge, Harvard U. P. a. Belknap, 2005, VIII-541 p.

5904. JACKS (Davis S.). Immigrant stocks and trade flows, 1870–1913. *Journal of European economic history*, 2005, 34, 3, p. 625-652.

5905. JACKSON (Gordon). The British whaling trade: research in maritime history number 29. St. John's, International Maritime Economic History Association, 2005, XVI-293 p.

5906. KHAN (B. Zorina). The democratization of invention: patents and copyrights in American economic development, 1790–1920. New York, Cambridge U. P., 2005, XVII-322 p.

5907. KÖB (Edgar Helmut). Die Brahma-Brauerei und die Modernisierung des Getränkehandels in Rio de Janeiro 1888 bis 1930. Stuttgart, Franz Steiner Verlag, 2005, 239 p.

5908. KOTILAINE (Jamo T.). Russia's foreign trade and economic expansion in the seventeenth century. Windows on the world. Leiden a. Boston, Brill Academic Publishers, 2005, XVII-611 p.

5909. LANDSMAN (Mark). Dictatorship and demand. The politics of consumerism in East Germany. Cambridge a. London, Harvard U. P., 2005, XII-296 p. (Harvard historical studies, 147).

5910. LAZĂR (Gheorghe). Negustori mecena în Țara Românească (Secolul al XVIII-lea). [The merchants "Maecenas" in Wallachia (the 18th century)]. *Studii și materiale de istorie medie. Academia Română*, 2005, 23, p. 159-168.

5911. MAC MILLAN (Hugh). An African trading empire: the story of Susman Brothers & Wulfsohn, 1901–2005. New York, I. B.Tauris, 2005, IX-492 p.

5912. MENTZ (Soeren). The English gentleman merchant at work. Madras and the city of London, 1660– 1740. København, Museum Tusculanum Press, 2005, 304 p.

5913. METTELE (Gisela). Kommerz und fromme Demut. Wirtschaftsethik und Wirtschaftspraxis im "Gefühlspietismus". *Vierteljahresschrift für Sozial- und Wirtschaftsgeschichte*, 2005, 92, 3, p. 301-321.

5914. O'CONNELL (Sean), REID (Chris). Working-class consumer credit in the UK, 1925–1960: the role of the check trader. *Economic history review*, 2005, 58, 2, p. 378-405.

5915. PAPADIA (Elena). La Rinascente. Bologna, Il Mulino, 2005, 166 p.

5916. PIKE (R.). Partnership companies in sixteenth-century transatlantic trade: the De la Fuente family of Seville. *Journal of European economic history*, 2005, 34, 1, p. 215-244.

5917. PONS PONS (Jerònia). Large American corporations in the Spanish life insurance market 1880–

1922. *Journal of European economic history*, 2005, 34, 2, p. 467-482.

5918. POOLE (Anthony). A market town and its surrounding villages. Cranbrook, Kent in the later seventeenth century. Chichester, Phillimore, 2005, XX-220 p.

5919. ROSOLINO (Riccardo). Crimes contre le marché, crimes contre dieu. Le juste prix dans la Sicile du XVIIe siècle. *Annales*, 2005, 60, 6, p. 1245-1273.

5920. SEGERS (Yves). Economic growth and living standards. Private consumer expenditure and food consumption in nineteenth century Belgium. *Journal of European economic history*, 2005, 34, 2, p. 513-538.

5921. SPECK (Mary). Closed-door imperialism: the politics of Cuban-U.S. trade, 1902–1933. *Hispanic American historical review*, 2005, 85, 3, p. 449-484.

5922. STITZIEL (Judd). Fashioning socialism: clothing, politics, and consumer culture in East Germany. New York, Berg, 2005, XII-260 p.

5923. TAGLIACOZZO (Eric). Secret trades, porous borders: smuggling and states along a Southeast Asian frontier, 1865–1915. New Haven, Yale U. P., 2005, XIV-437 p.

5924. UZUN (Ahmet). Osmanlı Devleti'nde Şehir Ekonomisi ve İaşe. (Economie urbaine et approvisionnement dans l'Empire Ottoman). *Türkiye Araştırmaları Literatür Dergisi*, 2005, 3, 6, p. 161-210.

5925. WELCH (Evelyn). Shopping in the Renaissance. Consumer cultures in Italy 1400–1600. New Haven, Yale U. P., 2005, IX-403 p.

§ 4. Addendum 2004.

5926. MITSUZONO (Isamu). Senzen-ki nihon ni okeru geppu hambai no keisei to tenkai. (The formation and development of monthly installment sales in pre-1941 Japan). *Historical review on manners and customs*, 2004, 28, p. 27-47.

Cf. n°s 4339, 6677, 6744, 7200

§ 5. Agriculture and agricultural problems.

5927. ABAKA (Edmund). "Kola is god's gift". Agricultural production, export initiatives and the kola industry of Asante and the Gold Coast, c. 1820–1950. Athens, Ohio U. P., 2005, XV-173 p. (Western African studies).

5928. ALSTON (Lee J.), FERRIE (Joseph P.). Time on the ladder: career mobility in agriculture, 1890-1938. *Journal of economic history*, 2005, 65, 4, p. 1058-1081.

5929. BANZATO (Guillermo). La expansión de la frontera bonaerense: posesión y propiedad de la tierra en Chascomús, Ranchos y Monte, 1780–1880. Buenos Aires, Universidad Nacional de Quilmes Editorial, 2005, 224 p. (ill.). (Colección Convergencia. Entre memoria y sociedad).

5930. BĂRBULESCU (Constantin). The contribution of the sanitary legislation to the modernization of the rural world in Romania at the end of the 19th century and the beginning of the 20th century. *Anuarul Institutului de Istorie "A.D. Xenopol" al Academiei Române*, 2005, 42, p. 175-180.

5931. BONHOMME (Brian). Forests, peasants, and revolutionaries: forest conservation and organization in Soviet Russia, 1917–1929. New York, Columbia U. P., 2005, 252 p. (East European monographs, 654).

5932. BROWN (Jonathan). Farming in Lincolnshire, 1850–1945. Lincoln, History of Lincolnshire Committee, 2005, XII-295 p.

5933. BUCHELI (Marcelo). Bananas and business: the United Fruit Company in Colombia, 1899–2000. New York U. P., 2005, XI-241 p.

5934. CAMERON (Ewen A.). Communication or separation? Reactions to Irish land agitation and legislation in the Highlands of Scotland, c.1870–1910. *English historical review*, 2005, 120, 487, p. 633-666.

5935. CUNFER (Geoff). On the great plains: agriculture and environment. College Station, Texas A&M U. P., 2005, IX-292 p.

5936. DAL LAGO (Enrico). Agrarian elites. American slaveholders and southern Italian landowners, 1815–1861. Baton Rouge, Louisiana State U. P., 2005, XVIII-372 p.

5937. DEKEL-CHEN (Jonathan L.). Farming the red land: Jewish agricultural colonization and local Soviet power, 1924–1941. New Haven, Yale U. P., 2005, XVIII-366 p.

5938. DOOLITTLE (Amity A.). Property and politics in Sabah, Malaysia: native struggles over land rights. Seattle, University of Washington Press, 2005, X-224 p. (Culture, Place, and Nature).

5939. DRONIN (Nikolai M.), BELLINGER (Edward G.). Climate dependence and food problems in Russia 1900–1990: the interaction of climate and agricultural policy and their effect on food problems. Budapest a. New York, Central European U. P., 2005, XVII-366 p.

5940. FEDERICO (Giovanni). Feeding the world: an economic history of agriculture, 1800–2000. Princeton, Princeton U. P., 2005, XIV-388 p. – IDEM. Not guilty? Agriculture in the 1920s and the Great Depression. *Journal of economic history*, 2005, 65, 4, p. 949-976. – IDEM. Seta, agricoltura e sviluppo economico in Italia. *Rivista di storia economica*, 2005, 21, 2, p. 123-154.

5941. FOLLETT (Richard). The sugar masters. Planters and slaves in Louisiana's cane world, 1820-1860. Baton Rouge, Louisiana State U. P., 2005, VIII-290 p.

5942. GAVRILĂ (Irina). Baze de date istorice. Marea proprietate funciară potrivit matricolelor nominale ale locuitorilor împroprietăriti prin legea rurală din 1864. (Historical databases. The big land property, according to the number of inhabitants put in possession of land in 1864). București, Editura Oscar Print, 2005, 150 p.

5943. Handwörterbuch zur ländlichen Gesellschaft in Deutschland. Hrsg. v. Stephan BEETZ, Kai BRAUER und Claudia NEU. Wiesbaden, Verlag für Sozialwissenschaften, 2005, 258 p.

5944. HARWOOD (Jonathan). Technology's dilemma: agricultural colleges between science and practice in Germany, 1860–1934. Oxford, Lang, 2005, 288 p.

5945. HIGMAN (B. W.). Plantation Jamaica, 1750–1850. Capital and control in a colonial economy. Kingston, University of the West Indies Press, 2005, XIV-386 p.

5946. JECH (Karel). Die Repressionen gegen die Großbauernschaft während der Kollektivisierung der tschechoslowakischen Landwirtschaft. *In*: Sozialgeschichtliche Kommunismusforschung. Tschechoslowakei, Polen, Ungarn und DDR 1948–1968. Ed. Christiane BRENNER. München, 2005, p. 319-336.

5947. KALC (Aleksej). Vinogradništvo in trgovina z vinom na Tržaškem v 18. stoletju kot področje spora med "tradicionalnim" in "inovativnim". (Wine growing and selling in the Trieste Area in the 18th century as a sphere of conflict between "the traditional" and "the innovative"). *Annales, Series historia et sociologia*, 2005, 15, 2, p. 291-308.

5948. KARNEEV (Andrej N.), KOZYREV (Vitalij A.), PISAREV (Aleksandr A.). Vlast' i derevnja v respublikanskom Kitae, 1911–1949. (Village and the state power in republican China, 1911–1949). Ed. Arlet V. MELIKSETOV. Mosk. gos. un-t im. M.V. Lomonosova, In-t stran Azii i Afriki; Tamkan. un-t, In-t rus. issled. Moskva, Sovero-print, 2005, XXIX-532 p. (bibl. p. 479-504; pers. ind. p. 505-511). [English summary]

5949. KLUGE (Ulrike). Agrarwirtschaft und ländliche Gesellschaft im 20. Jahrhundert. München, Oldenbourg, 2005, XI-158 p. (Enzyklopädie deutscher Geschichte, 73).

5950. MANFREDINI (Matteo), BRESCHI (Marco). Coresident and non-coresident kin in a nineteenth-century Italian rural community. *Annales de démographie historique*, 2005, 109, p. 157-172.

5951. MEDUSHEVSKIJ (Andrej N.). Proekty agrarnyh reform v Rossii: XVIII nachalo XXI vcka. (Projects of agrarian reforms in Russia, the 18th–the early 21st century). RAN, In-t rossijskoj istorii. Moskva, Nauka, 2005, 639 p. (bibl. p. 616-630; pers. ind. p. 631-639). [English summary]

5952. MUKHERJEE (Mridula). Colonializing agriculture: the myth of Punjab exceptionalism. New delhi a. London, Sage Publications, 2005, XXVII-209 p. (Sage series in modern Indian history, 9).

5953. MULGAN (Aurelia George). Where tradition meets change: Japan's agricultural politics in transition. *Journal of Japanese studies*, 2005, 31, 2, p. 261-298.

5954. MUREŞAN (Florin Valeriu). Satul românesc din nord-estul Transilvaniei la mijlocul secolului al XVIII-lea. (The Romanian village in North-Eastern Transylvania in the mid 18th century). Cluj-Napoca, Editura Centrului de Studii Transilvane, 2005, 446 p.

5955. NADAU (Thierry). Itinéraires marchands du goût moderne: Produits alimentaires et modernisation rurale en France et en Allemagne (1870–1940). Textes posthumes rassemblés et édités par Marie-Emmanuelle CHESSEL et Sandrine KOTT; préface d'Albert A. BRODER; postface de Martin BRUEGEL. Paris, Editions de la Maison des sciences de l'homme, 2005, XXI-300 p.

5956. NATION (Richard F.). At home in the Hoosier Hills: agriculture, politics, and religion in Southern Indiana, 1810–1870. Bloomington, Indiana U. P., 2005, XI-274 p. (Midwestern history and culture).

5957. NISSEN (Mogens R.). Til fælles bedste. Det danske landbrug under besættelsen. (Pour l'intérêt commun. L'agriculture danoise pendant l'occupation). København, Lindhardt og Ringhof, 2005, 423 p. (ill.).

5958. NORDIN (Dennis S.), SCOTT (Roy V.). From prairie farmer to entrepreneur: the transformation of Midwestern agriculture. Bloomington, Indiana U. P., 2005, XVI-356 p.

5959. OTIMAN (Păun Ion). Viaţa rurală a României în secolul al XX-lea. (Rural life in Romania in the 20th century). Timişoara, Editura Mirton, 2005, 131 p.

5960. PETRANSKÝ (Ivan A.). Pôdohospodárske majetky katolíckej cirkvi a pozemková reforma na Slovensku v rokoch 1945–1950. (Die Bodenwirtschaftliche Eigentümer der katholische Kirche und die Bodenreform in der Slowakei in den Jahren 1945–1950). *Historický časopis*, 2005, 53, 3, p. 487-504. [Deutsche Zsfassung].

5961. PIERCE (Steven). Farmers and the state in colonial Kano: land tenure and the legal imagination. Bloomington, Indiana U. P., 2005, XII-262 p.

5962. SCHÖNE (Jens). Frühling auf dem Lande? Die Kollektivierung der DDR-Landwirtschaft. Berlin, Links Verlag, 2005, 332 p.

5963. SITTON (Thad), CONRAD (James H.). Freedom colonies. Independent Black Texans in the time of Jim Crow. Austin, University of Texas Press, 2005, 248 p. (Jack and Doris Smothers series in Texas history, life, and culture, 15).

5964. SMOUT (T. Christopher), MAC DONALD (Alan R.), WATSON (Fiona). A history of the native woodlands of Scotland, 1500–1920. Edinburgh, Edinburgh U. P., 2005, XIII-434 p.

5965. SOLURI (John). Banana cultures. Agriculture, consumption, and environmental change in Honduras and the United States. Austin, University of Texas Press, 2005, XIII-321 p.

5966. SUMMERHILL (Thomas). Harvest of dissent: agrarianism in nineteenth-century New York. Urbana a. Chicago, University of Illinois Press, 2005, XI-287 p.

5967. Swing unmasked. The agricultural riots of 1830 to 1832 and their wider implications. Ed. by Michael

HOLLAND. Milton Keynes, FACHRS publications, 2005, IX-312 p.

5968. Țărănimea și puterea. Procesul de colectivizare a agriculturii în România (1949–1962). (Peasantry and the power. Agriculture collectivization process in Romania, 1949–1962). Editori Dorin DOBRINCU și Constantin IORDACHI, cuvânt înainte de Gail KLIGMAN și Katherine VERDERY. Iași, Editura Polirom, 2005, 502 p.

5969. TIKHONOV (Ju.A.). Dvorjanskaja usad'ba i krest'janskij dvor v Rossii XVII i XVIII vekov: sosushchestvovanie i protivostojanie. (Noble's manorial house and peasant's court in 17th and 18th century Russia: Coexistence and confrontation). Moskva, Letnij sad, 2005, 447-XVI p. (ill.; portr.; plates; ind. p. 437-447).

5970. TRAȘCĂ (Ottmar). Aspecte privind primii ani ai colectivizării agriculturii în România, 1949–1952. Studiu de caz: regiunea Cluj. (Aspects during the first years of the Collectivization of Agriculture in Romania, 1949–1952. Study case: The Region Cluj). *Anuarul Institutului de Istorie "George Barițiu". Academia Română*, 2005, 44, p. 385-406.

5971. VALENCIA (Marta). Tierras públicas, tierras privadas: Buenos Aires, 1852–1876. Buenos Aires, Editorial de la Universidad Nacional de La Plata, Asociación Amigos del Archivo Histórico de la Provincia de Buenos Aires, Instituto Cultural, Gobierno de la Provincia de Buenos Aires, 2005, 358 p. (ill., maps). (Colección sociales. Serie Historia argentina, 2).

5972. VIVES RIERA (Antoni). La resistencia de la Mallorca rural al proceso de modernización durante II República y el primer franquismo. *Historia social*, 2005, 52, p. 73-88.

5973. VOLANTO (Keith J.). Texas, cotton, and the New Deal. College Station, Texas A&M U. P., 2005, XV-194 p. (Sam Rayburn series on rural life, 7).

5974. WILSON (Jon E.). 'A Thousand Countries To Go To': peasants and rulers in late eighteenth-century Bengal. *Past and present*, 2005, 189, p. 81-109.

5975. Zemědělské družstevnictví. Kolektivizace zemědělství. Podmínky pro vznik JZD – 1947. (Agricultural cooperatives. The collectivization of agriculture. Conditions for the establishment of the Standard Farming Co-operative [JZD]). Ed. Jana PŠENIČKOVÁ. Praha, Národní archiv, 2005, 382 p. (Edice dokumentů z fondů Národního archivu).

5976. ZIMMERMAN (Andrew). A German Alabama in Africa: the Tuskegee expedition to German Togo and the transnational origins of West African cotton growers. *American historical review*, 2005, 110, 5, p. 1362-1398.

Cf. n°s 3593, 4461

§ 6. Money and finance.

5977. ARNOLDUS (Doreen), DANKERS (Joost). Management consultancies in the Dutch banking sector, 1960s and 1970s. *Business history*, 2005, 47, 3, p. 553-568.

5978. BÉGUIN (Katia). La circulation des rentes constituées dans la France du XVIIe siècle. Une approche de l'incertitude économique. *Annales*, 2005, 60, 6, p. 1229-1244.

5979. BLAUFARB (Rafe). Vers une histoire de l'exemption fiscale nobiliaire. La Provence des années 1530 à 1789. *Annales*, 2005, 60, 6, p. 1203-1228.

5980. BOEHME (Olivier). The involvement of the Belgian Central Bank in the Katanga secession, 1960–1963. *African economic history*, 2005, 33, p. 1-29.

5981. BOLDIZZONI (Francesco). Finanza pubblica nel Regno di Napoli fra Cinquecento e Seicento. *Rivista storica italiana*, 2005, 117, 3, p. 1037-1049. – IDEM. La rivoluzione dei prezzi rivisitata: moneta ed economia reale in Alta Italia. *Rivista storica italiana*, 2005, 117, 3, p. 1002-1036.

5982. BÖLÜKBAŞI (Ö.Faruk). Tezyid-i varidat ve tenkih-i masarifat: II.Abdülhamid döneminde mali idare. (Augmentation des revenus et abolition des dépences: administration financière au temps de Abdülhamid II). İstanbul, Osmanlı Bankası Arşiv ve Araştırma Merkezi, 2005, 151 p.

5983. BRION (René), MOREAU (Jean-Louis). La politique monétaire belge dans une Europe en reconstruction (1944–1958). [S. l.], Banque nationale de Belgique, 2005, 550 p. (Nationale Bank van België 1939–1971, 2).

5984. BURHOP (Carsten). Die Vergütung des Führungspersonals deutscher Großbanken, 1871–1913. *Vierteljahresschrift für Sozial- und Wirtschaftsgeschichte*, 2005, 92, 3, p. 281-300.

5985. CAILLOU (François). Une administration royale d'Ancien Régime. Le bureau des finances de Tours. Tours, Presses universitaires François Rabelais, 2005, 2 vol., 496 p., 436 p. (Travaux historiques, 3).

5986. CARNEVALI (Francesca). Europe's advantage: banks and small firms in Britain, France, Germany, and Italy since 1918. Oxford, Oxford U. P., 2005, XI-228 p.

5987. CASSIERS (Isabelle), LEDENT (Philippe). Politique monétaire et croissance économique en Belgique à l'ère de Bretton Woods (1944–1971). [S. l.], Banque nationale de Belgique, 2005, 224 p. (ill.). (Nationale Bank van België 1939–1971, 4).

5988. CLERICI (Luca). Pagamenti in natura, velocità di circolazione della moneta e rivoluzione dei prezzi. *Rivista di storia economica*, 2005, 21, 2, p. 155-180.

5989. COSGEL (Metin M.), MICELI (Thomas J.). Risk, transaction costs, and tax assignment: government finance in the Ottoman Empire. *Journal of economic history*, 2005, 65, 3, p. 806-821.

5990. DALE (Richard S.), JOHNSON (Johnnie E. V.), TANG (Leilei). Financial markets can go mad: evidence of irrational behaviour during the South Sea Bubble. *Economic history review*, 2005, 58, 2, p. 233-271.

5991. DAROVEC (Darko). Fiscal policy in Venetian Istria in modern age. *Megatrend review*, 2005, 2, 2, p. 63-85.

5992. DENZEL (M. A.), GERHARD (H. J.). Global and local aspects of pre-industrial inflations: new research on inflationary processes in XVIIIth-century Central Europe. *Journal of European economic history*, 2005, 34, 1, p. 149-186.

5993. DOĞRUEL (Fatma). Türkiye'de enflasyonun tarihi. (Histoire de l'inflation en Turquie). İstanbul, Türkiye Cumhuriyet Merkez Bankası; Tarih Vakfı, 2005, XIII-226 p.

5994. DOHNA (Jesko Zu). Die "jüdischen Konten" der Fürstlich Castell'schen Credit-Cassen und des Bankhauses Karl Meyer KG. Neustadt an der Aisch, Degener, 2005, 144 p. (Veröffentlichungen der Gesellschaft für fränkische Geschichte, Neujahrsblätter, 45).

5995. DUDA (Igor). Tehnika narodu! Trajna dobra, potrošnja i slobodno vrijeme u socijalističkoj Hrvatskoj. (Technology to the people! Durables, consumption and leisure in Socialist Croatia). *Časopis za suvremenu povijest*, 2005, 37, 2, p. 371-392.

5996. FISHBACK (Price V.), HORRACE (William C.), KANTOR (Shawn). Did New Deal grant programs stimulate local economies? A study of Federal grants and retail sales during the Great Depression. *Journal of economic history*, 2005, 65, 1, p. 36-71.

5997. FLANDREAU (Marc), JOBST (Clemens). The ties that divide: a network analysis of the International Monetary System, 1890–1910. *Journal of economic history*, 2005, 65, 4, p. 977-1007.

5998. GANTMAN (Ernesto R.). Capitalism, social privilege and managerial ideologies. Aldershot, Ashgate, 2005, VIII-185 p.

5999. GIEDEMAN (Daniel C.). Branch banking restrictions and finance constraints in early-twentieth-century America. *Journal of economic history*, 2005, 65, 1, p. 129-151.

6000. HALLON (Ľudovít). Úloha Milana Hodžu v komerčnom bankovníctve Slovenska v rokoch 1918–1938. (Milan Hodža and his role in the Slovak Commercial Banking Sector from 1918 until 1938). *Historický časopis*, 2005, 53, 1, p. 57-70. [Deutsche Zsfassung].

6001. HANLEY (Anne G.). Native capital. Financial institutions and economic development in Sao Paulo, Brazil, 1850–1920. Stanford, Stanford U. P., 2005, XVIII-286 p. (Social science history).

6002. HICKSON (Charles R.), TURNER (John D.). The genesis of corporate governance: nineteenth-century Irish joint-stock banks. *Business history*, 2005, 47, 2, p. 174-189.

6003. HICKSON (Kevin). The IMF [International Monetary Fund] crisis of 1976 and British politics. London, New York, I. B. Tauris, 2005, XI-259 p. (International library of political studies, 3).

6004. Historia monetaria contemporánea de Bolivia: siete momentos capitales en los 77 años de historia del Banco Central de Bolivia. La Paz, Departamento de Comunicación Institucional del Banco Central de Bolivia, 2005, 333 p. (ill.).

6005. HREINSSON (Einar). 'Noblesse de robe' in a classless society. The making of an Icelandic elite in the age of absolutism. *Scandinavian journal of history*, 2005, 30, 3-4, p. 225-237.

6006. HUYBENS (Elisabeth), LUCE JORDAN (Astrid), PRATAP (Sangeeta). Financial market discipline in early-twentieth-century Mexico. *Journal of economic history*, 2005, 65, 3, p. 757-778.

6007. Impôt (L') des campagnes. Fragile fondement de l'État dit moderne (XVe–XVIIIe siècle), colloque tenu à Bercy les 2 et 3 décembre 2002. Sous la dir. Antoine FOLLAIN et Gilbert LARGUIER. Paris, Comité pour l'histoire économique et financière de la France, 2005, 660 p.

6008. JONES (Geoffrey). Multinationals and global capitalism. From the nineteenth to the twenty-first century. Oxford, Oxford U. P., 2005, XL-340 p.

6009. KIM (Namsuk), WALLIS (John Joseph). The market for American state government bonds in Britain and the United States, 1830–1843. *Economic history review*, 2005, 58, 4, p. 736-764.

6010. KÖHLER (Ingo). Die "Arisierung" der Privatbanken im Dritten Reich: Verdrängung, Ausschaltung und die Frage der Wiedergutmachung. München, Verlag C. H. Beck, 2005, 602 p.

6011. KREUTZMÜLLER (Christoph). Händler und Handlungsgehilfen. Der Finanzplatz Amsterdam und die deutschen Großbanken, 1918–1945. Stuttgart, Steiner, 2005, 349 p.

6012. KUČERA (Jaroslav). Der zögerliche Expansionist. Die Commerzbank in den böhmischen Ländern 1938–1945. *Bankhistorisches Archiv*, 2005, 31, 1, p. 33-56.

6013. LAZAREVIĆ (Žarko), PRINČIČ (Jože). Bančniki v ogledalu časa: življenjske poti slovenskih bančnikov v 19. in 20. stoletju. (Bankers in the mirror of time: lives of Slovene bankers in the 19th and 20th centuries). Ljubljana, ZBS, Združenje bank Slovenije, 2005, 181 p.

6014. London and Paris as international financial centres in the twentieth century. Ed. by Youssef CASSIS et Éric BUSSIÈRE. Oxford, Oxford U. P., 2005, 367 p.

6015. MAC GOWEN (Randall). The Bank of England and the policing of forgery 1797–1821. *Past and present*, 2005, 186, p. 81-116.

6016. MAISSEN (Thomas). Verweigerte Erinnerung. Nachrichtenlose Vermögen und Schweizer Weltkriegsdebatte, 1989–2004. Zürich, Verlag Neue Zürcher Zeitung, 2005, 729 p.

6017. MATIS (Herbert). Die Schwarzenberg-Bank: Kapitalbildung und Industriefinanzierung in den habsburgischen Erblanden, 1787–1830. Wien, Verlag der Österreichischen Akademie der Wissenschaften, 2005, 452 p. (ill.). (Sitzungsberichte / Österreichische Akademie der Wissenschaften. Philosophisch-Historische Klasse, 731).

6018. Mesurer la monnaie. Banques centrales et construction de l'autorité monétaire (XIXe–XXe siècle). Sous la dir. de Olivier FEIERTAG. Paris, Albin Michel, 2005, 284 p.

6019. MITCHENER (Kris James). Bank supervision, regulation, and instability during the Great Depression. *Journal of economic history*, 2005, 65, 1, p. 152-185.

6020. MONTAGNE (Sabine). Pouvoir financier "vs" pouvoir salarial. Les fonds de pension américains: contribution du droit à la légitimité financière. *Annales*, 2005, 60, 6, p. 1299-1325.

6021. MOOIJ (Joke). Corporate culture of central banks: lessons from the past. *Journal of European economic history*, 2005, 34, 1, p. 11-42.

6022. OBSTFELD (Maurice), TAYLOR (Alan M.). Global capital markets. Integration, crisis and growth. Cambridge, Cambridge U. P., 2005, XVIII-350 p.

6023. Penuria sin fin: Historia de los impuestos en México, siglos XVIII–XX. Ed. por Luis ABOITES AGUILAR y Luis JÁUREGUI. México, Instituto de Investigaciones Dr. José María Luis Mora, 2005, 310 p. (Historia económica).

6024. PETRÁŠ (Jiří). Peněžní reforma 1953. (The financial reform of 1953). *Sborník Archivu ministerstva vnitra*, 2005, 3, p. 141-171.

6025. POLLARD (John F.). Money and the rise of the modern papacy: financing the Vatican, 1850–1950. Cambridge, Cambridge U. P., 2005, XX-265 p.

6026. TONIOLO (Gianni). Central bank cooperation at the bank for international settlements, 1930–1973. With the assistance of Piet CLEMENT. New York, Cambridge U. P., 2005, XXII-729 p.

6027. TORRES SÀNCHEZ (Rafael). The failure of the Spanish crown's fiscal monopoly over tobacco in Catalonia during the XVIIIth Century. *Journal of European economic history*, 2005, 34, 3, p. 721-760.

6028. TRISTRAM (Frédéric). Une fiscalité pour la croissance: la direction générale des impôts et la politique fiscale en France de 1948 à la fin des années 1960. Paris, Ministère de l'economie, des finances et de l'industrie, 2005, IX-740 p.

6029. ULLMANN (Hans-Peter). Der deutsche Steuerstaat: Geschichte der öffentlichen Finanzen vom 18. Jahrhundert bis heute. München, Beck, 2005, 269 p. (Beck'sche Reihe, 1616).

6030. VAN DER WEE (Herman), VERBREYT (Monique). Oorlog en monetaire politiek: de Nationale Bank van België, de Emissiebank te Brussel en de Belgische regering, 1939–1945. [S. l.], Nationale Bank van België, 2005, 760 p. (ill.). (Nationale Bank van België 1939–1971, boekdeel 1).

6031. VANOLI (André). A history of national accounting. Amsterdam, Ios Press, 2005, XX-522 p.

6032. WILKINS (Mira). The history of foreign investment in the United States, 1914–1945. Cambridge, Harvard U. P., 2005, XXVI-980 p. (Harvard studies in business history, 43).

6033. WOOD (John H.). A history of Central Banking in Great Britain and the United States. Cambridge, Cambridge U. P., 2005, XV-439 p.

6034. WRIGHT (Robert E.). The first Wall Street: Chestnut Street, Philadelphia, and the birth of american finance. Chicago, University of Chicago Press, 2005, VII-210 p.

6035. YOUNG-IOB (Chung). Korea under siege, 1876–1945: capital formation and economic transformation. New York, Oxford U. P., 2005, X-390 p.

Cf. n°s 144, 4339, 4364, 5576, 6990

§ 7. Demography and urban history.

* 6036. İNALCIK (Halil), ARI (Bülent). Türk-İslam-Osmanlı Şehirciliği ve Halil İnalcık'ın Çalışmaları. (Urbanisme turco-ottomano-musulman et les recherches de Halil İnalcık). *Türkiye Araştırmaları Literatür Dergisi*, 2005, 3, 6, p. 9-26.

* 6037. ÖZ (Mehmet). Osmanlı Klasik Döneminde Anadolu Kentleri. (Villes anatoliennes à l'époque classique ottomane). *Türkiye Araştırmaları Literatür Dergisi*, 2005, 3, 6, p. 27-56.

** 6038. EBEL (Kathryn A.). Osmanlı Şehir Tarihinin Görsel Kaynakları. (Sources visuelles de l'histoire de ville Ottomane). *Türkiye Araştırmaları Literatür Dergisi*, 2005, 3, 6, p. 457-486.

6039. ABOY (Rosa). Viviendas para el pueblo: Espacio urbano y sociabilidad en el barrio Los Perales, 1946–1955. Buenos Aires, Fondo de Cultura Económica de Argentina y Universidad de San Andrés, 2005, 195 p. (Sección de Obras Historia).

6040. AHLBERG (Nils). Stadsgrundningar och planförändringar: svensk stadsplanering 1521–1721. (Fondations de villes et modifications de plans urbains: la planification urbaine en Suède 1521–1721). Uppsala, Swedish University of Agricultural Sciences, 2005, 860 p. (Acta Universitatis agriculturae Sueciae, 2005:94). [English summary].

6041. ARLI (Alim). Cumhuriyet Döneminde Türkiye'de Şehirleşme ve Gecekondu Araştırmaları. (Recherches sur l'urbanisme et les bidonvilles en Turquie à l'époque de la République). *Türkiye Araştırmaları Literatür Dergisi*, 2005, 3, 6, p. 257-281.

6042. ARMAĞAN (A. Latif). XVI. Yüzyılda Antalya. (Antalya in the XVIth century). *Tarih Araştırmaları Dergisi (Ankara Üniversitesi Dil Ve Tarih Coğrafya Fakültesi)*, 2005, 24, 38, p. 93-112.

6043. ATTANÉ (Isabelle). Une Chine sans femmes? Paris, Perrin 2005, 391 p.

6044. AYDIN (Suavi) [et al.]. Küçük Asya'nın bin yüzü: Ankara. (Mille faces de l'Asie Mineure: Ankara). Ankara, Dost, 2005, 670 p.

6045. BALLENT (Anahí). Las huellas de la política: vivienda, ciudad, peronismo en Buenos Aires, 1943–1955. Buenos Aires, Universidad Nacional de Quilmes, 2005, 273 p. (Las ciudades y las ideas).

6046. BARCELLA (Paolo). Religione e città: Silvio D'Amico in viaggio tra New York e Chicago. *Storia urbana*, 2005, 28, 109, 4, p. 5-18.

6047. BECKETT (John). City status in the British Isles, 1830–2002. Aldershot, Ashgate, 2005, X-202 p.

6048. BENTI (Getahun). The dynamics of migration to Addis Ababa (Ethiopia) and the overurbanization of the city, c.1941–c.1974. Ann Arbor, UMI Dissertation Services, 2005, [s. p.]. (maps).

6049. BETKER (Frank). "Einsicht in die Notwendigkeit". Kommunale Stadtplanung in der DDR und nach der Wende, 1945–1994. Stuttgart, Steiner, 2005, 412 p. (Beiträge zur Stadtgeschichte und Urbanitätsforschung, 3).

6050. BOUDJAABA (Fabrice). Parenté, alliance et marché dans la France rurale traditionnelle. Essai d'application de l'analyse de réseaux au marché foncier et immobilier de Saint-Marcel (Normandie) 1760–1824. *Annales de démographie historique*, 2005, 109, p. 33-59.

6051. BRUEGMANN (Robert). Sprawl: a compact history. Chicago, University of Chicago Press, 2005, 301 p.

6052. BRUNBORG (Helge), TABEAU (Ewa). Demography of conflict and violence: an emerging field. *European journal of population*, 2005, 21, 2-3, p. 131-144.

6053. BUCKLER (Julie A.). Mapping St. Petersburg. Imperial text and cityscape. Princeton, Princeton U. P., 2005, 364 p.

6054. BURTON (Andrew). African underclass: urbanisation, crime and colonial order in Dar es Salaam, 1919–1961. Oxford, British Institute in Eastern Africa in association with James Currey, 2005, 320 p. (ill., maps). (Eastern African studies).

6055. CACCIA (Patrizia). America e Italia: mode e modelli urbani degli anni Venti. *Storia urbana*, 2005, 28, 109, 4, p. 19-34.

6056. ČAPKA (František), SLEZÁK (Lubomír), VACULÍK (Jaroslav). Nové osídlení pohraničí českých zemí po druhé světové válce. (Die neue Besiedlung des Grenzgebietes der tschechischen Länder nach dem Zweiten Weltkrieg). Brno, Cerm a Akademické nakl., 2005, 359 p. (ill., tb., mp.).

6057. CARVAIS (Robert). L'apocalisse urbana. La Parigi di ancien régime attraverso i verbali della police du bâtiment. *Storia urbana*, 2005, 28, 108, 3, p. 87-104.

6058. CATHERIN-QUIVET (Agnès). Évolution de la population âgée en institution et politiques mises en œuvre (1962–2004). *Annales de démographie historique*, 2005, 110, p. 185-219.

6059. CHEN (Shuang), CAMPBELL (Cameron), LEE (James). Vulnerability and resettlement: mortality differences in northeast China by place of origin, 1870-1912, comparing urban and rural migrants. *Annales de démographie historique*, 2005, 110, p. 47-80.

6060. Cidades e espaços urbanos. Organização de Magda PINHEIRO e Frédéric VIDAL. *Ler historia*, 2005, 48, p. 1-226.

6061. COLLINS (Christiane Crasemann). Werner Hegemann and the search for universal urbanism. New York, W. W. Norton, 2005, 417 p.

6062. CONFORTI (Claudia). La città del tardo Rinascimento. Roma e Bari, Laterza, 2005, VII-160 p. (ill., maps). (Storia della città, 7).

6063. DENNIS (Richard). La regolamentazione degli edifici per appartamenti a Toronto e Londra: una preistoria. *Storia urbana*, 2005, 28, 108, 3, p. 15-38.

6064. DHAR CHAKRABARTI (Prabodh G.). Le due città. Il rapporto tra città formale e città informale a Delhi, 1960–2000. *Storia urbana*, 2005, 28, 108, 3, p. 105-122.

6065. European cities, youth and the public sphere in the twentieth century. Ed. by Axel SCHILDT and Detlef SIEGFRIED. Aldershot a. Burlington, Ashgate, 2005, XII-162 p. (Historical urban studies).

6066. FALK (Barbara). Sowjetische Städte in der Hungersnot 1932/33. Staatliche Ernährungspolitik und städtisches Alltagsleben. Köln, Weimar u. Wien, Böhlau, 2005, VIII-445 p. (Beiträge zur Geschichte Osteuropas, 38).

6067. FAWAZ (Mona). La costruzione di un sobborgo informale: imprenditori e stato a Hayy el Sellom, Beirut, 1950-2000. *Storia urbana*, 2005, 28, 108, 3, p. 123-138.

6068. FOGELSON (Robert M.). Bourgeois nightmares: suburbia, 1870–1930. New Haven, Yale U. P., 2005, 264 p.

6069. FOOT (John). Dentro la città irregolare. Una rivisitazione delle coree milanesi, 1950–2000. *Storia urbana*, 2005, 28, 108, 3, p. 139-156-

6070. FÜLBERTH (Andreas). Tallinn – Riga – Kaunas. Ihr Ausbau zu modernen Hauptstädten 1920–1940. Köln, Weimar u. Wien, Böhlau, 2005, 395 p.

6071. GARB (Margaret). City of American dreams: a history of home ownership and housing reform in Chicago, 1871–1919. Chicago, University of Chicago Press, 2005, XV-261 p.

6072. GONZALEZ (Evelyn). The Bronx. New York, Columbia U. P., 2005, XII-263 p. (Columbia history of urban life).

6073. GRANGE (Cyril). Les réseaux matrimoniaux intra-confessionnels de la haute bourgeoisie juive à Paris à la fin du XIXe siècle. *Annales de démographie historique*, 2005, 109, p. 131-156.

6074. GREEN (Alan), MAC KINNON (Mary), MINNS (Chris). Conspicuous by their absence: French Canadians and the settlement of the Canadian west. *Journal of economic history*, 2005, 65, 3, p. 822-849.

6075. HACKER (J. David), HAINES (Michael R.). American Indian mortality in the late nineteenth century: the impact of Federal assimilation policies on a vulnerable population. *Annales de démographie historique*, 2005, 110, p. 17-29.

6076. HANNA (Nelly). Survey of urban history of Arab cities in the Ottoman period. *Türkiye Araştırmaları Literatür Dergisi*, 2005, 3, 6, p. 57-88.

6077. HANSSEN (Jens). Fin de siècle Beirut. The making of an Ottoman capital. New York, Clarendon Press of Oxford U. P., 2005, VIII-307 p.

6078. HAUTANIEMI LEONARD (Susan), GUTMANN (Myron P.). Isolated elderly in the U.S. great plains. The roles of environment and demography in creating a vulnerable population. *Annales de démographie historique*, 2005, 110, p. 81-108.

6079. HENZ (Ursula), THOMSON (Elizabeth). Union stability and stepfamily fertility in Austria, Finland, France and West Germany. *European journal of population*, 2005, 21, 1, p. 3-29.

6080. HJORTH (Birte). Dragør i 1700-tallet. Et maritimt bysamfund. (Dragør au 18e siècle. Une société urbaine maritime). Dragør, Dragør lokalarkive, 2005, 294 p. (ill., cartes). (Dansk Center for Byhistorie).

6081. Inszenierter Stolz. Stadtrepräsentationen in drei deutschen Gesellschaften 1935–1975. Hrsg. v. Adelheid von SALDERN; unter Mitarbeit von Lu SEEGERS. Stuttgart, Steiner, 2005, 498 p. (Beiträge zur Stadtgeschichte und Urbanisierungsforschung, 2).

6082. ISENBERG (Alison). Downtown America: a history of the place and the people who made it. Chicago, University of Chicago Press, 2005, XVIII-441 p. (Historical studies of Urban America).

6083. JAGER (Markus). Der Berliner Lustgarten. Gartenkunst und Stadtgestalt in Preußens Mitte. München u. Berlin, Deutscher Kunstverlag, 2005, 365 p. (Kunstwissenschaftliche Studien, 120).

6084. KINSBRUNER (Jay). The colonial Spanish-American city: urban life in the age of Atlantic capitalism. Austin, University of Texas Press, 2005, XIII-182 p.

6085. Klassiske købstad (Den). (La ville à l'époque classique). Red. Søren Bitsch CHRISTENSEN. Aarhus, Aarhus Universitetsforlag, 2005, 417 p. (Danske bystudier, 2).

6086. KOSTELECKÝ (Tomáš). Political behavior in metropolitan areas in the Czech Republic between 1990 and 2002. Patterns, trends, and the relation to suburbanization and its socio-spatial patterns. Praha, Sociologický ústav AV ČR, 2005, 104 p. (tb., graf.). (Sociological Studies, 05/02).

6087. LEMERCIER (Claire). Analyse de réseaux et histoire de la famille: une rencontre encore à venir? *Annales de démographie historique*, 2005, 109, p. 7-31.

6088. Lima en el siglo XVI. Directora de investigación Laura GUTIÉRREZ ARBULÚ. Lima, Pontificia Universidad Católico del Perú e Instituto Riva-Agüero, 2005, 863 p. (Publicación del Instituto Riva-Agüero, 225).

6089. LOHSE (York). Mexiko-Stadt im 18. Jahrhundert: das Bild einer kolonialen Metropole aus zeitgenössischer Perspektive. Frankfurt am Main u. Oxford, P. Lang, 2005, 446 p. (ill., maps). (Europäische Hochschulschriften. Reihe III, Geschichte und ihre Hilfswissenschaften, 1016 = Publications universitaires européennes. Série III, Histoire, sciences auxiliaires de l'histoire, 1016 = European university studies. Series III, History and allied studies, 1016).

6090. LOUSADA (Maria Alexandre). Una nuova grammatica per lo spazio urbano: la polizia e la città a Lisbona, 1760–1833. *Storia urbana*, 2005, 28, 108, 3, p. 67-86.

6091. LUCONI (Stefano). La città della Grande depressione: Manlio Morgagni a New York, 1932. *Storia urbana*, 2005, 28, 109, 4, p. 35-50.

6092. MARTIN GARCIA (Alfredo). Demografía y comportamientos demográficos en la Galicia moderna, la villa de Ferrol y su tierra, siglos XVI–XIX. León, Universidad de León, Secretariado de publicaciones, 2005, 356 p.

6093. Město a voda. Praha, město u vody. Sborník příspěvků z 22. vědecké konference. (Town and water. Prague, the town near water). Ed. Olga FEJTOVÁ, Václav LEDVINKA, Jiří PEŠEK. Praha, Scriptorium, 2005, 577 p. (Documenta Pragensia, 24).

6094. MIER Y TERÁN ROCHA (Lucía). La primera traza de la Ciudad de México, 1524–1535. México D.F., Universidad Autónoma Metropolitana, Fondo de Cultura Económica, 2005, 2 vol., [s. p.]. (ill., maps, plans). (Sección de obras de historia).

6095. Monde caraïbe (Le). Défis et dynamiques. 1. Visions identitaires, diasporas, configurations culturelles. Actes du colloque international, bordeaux, 2–7 juin 2003. Sous la dir. de Christian LERAT. Pessac, Maison des sciences del'homme d'Aquitaine, 2005, 579 p.

6096. MÜLLER (Marco). Die Sozial- und Wirtschaftsgeschichte der Stadt Rastatt: 1815–1890. Heidelberg, Ubstadt-Weiher u. Basel, Verl. Regionalkultur, 2005, 464 p. (ill.). (Stadtgeschichtliche Reihe / Stadt Rastatt, 8).

6097. MÜLLER (Peter). Symbolsuche: die Ost-Berliner Zentrumsplanung zwischen Repräsentation und Agitation. Berlin, Gebr. Mann, 2005, 344 p. (ill., maps, plans). (Berliner Schriften zur Kunst, 19).

6098. Municipalités mediterranéennes: les réformes urbaines ottomanes au miroir d'une histoire compareée (Moyen-Orient, Maghreb, Europe méridionale). Sous la dir. de Nora LAFI. Berlin, Klaus Schwarz Verlag, 2005, 373 p. (Zentrum Moderner Orient, 21).

6099. Naselenie Rossii v XX veke: Istoricheskie ocherki. (The population of Russia in the 20th century: Historical Essays). Vol. 3. Book 1. 1960–1979 gg. Ed. Jurij A. POLJAKOV, V.B. ZHIROMSKAJA, V.A. ISUPOV. Moskva, ROSSPEN, 2005, 304 p. (plates; bibl. p. 279-299; pers. ind. p. 300-303).

6100. NEMES (Robert). The once and future Budapest. De-Kalb, Northern Illinois U. P., 2005, XI-247 p.

6101. NUZZO (Mariella). Lo spazio sascro della Terza Roma: dinamiche di insediamento, aspetti devozionali e caratteri formali degli edifici religiosi dei nuovi quartieri della capitale tra Leone XIII e Pio X (1878–1914). *Mélanges de l'École française de Rome: Italie et mediterranée (MEFRIM)*, 2005, 117, 2, p. 497-518.

6102. Other New York (The). The American revolution beyond New York City, 1763–1787. Ed. by Joseph S. TIEDEMANN and Eugene R. FINGERHUT. Albany, State University of New York Press, 2005, XI-246 p. (An American region, studies in the Hudson Valley).

6103. OVERUD (Johanna). I beredskap med Fru Lojal: behovet av kvinnlig arbetskraft i Sverige under andra världskriget. (Le besoin de main-d'œuvre féminine en Suède pendant la seconde Guerre mondiale). Stockholm, Almqvist & Wiksell International, 2005, 245 p. (Stockholm studies in history, 76).

6104. PARSON (Don). Making a better world. Public housing, the red scare, and the direction of modern Los Angeles. Minneapolis, University of Minnesota Press, 2005, XX-289 p.

6105. PIKE (David L.). Subterranean cities. The world beneath Paris and London, 1800–1945. Ithaca, Cornell U. P., 2005, XVIII-355 p.

6106. PLATT (Harold L.). Shock cities. The environmental transformation and reform of Manchester and Chicago. Chicago a. London, University of Chicago Press, 2005, XVI-628 p.

6107. REVELL (Keith D.). Building gotham: civic culture and public policy in New York City, 1898–1938. Baltimore, Johns Hopkins U. P., 2005, X-327 p.

6108. SCHÜRMANN (Sandra). Dornröschen und König Bergbau. Kulturelle Urbanisierung und bürgerliche Repräsentation am Beispiel der Stadt Recklinghausen, 1930–1960. Paderborn, Schöningh, 2005, 325 p. (Forschungen zur Regionalgeschichte, 52).

6109. SELIGMAN (Amanda I.). Block by block: neighborhoods and public policy on Chicago's West Side. Chicago, University of Chicago Press, 2005, XIII-301 p. (Historical studies of urban America).

6110. SITTON (Tom). Los Angeles transformed. Fletcher Bowron's urban reform revival, 1938–1953. Albuquerque, University of New Mexico Press, 2005, XVI-256 p.

6111. SLABOTINSKÝ (Radek). Pod českou správou v nové republice. Karlovy Vary v prvním poválečném roce (květen 1945–červenec 1946). (Unter der tschechischen Verwaltung in der neuen Republik. Karlsbad im ersten Jahr nach dem Weltkrieg). Karlovy Vary, Krajské muzeum, 2005, 110 p. (ill.).

6112. Social (The) and cultural demographics of minorities. Ed. By Michel ORIS, Guy BRUNET and Alain BIDEAU. *History of the family*, 2005, 10, 1, p. 1-98. [Contents: ORIS (Michel), BRUNET (Guy), BIDEAU (Alain). The social and cultural demography of minorities: Introduction (p. 1-5). – PLAKANS (Andrejs). Minority nationalities in the Russian Baltic provinces: The 1881 Baltic census (p. 7-20). – GRANGE (Cyril). The choice of matrimonial witnesses by Parisian Jews: Integration into greater society and socio-professional cohesion (1875–1914) (p. 21-44). – MARTIN (Elisa), GAMELLA (Juan F.). Marriage practices and ethnic differentiation: The case of Spanish Gypsies (1870–2000) (p. 45-63). – ODILON NADALIN (Sergio), BIDEAU (Alain). How German Lutherans became Brazilians: A methodological essay (p. 65-85). – RYGIEL (Philippe). What became of the second generation? The children of European immigrants in France between the world wars (p. 87-98)].

6113. SOLIS (Julia). New York underground: the anatomy of a city. New York a. London, Routledge, 2005, VII-251 p. (ill., maps).

6114. SPEARS (Timothy B.). Chicago Dreaming. Midwesterners and the city, 1871–1919. Chicago, University of Chicago Press, 2005, XXIII-322 p.

6115. Stadtverwaltung im Nationalsozialismus. Systemstabilisierende Dimensionen kommunaler Herrschaft. Hrsg. v. Sabine MECKING und Andreas WIRSCHING. Paderborn, München u. Wien, Schöningh, Landschaftsverband Westfalen-Lippe, 2005, VII-418 p. (Forschungen zur Regionalgeschichte, 53).

6116. TABEAU (Ewa), BIJAK (Jakub). War-related deaths in the 1992–1995 armed conflicts in Bosnia and Herzegovina: a critique of previous estimates and recent results. *European journal of population*, 2005, 21, 2-3, p. 187-215.

6117. TOFTEGAARD JENSEN (Jens), NORDSKOV (Jeppe). Købstadens metamorfose. Byudvikling og byplanlægning i Århus 1800–1920. (La métamorphose de la ville. Développement urbain et planification urbaine à Århus 1800–1920). Århus, Aarhus Universitets Forlag, 2005, 213 p. (Dansk Center for Byhistorie).

6118. TREJO BARAJAS (Dení). Declinación y crecimiento demográfico en Baja California, siglos XVIII y XIX. Una perspectiva desde los censos y padrones locales. *Historia mexicana*, 2005, 54, 3, p. 761-831.

6119. UMBACH (Maiken). A tale of second cities: autonomy, culture, and the law in Hamburg and Barcelona in the late nineteenth century. *American historical review*, 2005, 110, 3, p. 659-692.

6120. Villes en crise? Les politiques municipales face aux pathologies urbaines (fin XVIIIe–fin XXe siècle). Sous la dir. de Yannick MAREC. Paris, Creaphis, 2005, 763 p.

6121. VORMS (Charlotte). La pratica e la regola. Gestione e controllo pubblico dell'urbanizzazione non pianificata nella periferia di Madrid (1860–1931). *Storia urbana*, 2005, 28, 108, 3, p. 39-52.

6122. WALTER (Richard J.). Politics and urban growth in Santiago, Chile, 1891–1941. Stanford, Stanford U. P., 2005, XVII-319 p.

§ 7. Addenda 2002–2003.

6123. BÜHL-GRAMER (Charlotte). Nürnberg 1850 bis 1892. Stadtentwicklung im Zeichen von Industrialisierung und Urbanisierung. Nürnberg, Stadtarchiv Nürnberg, 2003, XVI-712 p. (Nürnberger Werkstücke zur Stadt- und Landesgeschichte, 62).

6124. PLANT (Margaret). Venice: fragile city, 1797–1997. New Haven, Yale U. P., 2002, X-550 p.

Cf. n°s 3500, 4707, 5969, 6127, 6854

§ 8. Social history.

* 6125. FERGUSON (Priscilla). Eating orders: markets, menus, and meals. *Journal of modern history*, 2005, 77, 3, p. 679-700.

* 6126. SANFILIPPO (Matteo). Problemi di storiografia dell'emigrazione italiana. Viterbo, Sette Città, 2005, 389 p. (Biblioteca, 3).

* 6127. YAŞAR (Ahmet). Osmanlı Şehir Mekânları: Kahvehane Literatürü. (Milieux urbains ottomans: littérature sur les cafés). *Türkiye Araştırmaları Literatür Dergisi*, 2005, 3, 6, p. 211-235.

** 6128. KURT (Yılmaz). Çukurova tarihinin kaynakları II. 1547 tarihli Adana Sancağı mufassal tahrir defteri. III. 1572 tarihli Adana Sancağı mufassal tahrir defteri. (Sources de l'histoire de Çukurova II: Registre daté de 1547 du sandjak d'Adana. III. Registre daté de 1572 du sandjak d'Adana. Ankara, Türk Tarih Kurumu, 2005, 2 vol., XXXVII-854 p., CXXIV-946 p.

** 6129. Quellen zur Alltagsgeschichte der Deutschen 1815–1870. Hrsg. v. Hartwig BRANDT und Ewald GROTHE. Darmstadt, Wissenschaftliche Buchgesellschaft, 2005, XII-234 p. (Ausgewählte quellen zur deutschen Geschichte der Neuzeit, Freiherr-vom Stein-Gedächtnisausgabe, 44).

** 6130. Quellensammlung zur Geschichte der deutschen Sozialpolitik. Begr. von Peter RASSOW [ct al.] Im Auftr. der Historischen Kommission der Akademie der Wissenschaften und der Literatur, Mainz hrsg. von Hansjoachim HENNING [et al.]. Abt. 3. Ausbau und Differenzierung der Sozialpolitik seit Beginn des neuen Kurses (1890–1904). Band 3. Arbeiterschutz. Bearb. von Wolfgang AYASS. Darmstadt, WBG, 2005, XLIV-640 p. (ill.).

** 6131. Turci v Uhorsku. Časť 1. Život v Uhorskom kráľovstve počas tureckých vojen od tragickej bitky pri Moháči až do Bratislavského snemu. (Die Osmanen in Ungarn. Teil. 1. Das Leben in ungarischen Königreich während den türkischen Kriege seit der tragische Schlacht bei Mohács bis zum pressburger Landtag). Red. Viliam ČIČAJ. Ed. Pavel DVOŘÁK. Autor. Michal BADA, Viliam ČIČAJ, Michal DUCHOŇ [et. al.]. Bratislava, Literárne informačné centrum, 2005, 365 p. (Pramene k dejinám Slovenska a Slovákov, 7/1).

6132. 13 reformer af den danske velfærdsstat. (Treize réformes de l'Etat-providence danois). Red. Jørn Henrik PETERSEN, Klaus PETERSEN. Odense, Syddansk Universitetsforlag, 2005, 240 p. (University of Southern Denmark studies in history and social studies, 300).

6133. ABREU (Laurinda). Un destin exceptionnel: les enfants abandonnés au travail (Évora, 1650–1837). *Annales de démographie historique*, 2005, 110, p. 165-183.

6134. ALLEGRA (Luciano). Un modèle de mobilité sociale préindustrielle. Turin à l'époque napoléonienne. *Annales*, 2005, 60, 2, p. 443-474.

6135. ANDERSEN NEXØ (Sniff). Det rette valg. Dansk abortpolitik i 1930'erne og 1970'erne. (Le juste choix). La politique danoise de l'avortement dans les années 1930 et 1970). København, Københavns universitet, 2005, 318 p.

6136. ARRIZABALAGA (Marie-Pierre). Basque women and urban migration in the 19th century. *History of the family*, 2005, 10, 2, p. 99-117.

6137. ASSUNÇAO (Matthias Röhrig). Capoeira. The history of an Afro-Brazilian martial art. New York, Routledge, 2005, XIII-267 p. (Sport in the global society).

6138. ASTORGA (Pablo), BERGES (Ame R.), FITZGERALD (Valpy). The standard of living in Latin America during the twentieth century. *Economic history review*, 2005, 58, 4, p. 765-796.

6139. AZUMA (Eiichiro). Between two empires: race, history, and transnationalism in Japanese America. New York, Oxford U. P., 2005, XII-306 p.

6140. BARGAOUI (Sami). Des Turcs aux Hanafiyya. La construction d'une catégorie "métisse" à Tunis aux XVIIe et XVIIIe siècles. *Annales*, 2005, 60, 1, p. 209-228.

6141. BARNOW (David), GIMPEL (Madeleine), VAN RAAY (Rob). De hongerwinter 1944–1945. Kampen, Ten Have, 2005, 129 p.

6142. BARTH (Thomas). Adelige Lebenswege im Alten Reich. Der Landadel der Oberpfalz im 18. Jahrhundert. Regensburg, Pustet, 2005, 696 p.

6143. BASCUÑÁN AÑOVER (Oscar). Delincuencia y desorden social en la España agraria: La Mancha, 1900–1936. *Historia social*, 2005, 51, p. 111-138.

6144. BAUERKÄMPER (Arnd). Die Sozialgeschichte der DDR. München, Oldenbourg, 2005, X-148 p. (Enzyklopädie deutscher Geschichte, 76). – IDEM. Ein asymmetrisches Verhältnis. Gesellschaftliche und kulturelle Kontakte zwischen Großbritannien und der DDR von den Sechziger- zu den Achtzigerjahren. *Archiv für Sozialgeschichte*, 2005, 45, p. 43-58.

6145. BECQUEMIN (Michèle). Protection de l'enfance et placement familial. La Fondation Grancher. De l'Hygiénisme à la suppléance parentale. Paris, Éditions Petra, 2005, 259 p.

6146. BELLAMY (John). Strange, inhuman deaths. Murder in Tudor England. Westport, Praeger, 2005, 209 p.

6147. BENADUSI (Lorenzo). Il Nemico dell'uomo nuovo. L'omosessualità nell'esperimento totalitario fascista. Pref. di Emilio GENTILE. Milano, Feltrinelli, 2005, 430 p.

6148. BENSON (John). Affluence and authority. A social history of 20th century Britain. London a. Hodder, Arnold, 2005, XVI-240 p.

6149. BERG (Maxine). Luxury and pleasure in eighteenth-century Britain. Oxford, Oxford U. P., 2005, 373 p.

6150. Berlin – Paris, 1900–1993. Begegnungsorte, Wahrnehmungsmuster, Infrastrukturprobleme im Vergleich. Hrsg. Hans Manfred BOCK und Ilja MIECK. Bern, Lang, 2005, 384 p. (Convergences, 12).

6151. BERRY (Mary Frances). My face is black is true: Callie House and the struggle for ex-slave reparations. New York, Alfred A. Knopf, 2005, XIV-314 p.

6152. Beyond black and red. African-native relations in colonial Latin America. Ed. by Matthew RESTALL. Albuquerque, University of New Mexico Press, 2005, XV-303 p. (Diálogos).

6153. BJELOPERA (Jerome P.). City of clerks: office and sales workers in Philadelphia, 1870–1920. Urbana a. Chicago, University of Illinois Press, 2005, IX-208 p. (The Working Class in American history).

6154. BLIN (Thierry). Les sans-papiers de Saint-Bernard. Mouvement social et action organisée. Paris, L'Harmattan, 2005, 185 p.

6155. BOLLING (Hans). Sin egen hälsas smed: idéer, initiativ och organisationer inom svensk motionsidrott 1945–1981. (Forger sa propre santé. Idées, initiatives et organisations dans le sport suédois de 1945 à 1981). Stockholm, Acta Universitatis Stockholmiensis, 2005, 342 p. (Stockholm studies in history, 81). [English summary].

6156. BONIN (Pierre). Bourgeoisie et habitanage dans les villes du Languedoc sous l'ancien régime. Préf. de Albert RIGAUDIÈRE. Aix-en-Provence, Presses Universitaires d'Aix-Marseille, 2005, 584 p. (Collection d'Histoire du Droit; Série "Thèses et Travaux," 5).

6157. BORSAY (Anne). Disability and social policy in Britain since 1750. Basingstoke, Palgrave Macmillan, 2005, 306 p.

6158. BRÜGGEMEIER (Franz-Josef). Eine virtuelle Gemeinschaft. Deutschland und die Fußballweltmeisterschaft 1954. *Geschichte und Gesellschaft*, 2005, 31, 4, p. 610-635.

6159. BUKOWCZYK (John J.) [et al.]. Permeable border: the Great Lakes Basin as transnational region, 1650–1990. Pittsburgh: University of Pittsburgh Press, 2005, XII-298 p.

6160. BUREŠOVÁ (Jana). Žena na přelomu 19. a 20. století. Společenské postavení českých žen a souvislosti jeho proměn. (Woman in the end of the 19th century and the beginning of the 20th century. Change of the social position of Czech women). *Acta historica Neosoliensia*, 2005, 8, p. 201-216.

6161. Bürgertum nach 1945. Hrsg. v. Manfred HETTLING und Bernd ULRICH. Hamburg, Hamburger Edition, 2005, 438 p.

6162. BUSE (Dieter K.). Finding the nation in Bremen: the Lower Class and women after Napoleonic occupation. *Canadian journal of history*, 2005, 40, 1, p. 1-22.

6163. Business (The) of Dependency: Governments, Firms and the Consumption of Addictive Products. Editors: Matthias KIPPING, Lina GÁLVEZ MUÑOZ. *Business history*, 2005, 47, 3, p. 331-465. [Contents: KIPPING (Matthias), GÁLVEZ MUÑOZ (Lina). The Business of dependency: an introduction (p. 331-336). – CORLEY (T.A.B.). UK Government regulation of medicinal drugs, 1890–2000 (p. 337-351). – SLINN (Judy). Price controls or control through prices? Regulating the cost and consumption of prescription pharmaceuticals in the UK, 1948–1967 (p. 352-366). – SIMPSON (James). Too little regulation? The British market for sherry, 1840–1890 (p. 367-382). – PENNOCK (Pamela E.), KERR (K Austin). In the shadow of prohibition: domestic American alcohol policy since 1933 (p. 383-400). – GÁLVEZ MUÑOZ (Lina). Regulating an addictive product: the Spanish Government, brand advertising and tobacco business (1880s to 1930s) (p. 401-420). – COURTWRIGHT (David T.). 'Carry on smoking': public relations and advertising strategies of American and British tobacco companies since 1950 (p. 421-433). – MISKELL (Peter). Seduced by the silver screen: film addicts, critics and cinema regulation in Britain in the 1930s and 1940s (p. 433-448). – KIPPING (Matthias), SAINT-MARTIN (Denis). Between regulation, promotion and consumption: government and management consultancy in Britain (p. 449-465)].

6164. CALONACI (Stefano). Dietro lo scudo incantato : i fedecommessi di famiglia e il trionfo della borghesia fiorentina (1400ca–1750). Grassina, Le Monnier università, 2005, VI-299 p. (ill.). (Le Monnier università. Storia).

6165. CARLETON (Gregory). Sexual revolution in bolshevik Russia. Pittsburgh, University of Pittsburgh Press, 2005, X-272 p. (Russian and East European studies).

6166. CARR (Jacqueline Barbara). After the siege. A social history of Boston, 1775–1800. Boston, Northeastern U. P., 2005, XV-318 p.

6167. CHARLE (Christophe). "Les sociétés impériales" d'hier à aujourd'hui. Quelques propositions pour repenser l'histoire du second XXe siècle en Europe. *Journal of modern European history*, 2005, 3, 2, p. 123-139.

6168. CHIBRAC (Lucienne). Les pionnières du travail social auprès des étrangers. Le Service social d'aide aux émigrants, des origines à la Libération. Rennes, Éditions ENSP, 2005, 303 p.

6169. CIRIACONO (Salvatore). Migration, minorities and technology. Transfer in Early Modern Europe. *Journal of European economic history*, 2005, 34, 1, p. 43-64.

6170. COENEN (Craig R.). From Sandlots to the Super Bowl: the National Football League, 1920–1967. Knoxville, University of Tennessee Press, 2005, XII-342 p. (Sports and Popular Culture).

6171. COOK (Hera). The English sexual revolution: technology and social change. *History workshop*, 2005, 59, p. 109-128.

6172. COSTEA (Ionuț). Solam virtutem et nomen bonum: nobilitate, etnie, regionalism în Transilvania princiară: (sec. XVII). (Nobility, ethnic, regionalism in princely Transylvania XVII century). Cluj-Napoca, Editura Argonaut, 2005, 310 p.

6173. COWAN (Brian). The social life of coffee: the emergence of the British coffeehouse. New Haven, Yale U. P., 2005, XII-364 p.

6174. CROSBY (Emilye). A little taste of freedom: the Black freedom struggle in Claiborne County, Mississippi. Chapel Hill, University of North Carolina Press, 2005, XV-354 p. (The John Hope Franklin series in African American history and culture).

6175. CROSS (Gary S.), WALTON (John K.). The playful crowd: pleasure places in the twentieth century. New York, Columbia U. P., 2005, VIII-308 p.

6176. CROWSTON (Clare Haru). L'apprentissage hors des corporations. Les formations professionelles alternatives à Paris sous l'Ancien Régime. *Annales*, 2005, 60, 2, p. 409-441.

6177. CURRELL (Susan). The march of spare time: the problem and promise of leisure in the Great Depression. Philadelphia, University of Pennsylvania Press, 2005, 235 p.

6178. DA ORDEN (María Liliana). Inmigración española, familia y movilidad social en la Argentina moderna: una mirada desde Mar del Plata, 1890–1930. Buenos Aires, Editorial Biblos, 2005, 206 p. (ill., maps). (Colección La Argentina plural).

6179. DA ROSA (Gilson Justino). Imigrantes alemães, 1824–1853. Porto Alegre, Est Edições, 2005, 327 p. (map).

6180. DAVIES (Russell). Hope and heartbreak. A social history of Wales and the Welsh, 1776–1871. Cardiff, University of Wales Press, 2005, 559 p.

6181. DAWSON (Melanie). Laboring to play. Home entertainment and the spectacle of middle-class cultural life, 1850–1920. Tuscaloosa, University of Alabama Press, 2005, X-257 p.

6182. DAYAN-HERZBRUN (Sonia). Femmes et politique au Moyen-Orient. Paris, L'Harmattan, 2005, 157 p.

6183. DE GRAZIA (Victoria). Irresistible empire. America's advance through twentieth-century Europe. Cambridge, Belknapp Press of Harvard U. P., 2005, 586 p.

6184. DELYSER (Dydia). Ramona memories. Tourism and the shaping of Southern California. Minneapolis, University of Minnesota Press a. Santa Fe, Center for American places, 2005, XXIII-256 p.

6185. Désengagement militant (Le). Sour la dir. de Olivier FILLIEULE. Paris, Belin, 2005, 319 p.

6186. Désir (Le) et le goût. Une autre histoire (XIIIe–XVIIIe siècles). Sous la dir. de Odile REDON, Line SALLMANN et Sylvie STEINBERG. Saint-Denis, Presses universitaires de Vincennes, 2005, 403 p. (Temps et espaces).

6187. DIGBY (Anne). Early black doctors in South Africa. *Journal of African history*, 2005, 46, 3, p. 427-454.

6188. DINÇ (Güven). Şer'iyye Sicillerine Göre XIX. Yüzyılın Ortalarında Antalya'da Ailenin Sosyo-Ekonomik Durumu. (Structure socio-économique de la famille à Antalya d'après les registres tribunaux à la moitié du 19e siècle). *OTAM Ankara Üniversitesi Osmanlı Tarihi Araştırma Ve Uygulama Merkezi Dergisi*, 2005, 17, p. 103-130.

6189. Domestic servants in comparative perspective. Ed. by Antoinette FAUVE-CHAMOUX and Richard WALL. *History of the family*, 2005, 10, 4, p. 345-490. [Contents: FAUVE-CHAMOUX (Antoinette), WALL (Richard). Domestic servants in comparative perspective: Introduction (p. 345-354). – NAGATA (Mary Louise). One of the family: Domestic service in early modern Japan (p. 355-365). – MAC ISAAC COOPER (Sheila). Service to servitude? The decline and demise of life-cycle service in England (p. 367-386). – VIAZZO (Pier Paolo), AIME (Marco), ALLOVIO (Stefano). Crossing the boundary: peasants, shepherds, and servants in a western Alpine community (p. 387-405). – SARTI (Raffaella). The true servant: self-definition of male domestics in an Italian city (Bologna, 17th–19th centuries) (p. 407-433). – LEE (Robert). Domestic service and female domestic servants: a port-city comparison of Bremen and Liverpool, 1850–1914 (p. 435-460). – SWAIN (Shurlee). Maids and mothers: domestic servants and illegitimacy in 19th-century Australia (p. 461-471). – HIONIDOU (Violetta). Domestic service on three Greek islands in the later 19th and early 20th centuries (p. 473-489)].

6190. DONNER (Sandra). Von Höheren Töchtern und gelehrten Frauenzimmern. Mädchen- und Frauenbildung im 19. Jahrhundert. Dargestellt an den Schloßanstalten Wolfenbüttel. Frankfurt am Main, Lang, 2005, 275 p. (Europäische Hoschulschriften 3, Geschichte und ihre Hilfswissenschaften, 1006).

6191. DRIBE (Martin), STANFORS (Maria). Leaving home in post-war Sweden. A micro-level analysis of leaving the parental home in three birth cohorts. *Scandinavian economic history review*, 2005, 53, 2, p. 30-49.

6192. EDWARDS (Clive). Turning houses into homes: a history of the retailing and consumption of domestic furnishings. Aldershot, Ashgate, 2005, VIII-294 p.

6193. EHRICK (Christine). The shield of the weak: feminism and the state in Uruguay, 1903–1933. Albuquerque, University of New Mexico Press, 2005, XII-282 p.

6194. Eigener Sache (In). Frauen vor den höchsten Gerichten des Alten Reiches. Hrsg. v. Siegrid WESTPHAL. Köln, Weimar u. Wien, Böhlau, 2005, VI-273 p.

6195. EL-ROUAYHEB (Khaled). Before homosexuality in the Arab-Islamic world, 1500–1800. Chicago, University of Chicago Press, 2005, X-210 p.

6196. Entre la familia, la sociedad y el Estado: niños y jóvenes en América Latina (siglos XIX–XX). Ed. por Barbara POTTHAST y Sandra CARRERAS. Frankfurt am Main, Vervuert Verlagsgesellschaft, 2005, 403 p.

6197. ERICSSON (Kjersti), SIMONSEN (Eva). Children of World War II. The hidden enemy legacy, Oxford, Berg, 2005, 296 p.

6198. ESCOBARI DE QUEREJAZU (Laura). Caciques, yanaconas y extravagantes. la sociedad colonial en Charcas s. XVI–XVIII. Lima, IFEA, Instituto Francés de Estudios Andinos y La Paz, Plural Editores, 2005, 307 p. (ill., maps). (Travaux de l'Institut français d'études andines, 208).

6199. ESTES (Steve). I am a man! Race, manhood, and the civil rights movement. Chapel Hill, University of North Carolina Press, 2005, X-239 p.

6200. EVANS (Tanya). "Unfortunate objects". Lone mothers in eighteenth-century London. London, Palgrave Macmillan, 2005, X-279 p.

6201. Excluir para ser: Procesos identitarios y fronteras sociales en la América hispánica (XVII–XVIII). Ed. por Christian BÜSCHGES y Frédérique LANGUE.
Madrid, Iberoamericana y Frankfurt am Main, Vervuert / AHILA, 2005, 173 p. (Estudios AHILA de Historia Latinoamericana).

6202. Exil v Praze a Československu 1918–1938. (Exile in Prague and Czechoslovakia 1918–1938). Praha, Pražská edice, 2005, 213 p. (photogr.).

6203. Family transmission in Eurasian perspective. Edited by Antoinette FAUVE-CHAMOUX and Marie-Pierre ARRIZABALAGA. *History of the family*, 2005, 10, 3, p. 183-344. [Contents: FAUVE-CHAMOUX (Antoinette), ARRIZABALAGA (Marie-Pierre). Family transmission in Eurasian perspective: Introduction (p. 183-193). – HAYAMI (Akira), OKADA (Aoi). Population and households dynamics: a mountainous district in northern Japan in the Shûmon Aratame Chô of Aizu, 1750–1850 (p. 195-229). – FAUVE-CHAMOUX (Antoinette). A comparative study of family transmission systems in the central Pyrenees and northeastern Japan (p. 231-248). – BONNAIN (Rolande). Household mind and the ecology of the central Pyrenees in the 19th century: fathers, sons, and collective landed property (p. 249-270). – ARRIZABALAGA (Marie-Pierre). Succession strategies in the Pyrenees in the 19th century: the Basque case (p. 271-292). – DRIBE (Martin), LUNDH (Christer). Gender aspects of inheritance strategies and land transmission in rural Scania, Sweden, 1720–1840 (p. 293-308). – FERTIG (Christine), LÜNNEMANN (Volker), FERTIG (Georg). Inheritance, succession, and familial transfer in rural Westphalia, 1800–1900 (p. 309-326). – CRAIG (Béatrice). Farm transmission and the commercialization of agriculture in northern Maine in the second half of the 19th century (p. 327-344)].

6204. Famine and disease in Ireland. Ed. by L.A. CLARKSON and E. Margaret CRAWFORD. London, Pickering & Chatto, 2005, 5 vol., XXXVIII-2320 p.

6205. FEDELE (Santi). La massoneria italiana nell'esilio e nella clandestinità: 1927–1939. Milano, F. Angeli, 2005, 201 p. (Temi di storia, 63).

6206. FEHRENBACH (Heide). Race after Hitler. Black occupation children in postwar Germany and America. Princeton a. Oxford, Princeton U. P., 2005, XIII-263 p.

6207. FELLER (Élise). Histoire de la vieillesse en France, 1900–1960. Du vieillard au retraité. Paris, Éditions Seli Arslan, 2005, 352 p.

6208. FISCHER (Jasna). Družba, gospodarstvo, prebivalstvo: družbena in poklicna struktura prebivalstva na slovenskem ozemlju od druge polovice 19. stoletja do razpada habsburške monarhije. (Society, economy, population: social and occupational structure of the population of the Slovene territory from the second half of the 19th century to the disintegration of the Habsburg monarchy). Ljubljana, Inštitut za novejšo zgodovino, 2005, 247 p. (Zbirka Ekonomska knjižnica).

6209. FITZPATRICK (Sheila). Tear off the masks! Identity and imposture in twentieth-century Russia. Princeton, Princeton U. P., 2005, XII-332 p.

6210. FORD (Caroline). Divided houses. Religion and gender in modern France. Ithaca a. London, Cornell U. P., 2005, 170 p.

6211. FORT GODSHALK (David). Veiled visions: The 1906 Atlanta race riot and the reshaping of American race relations. Chapel Hill, University of North Carolina Press, 2005, XVI-365 p.

6212. FRANK (Andrew K.). Creeks and Southerners: biculturalism on the early American frontier. Lincoln, University of Nebraska Press, 2005, XI-192 p. (Indians of the southeast).

6213. FRANK (Denis). Staten, företagen och arbetskraftsinvandringen: en studie av invandringspolitiken i Sverige och rekryteringen av utländska arbetare 1960–1972. (L'Etat, les entreprises et l'immigration de main d'œuvre. Etude de la politique d'immigration en Suède et le recrutement de travailleurs étrangers de 1960 à 1972). Växjö, Växjö U. P., 2005, 256 p. (Acta Wexionensia. Samhällsvetenskap, 67).

6214. FROIDE (Amy M.). Never Married. Singlewomen in Early Modern England. New York, Oxford U. P., 2005, VII-246 p.

6215. FUCHS (Rachel G.). Gender and poverty in nineteenth-century Europe. Cambridge, Cambridge U. P., 2005, XI-267 p.

6216. GARDNER (Martha). The qualities of a citizen. Women, immigration and citizenship, 1870–1965. Princeton, Princeton U. P., 2005, VIII-271 p.

6217. Gender and slave emancipation in the Atlantic world. Ed. by Pamela SCULLY and Diana PATON. Durham, Duke U. P., 2005, VI-376 p.

6218. GHAZELEH (Pascale). L'economia dei diritti. Oggetti, legami e strategie di possesso al Cairo all'inizio del XIX secolo. *Quaderni storici*, 2005, 3, p. 777-800.

6219. GIBELLI (Antonio). Il popolo bambino. Infanzia e nazione dalla Grande guerra a Salò. Torino, Einaudi, 2005, 412 p.

6220. GIFFIN (William W.). African Americans and the color line in Ohio, 1915–1930. Columbus, Ohio State U. P., 2005, 312 p.

6221. GINZBERG (Lori D.). Untidy origins. A story of woman's rights in antebellum New York. Chapel Hill, University of North Carolina Press, 2005, XIV-222 p.

6222. GLASER (Clive). Managing the sexuality of urban youth: Johannesburg, 1920s–1960s. *International journal of African historical studies*, 2005, 38, 2, p. 301-328.

6223. GLOOTZ (Tanja Anette). Alterssicherung im europäischen Wohlfahrtsstaat: Etappen ihrer Entwicklung im 20. Jahrhundert. Frankfurt am Main, Campus, 2005, 300 p. (Campus-Forschung, 885).

6224. GOMES (Flávio dos Santos). A hidra e os pântanos: mocambos, quilombos e comunidades de fugitivos no Brasil (séculos XVII–XIX). São Paulo, UNESP, Polis, 2005, 462 p.

6225. GÓMEZ BRAVO (Gutmaro). La violencia y sus dinámicas: crimen y castigo en el siglo XIX español. *Historia social*, 2005, 51, p. 93-110.

6226. GORDON (Daniel A.). The back door of the nation state: expulsions of foreigners and continuity in twentieth-century France. *Past & present*, 2005, 186, p. 201-232.

6227. GOURDON (Vincent). Aux cœurs de la sociabilité villageoise: une analyse de réseau à partir du choix des conjoints et des témoins au mariage dans un village d'Île-de-France au XIXe siècle. *Annales de démographie historique*, 2005, 109, p. 61-94.

6228. GRANT (Oliver). Migration and inequality in Germany, 1870–1913. Oxford, Clarendon Press, 2005, VII-406 p.

6229. GREEN (Nancy L.). The politics of exit: reversing the immigration paradigm. *Journal of modern history*, 2005, 77, 2, p. 263-289.

6230. GREENE (Christina). Our separate ways: women and the black freedom movement in Durham, North Carolina. Chapel Hill, University of North Carolina Press, 2005, XVIII-366 p.

6231. GREGORY (James N.). The southern diaspora: how the great migrations of black and white southerners transformed America. Chapel Hill, University of North Carolina Press, 2005, XIV-446 p.

6232. GRIFFIN (Emma). England's revelry. A history of popular sports and pastimes, 1660–1830. Oxford, Oxford U. P., 2005, 295 p.

6233. GRØNGAARD JEPPESEN (Torben). Danske i USA 1850–2000: en demografisk, social og kulturgeografisk undersøgelse af de danske immigranter og deras efterkommere. (Les Danois aux Etats-Unis 1850–2000: étude démographique, sociale, géographique et culturelle des immigrants danois et de leurs descendants). Odense, Odense Bys Museer, 2005, 445 p.

6234. GURNEY (Peter). The battle of the consumer in postwar Britain. *Journal of modern history*, 2005, 77, 4, p. 956-987.

6235. HÄDER (Sonja). Selbstbehauptung wider Partei und Staat. Zum Einfluss westlicher Jugendkulturen auf die Jugend des Ostblocks. *Archiv für Sozialgeschichte*, 2005, 45, p. 449-476.

6236. HALPERÍN DONGHI (Tulio). La formación de la clase terrateniente bonaerense. Buenos Aires, Prometeo, 2005, 210 p. (ill.). (Colección de historia argentina).

6237. HANISCH (Ernst). Männlichkeiten. Eine andere Geschichte des 20. Jahrhunderts. Wien, Köln u. Weimar, Böhlau, 2005, 459 p.

6238. HERZOG (Dagmar). Sex after fascism. Memory and morality in twentieth-century Germany. Princeton, Princeton U. P., 2005, 361 p.

6239. HESSINGER (Rodney). Seduced, abandoned, and reborn. Visions of youth in middle-class America, 1780–1850. Philadelphia, University of Pennsylvania Press, 2005, 255 p. (Early American studies).

6240. HEUER (Jennifer Ngaire). The family and the nation. Gender and citizenship in Revolutionary France, 1789–1830. Ithaca a. London, Cornell U. P., 2005, X-256 p.

6241. Historia de la vida cotidiana en México. T. 3. Siglo XVIII: entre tradición y cambio. Coordinadora Pilar GONZALBO AIZPURU. T. 4. Bienes y vivencias, el siglo XIX. Coordinadora Anne STAPLES. México D.F., El Colegio de México, Fondo de Cultura Económica, 2005, 2 vol., 592 p., 615 p. (ill.).

6242. História em cousas miúdas: capítulos de história social da cronica no Brasil. Organizadores: Sidney CHALHOUB, Margarida de Souza NEVES e Leonardo Affonso de Miranda PEREIRA. Campinas, Editora da UNICAMP, 2005, 590 p. (ill.).

6243. Historia social y fútbol. El profesionalismo deportivo y sus alcances sociales. Intr. de Julio D. FRYDENBERG. *Entrepasados*, 2005, 27, [s. p.]. [Contiene: LANFRANCHI (Pierre). Los artistas del fútbol sudamericano en Europa, 1924–1940. – DIETSCHY (Paul). El "calcio" y el régimen. El fútbol italiano durante el "ventennio" fascista. – FREIRE RODRIGUES (Francisco Xavier). Profesionalizacion y modernizacion en el futbol brasileño. – FRYDENBERG (Julio D.). La profesionalización del fútbol argentino: entre una huelga de jugadores y la reestructuración del espectáculo].

6244. HÖIJER (Birgitta), RASMUSSEN (Joel). Medborgare om våldsdåd: reaktioner efter mordet på Anna Lindh och andra dåd. (Les citoyens et la violence. Réactions après l'assassinat d'Anna Lindh). Stockholm, Krisberedskapsmyndigheten, 2005, 120 p. (KBM:s temaserie, 2005:10).

6245. HOULBROOK (Matt). Queer London. Perils and pleasures in the sexual metropolis, 1918–1957. Chicago a. London, University of Chicago Press, 2005, 384 p.

6246. HUNN (Karin). "Nächstes Jahr kehren wir zurück ...": die Geschichte der türkischen "Gastarbeiter" in der Bundesrepublik Göttingen, Wallstein, 2005, 598 p. (Moderne Zeit, 11).

6247. Immigrants in Tudor and early Stuart England. Ed. by Nigel GOOSE and Lien LUU). Brighton, Sussex Academic Press, 2005, 263 p.

6248. INGERSOLL (Thomas N.). To intermix with our white brothers: Indian mixed bloods in the United States from earliest times to the Indian removal. Albuquerque, University of New Mexico Press, 2005, XXI-450 p.

6249. JACKSON (Brenda K.). Domesticating the West. The recreation of the nineteenth-century American middle class. Lincoln, University of Nebraska Press, 2005, XIII-180 p.

6250. JACOBS (Auke P.), D'ESPOSITO (Francesco). Migratory movements between Spain and the New World and the "Leyes Nuevas": passengers in both directions in 1543–1544. *Journal of European economic history*, 2005, 34, 2, p. 483-512.

6251. JAREB (Jere). Prilog povijesti hrvatskog iseljeništva u Sjedinjenim Američkim Državama 1941–1947. (Contribution for the history of Croatian immigrants in the U.S.A., 1941–1947). *Časopis za suvremenu povijest*, 2005, 37, 1, p. 37-70.

6252. JOERGENS (Bettina). Männlichkeiten. Deutsche Jungenschaft, CVJM und Naturfreundejugend in Minden, 1945–1955. Potsdam, Verlag für Berlin-Brandenburg, 2005, 603 p. (Potsdamer Studien, 17).

6253. JOHANSSON (Christina). Välkomna till Sverige? Svenska migrationspolitiska diskurser under 1900-talets andra hälft. (Bienvenus en Suède? Les discours sur la politique d'immigration en Suède dans la seconde moitié du 20e siècle). Malmö, Bokbox, 2005, 282 p. (Linköping studies in arts and science, 344). [English summary].

6254. JOHNSON (David K.). The lavender scare. The Cold War persecution of gays and lesbian in the Federal Government. Chicago, University of Chicago Press, 2005, XI-227 p.

6255. JUNHO ANASTÁSIA (Carla Maria). A geografia do crime. Violência nas Minas setecentistas. Belo Horizonte, Ed. Ufmg, 2005, 173 p.

6256. KALMIJN (Matthijs). The effects of divorce on men's employment and social security histories. *European journal of population*, 2005, 21, 4, p. 347-366.

6257. KAPLAN (Morris B.). Sodom on the Thames: sex, love, and scandal in wilde times. Ithaca, Cornell U. P., 2005, 314 p.

6258. KAY (Gwen). Dying to be beautiful. The fight for safe cosmetics. Columbus, Ohio State U. P., 2005, XII-190 p. (Women, gender, and health).

6259. KENNEDY (Cynthia M.). Braided relations, entwined lives: the women of Charleston's urban slave society. Bloomington, Indiana U. P., 2005, XII-311 p. (Blacks in the diaspora).

6260. KENT (E. J.). Masculinity and male witches in old and New England, 1593–1680. *History workshop*, 2005, 60, p. 69-92.

6261. KOÇ (Yunus). Osmanlıda Toplumsal Dinamizmden Celali İsyanlarına Giden Yol ya da İki Belgeye Tek Yorum. (Voie qui mène du dynamisme social otoman aux révoltes des Jelali ou une interprétation sur deux documents). *Bilig Türk Dünyası Sosyal Bilimler Dergisi / Journal of social sciences of the Turkish world*, 2005, 35, p. 229-245.

6262. KONOW (Jan von). Sveriges adels historia. (Histoire de la noblesse suédoise). Karlskrona, Abrahamson, 2005, 351 p.

6263. KOVAŘÍK (David). "V zájmu ochrany hranic". Přesídlení obyvatel ze zakázaného a hraničního pásma (1951–1952). ("In the interest of protecting the borders". Resettlement of the inhabitants from the border zone and forbidden zone, 1951–1952). *Soudobé dějiny*, 2005, 12, 3/4, p. 686-707. – IDEM. Vysídlení Němců z okresu Jindřichův Hradec 1945–1948. (Die Aussiedlung der Deutschen aus dem Bezirk Jindřichův Hradec (Neuhaus) in den Jahren 1945–1948). *Jihočeský sborník historický*, 2005, 74, p. 219-234.

6264. KRAINZ (Thomas A.). Delivering aid: implementing progressive era welfare in the American West. Albuquerque, University of New Mexico Press, 2005, XIV-325 p.

6265. KREUTZER (Susanne). Vom "Liebesdienst" zum modernen Frauenberuf: die Reform der Krankenpflege nach 1945. Frankfurt am Main, Campus-Verl., 2005, 306 p. (Reihe Geschichte und Geschlechter, 45).

6266. KUHN (Axel), SCHWEIGARD (Jörg). Freiheit oder Tod! Die deutsche Studentenbewegung zur Zeit der Französischen Revolution. Köln, Weimar u. Wien, Böhlau, 2005, X-481 p. (Stuttgarter historische Forschungen, 2).

6267. KULU (Hill). Migration and fertility: competing hypotheses re-examined. *European journal of population*, 2005, 21, 1, p. 51-87.

6268. LARUELLE (Marlène), PEYROUSE (Sébastien). Les russes du Kazakhstan. Identités nationales et nouveaux états dans l'espace post-soviétique. Paris, Maisonneuve & Larouse, Institut français d'études sur l'Asie central, 2005, 354 p.

6269. LATTE ABDALLAH (Stéphanie). Subvertir le consentement: itinéraires des femmes des camps de réfugiés palestiniens en Jordanie (1948–2001). *Annales*, 2005, 60, 1, p. 53-89.

6270. LAURIE (Bruce). Beyond Garrison. Antislavery and social reform. Cambridge, New York a. Melbourne, Cambridge U. P., 2005, XXI-340 p.

6271. LIEBSCHER (Daniela). La Obra Nacional Dopolavoro fascista y la NSGemeinschaft "Kraft durch Freude". Las relaciones entre las políticas sociales italiana y alemana desde 1925 a 1939. *Historia social*, 2005, 52, p. 129-146.

6272. LILTI (Antoine). Le monde des salons. Sociabilité et mondanité à Paris au XVIIIe siècle. Paris, Fayard, 2005, 568 p.

6273. LINDENBERGER (Thomas). "Asoziale Lebensweise". Herrschaftslegitimation, Sozialdisziplinierung und die Konstruktion eines "negativen Milieus" in der SED-Diktatur. *Geschichte und Gesellschaft*, 2005, 31, 2, p. 227-254.

6274. LINDENMEYR (Kriste). The greatest generation grows up: American childhood in the 1930s. Chicago, Ivan R. Dee, 2005, XIII-304 p.

6275. LONG (Jason). Rural-urban migration and socioeconomic mobility in Victorian Britain. *Journal of economic history*, 2005, 65, 1, p. 1-35.

6276. LORENZETTI (Luigi), MERZARIO (Raul). Il Fuoco acceso. Famiglie e migrazioni alpine nell'Italia d'età moderna. Roma, Donzelli editore, 2005, 194 p.

6277. LUCASSEN (Léo). The immigrant threat. The integration of old and new migrants in Western Europe since 1850. Urbana a. Chicago, University of Illinois Press, 2005, 277 p. (Studies of world migrations).

6278. LUNDSTRÖM (Catarina). Fruars makt och omakt: kön, klass och kulturarv 1900–1940. (Pouvoir et "non-pouvoir" des femmes: sexe, classe et héritage culturel, 1900–1940). Umeå, Institutionen för historiska studier, Umeå universitet, 2005, 303 p. (ill.). (Skrifter från Institutionen för historiska studier, 11). [English summary].

6279. MAC CARTHY (Angela). Irish migrants in New Zealand, 1840–1937: 'the desired haven'. Woodbridge, Boydell Press, 2005, XI-314 p. (ill., maps). (Irish historical monographs series).

6280. MAC MANUS (Sheila). The line which separates. Race, gender, and the making of the Alberta-Montana borderlands. Lincoln, University of Nebraska Press, 2005, XXIII-236 p. (Race and ethnicity in the American west).

6281. MAC TAVISH (Lianne). Childbirth and the display of authority in early modern France. Burlington, Ashgate, 2005, XIV-257 p. (Women and gender in the early modern world).

6282. MAC WILLIAMS (James E.). A revolution in eating. How the quest for food shaped America. New York, Columbia U. P., 2005, 386 p. (Arts and traditions of the table, perspectives on culinary history).

6283. MACHONIN (Pavel). Česká společnost a sociologické poznání. Problémy společenské transformace a modernizace od poloviny šedesátých let 20. století do současnosti. (Czech society and sociological knowledge). Praha, ISV nakl., 2005, 286 p.

6284. MAGER (Anne). 'One Beer, One Goal, One Nation, One Soul': South African Breweries, heritage, masculinity and nationalism 1960–1999. *Past and present*, 2005, 188, p. 163-194.

6285. MAHNKE-DEVLIN (Julia). Britische Migration nach Russland im 19. Jahrhundert. Integration – Kultur – Alltagsleben. Wiesbaden, Harrassowitz, 2005, 297 p. (Veröffentlichungen des Osteuropa-Institutes München, 69. Reihe Geschichte 9).

6286. MAHON (Rianne). Child care as citizenship right? Toronto in the 1970s and 1980s. *Canadian historical review*, 2005, 86, 2, p. 285-316.

6287. Marchés, migrations et logiques familiales dans les espaces français, canadien et suisse, XVIIIe–XXe siècles. Ed. par Luigi LORENZETTI, Anne-Lise HEADKÖNIG et Joseph GOY. Berne, Peter Lang, 2005, 321 p.

6288. MARK (James). Remembering rape: divided social memory and the Red Army in Hungary 1944–1945. *Past and present*, 2005, 188, p. 133-161.

6289. MILES (Tiya). Ties that bind. The story of an Afro-Cherokee family in slavery and freedom. Berkeley a. Los Angeles, University of California Press, George Gund Foundation input in African American studies, 2005, XIX-306 p. (American crossroads).

6290. MIRA ABAD (Alicia). Mujer, trabajo, religión y movilización social en el siglo XIX: modelos y paradojas. *Historia social*, 2005, 53, p. 85-102.

6291. Mişcarea naturală a populaţiei între 1901–1910. Transilvania. Vol. I-II. (The natural movement of the population between 1901–1910. Transylvania). Coordonatori Traian ROTARIU, Maria SEMENIUC şi Elemér MEZEI. Cluj-Napoca, Editura Presa Universitară Clujeană, 2005, 2 vol., 597 p., 299 p.

6292. MITCHELL (Pablo). Coyote nation: sexuality, race, and conquest in modernizing New Mexico, 1880–1920. Chicago, University of Chicago Press, 2005, XV-235 p. (Worlds of desire: the Chicago series on sexuality, gender, and culture).

6293. MOEHLING (Carolyn M.). "She has suddenly become powerful": youth employment and household decision making in the early twentieth century. *Journal of economic history*, 2005, 65, 2, p. 414-438.

6294. MORMINO (Gary R.). Land of sunshine, state of dreams. A social history of modern Florida. Gainesville, University Press of Florida, 2005, XVII-457 p. (The Florida history and culture series).

6295. MORRISSEY (Susan K.). Drinking to death: suicide, vodka and religious burial in Russia. *Past and present*, 2005, 186, p. 117-146.

6296. "Morts d'inanition": famine et exclusions en France sous l'Occupation. Sous la dir. Isabelle von BUELTZINGSLOEWEN. Rennes, Presses Universitaires de Rennes, 2005, 305 p. (Histoire).

6297. Mothers of the municipality: women, work, and social policy in post-1945 Halifax. Ed. by Judith FINGARD and Janet GUILDFORD. Buffalo, University of Toronto Press, 2005, VIII-318 p.

6298. MUCHEMBLED (Robert). L'orgasme et l'Occident. Une histoire du plaisir du XVIe siècle à nos jours. Paris, Ed. du Seuil, 2005, 382 p.

6299. MÜLLER (Ralf C.). Franken im Osten. Art, Umfang, Struktur und Dynamik der Migration aus dem lateinischen Westen in das Osmanische Reich des 15./16. Jahrhunderts auf der Grundlage von Reiseberichten. Leipzig, Eudora, 2005, 571 p.

6300. MÜLLER (Tanja). The making of elite women. Revolution and nation building in Eritrea. Leyden a. Boston, E. J. Brill, 2005, 299 p. (Afrika-Studiecentrum series, 4).

6301. MUNNO (Cristina). Prestige, integration, parentèle: les réseaux de parrainage dans une communauté de Vénétie (1834–1854). *Annales de démographie historique*, 2005, 109, p. 95-130.

6302. NAGEL (Silke). Ausländer in Mexiko: die Kolonien der deutschen und US-amerikanischen Einwanderer in der mexikanischen Hauptstadt, 1890–1942. Frankfurt am Main, Vervuert, 2005, XII-428 p. (Berliner Lateinamerika-Forschungen).

6303. NAJMABADI (Afsaneh). Women with mustaches and men without beards. Gender and sexual anxieties of Iranian modernity. Berkeley a. Los Angeles, University of California Press, 2005, XIV-363 p.

6304. NÄSLUND (Sture). Välfärd under Polstjärnan: nordbornas väg till tryggheten. (La société de bien-être sous l'étoile polaire: la voie scandinave vers l'Etat-providence). Stockholm, Hjalmarson & Högberg, 2005, 273 p.

6305. NEČAS (Ctibor). Romové na Moravě a ve Slezsku 1790–1945. (The Gypsies [Roma] in Moravia and Silesia, 1790–1945). Brno, Matice moravská, 2005, 475 p. (photogr., tb.). (Knižnice Matice moravské, 15).

6306. NEŠPOR (Zdeněk R.). České migrace 19. a 20. století a jejich dosavadní studium. (The study of Czech migration in the nineteenth and twentieth centuries). *Soudobé dějiny*, 2005, 12, 2, p. 245-284.

6307. NORRBY (Göran). Adel i förvandling: adliga strategier och identiteter i 1800-talets borgerliga samhälle. (Une noblesse en mutation: stratégies et identités aristocratiques dans la société bourgeoise du 19e siècle). Uppsala, Acta Universitatis Upsaliensis, 2005, 380 p. (ill.). (Studia historica Upsaliensia, 217). [English summary].

6308. NUMHAUSER (Paulina). Mujeres indias y señores de la coca: Potosí y Cuzco en el siglo XVI. Madrid, Cátedra, 2005, 407 p. (Historia. Serie menor).

6309. NURMIAINEN (Jouko). Northern European elites in historical perspectives. *Scandinavian journal of history*, 2005, 30, 3-4, p. 217-224.

6310. NUTINI (Hugo G.). Social stratification and mobility in central Veracruz. Austin, University of Texas Press, 2005, XI-178 p.

6311. OGILVIE (Sheilagh). Communities and the 'Second Serfdom' in early modern Bohemia. *Past and present*, 2005, 187, p. 69-119.

6312. OLIVIER (Cyril). Le vice ou la vertu, Vichy et les politiques de la sexualité. Toulouse, Presses Universitaires du Mirail, 2005, 311 p.

6313. OLTMER (Jochen). Migration und Politik in der Weimarer Republik: mit 18 Tabellen. Göttingen, Vandenhoeck & Ruprecht, 2005, 564 p. (ill.).

6314. ORLECK (Annelise). Storming Caesars Palace: how Black mothers fought their own war on poverty. Boston, Beacon, 2005, 368 p.

6315. OTTERNESS (Philip). Becoming German. The 1709 Palatine migration to New York. Ithaca, Cornell U. P., 2005, XIII-235 p.

8. SOCIAL HISTORY

6316. ÖZDEMIR (Hikmet). Salgın hastalıklardan ölümler: 1914−1918. (Morts par les maladies épidémiques). Ankara, Türk Tarih Kurumu, 2005, 444 p.

6317. ÖZTÜRK (Serdar). Cumhuriyet Türkiye'sinde kahvehane ve iktidar (1930−1945). [Café et pouvoir dans la Turquie de la République (1930−1945)]. İstanbul, Kırmızı, 2005, 546 p.

6318. PAS (Nicolas). Images d'une révolte ludique: le mouvement néerlandais Provo en France dans les années soixante. *Revue historique*, 2005, 634, p. 343-374.

6319. PÉREZ (Louis A. Jr.). To die in Cuba. Suicide and society. Chapel Hill, University of North Carolina Press, 2005, XIV-463 p. (E. Eugene and Lillian Youngs Lehman series).

6320. PICKUS (Noah). True faith and allegiance: immigration and American civic nationalism. Princeton, Princeton U. P., 2005, XIII-257 p.

6321. PIERCE (Richard B.). Polite protest: the political economy of race in Indianapolis, 1920–1970. Bloomington, Indiana U. P., 2005, X-155 p.

6322. PIVATO (Stefano). Bella ciao. Canto e politica nella storia d'Italia. Roma e Bari, Laterza, 2005, XII-378 p.

6323. Pobreza e historia en Costa Rica: determinantes estructurales y representaciones sociales del siglo XVII a 1950. Ed. por Ronny J. VIALES HURTADO. San José, Editorial de la Universidad de Costa Rica, CIHAC, Posgrado Centroamericano en Historia, Universidad de Costa Rica, 2005, 328 p. (ill.). (Colección Nueva historia).

6324. POKORNÝ (Jiří). Die Betriebsklubs in der Tschechoslowakei 1945–1968.: Zur Organisation sozialistischer Erziehung, Kultur und Erholung der Arbeiterschaft. *In*: Sozialgeschichtliche Kommunismusforschung. Tschechoslowakei, Polen, Ungarn und DDR 1948–1968. Ed. Christiane BRENNER. München, 2005, p. 263-275.

6325. POOLEY (Colin G.), TURNBULL (Jean), ADAMS (Mags). "...everywhere she went I had to tag beside her": family, life course, and everyday mobility in England since the 1940s. *History of the family*, 2005, 10, 2, p. 119-136.

6326. POUILLARD (Véronique). La publicité en Belgique, 1850–1975. Des courtiers aux agences internationales. Bruxelles, Académie Royale de Belgique, 2005, 509 p.

6327. Power and the people. A social history of Central European politics, 1945–1956. Ed. by Eleonore BREUNING, Jill LEWIS and Gareth PRITCHARD. Manchester, Manchester U. P., 2005, XXVII-292 p.

6328. PREČAN (Vilém). Dějiny exilu a emigrace. Prameny, výzkumné instituce a projekty. (History of exile and emigration. Sources, research institutions and projects). *Česko-slovenská historická ročenka*, 2005, p. 125-130.

6329. PREIN (Philipp). Bürgerliches Reisen im 19. Jahrhundert: Freizeit, Kommunikation und soziale Grenzen. Münster, LIT, 2005, VIII-297 p. (Kulturgeschichtliche Perspektiven, 3).

6330. PREMO (Bianca). Children of the Father King: youth, authority, and legal minority in colonial Lima. Chapel Hill, University of North Carolina Press, 2005, XIII-350 p. (ill., map).

6331. Přeshraniční spolupráce na východních hranicích České republiky. Růžový obláček a hrana reality. (Cross-bordier collaboration on eastern border of the Czech Republic). Ed. Vít DOČKAL. Brno, Masarykova univerzita, 2005, 140 p. (Studie, 35).

6332. Problémům menšin (K) v Československu v letech 1945–1989. Sborník studií. (Zu den Problemen der nationalen Minderheiten in der Tschechoslowakei in den Jahren 1945–1989). Ed. Helena NOSKOVÁ [et al.]. Praha, Ústav pro soudobé dějiny AV ČR, 2005, 267 p. (Studijní materiály Ústavu pro soudobé dějiny AV ČR).

6333. Prostitution (La) à Paris. Sous la dir. de Marie-Élisabeth HANDMAN et Janine MOSSUZ-LAVAU. Paris, Éditions de La Martinière, 2005, 414 p.

6334. RAINHORN (Judith). Paris, New York: des migrants italiens: années 1880-années 1930. Paris, CNRS éditions, 2005, 233 p. (ill., maps). (CNRS histoire).

6335. RILEY (James C.). Poverty and life expectancy. The Jamaica paradox. Cambridge, Cambridge U. P., 2005, 235 p.

6336. RINEY-KEHRBERG (Pamela). Childhood on the farm: work, play, and coming of age in the Midwest. Lawrence, University Press of Kansas, 2005, XI-300 p.

6337. ROBERTSON (Stephen). Crimes against children: sexual violence and legal culture in New York City, 1880–1960. Chapel Hill, University of North Carolina Press in association with the American Society for Legal History, 2005, XII-337 p. (Studies in legal history).

6338. ROBINSON (Armstead L.). Bitter fruits of bondage. The demise of slavery and the collapse of the Confederacy, 1861–1865. Charlottesville, University of Virginia Press, 2005, XVIII-352 p. (Carter G. Woodson Institute series).

6339. ROEDIGER (David R.). Working toward whiteness: how America's immigrants became white; the strange journey from Ellis Island to the suburbs. New York, Basic Books, 2005, VII-339 p.

6340. ROSS (Chad). Naked Germany. Health, race and the nation. Oxford a. New York, Berg, 2005, 239 p.

6341. ROTH (Ralf). Amerika – Deutschland. Folgen einer transatlantischen Migration. *Historische Zeitschrift*, 2005, 281, 3, p. 621-657.

6342. ROTHMAN (Adam). Slave country. American expansion and the origins of the deep South. Cambridge, Harvard U. P., 2005, XI-296 p.

6343. RUDY (Jarrett). The freedom to smoke: tobacco consumption and identity. Ithaca, McGill-Queen's U. P., 2005, X-232 p. (Studies on the history of Quebec, 18).

6344. SAFLEY (Thomas Max). Children of the laboring poor: expectation and experience among the orphans of Early Modern Augsburg. Boston, Brill, 2005, XVI-493 p. (Studies in Central European Histories, 38).

6345. SALERNO (Beth A.). Sister societies. Women's antislavery organizations in antebellum America. Dekalb, Northern Illinois U. P., 2005, X-233 p.

6346. SALLES (Vicente). O negro no pará sob o regime da escravidao. Belém, Instituto de artes do Pará, 2005, 372 p. (Programa raizes).

6347. SANBORN (Joshua A.). Unsettling the empire: violent migrations and social disaster in Russia during World War I. *Journal of modern history*, 2005, 77, 2, p. 290-324.

6348. SANDAGE (Scott A.). Born losers. A history of failure in America. Cambridge, Harvard U. P., 2005, X-362 p.

6349. SCHLEGEL-VOß (Lil-Christine). Alter in der "Volksgemeinschaft". Zur Lebenslage der älteren Generation im Nationalsozialismus. Berlin, Duncker & Humblot, 2005, 327 p. (Schriften zur Wirtschafts- und Sozialgeschichte, 80).

6350. SCHMIDT (C. Bettina). Jugendkriminalität und Gesellschaftskrisen. Umbrüche, Denkmodelle und Lösungsstrategien im Frankreich der Dritten Republik 1900–1914. Stuttgart, Steiner, 2005, 426 p. (Vierteljahrschrift für Sozial- und Wirtschaftsgeschichte, Beihefte, 182).

6351. SCHMIDTMANN (Christian). Katholische Studierende 1945–1973: eine Studie zur Kultur- und Sozialgeschichte der Bundesrepublik Deutschland. Paderborn, Schöningh, 2005, 535 p. (Veröffentlichungen der Kommission für Zeitgeschichte / B, 102).

6352. SCHOEN (Johanna). Choice and coercion. Birth control, sterilization, and abortion in public health and welfare. Chapel Hill, University of North Carolina Press, 2005, XIV-331 p. (Gender and American culture).

6353. SCHUBERT (Stefan). Saisonarbeit am Kanal: Rekrutierung, Arbeits- und Lebensverhältnisse ausländischer Arbeitskräfte beim Bau des Mittellandkanals im Osnabrücker Land 1910–1916. Frankfurt am Main, IKO, 2005, 364 p.

6354. SCHULTZ HANSEN (Hans). Hjemmetysken i Nordslesvig 1840–1867: den slesvig-holstenske bevægelse. Bd. 1. 1840–1850. Bd. 2. 1850–1867. (Les Allemands de l'intérieur dans le Schleswig du nord 1840–1867: le mouvement du Schleswig-Holstein. Tomes 1. 1840–1850. Tome 2. 1850–1867). Aabenraa, Historisk Samfund for Sønderjylland, 2005, 2 vol., 536 p., 494 p. (Skrifter udg. af Historisk Samfund for Sønderjylland, 93).

6355. SCHULZ (Knut), SCHUCHARD (Christiane). Handwerker deutscher Herkunft und ihre Bruderschaften im Rom der Renaissance. Darstellung und ausgewählte Quellen. Roma, Freiburg u. Wien, Herder, 2005, 712 p. (Römische Quartalschrift für Christliche Altertumskunde und Kirchengeschichte, Suppl., 57).

6356. SCHUSTER (Klaus). Wirtschaftliche Entwicklung, Sozialstruktur und biologischer Lebensstandard in München und dem südlichen Bayern im 19. Jahrhundert. St. Katherinen, Scripta Mercaturae, 2005, V-287 p. (Studien zur Wirtschafts- und Sozialgeschichte, 24).

6357. SCOTT (Rebecca J.). Degrees of freedom. Louisiana and Cuba after slavery. Cambridge, Belknap Press of Harvard U. P., 2005, 365 p.

6358. SGRAZZUTTI (Jorge P.), ROLDÁN (Diego). Tiempo libre y disciplinamiento en las clases obreras italiana y alemana durante el período de entreguerras. Dopolavoro y Kraft durch Freude: un análisis comparativo. *Historia social*, 2005, 52, p. 109-128.

6359. SHIPPER (Apichai W.). Criminals or victims? The politics of illegal foreigners in Japan. *Journal of Japanese studies*, 2005, 31, 2, p. 299-327.

6360. SIDDALI (Silvana R.). From property to person. Slavery and the confiscation acts, 1861–1862. Baton Rouge, Louisiana State U. P., 2005, X-298 p. (Conflicting worlds, new dimensions of the American Civil War).

6361. SIEVENS (Mary Beth). Stray wives: marital conflict in early national New England. New York, New York U. P., 2005, XII-171 p.

6362. SIMON (Bryant). Boardwalk of dreams. Atlantic City and the face of urban America. New York, Oxford U. P., 2005, XII-285 p.

6363. SIRINELLI (Jean-François). Die Babyboomer und der Mai 1968 in Frankreich. *Vierteljahrshefte für Zeitgeschichte*, 2005, 53, 4, p. 527-546.

6364. SIRKECI (İbrahim). Irak'tan Türkmen göçleri ve göç eğilimleri. (Turkmen immigration from Iraq and migration tendencies). Ankara, Global Strateji Enstitüsü = Global Strategy Institute, 2005, 88 p.

6365. SMITH (Susan L.). Japanese American midwives: culture, community, and health politics, 1880–1950. Urbana a. Chicago, University of Illinois Press, 2005, IX-280 p. (The Asian American experience).

6366. Societatea românească între modern și exotic, văzută de călători străini (1800–1847). (The Romanian society between modernity and exotic, viewd by foreign travellers, 1800–1847). Coordonatori Ileana CĂZAN și Irina GAVRILĂ. București, Editura Oscar Print, 2005, [s. p.].

6367. Sozialstaatlichkeit in der DDR. Sozialpolitische Entwicklungen im Spannungsfeld von Diktatur und Gesellschaft 1945/49–1989. Hrsg. v. Dierk HOFFMANN und Michael SCHWARTZ. München, Oldenbourg, 2005,

197 p. (Schriftenreihe der Vierteljahrshefte für Zeitgeschichte, Sondernummer).

6368. SPIRE (Alexis). Etrangers à la carte: l'administration de l'immigration en France, (1945–1975). Paris, B. Grasset, 2005, 402 p.

6369. SPRINGER (Kimberly). Living for the revolution. Black feminist organizations, 1968–1980. Durham, Duke U. P., 2005, X-228 p.

6370. STANĚK (Tomáš), ARBURG (Adrian von). Organizované divoké odsuny? Úloha ústředních státních orgánů při provádění "evakuace" německého obyvatelstva (květen až září 1945). (Organized spontaneous "transfers"? The role of the central state organs in "evacuating" the German population from May to September 1945). Tomo 1. Předpoklady a vývoj do konce května 1945. *Soudobé dějiny*, 2005, 12, 3/4, p. 465-533.

6371. STARGARDT (Nicholas). Witnesses of war. Children's lives under the nazis. London, Jonathan Cape, 2005, XVI-509 p.

6372. STEFANSKI (Valentina Maria). Nationalsozialistische Volkstums- und Arbeitseinsatzpolitik im Regierungsbezirk Kattowitz 1939–1945. *Geschichte und Gesellschaft*, 2005, 31, 1, p. 38-67.

6373. STOBART (Jon). Information, trust and reputation: shaping a merchant elite in early 18th-century England. *Scandinavian journal of history*, 2005, 30, 3-4, p. 298-307.

6374. STRANGE (Julie-Marie). Death, grief and poverty in Britain, 1870–1914. Cambridge, New York a. Melbourne, Cambridge U. P., 2005, X-294 p.

6375. Svenskt i Finland – finskt i Sverige. 1. Dialog och särart: människor, samhällen och idéer från Gustav Vasa till nutid. (Suédois en Finland – Finlandais en Suède. 1. Dialogue et spécificités: hommes, sociétés et idées de Gustave Vasa à aujourd'hui). Red. Gabriel BLADH, Christer KUVAJA. Helsinki, Svenska litteratursällskapet i Finland, Stockholm, Atlantis, 2005, 448 p. (Skrifter utgivna av Svenska litteratursällskapet i Finland, 682:1). [English summary].

6376. TAYLOR (Amy Murrell). The divided family in Civil War America. Chapel Hill, University of North Carolina Press, 2005, XIV-319 p. (Civil War America).

6377. TAYLOR (Matthew). The leaguers. The making of professional football in England, 1900–1939. Liverpool, Liverpool U. P., 2005, XXIV-320 p.

6378. TAYLOR (Nikki M.). Frontiers of freedom. Cincinnati's black community, 1802–1868. Athens, Ohio U. P., 2005, XVII-315 p. (Ohio University press series on law, society, and politics in the midwest).

6379. TAYLOR ALLEN (Ann). Feminism and motherhood in Western Europe, 1890–1970: the maternal dilemma. New York, Palgrave Macmillan, 2005, XI-354 p.

6380. TERPSTRA (Nicholas). Abandoned children of the Italian Renaissance: orphan care in Florence and Bologna. Baltimore, Johns Hopkins U. P., 2005, XII-349 p. (Johns Hopkins University studies in historical and political science, 123rd series, 4).

6381. THOMPSON (Paul). Feminism discreet but not forgotten: Papers from Marion Thompson, 1909–95. *History workshop*, 2005, 60, p. 152-169.

6382. TIMOSCHENKO (Tatjana). Die Verkäuferin im Wilhelminischen Kaiserreich. Etablierung und Aufwertungsversuche eines Frauenberufs um 1900. Frankfurt am Main, Lang, 2005, 177 p.

6383. TODD (Selina). Young women, work, and family in England, 1918–1950. Oxford, Oxford U. P., 2005, 272 p.

6384. TRACY (Sarah W.). Alcoholism in America. From reconstruction to prohibition. Baltimore, Johns Hopkins U. P., 2005, XXIII-357 p.

6385. VAN POPPEL (Frans), JONKER (Marianne), MANDEMAKERS (Kees). Differential infant and child mortality in three Dutch regions, 1812–1909. *Economic history review*, 2005, 58, 2, p. 272-309.

6386. VAN TILBURG (Marja). Gender and old age. Seniority in Dutch conduct literature of the early modern and modern period, 20th International Congress of Historical Sciences-proceedings. Sydney, CISH, 2005, [s. p.]. – EADEM. The attraction of Tahiti. Gender in late 18th century French texts on the Pacific", 20th International Congress of Historical Sciences-proceedings. Sydney, CISH 2005, [s. p.]. – EADEM. The seduction of the South Sea. Gender in late eighteenth century French travelogues on the Pacific. *In*: Identité et altérité. Analyses discursives de la communication touristique. Ed. par F. BAIDER, M. BURGER et D. GOUTSOS. Paris, L'Harmattan France 2005, p. 67-82. – EADEM. Where has 'the old, wise woman' gone…? Gender and age in early modern and modern advice literature. *In*: The prime of their lives. Wise old women in pre-industrial society. Ed. by A.B. MULDER-BAKKER and R. NIP. Leuven, Peeters 2005, p. 149-169. (Groningen studies in cultural change series).

6387. VERNON (James). The ethics of hunger and the assembly of society: the techno-politics of the school meal in modern Britain. *American historical review*, 2005, 110, 3, p. 693-725.

6388. VOLPI (Alessandro). Viareggio laica: la Massoneria in provincia, 1848–1925. Pisa, ETS, 2005, 103 p.

6389. WADAUER (Sigrid). Die Tour der Gesellen. Mobilität und Biographie im Handwerk vom 18. bis zum 20. Jahrhundert. Frankfurt am Main u. New York, Campus, 2005, 418 p. (Historische Sozialwissenschaft, 30).

6390. WAGNER (David). The poorhouse. America's forgotten institution. Lanham, Rowman and Littlefield, 2005, XI-179 p.

6391. WAGNER (Patrick). Bauern, Junker und Beamte: lokale Herrschaft und Partizipation im Ostelbien des 19. Jahrhunderts. Göttingen, Wallstein, 2005, 623 p. (Moderne Zeit, 9).

6392. WILD (Mark). Street meeting. Multiethnic neighborhoods in early twentieth-century Los Angeles. Berkeley, University of California Press, 2005, XI-298 p.

6393. WILLS (Abigail). Delinquency, masculinity and citizenship in England 1950–1970. *Past and present*, 2005, 187, p. 157-185.

6394. WOLFF (Charlotta). The Swedish aristocracy and the French enlightenment circa 1740–1780. *Scandinavian journal of history*, 2005, 30, 3-4, p. 259-270.

6395. WONG (John Chi-Kit). Lords of the rinks: the emergence of the National Hockey League, 1875–1936. Toronto, University of Toronto Press, 2005, VIII-235 p.

6396. WOODALL (Ann M.). What price the poor? William Booth, Karl Marx and the London residuum. Aldershot, Ashgate, 2005, XII-233 p.

6397. YANGWEN (Zheng). The social life of opium in China. Cambridge, New York a. Melbourne, Cambridge U. P., 2005, XIII-241 p.

6398. YATES (JoAnne). Structuring the information age: life insurance and technology in the twentieth century. Baltimore, Johns Hopkins U. P., 2005, X-351 p.

6399. ZAKHARINE (Dmitri). Von Angesicht zu Angesicht. Der Wandel direkter Kommunikation in der ost- und westeuropäischen Neuzeit. Konstanz, UVK, 2005, 689 p. (Historische Kulturwissenschaft, 7).

6400. Życie jest wszędzie-: ruchy społeczne w Polsce i Rosji do II wojny światowej : zbiór materiałów z konferencji 16–17 września 2003 r., Warszawa. (Life is everywhere: social movements in Poland and Russia to World War II: a collection of materials from the conference 16–17 September 2003). Pod redakcją Anny BRUS. Warszawa, Wydawn. Neriton, Instytut Historii PAN, 2005, 360 p.

§ 8. Addenda 2001–2004.

6401. CINOTTO (Simone). Una famiglia che mangia insieme: Cibo ed etnicità nella comunità italoamericana di New York, 1920–1940. Torino, Otto Editore, 2001, X-458 p.

6402. Deutsche Kriegsgesellschaft (Die) 1939 bis 1945: Ausbeutung, Deutungen, Ausgrenzung. Hsrg. v. Jörg ECHTERNKAMP. Müunchen, Deutsche Verlags-Anstalt, 2004, 1128 p.

6403. FLANDRIN (Jean-Louis). L'Ordre des mets. Paris, Odile Jacob, 2002, 278 p.

6404. MATSUZAWA (Yusaku). Ishin-ki tyokkatsu ken ni okeru kyūjutsu to bikō tyochiku. (Poor relief and emergency funds in the early Meiji period, 1869–1871). *Socio-economic history*, 2004, 70, 4, p. 71-92 [Eng. Summary].

6405. Zwischen "nationaler Revolution" und militärischer Aggression: Transformationen in Kirche und Gesellschaft während der konsolidierten NS-Gewaltherrschaft, 1934–1939. Hrsg. v. Gerhard BESIER und Elisabeth MÜLLER-LUCKNER. München, R. Oldenbourg Verlag, 2001, XXVII-276 p.

Cf. n^{os} *108-128, 771, 775, 3697, 3747, 3771, 3806, 3808, 3818, 3873, 3897, 4054, 4109, 4234, 4297, 4313, 4362, 4366, 4377, 4422, 4464, 4483, 4492, 4499, 4563, 4642, 4791, 4819, 5757, 5909, 5922, 5969, 6425*

§ 9. Working-class movement and socialism.

6406. ALEXANDER (Robert Jackson). A history of organized labor in Bolivia. With the collaboration of Eldon M. PARKER. Westport a. London, Praeger Publishers, 2005, XI-197 p. – IDEM. A history of organized labor in Uruguay and Paraguay. With the collaboration of Eldon M. PARKER. Westport, Praeger Publishers, 2005, X-164 p.

6407. Apogée (L') des syndicalismes en Europe occidentale, 1960–1985. Sous la direction de Michel PIGENET, Patrick PASTURE et Jean-Louis ROBERT. Paris, Publications de la Sorbonne, 2005, 282 p. (Série Internationale).

6408. Arbeiter im Staatssozialismus. Ideologischer Anspruch und soziale Wirklichkeit. Hrsg. v. Peter HÜBNER, Christoph KLESSMANN und Klaus TENFELDE. Köln, Weimar u. Wien, Böhlau, 2005, 515 p. (Zeithistorische Studien, 31).

6409. Asian labor in the wartime Japanese empire: unknown histories. Ed. by Paul H. KRATOSKA. Armonk, M. E. Sharpe, 2005, XX-433 p.

6410. AUSTIN (Gareth). Labour, land and capital in Ghana: from slavery to free labour in Asante, 1807–1956. Rochester, University of Rochester Press, 2005, XXIV-589 p. (Rochester studies in African history and the diaspora).

6411. BAILKIN (Jordanna). Making faces: tattooed women and colonial regimes. *History workshop*, 2005, 59, p. 33-56.

6412. BANERJEE (Dipankar). Labour movement in Assam: a study of non-plantation workers' strikes till 1939. New Delhi, Anamika Publishers & Distributors, 2005, 234 p.

6413. BARNES (John). Gentleman crusader: Henry Hyde Champion in the early Socialist Movement. *History workshop*, 2005, 60, p. 116-138.

6414. BARTON (Susan). Working-class organisations and popular tourism, 1840–1970. Manchester a. New York, Manchester U. P., 2005, XII-237 p.

6415. BEAVEN (Brad). Leisure, citizenship and working-class men in Britain, 1850–1945. Manchester a. New York, Manchester U. P., 2005, XII-258 p.

6416. BELL (Alan R.), MARTIN (Janette), MAC CAUSLAND (Sigrid). Labour's memory: a comparison of labour history archives in Australia, England, Wales and Scotland. *Labour history*, 2005, 88, p. 25-44.

6417. BERGER (Stefan), PATMORE (Greg). Comparative labour history in Britain and Australia. *Labour history*, 2005, 88, p. 9-24.

6418. BROOKE (Stephen). The body and socialism: Dora Russell in the 1920s. *Past and present*, 2005, 189, p. 147-177.

6419. CARLILE (Lonny E.). Divisions of labor. Globality, ideology, and war in the shaping of the Japanese labor movement. Honolulu, University of Hawai'i Press, 2005, X-292 p.

6420. COLLEY (Linda). How secure was that public service job? Redundancy in the Queensland public service 1859–1993. *Labour history*, 2005, 89, p. 141-158.

6421. CRUZ (Rafael). El órgano de la clase obrera: los significados de movimiento obrero en la España del siglo XX. *Historia social*, 2005, 53, p. 155-174.

6422. DEERY (Phillip), REDFERN (Neil). No lasting peace? Labor, Communism and the Cominform: Australia and Great Britain, 1945–1950. *Labour history*, 2005, 88, p. 63-86.

6423. DETER (Gerhard). Handwerk vor dem Untergang? Das westfälische Kleingewerbe im Spiegel der preußischen Gewerbetabellen 1816–1861. Stuttgart, Steiner, 2005, 160 p. (Studien zur Gewerbe- und Handelsgeschichte, 25).

6424. DUMONT (Dora M.). Workers in Risorgimento Bologna. *Canadian journal of history*, 2005, 40, 1, p. 23-44.

6425. Femmes, féminismes et socialismes dans l'espace germanophone après 1945. Sous la dir. de Corinne BOUILLOT et Paul PASTEUR. Paris, Belin, 2005, 240 p. (Europes centrales).

6426. FRAGER (Ruth A.), PATRIAS (Carmela). Discounted labour. women workers in Canada, 1870–1939. Toronto, University of Toronto Press, 2005, 189 p. (Themes in Canadian history).

6427. FUCHS (Rachel G.), THOMPSON (Victoria E.). Women in nineteenth-century Europe. Basingstoke, Palgrave, 2005, 216 p. (European culture and society series).

6428. FULBROOK (Mary). The people's state: East German society from Hitler to Honecker. New Haven, Yale U. P., 2005, XIV-350 p.

6429. GARCA-BRYCE (Iigo). From artisan to worker: the language of class during the age of liberalism in Peru, 1858–1879. *Social history*, 2005, 30, 4, p. 463-480.

6430. Generationen in der Arbeiterbewegung. Hrsg. v. Klaus SCHÖNHOVEN und Bernd BRAUN. München, Oldenbourg, 2005, 269 p. (Schriftenreihe der Stiftung Reichspräsident-Friedrich-Ebert Gedenkstätte, 12).

6431. HACHTMANN (Rüdiger). Chaos und Ineffizienz in der Deutschen Arbeitsfront. Ein Evaluierungsbericht aus dem Jahr 1936. *Vierteljahrshefte für Zeitgeschichte*, 2005, 53, 1, p. 43-78.

6432. HERON (Craig), PENFOLD (Steve). The workers' festival. a history of Labour Day in Canada. Toronto, University of Toronto Press, 2005, XVIII-340 p.

6433. HRUBÝ (Karel). Radikální socialismus a jeho návraty. (Radical Socialism and Its Recurrences). *Soudobé dějiny*, 2005, 12, 2, p. 334-342.

6434. HÜRTGEN (Renate). Zwischen Disziplinierung und Partizipation. Vertrauensleute des FDGB [Freier Deutscher Gewerkschaftsbund] im DDR-Betrieb. Köln, Weimar u. Wien, Böhlau, 2005, 353 p.

6435. Invisible woman (The). Aspects of women's work in eighteenth-century Britain. Ed. by Isabelle BAUDINO, Jacques CARRÉ and Cécile RÉVAUGER. Aldershot, Ashgate, 2005, XII-188 p.

6436. JONES (William P.). The tribe of Black Ulysses. African American lumber workers in the Jim Crow South. Urbana a. Chicago, University of Illinois Press, 2005, XV-235 p. (Working class in American history).

6437. KIR'JANOV (Jurij I.). Sotsial'no-politicheskij protest rabochikh Rossii v gody Pervoj mirovoj vojny (ijul' 1914–fevral' 1917 g.). (The social and political protest of workers in Russia during the First World War, July 1914–February 1917). Appendix by I.M. PUSHKAREVA. RAN, In-t rossijskoj istorii. Moskva, IRI RAN, 2005, 217 p. (bibl. incl.; pers. ind. p. 214-216).

6438. KIRBY (Peter). Debate: how many children were 'Unemployed' in Eighteenth- and Nineteenth-Century England?. *Past and present*, 2005, 187, p. 187-202. – CUNNINGHAM (Hugh). Reply. *Past and present*, 2005, 187, p. 203-215.

6439. KIRK (Neville). The conditions of royal rule: Australian and British socialist and labour attitudes to the monarchy, 1901–1911. *Social history*, 2005, 30, 1, p. 64-88.

6440. KITTNER (Michael). Arbeitskampf: Geschichte, Recht, Gegenwart. München, Beck, 2005, XXIV-783 p. (ill.).

6441. KOENKER (Diane P.). Republic of labor: Russian printers and Soviet socialism, 1918–1930. Ithaca, Cornell U. P., 2005, XII-343 p.

6442. LAMBERT (Josiah B.). "If the workers took a notion". The right to strike and American political development. Ithaca, ILR Press, 2005, X-259 p.

6443. LANGKAU-ALEX (Ursula). Deutsche Volksfront 1932–1939: zwischen Berlin, Paris, Prag und Moskau. Band 3. Dokumente zur Geschichte des Ausschusses zur Vorbereitung einer deutschen Volksfront, Chronik und Verzeichnisse. Berlin, Akademie, 2005, 544 p. [Band 1, 2. Cf. Bibl. 2004 n° 6744.]

6444. LEE (R. Alton). Farmers versus wage earners. Organized labor in Kansas, 1860–1960. Lincoln, University of Nebraska Press, 2005, XIV-340 p.

6445. LEIKIN (Steve). Practical utopians. American workers and the cooperative movement in the Gilded age. Detroit, Wayne State U. P., 2005, XXI-233 p.

6446. LICHTENSTEIN (Alex). Making Apartheid work: African trade unions and the 1953 Native labour (settlement of disputes) act in South Africa. *Journal of African history*, 2005, 46, 2, p. 293-314.

6447. LIEBIG SCHLABER (Gerret). Sozialpolitik im Schleswiger Land, 1840–1880. Neumünster, Wachholtz, 2005, 432 p. (Wirtschafts- und Sozialgeschichte Schleswig-Holsteins, 39).

6448. LUIS CHINEA (Jorge). Race and labor in the Hispanic Caribbean. The West Indian immigrant worker experience in Puerto Rico, 1800–1850. Gainesville, University Press of Florida, 2005, XV-227 p. (New directions in Puertorican studies).

6449. MAC IVOR (Arthur), WRIGHT (Christopher). Managing labour: UK and Australian employers in comparative perspective, 1900–1950. *Labour history*, 2005, 88, p. 45-62.

6450. MAC TAVISH (Duncan). Business and public management in the U.K., 1900–2003. Aldershot, Ashgate, 2005, VIII-274 p.

6451. MATSUMURA (Takao). Igirisu no tetsudo sogi to saiban: Tafuveiru hanketsu no rodo-shi. (Rail strikes and their litigation in Britain: a labour history of the Taff Vale railway company decision). Kyoto, Minerva Shobo, 2005, 295 p.

6452. MICHELS (Tony). A fire in their hearts. Yiddish socialists in New York. Cambridge, Harvard U. P., 2005, VIII-335 p.

6453. MINCHIN (Timothy J.). "Don't Sleep with Stevens!" The J.P. Stevens campaign and the struggle to organize the South, 1963–80. Foreword by John David SMITH. Gainesville, University Press of Florida, 2005, XVI-239 p. (New perspectives on the history of the South).

6454. MITTELSTADT (Jennifer). From welfare to workfare. the unintended consequences of liberal reform, 1945–1965. Chapel Hill, University of North Carolina Press, 2005, XIII-267 p. (Gender and American culture).

6455. MURPHY (Kevin). Revolution and counterrevolution: class struggle in a Moscow metal factory. New York, Berghahn Books, 2005, XIV-234 p. (International studies in social history, 6).

6456. NUTTALL (Jeremy). Labour revisionism and qualities of mind and character, 1931–1979. *English historical review*, 2005, 120, 487, p. 667-694.

6457. PARMET (Robert D.). The master of Seventh Avenue: David Dubinsky and the American Labor Movement. New York, New York U. P., 2005, XI-436 p.

6458. PORRINI (Rodolfo). La nueva clase trabajadora uruguaya (1940–1950). Montevideo, Universidad de la República, Facultad de Humanidades y Ciencias de la Educación, Departamento de Publicaciones, 2005, 366 p. (Serie Tesis de posgrado en humanidades, 4).

6459. SALMOND (John A.). Southern struggles. the Southern labor movement and the civil rights struggle. Gainesville, University Press of Florida, 2005, XIV-212 p. (New perspectives on the history of the south).

6460. SMITH (Justin Davis), OPPENHEIMER (Melanie). The Labour Movement and Voluntary Action in the UK and Australia: a comparative perspective. *Labour history*, 2005, 88, p. 105-120.

6461. Socialistes (Les) et la ville, 1890–1914. *Cahiers Jaurès*, 2005, 177-178, 112 p.

6462. VAN DER LINDEN (Marcel). Reading ethnography as labour history: the example of the Iatmul, East Sepik Province, Papua New Guinea. *Labour history*, 2005, 89, p. 197-214. – IDEM. The importance of antilabour history. *In*: 'Extreme Right' (The) in twentieth century Australia [Cf. n° 3610], p. 121-124.

6463. VARGAS (Zaragosa). Labor rights are civil rights: Mexican American workers in twentieth-century America. Princeton, Princeton U. P., 2005, XVI-375 p.

6464. VENUGOPAL REDDY (Kanchi). Working class and freedom struggle: Madras presidency, 1918–1922. New Delhi, Mittal Publications, 2005, XI-95 p. (maps).

6465. WEHRLE (Edmund F.). Between a river and a mountain: the AFL-CIO [American Federation of Labor and Congress of Industrial Organizations] and the Vietnam War. Ann Arbor, University of Michigan Press, 2005, VIII-304 p.

6466. WERUM (Stefan Paul). Gewerkschaftlicher Niedergang im sozialistischen Aufbau: der Freie Deutsche Gewerkschaftsbund (FDGB) 1945 bis 1953. Göttingen, Vandenhoeck & Ruprecht, 2005, 861 p. (Schriften des Hannah-Arendt-Instituts für Totalitarismusforschung, 26).

6467. ZIPF (Karin L.). Labor of innocents: forced apprenticeship in North Carolina, 1715–1919. Baton Rouge, Louisiana State U. P., 2005, XI-207 p.

§ 9. Addenda 1995–1998.

6468. BABIANO (José). Emigrantes, cronómetros y huelgas: un estudio sobre el trabajo y los trabajadores durante el franquismo, Madrid, 1951–1977. Madrid, Siglo XXI de España Editores, Fundación 1° de Mayo, 95, X-372 p. (Historia).

6469. MOLINERO (Carme), YSÀS (Pere). Productores disciplinados y minorías subversivas: clase obrera y conflictividad laboral en la España franquista. Madrid, Siglo XXI, 98, XIII-281 p. (ill.). (Historia).

Cf. n°s 3605, 3617, 3659, 3761, 4091, 4184, 4344, 4362, 4405, 4584, 4614, 4619

O

MODERN LEGAL AND CONSTITUTIONAL HISTORY

§ 1. General. 6470-6501. – § 2. History of constitutional law. 6502-6526. – § 3. Public law and institutions. 6527-6548. – § 4. Civil and penal law. 6549-6613. – § 5. International law. 6614-6618.

§ 1. General.

* 6470. HALLEBEEK (Jan). Bericht über die kirchenrechtsgeschichtliche Forschung im Gebiet der Niederlande und Belgiens. *Zeitschrift der Savigny-Stiftung für Rechtsgeschichte Kanonistische Abteilung*, 2005, 91, p. 421-445.

* 6471. ISHIBE (Masasuke). Neuere deutsche Rechtsgeschichte in Japan. 1. Teil. Von 1880 bis 1980. *Zeitschrift für Neuere Rechtsgeschichte*, 2005, 27, 1/2, p. 62-76.

* 6472. PETRÁS (René), FALADA (David). Die tschechische rechtshistorische Literatur in den Jahren 1984 bis 2003. *Zeitschrift für Neuere Rechtsgeschichte*, 2005, 27, 3/4, p. 229-253.

* 6473. SIPPEL (Harald). Neuere Untersuchungen zur kolonialen Rechts- und Verwaltungsgeschichte ehemaliger britischer und deutscher Überseegebiete. *Zeitschrift für Neuere Rechtsgeschichte*, 2005, 27, 1/2, p. 77-92.

* 6474. TRAMPUS (Antonio). La genesi e la circolazione della Scienza della Legislazione: saggio bibliografico. *Rivista storica italiana*, 2005, 117, 1, p. 309-359.

6475. AZEVEDO ALEXANDRINO FERNANDES (Joao Manuel). Die Theorie der Interpretation des Gesetzes bei Francisco Suárez. Frankfurt am Main, Lang, 2005, 172 p. (Rechtshistorische Reihe, 303).

6476. DEFLERS (Isabelle). Lex und ordo. Eine rechtshistorische Untersuchung der Rechtsauffassung Melanchthons. Berlin, Duncker & Humblot, 2005, 318 p. (Schriften zur Rechtsgeschichte, 121).

6477. ERDŐ (Péter). Die Forschung der Geschichte des kanonischen Rechts: ein Dialog zwischen Theologie und Rechtsgeschichte. *Zeitschrift der Savigny-Stiftung für Rechtsgeschichte Kanonistische Abteilung*, 2005, 91, p. 1-16.

6478. Gesetzliches Unrecht : rassistisches Recht im 20. Jahrhundert. Herausgegeben im Auftrag des Fritz Bauer Instituts von Micha BRUMLIK, Susanne MEINL und Werner RENZ. Frankfurt am Main, Campus, 2005, 244 p. (ill.). (Jahrbuch zur Geschichte und Wirkung des Holocaust, 2005).

6479. Giuristi (I) e la crisi dello Stato liberale (1918–1925). A cura di Pier Luigi BALLINI. Venezia, Istituto veneto di Scienze, Lettere ed Arti, 2005, 184 p.

6480. GÜN (Doğan). Überlegungen zum Rechtswesen im Klassischen Osmanischen Staat. *OTAM Ankara Üniversitesi Osmanlı Tarihi Araştırma Ve Uygulama Merkezi Dergisi*, 2005, 17, p. 279-294.

6481. Hans Kelsen. Staatsrechtslehrer und Rechtstheoretiker des 20. Jahrhunderts. Hrsg. v. Stanley L. PAULSON und Michael STOLLEIS. Tübingen, Mohr (Siebeck), 2005, XI-392 p. (Grundlagen der Rechtswissenschaft, 3).

6482. HARKE (Jan Dirk). Vorenthaltung und Verpflichtung. Philosophische Ansichten der Austauschgerechtigkeit und ihr rechtshistorischer Hintergrund. Berlin, Duncker & Humblot, 2005, 105 p. (Schriften zur Rechtstheorie 225).

6483. History of law in Japan since 1868. Ed. by Wilhelm RÖHL. Leiden a. Boston, Brill, 2005, VI-848 p. (Handbook of oriental studies. Section five, Japan, 12 = Handbuch der Orientalistik, 12).

6484. HLEDÍKOVÁ (Zdeňka), JANÁK (Jan), DOBEŠ (Jan). Dějiny správy v českých zemích. Od počátků státu po současnost. (The history of legislation in the Czech lands). Praha, Lidové noviny, 2005, 568 p. (map).

6485. Honor, status, and law in modern Latin America. Ed. by Sueann CAULFIELD, Sarah C. CHAMBERS and Lara PUTNAM. Durham, Duke U. P., 2005, VIII-331 p.

6486. JANSSEN (Albert). Die bleibende Bedeutung des Genossenschaftsrechts Otto von Gierkes für die Rechtswissenschaft. *Zeitschrift der Savigny-Stiftung für*

Rechtsgeschichte Germanistische Abteilung, 2005, 122, p. 352-366.

6487. Juristes (Les) et l'argent. Le coût de la justice et l'argent des juges du XIVe au XIXe siècle. Sous la dir. de Benoit GARNOT. Dijon, Editions universitaires de Dijon, 2005, 251 p. (Sociétés).

6488. Lois, justice, coutume: Amérique et Europe latines, (16e–19e siècle). Sous la direction de Juan Carlos GARAVAGLIA, Jean-Frédéric SCHAUB. Paris, Ecole des hautes études en sciences sociales, 2005, 317 p. (ill.). (Recherches d'histoire et de sciences sociales, 99).

6489. MIDDELBERG (Mathias). Judenrecht, Judenpolitik und der Jurist Hans Calmeyer in den besetzten Niederlanden 1940–1945. Göttingen, V & R Unipress [mit Universitätsverlag Osnabrück], 2005, 420 p. (ill., facsims.). (Osnabrücker Schriften zur Rechtsgeschichte, 5).

6490. MOORMAN VAN KAPPEN (Olav). Zur Holländischen Erklärung der Menschen- und Bürgerrechte von 1795. *Zeitschrift der Savigny-Stiftung für Rechtsgeschichte Germanistische Abteilung*, 2005, 122, p. 317-326.

6491. NEUMAIER (Helmut). "Daß wir kein anderes Haupt oder von Gott eingesetzte zeitliche Obrigkeit haben". Ort Odenwald der fränkischen Reichsritterschaft von den Anfängen bis zum Dreißigjährigen Krieg. Stuttgart, Kohlhammer, 2005. XXVI-258 p. (Veröffentlichungen der Kommission für geschichtliche Landeskunde in Baden-Württemberg, R. B, Forschungen, 161).

6492. PESANTE (Maria Luisa). Contro l'uguaglianza civile: discorsi inglesi sulla gerarchia nella seconda metá del Settecento. *Rivista storica italiana*, 2005, 117, 2, p. 448-493.

6493. SCHNEIDER (Christina). Die SS und "das Recht". Eine Untersuchung anhand ausgewählter Beispiele. Frankfurt am Main, Berlin, Bern, Bruxelles, New York, Oxford u. Wien, Lang, 2005, 278 p. (Rechtshistorische Reihe, 322).

6494. SOLEIL (Sylvain). La réception du modèle juridique français entre discours et réalité depuis la Révolution. *Tijdschrift voor rechtsgeschiedenis*, 2005, 73, 1-2, p. 171-182.

6495. SOMMA (Alessandro). I giuristi e l'Asse culturale Roma-Berlino: economia e politica nel diritto fascista e nazionalsocialista. Frankfurt am Main, Klostermann, 2005, XVI-791 p. (Das Europa der Diktatur, 8. Studien zur europäischen Rechtsgeschichte, 195).

6496. STORA-LAMARRE (Annie). La République des faibles. Les origines républicaines du droit républicain. Préf. de Michelle PERROT. Paris, Armand Colin, 2005, 220 p. (Histoire à l'œuvre).

6497. Sur la portée sociale du droit. Usage et légitimité du registre juridique. Actes du colloque de 14 et 15 novembre 2002. Sous la dir. de Liora ISRAËL, Guillaume SACRISTE, Antoine VAUCHEZ et Laurent WILLEMEZ. Paris, Presses universitaires de France, 2005, 395 p. (Collection du Curapp).

§ 1. Addenda 1999–2001.

6498. BERTHOLD (Lutz). Carl Schmitt und der Staatsnotstandsplan am Ende der Weimarer Republik. Berlin, Duncker und Humblot, 99, 94 p.

6499. GROSS (Raphael). Carl Schmitt und die Juden: Eine deutsche Rechtslehre. Frankfurt am Main, Suhrkamp, 2000, 442 p.

6500. QUARITSCH (Helmut). Carl Schmitt: Antworten in Nürnberg. Berlin, Duncker und Humblot, 2000, 153 p.

6501. SEITZER (Jeffrey). Comparative history and legal theory: Carl Schmitt in the first German democracy. Westport, Greenwood, 2001, XX-165 p.

Cf. n°s 724-750, 2752, 4501, 5773

§ 2. History of constitutional law.

** 6502. Quellen zu den Reformen in den Rheinbundstaaten. 7. Württemberg, 1797–1816/19. Quellen und Studien zur Entstehung des modernen Württembergischenstaates. Bearbeitet von Ina Ulrike PAUL. München, Oldenbourg, 2005, 2 vol., XIV-1424 p.

** 6503. Verfassungsdokumente Österreichs, Ungarns und Liechtensteins 1791–1849. Hrsg. v. Ilse REITER, András CIEGER und Paul VOGT = Ausztria, Magyarország és Liechtenstein alkotmányereju dokumentumai 1791–1849. Szerkesztette Ilse REITER, András CIEGER, Paul VOGT = Constitutional documents of Austria, Hungary and Liechtenstein 1791–1849. Ed. by Ilse REITER, András CIEGER and Paul VOGT. München, K.G. Saur, 2005, 352 p. (Constitutions of the world from the late 18th century to the middle of the 19th century. Europe, 2 = Verfassungen der Welt vom späten 18. Jahrhundert bis Mitte des 19. Jahrhunderts. Europa, 2).

** 6504. WILLIAMS (John Matthew). The Australian constitution: a documentary history. Carlton, Melbourne U. P., 2005, XV-1288 p. (ill., ports.).

6505. CARTIER (Emmanuel). La transition constitutionnelle en France (1940–1945): la reconstruction révolutionnaire d'un ordre juridique "républicain". Préf. de Michel VERPEAUX. Paris, L.G.D.J., 2005, XVI-665 p. (Bibliothèque constitutionnelle et de science politique, 126).

6506. Culture costituzionali a confronto. Europa e Stati Uniti dall'età delle rivoluzioni all'età contemporanea. A cura di Fernanda MAZZANTI PEPE. Genova, Name, 2005, 451 p.

6507. Debate (The) over the Constitutional Revolution of 1937. [AHR Forum]. Introduction by Alan BRINKLEY. *American historical review*, 2005, 110, 4, p. 1046-1115. [Contents: KALMAN (Laura). The Constitution, the Supreme Court, and the New Deal. – LEUCHTENBURG (William E.). Comment on Laura

Kalman's article. – WHITE (G. Edward). Constitutional change and the New Deal].

6508. DIPPEL (Horst). Modern constitutionalism, An introduction to a history in need of writing. *Tijdschrift voor rechtsgeschiedenis*, 2005, 73, 1-2, p. 153-170.

6509. FUNK (René). Die Wahlprüfung der volksgewählten Abgeordneten der Volksvertretungen im Frühkonstitutionalismus. Eine Untersuchung der Wahlprüfung in den Kammern der Abgeordneten des Großherzogtums Baden, des Königreichs Württemberg und des Großherzogtums Hessen. Frankfurt am Main, Lang, 2005, XLVI-258 p. (Rechtshistorische Reihe, 304).

6510. GRONSKÝ (Ján). Komentované dokumenty k ústavním dějinám Československa. (Commented Documents to the Constitutional History of the Czechoslovakia). Tomo 1. 1914–1945. Praha, Karolinum, 2005, 584 p.

6511. GROTHE (Ewald). Zwischen Geschichte und Recht. Deutsche Verfassgeschichtsschreibung, 1900–1970. München, Oldenbourg, 2005, 486 p. (Ordnungssysteme, Studien zur Ideengeschichte der Neuzeit, 16).

6512. HECKER (Michael). Napoleonischer Konstitutionalismus in Deutschland. Berlin, Duncker & Humblot, 2005, 204 p. (Schriften zur Verfassungsgeschichte, 72).

6513. HULSEBOSCH (Daniel J.). Constituting Empire: New York and the transformation of constitutionalism in the Atlantic World, 1664–1830. Chapel Hill, University of North Carolina Press in association with the American Society for Legal History, 2005, 494 p. (Studies in legal history).

6514. JOHNSON (Calvin H.). Righteous anger at the wicked states: the meaning of the founders' Constitution. New York, Cambridge U. P., 2005, XV-294 p.

6515. MALÝ (Karel). Die Böhmische Konföderation und die Verneuerte Landesordnung – zwei böhmische Verfassungsgestaltungen zu Beginn des 17. Jahrhunderts. *Zeitschrift der Savigny-Stiftung für Rechtsgeschichte Germanistische Abteilung*, 2005, 122, p. 285-300.

6516. MANTL (Wolfgang). Der österreichische Rechtsstaat zwischen habsburgischer Tradition und europäischer Zukunft. *Zeitschrift der Savigny-Stiftung für Rechtsgeschichte Germanistische Abteilung*, 2005, 122, p. 367-380.

6517. MÜLLER (Matthias). Gesellschaftlicher Wandel und Rechtsordnung. Die Zürcher Restauration, 1814–1831, und die Entstehung des bürgerlichen Staates. Zürich, Schulthess, 2005, 273 p. (Zürcher Studien zur Rechtsgeschichte, 54).

6518. O'NEILL (Johnathan). Originalism in American law and politics. A constitutional history. Baltimore, Johns Hopkins U. P., 2005, X-281 p. (Johns Hopkins series in constitutional thought).

6519. PANTOJA MORÁN (David). El supremo poder conservador: el diseño institucional en las primeras constituciones mexicanas. México D.F., El Colegio de México y Zamora, El Colegio de Michoacán, 2005, XV-572 p. (ill.).

6520. ROBERTSON (Lindsay G.). Conquest by law: how the discovery of America dispossessed indigenous peoples of their lands. New York, Oxford U. P., 2005, XIII-239 p.

6521. ROLIN (Jan). Der Ursprung des Staates. Die naturrechtlich-rechtsphilosophische Legitimation von Staat und Staatsgewalt im Deutschland des 18. und 19. Jahrhunderts. Tübingen, Mohr Siebeck, 2005, XII-298 p. (Grundlagen der Rechtswissenschaft, 4).

6522. SKACH (Cindy). Borrowing constitutional designs. Constitutional law in Weimar Germany and the French Fifth Republic. Princeton a. Oxford, Princeton U. P., 2005, XIII-151 p.

6523. STANOMIR (Ioan). Libertate, lege şi drept. O istorie a constituţionalismului românesc. (Freedom, law and justice. A history of the Romanian constitutionalism). Iaşi, Editura Polirom, 2005, 240 p.

6524. UERTZ (Rudolf). Vom Gottesrecht zum Menschenrecht: das katholische Staatsdenken in Deutschland von der Franzosischen Revolution bis zum II° Vatikanischen Konzil (1789–1965). Paderborn, Schöningh, 2005, 552 p. (Politik und. Kommunikationswissenschaftliche Veroffentlichungen der Gorres-Gesellschaft, 25).

6525. WALLIS (John Joseph). Constitutions, corporations, and corruption: American states and constitutional change, 1842 to 1852. *Journal of economic history*, 2005, 65, 1, p. 211-256.

6526. WALTER (Robert). Hans Kelsen als Verfassungsrichter. Wien, Manz, 2005, III-92 p. (Schriftenreihe des Hans Kelsen-Instituts, 27).

Cf. n°s 700-723, 4121, 4300

§ 3. Public law and institutions.

** 6527. BAUER (Volker). Repertorium territorialer Amtskalender und Amtshandbücher im Alten Reich, Adressen-, Hof-, Staatskalender und Staatshandbücher des 18. Jahrhunderts. Band 4. Repertorium reichischer Amtskalender und Amtshandbücher. Periodische Personalverzeichnisse des Alten Reiches und seiner Institutionen. Frankfurt am Main, Klostermann, 2005. X-480 p. (Studien zur europäischen Rechtsgeschichte, 196).

6528. BRYSON (W.H.). The prerogative of the sovereign in Virginia: Royal law in a Republic. *Tijdschrift voor rechtsgeschiedenis*, 2005, 73, 3-4, p. 371-384.

6529. BUSCH (Jörg W.). Administratio in der frühen Stauferzeit – Ein abgebrochener Versuch politischer Begriffsbildung. *Zeitschrift der Savigny-Stiftung für Rechtsgeschichte Germanistische Abteilung*, 2005, 122, p. 42-86.

6530. ENGERMAN (Stanley L.), SOKOLOFF (Kenneth L.). The evolution of suffrage institutions in the New

World. *Journal of economic history*, 2005, 65, 4, p. 891-921.

6531. Funciones, procedimientos y escenarios: un análisis de poder legislativo en América Latina. Ed. por Manuel ALCÁNTARA SÁEZ, Mercedes GARCÍA MONTERO y Francisco SÁNCHEZ LÓPEZ. Salamanca, Ediciones Universidad de Salamanca, 2005, 307 p.

6532. FURUKAWA (Takahisa). Showa-senchūki no gikai to gyosei. (The Diet and administration in Japan during Showa-wartime). Tokyo, Yoshikawa Kobunkan, 2005, 306 p.

6533. HEIDER (Matthias). Die Konzessionsverträge der Stadt Lüdenscheid über leitungsgebundene Versorgungsgüter und die Entwicklung der städtischen Versorgungsbetriebe zwischen 1856 und 1945. Zugleich ein Beitrag über den Ausbau der kommunalen Leistungsverwaltung in Preußen. Berlin, Duncker & Humblot, 2005, 274 p. (Schriften zur Rechtsgeschichte, 119).

6534. KARASOVÁ (Helena). Konstrukce monstrprocesů Státní bezpečností. (The State Security construction of monstertrials). *In*: Moc verzus občan. Úloha represie a politického násilia v komunizme. Ed. Pavel ŽÁČEK. Bratislava, 2005, p. 150-154.

6535. KLEMP (Stefan). "Nicht ermittelt": Polizeibataillone und die Nachkriegsjustiz; ein Handbuch. Essen, Klartext-Verl., 2005, 503 p. (Schriften / Villa ten Hompel, 5).

6536. KOCH (Arnd). Die Rückkehr der "Volksgerichte" – Das bayerische Schwurgericht der Nachkriegszeit. *Zeitschrift der Savigny-Stiftung für Rechtsgeschichte Germanistische Abteilung*, 2005, 122, p. 242-262.

6537. KOČOVÁ (Kateřina). Druhá retribuce. Činnost mimořádných lidových soudů v roce 1948. (The Second Retribution. The Work of the Extraordinary People's Courts in 1948). *Soudobé dějiny*, 2005, 12, 3/4, p. 586-625.

6538. Korporativismus in den südeuropäischen Diktaturen: [Tagung in Blankensee vom 24. bis 27. Juni 2002] = Il corporativismo nelle dittature sudeuropee / [Berlin-Brandenburgische Akademie der Wissenschaften]. Hrsg. von Aldo MAZZACANE, Alessandro SOMMA und Michael STOLLEIS. Frankfurt am Main, Klostermann, 2005, VIII-421 p. (Das Europa der Diktatur, 6. Studien zur europäischen Rechtsgeschichte, 185).

6539. LILLA (Joachim). Der Preußische Staatsrat 1921–1933. Eine biographisches Handbuch. Mit einer Dokumentation der im "Dritten Reich" berufenen Staatsräte. Düsseldorf, Droste, 2005, 330 p. (Geschichte des Parlamentarismus und der politischen Parteien, 13).

6540. MÖLLER (Caren). Medizinalpolizei. Die Theorie des staatlichen Gesundheitswesens im 18. und 19. Jahrhundert. Frankfurt am Main, Klostermann, 2005, IX-340 p.

6541. OWENS (J. B.). "By my absolute authority": justice and the Castilian commonwealth at the beginning of the first global age. Rochester, University of Rochester Press, 2005, XIX-371 p. (Changing perspectives on Early Modern Europe).

6542. PATTERSON (Catherine). Quo Warranto and Borough Corporations in early Stuart England: royal prerogative and local privileges in the Central Courts. *English historical review*, 2005, 120, 488, p. 879-906.

6543. Poder legislativo (El) en América Latina a través de sus normas. Ed. por Manuel ALCÁNTARA SÁEZ, Mercedes GARCÍA MONTERO y Francisco SÁNCHEZ LÓPEZ. Salamanca, Ediciones Universidad de Salamanca, 2005, 476 p.

6544. RAITHEL (Thomas). Das schwierige Spiel des Parlamentarismus. Deutscher Reichstag und französische Chambre des Députés in den Inflationskrisen der 1920 Jahre. München, Oldenbourg, 2005, IX-631 p.

6545. ROUSSELET-PIMONT (Anne). Le chancelier et la loi au XVIe siècle. D'après l'oeuvre d'Antoine Duprat, de Guillaume Poyet et de Michel de L'Hospital. Paris, de Boccard, 2005, 635 p. (Romanité et modernité du droit).

6546. SCHWIEGER (Christopher). Volksgesetzgebung in Deutschland. Der wissenschaftliche Umgang mit plebiszitärer Gesetzgebung auf Reichs- und Bundesebene in Weimarer Republik, Drittem Reich und Bundesrepublik Deutschland (1919–2002). Berlin, Duncker & Humblot, 2005, 422 p. (Tübinger Schriften zum Staats- und Verwaltungsrecht, 71).

6547. STOLZI (Irene). La progettazione corporativa: poteri "pensati" e poteri "esercitati" nell'Italia fascista. *Zeitschrift für Neuere Rechtsgeschichte*, 2005, 27, 1/2, p. 93-104.

6548. WULFF-KUCKELSBERG (Susanne). Procureurs – accusateurs – commissaires. Die ursprünglichen Funktionsträger der Staatsanwaltschaft. Frankfurt am Main, Lang 2005, 463 p. (Europäische Hochschulschriften 2, 4092).

§ 4. Civil and penal law.

** 6549. Briefe deutscher Strafrechtler an Karl Josef Anton Mittermaier 1832–1866. Hrsg. und bearb. Von Lieselotte JELOWIK. Frankfurt am Main, Klostermann, 2005, X-420 p. (Studien zur europäischen Rechtsgeschichte 188, Juristische Briefwechsel des 19. Jahrhunderts).

** 6550. COX (J. C. M.). Repertorium van de stadsrechten in Nederland, "quod vulgariter statreghte nuncupatur". Den Haag, VNG Uitgeverij, 2005, 304 p. (Werken der Stichting tot Uitgaaf der Bronnen van het Oud-Vaderlandse Recht, 33).

** 6551. Netzwerk (Das) der "Gefängnisfreunde", 1830–1872. Karl Josef Anton Mittermaiers Briefwechsel mit europäischen Strafvollzugsexperten. Herausgegeben und bearbeitet von Lars Hendrik RIEMER. Frankfurt am Main, Klostermann, 2005, 2 vol., XIV-XXX-1908 p. (Studien zur europäischen Rechtsgeschichte 192, 1, 2. Juristische Briefwechsel des 19. Jahrhunderts).

4. CIVIL AND PENAL LAW

** 6552. Repertorium der Policeyordnungen der Frühen Neuzeit. Hrsg. v. Karl HÄRTER und Michael STOLLEIS. Band 6. Reichsstädte 2. Köln. Hrsg. v. Klaus MILITZER. Frankfurt am Main, Klostermann, 2005, 2 vol., IX-V-1474 p. (Studien zur europäischen Rechtsgeschichte; Veröffentlichungen des Max-Planck-Instituts für euroäische Rechtsgeschichte, Frankfurt am Main, 191, 1-2).

6553. ALBRECHT (Matthias). Die Methode der preußischen Richter in der Anwendung des Preußischen Allgemeinen Landrechts von 1794. Eine Studie zum Gesetzesbegriff und zur Rechtswanwendung im späten Naturrecht. Frankfurt am Main, Lang, 2005, 243 p. (Schriften zur Preußischen Rechtsgeschichte, 2).

6554. ALLEN (Robert). Les tribunaux criminals sous la Révolution et l'Empire, 1792–1811. Rennes, Presses universitaires de Rennes, 2005, 318 p. (Histoire).

6555. BANNER (Stuart). How the Indians lost their land. Law and power on the frontier. Cambridge, Belknap Press of Harvard U. P., 2005, 344 p.

6556. BECKER (Martin). Arbeitsvertrag und Arbeitsverhältnis während der Weimarer Republik und in der Zeit des Nationalsozialismus. Frankfurt am Main, Klostermann, 2005, XVI-627 p. (Juristische Abhandlungen, 44).

6557. BECKER (Peter). Dem Täter auf der Spur. Eine Geschichte der Kriminalistik. Darmstadt, Primus Verlag, 2005, 288 p.

6558. BERGENHEIM (Åsa). Brottet, offret och förövaren: vetenskapens och det svenska rättsväsendets syn på sexuella övergrepp mot kvinnor och barn 1850–2000. (Le crime, la victime et le criminel. L'attitude de la science et de la justice suédoise vis-à-vis des agressions sexuelles contre les femmes et les enfants 1850–2000). Stockholm, Carlsson, 2005, 479 p.

6559. BLANKE (Sandro). Soziales Recht oder kollektive Privatautonomie? Hugo Sinzheimer im Kontext nach 1900. Tübingen, Mohr Siebeck, 2005, XII-238 p. (Beiträge zur Rechtsgeschichte des 20. Jahrhunderts, 46).

6560. BREWER (Holly). By birth or consent. Children, law, and the Anglo-American revolution in authority. Chapel Hill, University of North Carolina Press, 2005, 390 p.

6561. CVIRN (Janez). Boj za sveti zakon: prizadevanja za reformo poročnega prava od 18. stoletja do druge svetovne vojne. (Struggle for the holy matrimony: endeavours to reform the marriage law from the 18th century to WWII. Ljubljana, Zveza zgodovinskih društev Slovenije, 2005, 111 p. (Zbirka Zgodovinskega časopisa, 30).

6562. CZELK (Andrea). "Privilegierung" und Vorurteil. Positionen der bürgerlichen Frauenbewegung zum "Unehelichenrecht" und zur Kindstötung im Kaiserreich. Köln, Böhlau, 2005, XIV-260 p. (Rechtsgeschichte und Geschlechterforschung).

6563. DANIELS (Heinrich Gottfried Wilhelm). Kurkölnisches Landrecht. Eine Vorlesungsnachschrift (Universitäts- und Landesbibliothek Bonn S 1457). Herausgegeben und bearbeitet von Christoph BECKER. Köln, Böhlau, 2005, LXVIII-318 p. (Rechtsgeschichtliche Schriften, 19).

6564. DEERE (Carmen Diana), LEÓN (Magdalena). Liberalism and married women's property rights in nineteenth-century Latin America. *Hispanic American historical review*, 2005, 85, 4, p. 627-678.

6565. DENCKER (Friedrich). Täterschaft und Beihilfe bei NS-Gewaltverbrechen. *Zeitschrift für Neuere Rechtsgeschichte*, 2005, 27, 1/2, p. 49-61.

6566. DESCHAMPS (Olivier). Les origines de la responsabilité pour faute personelle dans le Code civil de 1804. Préf. de Anne LEFEBVRE-TEILLARD. Paris, Librairie générale de droit et de jurisprudence, 2005, XIII-562 p. (Bibliothèque de droit privé, 436).

6567. Deutsches Sachenrecht in polnischer Gerichtspraxis. Das BGB-Sachenrecht in der polnischen höchstrichterlichen Rechtsprechung in den Jahren 1920–1939. Tradition und europäische Perspektive. Hrsg. v. Wojciech DAIJCZAK und Hans-Georg KNOTHE. Berlin, Duncker & Humblot 2005, 378 p. (Schriften zur europäischen Rechts- und Verfassungsgeschichte, 49).

6568. DI MARTINO (Paolo). Approaching disaster: personal bankruptcy legislation in Italy and England, c.1880–1939. *Business history*, 2005, 47, 1, p. 23-43.

6569. DOLL (Natascha). Recht, Politik und "Realpolitik" bei August Ludwig von Rochau, 1810–1873. Ein wissenschaftsgeschichtlicher Beitrag zum Verhältnis von Politik und Recht im 19. Jahrhundert. Frankfurt am Main, Klostermann, 2005, IX-205 p. (Studien zur europäischen Rechtsgeschichte, 189).

6570. EDIGATI (Daniele). La tecnicizzazione della giustizia penale: il magistrato degli Otto di guardia e balia nella Toscana medicea del primo Seicento. *Archivio storico italiano*, 2005, 163, 3, p. 485-530.

6571. Éduquer et punir. la colonie agricole et pénitentiaire de Mettray, 1839-1937. Sous la dir. de Luc FORLIVESI, Georges-François POTTIER et Sophie CHASSAT. Rennes, Presses universitaires de Rennes, 2005, XXIV-256 p.

6572. EISFELD (Jens). Die Scheinehe in Deutschland im 19. und 20. Jahrhundert. Tübingen, Mohr Siebeck, 2005, 300 p. (Beiträge zur Rechtsgeschichte des 20. Jahrhunderts, 45).

6573. FLAMM (Michael W.). Law and order. Street crime, civil unrest, and the crisis of liberalism in the 1960s. New York, Columbia U. P., 2005, XIII-294 p.

6574. FLOßMANN (Ursula). Österreichische Privatrechtsgeschichte. Wien, Springer, 2005, XXIII-359 p. (Springers Kurzlehrbücher Rechtswissenschaft).

6575. FRANZIUS (Christine). Bonner Grundgesetz und Familienrecht. Die Diskussion um die Gleichberechtigung von Mann und Frau in der westdeutschen Zivilrechtslehre der Nachkriegszeit, 1945–1957. Frankfurt am Main, Klostermann, 2005, XII-202 p. (Studien zur europäischen Rechtsgeschichte, 178).

6576. GARRÉ (Roy). Consuetudo. Das Gewohnheitsrecht in der Rechtsquellen- und Methodenlehre des späten ius commune in Italien, 16.–18. Jahrhundert. Frankfurt am Main, Klostermann, 2005, XIV-288 p. (Studien zur europäischen Rechtsgeschichte, 183).

6577. GÓMEZ BRAVO (Gutmaro). Crimen y castigo. Cárceles, justicia y violencia en la España del siglo XIX. Madrid, Los libros de la catarata, 2005, 319 p.

6578. GUILLET (David). Customary law and the nationalist project in Spain and Peru. *Hispanic American historical review*, 2005, 85, 1, p. 81-114.

6579. GULCZYNSKI (Andrzej). Das Napoleonische Gesetzbuch (Code civil) und sein Einfluss auf die Stabilisierung des Familiennamens in den polnischen Gebieten. *Zeitschrift für Neuere Rechtsgeschichte*, 2005, 27, 1/2, p. 28-48.

6580. HÄRTER (Karl). Policey und Strafjustiz in Kurmainz. Gesetzgebung, Normdurchsetzung und Sozialkontrolle im frühneuzeitlichen Territorialstaat. Frankfurt am Main, Vittorio Klostermann, 2005, 2 vol., 1247 p. (Studien zur europäischen Rechtsgeschichte, 190).

6581. HARTMANN (Philip). Das Recht der vertraglichen Erbfolgeregelung in der neuern deutschen Privatrechtsgeschichte. Berlin, Duncker & Humblot, 2005, 411 p. (Schriften zur Rechtsgeschichte, 123).

6582. HULÍK (Milan). Právní pohled na násilný odpor vůči komunistickému režimu v 50. letech. (Legal opinion to the violent resistance in the fifties). *In*: Moc verzus občan. Úloha represie a politického násilia v komunizme. Ed. Pavel ŽÁČEK. Bratislava, 2005, p. 69-76.

6583. J. von Staudingers Kommentar zum Bürgerlichen Gesetzbuch mit Einführungsgesetz und Nebengesetzen. Eckpfeiler des Zivilrechts. Red. Michael MARTINEK. Berlin, Sellier de Gruyter, 2005, XI-1183 p.

6584. JANSEN (Nils). "Tief ist der Brunnen der Vergangenheit". Funktion, Methode und Ausgangspunkt historischer Fragestellungen in der Privatrechtsdogmatik. *Zeitschrift für Neuere Rechtsgeschichte*, 2005, 27, 3/4, p. 202-228.

6585. Justice (La) en Algérie, 1830–1962. [Colloque tenu les 22 et 23 octobre 2002 à la Bibliothèque nationale de France]. Pour l'Association française pour l'histoire de la justice. Paris, Association française pour l'histoire de la justice, Documentation française, 2005, 366 p. (Histoire de la justice, 16).

6586. Justice et argent. Les crimes et les peines pécuniaires du XIIIe au XXIe siècle. Sous la dir. de Benoit GARNOT. Dijon, Editions universitaires de Dijon, 2005, 336 p. (Société).

6587. KAISER (Christian). Kündigungsschutz ohne Prinzip. Der Weimarer Entwurf eines Arbeitsvertragsgesetzes und seine Bezüge zum heutigen Recht. Tübingen, Mohr Siebeck, 2005, XV-401 p. (Beiträge zur Rechtsgeschichte des 20. Jahrhunderts, 47).

6588. KORZILIUS (Sven). "Asoziale" und "Parasiten" im Recht der SBZ/DDR: Randgruppen im Sozialismus zwischen Repression und Ausgrenzung. Köln, Böhlau Verlag, 2005, IX-743 p. (Arbeiten zur Geschichte des Rechts in der DDR, 4).

6589. MACKEY (Thomas C.). Pursuing johns. Criminal law reform, defending character, and New York City's Committee of Fourteen, 1920–1930. Columbus, Ohio State U. P., 2005, X-297 p. (History of crime and criminal justice).

6590. MESSERSCHMIDT (Manfred). Die Wehrmachtjustiz, 1933–1945. Paderborn, Ferdinand Schöningh, 2005, XIV-511 p.

6591. MIDDLETON (Stephen). The black laws: race and the legal process in early Ohio. Athens, Ohio U. P., 2005, XI-363 p. (Law, society, and politics in the Midwest, 4).

6592. MIETH (Stefan). Die Entwicklung des Denkmalrechts in Preußen 1701–1947. Frankfurt am Main, Lang, 2005, XXIV-343 p. (Rechtshistorische Reihe, 309).

6593. MORRIS (Douglas G.). Justice imperiled. The anti-nazi lawyer Max Hirschberg in Weimar Germany. Ann Arbor, University of Michigan Press, 2005, XIV-443 p. (Social history, popular culture, and politics in germany).

6594. Napoleons nalatenschap. Tweehondered jaar Burgerlijk Wetboek in België = Un héritage Napoléonien. Bicentenaire du Code civil en Belgique. Ed. par Dirk HEIRBAUT et Georges MARTYN. Mechelen, Kluwer, 2005, 446 p.

6595. PEREIRA (Anthony W.). Political (in)justice: authoritarianism and the rule of law in Brazil, Chile and Argentina. Pittsburgh, University of Pittsburgh Press, 2005, XVI-262 p.

6596. PETRAK (Marko). Rimska pravna pravila kao izvor suvremenog hrvatskog obiteljskog prava. (Rules of Roman Law as a source of contemporary Croatian family law). *Zbornik Pravnog fakulteta u Zagrebu*, 2005, 55, 3-4, p. 597-627.

6597. QUAISSER (Friederike). Mietrecht im 19. Jahrhundert. Ein Vergleich der mietrechtlichen Konzeptionen im Allgemeinen Landrecht für die preußischen Staaten von 1794 und dem gemeinen Recht. Frankfurt am Main, Lang, 2005, XXXII-209 p. (Rechtshistorische reihe, 319).

6598. ROBERTS (Richard). Litigants and households: African disputes and colonial courts in the French Soudan, 1895–1912. Portsmouth, Heinemann, 2005, XII-309 p. (Social history of Africa).

6599. SAK (İzzet). Şer'iye sicillerinde bulunan Konya vakfiyeleri (1650–1800). [Les actes de vakf trouvant dans les registres judiciaires de Konya (1650–1800)]. Konya, Kömen, 2005, X-226 p.

6600. SCHUBERT (Werner). Die Anfänge eines modernen Verkehrsrechts im Radfahrrecht um 1900 – Von den regionalen und einzelstaatlichen polizeilichen Radfahrordnungen bis zu den reichseinheitlichen "Grundzügen, betreffend den Radfahrverkehr" vom April 1907. *Zeitschrift der Savigny-Stiftung für Rechtsgeschichte Germanistische Abteilung*, 2005, 122, p. 195-241.

6601. ŞENTOP (Mustafa). Osmanlı yargı sistemi ve Kazaskerlik. (Système judiciaire et le Kazaskerlik dans l'Empire Ottoman). İstanbul, Klasik, 2005, XII-162 p.

6602. SLYSTSCHENKOW (Wladimir A.). Der Entwurf eines russischen Zivilgesetzbuchs von 1905. Einflüsse westeuropäischer Rechtstraditionen. *Zeitschrift für Neuere Rechtsgeschichte*, 2005, 27, 3/4, p. 189-201.

6603. STOLTE (Stefan). Versandhandel und Verbraucherschutz. Entstehung und Genese in rechtshistorischer Perspektive. Köln, Weimar u. Wien, Böhlau, 2005, XIII-304 p. (Forschungen zur Neueren Privatrechtsgeschichte, 30).

6604. STOPANI (Antonio). La memoria dei confini. Giurisdizione e diritti comunitari in Toscana (XVI–XVIII secolo). *Quaderni storici*, 2005, 1, p. 73-96.

6605. UNGER (Dagmar). Adolf Wach, 1843–1926, und das liberale Zivilprozeßrecht. Berlin, Duncker & Humblot, 2005, 394 p. (Schriften zur Rechtsgeschichte, 120).

6606. VAN DE MORTEL (Johannes Petrus Maria). Criminaliteit, rechtspleging en straf in het Hollandse drostambt, Heusden, 1615–1714. Tilburg, Stichting zuidelijk historisch contact, 2005, 296 p. (Bijdragen tot de geschiedenis van het zuiden van Nederland, derde reeks).

6607. VOTHKNECHT (Michael). Das Recht der Arbeitsverpflichtungen. Eine rechtshistorische Untersuchung der Normsetzung und ihrer Entwicklung seit dem Ersten Weltkrieg. Berlin, Berliner Wissenschafts-Verlag, 2005, 724 p. (Berliner Juristische Universitätsschriften, Grundlagen des Rechts, 34).

6608. WEITZ (Mark A.). The Confederacy on trial: the piracy and sequestration cases of 1861. Lawrence, University Press of Kansas, 2005, XI-219 p. (Landmark law cases and American Society).

6609. WHITAKER (Matthew C.). Race work: the rise of civil rights in the urban West. Lincoln, University of Nebraska Press, 2005, XIV-382 p. (Race and ethnicity in the American West).

6610. WICKE (Christina). Kodifikationsbestrebungen und Wissenschaft in Hessen-Darmstadt im vorkonstitutionellen Zeitalter. Frankfurt am Main, Lang, 2005, XVII-273 p. (Rechtshistorische Reihe, 311).

6611. WIGGENHORN (Harald). Verliererjustiz – Die Leipziger Kriegsverbrecherprozesse nach dem Ersten Weltkrieg. Baden-Baden, Nomos, 2005, IV-548 p. (Studien zur Geschichte des Völkerrechts, 10).

6612. WOOD (Elizabeth A.). Performing justice. Agitation trials in early Soviet Russia. Ithaca, Cornell U. P., 2005, VIII-301 p.

6613. ZENTZ (Frank). Das amerikanische Strafverfahren als Element der Besatzungspolitik in Deutschland: Erziehung zur Demokratie durch den Court of Appeals 1948–1955. Frankfurt am Main u. Oxford, P. Lang, 2005, 286 p. (ill.). (Rechtshistorische Reihe, 318).

Cf. n°s 4366, 4445, 4471, 4696, 5006

§ 5. International law.

6614. FENRICK (William J.). International humanitarian law and combat casualties. *European journal of population*, 2005, 21, 2-3, p. 167-186.

6615. GOTO (Harumi). Ahen to Igirisu teikoku: Kokusai kisei no takamari. 1906–1943 nen. (The international control of opium and the British Empire, 1906–1943). Tokyo, Yamakawa Shuppansha, 2005, 245 p.

6616. Legislative history (The) of the International Criminal Court. Vol. 1. Introduction, analysis, and integrated text of the statute, elements of crimes and rules of procedure and evidence. Vol. 2. An article-by-article evolution of the statute from 1994–1998. Vol. 3. Summary records of the 1998 diplomatic conference. Ed. by M. Cherif BASSIOUNI. Ardsley, Transnational Publishers, 2005, 3 vol., LVIII-546 p., LVIII-546 p., VII-679 p. (col. ill., col. ports.).

6617. ROHT-ARRIAZA (Naomi). The Pinochet effect: transnational justice in the age of human rights. Philadelphia, University of Pennsylvania Press, 2005, XIII-256 p.

6618. ZIEGLER (Karl-Heinz). Zum 'gerechten Krieg' im späteren Mittelalter und in der Frühen Neuzeit – vom Decretum Gratiani bis zu Hugo Grotius. *Zeitschrift der Savigny-Stiftung für Rechtsgeschichte Romanistische Abteilung*, 2005, 122, p. 177-194.

P

HISTORY OF INTERNATIONAL RELATIONS

§ 1. General. 6619-6702. – § 2. History of colonization and decolonization (a. General; b. Asia; c. Africa; d. America; e. Oceania). 6703-6867. – § 3. From 1500 to 1789 (a. General; b. 1500–1648; c. 1648–1789). 6868-6914. – § 4. From 1789 to 1815. 6915-6943. – § 5. From 1815 to 1910. 6944-7005. – § 6. From 1910 to 1935. The First World War. 7006-7107. – § 7. From 1935 to 1945. The Second World War (a. General; b. Diplomacy. Economy; c. Military operations; d. Resistance). 7108-7308. – § 8. From 1945 (a. General; b. 1945–1956; c. From 1956). 7309-7561.

§ 1. General.

* 6619. MARSHALL (Julie G.). Britain and Tibet, 1765–1947: a select annotated bibliography of British relations with Tibet and the Himalayan states including Nepal, Sikkim, and Bhutan. Foreword by Alastair LAMB. London a. New York, RoutledgeCurzon, 2005, 656 p.

** 6620. Bălgarija v sekretnija archiv na Stalin: ot pravitelstvoto na Kimon Georgiev do smărtta na Stalin; dokumenti, pisma, noti, telegrami, protokoli, stenogrami. (La Bulgarie dans les archives secrete de Stalin. Depuis le gouvernment de Kimion Georgiev jusque à la mort de Stalin). Konsultant Dimităr GENCHEV; prevod Ekaterina ABRAMOVA, Marianna POPOVA. Sofija, Izd-vo RT Agencija, 2005, 382 p. (ill.). (Khroniki na XX vek).

** 6621. Correspondances (Les) des consuls du Royaume du Danemark dans les Etats du Maghreb au cours des XVIIIe et XIXe siècles. Éd. par Ali CHENOUFI. Tunis, Centre de publication universitaire, 2005, 510 p.

** 6622. Quellen zu den deutsch-tschechischen Beziehungen 1848 bis heute. Hrsg. v. Manfred ALEXANDER. Darmstadt, Wissenschaftliche Buchgesellschaft, 2005, XXIX-231 p. (Quellen zu den Beziehungen Deutschlands zu seinen Nachbarn im 19. und 20. Jahrhundert, 12).

** 6623. Records of Syria, 1918–1973. Vol. 1. 1918–1920. Vol. 2. 1920–1922. Vol. 3. 1923–1925. Vol. 4. 1926–1931. Vol. 5. 1931–1936. Vol. 6. 1937–1939. Vol. 7. 1940–1942. Vol. 8. 1943–1945. Vol. 9. 1945–1948. Vol. 10. 1949–1451. Vol. 11. 1952–1955. Vol. 12. 1956–1959. Vol. 13. 1959–1962. Vol. 14. 1962–1963. Vol. 15. 1964–1973. Research editor, Jane PRIESTLAND; preface by Patrick SEALE. Slough, Archive Editions, 2005, 15 vol., XXXII-901 p., XXXIII-907 p., XXXIV-846 p., XXXII-812 p., XXVIII-779 p., XXVI-791 p., XXXIV-856 p., XXXVI-717 p., XXIX-650 p., XXVIII-736 p., XXVII-766 p., XXXI-819 p., XXX-853 p., XXIX-774 p., XXXV-873 p. (maps).

** 6624. Relațiile Romano-Chineze, 1880–1974: documente. Coordonator: Ambasador Romulus Ioan BUDURA. București, Ministerul Afacerilor Externe, Arhivele Nationale 2005, 1288 p. (ill.).

6625. ANDERSON (Irvine H.). Biblical interpretation and Middle East policy. The promised land, America, and Israel, 1917–2002. Gainesville, University Press of Florida, 2005, X-187 p.

6626. Anglo-Iranian relations since 1800. Ed. by Vanessa MARTIN. London, Routledge, 2005, 192 p. (Royal Asiatic Society books).

6627. ARDAY (Lajos). Az Egyesült Királyság és Magyarország: Nagy-Britannia és a magyar – angol kapcsolatok a 20. században. (The United Kingdom and Hungary: Great Britain and Hungarian-English relations in the 20th century). Budapest, Mundus Magyar Egyetemi Kiadó, 2005, 504 p. (map). (Az Antall József Emlékbizottság és Baráti Társaság évkönyvei).

6628. AZZOU (El-Mostafa). Les États-Unis et le statut international du Maroc (1906–1956). *Guerres mondiales et conflits contemporains*, 2005, 53, 219, p. 103-112.

6629. BASHKURTI (Lisen). Diplomacia shqiptare. 1. Deri ne vitin 1945. (Albanian Diplomacy, until year 1945). Tiranë, Geer, 2005, 392 p.

6630. BATAKOVIC (Dusan T.). Les frontières balkaniques au 20e siècle. *Guerres mondiales et conflits contemporains*, 2005, 53, 217, p. 29-46.

6631. BENALI (Abdelkader), OBDEIJN (Herman). Marokko door Nederlandse ogen 1605–2005. Amsterdam, De Arbeiderspers, 2005, 255 p. (ill., maps).

6632. BLACK (Jeremy). Introduction to global military history: 1775 to the present day. London a. New York, Routledge, 2005, XIX-294 p.

6633. BOSOER (Fabián). Generales y embajadores: una historia de las diplomacias paralelas en la Argentina. Barcelona, Vergara, Grupo Zeta, 2005, 476 p. (Biografía e historia).

6634. British naval strategy east of Suez, 1900–2000: influences and actions. Ed. by Greg KENNEDY. London, Routledge, 2005, X-288 p. (maps). (Cass series-naval policy and history, 27).

6635. CARROLL (Francis M.). The American presence in Ulster: a diplomatic history, 1796–1996. Washington, Catholic University of America Press, 2005, XI-281 p.

6636. CASTILLO (Nelson). Presencia de Estados Unidos en la República Dominicana. Santo Domingo, Editora Manatí, 2005, VIII-569 p.

6637. CITINO (Robert Michael). The German way of war: from the Thirty Years' War to the Third Reich. Lawrence, University Press of Kansas, 2005, XIX-428 p. (ill., maps, ports.). (Modern war studies).

6638. Colonial Hong Kong and modern China: interaction and reintegration. Ed. by LEE Pui-tak. Hong Kong, Hong Kong U. P., 2005, XII-295 p. (ill.).

6639. Confluenţe româno-finlandeze. Trei secole de contacte. 85 de ani de relaţii diplomatice. (Romanian-Finnish confluence. Three centuries of connection. 85 years of diplomatic relationship). Coordonator Alexandru POPESCU. Bucureşti, Editura Institutului Cultural Român, 2005, 312 p.

6640. CONSTANTIN (Ion). Din istoria Poloniei şi a relaţiilor româno-polone. (From the history of Poland and Romanian-Polish relations). Cuvânt înainte de Georgeta FILITTI. Bucureşti, Editura Biblioteca Bucureştilor, 2005, 309 p.

6641. CORDELL (Karl), WOLFF (Stephan). Germany's foreign policy towards Poland and the Czech Republic: Ostpolitik revisited. London a. New York, Routledge, 2005, XI-183 p. (Routledge advances in European politics, 28).

6642. Corse (La) et l'Angleterre: XVIe–XIXe siècle: sixièmes journées de Bonifacio, juillet 2004, organisées par la mairie de Bonifacio. Ajaccio, A. Piazzola, 2005, 206 p.

6643. DESPRADEL (Alberto). El consulado de Belladere en las relaciones dominicohaitianas. Santo Domingo, Editora Manatí, 2005, 607 p.

6644. DILKS (David). The great dominion. Winston Churchill in Canada, 1900–1954. With the assistance of Richard DILKS. Toronto, Thomas Allen, 2005, XX-472 p.

6645. España y Estados Unidos en el siglo XX. Ed. por Lorenzo DELGADO y M. Dolores ELIZALDE. Madrid, Consejo Superior de Investigaciones Científicas, 2005, 362 p. (ill., ports.). (Biblioteca de Historia, 57).

6646. Exile armies. Ed. by Matthew BENNETT and Paul LATAWSKI. Basingstoke, Palgrave Macmillan, 2005, XIII-187 p.

6647. FERGUSON (Niall). Colossus: the rise and fall of the American empire. New York a. London, Penguin Press, 2005, XXIX-386 p. (ill.).

6648. FERMANDOIS H. (Joaquín). Mundo y fin de mundo: Chile en la política mundial, 1900–2004. Santiago, Ediciones Universidad Católica de Chile, 2005, 638 p. (Biblioteca bicentenario, 54).

6649. FERREIRA DA CUNHA (Paulo). Lusofilias: identidade portuguesa e relações internacionais. Porto, Edições Caixotim, 2005, 182 p. (Ideias).

6650. Foreign Office (The) and British diplomacy in the twentieth century. Ed. by Gaynor JOHNSON. London, Routledge, 2005, XIII-236 p.

6651. FRANCE (John). Close order and close quarter: the culture of combat in the West. *International history review*, 2005, 27, 3, p. 498-517.

6652. Franquismo, política exterior y memoria histórica. Editor de este número/zenbaki honen argitaratzaileak: Juan Carlos PEREIRA, Ricardo MIRALLES. *Historia contemporanea* (Universidad del Pais Vasco), 2005, 30, 1, 376 p.

6653. GELVIN (James L.). The Israel-Palestine conflict: one hundred years of war. Cambridge, Cambridge U. P., 2005, X-294 p. (ill., maps).

6654. GIORDANO (Giancarlo). Aspetti e momenti di storia diplomatica dell'Italia contemporanea. Roma, Aracne, 2005, 249 p. (Scienze politiche e sociali, 68).

6655. GOBAT (Michel). Confronting the American dream: Nicaragua under U.S. imperial rule. Durham: Duke U. P., 2005, XIII-373 p. (American encounters/global interactions).

6656. Goda grannar eller morska motståndare? Sverige och Norge från 1814 till idag. (Bons voisins ou adversaires hardis? La Suède et la Norvège de 1814 à nos jours). Stockholm, Carlsson, 2005, 254 p. (Projekt 1905).

6657. GÖKAY (Bülent). Soviet Eastern policy and Turkey, 1920–1991. London, Routledge, 2005, XIV-184 p. (Routledge studies in the history of Russia and Eastern Europe).

6658. GONĚC (Vladimír). K polskému faktoru v česko-slovenském vztahu. (The Polish Factor in Czech-Slovak Relations). *Česko-slovenská historická ročenka*, 2005, p. 227-258.

6659. Grenzregionen der Habsburgermonarchie im 18. und 19. Jahrhundert: ihre Bedeutung und Funktion aus der Perspektive Wiens. Hrsg. v. Hans-Christian MANER. Münster, Lit, 2005, 247 p. (map). (Mainzer Beiträge zur Geschichte Osteuropas, 1).

6660. HAMETZ (Maura). Making Trieste Italian, 1918–1954. Rochester, Boydell Press, 2005, XI-204 p. (Royal historical society studies in history, n.s.).

6661. HARBUĽOVÁ (Ľubica). Ruský diplomat Georgij Nikolajevič Garin-Michajlovskij a jeho pôsobenie na Slovensku. (Der russische Diplomat Georgij Nikolajevič Garin-Michajlovskij und sein Wirken in der Slowakei). *Historický časopis*, 2005, 53, 2, p. 355-364. [Deutsche Zsfassung].

6662. Histoire de la diplomatie française. Prés. de Dominique DE VILLEPIN. Paris, Perrin, 2005, 1050 p. (ill., maps).

6663. HOUŽVIČKA (Václav). Návraty sudetské otázky. (The Return of the Sudeten-German question). Praha, Karolinum, 2005, 543 p.

6664. Hrvatsko-bugarski odnosi u 19. i 20. stoljeću. Zbornik radova međunarodnog znanstvenog skupa, Zagreb, 27.–28. X 2003). (Le relazioni croato-bulgare nel XIX e XX secolo. Atti del convegno scientifico internazionale). Glavni i odgovorni urednik Josip BRATULIĆ. Zagreb, Hrvatsko-bugarsko društvo, 2005, 382 p. (ill.).

6665. INQUIMBERT (Anne-Aurore). Considérations sur les attachés militaires. *Revue d'histoire diplomatique*, 2005, 119, 2, p. 151-164.

6666. Japanese-German relations, 1895–1945: war, diplomacy and public opinion. Ed. by Christian W. SPANG and Rolf-Harald WIPPICH. New York, Routledge, 2005, 240 p.

6667. LIEBER (Keir). War and the engineers. The primacy of politics over technology. Ithaca, Cornell U. P., 2005, 236 p. (Cornell studies in security affairs).

6668. LOUCAS (Ioannis). La question d'Orient et la géopolitique de l'espace européen du Sud-Est. *Guerres mondiales et conflits contemporains*, 2005, 53, 217, p. 17-28.

6669. LUKS (Leonid). Der russische "Sonderweg"? Aufsätze zur neuesten Geschichte Russlands im europäischen Kontext. Stuttgart, Ibidem-Verlag, 2005, 433 p. (Soviet and post-soviet politics and society, 16).

6670. LYNN (John A.). Discourse, reality, and the culture of combat. *International history review*, 2005, 27, 3, p. 475-480.

6671. MATEEVA (Marija Atanasova). Istorija na diplomaticheskite otnoshenija na Bălgarija. (Storia delle relazioni diplomatiche della Bulgaria). Pod redakcijata na Petăr KONSTANTINOV. Sofija, Bălgarski bestselăr, 2005, 777 p. (ill.).

6672. MENÉNDEZ (Mario). Cuba, Haïti et l'interventionnisme américain: un poids, deux mesures. Paris, CNRS éditions, 2005, 178 p. (maps). (CNRS histoire).

6673. MIROIU (Andrei). Balanță și hegemonie: România în politica mondială, 1913–1989. (Equilibrio e egemonia: la Romania nella politica mondiale, 1913–1989). București, Tritonic, 2005, 206 p.

6674. Národnostná otázka v strednej Európe v rokoch 1848–1938. = Die Nationalitätenfrage in Mitteleuropa in den Jahren 1848–1938. Ed. Peter ŠVORC, Ľubica HARBUĽOVÁ, Karl SCHWARZ. Prešov, Vydavateľstvo Universum, 2005, 298 p. [Deutsche Zsfassung].

6675. Nazione, interdipendenza, integrazione: le relazioni internazionali dell'Italia, 1917–1989. A cura di Federico ROMERO e Antonio VARSORI. Roma, Carocci, 2005, 270 p. (Studi storici Carocci; 107).

6676. Norsk-svenske relasjoner i 200 år. (Norwegian-Swedish relations in 200 years). Redaktører, Øystein SØRENSEN og Torbjörn NILSSON. Oslo, Aschehoug, 2005, 254 p.

6677. O'BRIEN (John B.). Studies in Irish, British and Australian relations, 1916-1963: trade, diplomacy and politics. Ed. by Anne E. O'BRIEN. Dublin, Four Courts Press, 2005, 164 p.

6678. Osmanische Reich (Das) und die Habsburgermonarchie: Akten des internationalen Kongresses zum 150-jährigen Bestehen des Instituts für Österreichische Geschichtsforschung, Wien, 22.–25. September 2004. Hrsg. v. Marlene KURZ [et al.]. Wien, R. Oldenbourg Verlag, 2005, 650 p. (ill., maps). (Mitteilungen des Instituts für Österreichische Geschichtsborschung. 48. Ergänzungsband).

6679. Osmanismus, Nationalismus und der Kaukasus: Muslime und Christen, Türken und Armenier im 19. und 20. Jahrhundert. Hrsg. v. Fikret ADANUR und Bernd BONWETSCH. Wiesbaden, Reichert, 2005, VIII-327 p. (ill.). (Kaukasienstudien, 9).

6680. PAPP (Daniel S.), JOHNSON (Loch), ENDICOTT (John). American foreign policy: history, politics, and policy. New York a. London, Pearson Longman, 2005, XVI-542 p. (ill., maps).

6681. PATRIARCA (Marco Antonio). Due secoli di politica estera americana: vocazione, realtà e disincanto. Postfazione di Luciano PELLICANI. Soveria Mannelli, Rubbettino, 2005, 256 p. (Collana di studi diplomatici, 5).

6682. PERDUE (Peter C.). China marches west. The Qing conquest of Central Eurasia. Cambridge, Belknap Press of Harvard U. P., 2005, XX-725 p.

6683. PIERRI (Bruno). Guerra fredda e illusioni imperiali: la Gran Bretagna, gli Stati Uniti e i rapporti con l'Egitto (1848–1954). Galatina, Congedo, 2005, 304 p. (Pubblicazioni del Dipartimento di studi storici dal Medioevo all'età contemporanea, 82. Saggi e ricerche, 75).

6684. Polen & Sverige 1919–1999. Red. Harald RUNBLOM & Andrzej Nils UGGLA. Uppsala, Centrum för multietnisk forskning, 2005, 252 p. (ill.). (Uppsala multiethnic papers, 48).

6685. RĂCEANU (Mircea). Cronologie comentată a relațiilor româno-americane, de la începuturi până la 1989. (Comments on the chronology of the Romanian-American relations from the beginnings up to 1989). București, Editura Silex, 2005, 352 p.

6686. RASPOPOVIĆ (Radoslav M.). Crna Gora i Rusija: ogledi i eseji. (Montenegro e Russia, dati e saggi). Beograd, Službeni list Srbije i Crne Gore i Podgorica, Istorijski institut Crne Gore, 2005, 432 p. (ill.).

6687. ROBBINS (Keith). Britain and Europe, 1789–2005. London, Hodder Arnold, 2005, XVIII-350 p. (maps). (Britain and Europe).

6688. Rôle (Le) et la place des petits pays en Europe au XXe siecle = Small countries in Europe, their role and place in the XXth century. Sous la dir. de Gilbert TRAUSCH. Baden-Baden, Nomos et Bruxelles, Bruylant, 2005, 531 p. (Veröffentlichungen der Historiker-Verbindungsgruppe bei der Kommission der Europäischen Gemeinschaften, 6 = Publications du Groupe de Liaison des Professeurs d' Histoire Contemporaine auprès des Communautés Européennes, 6).

6689. RUCHNIEWICZ (Krzysztof). Zögernde Annäherung: Studien zur Geschichte der deutsch-polnischen Beziehungen im 20. Jahrhundert. Dresden, Thelem, 2005, 337 p. (Mitteleuropa-Studien, 7).

6690. Ruolo geopolitico (Il) della diplomazia e dei consolati: la codifica asburgica nel Manuale Piskur. A cura di Maria Paola PAGNINI e Aldo COLLEONI. Trieste, Edizioni Università di Trieste, 2005, 335 p.

6691. SCHÖLLGEN (Gregor). Jenseits von Hitler: die Deutschen in der Weltpolitik von Bismarck bis heute. Berlin, Propyläen, 2005, 399 p. (Schriftenreihe / Bundeszentrale für politische Bildung, 490).

6692. SINGH (Nirula). India and the Soviet Union 1917 to 1947. New Delhi, APH Publishing Corporation, 2005, 154 p.

6693. SMITH (Joseph). The United States and Latin America: a history of American diplomacy, 1776–2000. London a. New York, Routledge, 2005, VII-208 p. (maps). (International relations and history series).

6694. STOLER (Mark A.). War and diplomacy: or, Clausewitz for diplomatic historians. *Diplomatic history*, 2005, 29, 1, p. 1-26.

6695. SUCIU (Ioan Alexandru). România şi Comisia Europeană a Dunării (1856–1948). (Romania and the European Commission of the Danube, 1856–1948). Constanţa, Editura Ex Ponto, 2005, 380 p.

6696. VAN LAAK (Dirk). Über alles in der Welt: deutscher Imperialismus im 19. und 20. Jahrhundert. München, C.H. Beck, 2005, 228 p. (Beck'sche Reihe, 1650).

6697. WANG (Dong). China's unequal treaties: narrating national history. Lanham, Lexington Books, 2005, X-179 p. (AsiaWorld).

6698. War (From) to Cold War: Anglo-Finnish relations in the 20th century. Ed. by Juhana AUNESLUOMA. Helsinki, SKS/Finnish Literature Society, 2005, 201 p. (Studia historica, 72).

6699. When states kill: Latin America, the U.S. and technologies of terror. Ed. by Cecilia MENJÍVAR and Néstor RODRÍGUEZ. Austin, University of Texas Press, 2005, X-374 p.

§ 1. Addenda 1999–2004.

6700. Deutschland und Polen in schweren Zeiten 1933–1990. Alte Konflikte – neue Sichtweisen. Niemcy i Polska w trudnych latach 1933–1990. Nowe spojrzenie na dawne konflikty. Hrsg. v. Bernd MARTIN und Arkadiusz STEMPIN. Unt. Mitarb. v. Bożena GÓRCZYŃSKA-PRZYBYŁOWICZ. Mit ein. Geleitwort v. Czesław MADAJCZYK. Freiburg, Rombach u. Poznań, Instytut Historii UAM, 2004, 278 p.

6701. DOGO (Marco). Storie balcaniche: popoli e stati nella transizione alla modernità. Gorizia, Libreria editrice goriziana, 99, 173 p. (ill.). (I leggeri, 13).

6702. NISHIDA (Toshihiro). Shidehara Kijuro no kokusai ninshiki: daiichijisekaitaisengo no tenkanki wo chushin to shite. (Perception of the international relations of Shidehara Kijuro between the two World Wars). *Kokusai-Seiji*, 2004, p. 91-106.

Cf. n°s 3515-3569, 3742, 3756, 4039, 4168, 4499, 4621, 5079, 5111

§ 2. History of colonization and decolonization.

a. General

* 6703. MANN (Gregory). Locating colonial histories: between France and West Africa. *American historical review*, 2005, 110, 2, p. 409-434.

6704. ALTMANN (Gerhard). Abschied vom Empire: die innere Dekolonisation Großbritanniens 1945–1985. Göttingen, Wallstein, 2005, 461 p. (Moderne Zeit, 8).

6705. BEASLEY (Edward). Empire as the triumph of theory: imperialism, information, and the colonial society of 1868. London, Routledge, 2005, XIV-216 p. (Cass series-British foreign and colonial policy).

6706. BELMESSOUS (Saliha). Assimilation and racialism in seventeenth and eighteenth-century French colonial policy. *American historical review*, 2005, 110, 2, p. 322-349.

6707. BÉNOT (Yves.). Massacres coloniaux: 1944–1950, la IVe République et la mise au pas des colonies françaises. Paris, La Découverte, 2005, XV-202 p. (La Découverte poche. Sciences humaines et sociales, 107).

6708. British Empire (The) in the 1950s: retreat or revival? Ed. by Martin LYNN. New York a. Basingstoke, Palgrave Macmillan, 2005, XVII-242 p.

6709. BURIN (Eric). Slavery and the peculiar solution: a history of the American Colonization Society. Gainesville, University Press of Florida, 2005, XIV-223 p. (Southern dissent).

6710. COLEMAN (Deirdre). Romantic colonization and British anti-slavery. Cambridge, Cambridge U. P., 2005, XV-273 p. (Cambridge studies in Romanticism, 61).

2. HISTORY OF COLONIZATION AND DECOLONIZATION

6711. Connaissances et pouvoirs: les espaces impériaux (XVIe–XVIIIe siècles): France, Espagne, Portugal. Sous la dir. de Charlotte DE CASTELNAU-L'ESTOILE et François REGOURD. Pessac, Presses universitaires de Bordeaux, 2005, 412 p. (ill., maps). (La mer au fil des temps).

6712. Culturas imperiales: experiencia y representación en América, Asia y África. Ed. por Ricardo SALVATORE. Rosario, Beatriz Viterbo Editora, 2005, 384 p. (Estudios Culturales).

6713. Descobrimentos portugueses (Os) no mundo de língua inglesa (1880–1972) = The Portuguese discoveries in the English-speaking world (1880–1972). Coordenação de Maria Teresa PINTO COELHO; prefácio de John DARWIN. Lisboa, Edições Colibri, 2005, 248 p. (Relações Luso-Britânicas, 1).

6714. ELTIS (David), LEWIS (Frank D.), RICHARDSON (David). Slave prices, the African slave trade, and productivity in the Caribbean, 1674–1807. *Economic history review*, 2005, 58, 4, p. 673-700.

6715. Esclavage (L'), la colonisation, et après: France, Etats-Unis, Grande-Bretagne. Sous la dir. de Patrick WEIL et Stéphane DUFOIX. Paris, Presses universitaires de France, 2005, VIII-627 p. (ill.).

6716. ETEMAD (Bouda). De l'utilité des empires: colonisation et prospérité de l'Europe, XVIe–XXe siècle. Paris, A. Colin, 2005, 334 p. (maps).

6717. Fracture coloniale (La): la société française au prisme de l'héritage colonial. Sous la dir. de Pascal BLANCHARD, Nicolas BANCEL et Sandrine LEMAIRE. Paris, La Découverte, 2005, 310 p. (Cahiers libres). [Cf. n° <sélection> 430.]

6718. FRADERA (Josep Maria). Colonias para después de un imperio. Barcelona, Edicions Bellaterra, 2005, 751 p. (Serie General universitaria, 45).

6719. Germany's colonial pasts. Ed. by Eric AMES, Marcia KLOTZ and Lora WILDENTHAL; foreword by Sander L. GILMAN. Lincoln a. London, University of Nebraska Press, 2005, XXI-255 p.

6720. HALL (Gwendolyn Midlo). Slavery and African ethnicities in the Americas: restoring the links. Chapel Hill, University of North Carolina Press, 2005, XXI-225 p. (ill., maps).

6721. Interpreting Spanish colonialism: empires, nations, and legends. Ed. by Christopher SCHMIDT-NOWARA and John M. NIETO-PHILLIPS. Albuquerque, University of New Mexico Press, 2005, IX-269 p. (ill., maps).

6722. Italian colonialism. Ed. by Ruth BEN-GHIAT and Mia FULLER. New York a. Basingstoke, Palgrave Macmillan, 2005, XXII-266 p. (ill., maps). (Italian and Italian American studies).

6723. LAIDLAW (Zoë). Colonial connections, 1815–1845: patronage, the information revolution and colonial government. Manchester a. New York, Manchester U. P., 2005, XII-241 p. (Studies in imperialism).

6724. Loi (La) de 1905 et les colonies. Sous la dir. de Jean-Marc REGNAULT. *Outre-mers. Revue d'histoire*, 2005, 93, 348-349, p. 5-135.

6725. "Macht und Anteil an der Weltherrschaft": Berlin und der deutsche Kolonialismus. Hrsg. v. Ulrich VAN DER HEYDEN und Joachim ZELLER. Münster, Unrast, 2005, 288 p. (ill.).

6726. MARSHALL (Peter J.). The making and unmaking of Empires: Britain, India and America, c. 1750–1783. Oxford, Oxford U. P., 2005, IX-398 p.

6727. NOUSCHI (André). Les armes retournées: colonisation et décolonisation françaises: essai. Paris, Belin, 2005, 447 p. (Histoire et société).

6728. PITTS (Jennifer). A turn to empire: the rise of imperial liberalism in Britain and France. Princeton, Princeton U, P., 2005, XII-382 p.

6729. RODRIGUES (Jaime). De costa a costa: escravos, marinheiros e intermediários do tráfico negreiro de Angola ao Rio de Janeiro, 1780–1860. São Paulo, Companhia das Letras, 2005, 420 p. (ill., maps).

6730. Settler colonialism in the twentieth century: projects, practices, legacies. Ed. by Caroline ELKINS and Susan PEDERSEN. New York, Routledge, 2005, XIII-303 p. (ill., maps).

6731. SPEITKAMP (Winfried). Deutsche Kolonialgeschichte. Stuttgart, Philipp Reclam, 2005, 207 p. (maps). (Universal-Bibliothek, 17047).

6732. THOMAS (Martin). The French empire between the wars: imperialism, politics and society. Manchester, Manchester U. P., 2005, XVII-408 p. (maps). (Studies in imperialism).

6733. THOMPSON (Andrew). The empire strikes back? The impact of imperialism on Britain from the mid-nineteenth century. London a. New York, Pearson Longman, 2005, XVII-374 p.

6734. United Kingdom overseas territories (The). Ed. by David KILLINGRAY and David TAYLOR. London, Institute of Commonwealth Studies, University of London, 2005, IV-146 p. (Occasional paper of the OSPA Research Project at the Institute of Commonwealth Studies, University of London, 3).

6735. WEBSTER (Wendy). Englishness and empire, 1939–1965. Oxford a. New York, Oxford U. P., 2005, VIII-253 p.

6736. WINTER (Karin). Österreichische Spuren in der Südsee: die Missionsreise von S.M.S. Albatros in den Jahren 1895–1898 und ihre ökonomischen Hintergründe. Wien, NWV, Neuer Wissenschaftlicher Verlag, 2005, 285 p. (ill., ports.). (Geschichte).

a. Addendum 2003

6737. Phantasiereiche: Zur Kulturgeschichte des deutschen Kolonialismus. Hrsg. v. Birthe KUNDRUS. Frankfurt am Main, Campus Verlag, 2003, 328 p.

Cf. n°s 211, 409, 669, 752, 3516, 3565, 4941

b. Asia

6738. ASADA (Shinji). Kōshūwan soshakuchi ni okeru "chūgoku-jin" (1897–1914): doitsu shokuminchi-hō to shokuminchi seisaku no kanren kara. (The "Chinese" in the German Leasehold, Kiaochow [1897–1914]: between the German colonial law and colonial policy). *Rekishigaku Kenkyū*, 2005, 797, p. 1-17, 64.

6739. BERTRAND (Romain). État colonial, noblesse et nationalisme à Java: la tradition parfaite. Paris, Karthala, 2005, 800 p. (maps). (Recherches internationales).

6740. BOWEN (H. V.). The business of empire. The East Indian Company and imperial Britain, 1756–1833. Cambridge, Cambridge U. P., 2005, 304 p.

6741. British policy and Indian nationalism (1858–1919). Ed. by B.R. VERMA and S.R. BAKSHI. New Delhi, Ajay Verma for Commonwealth Publishers, 2005, 483 p.

6742. BROBST (Peter John). The future of the great game. Sir Olaf Caroe, India's independence, and the defense of Asia. Akron, University of Akron Press, 2005, XX-199 p.

6743. COOPER (Randolf G. S.). Culture, combat, and colonialism in eighteenth- and nineteenth-century India. *International history review*, 2005, 27, 3, p. 534-549.

6744. DASGUPTA (Biplab). European trade and colonial conquest. Vol. 1. London, Anthem, 2005, VIII-398 p. (Anthem South Asian studies).

6745. DESHPANDE (Anirudh). British military policy in India, 1900–1945: colonial constraints and declining power. New Delhi, Manohar, 2005, 223 p.

6746. GAUDART DE SOULAGES (Michel), RANDA (Philippe). Les dernières années de l'Inde française. Préf. de Douglas GRESSIEUX. Paris, Dualpha, 2005, 446 p. (ill., map). (Vérités pour l'histoire).

6747. JASANOFF (Maya). Edge of empire: Conquest and collecting in the East, 1750-1850. London, Fourth Estate, 2005, 404 p.

6748. MAC MILLAN (Richard). The British occupation of Indonesia, 1945–1946. London, RoutledgeCurzon, 2005, XII-248 p. (ill.). (Royal Asiatic Society books).

6749. MARK (Chi-Kwan). Defence or decolonisation? Britain, the United States, and the Hong Kong question in 1957. *Journal of imperial and Commonwealth history*, 2005, 33, 1, p. 51-72.

6750. METCALF (Thomas R.). Forging the Raj: essays on British India in the heyday of empire. New Delhi, Oxford U. P., 2005, VI-317 p. (ill.).

6751. MORLAT (Patrice). Indochine années vingt: le rendez-vous manqué (1918–1928): la politique indigène des grands commis au service de la mise en valeur. Paris, Les Indes savantes, 2005, 553 p. (ill., maps).

6752. NGUYEN-MARSHALL (Van). The moral economy of colonialism: subsistence and famine relief in French Indo-China, 1906–1917. *International history review*, 2005, 27, 2, p. 237-258.

6753. PEERS (Douglas M.). Colonial knowledge and the military in India, 1780–1860. *Journal of imperial and Commonwealth history*, 2005, 33, 2, p. 157-180.

6754. RUSCIO (Alain), TIGNÈRES (Serge). Dien Bien Phu, mythes et réalités: cinquante ans de passions françaises, 1954–2004. Paris, Indes savantes, 2005, 413 p.

6755. TAN (Tai Yong.). The garrison state: the military, government and society in colonial Punjab, 1849-1947. New Delhi a. London, Sage, 2005, 333 p. (maps). (Sage series in modern Indian history, 8).

6756. TRAVERS (Robert). Ideology and British expansion in Bengal, 1757–1772. *Journal of imperial and Commonwealth history*, 2005, 33, 1, p. 7-27.

6757. TURPIN (Frédéric). De Gaulle, les gaullistes et l'Indochine: 1940–1956. Paris, Indes savantes, 2005, 666 p.

6758. VAN NIEL (Robert). Java's Northeast coast 1740–1840: a study in colonial encroachment and dominance. Leiden, Research School CNWS, Leiden University, 2005, XIV-424 p. (maps). (CNWS publications, 137. Studies in overseas history, 6).

6759. VENUH (Neivetso). British colonization and restructuring of Naga polity. New Delhi, Mittal, 2005, XI-114 p. (map).

Cf. n°s 4150, 5216, 6638, 7562-7853

c. Africa

** 6760. British documents on the end of empire. Series B. Vol. 9. Central Africa. General editor Stephen Richard ASHTON; editor Philip MURPHY. Part 1. Closer association 1945–1958. Part 2. Crisis and dissolution 1959–1965. London, TSO, 2005, 2 vol., CXXVIII-448 p., XL-602 p.

** 6761. Commandant (Le) en tournée: une administration au contact des populations en Afrique noire coloniale. Sous la dir. de Francis SIMONIS. Paris, Seli Arslan, 2005, 287 p.

** 6762. LOBO ANTUNES (António). D'este viver aqui neste papel descripto: cartas da guerra. Organização Maria José LOBO ANTUNES e Joana LOBO ANTUNES. Lisboa, D. Quixote, 2005, 431 p. (ill., facsims., map, ports.).

** 6763. PROST (Antoine). Carnets d'Algérie. Préf. de Pierre VIDAL-NAQUET. Paris, Tallandier, 2005, 196 p. (ill., maps). (Archives contemporaines).

** 6764. "S'ist ein übles Land hier": zur Historiographie eines umstrittenen Kolonialkrieges: Tagebuchaufzeichnungen aus dem Herero-Krieg in Deutsch-Südwestafrika 1904 von Georg Hillebrecht und Franz Ritter von Epp. Hrsg. v. Andreas E. ECKL. Köln, R.

Köppe, 2005, 302 p. (ill., map, ports.). (History, cultural traditions and innovations in Southern Africa, 22).

———

6765. ACKERSON (Wayne). The African Institution (1807–1827) and the antislavery movement in Great Britain. Lewiston a. Lampeter, E. Mellen Press, 2005, IV-246 p (Studies in British history, 77).

6766. ADESOJI (Abimbola O.). Colonialism and intercommunity relations: the Ifon-Ilobu example. *History in Africa*, 2005, 32, p. 1-19.

6767. Africa and the Americas: interconnections during the slave trade. Ed. by José C. CURTO and Renée SOULODRE-LAFRANCE. Trenton, Africa World Press, 2005, VI-338 p.

6768. AGERON (Charles Robert). Le gouvernement du général Berthezène à Alger en 1831. Saint-Denis, Bouchène, 2005, 258 p.

6769. ANOUMA (René-Pierre). Aux origines de la nation ivoirienne, 1893–1946. Vol. 1. Conquêtes coloniales et aménagements terriotriaux, 1893–1920. Vol. 2. Corset colonial et prise de conscience, 1920–1946. Paris, L'Harmattan, 2005, 2 vol., 282 p., 386 p. (ill.). (Etudes africaines).

6770. ARNALTE (Arturo). Richard Burton, consul en Guinea Española: una visión europea de Africa en los albores de la colonización. Madrid, La Caterata, 2005, 189 p. (ill., maps).

6771. BIALUSCHEWSKI (Arne). Pirates, slavers, and the indigenous population in Madagascar, c. 1690–1715. *International journal of African historical studies*, 2005, 38, 3, p. 401-426.

6772. BRANCH (Daniel). Imprisonment and colonialism in Kenya, c. 1930–1952: escaping the carceral archipelago. *International journal of African historical studies*, 2005, 38, 2, p. 239-266.

6773. BRANCHE (Raphaëlle). La guerre d'Algérie: une histoire apaisée? Paris, Ed. du Seuil, 2005, 445 p. (Histoire en débats. Points. Histoire, 351).

6774. CISSOKO (Sékéné Mody). Un combat pour l'unité de l'Afrique de l'ouest: la Fédération du Mali (1959–1960). Dakar, Nouvelles Éditions africaines du Sénégal, 2005, 257 p.

6775. DAVID (Thomas), ETEMAD (Bouda), SCHAUFELBUEHL (Janick Marina). La Suisse et l'esclavage des noirs. Lausanne, Editions Antipodes et Société d'Histoire de la Suisse romande, 2005, 184 p. (Histoire.ch).

6776. DEL BOCA (Angelo). Italiani, brava gente? Un mito duro a morire. Vicenza, Neri Pozza, 2005, 318 p. (Colibrì).

6777. DIALLO (Mamadou Dian Cherif). Répression et enfermement en Guinée: le pénitencier de Fotoba et la prison centrale de Conakry de 1900 à 1958. Préf. de Catherine COQUERY-VIDROVITCH. Paris, L'Harmattan, 2005, 673 p. (ill., maps). (Etudes africaines).

6778. DION (Michèle). Quand La Réunion s'appelait Bourbon: (XVIIe–XVIIIe siècle). Paris, L'Harmattan, 2005, 222 p. (map). (Populations).

6779. DORÉ (Jean-Marie). La résistance contre l'occupation coloniale en région forestière: Guinée 1800–1930. Paris, L'Harmattan, 2005, 311 p. (Etudes africaines).

6780. DUMOULIN (Michel). Léopold II: un roi génocidaire? Bruxelles, Académie Royale de Belgique, 2005, 122 p. (Mémoire de la Classe des lettres. Collection in 8°, 3e série, 37).

6781. DURPAIRE (François). Les Etats-Unis ont-ils décolonisé l'Afrique noire francophone? Paris, L'Harmattan, 2005, 356 p. (Etudes africaines).

6782. ELKINS (Caroline). Imperial reckoning. The untold story of Britain's gulag in Kenya. New York, Henry Holt, 2005, XVI-475 p.

6783. ERGO (André-Bernard). Des bâtisseurs aux contempteurs du Congo belge: l'odyssée coloniale. Paris, L'Harmattan, 2005, 310 p.

6784. EVE (Prosper). Les sept dernières années du régime colonial à La Réunion: 1939–1946. Paris, Karthala et Saint-Denis de La Réunion, Université de La Réunion, 2005, 255 p. (Collection Tropiques).

6785. FERREIRA (Roquinaldo Amaral). Transforming Atlantic slaving: trade, warfare and territorial control in Angola, 1650–1800. Ann Arbor, UMI Dissertation Services 2005, XIV-260 p.

6786. Genozid und Gedenken: namibisch-deutsche Geschichte und Gegenwart. Hrsg. v. Henning MELBER. Frankfurt am Main, Brandes & Apsel, 2005, 208 p.

6787. GLASER (Antoine), SMITH (Stephen). Comment la France a perdu l'Afrique. Paris, Calmann-Lévy, 2005, 278 p. (ill., maps).

6788. GRANT (Kevin). A civilised savagery. Britain and the new slaveries in Africa, 1884–1926. New York, Routledge, 2005, XII-223 p.

6789. GUSTAFSSON (Kalle). The trade in slaves in Ovamboland, ca. 1850–1910. *African economic history*, 2005, 33, p. 31-68.

6790. JONES (Hilary). Citizens and subjects: Métis society, identity and the struggle over colonial politics in Saint Louis, Senegal, 1870–1920. Ann Arbor, UMI Dissertation Services, 2005, XII-280 p.

6791. L'ANGE (Gerald). The white Africans: from colonisation to liberation. Johannesburg, Jonathan Ball Publishers, 2005, XXX-524 p. (ill., maps, ports.).

6792. LABAND (John). The Transvaal rebellion: the First Boer War, 1880–1881. Harlow, Pearson Longman, 2005, XII-264 p.

6793. LAHOUEL (Badra). La résistance algérienne, 1830–1962. Oran, Dar El Gharb, 2005, 185 p.

6794. LAWRANCE (Benjamin N.). Bankoe v. Dome: traditions and petitions in the Ho-Asogli amalgamation,

British mandated Togoland, 1919–39. *Journal of African history*, 2005, 46, 2, p. 243-267.

6795. LE COUR GRANDMAISON (Olivier). Coloniser, exterminer: sur la guerre et l'état colonial. Paris, Fayard, 2005, 365 p.

6796. LEFEUVRE (Daniel). Chère Algérie: la France et sa colonie, 1930–1962. Préf. de Jacques MARSEILLE. Paris, Flammarion, 2005, 512 p. (ill.).

6797. MADLEY (Benjamin). From Africa to Auschwitz: how German South West Africa incubated ideas and methods adopted and developed by the Nazis in Eastern Europe. *European history quarterly*, 2005, 35, 3, p. 429-464.

6798. MIETTINEN (Kari). On the way to whiteness. Christianization, conflict and change in colonial Ovamboland, 1910–1965. Helsinki, Suomalaisen Kirjallisuuden Seura, 2005, 370 p. (Bibliotheca historica, 92).

6799. MURRAY (Colin), SANDERS (Peter). Medicine murder in colonial Lesotho. Edinburgh, Edinburgh U. P., 2005, XVI-493 p. (International African library, 31).

6800. NUNES (António Pires). Angola 1961: [da Baixa do Cassange a Nambuangongo]. Lisboa, Prefácio, 2005, 180 p. (ill., maps, ports.). (História militar. Guerras e combates, 4).

6801. NZENGUET IGUEMBA (Gilchrist Anicet). Colonisation, fiscalité et mutations au Gabon, 1910–1947. Préf. de Colette DUBOIS; avant-propos de Pierre N'DOMBI. Paris, L'Harmattan, 2005, 467 p. (ill., maps). (Etudes africaines).

6802. POLLARD (Lisa). Nurturing the nation. The family politics of modernizing, colonizing, and liberating Egypt, 1805–1923. Berkeley a. Los Angeles, University of California Press, 2005, XV 287 p.

6803. RANDAZZO (Antonella). In Affrica andammo: gli orrori negati dell'Africa italiana. Acireale, Bonanno, 2005, 155 p. (ill.). (Storia e politica, 15).

6804. REINWALD (Brigitte). Reisen durch den Krieg: Erfahrungen und Lebensstrategien westafrikanischer Weltkriegsveteranen. Berlin, Klaus Schwarz Verlag, 2005, 444 p. (Studien, 18).

6805. SALIH (Kamal O.). British colonial military recruitment policy in the Southern Kordofan region of Sudan, 1900–1945. *Middle Eastern studies*, 2005, 41, 2, p. 169-192.

6806. SCHAFFER (Matt). Bound to Africa: the Mandinka legacy in the new world. *History in Africa*, 2005, 32, p. 321-369.

6807. SCHNEIDER-WATERBERG (H. R.). Der Wahrheit eine Gasse: Anmerkungen zum Kolonialkrieg in Deutsch-Südwestafrika 1904. Swakopmund, Gesellschaft für Wissenschaftliche Entwicklung, 2005, 167 p. (ill.).

6808. SUDRY (Yves). Guerre d'Algérie: les prisonniers des Djounoud. Paris, L'Harmattan, 2005, 216 p. (ill., maps). (Histoire et perspectives méditerranéennes).

6809. THÉNAULT (Sylvie.). Histoire de la guerre d'indépendance algérienne. Paris, Flammarion, 2005, 303 p. (ill.).

6810. VANSINA (Jan). Ambaca society and the slave trade c. 1760–1845. *Journal of African history*, 2005, 46, 1, p. 1-27.

6811. VERA CRUZ (Elizabeth Ceita). O estatuto do indigenato: Angola: a legalização de discriminação na colonização portuguesa. Editores José Manuel DA NÓBREGA e Nuno PÁDUA DE MORA. Lisboa, Novo Imbonderio, 2005, 175 p. (Estudos e documentos).

6812. VIJGEN (Ingeborg). Tussen mandaat en kolonie: Rwanda, Burundi en het Belgische bestuur in opdracht van de Volkenbond (1916–1932). Leuven, Acco, 2005, 279 p. (ill., maps, ports.).

6813. Voci dall'Eritrea: ebrei e colonialismi tra XIX e XX secolo. A cura di Marco CAVALLARIN. Torino, L'Harmattan Italia, 2005, 129 p. (ill.). (Il politico e la memoria).

6814. Warri City and British colonial rule in Western Niger Delta. Ed. by Peter P. EKEH. Buffalo, Urhobo Historical Society, 2005, IV-XXX-295 p. (ill., maps). (Urhobo Historical Society monograph, 1).

6815. WILLIS (Justin). Hukm: the creolization of authority in condominium Sudan. *Journal of African history*, 2005, 46, 1, p. 29-50.

6816. WOOD (J. R. T.). So far and no further! Rhodesia's bid for independence during the retreat from empire 1959–1965. Johannesburg, 30 Degrees South Publishers, 2005, 533 p (ill., maps).

6817. YAZIDI (Béchir). La politique coloniale et le domaine de l'Etat en Tunisie: de 1881 jusqu'à la crise des années trente. Tunis, Editions Sahar et Mannouba, Faculté des lettres des arts et des humanités, 2005, 406 p.

Cf. n°s 648, 3575, 3577, 3689, 3727, 4255, 4257, 4282, 4634, 4666, 4673, 6585, 7402, 7489, 7854-7873

d. America

* 6818. MOSCOSO (Francisco). Bibliografía de la conquista y colonización de Puerto Rico: siglos XV-XVII (1492–1650). San Juan, Universidad de Puerto Rico, Recinto de Río Piedras, Departamento de Historia, 2005, III-42 p.

** 6819. Colonie française (Une) au Paraguay: la Nouvelle-Bordeaux. Présenté par Guido RODRÍGUEZ ALCALÁ et Luc CAPDEVILA. Paris, L'Harmattan, 2005, 124 p. (ill., maps). (Recherches Amériques latines).

** 6820. Documentos para el estudio de la cultura política de la transición: juras, poderes e instrucciones: Nueva España y la Capitanía General de Guatemala, 1808–1820. Compilación y estudio introductorio Beatriz ROJAS. México D.F., Instituto Mora, 2005, 524 p. (map). (Historia política).

** 6821. [FEDOROVA (T.S.), SPIRIDONOVA (L.I.).]. Rossijsko-amerikanskaja kompanija i izuchenie Tik-

2. HISTORY OF COLONIZATION AND DECOLONIZATION

hookeanskogo severa, 1815–1841: Sbornik dokumentov. (The Russian-American Company and the study of the Pacific North, 1815–1841: Documents). Ed. Nikolaj N. BOLKHOVITINOV. RAN. Otd. istoriko-filol. nauk; Fed. arkh. agenstvo, Ros. gos. arkhiv voen.-morsk. flota. Moskva, Nauka, 2005, 459 p. (ind. p. 439-457). (Issledovanja russkikh na Tikhom okeane v XVIII – pervoj polovine XIX v.).

6822. ANDRÉS-GALLEGO (José). La esclavitud en la América española. Madrid, Ediciones Encuentro – Fundación I. Larramendi, 2005, 415 p.

6823. ANDRIVON MILTON (Sabine). La Martinique et la Grande Guerre. Paris, L'Harmattan, 2005, 406 p. (maps).

6824. BÉNAT-TACHOT (Louise), LAVALLÉ (Bernard). L' Amérique de Charles Quint. Avant-propos de Jean VILAR. Pessac, Presses universitaires de Bordeaux, 2005, 248 p. (Parcours universitaires).

6825. BERKIN (Carol). Revolutionary mothers. Women in the struggle for America's independence. New York, Alfred A. Knopf, 2005, XVIII-194 p.

6826. BOSHER (J. F.). Vancouver island in the empire. *Journal of imperial and Commonwealth history*, 2005, 33, 3, p. 349-368.

6827. Creation (The) of the British Atlantic world. Ed. by Carole SHAMMAS and Elizabeth MANCKE. Baltimore, Johns Hopkins U. P., 2005, VI-400 p.

6828. DEL RÍO (María de las Mercedes). Etnicidad, territorialidad y colonialismo en los Andes: tradición y cambio entre los Soras de los siglos XVI y XVII (Bolivia). La Paz, Instituto de Estudios Bolivianos, ASDI y Lima, Institut français d'études andines, 2005, 341 p. (ill.). (Colección Cuarto centenario de la fundación de Oruro. Travaux de l'Institut français d'études andines, 212).

6829. DIXON (David). Never come to peace again: Pontiac's uprising and the fate of the British Empire in North America. Norman, University of Oklahoma Press, 2005, XVII-353 p. (Campaigns and commanders).

6830. ERICKSON (Lesley). Constructed and contested truths: aboriginal suicide, law, and colonialism in the Canadian West(s), 1823–1927. *Canadian historical review*, 2005, 86, 4, p. 595-618.

6831. FEDOSOV (Denis G.). Andskie strany v kolonial'nuju epokhu: Religioznaja i khudozhestvennaja kartina mira. (The Andes Countries in the colonial age: The religious and artistic image of the world). RAN, Feder. agenstvo po kul'ture i kinematografii RF, Gos. in-t iskusstvoznanija. Moskva, URSS – KomKniga, 2005, 247 p. (ill.; ind. p. 1867-190).

6832. FIGUEROA (Luis A.). Sugar, slavery and freedom in nineteenth-century Puerto Rico. Chapel Hill, University of North Carolina Press, 2005, 290 p.

6833. French colonial Louisiana and the Atlantic World. Ed. by Bradley G. BOND. Baton Rouge, Louisiana State U. P., 2005, XXI-322 p.

6834. GUERRERO RINCÓN (Amado Antonio), PÁEZ MARTÍNEZ (Laritza). Poblamiento y conflictos territoriales en Santander. Bucaramanga, Universidad Industrial de Santander, Escuela de Historia, 2005, 162 p. (ill., folded maps).

6835. HARVEY (Louis-Georges). Le printemps de l'Amérique française: américanité, anticolonialisme, et républicanisme dans le discours politique québécois, 1805–1837. Montréal, Boréal, 2005, 296 p.

6836. HINZ (Felix). "Hispanisierung" in Neu-Spanien 1519–1568: Transformation kollektiver Identitäten von Mexica, Tlaxkalteken und Spaniern. Hamburg, Verlag Dr. Kovač, 2005, 3 vol., 866 p. (ill.). (Studien zur Geschichtsforschung der Neuzeit, 45).

6837. HORNSBY (Stephen J.). British Atlantic, American frontier. Spaces of power in early modern British America. Hanover, University press of New England, 2005, XV-307 p.

6838. LÓPEZ (Adalberto). The colonial history of Paraguay: the revolt of the Comuneros, 1721–1735. New Brunswick a. London, Transaction Publishers, 2005, 213 p. (maps).

6839. MARTIRÉ (Eduardo). Las audiencias y la administración de justicia en las Indias. Madrid, Universidad Autónoma de Madrid, 2005, 286 p. (Colección de estudios, 104).

6840. METCALF (Alida C.). Go-betweens and the colonization of Brazil, 1500–1600. Austin, University of Texas Press, 2005, XIV-375 p. (ill., maps).

6841. Modos de governar: idéias e práticas políticas no Império português, séculos XVI–XIX. Organizadoras, Maria Fernanda BICALHO e Vera Lúcia AMARAL FERLINI. São Paulo, Alameda, 2005, 445 p.

6842. OOSTINDIE (Gert). Paradise overseas: the Dutch Caribbean: colonialism and its transatlantic legacies. Oxford, Macmillan Caribbean, 2005, XI-204 p. (map). (Warwick University Caribbean studies).

6843. PACHECO DÍAZ (Argelia). Una estrategia imperial. El situado de Nueva España a Puerto Rico, 1765–1821. México, Instituto Mora, 2005, 91 p.

6844. PULSIPHER (Jenny Hale). Subjects unto the same king. Indians, English, and the contest for authority in colonial New England. Philadelphia, University of Pennsylvania Press, 2005, 361 p.

6845. Revolución, independencia y las nuevas naciones de América. Coordinador Jaime E. RODRÍGUEZ O. Madrid, Fundación MAPFRE TAVERA, 2005, 614 p. (ill., charts, ports.). (Publicaciones del programa Iberoamérica, 9).

6846. SAETHER (Steinar A.). Identidades e independencia en Santa Marta y Riohacha, 1750–1850. Bogotá, Instituto Colombiano de Antropología e Historia, 2005, 300 p. (Colección año, 200).

6847. SARMIENTO GUTIÉRREZ (Julio). El Perú y la dominación hispánica: Cajamarca, conquista y colonia. Cajamarca, Universidad Nacional de Cajamarca, 2005, 180 p. (ill., maps).

6848. SARSON (Steven). British America, 1500–1800: creating colonies, imagining an empire. London, Hodder Arnold, 2005, XIX-332 p. (maps).

6849. SCHIAVINATTO (Iara Lis). A independência do Brasil: modos de lembrar e esquecer = La independencia de Brasil: formas de recordar y olvidar. Madrid, Fundación Mapfre y Aranjuez, Ediciones Doce Calles, 2005, 268 p. (Prisma histórico, 4. Publicaciones del programa Iberoamérica, 11. Iberoamérica: 200 años de convivencia independiente, 11).

6850. Soldados del rey: el ejército borbónico en América colonial en vísperas de la independencia. Ed. por Allen J. KUETHE y Juan MARCHENA FERNÁNDEZ. Castelló de la Plana, Universitat Jaume I, 2005, 282 p. (ill., map). (Col·lecció Amèrica, 4).

6851. SWEET (Julie Anne). Negotiating for Georgia: British-Creek relations in the trustee era, 1733–1752. Athens a. London, University of Georgia Press, 2005, X-267 p. (ill., maps).

6852. TEJERA (Eduardo J.). Causas de dos Américas: modelo de conquista y colonización hispano e inglés en el nuevo mundo. Madrid, Dykinson y Fundación Carlos III, 2005, 522-IV-IV p. (ill., maps).

6853. VELASCO GODOY (María de los Angeles). La historia de un cambio en el Valle de Ixtlahuaca: la formación de un pueblo colonial. Toluca, Universidad Autónoma del Estado de México, 2005, 344 p. (ill., maps). (Colección Humanidades. Serie Historia).

6854. VIDAL (Laurent). Mazagao, la ville qui traversa l'Atlantique. Du Maroc à l'Amazonie 1769–1783. Paris, Aubier, 2005, 314 p. (Coll. historique).

6855. WARREN (J. Benedict). Estudios sobre el Michoacán colonial: los inicios. Presentación por Gerardo SÁNCHEZ DÍAZ. Morelia, Universidad Michoacana de San Nicolás de Hidalgo, Instituto de Investigaciones Históricas, Fimax Publicistas Editores, 2005, 199 p. (ill.). (Colección Historia nuestra, 23).

6856. WEBER (David J.). Bárbaros: Spaniards and their savages in the Age of Enlightenment. New Haven a. London, Yale U. P., 2005, 466 p. (ill.).

6857. WEEKS (Charles A.). Paths to a middle ground. The diplomacy of Natchez, Boukfouka, Nogales, and San Fernando de las Barrancas, 1791–1795. Tuscaloosa, University of Alabama Press, 2005, X-292 p.

6858. WOOD (Betty). Slavery in colonial America, 1619–1776. Lanham a. Oxford, Rowman & Littlefield Publishers, 2005, XIV-131 p. (ill., map). (The African American history series).

Cf. n°s 644, 3586, 4112, 4663, 5845, 5893, 5941, 6767, 6875, 6900, 6921, 6930, 7874-7888

e. Oceania

6859. BLAIS (Hélène). Voyage au grand océan. Géographies du Pacifique et colonisation, 1815–1845. Paris, Éd. du CTHS, 2005, 351 p.

6860. CLAY (Brenda Johnson). Unstable images: colonial discourse on New Ireland, Papua New Guinea, 1875–1935. Honolulu, University of Hawaii Press, 2005, XIII-324 p. (ill., 1 map).

6861. CLENDINNEN (Inga). Dancing with strangers: Europeans and Australians at first contact. Cambridge, Cambridge U. P., 2005, 324 p. (ill., maps).

6862. DENOON (Donald). A trial separation: Australia and the decolonisation of Papua New Guinea. Canberra, Pandanus Books, 2005, XV-228 p. (maps).

6863. EVANS (Julie). Edward Eyre, race and colonial governance. Dunedin, University of Otago Press, 2005, 195 p. (ill., maps). (Otago history series).

6864. HARRIS (Richard). Making leeway in the Leewards, 1929–1951: the negotiation of colonial development. *Journal of imperial and Commonwealth history*, 2005, 33, 3, p. 393-418.

6865. Imperial communication: Australia, Britain, and the British Empire, c. 1830–1850. Ed. by Simon J. POTTER. London, University of London, King's College, Menzies Centre for Australian Studies, 2005, IV-121 p.

6866. KRUG (Alexander). "Der Hauptzweck ist die Tötung von Kanaken": die deutschen Strafexpeditionen in den Kolonien der Südsee 1872–1914. Tönning, Der Andere Verlag, 2005, VI-444 p. (maps).

6867. PICKLES (Katie). A link in 'the great chain of Empire friendship': the Victoria League in New Zealand. *Journal of imperial and Commonwealth history*, 2005, 33, 1, p. 29-50.

Cf. n°s 7889-7894

§ 3. From 1500 to 1789.

a. General

6868. CERNOVODEANU (Paul). O familie de diplomați români din Transilvania la cumpăna dintrea veacurile XVII și XVIII: Corbea din Șcheii Brașovului. (A family of Romanian diplomats from Transylvania at the end of the 17th century and the beginning of the 18th century: Corbea from Șcheii Brașovului). *Studii și Materiale de Istorie Medie. Academia Română*, 2005, 23, p. 145-158.

6869. ISOPESCU (Sergiu). Alianțele militare ale principatelor carpato-dunărene în secolul al XVII-lea. (Military alliances of the Carpathian-Danubian principalities in the 17th century). *Revista de istorie militară*, 2005, 3, p. 43-51.

6870. MATAR (Nabil I.). Britain and Barbary, 1589–1689. Gainesville, University Press of Florida, 2005, XIII-241 p.

b. 1500-1648

** 6871. Lettere di Vincenzo Priuli capitano delle galee di Fiandra al doge di Venezia, 1521-1523. A cura di Francesca ORTALLI; appendice ed indice a cura di Bianca LANFRANCHI STRINA. Venezia, Comitato per la pubblicazione delle fonti relative alla storia di Venezia, 2005, XLVIII-146 p. (ill.). (Fonti per la storia di Venezia. Archivi pubblici. Sez. I).

** 6872. Oranda shokancho nikki yakubun henno 10. [Nihon kankei kaigai shiryo]. (Diaries kept by the heads of the Dutch Factory in Japan. (Japanese translation, vol. 10). [historical documents in foreign languages relating to Japan]). Ed. Tokyo daigaku shiryo hensanjo (Historiographical Institute of the University of Tokyo). Tokyo, University of Tokyo Press, 2005, 276 p.

6873. ARFAIOLI (Maurizio). The Black Bands of Giovanni: infantry and diplomacy during the Italian wars (1526-1528). Pisa, Edizioni Plus-Pisa U. P., 2005, XVII-204 p. (ill., maps).

6874. BARTZ (Christian). Köln im Dreissigjährigen Krieg: die Politik des Rates der Stadt (1618-1635); vorwiegend anhand der im Historischen Archiv der Stadt Köln. Frankfurt am Main u. Oxford, P. Lang, 2005, 375 p. (maps). (Militärhistorische Untersuchungen, 6).

6875. Charles Quint, empereur d'Allemagne et roi d'Espagne, quelques aspects de son règne. Textes réunis par Marie-Catherine BARBAZZA. Montpellier, E.T.I.L.A.L., 2005, 325 p. (ill., maps). (Espagne médiévale et moderne, 6).

6876. Diplomate (Le) au travail. Entscheidungsprozesse, Information und Kommunikation im Umkreis des Westfälischen Friedenskongresses. Hrsg. v. Rainer BABEL. München, Oldenbourg, 2005, 217 p. (Pariser historische Studien, 65).

6877. DZIUBIŃSKI (Andrzej). Stosunki dyplomatyczne polsko-tureckie w latach 1500-1572 w kontekście międzynarodowym. (Relazioni diplomatiche polacco-turche nel periodo 1500-1572, nel contesto internazionale). Wrocław, Fundacja na Rzecz Nauki Polskiej, 2005, 309 p. (maps). (Monografie Fundacji na Rzecz Nauki Polskiej).

6878. ESTEBAN ESTRÍNGANA (Alicia). Madrid y Bruselas: relaciones de gobierno en la etapa postarchiducal (1621-1634). Leuven, Leuven U. P., 2005, 375 p. (ill.). (Avisos de Flandes, 10).

6879. ITO (Koji). Vestfålen-jōyaku to shinsei Rōma teikoku: Doitsu teikoku syoko toshite no Swêden. (Peace of Westphalia and the Holy Roman Empire: Sweden as a member of German empire). Fukuoka, Kyushu U. P., 2005, 214 p.

6880. KOBZAREVA (Elena I.). Shvedskaja okkupatsija Novgoroda v period Smuty XVII veka. (The Swedish occupation of Novgorod during the 17th-century Time of Troubles). RAN, In-t rossijskoj istorii. Moskva, IRI RAN, 2005, 454 p. (bibl. p. 390-440; pers. ind. p. 442-453).

6881. KUMRULAR (Özlem). El duelo entre Carlos V y Solimàn el Magnífico (1520-1535). İstanbul, Isis, 2005, 336 p. (Quadernos del Bósforo, 4).

6882. LEVIN (Michael J.). Agents of empire. Spanish ambassadors in sixteenth-century Italy. Ithaca, Cornell U. P., 2005, VIII-228 p.

6883. MAC DERMOTT (James). England and the Spanish Armada: the necessary quarrel. New Haven a. London, Yale U. P., 2005, XVI-411 p. (ill.).

6884. MACZKIEWITZ (Dirk). Der niederländische Aufstand gegen Spanien (1568-1609): eine kommunikationswissenschaftliche Analyse. Münster, Waxmann, 2005, 366 p. (ill., maps). (Studien zur Geschichte und Kultur Nordwesteuropas, 12).

6885. MICHON (C.). Étude sur le personnel diplomatique sous les règnes de François Ier et Henri VIII. Revue d'histoire diplomatique, 2005, 119, 3, p. 193-206.

6886. ÖHMAN (Jenny). Der Kampf um den Frieden: Schweden und der Kaiser im dreissigjährigen Krieg. Wien, Öbv & hpt, 2005, 288 p. (ill., maps, ports.). (Militärgeschichtliche Dissertationen österreichischer Universitäten, 16).

6887. OSTERRIEDER (Markus). Das wehrhafte Friedensreich: Bilder von Krieg und Frieden in Polen-Litauen (1505-1595). Wiesbaden, Reichert, 2005, 330 p. (Imagines Medii Aevi, 20).

6888. PEHLE (Hans). Der "Rheinübergang" des Schwedenkönigs Gustav II. Adolf: ein Ereignis im Dreißigjärigen Krieg. Riedstadt, Forum, 2005, 219 p. (ill., maps, ports.).

6889. REINBOLD (Markus). Jenseits der Konfession: die frühe Frankreichpolitik Philipps II. von Spanien, 1559-1571. Ostfildern, Thorbecke, 2005, 280 p. (Beihefte zu Francia, 61).

6890. RIECK (Anja). Frankfurt am Main unter schwedischer Besatzung, 1631-1635: Reichsstadt, Repräsentationsort, Bündnisfestung. Frankfurt am Main u. New York, P. Lang, 2005, 383 p. (ill.). (Europäische Hochschulschriften. Reihe III, Geschichte und ihre Hilfswissenschaften, 1011 = Publications universitaires européennes. Série III, Histoire, sciences auxiliaires de l'histoire, 1011 = European university studies. Series III, History and allied sciences, 1011).

6891. SÁENZ-CAMBRA (Concepción). Scotland and Philip II: 1580-1598. Sevenoaks, Amherst, 2005, XVII-232 p. (ill.).

6892. Tudor England and its neighbours. Ed. by Susan DORAN and Glenn RICHARDSON. Basingstoke a. New York, Palgrave Macmillan, 2005, XII-277 p. (Themes in focus).

6893. VOLPINI (Paola). Una storia di spie tra Ferdinando I di Toscana e Filippo II di Spagna (fine secolo

XVI). *Archivio storico italiano*, 2005, 163, 2, p. 229-258.

6894. WBG Deutsch-Französische Geschichte. Hrsg. v. Deutschen Historischen Institut in Paris u. d. Ecole des Hautes Etudes en Sciences Sociales. Band 3. BABEL (Rainer). Deutschland und Frankreich im Zeichen der habsburgischen Universalmonarchie, 1500–1648. Darmstadt, Wissenschaftliche Buchgesellschaft, 2005, 256 p. (ill., maps).

Cf. n° 4506

c. 1648–1789

** 6895. Embaixada de D. João V de Portugal ao imperador Yongzheng da China (1725–1728). Coordenação, António VASCONCELOS DE SALDANHA; leitura, estudo e anotação, Mariagrazia RUSSO; tradução e notas chinesas, Jin Guo PING. Lisboa, Fundação Oriente, 2005, 337 p..

** 6896. Rodney papers (The): selections from the correspondence of Admiral Lord Rodney. Ed. by David SYRETT. Vol. 1. 1741–1763. Aldershot, Ashgate, 2005, 540 p. (Publications of the Navy Records Society, 148).

** 6897. VAUGELADE (Daniel). La question américaine au 18ème siècle: au travers de la correspondance du duc Louis Alexandre de La Rochefoucauld (1743–1792). Paris, Publibook, 2005, 362 p. (ill.).

6898. BESPALOV (Aleksandr V.). Bitvy Velikoj Severnoj vojny, 1700–1721. (Battles of the Great Northern War, 1700–1721). Moskva, Rejtar', 2005, 251 p. (ill., maps; bibl. p. 242-248).

6899. BLACK (Jeremy). Hanover and British foreign policy 1714–1760. *English historical review*, 2005, 120, 486, p. 303-339. – IDEM. The continental commitment: Britain, Hanover, and interventionism 1714–1793. New York, Routledge, 2005, XIV-214 p.

6900. CARRER (Philippe). La Bretagne et la guerre d'indépendance américaine. Rennes, Portes du large, 2005, 237 p. (ill., map).

6901. CÉNAT (Jean-Philippe). Le ravage du Palatinat: politique de destruction, stratégie de cabinet et propagande au début de la guerre de la Ligue d'Augsbourg. *Revue historique*, 2005, 633, p. 97-132.

6902. Dansk udenrigspolitiks historie. 2. JESPERSEN (Knud J. V.), FELDBAEK (Ole). Revanche og neutralitet 1648–1814. København, Gyldendal Leksikon, 2005, 808 p. (ill., maps, ports.).

6903. DULL (Jonathan R.). The French Navy and the Seven Years' War. Lincoln, University of Nebraska Press, 2005, XIII-445 p. (maps). (France overseas).

6904. ECKERSLEY (Peter). Le difficile retour en grâce de Bruant des Carrières, ancien commis de Fouquet et agent de Louis XIV à Liège (1667–1676). *Revue d'histoire diplomatique*, 2005, 119, 3, p. 277-291.

6905. GUS'KOV (Andrej G.). Velikoe posol'stvo Petra I: Istochnikovedcheskoe issledovanie. (The Great Embassy of Peter the Great [1697–1698]: a source-critical study). RAN, In-t rossijskoj istorii. Moskva, IRI RAN, 398 p. (plates; bibl. p. 222-231; ind. p. 373-397).

6906. HASQUIN (Hervé). Louis XIV face à l'Europe du Nord: l'absolutisme vaincu par les libertés. Bruxelles, Editions Racine, 2005, 334 p. (ill., maps). (Les racines de l'histoire).

6907. JEANMOUGIN (Bertrand). Louis XIV à la conquête des Pays-Bas espagnols: la guerre oubliée, 1678–1684. Préface de Lucien BÉLY. Paris, Economica, 2005, XII-235 p. (maps). (Campagnes & stratégies. Les grandes batailles).

6908. KLONOWSKI (Martin). Im Dienst des Hauses Hannover: Friedrich Christian Weber als Gesandter im Russischen Reich und in Schweden 1714–1739. Husum, Matthiesen, 2005, 219 p. (ill., maps). (Historische Studien, 485).

6909. MATHIS (Rémi). De la négociation à la relation d'ambassade Simon Arnauld de Pomponne et sa deuxième ambassade de Suède (1671). *Revue d'histoire diplomatique*, 2005, 119, 3, p. 263-276.

6910. Personalunionen (Die) von Sachsen-Polen, 1697–1763, und Hannover-England, 1714–1837: ein Vergleich. Hrsg. v. Rex REXHEUSER. Wiesbaden, Harrassowitz, 2005, VII-495 p. (Quellen und Studien / Deutsches Historisches Institut Warschau, 18).

6911. SÉRÉ (Daniel). La paix des Pyrénées ou la paix du roi: le rôle méconnu de Philippe IV dans la restauration de la paix entre l'Espagne et la France (1659). *Revue d'histoire diplomatique*, 2005, 119, 3, p. 243-262.

6912. TOUZARD (Anne-Marie). Le drogman Padery: émissaire de France en Perse, 1719–1725. Paris, P. Geuthner, 2005, 318 p. (ill., maps).

6913. Tussen Munster en Aken: de Nederlandse Republiek als grote mogendheid (1648–1748). Red. Simon GROENVELD, Maurits EBBEN en Raymond FAGEL. Maastricht, Shaker Publishing, 2005, 75 p. (Publicaties van de Vlaams-Nederlandse Vereniging voor Nieuwe Geschiedenis, 2).

6914. WAQUET (Jean-Claude). François de Callières. L'art de négocier en France sous Louis XIV. Paris, Editions Rue d'Ulm, Presse de l'Ecole normale superieure, 2005, 288 p.

Cf. n°s 6918, 6940

§ 4. From 1789 to 1815.

** 6915. Bürger, Bauern und Soldaten: Napoleons Krieg in Thüringen 1806 in Selbstzeugnissen: Briefe, Berichte, Erinnerungen. Hrsg. v. Birgitt HELLMANN. Weimar, Hain, 2005, 327 p. (ill., maps, ports.). (Bausteine zur Jenaer Stadtgeschichte, 9).

** 6916. Sachsen, der Rheinbund und die Exekution der Sachsen betreffenden Entscheidungen des Wiener

Kongresses (1803–1816). Edition von Dokumenten des Sächsischen Hauptstaatsarchivs Dresden 1.-116. Ausgewählt, übertragen und kommentiert von Rudolf JENAK. Neustadt an der Aisch, Ph. C.W. Schmidt, 2005, 455 p. (ill.).

** 6917. SANTACARA (Carlos). La Guerra de Independencia vista por los británicos: 1808–1814. Boadilla del Monte, A. Machado Libros, 2005, 820 p. (ill., maps). (Papeles del tiempo, 7).

6918. BÉGAUD (Stéphane), BELISSA (Marc), VISSER (Joseph). Aux origines d'une alliance improbable: le réseau consulaire français aux Etats-Unis, (1776– 1815). Bruxelles et Oxford, P.I.E.-P. Lang, 2005, 302 p. (ill., col. maps). (Diplomatie et histoire).

6919. BERTSCH (Daniel). Anton Prokesch von Osten, 1795–1876: ein Diplomat Österreichs in Athen und an der Hohen Pforte: Beiträge zur Wahrnehmung des Orients im Europa des 19. Jahrhunderts. München, R. Oldenbourg, 2005, 754 p. (Südosteuropäische Arbeiten, 123).

6920. BEZOTOSNYJ (Viktor M.). Razvedka i plany storon v 1812 godu. (Intelligence and the plans of the French and the Russians in 1812). Gos. ist. muzej. Moskva, ROSSPEN, 2005, 287 p. (ill.; portr.; bibl. p. 237-278; pers. ind. p. 280-286).

6921. BROWN (Gordon S.). Toussaint's clause: the founding fathers and the Haitian revolution. Jackson, University Press of Mississippi, 2005, XI-321 p. (map).

6922. CIOBANU (Veniamin). Statele nordice și problema orientală (1792–1814). Iași, Editura Junimea, 2005, 295 p. (Bibliotheca historiae universalis, 10).

6923. CONNELLY (Owen). The wars of the French Revolution and Napoleon, 1792–1815. Milton Park, Abingdon a. New York, Routledge, 2005, IX-270 p.

6924. DELOS-HOURTOULE (Sarah). Les origines des garanties disciplinaires des hauts fonctionnaires des relations extérieures sous le Consulat et l'Empire, ainsi que leurs limites à travers l'exemple de la destitution du Consul général aux États-Unis. *Revue d'histoire diplomatique*, 2005, 119, 2, p. 101-134.

6925. GARNIER (Jacques). Austerlitz: 2 décembre 1805. Propos liminaire par Jean TULARD. Paris, Fayard, 2005, 457 p. (maps).

6926. HANTRAYE (Jacques). Les Cosaques aux Champs-Elysées: l'occupation de la France après la chute de Napoléon. Paris, Belin, 2005, 303 p. (Histoire & société).

6927. HILL (Peter P.). Napoleon's troublesome Americans: Franco-American relations, 1804–1815. Dulles, Brassey's, 2005, 272 p.

6928. LAMBERT (Frank). The Barbary wars: American independence in the Atlantic World. New York, Hill and Wang, 2005, 228 p.

6929. MIQUEL (Pierre). Austerlitz. Paris, Albin Michel, 2005, 458 p. (ill., maps, ports.).

6930. MOREL (Marco). A Revoluçao do Haiti e o Império do Brasil: intermediaçoes e rumores. *Anuario de estudios bolivarianos*, 2005, 12, p. 189-212.

6931. Napoléon et l'Europe: regards sur une politique: actes du colloque organisé par la Direction des archives du Ministère des affaires étrangères et la Fondation Napoléon, 18 et 19 novembre 2004. Coordonné par Thierry LENTZ. Paris, Fayard, 2005, 445 p.

6932. ORLOV (Aleksandr A.). Sojuz Peterburga i Londona: Rossijsko-britanskie otnoshenija v epokhu napoleonovskikh vojn. (The union of Saint-Petersburg and London: The relations between Russia and Britain in the time of Napoleon Wars). Moskva, Progress-Traditsija, 2005, 367 p. (ill.; facs.; bibl. incl.; pers. ind. p. 351-366).

6933. ORTÍZ SOTELO (Jorge). Perú y Gran Bretaña: política y economía (1808–1839), a través de los informes navales británicos. Lima, Asociación de Historia Marítima y Naval Iberoamericana e Instituto de Estudios Internacionales, Pontificia Universidad Católica del Perú, 2005, 293 p. (ill.).

6934. Paix d'Amiens (La): actes du colloque: Amiens, 24 et 25 mai 2002. Sous la direction de Nadine-Josette CHALINE. Amiens, Encrage, 2005, 302 p. (ill., maps). (Hier, 21).

6935. Popular resistance in the French wars: patriots, partisans and land pirates. Ed. by Charles J. ESDAILE. Basingstoke, Palgrave Macmillan, 2005, XIII-233 p.

6936. Portraits croisés de Thomas Jefferson et Napoléon Bonaparte: La cession de la Louisiane: Actes du colloque de la Fondation Singer-Polignac. Sous la direction de Jean-François LEMAIRE. Paris, Editions SPM, 2005, 154 p.

6937. RAPONI (Nicola). Il mito di Bonaparte in Italia: atteggiamenti della società milanese e reazioni nello Stato romano. Roma, Carocci, 2005, 200 p. (ill., ports.). (Studi storici Carocci, 78).

6938. RASPOPOVIĆ (Radoslav M.). Ruski konzulat u Kotoru i Crna Gora 1804–1806. (Russian Consulate in Kotor and Montenegro 1804–1806). Podgorica, Istorijski institut Crne Gore, 2005, 216 p. (Edicija Posebna izdanja).

6939. RIOTTE (Torsten). Hannover in der britischen Politik (1792–1815): dynastische Verbindung als Element außenpolitischer Entscheidungsprozesse. Münster, Lit, 2005, XIV-237 p. (ill.). (Historia profana et ecclesiastica, 13).

6940. RODRÍGUEZ GONZÁLEZ (Agustín Ramón). Trafalgar y el conflicto naval anglo-español del siglo XVIII. San Sebastián de los Reyes, Actas, 2005, 459 p. (ill.).

6941. SCHNEID (Frederick C.). Napoleon's conquest of Europe: the war of the Third Coalition. Westport a. London, Praeger Publishers, 2005, XII-192 p. (maps). (Studies in military history and international affairs).

6942. SPAANS (K. W.). 'Un pays absolument artificiel': Beierse diplomaten over politieke en maatschappelijke veranderingen in Nederland in de Bataafs-Franse periode (1795–1810). Amsterdam, De Bataafsche Leeuw, 2005, 103 p. (ill., maps).

6943. Suisse (La) de la Médiation dans l'Europe napoléonienne (1803–1814) =: Die Schweiz unter der Mediationsakte in Napoleons Europa (1803–1814): actes du colloque de Fribourg (journée du 10 octobre 2003). Éd. par Mario TURCHETTI. Fribourg, Academic Press, 2005, VI-151 p. (Ars historica & politica, 2. Réligion, politique, société en Suisse = Religion, Politik, Gesellschaft in der Schweiz, 36).

Cf. n°s 3630, 6899, 6902, 6964, 6966, 6970, 7003

§ 5. From 1815 to 1910.

** 6944. Akten zur Geschichte des Krimkriegs. Ser. 3. Englische Akten zur Geschichte des Krimkriegs. Band 1. 20. November 1852 bis 10. Dezemebr 1853. Hrsg. v. Winfried BAUMGART. München, R. Oldenbourg, 2005, 750 p.

** 6945. American Civil War (The) through British eyes: dispatches from British diplomats. Vol. 2. April 1862–February 1863. Compiled by James J. BARNES and Patience P. BARNES. Kent, Kent State U. P. a. London, Caliban Books, 2005, 353 p.

** 6946. CLAUDEL (Paul). Correspondance consulaire de Chine (1896–1909). Introduction par Jacques HOURIEZ; établissement du texte et annotation par Andrée HIRSCHI; index par Nicole BERTIN et Maryse BAZAUD. Besançon, Presses Universitaires de Franche-Comté, 2005, 377 p. (Annales littéraires de l'Université de Franche-Comté, 780. Série Centre Jacques-Petit, 106).

** 6947. Correspondence (The) of Sir Ernest Satow while he was British Minister in Japan (1895–1900) from the Satow Papers held at the National Archives, Kew, London. Published in full for researchers with notes by Ian RUXTON. Kyūshū, Ian Ruxton, 2005, [s. p.].

** 6948. Correspondence between the Foreign Office and the British embassy and consulates in the Ottoman empire 1820–1833: Foreign Office 78/97-221: a descriptive list. Vol. 2. Compiled by Eleftherios PREVELAKIS and Helen GARDIKAS-KATSIADAKIS. Athēna, Akadēmia Athēnōn, 2005, 614 p. (Mnēmeia tēs Hellēnikēs historias, 14).

** 6949. Germanski dokumenti za politikata na Germanija i evropskite Golemi sili vo Makedonija (1904–1910). (Documenti tedeschi sulla politica della Germania e delle grandi potenze europee in Macedonia). Izbor, predgovor i redakcija Gjorgji STOJČEVSKI; prevod Petar STOJČEVSKI i Gjorgji STOJČEVSKI. Skopje, Državen arhiv na Republika Makedonija, 2005, 292 p. (ill.).

** 6950. Ruolo geopolitico (Il) dei consolati a Trieste: istruzioni pubbliche e segrete di Casa Savoia al consolato del Regno di Sardegna. A cura di Maria Paola PAGNINI e Aldo COLLEONI. Trieste, Edizioni Università di Trieste, 2005, 392 p. (ill., facsims.).

** 6951 Türkiye'nin parçalanması ve İngiliz politikası (1900–1920): İngiliz Devlet Arşivi gizli belgeleri. [Démembrement de la Turquie et la politique anglaise (1900–1920): documents secrets dans les Archives Anglaises]. Ed. Ö. Andaç UĞURLU. İstanbul, Örgün, 2005, 663 p.

** 6952. VALENZUELA LAFOURCADE (Mario). Cartas en sótano de embajada: Arturo Alessandri Palma, 1908–1909, con el cónsul de Chile en Buenos Aires, Carlos Henríquez Argomedo. Santiago de Chile, Tajamar, 2005, 146 p.

6953. ADAMS (Iestyn). Brothers across the ocean: British foreign policy and the origins of the Anglo-American 'special relationship', 1900–1905. London a. New York, Tauris Academic Studies, 2005, 282 p. (Library of international relations, 24).

6954. AYDIN (Mithat). Balkanlarda isyan: Osmanlı-İngiliz rekabeti Bosna-Hersek ve Bulgaristan'daki ayaklanmalar (1875–1876). [Révolte en Balkans: concurrence turco-anglaise, soulèvements en Bosnie-Herzégovine et Bulgarie (1875–1876)]. İstanbul, Yeditepe, 2005, 208 p.

6955. Battle (The) of Adwa: reflections on Ethiopia's historic victory against European colonialism. Ed. by Paulos MILKIAS and Getachew METAFERIA. New York, Algora, 2005, XV-320 p.

6956. BERDINE (Michael D.). The accidental tourist, Wilfrid Scawen Blunt, and the British invasion of Egypt in 1882. New York a. London, Routledge, 2005, XXIII-305 p. (Middle East studies, history, politics & law).

6957. BLAUFARB (Rafe). Bonapartists in the borderlands: French exiles and refugees on the Gulf Coast, 1815–1835. Tuscaloosa, University of Alabama Press, 2005, XIX-302 p. (ill., maps).

6958. BURGAUD (Stéphanie). La politique de Gortchakov face à la Prusse à l'été 1866 à la lumière des archives russes. *Revue d'histoire diplomatique*, 2005, 119, 2, p. 135-150.

6959. Conselho (O) de Estado e a política externa do Império: consultas da Seção dos Negocios Estrangeiros. 1858–1862. Fundação Alexandre de Gusmão, Centro de História e Documentação Diplomática. Rio de Janeiro, FUNAG, 2005, 450 p.

6960. CORZO GONZÁLEZ (Diana). La política exterior mexicana ante la nueva doctrina Monroe, 1904–1907. México D.F., Instituto Mora, 2005, 123 p. (Historia internacional).

6961. DAUGE (Louis). L'Entente cordiale – de l'événement au concept. *Revue d'histoire diplomatique*, 2005, 119, 2, p. 97-100.

6962. EKINCI (İlhan). Kızıldeniz'in Güneyinde Rekabet- Şeyh Said ve Fersan Adaları Meselesi- (1 harita ile birlikte). [Competition South of the Red Sea – Sheikh Said and the Farasan Islands question (with 1 mape)]. *Belleten*, 2005, 69, 255, p. 567-598.

6963. ENGIN (Vahdettin). II. Abdülhamid ve dış politika. (II. Abdülhamid et la politique extérieure). İstanbul, Yeditepe, 2005, 323 p.

6964. FIEBIG (Eva Susanne). Hanseatenkreuz und Halbmond: die hanseatischen Konsulate in der Levante im 19. Jahrhundert. Marburg, Tectum, 2005, 330 p. (ill., maps).

6965. FRAUDET (Xavier). Politique étrangère française en mer Baltique (1871–1914): de l'exclusion à l'affirmation. Stockholm, Stockholms universitet, 2005, 312 p. (maps (Acta Universitatis Stockholmiensis, 77).

6966. GAULTIER-KURHAN (Caroline). Mehemet Ali et la France: 1805–1849. Histoire singulière du Napoléon de l'orient. Paris, Maisonneuve et Larose, 2005, 267 p.

6967. GEYER (Dietrich). Russian imperialism: the interaction of domestic and foreign policy, 1860–1914. New Haven, Yale U. P., 2005, 385 p.

6968. GIBB (Paul). Unmasterly inactivity? Sir Julian Pauncefote, Lord Salisbury, and the Venezuela boundary dispute. *Diplomacy & statecraft*, 2005, 16, 1, p. 23-55.

6969. GOTTSMANN (Andreas). Venetien, 1859–1866: österreichische Verwaltung und nationale Opposition. Wien, Verlag der Österreichischen Akademie der Wissenschaften, 2005, 607 p. (ill., map). (Zentraleuropa-Studien, 8).

6970. HELLYER (Robert I.). The missing pirate and the pervasive smuggler: regional agency in coastal defence, trade, and foreign relations in nineteenth-century Japan. *International history review*, 2005, 27, 1, p. 1-24.

6971. KANG (Woong Joe). The Korean struggle for international identity in the foreground of the Shufeldt Negotiation, 1866–1882. Lanham, University Press of America, 2005, VII-225 p.

6972. KIM (Seung-young). Russo-Japanese rivalry over Korean buffer at the beginning of the 20th century and its implications. *Diplomacy & statecraft*, 2005, 16, 4, p. 619-650.

6973. KNEŽEVIĆ (Saša). Velika Britanija i aneksiona kriza. (Great Britain and the annexation crisis). Podgorica, Istorijski institut Crne Gore, 2005, 356 p. (Posebna izdanja).

6974. MARTIN (Vanessa). The role of the "karguzar" in the foreign relations of state and society of Iran from the mid-nineteenth century to 1921. 1. Diplomatic relations. *Journal of the Royal Asiatic Society*, 2005, 3, p. 261-279.

6975. MATVEEV (Vladimir D.). Negostepriimnaja Tavrida: (K uchastiju Sardinskogo korolevstva i ego ekspeditsionnogo korpusa v Krymskoj vojne v 1855–1856 gg.). (The unhospitable Crimea: The participation of the Kingdom of Sardinia and its expeditionary corps in the Crimean War of 1855–1856). Sevastopol', EKOSI-Gidrofizika, 2005, 215 p. (ill.; maps; facs.; bibl. p. 212). [Ukrainian and Italian summaries]

6976. MAURER (Golo). Preußen am Tarpejischen Felsen: Chronik eines absehbaren Sturzes: die Geschichte des Deutschen Kapitols in Rom 1817–1918. Regensburg, Schnell + Steiner, 2005, 320 p. (ill., map).

6977. MIRANDA CASTAÑÓN (Edmundo). Eramos una bola perdida en el desierto: tres ensayos críticos sobre la guerra del Pacífico y las relaciones boliviano-chilenas. La Paz, Bolivia, C&C Editores, 2005, 142 p.

6978. Missão Varnhagen (A) nas Repúblicas do Pacífico, 1863 a 1867. V. 1. 1863 a 1865. V. 2. 1866 a 1867. Centro de História e Documentação Diplomática. Brasília, FUNAG, Fundação Alexandre de Gusmao e Rio de Janeiro, Centro de História e Documentação Diplomática, 2005, 2 vol., 592 p., 508 p.

6979. MONICO (Reto). Suisse-Portugal: regards croisés, (1890–1930). Genève, Société d'histoire et d'archéologie de Genève, 2005, 561 p. (ill.). (Hors-série).

6980. MONTIEL ARGÜELLO (Alejandro). Nicaragua y Costa Rica en la Constituyente de 1823. Managua, Fundación Uno, 2005, XII-270 p. (Colección cultural de Centro América, 9 Fuentes históricas).

6981. NIERI (Rolando). Sonnino, Guicciardini e la politica estera italiana, 1899–1906. Pisa, ETS, 2005, 217 p. (Studi di storia, 5).

6982. PAINE (S. C. M.). The Sino-Japanese War of 1894–1895: perceptions, power and primacy. Cambridge, Cambridge U. P., 2005, XI-412 p. (maps).

6983. PALACE (Wendy). The British Empire and Tibet, 1900–1922. London, RoutledgeCurzon, 2005, 194 p. (ill.). (RoutledgeCurzon studies in the modern history of Asia, 25).

6984. QUAN (Hexiu). Ri E zhanzheng dui jindai Zhong Han guanxi de yinxiang. (The impact of the Russo-Japanese war on Sino-Korean relations). *Jindai shi yanjiu*, 2005, 6, p. 117-141.

6985. QUINTANA GARCÍA (José Antonio). Venezuela y la independencia de Cuba, 1868–1898. Havana, P. de la Torriente Editorial, 2005, 223 p. (ill.).

6986. RASTOVIĆ (Aleksandar). Velika Britanija i Srbija, 1903–1914. (Great Britain and Serbia, 1903–1914). Beograd, Istorijski institut, 2005, 519 p. (ill.). (Posebna izdanja, 50).

6987. Russisch-Japanische Krieg (Der), 1904/05. Hrsg. v. Josef KREINER. Göttingen, V & R Unipress u. Bonn, Bonn U. P., 2005, 186 p.

6988. Russo-Japanese war (The) in global perspective. World war zero. Vol. 1. Ed. by John W. STEINBERG. Boston, Brill, 2005, XXIII-671 p. (History of warfare, 29).

6989. RYBACHENOK (Irina S.). Rossija i Pervaja konferentsija mira 1899 goda v Gaage. (Russia and the first peace conference in the Hague, 1899). RAN. In-t rosijskoj istorii. Moskva, ROSSPEN, 2005, 391 p. (ill.; portr.; facs.; plates; bibl. incl.; pers. ind. p. 368-391).

6990. SEXTON (Jay). Debtor diplomacy: finance and American foreign relations in the Civil War era, 1837–1873. Oxford, Clarendon, 2005, IX-287 p. (ill.). (Oxford historical monographs).

6991. SFETAS (Spyridon). Aspects of Greek-Bulgarian political relations after the Congress of Berlin – the failure of the Greek-Bulgarian rapprochement 1878–1900. *Balkan Studies*, 2005, 4, p. 41-52.

6992. SHAHNAVAZ (Shahbaz). Britain and the opening up of South-West Persia, 1880–1914: a study in imperialism and economic dependence. London a. New York, RoutledgeCurzon, 2005, X-290 p. (ill., maps).

6993. SIEG (Dirk). Die Ära Stosch: die Marine im Spannungsfeld der deutschen Politik 1872 bis 1883. Bochum, Winkler, 2005, 565 p. (ill., map). (Kleine Schriftenreihe zur Militär- und Marinegeschichte, 11).

6994. TURAN (Ömer), EVERED (Kyle T.). Jadidism in South-eastern Europe: the influence of Ismail Bey Gaspirali among Bulgarian Turks. *Middle Eastern studies*, 2005, 41, 4, p. 481-502.

6995. VERNASSA (Maurizio). All'ombra del Bardo: presenze toscane nella Tunisia di Ahmed Bey: 1837–1855. Pisa, PLUS-Pisa U. P., 2005, 232 p. (ill., map). (Studi pisani, 13).

6996. VIAENE (Vincent). The papacy and the new world order: Vatican diplomacy, catholic opinion and international politics at the time of Leo XIII 1878–1903. Bruxelles, Institut historique belge de Rome = Belgisch historisch Instituut te Rome, 2005, 516 p. (ill.). (Bibliothèque de l'Institut historique belge de Rome, 57).

6997. VILLEGAS REVUELTAS (Silvestre). Deuda y diplomacia: la relación México-Gran Bretaña, 1824–1884. México D.F., Universidad Nacional Autónoma de México, 2005, 280 p. (Serie Historia moderna y contemporánea, 42).

6998. VINOGRADOV (Vladelen N.). Balkanskaja epopeja knjazja A.M. Gorchakova. (The Balkanic epopee of Prince Aleksander Gorchakov). RAN, In-t slavjanovedenija. Moskva, Nauka, 2005, 301 p. (portr.; bibl. incl.; pers. ind. p. 295-300).

6999. VOILLERY (P.). La France, l'Hôtel Lambert, l'Empire Ottoman et les Bulgares. De la Monarchie de Juillet à la guerre de Crimée, 1830–1856. *Balkan Studies*, 2005, 2, p. 69-86.

7000. VOJVODIC (Mihailo). Pregovori srbije i grcke o makedoniji 1890–1893. (Négociations, la Serbie et la Grèce sur la Macédoine 1890–1893). *Vardarski zbornik*, 2005, 4, p. 13-33.

7001. WETZEL (David). Duell der Giganten: Bismarck, Napoleon III. und die Ursachen des Deutsch-Französischen Krieges 1870–1871. Paderborn, Ferdinand Schöningh, 2005, 239 p. (Wissenschaftliche Reihe [Otto-von-Bismarck-Stiftung], 7).

7002. YOKOTE (Shinji). Nichiro senso-shi: 20seiki saisho no taikoku-kan kyōso. (The history of Russo-Japanese War: the first conflict between the great powers in the 20th century). Tokyo, Chuokoron Shinsha, 2005, 212 p. (Chuko Shinsho).

7003. YURDAKUL (İlhami). XIX. Yüzyılda Osmanlı-Açe İlişkileri Osmanlı Hilafetinin Güney Asya'da Dinî-Siyasî Nüfuzu. (Relations entre l'Empire Ottoman et l'Aceh au 19e siècle: influence religieuse et politique de l'Empire Ottoman en Asie du sud). *Türk Kültürü İncelemeleri Dergisi / Journal of Turkish cultural studies*, 2005, 13, p. 19-48.

7004. ZOLOTAREV (Vladimir A.). Protivoborstvo imperij: vojna 1877–1878 gg., apofeoz vostochnogo krizisa. (The rivalry of the Russian and Turkish empires: The war of 1877–1878 and the apotheosis of the Eastern Crisis). Moskva, Animi Fortitudo, 2005, 567 p. (ill.; portr.; maps; bibl. p. 446-553; ind. p. 554-564). (Voennaja istorija gosudarstva Rossijskogo).

§ 5. Addendum 2004.

7005. MORITA (Yoshihiko). Bakumatsu ishinki no tai-Shin seisaku to Nisshin-syūkō-jōki: Nihon, Chūka-teikoku, Seiyō-shakai no sankaku kankei to higashi-ajia chitsujo no nijūsei, 1862–1871. (Sino-Japanese amity treaty and the east Asian world order: the triangle relationship and dual visions of Japan, 1862–1871). *Kokusai Seiji*, 2004, 139, p. 29-44 [Eng. Summary].

Cf. n°s 3630, 4281, 4310, 4460, 4520, 4678, 6821, 6916, 6919, 6930, 6933, 7018, 7022, 7048, 7049, 7074, 7099, 7102, 7224, 7562

§ 6. From 1910 to 1935. The First World War.

* 7006. VAN HARTESVELDT (Fred R.). The battles of the British Expeditionary Forces, 1914–1915: historiography and annotated bibliography. Westport a. London, Praeger, 2005, 195 p. (Bibliographies of battles and leaders, 25).

** 7007. BABITS (István). "Áll az idő és máll a tér": Babits István levelei a keleti frontról és a hadifogságból, 1915–1920. ("Time has stopped and space is weathering": letters of István Babits form the Eastern front line and from the prison of war, 1915–1920). A kötetet összeállította, a szöveget gondozta, a német nyelvű leveleket fordította, a jegyzeteket és a bevezetőt írta BUDA Attila. Budapest, Akadémiai Kiadó, 2005, 737 p. (ill., ports., facsims.). (Babits könyvtár, 13).

** 7008. Bălgarija v Părvata svetovna vojna: germanski diplomatičeski dokumenti: sbornik dokumenti v dva toma, 1913–1918. T. 2. 1916–1918 g. (La Bulgarie dans la Première guerre mondiale. Documents diplomatiques allemands. Recueil de documents, 1913–1918. T. 2. 1916–1918). Naučen redaktor Cvetana TODOROVA. Sofija, Glavno Upravlenie na Archivite pri Ministerskija Săvet, 2005, 744 p. (Archivite govorjat, 39).

** 7009. Car Boris III v britanskata diplomatičeska korespondencija (1919–1941). Tom 1. 1919–1934. (Zar

Boris nella corrispondenza diplomatica britannica, 1919–1941. Tomo 1. 1919–1934). Săstavitel Dimităr MITEV. Sofija, Univ. izd-vo "Sv. Kl. Okhridski", 2005, 319 p.

** 7010. Chile en los archivos soviéticos, 1922–1991. Ed. por Olga ULIANOVA y Alfredo RIQUELME SEGOVIA. T. 1. Komintern y Chile, 1922–1931. Santiago, Dirección de Bibliotecas, Archivos y Museos y LOM Ediciones, 2005, 800 p. (Fuentes para la historia de la república, 23).

** 7011. GANEV (Venelin). Dnevnik 1919 g. Njojskijat miren dogavo. (Journal, 1919. Le Contrat de paix de Neuilly). Săstaviteli Cočo BILJARSKI, Liljana VANOVA. Sofija, "Sineva", 2005, 270 p. (Biblioteka "Bălgarska pamet").

** 7012. Kaiser Wilhelm II. als Oberster Kriegsherr im Ersten Weltkrieg. Quellen aus der militärischen Umgebung des Kaisers, 1914–1918. Bearb. und eingeleitet von Holger AFFLERBACH. München, Oldenbourg, 2005, XII-1051 p. (Deutsche Geschichtsquellen des 19. und 20. Jahrhunderts, 64).

** 7013. Kominternăt i Bălgarija: (mart 1919–septemvri 1944 g.); dokumenti. Glavno Upravlenie na Archivite pri Ministerskija Săvet. Săst.: Luiza REVJAKINA. Sofija, Glavno upravlenie na archivite pri Ministerskija săvet na Republika Bălgarija, 2005, 2 vol., 1308 p. (Archivite govorjat, 36, 37).

** 7014. Landsturmmann (Ein) im Himmel: Flandern und der Erste Weltkrieg in den Briefen von Herman Nohl an seine Frau. Zusammengetragen, eingeleitet und erläutert von Walter THYS. Leipzig, Leipziger Universitätsverlag, 2005, 359 p. (ill.).

** 7015. Lettre (Une) par jour. Correspondance de Joannès Berger, poilu forézien, avec sa famille, 1913–1919. Tome I. D'Epinal, Vosges, à Legé, Loire-inférieure, novembre 1913 à septembre 1915. Ed. par Gérard BERGER; publié par le centre de recherche en histoire. Saint-Etienne, Publications de l'université de Saint-Etienne, 2005, 397 p. (Bibliothèque du CERHI, 2).

** 7016. Mission militaire française auprès de la République Tchécoslovaque 1919–1939. Edition documentaire. Sér. 1. 1919–1925. Vol. 2 B. 1921–1925, Rapports courants. Praha, Vojenský historický ústav, 2005, 184 p. (photogr.).

** 7017. Österreich-Ungarn und Armenien 1912–1918: Sammlung diplomatischer Aktenstücke. Zusammengestellt und eingeleitet von Artem OHANDJANIAN. Jerewan, Institut-Museum für Armenischen Genozid der Akademie der Wissenschaften Armeniens, 2005, 521 p.

7018. ABLOVA (Nadezhda E.). KVZhD i rossijskaja emigracija v Kitae: mezhdunar. i polit. aspekty istorii (pervaja polovina XX v.). (The Chinese Eastern Railway and the Russian diaspora in China: International and political aspects of history, the first half of the 20th century). RAN, In-t Dal'nego Vostoka. Moskva, Russkaja panorama, 2005, 430 p. (bibl. p. 402-420; pers. ind. p. 421-430).

7019. ADA (Serhan). Türk-Fransız ilişkilerinde Hatay sorunu (1918–1939). [Le Problème de Hatay dans les relations turco-françaises (1918–1939)]. İstanbul, İstanbul Bilgi Üniversitesi, 2005, XVI-266 p.

7020. AIRAKSINEN (Tiina Helena). Love your country on Nanjing Road: the British and the May Fourth Movement in Shanghai. Helsinki, Renvall Institute, 2005, 262 p. (ill., maps). (Renvall Institute publication, 19).

7021. ANTIER (Chantal). 1915. La France en chantier. *Guerres mondiales et conflits contemporains*, 2005, 53, 219, p. 53-62.

7022. AYDOĞAN (Erdal). İttihat ve Terakkî'nin doğu politikası (1908–1918). [(La politique orientale de l'Union et Progrès (1908–1918)]. İstanbul, Ötüken, 2005, 400 p.

7023. Balkan Wars (The). Ed. by Valery KOLEV and Christina KOULOURI. Thessaloniki, Center for Democracy and Reconciliation in Southeast Europe, 2005, 135 p. (ill.). (Teaching modern Southeast European History: alternative educational materials, workbook 3).

7024. BAYCROFT (Timothy). The Versailles settlement and identity in French Flanders. *Diplomacy & statecraft*, 2005, 16, 3, p. 589-602.

7025. BECKER (Jean-Jacques). Réflexions sur la guerre en 1915 sur le front occidental (d'après les notes des généraux Fayolle et Haig). *Guerres mondiales et conflits contemporains*, 2005, 53, 219, p. 5-14.

7026. BENVINDO (Bruno). Des hommes en guerre: les soldats belges entre ténacité et désillusion, 1914–1918. Bruxelles, Archives générales du Royaume, 2005, 185 p. (Etudes sur la Première Guerre mondiale = Studies over de Erste Wereldoorlog, 12).

7027. BEST (Antony). "Our respective empires should stand together": the Royal dimension in Anglo-Japanese relations, 1919–1941. *Diplomacy & statecraft*, 2005, 16, 2, p. 259-279.

7028. BESTRY (Jerzy). Służba konsularna Drugiej Rzeczypospolitej w Czechosłowacji. (Il servizio Consolare della Seconda Repubblica in Cecoslovacchia). Wrocław, Wydawnictwo Uniwersytetu Wrocławskiego, 2005, 221 p. (ill., maps). (Acta Universitatis Wratislaviensis, 2767).

7029. BŘACH (Radko). Dohoda z 18. února 1919 o podřízení čs. ozbrojených sil vrchnímu velení maršála Foche. (Agreement on the subordination of Czechoslovak Armed Forces to the High Command of Marshall Foch, signed on February 18, 1919). *Historie a vojenství*, 2005, 54, 1, p. 30-39. – IDEM. Generál Pellé a československo-polský spor o Těšínsko. (Generál Pellé and the dispure between Czechoslovakia and Poland over the Cieszyn region). *Historie a vojenství*, 2005, 54, 3, p. 33-42.

7030. BRAVO VALDIVIESO (Germán). La Primera Guerra Mundial en la costa de Chile: una neutralidad que no fue tal. Viña del Mar, Ediciones Altazor, 2005, 329 p.

7031. BROOKS (John). Dreadnought gunnery and the battle of Jutland. The question of fire control. London a. New York, Routledge, 2005, XIV-321 p.

7032. BUDINOV (Svetoslav). Balkanskite vojni 1912–1913. (Les Guerres Balkaniques 1912–1913). Sofija, Universitetsko izdatelstvo, 2005, 252 p.

7033. Canada and the First World War: essays in honour of Robert Craig Brown. Ed. By David Clark MAC KENZIE. Toronto a. Buffalo, University of Toronto Press, 2005, XII-452 p. (ill.).

7034. CASTELLAN (Georges). Les Balkans, poudrière du 20e siècle. *Guerres mondiales et conflits contemporains*, 2005, 53, 217, p. 5-16.

7035. CLAVIN (Patricia). Reparations in the long run. *Diplomacy & statecraft*, 2005, 16, 3, p. 515-530.

7036. COMȘA (Ioan). Cronologia relațiilor româno-americane. Contacte "relații culturale, economice, consulare și diplomatice 1913–1944. (The chronology of the Romanian – American relations. Contacts, cultural, economical, consular and diplomatic relations). *Revista Româno-Americană*, 2005, 10-11, p. 59-64.

7037. DEMIRCI (Sevtap). Strategies and struggles: British rethoric and Turkish response: the Lausanne Conference (1922–1923). İstanbul, Isis Press, 2005, 218 p.

7038. DOUGHTY (Robert A.). Pyrrhic victory. French strategy and operations in the Great War. Cambridge, Belknap Press of Harvard U. P., 2005, XIII-578 p.

7039. DUDDEN (Alexis). Japan's colonization of Korea: discourse and power. Honolulu, University of Hawai'i Press, 2005, X-215 p. (ill.). (Studies of the Weatherhead East Asian Institute).

7040. EHRENPREIS (Petronilla). Kriegs- und Friedensziele im Diskurs: Regierung und deutschsprachige Öffentlichkeit Österreich-Ungarns während des Ersten Weltkriegs. Innsbruck, StudienVerlag, 2005, 513 p. (Wiener Schriften zur Geschichte der Neuzeit, 3).

7041. FELDMAN (Gerald D.). The reparations debate. *Diplomacy & statecraft*, 2005, 16, 3, p. 487-498.

7042. FIÇORRI (Ramiz). Ndërhyrja italiane në Shqipëri (1925–1939). (Ingerenze italiane in Albania). Tiranë, Koçi, 2005, 285 p.

7043. FISCHER (Conan). The human price of reparations. *Diplomacy & statecraft*, 2005, 16, 3, p. 499-514.

7044. FOLEY (Robert T.). German strategy and the path to Verdun. Erich von Falkenhayn and the development of attrition, 1870–1916. Cambridge, New York a. Melbourne, Cambridge U. P., 2005, XIV-301 p. (Cambridge military histories).

7045. FOTAKIS (Zisis). Greek naval strategy and policy, 1910–1919. London a. New York, Routledge, 2005, XV-223 p. (maps).

7046. FRIED (Marvin Benjamin). Brockdorff-Rantzau and the struggle for a just peace. *Diplomacy & statecraft*, 2005, 16, 2, p. 403-416.

7047. GALLUPPI (Massimo). Rivoluzione, controrivoluzione e politica di potenza in Cina, 1914–1931. Napoli, L'Orientale, 2005, 85 p.

7048. GENOV (Georgi Petrov). Bălgarskata vănšna politika i Makedonskijat văpros. Č. 2. Bălgarskata vănshna politika sled Ilindenskoto văstanie. Č. 3. Balkanskijat săjuz, 1912 g.; Balkanskata vojna, 1912–1913. (La politique extérieure bulgare et la Question macédonienne. P. 2. La politique bulgare extérieure après le Soulèvement d'Ilinden. P. 3. L'Union Balkanique 1912. La Guerre Balkanique, 1912–1913). Sofija, Vano Nedkov, 2005, 2 vol., 275 p., 256 p. (Političeska i diplomatičeska istorija na Bălgarija, 19, 20).

7049. Georges Clémenceau et le monde anglo-saxon: Actes du colloque international, palais du Luxembourg, bibliothèque nationale de France, 27–28 novembre 2004. Sous la direction de Sylvie BRODZIAK et Michel DROUIN. La Crèche, Geste, 2005, 220 p.

7050. GÉRARD-PLASMANS (Delphine). La présence française en Egypte entre 1914 et 1936: de l'impérialisme à l'influence et de l'influence à la coopération. Darnétal, Editions darnétalaises S.A.R.L., 2005, 604 p. (ill.). (Actualité de l'histoire).

7051. GREENHALGH (Elizabeth). Victory through coalition. Britain and France during the First World War. Cambridge, New York a. Melbourne, Cambridge U. P., 2005, XVI-304 p.

7052. GROHMANN (Carolyn). From Lothringen to Lorraine: expulsion and voluntary repatriation. *Diplomacy & statecraft*, 2005, 16, 3, p. 571-587.

7053. HORNE (John). Entre expérience et mémoire: les soldats français de la Grande Guerre. *Annales*, 2005, 60, 5, p. 903-919.

7054. HOSFELD (Rolf). Operation Nemesis: Die Türkei, Deutschland und der Völkermord an den Armeniern. Köln, Kiepenheuer & Witsch, 2005, 351 p. (ill.).

7055. Illusions de puissance, puissance de l'illusion: historiographies et histoire de l'Europe centrale dans les relations internationales de l'entre-deux-guerres. Éditeur du volume, Traian SANDU; préface d'Elisabeth DU RÉAU; conclusions par Antoine MARÈS. Paris, L'Harmattan, 2005, 292 p. (Cahiers de la nouvelle Europe, 2).

7056. JANSEN (Anscar). Der Weg in den Ersten Weltkrieg: das deutsche Militär in der Julikrise 1914. Marburg, Tectum, 2005, 568 p.

7057. JAZHBOROVSKAJA (Inessa S.), PARSADANOVA (Valentina S.). Rossija i Pol'sha: Sindrom vojny 1920 g. (Russia and Poland: The syndrome of the war of 1920). 1914–1917–1920–1987–2004. Moskva, Academia, 2005, 404 p. (bibl. incl.).

7058. JEANNESON (Stanislas). French policy in the Rhineland. *Diplomacy & statecraft*, 2005, 16, 3, p. 475-486.

7059. JEFFERY (Keith). "Hut ab," "Promenade with Kamerade for Schokolade," and the Flying Dutchman: British soldiers in the Rhineland, 1918–1929. *Diplomacy & statecraft*, 2005, 16, 3, p. 455-473.

7060. KARAGANEV (Rumen). Bălgarija i nejnata insufficientia pulmonum ili nacionalnata kauza za izlaz na Bjalo more 1919–1941. (Bulgarie et son Insufficientia Pulmonum ou bien la cause de débouché maritime à la Mer Egée 1919–1941). Sofija, Gutenberg, 2005, 390 p.

7061. KOLB (Eberhard). Der Frieden von Versailles. München, Beck, 2005, 120 p. (ill., maps). (Beck'sche Reihe, 2375).

7062. KÖSSLER (Reinhart). In search of survival and dignity: two traditional communities in southern Namibia under South African rule. Windhoek, Gamsberg Macmillan, 2005, XVI-374 p. (ill.).

7063. KOSTOV (Aleksandăr). Razvitie na Bălgaro-Belgijskite otnošenija meždu dvete svetovni vojni. (Lo sviluppo delle relazioni tra Bulgaria e Belgio nel periodo interbellico). Sofija, Art Media Communications, 2005, 164 p. (ill.).

7064. KOVAČ (Miro). Francuska i hrvatsko pitanje 1914.–1929. (La Francia e la questione croata). Zagreb, Dom i svijet, 2005, 363 p. (Biblioteka Povjesnica).

7065. KURAL (Václav). "Memorandum Hergl" a formování strategických sudetoněmeckých (a německých) cílů v české otázce. ("Memorandum Hergl" und Formierung der strategischen sudetendeutschen (und deutschen) Ziele in der böhmischen Frage). *In*: Historie okupovaného pohraničí 1938-1945. Tomo 10. Ed. Zdeněk RADVANOVSKÝ. Ústí n.L., 2005, p. 237-256.

7066. LA MOAL (Frédéric). L'année 1915 dans les relations franco-italiennes: l'année de la rupture? *Guerres mondiales et conflits contemporains*, 2005, 53, 220, p. 5-22.

7067. LATOUR (Francis). Les relations entre le Saint-Siège et la Sublime Porte à l'èpreuve du génocide des chrètiens d'orient, pendant la grande guerre. *Guerres mondiales et conflits contemporains*, 2005, 53, 219, p. 31-44.

7068. LERHIS (Ainārs). Latvijas Republikas ārlietu dienests 1918–1941. (Foreign service of the Republic of Latvia, 1918–1941). Rīga, Latvijas vēstures institūta apgāds, 2005, 326 p. (ill.).

7069. LIUȘNEA (Mihaela-Denisia). România și încercari de constituire a unui "Locarno oriental". (Romania and the attempts of creation of an "Oriental Locarno"). București, Editura Cartea Universitară, 2005, 202 p.

7070. LOJKÓ (Miklós). Meddling in Middle Europe: Britain and the 'lands between' 1919–1925. Budapest a. New York, Central European U. P., 2005, X-377 p.

7071. LÜDKE (Tilman). Jihad made in Germany: Ottoman and German propaganda and intelligence operations in the First World War. Münster, Lit, 2005, IX-251 p. (Studien zur Zeitgeschichte des Nahen Ostens und Nordafrikas, 12).

7072. Małopolska i Podhale w latach wielkiej wojny 1914–1918: materiały z ogólnopolskiej konferencji naukowej zorganizowanej z okazji 90. rocznicy wybuchu I wojny światowej, Nowy Targ, 27–29 sierpnia 2004 r. (Małopolska e Podhale negli anni della grande guerra 1914–1918: materiali della conferenza scientifica nazionale organizzata in occasione del 90 anniversario dello scoppio della Prima guerra mondiale, Nowy Targ, 27–29 agosto 2004). Pod redakcją Roberta KOWALSKIEGO. Nowy Targ, Wydawn., Nakładem Polskiego Towarzystwa Historycznego, 2005, [s. p.]. (ill., ports., facsims.). (Prace Komisji Historii Wojskowości, 4).

7073. MATERSKI (Wojciech). Na widecie: II Rzeczpospolita wobec Sowietów 1918–1943. Warszawa, Instytut Studiów Politycznych PAN i "Rytm", 2005, 776 p. (ill., maps, ports., facsims.).

7074. MITCHINSON (K. W.). Defending Albion: Britain's Home Army 1908–1919. Basingstoke, Palgrave Macmillan, 2005, XII-262 p. (ill.). (Studies in military and strategic history).

7075. MOORHOUSE (Roger). "The Sore That would Never HeaL": the genesis of the Polish corridor. *Diplomacy & statecraft*, 2005, 16, 3, p. 603-613.

7076. NACHTIGAL (Reinhard). Kriegsgefangenschaft an der Ostfront 1914 bis 1918: Literaturbericht zu einem neuen Forschungsfeld. Frankfurt am Main u. Oxford, P. Lang, 2005, 162 p.

7077. NEIBERG (Michael S.). Fighting the Great War: a global history. Cambridge a. London, Harvard U. P., 2005, XX-395 p. (ill., maps).

7078. NIPPOLD (Otfried). "Die Wahrheit über die Ursachen des Europäischen Krieges": Japan, der Beginn des Ersten Weltkriegs und die völkerrechtliche Friedenswahrung. Hrsg. v. Harald KLEINSCHMIDT und eingeleitet von Akio NAKAI. München, Iudicium, 2005, LVII-250 p.

7079. O'RIORDAN (Elspeth). The British zone of occupation in the Rhineland. *Diplomacy & statecraft*, 2005, 16, 3, p. 439-454.

7080. OSTENC (Michel). 1915. L'Italie en guerre. *Guerres mondiales et conflits contemporains*, 2005, 53, 219, p. 15-30.

7081. PAYASLIAN (Simon). United States policy toward the Armenian question and the Armenian genocide. New York a. Basingstoke, Palgrave Macmillan, 2005, XII-268 p.

7082. PEHR (Michal). Mnichovská tragédie očima prostých lidí. Co psali občané prezidentu republiky na podzim 1938. (Die Münchner Tragödie mit den Augen einfacher Leute gesehen. Was Staatsbürger dem Präsidenten der Republik im Herbst 1938 schrieben). *Moderní dějiny*, 2005, 13, p. 235-243.

7083. POPOV (Zheko). Aspects of the Balkan politics of Romania at the beginning of the 20th century (1912). *Balkan Studies*, 2005, 3, p. 51-76.

7084. POPOVIĆ (Nikola B.). Srbi u građanskom ratu u Rusiji 1918–1921. (Serbs in the Russian Civil War). Beograd, Institut za savremenu istoriju, 2005, 318 p. (Biblioteka "Studije i monografije").

7085. PRIOR (Robin), WILSON (Trevor). The Somme. New Haven a. London, Yale U. P., 2005, 358 p.

7086. ROMSICS (Ignac). Der Friedensvertrag von Trianon. Herne, Schäfer, 2005, 224 p. (Studien zur Geschichte Ungarns, 6).

7087. Roumanie (La) et la Grande Guerre. Ouvrage édité par Dumitru IVĂNESCU et Sorin D. IVĂNESCU. Iaşi, Editura Junimea, 2005, 343 p. (Historia magistra vitae).

7088. RUSCONI (Gian Enrico). L'azzardo del 1915: come l'Italia decide la sua guerra. Bologna, Il Mulino, 2005, 199 p. (Intersezioni, 273).

7089. ŠAMBERGER (Zdeněk). K problematice vývoje česko-(sudeto)německých vztahů před rokem 1938. (Zur Problematik der Entwicklung der tschechisch-(sudeten)deutschen Beziehungen vor 1938). *In*: Historie okupovaného pohraničí 1938–1945. Tomo 10. Ed. Zdeněk RADVANOVSKÝ. Ústí n.L., 2005, p. 7-99.

7090. SCHICKEL (Matthias). Zwischen Wilson und Lenin: die Anfänge der globalen Blockbildung in den Jahren 1917–1919: dargestellt am Beispiel des amerikanischen Diplomaten William Christian Bullitt. Hamburg, Kovač, 2005, 380 p. (Schriftenreihe Studien zur Zeitgeschichte, 45).

7091. SEMERDŽIEV (Petăr). Ruskata imperija i Săvetskijat Săjuz v sădbata na Bălgarija. T. 2. Săvetsko-Bălgarskite otnošenija 1918–1943. (L'empire Russe et l'URSS dans le destin de la Bulgarie. T. 2. Relations soviètiques-bulgares 1918–1943). Farmington Hill, Inst. za Istorijata na Bălgarskata Emigracija v Severna Amerika "Ilija Todorov Gadžev", 2005, 504 p.

7092. SHARP (Alan). The enforcement of the Treaty of Versailles, 1919–1923. *Diplomacy & statecraft*, 2005, 16, 3, p. 423-438.

7093. SPÂNU (Alin). Spionaj şi contraspionaj la graniţa de est a României, 1920. (Espionage and counter-espionage at the Eastern border of Romania, 1920) *Arhivele totalitarismului*, 2005, 13, 48-49, p. 185-191.

7094. STEINER (Zara S.). The lights that failed: European international history, 1919–1933. Oxford, Oxford U. P., 2005, XV-938 p. (maps). (Oxford history of modern Europe).

7095. SUTTIE (Andrew). Rewriting the First World War: Lloyd George, politics and strategy, 1914–1918. Basingstoke, Palgrave Macmillan, 2005, IX-282 p.

7096. TELI DIBRA (Pranvera). Shqipëria dhe diplomacia angleze 1919-1927: nën dritën e burimeve arkivore angleze. (Albania and English diplomacy 1919-1927 [on the light of English archival sources]). Tiranë, "Neraida", 2005, 485 p.

7097. TENFELDE (Klaus). Disarmament and big business: the case of Krupp, 1918–1925. *Diplomacy & statecraft*, 2005, 16, 3, p. 531-549.

7098. TOMÁŠEK (Dušan). Nevyhlášená válka. Boje o Slovensko 1918–1920. (Der unerklärte Krieg. Die Kämpfe um der Slowakei in den Jahren 1918-1920). Praha, Epocha, 2005, 260 p. (photogr.). (Polozapomenuté války, 3).

7099. UGALDE (Luis). Gomecismo y la política panamericana de Estados Unidos. Caracas, Universidad Católica Andrés Bello, 2005, 121 p. (Colección Histórica, 7).

7100. WEBSTER (Andrew). Making disarmament work: the implementation of the international disarmament provisions in the League of Nations covenant, 1919–1925. *Diplomacy & statecraft*, 2005, 16, 3, p. 551-569.

7101. WEIGL (Michael). Das Bayernbild der Repräsentanten Österreichs in München 1918–1938. die diplomatische und konsularische Berichterstattung vor dem Hintergrund der bayerisch-österreichischen Beziehungen. Frankfurt am Main u. Oxford, P. Lang, 2005, 462 p. (Europäische Hochschulschriften. Reihe III, Geschichte und ihre Hilfswissenschaften, 1013 = Publications universitaires européennes. Série III, Histoire, sciences auxiliaires de l'histoire, 1013 = European university studies. Series III, History and allied studies, 1013).

7102. WINKLER (Henry Ralph). British Labour seeks a foreign policy, 1900–1940. New Brunswick a. London, Transaction Publishers, 2005, XIII-207 p.

7103. World War I. Ed. by Michael NEIBERG. Aldershot, Ashgate, 2005, XXIV-604 p. (International library of essays in military history).

7104. XU (Guoqi). China and the great war: China's pursuit of a new national identity and internationalization. New York a. Oxford, Cambridge U. P., 2005, XIV-316 p. (ill., ports.). (Studies in the social and cultural history of modern warfare).

7105. ZEIDNER (Robert Farrer). The tricolor over the Taurus: the French in Cilicia and vicinity, 1918–1922. İstanbul, Turk Tarih Kurumu Basimevi, 2005, XV-368 p. (Türk tarih kurumu yayınları, 105 16 dizi).

§ 6. Addenda 1993–1998.

7106. Concise history (A) of the Balkan Wars, 1912–1913. Athenae, Hellenic Army General Staff, Army History Directorate, 98, 385 p. (ill., maps).

7107. Other Balkan wars (The): a 1913 Carnegie Endowment inquiry in retrospect. Report of the International Commission to inquire into the causes and conduct of the Balkan Wars. With a new introduction and reflections on the present conflict by George F. KENNAN. Washington, Carnegie Endowment for International Peace, 93, IX-413 p. (ill., maps). (A Carnegie Endowment book).

Cf. n⁰ˢ 326, 327, 3570, 3640, 3645, 3967, 4069, 4084, 4224, 4280, 4395, 4460, 4520, 4675, 4678, 4710, 4769, 5034, 6437, 6823, 6951, 6965, 6967, 6973, 6974, 6976, 6979, 6983, 6986, 6992, 6994, 7117, 7120, 7132, 7136, 7173, 7196, 7199, 7205, 7208, 7224, 7232, 7238, 7240, 7286

§ 7. From 1935 to 1945. The Second World War.

a. General

** 7108. BERGSTRØM (Vilhelm). En borger i Danmark under krigen: dagbog 1939–1945. (Journal d'un citoyen danois pendant la seconde guerre mondiale). Bd 1. Bd 2. Ed. John T. LAURIDSEN. København, Gads forlag, 2005, 1192 p.

** 7109. Buch Hitler (Das): Geheimdossier des NKWD für Josef W. Stalin, zusammengestellt aufgrund der Verhörprotokolle des Persönlichen Adjutanten Hitlers, Otto Günsche, und des Kammerdieners Heinz Linge, Moskau 1948/49. Hrsg. v. Henrik EBERLE und Matthias UHL. Bergisch Gladbach, G. Lübbe, 2005, 672 p. (ill.).

** 7110. CELOVSKY (Boris). Germanisierung und Genozid: Hitlers Endlösung der tschechischen Frage; deutsche Dokumente 1933–1945. Dresden, Neisse u. Brno, Stilus, 2005, 431 p. (ill., ports.).

** 7111. Lithuania under German occupation, 1941–1945: despatches from US Legation in Stockholm. Compiled and edited by Thomas REMEIKIS. Vilnius, Vilnius U. P., 2005, 731 p.

** 7112. Okupowany Poznań i Wielkopolska w niemieckich fotografiach i dokumentach, 1939–1941: ze zbiorów Instytut Zachodniego w Poznaniu. (L'occupazione di Poznań and Wielkopolska in German photos and documents). Redakcja merytoryczna: Maria RUTOWSKA i Maria TOMCZAK. Poznań, Instytut Zachodni, 2005, 109 p. (ill., facs.). (Documenta occupationis, 15).

** 7113. Polacy i ukraińcy pomiędzy dwoma systemami totalitarnymi 1942–1945. (Poles and ukrainians among two totalitarian systems 1942–1945). Komitet redakcyjny Serhij BOHUNOW [et al.]. Warszawa i Kijów, Instytut Pamięci Narodowej--Komisja Ścigania Zbrodni przeciwko Narodowi Polskiemu, 2005, 2 vol., [s. p.]. (Polska i Ukraina w latach trzydziestych-czterdziestych XX wieku, 4).

** 7114. Wrzesień 1939: radzieckie zagrożenie Rzeczypospolitej w dokumentach, relacjach i wspomnieniach. (September 1939. Die sowjetische Bedrohung der Republik in Dokumenten, Berichten und Erinnerungen). Wybór i opracowanie Wojciech WŁODARKIEWICZ. Warszawa, "Neriton", 2005, 283 p.

7115. AMERSFOORT (Hermanus). "Ik had mijn Roode-Kruis band afgedaan": oorlogsrecht en gedragingen van Nederlandse en Duitse militairen in gevecht, mei 1940. Den Haag, Sdu Uitgevers, 2005, Den Haag, Sdu Uitgevers, 2005, 208 p. (ill.).

7116. BALTA (Sebastian). Rumänien und die Großmächte in der Ära Antonescu, 1940–1944. Stuttgart, F. Steiner, 2005, 540 p. (map). (Quellen und Studien zur Geschichte des östlichen Europa, 69).

7117. BARNES (James J.), BARNES (Patience P.). Nazis in pre-war London, 1930–1939: the fate and role of German party members and British sympathizers. Brighton, Sussex Academic Press, 2005, X-283 p.

7118. BAYLY (Christopher Alan), HARPER (Tim). Forgotten armies: the fall of British Asia, 1941–1945. Cambridge, Harvard U. P., 2005, XXXIII-555 p. (ill., maps, ports.).

7119. BOĆKOWSKI (Daniel). Na zawsze razem: Białostocczyzna i Łomżyńskie w polityce radzieckiej w czasie II wojny światowej (IX 1939–VIII 1944). (Together forever: Białostocczyzna and Lomzynski in Soviet policy during the Second World War). Warszawa, Wydawn. Neriton, Instytut Historii PAN, 2005, 331 p. (map).

7120. BRADSHAW (John Tancred Landon). British imperial strategy, King Abdullah and the Jewish Agency, 1921–1951. London, School of Oriental and African studies, 2005, 291 p.

7121. BUNDGÅRD CHRISTENSEN (Claus). Danmark besat: krig og hverdag 1940–1945. (Le Danemark occupé: guerre et vie quotidienne 1940–1945). København, Høst, 2005, 844 p.

7122. BURGWYN (H. James). Empire on the Adriatic: Mussolini's conquest of Yugoslavia, 1941–1943. New York, Enigma, 2005, XIX-385 p. (ill).

7123. BUZATU (Gheorghe). Hitler, Stalin, Antonescu. Ploieşti, Editura Societăţii Culturale Ploeşti-Mileniul III, 2005, 579 p.

7124. CAMERON (Craig M.). Race and identity: the culture of combat in the Pacific War. *International history review*, 2005, 27, 3, p. 550-567.

7125. CAREW (Michael G.). The power to persuade: FDR [Franklin Delano Roosevelt], the newsmagazines, and going to war, 1939–1941. Lanham a. Oxford, University Press of America, 2005, X-247 p. (ill.).

7126. CORNI (Gustavo). Il sogno del "grande spazio". Le politiche d'occupazione nell'Europa nazista. Roma e Bari, Laterza, 2005, 276 p.

7127. CÜPPERS (Martin). Wegbereiter der Shoah: die Waffen-SS, der Kommandostab Reichsführer-SS und die Judenvernichtung 1939–1945. Darmstadt, Wissenschaftliche Buchgesellschaft, 2005, 464 p. (ill.). (Veröffentlichungen der Forschungsstelle Ludwigsburg der Universität Stuttgart, 4).

7128. DALLAS (Gregor). 1945: the war that never ended. New Haven, Yale U. P., 2005, XX-739 p. (ill., maps).

7129. DEZHGIU (Muharrem). Shqipëria nën pushtimin italian, 1939–1943. (Albania under Italian occupation, 1939–1943). Tiranë, Akademia e Shkencave, Instituti i Historisë, 2005, 467 p.

7130. DI CAPUA (Giovanni). Il biennio cruciale (luglio 1943–giugno 1945): l'Italia di Charles Poletti. Presentazione di Giancarlo GALLI; prefazione di Giuseppe GARGANI. Soveria Mannelli, Rubbettino, 2005, 458 p.

7131. DINARDO (Richard L.). Germany and the Axis powers: from coalition to collapse. Foreword by Dennis SHOWALTER. Lawrence, University Press of Kansas, 2005, XIV-282 p. (ill., map, ports.). (Modern war studies).

7132. DOST (Pinar). Mustafa Kemal Atatürk "le bon dictateur": une construction politique d'une figure désirée dans la France des années 30. *Revue d'histoire diplomatique*, 2005, 119, 4, p. 361-378.

7133. Entrechtung, Vertreibung, Mord: NS-Unrecht in Slowenien und seine Spuren in Bayern 1941–1945. Hrsg. v. Gerhard JOCHEM und Georg SEIDERER; herausgegeben im Auftrag des Stadtarchivs Nürnberg und der Stiftung "Nürnberg-Stadt des Friedens und der Menschenrechte" in Zusammenarbeit mit der Slowenischen Vereinigung der Okkupationsopfer 1941–1945, Kranj. Berlin, Metropol, 2005, 348 p. (ill., maps).

7134. FIEDOR (Karol). Polska i Polacy w polityce Trzeciej Rzeszy 1933–1939. (La Polonia e i polacchi nella politica del Terzo Reich, 1933–1939). Łódź, Wyższa Szkoła Studiów Międzynarodowych, 2005, 288 p.

7135. Français (Les) et la guerre d'Espagne: actes du colloque tenu a Perpignan les 28, 29, et 30 Septembre 1989. Ed. par Jean SAGNES et Sylvie CAUCANAS; introduction par Bartolomé BENNASSAR; cloture du colloque par Pierre VILAR. Perpignan, Presses Universitaires de Perpignan, 2005, 437 p. (Collection Études).

7136. France (La) et l'URSS dans l'Europe des années 30. Sous la dir. de Mikhail Matveevich NARINSKI [et al.]; avec la collaboration de Christophe RÉVEILLARD. Paris, Presses de l'Université de Paris-Sorbonne, 2005, 192 p. (Mondes contemporains).

7137. GARRAUD (Philippe). La politique française de réarmement de 1936 à 1940: priorités et contraintes. *Guerres mondiales et conflits contemporains*, 2005, 53, 219, p. 87-102. – IDEM. La politique française de réarmement de 1936 à 1940: une production tardive mais massive. *Guerres mondiales et conflits contemporains*, 2005, 53, 220, p. 97-114.

7138. HASEGAWA (Tsuyoshi). Racing the enemy. Stalin, Truman, and the surrender of Japan. Cambridge, Belknap Press of Harvard U. P., 2005, IX-382 p.

7139. Historie okupovaného pohraničí 1938–1945. (Die Geschichte des Okkupationsgrenzgebiets 1938–1945). Tomo 10. Ed. Zdeněk RADVANOVSKÝ. Ústí nad Labem, Albis international, 2005, 256 p.

7140. KALLIS (Aristotle A.). Nazi propaganda and the Second World War. Basingstoke, Palgrave Macmillan, 2005, XI-294 p.

7141. KIRCHHOFF (Hans). Et menneske uden pas er ikke noget menneske: Danmark i den internationale flygtningepolitik 1933–1939. (Un sans-papier n'est pas un être humain. Le Danemark dans la politique internationale des réfugiés 1933–1939). Odense, Syddansk Universitetsforlag, 2005, 255 p. (Dansk flygtningepolitik 1933–1945).

7142. KNABE (Hubertus). Tag der Befreiung? Das Kriegsende in Ostdeutschland. Berlin, Propyläen, 2005, 388 p.

7143. KOERNER (Francis). Deux défenseurs de l'indépendance de la Finlande: Alexandre Varenne et Jacques Bardoux (décembre 1939–mars 1940). *Guerres mondiales et conflits contemporains*, 2005, 53, 218, p. 3-14.

7144. Konec soužití Čechů a Němců v Československu. Sborník k 60. výročí ukončení 2. světové války. (Interpretations of World War II and Asymmetric Relations between Czechs and Germans in Europe). Ed. Hynek FAJMON, Kateřina HLOUŠKOVÁ. Brno, Centrum pro studium demokracie a kultury, 2005, 115 p. (photogr.).

7145. Kriegsgefangene des Zweiten Weltkrieges: Gefangennahme, Lagerleben, Rückkehr: zehn Jahre Ludwig Boltzmann-Institut für Kriegsfolgen-Forschung. Hrsg. v. Günter BISCHOF, Stefan KARNER und Barbara STELZL-MARX; unter Mitarbeit von Edith PETSCHNIGG. Wien, Oldenbourg, 2005, 599 p. (ill.). (Kriegsfolgen-Forschung, 4).

7146. KUNZ (Norbert). Die Krim unter deutscher Herrschaft, 1941–1944. Germanisierungsutopie und Besatzungsrealität. Darmstadt, Wissenschaftliche Buchgesellschaft, 2005, 448 p. (Veröffentlichungen der Forschungsstelle Ludwigsburg der Universität Stuttgart, 5).

7147. LIDEGAARD (Bo). Kampen om Danmark 1933–1945. (La lutte pour le Danemark 1933–1945). København, Gyldendal, 2005, 629 p.

7148. LOWER (Wendy). Nazi empire-building and the Holocaust in Ukraine. Chapel Hill, University of North Carolina Press, 2005, XVIII-307 p. (ill., maps).

7149. LUND (Joachim). Hitlers spisekammer: Danmark og den europæiske nyordning 1940–1943. (Le garde-manger d'Hitler: le Danemark et l'Ordre nouveau européen 1940–1943). København, Gyldendal, 2005, 405 p.

7150. MARTIN (Benjamin F.). France in 1938. Baton Rouge, Louisiana State U. P., 2005, X-252 p. (ill.).

7151. MATHOT (René). Au ravin du loup: Hitler en Belgique et en France, mai–juin 1940. Bruxelles, Racine, 2005, 321 p. (ill., maps, ports.).

7152. MATTIOLI (Aram). Experimentierfeld der Gewalt: der Abessinienkrieg und seine internationale Bedeutung 1935–1941. Mit einem Vorwort von Angelo DEL BOCA. Zürich, Orell Füssli, 2005, 239 p. (ill., maps). (Kultur, Philosophie, Geschichte, 3).

7153. MÜLLER (Rolf-Dieter). Der letzte deutsche Krieg, 1939–1945. Stuttgart, Klett-Cotta, 2005, 415 p. (ill., maps).

7154. MUNHOLLAND (Kim). Rock of contention. Free French and Americans in war in New Caledonia, 1940–1945. New York, Berghahn Books, 2005, XI-251 p.

7155. NEVILLE (Peter). The origins of the Second World War revisited. *European history quarterly*, 2005, 35, 4, p. 569-582.

7156. RADOWITZ (Sven). Schweden und das "Dritte Reich" 1939–1945: die deutsch-schwedischen Beziehungen im Schatten des Zweiten Weltkrieges. Hamburg, Reinhold Krämer, 2005, 624 p. (Beiträge zur deutschen und europäischen Geschichte, 34).

7157. Répression nazie (La) en Basse-Normandie pendant la Seconde Guerre mondiale: actes du colloque du 25 avril 2004. Caen, Centre de recherche d'histoire quantitative, 2005, 143 p. (ill., maps). (Seconde Guerre mondiale. Université de Caen. Centre de recherche d'histoire quantitative, 3).

7158. ŘÍHOVÁ (Romana). Memorandum "Německo-česká otázka v rámci celoněmecké politiky" a otázka jeho autorství. (Das Memorandum "Deutsch-tschechische Frage im Rahmen der alldeutschen Politik" und die Frage seiner Autorschaft). *In*: Historie okupovaného pohraničí 1938–1945. Tomo 10. Ed. Zdeněk RADVANOVSKÝ. Ústí n.L., 2005, p. 187-236.

7159. ROCHAT (Giorgio). Le guerre italiane 1935–1943: dall'Impero d'Etiopia alla disfatta. Torino, G. Einaudi, 2005, XVI-460 p. (maps). (Einaudi tascabili, 5. Storia).

7160. RODOGNO (Davide). Italiani brava gente? Fascist Italy's policy toward the Jews in the Balkans, April 1941–July 1943. *European history quarterly*, 2005, 35, 2, p. 213-240.

7161. Rossija v XX veke. Vojna 1941–1945 gg.: sovremennye podhody. (Russia in the 20th century: The war of 1941-1945: Recent approaches: [Articles]). Ed. Andrej N. SAKHAROV, Ljudmila P. KOLODNIKOVA. RAN, In-t rossijskoj istorii, etc.. Moskva, Nauka, 2005, 567 p. (bibl. incl.). [English summary]

7162. SANFORD (George). Katyn and the Soviet massacre of 1940: truth, justice and memory. London, Routledge, 2005, XV-250 p. (ill., maps).

7163. SCHECK (Raffael). "They Are Just Savages": German massacres of black soldiers from the French Army in 1940. *Journal of modern history*, 2005, 77, 2, p. 325-344.

7164. SODE-MADSEN (Hans). Reddet fra Hitlers helvede: Danmark og De Hvide Busser 1941–1945. (Sauvés de l'enfer hitlérien. Le Danemark et les Autobus blancs 1941–1945). København, Aschehoug, 2005, 299 p.

7165. STOLER (Mark A.). Allies in war: Britain and America against the Axis powers, 1940–1945. London, Hodder Arnold, 2005, XXV-292 p. (ill., maps).

7166. THOMPSON (John A.). Conceptions of national security and American entry into World War II. *Diplomacy & statecraft*, 2005, 16, 4, p. 671-697.

7167. UEBERSCHÄR (Gerd R.), MÜLLER (Rolf-Dieter). 1945: das Ende des Krieges. Darmstadt, Wiss. Buchges., 2005, 240 p. (ill.).

7168. UHLÍŘ (Jan B.), KAPLAN (Jan). Praha ve stínu hákového kříže. (Prague in the shadow of the swastika). Praha, Ottovo nakladatelství, 2005, 199 p. (photogr.).

7169. "Unsere Opfer zählen nicht": die Dritte Welt im Zweiten Weltkrieg. Rheinisches Journalistinnenbüro; herausgegeben von Recherche International; Recherche, Redaktion und Text: Birgit MORGENRATH und Karl RÖSSEL. Berlin, Assoziation A, 2005, 444 p. (ill., maps).

7170. Verbrechen der Wehrmacht: Bilanz einer Debatte. Hrsg. v. Christian HARTMANN, Johannes HÜRTER und Ulrike JUREIT. München, Beck, 2005, 230 p. (maps). (Beck'sche Reihe, 1632).

7171. WEINBERG (Gerhard L.). Visions of victory: the hopes of eight World War II leaders. Cambridge, Cambridge U. P., 2005, XXIV-292 p. (ill., maps).

7172. World (A) at total war. Global conflict and the politics of destruction, 1937–1945. Ed. by Roger CHICKERING, Stig FÖRSTER and Bernd GREINER. Cambridge, New York a. Port Melbourne, Cambridge U. P., 2005, X-392 p.

7173. YU (Xue). Buddhism, war, and nationalism: Chinese monks in the struggle against Japanese aggressions, 1931–1945. New York a. London, Routledge, 2005, XIII-278 p. (East Asia, history, politics, sociology, culture).

7174. ZAMIR (Meir). An intimate alliance: the joint struggle of General Edward Spears and Riad al-Sulh to oust France from Lebanon, 1942–1944. *Middle Eastern studies*, 2005, 41, 6, p. 811-832.

7175. ZEIDLER (Manfred). Das "kaukasische Experiment". Gab es eine Weisung Hitlers zur deutschen Besatzungspolitik im Kaukasus? *Vierteljahrshefte für Zeitgeschichte*, 2005, 53, 3, p. 475-500.

7176. ŻERKO (Stanisław). Niemiecka polityka zagraniczna: 1933–1939. (German foreign policy: 1933–1939). Poznań, Instytut Zachodni, 2005, 500 p. (ill.). (Studium niemcoznawcze Instytutu Zachodniego, 81).

Cf. n°s 3897, 3939, 4008, 4084, 4274, 4280, 4334, 4655, 4710, 5669, 6797, 7018, 7039, 7042, 7050, 7055, 7062, 7073, 7099, 7363, 7425

b. Diplomacy. Economy

** 7177. Československo-francouzské vztahy v diplomatických jednáních 1940–1945. (The Czechoslovak-French relations in the diplomatic agendas). Ed. Jan NĚMEČEK, Helena NOVÁČKOVÁ, Ivan ŠŤOVÍČEK, Jan KUKLÍK. Praha, Historický ústav AV ČR, 2005, 647 p.

** 7178. Correspondencia de um diplomata do III Reich: Veiga Simões, ministro acreditado em Berlin de

1933 a 1940. Introdução, selecção e organização de Lina Alves MADEIRA. Coimbra, Mar da Palavra, 2005, 257 p. (ill.). (Colecção Barca de cronos, 1).

** 7179. Documents diplomatiques francais 1932–1939. Ministère des Affaires Etrangères, Commission de Pubblication des Documents Diplomatiques Français. 2. 1936–1939. Tome 2. 1936. 1 avril–18 juillet 1936. Bruxelles, P. Lang, 2005, LXVIII-763 p.

** 7180. My dear Mr. Stalin: the complete correspondence between Franklin D. Roosevelt and Joseph V. Stalin. Edited, with commentary, by Susan BUTLER; foreword by Arthur M. SCHLESINGER, Jr. New Haven a. London, Yale U. P., 2005, XIX-361 p.

** 7181. Polskie dokumenty dyplomatyczne: 1939 styczeń–sierpień. (Polish diplomatic documents: 1939 Jan.–August). Redaktor, Stanisław ŻERKO; współpraca, Piotr DŁUGOŁĘCKI. Warszawa, Polski Instytut Spraw Międzynarodowych, 2005, LXV-896 p. (Polskie dokumenty dyplomatyczne).

** 7182. Portugal visto pelos Nazis: documentos 1933–1945. Seleção e organização António LOUÇÃ. Lisboa, Fim de Seculo, 2005, 428 p. (ill., ports.).

** 7183. Viedenská arbitráž 2. november 1938. Dokumenty. Zv. 3. Rokovania. (3. november 1938–4. apríl 1939). (Der Wiener Schiedsspruch vom 2. November 1938. Dokumente. Band 3.) Ed. Ladislav DEÁK. Martin, Matica slovenská, 2005, 454 p.

7184. ADAMSON (Michael R.). "Must We Overlook All Impairment of Our Interests?" Debating the foreign aid role of the Export-Import Bank, 1934–1941. *Diplomatic history*, 2005, 29, 4, p. 589-623.

7185. BECKER (Steffen). Von der Werbung zum "Totaleinsatz": die Politik der Rekrutierung von Arbeitskräften im "Protektorat Böhmen und Mähren" für die deutsche Kriegswirtschaft und der Aufenthalt tschechischer Zwangsarbeiter und -arbeiterinnen im Dritten Reich 1939–1945. Berlin, Dissertation.de, 2005, 534 p.

7186. BERCUSON (David Jay), HERWIG (Holger H.). One Christmas in Washington: Churchill and Roosevelt forge the Grand Alliance. London, Weidenfeld & Nicolson, 2005, 320 p. (ill., ports.).

7187. BERGGREN (Jan). Världskrig, kommunism och nazism i själva verket: aktörer, åsikter och aktioner inom statlig förvaltning under andra världskriget: exempel Tullverket. (Guerre mondiale, communisme et nazisme. Acteurs, opinions et actions dans l'administration suédoise pendant la seconde Guerre mondiale: exemple de l'administration des Douanes). Stockholm, Carlsson, 2005, 248 p. (ill.). [English summary].

7188. BERSKA (Barbara). Kłopotliwy sojusznik: wpływ dyplomacji brytyjskiej na stosunki polsko-sowieckie w latach 1939–1943. (L'ingombrante alleato: l'impatto della diplomazia britannica nelle relazioni polacco-sovietiche negli anni 1939–1943). Kraków, Księgarnia Akademicka, 2005, 268 p.

7189. BOISDRON (Matthieu). Le projet de pacte oriental (février 1934–mai 1935). *Guerres mondiales et conflits contemporains*, 2005, 53, 220, p. 23-44.

7190. BÖSL (Elsbeth), KRAMER (Nicole), LINSINGER (Stephanie). Die vielen Gesichter der Zwangsarbeit: "Ausländereinsatz" im Landkreis München 1939–1945. Mit einem Vorwort von Hans Günter HOCKERTS. München, K.G. Saur, 2005, 195 p. (ill., maps).

7191. BRANDEBORG JENSEN (Ole). Besættelsetidens økonomiske og erhvervsmæssige forhold: studier i de økonomiske relationer mellem Danmark og Tyskland 1940–1945. (Les conditions économiques de l'occupation: études des relations économiques entre le Danemark et l'Allemagne, 1940–1945). Odense, Syddansk universitetsforlag, 2005, 449 p. (University of Southern Denmark studies in history and social sciences, 309).

7192. BRECHENMACHER (Thomas). Teufelspakt, Selbsterhaltung, universale Mission? Leitlinien und Spielräume der Diplomatie des Heiligen Stuhls gegenüber dem nationalsozialistischen Deutschland (1933–1939) im Lichte neu zugänglicher vatikanischer Akten. *Historische Zeitschrift*, 2005, 280, 3, p. 591-645.

7193. BROOK (Timothy). Collaboration: Japanese agents and local elites in wartime China. Cambridge, Harvard U. P., 2005, 288 p.

7194. BRYCE (Robert B.). Canada and the cost of World War II. The international operations of Canada's Department of finance, 1939–1947. Ed. by Matthew J. BELLAMY. Montreal, McGill-Queen's U. P., 2005, XVI-392 p.

7195. CHAPNICK (Adam). The middle power project: Canada and the founding of the United Nations. Vancouver, UBC Press, 2005, XIV-210 p.

7196. DIETERICH (Renate). Germany's relations with Iraq and Transjordan from the Weimar Republic to the end of the Second World War. *Middle Eastern studies*, 2005, 41, 4, p. 463-479.

7197. DOENECKE (Justus D.), STOLER (Mark A.). Debating Franklin D. Roosevelt's foreign policies, 1933–1945. Lanham, Rowman & Littlefield Publishers, 2005, VI-238 p. (Debating twentieth-century America).

7198. DUNTIIORN (David J.). The Paris conference on Tangier, August 1945: the British response to Soviet interest in the "Tangier question". *Diplomacy & statecraft*, 2005, 16, 1, p. 117-137.

7199. DUTTON (David). Sir Austen Chamberlain and British foreign policy 1931–1937. *Diplomacy & statecraft*, 2005, 16, 2, p. 281-295.

7200. EICHHOLTZ (Dietrich). Deutsche Politik und rumänisches Öl (1938–1941): eine Studie über Erdölimperialismus. Leipzig, Leipziger Universitätsverlag, 2005, 68 p. (map).

7201. Europa (Das) des "Dritten Reichs": Recht, Wirtschaft, Besatzung. Hrsg. v. Johannes BÄHR und Ralf BANKEN. Frankfurt am Main, Klostermann, 2005, VI-

288 p. (Europa der Diktatur, 5. Studien zur europäischen Rechtsgeschichte, 181).

7202. Fascismo e franchismo: relazioni, immagini rappresentazioni. A cura di Giuliana DI FEBO e Renato MORO. Soveria Mannelli, Rubbettino, 2005, XVIII-507 p. (Saggi, 163).

7203. FEICHTLBAUER (Hubert). Zwangsarbeit in Österreich 1938–1945. Wien, Österreichischer Versöhnungsfonds u. Braintrust, 2005, 335 p. (ill., maps, ports.).

7204. GERRARD (Craig). The Foreign Office and Finland, 1938–1940: diplomatic sideshow. London, Frank Cass, 2005, VII-190 p. (Cass contemporary security studies series).

7205. GLANTZ (Mary E.). FDR [Franklin Delano Roosevelt] and the Soviet Union: the President's battles over foreign policy. Lawrence, University Press of Kansas, 2005, VIII-253 p. (Modern war studies).

7206. GLUR (Stefan). Vom besten Pferd im Stall zur persona non grata: Paul Ruegger als Schweizer Gesandter in Rom 1936–1942. Bern u. Oxford, Lang, 2005, 352 p. (Geist und Werk der Zeiten, 100).

7207. HALL (Thomas). "Mere drops in the ocean": the politics and planning of the contribution of the British Commonwealth to the final defeat of Japan, 1944–1945. *Diplomacy & statecraft*, 2005, 16, 1, p. 93-115.

7208. HORN (Martin). Money in wartime: France's financial preparations for the two World Wars. *International history review*, 2005, 27, 4, p. 709-753.

7209. IVONE (Diomede). Raffaele Guariglia tra l'ambasceria a Parigi e gli ultimi "passi" in diplomazia: 1938–1943. Napoli, Editoriale scientifica, 2005, 284 p. (Ricerche storiche, 6).

7210. JOHNSON (Gaynor). Sir Eric Phipps, the British government, and the Appeasement of Germany, 1933–1937. *Diplomacy & statecraft*, 2005, 16, 4, p. 651-669.

7211. KAMIŃSKI (Marek K.). Edvard Beneš kontra gen. Władysław Sikorski: polityka władz czechosłowackich na emigracji wobec rządu polskiego na uchodźstwie 1939–1943. (Edvard Beneš contro il generale Wladyslaw Sikorski: la politica delle autorità cecoslovacche in esilio verso il governo polacco in esilio 1939–1943). Warszawa, Wydawn. Neriton, Instytut Historii PAN, 2005, 383 p.

7212. KIT-CHING (Chan Lau). Symbolism as diplomacy: the United States and Britain's China policy during the first year of the Pacific War. *Diplomacy & statecraft*, 2005, 16, 1, p. 73-92.

7213. KOCHAVI (Arieh J.). Confronting captivity. Britain and the United States and their POWs in Nazi Germany. Chapel Hill a. London, University of North Carolina Press a. Toronto, Scholarly book services, 2005, X-382 p.

7214. LUSSY (Hanspeter), LÓPEZ (Rodrigo). Finanzbeziehungen Liechtensteins zur Zeit des Nationalsozialismus: Studie im Auftrag der Unabhängigen Historikerkommission Liechtenstein Zweiter Weltkrieg. Vaduz, Historischer Verein für das Fürstentum Liechtenstein u. Zürich, Chronos, 2005, 2 vol., 819 p. (ill.). (Veröffentlichungen der Unabhängigen Historikerkommission Liechtenstein Zweiter Weltkrieg, 3).

7215. MARXER (Veronika), RUCH (Christian). Liechtensteinische Industriebetriebe und die Frage nach der Produktion für den deutschen Kriegsbedarf 1939–1945: Studie im Auftrag der Unabhängigen Historikerkommission Liechtenstein Zweiter Weltkrieg. Vaduz, Historischer Verein für das Fürstentum Liechtenstein u. Zürich, Chronos, 2005, 153 p. (Veröffentlichungen der Unabhängigen Historikerkommission Liechtenstein Zweiter Weltkrieg, 2).

7216. MESSENGER (David A.). Rival faces of France: refugees, would-be allies, and economic warfare in Spain, 1942–1944. *International history review*, 2005, 27, 1, p. 25-46.

7217. MORADIELLOS (Enrique). Franco frente a Churchill: España y Gran Bretaña en la Segunda Guerra Mundial (1939–1945). Barcelona, Ediciones Península, 2005, 479 p. (Atalaya, 215).

7218. MORI (Shigeki). Matsuoka gaikō to nisso kokkō chousei: seiryoku kinkō senryaku no kansei. (Matsuoka's diplomacy and Japanese-Soviet relations: demise of the balance of power strategy). *Rekishigaku Kenkyu*, 2005, 801, p. 1-18.

7219. NASH (Philip). America's first female chief of mission: Ruth Bryan Owen, minister to Denmark, 1933–1936. *Diplomacy & statecraft*, 2005, 16, 1, p. 57-72.

7220. PELIN (Mihai). Diplomație de război. România – Italia 1939–1945. (War diplomacy. Romania-Italy 1939–1945). București, Editura Elion, 2005, 344 p.

7221. PELTOVUORI (Risto). Suomi saksalaisin silmin 1933–1939: Lehdistön ja dipolmatian näkökulmia. (Finland through German eyes 1933–1939: the views of journalism and diplomacy). Helsinki, Suomalaisen Kirjallisuuden Seura, 2005, 306 p. (Historiallisia Tutkimuksia, 223).

7222. PENTER (Tanja). Zwangsarbeit – Arbeit für den Feind. Der Donbass unter deutscher Okkupation (1941–1943). *Geschichte und Gesellschaft*, 2005, 31, 1, p. 68-100.

7223. Polscy robotnicy przymusowi w Trzeciej Rzeszy. (Polish conscript workers in the Third Reich). Pod redakcją Włodzimierza BONUSIAKA. Rzeszów, Wydawn. Uniwersytetu Rzeszowskiego, 2005, 287 p. (ill.).

7224. PREPARATA (Guido Giacomo). Conjuring Hitler: how Britain and America made the Third Reich. London, Pluto Press, 2005, XIX-311 p. (ill.).

7225. PRÉVOST (Philippe). Le temps des compromis: mai–décembre 1940. Préf. de F.-G. DREYFUS. Paris, Centre d'études contemporaines, 2005, 210 p.

7226. RUOTSILA (Markku). Churchill and Finland: a study in anticommunism and geopolitics. London a. New York, Routledge, 2005, 199 p.

7227. SAMARANI (Guido). Shaping the future of Asia: Chiang Kai-shek, Nehru and China-India relations during the second world war period. Lund, Centre for East and South-East Asian Studies, Lund University, 2005, 20 p. (Working papers in contemporary Asian studies, 11).

7228. ŠAMBERGER (Zdeněk). Likvidační protičeský dokument. Na okraj nacistické koncepce plánu "Grundplanung". (Ein antitschechisches Liquidationsdokument. Am Rande der nazistischen Konzeption "Grundplanung"). *Právněhistorické studie*, 2005, 37, p. 157-205.

7229. SEYDI (Süleyman), MOREWOOD (Steven). Turkey's application of the Montreux Convention in the Second World War. *Middle Eastern studies*, 2005, 41, 1, p. 79-101.

7230. SMETANA (Vít). In the shadow of Munich. British policy towards Czechoslovakia from 1938 to 1942. Praha, Faculty of Social Sciences – Charles University, 2005, 320 p. [Ph.D. Thesis].

7231. SOMMELLA (Valentina). Un' allenza difficile: Churchill, De Gaulle, e Roosevelt negli anni della guerra. Roma, Aracne, 2005, 209 p. (Scienze politiche e sociali, 67).

7232. STONE (Glyn). Spain, Portugal and the Great Powers, 1931–1941. Basingstoke, Palgrave Macmillan, 2005, XIII-316 p. (maps). (Making of the 20th century).

7233. SZUBTARSKA (Beata). Ambasada polska w ZSRR w latach 1941–1943. (Polish embassy in the Soviet Union in the years 1941–1943). Warszawa, Wydawn. DiG, 2005, 205 p. (ill.).

7234. Totaler Arbeitseinsatz für die Kriegswirtschaft: Zwangsarbeit in der deutschen Binnenschifffahrt 1940–1945: Erinnerungen, Dokumente, Studien. Hrsg. v. Eckhard SCHINKEL. Münster, Landschaftsverband Westfalen-Lippe, 2005, 237 p. (ill., ports.). (Quellen und Studien / Westfälisches Industriemuseum, 11).

7235. TRAŞCĂ (Ottmar). Relaţiile româno-maghiare şi problema Transilvaniei, 1940–1944. (The Romanian-Magyar relations and the question of Transylvania, 1940–1944). *Anuarul Institutului de istorie "A.D. Xenopol"*. Academia Română, 2005, 42, p. 377-408.

7236. UNGUREANU (George). Chestiunea Cadrilaterului. Interese româneşti şi revizionism bulgar (1938–1940). (The question of the Quadrilater. Romanian interests and Bulgarian revisionism, 1938–1940). Bucureşti, Editura Ars Docendi, 2005, 157 p.

7237. VAYSSIÈRE (Bertrand). Le Manifeste de Ventotene (1941): acte de naissance du fédéralisme européen. *Guerres mondiales et conflits contemporains*, 2005, 53, 217, p. 69-76.

7238. VESELÝ (Zdeněk). Edvard Beneš – Československo – Evropa. (Edvard Beneš – Czechoslovakia – Europe). Praha, Professional Publ., 2005, 164 p. (ill.).

7239. VILANOVA (Francesc). El franquismo en guerra: de la destrucción de Checoslovaquia a la batalla de Stalingrado. Prólogo de Pere YSÀS. Barcelona, Ediciones Península, 2005, 253 p. (Atalaya, 210).

7240. VLASOV (Leonid V.), VLASOVA (Marina A.). Mannergejm i Pol'sha. (Carl Gustaf Emil Mannerheim and Poland). Sankt-Peterburg, Russkaja voennaja entsiklopedija, 2005, 223 p. (ill.; portr.; bibl. p. 219-222). (Rozhden dlja sluzhby tsarskoj... Voennye biografii).

7241. WIGG (Richard). Churchill and Spain: the survival of the Franco regime, 1940–1945. London a. New York, Routledge, 2005, 211 p. (Routledge/Cañada Blanch studies on contemporary Spain).

b. Addendum 1999

7242. MORADIELLOS (Enrique). The Allies and the Spanish Civil War. *In*: Spain and the Great Powers in the twentieth century. London, Routledge, 99, p. 96-126. [Cf. Bibl. 99, n° 7294]

Cf. n°s 3680, 4460, 5792, 5867, 6030, 7009, 7013, 7019, 7027, 7028, 7036, 7060, 7063, 7068, 7091, 7101, 7102, 7310, 7371, 7377, 7380, 7437, 7439, 7446

c. Military operations

** 7243. GELFAND (Wladimir). Deutschland-Tagebuch, 1945–1946. Aufzeichnungen eines Rotarmisten. Aus dem Russischen von Anja LUTTER und Hartmut SCHRÖDER; ausgewählt und kommentiert von Elke SCHERSTJANOI. Berlin, Aufbau-Verlag, 2005, 357 p.

** 7244. Guy Liddell diaries (The): MI5's director of counter-espionage in World War II. Vol. 1. 1939–1942. Vol. 2. 1942–1945. Ed. by Nigel WEST. London a. New York, Routledge, 2005, 2 vol., XVII-329 p., XVII-314 p.

** 7245. Italiener (Die) an der Ostfront 1942/43. Dokumente zu Mussolinis Krieg gegen die Sowjetunion. Herausgegeben und eingeleitet von Thomas SCHLEMMER; Übersetzung der Dokumente aus dem Italienischen von Gerhard KUCK. München, Oldenbourg, 2005, VII-291 p. (Schriftenreihe der Vierteljahrshefte für Zeitgeschichte, 91).

** 7246. METZ (Paul). Mussertman aan het oostfront: oorlogsdagboek 1941–1942. Bezorgd en ingeleid door Gerard GROENEVELD. Nijmegen, Vantilt, 2005, 128 p. (ill.).

7247. ARNOLD (Klaus Jochen). Die Wehrmacht und die Besatzungspolitik in den besetzten Gebieten der Sowjetunion: Kriegführung und Radikalisierung im "Unternehmen Barbarossa". Berlin, Duncker & Humblot, 2005, 579 p. (ill., maps). (Zeitgeschichtliche Forschungen, 23).

7248. BELL (Roger John), BRAWLEY (Sean), DIXON (Chris). Conflict in the Pacific 1937–1951. Cambridge a. Port Melbourne, Cambridge U. P., 2005, 235 p. (ill., maps, ports.). (Cambridge senior history series).

7249. BOWD (Rueben Robert Ernest). A basis for victory: the Allied Geographical Section, 1942–1946. Canberra, Strategic and Defence Studies Centre, Australian National University, 2005, XXIII-168 p. (ill., maps, ports.). (Canberra papers on strategy and defence, 157).

7250. BREITMAN (Richard) [et al.]. US intelligence and the Nazis. New York, Cambridge U. P., 2005, 495 p.

7251. CIOBANU (Nicolae), MATEI (Aurel). Principalele acțiuni militare desfășurate de armata română în campania de est (22 iunie 1941–23 august 1944). (The major military actions of the Romanian army during the East Campaign, June 22, 1941–August 23, 1944). Râmnicu Vâlcea, Editura Almarom, 2005, 202 p.

7252. COUTU (Éric). Le quartier général des opérations combinées et l'expédition canado-britannique au Spitzberg (août 1941). *Guerres mondiales et conflits contemporains*, 2005, 53, 220, p. 45-70.

7253. DI GIUSTO (Stefano). Operationszone Adriatisches Künstenland: Udine, Gorizia, Trieste, Pola, Fiume e Lubiana durante l'occupazione tedesca, 1943–1945. Udine, Istituto friulano per la storia del movimento di liberazione, 2005, 807 p. (ill.).

7254. DÍAZ BENÍTEZ (Juan José). Voluntarios de la zona aérea de Canarias y África occidental en la Wehrmacht. *Historia social*, 2005, 53, p. 47-64.

7255. Esercito italiano (L'): dall'armistizio alla Guerra di Liberazione: 8 settembre 1943–25 aprile 1945. A cura di Filippo CAPPELLANO e Salvatore ORLANDO. Roma, Stato maggiore dell'esercito, Ufficio storico, 2005, 214 p. (ill., maps, ports.).

7256. FORD (Christopher A.), ROSENBERG (David A.), BALANO (Randy C.). The admirals' advantage: U.S. Navy operational intelligence in World War II and the Cold War. Annapolis, Naval Institute Press, 2005, XXI-219 p.

7257. FORD (Douglas). Britain's secret war against Japan. London, Routledge, 2005, 258 p.

7258. FRIEDL (Jiří). Na jedné frontě. Vztahy československé a polské armády (Polskie siły zbrojne) za druhé světové války. (For the same cause. Czechoslovak-Polish military relations during the Second World War). Praha, Ústav pro soudobé dějiny AV ČR, 2005, 384 p. (Sešity Ústavu pro soudobé dějiny, 40).

7259. GANZENMÜLLER (Jörg). Das belagerte Leningrad, 1941–1944: die Stadt in den Strategien von Angreifern und Verteidigern. Paderborn, Ferdinand Schöningh, 2005, X-412 p. (Krieg in der Geschichte, 22).

7260. GLANTZ (David M.). Colossus reborn: the Red Army at war: 1941–1943. Kansas, University Press of Kansas, 2005, XIX-807 p. (ill., maps). (Modern war studies).

7261. GRANIER (Pierre). Les soldats oubliés de la 1re D.F.L. [Première Division française libre]: le pied-la route sous la croix de Lorraine. Toulon, Presses du midi, 2005, 414 p. (ill., map).

7262. GRIBAUDI (Gabriella). Guerra totale. Tra bombe alleate e violenze naziste. Napoli e il fronte meridionale 1940–1944. Torino, Bollati Boringhieri, 2005, 657 p.

7263. GRUNDEN (Walter E.). Secret weapons and World War II: Japan in the shadow of big science. Lawrence, University Press of Kansas, 2005, XI-335 p. (ill., ports.). (Modern war studies).

7264. HAAEST (Erik). Forrædere. Frikorps Danmarkfolk, om drømmen og virkeligheden. (Les traîtres. Les membres du Frikorp danois, du rêve à la réalité). Højbjerg, Bogan, 2005, 180 p. (ill., cartes).

7265. HERSHBERG (James G.). Reading and warning the likely enemy: China's signals to the United States about Vietnam in 1965. *International history review*, 2005, 27, 1, p. 47-85.

7266. KIRKPATRICK (Charles Edward). An unknown future and a doubtful present: writing the victory plan of 1941. New York, University Press of the Pacific, 2005, 158 p.

7267. KOKOŠKA (Stanislav). Praha v květnu 1945. Historie jednoho povstání. (Prague in May of 1945. History of the uprising). Praha, Lidové noviny, 2005, 277 p. (photogr.). (Válka a mír v moderní době).

7268. KOPECKÝ (Milan). Závěrečné boje 1. československé samostatné tankové brigády v Ostravské operaci (21. dubna–5. května 1945). (Concluding engagements of the 1st Czechoslovak Independent Tank Brigade during the Ostrava operation). *Historie a vojenství*, 2005, 54, 2, p. 76-88.

7269. KUDRNA (Ladislav). Českoslovenští letci v německém zajetí 1940–1945. (Czechoslovak Airmen in the German captivity). Praha, Naše vojsko, 2005, 437 p. (photogr.).

7270. Life and death in besieged Leningrad, 1941–1944. Ed. by John BARBER and Andrei DZENISKEVICH. Basingstoke, Palgrave Macmillan, 2005, XXVII-243 p. (ill.). (Studies in Russian and East European history and society).

7271. MAREK (Jindřich). Bojová činnost 2. polské armády na severu Čech v květnu 1945. (Action taken in combat by the 2nd Polish Army in the north of Bohemia in 1945). *Historie a vojenství*, 2005, 54, 2, p. 19-36.

7272. MAWDSLEY (Evan). Thunder in the East. The Nazi-Soviet war, 1941–1945. London, Hodder Arnold, 2005, XXVI-502 p.

7273. MOREMAN (T. R.). The jungle, the Japanese and the British Commonwealth armies at war, 1941–1945: fighting methods, doctrine and training for jungle warfare. London, Frank Cass, 2005, XII-277 p. (maps). (Cass series-military history and policy, 4).

7274. MORENO JULIÀ (Xavier). La División Azul: sangre española en Rusia, 1941–1945. Barcelona, Crítica, 2005, XIX-553 p. (ill., maps, ports.). (Crítica contrastes).

7275. MOREWOOD (Steven). The British defence of Egypt, 1935–1940: conflict and crisis in the eastern Mediterranean. London a. New York, Frank Cass, 2005, XVII-274 p (Cass series: military history and policy, 3).

7276. MÜLLER (Zdeněk), FLODROVÁ (Milena), BUDÍK (Miloš), SCHILDBERGER (Vlastimil). Bomby nad Brnem. Zpráva o leteckém bombardování města Brna v letech 1944 a 1945. (Bombing on Brno, 1944–1945). Brno, Expo data, 2005, 254 p. (photogr.).

7277. MURPHY (David E.). What Stalin knew: the enigma of Barbarossa. New Haven, Yale U. P., 2005, XXII-310 p.

7278. NEDBÁLEK (František). Železniční transporty a pochody smrti vězňů koncentračních táborů a válečných zajatců přes české země (zima a jaro 1945). (Die Zugtransporte und Todesmärsche der Häftlinge und Kriegsgefangenen über die böhmischen Länder). Ed. Václav KURAL. Ústí nad Labem, Albis international, 2005, 165 p. (mp). (Historie okupovaného pohraničí 1938–1945, 9).

7279. NEITZEL (Sönke). Abgehört: deutsche Generäle in britischer Kriegsgefangenschaft, 1942–1945. Berlin, Propyläen, 2005, 638 p. (ill.).

7280. OCHEA (Lionede). Serviciul Special de Informații al României pe Frontul de Sud, 1940–1944. (The Special intelligence service of Romania on the South front 1940–1944). Pitești, Editura Tipart, 2005, 595 p.

7281. PELIN (Mihai). Raidul escadrei trădate. Bombardamente asupra României 1941–1944. (The raid of the betrayed squadron. Bombing Romania 1941–1944). București, Editura Elion, 2005, 295 p.

7282. Populations civiles (Les) face au débarquement et à la bataille de Normandie: colloque international, Mémorial de Caen, 25–27 mars 2004. Textes rassemblés et édités par Bernard GARNIER [et al.]. Caen, Université de Caen, centre de recherche d'histoire quantitative et Caen, Mémorial de Caen, 2005, 320 p. (ill., maps, ports.). (Seconde guerre mondiale, 5).

7283. RAJLICH (Jiří). Českoslovenští příslušníci Royal Air Force v německém zajetí v letech 1940–1945. (Czechoslovak members of the Royal Air Force in German captivity during 1940-45). Tomo 1-2. Historie a vojenství, 2005, 54, 2-3, p. 60-75, p. 61-71.

7284. ROTHWELL (Victor). War aims in the Second World War. The war aims of the major belligerents, 1939–1945. Edinburgh, Edinburgh U. P., 2005, 244 p.

7285. SCHRIJVERS (Peter). The unknown dead: civilians in the Battle of the Bulge. Lexington, University Press of Kentucky, 2005, XVIII-430 p. (ill., map).

7286. SOYBEL (Phyllis L.). A necessary relationship: the development of Anglo-American cooperation in naval intelligence. Westport, Praeger, 2005, XVI-172 p.

7287. Sur les chemins de la Libération: Dunkerque, 1944–1945. Ed. par Patrick ODDONE; avant-propos de Bruno BÉTHOUART; préface de Michel DELEBARRE. Villeneuve d'Ascq, Presses universitaires du Septentrion, 2005, 333 p. (ill., maps). (Documents et témoignages).

7288. TAMAYAMA (Kazuo). Railwaymen in the war: tales by Japanese railway soldiers in Burma and Thailand, 1941–1947. Basingstoke, Palgrave Macmillan, 2005, XVI-286 p. (ill.).

7289. VÁŇA (Josef), RAIL (Jan). Českoslovenští letci ve Francii 1939–1940. (The Czechoslovak Airmen in the France, 1939–1940). Praha, Ministerstvo obrany ČR, 2005, 109 p. (photogr.).

7290. WESTERMANN (Edward B.). Hitler's police battalions. Enforcing racial war in the East. Lawrence, University Press of Kansas, 2005, XV-329 p. (ill.). (Modern war studies).

Cf. n°s 3693, 3850

d. Resistance

* 7291. REID (Donald). Resistance and its discontents: affairs, archives, avowals, and the Aubracs. *Journal of modern history*, 2005, 77, 1, p. 97-137.

** 7292. AUSTAD (Torleiv). Kirkelig motstand: dokumenter fra den norske kirkekamp under okkupasjonen 1940–1945, med innledninger og kommentarer. Kristiansand, Høyskoleforlaget, 2005, 298 p.

** 7293. Polskie Państwo podziemne wobec komunistów polskich (1939–1945): wypisy prasy konspiracyjnej. (Der polnische Untergrundstaat zu den polnischen Kommunisten, 1939–1945). Auszüge der Untergrundpresse). Wyboru dokonał i opracował Karol SACEWICZ. Olsztyn, Wydawn. Uniwersytetu Warmińsko-Mazurskiego w Olsztynie, 2005, 421 p.

7294. CAPORALE (Riccardo). La banda Carità: storia del reparto servizi speciali (1943–1945). Prefazione di Dianella GAGLIANI. Lucca, Edizioni S. Marco litotipo, 2005, 432 p. (ill.).

7295. CHAUBIN (Hélène). Corse des années de guerre: 1939–1945. La Résistance en Corse. Paris, Tirésias, AERI, 2005, 117 p. (ill., maps, ports.). (Histoire en mémoire).

7296. COLLIN (Claude). Les italiens dans la MOI [Main-d'œuvre immigrée] et les FTP-MOI [Francs-tireurs et partisan-Main-d'œuvre immigrée] à Lyon et à Grenoble. *Guerres mondiales et conflits contemporains*, 2005, 53, 218, p. 67-84.

7297. DE BOISFLEURY (Bernard). L'armée en résistance, France, 1940–1944. Préface de Henri AMOUROUX. Fontenay-aux-Roses, L'esprit du livre, 2005, 718 p. (map). (Histoire & mémoires combattantes).

7298. Frauen aus Deutschland in der französischen Résistance: eine Dokumentation. Hrsg. v. Ulla PLENER. Berlin, Edition Bodoni, 2005, 221 p. (ill., map, ports.). (Arbeiterbewegung: Forschungen, Dokument, Biografien).

7299. HULL (Eugeniusz). Obraz cywilnych struktur państwa podziemnego 1939–1944 w opisie wspomnieniowym. (Image of civil structure of underground state in recollective description 1939–1944). Olsztyn, Wydawnictwo Uniwersytetu Warmińsko-Mazurskiego, 2005, 293 p. (ill., col. maps).

7300. ISRAËL (Liora). Robes noires, années sombres. Avocats et magistrats en Résistance pendant la Seconde Guerre Mondiale. Paris, Fayard, 2005, 547 p.

7301. JOHANSSON (Anders). Den glömda armén: Norge-Sverige 1939–1945. (L'armée oubliée: Norvège-Suède, 1939–1945). Rimbo, Fischer & Co, 2005, 516 p. (ill.).

7302. KITSON (Simon). Vichy et la chasse aux espions nazis. 1940–1942, complexités de la politique de collaboration. Paris, Éditions Autrement, 2005, 268 p.

7303. LALIEU (Olivier). La zone grise? La Résistance française à Buchenwald. Préface de Jorge SEMPRUN. Paris, Tallandier, 2005, 441 p. (map).

7304. MAREK (Jindřich). Vojenská odbojová skupina "Bartoš" a její přípravy Pražského povstání. (The "Bartoš" Military Resistance Group and its preparations for the Prague uprising). *Historie a vojenství*, 2005, 54, 4, p. 74-89.

7305. MIANNAY (Patrice). Dictionnaire des agents doubles dans la Résistance. Paris, Cherche midi, 2005, 352 p. (Collection Documents).

7306. Organy bezpieczeństwa i wymiar sprawiedliwości Polskiego Państwa Podziemnego. (Gli organi di sicurezza e giustizia dello stato polacco clandestino. Pod redakcją Waldemara GRABOWSKIEGO. Warszawa, Instytut Pamięci Narodowej-Komisja Ścigania Zbrodni przeciwko Narodowi Polskiemu, 2005, 168 p. (Konferencje IPN, 26).

7307. PRŮCHA (Václav), KALINOVÁ (Lenka). Koncepce budoucí hospodářské a sociální politiky v československém odboji za druhé světové války. (Programmatic principles of the economic and social policy in the Czechoslovak Resistance Movement during the Second World War). *Acta Oeconomica Pragensia*, 2005, 13, 3, p. 81-108.

7308. Services publics (Les) et la résistance en zone interdite et en Belgique (1940–1944): colloque organisé à Bondues par le Centre de Recherche sur l'Histoire de l'Europe du Nord-Ouest et la ville de Bondues le 31 janvier 2004. Sous la dir. de Robert VANDENBUSSCHE. Lille, Centre de Recherche sur l'Histoire de l'Europe du Nord-Ouest et CEGES, 2005, 216 p. (ill., map). (Centre de Recherche sur l'Histoire de l'Europe du Nord-Ouest, 35).

Cf. n°s 3897, 3903, 4022, 5263, 5683

§ 8. From 1945.

a. General

** 7309. Bulgarian intelligence and security services in the Cold War years. Sofija, Academic Publishing House, 2005, CD-ROM.

** 7310. DE MENESES (Filipe Ribeiro). Correspondência diplomática irlandesa sobre Portugal, o Estado Novo e Salazar, 1941–1970. Lisboa, Ministério dos Negócios Estrangeiros, 2005, 551 p. (Colecção Biblioteca diplomatica. 3 série A).

** 7311. 'Dear Joe': Sir Alvary Frederick Gascoigne, G.B.E. (1893–1970): a British diplomat in Hungary after the Second World War: a collection of documents from the British Foreign Office. Compiled by Éva HARASZTI-TAYLOR. Nottingham, Astra Press, 2005, XXI-250 p.

** 7312. Europa (L') da Togliatti a Berlinguer: testimonianze e documenti: 1945–1984. A cura di Mauro MAGGIORANI e Paolo FERRARI; postfazione di Giorgio NAPOLITANO. Bologna, Il Mulino, 2005, 350 p. (Storia del federalismo e dell'integrazione europea).

** 7313. GODEŠA (Bojan). Jozef Tiso a Anton Korošec – vzťahy medzi Slovákmi a Slovincami. (Jozef Tiso and Anton Korošec – the Relations between the Slovaks and the Sloviens). *Historický časopis*, 2005, 53, 2, p. 365-379.

** 7314. MASTROLILLI (Paolo), MOLINARI (Maurizio). L'Italia vista dalla CIA: 1948–2004. Roma e Bari, Laterza, 2005, XX-365 p. (ill.). (Robinson. Letture).

** 7315. Nepal-India, Nepal-China relations: documents 1947–June 2005. Ed. by Avtar Singh BHASIN. New Delhi, Geetika Publishes, 2005, 5 vol., 161-3584 p. (ill., folded map).

** 7316. Polska w stosunkach międzynarodowych 1945–1989: wybór dokumentów. (La Polonia nelle relazioni internazionali 1945–1989: selezione di documenti). Wybór i opracowanie: Justyna ZAJĄC. Warszawa, Wydawnictwa Uniwersytetu Warszawskiego, 2005, 322 p.

** 7317. Vorläufiges Findbuch zur Abteilung X: "Internationale Verbindungen" des Ministeriums für Staatssicherheit der DDR. Abteilung Archivbestände der BStU (Hrsg.); bearbeitet von Marko POLLACK und Doreen BOMBITZKI. Münster, Lit, 2005, 335 p. (ill.). (Archiv zur DDR-Staatssicherheit, 8).

7318. Auswärtige Repräsentationen: deutsche Kulturdiplomatie nach 1945. Hrsg. v. Johannes PAULMANN. Köln, Böhlau, 2005, VI-314 p. (ill.).

7319. BÍLEK (Jiří). Vznik organizace Varšavské smlouvy a československá armáda. (The establishment of the Warsaw Pact and the role played by the Czechoslovak Army). *Historie a vojenství*, 2005, 54, 3, p. 20-32.

7320. BORHI (László). Magyarország a hidegháborúban: a Szovjetunió és az Egyesült Államok között, 1945-1956. (Hungary in the Cold War: being between the Soviet Union and The United States, 1945–1956). Budapest, Corvina, 2005, 360 p. (ill.). (Corvina Faktum).

7321. BORRING OLESEN (Thorsten), VILLAUME (Poul). I blokopdelingens tegn 1945–1972. Dansk udenrigspolitiks historie. Bd. 5. (Sous le signe de la division entre les blocs 1945–1972. Histoire de la politique étrangère danoise. Tome 5). Red. Carsten DUE-NIELSEN, Ole FELDBÆK, Nikolaj PETERSEN. København, Danmarks Nationalleksikon, 2005, 808 p.

7322. BRAGG (Christine). Vietnam, Korea and US foreign policy 1945–1975. Oxford, Heinemann, 2005, 192 p. (Heinemann advanced history).

7323. CHIRIȚOIU (Mircea). Între David și Goliath: România și Iugoslavia în balanța războiului rece. (Tra Davide e Golia: Romania e Jugoslavia nell'equilibrio della Guerra fredda). Iași, Casa Editorială Demiurg, 2005, 488 p. (Colecția Românii în istoria universală = The Romanians in world history, 112).

7324. CHMIELEWSKI (Paweł). Dyplomacja sowiecka w Radzie Bezpieczeństwa ONZ: wobec zadań utrzymania pokoju i bezpieczeństwa międzynarodowego u progu "zimnej wojny". (La diplomazia sovietica nel Consiglio di Sicurezza delle Nazioni Unite: i compiti di mantenimento della pace e della sicurezza internazionale alle soglie della "guerra fredda"). Łódź, Wydawn. Uniwersytetu Łódzkiego, 2005, 2 vol., [s. p.]. (Rozprawy habilitacyjne Uniwersytetu Łódzkiego).

7325. COHEN (Warren I.). America's failing empire: U.S. foreign relations since the Cold War. Malden, Blackwell Pub., 2005, 204 p. (America's recent past, 1).

7326. Cold War (The): a military history. Ed. by Robert COWLEY. New York, Random House, 2005, XVI-478 p. (ill., maps).

7327. Danmark under den kolde krig: den sikkerhedspolitiske situation, 1945–1991. (Le Danemark pendant la guerre froide. La politique de sécurité 1945–1991. Tomes 1-4). Bd. 1. 1945–1962. Bd. 2. 1963–1978. Bd. 3. 1979–1991. Bd. 4. Konklusion og perspektiver, Bilag. Dansk institut for internationale studier. København, DIIS, 2005, 4 vol., 721 p., 734 p., 636 p., 259 p. (ill., maps, ports.).

7328. DDR (Die) in Europa: zwischen Isolation und Öffnung. Hrsg. v. Heiner TIMMERMANN. Münster, Lit, 2005, 554 p. (Dokumente und Schriften der Europäischen Akademie Otzenhausen, 140).

7329. DOMENICO (Roy Palmer). "For The Cause of Christ Here in Italy": America's Protestant challenge in Italy and the cultural ambiguity of the Cold War. *Diplomatic history*, 2005, 29, 4, p. 625-654.

7330. FANZUN (Jon A.). Die Grenzen der Solidarität: schweizerische Menschenrechtspolitik im Kalten Krieg. Mit einem Vorwort von Walter KÄLIN. Zürich, Neue Zürcher Zeitung, 2005, XV-462 p.

7331. Freundschaftsbande und Beziehungskisten: die Afrikapolitik der DDR und der BRD gegenüber Mosambik. Hrsg. v. Hans-Joachim DÖRING und Uta RÜCHEL. Frankfurt am Main, Brandes & Apsel, 2005, 213 p. (ill.).

7332. GADDIS (John Lewis). The Cold War: a new history. New York, Penguin Press, 2005, XV-333 p. (ill.).

7333. GALEAZZI (Marco). Togliatti e Tito: tra identità nazionale e internazionalismo. Roma, Carocci, 2005, 271 p. (Studi storici Carocci, 88. Storia internazionale del XX secolo, 6).

7334. GEHLER (Michael). Österreichs Außenpolitik der Zweiten Republik: von der alliierten Besatzung bis zum Europa des 21. Jahrhunderts. Innsbruck, Studien-Verlag, 2005, 2 vol., 1292 p. (ill., maps).

7335. HAHN (Peter L.). Crisis and crossfire: the United States and the Middle East since 1945. Washington, Potomac Books, 2005, XIX-223 p. (ill., maps). (Issues in the history of American foreign relations).

7336. HOLLMANN (Anna). Die Schweizer und Europa: Wilhelm Tell zwischen Bern und Brüssel. Baden-Baden, Nomos, 2005, 151 p.

7337. HUGHES (R. Gerald). "Possession is nine tenths of the law": Britain and the boundaries of Eastern Europe since 1945. *Diplomacy & statecraft*, 2005, 16, 4, p. 723-747.

7338. HURST (Steven). Cold War US foreign policy: key perspectives. Edinburgh, Edinburgh U. P., 2005, 188 p.

7339. Images aux frontières. Représentations et constructions sociales et politiques. Palestine, Jordanie (1948-2000). Sous la dir. de Stéphanie LATTE ABDALLAH. Beyrouth, Ifpo, 2005, 371 p.

7340. India-Pakistan conflict (The): an enduring rivalry. Ed. by T.V. PAUL. Cambridge, Cambridge U. P., 2005, XIV-273 p. (ill).

7341. Italia e Germania 1945–2000: la costruzione dell'Europa. A cura di Gian Enrico RUSCONI e Hans WOLLER; premessa di Giorgio CRACCO. Bologna, Il Mulino, 2005, XIII-525 p.

7342. JEŘÁBEK (Vojtěch). Českoslovenští uprchlíci ve studené válce. Dějiny American Fund for Czechoslovak Refugees. (Czechoslovak Refugees in the Cold War. The history of the American Fund for Czechoslovak Refugees). Brno, Stilus; Ústav pro soudobé dějiny AV ČR, 2005, 296 p. (photogr.). (Prameny a studie k dějinám československého exilu 1948–1989, 8).

7343. KRONVALL (Olof), PETERSSON (Magnus). Svensk säkerhetspolitik i supermakternas skugga 1945–1991. (La politique de sécurité de la Suède à l'ombre des grandes puissances 1945–1991). Stockholm, Santérus, 2005, 213 p. (pl.).

7344. LAMMERS (Karl Christian). Hvad skal vi gøre ved tyskerne bagefter. Det dansk-tyske forhold efter 1945. (Que faire avec les Allemands après. Les relations germano-danoises après 1945). København, Schønberg, 2005, 322 p.

7345. LE DORH (Marc). Les démocrates chrétiens français face à l'Europe: mythes et réalités. Paris, L'Harmattan, 2005, 559 p. (Logiques juridiques).

7346. MAC SHERRY (J. Patrice). Predatory states. Operation Condor and covert war in Latin America. Lanham, Rowman & Littlefield, 2005, XXX-284 p.

7347. MILLER (Rory). Ireland and the Palestine question: 1948–2004. Dublin a. Portland, Irish Academic Press, 2005, XI-266 p.

7348. MITCHELL (Otis C.). The cold war in Germany: overview, origins, and intelligence wars. Lanham, University Press of America, 2005, 256 p.

7349. MÖLLER (Kay). Die Aussenpolitik der Volksrepublik China 1949–2004: eine Einführung. Wiesbaden, VS Verlag für Sozialwissenschaften, 2005, 280 p. (ill.). (Lehrbuch (Studienbücher Aussenpolitik und internationale Beziehungen).

7350. NASHEL (Jonathan). Edward Lansdale's Cold War. Amherst, University of Massachusetts Press, 2005, XII-278 p. (ill., ports.). (Culture, politics, and the cold war).

7351. Neutralität: Chance oder Chimäre? Konzepte des dritten Weges für Deutschland und die Welt 1945–1990. Hrsg. v. Dominik GEPPERT und Udo WENGST; herausgegeben im Auftrag des Deutschen Historischen Instituts London und des Instituts für Zeitgeschichte München-Berlin. München, R. Oldenbourg, 2005, 304 p.

7352. Other Germany (The). Perceptions and influences in British-East German relations, 1945–1990. Hrsg. v. Stefan BERGER und Norman LAPORTE. Augsburg, Wissner-Verlag, 2005, 343 p. (Beiträge zur England-Forschung: Schriftenreihe des Arbeitskreis Deutsche England-Forschung, 52).

7353. PAN (Liang). The United Nations in Japan's foreign and security policymaking, 1945–1992: national security, party politics, and international status. Cambridge a. London, Harvard University Asia Center, 2005, XV-384 p. (Harvard East Asian monographs, 257).

7354. Pol'sha-SSSR. 1945–1989: Izbrannye politicheskie problemy, nasledie proshlogo: Kollektivnaja monografija. (Poland and the USSR, 1945–1989: Selected political problems and the heritage of the past: A collective monograph). RAN, In-t rossijskoj istorii; Polska Akademia nauk, In-t historii. Ed. Eugeniusz DURACINSKI, Andrej N. SAKHAROV. Moskva, IRI RAN, 2005, 446 p. (plates; bibl. incl.).

7355. Relations internationales et stratégie de la guerre froide à la guerre contre le terrorisme. Centre de recherches en histoire internationale et atlantique, Université de Nantes; textes réunis par Frédéric BOZO. Rennes, Presses universitaires de Rennes, 2005, 165 p. (map). (Enquêtes & documents, 31).

7356. SCHLIENGER (Micheline). Vers l'éclatement du conflit franco-vietnamien, Paris, Mare et Martin, 2005, 304 p.

7357. STAMOVA (Marijana). Albanskijat vǎpros na Balkanite (1945–1981). (Le problème Albanais aux Balkans, 1945–1981). Veliko Tǎrnovo, Faber, 2005, 362 p. (maps).

7358. STANCIU (Cezar). România stalinistă în relațiile internaționale. (Stalinist Romania in international relations). Studii și Materiale de Istorie Contemporană. Academia Română, 2005, 4, p. 73-88.

7359. STOKES (Doug). America's other war: terrorizing Colombia. London a. New York, Zed, 2005, 147 p.

7360. THORNTON (Martin). Times of heroism, times of terror: American presidents and the Cold War. Westport, Praeger Publishers, 2005, XII-175 p.

7361. Traité (Le) de l'Elysée et les relations franco-allemandes: 1945–1963–2003. Dirigé par Corinne DEFRANCE et Ulrich PFEIL; préface de Claudie HAIGNERÉ et Peter MÜLLER. Paris, CNRS éditions, 2005, 268 p. (CNRS histoire).

7362. VAN DER HEYDEN (Ulrich). Zwischen Solidarität und Wirtschaftsinteressen: die "geheimen" Beziehungen der DDR zum südafrikanischen Apartheidregime. Münster, Lit, 2005, 181 p. (DDR und die Dritte Welt, 7).

7363. Vie parallele: Italia e Germania 1944-2004 = Parallele Wege: Italien und Deutschland 1944–2004. A cura di Renato CRISTIN. Frankfurt am Main e Oxford, P. Lang, 2005, 163 p. (Italien in Geschichte und Gegenwart, 23).

7364. VIGEZZI (Brunello). The British Committee on the Theory of International Politics, 1954–1985. The rediscovery of history. Milano, Edizioni Unicopli, 2005, XXVIII-440 p.

7365. WEGENER FRIIS (Thomas). Den usynlige front. (Le front invisible [L'espionnage de la RDA au Danemark pendant la guerre froide]). København, Lindhardt og Ringhof, 2005, 433 p.

7366. WESTAD (Odd Arne). The global Cold War: third world interventions and the making of our times. Cambridge, Cambridge U. P., 2005, XIV-484 p. (ill., maps).

7367. WESTERN (Jon). Selling intervention and war. The presidency, the media, and the American public. Baltimore a. London, Johns Hopkins U. P., 2005, XI-305 p.

Cf. n[os] 3573, 3953, 4460, 4543, 4662, 4781, 7057, 7256

b. 1945–1956

** 7368. British documents on foreign affairs: reports and papers from the Foreign Office confidential print. General eds. Paul PRESTON and Michael PARTRIDGE. Pt. 5. From 1951 through 1956. Ser. A. The Soviet Union and Finland, 1951. Ed. Daniel KOWALSKY. Vol. 1. The Soviet Union and Finland, 1951. Ser. B. Near and Middle East 1951. Ed. by Bülent GÖKAY. Vol. 1. Afghanistan, Persia and Iraq, 1951. Vol. 2. Turkey, Jordan, Arabia, Lebanon, Israel and Syria, 1951. Ser. C. North America 1951. Ed. by Peter BOYLE.

Vol. 1. United States, 1951. Ser. D. Latin America 1951. Ed. by James DUNKERLEY. Vol. 1. Latin America, 1951. Ser. E. Asia, 1951. Ed. by Anthony BEST. Vol. 1. China, Japan and Korea, 1951. Vol. 2. Siam, Burma, South-East Asia and the Far East, Indo-China, Indonesia, Nepal and The Philippines, 1951. Ser. F. Europe 1951. Ed. by Piers LUDLOW. Vol. 1. Germany, France and Western Europe, 1951. Vol. 2. Poland, Bulgaria, Czechoslovakia, Hungary, Roumania, Yugoslavia and Albania and Eastern Europe (General), 1951. Vol. 3. Austria, Belgium and Luxembourg, Denmark, Iceland, The Netherlands, Norway and Sweden, 1951. Vol. 4. Greece, Italy and Trieste, Portugal, Spain, and The Vatican, 1951. Ser. G. Africa 1951. Ed. By Peter WOODWARD. Vol. 1. Ethiopia, Egypt, Morocco, Libya and Africa (General), 1951. Bethesda, LexisNexis, 2005, 12 vol., XIV-401 p., XIII-392 p., XIV-420 p., XIII-406 p., XVI-586 p., XVI-534 p., XVI-418 p., XIV-555 p., XIV-456 p., XIV-362 p., XIV-368 p., XV-341 p.

** 7369. Documenti diplomatici italiani (I). Undicesima serie. Vol. 1. 1948–1953. 8 maggio–31 dicembre 1948. Ministero degli Affari Esteri, Commissione per la Pubblicazione dei Documenti Diplomatici. Roma, Istituto poligrafico e zecca dello stato, 2005, LXXI-1253 p.

** 7370. Eden-Eisenhower correspondence (The), 1955–1957. Ed. by Peter G. BOYLE. Chapel Hill, University of North Carolina Press, 2005, 230 p. (ill.).

** 7371. Iratok a magyar – szovjet kapcsolatok történetéhez, 1944. október–1948. június: dokumentumok. (Writings about the history of Hungarian- Soviet relationship, 1944, October–1948, July: Documentaries). Szerkesztette VIDA István. Budapest, Gondolat Kiadó, 2005, 384 p. (Iratok a magyar diplomácia történetéhez).

** 7372. LEDOVSKIJ (A.M.). SSSR, SShA i kitajskaja revoljutsija glazami ochevidtsa. (The USSR, the USA and the Chinese Revolution as seen by an eye-witness, 1946–1949). RAN, In-t Dal'nego Vostoka. Moskva, IDV, 2005, 189 p.

** 7373. Midst (In the) of events: the Foreign Office diaries and papers of Kenneth Younger, February 1950 October 1951. Ed. by Geoffrey WARNER. London a. New York, Routledge, 2005, XV-123 p. (British politics and society).

** 7374. Politik (Die) der Sowjetischen Militäradministration in Deutschland (SMAD): Kultur, Wissenschaft und Bildung 1945–1949: Ziele, Methoden, Ergebnisse, Dokumente aus russischen Archiven. Im Auftrag der Gemeinsamen Kommission zur Erforschung der jüngeren Geschichte der deutsch-russischen Beziehungen; hrsg. v. Horst MÖLLER und Alexandr O. TSCHUBARJAN; in Zusammenarbeit mit Wladimir P. KOSLOW, Sergei W. MIRONENKO und Hartmut WEBER; verantwortliche Bearbeiter, Jan FOITZIK, Natalja P. TIMOFEJEWA. München, K. G. Saur, 2005, 468 p. (Texte und Materialien zur Zeitgeschichte, 15).

** 7375. Rote Armee (Die) in Österreich: sowjetische Besatzung, 1945–1955 = Krasnaja Armija v Avstrii: sovetskaja okkupacija, 1945–1955. Band 1. Beiträge. Hrsg. v. Stefan KARNER und Barbara STELZL-MARX. Band 2. Dokumente. Hrsg. v. Stefan KARNER; Barbara STELZL-MARX; Alexander TSCHUBARJAN. Graz, Verein zur Förderung der Forschung von Folgen nach Konflikten und Kriegen u. Wien, Oldenbourg, 2005, 2 vol., 888 p., 979 p.

** 7376. Sowjetische Politik in Österreich 1945–1955: Dokumente aus russischen Archiven = Sovetskaja politika v Avstrii 1945–1955 gg. Österreichische Akademie der Wissenschaften, Philosophisch-Historische Klasse, Historische Kommission. Hrsg. v. Wolfgang MUELLER, Arnold SUPPAN, Norman M. NAIMARK und Gennadij BORDJUGOV. Wien, Verlag der Österreichische Akademie der Wissenschaften, 2005, 1119 p. (Fontes rerum Austriacarum: Abteilung 2, Diplomataria et acta, 93).

** 7377. Vojna in mir na Primorskem: od kapitulacije Italije leta 1943 do Londonskega memoranduma leta 1954. (Guerra e pace nel Litorale: dalla capitolazione dell'Italia nel 1943 al memorandum di Londra nel 1954). Zbrali in uredili Jože PIJEVEC, Gorazd BAJC, Borut KLABJAN. Koper, Založba Annales, 2005, 511 p. (ill.).

7378. ALDOUS (Richard). Macmillan, Eisenhower and the Cold War. Dublin, Four Courts Press, 2005, 205 p.

7379. ALMEIDA (B. Hamilton). Sob os olhos de Perón: o Brasil de Vargas e as relações com a Argentina. Rio de Janeiro, Record, 2005, 335 p. (ill., facsims., ports.).

7380. ALONSO VÁZQUEZ (Francisco Javier). La alianza de dos generalísimos: relaciones diplomáticas Franco-Trujillo. Santo Domingo, Fundación García Arévalo, 2005, XXII-744 p. (ill.). (Serie Documental, 6).

7381. ANTONOVA (Liudmila). La politique étrangère de l'Union Soviétique d'après les notes internes du Présidium du Comité central du Parti communiste de l'URSS (1954–1956). *Revue d'histoire diplomatique*, 2005, 119, 1, p. 19-42.

7382. ARA (Angelo). Lo Staatsvertrag austriaco nelle fonti diplomatiche italiane. *Rivista storica italiana*, 2005, 117, 3, p. 741-779.

7383. BÉZIAS (Jean Rémy). Prélude au Conseil de l'Europe: la déclaration de Georges Bidault à La Haye (19 juillet 1948). *Guerres mondiales et conflits contemporains*, 2005, 53, 220, p. 115-128.

7384. BRITS (J. P.). Tiptoeing along the apartheid tightrope: the United States, South Africa, and the United Nations in 1952. *International history review*, 2005, 27, 4, p. 754-779.

7385. BYSTROVA (Nina E.). SSSR i formirovanie voenno-blokovogo protivostojanija v Evrope (1945–

1955 gg.). (The USSR and the formation of the face-off of military blocks in Europe, 1945–1955). RAN, In-t rossijskoj istorii. Book 1. Na puti k konfrontatsii, 1945–1948 gg. (On the way to confrontation, 1945–1948). Book 2. Pik konfrontatsii i kodifikatsija blockov, 1949–1955 gg. (The peak of confrontation and the codification of the blocks, 1949–1955). Moskva, IRI RAN, 2005, 541 p. (bibl. p. 217-236, 504-523; pers. ind. p. 527-534).

7386. CASEY (Steven). Selling NSC-68: the Truman administration, public opinion, and the politics of mobilization, 1950–1951. *Diplomatic history*, 2005, 29, 4, p. 655-690.

7387. COLLADO-SEIDEL (Carlos). España, refugio nazi. Madrid, Temas de Hoy, 2005, 349 p. (Coleccion historia).

7388. CONWAY-LANZ (Sahr). Beyond No Gun Ri: refugees and the United States military in the Korean war. *Diplomatic history*, 2005, 29, 1, p. 49-81.

7389. DIMITRIJEVIĆ (Bojan B.). Od Staljina do Atlantskog pakta: armija u spoljnoj politici Titove Jugoslavije 1945–1958. (Da Stalin al Patto Atlantico: l'esercito nella politica estera della Jugoslavia di Tito). Beograd, Službeni list SCG, 2005, 434 p. (ill.).

7390. DUSSAULT (Éric). La dénazification de l'Autriche par la France: la politique culturelle de la France dans sa zone d'occupation, 1945–1955. Sainte-Foy, Presses de l'Université Laval, 2005, 122 p. (ill., carte).

7391. FEJÉRDY (Gergely). La question de la représentation des intérêts français en Hongrie à la fin et après la Seconde guerre mondiale. *Revue d'histoire diplomatique*, 2005, 119, 2, p. 165-185.

7392. GIORGI (Luigi). Giuseppe Dossetti e la politica estera italiana, 1945–1956: metodo, prospettive, sviluppo. Cernusco sul Naviglio, Scriptorium, 2005, 305 p. (ill.). (Saggi).

7393. GIRZYŃSKI (Zbigniew). Polska-Francja, 1945–1950. Toruń, Mado, 2005, 283 p. (map).

7394. GRANVILLE (Johanna). "Caught with Jam on Our Fingers": Radio Free Europe and the Hungarian Revolution of 1956. *Diplomatic history*, 2005, 29, 5, p. 811-839.

7395. Gunst (Die) des Augenblicks: neuere Forschungen zu Staatsvertrag und Neutralität. Hrsg. v. Manfried RAUCHENSTEINER und Robert KRIECHBAUMER. Wien, Böhlau, 2005, 564 p. (Schriftenreihe des Forschungsinstitutes für politisch-historische Studien der Dr.-Wilfried-Haslauer-Bibliothek, 24).

7396. HUGHES (R. Gerald). Unfinished business from Potsdam: Britain, West Germany, and the Oder-Neisse line, 1945–1962. *International history review*, 2005, 27, 2, p. 259-294.

7397. HUMBLOT (Guillaume). La puissance militaire soviétique, vue par les attachés militaires français à Moscou (1945–1953). *Guerres mondiales et conflits contemporains*, 2005, 53, 218, p. 101-114.

7398. IOANID (Radu). The ransom of the Jews. The story of the extraordinary secret bargain between Romania and Israel. Chicago, Ivan R. Dee, 2005, XVIII-217 p.

7399. Israel-Österreich: von den Anfängen bis zum Eichmann-Prozess 1961. Hrsg. v. Sabine FALCH und Moshe ZIMMERMANN. Innsbruck, Studien, 2005, 241 p. (Österreich-Israel-Studien, 3).

7400. JONES (Matthew). A "segregated" Asia? Race, the Bandung Conference, and Pan-Asianist fears in American thought and policy, 1954–1955. *Diplomatic history*, 2005, 29, 5, p. 841-868.

7401. KARP (Candace). Missed opportunities: US diplomatic failures and the Arab-Israeli conflict, 1947–1967. Claremont, Regina Books, 2005, XI-309 p. (maps).

7402. KENT (John). United States reactions to empire, colonialism, and cold war in Black Africa, 1949–1957. *Journal of imperial and Commonwealth history*, 2005, 33, 2, p. 195-220.

7403. KISATSKY (Deborah). The United States and the European right, 1945–1955. Columbus, Ohio State U. P., 2005, XIV-237 p.

7404. KNIGHT (Amy). How the Cold War began. The Gouzenko affair and the hunt for Soviet spies. Toronto, McLelland & Stewart, 2005, IX-358 p.

7405. Konrad Adenauer und Frankreich, 1949–1963: Stand und Perspektiven der Forschung zu den deutsch-französischen Beziehungen in Politik, Wirtschaft und Kultur. Hrsg. v. Klaus SCHWABE. Bonn, Bouvier, 2005, VIII-266 p. (Rhöndorfer Gespräche, 21).

7406. KUČERA (Jaroslav). Žralok nebude nikdy tak silný. Československá zahraniční politika vůči Německu 1945–1948. (Der Hai wird nie so stark sein. Tschechoslowakische Deutschlandpolitik 1945–1948). Praha, Argo, 2005, 198 p. (Historické myšlení, 24).

7407. KWIATKIEWICZ (Piotr). Mocarstwa wobec Iraku w latach 1945–1967. (Powers against Iraq in the years 1945–1967). Toruń, Wydawnictwo Adam Marszałek, 2005, 369 p.

7408. LEVINE (Alan J.). Stalin's last war: Korea and the approach to World War III. Jefferson a. London, McFarland, 2005, VIII-320 p.

7409. LIEBREICH (Fritz). Britain's naval and political reaction to the illegal immigration of Jews to Palestine, 1945–1948. London, Routledge, 2005, XXIII-370 p. (ill.).

7410. LONG (Stephen John). Containing liberation: the US cold war strategy towards Eastern Europe and the Hungarian revolution of 1956. Birmingham, University of Birmingham, 2005, 102 p.

7411. MAC KENZIE (Brian Angus). Remaking France: Americanization, public diplomacy, and the Marshall

Plan. New York a. Oxford, Berghahn Books, 2005, XII-259 p. (ill.). (Explorations in culture and international history, 2).

7412. MACHINANDIARENA DE DEVOTO (Leonor). Las relaciones con Chile durante el peronismo, 1946–1955. Buenos Aires, Lumière, 2005, 669 p.

7413. MILLETT (Allan R.). The war for Korea, 1945–1950: a house burning. Lawrence, University Press of Kansas, 2005, XVIII-348 p. (ill., maps). (Modern war studies).

7414. MUELLER (Wolfgang). Die sowjetische Besatzung in Österreich 1945–1955 und ihre politische Mission. Wien, Böhlau, 2005, 300 p. (ill., ports.).

7415. National security legacy (The) of Harry S. Truman. Truman Legacy Symposium (1st: 2003: Key West, Fla.). Ed. by Robert P. WATSON, Michael J. DEVINE and Robert J. WOLZ. Kirksvikke, Truman State U. P., 2005, XVI-198 p. (ill.). (Truman legacy series, 1).

7416. ONOZAWA (Toru). Formation of American regional policy for the Middle East, 1950–1952: the Middle East command concept and its legacy. *Diplomatic history*, 2005, 29, 1, p. 117-148.

7417. Österreichische Staatsvertrag 1955 (Der): internationale Strategie, rechtliche Relevanz, nationale Identität = The Austrian State Treaty 1955: international strategy, legal relevance, national identity. Hrsg.v. Arnold SUPPAN, Gerald STOURZH und Wolfgang MUELLER. Wien, Verlag der Österreichischen Akademie der Wissenschaften, 2005, 1019 p (Archiv für österreichische Geschichte, 140).

7418. Reports (The) of the last British consul in Trabzon, 1949–1956: a foreigner's perspective on a region in transformation. Ed. Christopher HARRIS. İstanbul, Isis, 2005, 389 p.

7419. REŽEK (Mateja). Med resnicnostjo in iluzijo: slovenska in jugoslavanska politika v desetletju po sporu z Informbirojem 1948–1958. (Between reality and illusion: Slovenian and Yugoslav policy in the decade after conflict with Cominform 1948–1958). Ljubljana, Modrijan, 2005, 224 p. (ill., port.).

7420. RUANE (Kevin). SEATO, MEDO, and the Baghdad pact. Anthony Eden, British foreign policy and the collective defense of Southeast Asia and the Middle East, 1952–1955. *Diplomacy & statecraft*, 2005, 16, 1, p. 169-199.

7421. SALE (Giovanni). De Gasperi, gli USA e il Vaticano: all'inizio della guerra fredda. Milano, Jaca book, 2005, XLIII-449 p. (Di fronte e attraverso, 694).

7422. SAUVAGE (Jean-Christophe). Les relations diplomatiques entre la France et le Saint-Siège sous la IVe république. *Revue d'histoire diplomatique*, 2005, 119, 1, p. 43-74.

7423. SHEN (Zhihua). Yijiuwuliu nian shiyue de weiji: Zhongguo de jiaosi yu yingxiang. "Poxiong shijian yu Zhongguo" yanjiu zhi yi. (The October 1956's crisis: an exploration of China's role and influence). *Lishi yanjiu*, 2005, 2, p. 119-143.

7424. SOL GOLDSTEIN (Cora). Before the CIA: American actions in the German fine arts (1946–1949). *Diplomatic history*, 2005, 29, 5, p. 747-778.

7425. STEININGER (Rolf). Der Staatsvertrag: Österreich im Schatten von deutscher Frage und Kaltem Krieg 1938–1955. Innsbruck, Studien Verlag, 2005, 198 p. (ill.).

7426. SWENSON-WRIGHT (John). Unequal allies? United States security and alliance policy toward Japan, 1945–1960. Stanford, Stanford U. P., 2005, XIV-349 p.

7427. TARLING (Nicholas). Britain, Southeast Asia, and the impact of the Korean War. Singapore, Singapore U. P., 2005, XIII-538 p.

7428. TERMIS SOTO (Fernando). Renunciando a todo: el régimen franquista y los Estados Unidos desde 1945 hasta 1963. Madrid, Universidad Nacional de Educación a Distancia y Biblioteca Nueva, 2005, 243 p. (Colección Historia Biblioteca Nueva).

7429. THOMAS (Markus). Finnland zwischen Frieden und Kaltem Krieg: die Aussenpolitik des Präsidenten Paasikivi, 1947–1955. Hamburg, Kovač, 2005, 231 p. (Schriftenreihe Studien zur Zeitgeschichte, 42).

7430. ȚÎRĂU (Liviu). Între Washington și Moscova: politicile de securitate națională ale SUA și URSS și impactul lor asupra României, 1945–1965. (Tra Washington e Mosca: le politiche di sicurezza nazionale di Stati Uniti ed Unione Sovietica e il loro impatto sulla Romania, 1945–1965). Cluj-Napoca, Editura Tribuna, 2005, 555 p.

7431. TREMBATH (Richard). A different sort of war: Australians in Korea, 1950–1953. Melbourne, Australian Scholarly, 2005, XXIV-266 p.

7432. TSANG (Steven Yui-Sang). The Cold War's odd couple: the unintended partnership between the Republic of China and the UK, 1950–1958. London, I. B. Tauris, 2005, XVIII-269 p.

7433. VAN METER (Robert H.). Secretary of state Marshall, General Clay, and the Moscow council of foreign ministers meeting of 1947: a response to Philip Zelikow. [Cf. Bibl. 1997, n° 7706.] *Diplomacy & statecraft*, 2005, 16, 1, p. 139-167.

7434. VAUGHAN (James R.). The failure of American and British propaganda in the Arab Middle East, 1945–1957: unconquerable minds. New York, Palgrave Macmillan, 2005, VIII-316 p. (Cold War history series).

7435. Velike sile i male države u hladnom ratu 1945–1955: slučaj Jugoslavije: zbornik radova sa Međunarodne naučne konferencije, Beograd, 3–4. novembar 2003 = Great Powers and small countries in Cold War, 1945–1955: issue of ex-Yugoslavia: proceedings of the International Scinetific Conference, Belgrade,

November 3rd–4th, 2003. Urednik Ljubodrag DIMIĆ. Beograd, Katedra za istoriju Jugoslavije Filozofskog fakulteta, 2005, 400 p. (ill.).

7436. VILLATOUX (Paul), VILLATOUX (Marie-Catherine). La République et son armée face au péril subversif: guerre et action psychologiques en France, 1945–1960. Préface de Maurice VAÏSSE. Paris, Indes savantes, 2005, 694 p.

7437. Vojna in mir na Primorskem: od kapitulacije Italije leta 1943 do Londonskega memoranduma leta 1954. (War and peace in the Primorska region: from the capitulation of Italy in 1943 to the Memorandum of London in 1954). Ed. Jože PIRJEVEC, Goradz BAJC, Borut KLABJAN. Koper, Univerza na Primorskem, Znanstveno-raziskovalno središče, Založba Annales, Zgodovinsko društvo za južno Primorsko, 2005, 511 p. (Knjižnica Annales Majora).

7438. Wandel und Integration: deutsch-französische Annäherungen der fünfziger Jahre = Mutations et integration: les rapprochements franco-allemands dans les annees cinquante. Hrsg. v. Hélène MIARD-DELACROIX und Rainer HUDEMANN. München, Oldenbourg, 2005, 463 p. (ill.).

7439. WOLSZA (Tadeusz). Za żelazną kurtyną: Europa Środkowo-Wschodnia, Związek Sowiecki i Józef Stalin w opiniach polskiej emigracji politycznej w Wielkiej Brytanii 1944/1945–1953. (Behind the Iron Curtain: Central and Eastern Europe, Soviet Union and Joseph Stalin in the opinions of the Polish political exile in the UK 1944/1945–1953). Warszawa, Instytut Historii PAN, 2005, 286 p. (ill.).

7440. YUSTE DE PAZ (Miguel Ángel). La II República Española en el exilio en los inicios de la Guerra Fría (1945–1951). Madrid, Fundación Universitaria Española, 2005, 336 p. (ill.). (Publicación de la Fundación Universitaria Española, 8. Archivo II República en el exilio).

7441. ZHANG (Hong Bo). Nihon-gun no Shansei zanryū ni miru sengo-shoki chū-niti kankei no keisei. (Presence of Japanese-military in post-war Shanxi, 1945). *Hitotsubashi Ronsō*, 2005, 134, 2, p. 125-146.

7442. ZNAMIEROWSKA-RAKK (Elżbieta). Federacja Słowian południowych w polityce Bułgarii po II wojnie światowej: korzenie, próby realizacji, upadek. (La Federazione degli Slavi del sud nella politica della Bulgaria dopo la Seconda guerra mondiale: radici, tentativi di attuazione, caduta). Warszawa, Wydawnictwo "Neriton", 2005, 492 p. (maps).

7443. ZUREK (Robert). Zwischen Nationalismus und Versöhnung: die Kirche und die deutsch-polnischen Beziehungen 1945–1956. Köln, Weimar u. Wien, Böhlau, 2005, XIII-413 p. (Forschungen und Quellen zur Kirchen und Kuturgeschichte Ostdeutschlands, 35).

b. Addendum 2004

7444. MARX (Thomas Christoph). Zwischen Schwert und Schild. Die US-Streitkräfte in Deutschland 1953 bis 1963 und die Umsetzung der Militärstrategie der USA. Münster, Lit, 2004, XXI-502 p. (Forschungen zur Geschichte der Neuzeit. Marburger Beiträge, 8).

Cf. n^{os} *3680, 4262, 4434, 6613, 7120, 7194, 7195, 7248, 7446, 7452, 7462, 7464, 7472, 7473, 7481, 7492, 7501, 7505, 7508, 7509, 7514, 7517, 7519, 7527, 7532, 7543, 7549, 7552, 7560*

c. From 1956

** 7445. Akten zur auswärtigen Politik der Bundesrepublik Deutschland. 1974. Bd. 1. 1. Januar bis 30. Juni 1974. Bd. 2. 1. Juli bis 31. Dezember 1974. Wissenschaftlicher Leiterin: Ilse Dorothee PAUTSCH; Bearbeiter: Daniela TASCHLER, Fabian HILFRICH und Michael PLOETZ. München, R. Oldenbourg, 2005, 2 vol., LXXVIII-1805 p.

** 7446. BOSSUAT (Gérard). Faire l'Europe sans défaire la France: 60 ans de politique d'unité européenne des gouvernements et des présidents de la République française (1943–2003). Bruxelles, P.I.E. et Oxford, P. Lang, 2005, 630 p. (Euroclio. Etudes et documents, 30).

** 7447. Documents on Australian foreign policy. Australia and the formation of Malaysia: 1961–1966. Ed. by Moreen DEE. Barton, A.C.T., Dept. of Foreign Affairs and Trade, 2005, XXXII-651 p. (maps).

** 7448. Foreign relations of the United States, 1964–1968. General editor Edward C. KEEFER. Vol. 32. Dominican Republic; Cuba; Haiti; Guyana. Ed. by Daniel LAWLER and Carolyn YEE. Washington, United States Government Printing Office, 2005, XXXV-992 p. (Department of State publication, 11173).

** 7449. Foreign relations of the United States, 1969–1976. General editor, Edward C. KEEFER. Vol. 11. South Asia crisis, 1971. Ed. by Louis J. SMITH. Washington, United States Government Printing Office, 2005, XXXIII-900 p. (Department of State publication, 11199).

** 7450. FROMENT-MEURICE (Henri). Journal d'Asie. Chine-Inde-Indochine-Japon, 1969–1975. Paris, L'Harmattan, 2005, 488 p.

** 7451. Macmillan-Eisenhower correspondence (The), 1957–1969. Ed. by E. Bruce GEELHOED and Anthony O. EDMONDS; with the assistance of Michael DAVISON. Basingstoke, Palgrave Macmillan, 2005, IX-436 p.

** 7452. Polityka i dyplomacja polska wobec Niemiec. T. 1. 1945–1970. (Politica e diplomazia polacca verso la Germania. Wstęp, wybór i opracowanie dokumentów Mieczysław TOMALA. Warszawa, Dom Wydawniczy "Elipsa", 2005, 434 p.

** 7453. Polskie dokumenty dyplomatyczne: 1972. (Documenti diplomatici polacchi: 1972). Redaktor Włodzimierz BORODZIEJ; współpraca Piotr DŁUGOŁĘCKI. Warszawa, Polski Instytut Spraw Międzynarodowych, 2005, XXXVIII-787 p.

7454. Abgrund (Vor dem). Die Streitkräfte der USA und der UdSSR sowie ihrer deutschen Bündnispartner in der Kubakrise. Hrsg. v. Dimitrij N. FILIPPOVYCH und Matthias UHL. München, Oldenbourg, 2005, XIV-265 p. (Schriftenreihe der Vierteljahrshefte für Zeitgeschichte).

7455. ALI (S. Mahmud). US-China Cold War collaboration, 1971–1989. London a. New York, Routledge, 2005, XV-276 p. (ill.). (Routledge studies in the modern history of Asia, 31).

7456. ANDERSON (David L.). The Vietnam War. Basingstoke, Palgrave Macmillan, 2005, IX-152 p. (map). (Twentieth-century wars).

7457. ASHTON (Nigel J.). 'A "Special Relationship" sometimes in spite of ourselves': Britain and Jordan, 1957–1973. *Journal of imperial and Commonwealth history*, 2005, 33, 2, p. 221-244. – IDEM. Harold Macmillan and the "Golden Days" of Anglo-American relations revisited, 1957–1963. *Diplomatic history*, 2005, 29, 4, p. 691-723.

7458. ASHTON (Stephen Richard). Mountbatten, the royal family, and British influence in post-independence India and Burma. *Journal of imperial and Commonwealth history*, 2005, 33, 1, p. 73-92.

7459. BAEV (Jordan). Bulgarian policy towards Greek military junta regime, 1967–1974. *Balkan Studies*, 2005, 4, p. 147-156.

7460. BAIN (Mervyn J.). Cuba–Soviet relations in the Gorbachev era. *Journal of Latin American studies*, 2005, 37, 4, p. 769-791.

7461. BALAWYDER (Aloysius). On the road to freedom: Canadian-East European relations, 1963–1991. Boulder, East European Monographs, 2005, VIII-166 p. (map). (East European monographs, 669).

7462. BANU (Florian). Activitatea Direcției de Informații externe a Securității – între atribuțiile oficiale și acțiunile reale (1951–1989). (The activity of Foreign Intelligence Division of the Securitate – between official responsabilities and actual actions, 1951–1989). *Cetatea Biharei. Revistă de Cultură și Istorie Militară*, 2005, 1, p. 99-111.

7463. BAULON (Jean Philippe). Mai 68 et la réconciliation franco-américaine. Les vertus diplomatiques d'une crise intérieure. *Guerres mondiales et conflits contemporains*, 2005, 53, 218, p. 115-132.

7464. BIALER (Uri). Cross on the star of David: the Christian world in Israel's foreign policy, 1948–1967. Bloomington, Indiana U. P., 2005, XIV-240 p. (ill). (Indiana series in Middle East studies).

7465. BIŚVĀSA (Sukumāra). Japan-Bangladesh relations, 1972–1990. Dhaka, Mowla Brothers, 2005, XV-215 p. (ill.).

7466. BLACKWELL (Stephen). The British intervention in Jordan, 1958. London, Routledge, 2005, 224 p.

7467. BOHNING (Don). The Castro obsession: U.S. covert operations against Cuba, 1959–1965. Dulles, Brassey's, 2005, 320 p. (ill.).

7468. BOUSSELHAM (Abdelkader). Regards sur la diplomatie algérienne. Préf. by Mohamed BEDJAOUI. Algers, Casbah, 2005, 396 p.

7469. BOZO (Frédéric). Mitterrand, la fin de la guerre froide et l'unification allemande: de Yalta à Maastricht. Paris, Jacob, 2005, 519 p.

7470. BRAZINSKY (Gregg Andrew). From pupil to model: South Korea and American development policy during the early Park Chung Hee era. *Diplomatic history*, 2005, 29, 1, p. 83-115.

7471. BRENCHLEY (Frank). Britain, the Six Day War and its aftermath. London, I.B. Tauris, 2005, XXIV-184 p. (International library of twentieth century history, 3).

7472. COHEN (Michael Joseph). Strategy and politics in the Middle East, 1954–1960: defending the northern tier. London, Frank Cass, 2005, XIII-272 p. (maps).

7473. DAUER (Richard P.). A North-South mind in an East-West world: Chester Bowles and the making of United States Cold War foreign policy, 1951–1969. Westport a. London, Praeger, 2005, XII-273 p. (Contributions to the study of world history, 110).

7474. DEMKER (Marie). Dans på slak lina: Sverige och den grekiska diktaturen 1967–1974. (Sur la corde raide: la Suède et la dictature grecque 1967–1974). Göteborg, Statsvetenskapliga institutionen, Göteborgs universitet, 2005, 52 p. (Arbetsrapport / Forskningsprogrammet Sverige under kalla kriget, 15).

7475. DOBSON (Alan P.). The Reagan administration, economic warfare, and starting to close down the Cold War. *Diplomatic history*, 2005, 29, 3, p. 531-556.

7476. DOCKRILL (Saki Ruth). The end of the Cold War era: the transformation of the global security order. London, Hodder Arnold, 2005, XXIV-280 p.

7477. DRUKS (Herbert). John F. Kennedy and Israel. Westport, Praeger Security International, 2005, IX-183 p. (ill.).

7478. DURAND (Pierre Michel). Le Peace corps en Afrique française dans les années 1960. Histoire d'un succès paradoxal. *Guerres mondiales et conflits contemporains*, 2005, 53, 217, p. 91-104.

7479. DUVAL (Marcel). Un secret bien gardé: la bombe H française. *Revue d'histoire diplomatique*, 2005, 119, 4, p. 379-395.

7480. EKENGREN (Ann-Marie). Olof Palme och utrikespolitiken: Europa och Tredje världen. (Olof Palme et la politique étrangère: l'Europe et le Tiers-monde). Umeå, Boréa, 2005, 275 p.

7481. EL MÉCHAT (Samya). Les relations franco-tunisiennes: histoire d'une souveraineté arrachée, 1955–

1964. Paris, L'Harmattan, 2005, 254 p. (Histoire et perspectives méditerranéennes).

7482. ENGEL (Jeffrey A.). Of fat and thin communists: diplomacy and philosophy in Western economic warfare strategies toward China (and tyrants, broadly). *Diplomatic history*, 2005, 29, 3, p. 445-474.

7483. Etats-Unis (Les) et la fin de la guerre froide. Sous la dir. de Pierre MELANDRI et Serge RICARD. Paris, L'Harmattan, 2005, 213 p. (L'aire anglophone).

7484. FARBER (David). Taken hostage. The Iran hostage crisis and America's first encounter with radical Islam. Princeton a. Oxford, Princeton U. P., 2005, VIII-212 p.

7485. FILIU (Jean-Pierre). Mitterrand et la Palestine: l'ami d'Israël qui sauva par trois fois Yasser Arafat. Paris, Fayard, 2005, 368 p.

7486. FREEDMAN (Lawrence). The official history of the Falklands Campaign. Vol. 1. The origins of the Falklands war. Vol. 2. War and diplomacy. London, Routledge, Taylor & Francis Group, 2005, 2 vol., XIV-253 p., XXX-849 p. (ill., maps). (Government official history series).

7487. GANSER (Daniele). Nato's secret armies. Operation Gladio and terrorism in Western Europe. London a. New York, Frank Cass, 2005, XX-315 p.

7488. GAT (Moshe). The great powers and the water dispute in the Middle East: a prelude to the Six day war. *Middle Eastern studies*, 2005, 41, 6, p. 911-935.

7489. GEORGE (Edward). The cuban intervention in Angola, 1965–1991: from Che Guevara to Cuito Cuanavale. London a. New York, Frank Cass, 2005, XIV-354 p. (maps). (Cass military studies series).

7490. GOH (Evelyn). Constructing the U.S. rapprochement with China, 1961–1974: from "red menace" to "tacit ally". Cambridge a. New York, Cambridge U. P., 2005, XIV-299 p. – EADEM. Nixon, Kissinger, and the "Soviet Card" in the U.S. opening to China, 1971–1974. *Diplomatic history*, 2005, 29, 3, p. 475-502.

7491. GÖKTEPE (Cihat). The Cyprus crisis of 1967 and its effects on Turkey's foreign relations. *Middle Eastern studies*, 2005, 41, 3, p. 431-444.

7492. GRANIERI (Ronald J.). The ambivalent alliance. Konrad Adenauer. The CDU/CSU, and the West, 1949–1966. New York a. Oxford, Berghahn, 2005, XVI-250 p. (Monographs in German history, 9).

7493. Greek-Turkish relations in an era of détente. Ed. by Ali ÇARKOĞLU and Barry RUBIN. London a. New York, Routledge, 2005, VIII-166 p. (ill.). (Turkish studies, 5, 1).

7494. GÜRBEY (Gülistan). Aussenpolitik in defekten Demokratien: gesellschaftliche Anforderungen und Entscheidungsprozesse in der Türkei 1983–1993. Frankfurt am Main u. New York, Campus, 2005, 396 p. (ill.). (Studien der Hessischen Stiftung Friedens- und Konfliktforschung, 46).

7495. HASLAM (Jonathan). The Nixon administration and the death of Allende's Chile: a case of assisted suicide. London a. New York, Verso, 2005, XVI-255 p.

7496. HASLINGER (Peter). Eine Option in Richtung Europa? Österreich im außenpolitischen Kalkül der ungarischen Kommunisten 1956–1989. *Archiv für Sozialgeschichte*, 2005, 45, p. 59-76.

7497. HAYCRAFT (William Russell). Unraveling Vietnam: how American arms and diplomacy failed in Southeast Asia. Jefferson, McFarland & Co., 2005, VII-263 p. (map).

7498. HOLT (Andrew). Lord Home and Anglo-American relations, 1961–1963. *Diplomacy & statecraft*, 2005, 16, 4, p. 699-722.

7499. Israel und Deutschland: dorniger Weg zur Partnerschaft: die Botschafter berichten über vier Jahrzehnte diplomatische Beziehungen, 1965–2005. Hrsg. v. Asher BEN NATAN und Niels HANSEN. Köln, Böhlau, 2005, VI-301 p. (ill.).

7500. KASSIANIDES (Yoann). La politique étrangère américaine à Chypre, (1960–1967). Paris, L'Harmattan, 2005, 204 p. (Histoire et perspectives méditerranéennes).

7501. København-Bonn erklæringerne 1955–2005. De dansk-tyske mindretalserklæringers baggrund, tilblivelse og virkning. (Les déclarations de Copenhague et Bonn 1955–2005. Arrière-plan, mise en œuvre et résultats des accords germano-danois sur les minorités). Red. Jørgen KÜHL. Aabenraa, Institut for Grænseregionsforskning, 2005, 668 p. (ill., cartes).

7502. KOCHAVI (Noam). Insights abandoned, flexibility lost: Kissinger, Soviet Jewish emigration, and the demise of Détente. *Diplomatic history*, 2005, 29, 3, p. 503-530.

7503. KÜNZLI (Jörg). Zwischen Recht und Politik: der rechtliche Handlungsspielraum der schweizerischen Südafrikapolitik (1976–1994). Zürich, Chronos, 2005, 415 p.

7504. LAWRENCE (Mark Atwood). Assuming the burden: Europe and the American commitment to war in Vietnam. Berkeley a. Los Angeles, University of California Press, 2005, XII-358 p. (From Indochina to Vietnam: revolution and war in a global perspective, 1).

7505. MADHOK (Shakti). Sino-Mongolian relations, 1949–2004. New Delhi, Reliance Publishing House, 2005, IX-246 p.

7506. MASTNY (Vojtech). Was 1968 a strategic watershed of the Cold War?. *Diplomatic history*, 2005, 29, 1, p. 149-177.

7507. MATĚJKA (Zdeněk). Jednání o rozpuštění Varšavské smlouvy. (Negotiations on the dissolution of the Warsaw Pact). *Historie a vojenství*, 2005, 54, 3, p. 4-19.

7508. MAWBY (Spencer). British policy in Aden and the protectorates 1955–1967: forwards and backwards. New York, Routledge, 2005, XI-210 p. (ill.).

7509. MIDTGAARD (Kristine). Småstat, magt og sikkerhed: Danmark og FN 1949-1965. (Petit Etat, puissance et sécurité: le Danemark et l'ONU 1949–1965). Odense, Syddansk Universitetsforlag, 2005, 390 p. (ill.). (University of Southern Denmark studies in history and social sciences, 296).

7510. Między Październikiem a Grudniem: polityka zagraniczna doby Gomułki. (Tra ottobre e dicembre: la politica estera al tempo di Gomułka). Pod redakcją Krzysztofa RUCHNIEWICZA, Bożeny SZAYNOK i Jakuba TYSZKIEWICZA. Toruń, Wydawnictwo Adam Marszałek, 2005, 218 p. (Wrocławskie studia z polityki zagranicznej, 2).

7511. MIGEV (Vladimir). Prazhkata prolet '68 i Bălgarija. (Le printemps '68 de Prague et la Bulgarie). Sofija, Iztok-Zapad, 2005, 307 p.

7512. MOURLANE (Stéphane). La guerre d'Algérie dans les relations franco-italiennes (1958–1962). *Guerres mondiales et conflits contemporains*, 2005, 53, 217, p. 77-90.

7513. MÜFTÜLER-BAC (Meltem), GÜNEY (Aylin). The European Union and the Cyprus problem 1961–2003. *Middle Eastern studies*, 2005, 41, 2, p. 281-293.

7514. MUSCHIK (Alexander). Die beiden deutschen Staaten und das neutrale Schweden: eine Dreiecksbeziehung im Schatten der offenen Deutschlandfrage 1949–1972. Münster, Lit, 2005, 298 p. (Nordische Geschichte, 1).

7515. NEU (Charles E.). America's lost war: Vietnam, 1945–1975. Wheeling, Harlan Davidson, 2005, XIX-272 p. (ill.). (The American history series).

7516. ÖKE (Mim Kemal). Kılıç ve ney: 21. yüzyılda uluslararası ilişkiler ve Türk dış politikası. (L'épée et la flûte: Relations internationales et la politique extérieur turque au 21ème siècle). İstanbul, Etkileşim, 2005, 336 p.

7517. OLIVEIRA (Ione). Außenpolitik und Wirtschaftsinteresse in den Beziehungen zwischen Brasilien und der Bundesrepublik Deutschland 1949–1966. Frankfurt am Main u. Oxford, Lang, 2005, 337 p. (Moderne Geschichte und Politik, 19).

7518. OLSCHOWSKY (Burkhard). Einvernehmen und Konflikt: das Verhältnis zwischen der DDR und der Volksrepublik Polen 1980–1989. Osnabrück, Fibre, 2005, 690 p. (Veröffentlichungen der Deutsch-Polnischen-Gesellschaft Bundesverband, 7).

7519. OLTEANU (Constantin), DUȚU (Alesandru), ANTIP (Constantin). România și Tratatul de la Varșovia. Istoric. Mărturii. Documente. Cronologie. (Romania and the Warsaw Pact. History. Testimonies. Documents. Chronology). București, Editura Pro Historia, 2005, 334 p.

7520. OREKHOV (Aleksandr M.). Sovetskij Sojuz i Pol'sha v gody ottepeli: iz istorii sovetsko-pol'skikh otnoshenij. (The Soviet Union and Poland during the Khrushchev's Thaw). Moskva, Indrik, 2005, 327 p. (bibl. p. 310-320; pers. ind. p. 321-327).

7521. ORTUÑO ANAYA (Pilar). Los socialistas europeos y la transición española (1959–1977). Madrid, Marcial Pons Historia, 2005, 283 p.

7522. PANASPORNPRASIT (Chookiat). US-Kuwaiti relations, 1961–1992: an uneasy relationship. London, Routledge, 2005, VII-196 p.

7523. PANKOVITS (József). Fejezetek a magyarolasz politikai kapcsolatok történetéből, 1956–1977. (Chapters about the history of the Hungarian- Italian political relations, 1956–1977). Budapest, Gondolat Kiadó, 2005, 164 p. (ill.). (Magyarország és a világ.. Diplomáciatörténet).

7524. PETERSEN (Tore T.). The decline of the Anglo-American Middle East, 1961–1969: a willing retreat. Brighton, Sussex Academic Press, 2005, XI-181 p.

7525. PFISTER (Roger). Apartheid South Africa and African states: from pariah to middle power, 1961–1994. London, I. B. Tauris, 2005, XVI-248 p. (ill., maps). (International library of African studies, 14).

7526. PRIMICERI (Emanuela). Il sequestro dell' "Achille Lauro" e il governo Craxi: relazioni internazionali e dibattito politico in Italia. Manduria, P. Lacaita, 2005, 172 p. (Biblioteca di storia contemporanea, 50).

7527. RABE (Stephen G.). US intervention in British Guiana. a Cold War story. Chapel Hill, University of North Carolina Press, 2005, 240 p. (The new Cold War history).

7528. RABEL (Roberto Giorgio). New Zealand and the Vietnam war: politics and diplomacy. Auckland, Auckland U. P., 2005, XI-443 p. (ill., ports.).

7529. Reinterpreting the end of the Cold War: issues, interpretations, periodizations. Ed. by Silvio PONS and Federico ROMERO. London, Frank Cass, 2005, VIII-237 p. (Cass series- Cold War history, 6).

7530. România și "Primăvara de la Praga". (La Romania e la 'Primavera di Praga'). Ediție îngrijită și cuvânt înainte de Dan CĂTĂNUȘ. București, Institutul Național pentru Studiul Totalitarismului, 2005, 101 p. (ill.). (Colecția dezbateri).

7531. ROMERO (María Teresa). Venezuela en defensa de la democracia, 1958–1998: el caso de la doctrina Betancourt. Caracas, Fundación para la Cultura Urbana, 2005, XXVI-262 p. (Fundación para la cultura urbana, 41).

7532. ROSENBERG (Victor). Soviet-American relations, 1953–1960: diplomacy and cultural exchange during the Eisenhower presidency. Jefferson a. London, McFarland & Co., 2005, VII-324 p.

7533. RYE OLSEN (Gorm). Danmark, 11. september og den fattige verden. (Le Danemark, le 11 septembre et le tiers-monde). København, Columbus, 2005, 128 p. (ill.).

7534. SCHMITZ (David F.). The Tet Offensive: politics, war, and public opinion. Lanham, Rowman & Littlefield, 2005, XVII-183 p. (ill.). (Vietnam-America in the war years).

7535. SCOTT-SMITH (Giles). Mending the "Unhinged Alliance" in the 1970s: transatlantic relations, public diplomacy, and the origins of the European Union visitors program. *Diplomacy & statecraft*, 2005, 16, 4, p. 749-778.

7536. SHAW (John M.). The Cambodian campaign: the 1970 offensive and America's Vietnam War. Lawrence, University Press of Kansas, 2005, XIV-222 p. (ill., maps). (Modern war studies).

7537. SINGH (Karminder Dhillon). Malaysian foreign policy in the Mahathir era, 1981–2003. Ann Arbour, UMI Dissertation Services, 2005, XVI-446 p.

7538. SKÁLOVÁ (Ivana). Podíl Bulharska na potlačení Pražského jara 1968. (The role of the Bulgaria in the 1968 intervention in Czechoslovakia). Praha, Univerzita Karlova – Filozofická fakulta, 2005, 93 p. (Sešity Ústavu českých dějin UK FF. Řada B, 2).

7539. SOLAÚN (Mauricio). U.S. intervention and regime change in Nicaragua. Lincoln, University of Nebraska Press, 2005, XII-391 p.

7540. Sovětská okupace Československa a její oběti. (Soviet occupation of Czechoslovakia and its victims). Ed. Hynek FAJMON. Brno, Centrum pro studium demokracie a kultury, 2005, 131 p.

7541. STERN (Lewis M.). Defense relations between the United States and Vietnam: the process of normalization, 1977–2003. Jefferson a. London, McFarland & Co., 2005, V-291 p.

7542. SZABÓ (Csaba). A Szentszék és a Magyar Népköztársaság kapcsolatai a hatvanas években. (The relatishionship of the Holy See and the People's Republic of Hungary int he Sixties). Budapest, Szent István Társulat, Magyar Országos Levéltár, 2005, 506 p.

7543. SZALONTAI (Balazs). Kim Il Sung in the Khrushchev era: Soviet-DPRK [Democratic people's republic of Korea] relations and the roots of North Korean despotism, 1953–1964. Washington, Woodrow Wilson Center Press a. Stanford, Stanford U. P., 2005, XXIII-343 p. (Cold War International History Project series).

7544. SZEPTYCKI (Andrzej) Francja czy Europa? Dziedzictwo generała de Gaulle'a w polityce zagranicznej V Republiki. (France or Europe? Heritage of General de Gaulle's foreign policy in the V Republic). Warszawa, Wydawnictwo Naukowe Scholar, Fundacja Studiów Międzynarodowych, 2005, 379 p. (ill.).

7545. TEBINKA (Jacek). Nadzieje i rozczarowania: polityka Wielkiej Brytanii wobec Polski 1956–1970. (Le speranze e le delusioni. La politica della Gran Bretagna verso la Polonia, 1956–1970). Warszawa, Neriton, Instytut Historii PAN, 2005, 433 p.

7546. TELLO DÍAZ (Carlos). El fin de una amistad: la relación de México con la Revolución cubana. México D.F., Planeta, 2005, 204 p. (ill.).

7547. Tio år i EU: vad vet vi och vad vill vi? (La Suède depuis dix ans dans l'Union européenne: quels enseignements et quel avenir?) Ed. par Fredrik BYSTEDT et Ylva NILSSON. Stockholm, SNS förlag, 2005, 111 p.

7548. TVEIT (Odd Karsten). Krig & diplomati: Oslo-Jerusalem 1978–1996. Oslo, Aschehoug, 2005, 761 p. (ill., maps).

7549. WAGNER (Christian). Die "verhinderte" Großmacht? Die Außenpolitik der Indischen Union, 1947–1998. Baden-Baden, Nomos, 2005, 373 p. (Internationale Politik und Sicherheit, 56).

7550. WALKER (Jonathan). Aden insurgency: the savage war in South Arabia 1962–1967. Staplehurst, Spellmount, 2005, XX-332 p. (ill., maps, ports.).

7551. WATERS (Robert), DANIELS (Gordon). The world's longest general strike: the AFL-CIO, the CIA, and British Guiana. *Diplomatic history*, 2005, 29, 2, p. 279-307.

7552. WEGENER FRIIS (Thomas), LINDEROTH (Andreas). DDR og Norden: østtysk-nordiske relationer 1949–1989. (La RDA et la Scandinavie: les relations entre l'Allemagne de l'Est et les pays nordiques 1949–1989). Odense, Syddansk Universitetsforlag, 2005, 363 p. (University of Southern Denmark studies in history and social sciences, 277).

7553. WIEGREFE (Klaus). Das Zerwürfnis: Helmut Schmidt, Jimmy Carter und die Krise der deutschamerikanische Beziehungen. Berlin, Propyläen, 2005, 523 p.

7554. Willy Brandt und Frankreich. Hrsg. v. Horst MÖLLER und Maurice VAÏSSE. München, R. Oldenbourg, 2005, X-286 p. (Schriftenreihe der Vierteljahrshefte für Zeitgeschichte.. Sondernummer).

7555. Wir haben Spuren hinterlassen! Die DDR in Mosambik. Erlebnisse, Erfahrungen und Erkenntnisse aus drei Jahrzehnten. Hrsg. v. Matthias VOß. Münster, Lit, 2005, 604 p. (Die DDR und die Dritte Welt, 6).

7556. WOLVAARDT (Pieter). A diplomat's story: apartheid and beyond, 1969–1998. Alberton, Poole, Galago, 2005, 336 p. (ill.).

7557. XIA (Yafeng). Negotiating at cross-purposes: Sino-American ambassadorial talks, 1961–1968. *Diplomacy & statecraft*, 2005, 16, 2, p. 297-329.

7558. YEILBURSA (Behçet Kemal). The 'Revolution' of 27 May 1960 in Turkey: British policy towards Turkey. *Middle Eastern studies*, 2005, 41, 1, p. 121-151.

7559. ZAVACKÁ (Marína). Kto žije za ostnatým drôtom? Oficiálna zahraničnopolitická propaganda na Slovensku, 1956–1962. Teórie, politické smernice a spoločenská prax. (Wer lebt hinter den Stacheldraht? Offizielle aussenpolitische Propaganda auf der Slowakei 1956–1962. Theorien, politische Anleitungen und die gesellschaftliche Praxis). Bratislava, Ústav politických vied Slovenskej akadémie vied, 2005, 168 p.

c. Addenda 2001–2004

** 7560. Documents diplomatiques suisses = Diplomatische Dokumente der Schweiz. Commission pour la Publication de Documents Diplomatiques Suisses. Vol. 20. 1.IV.1955-28.II.1958. Directeur de la recherche: Antoine FLEURY. Coresponsable: Mauro CERUTTI. Zürich, Chronos Verl., 2004, CIV-423 p.

7561. BAR-JOSEPH (Uri). Ha-Tsofeh she-nirdam: haftaat Yom ha-Kipurim u-mekoroteha. (The watchman fell asleep: the surprise of Yom Kippur and its sources). Lod, Zemorah-Bitan, 2001, 532 p.

Cf. n°s 3751, 3752, 3767, 3957, 4253, 4381, 4434, 4715, 4744, 7370, 7378, 7380, 7389, 7396, 7399, 7401, 7402, 7405, 7407, 7419, 7422, 7426, 7428, 7430, 7432, 7434, 7436, 7438, 7444

R

ASIA

§ 1. General. 7562-7563. – § 2. Western and central Asia. 7564-7581. – § 3. South Asia and Southeast Asia. 7582-7595. – § 4. China. 7596-7801. – § 5. Japan (esp. before 1868). 7802-7849. – § 6. Korea. 7850-7853.

§ 1. General.

7562. Istorija Vostoka: V 6 tomakh. (A history of the East). Vol. 4. Vostok v novoe vremja (konets XVI-II–nachalo XX v.). (The East in the Modern Time, the late 18th–the early 20th century). Book. 2. Ed. Rostislav B. RYBAKOV, Leonid B. ALAEV [et al.]. Moskva, Vostochnaja literatura, 2005, 574 p. (maps; bibl. p. 526-539; ind. p. 540-571).

7563. Time, temporality, and imperial transition. East Asia from Ming to Qing. Ed. by Lynn A. STRUVE. Honolulu, University of Hawai'i Press, 2005, X-300 p. (Asian interactions and comparisons).

Cf. n°s 6738-6759

§ 2. Western and central Asia.

7564. Antropoekologija Tsentral'noj Azii. (Anthropological ecology of Central Asia). Ed. Tat'jana I. ALEKSEEVA. Mosovskij gos. un-t im. M.V. Lomonosova, NII i muzej antropologii; RAN, In-t arkheologii. Moskva, Nauchnyj mir, 2005, 326 p. (ill.; portr.; plates; bibl. p. 315-326). [English summary]

7565. GÜL (Muammer). XIII. ve XIV. yüzyıllarda Doğu ve Güneydoğu Anadolu'da Moğol hâkimiyeti. (Domination mongole en Anatolie orientale aux 13e et 14e siècles). İstanbul, Yeditepe, 2005, 248 p.

7566. HANEDA (Masashi). Isurâmu sekai no sōzō. (Creating the notion of Islamic world). Tokyo, University of Tokyo Press, 2005, 316 p.

7567. ISHIHAMA (Yumiko). Study on Qianlong as Cakravartin, a manifestation of Bodhisattva Mañjuśrī, Tangka. *Bulletin of Waseda Institute for Mongolian Studies*, 2005, 2, p. 19-39.

7568. JACKSON (Peter). The Mongols and the West, 1221–1410. Harlow, Longman, 2005, XXXIV-414 p.

7569. KAWAHARA (Yayoi). Kōkando-han koku ni okeru Marugiran no tora tachi: kyōdankei no seija ichizoku ni kansuru ichikōsatsu. (Margilian's Tūras in the Khoqand Khanate). *Annals of Japan Association for Middle East Studies*, 2005, 20, 2, p. 269-294.

7570. KISAICHI (Masatoshi). Three renowned 'Ulamā' families of Tlemcen: the Maqqarī, the Marzūqī and the 'Uqbānī. *Journal of Sophia Asian studies*, 2005, 22, p. 121-137.

7571. KYCHANOV (Evgenij I.), MEL'NICHENKO (Boris N.). Istorija Tibeta: S drevnejshikh vremen do nashikh dnej. (A history of Tibet from the earliest time to nowadays). Moskva, Vostochnaja literatura, 2005, 351 p. (maps; bibl. p. 319-326; pers. ind. p.327-347).

7572. MORIMOTO (Kazuo). Towards the formation of Sayyido-Sharifology: questioning accepted facts. *Journal of Sophia Asian studies*, 2005, 22, p. 1-13.

7573. MORIYAMA (Teruaki). The vilification and the reverence: particularities of the Syrian Muslim's view on 'Alī in the 9th and 12th centuries reflected in the accounts of the History of Damascus. *Journal of Sophia Asian studies*, 2005, 22, p. 15-30.

7574. NODA (Jin). Ro Shin no hazama no Kazan hân koku: surutan to Shintyō tono kankei wo tyūsin ni. (The Kazakh khanate between the Russian and Qing Empires: mainly on the relations of Kazakh sultans with the Qing dynasty). *Toyo Gakuho*, 2005, 87, 2, p. 29-59 [Eng. Summary].

7575. ÖZDEMIR (H. Ahmet). Moğol istîlâsı ve Abbâsî Devleti'nin yıkılışı: Cengiz ve Hülâgû dönemleri (612–656/1216–1258). [Invasion des Mongoles et renversement de l'Etat Abbaside: époques de Cengiz et de Hulâgû (612–656/1216–1258)]. İstanbul, İz, 2005, 335 p.

7576. SHIMIZU (Kazuhiro). Gunji dorei, Kanryō, Minsyū: abbâsu-tyō kaitaiki no iraku syakai. (Slave soldiers, bureaucrats and the people: Iraqi society during the disintegration of the 'Abbāsid dynasty). Tokyo, Yamakawa Shuppansya, 2005, 247 p.

7577. SUZUKI (Kosetsu). Tokketsu Ashina-shima keifu kō: tokketsu daiichi han koku no han keifu to tōdai orudosu no tokketsu shūdan. (On the genealogical line of Türk's Ashina Simo: the royal genealogy of the first Türkic Qaɣunate and the Ordos region during the Tang period). *Toyo Gakuho*, 2005, 87, 1, p. 37-68 [Eng. Summary].

7578. TACHIBANA (Makoto). Bogudo-han seiken no uchi mongoru tōgō no kokoromi: shirīngoru dōmei wo jirei toshite. (Approaches of Bagd Khaan's government to unifying Inner Mongolia: the case of the Shiliin Gol League). *Toyo Gakuho*, 2005, 87, 3, p. 63-94 [Eng. Summary].

7579. TONAGA (Yasushi). Sufi Saints and Non-Sufi Saints in early Islamic history. *Journal of Sophia Asian studies*, 2005, 22, p. 1-13.

7580. Voennoe delo nomadov Tsentral'noj Azii v Sjan'bijskuju epokhu: Sb. nauch. tr. (The art of war of the nomads of Central Asia in the Xianbei epoch: Articles). Ed. Julij S. KHUDJAKOV, Sergej G. SKOBELEV. Min-vo obrazovanija i nauki RF, Feder. tselevaja programma "Integratsija"; Novosibirskij gos. un-t. Novosibirsk, NGU, 2005, 231 p. (bibl. incl.).

7581. YOKKAICHI (Yasuhiro). Jaruguchi kō: mongoru-teikoku no jūsōteki kokka kōzō oyobi bumpai shisutemu tono kakawari kara. (Study on the Ĵarɣuči, from the standpoint of political structure of the Mongol Empire and distribution system of benefits). *Shigaku Zasshi*, 2005, 114, 4, p. 1-30 [Eng. Summary].

Cf. n[os] 570, 688, 1104, 1328

§ 3. South Asia and Southeast Asia.

7582. BARUA (Pradeep P.). The state at war in South Asia. Lincoln, University of Nebraska Press, 2005, XVI-437 p.

7583. FUJITA (Kōichi). Banguradeshu nōson kaihatsu no naka no kaisō-hendō: Hinkon sakugen no tame no kiso-kenkyū. (Hierarchical changes in the development of Bangladeshi farm villages today: basic studies for poverty reduction). Kyōto, Kyōto U. P., 2005, 287 p.

7584. GUMISAWA (Hideo). "Gotong Royong" gainen no tanjō to henyō: Shokuminchi-ki makki kara Sukaruno-ki made. (Invention and transformation of the "Gotong Royong" concept in Indonesia: from the late colonial period to the Sukarno period). *Ajia Keizai*, 2005, 45, 6, p. 2-29.

7585. IMAI (Keiko) Nihon senryō-ka no Maraya, 1941 nen-1945 nen. (The Japanese occupation of Malaya, 1941–1945). Tōkyō, Kōjinsha, 2005, 410 p.

7586. IWASAKI (Ikuo). Singapōru-kokka no kenkyū: Chitsujo to seichō no seidoka kinō akutâ. (The studies of Singapore state today: institutionalization, function and actor of 'order and growth'). Tōkyō, Fūkyōsha, 2005, 366 p.

7587. KATAOKA (Kei). (Critical edition of the Isvarasiddhi section of Bhatta Jayanta's Nyayamanjiri). *Kiyō*, 2005, 148, p. 57-110.

7588. KITAGAWA (Takako). Komupōt-rijikanku ni okeru kōkyō-seido no dōnyū: rijikanfu-teiki-hōkokusho. (Introduction of public education system into the circonscription résidentielle de Kampot in Cambodia). *Tōnan Ajia*, 2005, 34, p. 80-101 [Eng. Summary].

7589. KOIKE (Makoto). Higashi-Indoneshia no ie-shakai: Sumba no sinzoku to girei. (The family in Eastern Indonesia: kinship and ritual in Sumba). Kyōto, Kōyō-shobō, 2005, 281 p.

7590. KURIHARA (Hirohide). Komintern-system to Indoshina kyōsantō. (The Indochinese Communist Party in the Comintern system). Tokyo, University of Tokyo Press, 2005, 319 p.

7591. LAL (Ruby). Domesticity and power in the early Mughal world. Cambridge, Cambridge U. P., 2005, XV-241 p. (ill.). (Cambridge studies in Islamic civilization).

7592. NAKAJIMA (Takeshi). Nashonarizumu to shūkyō: Gendai Indo no Hindū Nashonarizumu-undō. (Nationalism and religion: Hindu Nationalism in modern India). Yokohama, Shunpūsha, 2005, 380 p.

7593. OTA (Shoichi). 1940 nendai futsu-ryō Indoshina no kōkyō jigyō seisaku: Dokū no seisaku to toshi kenchiku. (The politics of public works in French Indochina in the 1940s: Jean Decoux's policy, towns and architecture). *Kiyō*, 2005, 147, p. 117-158.

7594. SATŌ (Hiroshi). Minami-Ajia ni okeru Minoritie to nanmin: kokumin kokka keisei-ki ni okeru Tōzai-Bengaru. (Minorities and refugees in South Asia: Partition of Bengal and the nation-state formation). *Ajia Keizai*, 2005, 46, 1, p. 2-34.

7595. UEDA (Tomoaki). 19 seiki no Ei-In keizai kankei to nashonarizumu: M.G.Rânadê no keizai-shisō to Igirisu-kan. (Indo-British economic relations and nationalism in the 19th century: M. G. Ranade's economic thought and his view on the British Empire). *Hōgaku Ronsō*, 2005, 157, 3, p. 54-75; 5, p. 78-97.

Cf. n°s 953, 1104, 4361, 4362

§ 4. China.

** 7596. PAN (Guangdan). Zhongguo minzu shiliao huibian. (A collection of historical sources on Chinese nationalities). Tianjin, Tianjing guji chubanshe, 2005, 17, 487 p.

** 7597. [UL'JANOV [Ulyanov] (Mark Ju.).] Bambukovye annaly: Drevnij tekst (Gu Ben' Chzhu Shu Tszi Njan'). (The Bamboo Annals [Gu ben zhu shu ji nian]: The ancient text). Edited, translated, with introduction, commentary and appendix by Mark Ju. UL'JANOV, with participation of Dega V. DEOPIK and A.I. TARKINA. MOSOVSKIJ gos. un-t im. M.V. Lomonosova, In-t stran Azii i Afriki, In-t prakticheskogo vostokovedenija. Moskva, Vostochnaja literatura, 2005, 311 p. (plates; bibl. p. 276-284; ind. p. 285-309). [Table of contents in English; text partly in Chinese]

7598. ATWILL (David G.). The Chinese sultanate: Islam, ethnicity, and the Panthay Rebellion in southwest China, 1856–1873. Stanford, Stanford U. P., 2005, XII-264 p. (ill., maps).

7599. BAI (Limin). Shaping the ideal child: children and their primers in late imperial China. Hong Kong, Chinese U. P., 2005, XXIV-311 p.

7600. Bainian sangcang. Zhongguo Guomindang shi. (History of the Chinese National Party). Edited by Jiaqi MAO. Xiamen, Lujiang chubanshe, 2005, 24, 1537 p.

7601. BELLO (David Anthony). Opium and the limits of empire. Drug prohibition in the Chinese interior, 1729–1850. Cambridge, Harvard University Asia Center, 2005, XXI-361 p. (Harvard East Asian monographs, 241).

7602. BELSKY (Richard). Localities at the center: native place, space, and power in late imperial Beijing. Cambridge, Harvard U. P., 2005, XII-318 p. (Harvard East Asian monographs, 258).

7603. BENITE (Zvi Ben-Dor). The dao of Muhammad. A cultural history of Muslims in late imperial China. Cambridge, Harvard U. P., 2005, XII-280 p. (Harvard East Asian monographs, 248).

7604. BIAN (Morris L.). The making of the state enterprise system in modern China. The dynamics of institutional change. Cambridge, Harvard U. P., 2005, XI-331 p.

7605. CARROLL (John M.). Edge of empires. Chinese elites and British colonials in Hong Kong. Cambridge, Harvard U. P., 2005, XII-260 p.

7606. CHANG (Jincang). Ershiji gushi yanjiu fansi lu. (Rethinking the study of ancient history in the twentieth century). Beijing, Zhongguo shehui kexue chubanshe, 2005, 2, 3, 4, 354 p.

7607. CHANG (Jung), HALLIDAY (Jon). Mao: the unknown story. London, Jonathan Cape, 2005, XVI-814 p. (ill., maps, ports.).

7608. CHE (Wenming). Zhongguo shemiao juchang. (The temple theatres of China). Beijing, Wenhua yishu chubanshe, 2005, 4, 8, 272 p.

7609. CHEN (Baoliang). Mingdai ruxue shengyuan yu difang shehui. (Confucian literati and local society in Ming times). Beijing, Zhongguo shehui kexue chubanshe, 2005, 5, 568 p.

7610. CHEN (Dasheng). Zheng He yu Maliujia. (Zheng He and Malacca). Singapore, Guoji Zheng He xuehui, 2005, 104 p.

7611. CHEN (Hu). Changzheng riji. (A journal of the Long March). Beijing, Chang'an chubanshe, 2005, 4, 486 p.

7612. CHEN (Jiuru). Liu Mingzhuan yu jindai Taiwan chaye de fazhan. (Liu Mingzhuan and the development of tea industry in Taiwan). *Lishi dang'an*, 2005, 1, p. 57-62.

7613. CHEN (Lizhu). Wei zi fengjian kao. (A note on Wei Zi's seignory). *Lishi yanjiu*, 2005, 6, p. 63-73.

7614. CHEN (Xiaoqing), LI (Jifeng), ZHU (Lexian). Yige shidai de ceying: Zhongguo 1931–1945. (Profile of an epoch: China 1931–1945). Guilin, Guangxi shifan daxue chubanshe, 2005, 405 p.

7615. CHEN (Yan). Waijiao, wai ze he paixi: cong "Liang Yan zhengzheng" kan 20 shiji 20 niandai chuqi Beijing zhengfu de waijiao yunzuo. (Foreign policy, foreign loans and factions. The Beijing government's foreign relations at the beginning of the 1920s as seen in the political war between Liang Shizhi and Yan Huiqing). *Jindai shi yanjiu*, 2005, 1, p. 188-211.

7616. CHEN (Yanxiang). 1920 nian qianhou liangci zheng guoquan yundong de yixiang xingtai ji xingcheng yuanyin. (The disparate forms and causes of two national rights movement before and after 1920). *Jindai shi yanjiu*, 2005, 2, p. 38-77.

7617. CHIBA (Masashi). Shin-matsu rikken kaikakuka ni okeru kokka tōgō no saihen to tetsudō. (The interrelation of the reorganization of the state integration and the railway construction under the constitutional reform of the Late Qing Period). *Shigaku Zasshi*, 2005, 114, 2, p. 1-35 [Eng. Summary].

7618. DAI (Lu). Han-Da-Ping gongsi de gangtie xiaoshou yu woguo jindai gangtie shichang (1908–1927). (The Hankou-Daye-Pingxiang company's steel sales and China's modern steel market 1908–1927). *Jindai shi yanjiu*, 2005, 6, p. 47-82.

7619. Dalizhabu, Qingdai Chaher zhasa keqi kao. (A study of Aqay Jasay banner in the Qing). *Lishi yanjiu*, 2005, 5, p. 47-59.

7620. DENG (Ye). Fu Zuoyi zhengzhe zhuanxing guocheng zhong de shuangzhongxing. (The dual nature of Fu Zuoyi's political conversion). *Lishi yanjiu*, 2005, 5, p. 117-131. – IDEM. Lun Guo Gong Chongqing tanpan de zhengzhi xingzhi. (On the political nature of the Chongqing negotiations between the Nationalists and the Communist parties). *Jindai shi yanjiu*, 2005, 1, p. 30-64.

7621. DING (Guangxun). Lishi yu dang'an. (History and archives). Shanghai, Sanlian shudian, 2005, 2, 383 p.

7622. DU (Lihong). 1930 niandai de Beiping chengshi wuwu guanli gaige. (The reform of waste disposal management in Beijing in the 1930s). *Jindai shi yanjiu*, 2005, 5, p. 90-113.

7623. DUTTON (Michael). Policing Chinese politics. A history. Durham, Duke U. P., 2005, XIII-411 p. (Asia-Pacific, culture, politics, society).

7624. FANG (Huawen). 20 shiji Zhongguo fanyi shi. (A history of translation in twentieth century China). Xi'an, Xibei daxue chubanshe, 2005, 3, 2, 651 p.

7625. FANG (Jun). 1931–1945 qinli Riben qin Hua zhanzheng de zuihou yipiren. (The last generation who experienced personally the Japanese invasion of China 1931–1945). Xi'an, Shanxi renmin chubanshe, 5, 483 p.

7626. Fengyun licheng: 1949–1978. (Through trials and hardships: 1949–1978). Ed. by Liping XIE, Jian HUANG and the Archives of the Shanghai Municipality Committee of the Chinese Communist Party). Shanghai, Shanghai guji chubanshe, 2005, 3, 609 p.

7627. FU (Chonglan). Zhongguo yunhe zhuan. (A history of canals in China). Taiyuan, Shanxi renmin chubanshe, 2005, 8, 297 p.

7628. FUNADA (Yoshiyuki). Gendai no meirei bunsho no kaidoku nitsuite. (On the promulgation of written edicts in the Yuan Period). *Tōyōshi Kenkyū*, 2005, 63, 4, p. 36-67 [Eng. Summary].

7629. GALENOVICH (Jurij M.). Smert' Mao Tszeduna. (The death of Mao Zedong). Moskva, Izografus, 2005, 671 p. (bibl. p. 638-666).

7630. GAO (Fu). Ye shiji: chuanshuo zhong de jindai Zhongguo. (Unofficial history: modern China in popular tales). Beijing, Zhongguo shehui kexue chubanshe, 2005, 4, 252 p.

7631. GAO (Xiuqing), ZHANG (Lipeng). Liumang de lishi. (The history of blackguard). Beijing, Zhongguo wenshi chubanshe, 2005, 220 p.

7632. GE (Fuping). Zhong Fa geng kuan an zhong de wuli zejuan wenti. (The issue of no-interest bonds in the Sino-French dispute over indemnities after the 1900 war). *Jindai shi yanjiu*, 2005, 2, p. 123-141.

7633. GOLDMAN (Merle). From comrade to citizen: the struggle for political rights in China. Cambridge, Harvard U. P., 2005, VIII-286 p.

7634. GUAN (Shaohong). Keju tingfei yu jindai xiangcun shizi. Yi Liu Dapeng, Zhu Shisan riji wei shijiao de bijiao kaocha. (The abolition of the imperial examination and modern rural intellectuals: a comparison of the personal diaries of Liu Dapeng and Zhu Shisan). *Lishi yanjiu*, 2005, 5, p. 84-99.

7635. GUAN (Yanbo). Zhongguo xinan minzu shehui shenghuo shi. (A history of social life of South-Western China ethnic groups). Har'erbin, Heilongjiang renmin chubanshe, 2005, 4, 266 p.

7636. GUO (Peigui). Mingdai wenguan yinxu zhidu kaolun. (An examination on the policy of favoring the offsprings of civil officers in the selection of officers during the Ming dynasty). *Lishi yanjiu*, 2005, 2, p. 42-58.

7637. GUO (Qitao). Ritual opera and mercantile lineage. The Confucian transformation of popular culture in late imperial Huizhou. Stanford, Stanford U. P., 2005, 366 p.

7638. GUO (Yisheng). Luo Ergang zhuan. (Biography of Luo Ergang). Guilin, Guangxi shifan dazue chubanshe, 2005, 3, 2, 111 p.

7639. HAN (Shufeng). Qin Han tuxing san lun. (Some remarks on criminal punishment in the Qin and Han dynasties). *Lishi yanjiu*, 2005, 3, p. 37-52.

7640. HAN (Xiaorong). Chinese discourses on the peasant, 1900–1949. Albany, State University of New York Press, 2005, XI-259 p. (Chinese philosophy and culture, SUNY series).

7641. HARA (Motoko). 'Nōhon' shugi to 'ōdo' no hassei: kodai Chūgoku no kaihatsu to kankyō. (Agricultural fundamentalism and the creation of China's 'yellow soil.' Study of development and the environment in ancient China). Tokyo, Kenkyu Shuppan, 2005, 506 p.

7642. HARRISON (Henrietta). The man awakened from dreams. One man's life in a north China village, 1857–1942. Stanford, Stanford U. P., 2005, VIII-207 p.

7643. HE (Xuyan). Xintuoye zai Zhongguo xingqi. Jianlun "xinjiao fengchao" zhongde xintuo gongsi. (The rise of trust industry in China). *Jindai shi yanjiu*, 2005, 4, p. 187-212.

7644. HE (Yanyan). "Guomin waijiao"beijing xia de Zhong Su jianjiao tanpan (1923–1924). (Sino-Soviet negotiations and people's diplomacy 1923–1924). *Jindai shi yanjiu*, 2005, 4, p. 237-254.

7645. HO (Virgil K. Y.). Understanding Canton: rethinking popular culture in the Republican period. New York, Oxford U. P., 2005, VI-510 p. (Studies on Contemporary China).

7646. HOU (Xudong). Zhongguo gudai ren "ming" de shiyong ji qi yiyi. Zunbei, zongshu yu zeren. (The use of "personal names" in ancient China and its meanings: social status, domination-subordination and responsability). *Lishi yanjiu*, 2005, 5, p.3-21.

7647. HSIUNG (Ping-Chen). A tender voyage. Children and childhood in late imperial China. Stanford, Stanford U. P., 2005, XVI-351 p.

7648. HU (Shaohua). Zhongguo nanfang minzu lishi wenhua tansuo. (Investigations on Chinese southern nationalities's history and culture). Beijing, Minzu chubanshe, 2005, 2, 2, 443 p.

7649. HUANG (Daoxuan). Yijiuerling – yijiusiling niandai Zhongguo dinan nongqu de ditu zhanyou. Jiantan dizhu, nongmin yu ditu geming. (The appropriation of land in South-Western China between 1930s and 1940s, with a discussion of landlords, peasants and the land revolution). *Lishi yanjiu*, 2005, 1, p. 34-53.

7650. HUANG (Dewen). Beijing waijiaotuan de fazhan ji qiyi tiaoyue liyi wei zhuti de yunzuo. (The development of the Peking diplomatic corp and its activity centered on the treaty interests). *Lishi yanjiu*, 2005, 3, p. 97-114.

7651. HUANG (Kuangzhong). Cong zhongyang yu difang guangxi hudong kan Song dai jiceng shehui yanbian. (The transformation of the base structure of society during the Song as seen from the interaction of central government and local society). *Lishi yanjiu*, 2005, 4, p. 100-117.

7652. HUKUI (Shigemasa). Kan-dai jukyō no shiteki kenkyū: jukyō no kangaku-ka wo meguru têsetsu no saikentō. (A historical study of Han confucianism: a reexamination of the established opinion concerning confucianism as official learning). Tokyo, Kyuko Shoin, 2005, 553 p.

7653. HUTERS (Theodore). Bringing the world home. Appropriating the West in late Qing and early republican China. Honolulu, University of Hawai'i Press, 2005, IX-370 p.

7654. IIJIMA (Wataru). Mararia to têkoku: shokuminchi igaku to higashi-Ajia no kōiki chitsujo. (The hidden history of Malaria: colonial / imperial medicine and an integrated regional order in the 20th-century East Asia). Tokyo, University of Tokyo Press, 2005, 441 p.

7655. IIYAMA (Tomoyasu). Kakyo/gakko seisaku no hensen kara mita Kin-dai shijin sō. (Local literati during the Jin period in relation to reforms of the civil service examination and school systems). *Shigaku Zasshi*, 2005, 114, 12, p. 1-34 [Eng. Summary].

7656. IP (Hung-yok). Intellectuals in revolutionary China, 1921–1949: leaders, heroes and sophisticates. London, RoutledgeCurzon, 2005, XII-328 p. (Chinese worlds).

7657. JIANG (Pei), XIONG (Yaping). Tielu yu Shijiazhuang chengshi de chuqi. 1905–1937 nian. (The railway and the rise of Shijiazhuang 1905–1937). *Jindai shi yanjiu*, 2005, 3, p. 170-197.

7658. JIN (Chongji). Kangzhan qianye Zhonggong zhongyang zhanlüe juece de xingcheng (Strategic decision making of the CCP Central Committee at the eve of the war of resistance). *Lishi yanjiu*, 2005, 4, p. 3-24.

7659. JIN (Yilin). Diyu guannian yu paixi chongtu: Yi er san shi niandai Guomindang Yueji lingyou wei zhongxin de kaocha. (The stress on the place of origin and factions inside the Nationalist Party: A study of the the Nationalist Party leaders from Guangdong in the 1920s and 1930s). *Lishi yanjiu*, 2005, 3, p. 115-128. – IDEM. Ning Yue duishi qianhou Yan Xishan de fan Jiang dao Zhang huodong. (Yan Xishan's activities opposing Chiang Kai-shek an Zhang Xueliang before and after the Nanjing-Guangdong conflict). *Jindai shi yanjiu*, 2005, 5, p. 50-89.

7660. JIN (Zhihuan). Zhongguo fangzhi jianshe gongsi de minyinghua yu gupiao faxing tansi. (The privatization of Chinese textile enterprises and the emission of stocks). *Jindai shi yanjiu*, 2005, 2, p. 176-211.

7661. Jindai yilai Tianjin chengshihua jingcheng shilu. (Historical records about the urbanization of Tianjin since the modern era). Edited by the Tianjin municipal archives. Tianjin, Tianjin renmin chubanshe, 2005, 10, 9, 717 p.

7662. KATZ (Paul R.). When valleys turned blood red. The Ta-pa-ni incident in colonial Taiwan. Honolulu, University of Hawai'i Press, 2005, XVI-313 p.

7663. KONNO (Jun). Kenkoku shoki Chūgoku shakai ni okeru sêji dōin to taishū undō: 'San-han' undō to Shanhai shakai (1951–1952). (Political mobilization and the mass campaign in the early period of the Chinese People's Republic: the 'Three anti campaign' and Shanghai society, 1951–1952). *Aziya Kenkyu*, 2005, 51, 3, p. 1-22 [Eng. Summary].

7664. KOTERA (Atsushi). Shi no sêritsu to dempa ni kansuru ichi-kōsatsu: kyōdō saishi no ba to no kankê wo chūsin ni. (On the formation of Chinese poetry and its diffution in relation to where communal ceremonies were held). *Shigaku Zasshi*, 2005, 114, 9, p. 36-59 [Eng. Summary].

7665. KUBO (Toru). Senkan-ki Chūgoku no mengyō to kigyō keiei. (The cotton industry and enterprise management in war interim period China). Tokyo, Kyuko Shoin, 2005, 320 p.

7666. LEI (Bing). "Gaixing de zuojia": Shizhang Li Jieren jiaosi rentong de kunjiong (1950–1962). (A writer's change of profession: Li Jieren's trouble in identifyng with his new mayoral role 1950–1962). *Lishi yanjiu*, 2005, 1, p. 20-33 .

7667. LI (Aili). Wan Qing Meiji shuiwusi yanjiu: yi Yue haiguan wei zhongxin. (A study on the American commissioners in late Qing, with the Guangdong customs as focus). Tianjin, Tianjin guji chubanshe, 2005, 3, 2, 302 p.

7668. LI (Houmu). Zhongguo jindai diyiwei haijun siling: Ding Ruchang. (China first modern naval commander: Ding Ruchang). Beijing, Xinhua chubanshe, 2005, 2, 5, 265 p.

7669. LI (Hu). "Lihu" xianyi. Cong Changsha Zoumalou Wu jian tanqi". (A discussion on the registration of government employees based on the Wu State bamboo slips of Zamalou at Changsha). *Lishi yanjiu*, 2005, 3, p. 53-68.

7670. LI (Li). Qingdai minfa yujing zhong "ye" de biaoda ji qi yiyi. (The expression and meaning of "ye" in Qing civil code). *Lishi yanjiu*, 2005, 4, p. 131-142.

7671. LI (Pozhong). Dongjin Nanchao Jiangdong wenhua ronghe. (The cultural integration in the Jiangdong area during the Eastern Jin and Southern dynasties). *Lishi yanjiu*, 2005, 6, p. 91-107.

7672. LI (Qing). Qin Han Wei Jin Nan Bei chao shiqi jiazu, zongzu guanxi yanjiu. (A study of families and clan's relations during the Qin, Han, Wei, Jin and the Northern and Southern dynasties). Shanghai, Shanghai renmin chubanshe, 2005, 324 p.

7673. LI (Rong). Zhonghua minzu kang Ri zhanzheng shi (History of the war of resistance against Japan). Beijing, Zhongyang wenxian chubanshe, 2005, 4, 662 p.

7674. LI (Tianyi). Zhonggongdang shi lun cong. (Selected papers on the history of the Chinese Communist Party). Chansha, Yelu shushe, 2005, 1, 2, 335 p.

7675. LI (Wei). Zhongguo jindai fanyi yi. (A history of modern translation in China). Jinan, Qulu shushe, 2005, 2, 14, 4, 334 p.

7676. LI (Xizhu). Xingbie chongtu yu min chu zhengzhi minzhuhua de xiandu. Yi minchu nüzi can zhengquan an wei lie. (Gender conflicts and the limits of political democratization in the early Republican period). *Lishi yanjiu*, 2005, 4, p. 69-83.

7677. LI (Xuetong). Huanmie de meng: Weng Wenhao yu Zhongguo zaoqi gongyehua. (Broken dreams: Weng Wenhao and Chinese early industrialization). Tianjin, Tianjin guji chubanshe, 2005, 2, 276 p.

7678. LI (Yangfan). Zouchu wan Qing: shewai renwu ji Zhongguo de shijie guannian zhi yanjiu (Beyong the late Qing: foreign-related personalities and China's view of the world). Beijing, Beijing daxue chubanshe, 2005, 13, 454 p.

7679. LI (Yanjing). Difangzhi yu lüyou. (Local gazetteers and travels). Beijing, Fangzhi chubanshe, 2005, 3, 248 p.

7680. LIN (Shu-mei). 18sêki kōhan no Taiwan ijūmin shakai to dōshi huhō juken jiken: juken no shojōken to rimpo. (Immigrant society in Taiwan at the end of the 18th century: the case of unqualified Tongshih examinees). *Toyo Gakuho*, 2005, 87, 3, p. 33-61 [Eng. Summary].

7681. LIN (Xiaoting). Erzhan shi qi Zhong Ying guanxi zai tantao: yi Nan Ya wenti wei zhongxin. (Sino-British relations during the Second World War: focuzing on the problem of South-Asia). *Jindai shi yanjiu*, 2005, 4, p. 32-56.

7682. Lishi yu minzu: Zhongguo bianjiang de zhengzhi, shehui yu wenhua. (Ethno-history: politics, society and culture at China's frontiers). Ed. by Xianyou LUO. Beijing, Shehui kexue wenxian chubanshe, 2005, 4, 7, 2, 3, 4, 541 p.

7683. LIU (Dalin), HU (Hongxia). Yunyu yiyang: Zhongguo xing wenhua xiangzheng. (Chinese cultural symbols of sex). Chengdu, Sichuan renmin chubanshe, 2005, 14, 7, 246 p.

7684. LIU (Wei). "Jiaoxue tongyi" yu Kang Youwei de zaoqi jingxue luxiang ji qi zhuanxiang. Jianji Kang shi yu Liao Ping de xueshu jiuge. (Jiaoxue tongyi and Kang Youwei's approach to Jingxue: with reference to the alleged plagiarization of Kang by Liao Ping). *Lishi yanjiu*, 2005, 4, p. 49-68.

7685. LIU (Wusheng). Zhou Enlai yu Gongheguo zhongda lishi shijian. (Zhou Enlai and the most important events of the People's Republic). Beijing, Zhonghua shuju, 2005, 4, 3, 281 p.

7686. LIU (Xinghua). Heshen simi shenghuo quan gongkai. (A disclosure of the secret life of Heshen). Beijing, Zhongguo xiju chubanshe, 2005, 2, 5, 355 p.

7687. LIU (Yangdong). Hongdi jinzi: liushi niandai de Beijing haizi. (Golden characters on a red ground: Beijing children in the sixties). Beijing, Zhongguo qingnian chubanshe, 2005, 5, 315 p.

7688. LU (Bowen). Jiangshan meiren and wenxue: Zhongguo lishi gushi daguan. ("The throne and the beauty" and literature: Chinese historical tales). Taibei, Li Kenan, 2005, 1210 p.

7689. LU (Hanchao). Street criers. A cultural history of Chinese beggars. Stanford, Stanford U. P., 2005, XIV-269 p.

7690. LUAN (Chengxuan). Ming Qing Huizhou zongzu de yixing chengxu. (Passing the family properties to an outsider: cases in Huizhou during the Ming and the Qing). *Lishi yanjiu*, 2005, 3, p. 85-96.

7691. LUO (Zhitian). Yin xiangjin er qufen: "wenti yu zhuyi" zhi zheng xai renshi zhi yi. (Difference and similarity. A reconsideration on the dispute of "issues versus isms). *Jindai shi yanjiu*, 2005, 3, p. 44 -82. – IDEM. Zhengti gaizao he diandi gaige. "Wenti yu zhuyi" zhi zheng zai renshi zhi er. (Total transformation or gradual reform. A reflection of the debate between "issues versus isms"). *Lishi yanjiu*, 2005, 5, p. 100-116.

7692. MA (Zili). Zhong Tang wenren zhi shehui jiaose yu wenxue huodong. (Mid-Tang literati's social outlook and literary activities). Beijing, Zhongguo shehui kexue chubanshe, 2005, 2, 5, 2, 229 p.

7693. MAO (Haijian). "Gongzheng shangshu" kaozheng pu (yi). (A supplement to the evidential study of the 1895 joint petition of examination candidates to the emperor. Part 1). "Gongzheng shangshu" kaozheng pu (er). (A supplement to the evidential study of the 1895 joint petition of examination candidates to the emperor. Part 2). *Jindai shi yanjiu*, 2005, 3, p. 1-43; 4, p. 85-147. – IDEM. Qiushi de pianfang: wuxu bianfa qijian siyuan shimin shangshu zhong junshi waijiaolun. (The view on foreign relations and military affairs in the petitions of literati during the Hundred Days reforms). *Jindai shi yanjiu*, 2005, 1, p. 212-261.

7694. Mao Zedong zheyang xuexi lishi, zheyang pingdian lishi. (How Mao Zedong studies and commented history). Ed. by Xunchang SHENG, Weiwei OU and Yanghong SHENG. Beijing, Renmin chubanshe, 2005, 123, 320 p.

7695. MARUHASHI (Mitsuhiro). To-So henkakuki no guntai to chitsujo. (Military ceremonials and social order in the period of upheaval between the Tang and Song dynasties). *Tokyoshi Kenkyu*, 2005, 64, 3, p. 34-66 [Eng. Summary].

7696. MENG (Yanhong). Qin Han fadian tixi de yanbian. (The evolution of the system of the Corpus Iuris in the Qin and Han dynasties). *Lishi yanjiu*, 2005, 3, p. 19-36.

7697. MENG (Zhaochen). Zhongguo jindai xiaobao shi. (A history of modern Chinese tabloids). Beijing, Shehui kexue xianwen chunbanshe, 2005, 12, 4, 2, 673 p.

7698. MING (Fung Chi). Reluctant heroes: rickshaw pullers in Hong Kong and Canton, 1874–1954. Hong Kong, Hong Kong U. P., 2005, XX-216 p. (Royal Asiatic Society Hong Kong studies series).

7699. MO (Zigang). Guizhou qiye gongsi yanjiu (1939–1949). (A research on Guizhou industries). *Jindai shi yanjiu*, 2005, 1, p. 104-136.

7700. MURAI (Hiroshi). Minkoku jiki Shanghai no kōkoku to media. (Advertizing and the media in Shanghai during the Republic of China). *Shigaku Zasshi*, 2005, 114, 1, p. 1-33 [Eng. Summary].

7701. NIE (Yuanzi). Wenge "wuda lingxiu". Nie Yuanzi huiyilu (Memories of Nie Yuanzi). Hong Kong, Shidai guoji chubanshe, 2005, 17, 497 p.

7702. NISHIZATO (Kikō). Shin-matsu chū-ryū-nichi kankeishi no kenkyū. (Oriental research series 66: A study of relations between China, Ryukyu and Japan in late Qing period). Kyoto, Kyoto U. P., 2005, 848 p.

7703. OKAMURA (Hidenori). Chūgoku kodai ōken to saishi. (Kingship and ritual in ancient China). Tokyo, Gakuseisha, 2005, 498 p.

7704. PALMER (David A.). La fièvre du Qigong: guérison, religion et politique en Chine, 1949-1999. Paris, Ecole des hautes études en sciences sociales, 2005, 511 p. (Recherches d'histoire et de sciences sociales, 101).

7705. PAN (Guangzhe). "Shiwubao" he tade duzhe. (The "Shiwubao" and its readers). *Lishi yanjiu*, 2005, 5, p. 60-83.

7706. PARK (Sang-soo). 20 shiji san si shi niandai Zhonggong zai Shen Gan Ning bianqu yu Gelaohui guanxi lunsi. (On the relationship between the; Chinese Communist Party and the Gelaohui in Shen-Gan-Ning border area). *Jindai shi yanjiu*, 2005, 6, p. 142-170.

7707. PENG (Dachen). Wei Yuan and Xi xue Dong jian: Zhongguo zouxiang jindaihua de jiannan licheng. (Wei Yuan and the arrival of Western learning in the East: the difficult process of Chinese modernization). Changsha, Hunan shifan daxue chubanshe, 2005, 1, 511 p.

7708. QIN (Feng). Minguo Nanjing 1927-1949. (Nanking in the Republican period 1927-1949). Shanghai, Wenhui chubanshe, 2005, 2, 300 p.

7709. QIN (Guojing). Ming Qing dang'an xue. (Ming and Qing archival science). Beijing, Xueyuan chubanshe, 2005, 5, 14, 883 p.

7710. QIN (Yongzhang). Jindai Riben shentou Xizang shulun. (Japanese penetration in Tibet in modern times). *Jindai shi yanjiu*, 2005, 3, p. 144-169.

7711. Qingdai zouzhe huibian. Nongye, huanjing. (Collection of imperial memorials of the Qing period: agriculture and environment). Edited by the Centre of geographical sciences and natural resources of the Chinese Academy of Social Sciences and the First Historical Archives. Beijing, Shangwu yinshiguan, 2005, 12, 640 p.

7712. QIU (Fanzhen). Jingqi de bizhi gaige fang'an yu wan Qing bizhi wenti. (J.W. Jenks's program for currency reform and late Qing currency problems). *Jindai shi yanjiu*, 2005, 3, p. 117-143.

7713. REARDON-ANDERSON (James). Reluctant pioneers. China's expansion northward, 1644-1937. Stanford, Stanford U. P., 2005, XVII-288 p. (Studies of the Weatherhead East Asian Institute, Columbia university).

7714. SAGAWA (Eiji). Tōgi-Hokusê kakumei to Gisho no hensan. (The compilation of the Weishu and the revolutionary shift from the Eastern Wei to the Northern Qi). *Toyoshi-Kenkyu*, 2005, 64, 1, p. 37-64 [Eng. Summary].

7715. SANG (Bing). Minguo xuezhe de laopei. (The old generation of scholars in the Republican period). *Lishi yanjiu*, 2005, 6, p. 3-24.

7716. SAO (Qichuang). Yuandai duozu shiren wangluo zhongde shisheng guanxi. (The master-disciple relationship in the multi-ethnic networks during the Yuan dynasty). *Lishi yanjiu*, 2005, 1, p. 119-141.

7717. SATO (Hitoshi). Shin-matsu Min-sho no seisō ni okeru chiiki tairitsu no kōzu. (The structure of local political conflict in late Qing and early republican China). *Rekisigaku Kenkyu*, 2005, 806, p. 19-35.

7718. SHEN (Fuwei). Shisi zhi shiwu shiji zhong fachuan de Feizhou hancheng. (Chinese junks 's voyages to Africa in the Fourteenth and Fifteenth centuries). *Lishi yanjiu*, 2005, 6, p. 119-134.

7719. SHI (Jinpo). Xi Xia nongye zushui kao. Xi Xiawen nongye zushui wenshu yishi. (A study of agricultural taxes and rents during the Western Xia). *Lishi yanjiu*, 2005, 1, p. 107-118.

7720. SUI (Xiaozuo). Gong qinwang simi shenghuo quan gongkai. (A disclosure of the secret life of Prince Gong). Beijing, Zhongguo xiju chubanshe, 2005, 2, 6, 361 p.

7721. SUN (Lijuan). Qingdai shangye shehui de guize yu zhixu: cong beike ziliao jiedu Qing dai Zhongguo shangshi xiguan fa. (Rules and procedures of commercial societies during the Qing period). Beijing, Zhongguo shehui kexue chubanshe, 2005, 4, 3, 3, 317 p.

7722. TAGUCHI (Kojiro). Min-dai no sōryō to yomai. (Tribute grain and its surplus during the Ming dynasty). *Toyoshi Kenkyu*, 2005, 64, 3, p. 67-103 [Eng. Summary].

7723. TANII (Toshihito). Isshin ittoku kō: Shin-chō ni okeru seijiteki seitōsei no ronri. (A consideration of the phrase "Yixin yide": the logic of political legitimacy in the Qing Dynasty). *Tōyōshi Kenkyū*, 2005, 63, 4, p. 68-104 [Eng. Summary].

7724. TAO (Feiya). Bianyuan de lishi: Jidujiao yu jindai Zhongguo. (Christian religion and modern China). Shanghai, Guji chubanshe, 2005, 2, 13, 359 p.

7725. TONG (Zhiqiang). Guanyu Xin silu. (The New Fourth Army). Shanghai, Shanghai kexue jishu wenxian chubanshe, 2005, 9, 3, 435 p.

7726. TUTTLE (Gray). Tibetan Buddhists in the making of modern China. New York, Columbia U. P., 2005, XVIII-337 p.

7727. USUI (Sachiko). Kishū shōnin no kenkyū. (The study of Huizhou merchants). Tokyo, Kyuko Shoin, 2005, 556 p.

7728. WANG (Chunxia). "Pai Man" yu minzu zhuyi. (Anti-Manchuism and nationalism). Beijing, Shehui kexue wenxian chubanshe, 2005, 2, 3, 4, 11, 285 p.

7729. WANG (Fansen). Cong jinxue dao shixue de guodu: Liao Ping yu Meng Wentong de liezi. (The transition from the study of the Classics to the study of history: the case of Liao Ping and Meng Wentong). *Lishi yanjiu*, 2005, 2, p. 59-74.

7730. WANG (Jianlang). Beijing zhengfu canzhan wenti zai kaocha. (A new analysis on the problem of Beijing' government decision to enter the war). *Jindai shi yanjiu*, 2005, 4, p. 1-31.

7731. WANG (Jinlong). Nayidai fengliu: Songdai shidafu shenghuo. (The life of Song literati). Xianggang (Hong Kong), Heping tushu youxian gongsi, 2005, 315 p.

7732. WANG (Qingcheng). Wan Qing Huabei dingqi jishi shu de zengchang ji qi yiyi zhi yijie. (The increase in the number of regular markets in North China during the late Qing and an interpretation of its meaning). *Jindai shi yanjiu*, 2005, 6, p. 9-46.

7733. WANG (Qingjia). Xuechao yu jiaoshou: Kangzhan qianhou zhengzhi yu xueshu hudong de yige kaocha. (Professors and students in China during the Second World War: the political involvement of intellectuals in war time). *Lishi yanjiu*, 2005, 4, p. 25-48.

7734. WANG (Wenguang). Zhongguo minzu fazhan shi. (History of development of Chinese nationalities). Beijing, Minzu chubanshe, 2005, 10, 878 p.

7735. WANG (Xiangyuan). Riben dui Zhongguo de wenhua qinlüe. Xuezhe, wenhuaren de qin Hua zhanzheng. (Japan' cultural invasion of China. Scholars and intellectuals's war of invasion of China). Beijing, Kunlun chubanshe, 2005, 11, 383 p.

7736. WANG (Xiaohua), ZHANG (Qingjun). Minguo shi da junfa. (Ten great warlords of the Republic). Taibei, Linghuo wenhua shiye youxian gongsi, 2005, 303 p.

7737. WANG (Ying). Xin minzhu zhuyi geming shiqi xuanju zhidu yanjiu. (Researches on the electoral systems during the period of the revolution of the New Democracy). Beijing, Zhongguo shehui kexue chubanshe, 2005, 8, 4, 2, 285 p.

7738. WANG (Youming). Kangzhan shiqi Zhonggong de jian zu jian xi zhengce yu diquan biandong. Dui Shandong genjudi Junan xian de ge an fenxi. (The Chinese Communist Party's policy of rent and interest reduction during the war of resistance and the changes of land ownership: Junan county in Shandong base area). *Jindai shi yanjiu*, 2005, 6, p. 83-117.

7739. WANG (Yuquan). Wang Yuquan shilun ji. (Wang Yuquan's historical essays). Beijing, Zhonghua shuju, 2005, 2 vol., 7, 6, 1282 p.

7740. WANG (Zhongshu). Zhong Ri liang guo kaoguxue, gudaishi wenlunji (Historical essays on archeology and ancient history of China and Japan). Beijing, Kexue chubanshe, 2005, 7, 569 p.

7741. Wang Pengsheng yu Taiwan kang Ri zhishi. (Wang Pengsheng and the Taiwan patriots of the war against Japan). Ed. by Erjing CHEN. Taipei, Dadi chubanshe, 2005, 284 p.

7742. WATANABE (Miki). Shin ni taisuru ryu-nichi kankê no impei to hyochaku mondai. (Concealing Ryukyu-Japanese relations from Qing China and the problems of castaways). *Shigaku Zasshi*, 2005, 114, 1, p. 1-35 [Eng. Summary].

7743. Wo suo zhidao de wei Man zhengquan. (The puppet Manzhouguo as I knew it). Edited by Wen FEI. Beijing, Wenshi chubanshe, 2005, 3, 374 p.

7744. WRIGHT (David Curtis). From war to diplomatic parity in eleventh-century China. Sung's foreign relations with Kitan Liao. Boston, Brill, 2005, XI-287 p. (History of warfare, 33).

7745. WU (Tiaopo). Kangxi di yu Fojiao. (Kangxi Emperor and Buddhism). *Lishi dang'an*, 2005, 1, p. 38-44.

7746. WU (Yixiong). Yapian zhanzheng qian Yue haiguan shuifei wenti yu zhan hou haiguan shui ze tanpan. (Pre-Opium war tariff problems in Guangdong and the post-Opium war negotiation custom tariffs). *Lishi yanjiu*, 2005, 1, p. 54-74 .

7747. WU (Yongming). Li nian, zhidu yu shijian: Zhongguo sifa xiandaihua biange yanjiu 1912–1928. (Studies on the modernization of Chinese law). Beijing, Falü chubanshe, 2005, 2, 278 p.

7748. WU (Zengfeng), HAN (Chunying). Shi lun Liang Dingfen yu Zhang Zhidong de guanxi. (The relations between Liang Dingfen and Zhang Zhidong). *Lishi dang'an*, 2005, 1, p. 63-70.

7749. XIAN (Bo). Yandu de lishi. (The history of opium). Beijing, Zhongguo wenshi chubanshe, 2005, 219 p.

7750. XIAO (Zhizhi), SHI (Yantao). Huang Xing yu xinhai geming. (Huang Xing and the Republican revolution). Changsha, Yelu shushe, 2005, 2, 11, 276 p.

7751. XIAO (Zili). Minguo shiqi wusha zousi xianxiang tansi. (On the phenomenon of tungten ore smuggling in the Republican period). *Jindai shi yanjiu*, 2005, 4, p. 148-186.

7752. XIE (Zhonghou). Riben qinlüe Huabei zuixing shigao. (History of Japanese atrocities during the occupation of Northern China). Beijing, Zhongguo shehui kexue chubanshe, 2005, 3, 10, 7, 560 p.

7753. XING (Tie). Songdai jiating yanjiu. (Studies on the family in the Song period). Shanghai, Shanghai renmin chubanshe, 2005, 2, 347 p.

7754. XU (Chang). Nongjia zeren yu diquan yidong. yi 20 shiji 30 niandai qianqi Chang Jiang zhongxia yu diqu nongcun wei zhongxin. (Peasant household debts and the shift in land ownership, as seen in Lower Yangzi area in the early 1930s). *Jindai shi yanjiu*, 2005, 2, p. 78-122.

7755. XU (Xing). Zhong Han guanxi shi shang de zhongyao yi ye: zhanhou Tianjin ji Huabei Hanqiao de jizhong guanli yu qianfan. (An important page in the history of Sino-Korean relations: the centralized management and repatriation of Koreans in Tianjin and North-China after the war of resistance). *Jindai shi yanjiu*, 2005, 4, p. 57 -84.

7756. XU (Zhuoyun). Handai nongye: Zhongguo nongye jingji de qiyuan yu texing. (Han agriculture: the rise and characteristics of Chinese agriculture economy). Guilin, Guangxi shifan daxue chubanshe, 2005, 11, 10, 334 p.

7757. Xue yu huo de niandai: Zhongguo renmin wushi nian kang Ri jiu guo zhangzheng. (The age of blood and fire: the fifty-years patriotic war of Chinese people to resist Japan). Ed. by Yuqing MA, Yuan YU and Lingxia LI. Beijing, Zhongguo renmin chubanshe, 2005, 294 p.

7758. YAMAMOTO (Shin). Fukken sêbu kakumê konkyochi ni okeru shakai kozo to tochi kakumê. (Communist land reform and social structure in revolutionary west Fujian). *Toyo Gakuho*, 2005, 87, 2, p. 33-61 [Eng. Summary].

7759. YAN (Lieshang). Huiyu zhibian. Yan Lieshang lishi suibi. (Yan Lieshang's notes on history). Fuzhou, Fujian renmin chubanshe, 2005, 249 p.

7760. YANG (Kuisong). Jiang Jieshi, Zhang Xueliang yu Zhongdong lu shijian de jiaoshe. (Chiang Kai-shek, Zhang Xueliang and the negotiations over the Zhongdong railway incident). *Jindai shi yanjiu*, 2005, 1, p. 137-187. – IDEM. Yijiuerqi nian Nanjing Guomindang "qingdang" yundong yanjiu. (A study on the "party purification" movement in the Nationalist Party in Nanjing in 1927). *Lishi yanjiu*, 2005, 6, p. 42.

7761. YANG (Tianhong). Beiyang waijiao yu "zhiwai faquan" de chefei. Jiyu faquan huiyi suozai de lishi kaocha. (The Beiyang government foreing policy and the abolition of extraterritoriality). *Jindai shi yanjiu*, 2005, 3, p. 83-116.

7762. YANG (Tianshi). "Tong gongzuo" de biansi. (An exploration of the Kiri project). *Lishi yanjiu*, 2005, 2, p. 96-118.

7763. YANG (Zhenhong). Qin Han lübian erji fenlei shuo. Lun "ernian lüling" ershiqi zhong lü junshu jiuzhang. (The two leverl structure of the Qin Han code). *Lishi yanjiu*, 2005, 6, p. 74-90.

7764. YOSHIZAWA (Seichiro). Sêhoku kensetsu seisaku no shido: Nankin kokumin seifu ni okeru kaihatsu no mondai. (Start of the project to establish Northwestern China: problems about exploitation in Nanjing Nationalist Government). *Kiyo*, 2005, 148, p. 19-74.

7765. YOU (Ziyong). Mozhao, mochi yu Tang Wu dai de zhengwu yunxing. (Mozhao, mochi and the administrative operation of the Tang and the Five dynasties). *Lishi yanjiu*, 2005, 5, p. 32-46.

7766. YU (Shicun). Feichang dao: 1840–1999 Zhongguo huayu. (Exceptionality: Chinese discourse 1840–1999). Beijing, Shehui kexue chubanshe, 2005, 2, 250 p.

7767. YU (Wei). Wanzheng zhi yu fenlizhi. Songdai difang xingzheng quanli de zhuanyi. (Centralization versus devolution: the transfer of local administrative power during the Song). *Lishi yanjiu*, 2005, 4, p. 118-130.

7768. YU (Yongding). Yige xuezhe de sixiang guiji. (L'itinéraire d'un érudit). Textes réunis par WANG Wei. Beijing, Zhongxin chubanshe, 2005, 415 p.

7769. YUAN (Li). Zhengtu manman: wode Hongjun shengya. (My life in the Red Army). Edited and annotated by Yongmin YUAN and Shangqing WEN. Beijing, Zhongyang wenxian chubanshe, 2005, 2, 1, 381.

7770. Zaikabo to Chūgoku shakai. (Japanese owned cotton spinning mills in China and Chinese society). Ed. by Tokihiko MORI. Kyoto, Kyoto U. P., 2005, 232 p.

7771. ZARROW (Peter). China in war and revolution, 1895–1949. New York, Routledge, 2005, XVIII-411 p. (Asia's transformations).

7772. ZENG (Xiongsheng). Si Songdai "dao mai ershu" shuo (A discussion on crop rotation of rice and wheat in the Song dynasty). *Lishi yanjiu*, 2005, 1, p. 86-106.

7773. ZENG (Zhigong). Jiangxisheng Nanfeng Nuo wenhua. (La culture Nuo de Nanfeng, province du Jiangxi). Beijing, Zhongguo xiju chubanshe, 2005, 2 vol., 740 p.

7774. ZHANG (Fan). Tuizhaji" yu Xu Heng Liu Yin chu chu jin tui. Yuan dai rushi jinyu xintai zhi yiban. (The situation and mentality on the Yuan dynasty Confucisan scholars as seen in Tui Zhaji and Heng Liu and Liu Yin). *Lishi yanjiu*, 2005, 3, p. 69-84.

7775. ZHANG (Fuxiang). Shang wang minghao yu shang gu riming zhi yanjiu. (Appellations of the Shang dynasty kings and the practices of naming by birth date in ancient China). *Lishi yanjiu*, 2005, 2, p. 3-27.

7776. ZHANG (Guogang). Tangdai jiating xingtai de fuhexing tezheng. (Compound family as characteristic of the Tang family). *Lishi yanjiu*, 2005, 4, p. 84- 99.

7777. ZHANG (Haipeng). 20 shiji Zhongguo jindai shixue ke tixi wenti de tansuo. (Probing the structure of 20th Century field of modern Chinese history). *Jindai shi yanjiu*, 2005, 1, p. 1-29.

7778. ZHANG (Ruide). Yaozhi- Jiang Jieshi shou ling yanjiu. (Remote control: a study on Chiang Kai-shek's use of handwritten personal orders). *Jindai shi yanjiu*, 2005, 5, p. 27-49.

7779. ZHANG (Shouchang). Zhongguo nongmin yu jindai geming. (Chinese peasants and the modern revolution). Zhengzhou, Daxiang chubanshe, 2005, 2, 3, 411 p.

7780. ZHANG (Shude). Hongqiang dashi: Gongheguo lishi shijian de lailong qumai. (Origin and development of historical events of the People's Republic of China). Beijing, Zhongyang wenxian chubanshe, 2005, 2 vol., 2, 22, 774 p.

7781. ZHANG (Taiyuan). 20 shiji sanshi niandai de wenshi zhi zheng. (Literature vs. science conflict in the 1930s). *Jindai shi yanjiu*, 2005, 6, p. 171-204.

7782. ZHANG (Xianwen). Zhonghua Minguo shi. (A history of the Republic of China). Nanjing, Nanjing daxue chubanshe, 2005, 4 vol., [s. p.].

7783. ZHANG (Xueji). Huang Fu zhuan. (Huang Fu's biography). Beijing, Tuanjie chubanshe, 2005, 2, 257 p. – IDEM. Yuan Shikai mufu. (The personal advisers of Yuan Shikai). Beijing, Zhongguo guangbo dianshi chubanshe, 2005, 4, 391 p.

7784. ZHANG (Xuexu). Gu Denuo yu Minchu xianzheng wenti yanjiu. (Frank Johnson Goodnow and the costitutional question in the early republican period). *Jindai shi yanjiu*, 2005, 2, p. 142-175.

7785. ZHANG (Yinke), YU (Zhanghua). Shiji yanjiu Shiji yanjiujia. (Researches and researchers on the Shiji). Beijing, Huawen chubanshe, 2005, 3, 552 p.

7786. ZHANG (Yusheng). Shanshui wenmai. Zhang Yusheng lishi suibi. (Zhang Yusheng's notes on history). Fuzhou, Fujian renmin chubanshe, 2005, 3, 2, 381 p.

7787. ZHAO (Zhihua), JIN (Weiping). Zhonghua diguo dahang hai: Zheng He xia Xiyang. (The great sea travels of Chinese imperialism: Zheng He in the Western oceans). Kunming, Yunnan renmin chubanshe, 2005, 444 p.

7788. Zhaoshangju yu jindai Zhongguo yanjiu. (China merchants marine company and modern China). Ed. by Huili YI and Zheng HU. Beijing, Zhongguo shehui kexue chubanshe, 2005, 2, 694 p.

7789. ZHENG (Shiqu). Liang Qichao yu Xin wenhua yundong. (Liang Qichao and the New Culture Movement). *Jindai shi yanjiu*, 2005, 2, p. 1-37.

7790. ZHENG (Yijun). Zheng He quanzhuan. (Zheng He's biography). Beijing, Zhongguo qingnian chubanshe, 2005, 10, 446 p.

7791. ZHENG (Zelong). Junren congzheng: kang Ri zhanzheng shiqi de Li Hanhun. (Armyman in power. Li Hanhun during the war of resistance against Japan). Tianjin, Tianjin guji chubanshe, 2005, 3, 443 p.

7792. ZHENPING (Wang). Ambassadors from the islands of immortals. China-Japan relations in the Han-Tang period. Honolulu, University of Hawai'i Press, 2005, XIII-387 p. (Asian interactions and comparisons).

7793. Zhongguo Gongchandang tanpan shi. (History of the Chinese Communist Party's negotiations). Edited by Shenqing YANG. Beijing, Zhongyang wenxia chubanshe, 2005, 143, 4, 944 p.

7794. Zhongguo kang Ri zhanzheng shihua (An illustrated history of China's war against Japan). Edited by the Institute of History of Military History of the Academy of Military Sciences. Beijing, Junshi kexue chubanshe, 2005, 5 vol., [s. p.].

7795. ZHOU (Suiyuan). Wu sa qian Gongchandangren dui zhishi fenzi shehui jiaosi de tansuo. (Chinese communist explorations on the social role of intellectuals before the May Thirthieth Movement). *Lishi yanjiu*, 2005, 1, p. 1-19.

7796. ZHOU (Xin). Zhongguo nongju fazhan shi. (A history of the development of Chinese agricultural tools). Jinan, Zhongguo kexue jishu chubanshe, 2005, 14, 988 p.

7797. ZHU (Hu). Jiangnan ren zai Huabei. Cong wan Qing yizhen de xingqi kan difang shi lujing de kongjian juxian. (Jiangnan people in North China. Using the rise of charitable aid in late Qing to look at the spatial limitations of the practice of local history). *Jindai shi yanjiu*, 2005, 5, p. 114-148.

7798. ZHU (Ying). Minguo shiqi Jiangsu jianhang fenzheng yu sheng yihui bei hui'an. (The disputes over filatures and the assault on the provincial congress in Republican Jiangsu). *Lishi yanjiu*, 2005, 6, p. 25-41.

7799. ZHU (Yingui). Shilun Nanjing guomin zhengfu shiqi guojia ziben gufenzhi qiye xingcheng de tujing. (On the route to forming State capital joint stock companies during the Nanjing government period). *Jindai shi yanjiu*, 2005, 5, p.1-26.

7800. ZUO (Shuangwen). 1946 nian Shen Chong shijian: Nanjing zhengfu de duice. (The Shen Chong's incident in 1946: the Nanjing Nationalist government's policies). *Jindai shi yanjiu*, 2005, 1, p. 65-103.

§ 4. Addendum 2004.

7801. KUROIWA (Takashi). 'Gaku' to 'kyō'. Kaimin hōki ni miru Shin-dai musurimu-shakai no chiiki sō. (Xue and Jiao: diversity in Chinese Muslim society observed in the Muslim rebellion of 1862–1878). *Toyo Gakuho*, 2004, 86, 3, p. 99-133 [Eng. Summary].

Cf. n^{os} *352, 1104, 3649, 5948, 7018, 7372, 7571, 7580*

§ 5. Japan (before 1868).

7802. Akutō to nairan. (Villains and rebellion in medieval Japan). Ed. Akuto Kenkyukai. Tokyo, Iwata Shoin, 2005, 379 p.

7803. Edo no hiroba. (Public spaces in Edo Japan). Ed. by Nobuyuki YOSHIDA, Hiroaki NAGASHIMA and Takeshi ITO. Tokyo, University of Tokyo press, 2005, 228 p.

7804. FUJITA (Satoru). Kinsei kouki seijishi to taigai kankei. (International relations of the late Edo period). Tokyo, University of Tokyo press, 2005, 323 p.

7805. HASHIMOTO (Yu). Chūsei nihon no kokusaikankei: higashi ajia kōtsuken to gishimondai. (International relations of medieval Japan: the problems of diplomatic authenticity in eastern Asia). Tokyo, Yoshikawa Kobunkan, 2005, 355 p.

7806. Higashi azia kinsei toshi ni okeru shakai teki ketsugō: shomibun shokaisou no sonzai keitai. (Social unity in the cities of early modern East Asia: orders & status). Ed. by Tōru INOUE and Takashi TSUKADA. Osaka, Seibundo Shuppan, 2005, 326 p.

7807. HONGO (Masatsugu). Ritsuryō kokka bukkyō no kenkyū. (Studies in Buddism in ancient Japan). Kyoto, Hozokan, 2005, 344 p.

7808. HOWELL (David L.). Geographies of identity in nineteenth-century Japan. Berkeley a. Los Angeles, University of California Press, 2005, X-261 p.

7809. HUKUTO (Sanae). Heian ōchyō shakai no jendâ: Ie, ōken, seiai (Rekishi kagaku sensho). (Gender in the Heian court: family, royal authority and sexual love). Tokyo, Azekura Shobo, 2005, 350 p. (Series of historical science).

7810. IKEGAMI (Eiko). Bonds of civility. Aesthetic networks and the political origins of Japanese culture. New York, Cambridge U. P., 2005, XIV-460 p. (Structural analysis in the social sciences).

7811. IRUMADA (Nobuo). Kitanihon chūsei shakaishiron. (Historical studies in the society of medieval northern Japan). Tokyo, Yoshikawa Kobunkan, 2005, 348 p.

7812. KANAI (Kiyomitsu). Ippen hijirie shinkō. (Rethinking the picture scroll of reverend Ippen). Tokyo, Iwata Shoin, 2005, 355 p.

7813. Kentō-shi no mita Chūgoku to Nihon: Shinhakken SEI Shinsei boshi kara nani ga wakaru ka Asahi sensho. (How a Japanese mission to Tang-dynasty China saw China and Japan: What have been clarified by the new discovery of the Epitaph of SEI Shinsei). Ed. Senshu University & Northwestern University kyodo porojekuto. Tokyo, Asahi Shimbunsha Shuppan, 2005, 358 p. (Asahi selected books).

7814. Kinsei toshi no seiritsu sirīzu toshi kenchiku rekishi. (History of modern architecture 5: Early modern city and architecture). Ed. by Hiroyuki SUZUKI, Osamu ISHIYAMA, Takeshi ITO and Tuneto YAMAGISHI. Tokyo, University of Tokyo press, 2005, 412 p.

7815. Kisei hou no saikentou: rekishigaku to houshigaku no taiwa. (Early modern Japanese laws reconsidered: history and historical jurisprudence). Ed. by Satoru FUJITA. Tokyo, Yamakawa Shuppansha, 2005, 225 p.

7816. KOJIMA (Michihiro). Sengoku shokuhōki no toshi to chīki. (Towns and regions in Japan during the warring states and Oda-Toyotomi period). Tokyo, Seishi Shuppan, 2005, 355 p.

7817. KUBOTA (Masaki). Sengoku daimyō Imagawa shi to ryōgoku shihai. (The Daimyo Imagawa and the rule of their domains in the warring states period of Japan). Tokyo, Yoshikawa Kobunkan, 2005, 410 p.

7818. MAC NALLY (Mark). Proving the way. Conflict and practice in the history of Japanese nativism. Cambridge, Harvard University Asia Center, 2005, XIV-287 p. (Harvard East Asian monographs, 245).

7819. Midō Kanpaku-ki zen-tyūsyaku, Kankō 3nen and Kankō 7nen. (An annotated edition of Diaries of Mido Kanpaku in 1006 and 1010). Ed. by Yutaka YAMANAKA. Kyoto, Shibunkaku Shuppan, 2005, 2 vol., 208 p., 208 p.

7820. MIKAMI (Yoshitaka). Nihon kodai no kahei to shakai. (Currency and society in ancient Japan). Tokyo, Yoshikawa Kobunkan, 2005, 270 p.

7821. MIYAMOTO (Masaaki). Toshi kukan no kinseishi kenkyū. (A study of urban space in early modern Japan). Tokyo, Chuo Koron Bijutsu Shuppan, 2005, 718 p.

7822. MIZUNO (Tomoyuki). Muromachijidai kōbukankei no kenkyū. (Studies on the relation between imperial court and shogunate in the Muromachi period, Japan). Tokyo, Yoshikawa Kobunkan, 2005, 385 p.

7823. MORI (Yukio). Rokuharatandai no kenkyū. (A study of Kamakura Bakufu's Rokuhara Tandai, Kyoto). Tokyo, Zokugunshoruiju Kanseikai, 2005, 330 p.

7824. MURAI (Shosuke). Chūsei no kokka to zaichi shakai. (State and local society in medieval Japan). Tokyo, Azekura Shobo, 2005, 498 p. – IDEM. Higashi ajia no naka no nihon bunka. (Japanese culture in eastern Asia). Tokyo, Society for the promotion of the University of the Air, 2005, 375 p.

7825. NAKAYAMA (Tomihiro). Kinsei no keizai hatten to chihou shakai: geibi chihou no toshi to nouson. (Economic growth and local societies of Aki and Owari districts in the early modern period). Osaka, Seibundo Shuppan, 2005, 380 p.

7826. NAOKI (Kojiro). Kodai Kawachi seiken no kenkyū. (Studies in the political power in ancient Kawachi). Tokyo, Hanawa Shobo, 2005, 348 p.

7827. Nihon kodaishi kenkyū to siryō. (Researches on ancient Japan and historical documents). Ed. by Arikiyo SAEKI. Tokyo, Seishi Shuppan, 2005, 346 p.

7828. Nihonshi kōza. 6. 7. (Japanese history. Vol. 6. Social structure in the pre modern world. Vol. 7. The dismantling of pre modern society). Eds. Rekishigaku kenkyukai & Nihonshi kenkyukai. Tokyo, University of Tokyo press, 2005, 2 vol., 329 p., 329 p.

7829. NIUNOYA (Tetsuichi). Mibun, sabetsu to chūsei shakai. (Status and discrimination in medieval Japan). Tokyo, Hanawa Shobo, 2005, 438 p.

7830. NORITAKE (Yuichi). Sengoku daimyō ryōgoku no kenryokukōzō. (Power structure of the feudal territories in the Warring State Period, Japan). Tokyo, Yoshikawa Kobunkan, 2005, 375 p.

7831. OGA (Ikuo). Kinsei sanson shakai kouzou no kenkyū. (Studies in the mountainious societies of early modern Japan). Tokyo, Azekura Shobo, 2005, 378 p.

7832. Oranda shōkanchō no mita nihon: Titsingh ōfuku shokan shū. (The private correspondence of Isaac Titsingh, a Dutch captain of Nagasaki). Ed. by Korenori YOKOYAMA. Tokyo, Yoshikawa Kōbunkan, 2005, 528 p.

7833. Rettō no kodaishi. (Ancient history of the Japanese islands). Vol. 2-4. Ed. by Mahito UEHARA, Taichiro SHIRAISHI, Shinji YOSHIKAWA and Takehiko YOSHIMURA. Tokyo, Iwanami Shoten, 2005, 3 vol., 320 p., 296 p., 310 p.

7834. Rituryō-sei kokka to kodai shakai. (The state and society in ancient Japan). Ed. by Takehiko YOSHIMURA. Tokyo, Hanawa Shobo, 2005, 573 p.

7835. Saidai-ji koezu no sekai. (Historical landscapes in old maps of Saidai-ji, Nara). Ed. by Makoto SATO. Tokyo, University of Tokyo Press, 2005, 316 p.

7836. SAKAI (Kimi). Yume kara saguru chūsei. (Dream in medieval Japan). Tokyo, Kadokawa Shoten, 2005, 222 p.

7837. SASAKI (Junnosuke). Edo jidai ron. (Studies in the Edo era). Tokyo, Yoshikawa Kobunkan, 2005, 446 p.

7838. Sato Sojun sensei taikan kinen ronnbun-shu kanko-kai. Chikanobu-kyō-ki no kenkyū. (Studies in the Diary of Chikanobu-kyo: Festschrift for Professor Sojun Sato). Kyoto, Shibunkaku Shuppan, 2005, 590 p.

7839. Shōsō-in monjo ronshū. (Essays on the historical documents of Shoso-in, Nara). Ed. by Yoko NISHI and Eichi ISHIGAMI.Tokyo, Seisi Shuppan, 2005, 350 p.

7840. SUZUKI (Tetsuo). Chūsei Kantō no naikai sekai. (Worlds of inland sea in medieval Kanto, Japan). Tokyo, Iwata Shoin, 2005, 251 p.

7841. THAL (Sarah). Rearranging the landscape of the gods. The politics of a pilgrimage site in Japan, 1573–1912. Chicago, University of Chicago Press, 2005, XIV-409 p. (Studies of the Weatherhead East Asian Institute, Columbia University).

7842. TONO (Haruyuki). Nihon kodai shiryō-gaku. (Studies in the ancient Japanese historical documents). Tokyo, Iwanami Shoten, 2005, 333 p.

7843. TSUGEI (Yukio). Sekkan-ki kizoku shakai no kenkyū. (Studies in the aristocratic society in the age of regency in the Heian period). Tokyo, Hanawa Shobo, 2005, 378 p.

7844. WALKER (Brett L.). The lost wolves of Japan. Forew. by William CRONON. Seattle, University of Washington Press, 2005, XVIII-331 p. (Weyerhauser environmental books).

7845. WATANABE (Hisashi). Han chiiki no kozo to henyo: shinanonokuni matsushirohan chiiki no kenkyu. (The structure and transfiguration of Japanese clan society: Matsushiro, Shinano). Tokyo, Iwata Shoin, 2005, 380 p.

7846. YAMAZAKI (Kei). Kinsei bakuryo chiiki shakai no kenkyū. (A study of local societies in shogunate estate in early modern Japan). Tokyo, Azekura Shobo, 2005, 316 p.

7847. YUASA (Haruhisa). Chūsei tōgoku no chikishakaishi. (A history of regional society in medieval eastern Japan). Tokyo, Iwata Shoin, 2005, 441 p.

§ 5. Addenda 2004.

7848. IWABUCHI (Reiji). Edo bukechi no kenkyū. (Studies in samurai's residential area in Edo). Tokyo, Hanawa Shobo, 2004, 708 p.

7849. Kinsei gimin nenpyō. (A Japanese chronology of early modern self-sacrificing people). Ed. by Satoshi HOSAKA. Tokyo, Yoshikawa Kobunkan, 2004, 526 p.

Cf. n°s 217, 7792

§ 6. Korea.

7850. ISHIKAWA (Ryōta). Chōsen kaikō-go ni okeru kashō no tai-Shanghai bōeki: Tongshuntai-shiryō wo tsūjite. (Trade with Shanghai by Chinese merchants after the opening of Korean ports: viewed through the historical records of Tongshuntai). *Tōyōshi Kenkyū*, 2005, 63, 4, p. 21-56 [Eng. Summary].

7851. PARK (Hyun Ok). Two dreams in one bed: empire, social life, and the origins of the North Korean revolution in Manchuria. Durham, Duke U. P., 2005, XIX-314 p. (Asia-Pacific: culture, politics, and society).

7852. RIOTTO (Maurizio). Storia della Corea. Dalle origini ai giorni nostri. Milano, Bompiani, 2005, 408 p.

7853. ROKUTANDA (Yutaka). Chōsen shoki ni okeru denzê-koku no yusō/jōnō kigen: sōun-koku wo chūsin to site. (The term for paying the rice-field tax in the early Joseon dynasty, with a focus on shipment of grain by boat). *Toyoshi Kenkyū*, 2005, 64, 2, p. 35-64 [Eng. Summary].

Cf. n° 4262

S

AFRICA

(esp. to its colonization)

* 7854. MITCHELL (Peter), WHITELAW (Gavin). The archaeology of Southernmost Africa [South Africa, Lesotho and Swaziland] from c. 2000 bp to the early 1800s: a review of recent research. *Journal of African history*, 2005, 46, 2, p. 209-241.

** 7855. [TSYPKIN (G.).] Istorija Afriki v dokumentakh, 1870–2000. (The history of Africa in documents, 1870–2000). RAN, In-t vseobshchej istorii. Gen ed. Apollon B. DAVIDSON. Vol. 1. 1870–1918. Moskva, Nauka, 2005, 499 p. (pers. ind. p. 468-477).

7856. African archaeology: a critical introduction. Ed. by Ann B. STAHL. Oxford, Blackwell, 2005, XIV-490 p. (Blackwell studies in global archaeology).

7857. African family archive (An). The Lawsons of little Popo/Aneho (Togo), 1841–1938. Ed. by Adam JONES and Peter SEBALD. Oxford, Oxford U. P., 2005, 566 p. (Fontes historiae Africanae, 7).

7858. African urban spaces in historical perspective. Ed. by Steven J. SALM and Toyin FALOLA. Rochester, University of Rochester Press, 2005, XL, 395 p. (Rochester studies in African history and the diaspora, 21).

7859. ALLEN (William E.). Historical methodology and writing the Liberian past: the case of agriculture in the nineteenth century. *History in Africa*, 2005, 32, p. 21-39.

7860. ALPERN (Stanley Bernard). Did they or didn't they invent it? Iron in Sub-Saharan Africa. *History in Africa*, 2005, 32, p. 41-94.

7861. BECKERT (Sven). Von Tuskegee nach Togo. Das Problem der Freiheit im Reich der Baumwolle. *Geschichte und Gesellschaft*, 2005, 31, 4, p. 505-545.

7862. BONK (Jonathan J.). Ecclesiastical cartography and the problem of Africa. *History in Africa*, 2005, 32, p. 117-132.

7863. DEPELCHIN (Jacques). Silences in African history: between the syndromes of discovery and abolition. Dar Es Salaam, Mkuki na Nyota Publishers, 2005, XIX-256 p.

7864. DEVEAU (Jean-Michel). L'or et les esclaves: histoire des forts du Ghana du XVIe au XVIIIe siècle. Paris, Éds. UNESCO et Karthala, 2005, 330 p. (ill., maps). (Mémoire des peuples. Route de l'esclave).

7865. Encyclopedia of African history and culture. Ed. by Willie F. PAGE. Rev. ed. by R. Hunt DAVIS, Jr. New York, Facts On File, 2005, 5 vol., [s. p.]. (ill., maps).

7866. GIRAUT (F.). La nature, les territoires et le politique en Afrique du sud. *Annales*, 2005, 60, 4, p. 695-717.

7867. ILIFFE (John). Honour in African history. Cambridge, New York a. Port Melbourne, Cambridge U. P., 2005, XXIV-404 p.

7868. LEOPOLD (Mark). Inside West Nile: violence, history and representation on an African frontier. London, James Currey, 2005, X-180 p. (World Anthropology).

7869. LULAT (Y. G.-M.). A history of African higher education from antiquity to the present. Westport, Praeger, 2005, XII-624 p. (Studies in higher education).

7870. MAC CANN (James C.). Maize and grace: Africa's encounter with a New World crop, 1500–2000. Cambridge, Harvard U. P., 2005, XIII-289 p.

7871. MAC GAFFEY (Wyatt). Changing representations in Central African history. *Journal of African history*, 2005, 46, 2, p. 189-207.

7872. MAC INTOSH (Roderick J.). Ancient middle Niger: urbanism and the self-organizing landscape. New York, Cambridge U. P., 2005, XVI-261 p. (Case Studies in Early Societies).

7873. Writing African history. Ed. by John PHILLIPS. Woodbridge, University of Rochester Press, 2005, 552 p.

Cf. n°s 114s, 4634, 6760-6817, 7562

T

AMERICA

(esp. to its colonization)

7874. BONILLA (Heraclio). El futuro del pasado: las coordenadas de la configuración de los Andes. Lima, Editorial del Pedagógico San Marcos, Instituto de Ciencias Sociales y Humanidades, 2005, 1261 p. (Serie historia).

7875. COMBÈS (Isabelle). Etno-historias del Isoso: Chané y chiriguanos en el Chaco boliviano (siglos XVI a XX). La Paz, Programa de Investigación Estratégica en Bolivia y Lima, Instituto Francés de Estudios Andinos, 2005, 396 p. (Travaux de l'Institut français d'études andines, 189).

7876. Copán: the history of an ancient Maya kingdom. Ed. by E. Wyllys ANDREWS and William L. FASH. Santa Fe, School of American Research Press a. Oxford, James Currey, 2005, XVI-492 p. (School of American Research advanced seminar series).

7877. DENZER (Jörg). Die Konquista der Augsburger Welser-Gesellschaft in Südamerika (1528–1556): historische Rekonstruktion, Historiografie und lokale Erinnerungskultur in Kolumbien und Venezuela. München, Beck, 2005, 339 p. (Schriftenreihe zur Zeitschrift für Unternehmensgeschichte, 15).

7878. FOSTER (Lynn V.). Handbook to life in the ancient Maya world. New York a. Oxford, Oxford U. P., 2005, XIV-402 p. (ill., maps).

7879. GOLDSTEIN (Paul S.). Andean diaspora: the Tiwanaku colonies and the origins of South American empire. Gainesville, University Press of Florida, 2005, XX-403 p. (New world diasporas).

7880. HERRING (Adam). Art and writing in the Maya cities, a.d. 600–800: a poetics of line. New York, Cambridge U. P., 2005, XVI-316 p.

7881. KELLOGG (Susan). Weaving the past. A history of Latin America's indigenous women from the prehispanic period to the present. New York, Oxford U. P., 2005, X-338 p.

7882. LAVALLÉE (Danièle). Néolithisations en Amérique. Des prédateur semi-nomades aux sociétés complexes. *Annales*, 2005, 60, 5, p. 1035-1067.

7883. LIVI BACCI (Massimo). Conquista : la distruzione degli indios americani. Bologna, Il Mulino, 2005, 335 p. (ill.). (Biblioteca storica).

7884. Maw (In the) of the earth monster: Mesoamerican ritual cave use. Ed. by James E. BRADY and Keith M. PRUFER. Austin, University of Texas Press, 2005, VIII-438 p. (Linda Schele series in Maya and pre-Columbian studies).

7885. MOORE (Jerry D.). Cultural landscapes in the ancient Andes: archaeologies of place. Gainesville, University Press of Florida, 2005, XII-270 p.

7886. SAYRE (Gordon M.). The Indian chief as tragic hero: native resistance and the literatures of America, from Moctezuma to Tecumseh. Chapel Hill, University of North Carolina Press, 2005, X-357 p.

7887. SUGIYAMA (Saburo). Human sacrifice, militarism, and rulership: materialization of state ideology at the Feathered Serpent Pyramid, Teotihuacan. New York, Cambridge U. P., 2005, XVII-280 p. (New studies in archeology).

7888. WESSON (Cameron B.). Historical dictionary of early North America. Lanham a. Oxford, Scarecrow Press, 2005, XXXV-273 p. (maps). (Historical dictionaries of ancient civilizations and historical eras, 15).

Cf. n°s 113, 3554, 6818-6858

U

OCEANIA

(esp. to its colonization)

7889. ANGLEVIEL (Frédéric), ALDRICH (Robert). Images of the Pacific – images du Pacifique. *In*: 20ᵉ International congress of Historical Sciences. Programme. Sidney, University of Sydney, 2005, p. 161-162.

7890. ANGLEVIEL (Frédéric). Histoire de la Nouvelle-Calédonie : nouvelles approches, nouveaux objets. Paris, L'Harmattan, 2005, 350 p. (ill.). (Portes océanes).

7891. ATTWOOD (Bain). Telling the truth about Aboriginal history. Crows Nest, Allen and Unwin, 2005, VIII-264 p.

7892. FISCHER (Steven Roger). Island at the end of the world. The turbulent history of Easter Island. London, Reaktion Books, 2005, 304 p.

7893. French (The) and the Pacific world, 17th–19th centuries: discoveries, migrations and cultural exchanges. Ed. by Annick FOUCRIER. Aldershot, Ashgate Variorum, 2005, XLVI-342 p. (ill., maps). (Pacific world, 7).

7894. LE BORGNE (Jean). Nouvelle-Calédonie, 1945–1968: la confiance trahie. Paris, L'Harmattan, 2005, 598 p.

Cf. nᵒˢ 568, 6859-6867

INDEX OF NAMES[1]

A

AASERUD (Finn), 5432.
AASLESTAD (Katherine), 3917.
ABAKA (Edmund), 5927.
ABAY (Eşref), 1101, 1343.
ABAZOV (Rafis), 594.
ABBAMONTE (Giancarlo), 3066.
ABBASSI (Driss), 4671.
'Abbāsid, dinasty, 7576.
Abdülhamid II, Sultan of the Turks, 6963.
Abdullah, king of Jordan, 7120.
Abelardus (Petrus), 3278, 3293.
ABELLÁN (José Luis), 4552.
ABERTH (John), 2779.
Abetz (Otto), 3870.
ABILDGREN (Kim), 5887.
ABLOVA (Nadezhda E.), 7018.
ABOAL FERNÁNDEZ (Roberto), 1076.
ABOITES AGUILAR (Luis), 6023.
ABOY (Rosa), 6039.
ABRAMOVA (Ekaterina), 6620.
ABRAMS (Bradley F.), 3755.
ABRAMSON (Daniel M.), 5576.
ABRAMSON (Meri L.), 2585.
ABRAMZOM (Mikhail G.), 138, 1852.
ABREU (Laurinda), 6133.
Absalon, archbishop of Lund, 2828.
ABU NOWAR (Ma'an), 4249.
ABU-BAKER (Khawla), 4193.
ABUKHANFUSA (Kerstin), 4638.
ACAR (Rahim), 3257.
ACATRINEI (Filaret), 5619.
ACCATTINO (Paolo), 1418.
ACHARYA (Subrata Kumar), 2.
ACHEBE (Nwando), 4351.
ACKERMAN (Bruce), 4717.
ACKERSON (Wayne), 6765.
ACTON (Edward), 4459.
ADA (Serhan), 7019.
Adalbert von Prag, 3412.
ADAM (Thomas), 3918.

ADAM (Wolfgang), 71.
ADAMCZYK (Mieczysław), 4388.
ADAMOVSKY (Ezequiel), 328.
ADAMS (Iestyn), 6953.
Adams (John), 4778.
ADAMS (Julia), 5721.
ADAMS (Mags), 6325.
ADAMSON (Michael R.), 7184.
ADAMSON (Peter), 956.
ADAMTHWAITE (Anthony), 329.
ADANIR (Fikret), 6679.
ADDISON (Paul), 615, 4038.
Adenauer (Konrad), 7405, 7492.
ADENEY (Katharine), 4138.
ADESOJI (Abimbola O.), 6766.
ADLER (Laure), 5323.
ADLY (Emad), 1178.
ÅDNA (Jostein), 2277.
ADRIANSEN (Inge), 3806.
Aelfric, abbot of Eynsham, 3410.
AELRED OF RIEVAULX, SAINT, 2570, 2882.
Aemilius Sura, 1829.
Aeschines, 1592.
Aeschylus, 1534, 1598, 1614.
AFARY (Janet), 4158.
AFFLERBACH (Holger), 7012.
AFIGBO (Adiele), 4352.
AFRICANO (António de Freitas), 4415.
Agathocles, tiranno di Siracusa, 1459.
AGER (Sheila L.), 1151.
AGERON (Charles Robert), 6768.
Agesilaus, re di Sparta, 1469.
AGGIO (Regina), 5620.
AGGOUN (Lounis), 3573.
AGOSTINI (Lucia), 1044.
ÁGOSTON (Gábor), 4677.
AGRAWAL (Arun), 5362.
AGUIRRE (Carlos), 4366.
AGUIRRE SALVADOR (Rodolfo), 4383.
AGULHON (Maurice), 491.
AGUSTÍ (Carme), 4596.

AHLBERG (Nils), 6040.
AHLRICHS (Bernhard), 1529.
Ahmed, Bey of Constantine, 6995.
AHMIDA (Ali Abdullatif), 4273.
AHONEN (Kalevi), 5722.
AHRENS (Stefan), 5323.
AHUJA (M. L.), 4133.
AHUJA (Ravi), 5888.
AIGNER-FORESTI (Luciana), 1375.
AIKEN (Katherine G.), 5787.
AILLAUD (S.), 1078.
AILZADEH (Abbas), 1102.
AIME (Marco), 6189.
AIRAKSINEN (Tiina Helena), 7020.
AKAGÜL (Deniz), 4702.
AKAR (Ali), 152.
AKASOY (Anna), 2711, 3246.
AKEROYD (J. R.), 1109.
AKHRAS (Mahmūd), 4248.
AKKERMAN (Fokke), 5355.
AKMEŞE (Handan Nezir), 4678.
AKÖZ (Alaaddin), 4679.
AKSAN (Virginia H.), 27.
AKUJÄRVI (Johanna), 1530.
AKYILDIZ (Ali), 5723.
ALAEV (Leonid B.), 7562.
ALAI (Cyrus), 204.
ALAJOUTSIJÄRVI (Kimmo), 5788.
ALAM (Abu Yusuf), 4134.
ALAND (Barbara), 2282.
ALANIS ENCISO (Fernando), 4288.
Alanus de Insulis, 3259.
ALAURA (Silvia), 1252.
ALBA (Elisabetta), 1053.
ALBANESE (Umberto), 153.
ALBANI (Riccardo), 884.
ALBAREDA I SALVADÓ (Joaquim), 4553.
AL-BĀZYĀR (Muhammad ibn 'Abdāllah), 2711.
ALBERIGO (Giuseppe), 4826.
ALBERTI (Leon Battista), 3152.
Albertus Magnus, Sanctus, 3260.
ALBRECHT (Andrea), 5055.
ALBRECHT (Joachim), 3919.

1. The Slavonic and in particular the Russian names are given in their national form translitterated following the usual methods and are classified accordingly. Characters with diacritics, for instance ć, ś, č, š are considered as if ordinary c, s. the German modified vowels ä, ö, ø, ü are considered as if a, o, u. The names of Classical authors, Byzantine Kings and Emperors, Saints and Popes are indexed in their Latin form. Authors' names are given in capital letters.

ALBRECHT (Matthias), 6553.
ALBRECHT (Michael von), 5050.
ALBRECHT (Peter), 5280.
Albrecht IV, Herzog von Bayern-München, 2667.
ALCALÁ (César), 4554.
Alcalá Galiano (Antonio), 4602.
ALCALÁ GALVE (Angel), 5007.
ALCALÁ-ZAMORA Y QUEIPO DE LLANO (José), 4568.
ALCÁNTARA SÁEZ (Manuel), 6531, 6543.
Alcidamas, 1666.
ALDOUS (Richard), 7378.
ALDRICH (Robert), 7889.
ALDROVANDI (Cibele), 1344.
ALEKSEEV (Veniamin V.), 4464.
ALEKSEEVA (Tat'jana I.), 7564.
ALÉN GARABATO (M. Carmen), 166.
ALENT'EVA (Tat'jana V.), 4718.
Alessandri Palma (Arturo), 6952.
ALEXANDER (Alison), 622.
ALEXANDER (Manfred), 6622.
ALEXANDER (Robert Jackson), 6406.
Alexander Aphrodisiensis, 1418.
Alexander II, king of Scotland, 2642.
Alexander II, Nikolaevich, empereur de Russie, 4510.
Alexander of Hales, 3284.
Alexandre (Jeanne), 3901.
Alexandros III, ho Megas, king of Macedonia, 150, 1362, 1423, 1445, 1455, 1466, 1474, 1620, 1819, 1881.
ALFÖLDY (Géza), 1779.
Alfonso X, rey de Castilla y León, 3050.
Alfonso XIII, rey de España, 4555.
ALI (S. Mahmud), 7455.
ALIAGA (Joan), 3154.
ALJOVÍN DE LOSADA (Cristóbal), 3552, 3703, 4373.
ALKIER (Stefan), 2295.
ALLAN (Diana), 431.
ALLASON-JONES (Lindsay), 1993.
ALLEGRA (Luciano), 6134.
ALLÉGRET (Marc), 5515.
ALLEN (Barbara), 5357.
ALLEN (Elizabeth), 2978.
ALLEN (Philip M.), 595.
ALLEN (Robert C.), 762.
ALLEN (Robert), 6554.
ALLEN (William E.), 7859.
ALLEN (William), 1640.
Allende Gossens (Salvador), 3712, 7495.
ALLOVIO (Stefano), 6189.
Almagro Basch (Martin), 1008.

Almeida (António José de), 4427.
ALMEIDA (B. Hamilton), 7379.
ALMGREN (Birgitta), 4639.
ALONSO RUIZ (Begoña), 302.
ALONSO VÁZQUEZ (Francisco Javier), 7380.
ALPERN (Stanley Bernard), 7860.
ALPERS (Svetlana), 5592.
ALPERSON-AFIL (Nira), 1023.
Al-Sabah (Mubarak), Shaykh of Kuwait, 4263.
al-Shanfarā, 1311.
ALSTON (Lee J.), 5928.
ALT (Peter-André), 5480.
ALTAVISTA (Clara), 3156.
ALTERMAN (Eric), 4719.
ALTHANN (Robert), 878.
Althusius (Johannes), 710.
ALTMANN (Gerhard), 6704.
ALTMEYER (Claudia), 3261.
ALVAR (Carlos), 990.
ALVARENGA VENUTOLO (Ana Patricia), 3729.
ÁLVAREZ DE LA BORDA (Joel), 5789.
ÁLVAREZ DE SOTOMAYOR DOMÍNGUEZ (Agustín), 4583.
ÁLVAREZ HERNÁNDEZ (Arturo), 1826.
ÁLVAREZ-MON (Javier), 1345.
ALVIS (Robert E.), 4389.
ALY (Götz), 3920.
AMADO (Mª Teresa), 1574.
AMALVI (Christian), 254, 391.
AMAR (Hanania Alain), 4655.
AMARA (Michaël), 3645.
AMARAL FERLINI (Vera Lúcia), 6841.
AMARELLI (Francesco), 1927, 2283.
Amar-Suena, king of Ur, 1237.
AMATO (Eugenio), 1152, 1429, 2075, 2127, 2413, 2432.
AMATORI (Franco), 5803.
AMBOS (Claus), 587.
Ambrosius Mediolanensis, episcopus, Sanctus, 2211.
AMELANG (James S.), 545.
AMENGUAL I BATLE (Josep), 2433.
AMERINI (Fabrizio), 3262.
AMERISE (Marilena), 2284.
AMERSFOORT (Hermanus), 7115.
AMES (Eric), 6719.
AMHERDT (David), 1854, 2285.
AMICI (Angela), 2286.
AMIET (Pierre), 1203.
AMIGUES (Suzanne), 1371.
AMIN (Shahid), 596.
AMIRKHANOV (Khizri A.), 1037.
AMIR-MOEZZI (Mohammad Ali), 908.

AMITH (Jonathan D.), 5724.
Ammianus Marcellinus, 1795, 1796, 2058.
AMODIO (Mark), 3057.
AMORE (Maria Grazia), 1725.
AMOURETTI (Marie-Claire), 776.
AMOUROUX (Henri), 7297.
ANASTAKIS (Dimitry), 5790.
Anastasius I (Flavius), Byzantine emperor, 140, 2411.
ANASTASSIADOU-DUMONT (Méropi), 5473.
ANCA (Alexandru S.), 2434.
ANCEL (Jean), 4432.
ANCOS-GARCÍA (Pablo), 2957.
ANDĚL (Michael), 5297.
ANDENMATTEN (Bernard), 2780.
ANDENNA (Cristina), 3389.
ANDENNA (Giancarlo), 3376.
ANDERLE (Ádám), 5487.
ANDERMANN (Jens), 265.
ANDERSEN (Burton R.), 1248.
ANDERSEN (Elizabeth), 3102.
ANDERSEN (Ole Ejnar), 5621.
ANDERSEN (Per), 2621.
ANDERSEN (Steen), 5791.
ANDERSEN NEXØ (Sniff), 6135.
ANDERSON (Benedict), 3516.
ANDERSON (Betty Signe), 4250.
ANDERSON (David L.), 7456.
ANDERSON (David), 4253.
ANDERSON (Fred), 4721.
ANDERSON (Gary Clayton), 4722.
ANDERSON (Irvine H.), 6625.
ANDERSON (John Julian), 2979.
ANDERSON (Katharine), 5363.
ANDERSON (Kevin B.), 4158.
ANDERSON (Leslie E.), 4345.
ANDERSSON (Irene), 4654.
ANDÒ (Valeria), 1509.
Andocides, 1419.
ANDRÉ (Jean-Marie), 2117.
ANDREA DA BARBERINO, 2949.
Andreas Capellanus, 3295.
ANDREESCU (Ştefan), 597.
ANDREEV (Jurij V.), 1462.
ANDRÉN (Anders), 865.
ANDREOPOULOS (Andreas), 2435.
ANDRESEN (Gustavo de Mello Breyner), 4414.
ANDRÉS-GALLEGO (José), 6822.
ANDREWS (E. Wyllys), 7876.
ANDRIST (Patrick), 2287.
ANDRIVON MILTON (Sabine), 6823.
ANDUZE (Eric), 4680.
ANGELI (Lucia), 1040.
Angelico, fra, 3181.
ANGELINI (Margherita), 432.
ANGLEVIEL (Frédéric), 331, 886, 7889, 7890.

ANGLO (Sydney), 5344.
ANGOSTO VÉLEZ (Pedro Luis), 4555.
ANKERSMIT (Frank), 492.
Anna Comnena, 2480.
Anne, queen of Great Britain, 4029.
ANNOUSSAMY (David), 4135.
ANOKHINA (Tat'jana G.), 4458.
ANOUMA (René-Pierre), 6769.
ANSARI (Sarah), 4358.
ANSBACHER (Walter), 3365.
ANSELMI (Andrea), 212.
ANTIER (Chantal), 7021.
ANTIP (Constantin), 7519.
ANTIPENKO (Anton L.), 1531.
Antiphon, 1419.
ANTOINE (Jacques), 5058.
Antonescu (Ion), 7116, 7123.
Antoninus Pius (Titus Aurelius Fulvus Boionius Arrius), Roman emperor, 1791, 1928.
ANTONOVA (Liudmila), 7381.
ANTTONEN (Veikko), 887.
ANUŠAUSKAS (Arvydas), 4277.
Anu-uballit, 1210.
AOULAD TAHER (Mohamed), 1273.
AOUN (Marc), 2436.
APELT (Wolfgang), 4666.
Apollonius Dyscolus, 1420.
Apollonius Rhodius, 1605.
APPEL (Reinhard), 5282.
Appianus, 1799.
Appius Claudius Caecus, 1904.
APPLEGATE (Celia), 5622.
Apuleius (Lucius), 1800, 1801, 2123, 2360.
AQUINO-WEBER (Dorothee), 164.
ARA (Angelo), 7382.
ARA (Konomi), 5151.
ARAFAT (Alâ Al-dîn), 3823.
ARAGNO (Giuseppe), 4199.
ARAI (Yukio), 2781.
ARAMBURU-ZABALA HIGUERA (Javier), 1086.
ARANCIBIA CLAVEL (Patricia), 3700.
ARANDA JIMÉNEZ (Gonzalo), 1054.
ARANDA PÉREZ (Francisco José), 4582.
ARASSE (Daniel), 564.
Aratus, 1560.
ARAVANTINOS (Vassilis L.), 1388.
ARBURG (Adrian von), 6370.
ARCA (Andrea), 1007.
ARCE (Moisés), 4367.
ARCHER (Robert), 2980.
ARCHI (Alfonso), 1103, 1204.
Archilochus, 1542.

ARDA (Basak), 1389.
ARDAY (Lajos), 6627.
ARDUINI (Franca), 35.
ARENA (Antonella), 2055.
ARENA (Gaetano), 1857.
Arendt (Hannah), 5323.
ARFAIOLI (Maurizio), 6873.
ARFÉ (Gaetano), 4199.
Argan (Giulio Carlo), 854.
ARGE (Símun V.), 2729.
ARGERI (María F.), 3582.
ARGYRIOU (Astérios), 2438.
ARI (Bülent), 6036.
Aribo Scholasticus, 3235.
Ariosto (Ludovico), 5542, 5543.
Aristophanes, 1567, 1619, 1635.
ARISTOTELES, 1421, 1422, 1641, 1653, 1655, 1660, 1673, 1675, 1676, 1685, 2464, 3246.
Arius, monachus, 2336.
ARIZAGA BOLUMBURU (Beatriz), 3491.
ARLEGUI SUESCUN (José), 3136.
ARLENE (Allen), 1387.
ARLI (Alim), 6041.
ARLT (Thomas), 5593.
ARMADA PITA (Xosé-Lois), 1055.
ARMAĞAN (A. Latif), 6042.
ARMANI (Barbara), 4200.
ARMISEN-MARCHETTI (Mireille), 2069.
ARMSTRONG (Catherine), 92.
ARMSTRONG (Megan), 4926.
ARMSTRONG (Richard H.), 494.
ARMSTRONG (Robert), 4171.
Arnaldus de Villanova, 3258.
ARNALTE (Arturo), 6770.
ARNAUDIÈS (Alain), 1274.
ARNAUD-LINDET (Marie-Pierre), 1869.
ARNDT (Astrid), 251.
ARNDT (Richard T.), 5059.
ARNOLD (Dorothee), 3350.
ARNOLD (John), 3429.
ARNOLD (Klaus Jochen), 7247.
ARNOLDUS (Doreen), 5977.
ARNULFUS AURELIANENSIS, 2981.
Arnuwanda I, king of Hittites, 1257.
Aron (Raymond), 5113.
ARÓSTEGUI (Julio), 4609.
AR-RAWI (F.), 1205.
Arrianus, 1423.
ARRIZABALAGA (Marie-Pierre), 6136, 6203.
ARROYO REYES (Carlos), 4368.
ARSLAN (Ermanno A.), 2782.
ARTIGAS-MENANT (Geneviève), 4999.
ARTOLA GALLEGO (Miguel), 3517.
ARVIGO (Tiziana), 2973.

ASADA (Shinji), 6738.
ASBACH (Olaf), 3840.
ASCH (Ronald G.), 5156.
ASCHERI (Mario), 4230.
ASCHERI (Paola), 1532.
ASHKENAZI (S.), 1275.
ASHMAN ROWE (Elizabeth), 2720.
ASHTON (Nigel J.), 7457.
ASHTON (Stephen Richard), 6760, 7458.
ASIRVATHAM (Sulochana Ruth), 1533.
ASKER (Björn), 4652.
ASMIS (Elizabeth), 1858.
ASPINALL (Edward), 4154.
ASSAYAG (Jackie), 415.
Assemani (Simone), 148.
ASSUNÇAO (Matthias Röhrig), 6137.
Assurbanipal II, king of Assyria, 1240.
ASTARITA (Carlos), 2783.
Asterius Amasenus, episcopus, 2339.
ASTORGA (Pablo), 6138.
ASTROEM (Paul), 1738.
ASTUTI (Paola), 1024.
ASTUTO (Giuseppe), 4201.
ATABEKOVA (A. G.), 4458.
ATACK (Jeremy), 5793.
Atatürk Mustafa Kemal, 4698, 7132.
ATCHISON (Mary), 2953.
Athanasius Alexandrinus, Sanctus, 2213, 2287.
ATKINSON (Tim C.), 1042.
ATTAL (Frédéric), 821.
ATTANÉ (Isabelle), 6043.
ATTINGER (Pascal), 1206.
ATTWOOD (Bain), 7891.
ATWILL (David G.), 7598.
ATZBACH (Rainer), 2877.
AUBERT (Hippolyte), 5053.
Aubigné (Agrippa d'), 5489.
Aubin (Hermann), 449.
AUBOIN (Michel), 599.
Aubrac (Lucie), 7291.
Aubrac (Raymond), 7291.
AUDIGIER (François), 3841.
Aue (Hartmann von), 621.
AUE (Maximilian), 5360.
AUERBACH (Thomas), 3998.
AUFENANGER (Stephan), 5624.
AUFFARTH (Christoph), 3430.
AUGÉ (Axel Eric), 3906.
AUGELLO (Massimo M.), 5779.
Augustinus Aurelius, Sanctus, 2214, 2289, 2293, 2303, 2305, 2311, 2347, 2356.
Augustus (Gaius Iulius Caesar Octavianus), Roman emperor,

611, 1867, 1875, 1938, 1947, 1975, 2008, 2028, 2097, 2118, 2147, 2150, 2156, 2170, 2186.
AULINGER (Rosemarie), 4550.
Aulus Caecina Severus, 1961.
Aulus Gellius, 2084.
AUNESLUOMA (Juhana), 6698.
AUPIAIS (Grégory), 134.
Aurelianus (Lucius Domitius), Roman emperor, 1952.
Aurelius Victor (Sextus), 1873.
AURELL (Martin), 125, 2785.
AUROV (Oleg V.), 3488.
AUSTAD (Torleiv), 7292.
AUSTIN (Alvyn), 4932.
AUSTIN (Gareth), 6410.
AUWERS (J.-M.), 857.
AVANZA (Martina), 154.
AVDELA (Efi), 4100.
AVENARIUS (Alexander), 2439.
AVENARIUS (Martin), 1836, 1837.
Averroes (Ibn Rushd), 3256, 3265.
AVEZOU (Laurent), 391.
AVIANUS, 1810, 2983.
Avicenna, 2706, 3257.
AVNI (Gideon), 1276.
AVON (Dominique), 4911.
AVRAMOVIČ (Sima), 1480.
AXELSSON (Per), 560.
AYÁN VILA (Xurxo M.), 1076.
AYAß (Wolfgang), 6130.
AYDIN (Mithat), 6954.
AYDIN (Suavi), 6044.
AYDOĞAN (Erdal), 7022.
AYGÜN (Necmettin), 5889.
AYMES (Jean René), 5060.
AYTON (Andrew), 2650.
AYVAZIAN (Alina), 1207.
AZAR (Paola), 5745.
AZEVEDO ALEXANDRINO FERNANDES (Joao Manuel), 6475.
AZUMA (Eiichiro), 6139.
AZZOU (El-Mostafa), 6628.

B

Baader (Andreas), 3977.
BAADER (Gerhard), 5422.
BABEL (Rainer), 221, 6876, 6894.
BABEROWSKI (Jörg), 495, 5061.
BABIANO (José), 6468.
BABICHEVA (Majja E.), 5503.
BABITS (István), 7007.
BACCHETTI (Enrico), 2733, 2735.
Bacchylides, 1544, 1559, 1690.
BACCI (Michele), 3431.
Bach (Johann Sebastian), 5622.
BADA (Michal), 6131.
BADDACK (Cornelia), 443.
BADEL (Christophe), 1859.
BAECHLER (Christian), 5193.

BAEV (Jordan), 7459.
BAGGE (Sverre), 2586.
BAGHDIANTZ MAC CABE (Ina), 754.
BAGLIONI (Lapo), 1056.
BAGROV (Lev S.), 206.
BAGUS (Anita), 5364.
BAHAMONDE MAGRO (Angel), 4611.
BAHLCKE (Joachim), 3621.
BÄHR (Johannes), 7201.
BAI (Limin), 7599.
BAIDER (F.), 6386.
BAIKIE (James), 1153.
BAILKIN (Jordanna), 6411.
BAILYN (Bernard), 3519.
BAIN (Mervyn J.), 7460.
Bainton (Roland H.), 454.
BAIOCCHI (Gianpaolo), 3658.
BAJC (Gorazd), 7377, 7437.
BAKER (Chris), 4669.
BAKER (Janice E.), 671.
BAKER (Patrick), 1405.
BAKHLE (Janaki), 5625.
BAKKE (Odd Magne), 3137.
BAKKER (Gerben), 5626.
BAKSHI (Shiri Ram), 6741.
BALANO (Randy C.), 7256.
BALARD (Michel), 2444.
BALAWYDER (Aloysius), 7461.
BALD (Detlef), 3921, 4022.
BALDINI MOSCADI (Loretta), 2056.
Baldo degli Ubaldi, 2754.
BALDWIN (Matthew C.), 2290.
BALDZUHN (Michael), 5210.
BALIVET (Michel), 547.
BALL (Jennifer L.), 2440.
BALLANTYNE (Tony), 752.
BALLENT (Anahí), 6045.
BALLESTEROS GARCÍA (Víctor M.), 4289.
BALLINI (Pier Luigi), 6479.
BALOGLOU (Christos P.), 2389.
Balsamo (Luigi), 5543.
BALSLEV (Ole), 1431.
BALTA (Sebastian), 7116.
Balzac (Honoré de), 5527.
BALZAMO (Elena), 235.
BANCEL (Nicolas), 347, 6717.
BANDELLI (Gino), 1959.
BANDIERI (Susana), 600.
Bandini (Angelo Maria), 5143.
BANDLE (Oskar), 187.
BANERJEE (Dipankar), 6412.
BANFI (Antonio), 2291.
BANGOURA (Mohamed Tétémadi), 3699.
BANKEN (Ralf), 7201.
BANKIER (David), 5032.
BANNER (Stuart), 6555.
BANNERT (H.), 1534.

BANTI (Alberto Mario), 4202.
BANU (Florian), 7462.
BANZATO (Guillermo), 5929.
BAR (Doron), 1960, 2292.
BÄR (Gerald), 5627.
BARACCO (Luciano), 4346.
BARAGLI (Sandra), 829, 3159.
BARAHONA (Marvin), 4113.
BARAJAS (Rafael), 4290.
BARBAGLI (Marzio), 771.
BARBÀRA (Maria Antonietta), 2242.
BARBAZZA (Marie-Catherine), 6875.
BARBER (John), 7270.
Barberini, famiglia, 5614.
BARBIERI (Edoardo), 97, 98.
BARBOSA (Francisco J.), 4347.
BĂRBULESCU (Constantin), 548, 5930.
BĂRBULESCU (Mihai), 598, 2013.
BARBUTO (Gennaro Maria), 700.
BARCELLA (Paolo), 6046.
BARCELÓ (Pedro A.), 1860.
BARCHIESI (Alessandro), 1535, 1800.
BARCIA (Pedro Luis), 5483.
Bardoux (Jacques), 7143.
Barg (Mikhail A.), 436.
BARGAOUI (Sami), 6140.
BAR-GIL (Shlomo), 4190.
BARIĆ (Nikica), 3737.
BARILLARI (Sonia Maura), 2943.
BARIŞTA (Örçün), 5613.
BAR-JOSEPH (Uri), 7561.
BARKAN (Elazar), 496.
BARKER (Katherine), 3396.
BARNAY (Sylvie), 4860.
BARNES (James J.), 6945, 7117.
BARNES (John), 6413.
BARNES (Patience P.), 6945, 7117.
Barnes (Viola Florence), 450.
BARNEY (Stephen A.), 2954, 2995.
BARNOW (David), 4331, 6141.
BARNWELL (P. S.), 3432.
BARON (Beth), 3822.
BARON (Roman), 812.
BARONE (Francesca Prometea), 2236.
BARONI (Maria Franca), 2557.
Baronio (Cesare), 69.
BARR (Helen), 3309.
BARR (Niall), 4041.
BARRACCA (Steven), 4291.
BARRAGÁN MORIANA (Antonio), 4556.
BARRAGÁN ROMANO (Rossana), 3650.
BARRAL (Pierre), 601.
BARRATT BROWN (Michael), 4517.
BARRET (Sébastien), 3333.

BARRETT (Anthony A.), 1861, 1961.
BARRETT (David M.), 4723.
BARRIO GOZALO (Maximiliano), 3352.
BARRIOS GARCÍA (Angel), 2580.
BARROSO BERMEJO (Rosa), 1057.
BARROVECCHIO (Anne-Sophie), 485.
BARROW (Julia S.), 3397.
BARTAL (Israel), 5018.
BARTH (Boris), 822.
BARTH (Ferdinand), 2984.
BARTH (Thomas), 6142.
Bartholomaeus Anglicus, 3266.
BARTL (Anna), 18.
BARTLEY (Adam), 1536.
BARTMAN (Elizabeth), 2152.
BARTON (Susan), 6414.
BARTOSIEWICZ (László), 1045.
BARTOV (Omer), 5628.
BARTZ (Christian), 6874.
BARUA (Pradeep P.), 7582.
BARVÍKOVÁ (Hana), 5153.
BAR-YOSEF (Eitan), 5062.
BARZAZI (Antonella), 4926.
BASCH (Françoise), 3904.
BASCH (Victor), 3904.
BASCUÑÁN AÑOVER (Oscar), 6143.
BASDEVANT-GAUDEMET (Brigitte), 5211.
BASHKURTI (Lisen), 6629.
Basilius II, Byzantine emperor, 2463.
Basilius Magnus Caesariensis, episcopus, Sanctus, 2215.
BASKICI (Mehmet Murat), 5726.
BASSIOUNI (M. Cherif), 6616.
Basso (Lelio), 4222.
BASTIEN (Paul), 4810.
BASTOS DA SILVA (Jorge), 5484.
BATABYAL (Rakesh), 4136.
BATAILLON (Gilles), 4348.
BATAKOVIC (Dusan T.), 6630.
BATEMAN (Fred), 5793.
BATEN (Joerg), 5794.
BATTISTI (Eugenio), 5548.
BATTISTINI (Francesco), 5890.
BATUR (Afife), 5577.
Baudelaire (Charles), 976, 5506.
BAUDINO (Isabelle), 6435.
BAUDOU (Alban), 155.
BAUDRY (G. H.), 894.
BAUDUIN (Pierre), 2722.
BAUER (Volker), 6527.
BAUERKÄMPER (Arnd), 3937, 6144.
BAUJÍN (José Antonio), 5488.
BAULON (Jean Philippe), 7463.

BAUMBACH (Manuel), 1433.
BAUMGART (Winfried), 6944.
BAUMGARTNER (Emmanuéle), 2878.
BAUSI (Francesco), 5344.
BAUTIER (Anne-Marie), 3026.
BAUTISTA GARCÍA (Cecilia Adriana), 4861.
BAUTZ (Friedrich Wilhelm), 889.
BAUTZ (Traugott), 889.
BAVAJ (Riccardo), 3922, 3940.
BAVANT (B.), 1118.
BAYCROFT (Timothy), 7024.
BAYLE (Pierre), 5052, 5324.
BAYLY (Christopher Alan), 7118.
BAZAUD (Maryse), 6946.
BAZZANA (André), 3265.
BAZZUCCHI (Giorgio), 3153.
BEACHY (Robert), 3923.
BEAGON (Mary), 1828.
BEAL (Peter), 47.
BEARD (Mary), 2172.
BEARMAN (P. J.), 900.
BEARZOT (Cinzia), 602.
BEASLEY (Edward), 6705.
Beauharnais, family, 128.
BEAUNE (Colette), 3086.
BEAUREPAIRE (Pierre-Yves), 693.
BEAVEN (Brad), 6415.
BEBIANO (Rui), 4416.
BECCARIA (Gian Luigi), 178.
BECHERUCCI (Isabella), 5523.
BECK (Hans), 1962.
BECK (Nora), 3160.
BECK (Ralf), 3924.
BECKER (Christoph), 6563.
BECKER (Eve-Marie), 332, 2377.
BECKER (Jean Jacques), 3842, 3864, 5856, 7025.
BECKER (Martin), 6556.
BECKER (Peter), 6557.
BECKER (Steffen), 7185.
BECKER (Thomas), 549.
BECKERT (Sven), 7861.
BECKETT (John), 6047.
BECKMANN (Martin), 2153.
BECQUEMIN (Michèle), 6145.
BECQUET LAVOINNE (Claude), 3925.
Bede the Venerable, saint, 3433.
BEDINGFIELD (M. Bradford), 3455.
BEDJAOUI (Mohammed), 7468.
BEDON (Robert), 2003.
BEDOS-REZAK (Brigitte Miriam), 2824.
BEDOUELLE (Guy), 4809.
BEDUHN (Jason D.), 2293.
BEER (Harald), 923.
BEETZ (Stephan), 5943.
BÉGAUD (Stéphane), 6918.
BÉGIN (Natalie), 4390.

BEGLEY (Ronald B.), 3145.
BÉGUIN (Katia), 5978.
BEHRENDS (Jan C.), 3520, 5056.
BEHRENDT (Ralf), 3912.
Behring (Emil von), 5417.
BEIDERBECK (Friedrich), 3843.
BEIDLER (Peter G.), 3044.
BEIER-DE HAAN (Rosmarie), 303.
BEIK (William), 3844.
BEISER (Frederick), 5352.
BEIT-ARIEH (Itzhaq), 1062.
BÉJAR (María Dolores), 3583.
BEJKO (Lorenc), 1725.
BEKAR (C. T.), 5752.
BÉLAND (Daniel), 4724.
BELARDELLI (Giovanni), 5063.
BELFER-COHEN (Anna), 1041.
BELFIELD (Gervase), 3570.
BELGE (Murat), 5064.
BELISSA (Marc), 6918.
BELJAKOVA (Elena V.), 2752.
BELL (Alan R.), 6416.
BELL (Duncan S. A.), 3521.
BELL (Maureen), 55.
BELL (Roger John), 7248.
BELLAMY (John), 6146.
BELLAMY (Matthew J.), 7194.
BELLENGER (Dominic Aidan), 3353.
BELLINGER (Edward G.), 5939.
BELLITTO (Christopher M.), 3334.
BELLO (David Anthony), 7601.
BELLÓN (Juan Pedro), 1070.
BELMESSOUS (Saliha), 6706.
BELSKY (Richard), 7602.
Belzu (Manuel Isidoro), 3655.
Bembo (Pietro), 295.
BEN ABDALLAH (Z. Benzina), 1780.
BEN NÉFISSA (Sarah), 3823.
BEN YOM TOV (David), 2676.
BENABDELLAH (Saïd), 725.
BENADUSI (Lorenzo), 6147.
BENALI (Abdelkader), 6631.
BENATI (Daniele), 3155.
BÉNAT-TACHOT (Louise), 6824.
BENATTI (Corrado), 2563.
BENAVIDES HINOJOSA (Artemio), 4285, 4320.
BENDA WEBER (Isabella), 1119.
BENDALL (Lisa M.), 1390.
BEN-DOV (Jonathan), 1208.
BENEDETTI (Francesco), 1537.
Benedictus XIII, Papa, 2521.
BÉNÉÏ (Véronique), 415.
Beneš (Edvard), 7238.
BENESS (J. Lea), 1862.
BENFERHAT (Yasmina), 2118.
BEN-GHIAT (Ruth), 6722.
BENGTSSON (Tommy), 762.
Ben-Gurion (David), 4194.

BENITE (Zvi Ben-Dor), 7603.
BENNASSAR (Bartolomé), 7135.
BEN-NATAN (Asher), 7499.
BENNE (Christian), 5348.
BENNEMA (Cornelis), 2294.
BENNETT (Matthew), 6646.
BENOIST (Stéphane), 1863.
BÉNOT (Yves.), 6707.
BENREKASSA (Georges), 558.
BENSON (Eugene), 989.
BENSON (John), 6148.
BENTI (Getahun), 6048.
BENTIVOGLIO (Enzo), 3152.
BENTLEY (Michael), 333.
BENTLEY (R. Alexander), 1042.
BEN-TOR (Amnon), 1277.
BENVENUTI PAPI (Anna), 2940, 3418.
BENVINDO (Bruno), 7026.
BENZ (Wolfgang), 4001.
Beomondo d'Altavilla, 2645.
BEOZZO (Jose Oscar), 4862.
BERAN (Jiří), 5365.
BERARDI (Maria Rita), 2532.
BERBEN (Tobias), 5326.
BERCÉ (Yves-Marie), 5797.
BERCHMAN (Robert M.), 2246.
BERCUSON (David Jay), 7186.
BERDINE (Michael D.), 6956.
BÉRENGER (Jean), 4116.
BERENSON (Edward), 334.
BERG (Manfred), 4725.
BERG (Maxine), 6149.
BERGDOLT (Klaus), 5169.
BERGEMANN (Johannes), 1726.
BERGEN (Doris), 3932.
BERGENHEIM (Åsa), 6558.
BERGER (Albercht), 2383.
BERGER (Denis), 4560.
BERGER (Gérard), 7015.
BERGER (Joannès), 7015.
BERGER (Klaus), 2253.
BERGER (Stefan), 335, 4042, 6417, 7352.
BERGER BUSSE (Anna Maria), 3225.
BERGES (Ame R.), 6138.
BERGGREN (Jan), 7187.
BERGHAHN (Klaus L.), 4018.
BERGHAHN (Volker), 3932, 5170.
Bergman (Robert P.), 3490.
BERGQUIST (Lars-Göran), 551.
BERGSTRØM (Vilhelm), 7108.
BÉRIOU (Nicole), 3265.
Berkeley (George), 957.
BERKHOFER (Robert F.), 2627.
BERKIN (Carol), 6825.
BERKOWITZ (Roger), 5342.
Berlin (Isaiah), 535.
BERLING ÅSELIUS (Ebba), 4640.

Berlinguer (Enrico), 7312.
BERMAN (Constance H.), 3317.
BERMEJO BARRERA (José Carlos), 497.
BERNABÉ (Albertus), 1439.
BERNAL (Antonio-Miguel), 3522.
BERNARD (Antonia), 4540.
BERNARD (G. W.), 4961.
Bernard (Jean Frédéric), 4962.
BERNARD (Mathias), 3899.
Bernard de Clairvaux, 3139.
BERNARDI (Philippe), 3221.
BERNDT (Rainer), 2985, 3306.
BERNHARD (Michael H.), 3927.
BERNHARD (Patrick), 3928.
Bernini (Pietro), 5560.
BERNS (Thomas), 701.
BERNSTEIN (Gail Lee), 109.
BERNSTEIN (Serge), 3860.
BEROIZ (Marcelino), 2519.
BERR (Karsten), 5334.
BERRENDONNER (Clara), 1864.
BERRÍOS BERRÍOS (Alexi), 4795.
BERROCAL (María Cruz), 1008.
BERRY (Craig), 3104.
BERRY (Mary Frances), 6151.
BERSKA (Barbara), 7188.
BERSTAN (R.), 1078.
BERSTEIN (Serge), 3839, 3877.
BERTAU (Karl), 2986.
Berthezène (Pierre, baron), 6768.
BERTHIER (Annie), 25.
BERTHOLD (Lutz), 6498.
BERTI (Silvia), 4962.
BERTIN (Emiliano), 2622.
BERTIN (Nicole), 6946.
BERTINO (Magdalena), 5745.
BERTOLETTI (Nello), 2975.
BERTOLINI (Francesco), 1538, 1564.
BERTRAM (Jerome), 3354.
BERTRAM (Martin), 2771.
BERTRAND (Jean-Marie), 1527.
BERTRAND (Pascal-François), 5614.
BERTRAND (Romain), 6739.
BERTRAND-DAGENBACH (Cécile), 729.
BERTRANDY (François), 1785, 1865.
BERTSCH (Daniel), 6919.
BESHARA (Adel), 4270.
BESIER (Gerhard), 4846, 6405.
BEŠLIN (Branko), 4518.
BESPALOV (Aleksandr V.), 6898.
BESPJATYKH (Jurij N.), 4466.
BESSONE (Federica), 1926.
BEST (Anthony), 7368.
BEST (Antony), 7027.
BEST (Geoffrey), 4043.
BESTRY (Jerzy), 7028.

Betancourt (Rómulo), 7531.
BETHEA (David M.), 998.
BETHLEHEM (Ulrike), 2987.
BÉTHOUART (Bruno), 7287.
BETKER (Frank), 6049.
BETRÒ (Marilina), 1157.
BETTARINI (Luca), 1406.
BETTELLA (Patrizia), 2988.
BETTELS (Christian), 3231.
BETTETINI (Maria), 3275.
BETTHAUSEN (Peter), 488.
BETTI (Fabio), 3153.
BETZ (Hans Dieter), 874.
BEYME (Klaus von), 5549.
BÈZE (Théodore de), 5053.
BÉZIAS (Jean Rémy), 7383.
BEZNER (Frank), 2881.
BEZOTOSNYJ (Viktor M.), 6920.
BHANA (Surendra), 4541.
BHASIN (Avtar Singh), 7315.
BHATTACHARYA (Tithi), 5212.
BIAGINI (Lorenzo), 1539.
BIALER (Uri), 7464.
BIALUSCHEWSKI (Arne), 6771.
BIAN (Morris L.), 7604.
BIANCALANA (Joseph), 2753.
BIANCHIN (Lucia), 5262.
Bianchini (Francesco), 451, 5391.
BIANCONI (Daniele), 2441.
BIANQUIS (T.), 900.
BIARD (Joël), 3267.
BIÇAK (Ayhan), 955.
BICALHO (Maria Fernanda), 6841.
Bidault (Georges), 7383.
BIDEAU (Alain), 6112.
BIEBER (Florian), 4519.
BIEBER (Hans-Joachim), 5366.
BIELENSTEIN (Hans), 603.
BIENVENU (Jean-Marc), 2544.
BIÈRE-CHAUVEL (Delphine), 5594.
BIERINGER (Reimund), 2258.
BIERNATH (Andrea), 2296.
BIESEN (Sheri Chenin), 5629.
BIESTER (Björn), 293, 487.
BIGG (Charlotte), 5367.
BIGNAMI ODIER (Jeanne), 3422.
Bigod, Roger (I), Earl of Norfolk, 2639.
Bigod, Roger (II), Earl of Norfolk, 2639.
BIGONGIARI (Dino), 2989.
BIHARI (Mihály), 4117.
BIJAK (Jakub), 6116.
BÍLEK (Jiří), 3557, 3757, 3764, 7319.
BILGE (M. Sadık), 4681.
BILGE BAŞTÜRK (Mahmut), 1209.
BILJARSKI (Čočo V.), 7011.
Billinghurst (Guillermo Eduardo), 4372.
BINDER (Dieter A.), 3623.

BINDER (Harald), 3624.
BINDER (Timon), 2297.
BINGEN (Dieter), 4391.
BIONDI (Marino), 5504.
BIRK (Bonnie A.), 3435.
BIRKHAN (Helmut), 3054.
BIRLEY (Anthony Richard), 1866.
Birnbaum (Eleazar), 27.
BIRNBAUM (Pierre), 5019.
BIRTEK (Faruk), 4101.
BISANTI (Armando), 2947.
BISCHOF (Günter), 7145.
BISER (Eugen), 4022.
Bismarck (Otto von), 3930, 3951, 7001.
BISPINCK (Henrik), 336.
BISSO (Andrés), 3584.
BIŚVĀSA (Sukumāra), 7465.
BITSCH CHRISTENSEN (Søren), 6085.
BITTLESTONE (Robert), 1540.
BITTON (R.), 1275.
BITTON-ASHKELONY (Brouria), 2298.
BIVAR (A. David H.), 1346.
BIZZARRO (Salvatore), 605.
BJELOPERA (Jerome P.), 6153.
BLACK (Fiona A.), 75.
BLACK (Jeremy), 5065, 6632, 6899.
BLACK-VELDTRUP (Mechthild), 5879.
BLACKWELL (Stephen), 7466.
BLADH (Gabriel), 6375.
BLAIR (John), 3318.
BLAIS (André), 3694.
BLAIS (Hélène), 6859.
BLAIVE (Muriel), 3758.
BLAKELY (Ruth Margaret), 110.
BLANCHARD (Ian), 2786.
BLANCHARD (Pascal), 347, 6717.
BLANCKAERT (Claude), 558.
BLANCPAIN (Jean-Pierre), 3701.
BLANEY (Gerald), 3551.
BLANKE (Sandro), 6559.
BLÄNSDORF (Agnes), 360.
BLASCO HERRANZ (Inmaculada), 890.
Blasco Ibáñez (Vicente), 4603.
BLASIUS (Dirk), 3929, 5353.
BLATMAN (Daniel), 5263.
BLAUFARB (Rafe), 5979, 6957.
BLAŽEK (Petr), 3759, 3760, 3780.
BLEANEY (Heather), 880.
BLED (Jean-Paul), 3930.
BLEICH (David), 5258.
BLEIERE (Daina), 4265.
BLEJMAN (Saul), 3585.
BLESA I DUET (Isaïes), 4557.
BLICK (Sarah), 3157.
BLICKLE (Peter), 814.

BLIN (Thierry), 6154.
Bloch (Marc), 452, 520.
BLÖDORN (Andreas), 251.
BLŒDÉ (James), 3162.
BLÖMER (Michael), 1120.
BLOMKVIST (Nils), 3319.
BLOMQUIST (Thomas W.), 2787.
BLONDEAU (Chrystele), 2902.
BLOXHAM (Donald), 4682.
BLUE (Frederick J.), 4726.
BLUM (Edward J.), 4727.
BLUM (Wilhelm), 2390.
BLUME (Kenneth J.), 606.
BLUMENWITZ (Dieter), 4533.
Blunt (Wilfrid Scawen), 6956.
BLYSIDU (B.), 2425.
BOADAS I RASET (Joan), 2524.
Bobadilla (Nicolás), 4944.
BOBKOVA (Marina S.), 3553.
BOCCACCIO (Giovanni), 2954, 2995, 3069.
BOCCHIOLA (Massimo), 1867.
BOCHMANN (Klaus), 156.
BOCK (Hans Manfred), 5079, 6150.
BOCKMUEHL (Markus), 2375.
BOĆKOWSKI (Daniel), 7119.
BODEI (Remo), 5507.
Bodin (Jean), 701.
BODON (Giulio), 5550.
BOECKH (Katrin), 4709.
BOEDEKER (Deborah), 1510.
BOEHME (Olivier), 5980.
Boethius (Anicius Manlius Torquatus Severinus), 2221, 3312.
BOFFEY (Julia), 3056.
BOGDANOS (Matthew), 304.
BOGDANOV (Andrej P.), 4467.
BOGDANOV (V. N.), 4262.
BOHLER (Danièle), 359.
BÖHME-KAßLER (Katrin), 5368.
BOHNING (Don), 7467.
Bohr (Niels Henrik David), 5432.
BÖHRINGER (Hannes), 855.
BOHUNOV (Serhij), 7113.
BOIA (Lucian), 4433.
BOIARDO (Matteo Maria), 2951.
BOISDRON (Matthieu), 7189.
BOITO JÚNIOR (Armando), 3659.
BOIY (Tom), 1210.
BOJAR (Tomáš), 270.
BOJARCHENKOV (Vladislav V.), 337.
BOJKO (Nadezhda V.), 4460.
BOJOVIC (Bosko), 2788.
BÖKER (Hans Josef), 3163.
BÖKER (Hans), 3164.
BOKOKO BOKO (Djongele), 3821.
BOL (Renate), 1121.

Bol'shakova (T. F.), 2568.
BOLDIZZONI (Francesco), 5981.
Bolesław III Krzywousty, książę Polski, 2625.
Bolívar (Simón), 4796.
BOLKHOVITINOV (Nikolaj N.), 338.
BOLLING (Hans), 6155.
BOLLMANN (Anne M.), 2916.
BOLTANSKI (Christian), 253.
BOLTON (Brenda), 2597.
BOLTON (Charles C.), 5213.
BÖLÜKBAŞI (Ö. Faruk), 5982.
BOLVIG (Axel), 3165.
BOMBITZKI (Doreen), 7317.
BOMM (Werner), 2602.
BON (Bruno), 3026.
BONACINA (Giovanni), 5330.
BONAH (Christian), 5405.
BONAMENTE (Giorgio), 1814.
BOND (Bradley G.), 6833.
BONHOMME (Brian), 5931.
Bonifacius VIII, Papa, 3356.
BONILLA (Heraclio), 7874.
BONILLA (Walter R.), 3810.
BONIN (Pierre), 6156.
BONINCHI (Marc), 3845.
BÖNING (Holger), 5280.
BONINI (Celina), 3581.
BONK (Jonathan J.), 7862.
BONNAIN (Rolande), 6203.
BONNER (Elizabeth), 3846.
BONNER (Michael), 2437.
BONNET (Charles), 1154.
BONNET (Christian), 5354.
BONNET (Corinne), 339, 455.
BONNICHON (Philippe), 485.
BONUSIAK (Włodzimierz), 7223.
BONWETSCH (Bernd), 6679.
BOORMAN (Stanley), 56.
BOOTH (Paul), 2540.
Booth (William), 6396.
BOQUET (Damien), 2882.
BOR (Michael), 4044.
BORÁK (Mečislav), 3762.
BORDJUGOV (Gennadij A.), 7376.
Borges (Jorge Luis), 5540.
BORGES COELHO (João Paulo), 402.
BORGES FILHO (Nilson), 3660.
Borghese (Scipione), 4878.
BORGHETTI (Maria Novella), 469.
BORGOGNO (Alberto), 1424.
BORGOLTE (Michael), 877, 2917.
BORHI (László), 7320.
Boris III, Car na Bălgarija, 7009.
BORISOVA (L. V.), 4461.
Borjia Medina (Francisco de), 4918.
BORK (Robert Odell), 3125.
BÖRM (Henning), 471.
BORMANN (Lukas), 2217.

BORN (Robert), 850.
BORODZIEJ (Włodzimierz), 7453.
BOROS (Gábor), 5313.
BOROVKOVA (Ljudmila A.), 1104.
BORRING OLESEN (Thorsten), 7321.
Borromini (Francesco), 5596.
BORSAY (Anne), 6157.
BORTOLAMI (Sante), 2742.
BOS (Gerrit), 2676.
BOSCANI LEONI (Simona), 5066.
BOSCH (Aurora), 4612.
BÖSCH (Frank), 5795.
BOSCHI (Alberto), 5630.
BOSHER (J. F.), 6826.
BÖSL (Elsbeth), 7190.
BOSMA (Ulbe), 5264.
BOSOER (Fabián), 6633.
BOSSUAT (Gérard), 7446.
BOSTRÖM (Hans-Olof), 5564.
BOSWORTH (C. E.), 900.
BOSWORTH (R. J. B.), 4203.
BOTELHO (André), 3661.
BOTERMANN (Helga), 1995.
BOTEY SOBRADO (Ana María), 3730.
Botrel (Jean-François), 90.
BOTSMAN (Daniel V.), 4235.
BOTTACIN (Annalisa), 5508.
BÖTTCHER (Nicolaus), 3526.
BOTTER (Barbara), 1641.
BOTTIN (Francesco), 3269.
BOTTONI (Riccardo), 785.
BOTWE-ASAMOAH (Kwame), 4026.
BOUCHARLAT (Rémy), 1077.
BOUCHER (Jacqueline), 3905.
BOUCHERON (Patrick), 3166.
BOUCKAERT (Bruno), 3229.
BOUD'HORS (Anne), 1155, 1156.
BOUDENOT (Jean-Claude), 5448.
BOUDEWIJN SIRKS (Adriaan Johan), 1963.
BOUDJAABA (Fabrice), 6050.
BOUDON-MILLOT (Véronique), 1642, 2344.
BOUDOUHOU (Nouzha), 1278.
BOUILLOT (Corinne), 6425.
BOULANGER (Philippe), 236.
BOULHOSA (Patricia Pires), 2721.
BOUNEGRU (Octavian), 1996.
BOURDELAIS (Patrice), 5433.
BOURDIN (Philippe), 3899.
BOURDIN (Stéphane), 1997.
BOURGAIN (Pascale), 157.
BOURGEOIS (Ariane), 2154.
BOURGEOIS (Nicolas), 3373.
Bourguiba (Habib), 4671.
BOURIN (Monique), 2809.
BOURQUIN (Laurent), 633.

BOUSSELHAM (Abdelkader), 7468.
BOUTIER (Jean), 5124.
BOUZIDI (Abdelmadjid), 4672.
BOWD (Rueben Robert Ernest), 7249.
BOWDEN (Hugh), 1444.
BOWEN (H. V.), 6740.
Bowersock (G. W.), 1285.
BOWES (Kim), 2197.
Bowles (Chester), 7473.
BOWLES (Nigel), 4728.
BOWMAN (Alan K.), 610.
BOWMAN (Curtis), 5338.
Bowron (Fletcher), 6110.
BOYCE (D. George), 4186.
BOYER (Jean-Paul), 2760, 2785.
BOYER (Josef), 647.
BOYER (Régis), 2730.
Boyle (Leonard), 2597.
BOYLE (Peter G.), 7370.
BOYLE (Peter), 7368.
BOYNTON (Susan), 3374.
BOZO (Frédéric), 7355, 7469.
BRACCINI (Tommaso), 2442.
BŘACH (Radko), 7029.
BRACHER (Andreas), 3911.
BRACKMANN (Stephan), 1868.
BRADLEY (Keith), 1998.
BRADLEY (Richard), 1043.
BRADSHAW (David), 787.
BRADSHAW (John Tancred Landon), 7120.
BRADY (Ciaran), 4172.
BRADY (James E.), 7884.
BRAGG (Christine), 7322.
BRAKELMANN (Günter), 3916.
BRAMANI (Lidia), 5631.
BRAMBILLA (Elena), 784.
BRANCH (Daniel), 6772.
BRANCHE (Raphaëlle), 6773.
BRAND (Paul), 2740, 2746.
BRANDEBORG JENSEN (Ole), 7191.
BRANDENBURG (Philipp), 1420, 1541.
BRANDT (Hartwig), 6129.
Brandt (Willy), 7554.
BRANKAER (Johanna), 1279.
BRATOŽ (Rajko), 552.
BRATULIĆ (Josip), 6664.
Braudel (Fernand), 453.
BRÄUEL (Ulrich), 4863.
BRAUER (Kai), 5943.
BRÄUER (Uta Maria), 3192.
BRAUN (Bernd), 6430.
BRAUN (Karl-Heinz), 3336.
BRAUN (Lucien), 3167.
BRAUN (Manuel), 2884.
BRAUN-NIEHR (Beate), 21.
BRAVO VALDIVIESO (Germán), 7030.

BRAWLEY (Sean), 7248.
BRAY (Gerald Lewis), 3349.
BRAZINSKY (Gregg Andrew), 7470.
BREATNACH (Liam), 727.
BRECHENMACHER (Thomas), 4827, 7192.
Brecht (Bertolt), 5698.
BRECKENRIDGE (Keith), 4542.
BREDE (Christina), 5796.
BREDOW-KLAUS (Isabel von), 3168.
BREGNSBRO (Michael), 607.
BREISACH (Ernst), 542.
BREITBACH (Michael), 1481.
BREITENSTEIN (Mirko), 3376.
BREITMAN (Richard), 7250.
BRÉLAZ (Cédric), 1964.
Brelich (Angelo), 885, 893.
BREMM (Klaus-Jürgen), 3931.
BREMMER (Jan N.), 2128.
BRENCHLEY (Frank), 7471.
BRENDEL (Thomas), 3524.
BRENER (Ann), 2680.
BRENNAN (James R.), 5632.
BRENNER (Bernhard), 2651.
BRENNER (Christiane), 3762, 3781, 3784, 5099, 5946, 6324.
BŘEŇOVÁ (Věra), IV.
BRESC (Henri), 3086.
BRESCHI (Giancarlo), 2997.
BRESCHI (Marco), 5950.
BRESCIANI (Edda), 1157, 1158.
BREUER (Stefan), 3525.
BREUILLY (John), 3950.
BREUNING (Eleonore), 6327.
BREWER (Holly), 6560.
BREWSTER (Claire), 4292.
BREYNER (Thomaz de Mello, Conde de Mafra), 4414.
BRICAULT (Laurent), 1407.
BRICKHOUSE (Thomas C.), 1643.
BRIÇONNET (Guillaume), 2654.
BRIDENTHAL (Renate), 5089.
BRIGGS (Asa), 5369.
BRIGGS (Chris), 2789.
BRINGMANN (Klaus), 1280.
BRINKLEY (Alan), 6507.
BRINKLEY (Douglas), 4114.
BRINKMAN (Inge), 3575.
BRION (René), 5983.
BRIQUEL CHATONNET (Françoise), 1281.
BRISEÑO SENOSIAIN (Lillian), 4293.
BRISON (Jeffrey D.), 5551.
BRISSON (Luc), 1644.
Britannicus, 1941.
BRITNELL (R. H.), 2847, 2939.
BRITO (E.), 857.
BRITS (J. P.), 7384.

BROADBERRY (Stephen), 5734.
BROADHEAD (P. J.), 4963.
BROBST (Peter John), 6742.
BROCCIA (Giuseppe), 1542.
BROCK (Roger), 748.
Brockdorff-Rantzau (Ulrich, Graf von), 7046.
BRODER (Albert A.), 5955.
BRODKORB (Clemens), 4811.
BRODOCZ (André), 709.
BRODY (Lisa R.), 680.
BRODZIAK (Sylvie), 7049.
BROERS (Michael), 4204.
BROGIOLO (Gian Pietro), 3487, 3493, 3494.
BROHED (Ingmar), 4964.
BRONSTEIN (Judith), 3436.
BROOK (Timothy), 7193.
BROOKE (Stephen), 6418.
BROOKER (Robert E.), 3169.
BROOKS (Douglas A.), 91.
BROOKS (John), 7031.
BROSSEDER (Claudia), 5370.
BROTÁNKOVÁ (Helena), 4864.
BROTTIER (Laurence), 2299.
BROUILLET (Eugénie), 5067.
BROWN (Christopher Boyd), 4965.
BROWN (Gordon S.), 6921.
BROWN (Gregory S.), 5633.
BROWN (Jonathan), 5932.
BROWN (Keith M.), 4078.
BROWN (Michael), 2652.
Brown (Robert Craig), 7033.
BROWN (Sarah Annes), 2057.
BROWN (Virginia), 3.
Browne (Frank), 3614.
BROWNING (Don S.), 874.
BROZYNA (Martha A.), 2876, 2897.
Bruant des Carrières (Louis), 6904.
BRUBAKER (Leslie), 2443.
BRUCE (Yvonne), 3025.
BRUCH (Anne), 4205.
BRUCH (Rüdiger vom), 5174.
BRUCKER (Gene A.), 3170.
BRUCKMÜLLER (Ernst), 3623.
BRUEGEL (Martin), 5955.
BRUEGMANN (Robert), 6051.
BRUENING (Michael W.), 4966.
BRÜGGEMEIER (Franz-Josef), 3964, 6158.
BRUIT-ZAIDMAN (Louise), 1687.
BRULÉ (Pierre), 1688.
BRUMLIK (Micha), 6478.
BRUN (Jean-Pierre), 1999.
BRUNBORG (Helge), 6052.
BRUNDAGE (William Fitzhugh), 4729.
BRUNEL (Ghislain), 3171.
BRUNET (Guy), 6112.
BRUNET (Serge), 111.

Brüning (Heinrich), 3963.
BRUNIUS (Jan), 41.
Bruno (Giordano), 5325.
BRUNS (Peter), 3343.
BRUN-TRIGAUD (Guylaine), 177.
BRUS (Anna), 6400.
Brus, family, 110.
BRUSCAGLI (Riccardo), 2990.
BRUSH (Kathryn), 844.
BRUUN (Christer), 2173.
BRY (Paul), 1211.
BRYAN (Christopher), 2300.
BRYCE (Robert B.), 7194.
BRYCE (Trevor), 1253.
BRYSON (W. H.), 6528.
Bucer (Martin), 5011.
BUCHANAN-BROWN (John), 57.
BÜCHEL (Jochen), 5371.
BUCHELI (Marcelo), 5933.
BUCHET (Constantin), 4434.
BÜCHNER (Christine), 3270.
BÜCHSEL (Martin), 3205.
BUCHSTAB (Günter), 3913, 3981.
BUCHWALD-PELCOWA (Paulina), 59.
BUCKLER (Julie A.), 6053.
BUCKLEY (Ian M.), 1159.
BUCKLEY (Peter), 1159.
BUDA (Attila), 7007.
BUDD (Louis J.), 984.
BUDÍK (Miloš), 7276.
BUDINOV (Svetoslav), 7032.
BUDNITSKIJ (Oleg V.), 5020, 5034.
BUDURA (Romulus Ioan), 6624.
BUEL (Richard Jr.), 4730.
BUELTZINGSLOEWEN (Isabelle von), 6296.
BUENO RAMIREZ (Primitiva), 1057.
BUES (Almut), 3568.
Buffon (Georges Louis Leclerc, comte de), 5406.
BUFFON (Giuseppe), 4912.
BÜHL-GRAMER (Charlotte), 6123.
BUJOLI (Marina), 255.
BUKOWCZYK (John J.), 6159.
BULBILIA (Joseph), 892.
BULL (Marcus Graham), 3120.
BULLARD (Melissa Meriam), 482.
BULL-CHRISTIANSEN (Lene), 4806.
BULLINGER (Heinrich), 4960.
Bullitt (William Christian), 7090.
BUNDGÅRD CHRISTENSEN (Claus), 7121.
Buñuel (Luis), 835.
BÜNZ (Enno), 4975.
BURANELLI (Francesco), 1020.
BURCKHARDT (Jacob), 377, 531, 847, 5595.
BURDETT (Anita L. P.), 4168.
BURDIEL (Isabel), 4613.

BUREŠOVÁ (Jana), 6160.
BURGAUD (Stéphanie), 6958.
BURGER (M.), 6386.
BURGESS (Richard W.), 2058.
BURGESS (Richard), 2416.
BÜRGI (Michael), 818.
BURGIO (Alberto), 5341.
BURGMANN (Ludwig), 2388.
BURGSMÜLLER (Anne), 2249.
BURGUIÈRE (André), 452.
BURGWYN (H. James), 7122.
BURHOP (Carsten), 5727, 5984.
BURIDANT (Jérôme), 5797.
BURIN (Eric), 6709.
BURITY (Joanildo A.), 3677.
BURKARD (Dominik), 4865, 4898.
BURKE (Brendan), 1391.
BURKE (Peter), 498.
BURKHARDT (Richard W., Jr.), 5372.
BURMAN (Thomas E.), 3470.
BURNAND (Yves), 2000.
BURNBURY (Judith), 1201.
BURNETT (Andrew), 136.
BURNETT (Anne Pippin), 1436.
BURNETT (Charles S. F.), 158, 2932.
BURNETT (Charles), 2676.
BURNS (Grant), 973.
BURNS (Kathryn), 340.
BURROUGHS (William J.), 1009.
BURROWS (Daron Lee), 2991.
BURSTIN (Haim), 3847.
BURT (Caroline), 2623.
BURTON (Andrew), 6054.
BURTON (Antoinette), 277, 752.
Burton (Richard Francis sir), 227, 6770.
BURTON (T. L.), 3091.
BURY (Emmanuel), 202.
BURZACCHINI (Gabriele), 1543.
BUȘĂ (Daniela), 5891.
BUSBY (Keith), 3090.
BUSCH (Jörg W.), 6529.
BÜSCHGES (Christian), 6201.
BUSE (Dieter K.), 6162.
BUSEMANN (Jan Dirk), 4898.
BUSHNELL (Rebecca), 1514.
BUSINO (Giovanni), 499.
BUSSIÈRE (Éric), 6014.
BUSUTTIL (James), 891.
BUTLER (Susan), 7180.
BUTTERWICK (Richard), 4392.
BUTTERWORTH (Susan), 4342.
BÜTTNER (Heinrich), 2542.
BUZATU (Gheorghe), 7123.
BUZHILOVA (Aleksandra P.), 609.
BUZI (Paola), 22.
BYERLY (Carol R.), 5373.

BYRNES (Joseph F.), 4866.
BYSTEDT (Fredrik), 7547.
BYSTRICKÝ (Jozef), 4528.
BYSTROVA (Nina E.), 7385.

C

CABADA (Ladislav), 3763.
CABANAS (Nathalie), 5855.
CABANEL (Patrick), 4916.
CABERLIN (Luigi), 2742.
CABEZA SAN DEOGRACIAS (José), 5634.
CABOURET (Bernadette), 1869.
CABRERA (Miguel A.), 321.
CACCIA (Patrizia), 6055.
CADEDDU (Antonio), 5374.
CADENA INOSTROZA (Cecilia), 4295.
Cadier (Léon), 12.
CADIOU (François), 500.
CADOUX (T. J.), 1870.
Caesar (Gaius Iulius), 1811, 1878, 1891, 1907, 1939, 1945, 1954, 1988, 2139.
CAFFIERO (Marina), 4850.
CAGNAZZI (Silvana), 1445.
CAHILL (Nicholas), 1122.
CAILLE (Jacqueline), 2992.
CAILLOU (François), 5985.
CAIN (Louis P.), 5798.
CAINE (Barbara), 112, 345.
CAIRE (Emmanuèle), 1897.
CAIRNS (Douglas L.), 1544.
ÇAKIRHAN (Nail), 5578.
CALAME (Claude), 1545.
CALAMENT (Florence), 1155.
CALBOLI (Gualtiero), 159.
CALCANI (Giuliana), 553.
CALCANTE (Cesare Marco), 1546.
CALDWELL (Peter C.), 5353.
CALDWELL AMES (Christine), 860.
Caligula (Gaius Iulius Caesar Germanicus), Roman emperor, 1868, 1953.
CALIÒ (Tommaso), 4867.
CALKIN (Siobhain Bly), 2885.
CALLAHAN (Christopher J.), 3228.
Callimachus, 1548, 1566.
Calmeyer (Hans), 6489.
CALONACI (Stefano), 6164.
CALVET (Louis Jean), 175.
CALVI REZIA (Gabriella), 1044.
Calvin (Jean), 5007.
CALZADILLA (Pedro Enrique), 209.
CAMARA (Mohamed Saliou), 4108.
CAMARERO (Hernán), 3599.
CAMASSA (Giorgio), 1689.
CAMERON (Averil), 610.
CAMERON (Craig M.), 7124.
CAMERON (Ewen A.), 5934.
CAMERON (Hamish), 1871.

CAMEROTTO (Alberto), 1547.
CAMIC (Charles), 486.
CAMOGLI (Pablo), 3586.
CAMPANINI (Massimo), 3824.
CAMPBELL (Bruce M. S.), 2791.
CAMPBELL (Cameron), 6059.
CAMPBELL (Fergus), 4173.
CAMPBELL (Gwyn), 4282.
CAMPBELL (Peter R.), 3883.
CAMPITELLI (Alberta), 318.
CAMPOPIANO (Michele), 2835.
Camporeale (Salvatore I.), 482.
CAMY (Olivier), 732.
CANALI (Luca), 2247.
CANCIK (Hubert), 608.
CANDAR (Gilles), 3864.
CÀNDITO (Cristina), 5596.
CÁNEPA (Mercedes Maria Loguércio), 3663.
CANFORA (Davide), 5068, 5509.
CAÑIZAR PALACIOS (J. L.), 2445.
CANNING (Charlotte M.), 5635.
CANTA (Caterina Chiara), 4868.
CANTARELLA (Eva), 1483.
CANTARELLA (Glauco Maria), 3355.
Cantimori (Delio), 454.
CANTÓN (Darío), 3587.
Canute, king of Denmark, England and Norway, 2728.
CAPASSO (Mario), 23.
CAPDEVILA (Luc), 6819.
CAPITANI (Ovidio), 3356.
ČAPKA (František), 6056.
ČAPKOVÁ (Kateřina), 3527.
ČAPLOVIČ (Miloslav), 4528.
CAPOCCIA (Giovanni), 3528.
CAPOGROSSI COLOGNESI (Luigi), 486.
CAPOMACCHIA (Anna Maria G.), 893.
CAPORALE (Riccardo), 7294.
CAPPELLANO (Filippo), 7255.
CAPPELLETTI (Lorenzo), 4855.
CAPPOZZO (Mario), 1160.
CAPRIGLIONE (Jolanda), 1645.
CAPRISTO (Annalisa), 294.
CAPROŞU (Ioan), 5871.
CARANDE HERRERO (Rocio), 1780.
CARATINI (Roger), 3437.
Caravaggio (Michelangelo Merisi da), 5601.
CARAVALE (Giorgio), 4914.
CARAVALE (Mario), 728.
CARBONE (Gabriella), 1373.
CARBONETTI VENDITTELLI (Cristina), 1.
CÁRCEL ORTI (María Milagros), 2522, 2541.
CAREW (Michael G.), 7125.
Carías Andino (Tiburcio), 4114.

Carità (Mario), 7294.
ÇARKOĞLU (Ali), 7493.
CARLAW (Kenneth I.), 5752.
CARLETON (Gregory), 6165.
CARLILE (Lonny E.), 6419.
CARLUCCI (Nadia), 2060.
CARMICHAEL (Peter S.), 4731.
CARMICHAEL (Tim), 3832.
CARNEVALI (Francesca), 5986.
CARNEY (Elizabeth Donnelly), 1123.
CARNOVALE (Vera), 3580.
Caroe (Olaf Kirkpatrick), 6742.
CARON (François), 5799.
CARON (Jean-Claude), 3899.
CAROTTI (Carlo), 5636.
CARPINATO (Caterina), 2404.
CARQUÉ (Bernd), 3186.
CARR (Graham), 341.
CARR (Jacqueline Barbara), 6166.
CARRÀ (Ettore), 4206.
CARRASCO (Juan), 2519.
CARRATTIERI (Mirco), 432.
CARRAZ (Damien), 3375.
CARRÉ (Jacques), 6435.
CARRER (Philippe), 6900.
CARRERA DAMAS (Germán), 4796.
CARRERA RAMIREZ (F.), 1072.
CARRERAS (Juan Jose), 5703.
CARRERAS (Sandra), 6196.
CARRERO SANTAMARÍA (Eduardo), 3172.
CARRILLO (Ana María), 5375.
CARROLL (Francis M.), 6635.
CARROLL (John M.), 7605.
CARRUBA (Onofrio), 1254.
Carter (James Earl 'Jimmy'), 7553.
CARTER (Lionel), 4132.
CARTER (Michael), 1511.
CARTIER (Emmanuel), 6505.
CARTWRIGHT (David E.), 958.
CARVAIS (Robert), 6057.
CARVALHO (Maria Manuela), 501.
CARVALHO DA SILVA (Paulo José), 5376.
CARVER (M. O. H.), 2938.
CASALI (Luciano), 4558.
CASALS (Xavier), 4559.
CASAMENTO (Alfredo), 2061.
CASANOVA (Julián), 4614.
CASAS SÁNCHEZ (José Luis), 4567.
CASASSAS I YMBERT (Jordi), 4598.
CASAÚS ARZÚ (Marta), 5137.
CASAVOLA (Franco), 1971.
CASCÓN DORADO (Antonio), 1825.
CASELLA (Mario), 4207.
CASELLAS I SERRA (Lluís-Esteve), 2524.
CASELLI (Gian Paolo), 5728.
CASEY (Steven), 7386.

CASIMIRO (Isabel), 402.
CASINOS MORA (Francisco Javier), 1965.
CASPARD (Pierre), 5234.
CASPERS (Christiaan), 1548.
CASSAGNES-BROUQUET (Sophie), 3173.
CASSEN (Serge), 1050.
CASSIERS (Isabelle), 5987.
CASSIMATIS (Hélène), 1727.
CASSIO (Albio Cesare), 1549.
Cassiodorus (Flavius Magnus Aurelius), 1802, 2286.
Cassiodorus, Senator, 2198.
CASSIRER (Ernst), 5326, 5327.
CASSIS (Youssef), 5736, 6014.
Cassius Dio, 2066.
CASTALDI (Lucia), 3105, 3369.
CASTELLAN (Georges), 7034.
CASTELLI (Elizabeth), 2301.
CASTELLVELL (Ventura), 2734.
CASTIGLIONE (Caroline), 4208.
CASTIGLIONI (Rossana), 4791.
CASTILLO (Alicia), 1010.
CASTILLO (Nelson), 6636.
CASTILLO (R.), 2446.
CASTRO (Ivo), 160.
CASTRO-GÓMEZ (Santiago), 5377.
CASTRONOVO (Valerio), 5872.
CATALDI (Silvio), 750.
CATANA (Leo), 5325.
CĂTĂNUȘ (Ana-Maria), 4454.
CĂTĂNUȘ (Dan), 7530.
CÁTEDRA (Pedro M.), 2993.
CATELLA (Alceste), 3425.
CATHERIN-QUIVET (Agnès), 6058.
Catilina (Lucius Sergius), 1811, 1870.
Cato (Marcus Porcius), 1916.
Cato (Marcus Porcius), maior, 2120.
CATRINA (Constantin), 832.
Cattaneo (Enrico), 2378.
CATTARUZZA (Marina), 3560, 4223.
CATTIN (Florence), 164.
CATTIN (Giulio), 94.
Catullus (Gaius Valerius), 1803, 2093.
CAUCANAS (Sylvie), 7135.
CAUCHIES (Jean-Marie), 2918.
CAULFIELD (Sueann), 6485.
CAUNE (Andris), 4269.
CAVALIER (Laurence), 1124.
CAVALLARI (Cinzia), 2887.
CAVALLARIN (Marco), 6813.
CAVALLO (Guglielmo), 1, 4.
CAVIERES FIGUEROA (Eduardo), 3703.
CAWKWELL (George), 1446.
CAYTON (Andrew), 4721.
CAYUELA (Anne), 60.

CĂZAN (Ileana), 6366.
CAZANAVE (Caroline), 3005.
CAZELLES (Brigitte), 2994.
CAZZANIGA (Gian Mario), 969, 5502.
CAZZELLA (Alberto), 1058.
CEAMANOS LLORENS (Roberto), 342.
Ceaușescu (Nicolae), 4454.
CEGLAR (Stanislaus), 3249.
CELENZA (Christopher S.), 482.
CELEYRETTE (Jean), 3267.
Céline (Louis-Ferdinand), 5532.
CELOVSKY (Boris), 7110.
CENARRO LAGUNAS (Angela), 4615.
CÉNAT (Jean-Philippe), 6901.
CENGARLE (Federica), 2855.
CENNI (Francesca), 34.
CENSER (Jack), 256.
CENTANNI (Monica), 842.
CENTENERO DE ARCE (Domingo), 4565.
CERDEÑO (M.ª Luisa), 1010.
CERESA (Massimo), 17.
CERISIER (Alban), 5516.
CERNADAS (Jorge), 3581.
CERNOVODEANU (Paul), 6868.
CEROVA (Ylli), 1725.
CERRI (Giovanni), 2080.
CERUTI (Leonidas), 3588.
CERUTTI (Mauro), 7560.
CERVANTES (Fernando), 5485.
Cervantes Saavedra (Miguel de), 230, 990, 1005, 5483, 5485-5488, 5491, 5492, 5495.
CERVERA GIL (Javier), 4611.
CESARETTI (Paolo), 2447.
CESERANI (Giovanna), 473.
CESEREANU (Ruxandra), 5069.
ÇETINSAYA (Gökhan), 4683.
ÇEVIK (Özlem), 1011, 1101, 1105.
CHADWICK (Henry), 3320.
CHAGNIOT (Jean), 128.
CHALHOUB (Sidney), 6242.
CHALINE (Nadine-Josette), 6934.
CHALUS (Elaine), 4046.
CHAMBERLAIN (Andrew), 1069.
Chamberlain (Austen), 7199.
CHAMBERLAIN (Neville), 4035.
CHAMBERS (Sarah C.), 6485.
CHAMBERT (Régine), 2001.
CHAMBON (Pascal), 3848.
Champion (Henry Hyde), 6413.
CHAMPION (Pierre), 3905.
CHAMPLIN (Edward), 2062.
CHANCEY (Mark A.), 2302.
CHANDLER (Alfred D. Jr.), 660, 5800.
CHANG (Jincang), 7606.
CHANG (Jung), 7607.

CHANIOTIS (Angelos), 1106, 1402, 1404, 1484.
CHAPMAN (Jane), 5265.
CHAPMAN (John), 1045.
CHAPNICK (Adam), 7195.
CHAPPAZ (Jean-Luc), 1161.
Chariton Aphrodisiensis, 1424.
CHARLE (Christophe), V, 6167.
Charlemagne v. Karl I der Große, röm.-deutscher Kaiser, König der Franken.
Charles de Valois-Bourgogne, dit Charles le Téméraire, 2675.
Charles I, king of England, 4049, 4064.
Charles II, king of England, 4061.
Charles II, le Chauve, roi de France, 2616.
CHARMASSON (Thérèse), 5209.
CHARNOW (Sally Debra), 5637.
CHARPIN (Dominique), 1202, 1212.
CHARPIN-PLOIX (Marie-Lucie), 2241.
CHARTIER (Roger), 981.
CHARVOLIN (Florian), 777.
CHASE (Martin), 2572, 2718.
CHASE (Michael), 2303.
Chase (Salmon Portland), 4765.
CHASSAT (Sophie), 6571.
Chateaubriand (François-René de), 5500.
CHATEL (Élisabeth), 5777.
CHATTERJEE (Kumkum), 343.
CHAUBIN (Hélène), 7295.
CHAUCER (Geoffrey), 2954, 2955, 2995, 2998, 3023, 3027, 3042, 3046, 3064, 3091, 3092, 3109.
CHAUMARTIN (François-Régis), 1840.
CHAUMONT (Michel), 167.
CHAUSSON (François), 2173.
CHAUVARD (Jean-François), 5892.
CHAVARRÍA ARNAU (Alexandra), 3494.
CHAVES (Liliana), 3589.
CHAVES PALACIOS (Julián), 4616.
CHÁVEZ (Fermín), 630.
CHÁVEZ SOTO (Tania Lilia), 248.
CHE (Wenming), 7608.
CHEKANOVA (Nina V.), 1872.
CHEN (Baoliang), 7609.
CHEN (Dasheng), 7610.
CHEN (Erjing), 7741.
CHEN (Hu), 7611.
CHEN (Jiuru), 7612.
CHEN (Lizhu), 7613.
CHEN (Shuang), 6059.
CHEN (Xiaoqing), 7614.
CHEN (Yan), 7615.
CHEN (Yanxiang), 7616.

CHENERY (Carolyn), 1069.
CHEN-MORRIS (Raz), 5378.
CHENOUFI (Ali), 6621.
CHERNETSOV (Aleksej V.), 2599.
CHERNOBAEV (Anatolij A.), 322.
CHERNOV (Sergej Z.), 3500.
CHERNYKH (Evgenij N.), 493, 1063.
CHERUBINI (Paolo), 28, 2218.
CHERVONOV (P. D.), 3488.
CHESSEL (Marie-Emmanuelle), 5955.
CHEVALIER (Bernard), 2654.
CHEVREAU (Emmanuelle), 1966.
CHEWNING (Susannah Mary), 2888.
CHEYNET (Jean-Claude), 2421.
CHIABÀ (Monica), 1959.
CHIABURU (Elena), 5266.
Chiang (Kai-shek), 7171, 7227, 7659, 7760, 7778.
CHIAPPA (Bruno), 2750.
CHIARA (Frugoni), 53.
Chiara di Assisi, santa, 3406.
CHIASSON (Charles C.), 1550.
CHIBA (Masashi), 7617.
CHIBRAC (Lucienne), 6168.
CHICKERING (Roger), 7172.
CHIEFFO RAGUIN (Virginia), 3482.
CHIESA (Federica), 1760.
CHIESA (Paolo), 3105.
CHILDS (William R.), 5801.
CHILUNDO (Arlindo Gonçalo), 402.
CHINCA (Mark), 3062.
CHIPMAN (Donald E.), 113.
CHIRHART (Ann Short), 5214.
CHIRIȚOIU (Mircea), 7323.
CHIRKOV (Sergej V.), 344.
CHIRONI (Giuseppe), 279.
CHIŞ (Silvius Ovidiu), 2155.
CHISHTI (S. M. A. W.), 4137.
CHISICK (Harvey), 786.
Chisthenes, 1465.
CHITTOLINI (Giorgio), 2655, 2855.
Chizhov (Fedor), 5853.
CHMIELEWSKI (Paweł), 7324.
CHOAT (Malcolm), 1162.
CHOI (Suk Wan), 4258.
CHOQUE CANQUI (Roberto), 3652.
CHOUEIRI (Youssef M.), 618.
CHOWNING (Margaret), 4869.
Chrétien de Troyes, 614.
CHRIST (Karl), 1873.
CHRIST (Michel), 4560.
CHRISTCHEV (T.), 163.
CHRISTENSSON (Jakob), 5152.
CHRISTIANSEN (Palle Ove), 3807.
CHRISTIN (Olivier), 4968.
Christina of Markyate, Saint, 29, 3407.

Christine de Pisan, 3282, 3435.
CHRISTOF (Eva), 1125.
CHRISTOL (Michel), 1967.
CHROBAK (Werner Johann), 3364.
Chrodegang, saint, 3354.
Chruščëv (Nikita Sergeevič), 4470, 7520.
CHUDINOV (Aleksandr V.), 3859.
CHUKU (Gloria), 5729.
Churchill (Winston Leonard Spencer), 4038, 4043, 6644, 7171, 7186, 7217, 7226, 7231, 7241.
CHYTRY (Josef), 5880.
CIAMMARUCONI (Clemente), 5117.
CIARAMELLI (Giancarlo), 61.
CIBU (Simina), 2129.
ČIČAJ (Viliam), 6131.
CICCUTO (Marcello), 3152.
ÇIÇEK (Kemal), 4684.
Cicero (Marcus Tullius), 1804-1808, 1854, 1858, 1906, 2072, 2073, 2098, 2111, 2126.
CIEGER (András), 6503.
CIENFUEGOS GARCÍA (Juan José), 2244.
CIGLIANO (Giovanna), 4468.
CÍLEK (Roman), 3557.
CILIBERTO (Michele), 5311.
CIMBALA (Paul A.), 588.
CINGANO (Ettore), 1551.
CINOTTO (Simone), 6401.
CIOBANU (Nicolae), 7251.
CIOBANU (Veniamin), 6922.
CIOBANU (Vitalie), 4324.
CIOC (Mark), 3964.
CIOCIOLA (C.), 99.
Cioran (Emil), 4438.
CIRIACONO (Salvatore), 6169.
CIRILLO (Olga), 2063.
CISSOKO (Sékéné Mody), 6774.
CITINO (Robert Michael), 6637.
CIVES (Simona), 22.
CLAEYS (Gregory), 962.
CLAISSE (Stéphanie), 3639.
CLANCIER (Philippe), 1213.
CLARK (Elizabeth A.), 896.
CLARK (Linda S.), 2919.
CLARK (Michael D.), 5071.
CLARK (Suzannah), 3227.
Clarke (H. B.), 225.
CLARKE (William), 4030.
CLARKSON (Leslie), 6204.
CLASSEN (Albrecht), 3138.
CLASSEN (Steven D.), 5639.
CLAUDE (Jean), 5515.
CLAUDEL (Paul), 6946.
Claudius Quadrigarius (Quintus), 1809.
Clausewitz (Karl von), 6694.
CLAVIN (Patricia), 7035.

CLAWSON (Patrick), 4159.
CLAY (Brenda Johnson), 6860.
Clay (Lucius D.), 7433.
CLAYSON (Hollis), 5573.
Clémenceau (Georges), 7049.
Clemens VII, Papa, 4853.
CLÉMENT (Jean Louis), 3849.
CLÉMENT (Michèle), 5312.
CLEMENT (Piet), 6026.
CLEMENTE RAMOS (Julián), 2593.
CLENDINNEN (Inga), 6861.
Cleopatra VII, queen of Egypt, 1181.
CLERICI (Luca), 5988.
CLOUGH (Cecil H.), 2653, 2675.
CLOUZOT (Martine), 3241.
COARELLI (Filippo), 1512, 1874.
COATES (David), 3617.
COBBOLD (G. B.), 1849.
COBO ROMERO (Francisco), 4561.
COCCHI GENICK (Daniela), 1012.
COCHELIN (Isabelle), 3374.
COCLANIS (Peter A.), 5725.
COCO (Virginia Giuseppina), 2305.
CODATO (Piero), 5588.
Cody (William), 5164.
COELHO (Maria Teresa Pinto), 6713.
COENEN (Craig R.), 6170.
COHEN (David William), 4254.
COHEN (David), 1482, 1485.
COHEN (Eran), 1214.
COHEN (Gerson David), 2679.
Cohen (Gustave), 452.
COHEN (Mark R.), 2682.
COHEN (Michael Joseph), 7472.
COHEN (Shaye J. D.), 5021.
COHEN (Susan Sarah), 589.
COHEN (Warren I.), 7325.
COHEN (Yoram), 1215, 1255.
COINTET (Michèle), 3850.
COLARIZI (Simona), 4209.
COLCLOUGH (David), 4047.
COLDAGELLI (Umberto), 481.
COLE (Shawn), 5893.
COLEMAN (Deirdre), 6710.
COLEMAN (Heather J.), 4969.
COLIN MUSET, 3228.
COLLADO-SEIDEL (Carlos), 7387.
COLLEONI (Aldo), 6690, 6950.
COLLEY (Linda), 6420.
COLLIN (Claude), 7296.
Collingwood (Robin George), 535.
COLLINS (Christiane Crasemann), 6061.
COLLINS (Jeffrey R.), 5337.
COLLINS (John J.), 1282.
COLLINS (Matthew), 1045, 1069.
COLLINS (Peter), 897.
COLLINS (Robert O.), 4631.
COLLOMBERT (Philippe), 1163.

COLLOUD-STREIT (Marlis), 1646.
COLOMBANI (Marie-Françoise), 5267.
COLOMO (Daniela), 1552.
COLON (Germà), 5486.
COLONESE (André Carlo), 1025.
COLOT (Blandine), 2307.
COLVILLE (John Rupert, Sir.), 4031.
COMBA (Michele), 5579.
COMBÈS (Isabelle), 7875.
COMELLA (Annamaria), 2131.
COMET (Georges), 776.
Commodus (Marcus Aurelius), Roman emperor, 2170.
COMPAGNON (Antoine), 5510.
COMPANY COMPANY (Concepción), 3101.
COMPANY I CLIMENT (Ximo), 3154.
COMȘA (Ioan), 7036.
CONAGHAN (Catherine M.), 4369.
CONDRAU (Flurin), 5802.
CONFESSORE (Ornella), 4870.
CONFORTI (Claudia), 6062.
CONGAR (Yves), 4829.
CONGOURDEAU (Marie-Hélène), 2448.
CONINGHAM (R. A. E.), 1084.
CONLON (Pierre M.), 5051.
CONNELL (William J.), 4871.
CONNELLY (John), 5202.
CONNELLY (Owen), 6923.
CONNOLLY (Peter), 2004.
CONOLLY (Leonard W.), 989.
CONRAD (James H.), 5963.
CONRAD-MARTIUS (Hedwig), 5340.
CONSANI (Carlo), 1553.
CONSTABLE (Giles), 4871.
Constant (Benjamin), 538.
CONSTANTIN (Ion), 6640.
CONSTANTINEAU (Philippe), 749.
CONSTANTINESCU (Jean Leontin), 5268.
CONSTANTINOU (Stavroula), 2449.
Constantinus (Lucius Flavius Valerius), Roman emperor, 1817, 1900, 1985, 1988, 2071, 2194, 2284.
Constantinus Manasses, 2484.
Constantius II, Roman emperor, 1796.
CONTAMINE (Philippe), 2587, 3130.
CONTE (Domenico), 5327.
CONTINI (Alessandra), 4230.
CONWAY-LANZ (Sahr), 7388.
CONZE (Eckart), 3933, 4198.
CONZE (Vanessa), 3530.
COOK (Chris), 624, 625.

COOK (Gordon), 1069.
COOK (Hera), 6171.
COOK (James Wyatt), 986.
Cook (James), 242.
COOK (William Robert), 3174.
COOKE (Ashley), 1159.
COOPER (Alan), 2627.
COOPER (Catherine Fales), 3417.
COOPER (David), 3087.
COOPER (Duff, Viscount Norwich), 4033.
COOPER (Randolf G. S.), 6743.
COOPEY (Richard), 5894.
COPANI (Fabio), 1554.
COPENHAVER (Brian P.), 482.
COPLAND (Ian), 4139.
COPLEY (M. S.), 1078.
COPPENS (Chris), 5293.
COPPOLINO (Nina Carmel), 2065.
COPSEY (Nigel), 4045.
Corbea, family, 6868.
CORBIER (Mireille), 1778.
CORBIN (Alain), 564.
Corbin (Henry), 908, 925.
CORCELLA (Aldo), 1555.
CORDASCO (Pasquale), 1.
CORDELL (Karl), 6641.
CORDESMAN (Anthony H.), 4514.
CORDIER (Pierre), 2005, 2066.
CORDOBA (Joaquín María), 1079.
CORDOVA (Ferdinando), 4210.
CORFIS (Ivy A.), 2957.
CORK (Edward), 1059.
CORKIN (Stanley), 5640.
CORLEY (T. A. B.), 6163.
CORNEJO (Francisco J.), 5597.
CORNELISSEN (Christoph), 355.
CORNI (Gustavo), 4198, 7126.
CORRADI (Michele), 1647.
CORREA (Sofía), 3704.
CORREALE (Robert M.), 3092.
CORRIENTE (Federico), 1283.
CORRITORE (Renzo P.), 62.
CORRY (Jennifer M.), 2996.
CORSELLIS (John), 4534.
CORSTEN (Thomas), 1404.
CORTASSA (Guido), 2397.
CORTELLESSA (Andrea), 5531.
CORTÉS (Jorge), 3653.
CORTÉS GABAUDAN (Francisco), 1556.
CORTÉS PEÑA (Antonio Luis), 4562.
CORTI (Maria), 2997.
CORVOL (Andrée), 5797.
CORZO GONZÁLEZ (Diana), 6960.
COȘA (Anton), 4872.
COSGEL (Metin M.), 5989.
COSKI (John M.), 257.
COSKUN (Altay), 1933.
COȘKUN (Rıdvan), 5598.

COSMA (Călin), 2002.
COSMA (Ela), 4437.
COSMA (Viorel), 832.
COSMACINI (Giorgio), 788.
COSME (Pierre), 1875.
COSTA (Desmond), 1432.
COSTANTINI (Emanuela), 4438.
COSTANZA (Salvatore), 2006, 2450.
COSTEA (Ionuț), 6172.
COSTIGAN (Richard F.), 4830.
COTTA RAMOSINO (Luisa), 1829.
COTTIER (Jean-François), 4809.
COTTLE (Drew), 3609.
COUHADE-BEYNEIX (Cynthia), 2156.
COULET (Noël), 2785.
COULOMB (Clarisse), 500.
COULON (Virginia), 974.
COULTON (J. J.), 1126.
COURTHÈS (Eric), 214.
COURTINE (Jean-Jacques), 564.
COURTWRIGHT (David T.), 6163.
COUTELLE (Éric), 2067.
COUTU (Éric), 7252.
COUVENHES (Jean-Christophe), 1515.
COVELL (Maureen), 595.
COWAN (Brian), 6173.
COWEN (Painton), 3175.
COWLEY (Robert), 7326.
COX (Catherine S.), 2998.
COX (Iris), 5347.
COX (J. C. M.), 6550.
CRAIG (Béatrice), 6203.
CRAIG (Edward), 960.
CRAIG (Geoffrey), 1069.
CRAIG (Oliver E.), 1045.
CRAIG (Oliver), 1069.
CRAIG BRITTAIN (Christopher), 898.
Cratylus, 1546.
CRAWFORD (E. Margaret), 6204.
CRAWFORD (Elisabeth T.), 5197.
CRAWFORD (Jon G.), 4174.
Craxi (Benedetto 'Bettino'), 4209, 7526.
CREAZZO (Tiziana), 2451.
CREIGHTON (Margaret S.), 4732.
CREIGHTON (Oliver Hamilton), 3492.
CREMANTE (Renzo), 5505.
CREPAS (Nicola), 5803.
CRÉPIN (Annie), 3851.
CRÉPIN (Marie-Yvonne), 730.
CREVATIN (Giuliana), 3152.
CRIADO (Cecilia), 1574.
CRIADO BOADO (Felipe), 1076.
CRISCENTI (Luca), 4197.
CRISTIANI TESTI (Maria Laura), 3176.

CRISTIN (Renato), 7363.
CRISTOFOLI (Roberto), 2007.
CRISTÓVÃO (Fernando), 4427.
Critobulus of Imbros, 2385.
CRIVELLO (Fabrizio), 3177, 3425.
CROCE (Benedetto), 502, 519, 5327.
CROISILLE (Jean-Michel), 2157.
CROIX (Alain), 5092.
CROKE (Brian), 2452.
CRONON (William), 7844.
CROOK (David), 2737.
CROPP (Martin), 1557.
CROSBY (Emilye), 6174.
CROSS (Gary S.), 6175.
CROSS (M. Claire), 3432.
CROSS (Richard), 3271.
CROUCH (David), 2797.
CROUSAZ (Karine), 5269.
CROUTSCH (Christophe), 1050.
CROUZET (Denis), 3854.
Crow (Jim), 4755, 5963, 6436.
CROWE (Benjamin D.), 503.
CROWSTON (Clare Haru), 6176.
CROWTHER (Charles), 1127.
CRUM (Roger J.), 5555.
CRUZ (Consuelo), 3731.
CRUZ (Rafael), 6421.
CRUZ BARNEY (Oscar), 5895.
CRUZ DE CARLOS (María), 302.
CRUZ E SILVA (Teresa), 402.
CRUZ ROMEO (M.), 4613.
CSUKOVITS (Enikő), 751.
Cu'ò'ng Đê, 4803.
CUDJOE (Richard V.), 1486.
CUELLA ESTEBAN (Ovidio), 2521.
CUEVA SEVILLANO (Alfonso), 631.
CUFF (Timothy), 5730.
Čulen (Martin), 4532.
CULLEN (L.), 469.
CUMMINGS (Owen F.), 2275.
Cumont (Franz), 455.
CUNFER (Geoff), 5935.
CUNLIFFE (Barry), 1094.
CUNNINGHAM (Hugh), 6438.
CUOMO (Valentina), 2453.
CÜPPERS (Martin), 7127.
CURL (James Stevens), 833.
CURP (T. David), 4393.
CURRAN (James), 4048.
CURRAN (John), 1876.
CURRELL (Susan), 6177.
CURRIE (Bruno), 1558.
CURRY (Ann), 2746.
CURSENTE (Benoît), 434, 2523.
CURSIETTI (Mauro), 2949.
CURTA (Florin), 2588, 2883.
CURTHOYS (Ann), 345.
CURTIS (John), 1080.
CURTO (José C.), 6767.

CUSHING (Kathleen G.), 3357.
CUSSEN (Celia L.), 4873.
CUST (Richard), 4049.
CVIRN (Janez), 3625, 6561.
Cyrillus Alexandrinus, 2222.
Cyrillus Scythopolitanus, 2370.
CYRINO (Monica Silveira), 5641.
CZELK (Andrea), 6562.
CZIGÁNY (István), 4125.

D

D'ADAMI (Luisa Maria), 2798.
D'AGOSTINO (Bruno), 1761.
D'AGOSTINO (Franco), 1205, 1216.
D'Ailly (Pierre), 3464.
D'ALESSIO (Giovan Battista), 1559.
D'Amico (Silvio), 6046.
D'ANGELO (Augusto), 4212.
D'ANGOUR (Armand), 1597.
D'AUTILIA (Gabriele), 258, 4197.
D'AVRAY (David L.), 2755, 2891.
D'ELIA (Gianni), 5511.
D'ERCOLE (Maria Cecilia), 1752.
D'ESPOSITO (Francesco), 6250.
D'ORIANO (Rubens), 1753.
DA COSTA KAUFMANN (Thomas), 5599.
DA NÓBREGA (José Manuel), 6811.
DA ORDEN (María Liliana), 6178.
DA ROSA (Gilson Justino), 6179.
DABDAB TRABULSI (José Antonio), 348.
DACHER (Michèle), 114.
DAFTARY (Farhad), 2696.
DAGLI ORTI (Alfredo), 2447.
DAGUET-GAGEY (Anne), 2158.
DAHL (Gunnar), 1877.
DAHLHEIM (Werner), 1878.
DAHLMANN (Dittmar), 5884.
DAI (Lu), 7618.
DAJCZAK (Wojciech), 6567.
DAJNOWICZ (Małgorzata), 4394.
DAL LAGO (Enrico), 5936.
DALARUN (Jacques), 3377.
DALE (Gareth), 3935.
DALE (Richard S.), 5990.
DALEWSKI (Zbigniew), 2625.
DALLA MASSARA (Tommaso), 1968.
DALLA VECCHIA (Patrizia), 94.
DALLAS (Gregor), 7128.
DALLEY (Stephanie), 1217.
DALMON (Laurence), 2308.
DALTON (Margaret Stieg), 5552.
DALY (Peter Maurice), 130.
DAMERON (George Williamson), 3439.
DAMIANAKI (Chrysa), 5140.
DAMMANN (Clas), 5642.
DAMOUSI (Joy), 5380.

DANDAMAEVA (Mariam M.), 1110.
DANDO-COLLINS (Stephen), 1969.
DANEK (Georg), 1690.
DANGL (Vojtech), 627.
DANGLER (Jean), 2892.
DANIEL (Pete), 5381.
DANIELE (Chiara), 5331.
DANIELEWICZ (Jerzy), 1560.
DANIELS (Gordon), 7551.
DANIELS (Heinrich Gottfried Wilhelm), 6563.
DANILOV (Viktor N.), 349.
DANKER (Uwe), 3936.
DANKERS (Joost), 5977.
DANTE ALIGHIERI, 2622, 2899, 2958, 2984, 2989, 2998, 3096, 3116, 3123, 5149.
DANTHIEUX (Dominique), 3852.
DANTO (Elizabeth Ann), 5382.
DANZI (Massimo), 295.
DANZIGER (Danny), 1879.
Darby (Douglas Evelyn), 3609.
DARD (Olivier), 3853, 5765.
DARDANO (Paola), 1256.
Darius III, king of Persia, 1362.
DARK (Ken), 2159.
DARNTON (Robert), 5298.
DAROVEC (Darko), 5991.
DARTHOU (Sonia), 1691.
Darwin (Charles), 5467.
DARWIN (John), 6713.
DAS NEVES TEMBE (Joel), 402.
DASGUPTA (Biplab), 6744.
DAUB (Susanne), 2999.
DAUBRESSE (Sylvie), 3854.
DAUCHY (Serge), 740.
DAUDIN (Guillaume), 5897.
DAUER (Richard P.), 7473.
DAUGE (Louis), 6961.
DAUM (Werner), 5270.
DAUNTON (Martin J.), 5435.
DAUPHINAIS (Michael), 3302.
DAVARI (Mahmood T.), 4160.
DÁVID (Géza), 4118.
DAVID (Thomas), 6775.
David, king of Israel, 2219, 2367.
DAVIDS (Mila), 5804.
DAVIDSON (Apollon B.), 7855.
DAVIDSON (Christopher), 4714.
DAVIDSON (Clifford), 3000.
DAVIDSON (Herbert Alan), 2683.
DAVIES (John), 1487.
DAVIES (Paul), 1026.
DAVIES (Russell), 6180.
DAVIES (Sarah), 4505.
DAVIS (Eric), 350.
DAVIS (Norman), 2555.
DAVIS (R. Hunt), 7865.
DAVIS (Richard W.), 4079.
DAVIS (Timothy), 5778.
DAVISON (Michael), 7451.

DAWSON (Graham), 351.
DAWSON (Mark S.), 5643.
DAWSON (Melanie), 6181.
DAXNER (Michael), 3523.
DAY (Juliette), 2410.
DAYAN-HERZBRUN (Sonia), 6182.
DAYIOĞLU (Ali), 3681.
DE ALMEIDA BALDISSERA (Marli), 3664.
De Arrizubieta (Martín), 4595.
DE BAETS (Antoon), 504.
DE BALBÍN BEHRMANN (Rodrigo), 1057.
DE BENOIST (Alain), 3835.
DE BOISFLEURY (Bernard), 7297.
DE CALLATAY (François), 1516.
De Callières (François), 6914.
DE CARLO (Antonella), 1970.
DE CASTELNAU-L'ESTOILE (Charlotte), 6711.
DE CASTRO GOMES (Angela), 3672.
DE CAZANOVE (Olivier), 2130.
DE CEVINS (Marie-Madeleine), 3141.
DE CONINCK (Pieter), 4332.
DE COURCELLES (Dominique), 5476.
DE DAINVILLE BARBICHE (Ségolène), 4874.
DE DOMINICIS (Alessandro), 1058.
DE FLEURQUIN (L.), 857.
DE FREIXO (Adriano), 3666.
DE GALBERT (Geoffroy), 1880.
De Gasperi (Alcide), 4198, 7421.
De Gaulle (Charles), 3841, 3850, 5713, 6757, 7171, 7231, 7544.
DE GIORGI (Laura), 352.
DE GRAZIA (Victoria), 6183.
DE HAAS (Claire de), 2543.
DE HOZ (Javier), 1754.
DE KREY (Gary S.), 4050.
DE LA FLOR (Fernando R.), 5072.
De la Fuente, familia, 5916.
DE LA IGLESIA DUARTE (José Ignacio), 2629, 2794.
DE LA PUENTE (Cristina), 2695.
DE LA RONCIÈRE (Charles Marie), 2799.
DE LA ROSA (Jesús), 3811.
DE LA TORRE (Renée), 4318.
DE LA VAISSIÈRE (Étienne), 753.
De Luca (Giuseppe), 456.
DE LUCIA BROLLI (Maria Anna), 2160.
DE LUNA (Giovanni), 555, 4197.
DE MADARIAGA (Isabel), 4469.
DE MAMBRO SANTOS (Ricardo), 5553.
DE MARCO (Miguel Angel), 3586.
De Martino (Ernesto), 884.

DE MARTINO (Stefano), 1257.
DE MATOS (Artur Teodoro), 211.
DE MENESES (Filipe Ribeiro), 7310.
DE NIE (Giselle), 3214.
DE PALMA (Luigi Michele), 353.
De Palma (Pedro), 3196.
DE PEDRO MICHÓ (María Jesús), 1060.
DE POLIGNAC (François), 1081.
DE PRÉMARE (Alfred-Louis), 2697.
DE PRIVITELLO (Luciano), 3586.
DE QUEIRÓS MATTOSO (Katia M.), 3667.
DE RIJK (Lambertus Marie), 3253.
DE ROBERTIS (Domenico), 2958.
DE ROMILLY (Jacqueline), 480, 1448.
DE ROTEN (Philippe), 2309.
DE SCHAEPDRIJVER (Sophie), 3640.
DE SENARCLENS (Vanessa), 490.
DE SOUZA (Ismara Izepe), 3665.
De Thérines (Jacques), 3328.
DE VEGA (Mercedes), 4298.
DE VENUTO (Liliana), 5073.
DE VILLEPIN (Dominique), 6662.
DE VOS (Luc), 3726.
DE VRIES (Kelly), 3124, 3132.
DEÁK (Ladislav), 7183.
DEAR (Ian), 674.
DEBENEDETTI (Giacomo), 5512.
DĘBNICKI (Krzysztof), 4329.
DEBORD (Pierre), 1692.
Debré (Michel), 3877.
DEBRIS (Cyrille), 115.
DECHERNEY (Peter), 5644.
DECKER (Wolfgang), 1370.
DECOT (Rolf), 4837.
Decoux (Jean), 7593.
DEDNER (Burghard), 5081.
DEE (James H.), 2068.
DEE (Moreen), 7447.
DEER (Cécile), 5216.
DEERE (Carmen Diana), 6564.
DEERY (Phillip), 6422.
DEFFAYET (Laurence), 4875.
DEFLERS (Isabelle), 6476.
DEFOSSE (Pol), 790.
DEFRANCE (Corine), 7361.
DEGL'INNOCENTI (Antonella), 3405.
DEGL'INNOCENTI (Maurizio), 4213.
DEGOLS (Renaud), 637.
DEGÓRSKI (Bazyli), 2225.
Degrelle (Léon), 3641.
Dehio (Georg), 488.
DEHN-NIELSEN (Henning), 628.
DEKEL-CHEN (Jonathan L.), 5937.
DEKESEL (Christian E.), 137.

DEKEYZER (Brigitte), 39, 3197.
DEKKER (Kees), 3079.
DEKONINCK (Ralph), 259.
DEL BOCA (Angelo), 6776, 7152.
DEL CAMPO (Hugo), 3590.
DEL CERRO (María del Carmen), 1079.
DEL FREO (Maurizio), 1388, 1392.
DEL LLANO IBÁÑEZ (Ramón), 4299.
DEL OLMO LETE (Gregorio), 1107, 1284.
DEL RÍO (María de las Mercedes), 6828.
DEL VAL VALDIVIELSO (María Isabel), 2657.
DELAPLACE (Christine), 3351.
Delaunay (Robert), 5594.
DELCARO (Dino), 1050.
DELEBARRE (Michel), 7287.
DELFINO (Susanna), 5821.
DELFOUR (Christine), 3654.
DELGADO (Lorenzo), 6645.
DELGADO (Mariano), 5495.
DELGADO BARÓN (Mariana), 3719.
DELGADO JARA (Inmaculada), 2254.
DELL'ORO (F.), 3425.
DELLA CORTE (Federico), 2972.
DELLA FINA (Giuseppe M.), 1762, 1771.
DELMAIRE (Roland), 1985.
DELORT (Robert), 778.
DELOS-HOURTOULE (Sarah), 6924.
DELPAL (Bernard), 4820.
DELPRAT (François), 3654.
DELSALLE (Paul), 629.
DELTOMBE (Thomas), 5074.
DELYSER (Dydia), 6184.
DEMANDT (Alexander), 471.
DEMANGE (Jean-François), 129.
DEMARS-SION (Véronique), 740.
DEMIDCHIK (Arkadij E.), 1164.
DEMİR (Hüseyin), 5217.
DEMIR (Remzi), 961.
DEMIRCI (Sevtap), 7037.
DEMIRYÜREK (Mehmet), 3750.
DEMKER (Marie), 4357, 7474.
Democritus, 1660.
Demosthenes, 1425.
DEMURA (Akira), 3323.
DEMURGER (Alain), 3375.
DEN BOEFT (Jan), 1795.
DEN HENGST (Daniel), 1795.
Denaux (A.), 2258.
DENCH (Emma), 1881.
DENCKER (Friedrich), 6565.
DENDORFER (Jürgen), 2630.
DENERY (Dallas George), 2893.
DENG (Ye), 7620.
DENIZE (Eugen), 4439, 4440.

DENNINGMAN (Susanne), 1648.
DENNIS (George T.), 2576.
DENNIS (Richard), 6063.
DENOËL (Charlotte), 52.
DENOON (Donald), 6862.
DENZEL (M. A.), 5992.
DENZER (Jörg), 7877.
DEOPIK (Dega V.), 7597.
DEPELCHIN (Jacques), 7863.
DEPLOIGE (Jeroen), 2920.
DÉROCHE (François), 25.
DERORI (Ze'ev), 4191.
DEROUX (Carl), 2112.
DES BOSCS-PLATEAUX (Françoise), 2008.
DES BOUVRIES (Synnove), 1693.
Descartes (René), 5326.
DESCHAMPS (Olivier), 6566.
DESCIMON (Robert), 4876.
DESCOLA (Philippe), 556.
DESGRAVES (Louis), 64.
DESHPANDE (Anirudh), 6745.
DESLANDES (Paul R.), 5176.
DESMULLIEZ (Janine), 1789.
DESPOIX (Philippe), 215.
DESPORTES (Marc), 5805.
DESPORTES (Pierre), 3346.
DESPRADEL (Alberto), 6643.
DESSÌ (Rosa Maria), 2928.
DESSÌ SCHMID (Sarah), 5327.
DESTANI (Bejtullah D.), 3570.
DESTRÉE (Pierre), 1649, 1678.
DESVOIS (Jean-Michel), 90.
DESY (Ph.), 2070.
DETER (Gerhard), 6423.
DETIENNE (Marcel), 354, 557.
DETTMAR (Kevin J. H.), 787.
DEUS (Valdemar de), 2739.
DEVEAU (Jean-Michel), 7864.
DEVINE (Michael J.), 7415.
DEVLET (Ekaterina G.), 1013.
DEVLET (Marianna A.), 1013.
DEZHGIU (Muharrem), 7129.
DHAR CHAKRABARTI (Prabodh G.), 6064.
DHILLON (Kirpal S.), 4140.
DHUGA (Umit Singh), 1561.
DI BENEDETTO (Vincenzo), 1562.
DI CAPUA (Giovanni), 7130.
DI DONATO (Riccardo), 1563.
DI FAZIO (Massimiliano), 1763.
DI FEBO (Giuliana), 7202.
DI GIACINTO (Valter), 5731.
DI GIUSEPPE (Helga), 1764.
DI GIUSTO (Stefano), 7253.
DI MACCO (Michela), 854.
DI MARCO (Massimo), 1417.
DI MARTINO (Paolo), 6568.
DI MURO (Giovanni F.), 3641.
DI PAOLA (Lucietta), 2195.
DI RIENZO (Daniele), 2224.

DI SEGNI (Leah), 1285.
DI VENOSA (Elena), 2894.
DIALLO (Mamadou Dian Cherif), 6777.
DIAMOND (Hanna), 3897.
DÍAZ ARIAS (David), 3732.
DÍAZ BENÍTEZ (Juan José), 7254.
DÍAZ DE BUSTAMANTE (J. M.), 3070.
DÍAZ DE CERIO (Mercedes), 1650.
DÍAZ DE DURANA (José Ramón), 2591.
DÍAZ Y DÍAZ (Manuel C.), 19, 3070.
DIČEV (Borislav), 5669.
DICK (Michael B.), 1218.
DICKMANN (Jens-Arne), 2161.
DICKSTEIN-BERNARD (C.), 3370.
DIECKMANN (Sandra), 3002.
DIEDERICH (Toni), 323.
DIEDRICH (Torsten), 4013.
DIEHL (Markus Albert), 5732.
DIEHL (Paula), 260.
DIETERICH (Renate), 7196.
DIETSCHY (Paul), 6243.
DIETZ (Günter), 1565.
DIETZ (Gunther), 2698.
DIETZ (Maribel), 3378.
DIEZ DE VELASCO (Francisco), 862.
DIGBY (Anne), 6187.
DIGGLE (James), 1540.
DILAN (Hasan), 4675.
DILKS (David), 6644.
DILKS (Richard), 6644.
DILLERY (John), 1694.
DILLON (Matthew), 1855.
Dilthey (Wilhelm), 503, 5328.
DIM (Emmanuel Uchenna), 2268.
DIMASHQĪ (Shams al-Dīn Muhammad ibn Abī Tālib), 2693.
DIMIĆ (Ljubodrag), 7435.
DIMITRIJEVIĆ (Bojan B.), 7389.
DIMMER (C.), 4029.
DIMOV (Dimcho N.), 3682.
DINARDO (Richard L.), 7131.
DINÇ (Güven), 6188.
DINET-LECOMTE (Marie-Claude), 4915.
DING (Guangxun), 7621.
Ding (Ruchang), 7668.
DINI (Mario), 1024.
Dio Chrysostomus, 1604, 1684.
Diodorus Siculus, 1423.
Diodorus Tarsensis, 2223.
Diogenes Laertius, 1666.
DION (Michèle), 6778.
Dionysius Alexandrinus, 1152.
Dionysius Halicarnassensis, 1546, 1616.

Dionysius Oenoandensis, 1672.
DIPPEL (Horst), 6508.
DIPPER (Christof), 5075.
DIRLMEIER (Ulf), 2800.
Disney (Walt), 5705.
Disraeli (Benjamin, Earl of Beaconsfield), 4068.
DISTEL (Barbara), 4001.
DITTMANN (Gudrun), 5645.
DIXON (Chris), 7248.
DIXON (David), 6829.
DLUBEK (Rolf), 5346.
DŁUGAICZYK (Martina), 4333.
DŁUGAJCZYK (Edward), 4395.
DŁUGOŁĘCKI (Piotr), 7181, 7453.
DMITRIEV (Sviatoslav), 1128.
Dmowski (Roman), 4393.
DO RÊGO (André Heráclio), 3667.
DOANE (Alger Nicolaus), 2950.
DOBEŠ (Jan), 6484.
DOBIAS-LALOU (Catherine), 1408.
DOBLES (Ignacio), 3733.
DOBOZY (Maria), 3003.
DOBRE (Dumitru), 4436.
DOBRINCU (Dorin), 5968.
DOBROVOL'SKAJA (Maria V.), 1014.
DOBSON (Alan P.), 7475.
DOCHHORN (Jan), 1286.
DOČKAL (Vít), 6331.
DOCKRILL (Saki Ruth), 7476.
DOCTER (Roald Fritjof), 1883.
DODD (David B.), 1696.
DODD (Lawrence C.), 4345.
DODD (Thomas J.), 4114.
DOENECKE (Justus D.), 7197.
DOERKSEN (Clifford J.), 5646.
DOGO (Marco), 6701.
DOĞRUEL (Fatma), 5993.
DOHNA (Jesko Zu), 5994.
DOIGNON (Jean), 2234.
DOLBEAU (François), 3408, 3476.
DOLHNIKOFF (Miriam), 3668.
DOLL (Natascha), 6569.
DOMAGAŁA (Tadeusz), 2738.
DOMARADZKI (Kamil), 2454.
DOMENICO (Roy Palmer), 7329.
Domenico, santo, 3379.
DOMINGUEZ PÉREZ (Juan Carlos), 2009.
Domitianus (Titus Flavius), Roman emperor, 1949, 1990.
DOMPNIER (Bernard), 5682.
DONADIEU-RIGAUT (Dominique), 261.
Donatello, 5555.
DONATI (Claudio), 816.
Donatus (Aelius), 1835.
Doni (Anton Francesco), 5544.
DÖNMEZ (Şevket), 1082.
DONNER (Sandra), 6190.

DONOUGHUE (Bernard), 4032.
DOOLITTLE (Amity A.), 5938.
DOR (Fabrice), 3202.
DORAN (Susan), 6892.
DORÉ (Jean-Marie), 6779.
DÖRING (Hans-Joachim), 7331.
Döring (Mathias), 3402.
DORION (Louis-André), 1651.
DORMEYER (Detlev), 2255.
DORNIER (Carole), 506.
DOSI (Giovanni), 5880.
DOSKOČIL (Zdeněk), 3767.
Dossetti (Giuseppe), 7392.
DOST (Pinar), 7132.
DOUDET (Estelle), 2922.
DOUGHERTY (Jack), 5218.
DOUGHTY (Robert A.), 7038.
DOUGLAS (Michele T.), 1042.
DOUGLAS (Roy), 4051.
Douglas, family, 2652.
Douglas-Home (Alec), 7498.
DOUKELLIS (Panagiotis), 1449.
DOUTHWAITE (Julia V.), 5094.
DOUZOU (Laurent), 357.
DOVERE (Elio), 1971.
DOWNEY (Susan B.), 1287.
DOWNING (Eric), 5514.
DOYLE (Matthew A.), 3139.
Drach (Peter), 80.
Dracontius (Blossius Aemilius), 2361.
DRÄGER (Paul), 1817, 2071.
DRAGO (Corinna), 1.
DRAGONAS (Thalia G.), 4101.
DRAKE (David), 3855.
DRÁPALA (Milan), 5271.
DRAY (Philip), 5383.
DREESBACH (Anne), 305.
DREISBACH (Kai), 1001.
DREMEL (Erik), 5647.
DRÉVILLON (Hervé), 3856.
DREYER (Elizabeth), 3441.
Dreyfus (Alfred), 416, 3904.
DREYFUS (F.-G.), 7225.
DREYFUS (Hubert L.), 959.
DRIBE (Martin), 762, 6191, 6203.
DRIJVERS (Jan Willem), 1795.
DRINOT (Paulo), 4375.
DRONIN (Nikolai M.), 5939.
DROUIN (Michel), 7049.
DROUX (Xavier), 1165.
DROZDEK (Adam), 1652.
DRUKS (Herbert), 7477.
DU (Lihong), 7622.
DU RÉAU (Elisabeth), 7055.
DUBBER (Markus Dirk), 4733.
DUBERT (Isidro), 117.
Dubinsky (David), 6457.
DUBLIN (Thomas), 5806.
DUBOIS (Laurent), 1403, 1409.
DUBOIS (Sébastien), 3642.

DUBOIS (Thomas David), 721.
DUBOURG GLATIGNY (Pascal), 5384, 5580.
DUBRAY (Jean-Marc), 637.
DUBROVSKIJ (Aleksandr M.), 358.
DUCH I PLANA (Montserrat), 4563.
DUCHET (Claude), 5049.
DUCHET (Michèle), 558.
DUCHOŇ (Michal), 6131.
DUCLOS (France), 229.
DUDA (Igor), 5995.
DUDD (S. N.), 1078.
DUDDEN (Alexis), 7039.
DUE-NIELSEN (Carsten), 7321.
DUFAYS (Jean-Michel), 790.
DUFEK (Pavel), 5807.
DUFFY (Seán), 664.
DUFOIX (Stéphane), 6715.
DUFOUR (Alain), 5053.
DUFOUR (Jean), 2560.
DUFOUR (Xavier), 864.
DUGAC (Željko), 5076.
DUGAN (John), 2072.
DUGGAN (Anne J.), 2597.
DUGGAN (Hoyt N.), 3036.
DUHOUX (Yves), 1393.
DULAEY (Martine), 2311.
DULL (Jonathan R.), 6903.
DULONG (Renaud), 506.
DULOVIČ (Erik), 4529.
DUMARÇAY (Jacques), 791.
DUMÉZIL (Bruno), 3324.
Dumézil (Georges), 457.
DUMMER (Jürgen), 2382.
DUMOLYN (Jan), 2801.
DUMONT (Dora M.), 6424.
DUMONT (Georges Henri), 3643.
DUMOULIN (Michel), 3646, 3648, 6780.
DUMVILLE (David. N.), 2571.
DUNAWAY (Finis), 755.
DUNCAN (Jason K.), 4877.
DUNCAN (Thomas Gibson), 619.
DUNCAN-JONES (Richard P.), 2010.
DUNKERLEY (James), 3650, 7368.
Dunlap (William), 5562.
DUNLOP (D. M.), 3246.
DUNLOP (Eileen), 2626.
DUNN (Geoffrey B.), 2312.
DUNN (James G.), 2273.
DUNN (John P.), 3825.
Duns Scotus (Iohannes), 3254, 3271, 3300.
DUNTHORN (David J.), 7198.
DUPLESSIS (Robert S.), 5898.
DUPLOUY (Alain), 1517.
DUPONT (Christine A.), 5556.
DUPONT (Jacques), 5506.
Duprat (Antoine), 6545.
DURACINSKI (Eugeniusz), 7354.

DURÁN ALCALÁ (Francisco), 4567.
DURÁN TAPIA (R.), 2473.
DURAND (Jean Dominique), 4916.
DURAND (Jean Marie), 1219.
DURAND (Pierre Michel), 7478.
DURBEC (Yannick), 1566.
ĎURČANSKÝ (Marek), 5153.
DURNFORD (S. P. B.), 1109.
DURPAIRE (François), 6781.
DÜRR (Renate), 218, 4970.
DURU (Refik), 1015.
DURUKAN (Murat), 1129.
DUSHIN (Oleg E.), 2895.
DUSSAULT (Eric), 7390.
DUTSCH (Dorota), 2119.
Dutschke (Rudi), 3977.
DUTTON (David), 7199.
DUTTON (Marsha L.), 2570.
DUTTON (Michael), 7623.
DUŢU (Alesandru), 7519.
DUVAL (Marcel), 7479.
DUVAL (Yves-Marie), 2313.
DUYRAT (Frédérique), 1130.
DVOŘÁK (Jiří), 3768.
DVOŘÁK (Pavel), 6131.
DVOŘÁK (Tomáš), 5808.
Dyakonov (Igor' M.), 1110.
DYAS (Dee), 3428.
DYER (Christopher), 2802, 2868.
DYHOUSE (Carol), 5219.
DYSINGER (Luke), 2230.
DZENISKEVICH (Andrei), 7270.
DZIUBIŃSKI (Andrzej), 6877.

E

EALHAM (Chris), 4564, 5155.
EATON (Richard M.), 756.
EATOUGH (Geoffrey), 195.
EBBEN (Maurits), 6913.
EBBINGHAUS (Susanne), 1695.
EBEL (Kathryn A.), 6038.
EBENBAUER (Alfred), 2965.
EBERLE (Henrik), 286, 7109.
EBIED (Rifaat. Y.), 2693.
EBNETER (Thomas), 3272.
ECHAVARRÍA (Martín F.), 3273.
ECHTERNKAMP (Jörg), 3939, 6402.
ECK (Werner), 1943, 1972.
ECKARDT (Hella), 2162.
ECKART (Wolfgang Uwe), 792.
ECKEL (Jan), 475.
ECKERSLEY (Peter), 6904.
ECKHART (Hans Wilhelm), 5.
Eckhart, meister, 3261, 3270, 3292, 3299.
ECKL (Andreas E.), 6764.
EDDEN (Valerie), 3428.
Eden (Anthony, Earl of Avon), 7370, 7420.
EDEN (Kathy), 1653.
EDER (Franz X.), 825.

EDGERTON (David), 4052.
EDIGATI (Daniele), 6570.
EDIGER (Volkan Ş.), 5809.
EDMONDS (Anthony O.), 7451.
EDMONDS (Mark), 1043.
EDMONDSON (Jonathan C.), 1884.
Edmont (Edmond), 177.
EDOUARD (Sylvène), 262.
EDROIU (Nicolae), 3495.
Edward I, king of England, 2623, 2746.
Edward II, king of England, 2671, 2746.
Edward III, king of England, 2746, 2848.
Edward IV, king of England, 2746.
EDWARDS (Anthony Stockwell Garfield), 47, 3056.
EDWARDS (Clive), 6192.
EDWARDS (Gwynne), 835.
EDWARDS (John), 3442.
EDWARDS (Nancy), 2538.
EGAN-BUFFET (Máire), 3224.
Egeria, 2237.
EGGERT (Heinz-Ulrich), 5220.
EGO (Beate), 2353.
EHRENBERG (Victor), 5329.
EHRENPREIS (Petronilla), 7040.
EHRESMANN (Nina), 5600.
EHRICK (Christine), 6193.
EIBL (Karl), 5080.
EICH (Peter), 1885.
EICHHOLTZ (Dietrich), 7200.
EIKELMANN (Manfred), 3102.
EINAR SKÚLASON, 2572, 2572, 2718.
EINONEN (Piia), 4641.
Einstein (Albert), 5450.
EISENBERG (Christiane), 5077.
EISENHOWER (Dwight David), 7370, 7378, 7451, 7532.
EISENHOWER (John S. D.), 635.
EISENSTAEDT (Jean), 5385.
EISFELD (Jens), 6572.
EITRHEIM (Øyvind), 5899.
EKEH (Peter P.), 6814.
EKENGREN (Ann-Marie), 7480.
EKINCI (İlhan), 6962.
EKSHTUT (Semen A.), 5557.
EL KENZ (David), 757.
EL MÉCHAT (Samya), 7481.
ELDAMATY (Mamdough), 1166.
ELDEM (Edhem), 559.
Eleanor de Aquitaine, queen, 3120.
ELENA (Eduardo), 3591.
ELEONORE VON ÖSTERREICH, 2959.
ELGEMYR (Göran), 5648.
ELIADE (Mircea), 906, 910, 912, 925, 4438.
Elias (Norbert), 458.

ELIASON (Eric), 3036.
ELIASSON (Per), 779.
Elizabeth I, queen of England and Ireland, 4073.
ELIZALDE (María Dolores), 6645.
ELKINS (Caroline), 6730, 6782.
ELLIOTT (Gregory), 4489.
ELLIS (Faron), 3691.
ELLIS (Joseph J.), 4734.
ELLIS (Roger), 3428.
ELLIS (Steve), 613.
ELMAN (Benjamin A.), 5386.
EL-MASRI (Yahia), 1167.
ELMQUIST JØRGENSEN (Kasper), 5735.
ELO (Kimmo), 3941.
EL-ROUAYHEB (Khaled), 6195.
ELSNER (Jaś), 873, 2163.
ELTIS (David), 6714.
ELTON (Hugh), 2416.
ELTZ (Erwein), 4550.
EMANUELSSON (Anders), 2803.
EMBERGER (Peter), 1811.
EMBLETON (Ron), 1988.
EMICH (Birgit), 4214.
EMINGER (Stefan), 5810.
Emma, queen (consort of Canute I, king of England), 2726.
EMMEL (Stephen), 1168.
Empedocles, 1667.
ENACHE (George), 4441.
ENCREVÉ (André), 4971.
ENDICOTT (John), 6680.
ENDO (Yasuo), 5151.
ENE (Sorin), 5222.
ENEEVA (Natal'ja T.), 3555.
ENENKEL (Karl. A. E.), 63.
ENGBERG (Jens), 5078.
ENGEHAUSEN (Frank), 3942.
ENGEL (Gisela), 218.
ENGEL (Jeffrey A.), 7482.
ENGELBERG (Meinrad von), 5581.
ENGELHARD (Kristina), 5338.
ENGELS (Friedrich), 5346.
ENGEN (Darel Tai), 1488.
ENGERMAN (Stanley L.), 6530.
ENGIN (Vahdettin), 6963.
ENGLER (Steven), 909.
ENGLISH (Brigitte), 107.
ENGLISH (Edward D.), 638.
ENGLISH (Mary), 1567.
ENGSTROM (Eric J.), 5390.
Ennodius (Magnus Felix), episcopus, 2224.
ENOKI (Kazue), 5811.
Enrich, familia, 5814.
ENRÍQUEZ FERNÁNDEZ (Javier), 2732.
Epictetus, 1426.
Epitadeus, 1480.
EPKING (Simone), 3178.

EPP (Franz, Ritter von), 6764.
ERASMUS ROTERODAMUS (Desiderius), 4339, 4809, 5269.
ERATH-KOINER (Gabriele), 1125.
ERBEN (Dietrich), 5574.
ERDEM (İlhan), 2804.
ERDKAMP (Paul P. M.), 2011.
ERDMANN (Karl Dietrich), 360.
ERDŐ (Péter), 6477.
Erenburg (Il'ja), 5517.
ĒRGLIS (Dzintars), 4269.
ERGO (André-Bernard), 6783.
Erhard (Ludwig), 3989.
ERICKSON (Brice), 1728.
ERICKSON (Lesley), 6830.
ERICSSON (Kjersti), 6197.
Erigena (Johannes Scotus), 3304.
Erik XIV, king of Sweden, 4650.
ERKENS (Franz-Reiner), 3447.
ERKER (Paul), 5812.
ERLANDSEN (Solveig K.), 5899.
ERMACHENKO (Igor' O.), 807.
ERMAKOVA (Ljudmila M.), 217.
ERNST (Peter), 165.
EROĞLU (Cengiz), 4685.
ERRERA (Michel), 1050.
ERSKINE (Andrew), 639.
ERTL (Thomas), 362.
Erxleben (Johann Christian Polycarp), 5416.
Erzurumlu İbrahim Hakkı, 5223.
ESAKOV (Vladimir D.), 4471.
ESCH (Arnold), 3179.
ESCHER (Monika), 3512.
ESCOBAR OHMSTEDE (Antonio), 5900.
ESCOBARI DE QUEREJAZU (Laura), 6198.
ESDAILE (Charles J.), 6935.
ESENWEIN (George Richard), 4566.
ESPAGNE (Michel), 363.
ESPAHANGIZI (Kijan Malte), 5387.
ESPELI (Harald), 5128.
ESPINEL (Andrés D.), 1169.
ESTEBAN ESTRÍNGANA (Alicia), 6878.
ESTEFANÍA (Dulce), 1574.
ESTES (James M.), 4972.
ESTES (Kenneth W.), 4605.
ESTES (Steve), 6199.
ESTÈVES (Aline), 2074.
ETEMAD (Bouda), 6716, 6775.
ETHERINGTON (Norman), 4941.
ETINGOF (Ol'ga E.), 3222.
ETZEMÜLLER (Thomas), 4642.
EUGENIUS TOLETANUS, 3247.
Eugippius, 2225.
EURAQUE (Dario A.), 3544.
Euripides, 1427.
Euripides, 1428, 1600, 1603, 1638.

Eusebius Caesariensis, 2226-2229.
EUSTACHE LE MOINE, 2960.
Eustathius Thessalonicensis, episcopus, 2393.
Eustratius Constantinopolitanus, 2387.
Eutropius, 2058.
Evagrius Ponticus, 2230.
Evagrius, episcopus, 2343.
EVANGELISTI (Paolo), 2659.
EVANS (Angela Care), 2938.
EVANS (James A.), 2455.
EVANS (James R.), 4520.
EVANS (Jane), 1069.
EVANS (Julie), 6863.
EVANS (Richard J.), 3943.
EVANS (Tanya), 6200.
EVE (Prosper), 6784.
EVERED (Kyle T.), 6994.
EVERSHED (R. P.), 1078.
EVORA (José Silva), 3698.
EVSEENKO (Timur P.), 1374.
EWALD (Owen M.), 1306.
Eyre (Edward John), 6863.

F

FABBRI (Cristina), 1040.
FABER (Martin), 4878, 4926.
FÁBREGAS VALCARCE (R.), 1072.
FABRIS (Cécile), 3140.
FAEHNDRICH (Jutta), 5326.
FAGEL (Raymond), 6913.
FAGERFJÄLL (Ronald), 5737.
FAGOAGA HERNÁNDEZ (Ricardo A.), 5900.
FAHLBUSCH (Michael), 371.
FÄHNDERS (Walter), 207.
FÄHNDRICH (Sabine), 2164.
FÄHNRICH (Heinz), 196.
FAHY (Conor), 96, 5542, 5543.
Faianus Plebeius (Publius), 1874.
FAIRBAIRN (Andrew), 1111.
FAIRBURN (Miles), 4053, 4343, 4344.
FAIST (Thomas), 780.
FAJMON (Hynek), 7144, 7540.
Fakkenhayn (Erich von), 7044.
FALADA (David), 6472.
FALASCA-ZAMPONI (Simonetta), 5575.
FALCH (Sabine), 7399.
FALCÓN (Romana), 4297, 4302.
FALILEEV (Aleksandr I.), 3274.
FALK (Barbara), 6066.
FALK (Beatrice), 3910.
FALKENBERG (Karin), 5649.
FALOLA (Toyin), 4352, 7858.
FAMERÉE (J.), 857.
FANG (Huawen), 7624.
FANG (Jun), 7625.

FANIZZA (Lucia), 1973.
FANOUS (Samuel), 3407.
FANTUZZI (Marco), 1568.
FANZUN (Jon A.), 7330.
FARAGÓ-GÜNTHER (Zita), 4957.
FARAGUNA (Michele), 1489.
FARAJ (Ali H.), 1220.
FARAONE (Chris A.), 1288.
FARAONE (Christopher A.), 1696.
FARBER (David), 7484.
FARBER (Walter), 1221.
FARENGA (Paola), 67, 5272.
FARGE (Arlette), 167.
FARKAS (Paul), 4686.
FARMHOUSE (Alberto Paulo), 3247.
FARNUM (Jerome H.), 1886.
Farquhar, family, 118.
FARQUHARSON (Geoffrey), 118.
FARRAUDIÈRE (Yvette), 4110.
FARRENKOPF (Michael), 5879.
FASCE (Ferdinando), 219.
FASH (William L.), 7876.
FATHI (Habiba), 5046.
FATTI (Federico), 2456.
FATTORINI (Emma), 4879.
Faupel (Wilhelm), 4595.
FAURE (Olivier), 4820, 5433.
FAUSER (Annegret), 5650.
FAUSER (Markus), 71.
FAUST (Avraham), 1289.
FAUTH (Tim), 3793.
FAUVE-CHAMOUX (Antoinette), 6189, 6203.
FAVARO (Orietta), 3607.
FAVIER (Jean), 2559, 2560, 2620.
FAVINO (Federica), 5389.
Favorinus, 1429.
FAVREAU (Robert), 2544.
FAWAZ (Mona), 6067.
FAXAS (Laura), 3812.
FAYE (Emmanuel), 5335.
Fayolle (Marie Émile), 7025.
FAZLIOĞLU (İhsan), 507.
FAZLIOĞLU (Şükran), 5223.
FEAR (Jeffrey R.), 5813.
Febvre (Lucien), 452.
FECI (Simona), 4926.
FEDELE (Santi), 6205.
FEDELI (Paolo), 1832.
FEDERICO (Giovanni), 5940.
FEDOROV (Sergej E.), 4054.
FEDOROVA (T. S.), 6821.
FEDOSOV (Denis G.), 6831.
FEENSTRA (Robert), 733.
FEHÉR (Bence), 1781.
FEHRENBACH (Heide), 6206.
FEHSE (Monika), 2805.
FEI (Wen), 7743.
FEICHTLBAUER (Hubert), 7203.
FEIERTAG (Olivier), 6018.

FEIGELSON (Kristian), 5651.
FEIN (Susanna), 3044.
FEINSTEIN (Charles H.), 5738.
FEISS (Hugh), 3416.
FEITL (István), 4126.
FEJÉRDY (Gergely), 7391.
FEJTOVÁ (Olga), 6093.
FELBER (Andreas), 5652.
FELD (Karl), 1887.
FELDBÆK (Ole), 6902, 7321.
FELDMAN (Gerald D.), 3997, 7041.
FELDMAN (Glenn), 4845.
FELICE (Domenico), 5347.
FELICE (Emanuele), 5739.
FELICI (Candida), 5653.
FELICI (Lucia), 454.
Felipe II, rey de España, 262, 6889, 6891, 6893.
Felipe III, rey de España, 4593.
Felipe IV, rey de España, 4568, 6911.
FÉLIX BALLESTA (Mª. Angeles), 4551.
FELLE (Antonio Enrico), 2314.
FELLER (Élise), 6207.
FELLER (Laurent), 2834.
FELLOWS-JENSEN (Gillian), 24.
FEND (Michael), 5696.
FENECH (Dominic), 4284.
FENERICK (José Adriano), 5654.
FENNO (Jonathan), 1569.
FENOALTEA (Stefano), 5740.
FENOGLIO (Silvia), 1570.
FÉNOT (Anne), 4707.
FENRICK (William J.), 6614.
FENTRESS (Elizabeth), 3513.
FENWICK (Carolyn C.), 2558.
FÉRAL (Thierry), 4655.
FERDIÈRE (Alain), 1888.
Ferdinand I, röm.-deutscher Kaiser, 3536.
Ferdinando I de' Medici, granduca di Toscana, 6893.
FERES (João), 364.
FERGUSON (Niall), 694, 6647.
FERGUSON (Priscilla), 6125.
Ferid Pacha, 5229.
FERMANDOIS H. (Joaquín), 6648.
FERNANDES (Filipe S.), 4417.
FERNÁNDEZ ÁLVAREZ (Maria Pilar), 2719.
FERNÁNDEZ CONDE (Francisco Javier), 3443.
FERNÁNDEZ FLORES (Álvaro), 1083.
FERNÁNDEZ LLAMAZARES (José), 3444.
FERNÁNDEZ LUZÓN (Antonio), 5178.
FERNÁNDEZ MARTÍNEZ (Concepción), 1780.

FERNÁNDEZ SANDE (Manuel), 5655.
FERNÁNDEZ TRABAL (Josep), 2806.
FERNÁNDEZ VALVERDE (Juan), 1821.
FERONE (Claudio), 1782.
FERRAR (Marcus), 4534.
FERRARA (Enzo), 1729.
FERRARI (Franco), 1654.
Ferrari (Giuseppe), 4205.
FERRARI (Paolo), 7312.
FERRARI (Silvia), 1027.
FERRARY (Jean-Louis), 1783.
FERREIRA (Jorge), 3669.
FERREIRA (Manuel Pedro), 3226.
FERREIRA (Roquinaldo Amaral), 6785.
FERREIRA DA CUNHA (Paulo), 6649.
FERREIRA DA SILVA (Alvaro), 5744.
FERREIRO (Alberto), 3445.
FERRER I ALÓS (Llorenç), 5814.
FERRER MALLOL (María Teresa), 2660, 2844.
FERRERA (Carlos), 640.
FERRETTI (Maria), 365.
FERRI (Andrea), 3347.
FERRIE (Joseph P.), 5928.
FERRO (Marc), 366.
FERRO (Teresa), 4935.
FERSTLER (Howard), 836.
FERTIG (Christine), 6203.
FERTIG (Georg), 561, 6203.
FERTL (Evelyn), 2012.
FERZIGER (Adam S.), 5023.
FESTUGIÈRE (André Jean), 2250.
Festus, 2058.
FEUCHTWANG (Stephan), 367.
FEUERLE (Mark), 3276.
FEYEL (Gilles), 789.
FIÇORRI (Ramiz), 7042.
FIDALGO (Elvira), 3009.
FIDORA (Alexander), 3246.
FIEBIG (Eva Susanne), 6964.
FIEDLER (Peter), 2315.
FIEDOR (Karol), 7134.
Field (Arthur Nelson), 3615.
FIELDHOUSE (Edward), 4083.
FIELDHOUSE (Roger), 4055.
FIETZE (Katharina), 2807.
FIGUEROA (Luis A.), 6832.
Filangeri (Gaetano), 6474.
FILER (Joyce M.), 1170.
FILGUERAS (Fernando), 3660.
FILIOS (Denise K.), 3007.
FILIP-FRÖSCHL (Johanna), 1958.
FILIPPI (Omar), 1033.
FILIPPOVYCH (Dimitrij N.), 7454.
FILITTI (Georgeta), 6640.

FILIU (Jean-Pierre), 7485.
FILLIEULE (Olivier), 6185.
FILLION (Réal), 721.
FINADEEV (Aleksandr P.), 276.
FINCH (Henry), 5780.
FINDLEY (Carter Vaughn), 641.
FINE (Agnès), 186.
FINE (Lisa M.), 5815.
FINGARD (Judith), 6297.
FINGER (Heinz), 43.
FINGERHUT (Eugene R.), 6102.
FINGS (Karola), 3944.
FINK (Gerhard), 1850.
FINKELBERG (Margalit), 1394.
FINKELSTEIN (David), 68.
FINKELSTEIN (Israel), 1290.
Finley (Moses), 396.
FINOCCHIARO (Giuseppe), 69.
FINZSCH (Norbert), 4735.
FIORILLA (Maurizio), 3008.
FIRPO (Giulio), 464, 1291.
FIRPO (Massimo), 330, 4880.
FISCHER (Conan), 7043.
FISCHER (David Hackett), 703.
FISCHER (Jasna), 4539, 6208.
FISCHER (Nick), 3611.
FISCHER (Steven Roger), 7892.
FISCHER (Svante), 1889.
FISHBACK (Price V.), 5996.
FISHER (Louis), 4736.
FISHWICK (Duncan), 2133.
FISIAK (Jacek), 3100.
FITOUSSI (Michèle), 5267.
FITZGERALD (Robert), 5901.
FITZGERALD (Valpy), 6138.
FITZPATRICK (Sheila), 6209.
FLACH (Dieter), 1916, 2120.
FLACHENECKER (Helmut), 816, 3148.
Flamininus (Titus Quinctius), 1924.
FLAMM (Michael W.), 6573.
FLANDREAU (Marc), 5997.
FLANDRIN (Jean-Louis), 6403.
FLASH (Kurt), 463.
Flaubert (Gustave), 5334.
FLAVIO BIONDO, 2961, 5054.
Flavius Iosephus, 1286, 1298, 1299, 1306, 1312, 1319, 1327, 1335, 1583.
FLEDELIUS KNAP (Torben), 3802.
FLEMING (Andrew), 1016.
FLEMING (Neil C.), 642, 4175.
FLEMING (Patricia), 75.
FLEMING (S. J.), 1347.
FLENSTED-JENSEN (Pernille), 1476.
FLETCHER (Stella), 3353.
FLEURIMOND (Wiener Kerns), 4111.
FLEURY (Antoine), 7560.
FLIGHT (Colin), 2808.

FLINT (Peter W.), 2267.
FLODROVÁ (Milena), 7276.
FLOGAITIS (Spyridon), 702.
FLORES (Marcello), 3560.
FLORESCU (Nicolae), 5268.
Florinsky (Mikhail), 338.
Florus (Lucius Annaeus), 1811, 1812.
FLORYSZCZAK (Silke), 2231, 3359.
FLOßMANN (Ursula), 6574.
FLÜGEL (Wolfgang), 4973.
FLÜGGE (Lars), 70.
Flynn (John T.), 4757.
FOBELLI (Maria Luigia), 2409.
FOCANT (Guy), 3202.
FOCARDI (Filippo), 368.
Foch (Ferdinand), 7029.
Focillon (Henri), 459, 489.
FODOREAN (Florin), 2013.
FOEHR-JANSSENS (Yasmina), 3115.
FOGELSON (Robert M.), 6068.
FOITZIK (Jan), 3782, 7374.
FOLDA (Jaroslav), 3180.
FOLEY (John Miles), 982.
FOLEY (Robert T.), 7044.
FOLI (John Miles), 1571.
FOLLAIN (Antoine), 6007.
FOLLET (Simone), 1410.
FOLLETT (Richard), 5941.
FÖLLMER (Moritz), 3945, 3978.
Fomenko (Anatoly), 422.
FONQUERNIE (Laurent), 5855.
FONSECA (Cosimo Damiano), 2917.
FONT (Márta), 2589.
FONTANA I LÀZARO (Josep), 4570.
FONTES (Manuel da Costa), 3004.
FONTES-BARATTO (Anna), 3072.
FOOT (John), 6069.
FOOT (Michael Richard Daniell), 674.
FORCADE (Olivier), 704.
FORD (Alan), 4818.
FORD (Caroline), 6210.
FORD (Christopher A.), 7256.
FORD (Douglas), 7257.
Ford (Henry), 5868.
FORD (L. A.), 1084.
FORLANINI (Massimo), 1258.
FORLIVESI (Luc), 6571.
Forman (Simon), 5410.
FORMICOLA (Crescenzo), 2134.
FORNI (Kathleen), 2955.
FORNO (Mauro), 5273.
FORONDA (François), 2656.
FORREST (Ian), 3446.
FORSDYKE (Sara), 1450, 1518.
FORSSBERG (Anna Marie), 4643.
FORSSNER (Thorvald), 168.
FORSTER (Ralf), 5657.

FÖRSTER (Stig), 7172.
FORSYTH (Elliott Christopher), 5489.
FORSYTHE (Gary), 1890.
FORT GODSHALK (David), 6211.
FORTE (Angelo D. M.), 2723.
FORTESCUE (William), 3857.
FOSTER (Lynn V.), 7878.
FOSTER (Stephen C.), 827.
FOTAKIS (Zisis), 7045.
Foucault (Michel), 460, 4158.
FOUCHÉ (Pascal), 65.
FOUCRIER (Annick), 7893.
FOUILLOUX (Étienne), 4917.
FOURACRE (Paul), 668.
Fourmont (Michel), 1410.
FOURNIER (Julien), 1411.
FOWLER (Richard), 141.
FOWLER-MAGERL (Linda), 2736.
FOX (Robert), 5192.
FOYSTER (Elizabeth), 119.
FRACKOWIAK (Johannes), 3946.
FRADERA (Josep Maria), 6718.
FRAESDORFF (David), 251.
FRAGER (Ruth A.), 6426.
FRAGNITO (Gigliola), 169.
FRAME (Robin), 2939.
FRANCE (John), 2590, 6651.
Francesco da Prato, 3262.
FRANCFORT (Didier), 5658.
FRANCFORT (Henri-Paul), 1077.
FRANCO (Marina), 3580.
FRANCO BAHAMONDE (Francisco), 4549, 4559, 4585, 4590, 4601, 4621, 7217, 7239, 7241, 7380.
FRANCO LANAO (Elena), 4571.
FRANÇOIS (Michel), 3905.
François Ier, roi de France, 6885.
FRANK (Alison Fleig), 5816.
FRANK (Andrew K.), 6212.
Frank (Anne), 4331.
FRANK (Denis), 6213.
FRANK (Günter), 4988.
FRANK (Richard M.), 2699.
FRANKEMÖLLE (Hubert), 2256.
Franklin (Benjamin), 5166, 5383.
Franklin (Clarence LaVaughn), 5004.
FRANSES (Henri), 2458.
FRANZ (Corinna), 647.
FRANZ (Eckhart G.), 120.
FRANZ (Kathleen), 5817.
FRANZINELLI (Mimmo), 785.
FRANZINI (Antoine), 2661.
FRANZIUS (Christine), 6575.
FRASCHETTI (Augusto), 1891, 2076.
FRASER (James Earle), 1892.
FRASER (P. M.), 170.
FRASER (Steve), 4771.
FRASSO (Giuseppe), 97.

FRATANTUONO (Lee), 2135.
FRATEANTONIO (Christa), 2132.
FRATI (Marco), 3126.
FRAUDET (Xavier), 6965.
FRAZIER (Alison Knowles), 5490.
FREDERICK (Julia C.), 4799.
FREEDMAN (Lawrence), 7486.
FREEDMAN (Paul), 2809.
FREELAND (Jane Patricia), 2570.
FREGUGLIA (Margherita), 1035.
FREI (Norbert), 369.
FREIGANG (Christian), 2902.
FREIRE RODRIGUES (Francisco Xavier), 6243.
FREIST (Dagmar), 5156.
FREITÄGER (Andreas), 5177.
FRENCH (David), 4056.
FRENCH (Michael), 5902.
FRENZ (Thomas), 9.
Freud (Sigmund), 494, 5380, 5382.
FREUDENBURG (Kirk), 2059.
FREVERT (Ute), 399.
FREY (Jörg), 2310.
FREYDANK (Helmut), 1131.
Freyre (Gilberto), 461.
FRIED (Marvin Benjamin), 7046.
FRIEDEBURG (Robert von), 3947.
FRIEDL (Christian), 2628.
FRIEDL (Jiří), 7258.
FRIEDMAN (Edward), 3713.
FRIEDMAN (Rebecca), 5179.
Friedrich II, röm.-deutscher Kaiser, 2628.
Friedrich III, röm.-deutscher Kaiser, 2666.
FRIISBERG (Claus), 3795.
FRIMMOVÁ (Eva), 226.
FRINGS (Irene), 2077.
FRINGS (Jutta), 5547.
FRIOLI (Donatella), 3405.
FRITZEN (Florentine), 3940.
FRÖHLICH (Elke), 3915.
FRÖHLICH (Pierre), 1447.
FROIDE (Amy M.), 6214.
FROLOVA (Nina A.), 138.
FROM (Peter), 4644.
FROMENT-MEURICE (Henri), 7450.
FROMMER (Benjamin), 3769.
Frontinus (Sextus Iulius), 1792, 2082.
Fronto (Marcus Cornelius), 2122.
FRONZAROLI (Pelio), 1292.
FROSCHAUER (Harald), 1171.
FROST (Frank J.), 1451.
FROTSCHER (Ann), 4359.
FROVA (Carla), 2754.
FRUGONI (Chiara), 2583.
FRÜHAUF (Martin), 2216.
Frutolf von Michelsberg, 104.
FRY (Geoffrey Kingdon), 4057.
FRYDENBERG (Julio D.), 6243.

FU (Chonglan), 7627.
Fu (Zuoyi), 7620.
FUBINI (Enrico), 5659.
FUCHS (Martina), 3536.
FUCHS (Rachel G.), 6215, 6427.
Fuentes (Carlos), 4292.
FÜHNER (Harald), 4334.
FUHRIMANN (Daniel), 5660.
FUHRMANN (Bernd), 2800.
FUJIKAWA (Takao), 4066.
Fujimori (Alberto), 4369.
FUJITA (Kōichi), 7583.
FUJITA (Satoru), 7804, 7815.
FÜLBERTH (Andreas), 6070.
FULBROOK (Mary), 6428.
FULCHER (Jane F.), 5661.
FULKERSON (Laurel), 2078.
FULLER (Dorian Q.), 1017.
FULLER (Mia), 6722.
FULLOLA I PERICOT (Josep M.^{a}), 1028.
FULTON (Helen), 3047.
FUNADA (Yoshiyuki), 7628.
FUNIGIELLO (Philip J.), 5392.
FUNK (René), 6509.
FURDELL (Elizabeth Lane), 3310.
FURE (Eli), 4355.
FURUKAWA (Takahisa), 6532.

G

GABACCIA (Donna R.), 781.
GABELKO (Oleg L.), 1132.
GABER (Ivor), 4048.
GABRIEL (Jiří), 5322.
GABRIEL MELÉNDEZ (A.), 5274.
GABUCCI (Ada), 643.
GABZDILOVÁ (Soňa), 3770.
GADDIS (John Lewis), 7332.
GADDIS (Michael), 2316.
GADE (Dietlind), 3011.
GAERTNER (Jan Felix), 1822.
GAGARIN (Michael), 1482, 1490, 1507.
GAGLIARDI (Isabella), 3448.
GAILLE-NIKODIMOV (Marie), 706.
GAILUS (Manfred), 4992.
GALAND-HALLYN (Perrine), 5131.
Galanti (Giuseppe Maria), 462.
GALAOR (Isabel), 3526.
GALASSO (Giuseppe), 468.
GALAVOTTI (Enrico), 4851.
GALEAZZI (Marco), 7333.
GALENOVICH (Jurij M.), 7629.
Galenus (Claudius), 1642, 1662, 1671.
GALFRÉ (Monica), 5275.
GALGANO (Robert C.), 4936.
Galileo Galilei, 5379, 5421.
GALINDO CÁRDENAS (Benjamín), 4303.
GALINIER (Martin), 457.

GALINSKY (Karl), 611.
GALLAGHER (Tom), 4442.
GALLÉ CEJUDO (R. J.), 1572.
GALLEGOS VÁZQUEZ (Federico), 2756.
GALLI (Giancarlo), 7130.
GALLICHAN (Gilles), 75, 5279.
GALLO (Klaus), 5662.
GALLUPPI (Massimo), 7047.
Gallus, Anonymus, 2625.
GALONNIER (Alan), 3259.
GALTIER MARTÍ (Fernando), 20.
GALUNOV (T.), 3683.
Galvão (Henrique), 4424.
GÁLVEZ MUÑOZ (Lina), 6163.
GAMBA (Aldo), 5388.
GAMBERO (Luigi), 3277.
GAMELLA (Juan F.), 6112.
Gandhi (Mohandas Karamchand), 4147, 4149, 4151, 4541.
GANDINI (Mario), 902.
GANDOULPHE (Pascal), 4573.
GANEV (Venelin), 7011.
GANSER (Daniele), 7487.
GANTMAN (Ernesto R.), 5998.
GANZENMÜLLER (Jörg), 7259.
GAO (Fu), 7630.
GAO (Xiuqing), 7631.
GARAVAGLIA (Juan Carlos), 644, 6488.
GARB (Margaret), 6071.
GARCA-BRYCE (Iigo), 6429.
GARCEA (Alessandro), 1804.
GARCÉS (María Antonia), 5491.
GARCÉS (Mario), 3706.
GARCIA (Dominique), 1085.
GARCÍA (Gervasio Luis), 370.
GARCÍA (María Elena), 4370.
GARCÍA (Prudencio), 4106.
GARCÍA ATIÉNZAR (Gabriel), 1046.
GARCÍA BORJA (Pablo), 1060.
GARCÍA DE CORTÁZAR (Fernando), 220.
GARCÍA DE CORTÁZAR Y RUIZ DE AGUIRRE (José Angel), 2591.
GARCÍA DE SALTOR (Irene), 3601, 3602.
GARCÍA FERNÁNDEZ (Ernesto), 2811.
GARCÍA FERNÁNDEZ (Hugo), 4574.
GARCÍA GARCÍA (Bernardo J.), 5703.
GARCÍA MONTERO (Mercedes), 6531, 6543.
GARCÍA SANJUÁN (Leonardo), 1061.
GARCÍA SEBASTIANI (Marcela), 3592.
GARCÍA UGARTE (Marta Eugenia), 4318.

GARCÍA-ARGÜELLES ANDREU (Pilar), 1028.
GARCIADIEGO DANTAN (Javier), 4286.
GARCÍA-SANZ MARCOTEGUI (Angel), 4575.
Gardano (Antonio), 79.
GARDIKA-KATSIADAKE (Elene), 6948.
GARDNER (Martha), 6216.
GARDNER NAKAMURA (Ellen), 5393.
GARGANI (Giuseppe), 7130.
GARGANO (Innocenzo), 3358.
Garibaldi (Giuseppe), 4227.
GARIBALDI (Patrizia), 1050.
Garin (Eugenio), 463.
Garin-Michajlovskij (Georgij Nikolajevič), 6661.
GARLAND (Lynda), 1855.
GARNAND (Brien), 1288.
GARNIER (Bernard), 7282.
GARNIER (Jacques), 6925.
GARNOT (Benoit), 6487, 6586.
GARNSEY (Peter), 610.
GAROFALO (Leo), 4375.
GARR (W. Randall), 1293.
GARRAUD (Philippe), 7137.
GARRÉ (Roy), 6576.
GARRETT (David T.), 4371.
GARRIGUES (Jean), 821.
GARSTANG (John), 1259.
GÄRTNER (Thomas), 2079.
GÄRTNER (Ursula), 1931, 2110.
GARUTI (Adriano), 4951.
GARUTI (Giovanni), 2248.
GARVIN (Tom), 4176.
GARZANITI (Marcello), 3418.
GARZYA (Antonio), 1573.
Gascoigne (Alvary Frederick), 7311.
GĄSOWSKI (Tomasz), 4386.
GAŠPARIČ (Jure), 3625.
GASPARRI (Stefano), 2792.
GASS-BOLM (Torsten), 5224.
GASSERT (Philipp), 3934, 3981.
GASTI (Fabio), 1564, 2095.
GAT (Moshe), 7488.
GATAGOVA (Ljudmila S.), 4462.
GATES (Charles), 1348.
GATHERU (R. Mugo), 4255.
GATRELL (Peter), 5741.
GATTI (Paolo), 3405.
Gatz (Erwin), 4811.
GAUCHER (Élisabeth), 3114.
GAUDART DE SOULAGES (Michel), 6746.
GAUDE-FERRAGU (Murielle), 2812.
GAUDELUS (Sébastien), 5663.
GAUGER (Jörg-Dieter), 5161.

GAUL (Volker), 3797.
GAULIN (Jean-Louis), 2564, 3485.
GAULME (François), 3906.
GAULTIER-KURHAN (Caroline), 6966.
GAUSEMEIER (Bernd), 5394.
GAUTHIER (Philippe), 1403.
GAUTIER (Claude), 508.
GAUVREAU (Michael), 4881.
GAVIRA MÁRQUEZ (María Concepción), 5818.
GAVRILĂ (Irina), 5942, 6366.
GAVRILOV (K. N.), 1037.
GAVRILOVA (Stella), 5276.
GAVRILOVIĆ (Ana), 4521.
GAWARECKI (Kathrin), 386.
GAYIBOR (Nicoué Lodjou), 648.
GAZIAUX (É.), 857.
GE (Fuping), 7632.
GÉAL (Pierre), 306.
GEARY (Laurence M.), 4183.
GEBAUER (Anne Birgitte), 1029.
GEBAUER (Annekatrin), 3948.
GEBHARDT (Volker), 846.
GEDDES (Jane), 29.
GEELHOED (E. Bruce), 7451.
GEHLER (Michael), 7334.
GEISEL (Christof), 3949.
GELEZ (Philippe), 903.
GELFAND (Wladimir), 7243.
GELICHI (Sauro), 2790.
GÉLIS (Jacques), 564.
GELJON (Albert-Kees), 2317.
GELLER (Jay Howard), 5024.
GELLER (Jay), 904.
GELLI (Licio), 278.
GELMAN (Jorge), 3593.
GELONCH (Josep), 4596.
GELVIN (James L.), 3533, 6653.
GELZER (Matthias), 1893.
GEMUEV (Izmail N.), 576.
GEMÜNDEN (Petra von), 2261.
GENCHEV (Dimităr), 6620.
GENET (Jean-Philippe), 2656.
GÉNETIOT (Alain), 837.
GENITO (Bruno), 1349.
GENOV (Georgi Petrov), 7048.
GENOVESE (Gianluca), 5327.
GENTILE (Emilio), 6147.
GENTILI (Anna Maria), 402.
GENTILI (Bruno), 2080.
GENTILI (Sonia), 3012.
GENTON (Hervé), 5053.
GENTRY (Francis G.), 621.
GEOFFROY (Marc), 3256.
GEORGAKIS (Angelo), 4102.
GEORGE (Edward), 7489.
GEORGE (Michele), 2014, 2040.
George Gordon, 4th Earl of Huntly and Chancellor of Scotland, 3846.

GEORGELIN (Hervé), 4687.
GEORGES (Stefan), 2711.
GEORGES (Tobias), 3278.
Georgiev (Kimon), 6620.
Georgius Gemistus Plethon, 2389-2391.
Georgius Pachymeres, 2392.
Georgius Xiphilinus, 2386.
GEORGOUDI (Stella), 899.
GEORGOULA (Electra), 1730.
GEPPERT (Dominik), 7351.
GERARD (Emmanuel), 3726.
Gérard Ier, évêque de Cambrai et d'Arras, comte du Cambrésis, 2527.
Gérard II, évêque de Cambrai et d'Arras, comte du Cambrésis, 2527.
GÉRARD-LIBOIS (Jules), 3726.
GÉRARD-PLASMANS (Delphine), 7050.
GERBER (Christine), 2318.
GERBER (Larry G.), 5819.
GERBER (Stefan), 5186.
GERHARD (H. J.), 5992.
GERHARDS (Albert), 2332.
GERHARDS (Jürgen), 171.
GERHARDT (Volker), 5338.
GERICKE (Hans Otto), 5820.
GERLACH (Christian), 5742.
GERLACH (Jens), 1433.
Gerlaksson (Brynolf), 4885.
GERLAUD (Bernard), 2407.
GERL-FALKOVITZ (Hanna-Barbara), 5340.
Germanicus Caesar, 1711.
GERMANY (Robert), 1575.
GERMOND (Philippe), 1172.
GERNET (Jacques), 5359.
GERRARD (Craig), 7204.
Gerson (Jean), 3331.
GERSON (Lloyd P.), 1655.
Gerstenmaier (Eugen), 3952.
GERSTLE (Gary), 4771.
GERVASONI (Marco), 4209.
GERWARTH (Robert), 3951.
GERZAGUET (Jean-Pierre), 2528.
GESCHE (Bonifatia), 2220.
GESKE (Norbert), 1452.
GETHMANN-SIEFERT (Annemarie), 963, 5334.
GEYER (Angelika), 2165.
GEYER (Dietrich), 6967.
Ghazālī, 2710.
GHAZELEH (Pascale), 6218.
GHELARDI (Maurizio), 847, 5595.
GHERMANI (Naïma), 263.
GHEZZI (Vania), 1491.
GHIATI (Claude), V.
GHINOIU (Ion), 546.
GIAGNACOVO (Maria), 2813.

GIANA (Luca), 4926.
GIANFREDA (Sandra), 5601.
GIANNOBILE (Sergio), 1173.
Giannone (Pietro), 464.
GIARDINA (Giancarlo), 1833.
GIAVARINI (Carlo), 2166.
GIBB (Paul), 6968.
Gibbon (Edward), 5330.
GIBELLI (Antonio), 6219.
GIBLIN (James L.), 4668.
GIDE (André), 5515, 5516.
GIEBEL (Marion), 1830.
GIEBELS (Lambert J.), 4155.
GIEDEMAN (Daniel C.), 5999.
GIESLER (Gerd), 5353.
GIFFIN (William W.), 6220.
Gil Carles (Fernando), 1008.
GIL ESTEBAN (Manuel), 1008.
GIL NOVALES (Alberto), 645, 5060.
GILBERT (Creighton), 3181.
GILBERT (Shirli), 5666.
GIL-CARLES ESTEBAN (Josè Manuel), 1008.
GILCHRIST (Roberta), 3381.
GILES (Katherine), 2868.
GILLAM (Robyn), 1174.
GILLESPIE (Michele), 5821.
GILLESPIE (Raymond), 76, 225.
GILLI (Patrick), 2814.
Gilliéron (Jules), 177.
GILLINGHAM (Paul), 4304.
GILLIVER (Catherine Mary), 1894.
GILLY (Adolfo), 4319.
Gilman (Sander L.), 6719.
GILMONT (Jean François), 4974.
GIMBUTAS (Marija), 1047.
GIMPEL (Madeleine), 6141.
GINAT (Rami), 4662.
GINTRAC (Cécile), 4707.
GINZBERG (Lori D.), 6221.
GINZBURG (Carlo), 431.
Gioacchino da Fiore, 3314.
Giolito de' Ferrari (Gabriele), 5293.
Giolito de' Ferrari (Giovanni), 5293.
GIOLO (Orsetta), 734.
GIORDANO (Giancarlo), 6654.
GIORGI (Luigi), 7392.
GIORGIERI (Mauro), 1260.
Giotto, 3160.
GIOVAGNOLI (Agostino), 4216, 4933.
Giovannini da Capuagnano (Girolamo), 5544.
GIOVINI (Marco), 3013.
GIPPERT (Wolfgang), 4396.
GIRARD (Philippe R.), 4112.
GIRARDET (Klaus Martin), 738.
Giraud (Henri), 3850.
GIRAUT (F.), 7866.

GIRVIN (Brian), 4179.
GIRZYŃSKI (Zbigniew), 7393.
Giscard d'Estaing (Valéry), 3839.
GITTOS (Helen), 3455.
GIULIETTI (Fabrizio), 4233.
GIULIODORI (Serena), 2757.
Giunigi (Paolo), 3156.
GIUNIPERO (Elisa), 4933.
GIUNTA (Claudio), 3014.
GIUROVICH (Sara), 1576.
Giustiniano Partecipazio, doge di Venezia, 2475.
GIVENS (Jean Ann), 3182.
GIVEN-WILSON (Chris), 2746.
Gladstone (William Ewart), 4068.
GLANTZ (David M.), 7260.
GLANTZ (Mary E.), 7205.
GLASER (Antoine), 6787.
GLASER (Clive), 6222.
GLASER (Hermann), 5679.
GLÄSER (Pascal), 3279.
GLASSHEIM (Eagle), 3771.
GLASSNER (Jean-Jacques), 1222.
GLAUCH (Sonja), 2986, 3017.
GLAZEBROOK (Allison), 1492.
GLAZER (Peter), 4576.
GŁĘBOCKI (Henryk), 4406.
GLENN (Jonathan A.), 2964.
GLICK (Thomas F.), 2700, 3290.
GLIGOR (Mihaela), 910.
GLIWITZKY (Christian), 1133.
GLOCK (Anne), 2132.
GLOOTZ (Tanja Anette), 6223.
Gluck (Christoph Willibald), 5719.
GLUR (Stefan), 7206.
GMITRUK (Janusz), 4388.
GNAD (Oliver), 647.
GNISS (Daniela), 3952.
GOBAT (Michel), 6655.
GOCKING (Roger S.), 4027.
GODART (Louis), 1388.
GODDEERIS (Idesbald), 3644, 3646.
GODDING (Philippe), 2648, 2745.
GODEA (Ioan), 562.
GODEŠA (Bojan), 7313.
GOEBBELS (Joseph), 3915.
GOELDEL (Denis), 3953.
Goethe (Johann Wolfgang von), 5328.
GOH (Evelyn), 7490.
GOICOLEA JULIÁN (Francisco Javier), 2820.
GOICOVIC DONOSO (Igor), 3707.
GÖKAY (Bülent), 6657, 7368.
GÖKTEPE (Cihat), 7491.
GOLD (Steven J.), 782.
GOLDAMMER (Yvonne), 2966.
GOLDBERG (Martine), 790.
GOLDBERG (Sander Michael), 2081.

GOLDER (Hilary), 3612.
GOLDHAUSEN (Marco), 1134.
GOLDMAN (Merle), 7633.
GOLDMAN (Noemí), 3608.
GOLDMAN (Wendy), 4472.
GOLDSCHMIDT (Arthur), 3827.
GOLDSCHMIDT (Nils), 4023.
GOLDSTEIN (Henri), 5043.
GOLDSTEIN (Ivo), 2459.
GOLDSTEIN (Paul S.), 7879.
GOLDSWORTHY (Adrian Keith), 1894.
GOLOVINA (Galina), 5346.
GOLTZ (Andreas), 471.
GOMES (Flávio dos Santos), 6224.
Gómez (Juan Vicente), 7099.
Gómez (Laureano), 3725.
GÓMEZ BRAVO (Gutmaro), 6225, 6577.
GÓMEZ HERNANZ (Juan), 1030.
GÓMEZ PALLARÈS (Joan), 1780.
GÖMÖRI (George), 5181.
Gomułka (Władysław), 7510.
GONĚC (Vladimír), 6658.
GONIS (Nikolaos), 1175.
GONZALBO AIZPURU (Pilar), 6241.
GONZÁLES (Osmar), 4372.
GONZÁLEZ (Aurelio), 3101.
GONZALEZ (Evelyn), 6072.
GONZÁLEZ (Lorenzo), 630.
GONZÁLEZ (M. Paula), 3580.
GONZÁLEZ ADÁNEZ (Noelia), 4058.
GONZÁLEZ CALLEJA (Eduardo), 4577.
GONZÁLEZ CUEVAS (Pedro Carlos), 705, 4617.
González de Hoces (Pedro), 2553.
GONZÁLEZ GADEA (Diego N.), 5492.
GONZÁLEZ LEANDRI (Ricardo), 5395.
GONZÁLEZ MARTÍNEZ (Carmen), 4618.
GONZALEZ RUIBAL (Alfredo), 1018.
GONZÁLEZ SALAZAR (Juan Manuel), 1261.
GONZÁLEZ SUBÍAS (José Luis), 5618.
GONZÁLEZ TRAVIESO (David), 3720.
GOOD (Deirdre), 2379.
GOODAY (Graeme), 5192.
Goodnow (Frank Johnson), 7784.
GOODRICH (Amanda), 4059.
GOODWIN (Tony), 139.
GOOSE (Nigel), 6247.
GOOSMAN (Stuart L.), 5667.
GOOT (Murray), 3613.
GOPNIK (Hilary), 1350.

Gorbachev (Mikhail S.), 7460.
Gorchakov (Aleksandr), 6998.
GÓRCZYŃSKA-PRZYBYŁOWICZ (Bożena), 6700.
GORDON (Daniel A.), 6226.
GORDON (Matthew), 2701.
GORDON (Robert), 5668.
GOREN-INBAR (N.), 1275.
GORI (Franco), 2214.
GORSKI (Philip S.), 486.
GORT JUANPERE (Ezequiel), 2743.
Gortchakov (Alexandre M.), 6958.
GOSCHLER (Constantin), 3954.
GOSEWINKEL (Dieter), 5773.
GOTHONI (René), 866.
GOTO (Harumi), 6615.
GOTOR (Miguel), 474.
GOTTSCHALL (Dagmar), 249.
GOTTSMANN (Andreas), 6969.
GOULD (Eliga H.), 3532.
GOULD (Jeffery L.), 3544.
GOULD (Lewis L.), 4715, 4737.
GOULLET (Monique), 3409.
GOURDON (Vincent), 6227.
GOUTSOS (D.), 6386.
GOUWENS (Kenneth), 4853.
Gouzenko (Igor), 7404.
Gower (John), 3058.
GOWING (Alain Michael), 1895.
GOY (Joseph), 6287.
GOY (Rudolf), 3248.
GOYON (Jean-Claude), 1176.
GRABAR (Oleg), 2702.
GRABER (Tom), 2658.
GRABMAYER (Johannes), *I*.
GRABOWSKI (Waldemar), 7306.
GRADL (Hans-Georg), 2320.
GRADMANN (Christoph), 5396.
GRADY (Frank), 3015.
GRAF (Friedrich Wilhelm), 486.
GRAF (Rüdiger), 3978.
GRÄFF (Andreas), 1223.
GRAGG (Gene), 1294.
GRAGNOLATI (Manuele), 2899.
GRAHAM (Helen), 4578.
GRAHAM (Patricia Albjerg), 5225.
GRAHAM-LEIGH (Elaine), 2815.
GRAINGE (Gerald), 1896.
GRAJETZKI (Wolfram), 1177.
Gramsci (Antonio), 5331.
GRAN (John W.), 4831.
GRAND'HENRY (Jacques), 2232.
GRANDIN (Greg), 373.
GRANGE (Cyril), 6073, 6112.
GRANGE (Daniel J.), 5083.
GRANIER (Pierre.), 7261.
GRANIERI (Ronald J.), 7492.
GRANT (Kevin), 6788.
GRANT (Lindy), 3183.
GRANT (Oliver), 6228.
GRANVILLE (Johanna), 7394.

GRATHOFF (Stefan), 3449.
GRAU (Andreas), 3955.
GRAUBARD (Stephen), 4738.
GRAY (Thomas), 2574.
Grayson, family, 126.
GRAZIOSI (Andrea), 280.
GRAZIOSI (Paolo), 1019.
GRDINA (Igor), 5084.
GREATREX (Geoffrey), 2416.
GREATREX (Joan), 2597.
GREAVES (Julian), 5822.
GREBE (Anja), 2908.
GREBING (Helga), 855.
GREEN (Alan), 6074.
GREEN (Dennis Howard), 3062.
GREEN (Karen), 3282.
GREEN (Nancy L.), 334, 6229.
GREEN (Peter), 1823.
GREENE (Christina), 6230.
GREENE (Jody), 5474.
GREENE (Kevin T.), 2167.
GREENHALGH (Elizabeth), 7051.
GREENSHIELDS (Malcolm R.), 907.
GREEVEN (Heinrich), 2264.
Grégoire (Henri), 4902.
GREGOR (A. James), 5085.
GREGOR (Neil), 3996.
Gregorius Antiochus, 2393, 2502.
Gregorius I Magnus, Papa, Sanctus, 2231, 3177, 3280, 3358.
Gregorius Nazianzenus, Sanctus, 2232.
Gregorius Nyssenus, Sanctus, 2233, 2317.
Gregorius VII, Papa, 3355.
Gregorius XIII, Papa, 5446.
GREGORY (Andrew), 2350.
GREGORY (James N.), 6231.
GREGORY (Jeremy), 3417.
GREGORY (Justina), 1513, 1577.
GREGORY (Timothy E.), 2422.
GREINER (Bernd), 7172.
GRELKA (Frank), 4710.
GRELLE (Francesco), 1973.
GRENIER (John), 4739.
Grénon (Louis), 5682.
GRESCHAT (Katharina), 3280.
GRESCHAT (Martin), 4976.
GRESSER (Georg), 3343.
GRESSIEUX (Douglas), 6746.
GRETHLEIN (Jonas), 1578.
GRETSCH (Mechthild), 3410.
GREVE (Gregory P.), 909.
GRÉVY (Jérôme), 3860.
GRIBAUDI (Gabriella), 7262.
GRIFFIN (Clive), 4813.
GRIFFIN (Emma), 6232.
GRIFONI CREMONESI (Renata), 1024.
GRIGSBY (John), 3016.

GRILLET (Bernard), 2250.
Grillius grammaticus, 1813.
GRIMAL (Nicolas), 1178.
Grimani (Giovanni), 4880.
GRIMBERT (Joan T.), 614.
GRIMSHAW (Mike), 905.
GRINDHEIM (Sigurd), 2322.
GRIOLET (Patrick), 175.
GROENEVELD (Gerard), 7246.
GROENVELD (Simon), 6913.
GROHMANN (Carolyn), 7052.
GRØNGAARD JEPPESEN (Torben), 6233.
GRONSKÝ (Ján), 6510.
GROS (Dominique), 732.
GROS (Pierre), 1903.
GROSS (Raphael), 6499.
GROßE KRACHT (Klaus), 374.
GROSSI (Paolo), 2758.
GROTHE (Ewald), 6129, 6511.
Grotius (Hugo), 5332.
GROTTANELLI (Cristiano), 906.
GROVE (Eric), 4060.
GROVE (Laurence), 264.
GRUBE (Ernst J.), 2703, 2704.
GRÜBEL (Monika), 800.
GRUENINGER (Donat), 3184.
GRUFFUDD AP MAREDUDD AP DAFYDD, 2963, 2976.
GRÜNBART (Michael), 2394, 2460.
GRUNDBERG (Malin), 563.
GRUNDEN (Walter E.), 7263.
GRUNER (Wolf), 3956.
GRUPP (Peter), 604.
GRÜTTNER (Michael), 5202.
GUADALUPI (Gianni), 2447.
GUALDO ROSA (Lucia), 3066.
GUALERZI (Saverio), 2136.
GUAN (Shaohong), 7634.
GUAN (Yanbo), 7635.
GUARAGNELLA (Pasquale), 5341.
GUARDINO (Peter F.), 4305.
Guariglia (Raffaele), 7209.
GUARINO (Giuseppe), 5781.
GUARNIZO (Luis E.), 765.
GUDIS (Catherine), 5086.
GUDMUNDSSON (Rögnvaldur), 3087.
GUELKE (Adrian), 4543.
GÜEMEZ PINEDA (Arturo), 4306.
GUENÉE (Bernard), 375.
Guercino, 5601.
GUERDAT (Marcel), 5298.
GUÉRIN (Jeanyves), 987.
GUERRA (Cesare), 61.
GUERRA (Lillian), 3743.
GUERRAGGIO (Angelo), 5397.
GUERRERO ARJONA (Melchor), 4579.
GUERRERO RINCÓN (Amado Antonio), 6834.
GUGLIELMETTI (Rossana), 3281.
GUHA (Ranajit), 695.
Guicciardini (Francesco), 6981.
GUICHARD (Michael), 1224.
GUICHARD (Pierre), 3265.
GUIDA (Francesco), 4443.
GUIDI (Elisa), 432.
GUIDI (Marco E. L.), 5779.
GUIDI (Patrizia), 2272.
GUIDI (Remo L.), 3411.
GUIDOBALDI (Maria Paola), 2179.
GUIDOT (Bernard), 3090.
GUILAINE (Jean), 1048.
GUILDERSON (T.), 1072.
GUILDFORD (Janet), 6297.
GUILLAMÓN (F. Javier), 4565.
GUILLAUMIN (Jean-Yves), 2082.
GUILLELMUS A SANCTO THEODORICO, 3249.
GUILLET (David), 6578.
GUINOT (Jean-Noël), 2323.
GUIRAUD (Hélène), 1853.
GUISSET (Jacqueline), 2918.
GÜL (Muammer), 7565.
GÜLALP (Haldun), 4688.
GULCZYNSKI (Andrzej), 6579.
GULJAEV (Valerij I.), 1351.
GULLA (Bob), 838.
GULLBEKK (Svein H.), 2816.
GULLÌ (Domenica), 1755.
GULLINO (Giuseppe), 2749.
GUMISAWA (Hideo), 7584.
GÜN (Doğan), 6480.
GÜNAY (Mustafa), 5314.
GÜNERGUN (Feza), 5398.
GÜNEŞ (Ahmet), 376.
GÜNEY (Aylin), 7513.
Günsche (Otto), 7109.
GÜNTHART (Romy), 2909.
GUO (Peigui), 7636.
GUO (Qitao), 7637.
GUO (Yisheng), 7638.
GUPTA (Bishnupriya), 5823.
GÜRBEY (Gülistan), 7494.
GUREVICH (Aron Ja.), 672.
GURNEY (Peter), 6234.
GUROCK (Jeffrey S.), 5025.
GUS'KOV (Andrej G.), 6905.
GUST (Wolfgang), 4676.
GUSTAFSSON (Kalle), 6789.
Gustav II Adolf, king of Sweden, 6888.
Gustav V, king of Sweden, 4645.
Gustav Vasa, 6375.
GUŢAN (Manuel), 4444.
GUTAS (Dimitri), 2699.
GUTIÉRREZ ARBULÚ (Laura), 6088.
Gutiérrez Cuevas (Teodomiro Augusto), 4368.
GUTIÉRREZ PALACIO (Javier), 5261.
GÜTING (Eberhard), 2264.
GUTMANN (Myron P.), 6078.
GUTTERMAN (David S.), 4832.
GUTZWILLER (Kathryn), 1579, 1606.
GUY (Jean-Claude), 2243.
GUYER (Paul), 5338.
GUYOTJEANNIN (Olivier), 8, 222.
GUZZO (Pietro Giovanni), 2179.
GWASSA (Gilbert Clement Kamana), 4666.
GWENAEL (Murphy), 4833.
GYGAX (Marc Domingo), 1135.
GZELLA (Holger), 1295.

H

HAAEST (Erik), 7264.
HAAR (Gerrie ter), 891.
HAAR (Ingo), 371.
HÅARD (Mikael), 5399.
HAARMANN (Lutz), 3957.
HAAS (Mark L.), 5087.
HAAS (Max), 3232.
HAAS (Stefan), 3958.
HAAS (Volkert), 1262.
HAASE (Marie-Luise), 5348.
HAASE (Richard), 1225, 1263.
HABINEK (Thomas Noel), 2015.
HÄCHLER (Stefan), 5361.
HACHTMANN (Rüdiger), 6431.
HACKEL (Steven W.), 4937.
HACKER (Barton C.), 590.
HACKER (J. David), 6075.
HACKER (Peter), 4279.
HADDOCK (David D.), 5798.
HADENIUS (Stig), 4645.
HÄDER (Sonja), 6235.
HADERMANN-MISGUICH (Lydie), 2461.
Hadewijch, 3441.
HADFIELD (Andrew), 76.
Hadrianus (Publius Aelius), Roman emperor, 1879, 1881, 1898, 2008.
HAECK (Tom), 2016.
HAEHLING (Raban von), 2321.
HAEMERS (Jelle), 2801.
HAERS (J.), 857.
Haffner (Sebastian), 3924.
HAGBERG (Johnny), 4885.
HAGEMAN (Maarten), 4335.
HAGEMAN (Mariëlle), 2929.
HAGENS (Graham), 1226.
HÄGG (Göran), 4646.
HAGNER (Donald A.), 2375.
HAHN (Hans-Joachim), 5670.
HAHN (Judith), 5422.
HAHN (Peter L.), 7335.
HAHN (Reinhard), 2959.

HAHN (Roger), 5400.
HAHN (Wolfgang R. O.), 140.
HAID (Oliver), 550.
HAIDACHER (Christoph), 3636.
Haig (Douglas), 7025.
HAIGNERÉ (Claudie), 7361.
HAIMSON (Leopold H.), 4473.
HAINES (Michael R.), 6075.
HALBWACHS (Maurice), 5333.
HALDON (John F.), 2423.
HALE (Charles R.), 3544.
HALEY (Evan W.), 1898.
HALL (Gwendolyn Midlo), 6720.
HALL (Jackie), 3394.
HALL (Leslie), 825.
HALL (Marcia B.), 831.
HALL (Marcus), 759.
HALL (Thomas), 7207.
HALLAMAA (Olli), 3255.
HALLAQ (Wael B.), 5026.
HALLEBEEK (Jan), 6470.
Haller (Albrecht von), 5361.
HALLET (Christopher H.), 2168.
HALLEWELL (Laurence), 5277.
HALLIDAY (Jon), 7607.
HALLIWELL (Stephen), 1580.
HALLON (Ľudovít), 6000.
HALLYN (Fernand), 5131.
HALPERÍN DONGHI (Tulio), 6236.
HALSTEAD (Paul), 1395.
HALTENHOFF (Andreas), 1932.
HAMAMOTO (Mami), 4952.
HAMAMSY (Chafika Soliman), 3826.
HAMBLIN (Jacob Darwin), 5401.
HAMEL (Mary), 3092.
HAMETZ (Maura), 6660.
Hamilton (Alexander), 4778.
HAMILTON (Keith), 281.
HAMILTON (Kenneth), 830.
HAMILTON (Louis I.), 3334.
HAMILTON (Sarah), 3440.
HAMMER (Dean), 1493.
HAMMERSLEY (Rachel), 3861.
HAMMERSTEIN (Notker), 5226.
HAMMOND (P. W.), 2817.
HAN (Chunying), 7748.
HAN (Shufeng), 7639.
HAN (Xiaorong), 7640.
HANDMAN (Marie-Élisabeth), 6333.
HANEDA (Masashi), 7566.
HANISCH (Ernst), 6237.
HANKE (Edith), 486.
HANLEY (Anne G.), 6001.
HANLEY (Wayne), 3862.
HANNA (Nelly), 6076.
HANNA (Ralph), 3018.
Hannibal, 1880, 1899.
Hanniballianus, rex regum et Pontiacarum gentium, 1919.

Hanschmidt (Alwin), 5221.
HANSEN (Dirk Uwe), 1433.
HANSEN (Mogens Herman), 1454, 1456, 1477.
HANSEN (Niels), 7499.
HANSEN (Peer Henrik), 3798.
HANSEN (Peter Allan), 1430.
HANSEN (Rasmus Thorning), 3049.
HANSEN-SCHABERG (Inge), 5240, 5250.
HANSON (Ann), 1376.
Hanson (Pauline), 3613.
HANSSEN (Jens), 6077.
HANSSON (Anders), 2818.
HANTRAYE (Jacques), 6926.
HANUŠ (Jiří), 793, 4834.
HAQQANI (Husain), 4360.
HAQUIN (A.), 857.
HARA (Motoko), 7641.
HARASZTI-TAYLOR (Éva), 7311.
HARBUĽOVÁ (Ľubica), 6661, 6674.
HÄRDELIN (Alf), 3325.
HARDER (Annette), 1803.
HARDIS (Arne), 3794.
Hardt (Michael), 721.
HARDTWIG (Wolfgang), 372, 5088, 5133.
HARDY-HÉMERY (Odette), 5824.
HARDYMENT (Christina), 3019.
HARF-LANCNER (Laurence), 2878.
HARINCK (George), 895.
HARISMENDY (Patrick), 4977.
HARKE (Jan Dirk), 736, 1974, 6482.
HARLAFTIS (Gelina), 754.
HARO TECGLEN (Eduardo), 4555.
HARPER (Tim), 7118.
HARPER-BILL (Christopher), 2594.
HARRIS (Bob), 4087.
HARRIS (Christopher), 7418.
HARRIS (Edward M.), 1494.
HARRIS (James), 4505.
HARRIS (Jonathan), 2426, 2430.
HARRIS (Julie A.), 3322.
HARRIS (Neil), 5493, 5543.
HARRIS (Richard), 6864
HARRIS (Rita), 4882.
HARRIS (Tim), 4061.
HARRIS (William V.), 2281, 2324.
HARRISON (Henrietta), 7642.
HARRISON (Mark), 5734.
HARRISON (Stephen J.), 2064.
HARRISS (Gerald L.), 2663.
HARROP (Paul), 2540.
HARROP (Sylvia A.), 2540.
HART (Joan), 848.
HART (Paul), 4307.
HART (Stephen M.), 983.
HÄRTER (Karl), 6552, 6580.
HARTMANN (Christian), 7170.
Hartmann (Johannes), 4858.

HARTMANN (Philip), 6581.
HARTMANNSGRUBER (Friedrich), 3908.
HARTMANN-VIRNICH (Andreas), 3221.
HARTOG (François), 510.
HARTUNG (Gerald), 5326.
HARTWICH (Wolf-Daniel), 5315.
HARVEY (Anthony), 3020.
HARVEY (L. P.), 5027.
HARVEY (Louis-Georges), 6835.
HARVEY (Paul), 4978.
HARWOOD (Jonathan), 5944.
HASEGAWA (Kinuko), 1784.
HASEGAWA (Tsuyoshi), 7138.
HASELBERGER (Lothar), 1731.
HASHIMOTO (Yu), 7805.
HASKIN (Jeanne M.), 3727.
HASLAM (Jonathan), 7495.
HASLETT (Stephen), 4053, 4344.
HASLINGER (Peter), 7496.
HASQUIN (Hervé), 6906.
HAßKAMP (Dorothee), 1495.
HATANO (Isamu), 5825.
HATELY-BROAD (Barbara), 411.
HATTON (Timothy J.), 5743.
Hattušili II, king of Hittites, 1254, 1258.
HATZFELD (Hélène), 3863.
HATZISTAVROU (Antony), 1656.
HATZIVASSILIOU (Evanthis), 3751.
HAUDRY (Jean), 2137.
HAUMANN (Heiko), 377.
HAUNSCHILD (Norbert), 4831.
HAUPT (Heinz-Gerhard), 399.
HAUSBERGER (Bernd), 3526.
HAUSBERGER (Karl), 3364.
HAUSBERGHER (Mauro), 72.
HAUSER (Marie-Christine), 5298.
HAUSMANN (Albrecht), 73.
HAUSNER (Isolde), 161.
HAUTANIEMI LEONARD (Susan), 6078.
HAVELANGE (Isabelle), V.
HAVERKAMP (Eva), 2677.
HAWTING (Gerald R.), 2640.
HAY (Gilbert), 2964.
HAY (William Anthony), 4062.
HAYAMI (Akira), 6203.
HAYCRAFT (William Russell), 7497.
HAYDEN (J. Michael), 907.
HAYE (Thomas), 2901.
HAYNES (Sybille), 1766.
HAYTON (Heather), 3104.
HAZAREESINGH (Sudhir), 378.
HE (Fan), 5776.
HE (Xuyan), 7643.
HE (Yanyan), 7644.
HEAD-KONIG (Anne-Lise), 6287.
HEAL (Felicity), 4979.

HEALY (Mark), 1899.
HEATHER (Peter J.), 2196.
HEATON-ARMSTRONG (Duncan), 3570.
HEBEISEN (Erika), 4883.
HECKER (Joel), 2684.
HECKER (Michael), 6512.
HEDGES (David), 402.
HEERS (Jacques), 2462.
HEEßEL (Nils P.), 1112.
HEFFERNAN (Kevin), 5671.
HEFFERNAN (Thomas J.), 3044, 3470.
HEFTY (Julia), 3960.
HEGEL (Georg Wilhelm Friedrich), 963, 966, 5334, 5338.
Hegemann (Werner), 6061.
HEGER-ETIENVRE (Marie-Jeanne), 4660.
HEHL (Ulrich von), 5182.
Heidegger (Martin), 959, 5334, 5335, 5338.
HEIDEMANN (Gudrun), 5517.
HEIDER (Matthias), 6533.
HEIL (Andreas), 1932.
HEIL (Matthäus), 1943.
HEIMANN-SEELBACH (Sabine), 2977.
HEINE NIELSEN (Thomas), 1475.
HEINICH (Nathalie), 5090.
Heinrich der Teichner, 3032.
HEINRICH VON DEM TÜRLIN, 2965.
HEINRICHS (W. P.), 900.
HEINZ (Werner), 2278.
HEINZE (Traudel), 1900.
HEIRBAUT (Dirk), 6594.
HEISE (Martin), 5391.
HEISTER (Hanns-Werner), 5665.
HEITSCH (Ernst), 1581.
HEITZ (Claudius), 4884.
HEKMA (Gert), 825.
HEKSTER (Olivier), 141.
Helagabalus (Marcus Aurelius Antoninus cognominatus), Roman emperor, 1815, 2136.
HÉLARD (André), 3904.
HELD (Winfried), 1136.
Helena (Iulia Flavia), mother of emperor Constantine, 2071.
Helena, Sancta, 1817.
HELFERS (James Peter), 2923.
HELLEGOUARC'H (Jacqueline), 485.
HELLER (Erich), 1843.
HELLIE (Richard), 721.
HELLMANN (Birgitt), 6915.
HELLYER (Marcus), 5402.
HELLYER (Robert I.), 6970.
Heloîse, 3293.
HELTEN (Leonhard), 3185.

HEMELRIJK (Emily A.), 1975, 2138.
HEMMES-HOOGSTADT (Annette C.), 3021.
HEMPTON (David), 4980.
HENARE (Amiria J. M.), 307.
HENDERSON (Frances), 4030.
HENDERSON (Peter), 3614.
HENGERER (Mark), 392.
HENNING (Andreas), 5602.
HENNING (Hansjoachim), 6130.
Henri II, roi de France, 3846.
Henri III, roi de France, 3905.
Henri IV, roi de France, 3843.
HENRICUS DE GANDAVO, 3250.
Henríquez Argomedo (Carlos), 6952.
HENRY (W. Ben), 1437.
Henry III, king of England, 2545, 2737.
Henry IV, king of England, 2746.
Henry V, king of England, 2746.
Henry VI, king of England, 2746.
Henry VII, king of England, 2746.
Henry VIII, king of England, 4961, 5003, 6885.
HENSCHEL (Christine), 223.
HENTSCHEL (Klaus), 5403.
HENZ (Ursula), 6079.
HENZE (Barbara), 3336.
Heraclius I (Flavius), Byzantine emperor, 140.
Heraclius, Byzantine emperor, 2490.
HERBERICHS (Cornelia), 2884.
Herbert de Clairvaux, 3033.
HERBERT MAC AVOY (Liz), 3450.
Herder (Johann Gottfried), 5336.
HERMAN (Bernard L.), 5582.
HERMAND (Jost), 4018.
HERMANN (Angela), 3915.
HERMANN (Jakobs), 2542.
HERMANN (Pernille), 3038.
HERMANN (Tomáš), 5404.
HERMANNSSON (Birgir), 4131.
HERMANSEN (Karsten), 3799.
Hermogenianus, 1971.
HERNÁNDEZ GUERRA (Liborio), 1902.
HERNÁNDEZ-GASCH (Jordi), 1086.
Herodes I, king of Iudaea, 1335.
Herodotus, 1550, 1555.
HEROLD (Paul A.), 16.
HERON (Carl), 1045.
HERON (Craig), 6432.
HERRERA (Carlos Miguel), 3599.
HERRERA (María Esther), 2968.
HERRERA NAVARRO (Javier), 5672.
HERRERÍN LÓPEZ (Angel), 4619.
HERRING (Adam), 7880.

HERRMANN (Horst), 1901.
HERRMANN (Ulrich G.), 5226.
HERRY (Simone), 5193.
HERSHBERG (James G.), 7265.
HERWIG (Holger H.), 7186.
HERZIG (Arno), 5015.
HERZOG (Dagmar), 6238.
HERZSTEIN (Robert E.), 5278.
HESBERG (Henner von), 1976, 2169.
Heshen, 7686.
Hesiodus, 1609, 1632.
HESS (Michael), 3622.
HESS (Volker), 5390.
HESSE (Hans), 3961.
HESSINGER (Rodney), 6239.
HEßLER (Martina), 795.
Hesychius Alexandrinus, 1430.
HETTEMA (T. L.), 940.
HETTLING (Manfred), 6161.
HEUCHERT (Volker), 136, 2017.
HEUER (Jennifer Ngaire), 6240.
HEVELONE-HARPER (Jennifer L.), 2325.
HEYEN (Elk Volkmar), 4981.
HEZSER (Catherine), 1296.
HICKEY (Raymond), 3294.
HICKSON (Charles R.), 6002.
HICKSON (Kevin), 4080, 6003.
HIDALGO DE CISNEROS AMESTOY (Concepción), 2732.
Hieron I, tiranno di Siracusa, 1459.
Hieron II, tiranno di Siracusa, 1413, 1459.
Hieronymus, Sanctus, 2058.
HIESTAND (Rudolf), 2546.
HIGGINS (David M.), 5886.
HIGHAM (Robert), 3492.
HIGMAN (B. W.), 5945.
Hilarius Pictaviensis, 2234.
HILBER (Kathrin), 4657.
Hildegard von Bingen, 3240.
Hildegard, saint, 3441.
HILFRICH (Fabian), 7445.
HILL (Peter P.), 6927.
HILL (Robert C.), 2223, 2326.
HILLEBRECHT (Georg), 6764.
HILLENBRAND (Robert), 2704.
HILLER (Jen), 1069.
HILLES (Sharon), 3041.
HILLS (Philip), 2083.
HIMMELBERG (Robert F.), 588.
HINDMARSH (D. Bruce), 4982.
HINGLEY (Richard), 2018.
HINKS (John), 92.
HINNELLS (John), 876.
HINTON (David Alban), 2819, 3396.
HINZ (Felix), 6836.
HIONIDOU (Violetta), 6189.
HIRATA (Elaine), 1344.

HIRSCH (Francine), 4474.
Hirschberg (Max), 6593.
HIRSCHI (Andrée), 6946.
HIRSCHI (Caspar), 5091.
HIRSCHINGER (Frank), 3962.
HIRSCHMANN (Frank G.), 3512.
HIRSCHMANN (Vera-Elisabeth), 2327.
HIRTE (Markus), 3360.
HIß (Reinhard), 2384.
HITCHNER (R. Bruce), 1113.
Hitler (Adolf), 286, 3908, 3920, 3969, 3983, 4018, 7109, 7123, 7151, 7171, 7175.
HJORTH (Birte), 6080.
HLAVIČKOVA (Zora), 3564, 5103.
HLEDÍKOVÁ (Zdeňka), 6484.
HLOUŠKOVÁ (Kateřina), 7144.
HO (Virgil K. Y.), 7645.
Hobbes (Thomas), 710, 5337.
HÖCHLI (Daniel), 707.
HOCHSTETTER (Dorothee), 5827.
HOCKERTS (Hans Günter), 7190.
HODEL (Paul-Bernard), 3420.
HODGES (Laura Fulkerson), 3023.
HODGSON (Dorothy L.), 4938.
HODKINSON (Stephen), 748.
HÖDL (Uta), *I*.
HODOS (George H.), 4119.
HODOS (Tamar), 1087.
HOEKSTRA (Hanneke), 4336.
HOFENEDER (Andreas), 2139.
HOFER (Sigrid), 5583.
HOFFMANN (Dierk), 6367.
HOFFMANN (Frank W.), 836.
HOFFMANN (Lars M.), 2431.
HOFFMANN (Léon-François), 5518.
HOFFMANN (Matthias Reinhard), 2328.
HOFFMANN (Peter), 472.
HOFFMEYER (Erik), 3800.
HOFMANN (Daniel), 3909.
HOFMANN (Heinz), 195.
HÖHLER (Sabine), 250.
HÖIJER (Birgitta), 6244.
HØJTE (Jakob Munk), 2170.
HOLDEN (Anthony J.), 2960.
HOLDENRIED (Anke), 3024.
Hölderlin (Friedrich), 5328.
HOLDSTEIN (Deborah), 5258.
HOLEC (Roman), 355.
HOLFORD-STREVENS (Leofranc), 2084.
HOLGUÍN (Sandie), 4580.
HOLL (Richard E.), 5782.
HOLLAND (Michael), 5967.
HOLLENDER (Elisabeth), 2678.
HOLLENSTEINER (Stephan), 3595.
HOLLMANN (Anna), 7336.
HOLLÝ (Karol), 2631.
HOLM (Tom), 4740.

HOLMA (Heikki), 5788.
HOLMES (Catherine), 2463.
HOLSCHER (Lucian), 4983.
HOLSINGER (Bruce), 381.
HOLT (Andrew), 7498.
HOLT (Frank L.), 1455.
HÖLTGEN (Karl Josef), 132.
HOLTHUSEN (Christoph), 4177.
HOLÝ (Martin), 812.
HOLZBERG (Niklas), 1798.
HOLZHEY (Helmut), 964.
HOMANN (Eckhard), 3307.
Homerus, 1540, 1565, 1590, 1613.
HÖMIG (Herbert), 3963.
HOMOTH-KUHS (Clemens), 1179.
HONCIUC BELDIMAN (Dana), 4445.
HONEMANN (Volker), 3231.
HONGO (Masatsugu), 7807.
HONNEFELDER (Ludger), 3260.
HONOLD (Marianne), 3283.
Honorius Augustodunensis, 104.
HONT (Istvan), 5903.
HÖPKEN (Constanze), 2171.
HOPKINS (Clare), 5183.
HOPKINS (Keith), 2172.
HOPPENKOTHEN (Dieter), 3569.
HOQUET (Thierry), 5406.
Horatius Flaccus (Quintus), 2083, 2085, 2091.
HORČÁKOVÁ (Václava), *IV*, 404.
HORN (Martin), 7208.
HORNE (Gerald), 4308.
HORNE (John), 7053.
HORNER (Patrick J.), 3044.
HORNSBY (Stephen J.), 6837.
Hornstein (Gabriel), 130.
HOROWITZ (Wayne), 1208.
HOROWSKI (Aleksander), 3284.
HORRACE (William C.), 5996.
HORROX (Rosemary), 2746.
HORTON (Carol A.), 4741.
HORTON TREMBLAY (Carol), 5876.
HORVATH (Robert), 4475.
HOSAKA (Satoshi), 7849.
HOSAKA (Takaya), 2329.
HOSFELD (Rolf), 7054.
HOSKIN (Philippa M.), 3345.
HOSSAM MUJTĀR AL ABADI, 30.
HOU (Xudong), 7646.
HOUBEN (Hubert), 2795, 2917.
HOULBROOK (Matt), 6245.
HOURIEZ (Jacques), 6946.
HOURIHANE (Colum), 3161.
HOURS (Bernard), 4917.
House (Callie), 6151.
HOUŽVIČKA (Václav), 6663.
HOVBAKKE SØRENSEN (Lars), 3801.
HOVERS (Erella), 1023, 1041.
HOWE (Eunice D.), 3188.
HOWELL (David L.), 7808.

HOWELL CHAPMAN (Honora), 1583.
HOWGEGO (Christopher), 136, 2019.
HOWLETT (D. R.), 163.
HOWNSLOW (David), 92.
HRADSKÁ (Katarína), 5093.
HREINSSON (Einar), 6005.
Hrotsvitha, 2947.
HRUBÝ (Karel), 6433.
HRUZA (Karel), 16.
HSIUNG (Ping-Chen), 7647.
HU (Hongxia), 7683.
HU (Shaohua), 7648.
HU (Zheng), 7788.
HUANG (Daoxuan), 7649.
HUANG (Dewen), 7650.
Huang (Fu), 7783.
HUANG (Jian), 7626.
HUANG (Kuangzhong), 7651.
Huang (Xing), 7750.
HUBER (Irene), 1114.
HUBERT (Marie-Clotilde), 157.
HÜBNER (Peter), 6408.
HÜBNER (Sabine), 2330.
HUCK (Oliver), 3233.
HUDEMANN (Rainer), 7438.
Hudson (Anne), 3309.
HUDSON (Benjamin T.), 2725.
HUFF (Dietrich), 1352.
Hugh of Saint-Victor, 2985, 3306.
HUGHES (R. Gerald), 7337, 7396.
HUGHES-EDWARDS (Mari), 3450.
HUGLO (Michel), 3234.
Hugo (Victor), 5518.
HUGONIOT (Christophe), 1519.
HUI (Victoria Tin-Bor), 708.
HUKUI (Shigemasa), 7652.
HUKUTO (Sanae), 7809.
HULAS (Magdalena), 812, 5101.
HULÍK (Milan), 6582.
HULL (Caroline Susan), 654.
HULL (Eugeniusz), 7299.
HULL (Isabel V.), 3965.
HÜLLEN (Werner), 174.
HULSEBOSCH (Daniel J.), 6513.
HÜLSEN-ESCH (Andrea von), 2905.
HUMBLOT (Guillaume), 7397.
Humboldt (Alexander von), 224.
Humboldts (Wilhelm von), 5254.
Hume (David), 466, 508.
HUMM (Michel), 1904.
HUMMEL (Karl Joseph), 4886.
HUMMER (Hans J.), 2613.
HUMPHREY (Caroline), 565.
HUNGER (Bettina), 4673.
HUNINK (Vincent), 2252.
HUNN (Karin), 6246.
HUNNER (Jon), 5407.
HUNT (Alan), 3396.
HUNT (Lynn), 256.

HUNT ORTIZ (Mark A.), 1055.
HUNTER (Ian), 3451.
HUNTER (Richard), 1582, 1584.
HUOT (Jean-Louis), 1227.
HÜRELBAATAR (Altanhuu), 565.
HURST (Henry R.), 1732.
HURST (J. D.), 2821.
HURST (Steven), 7338.
HURTADO OVIEDO (Víctor), 2549.
HÜRTER (Johannes), 7170.
HÜRTGEN (Renate), 6434.
Husák (Gustáv), 3767.
HUSCROFT (Richard), 2632.
HÜSMERT (Ernst), 5353.
Hussein, king of Jordan, 4252.
HUSSEY (Matthew T.), 2950.
HUTCHISON (Ann M.), 3309.
HUTERS (Theodore), 7653.
HUTTON (Ronald), 2822.
HUTTON (William), 1585.
HUYBENS (Elisabeth), 6006.
HVIID JENSEN (Anne), 4939.
Hyginus Gromaticus, 2082.
HYLDTOFT (Ole), 5828.
HYMAN (Paula E.), 324.
HYSA (Mahmud), 3571.

I

IACONO (Antonietta), 2974.
IACOVOU (Maria), 1088.
IAMBOR (Petru), 3495.
IANCU (Carol), 796.
IANNELLA (Cecilia), 2835.
IAROSSI (Maria Felicia), 198.
IBN DAUD (Abraham ben David, Halevi), 2679.
ICHIHARA (Koichi), 3496.
Iddin-Dagan, king of Larsa, 1234.
IDRIS (Amir H.), 4632.
IERODIAKONOU (Katerina), 2464.
Iesus Christus, 433, 1313, 2213, 2271, 2273, 2300, 2302, 2310, 2316, 2323, 2333, 2335, 2345, 2354, 2357, 2371, 4920.
IGERSHEIM (François), 5193.
IGGERS (Georg G.), 382.
IHRING (Peter), 240, 3076.
IHSE (Cecilia), 4647.
IIJIMA (Wataru), 7654.
IIYAMA (Tomoyasu), 7655.
IKEGAMI (Eiko), 7810.
IKEGAMI (Keiko), 3371.
IKEGAMI (Shunichi), 2600, 2601.
ILBERT (Robert), 4671.
ILIFFE (John), 7867.
ILNITCHI (Gabriela), 3235.
IMAI (Keiko), 7585.
IMLAY (Talbot), 3865.
INACHIN (Kyra T.), 3966.
İNALCIK (Halil), 6036.
INGEGNERI (Gabriele), 4913.

INGERSOLL (Thomas N.), 6248.
INGLEBERT (Hervé), 1903.
INGLESE (Alessandra), 1520.
INGLESE (Lionello), 1586.
INGRAM (Martin), 2778.
INNES (Matthew), 696.
Innocentius I, Papa, Sanctus, 2312.
Innocentius III, Papa, 3360.
Innocentius IV, Papa, 2761.
İnönü (Ismet), 4704.
INOUE (Tōru), 7806.
INQUIMBERT (Anne-Aurore), 6665.
INWOOD (Brad), 2121.
IOANID (Radu), 7398.
IOGNA-PRAT (Dominique), 2824.
Iohannes Antiochenus, 2235.
Iohannes Baptista, Sanctus, 2257.
Iohannes Chrysostomus, 2236, 2299, 2309, 2312, 2319, 2369.
Iohannes Lydus, 2396, 2468, 2498.
Iohannes Mauropus, 2397.
Iohannes Philoponus, 2398, 2399.
Iohannes VIII Palaeologus, Byzantine emperor, 2442.
Iohannes VIII, Papa, 2615, Papa, 3350.
Iohannes XXIII, Papa, 4851.
Iohannes, evangelista, Sanctus, 2259, 2265.
IOLY ZORATTINI (Pier Cesare), 5039.
IONCIOAIA (Florea), 384.
IONESCU (Mihai), 2020.
Ionescu (Nae), 4438.
IONESCU-GURĂ (Nicoleta), 4446.
IORDACHI (Constantin), 5968.
Iordanus de Saxonia, 3420.
Iorga (Nicolae), 467.
IORGULESCU (Vasile), 2592.
Iosephus (Flavius), 1818, 1884.
Ip (Hung-yok), 7656.
Ippen reverend, 7812.
IRBOUH (Hamid), 5558.
Irenaeus Lugdunensis, episcopus, Sanctus, 2378.
IRIE (Koji), 4648.
IRUMADA (Nobuo), 7811.
IRUROZQUI VICTORIANO (Marta), 3819.
IRWIN (Elizabeth), 1440, 1697.
IRWIN G. (Domingo), 4797.
Isabel I de Castilla 'la Católica', reina de Castilla y de León, 2657.
Isabella of Woodstock, 2848.
Isaeus, 1419.
Isaiah, 2268.
ISELLA (Dante), 5524.
ISENBERG (Alison), 6082.
ISENBERG (Andrew C.), 5830.

ISENMANN (Moritz), 4887.
ISETTI (Eugenia), 1050.
ISHIBE (Masasuke), 6471.
ISHIGAMI (Eichi), 7839.
ISHIHAMA (Yumiko), 7567.
ISHIKAWA (Ryōta), 7850.
ISHINABE (Masumi), 3189.
ISHIYAMA (Osamu), 7814.
ISKHAKOV (Salavat M.), 4496.
Ismail Bey Gaspirali, 6994.
Ismaïl, Khedive of Egypt and the Sudan, 3825.
Isocrates, 1419, 1552.
ISOPESCU (Sergiu), 6869.
ISRAËL (Liora), 6497, 7300.
ISRAËL (Stéphane), 5184.
ISRAEL (Uwe), 2825.
ISTRIA (Daniel), 3497.
ISUPOV (V. A.), 6099.
ITGENSHORST (Tanja), 1905.
ITIHARA (Kenji), 5346.
ITO (Koji), 6879.
ITO (Takeshi), 7803, 7814.
Iulianus (Flavius Claudius), Roman emperor, 1796, 1951, 2141.
Iustinianus I (Flavius), Byzantine emperor, 2400, 2401, 2409, 2420, 2445, 2455, 2490.
Iustinus (Iunianus Marcus), 1620, 1819.
Iustinus, Sanctus, 2238.
Ivan IV Groznyj, Vasilevič, Tsar of Russia, 4469.
IVANCHIK (Askol'd I.), 567.
IVĂNESCU (Dumitru), 7087.
IVĂNESCU (Sorin D.), 7087.
IVANIČ (Peter), 2826.
IVANOV (Jurij M.), 4476.
IVANOVA (Julija), 2946.
IVANOVA (Ljudmila A.), 568.
IVANTCHIK (Askold), 1353.
IVANTSOV (Igor' G.), 4477.
IVERSEN (Tore), 2762.
IVETIC (Egidio), 4522.
IVONE (Diomede), 7209.
IWABUCHI (Reiji), 7848.
IWASAKI (Ikuo), 7586.

J

JÄCKEL (Dirk), 2605.
JACKOB (Nikolaus), 1906.
JACKS (Davis S.), 5904.
JACKSON (Bernard S.), 1297.
JACKSON (Brenda K.), 6249.
JACKSON (Gordon), 5905.
JACKSON (Jerma A.), 5673.
JACKSON (Peter), 914, 7568.
JACKSON (Robert H.), 4940.
JACKSON (Tat'jana N.), 988.
JACKSON (William Keith), 651.
JACOB (Christian), 2021.

JACOBS (Auke P.), 6250.
JACOBS (Meg), 5746.
JACOBSEN (Nils), 3552.
JACOBY (David), 2465.
JACOBY (Sanford M.), 5831.
JACQUART (Danielle), 2932.
JACQUEMIN (Anne), 1698.
JACQUET-FRANCILLON (François), 5215.
Jacquier (François), 5389, 5580.
JAGEMANN (Norbert), 5227.
JAGER (Markus), 6083.
JAHN (Beate), 1228.
JAHR (Christoph), 5174.
JAIMOUKHA (Amjad M.), 652.
JAKOBI (Rainer), 1813.
JAKOVLEV (Aleksandr N.), 5016.
JAKUBEC (Ivan), 5747.
JAMAL (Amal), 4363.
JAMAR (Jo T. J.), 3344.
JAMBET (Christian), 908.
James VI and I, king of Scots and king of England, 4076.
JAMISON (Andrew), 5399.
JAMME (Armand), 2564, 3485, 4843.
JAMROZIAK (Emilia), 3382.
JANÁK (Dušan), 3762.
JANÁK (Jan), 6484.
JANATKOVÁ (Alena), 850.
JANCSÓ (István), 650.
JANDA (Richard D.), 172.
JANECZEK (Zdzisław), 4397.
JANIN (Hunt), 2705.
JANKOVSKAJA (Ninel' B.), 10.
JANNOT (Jean-René), 1767.
JANOSIK (MaryAnn), 838.
JANOWSKI (Berndt), 874.
JANSEN (Anscar), 7056.
JANSEN (Hans Heinrich), 3909.
JANSEN (Nils), 6584.
JANSEN (Ulrike), 1786.
JANSON (Henrik), 2724.
JANSSEN (Albert), 6486.
JANSSON (Maija), 4037.
JANUŽYTĖ (Audronė), 4276.
JAQUET (Chantal), 5355.
JARA HINOJOSA (Isabel), 3700.
JARAUSCH (Konrad H.), 3937.
JARDIN (Pierre), 3967.
JAREB (Jere), 6251.
JARITZ (Gerhard), 3361.
JARMAN (Robert L.), 3742.
JAROSZ (Dariusz), 4413.
JARVIS (Mark), 4063.
JASANOFF (Maya), 6747.
JASNOW (Richard), 1180.
JAUERNIG (Martha), *I*.
JAUMAIN (Serge), 3645.
Jaume d'Aragó, bisbe de València, 2541.

JÁUREGUI (Luis), 6023.
JAY (Martin), 798.
JAZHBOROVSKAJA (Inessa S.), 7057.
JEANMOUGIN (Bertrand), 6907.
JEANNESON (Stanislas), 7058.
JEATER (Diana), 5408.
JEBB (Richard C.), 1419.
JECH (Karel), 5946.
Jefferson (Thomas), 4717, 4778, 6936.
JEFFERY (Keith), 7059.
JEFFREYS (David), 1201.
JELLINEK (Georg), 5329.
JELOWIK (Lieselotte), 6549.
JENAK (Rudolf), 6916.
JENKINS (Brian), 3858.
JENKS (Andrew L.), 5559.
Jenks (Jeremiah Whipple), 7712.
JENNBERT (Kristina), 865.
JENNINGS (Eric T.), 3837.
JENSEN (Bent), 4478.
JENSEN (Kurt Villads), 3459.
JENSEN (Uffa), 5028.
JEŘÁBEK (Vojtěch), 7342.
JESPERSEN (Knud J. V.), 6902.
Jiang (Jieshi), 7659, 7778.
JIANG (Pei), 7657.
JIMENEZ (O. F. M.), 4918.
JIMÉNEZ ÁVILA (Javier), 1089.
JIMURA (Akiyuki), 3027.
JIN (Chongji), 7658.
JIN (Weiping), 7787.
JIN (Yilin), 7659.
JIN (Zhihuan), 7660.
JIRÁSEK (Zdeněk), 4398.
JIRGENS (Eckhard), 5674.
Joan of the Tower, 2848.
JOANNÈS (Sidonie), 3534.
João IV, Rei de Portugal, 4415.
João V, Rei de Portugal, 4425, 6895.
JOBST (Clemens), 5997.
JOCHEM (Gerhard), 7133.
JOCKENHÖVEL (Albrecht), 3127.
JOERGENS (Bettina), 6252.
Johan III, king of Sweden, 4650.
JOHANNES DE RUPESCISSA, 3422.
JOHANSEN (Anja), 3866.
JOHANSEN (Hans Christian), 5828.
JOHANSSON (Anders), 7301.
JOHANSSON (Christina), 6253.
John of Gaunt, 2848.
John of Salisbury, bishop of Chartres, 3281.
JOHNMAN (Lewis), 5832.
JOHNSON (Amy J.), 3827.
JOHNSON (Calvin H.), 6514.
JOHNSON (David K.), 6254.
JOHNSON (David), 992.
JOHNSON (Gaynor), 6650, 7210.

JOHNSON (Johnnie E. V.), 5990.
JOHNSON (Loch), 6680.
Johnson (Louis Arthur), 4750.
Johnson (Lyndon Baines), 4744, 4747, 4749, 5255.
JOHNSON (Marguerite), 1377.
JOHNSON (Richard Freeman), 3452.
JOHNSON (Timothy Scott), 2085.
JOHNSTON (Cristina), 794.
JOHNSTON (Dafydd), 3028.
JOHRENDT (Jochen), 3327.
JOLY (Bertrand), 653.
JOLY (Hervé), 282, 5792.
JOMINI (Marie-Noëlle), 2827.
JONES (Adam), 7857.
JONES (Christopher P.), 1435.
JONES (Clyve), 4079.
JONES (Geoffrey), 5833, 6008.
JONES (Harriet), 615.
JONES (Hilary), 6790.
JONES (Karen R.), 799.
JONES (Lindsay), 863.
JONES (Matthew), 7400.
JONES (Polly), 4470.
JONES (Prudence J.), 2086.
JONES (Rick), 2174.
JONES (Stephen F.), 3907.
JONES (William P.), 6436.
Jones (William Stanley), 5784.
JONKER (Marianne), 6385.
JONQUIÈRE (Tessel), 1298.
JONSSON (Alexander), 4649.
JOON-HAI LEE (Christopher), 569.
JORBA (Norma), 1780.
JORDAN (William Chester), 3328.
JORGE (Ana Maria C. M.), 3438.
JØRGENSEN (Torsten Bo.), 3361.
JORIO (Marco), 649.
JORLAND (Gérard), 135.
JORRAT (Jorge Raúl), 3587.
JOSEFSON (Kristina), 2828.
JOSEPH (Brian D.), 172.
Joseph II, röm.-deutscher Kaiser, 3627, 4930.
JOST (Madeleine), 1699.
JOTISCHKY (Andrew), 654.
JOUANNO (Corinne), 1587, 2386.
JOURDAN (Fabienne), 1657.
JOURNET (Charles), 5345.
JOYE (Frédéric), 5459.
JOYKUMAR SINGH (N.), 4141.
Juan de Austria, 262.
Juan de Prado, 5035.
JUÁREZ (Benito), 4285, 4322.
JUCKER (Michael), 2909.
JUD (Ursina), 4274.
Judah, ha-Levi, 2680.
Judith von Bayern, zweite Gemahlin Ludwigs des Frommen, 2614.

JUDT (Tony), 3535.
Juel (Niels), 3803.
Jugurtha, king of Numidia, 1908.
JULIÁ (Santos), 4628, 4629.
JULIAN OF NORWICH, 3423, 3467.
JULIAN, ARCHBISHOP OF TOLEDO, 2577, 2581.
JULIEN (Yvette), 1429.
JULLIEN DE POMMEROL (Marie-Henriette), 3422.
JUNGEL (Eberhard), 874.
Jünger (Ernst), 906.
JUNHO ANASTÁSIA (Carla Maria), 6255.
JUNQUERAS (Oriol), 4581.
Junta (Giunti), famiglia, 5295.
JUNTUNEN (Eveliina), 5603.
JUNZ (Helen B.), 3569.
JUREIT (Ulrike), 509, 7170.
JURGANOV (Andrej L.), 513.
JÜRGENSMEIER (Friedhelm), 4925.
JUSTESEN (Ole), 4025.
JUSTICE (Benjamin), 5228.

K

KABELE (Jiří), 3773.
KADIOĞLU (Musa), 1137.
KADRA (Haouaria), 1908.
KAESTLI (Tobias), 4656.
KAGAY (Donald J.), 2664.
KĀHSĀY (Barha), 3833.
KAHYA (Esin), 2706.
KAISER (Christian), 6587.
KAISER (Michael), 3986.
KAISER (Monika), 3909.
KAISER (Wolfram), 5095.
KAKRIDI (Christina), 2198.
KALAIDJIAN (Walter B.), 975.
KALC (Aleksej), 5947.
KALCK (Pierre), 655.
KALDELLIS (Anthony), 2396.
KALIFA (Dominique), 5096.
KALIFF (Anders), 661.
KALIMI (Isaac), 1300.
KÄLIN (Walter), 7330.
KALINOVÁ (Lenka), 7307.
KALITKINA (N. L.), 4458.
KALLAS (Marian), 4399.
KALLIS (Aristotle A.), 7140.
KALMAN (Laura), 6507.
KALMAR (Ivan Davidson), 403.
KALMIJN (Matthijs), 6256.
KALMYKOV (Nikolaj P.), 3655.
Kalu (Ogbu U.), 4354.
KAMBOUCHNER (Denis), 5215.
KAMIL (Neil), 4984.
KAMIŃSKI (Łukasz), 646, 3759.
KAMIŃSKI (Marek K.), 7211.
KAMRAVA (Mehran), 3537.
KANAAN (Claude Boueiz), 4271.
KANAI (Kiyomitsu), 7812.

KANDLER (Johannes), 3236.
KANE (George), 3029.
KANELLOPOULOS (Chrysanthos), 1398.
KANG (Woong Joe), 6971.
Kang (Youwei), 7684.
Kangxi Emperor, 7745.
KANJASCHIN (Jurij), 2829.
KANN (Mark E.), 4742.
KANNAPIN (Detlef), 5675.
KANT (Immanuel), 964, 5338.
KANTOR (Shawn), 5996.
KANTZIOS (Ippokratis), 1588.
KAPILA (Shruti), 5409.
KAPLAN (Jan), 7168.
Kaplan (Karel), 3774.
KAPLAN (Marion A.), 5029, 5030.
KAPLAN (Morris B.), 6257.
KAPRIEV (Georgi), 2466.
KAPS (Gabriele), 3030.
KAPTEYN (Johannes Marie Neele), 2977.
KARA (Mustafa), 915.
KARABEL (Jerome), 5185.
KARAGANEV (Rumen), 7060.
KARAHAN (Anne), 2467.
KARAKASIS (Evangelos), 2087.
Karamanoğlu İbrahim Bey II, 4679.
KARASIK (Avshalom), 1062, 1747.
KARASOVÁ (Helena), 6534.
KARASOY (Yakup), 108.
KARATEKE (Hakan T.), 4693.
KARAVASHKIN (Andrej Ju.), 513.
KARAYIANNIS (Anastassios D.), 2389.
KARILA-COHEN (Karine), 1700.
KARILA-COHEN (Pierre), 3867.
Karl I der Große, röm.-deutscher Kaiser, König der Franken, 2612.
Karl IV, röm.-deutscher Kaiser, 2931.
Karl V, röm.-deutscher Kaiser, 4812, 6824, 6875, 6881.
Karl IX, king of Sweden, 4650.
Karl XII, king of Sweden, 4644.
KARLEKAR (Malavika), 5615.
KARLSCH (Rainer), 3968.
KARN (Nicholas), 2539, 3345.
KARNEEV (Andrej N.), 5948.
KARNER (Stefan), 7145, 7375.
KÁRNIK (Zdeněk), 3761.
KARP (Candace), 7401.
KARPOV (Sergej P.), 2598.
Karpovich (Mikhail), 338.
KARRAS (Ruth Mazo), 2906.
KARTOUS (Peter), 131.
Kashtanov (Sergey M.), 2520.
KASSELL (Lauren), 5410.
KASSIANIDES (Yoann), 7500.

KASTER (Robert Andrew), 2022.
KASULKE (Christoph Tobias), 2122.
KATAOKA (Kei), 7587.
KATHE (Steffen R.), 5097.
KATOS (Demetrios), 2331.
KATZ (Kimberly), 4251.
KATZ (Paul R.), 7662.
KAUKO (Miriam), 5180.
Kaunitz-Rietberg (Wenzel Anton, Fürst), 3627.
KAVČIČ (Silvija), 5422.
KAWAHARA (Yayoi), 7569.
KAWAKITA (Minoru), 4066.
KAWAMI (Trudy S.), 1354.
KAY (Gwen), 6258.
KAYE (Alan S.), 1301.
KAYLAN (Muammer), 4689.
KAYNAR (Hakan), 810.
KAZBEKOVA (Elena V.), 2761.
KAZGAN (Gülten), 5749.
KEATING (Daniel A.), 3263.
KEAVENEY (Arthur), 1909.
KEAY (Simon J.), 2182.
KECHRIOTIS (Vangelis), 4690.
KECSKEMÉTI (Charles), 4116.
KEEFER (Edward C.), 7448, 7449.
KEEP (John), 4491.
KEETMAN (Jan), 1229.
KEHNEL (Annette), 3362.
KEHR (Paul Fridolin), 2546.
KEIPER (Gerhard), 604.
KEIPERT (Maria), 604.
KEITT (Andrew W.), 4836.
KEJŘ (Jiří), 3329.
KEKEWICH (Margaret Lucille), 2665.
KELÉNYI (György), 4120.
KELHOFFER (James A.), 2257.
KELLEHER (Margaret), 4183.
KELLER (Christian), 5329.
KELLER (Donald R.), 1325.
KELLER (Manfred), 3916.
KELLERHOFF (Sven Felix), 3969.
KELLEY (Donald R.), 383.
KELLOGG (Michael), 3970.
KELLOGG (Susan), 7881.
KELLY (Catriona), 4479.
KELLY (Christopher), 2468.
KELLY (Gavin), 1796.
KELLY (Stephen), 31.
KELLY (Susan E.), 2526.
Kelsen (Hans), 6481, 6526.
KEMP (Barry), 1201.
Kempe (Margery), 3467.
KEMPER (Michael), 739, 4480.
KEMPFER (Jacqueline), 4985.
KENISTON MAC INTOSH (Marjorie), 760.
KENNAN (George Frost), 7107.
KENNEDY (Cynthia M.), 6259.

KENNEDY (Dane), 227.
KENNEDY (Greg), 6634.
Kennedy (John Fitzgerald), 4715, 5255, 7477.
KENNEDY (Philip F.), 2712.
KENNEY (William Howland), 5676.
KENNY (Anthony John Patrick), 965.
KENT (E. J.), 6260.
KENT (John), 7402.
Kepler (Johannes), 5378.
KÉRAUTRET (Michel), 656.
KERBY-FULTON (Kathryn), 3112.
KEREN (Zvi), 3684.
KERIMOV (Abusaid K.), 570.
KÉRIVEN (Brigitte), V.
KERKVLIET (Benedict J.), 4802.
Kermavner (Dušan), 4536.
KERR (K Austin), 6163.
KERSTEN (Rikki), 4238.
KERTZER (David I.), 771.
KERVÉGAN (Jean-François), 966.
KESSLER (Edward), 2276.
KESSLER (Hans-Ulrich), 5560.
KESSLER (Martin), 5336.
KEYNES (Simon), 591.
KEYS (Angela), 3609.
Khaemhat, 1189.
KHALILIEH (Hassan S.), 2469.
KHAN (B. Zorina), 5906.
KHAN (Farid), 1360.
KHRISTOVA (Natalija), 3685.
KHUDJAKOV (Julij S.), 7580.
KHVOSTOVA (Ksenija V.), 2424.
KICK (Hermes Andreas), 1565.
KIENAST (Dietmar), 1457.
KIESER (Hans-Lukas), 4691.
Kiesinger (Kurt Georg), 3913, 3981.
KIESSLING (Rolf), 3148.
KIHORO (Wanyiri), 4256.
KIISKINEN (Elina), 3971.
KILIKOGLOU (Vassilis), 1087, 1115.
KILLINGRAY (David), 6734.
Kim (Il-saong), 7543.
KIM (Namsuk), 6009.
KIM (Seung-young), 6972.
KIM (YoungSook), 4236.
KIMMERLE (Ralph), 1496.
KIND (Friederike), 3520.
KING (Andy), 2574.
KING (Karen L.), 1658.
KINGSLAND (Sharon E.), 761.
KINLOCH TIJERINO (Frances), 4349.
KINSBRUNER (Jay), 6084.
KINTZINGER (Martin), 2633.
KIPKE (Rüdiger), 3756.
KIPPING (Matthias), 6163.

KIR'JANOV (Jurij I.), 6437.
KIRAT (Thierry), 5777.
KIRBY (Peter), 6438.
Kirby Smith (Edmund), 4767.
KIRCHHOFF (Hans), 7141.
KIRCHNER (Bill), 839.
KIRK (Neville), 6439.
KIRK (Thomas Allison), 4218.
KIRKPATRICK (Charles Edward), 7266.
KIRMIZI (Abdulhamit), 5229.
KIRON (Arthur), 77.
KIROS (Asmeret), 1033.
KISAICHI (Masatoshi), 7570.
KISATSKY (Deborah), 7403.
KISELEV (Nikolaj P.), 4481.
KISELEVA (L. I.), 32.
KISHLANSKY (Mark), 4064.
KISLUK (Eugene J.), 4400.
KISSANE (Bill), 4178.
KIBEL (Walter), 5050.
KIßENER (Michael), 3972.
Kissinger (Henry), 7490, 7502.
KITAGAWA (Takako), 7588.
KIT-CHING (Chan Lau), 7212.
KITSON (Simon), 3897, 7302.
KITTEL (Manfred), 5161.
KITTLESON (Roger Alan), 3670.
KITTNER (Michael), 6440.
KIVISTO (Peter), 775.
KLABJAN (Borut), 7377, 7437.
KLAPP (Otto), 971.
KLAPPER (Melissa R.), 5031.
KLAPP-LEHRMANN (Astrid), 971.
KLÁPŠTE (Jan), 2873.
KLASSEN (Lutz), 1050.
Klauck (Hans-Josef), 2212.
KLAUSA (Gustava Alice), 3914.
KLAUSCH (Hans-Peter), 3973.
KLECKER (Elisabeth), 226.
KLEIN (Thomas), 2948.
KLEINBERG (Aviad M.), 3413.
KLEINBERG (Ethan), 5335.
KLEINEN (Karin), 5187.
KLEINER (Diana E. E.), 1181.
KLEINSCHMIDT (Harald), 2907, 7078.
KLEISNER (Karel), 5404.
KLEISS (Wolfram), 1355.
KLEMP (Stefan), 6535.
KLENGEL (Horst), 1264.
KLEẞMANN (Christoph), 6408.
KLEYWEGT (Adrianus Jan), 1845.
KLIGMAN (Gail), 5968.
KLIMASMITH (Betsy), 5098.
KLIMMEK (Nikolai F.), 5338.
KLIMÓ (Árpád von), 5056.
KLINGENBERG (Georg), 1958.
KLINGER (Jörg W.), 1265.
KLINKENBERG (Emanuel S.), 3192.
KLINKHAMMER (Lutz), 356.

KLINKOTT (Hilmar), 1356.
KLONOWSKI (Martin), 6908.
KLOTZ (Marcia), 6719.
KLUETING (Edeltraud), 3383.
Klueva (N. G.), 4471.
KLUGE (Ulrike), 5949.
KLUGER (Helmuth), 2602.
KLUNDER (Nolanda), 3031.
KNABE (Hubertus), 7142.
KNAPÍK (Jiří), 5099.
KNAPP (Bernard A.), 1389.
KNAPP (Fritz Peter), 385, 2965.
KNAPP (Keith Nathaniel), 2830.
KNAPP (Marion), 5100.
KNAPP (Raymond), 5678.
KNAPPETT (Carl), 1087, 1115.
KNEŽEVIĆ (Saša), 6973.
KNIGGE (Volkhard), 309.
KNIGHT (Alan), 4294.
KNIGHT (Amy), 7404.
KNIGHTON (C. S.), 4029.
KNIGHTS (Mark), 4065.
KNIPP (David), 3158.
KNOCH (Stefan), 1977.
KNOTHE (Hans-Georg), 6567.
KNOTT (Kim), 916.
KNOX (Robert), 1360.
KÖB (Edgar Helmut), 5907.
KÖBLER (Gerhard), 2967.
KOBZAREVA (Elena I.), 6880.
KOÇ (Murat), 5519.
KOÇ (Yunus), 6261.
Koca-Memi, familie, 122.
KOCAMEMI (Fazıl Bülent), 122.
KOCH (Armin), 2614.
KOCH (Arnd), 6536.
KOCH (Bettina), 710.
KOCH (Hans Jürgen), 5679.
KOCH (Renée), 899.
Koch (Robert), 5396.
KOCHAVI (Arieh J.), 7213.
KOCHAVI (Noam), 7502.
KOCIAN (Jiří), 405, 3765, 3775, 5101.
KOCKA (Jürgen), 360.
KOCKEL (Titus), 5834.
KOCKEL (Valentin), 5391.
KOČOVÁ (Kateřina), 6537.
KOELB (Clayton), 5514.
KOENKER (Diane P.), 6441.
KOERNER (Francis), 7143.
KOFMAN (Jakov M.), 5022.
KOGAN (L. E.), 1110.
KOGAN (Leonid), 1302.
KÖHLER (Ingo), 6010.
KÖHLER (Kai), 5081.
KOHLRAUSCH (Martin), 3974.
KOHNLE (Armin), 657.
KOHUT (Zenon E.), 658.
KOIKE (Makoto), 7589.
KOIV (Mait), 1497.

Kojève (Alexandre), 5334, 5339.
KOJIMA (Michihiro), 7816.
KOKOŠKA (Stanislav), 3765, 7267.
KOKOŠKOVÁ (Zdeňka), 3765.
KOL'TSOV (Lev V.), 1031.
KOLAR-DIMITRIJEVIĆ (Mira), 5869.
KOLB (Anne), 1786.
KOLB (Eberhard), 7061.
KÖLBL (Angelika), 3032.
KOLEV (Valeri), 7023.
KOLL (Johannes), 3647.
KOLLER (Heinrich), 2666.
KOLMER (Lothar), 2618.
KOLOBU (Photeine), 2385.
KOLODNIKOVA (Ljudmila P.), 4464, 7161.
KOLOĞLU (Orhan), 5281.
KÖLZER (Theo), 11.
KOMIYA (Hitoshi), 4237.
KOMOLOVA (Nelli P.), 5102.
KOMPATSCHER GUFLER (Gabriela), 3033.
KONAKÇI (Erim), 1209.
KONDRATENKO (D. P.), 4482.
KONG (Lily), 917.
KÖNIG (Jason), 2088.
KÖNIGSEDER (Angelika), 4001.
KONJAVSKAJA (Elena L.), 2568.
KONNO (Jun), 7663.
KONOPPA (Claudia), 2240.
KONOW (Jan von), 6262.
Konrad von Megenberg, 3308.
KÖNSGEN (Ewald), 2948.
KONSTANTINOU (Evangelos), 2418.
KONSTANTINOV (Petăr Stefanov), 6671.
KONTANI (Ryoichi), 1357.
KOOPMANN (Helmut), 1002.
KOPEČEK (Lubomír), 3562.
KOPEČEK (Michal), 387, 3761, 5103.
KOPECKÝ (Milan), 7268.
KOPKE (Christoph), 5422.
KOPP-OBERSTEBRINK (Herbert), 5326.
KOPTEV (Alexandr), 1978.
KORECKY-KRÖLL (Katharina), 161.
KORELIN (Avenir P.), 4494.
KORELIN (Mikhail S.), 2199.
KORIEH (Chima J.), 4354.
KORMYSHEVA (Eleonora E.), 1182.
KOROLEVA (Nadezhda G.), 4512.
Korošec (Anton), 7313.
KORYN (Andrzej), 3542.
KORZILIUS (Sven), 6588.
KOS (Dušan), 3498.
Koselleck (Reinhart), 530.
KOSESKI (Adam), 4388.

KOSHELEVA (L. P.), 4462.
KOSLOWSKI (Stefan), 5356.
KOSMETATOU (Elisabeth), 1138.
KÖSOĞLU (Nevzat), 5520.
KÖSSLER (Reinhart), 7062.
KÖSSLER (Till), 647, 3976.
KOSTA (Jiří), 5750.
KOSTELECKÝ (Tomáš), 6086.
KOSTELENOS (G.), 5751.
KÖSTERS (Christoph), 4835.
KOSTO (Adam J.), 2627.
KOSTOV (Aleksandăr), 7063.
KOSTYRCHENKO (Gennadij V.), 5016.
KOTALA (Tomasz), 2470.
KOTERA (Atsushi), 7664.
KOTERSKI (Joseph W.), 3145.
KOTILAINE (Jamo T.), 5908.
KOTT (Sandrine), 5955.
KOUAMÉ (Thierry), 821, 3143.
KOULOURI (Christina), 7023.
KOUSSER (Rachel), 1733.
KOUWENBERG (N. J. C.), 1230.
KOVAČ (Miro), 7064.
KOVAL' (Vladimir Ju.), 2599.
KOVALEVA (L. A.), 274.
KOVALEVSKAJA (Vera B.), 1090.
KOVAŘÍK (David), 6263.
KOWALCZUK (Ilko-Sascha), 4013.
KOWALSKI (Robert), 7072.
KOWALSKY (Daniel), 4605, 7368.
KOWERSKI (Lawrence M.), 1589.
Koyré (Alexandre), 5340.
KOZÁRI (Monika), 4121.
KOZHOKIN (Evgenij M.), 3868.
KOZLOV (Vladimir P.), 7374.
KOZLOVA (N. V.), 1110.
KOZLOWSKI (Stefan), 1024.
KOZOLUPENKO (Dar'ja P.), 868.
KOZYREV (Vitalij A.), 5948.
KRAFFT (Otfried), 918, 3424.
KRAFT (John), 4814.
KRAGH (Helge), 801, 5411.
KRAGL (Florian), 2965, 3034.
KRAINZ (Thomas A.), 6264.
KRAJEWSKI (Kazimierz), 4412.
KRAKAU (Constanze), 4192.
KRAMER (Emil A.), 1848.
KRAMER (Ferdinand), 121.
KRAMER (Nicole), 7190.
KRANICH-HOFBAUER (Karin), 54.
KRANNICH (Torsten), 2333.
KRAPINGER (Gernot), 1838.
KRAPOTH (Hermann), 505.
KRATOSKA (Paul H.), 6409.
KRATZKE (Christine), 3394.
KRAUS (Jeremia), 3190.
Kraus (Karl), 5536.
KRAUS (Thomas R.), 2747.
KRAUSHAAR (Wolfgang), 3977.
KREINER (Josef), 6987.

KREJČÍ (Oskar), 5104.
KREKOVIČ (Eduard), 5123.
KREKOVIČOVÁ (Eva), 5123.
KREMER (Klaus), 3471.
KREMLIČKOVÁ (Ladislava), 3772.
KŘEN (Jan), 514, 3538.
KRENN (Michael L.), 5561.
KRĒSLIŅŠ (Uldis), 4267.
KREUL (Andreas), 5546.
KREUTZ (Bernhard), 3499.
KREUTZER (Susanne), 6265.
KREUTZMÜLLER (Christoph), 6011.
KREUZER (Georg), 3518.
KREY (Hans-Josef), 2667.
KRIECHBAUMER (Robert), 3619, 7395.
KRIECKHAUS (Andreas), 2023.
KRIGER (Colleen), 5835.
KRIKLER (Jeremy), 4544.
KRIM (Arthur J.), 266.
KRISHAN (Shri), 4142.
KRISTAL (Efraín), 979.
KRISTIANSEN (Kristian), 1064.
KRISTOL (Andres), 164.
KROES (Rob), 3558.
KRÖGER (Bernward), 4888.
KRÖGER (Martin), 604.
KROL (Aleksej A.), 1183.
KRÓLCZYK (Krzysztof), 1787.
KROLL (Frank-Lothar), 3980.
KROLL (John H.), 1122.
KROLL (Stephan), 1358.
KROLL (Thomas), 486.
KRON (Geoffrey), 1378, 2024.
KRONVALL (Olof), 7343.
KROPP (Alexander), 5584.
KRÜCK (M.-P.), 1590.
KRUG (Alexander), 6866.
KRÜGER (Gunnar), 5282.
KRUGER (Paul A.), 1303.
KRUGER (Steven F.), 2686.
KRUMBACHER (Karl), 2384.
KRUMMENACHER-SCHÖLL (Jörg), 4657.
KRÜPE (Florian), 592.
KRUSE (Holger), 2759.
KRUSE (Kevin M.), 4743.
KU (Su-Hun), 3191.
KUBERSKY-PIREDDA (Susanne), 5604.
KUBO (Toru), 7665.
KUBOTA (Masaki), 7817.
KUBOUCHI (Tadao), 3371.
KUČERA (Jaroslav), 6012, 7406.
KUČERA (Radek), 5836.
KUCK (Gerhard), 7245.
KÜÇÜK (Serhat), 810.
KUDRNA (Ladislav), 7269.
KUETHE (Allan J.), 6850.
KÜHL (Jørgen), 7501.

KUHN (Axel), 6266.
KUHN (Lambrecht), 3363.
KÜHN (Thomas), 4804.
KUHNS (Richard Francis), 3035.
KUHOFF (W.), 2471.
KUHR-KOROLEV (Corinna), 4483.
KUKLÍK (Jan), 7177.
KULAKOV (Arkadij A.), 4464.
KULENKAMPFF (Angela), 3627.
KULIKOWSKI (Michael), 2197.
KULKE (Hermann), 659.
KULLY (Rolf Max), 164.
KULU (Hill), 6267.
KUMANEV (G. A.), 4463.
KUMAR (Sunil), 2707.
KUMARI (Saroj), 4143.
KUMRULAR (Özlem), 6881.
KUNDRUS (Birthe), 6737.
KUNST (Christiane), 2025.
KUNŠTÁT (Miroslav), 387.
KUNZ (Andreas), 3979.
KUNZ (Norbert), 7146.
KUNZE (Rolf-Ulrich), 711.
KÜNZE-GÖTTE (Erika), 1734.
KÜNZLI (Jörg), 7503.
KÜPPER (Herbert), 737.
KURAL (Václav), 7065, 7278.
KURAN-BURÇOĞLU (Nudret), 813.
KURIHARA (Hirohide), 7590.
KURLAEV (Evgenij A.), 5837.
KURODA (Hideo), 414.
KUROIWA (Takashi), 7801.
KUROMIYA (Hiroaki), 4484.
KURT (Yılmaz), 6128.
KUSHNER (Ali Ekrem Bey), 4692.
KUSTER (Friederike), 5350.
KUT (Günay), 993.
KUVAJA (Christer), 6375.
KUWAHARA (Makiko), 571.
KUZINA (Inna N.), 2599.
KUZNECOVA (Tamara E.), 4485.
KUZNETSOV (A.), 4268.
KWIATKIEWICZ (Piotr), 7407.
KWIEK (Julian), 4386.
KWON (Jeong-Im), 5334.
KYCHANOV (Evgenij I.), 7571.
KYNOCH (Gary), 4545.
KYRIAKIDIS (Evangelos), 1065, 1396.

L

L'ANGE (Gerald), 6791.
L'Hospital (Michel de), 6545.
L'OCCASO (Stefano), 2547.
LA MOAL (Frédéric), 7066.
LA ROCCA (Adolfo), 2123.
LA ROCCA (Cristina), 2792.
La Rochefoucauld-d'Enville (Louis-Alexandre, duc de), 6897.
LA ROOIJ (Marinus), 3615.

LA VOPA (Anthony J.), 5338.
LABAND (John), 6792.
LABORDE (Denis), 505.
LABOW (Dagmar), 1818.
Labriola (Antonio), 468, 5341.
Labrousse (Camille-Ernest), 469.
LABROUSSE (Elisabeth), 5052, 5324.
LABUDA (Adam S.), 850.
ŁABUSZEWSKI (Tomasz), 4412.
LABUTINA (Tat'jana L.), 5106.
LACH (Jiří), 388.
LACKMANN (Heinrich), 4858.
LACONI (Sonia), 1809.
Lactantius (Lucius Caecilius Firmianus), 2239, 2307.
LĂCUSTĂ (Ioan), 4447.
LACY (Norris J.), 614, 3010.
LADA-RICHARDS (Ismene), 1591.
LADBROOK (Dave), 1026.
LADRÓN DE GUEVARA MELLADO (Pedro Luis), 5522.
LAEBEN-ROSÉN (Viktoria), 1910.
LAFEBER (Walter), 4744.
LAFI (Nora), 6098.
LAFON (Xavier), 2187.
LAFOND (Yves), 1701.
LAGATZ (Tobias), 4898.
LAGERLF NILSSON (Ulrika), 4889.
Lagidi, dinastia egiziana, 1151.
LAGOS (Katerina), 4103.
LAGROU (Pieter), 411, 5032.
LAHOUEL (Badra), 6793.
LAIDLAW (Zoë), 6723.
LAINS (Pedro), 5744, 5838.
LAIOU (Angeliki), 2472, 2511.
LAL (Ruby), 7591.
LALIEU (Olivier), 7303.
LALLA (Sebastian), 4988.
LALONDE (Gerald V.), 2334.
LALOUE (Christine), 3241.
LALOUETTE (Jacqueline), 3838, 3869.
Lamares, king of Egypt, 1200.
LAMARRA (Antonio), 5342.
LAMBAUER (Barbara), 3870.
LAMBERIGTS (M.), 857.
LAMBERT (David), 3638.
LAMBERT (Frank), 6928.
LAMBERT (Josiah B.), 6442.
LAMBERT (Laura), 797.
LAMBERT (Mayer), 1304.
LAMBERTS BENDROTH (Margaret), 4838.
LAMBIN (Gérard), 1434.
LAMMERS (Karl Christian), 7344.
LAMONDE (Yvan), 5279.
LAMOTHE (Kimerer L.), 919.
LAMPAKES (S.), 2425.
LAN'KOV (Andrei Nikolaevich), 4259.

LANCASTER (Lynne C.), 2175.
LANCEL (Serge), 2356.
LANCELLOTTI (Maria Grazia), 885.
LANDAVAZO (Marco Antonio), 4309.
LANDI (Fiorenzo), 4839.
LANDINO (Cristoforo), 3123.
LANDOLT (Patricia), 765.
LANDON (William J.), 5107.
LANDRY (Yves), 287.
LANDSKRON (Alice), 1359.
LANDSMAN (Mark), 5909.
LANDWEHR (Eva-Maria), 5391.
LANE (Fintan), O DRISCEOIL (Donal), 4184.
LANFRANCHI (Pierre), 6243.
LANFRANCHI STRINA (Bianca), 6871.
Lanfredini (Giovanni), 2530.
LANG (Peter Thaddäus), 3981.
LANGENBACHER (Ferdy), 4918.
LANGER (Ullrich), 977.
LANGERMANN (Y. Tzvi), 2676.
LANGKAU-ALEX (Ursula), 6443.
LANGLAND (William), 3029, 3036.
LANGLOIS (Claude), 4890.
LANGUE (Frédérique), 4797, 6201.
LANKOV (Andrei), 4260.
LANNI (Adriaan), 1498.
LANNUTTI (Maria Sofia), 3243.
Lansdale (Edward Geary), 7350.
LAPATIN (Kenneth D. S.), 1735.
LAPINI (Walter), 1592.
Laplace (Pierre Simon), 5400.
LAPORTE (Norman), 4042, 7352.
LAPTEVA (Ljudmila P.), 389.
LARA GUZMÁN (Marco), 3817.
LARGUIER (Gilbert), 6007.
LARKIN (Edward), 5521.
LARSEN (Erling Røed), 5783.
LARSSON (Esbjörn), 5188.
LARSSON (Lars-Olof), 4650.
LARUELLE (Marlène), 4486, 6268.
LASER (Günter), 1812.
LASOCHKO (L. S.), 275.
LASSAVE (Pierre), 995.
LASSONDE (Stephen), 5230.
LATAWSKI (Paul Chester), 6646.
LATEINER (Donald), 1593.
LATERZA (Giovanni), 5327.
LATOUR (Francis), 7067.
LATTE ABDALLAH (Stéphanie), 6269, 7339.
LATZIN (Ellen), 5108.
LAU (Thomas), 116.
LAUDANI (Simona), 4219.
LAUGKAU-ALEX (Ursula), 3982.
LAURAND (Valéry), 1659.
LAURENT (Sébastien), 704, 3871.
LAURIDSEN (John T.), 7108.
LAURIE (Bruce), 6270.

LAURINI (Gerardo), 5327.
LAURIOUX (Bruno), 2910.
LAURSEN (John Christian), 3451.
LAUSE (Mark A.), 4745.
LAUWERS (Michel), 2911, 3453.
LAVAGETTO (Mario), 5512.
LAVALLÉ (Bernard), 6824.
LAVALLÉE (Danièle), 7882.
LAVELLE (B. M.), 1458.
LAVILLE (Helen), 4781.
LAVOIX (Hélène), 3687.
LAVRENOV (Sergej Ja.), 4262.
LAWLER (Daniel), 7448.
LAWRANCE (Benjamin N.), 6794.
LAWRENCE (Christopher), 5412.
LAWRENCE (Marilyn), 3068.
LAWRENCE (Mark Atwood), 7504.
LAWRENCE (Stuart), 1594.
Lawrence of Durham, 2999.
Lawson, family, 7857.
LAWSON (George), 3539.
LAZĂR (Gheorghe), 5910.
LAZAR (Lance Gabriel), 4919.
LAZAREVIĆ (Žarko), 6013.
LAZZARO (Claudia), 5555.
LE BERRE (Yves), 177.
LE BLANC (Marianne), 5580.
LE BOHEC (Yann), 1911.
LE BORGNE (Jean), 7894.
LE COUR GRANDMAISON (Olivier), 6795.
LE DORH (Marc), 7345.
LE DU (Jean), 177.
LE GALL (Dina), 920.
LE MOINE (Frederic), 3872.
LE ROUX DE LINCY (Antoine-Jean-Victor), 37.
Le Seur (Thomas), 5389, 5580.
LE THIEC (Guy), 4812.
LE TOURNEAU (Dominique), 921.
LEACH (Elizabeth Eva), 3227.
LEANDRO ZUÑIGA (Vilma), 3733.
LEÃO (Delfim F.), 1499.
LEBECQ (Stéphane), 3433.
LEBOUCHER (Marc), 4821.
LEBRETON (Philippe), 123.
LECHNER (Stefan), 4220.
LECLANT (Jean), 632, 2900.
LECOMTE (Olivier), 1077, 1091.
LECUPPRE (Gilles), 2606.
LECUPPRE-DESJARDIN (Elodie), 2896.
LEDENT (Philippe), 5987.
LEDOVSKIJ (A. M.), 7372.
LEDUC (Claudine), 1702.
LEDVINKA (Václav), 6093.
LEE (Aquila H. I.), 2335.
LEE (Benjamin Todd), 1801.
LEE (James), 6059.
LEE (Mi-Kyoung), 1660.
LEE (Pui-tak), 6638.

LEE (R. Alton), 6444.
LEE (Robert), 6189.
LEE (Stephen J.), 4068.
Lee (Teng-hui), 4665.
LEE VAN COTT (Donna), 3540.
LEERSON (Benoît), 1184.
LEFEBVRE-TEILLARD (Anne), 3144, 6566.
LEFEUVRE (Daniel), 6796.
LEFÈVRE (Eckard), 2110.
LEFKA (Aikaterini), 1661.
LEFORT (Jacques), 2514.
LEGENDRE (Olivier), 3251.
LÉGLU (Catherine), 3120.
LEGUAI (André), 2668.
LEHANE (Brendan), 3454.
LEHMANN (Hartmut), 4992.
LEHMANN-BRAUNS (Sicco), 543.
LEHMLER (Caroline), 1459.
LEHTONEN (Tuomas M. S.), 3459.
LEI (Bing), 7666.
LEIBNIZ (Gottfried Wilhelm, Freiherr von), 5342.
LEIKIN (Steve), 6445.
LEINKAUF (Thomas), 3285.
LEISCH-PROST (Edith), 3569.
LEISER (Gary), 5413.
LEITERITZ (Christiane), 5480.
LEITNER (Philipp), 2089.
LEIVA (Fernando), 3705.
LEIVAM (Sebastián), 3706.
LEIVERKUS (Yvonne), 3501.
LEJON (Kjell O.), 3330.
LEKAN (Thomas), 5082.
LEMAIRE (Jean-François), 6936.
LEMAIRE (Sandrine), 347, 6717.
LEMAÎTRE (Jean Loup), 33.
LEMAÎTRE (Nicole), 111.
Lemass (Seán), 4179.
LEMAY (Benoît), 3983.
LEMBI (Gaia), 1299.
LEMBKE (Janet), 1851.
LEMERCIER (Claire), 515, 6087.
LEMERLE (Frédérique), 5585.
LEMOINE (Thierry), 3639.
LEMOINE (Yves), 453.
LEMONDE (Anne), 500.
LEMOV (Rebecca), 5414.
LENDON (Jon E.), 1379.
LENGWILER (Martin), 5189.
LENGYELOVÁ (Tünde), 751.
LENIAUD (Jean-Michel), 5586.
Lenin (Vladimir Ilič Uljanov), 4498.
LENSKI (Noel), 2194.
LENTZ (Thierry), 6931.
Leo Diaconus, 2402.
Leo I Magnus, Papa, Sanctus, 2333.
Leo I, Byzantine emperor, 2452, 2471.

Leo the Deacon, 2576.
Leo XIII, Papa, 4855, 6996.
Leo, metropolitan of Chalcedon, 2451.
LÉON (Antoine), 5231.
LEÓN (Leonardo), 3708.
LEÓN (Magdalena), 6564.
LEÓN LIRA (Matías), 5283.
LEONARD (Amy), 4891.
LEONARDI (Claudio), 3434.
LEONARDI (Lino), 3037.
LEONE (Alfonso), 3291.
LEONOVA (Elena V.), 1037.
Leontius Scholasticus, 2403.
Leopardi (Giacomo), 5511, 5522.
LEOPOLD (Mark), 7868.
Leopold II, king of the Belgians, 6780.
LEPORE (Jill), 4746.
Leporius, monachus, 2333.
LEPP (Claudia), 4987.
LEPPIN (Volker), 5336.
LERAT (Christian), 6095.
Lerdo de Tejada (Sebastián), 4290.
LERHIS (Ainārs), 7068.
LERNER (Mitchell B.), 4749.
Leroy-Beaulieu (Pier), 3891.
LEȘCU (Anatol), 4488.
LESLIE (William Bruce), 5190.
Lessing (Gotthold Ephraim), 5328.
LÉTRILLIART (Philippe), 3744.
LETZNER (Wolfram), 2176.
LEUCHTENBURG (William E.), 4747, 6507.
LEUGERS (Antonio), 3926.
LEUSCHNER (Eckhard), 5605.
LEUSTEAN (Lucian N.), 922.
LEUȘTEAN (Lucian), 5232.
LEV (Yaacov), 2708.
LEVEAU (Rémy), 5037.
LEVERING (Matthew), 3302.
LEVÍN (Florencia), 3580.
LEVIN (Michael J.), 6882.
LEVIN (Miriam R.), 5415.
LEVINA (Elena S.), 4471.
Levinas (Emmanuel), 5343.
LEVINE (Alan J.), 7408.
LEVINE (Bruce), 4748.
LEVINE (Joseph M.), 383.
LEVINE (Lee I.), 1305.
LEVINE (Mark), 228.
LEVINSON (Irving W.), 4310.
LEVIS SULLAM (Simon), 3560.
LÉVISON (John R.), 1306.
LÉVY (Antoine), 2210.
LÉVY-DUMOULIN (Olivier), 390.
LEWIN (Moshe), 4489.
LEWIS (A. D. E.), 5347.
LEWIS (Barry James), 2963, 2976.
LEWIS (Frank D.), 6714.
Lewis (Jill), 6327.

LEWIS (Mary S.), 79.
LEWIS (Stephen E.), 4311.
LEWY (Guenter), 4694.
LEYDER (Dirk), 5233.
LEYGUARDA DOMÍNGUEZ (Manuel), 3372.
LEYSER (Henrietta), 3407.
LI (Aili), 7667.
Li (Hanhun), 7791.
LI (Houmu), 7668.
LI (Hu), 7669.
LI (Huaiyin), 3714.
Li (Jieren), 7666.
LI (Jifeng), 7614.
LI (Li), 7670.
LI (Lingxia), 7757.
LI (Pozhong), 7671.
LI (Qing), 7672.
LI (Rong), 7673.
LI (Tianyi), 7674.
LI (Wei), 7675.
LI (Xizhu), 7676.
LI (Xuetong), 7677.
LI (Yangfan), 7678.
LI (Yanjing), 7679.
Liang (Dingfen), 7748.
Liang (Qichao), 7789.
Liang (Shizhi), 7615.
Liao (Ping), 7684, 7729.
LIBSEKAL (Yosief), 1033.
LICCIARDELLO (Pierluigi), 3414.
LICHT (Walter), 5806.
LICHTBLAU (Karin), 3054.
LICHTENBERG (Georg Christoph), 5416, 5431.
LICHTENSTEIN (Alex), 6446.
Licinius Rufinus (Marcus Gnaeus), 1980.
LIDA (Miranda), 4892.
LIDDELL (Guy Maynard), 7244.
LIDDY (Christian Drummond), 2669, 2847.
LIDEGAARD (Bo), 7147.
LIE (Henry), 2178.
LIEBER (Keir), 6667.
LIEBERG (Godo), 1788.
Liébert, évêque de Cambrai et d'Arras, comtes du Cambrési, 2527.
LIEBIG SCHLABER (Gerret), 6447.
LIEBREICH (Fritz), 7409.
LIEBSCHER (Daniela), 6271.
LIENHARD (Luc), 5361.
LIERMANN (Christiane), 477.
LIETZ (Gero), 180.
LIGHT (Ivan), 782.
LIGUORI (Guido), 5331.
LIIVIK (Olev), 3830.
LILLA (Joachim), 6539.
LILLIE (Amanda), 3502.
LILTI (Antoine), 6272.

LIN (Shu-mei), 7680.
LIN (Xiaoqing Diana), 5191.
LIN (Xiaoting), 7681.
Lincoln (Abraham), 4765, 4780.
LINCOLN (Bruce), 896, 924, 935, 946.
LIND (Amy), 3818.
LINDENBERGER (Thomas), 6273.
LINDENMEYER (Kriste), 6274.
LINDER (Jan), 4356.
LINDEROTH (Andreas), 7552.
Lindh (Anna), 6244.
LINDNER (Stephan H.), 5839.
LINDSAY (Denise), 3913.
LING (Lesley), 2177.
LING (Roger), 2177.
Linge (Heinz), 7109.
LINGELBACH (Gerhard), 2748.
LINIGER-GOUMAZ (Max), 3821.
LINK (Stephan), 3984.
LINKE (Bernhard), 1703, 1913.
LINSINGER (Stephanie), 7190.
LINTON (Derek S.), 5417.
Lionel of Antwerp, 2848.
LIPHARDT (Elizaveta), 4490.
LIPPE (Marcus Chr), 5680.
LIPSEY (Richard G.), 5752.
Lipsius (Justus), 3881.
Liszt (Franz), 830.
LITTLE (David), 4994.
LITVIN (Alter), 4491.
LIU (Dalin), 7683.
Liu (Dapeng), 7634.
Liu, family, 7642.
Liu (Mingzhuan), 7612.
LIU (Wei), 7684.
LIU (Wusheng), 7685.
LIU (Xinghua), 7686.
LIU (Yangdong), 7687.
Liu (Yin), 7774.
LIUȘNEA (Mihaela-Denisia), 7069.
Liutprando da Cremona, 3013.
Liuzzo (Viola), 4752.
LIVELEY (Genevieve), 1824.
LIVESEY (Steven John), 3290.
LIVI (Massimiliano), 325.
LIVI BACCI (Massimo), 7883.
Livius (Titus), 1820.
LJUBART (Margarita K.), 3873.
LLANO ISAZA (Rodrigo), 3721.
LLERAS DE LA FUENTE (Carlos), 3718.
LLERAS RESTREPO (Carlos), 3718.
LLOYD (Geoffrey E. R.), 803.
LLOYD (Joan Barclay), 3385.
LLOYD (Rosemary), 976.
Lloyd George (David), 7095.
LLOYD-JONES (Hugh), 1380.
LLOYD-JONES (Stewart), 4418.
LLUCH BRAMON (Rosa), 2831.
LLULL (Ramon), 3252.

LO CASCIO (Elio), 763.
LO RE (Salvatore), 5606.
LOBO ANTUNES (António), 6762.
LOBO ANTUNES (Joana), 6762.
LOBO ANTUNES (Maria José), 6762.
LOCANTO (Massimiliano), 3243.
LOCHARD (Eric-Olivier), 5052.
LOCHRIE (Karma), 2912.
LOCKHART FLEMING (Patricia), 5279.
LODS (Adolphe), 1307.
LODY (Raul Giovanni da Motta), 310.
LÖFFLER (Ursula), 3985.
LOFSTROM M. (William Lee), 124.
LOGAN (Francis Donald), 2763, 3348.
LOGUÉRCIO CÁNEPA (Mercedes Maria), 3671.
LÖHR (Winrich Alfried), 2336.
LOHSE (York), 6089.
LOISEAUX (Olivier), 229.
LOJKÓ (Miklós), 7070.
LOMBARDI (Chiara), 3040.
LOMONACO (Fabrizio), 5316.
Londonderry (Charles Stewart Henry Vane-Tempest-Stewart, Marquess), 4175.
LONG (Jason), 6275.
LONG (Stephen John), 7410.
Longhi (Roberto), 852, 853.
Longus, 1424, 1636.
LONNET (Antoine), 1308.
LONZA (Nella), 2744.
LÓPEZ (Adalberto), 6838.
LOPEZ (Brice), 2050.
LÓPEZ (Cristina del Carmen), 3601, 3602.
LÓPEZ (Mario Justo), 3603.
LÓPEZ (Rodrigo), 7214.
LÓPEZ (Sinesio), 4373.
LÓPEZ ELUM (Pedro), 3237.
LÓPEZ MARTÍN (Damián), 3581.
LÓPEZ MONDÉJAR (Publio), 5616.
LÓPEZ RODRÍGUEZ (Carlos), 2832.
LÓPEZ-CALVO (Ignacio), 3813.
LÓPEZ-RUIZ (Carolina), 1288.
Lorentz (Hendrik Antoon), 5448.
LORENZ (Federico), 3580.
Lorenz (Konrad), 5372.
LORENZ (Sönke), 3380.
Lorenzetti (Ambrogio), 3166.
LORENZETTI (Luigi), 6276, 6287.
Lorenzo de' Medici, 35.
LORUSSO (Vito), 1662.
LORY (Bernard), 3541.
LORY (Pierre), 908, 925.
LÖSER (Freimut), 3113.
LÖSER (Philipp), 5201.
LÖSER (Werner), 2216.

LOSKANT (Alexander), 5681.
ŁOSSOWSKI (Piotr), 3542.
LOTZ (Almuth), 2200.
LOUÇÃ (António), 7182.
LOUCAS (Ioannis), 6668.
Louis XIV, roi de France, 485, 3844, 3856, 5574, 6904, 6906, 6907.
Louis XVI, roi de France, 255.
LOUKAKI (Marina), 2386.
LOUKOPOULOU (Louisa D.), 1412.
LOUNGHIS (T.), 2425.
LOUSADA (Maria Alexandre), 6090.
LOUTH (Andrew), 926.
LOVATT (Helen), 2026.
LOVE (Nicholas), 3426.
LOVELL (Margaretta M.), 5607.
LOW (Polly), 1460.
LOWE (Jeremy), 2913.
LOWER (Michael), 2634.
LOWER (Wendy), 7148.
Löwith (Karl), 5334.
LOWNEY (Chris), 2914.
LOYER (Emmanuelle), 5109.
Loyola (Carlos M.), 4299.
LOZANO NAVARRO (Julián J.), 4920.
LU (Bowen), 7688.
LU (Hanchao), 7689.
LUAN (Chengxuan), 7690.
LÜBBE (Hermann), 4846.
LUC (Jean-Noël), 3836, 5234.
Luca, evangelista, Sanctus, 2258, 2320, 2366.
Lucanus (Marcus Annaeus), 2061, 2109.
LUCARINI (Carlo Martino), 1595.
LUCAS (A. A.), 5418.
LUCASSEN (Léo), 6277.
LUCCHESI (Enzo), 1116, 1309.
Luce (Henry Robinson), 5278.
LUCE JORDAN (Astrid), 6006.
LUCHNER (K.), 2251.
LUCIANO DI SAMOSATA, 3094.
Lucianus Samosatensis, 1431, 1432, 1433, 1536, 1591, 1617.
Lucius of England, 4979.
LUCKHARDT (C. Grant), 5360.
LUCONI (Stefano), 6091.
Lucretius Carus (Titus), 2124.
LÜDKE (Tilman), 7071.
LUDLOW (Piers), 7368.
LUDOVA (Ludmilla), 3569.
LUDWIG (Walther), 5169.
Ludwig der Fromme, 11.
Ludwig IV, röm.-deutscher Kaiser, 2651.
LUHTALA (Anneli), 2201.
LUIK (Martin), 1914.
LUIS CHINEA (Jorge), 6448.

LUKAS-GÖTZ (Elisabeth), 121.
LUKOWSKI (Jerzy), 697.
LUKS (Leonid), 6669.
LULAT (Y. G.-M.), 7869.
Lumumba (Patrice), 3726, 3728.
LUND (Allan A.), 1381.
LUND (Joachim), 7149.
LUNDH (Christer), 6203.
LUNDQUIST (Jennifer H.), 4350.
LUNDSTRÖM (Catarina), 6278.
LUNEAU (Aurélie), 5683.
LÜNNEMANN (Volker), 6203.
Luo (Ergang), 7638.
LUO (Xianyou), 7682.
LUO (Zhitian), 7691.
LUOMANEN (Petri), 861.
LUPI (Regina), 4893.
Lupus di Ferrières, 2569, 2616.
LUQUE BALLESTEROS (Antonio), 4583.
LURAGHI (Nino), 1596.
LURIA (Keith P.), 4815.
LUSIGNAN (Serge), 8.
LUSSY (Hanspeter), 7214.
LUST (J.), 857.
LUTARD-TAVARD (Catherine), 4523.
Luther (Martin), 4972.
LÜTKEMEYER (Sabine), 2090.
LUTTER (Anja), 7243.
LUTTER (Christina), 3386.
LUTZ (Eckart Conrad), 3039.
LUTZ (Helma), 386.
LUU (Lien Bich), 5840, 6247.
LUX (Rüdiger), 2219.
LUXFORD (Julian M.), 3193.
LUX-STERRITT (Laurence), 4921.
LUZ (Christine), 1597.
LUZZI (Serena), 5110.
Lycophron, 1434.
Lydgate (John), 2926, 3053.
LYNCH (David), 4180.
LYNE (R. O. A. M.), 2091.
LYNN (John A.), 6670.
LYNN (Martin), 6708.
LYONS (Maura), 5562.
Lysias, 1419.

M

MA (Yuqing), 7757.
MA (Zili), 7692.
MAAS (Harro), 5784.
MAAS (Michael Robert), 2420.
MABEKO-TALI (Jean-Michel), 3576.
MAC CAFFERTY (John), 4818.
MAC CANN (James C.), 7870.
MAC CARTHY (Angela), 6279.
MAC CARTNEY (Helen B.), 4069.
MAC CAUSLAND (Sigrid), 6416.
MAC CLEERY (Alistair), 68.

MAC CLELLAND (Charles), 5172.
MAC CLENDON (Charles B.), 3194.
MAC CONNELL (Stuart Charles), 283.
MAC CULLY (Chris), 3041.
MAC CUSKER (John J.), 5284.
MAC CUTCHEON (Russel T.), 927.
MAC DERMOTT (James), 6883.
MAC DONALD (Alan R.), 5964.
MAC DONALD (Alasdair A.), 3079.
MAC DONALD (David), 4473.
MAC DONALD (Helen), 5419.
MAC DONOUGH (Terrence), 4187.
MAC FARLAND (Keith D.), 4750.
MAC FARLANE (John), 3692.
MAC GAFFEY (Wyatt), 7871.
MAC GEE (Owen), 4181.
MAC GINN (Bernard), 3456.
MAC GOWEN (Randall), 6015.
MAC GRATH (Stamatina), 2576.
MAC GUIRE (Brian Patrick), 804, 3331.
MAC INTOSH (Marjorie Keniston), 2833.
MAC INTOSH (Roderick J.), 7872.
MAC INTOSH TURFA (Jean), 1768.
MAC ISAAC COOPER (Sheila), 6189.
MAC IVOR (Arthur), 6449.
MAC KAY (George), 5684.
MAC KECHNIE (Paul), 1461.
MAC KENNA (Antony), 4999.
MAC KENZIE (Brian Angus), 7411.
MAC KENZIE (David Clark), 7033.
MAC KENZIE (Donald Francis), 55.
MAC KINNON (Mary), 6074.
MAC LEAN (Iain), 4070.
MAC LEOD (Frederick G.), 2271.
MAC MAHON (Ardle), 2041.
MAC MANUS (Sheila), 6280.
MAC MILLAN (Alistair), 4070.
MAC MILLAN (Hugh), 5911.
MAC MILLAN (Richard), 6748.
MAC MULLIN (Ernan), 5379.
MAC NALLY (Mark), 7818.
MAC NAMARA (Patrick), 4894.
MAC NEIL (Lynda), 1598.
MAC PHERRAN (Mark L.), 1663.
MAC PHILLIPS (Kevin), 4071.
MAC ROBIE (Alan), 651.
MAC SHAMHRÁIN (Ailbhe), 664.
MAC SHEFFREY (Shannon), 4816.
MAC SHERRY (J. Patrice), 7346.
MAC TAVISH (Duncan), 6450.
MAC TAVISH (Lianne), 6281.
MAC TURK (Rory), 616, 3042.
MAC WILLIAMS (James E.), 6282.
MACHAČOVÁ (Jana), 5111.
MACHADO, (Maria das Dores Campos), 3677.

MACHIAVELLI (Niccolò), 525, 700, 2389, 5107, 5344.
MACHINANDIARENA DE DEVOTO (Leonor), 7412.
MÄCHLER (Stefan), 4658.
MACHONIN (Pavel), 6283.
MACHOVEC (Milan), 3776.
MACINNES (Allan I.), 4072.
MAČIULIS (Dangiras), 4278.
MACKEY (Thomas C.), 6589.
MACMILLAN (Harold), 7378, 7451, 7457.
MAČUHA (Maroš), 4526.
MACURA (Miloš), 4521.
MACZKIEWITZ (Dirk), 6884.
MADAJCZYK (Czesław), 6700.
MADDOX (Gregory H.), 4668.
MADDOX (Lucy), 5112.
MADEIRA (Lina Alves), 7178.
MADER (Gottfried), 1815.
MADESANI (Angela), 5685.
MADHOK (Shakti), 7505.
MADIGAN (Kevin), 2337.
MADISON (James), 4716.
MADLEY (Benjamin), 6797.
MADSEN (Lennart S.), 3806.
MAFFI (Alberto), 1500.
MAGARIÑOS (Gustavo), 5753.
MAGEE (Peter), 1092, 1360.
MĀGELE (Semra), 1139.
MAGER (Anne), 6284.
MAGGIORANI (Mauro), 7312.
MAGIDOV (Vladimir M.), 284.
MAGISTRALE (Francesco), 1.
MAGNESS (Jodi), 1310.
MAGNIEN SIMONIN (Chatherine), 359.
Magnus Maximus, Roman emperor, 2211.
MAGODE (José), 4326.
Magritte (René), 273.
Mahathir (bin Mohamad), 7537.
MÄHLERT (Ulrich), 309, 3754.
MAHNKE-DEVLIN (Julia), 6285.
MAHON (Rianne), 6286.
MAIDA (Bruno), 5291, 5302.
MAIER (Gideon), 2202.
Maignan (Emmanuel), 5596.
MAIHOLD (Harald), 2764.
MAÍLLO FERNÁNDEZ (José Manuel), 1032.
MAILLOUX (Anne), 2760.
Maimonides (Moses), 2681, 2683, 2691, 2692, 5039.
MAIORINO (Marco), 9.
MAISSEN (Thomas), 5285, 6016.
MAIZTEGUI CASAS (Lincoln R.), 4792.
MAJERUS (Benoît), 3645.
MAJSKIJ [Mayský] (Ivan M.), 4460.

MAKAROV (Nikolaj A.), 2599.
MAKAROVA (Irina F.), 572.
MÄKELER (Hendrik), 80.
Makhno (Nestor), 4503.
MÄKINEN (Martti), 3294.
MAKOWSKI (Bronisław), 3542.
MAKOWSKI (Elizabeth M.), 3457.
MAL'KOVSKAJA (Tat'jana N.), 4492.
MALAISE (Michel), 1185.
MALAMUT (Élisabeth), 2444.
MALANIMA (Paolo), 763.
MALATO (Enrico), 99, 972.
MALAVASI (Massimiliano), 972.
Mâle (Émile), 470.
MALFITANA (Daniele), 1736.
MALGERI (Francesco), 4828.
MALIS (Christian), 5113.
MALITZ (Jürgen), 1915.
MALKIEWICZ (Andrzej), 4398.
MALLETTE (Karla), 3043.
MALLINCKRODT (Rebekka Von), 4922.
MALLINSON (William), 3752.
MALLON (Florencia E.), 3709.
MALONEY (Sean M.), 3693.
MALSCH (Gabriele), 5328.
MALTARICH (Bill), 5114.
MALTEZ (José Adelino), 4415, 4419.
MALTOMINI (Francesca), 1599.
MALÝ (Karel), 5286, 6515.
MALYSHEVA (Svetlana Ju.), 5115.
Mamilius Sura, 1829.
MAN'KOVA (Irina L.), 5837.
MANCKE (Elizabeth), 4751, 6827.
MANCURTI (Francesco Maria), 3347.
MÄND (Anu), 2915.
Mandela (Nelson), 4546.
MANDELBROTE (Scott), 5440.
MANDEMAKERS (Kees), 6385.
MANDOLESI (Alessandro), 1020.
MANDUMBINE (Marcos Agostinho), 4327.
MANER (Hans-Christian), 6659.
MANFERLOTTI (Stefano), 5494.
MANFREDINI (Matteo), 5950.
MANGANARO (Giacomo), 1413.
MANGANI (Elisabetta), 1397.
MANI (Braj Ranjan), 4144.
MANIKOWSKA (Halina), 812.
MANN (Alastair J.), 4078.
MANN (Gregory), 6703.
MANN (Michael), 3543.
MANN (Nicholas), 158.
Mann (Thomas), 1002, 5534, 5717.
Mannerheim (Carl Gustaf Emil), 7240.
MANNOVÁ (Elena), 2439, 5123.
MANOR (Dalia), 267.

MANRIQUE ANTÓN (Teodoro), 2719.
MANSERGH (Danny), 4182.
MANSFELD (Jaap), 1664.
MANSINGH (Surjit), 662.
MANSON (Enrique), 630.
MANSOUR (Wisam), 1311.
Mantegna (Andrea), 5593.
MANTELLI (Brunello), 285.
MANTL (Wolfgang), 6516.
MANTOVI (Dario), 5523.
MANTZOURANI (Eleni), 1398.
Manuel I Comnenus, Byzantine emperor, 2504.
MANUWALD (Bernd), 1600, 1665.
MANZOLI (Giacomo), 5623.
MANZONI (Alessandro), 5523, 5524.
MAO (Haijian), 7693.
MAO (Jiaqi), 7600.
Mao (Zedong), 7607, 7629, 7694.
Maqqarī, family, 7570.
MAQUET (Julien), 3202.
MARAITE (Hubert), 3476.
MARANGHELLO (César), 5686.
MARAS (Sabrina), 1361.
MARASCO (Gabriele), 2338.
MARBODE, 2968.
MARCHAND (Jean-Jacques), 5344.
MARCHANT (Laurence Gérard), 3195.
MARCHENA (Juan), 644.
MARCHENA FERNÁNDEZ (Juan), 6850.
MARCHENA HIDALGO (Rosario), 3196.
MARCHENKO (Konstantin K.), 1462.
MARCHETTI (P.), 901.
MARCHETTO (Agostino), 4840.
MARCIAK (Dorothée), 5687.
MARCÍLIO (Maria Luiza), 5235.
MARCONE (Arnaldo), 462, 5116.
MARCOVICH (Miroslav), 2238.
MĂRCULEȚ (Vasile), 2474.
MARCUS (Kenneth H.), 5688.
Marcus Antonius, 1984.
Marcus Aurelius Antoninus, Roman emperor, 1796, 1935, 2122, 2153.
Marcus, evangelista, Sanctus, 2255, 2263, 2264.
Marcuse (Herbert), 5334.
MAREC (Yannick), 6120.
MAREK (Doris), 970.
MAREK (Jindřich), 7271, 7304.
MAREK (Pavel), 4841.
MARENBON (John), 3287.
MAREȘ (Alexandru), 805.
MARÈS (Antoine), 7055.
MAREŠ (Miroslav), 3777.

MARESCA (Paola), 851.
MARFANY (Joan-Llus), 181.
MARGAIRAZ (Michel), 5765.
Margaret, Queen, consort of Malcolm III, king of Scotland, 2626.
MARGINESCU (Giovanni), 1414.
MARGO (Robert A.), 5793.
MARI (Paolo), 3045.
Maria Magdalena, Sancta, 2379.
Maria Theresia von Österreich, regierende Erzherzogin von Österreich und Königin u. a. von Ungarn und Böhmen, 3627.
Maria, Iesus' mother, 2379, 2517.
MARIE (Laurent), 5689.
MARIETHOZ (François), 1066.
MARIETTI (Marina), 3072.
MARIN (Brigitte), 5124.
MARÍN (Dolors), 4584.
MARÍN (Manuela), 2695.
MARIN (Olivier), 3458.
MARIN (Șerban), 2475.
MARÍN GELABERT (Miquel Ángel), 393.
Marina, Sancta, 2334.
MARINO (Domenico), 1093.
MARINO (Emanuele Valerio), 5690.
MARINO (Natalia), 5690.
MARINO (Rosanna), 1841.
Marinus Falierus, 2404.
MARITAIN (Jacques), 5345.
Marius Mercator, 2240.
MARJANEN (Antti), 861.
MARK (Chi-Kwan), 6749.
MARK (James), 4122, 6288.
MARKARIAN (Vania), 4793.
MARKEVICH (Andrej M.), 5841.
MARKOV (Georgi G.), 3679.
MARKOWITZ (Rhonda), 838.
MARKS (J.), 1601.
MARKS (Raymond David), 1917.
MARKUS (Tomislav), 516.
MARMARA (Rinaldo), 182.
MAROTTI (Arthur F.), 4842.
MARQUES (Tiago Pires), 4420.
MÁRQUEZ MORA (Belén), 1030.
MARRANI (Giuseppe), 3037.
MARROW (James H.), 3197.
MARSDEN (Ben), 5420.
MARSEILLE (Jacques), 6796.
MARSHALL (Bill), 794.
Marshall (George C.), 7433.
Marshall (John), 4717.
MARSHALL (Julie G.), 6619.
MARSHALL (Peter J.), 6726.
MARSHALL (Peter), 1069.
Marsilio da Padova, 398, 710.
MARTELLO (Concetto), 3288.
Martí (José), 3743.

MARTI (Lionel), 1219, 1231.
Martialis (Marcus Valerius), 1821, 2053, 2107.
MARTIN (Benjamin F.), 7150.
MARTIN (Bernd), 6700.
MARTIN (Elisa), 6112.
MARTIN (Geoffrey), 2746.
MARTIN (Janette), 6416.
MARTIN (Jean-Marie), 2476, 2477, 2607.
MARTIN (Jean-Philippe), 3874.
MARTIN (JoAnn), 4312.
MARTIN (Laurent), 5287.
MARTIN (Marc), 5288.
MARTIN (Pierre), 3694.
MARTIN (Scott C.), 5896.
MARTIN (Therese), 3322.
MARTIN (Vanessa), 4161, 6626, 6974.
MARTÍN ASUERO (Pablo), 232.
MARTÍN DE SANTA OLALLA SALUDES (Pablo), 4585.
MARTIN GARCIA (Alfredo), 6092.
MARTÍN RUBIO (Ángel David), 4586.
MARTINA (Rossella), 5327.
MARTINDALE (Charles Anthony), 2092.
MARTINEK (Michael), 6583.
MARTINELLI (Maria Clara), 1067.
MARTINET (Marie-Madeleine), 210.
MARTÍNEZ DE SAS (María Teresa), 4587.
MARTÍNEZ DÍEZ (Gonzalo), 2608.
MARTÍNEZ LAHIDALGA (Adela), 2732.
MARTÍNEZ LEAL (Juan), 4588.
MARTÍNEZ MARTÍN (Jesús A.), 4602.
MARTÍNEZ NAVARRETE (M.ª Isabel), 1008.
MARTÍNEZ NIETO (Roxana), 1721.
MARTÍNEZ PIZARRO (Joaquín), 2577, 2581.
MARTÍNEZ-VERNE (Teresita), 3814.
MARTINI (Aldo), 3425.
MARTINI (Fabio), 1025, 1033.
MARTINI (Mauro), 5118.
MARTINO (Gabriele), 1033.
MARTINS (Susana), 4421.
MARTIRÉ (Eduardo), 6839.
MARTY (Nicolas), 5843.
MARTYN (Georges), 6594.
MARUHASHI (Mitsuhiro), 7695.
MARUŠIČ (Branko), 4535.
MARX (Alfred), 2269.
MARX (Karl), 5346, 6396.
MARX (Thomas Christoph), 7444.

MARXER (Veronika), 7215.
Mary of Egypt, sancta, 3416.
MÂRZA (Iacob), 5236.
MARZAL (Manuel M.), 938.
MARZAL RODRÍGUEZ (Pascual), 4589.
MARZEL (Shoshana-Rose), 5525.
Marzūqī, family, 7570.
Masaryk (Jan), 3772.
MASARYK (Tomáš Garrigue), 3515.
MASCHIETTI (Stefano), 502, 5338.
MASCHKE (Günter), 5353.
MASER (Peter), 4986.
MASI (Michael), 3046.
MASLAN (Susan), 5691.
MASLOWSKI (Nicolas), 3564.
MASS (Edgar), 5347.
MASSA-PAIRAULT (Françoise-Hélène), 1769.
Massentius, 2166.
MASSENZIO (Marcello), 928.
MASSEY (Douglas S.), 4350.
MASSICARD (Elise), 4695.
MASSON (Pierre), 5515.
MASTANDREA (Paolo), 2958.
MASTINO (Attilio), 1757, 1758.
MASTNY (Vojtech), 7506.
MASTOROPOULOS (Geōrgios S.), 2478.
MASTROCINQUE (Attilio), 2141.
MASTROGREGORI (Massimo), *VIII*, 320.
MASTROLILLI (Paolo), 7314.
MASTROPAOLO (Alfio), 712.
MASUHARA (Ayako), 4156.
MASUZAWA (Tomoko), 869.
MÂȚĂ (Cezar), 4439.
MATAR (Nabil I.), 6870.
MATARD-BONUCCI (Marie-Anne), 183, 268, 5017.
MATEEVA (Marija Atanasova), 6671.
MATEI (Aurel), 7251.
MATĚJČEK (Jiří), 5111.
MATĚJKA (Zdeněk), 7507.
MATEOS (Abdón), 4620.
MATERNA (Ingo), 3910.
MATERSKI (Wojciech), 7073.
MATHEUS (Michael), 2810.
MATHIS (Rémi), 6909.
MATHON (G.), 894.
MATHOT (René), 7151.
MATIJEVIĆ (Zlatko), 3738, 3739.
MATINO (Giuseppina), 2411.
MATIS (Herbert), 6017.
MATKOVIĆ (Stjepan), 3738.
MATONTI (Frédérique), 3875.
MATOVINA (Timothy), 4895.
MATSUMURA (Takao), 6451.
Matsuoka (Yōsuke), 7218.

Matsuura, family, 109.
MATSUZAWA (Yusaku), 6404.
MATTÉONI (Olivier), 2668.
MATTERN (David B.), 4716.
Matthaeus, evangelista, Sanctus, 2260.
MATTHEE (Rudi), 806.
MATTHEW (Donald), 2635.
MATTHEWS (E.), 170.
MATTHEWS (Jill Julius), 5692.
MATTHIAE (Paolo), 1117.
MATTIOLI (Aram), 7152.
MATTONE (Antonello), 296.
MATTUSCH (Carol C.), 2178.
MATVEEV (Vladimir D.), 6975.
MATZ (Jean-Michel), 3141.
MATZUKES (C.), 2395.
MAUGANS DRIVER (Lisa D.), 2339.
MAUL (Stefan M.), 1232.
Maulbertsch (Franz Anton), 5599.
MAURACH (Gregor), 1602, 2093, 2094.
MAURER (Golo), 6976.
MAURÍCIO (Maria José), 4422.
MAUTNER (Thomas), 5332.
MAWBY (Spencer), 7508.
MAWDSLEY (Evan), 7272.
MAXEY (Trent Elliott), 4239.
Maximilian I, röm.-deutscher Kaiser, 3636.
Maximinus Thrax, Roman emperor, 1816.
Maximus Homologetes, 2241, 2405, 2482.
MAY (Gary), 4752.
MAYADE-CLAUSTRE (Julie), 2765.
MAYAUD (Pierre-Noël), 5421.
MAYER (Ines), 4004.
MAYER (Jochen Werner), 2027.
MAYER (Marc), 1814.
MAYER (Wendy), 2457.
MAYER (Werner R.), 1233.
MAYEUR (Jean Marie), 3876.
MAYEUR-JAOUEN (Catherine), 4817.
MAYORDOMO (Moisés), 2262.
MAZEL (Florian), 2836.
MAZÍN (Oscar), 663.
MAZLISH (Bruce), 660.
MAZOWER (Mark), 713, 783.
MAZZA (Mario), 2203.
MAZZACANE (Aldo), 6538.
MAZZANTI PEPE (Fernanda), 6506.
MAZZARA (Giuseppe), 1666.
Mazzini (Giuseppe), 4712.
MAZZOLENI (Gilberto), 929.
MBOGONI (Lawrence Ezekiel Yona), 4667.
MEARS (Natalie), 4073.
Mechthild of Magdeburg, 3313.
MECKING (Sabine), 6115.

MEDDA (Enrico), 1603.
MEDRI (Litta Maria), 851.
MEDUSHEVSKIJ (Andrej N.), 5951.
Megetius, 2432.
MEGILL (Allan), 383.
MEGINO RODRÍGUEZ (Carlos), 1667.
MEHLMAN (Jeffrey), 5119.
Mehmet Ali Paşa, governor of Egypt, 6966.
MEHTA (Jaswant Lal), 4145.
MEIER (Claudia Annette), 3198.
MEIKLE (Jeffrey L.), 5617.
MEINHARDT (Matthias), 770.
MEINL (Susanne), 6478.
MEL'NICHENKO (Boris N.), 7571.
MEL'NIKOVA (Alla S.), 144.
MEL'NIKOVA (Elena A.), 2406.
Melanchthon (Philipp), 4972, 4988, 6476.
MELANDER (Anders), 5844.
MÉLANDRI (Pierre), 7483.
MELBER (Henning), 6786.
MELCHOR GIL (Enrique), 1907.
MELERA-MORETTINI (Matteo), 5344.
MELHAOUI (Mohammed), 2709.
MELIKSETOV (Arlet V.), 5948.
MELLONI (Alberto), 4852.
MELTON (J. Gordon), 870.
MELVILLE (Gert), 3333, 3376, 3389.
MÉNAGER (Daniel), 5120.
MENANT (François), 2636.
MENANT (Sylvain), 485.
MENDELS (Doron), 1312.
MENDELSON (Jordana), 311.
Mendelssohn-Bartholdy (Felix), 5622.
MENDES (Nuno Canas), 3816.
MENDES DE ARAUJO (Manuel Garrido), 402.
MÉNDEZ (Cecilia), 4376.
MÉNDEZ BELTRÁN (Luz María), 5845.
MENEGUS BORNEMANN (Margarita), 4383.
MENÉNDEZ (Mario), 6672.
Meng (Wengtong), 7729.
MENG (Yanhong), 7696.
MENG (Zhaochen), 7697.
MENJÍVAR (Cecilia), 6699.
MENJOT (Denis), 2823.
MENNEKES (Ralf), 5587.
MENSCHING (Günther), 3307.
Menshikov (Aleksandr Danilovich), 4466.
MENTZ (Soeren), 5912.
Mercante (Domingo Alfredo), 3594.
MERGEL (Thomas), 4074.

MERI (Josef W.), 665.
MERKEL (Helmut), 2352.
MERMIER (Franck), 81.
MERTELSMANN (Olaf), 3831.
MERZARIO (Raul), 6276.
MESA-LAGO (Carmelo), 3745.
MESNER (Maria), 3626.
Messalinus, 2055.
MESSAMORE (Barbara J.), 3695.
MESSENGER (David A.), 7216.
MESSENT (Peter B.), 984.
MESSERSCHMIDT (Manfred), 6590.
META (Chiara), 5331.
METAFERIA (Getachew), 6955.
Metaxas (Ioannis), 4103.
METCALF (Alida C.), 6840.
METCALF (Thomas R.), 6750.
MÉTIVIER (Sophie), 2479.
METTAUER (Adrian), 58.
METTELE (Gisela), 5913.
Metternich (Klemens von), 3630.
METTLER (Suzanne), 4753.
METZ (Paul), 7246.
METZLER (Gabriele), 3987.
MEVIUS (Martin), 4123.
MEWS (C. J.), 3282, 3293.
MEYENDORFF (John), 2340.
MEYER (Ahlrich), 5033.
MEYER (Andrea), 2562.
MEYER (Annegret), 2259.
MEYER (Anneke), 3307.
MEYER (Beate), 5030.
MEYER (Fédéric), 4926.
MEYER (Heinz), 3266.
MEYER (Thomas), 3911.
MEYERS (Gretchen), 1775.
MEYNARD (Cécile), 5526.
MEZEI (Elemér), 6291.
Mezentius, rex Etruscorum, 1763.
MEZNÍK (Jaroslav), 3778.
MEZZADRA (Sandro), 394.
MIANNAY (Patrice), 7305.
MIARD-DELACROIX (Hélène), 7438.
MICAELLI (Claudio), 2341.
Micah, 1288.
MICCOLIS (Stefano), 5341.
MICELI (Thomas J.), 5989.
MICHAEL (Angelika), 3199.
Michael Hagiotheodoritus, 2502.
Michael III Anchialus, patriarch of Constantinople, 2504.
Michael Psellus, 2454.
MICHAILIDOU (Anna), 145.
MICHÁLEK (Slavomír), 3779, 5290.
MICHALOWSKI (Piotr), 1234.
MICHEL (Johann), 517.
MICHEL (Marco), 3988.
MICHELS (Eckard), 5121.
MICHELS (Stefanie), 409.
MICHELS (Tony), 6452.

MICHETTI (Laura Maria), 2160.
MICHON (C.), 6885.
MIDDELBERG (Mathias), 6489.
MIDDELL (Matthias), 3878.
MIDDLETON (Stephen), 6591.
MIDLARSKY (Manus I.), 3545.
MIDTGAARD (Kristine), 7509.
MIECK (Ilja), 6150.
MIEGGE (Mario), 764.
MIER Y TERÁN ROCHA (Lucía), 6094.
MIERNIK (Grzegorz), 4409.
MIERZEJEWSKI (Alfred C.), 3989.
MIESZKOWSKI (Sylvia), 5180.
MIETH (Stefan), 6592.
MIETTINEN (Kari), 6798.
MIGANI (Elena), 432.
MIGEOTTE (Léopold), 1501.
MIGEV (Vladimir), 7511.
MIGLIETTA (Massimo), 1313.
MIGUEL (Sonsoles), 4572.
MIHÁLIKOVÁ (Silvia), 4530.
MIKAMI (Yoshitaka), 7820.
MIKHAILOVA (Milena), 3055.
MIKHAJLOVA (Tatiana A.), 930.
MIKLAUTSCH (Lydia), 3048.
MILANI (Giuliano), 2637.
MILER (Susan Gilson), 813.
MILES (Margaret Ruth), 3332.
MILES (Tiya), 6289.
MILHAU (Marc), 2234.
MILIS (Ludovicus), 2920.
MILITAREV (Alexander), 1314.
MILITZER (Klaus), 2837, 6552.
MILKIAS (Paulos), 6955.
Mill (John Stuart), 5785.
MILLER (Catherine), 4633.
MILLER (John), 4989.
MILLER (Rory), 7347.
MILLER (Ruth Austin), 4696.
MILLER (William L.), 4040.
MILLETT (Allan R.), 7413.
MILLETT (Martin), 666.
MILLOT (Hélène), 5529.
MILLS (Robert), 2921.
MILLS (Sean), 3696.
MILLWARD (Robert), 5846.
MILNER (Stephen J.), 2784.
MILNOR (Kristina), 2028.
MILOVANOVIC (Nicolas), 269.
MILWARD (Alan S.), 5754.
MILZA (Pierre), 3877.
MIN (Kyuong-sik), 2260.
MIN (Pyong Gap), 636.
MINAGAWA (Taku), 3990.
MIÑAMBRES (Mª Teresa), 1574.
MINAUD (Gérard), 2029.
MINCHIN (Timothy J.), 6453.
MING (Fung Chi), 7698.
MINNA STERN (Alexandra), 5425.
MINNS (Chris), 6074.

MIÑO GRIJALVA (Manuel), 4314.
MINTSĒS (Geōrgios I.), 2480.
MIQUEL (Angel), 5693.
MIQUEL (Pierre), 6929.
MIR (Conxita), 4596.
MIRA ABAD (Alicia), 6290.
MIRA JODAR (Antonio José), 2838.
MIRALLES (Ricardo), 6652.
MIRANDA CASTAÑÓN (Edmundo), 6977.
MIRANDA GARCÍA (Jorge Luis), 248.
MIRANDOLA (Giorgio), 82.
MIRKOVIC (Miroslava), 1979.
MIROIU (Andrei), 6673.
MIRONENKO (Sergei V.), 7374.
MIRSKIJ (Stanislav V.), 4506.
MISIUK (Andrzej), 4401.
MISKELL (Peter), 5833, 6163.
MÍŠKOVÁ (Alena), 297.
MITCHELL (Leslie George), 4075.
MITCHELL (Otis C.), 7348.
MITCHELL (Pablo), 6292.
MITCHELL (Peter), 7854.
MITCHENER (Kris James), 5755, 6019.
MITCHINSON (K. W.), 7074.
MITEV (Dimităr), 7009.
MITEV (J.), 4124, 4423.
MITSUZONO (Isamu), 5926.
MITTAG (Achim), 380.
MITTELSTADT (Jennifer), 6454.
Mittermaier (Karl Josef Anton), 6549, 6551.
Mitterrand (François), 7469, 7485.
MIXA (Walter), 3365.
MIXON (Gregory), 4754.
MIYAMOTO (Masaaki), 7821.
MIZOGUCHI (Koji), 1095.
MIZUNO (Tomoyuki), 7822.
MKAIMA (Felista Elias), 4327.
MO (Zigang), 7699.
MOA (Pío), 4590.
MOADDEL (Mansoor), 3546.
MOALLEM (Minoo), 4162.
Moctezuma (Isabel), 113.
MODICA (Marilena), 4926.
MODORCEA (Grid), 840.
MOEHLING (Carolyn M.), 6293.
MOELLER (Robert), 3932.
MOHLE (Ingvar Brandvik), 2030.
MÖHLIG (Wilhelm J. G.), 4666.
MOHR (James C.), 5426.
MOHSEN-FINAN (Khadija), 5037.
MOKSHIN (Nikolaj F.), 574.
MOKSHINA (Elena N.), 574.
MOKSHINA (Julia N.), 575.
MOKYR (Joel), 5756.
MOLAS RIBALTA (Pere), XI.
MOLENDIJK (Arie L.), 931.
MOLES (John), 1604.

MOLICH (Georg), 800.
MOLINA (Iván), 626.
MOLINA (Manuel), 1235.
MOLINA APARICIO (Fernando), 4548.
MOLINA GONZÁLEZ (Fernando), 1054.
MOLINA JIMÉNEZ (Iván), 3734.
MOLINARI (Maurizio), 7314.
MOLINERO (Carme), 4591, 6469.
MÖLLER (Caren), 6540.
MÖLLER (Horst), 3991, 7374, 7554.
MÖLLER (Kay), 7349.
MOLODIN (Vjacheslav I.), 576.
Moltke (Helmut von), 3911.
MOM (Gijs), 5427.
MOMMSEN (Theodor), 471, 1985.
MOMMSEN (Wolfgang J.), 360.
MOMRAK (Kristoffer), 1315.
MONACA (Mariangela), 2142.
MONCHIZADEH (Anuscha), 2431.
MONCIATTI (Alessio), 3200.
MONDIO (Alessandra), 1745.
MONDRAGÓN CASTAÑEDA (Julio), 3722.
MONFERRER SALA (Juan Pedro), 216.
MONFRIN (Jacques), 2960.
MONG (Sai Kam), 3649.
MONGINI (Guido), 4923.
MONICO (Reto), 6979.
MONIÉ NORDIN (Jonas), 2670.
MONINA (Giancarlo), 4222.
MONNIER (Raymonde), 714.
MONOD (David), 5694.
Monsiváis (Carlos), 4292.
MONSON (Don Alfred), 3295.
MONTAGNE (Sabine), 6020.
Montaigne (Michel Eyquem de), 977.
MONTAÑA CONCHIÑA (Juan Luis de la), 2593.
MONTECCHI (Giorgio), 83, 99.
MONTEIRO DE CASTRO (Bernardo), 3050.
MONTELEONE (Ciro), 1806.
MONTERO CARTELLE (Enrique), 1821.
MONTERO RUIZ (Ignacio), 1055.
Montesquieu (Charles-Louis de Secondat, baron de), 490, 5347.
Montezuma II, Emperor of Mexico, 113.
MONTGOMERY (Janet), 1069.
MONTGOMERY (Scott Bradford), 3125.
MONTI (Maria Teresa), 5428.
MONTIEL ARGÜELLO (Alejandro), 6980.
MONTOITO (Eugénio), 4424.

MONTORSI (William), 3201.
MONTOYA (Rodrigo), 4378.
MONTRONI (Giovanni), 3547.
MOOIJ (Joke), 6021.
MOON (Krystyn R.), 5695.
MOORE (Andrew M. T.), 1236.
MOORE (Andrew), 3610, 3616.
MOORE (Bob), 411.
MOORE (David), 2638.
MOORE (Jerry D.), 7885.
MOORHOUSE (Roger), 7075.
MOORMAN VAN KAPPEN (Olav), 6490.
MOOS (Peter von), 3051.
MOOS (Thorsten), 4990.
MOOSA (Ebrahim), 2710.
MOOSLECHNER (Peter), 5852.
MORA (Clelia), 1145.
MORADIELLOS (Enrique), 7217, 7242.
MORAN (Bruce T.), 5429.
MORANDINI (Francesca), 3493.
MORCK (Randall K.), 5826.
MOREAU (Jean-Louis), 5983.
MOREL (Marco), 3673, 6930.
MORELLI (Serena), 12.
MOREMAN (T. R.), 7273.
MORENO (Hugo), 3596.
MORENO FERNÁNDEZ (Francisco), 184.
MORENO JULIÀ (Xavier), 7274.
MORENO SECO (Mónica), 4896.
MORENO SOLDEVILA (Rosario), 1821.
MORENT (Stefan), 3240.
MORENZONI (Franco), 3264.
MOREROD-FATTEBERT (Christine), 3422.
MORES (Francesco), 452.
MORESCHINI (Claudio), 1805, 2221.
MORETTI (Franco), 996.
MORETTI (Mauro), 484.
MOREWOOD (Steven), 7229, 7275.
Morgagni (Manlio), 6091.
MORGAN (Chad), 5847.
MORGAN (Francesca), 4755.
MORGAN (Gerald), 3052.
MORGAN (M. Gwyn), 1918.
MORGAN (Teresa), 2227.
MORGENRATH (Birgit), 7169.
MORI (Anatole), 1605.
MORI (Kenji), 5346.
MORI (Nobuyoshi), 2578.
MORI (Shigeki), 7218.
MORI (Tokihiko), 7770.
MORI (Yukio), 7823.
MORIGGI (Marco), 1220.
MORIMOTO (Kazuo), 7572.
MORIMOTO (Yoshiki), 2839.
MORITA (Yoshihiko), 7005.

MORIYAMA (Teruaki), 7573.
MÖRKE (Olaf), 4991.
MORLAT (Patrice), 6751.
MORMINO (Gary R.), 6294.
Moro (Aldo), 4212, 4216.
MORO (Renato), 7202.
Morone (Giovanni), 4880.
MOROZOV (Maksim A.), 2481.
Morozov (Pavlik), 4479.
MOROZOVA (Vera), 5346.
MORRIS (Colin), 3460.
MORRIS (Douglas G.), 6593.
MORRIS (Marc), 2639.
MORRIS (R. J.), 5757.
MORRIS (Sarah P.), 1521.
MORRISON (Jeffry H.), 4756.
MORRISON (Karl Frederick), 3214.
MORRISSEY (Susan K.), 6295.
MORRISSON (Cécile), 2514.
MORTENSEN (Andras), 2729.
MORTIMER (Ian), 2671.
MORTIMER (Nigel), 3053.
MORTON (Desmond), 3697.
MOSCA (Paul G.), 1316.
MOSCATI (Emma), 4196.
MOSCHOVITIS (Chris), 797.
MOSCOSO (Francisco), 6818.
MOSER (John E.), 4757.
MOSER (Marie Hélène), 2827.
MOSES (A. Dirk), 518.
MOSES (Dirk), 3932.
MOSETTI CASARETTO (Francesco), 2903.
MOSIG-WALBURG (Karin), 1919.
MOSKOVICH (Vol'f A.), 2685.
MOSOVSKIJ (A. I. Tarkina), 7597.
MOSSÉ (Claude), 1463.
MOSSUZ-LAVAU (Janine), 6333.
MOSTERT (Marco), 2929, 3214.
MOTRO (U.), 1275.
MOTTA (Franco), 4910.
MOTTE (André), 901.
MOULINET (Daniel), 3879.
MOULINIER (Laurence), 143.
Mountbatten (Louis), 7458.
MOURACADE (John), 1668.
MOURITSEN (Henrik), 2031.
MOURLANE (Stéphane), 7512.
MOURÓN FIGUEROA (Cristina), 2840.
MOUSAVI (Ali), 1068.
MOUSNIER (Mireille), 434.
MOUTAFTCHIEVA (Vera), 3548.
MOYA (Angel Omar), 3597.
MOYER-VINDING (Birgitte), 2731.
MOYN (Samuel), 395, 5343.
MOYNES (James), 664.
Mozart (Wolfgang Amadeus), 5631.
MOZET (Nicole), 5527.
Muahhari (Murtaża), 4160.

MUCHEMBLED (Robert), 6298.
MUCHNIK (Natalia), 5035.
MÜCKEL (Wenke), 185.
MUDROCH (Vilem), 964.
MUELLER (Wolfgang), 7376, 7414, 7417.
MUELLER-JOURDAN (Pascal), 2405, 2482.
MÜFTÜLER-BAC (Meltem), 7513.
MUGUETA (Iñigo), 2519.
MÜHLE (Eduard), 449.
MUIR (Tom), 3087.
MUKHERJEE (A. J.), 1078.
MUKHERJEE (Mridula), 5952.
MUKTA (Parita), 431.
MULDER-BAKKER (Anneke B.), 2904, 3461, 6386.
MULGAN (Aurelia George), 5953.
MÜLLER (Anett), 3992.
MÜLLER (Bertrand), 511.
MÜLLER (Christel), 1447.
MÜLLER (Dietmar), 4524.
Müller (Gerhard Friedrich), 472.
MÜLLER (Gerhard), 5208.
MÜLLER (Jürgen), 3993.
MÜLLER (Marco), 6096.
MÜLLER (Markus), 3994.
MÜLLER (Matthias), 6517.
MÜLLER (Miriam), 2841.
MÜLLER (Peter), 6097, 7361.
MÜLLER (Philipp), 5292.
MÜLLER (Ralf C.), 6299.
MÜLLER (Rolf-Dieter), 7153, 7167.
MÜLLER (Tanja), 6300.
MÜLLER (Zdeněk), 7276.
MÜLLER-LUCKNER (Elisabeth), 199, 5075, 6405.
MULLETT (Margaret), 2595.
MULLIGAN (William), 3995.
MULSOW (Martin), 5009.
MULVILLE (Jacqui), 1069.
Munabbin (Wahb b.), 2697.
MUNDING (Anne), 3915.
MÚNERA (Alfonso), 234.
MUNHOLLAND (Kim), 7154.
MUNIESA (Bernat), 4592.
MUÑIZ GRIJALVO (Elena), 2306.
MÜNKLER (Herfried), 715.
MUNNO (Cristina), 6301.
MUÑOZ ALTABERT (M. Lluïsa), 4593.
MUÑOZ RODRÍGUEZ (Julio D.), 4565.
MUNRO (John), 2842.
MUNTEAL FILHO (Oswaldo), 3666.
MUNTEAN (Ovidiu), 5122.
MUNZI (Luigi), 3066.
Murad II, Sultan of the Turks, 4679.
MURAI (Hiroshi), 7700.

MURAI (Shosuke), 684, 7824.
MURANO (Giovanna), 44.
MUREŞAN (Florin Valeriu), 5954.
MURGA ARMAS (Jorge), 4107.
MURGATROYD (Paul), 2096.
MURPHY (Brian), 4493.
MURPHY (David E.), 7277.
MURPHY (Gary), 4179.
MURPHY (Hugh), 5832.
MURPHY (James Jerome), 3146.
MURPHY (Kevin), 6455.
MURPHY (Philip), 6760.
MURRAY (Colin), 6799.
MURRAY (James M.), 2843.
MURRAY (Kevin), 173.
MURRAY (Oswyn), 680.
MUSCHIK (Alexander), 7514.
MUSIN (Aleksandr E.), 2596.
Musolff (Hans-Ulrich), 5221.
Mussolini (Benito), 4203, 4224, 5575, 7122, 7171.
MUSSON (Anthony), 726, 2766.
MUSTAKALLIO (Katariina), 3142.
MUSTÈ (Marcello), 519.
MUSTI (Domenico), 1382, 1770.
MUSULLAM (Adnan A.), 5036.
MUTLU (Şamil), 4942.
MUTSCHLER (Fritz-Heiner), 1932.
MUZZARELLI (Maria Giuseppina), 3462.
MYKING (John Ragnar), 2762.
MYLONOPOULOS (Jannis), 1402.

N

Nabokov (Vladimir Vladimirovich), 5517.
NACHTIGAL (Reinhard), 7076.
NACO DEL HOYO (Toni), 2032.
NADAL I LORENZO (Jordi), 1028.
NADASEN (Premilla), 4758.
NADAU (Thierry), 5955.
NADEL (Alan), 5697.
NADERER (Otto), 3628.
NADIS (Fred), 5430.
NAFISSI (Mohammad), 396.
NAGASHIMA (Hiroaki), 7803.
NAGATA (Mary Louise), 6189.
NAGEL (Anne Christine), 397.
NAGEL (Silke), 6302.
NAGOVITSYN (Aleksej E.), 872.
NAGY-ZEKMI (Silvia), 3705.
NAHUM (Benjamín), 5758.
NAIMARK (Norman M.), 7376.
NAJMABADI (Afsaneh), 6303.
NAKAI (Akio), 7078.
NAKAJIMA (Takeshi), 7592.
NAKAMURA (Shin'ichi), 1049.
NAKANO (Chièmi), 1156.
NAKAYAMA (Tomihiro), 7825.
NAM (Sang Ho), 4240.
NAMHILA (Ellen Ndeshi), 4328.

NANU (Veronica), 4436.
NAOKI (Kojiro), 7826.
Napoléon I, empereur de France, 378, 3862, 3902, 4204, 4231, 6594, 6923, 6927, 6931, 6936, 6937, 6941.
Napoléon III (Louis Bonaparte), empereur de France, 7001.
NAPOLITANO (Giorgio), 7312.
NAPPA (Christopher), 2097.
NARDUCCI (Emanuele), 2073, 2098.
NARINSKI (Mikhail Matveevich), 7136.
NASH (Gary B.), 4759.
NASH (Philip), 7219.
NASH (S. K.), 1347.
NASH (Suzanne), 5518.
NASHEL (Jonathan), 7350.
NÄSLUND (Sture), 6304.
NASRABADI (Behzad Mofidi), 1237.
NASTASI (Piero), 5397.
NASTI (Fara), 1980.
NATALI (Denise), 4169.
NATHANS (Benjamin), 5047.
NATION (Richard F.), 5956.
NATOLI (Claudio), 356.
NATTERMANN (Ruth), 446.
Naumkin (Vitaly V.), 2694.
NAUTA (Ruurd R.), 1803.
NAVICKAS (Katrina), 5848.
NAYROLLES (Jean), 841.
NDIAYE (Émilia), 2033.
NEAGOE (Stelian), 4449.
NEBESIO (Bohdan Y.), 658.
NEĆAK (Dušan), 4537.
NEČAS (Ctibor), 6305.
NEDBÁLEK (František), 7278.
NEDERMAN (Cary J.), 398, 3451.
NEESAM (Malcolm George), 2575.
NEFF (John R.), 4760.
Negri (Antonio), 721.
NEGRUZZO (Simona), 5237.
Nehru (Jawaharlal), 4147, 7227.
NEIBERG (Michael S.), 7077, 7103.
NEILS (Jenifer), 1737, 1742.
NEIRA FALEIRO (Concepción), 2408.
NEITZEL (Sönke), 7279.
NEJIME (Kenichi), 2924.
NELSON (Adam R.), 5238.
NELSON (Emmanuel Sampath), 991.
NELSON (Eric), 4924.
Nelson (Horatio), 4034.
NĚMEČEK (Jan), 7177.
NEMES (Robert), 6100.
NEMETH (Eduard), 1920, 1921.
NEMETI (Sorin), 1790, 2143.
NEMIROVSKIJ (Evgenij L.), 85.

NERI-ULTSCH (Daniela), 3880.
Nero Claudius Caesar, Roman emperor, 1901, 1915, 1941, 1969.
Nerses IV, patriarch of Armenia, 2504.
Neskhons, 1190.
NEŠPOR (Zdeněk R.), 6306.
NESSELRATH (Hans-Günter), 1464.
NESTORESCU-BĂLCEŞTI (Horia), 808.
NETHERTON (Robin), 3128.
NETZHAMMER (Nikolaus), 4859.
NETZHAMMER (Raymund), 4859.
NEU (Charles E.), 7515.
NEU (Claudia), 5943.
NEUBER (Wolfgang), 63.
NEUBERGER (Joan), 3566.
NEUBERT (Ehrhart), 3998.
NEUGEBAUER (Wolfgang), 3629.
NEUHEUSER (Hanns Peter), 544.
NEUJAHR (Matthew), 1362.
NEUMAIER (Helmut), 6491.
NEUMAYER (Christoph), 5852.
NEVES (Margarida de Souza), 6242.
NEVILLE (Ann), 1317.
NEVILLE (Cynthia J.), 2845.
NEVILLE (Peter), 7155.
NEVLING PORTER (Barbara), 1238.
NEWBY (L. J.), 3715.
NEWBY (Zahra), 2034.
NEWHAUSER (Richard), 2925.
NEWITT (Malyn), 669.
NEWMAN (Andrew J.), 4163.
NEWMAN (Frances Stickney), 2099.
NEWMAN (John Kevin), 2099.
NEWMAN (Mark), 4993.
NEWTON (Diana), 4076.
NG (On-Cho), 400.
NGANDO (B. Alfred), 4546.
NGUYEN-MARSHALL (Van), 6752.
NIAKAN (Lily), 1096.
NIBLETT (Rosalind), 3486.
NICCOLI (Ottavia), 5125.
Niccolò Cusano, 398, 3289, 3471.
NICHOLSON (Paul T.), 1201.
NICHOLSON (Peter), 3058.
Nicias, 1452.
NICO OTTAVIANI (Maria Grazia), 724, 2754.
NICOLAJ (Giovanna), 1.
NICOLAOU-KONNARI (Angel), 2624.
Nicolartius (Petrus), 4858.
NICOLÁS (Encarna), 4594.
NICOLAS (Nathalie), 3503.
NICOLÁS CHECA (Elena), 1030.
NICOLLIER (Béatrice), 5053.
NIDERST (Alain), 5499.

NIE (Yuanzi), 7701.
NIEDEREHE (Hans-Josef), 151.
NIEDERER (Monica), 3296.
NIEDERREITER (Zoltán), 1239.
NIEKERK (Carl), 5431.
NIELSEN (Morten), 3802.
NIEMINEN (May Kyi Win), 685.
NIERI (Rolando), 6981.
NIESNER (Manuela), 2687.
NIETO SORIA (José Manuel), 2656.
NIETO-PHILLIPS (John M.), 6721.
NIETZSCHE (Friedrich Wilhelm), 531, 5348.
NIEUS (Jean-François), 2641.
NIGHTINGALE (Pamela), 2846.
Nikarchos, 1210.
Nikolaj I Romanov, Tsar of Russia, 4501.
NIKOLAOU (Katerina), 2483.
NIKOLAOU (Theodor), 2342.
Nikophon, 1488.
NIKRMAJER (Leoš), 5239.
NILSSON (Ingela), 2484.
NILSSON (Torbjörn), 6656, 6676.
NILSSON (Ylva), 7547.
NINOMIYA (Hiroyuki), 520.
NIP (R.), 6386.
NIPPOLD (Otfried), 7078.
NISETICH (Frank), 1607.
NISHI (Yoko), 7839.
NISHIDA (Toshihiro), 6702.
NISHIKAWA (Makoto), 4241.
NISHIZATO (Kikō), 7702.
NISSE (Ruth), 3059.
NISSEN (Mogens R.), 5957.
NITOBURG (Eduard L.), 577.
NITSCHACK (Horst), 3662.
NIUNOYA (Tetsuichi), 7829.
NIWIŃSKI (Piotr), 4275.
NIXON (Lucia), 1743.
Nixon (Richard Milhous), 4728, 7490, 7495.
NIYOGI (Ruma), 2485.
NIZ (Faruk), 5126.
Nkrumah (Kwame), 4026, 4028.
NKWENGUE (Pierre), 3689.
NNAEMEKA (Obioma), 4354.
Noakes (Jeremy), 3996.
NOBLE (Thomas F. X.), 2204.
NODA (Jin), 7574.
NOËL (Erick), 128.
NOGUEIRA (Bernardo Guimarães Fisher de Sá), 13.
NOHL (Herman), 7014.
NOIRAY (Michel), 5696.
NOLAN (Maura), 2926.
NOLAN (Michael E.), 5127.
NOLLENDORFS (Valters), 4266.
NOLTE (Hans-Heinrich), 670.
NOLZEN (Armin), 3999, 4215.
Nonnus Panopolitanus, 2407.

NOORANI (Abdul Gafoor Abdul Majeed), 4146.
NORBERG (Arthur L.), 5849.
NORDIN (Dennis S.), 5958.
NORDSKOV (Jeppe), 6117.
NORITAKE (Yuichi), 7830.
NORRBY (Göran), 6307.
NORTH (John David), 3297.
NORTIER (Michel), 2559, 2620.
NORTMANNN (Ulrich), 738.
NORWICH (John Julius), 4033.
NOSKOVÁ (Helena), 6332.
NOTLEY (R. Steven), 2228.
NOUHOU (Alhadji Bouba), 4353.
NOUSCHI (André), 521, 6727.
NOVA (Giuseppe), 86.
NOVÁČKOVÁ (Helena), 7177.
Novalis, 5328.
Novikov (Michail M.), 5404.
NOVIKOVA (Ol'ga E.), 3881.
NOVOA MACKENNA (Andrea), 3700.
NOVOSELOV (Vasilij R.), 3882.
NOVOTNY (Jamie R.), 1240.
NOWAK (Jolanta), 4406.
NOWAKOWSKI BAKER (Denise), 3423.
NOYÉ (Ghislaine), 2476.
NUGIN (Raili), 3830.
NUMHAUSER (Paulina), 6308.
NUNES (António Pires), 6800.
NÚÑEZ SEIXAS (Xosé Manuel), 4595.
NÜNNING (Vera), 5475.
NUOVO (Angela), 5293.
NURMIAINEN (Jouko), 6309.
NURSER (John S.), 4994.
NUSS (Philippe), 2561.
NUTINI (Hugo G.), 6310.
NUTTALL (Jeremy), 6456.
NÜTZENADEL (Alexander), 5129.
NUZZO (Giorgio), 5731.
NUZZO (Mariella), 6101.
NWOKEJI (G. Ugo), 4354.
NYBERG (Kjell), 5788.
NYS (Karin), 1738.
NYS (Liesbet), 312.
NYSTRÖM (Esbjörn), 5698.
NZENGUET IGUEMBA (Gilchrist Anicet), 6801.

O

Ó CARRAGÁIN (Éamonn), 3060.
Ó CRÓINÍN (Dáibhí), 676.
O MURCHADHA (Diarmuid), 173.
O RIAIN (Pádraig), 173.
O'BRIEN (Denis), 1669.
O'BRIEN (Harriet), 2726.
O'BRIEN (John B.), 6677.
O'BRIEN (Paul), 4224.
O'CONNELL (Sean), 5894, 5914.

O'CONNOR (Sue), 1034.
O'DAY (Alan), 642, 4186.
O'DONNELL (Krista), 5089.
O'GORMAN (Francis), 985.
O'HARA (James J.), 2100.
O'MARA (Margaret Pugh), 5434.
O'NEILL (Johnathan), 6518.
O'NEILL (Kathleen), 3550.
O'RIORDAN (Elspeth), 7079.
O'SULLIVAN (Carolin), 2767.
O'SULLIVAN (Daniel E.), 3061.
O'SULLIVAN (James N.), 1443.
O'TOOLE (Thomas E.), 671.
OAKLEY (Christopher Arris), 4761.
OAKLEY (John H.), 1739.
OAKLEY (Stephen P.), 1820.
Oatis (William Nathan), 3779, 5290.
OBAID (Nawaf), 4514.
OBBINK (Dirk), 1608.
OBDEIJN (Herman), 6631.
OBERLÄNDER (Erwin), 4266.
OBERLY (James W.), 4762.
OBERMAIR (Hannes), 2552.
OBERPARLEITER (Veronika), 195.
OBORNI (Teréz), 3536.
OBSTFELD (Maurice), 6022.
OCAK (Ahmet Yaşar), 934.
OCHEA (Lionede), 7280.
OCITTI (Jim), 4708.
OCKINGA (Boyo G.), 1186.
OCONTRILLO (Eduardo), 3735.
ODART MORCHESNE, 8.
ODDONE (Patrick), 7287.
ODHIAMBO (E. S. Atieno), 4254.
ODILON NADALIN (Sergio), 6112.
ODONIS (Giraldus), 3253.
OEPEN (Joachim), 323.
OESTERLE (Günter), 361.
OGA (Ikuo), 7831.
OGBAR (Jeffrey O. G.), 4763.
OGILVIE (Sheilagh), 6311.
OGNYANOVA (I.), 3740.
OHANDJANIAN (Artem), 7017.
OHLMEYER (Jane), 4172.
ÖHMAN (Jenny), 6886.
OHNESORG (Aenne), 1740.
OIKONOMAKU (Konstantinou), 2392.
OIKONOMIDES (Nikolas), 2429.
OIZUMI (Akio), 3100.
OKADA (Aoi), 6203.
OKADO (Masakatsu), 4242.
OKAFOR (Eddie E.), 4943.
OKAMURA (Hidenori), 7703.
ÖKE (Mim Kemal), 7516.
OLASO (Vicent), 2550.
OLAUS MAGNUS, 235.
OLCOTT (Jocelyn), 4313.
OLEJNÍK (Milan), 3770.
OLENDER (Maurice), 578.

OLFF-NATHAN (Josiane), 5197.
OLICK (Jeffrey K.), 4000.
OLIVEIRA (Ione), 7517.
OLIVER OLMO (Pedro), 522.
OLIVERA (Mercedes), 4296.
OLIVIER (Cyril), 6312.
OLMO (Carlo), 5579.
OLOVSDOTTER (Cecilia), 2180.
OLSCHOWSKY (Burkhard), 7518.
OLSEN (Edward A.), 4261.
OLSEN (Olaf), 3809.
OLSON (Carl), 932.
OLSON (Linda), 3112.
OLSSEN (Erik), 4343.
OLTEANU (Constantin), 7519.
OLTMER (Jochen), 6313.
OMURA (Izumi), 5346.
OMURA (Sachihiro), 1111.
ONITSUKA (Hiroshi), 4243.
ÖNNERFORS (Alf), 1844.
ONOZAWA (Toru), 7416.
ONUF (Peter S.), 3532.
ONUK (Taciser), 5565.
ONUR (Bekir), 809.
OOSTINDIE (Gert), 6842.
OPFER-KLINGER (Björn), 4280.
Opicinus de Canistris, 3209.
OPINEL (Annick), 135, 5608.
OPITZ (Claudia), 458.
OPLL (Ferdinand), 3504.
OPP (James), 4995.
OPPENHEIMER (Melanie), 6460.
Oppianus, 1537.
OPRIŞ (Petre), 5850.
ORAM (Richard D.), 2642, 2723.
ORCHARD (Nicholas), 3468.
ORDEIG I MATA (Ramon), 2533.
OREKHOV (Aleksandr M.), 7520.
ORGHIDAN (I.), 4450.
Origenes, 2242, 2369.
ORIS (Michel), 6112.
ORLANDI (Antonella), 95.
ORLANDI (Giovanni), 3434.
ORLANDO (Salvatore), 7255.
ORLANDO (Sandro), 2971.
ORLECK (Annelise), 6314.
ORLOV (Aleksandr A.), 6932.
ORMROD (Mark), 2746.
ORMROD (W. M.), 2848.
OROPEZA (Lorena), 4764.
ORREGO PENAGOS (Juan Luis), 4379.
ORRILL (Robert), 523.
ORSI (Richard J.), 5851.
ORSINA (Giovanni), 4211.
ORSINI (Pasquale), 6, 2486.
ORTALLI (Francesca), 6871.
ORTALLI (Gherardo), 2733, 2735.
ORTEGA CANADELL (Rosa), *XI*.
ORTEGA LÓPEZ (Teresa María), 4561.

Ortega y Gasset (José), 5116.
ORTÍZ SOTELO (Jorge), 6933.
ORTOLÁ (Marie-Sol), 176.
ORTUÑO ANAYA (Pilar), 7521.
ORUÇ (Hatice), 2849.
ORYE (Lieve), 933.
OSBORNE (Robin), 1094, 1609.
OSBORNE (Thomas Michael), 3298.
OSÉS URRICELQUI (Merche), 2535.
OSHINSKY (David M.), 5436.
OSIEK (Carolyn), 2337.
OST (David), 4402.
OSTENC (Michel), 4221, 7080.
OSTERHAMMEL (Jürgen), 447, 822.
OSTERRIEDER (Markus), 6887.
OSTOS SALCEDO (Pilar), 2553.
ØSTREM (Eyolf), 2890.
OSWALD (Anne von), 5885.
OTA (Shoichi), 7593.
OTIMAN (Păun Ion), 5959.
OTRANTO (Giorgio), 2343.
OTRANTO (Rosa), 1610.
OTSUKI (Yasuhiro), 2487.
OTTERNESS (Philip), 6315.
Otto von Bamberg, 3412.
Otto von Freising, 104.
OU (Weiwei), 7694.
OUDIN-BASTIDE (Caroline), 5759.
OUELLETTE (Françoise-Romaine), 186.
Ouko (John Robert), 4254.
OUSTERHOUT (Robert G.), 2488.
OUYANG (Wen-chin), 983.
OVER (Kristen Lee), 3063.
OVERLAET (Bruno), 1363.
OVERUD (Johanna), 6103.
Ovidius Naso (Publius), 1822, 1823, 1824, 2057, 2077, 2079, 2090, 2096, 2101, 2116, 2127, 2149, 2981.
OVIEDO (José Miguel), 997.
Owen (Ruth Ryan), 7219.
OWEN (Thomas C.), 5853.
OWEN-CROCKER (Gale R.), 3128.
OWENS (J. B.), 6541.
OWUSU (Robert Yaw), 4028.
OWUSU-ANSAH (David), 673.
OY-MARRA (Elisabeth), 5609.
ÖZ (Mehmet), 6037.
ÖZDEMIR (H. Ahmet), 7575.
ÖZDEMIR (Hikmet), 6316.
ÖZDEN (Mehmet), 4698.
ÖZKUL (Ali Efdal), 5760.
OZOUF (Mona), 3884.
ÖZTÜRK (Serdar), 6317.
ÖZYETGIN (A. Melek), 2927.

P

Paasikivi (Juho Kusti), 7429.
PACHECO DÍAZ (Argelia), 6843.

PACHMANOVÁ (Martina), 5204.
Padery (Etienne), 6912.
Padre Pio da Pietralcina (Francesco Forgione), 4870.
PÁDUA DE MORA (Nuno), 6811.
PADUANO (Guido), 1427, 1611.
PÁEZ MARTÍNEZ (Laritza), 6834.
PAGDEN (Anthony), 252, 721.
PAGE (Willie F.), 7865.
PAGÈS I BLANCH (Pelai), 4606.
PAGNINI (Maria Paola), 6690, 6950.
PAHLOW (Louis), 746.
PAHTA (Päivi), 3294.
PAINE (S. C. M.), 6982.
Paine (Thomas), 5521.
PAKKALA-WECKSTRÖM (Mari), 3064.
PALACE (Wendy), 6983.
PALACIOS (Jesús), 4549.
PALACIOS CEREZALES (Diego), 4418.
PALAGIA (Olga), 1741.
PALAIA (Roberto), 5342.
PALAIMA (Thomas), 1399.
PALAU (Josep), 5038.
PALETSCHEK (Sylvia), 5206.
PÁLKA (Petr), 5045.
PALL (Francisc), 14.
PÄLL (Janika), 2419.
Palladio (Andrea), 5588.
Palladius Helenopolitanus, episcopus, 2331.
PALLANTZA (Elena), 1612.
PALLARES-BURKE (Maria Lúcia G.), 461.
PALMA DI CESNOLA (Arturo), 1035.
Palme (Olof), 7480.
PALMER (David A.), 7704.
PALMER (Steven), 626.
PALOMO (María Dolores), 4296.
PALTEMAA (Lauri), 3716.
PALUMBO STRACCA (Bruna M.), 1417.
PAN (Guangdan), 7596.
PAN (Guangzhe), 7705.
PAN (Liang), 7353.
PANASPORNPRASIT (Chookiat), 7522.
PÁNEK (Jaroslav), 404, 405, 812, 5101.
PANELLA (Claudio), 3594.
PANGRAZIO (Miguel Angel), 4365.
PANI (G. Giacomo), 3153.
PANI (Mario), 1981.
PANIAGUA (Javier), 4626.
PANICHKIN (Jurij N.), 4361.
PANITZ-COHEN (Nava), 1318.
PANKOVITS (József), 7523.
PANSTERS (Wil G.), 4294.

PANTELIS (Antoine), 702.
PANTOJA MORÁN (David), 6519.
PANZIG (Erik A.), 3299.
PAOLETTI (Orazio), 1765.
PAOLETTI (Paolo), 1241.
PAOLILLI (Antonio Luigi), 2850.
Paolino Veneto, 3484.
PAOLUCCI (Anne), 2989.
PAOLUCCI (Henry), 2989.
Paolus VI, Papa, 4585.
PAPACONSTANTINOU (Arietta), 1187.
PAPADIA (Elena), 5915.
PAPADOPOULOS (John K.), 1521.
PAPAILIAS (Penelope), 406.
PAPAIOANNOU (Sophia), 2101.
PAPAJÍK (David), 2851.
PAPANGELES (Theodoros D.), 1922.
PAPARELLA (Francesco D.), 3275.
PAPELL I TARDIU (Joan), 2534.
PAPP (Daniel S.), 6680.
PAPUC (Gheorghe), 2020.
PAQUIER (Serge), 5829.
PARADA (Alejandro E.), 5483.
PARADELA ALONSO (Nieves), 237.
PARADISO (Annalisa), 1923.
PARADOWSKI (Przemysław), 4403.
PARAS (Eric), 460.
PARAVICINI (Werner), 221, 2653, 2675, 2759, 2902, 3187.
PARAVICINI BAGLIANI (Agostino), 102.
PARENTE (Fausto), 1319.
PARENTE (Ulderico), 4944.
PARENTI (Stefano), 3387.
PARISELLA (Antonio), 4225.
PARISH (Helen L.), 3415.
PARISOLI (Luca), 3300.
PARISSAKI (Maria Gabriella), 1412.
Park (Chung Hee), 7470.
PARK (Hyun Ok), 7851.
PARK (Sang-soo), 7706.
PARKER (Bradley J.), 584.
PARKER (Eldon M.), 6406.
PARKER (Geoffrey), 612.
PARKER (Robert), 1502, 1704.
PARKER (Victor), 1871.
PARKER (Will), 3065.
PARKER PEARSON (Mike), 1069.
PARKS (Tim), 2852.
Parmenides, 1654.
PARMET (Robert D.), 6457.
PARPAROV (Fyodor), 286.
PARRA GARZÓN (Gabriela), 3579.
PARSADANOVA (Valentina S.), 7057.
PARSON (Don), 6104.
PARTRIDGE (Christopher H.), 867.
PARTRIDGE (Michael), 7368.
PÂRVULESCU (Adrian), 2102.

PAS (Nicolas), 6318.
PASCHOUD (François), 2103.
PASCOE (Louis B.), 3464.
PASERO (Nicolò), 2943.
PÄSLER (Ralf G.), 3113.
Pasolini (Pier Paolo), 5507, 5511, 5538.
PASSA (Enzo), 1670.
PASSINI (Michela), 459.
PASTA (Renato), 483.
Pasteur (Louis), 5374.
PASTEUR (Paul), 6425.
PASTORE (Stefania), 4927.
PASTURE (Patrick), 6407.
PATEL (Ismail Adam), 2427.
PATLAGEAN (Évelyne), 2428.
PATMORE (Greg), 3617, 6417.
PATON (Diana), 6217.
PATRIARCA (Marco Antonio), 6681.
PATRIARCA (Silvana), 4226.
PATRIAS (Carmela), 6426.
PATTERSON (Catherine), 6542.
PATURZO (Franco), 1772.
PAUL (Ina Ulrike), 6502.
PAUL (T. V.), 7340.
PAULET (Aurelie), 1188.
Paulinus Nolanus, Sanctus, 2244, 2245, 2285.
PAULMANN (Johannes), 7318.
PAULSON (Stanley L.), 6481.
Paulus Silentiarius, 2409.
Paulus, apostolus, Sanctus, 2318, 2322, 2357, 2359, 2373, 2376, 2377.
Pauncefote (Julian), 6968.
Pausanias, 1530, 1585, 1596, 1707, 1710.
PAUTSCH (Ilse Dorothee), 7445.
PAVLINCOVÁ (Helena), 5322.
PAVONE (Sabina), 4928.
PAYASLIAN (Simon), 7081.
PAYEN (Pascal), 407.
PAYK (Marcus M.), 3937.
PAYNE (S.), 1078.
PAYNE (Stanley G.), 4549.
Paz (Octavio), 4292.
PEARSON (Andrea G.), 3203.
PEARSON (David), 87.
PEARSON (Owen), 3572.
PEARSON (Richard), 1036.
PÉBARTHE (Christophe), 1415, 1465.
PECCHIOLI DADDI (Franca), 1266.
PECHA (Lukas), 1242.
PÉCHOIN (Daniel), 65.
PEDERSEN (Frederik), 2723.
PEDERSEN (Susan), 6730.
PEDIO (Alessia), 432.
PEDLEY (John Griffiths), 1705.
PEDLEY (Mary Sponberg), 238.

PEDROSO (Regina Célia), 3674.
PEDUCCI (Giulia), 1706.
PEERS (Douglas M.), 6753.
PEGUERO (Valentina), 3815.
PEHLE (Hans), 6888.
PEHR (Michal), 7082.
PELÁEZ (Manuel), 731.
PELIN (Mihai), 7220, 7281.
Pellé (Maurice), 7029.
PELLEGRINI (Maria), 2247.
PELLING (Margaret), 5440.
PELTOVUORI (Risto), 7221.
PÉNEAU (Corinne), 2573.
PENFOLD (Steve), 6432.
PENG (Dachen), 7707.
PENN (Shana), 4404.
PENNACCHIETTI (Fabrizio A.), 1320.
PENNINGTON (Brian K.), 935.
PENNOCK (Pamela E.), 6163.
PENSLAR (Derek J.), 403.
PENTER (Tanja), 7222.
PEPE (Luigi), 5437.
PEPELASIS MINOGLOU (Ioanna), 754, 5736.
PEPIN (Ronald E.), 3416.
PERA (Rossella), 142.
Perdiccas, 1466.
PERDUE (Peter C.), 6682.
PEREA CAVEDA (Alicia), 1097.
PEREDA (Felipe), 302.
PEREGO (Lucio G.), 1773.
PEREIRA (Anthony W.), 6595.
PEREIRA (Juan Carlos), 6652.
PEREIRA (Leonardo Affonso de Miranda), 6242.
PERESANI (Marco), 1027.
PERETZ (Daniel), 2035.
Pérez (Alonso), 60.
PÉREZ (Louis A. Jr.), 6319.
PÉREZ GONZÁLEZ (Carlos), 3071.
PÉREZ GONZÁLEZ (Silvia María), 2853.
PÉREZ LEDESMA (Manuel), 3517, 5137.
PÉREZ PONS (E.), 2688.
PÉREZ VILARIÑO (Álvaro), 1574.
PÉREZ-GONZÁLEZ (Alfredo), 1030.
PÉREZ-LÓPEZ (Jorge F.), 3745.
PERFETTI (Lisa Renée), 3067.
Pericles, 1463.
PERIFANO (Alfredo), 820.
PERLER (Dominik), 3286.
PERNES (Jiří), 3781, 3782.
Perón (Eva), 5708.
Perón (Juan Domingo), 7412.
PERONI (Adriano), 3207.
PEROTTI (Pier Angelo), 2104.
PEROVŠEK (Jurij), 4536, 4538.
Perpetua, Sancta, 2380.

PERRICCIOLI SAGGESE (Alessandra), 42.
PERRIN (Michel), 3433.
PERRONE (Raffaele), 1027.
PERROT (Michelle), 6496.
PERROUSSEAUX (Yves), 88.
PERRUS (Claude), 3072.
PERRY (Mary Elizabeth), 4897.
PERSAK (Krzysztof), 646, 3759.
PESANDO (Fabrizio), 1756.
PESANTE (Maria Luisa), 6492.
PESARESI (Roberto), 1982.
PESCATORE (Guglielmo), 5623.
PESCH (Otto Hermann), 4844.
PESCHLOW-BINDOKAT (Anneliese), 1140.
PEŠEK (Jiří), 89, 355, 6093.
PESIRI (Giovanni), 2615.
PETERS (Kate), 5294.
PETERSEN (Jørn Henrik), 6132.
PETERSEN (Klaus), 6132.
PETERSEN (Nikolaj), 7321.
PETERSEN (Tore T.), 7524.
PETERSMANN (Gerhard), 195.
PETERSON (Brent O.), 5528.
PETERSSON (Magnus), 7343.
PETIT (Caroline), 1671.
PETLEY (Julian), 4048.
PETOLETTI (Marco), 3069.
PETRAK (Marko), 6596.
PETRANSKÝ (Ivan A.), 5960.
Petrarca (Francesco), 3008, 3096.
PETRÁŠ (Jiří), 6024.
PETRÁŠ (René), 6472.
PÉTREQUIN (Anne-Marie), 1050.
PÉTREQUIN (Pierre), 1050.
PETRIE (Cameron), 1360.
PETROV (Bisser), 4104.
Petrucci (Ottaviano), 94.
PETRUF (Pavol), 4527.
PETRUKHIN (Vladimir Ja.), 2685.
Petrus, Sanctus, 2290.
PETSCHNIGG (Edith), 7145.
PETTAS (William A.), 5295.
Pettazzoni (Raffaele), 902, 929.
PETTEGREE (Andrew), 4996.
PETTI BALBI (Giovanna), 2609.
PETY (Dominique), 5049.
PETZL (Georg), 1672.
PETZOLDT (Leander), 550.
PEYROUSE (Sébastien), 6268.
PFAU (Marianne Richert), 3240.
PFAU (Michael William), 4765.
PFEIL (Ulrich), 7361.
PFEILSCHIFTER (Rene), 1924.
PFISTER (Roger), 7525.
PFISTERER (Stephan), 5854.
PFLEFKA (Sven), 3367.
PFLÜGER (Christine), 4002.
Phaedrus, 1810, 1825, 2062.

Philip II Augustus, roi de France, 2559, 2620.
Philippe III de Bourgogne 'Philippe le Bon, 2648, 2759.
Philipps des Kanzlers, 3315.
Philippus II, king of Macedonia, 150.
Philitas, 1636.
Phillips (John), 4079.
PHILLIPS (John), 7873.
PHILLIPS (Ruth B.), 313.
PHILLIPS (Seymour), 2746.
Philo Alexandrinus, 2317.
Philostorgius, 2338.
Philostratus, 1435.
PHILPOTT (William J.), 4084.
Phipps (Eric), 7210.
PHIRI (Bizeck Jube), 4805.
Phocas Augustus (Flavius), Byzantine emperor, 140.
PHONGPAICHIT (Pasuk), 4669.
Photius, 2506.
PIANA AGOSTINETTI (Paola), 1021.
PIAZZA (Alberto), 996.
PIAZZI (Lisa), 2124.
Picart (Bernard), 4962.
PICCATO (Pablo), 4287.
PICCHI (Michele), 456.
PICCINNO (Luisa), 62.
PICHOIS (Claude), 5506.
PICK (Daniel), 4227.
PICKLES (Katie), 6867.
PICKOWICZ (Paul G.), 3713.
PICKUS (Noah), 6320.
PIEPENBRINK (Karen), 2279.
PIERAU (Karl), 443.
PIERAZZO (Elena), 5544.
PIERCE (Richard B.), 6321.
PIERCE (Steven), 5961.
PIERI (Dino), 5504.
PIERI (Dominique), 2489.
Piero Della Francesca, 3189.
PIERRE (Antoine Fritz), 4111.
PIERRI (Bruno), 6683.
PIETRANTONIO (Vanessa), 5512.
PIETROBELLI (Antoine), 1642.
PIETRUSEWSKY (Michael), 1042.
PIETSCH (Andreas), 4898.
PIETSCH (Efthymia), 2412.
PIGENET (Michel), 6407.
PIGOTT (V. C.), 1347.
PIHLAJA (Pivi Maria), 5438.
PIKE (David L.), 6105.
PIKE (R.), 5916.
Pilati (Carlo Antonio), 5110.
PILHOFER (Peter), 1433, 2377.
PILLININI (Giovanni), 5296.
PILLON (Michel), 2036, 2490.
PILS (Holger), 4957.
PILTZ (Elisabeth), 2491.

PIMENTA (Fernando Tavares), 3577.
PIMENTEL (Cristina), 314.
PINCELLI (Agata), 2643.
Pindarus, 1436, 1437, 1554, 1558, 1559.
PINDSTRUP (Jacob), 3802.
PING (Jin Guo), 6895.
PINHEIRO (Magda), 6060.
PINI (Antonio Ivan), 3147.
PINI (Raffaella), 3204.
PINO (Cristina), 1189.
Pinochet Ugarte (Augusto), 3711, 3712, 6617.
PINTILIE (Florin), 4431.
PINTO (Anthony), 5855.
PINTO DA FONSECA (Francisco César), 3675.
PIOLOT (Laurent), 1707.
PIOTROVSKIJ (Mikhail B.), 32.
PIPER (Ernst), 4003.
PIPES (Richard), 4495.
PIPPIDI (Andrei), 467.
PIQUER FERRER (Esperança), 2525.
PIQUERAS (Josep A.), 4626.
PIQUERAS ARENAS (José Antonio), 3746.
PIRANDELLO (Andrea), 5513.
Pirandello (Luigi), 5341, 5513, 5530.
Pirandello (Stefano), 5513.
PIRILLO (Paolo), 2854.
PIRJEVEC (Jože), 4228, 7377, 7437.
PIRONTI (Gabriella), 1708.
PIROTTE (Dominique), 5339.
PISANI (Michael V.), 5699.
PISANO (Debora), 972.
PISAREV (Aleksandr A.), 5948.
Pisistratus, 1458, 1721.
PISO (Ioan), 1925.
PITRE (Brant), 2345.
PITTAU (Massimo), 188.
PITTIA (Sylvie), 1897.
PITTS (Jennifer), 6728.
Pius XI, Papa, 4857.
PIVATO (Stefano), 6322.
PIZZOLATO (Nicola), 5439.
PIZZORUSSO (Giovanni), 4945.
PLAKANS (Andrejs), 6112.
PLANT (Alisa), 4037.
PLANT (Margaret), 6124.
Plantagenets, kings of England, 2644.
PLATANIA (Marco), 5347.
PLATH (Christian), 4997.
Platina, 3188.
Plato, 1644, 1654, 1660, 1661, 1665, 1668, 1669, 1680, 1683, 3285.
PLATOVSKÁ (Marie), 5554.
PLATSCHEK (Johannes), 1807.

PLATT (Harold L.), 6106.
Plautus (Titus Maccius), 1826, 1827, 2108, 2119.
PLÉ (Bernhard), 3885.
PLENER (Ulla), 7298.
PLESHKOVA (Sof'ja L.), 3886.
Plinius Secundus (Gaius), maior, 1828-1830.
PLOETZ (Michael), 7445.
PLÖGER (Karsten), 3368.
PLOSCARU (Cristian), 4451.
Plumier (Charles), 5463.
PLUMIER (Jean), 2872.
PLUMMER (John F.), 3091.
PLUMMER (John), 3161.
Plutarchus, 1423, 1438, 1529, 1533, 1586, 1657.
POCOCK (John Greville Agard), 675, 811.
PODDAR (Prem), 992.
POGLIANO (Claudio), 579.
POHL (Walter), 2205.
PÓK (Attila), 408.
POKORNÝ (Jiří), 6324.
POKORNY (Michael), 5707.
Polanyi (Karl), 396.
POLASCHEGG (Andrea), 271.
Poletti (Charles), 7130.
POLFER (Michel), 1994.
Politi (Ambrogio Catarino), 4914.
POLITI (Janet), 1243.
POLJAKOV (Jurij A.), 6099.
POLJAKOV (Konstantin I.), 4634.
POLJAKOVA (Elena Ju.), 3549.
POLLACK (Marko), 7317.
Pollaiuolo (Antonio), 3223.
Pollaiuolo (Piero), 3223.
POLLARD (A. M.), 1084.
POLLARD (John F.), 6025.
POLLARD (Lisa), 6802.
POLLMANN (Karla), 2289.
POLOTTO (Federico), 698, 3598, 3608.
POLUKHINA (M. V.), 276.
POLUNOV (Alexander), 4497.
POLVANI (Anna Maria), 1267.
POMATA (Gianna), 379.
POMBENI (Paolo), 4198.
POMIAN (Krzysztof), 3758.
Pompeius (Sextus), 1923.
Pompeius Magnus (Cnaeus), 1620, 1819, 1893, 1945.
Pompeius Trogus, 1620.
POMPER (Philip), 721.
POMPILIO (Antonella), 5327.
POMPONIO (Francesco), 1216.
PON (Georges), 2544.
PONCET (Olivier), 291, 4843.
Poniatowska (Elena), 4292.
Pons, familia, 5814.
PONS (Nina), 1098.

PONS (Silvio), 7529.
PONS PONS (Jerònia), 5917.
PONSO (Marzia), 5338.
PONTANI (Filippo Maria), 1613.
Pontano (Giovanni Gioviano), 2974.
POOLE (Anthony), 5918.
POOLE (Federico), 1190.
POOLE (Hilary W.), 797.
POOLEY (Colin G.), 6325.
POP (Horea), 1920.
POP (Ovidiu H.), 4847.
POPA (Klaus), 4429.
POPE (Hugh), 4699.
POPESCU (Alexandru), 6639.
POPKES (Enno Edzard), 2346.
POPKIN (Jeremy D.), 410.
POPOV (Andrej V.), 4953.
POPOV (I. M.), 4262.
POPOV (Zheko), 7083.
POPOVA (Marianna), 6620.
POPOVA (Nina), III.
POPOVA (Olga), 2492.
POPOVIĆ (Nikola B.), 7084.
POPOVICI (Vlad), 4452.
Popper (Karl Raimund), 5349.
PORCARO (M.), 2347.
PORCELLI (Bruno), 189.
POROMBKA (Stephan), 208.
Porphyrius, 1657, 2246, 2303.
PORRINI (Rodolfo), 6458.
Porshnev (Boris F.), 3859.
PORSKROG RASMUSSEN (Carsten), 3806.
PORTE (Rémy), 5856.
PORTELLI (Alessandro), 431.
PORTER (David H.), 1614.
PORTER (Dilwyn), 5894.
PORTER (Joy), 978.
PORTER (Martin), 5134.
PORTES (Alejandro), 765.
PORTNOY (Alisse), 4766.
Posidippus, 1535, 1568, 1599, 1607, 1608, 1626, 1627, 1631, 1639, 1606.
POSKONINA (Ol'ga I.), 3554.
POSSENTI (Elisa), 3487.
POSTLER (Vicky), 5135.
Potocki (Ignacy), 4397.
POTT (Ute), 71.
POTTER (David Stone), 620.
POTTER (Simon James), 6865.
POTTHAST (August), 2579.
POTTHAST (Barbara), 6196.
POTTIER (Bruno), 1816.
POTTIER (Georges-François), 6571.
POTTS (Annie), 824.
POTTS (Daniel T.), 1364.
POUDERON (Bernard), 2344.
POUILLARD (Véronique), 6326.

POULAIN-GAUTRET (Emmanuelle), 3073.
POULAT (Émile), 4848.
POULIOT (François), 3301.
POULOT (Dominique), 315.
POULSON (Stephen C.), 4164.
POUNDS (Norman John Greville), 3506.
POUTRUS (Patrice G.), 5056.
POWELL (Martyn J.), 4185.
POWER (Jane), 3020.
POWITZ (Gerhardt), 46.
Poyet (Guillaume), 6545.
POZUELO RODRÍGUEZ (Felipe), 2536.
PRACA (Edwige), 5855.
PRADA RODRÍGUEZ (Julio), 4597.
PRADOS-TORREIRA (Teresa), 3747.
PRATAP (Sangeeta), 6006.
PRATSCH (Thomas), 2415, 2493.
PRAZ (Anne-Françoise), 5241.
PRAŽANOVÁ (Markéta), 5204.
PREČAN (Vilém), 6328.
PREIN (Philipp), 6329.
PREMO (Bianca), 6330.
PRENSA (Luis), 3238.
PREPARATA (Guido Giacomo), 7224.
PRESTON (Paul), 7368.
PRESTON BART (Philip), 2650.
PRESTWICH (Michael), 2644, 2939.
PRÊTEUX (Franck), 1709.
Preti (Mattia), 5601.
PREUSS (Monika), 5040.
PREVELAKIS (Eleutherios), 6948.
PRÉVÉLAKIS (Georges), 677.
Previtali (Giovanni), 853.
PRÉVOST (Philippe), 7225.
PREZIOSI (Ernesto), 4828.
PRICE (Douglas), 1029.
PRICE (Simon), 1743.
PRIESTLAND (Jane), 6623.
PRIETO MARTÍNEZ (M.ª Pilar), 1076.
PRIMICERI (Emanuela), 7526.
Primo de Rivera (Miguel), 4577.
Prince Gong, 7720.
PRINČIČ (Jože), 6013.
PRIOR (Charles W. A.), 4998.
PRIOR (Robin), 7085.
Priscianus, 2201.
PRITCHARD (Gareth), 4005, 6327.
PRIULI (Vincenzo), 6871.
PRIVITERA (G. Aurelio), 1615.
PROBST (Niels), 3803.
Probus (Marcus Valerius), 1831.
PROCACCI (Giuliano), 525.
PROCACCIOLI (Paolo), 3123, 5140.
PROCELLI (Enrico), 1006.
PROCHASSON (Christophe), 5351.
PRÓCHNIAK (Leszek), 4410.

Proclus, 2399, 2410.
Procopius Gazaeus, 2411, 2432.
PROFUMO (Maria Cecilia), 3507.
PRÖGER (Susanne), 970.
Prokesch von Osten (Anton, Graf), 6919.
Propertius (Sextus), 1832-1834, 2067.
PROSPERI (Adriano), 4819.
PROST (Antoine), 327, 6763.
Protagoras, 1647, 1652, 1660.
Proust (Marcel), 5512.
PROVENCE (Michael), 4663.
PROVIDENTI (Elio), 5530.
PRŮCHA (Václav), 7307.
Prudentius Clemens (Aurelius), 2247, 2248, 2362.
PRUFER (Keith M.), 7884.
PRUNIER (Gérard), 4635.
PRUSHANKIN (Jeffery S.), 4767.
Psellus (Michael), 2412.
PŠENIČKOVÁ (Jana), 5975.
Pseudo-Basilius, 2249.
Pseudo-Cassiodorus, 1835.
PSILOS (Christopher), 4281.
PSOMA (Selene), 1412, 1744.
Ptahhotep, Egyptian official, 1191.
PUCHNER (Martin), 5360.
PUGH (Martin), 4081.
PUIGGALI (Jacques), 1616.
PUJOL (Stéphane), 5477.
PUKELSHEIM (Friedrich), 3289.
Pulci (Luigi), 5493.
PULSIPHER (Jenny Hale), 6844.
PUPPI (Lionello), 5588.
PURCELL (Nicholas), 1879.
PURDEK (Imrich), 4528.
PURDIE (Rhiannon), 3089.
PURSEIGLE (Pierre), 326.
PUSHKAREVA (I. M.), 6437.
Pushkin (Aleksandr Sergeevich), 998.
PUTATURO DONATI MURANO (Antonella), 42.
PUTNAM (Lara), 6485.
PYTHON (Francis), 4659.
PYYSIÄINEN (Ilkka), 936.

Q

QIN (Feng), 7708.
QIN (Guojing), 7709.
QIN (Yongzhang), 7710.
QIU (Fanzhen), 7712.
QUACK (Joachim F.), 1141, 1191.
QUAGLIARIELLO (Gaetano), 4211.
QUAGLIONI (Diego), 2796.
QUAISSER (Friederike), 6597.
QUAN (Hexiu), 6984.
QUANCHI (Max), 239.
QUARANTOTTO (Diana), 1673.
QUARITSCH (Helmut), 6500.
QUAST (Bruno), 3074.
QUENET (Grégory), 5442.
QUÉNIART (Jean), 5092.
QUESTA (Cesare), 1827.
QUESTIER (Michael C.), 4036.
QUEZADA (Sergio), 4301.
QUINTA (Françoise), 5855.
QUINTA (Henri), 5855.
QUINTANA GARCÍA (José Antonio), 6985.
QUINTERO (Rafael), 3820.
Quintilianus (Marcus Fabius), 1838.
Quintus (Publius), 1807.
Quintus Curtius Rufus, 1423.
QUIROZ-PÉREZ (Lissel), 4380.
QUISBERT (Cristina), 3652.
Qutb (Sayyid), 5036.

R

RAAFLAUB (Kurt A.), 1510, 2047.
RABE (Stephen G.), 7527.
RABEL (Roberto Giorgio), 7528.
RABER (Christine), 5700.
RÁBIK (Vladimír), 3508.
Rabin (Yitzhak), 4189.
RABINOVICH (R.), 1275.
RABINOWITZ (Dan), 4193.
RABONI (Giovanni), 5531.
RĂCEANU (Mircea), 6685.
RACINE (Pierre), 5193.
Ráday (Pál), 4125.
RÁDAY-PESTHY (Pál Frigyes), 4125.
RADICI COLACE (Paola), 1745.
RADICIOTTI (Paolo), 1.
RADNER (Karen), 190.
RADOWITZ (Sven), 7156.
RADT (Stefan), 1442.
RADT (Wolfgang), 1142.
RADU (Sorin), 4453.
RADULESCU (Raluca), 2898.
RADVANOVSKÝ (Zdeněk), 7089, 7139, 7158.
RAEPSAET-CHARLIER (Marie-Thérèse), 2038.
RAFEL FONTANALS (Núria), 1055.
RAFFAELLI (Mauro), 245.
Raffaello Sanzio, 831, 5602.
RAFFARIN-DUPUIS (Anne), 5054.
RAFUSE (Ethan), 4720.
RAGGHIANTI (Renzo), 5141.
RAGUSA (Andrea), 4234.
RAHLF (Thomas), 443.
RAHMAN (Habibur), 4428.
RAHN (Werner), 3938.
RAICICH (Marino), 5242.
RAIL (Jan), 7289.
RAINER (Heinrich), 4960.
RAINHORN (Judith), 6334.
RAITHEL (Thomas), 6544.
RAJA (Maria Elisa), 3075.
RAJLICH (Jiří), 7283.
RAKHSHMIR (Pavel Ju.), 3630.
RAKOVÁ (Svatava), 404.
RAMBAUX (Claude), 2348.
RAMIRES (Alexandre), 4427.
RAMÍREZ SÁIZ (Juan Manuel), 4318.
Ramón Ramón (Antonio), 3707.
RAMOS (María Dolores), 4599.
RAMPAZZO (Natale), 1983.
Ramses II, king of Egypt, 1184.
Ramses III, king of Egypt, 1161.
RAMSEY (John T.), 1984.
Ranade (M. G.), 7595.
RANDA (Philippe), 6746.
RANDAZZO (Antonella), 6803.
RANFT (Andreas), 770.
RAO (Riccardo), 2768.
RAPETTI (Anna Maria), 3388.
RAPHAEL (Lutz), 412.
RAPONE (Leonardo), 356.
RAPONI (Nicola), 6937.
RAPOPORT (Yossef), 2714.
RAPP (Claudia), 2280.
RASCH (Gerhard), 191.
RASCH (Rudolf), 84.
RASCHLE (Christian R.), 2211.
RASHED (Marwan), 1421.
RASMUSSEN (Anne), 5405.
RASMUSSEN (Joel), 6244.
RASPOPOVIĆ (Radoslav M.), 6686, 6938.
RASSOW (Peter), 6130.
RASTOVIĆ (Aleksandar), 6986.
RATAJ (Jan), 5136.
RATH (Thomas), 4315.
Ratherius, episcopus veronensis, 3476.
RATHGEB (Eberhard), 4006.
RATHGENS (Sigrid), 5250.
RATHKOLB (Oliver), 3569.
RATHMANN (Michael), 1466.
Ratoldus, abbot of Corbie, 3468.
RAUCHENSTEINER (Manfried), 7395.
RAUDVERE (Catharina), 865.
RAUNIG (Walter), 2349.
RAUS (Edmund J. Jr.), 4768.
RAUSCHER (Frederick), 5338.
RAVEN (Maarten J.), 1192.
RAVIER (Xavier), 2523.
RAWE (Kai), 5857.
RAXHON (Philippe), 3726.
RAY (John D.), 1193.
RAYSKI (Adam), 3887.
RAZJMOU (Shakrokh), 1365.
RDVANOVSKÝ (Zdeněk), 7065.
READ (Christopher), 4498.
READMAN (Paul), 413.

Reagan (Ronald), 7475.
REAGIN (Nancy Ruth), 5089.
REANEY (Percy Hide), 192.
REARDON-ANDERSON (James), 7713.
RECCHIA (Alessandro), 4855.
RECCHIA (Giulia), 1058.
RECHT (Roland), 848.
REDDÉ (Michel), 1194.
REDFERN (Neil), 6422.
REDINGER (Matthew), 4316.
REDON (Odile), 6186.
REED (Annette Yoshiko), 2351.
REEMTSMA (Jan Philipp), 3977.
REES (Joachim), 241.
REESE (Roger R.), 4499.
REFF (Daniel T.), 937.
REGALADO (Nancy Freeman), 3068.
REGLERO DE LA FUENTE (Carlos Manuel), 2551.
REGNARD (Maude), 2872.
REGNAULT (Jean-Marc), 6724.
RÉGNIER (Philippe), 5049.
REGOLIOSI (Mariangela), 482.
REGOURD (François), 6711.
REHBERG (Karl-Siegbert), 233.
REICHARDT (Sven), 4215.
REICHERT (Hermann), 2970.
REICKE (Bo), 2274.
REID (Chris), 5914.
REID (Donald), 7291.
REID (John G.), 450.
REIF (Elizabeth), 387.
REIKAT (Andrea), 580.
Reinalter (Helmut), 967.
REINBOLD (Markus), 6889.
REINHARDT (Klaus), 3471.
REINHARDT (Volker), 116.
Reinhardt (Walther), 3995.
REINHARTZ (Dennis), 231.
REININGHAUS (Wilfried), 5879.
REINKOWSKI (Maurus), 739, 4693, 4700.
REINSCH (Diether Roderich), 2385, 2494.
REINWALD (Brigitte), 6804.
REIS (Maria de Fátima), 4425.
REIS PILAR (Maria), 2183.
REISS (Sheryl E.), 4853.
REITEMEIER (Arnd), 2856.
REITER (Ilse), 6503.
REITZAMMER (Laurialan), 1739.
REMEIKIS (Thomas), 7111.
REMEKE (Stefan), 4007.
RÉMOND (René), 741, 3888, 4821.
RÉMY (Bernard), 1791, 1928.
RENTON (Dave), 4045.
RENUCCI (Pierre), 1929.
RENWICK (John), 5358.
RENZ (Werner), 6478.

REPE (Božo), 4537.
REPINA (Lorina P.), 401, 436, 512, 807.
RESÉDENZ (Andrés), 3556.
RESTALL (Matthew), 6152.
REUDENBACH (Bruno), 3206.
RÉV (István), 5138.
RÉVAUGER (Cécile), 6435.
RÉVEILLARD (Christophe), 7136.
REVEL (Judith), 460.
REVELL (Keith D.), 6107.
REVJAKINA (Luiza), 7013.
REXOVÁ (Kristina), IV.
REY TRISTÁN (Eduardo), 4794.
REYDAMS-SCHILS (Gretchen J.), 2125.
REYERSON (Kathryn), 2992.
Reynaud (Paul), 3896.
REYNOLDS (Matthew), 5000.
REŽEK (Mateja), 7419.
RHEIN (Stefan), 4975.
RHODES (P. J.), 1383.
Riad al-Suhl, 7174.
RIASANOVSKY (Nicholas V.), 678.
RIBALTA I HARO (Jaume), 2769.
RIBÉMONT (Thomas), 416.
RICARD (Serge), 7483.
Ricardo (David), 5778.
RICCARDI (Andrea), 4849.
RICCI (Alessio), 3080.
RICCI (Andrea), 1134.
RICCI (Cecilia), 2039.
RICCI (Manuela), 5505.
RICCI (Roberto), 212.
RICCIARDI (Alberto), 2569, 2616.
RICCUCCI (Marina), 2951.
RICH (Jeremy), 5858.
RICHARD (François), 1985.
Richard de Saint-Victor, 3248.
Richard I, king of England, 2610.
Richard II, king of England, 2610, 2746, 2858, 3093.
Richard III, king of England, 2610, 2746.
Richard of St. Victor, 3272.
Richard of Wallingford, 3297.
RICHARDOT (Philippe), 2206.
RICHARDS (Michael), 5155.
RICHARDSON (Brian W.), 242.
RICHARDSON (David), 6714.
RICHARDSON (Gary), 2857.
RICHARDSON (Glenn), 6892.
RICHARDSON (Seth), 1244.
RICHTER (Amy G.), 5859.
RICHTER (Daniel), 1617.
RICHTER (Gerhard), 2353.
RICHTER (Karel), 3557.
RICHTER (Thomas), 1268.
RICHTER (Timm), 4954.
RICKETTS (Mac Linscott), 910.
Ricoeur (Paul), 511.

RICUPERATI (Giuseppe), 417, 451, 526, 5139.
RIDDERBOS (Bernhard), 834.
RIDGEON (Lloyd), 4165.
RIDLEY (Ronald T.), 2144.
RIDOLFI (Maurizio), 3531.
RIECK (Anja), 6890.
RIEDI (Eliza), 5243.
RIEDL (Peter Philipp), 5566.
RIEDO-EMMENEGGER (Christoph), 2354.
RIEGER (Barbara), 1370.
RIEGER (Bernhard), 5443.
RIEGER (Reinhold), 3303.
RIEKER (Jörg Rudolf), 2981.
RIEMER (Lars Hendrik), 6551.
RIES (Gerhard), 1143.
RIES (Julien), 871, 901, 942.
RIES (Klaus), 5208.
RIGAUDIÈRE (Albert), 2823, 6156.
RIGAUX (Dominique), 272.
RIGBY (Stephen Henry), 2858.
RIGHETTI (Marina), 3207.
RIGO (Georges), 1428.
ŘÍHOVÁ (Romana), 7158.
RILEY (James C.), 6335.
RIMPAU (Laetitia), 240, 3076.
Rinaldeschi (Antonio), 4871.
RINALDI (Alberto), 5761.
RINALDI (Massimo), 5444.
RINDE (Harald), 5128.
RINEY-KEHRBERG (Pamela), 6336.
RINOLDI (Paolo), 3095.
RIOBÓ (Carmen), 1574.
RÍOS ZÚÑIGA (Rosalinda), 4317.
RIOTTE (Torsten), 6939.
RIOTTO (Maurizio), 7852.
RIOUX (Jean-Pierre), 5092.
RÎPEANU (B. T.), 5656.
RIQUELME SEGOVIA (Alfredo), 7010.
RITT (Nikolaus), 3078.
RITTER (Gerhard Albert), 742.
RITTER (Markus), 5589.
RITTER (Nils C.), 1366.
RIVA (Corinna), 1099.
RIVERO (Yeidi M.), 5701.
RIVOIRE (Jran-Baptiste), 3573.
RIVOLTELLA (Massimo), 2106.
RIZZA (Alfredo), 1144.
RIZZARDI (Clementina), 2513.
RIZZO (Francesco Paolo), 2355.
ROACH (Andrew), 3465.
ROBB (John G.), 1026.
ROBBEN (Antonius C. G. M.), 3604.
ROBBINS (Keith), 6687.
ROBBINS (Rossell Hope), 3056.
ROBERSON (Houston Bryan), 5001.
ROBERT (Jean-Louis), 6407.
ROBERTO (Sebastiano), 5590.

ROBERTO (Umberto), 2235.
ROBERTS (Anna), 3081.
ROBERTS (Brynley F.), 2952.
ROBERTS (Jane Annette), 7.
ROBERTS (Priscilla Mary), 635.
ROBERTS (Richard), 6598.
ROBERTSON (John), 5317.
ROBERTSON (Lindsay G.), 6520.
ROBERTSON (Stephen), 6337.
Robertson (William), 473.
ROBIN (Régine), 511.
ROBINSON (Armstead L.), 6338.
ROBINSON (Cynthia), 3082, 3108.
ROBINSON (Damian), 2174.
ROBINSON (Joseph Armitage), 2380.
ROBSON (Eleanor), 1245.
ROBSON (John), 239.
ROCA (José Luis), 3650.
ROCA W. (Demetrio), 581.
ROCCHI (Fernando), 5860.
ROCCHI (Maria), 1710.
ROCHAT (Giorgio), 7159.
Rochau (August Ludwig von), 6569.
ROCHE (Julien), 3466.
ROCHE (Pierre), 5231.
ROCHE-PÉZARD (Fanette), 193.
RÖCKE (Werner), 994.
RÖCKELEIN (Hedwig), 3186.
ROD (Yann), 2827.
RODGER (N. A. M.), 4082.
RODLER (Klaus), 3254.
RODNEY (George Brydges Rodney, Baron), 6896.
RODOGNO (Davide), 7160.
RODRIGUES (Jaime), 6729.
RODRÍGUEZ (Néstor), 6699.
RODRÍGUEZ ADRADOS (Francisco), 194, 1774.
RODRÍGUEZ ALCALÁ (Guido), 6819.
RODRÍGUEZ AZOGUE (Araceli), 1083.
RODRÍGUEZ CASTILLO (Héctor), 3390.
RODRÍGUEZ CRUZ (Agueda María), 5194.
RODRÍGUEZ GÓMEZ (María Dolores), 216.
RODRÍGUEZ GONZÁLEZ (Agustín Ramón), 6940.
RODRÍGUEZ O. (Jaime E.), 4300, 6845.
RODRÍGUEZ ROSALES (Isolda), 5244.
RODSETH (Lars), 584.
ROECK (Bernd), 679.
ROEDIGER (David R.), 6339.
ROEGIERS (Patrick), 273.
ROEMER (Kenneth M.), 978.

ROEMER (Nils H.), 418.
ROFFAT (Sébastien), 5702.
ROGER (Philippe), 5142.
ROGERS (Rebecca), 5245.
ROGERSON (Margaret), 3404.
ROGOV (Evgenij Ja.), 1462.
ROGOVAJA (L. A.), 4462.
RÖHL (Wilhelm), 735, 6483.
ROHLÍKOVÁ (Slavěna), IV.
ROHLS (Jan), 5009.
ROHR (Christian), 2618.
ROHRSCHNEIDER (Michael), 3986.
ROHT-ARRIAZA (Naomi), 6617.
ROIG MIRANDA (Marie), 176.
ROISMAN (Hanna M.), 1441.
ROISMAN (Joseph), 1618.
ROJAS (Beatriz), 6820.
ROKUTANDA (Yutaka), 7853.
ROLAND (Martin), 3212.
ROLANDSEN (Øystein), 4636.
ROLDÁN (Darío), 3600.
ROLDÁN (Diego), 6358.
ROLIN (Jan), 6521.
ROLIŃSKI (Adam), 4406.
ROLL (David L.), 4750.
ROLLER (Duane W.), 1522.
ROMAN (Christopher), 3467.
ROMANATO (Gianpaolo), 4855.
ROMANELLI (Rita), 288.
ROMANINI (Angiola Maria), 3207.
ROMANO (Angelo), 5140.
ROMANO (Antonella), 5124, 5384.
ROMANO (Giovanni), 853.
Romano (Ruggiero), 474.
ROMANOV (Vladimir V.), 4769.
Romanus III Argyros, Byzantine emperor, 2470.
ROMEO MATEO (María Cruz), 4622.
ROMERO (Federico), 6675, 7529.
ROMERO (María Teresa), 7531.
ROMERO SAMPER (Miligrosa), 4600.
ROMM (James), 1423.
ROMSICS (Ignac), 7086.
Romulus, king of Rome, 1881, 1898.
RONCALLI (Marco), 4854.
RONCHESE (Gino), 3484.
RONCHI (Gabriella), 3095.
RONCHI (Veronica), 5533.
RONSIN (Samuel), 3783.
ROOS (Neil), 4547.
Roosevelt (Franflin Delano), 4750, 4747, 5762, 5782, 7125, 7171, 7180, 7186, 7197, 7205, 7231.
Roosevelt (Theodore), 5755.
ROREM (Paul), 3304.
ROSA (Mario), 5143, 5344.
ROSAS (Juan Manuel José Domingo Ortiz de), 3578, 3593.

ROSAS (Ruth Magali), 4374.
ROSAS LAURO (Claudia), 4377.
ROSCALLA (Fabio), 1467.
ROSE (Charles Brian), 1934.
ROSE (Pamela), 1201.
ROSE (R. S.), 3676.
ROSE (Susan), 2665.
ROSEN (Elliott A.), 5762.
ROSEN (Ralph M.), 1619.
ROSENAU (James N.), 721.
Rosenberg (Alfred), 4003, 4865.
ROSENBERG (David A.), 7256.
ROSENBERG (Samuel N.), 3228.
ROSENBERG (Victor), 7532.
ROSENDAAL (J. G. M. M.), 4337.
ROSENTHAL (Yemima), 4189.
ROSENWEIN (Barbara H.), 2930.
ROSER (Hannes), 5591.
ROSETH (Roger), 3255.
ROSIER-CATACH (Irène), 3287.
ROSILLO LÓPEZ (Cristina), 2042.
ROSINI (Martina), 1044, 1051.
ROSIVACH (Vincent J.), 1746.
ROSKAM (Geert), 1674.
Roskin (G. I.), 4471.
ROSLER (Andrés), 1675.
ROSOLINO (Riccardo), 5919.
ROSS (Chad), 6340.
ROSS (Margaret Clunies), 3083.
ROSSBACH (Sabine), 5144.
RÖSSEL (Karl), 7169.
RÖSSEL (Uta), 3912.
ROSSI (Elena), 1827.
ROSSI (Filli), 3493.
ROSSI (Guido), 1050.
ROSSI (Mariaclara), 3321.
ROSSI BELLOTTO (Carla), 3084.
ROSSIGNOLI (Benedetta), 1711.
Rossini (Gioacchino), 5680.
ROSSMAN (Jeffrey J.), 4500.
ROSSO (Maxime), 5318.
ROSSO (Paolo), 5195.
ROSSY (Michel), 1050.
ROST (Leonhard), 2274.
ROTARIU (Traian), 6291.
ROTH (Norman), 2689.
ROTH (Ralf), 5861, 6341.
ROTH (Ulrike), 2043.
ROTHENHÖFER (Peter), 2044.
ROTHERMUND (Dietmar), 4147.
Rothfels (Hans), 475.
ROTHMAN (Adam), 6342.
ROTHSCHILD (Joan), 5445.
ROTHSTEIN (Bret Louis), 3208.
ROTHWELL (Victor), 7284.
ROTMAN (Youval), 2715.
Rotteck (Karl von), 377.
ROTUNDO (Donatella), 456.
ROUAULT (Olivier), 1145.
ROUCHON-MOUILLERON (Véronique), 2564, 3485.

ROUET (Dominique), 3391.
ROUGÉ (Jean), 1985.
ROUHI (Leyla), 3108.
ROULIN (Jean-Marie), 5500.
ROUSE (Robert Allen), 3085.
ROUSSEAU (Frédéric), 4084.
Rousseau (Jean-Jacques), 1004, 1431, 5350.
ROUSSEAU (Vanessa), 943.
ROUSSELET-PIMONT (Anne), 6545.
ROUSSILLON (Alain), 243.
ROUSSIN (Philippe), 5532.
ROUSSOT (Thomas), 1935.
ROUX (Guy), 3209.
ROUX (Rhina), 4319.
ROWE (John Allen), 290.
ROWE (M. W.), 1072.
ROWE (William), 265.
ROYAN (Nicola), 3089.
ROŽAC-DAROVEC (Vida), 573.
ROZWADOWSKI (Helen M.), 244.
RUANE (Kevin), 7420.
RUANO-BORBALAN (Jean-Claude), 5253.
Rubens (Peter Paul), 5603.
RUBERG (Willemijn), 5145.
RUBIN (Aaron D.), 1321.
RUBIN (Anne Sarah), 4770.
RUBIN (Barry M.), 7493.
RUBIN (Michael), 4159.
RUBINCAM (Catherine), 1620, 1819.
RUBINSTEIN (Lene), 1503.
RUBIO (Gonzalo), 1322.
RUBLACK (Ulinka), 5002.
RÜBNER (Hartmut), 5862.
RUCH (Christian), 7215.
RÜCHEL (Uta), 7331.
RUCHNIEWICZ (Krzysztof), 5246, 6689, 7510.
RÜDEN (Peter von), 5664.
RUDERER (Stephan), 4957.
RÜDIGER (Axel), 718.
RUDOLPH (Ulrich), 3286.
RUDWICK (Martin J. S.), 103.
RUDY (Jarrett), 6343.
RUEDA (Carmen), 1070.
RUEDA RAMÍREZ (Pedro J.), 5299.
Ruegger (Paul), 7206.
RUFF (Mark Edward), 4899.
RUFFINI (Marco), 5446.
RUGE (Daniela), 309.
RUGGIERI (Vincenzo), 2495.
RUGGINI (Cristiana), 1058.
RUIZ (Julius), 4601.
RUIZ (Teofilo F.), 2604.
RUIZ CARNICER (Miguel Angel), 4623.
RUIZ DE ARBULO (Joaquín), 1055.
RUIZ IBÁÑEZ (José Javier), 4876.
RUIZ MONTERO (Consuelo), 1621.

RUIZ PÉREZ (Angel), 1712.
RUIZ PORTELLA (Javier), 4590.
RUIZ SALGUERO (Magda), 5447.
RUIZ ZAPATA (Blanca), 1030.
RUIZ-GÁLVEZ (Marisa), 1073.
RÚJULA LÓPEZ (Pedro), 4624.
RUMBLE (Alexander R.), 4029.
RUMENJAK (Nives), 3741.
RUNBLOM (Harald), 6684.
RUNDEL (Tobias), 1986.
RÜNITZ (Lone), 3804.
RUOTSILA (Markku), 7226.
RÜPKE (Jörg), 2132.
RUPPRECHT (Hans-Albert), 1195, 1504.
RUSAN (Romulus), 4435.
RUSCHENBUSCH (Eberhard), 1505.
RUSCIO (Alain), 6754.
RUSCONI (Gian Enrico), 7088, 7341.
RÜSEN (Jörn), 380.
RUSJAEVA (Anna S.), 1713.
RUSSELL (Andrew T.), 4083.
Russell (Dora), 6418.
Russell (John), 4086.
RUSSO (Daniel), 3210.
RUSSO (Flavio), 1936.
RUSSO (Luigi), 2645.
RUSSO (Mariagrazia), 6895.
RUSTOIU (Aurel), 1920, 2002.
RÜTH (Axel), 527.
RUTHERFORD (Ian), 873.
RUTHERFORD (Richard), 1384.
Rutilius Namatianus (Claudius), 1839, 2362.
RUTOWSKA (Maria), 7112.
RUXTON (Ian C.), 6947.
RUZHITSKAJA (Irina V.), 4501.
RYAN (Terry), 1377.
RYBACHENOK (Irina S.), 6989.
RYBAKOV (Rostislav B.), 7562.
RYBALKO (Natal'ja V.), 4506.
RYBÁŘOVÁ (Petra), 4531.
RYCHNER (Jacques), 5298.
RYCRAFT (Ann), 3432.
RYDELL (Robert W.), 3558.
RYDGREN (Jens), 4651.
RYE OLSEN (Gorm), 7533.
RYGIEL (Philippe), 6112.
RYRIE (Alec), 5003.
RZIHACEK-BEDÖ (Andrea), 3305.

S

SAALER (Sven), 528.
SABBAH (Guy), 2250.
SABIN (Paul), 5863.
SABOL (Miroslav), 5864.
SACCHETTI (Franco), 2972.
SACCHETTI (Giannozzo), 2973.
SACCHI (Osvaldo), 1987.
SACCONI (Anna), 1388.

SACEWICZ (Karol), 7293.
SACKS (David), 680.
SACRISTÁN (Cristina), 4287.
SACRISTE (Guillaume), 6497.
SADGROVE (Philip C.), 74.
SAEKI (Arikiyo), 7827.
Sáenz Peña (Roque), 3603.
SÁENZ ROVNER (Eduardo), 3748.
SÁENZ-CAMBRA (Concepción), 6891.
SAETHER (Steinar A.), 3723, 6846.
SÁEZ (Lawrence), 4138.
SAFLEY (Thomas Max), 6344.
SAFRAI (Zeev), 2228.
SAGARDOY (Teresa), 1010.
SAGAWA (Eiji), 7714.
SAGNES (Jean), 7135.
ŞAHIN (Mustafa), 1146.
SAIDEL (Benjamin Adam), 1323.
SAINT-HILAIRE (Janine Cels), 1937.
SAINT-MARTIN (Denis), 6163.
Saint-Pierre (Charles Irénée Castel de), 3840.
Saint-Simon (Henri, comte de), 5351.
SAK (İzzet), 6599.
SAKAI (Kimi), 7836.
SAKHAROV (Andrej N.), 7161, 7354.
SALAIS (Robert), 5777.
SĂLĂGEAN (Tudor), 3495.
SALAÜN (Marie), 5247.
Salazar (António de Oliveira), 4423, 7310.
SALAZAR (Christine F.), 608.
SALAZAR VERGARA (Gabriel), 3710.
SALDERN (Adelheid von), 6081.
SALE (Giovanni), 7421.
SALEM (Jean), 5354.
SALEMME (Carmelo), 2107.
SALERNO (Beth A.), 6345.
SALEWSKI (Michael), 4008.
SALIBA (Fabrice), 4084.
SALIH (Kamal O.), 6805.
Salih Zeki, 5398.
Salisbury (Robert Arthur Talbot Gascoyne-Cecil, 3rd Marquess of), 6968.
SALLES (Vicente), 6346.
SALLMANN (Line), 6186.
Sallustius Crispus (Gaius), 1865, 1871.
SALM (Steven J.), 7858.
SALMI (Hannu), 5704.
SALMON (Philip), 4079.
SALMOND (John A.), 6459.
SALMONI (Barak A.), 3827.
SALOMÓN CHÉLIZ (María del Pilar), 4822.

Salomon, king of Israel, 2219.
SALONEN (Kirsi), 3361.
SALVADORI (Emanuela), 2983.
SALVATICI (Silvia), 554.
SALVATORE (Nick), 5004.
SALVATORE (Ricardo), 6712.
Salvatorelli (Luigi), 476.
SALVUCCI (Richard), 5763.
SALY-GIOCANTI (Frédéric), 529.
SAMARANI (Guido), 352, 7227.
SAMB (Djibril), 4515.
ŠAMBERGER (Zdeněk), 7089, 7228.
SAMERSKI (Stefan), 4863.
SAMMOND (Nicholas), 5705.
SAMUELI (Jean-Jacques), 5448.
SAMUELS (Albert L.), 5248.
SAN FRANCISCO (Alejandro), 3702.
SANBORN (Joshua A.), 6347.
SÁNCHEZ (Alberto), 1070.
SÁNCHEZ DÍAZ (Gerardo), 6855.
SÁNCHEZ GARCÍA (Raquel), 4602.
SÁNCHEZ LÓPEZ (Francisco), 6531, 6543.
SÁNCHEZ MARTÍNEZ (Manuel), 2823.
SÁNCHEZ MOLINA (Ángel), 1060.
SÁNCHEZ RECIO (Glicerio), 4625.
SÁNCHEZ ROMÁN (José Antonio), 5865.
Sancho VII, "el Fuerte", rey de Navarra, 45.
SANDAGE (Scott A.), 6348.
SANDERS (Charles W. Jr.), 4772.
SANDERS (Peter), 6799.
SANDU (Traian), 7055.
SANFELIÚ (Luz), 4603.
SANFILIPPO (Matteo), 4945, 6126.
SANFORD (George), 7162.
SANG (Bing), 7715.
SÄNGER (P.), 2496.
SANGERMANO (Gerardo), 3291.
SANKEY (Margaret), 4085.
SANMARTÍN (Joaquín), 1324.
SANNIBALE (Maurizio), 1020.
SANSARIDOU-HENDRICKX (Thekla), 2497.
Santa Lucia da Foligno, 3377.
SANTACARA (Carlos), 6917.
SANTA-CRUZ (Arturo), 4381.
SANTAMARIA (Yves), 3889.
SANTAMARÍA GARCÍA (Antonio), 593.
SANTANACH I SUÑOL (Joan), 3252.
SANTANGELO (Federico), 2145.
SANTI (Isabel), 3600.
SANTIEMMA (Adriano), 944.
SANTORO (Marco), 66, 100.
SANTORO (Stefano), 5147.
SANTOS (Eduardo), 3718.
SANTOS (Hermínio), 4417.
SANTUCCI (Antonio A.), 5331.

SANTUCCI (Marco), 1523.
SAO (Qichuang), 7716.
SAPELLI (Giulio), 5533.
Sapieha (Jan), 4506.
Sappho, 1543, 1562.
SAPRYKIN (Sergej Ju.), 147.
SARAGUSTI (Idit), 1747.
SARAMANDU (Nicolae), 162.
SARDSHWELADSE (Surab), 196.
SARGENT (Michael G.), 3426.
SARILA (Narendra Singh), 4148.
SARKAR (Sumit), 419.
SARMIENTO GUTIÉRREZ (Julio), 6847.
Sarraut (Albert), 5158.
SARSON (Steven), 6848.
SARTI (Lucia), 1044, 1051.
SARTI (Raffaella), 6189.
SARTORE (Alberto Maria), 2556.
SARTORI (Marco), 1867.
SARTRE (Maurice), 1468.
SASAKI (Junnosuke), 7837.
Sasaki (Takayuki), 4241.
SASAKI (Takeshi), 682.
ŠAŠEL KOS (Marjeta), 1799.
SATLOW (Micheal L.), 945.
SATŌ (Hiroshi), 7594.
SATO (Hitoshi), 7717.
SATO (Makoto), 7835.
SATO (Shoichi), 2600.
Sato (Sojun), 7838.
SATO (Yoshiaki), 3483.
SATOW (Ernest Mason, Sir), 6947.
Sattler (Dieter), 5157.
SAÚD (S.), 1622.
SAUL (Nigel), 2610, 3217.
SAUNDERS (Robert), 4086.
SAUNT (Claudio), 126.
SAURON (Gilbert), 1903.
SAURON (Gilles), 2187.
SAUVAGE (Jean-Christophe), 7422.
SAUVY (Alfred), 5333.
SAVAGE (Stephen H.), 1325.
SAVINO (Eliodoro), 2207.
SAVOIE (Philippe), 5234.
SAVONA (Paolo), 5785.
SAVORELLI (Alessandro), 5141.
SAVY (Pierre), 2672.
SAXON (Gerald D.), 231.
SAXONBERG (Steven), 4127.
SAYAR (Ahmed Güner), 766.
SAYHI-PÉRIGOT (Béatrice), 968.
SAYRE (Gordon M.), 7886.
SCAFURO (Adele C.), 1506.
SCAHILL (John), 3100, 3371, 3404.
SCALES (Len), 716.
SCÀNDOLA (Mario), 1827.
SCARAMELLA (Pierroberto), 4929.
SCARTON (Elisabetta), 2530.
SCHAAL (Dirk), 5866.
SCHAARSCHMIDT (Rebecca), 5174.

SCHABAS (Margaret), 5786.
SCHABEL (Chris), 2624.
SCHÄFER (Christoph), 592.
SCHÄFER (Daniel), 5169.
SCHÄFER (Hans-Michael), 298.
SCHÄFER (Jörgen), 827.
SCHAFFER (Matt), 6806.
SCHAFFER (Simon), 5449.
SCHAFFRODT (Petra), 4957.
SCHALLER (Dieter), 2948.
SCHAMP (Jacques), 2498.
SCHAPIRA (Marie-Claude), 5529.
SCHAUB (Jean-Frédéric), 6488.
SCHAUFELBUEHL (Janick Marina), 6775.
SCHAUSBERGER (Franz), 3631.
SCHECK (Raffael), 7163.
SCHEDROVITSKIJ (Dmitrij V.), 1326.
SCHEID (John), 2146.
SCHEIDE (Carmen), 5884.
SCHEIDEL (Walter), 2045.
SCHEIDELER (Britta), 5450.
SCHEIN (Sylvia), 2646.
SCHELLER (Benjamin), 486.
SCHENDL (Herbert), 3078.
SCHENKE (Ludger), 2263.
SCHENKEL (Wolfgang), 1196.
SCHENNACH (Martin P.), 3620.
SCHEPENS (Guido), 1469.
SCHEPERS (Heinrich), 5342.
SCHEPERS (Judith), 4898.
SCHERSTJANOI (Elke), 7243.
SCHEUBLE (Robert), 3088.
SCHIAVINATTO (Iara Lis), 6849.
SCHIAVONE (Aldo), 743.
SCHICKEL (Matthias), 7090.
SCHIERA (Pierangelo), 5388.
SCHIERSNER (Dietmar), 4900.
SCHIFFRIN (André), 5516.
SCHIFFRIN (Jacques), 5516.
SCHILBRACK (Kevin), 946.
SCHILDBERGER (Vlastimil), 7276.
SCHILDT (Axel), 6065.
Schiller (Friedrich), 5352.
SCHIMANOWSKI (Gottfried), 1327.
SCHINKEL (Anders), 530.
SCHINKEL (Eckhard), 7234.
SCHIPANI (Sandro), 2400.
SCHIPPER (Friedrich T.), 316.
SCHLANGE-SCHÖNINGEN (Heinrich), 471, 1938.
SCHLARP (Karl-Heinz), 5764.
SCHLEGEL-VOß (Lil-Christine), 6349.
SCHLEIER (Hans), 448.
SCHLEIMER (Ute), 5259.
SCHLEMMER (Thomas), 4229, 7245.
SCHLESINGER (Arthur M.), 7180.

SCHLICHT (Markus), 3211.
SCHLIENGER (Micheline), 7356.
SCHLOESSER (Stephen), 4823.
SCHLÖGL (Rudolf), 814.
SCHLOSSER (Horst Dieter), 197.
SCHLOTHEUBER (Eva), 2931.
SCHLUCHTER (Wolfgang), 486, 767.
SCHLUEP (Christoph), 2357.
SCHLUP (Michel), 5298.
SCHLÜTER (Theodor C.), 5005.
SCHMAL (Stephan), 478.
SCHMEISER (Martin), 5207.
SCHMID (Alfred), 2147.
SCHMID (Josef), 3959.
SCHMIDT (Alexander), 5148.
SCHMIDT (Andrea), 2304.
SCHMIDT (Arno), 1676.
SCHMIDT (C. Bettina), 6350.
SCHMIDT (Elizabeth), 4109.
SCHMIDT (Francis), 899.
SCHMIDT (Gerhard), 3212.
SCHMIDT (Hans-Joachim), 3134.
SCHMIDT (Helmut), 3948, 7553.
SCHMIDT (Joël), 1939.
SCHMIDT (Manfred G.), 1786.
SCHMIDT (Michael), 1940, 2617, 5298.
SCHMIDT (Victor Michael), 3213.
SCHMIDT (Wilhelm R.), 970.
SCHMIDT-BURKHARDT (Astrit), 5535.
SCHMIDT-GLINTZER (Helwig), 380.
SCHMIDTMANN (Christian), 6351.
SCHMIDT-NOWARA (Christopher), 6721.
SCHMIEDER (Felicitas), 3509.
SCHMITT (Carl), 5334, 5353, 6498-6501.
SCHMITT (Hatto H.), 681.
SCHMITT (Jean Claude), 2902.
SCHMITZ (David F.), 7534.
SCHMITZ (Michael), 420.
SCHMITZ (Walter), 233.
SCHMITZER (Ulrich), 1941.
SCHMOECKEL (Mathias), 744.
SCHMUHL (Hans-Walter), 583.
SCHMUNDT (Hilmar), 208.
SCHNEEWIND (Jerome B.), 383.
SCHNEID (Frederick C.), 6941.
SCHNEIDER (Alejandro), 3605.
SCHNEIDER (Bernhard), 3336.
SCHNEIDER (Carolyn), 2213.
SCHNEIDER (Christina), 6493.
SCHNEIDER (Helmuth), 608.
SCHNEIDER (Herbert), 3239.
SCHNEIDER (Michael C.), 5867.
SCHNEIDER (Ulrich Johannes), 383, 5105.
SCHNEIDER (Ute), 5300.

SCHNEIDER-WATERBERG (H. R.), 6807.
SCHOBER (Richard), 3620, 3636.
SCHOCHOW (Werner), 299.
SCHOEN (Johanna), 6352.
Scholem (Gershom), 925.
SCHÖLLGEN (Gregor), 6691.
SCHOLLMEYER (Patrick), 2184.
SCHOLTEN (Clemens), 2398.
SCHOLTEN (Helga), 1246.
SCHOLTYSECK (Joachim), 4017.
SCHOLZ (Stephan), 4901.
SCHÖN (Amalie), 3914.
SCHÖN (Theodor), 3914.
Schönborn, Familie, 127.
SCHÖNE (Jens), 5962.
SCHÖNHOVEN (Klaus), 6430.
Schopenhauer (Arthur), 958.
SCHÖRNER (Günther), 1930.
SCHOTT (Christian-Erdmann), 4986.
SCHRATZ (Sabine), 4898.
SCHRAUT (Sylvia), 127.
SCHREIBER (Gebhard J.), 1247.
SCHREINER (Peter), 2384.
SCHREPFER (Susan R.), 768.
SCHRIJVERS (Peter), 7285.
SCHRÖDER (Christian), 2969.
SCHRÖDER (Gisela A.), 4954.
SCHRÖDER (Hartmut), 7243.
SCHRÖDER (Iris), 250.
SCHRÖDER (Stephen), 4017.
SCHUBERT (Charlotte), 1677.
SCHUBERT (Martin J.), 3001.
SCHUBERT (Stefan), 6353.
SCHUBERT (Werner), 5050, 6600.
SCHÜBL (Almar), 5196.
SCHUCHARD (Christiane), 6355.
SCHÜLE (Klaus), 3890.
SCHULIN (Ernst), 531.
SCHULTE (Hendrich), 2403.
SCHULTE (Jan Erik), 3975.
SCHULTZ HANSEN (Hans), 6354.
SCHULTZE (Andrea), 4946.
SCHULZ (Andreas), 815.
SCHULZ (Günther), 428, 4954.
SCHULZ (Knut), 6355.
SCHULZ (Matthias), 5451.
SCHULZ (Rainer), 5794.
SCHULZE (Christian), 2358.
SCHULZE (Thies), 5149.
SCHÜMANN (Nicola), 3518.
SCHÜRER (Markus), 3392.
SCHURIG (Sebastian), 2222.
SCHÜRMANN (Sandra), 6108.
SCHUSTER (Klaus), 6356.
SCHUSTER KESWANI (Priscilla), 1071.
SCHÜTRUMPF (Eckart), 1422.
SCHÜTTPELZ (Erhard), 999.
SCHÜTZ (Erhard), 372.

SCHUWER (Philippe), 65.
SCHWAB (Katherine A.), 1748.
SCHWABE (Astrid), 3936.
SCHWABE (Klaus), 7405.
SCHWAETZER (Harald), 3289.
SCHWANNINGER (Florian), 3632.
SCHWARTZ (Dov), 2690.
SCHWARTZ (Michael), 6367.
SCHWARZ (Karl W.), 5006.
SCHWARZ (Karl), 6674.
SCHWARZ (Peter), 3629.
SCHWARZBAUER (Fabian), 104.
SCHWEDT (Hernan H.), 4898.
SCHWEGEL (Andreas), 4009.
SCHWEIGARD (Jörg), 6266.
SCHWEIZER (Jürg), 2185.
SCHWENNINGER (Jean-Luc), 1069.
SCHWERDTFEGER (Regina Elisabeth), 4925.
SCHWERHOFF (Gerd), 3469.
SCHWIEGER (Christopher), 6546.
SCHWINDT (Jürgen Paul), 2105.
SCHWOB (Anton), 54.
SCIARRA (Elisabetta), 1623.
Scipio Aemilianus (Publius Cornelius), 1862.
Scipio Africanus (Publius Cornelius), 1917.
Scipio Barbatus, 1782.
SCIUTI RUSSI (Vittorio), 4902.
SCODEL (Ruth), 1624.
SCORDIA (Lydwine), 2859.
SCOT (Marie), 5198.
SCOTT (Allen J.), 5706.
SCOTT (Jamie S.), 4932.
SCOTT (Rebecca J.), 6357.
SCOTT (Roy V.), 5958.
SCOTTI (Francesca), 2401.
SCOTT-SMITH (Giles), 7535.
SCRIVANI-TIDD (Lisa), 838.
SCUCCIMARRA (Luca), 532.
SCULLION (Scott), 1385, 1714.
SCULLY (Pamela), 6217.
SCURLOCK (Jo Ann), 1248.
SEAFORD (Richard), 1715.
SEAGER (Robin), 1942.
SEALE (Patrick), 6623.
SEARLE (Alaric), 4010.
SEBALD (Peter), 7857.
SEBASTIÀ (Enric), 4626.
SEBILLOTTE CUCHET (Violaine), 1716.
SECK (Assane), 4515.
SECK (Mamadou), 4516.
SECONDY (Philippe), 3891.
SEDDON (Keith), 1426.
SEDGWICK (John), 5707.
SEDLIAKOVÁ (Alžbeta), X.
SEDOV (Aleksandr V.), 1328, 2694.
SEED (John), 466.
SEEGER (Tatyana), 5478.

SEEGERS (Lu), 6081.
SEEHER (Jürgen), 1269.
SEELBACH (Ulrich), 2977.
SEEMANN (Christoph), 3912.
SEESKIN (Kenneth), 2681, 2691.
SEGAL (Charles), 1800.
SEGAL (Howard P.), 5868.
SEGENNI (Simonetta), 1792.
SEGERS (Yves), 5920.
SEGEŠ (Dušan), 4527.
SEGEŠ (Vladimír), 2860.
SEGGERN (Jessica von), 4011.
SEGID (Amha), 1033.
SÉGINGER (Gisèle), 5334.
Segni (Lothar von), 3362.
SEGUIN (Maria Susana), 4999.
SEGURA (Mauricio), 5150.
SEGURA URRA (Félix), 2770.
SEHLMEYER (Markus), 465.
SEIBEL (Wolfgang), 3997.
SEIBERT (Hubertus), 2602, 2630.
SEIDENSTICKER (Bernd), 1625.
SEIDERER (Georg), 7133.
SEIFFERT (Wolfgang), 3957.
SEIGEL (Jerrold), 5319.
SEITA (Mario), 2108.
SEITTER (Walter), 2390.
SEITZER (Jeffrey), 6501.
SEKIGUCHI (Takehiko), 3393.
SELBMANN (Rolf), 5498.
SELDEN (Mark), 3713.
SELDERHUIS (Herman J.), 4988.
Seleukos of Seleukeia, astronomer, 1522.
SELF (Robert), 4035.
SELIGMAN (Amanda I.), 6109.
SELTMANN (Ingeborg), 3215.
SELUNSKAIA (Natal'ia), 4507.
SÉMELIN (Jacques), 719.
SEMENIUC (Maria), 6291.
SEMERANO (Giovanni), 198.
SEMERANO (Grazia), 1147.
SEMERDŽIEV (Petăr), 7091.
SEMPRUN (Jorge), 7303.
Seneca (Lucius Annaeus), 1840, 1841, 2070, 2094, 2121.
SENJAVSKIJ (Aleksandr S.), 4465.
SENS (Alexander), 1626.
SENS (Angelie), 5264.
SENSI (Mario), 4903.
ŞENTOP (Mustafa), 6601.
SEPE (Crescenzio), 4933.
Seqenenrê, king of Egypt, 1188.
SERDON (Valérie), 3130.
SÈRE (Bénédicte), 2933.
SÉRÉ (Daniel), 6911.
SERESSE (Volker), 4012.
SERIN (Ufuk), 2499.
SERKOV (Andrej I.), 4481.
SERNA (Pierre), 3892.
Serra (Renato), 5504, 5505.

SERVET (Miguel), 5007.
SERVICE (Robert), 4502.
SESÉ ALEGRE (José María), 4374.
Sestan (Ernesto), 477.
SETTESOLDI (Laura), 245.
SÉVENET (Jacques), 4904.
SÉVÉRAC (Pascal), 5355.
SEVERANCE (Ben H.), 4773.
Severus Alexandrinus, 2413.
SEVILLA SOLER (María Rosario), 5301.
SEVILLANO CALERO (Francisco), 4627.
SEVOST'JANOV (Grigorij N.), 4496.
SEWELL (William H. Jr.), 533.
SEXTON (Jay), 6990.
SEYDI (Süleyman), 7229.
Seydlitz-Kurzbach (Walther von), 3925.
SFEIR-KHAYAT (Jihane), 421.
SFETAS (Spyridon), 6991.
SGRAZZUTTI (Jorge P.), 6358.
SHA'BĀN (al-Sādiq), 4674.
Shackleton (Robert), 5347.
SHAFER (David A.), 3893.
SHAFFER (Kirwin R.), 3749.
SHAGAN (Ethan), 4967.
SHAHID (Kamran), 4149.
SHAHNAVAZ (Shahbaz), 6992.
Shakespeare (William), 5484, 5494, 5496.
SHALOM (Zakai), 4194.
SHAMMAS (Carole), 6827.
SHANAHAN (Rodger), 4272.
SHAPIRA (Dan), 2685.
SHAPIRO (Harold T.), 5199.
SHAPIRO (Linn), 523.
SHARE (Michael John), 2399.
SHARMA (Shalini), 4150.
SHARON (Ilan), 1747.
SHARP (Alan), 7092.
SHAUF (Scott), 2359.
SHAW (David Gary), 2861.
SHAW (Ian), 1197.
SHAW (John M.), 7536.
SHAW (Joseph W.), 1400.
SHAW (Mary C.), 1400.
SHAYA (Josephine), 1749.
SHCHAPOV (Jaroslav N.), 2752.
SHCHELCHKOV (Andrej A.), 3655.
SHCHERBAKOVA (E. I.), 3488.
Sheikh Said, 6962.
SHELDON (Kathleen E.), 769.
SHELDON (Rose Mary), 1944.
SHELL (Marc), 5453.
Shen (Chong), 7800.
SHEN (Fuwei), 7718.
SHEN (Zhihua), 7423.
SHENG (Xunchang), 7694.
SHENG (Yanghong), 7694.
SHENK (Gerald E.), 4774.

SHEPHERD (John J.), 947.
Shepilov (Dmitry), 4485.
SHEPPARD (Simon), 1945.
SHERRY (Vincent B.), 980.
SHI (Jinpo), 7719.
SHI (Qinghua), 289.
SHI (Yantao), 7750.
Shidehara (Kijuro), 6702.
SHILO (Margalit), 5041.
SHIMIZU (Kazuhiro), 7576.
SHIMIZU (Yuichiro), 4244.
SHIPPER (Apichai W.), 6359.
SHIRAISHI (Taichiro), 7833.
SHMIDT (Sigurd O.), 344, 422, 2520.
SHOJU (Keitaro), 2500.
SHORTER (Edward), 817.
SHOWALTER (Dennis), 7131.
SHPUZA (Gazmend), 3571.
SHUBIN (Aleksandr V.), 4503.
SHUVALOV (Petr V.), 2414.
SIANI-DAVIES (Peter), 4455.
SIARI-TENGOUR (Ouanassa), 444.
SIBANDA (Eliakim M.), 4807.
SIBILIO (Vito), 2501.
SICARD (Germain), 5211.
SICK (David H.), 2360.
SIDDALI (Silvana R.), 6360.
SIDEBOTTOM (Harry), 1946.
SIDER (David), 1627.
SIDERAS (Alexander), 2393, 2502.
Sidor (Karol), 4527.
SIDOROVICH (Ol'ga V.), 423.
SIDWELL (Keith C.), 195.
SIEBEN (Hermann Josef), 3335.
SIEBER (Dominik), 4947.
SIEBERS (Winfried), 241.
SIEBERTZ (Roman), 4166.
SIEG (Dirk), 6993.
SIEGENTHALER (Hansjörg), 717.
SIEGFRIED (Detlef), 6065.
SIELEZIN (Jan Ryszard), 4405.
SIENA (Kevin), 5456.
SIER (Kurt), 2110.
SIEVENS (Mary Beth), 6361.
SIEVERS (Joseph), 1299.
SIEWERT (Peter), 1375, 1470.
Sigismund III, king of Poland and Sweden, 4650.
SIGNES CODOÑER (Juan), 5057.
SIGNOLES (Aude), 424.
SIGNORI (Gabriela), 3216.
SILBER (Nina), 4775.
Silius Italicus (Tiberius Catius Asconius), 1917.
SILLIÈRES (Pierre), 1856.
Silone (Ignazio), 5154.
SILTALA (Juha), 3834.
SILVAS (Anna M.), 2215.
SILVESTRINI (Mara), 1051.
SILVESTRINI (Marina), 2046.

SIMÁNDI (Irén), 4128.
SIMBULAN (Dante C.), 4384.
SIMIS (Anita), 5638.
SIMKINS (Michael), 1988.
SIMMS (Anngret), 225.
SIMMS (R. Clinton), 1628.
SIMÕES (Alberto da Veiga), 7178.
SIMON (Anne), 3102.
SIMON (Bryant), 6362.
SIMON (Jacques), 3574.
SIMON (Jonathan), 5454.
Simon Arnauld de Pomponne, 6909.
Simon Magus, 3445.
ŠIMONČIĆ-BOBETKO (Zdenka), 5869.
SIMONIS (Francis), 6761.
SIMONS (Roswitha), 2361.
SIMONSEN (Eva), 6197.
SIMPSON (Christopher J.), 1947, 2186.
SIMPSON (James), 6163.
ŠIMŮNEK (Michal), 5455.
Sin-ah-usur, 1239.
SINDBÆK (Søren Michael), 2727.
SINGH (Karminder Dhillon), 7537.
SINGH (Nirula), 6692.
SINKOFF (Nancy), 5048.
SIPOS (Balázs), 4126.
SIPPEL (Harald), 6473.
SIRAISI (Nancy G.), 379.
SIRAT (Colette), 3256.
SIRINELLI (Jean-François), 683, 3839, 3877, 3896, 425, 5092, 5184, 6363.
SIRKECI (İbrahim), 6364.
SITTON (Thad), 5963.
SITTON (Tom), 6110.
SIVASUNDARAM (Sujit), 4948.
Sixtus IV, Papa, 3188.
SKACH (Cindy), 6522.
SKAGESTAD (Peter), 535.
SKÁLOVÁ (Ivana), 7538.
SKEATES (Robin), 1022.
SKINNER (Marilyn B.), 1386.
SKINNER (S. A.), 5008.
SKOBELEV (Sergej G.), 7580.
SKÖLD (Peter), 560.
SKOPAL (Pavel), 5677.
SKOU (Kaare R.), 3805.
SKUNCKE (Marie-Christine), 5289.
Skylitzes Matritensis, 2491.
SLABOTINSKÝ (Radek), 6111.
Slánský (Rudolf), 3782.
SLAVAZZI (Fabrizio), 2151.
SLAVIN (Michael), 48.
SLAWSON (Douglas J.), 5249.
ŚLESZYŃSKI (Wojciech), 4387.
SLEZÁK (Lubomír), 6056.
SLINN (Judy), 6163.
SLOANE (Barney), 3381.

SLOCUM (Kay Brainerd), 2935.
SŁOŃ (Marek), 2862.
SLOT (B. J.), 4263.
SLOTKIN (Richard), 4776.
SLUYTERMAN (Keetie E.), 5766.
SLYSTSCHENKOW (Wladimir A.), 6602.
SMAIL (Dan), 426.
SMALL (Jocelyn Penny), 1629.
SMALL (Melvin), 4777.
SMEETS (Jos), 4338.
SMETANA (Vít), 7230.
SMIL (Vaclav), 5457.
SMILANSKY (Uzy), 1062, 1747.
SMILJANSKAJA (Irina M.), 2694.
SMIRNOV (Valerij V.), 4464.
SMIRNOVA (Ol'ga P.), 1108.
SMITH (Angela K.), 4088.
SMITH (Chris), 838.
SMITH (Crosbie), 5420.
SMITH (David Lovatt), 4257.
SMITH (David M.), 2539.
SMITH (David Michael), 3345.
SMITH (Frederick H.), 5767.
SMITH (Harold Eugene), 685.
SMITH (Helen), 1069.
SMITH (Jay M.), 3894.
SMITH (Jeremy), 4504.
SMITH (John David), 4725, 6453.
SMITH (Joseph), 6693.
SMITH (Julia M. H.), 2936.
SMITH (Justin Davis), 6460.
SMITH (Louis J.), 7449.
SMITH (Marc), 1198.
SMITH (Nicolas D.), 1643, 1678.
SMITH (Paul), 3895.
SMITH (Philippa Mein), 686.
SMITH (Robert Douglas), 3132.
SMITH (Stephen), 6787.
SMITH (Susan L.), 6365.
SMOLINSKY (Heribert), 3336.
SMOUT (T. Christopher), 4039, 5964.
SMYSHLJAEV (Aleksandr L.), 1108.
SNEDDON (Shelagh), 2746.
SNELL (Daniel C.), 617.
SNOW (Joseph Thomas), 3004.
SOARES (Benjamin F.), 4283.
SOAVE (Sergio), 5154.
SOBĚHART (Radek), 5720.
Socrates, 1481, 1643, 1644, 1649, 1651, 1656, 1661, 1663, 1678, 1679.
SODE-MADSEN (Hans), 7164.
SODEN (Iain), 3510.
SODINI (Jean-Pierre), 2514.
Soekarno, 4154, 4155.
SOERGEL (Philip M.), 2934.
SÖĞÜT (Bilal), 1148.
ŠOKČEVIĆ (Dinko), 3738.
SOKOLOFF (Kenneth L.), 6530.

SOKOLOV (Andrej K.), 5841.
SOKOLOVA (Zoja P.), 576.
SOL (Thierry), 720.
SOL GOLDSTEIN (Cora), 7424.
SOLAK (Zbigniew), 4406.
SOLANO CHUQUIMIA (Franz), 3656.
SOLAÚN (Mauricio), 7539.
SOLCAN (Şarolta), 2863.
SÖLCH (Brigitte), 5391.
SOLDANI (Simonetta), 5242.
SOLEIL (Sylvain), 6494.
SOLER (Joëlle), 2362.
SOLÈRE (Jean-Luc), 3259.
SOLETTI (Elisabetta), 178.
SOLIMANO (Giannina), 1810.
SOLIS (Julia), 6113.
SOLL (Jacob), 427.
SÖLLNER (Alfred), 745.
Solon, 1440, 1499.
SOLÓRZANO TELECHEA (Jesús Ángel), 3491.
SOLURI (John), 5965.
SOMMA (Alessandro), 6495, 6538.
SOMMACAL (Matteo), 1021.
SOMMELLA (Valentina), 7231.
SOMMER (Michael), 1149.
SOMMERSTEIN (Alan H.), 1717.
SOMVILLE (Pierre), 1630.
SØNDERGAARD (Leif), 3049.
Sonnino (Giorgio Sidney), 6981.
SÖNTGEN (Beate), 855.
Sophocles, 1441, 1541, 1561, 1594, 1624.
SOPHOULIS (Pananos), 2503.
SORBY (Karol), 4664.
SORELLA (Antonio), 99.
SØRENSEN (Jesper), 948.
SØRENSEN (Øystein), 6656, 6676.
SORIA (Claudia), 5708.
SOROCHAN (Sergej B.), 1723.
SOROŞTINEANU (Valeria), 4955.
SORREL (Christian), 4905.
SOSIN (Joshua D.), 1416.
SOTINEL (Claire), 2363.
SOTO (Ángel), 3702.
SOTO CARMONA (Alvaro), 4604.
SOUFI (Fouad), 444.
SOULA (René), 3472.
SOULODRE-LAFRANCE (Renée), 6767.
SOURVINOU-INWOOD (Christiane), 1718.
SOUTHGATE (Beverley), 536.
SOYBEL (Phyllis L.), 7286.
SOYSAL (Oğuz), 1270.
Sozomen (Salminius Hermias), 2250.
SPAANS (K. W.), 6942.
SPADA (Gabriella), 1241.
SPADAVECCHIA (Anna), 5870.

SPADONI (CLAUDIO), 852.
SPAHLINGER (Lothar), 2126.
Spallanzani (Lazzaro), 5428.
SPANG (Christian W.), 6666.
SPÂNU (Alin), 4431, 7093.
SPANU (Pier Giorgio), 1757.
SPÄTI (Christina), 3559.
SPEAR (Valerie G.), 3395.
Spears (Edward), 7174.
SPEARS (Timothy B.), 6114.
SPECHT (Minna), 5250.
SPECK (Mary), 5921.
Speer (Albert), 5584.
SPEER (Andreas), 3292.
SPEIDEL (Michael Alexander), 1948.
SPEIRS (Ronald), 3950.
SPEITKAMP (Winfried), 6731.
SPENCER (Diana), 2109.
SPEVAK (Olga), 2237.
SPICER (Andrew), 3440.
SPICKERMANN (Wolfgang), 2140.
SPIEGEL (Gabrielle M.), 524.
SPIELVOGEL (Jörg), 2364.
SPIERLING (Karen E.), 5010.
SPIESER (Jean-Michel), 2444.
SPINELLI (María Estela), 3606.
SPINELLI (Mario), 2239.
SPINETO (Natale), 911.
SPINK (Walter M.), 687.
Spinoza (Benedictus de), 5355.
SPIRE (Alexis), 6368.
SPIRIDONOVA (L. I.)., 6821.
SPIRITOVÁ (Markéta), 3784.
SPIVEY (Nigel), 1471.
Splett (Karol Maria), 4863.
SPOERER (Mark), 5768.
SPONGBERG (Mary), 345.
SPRENKELS (Ralph), 3828.
SPRINGBORG (Peter), 24.
SPRINGER (Christian), 5709.
SPRINGER (Kimberly), 6369.
SPYRA (Ulrike), 3308.
SQUILLANTE (Marisa), 1839.
STABLEFORD (Brian M.), 1000.
STABLEFORD (Tom), 4459.
STÄCKER (Thomas), 137.
STADLER (Friedrich), 105.
STAFFORD (Emma J.), 1524, 1719.
STAHL (Ann B.), 7856.
STAHNISCH (Frank), 5423.
STAHULJAK (Zrinka), 3218.
STALEY (Lynn), 3093.
Stalin (Iosif Visarionovič Džugašvili), 286, 4472, 4478, 4484, 4491, 4500, 4502, 4505, 4511, 5061, 5651, 7109, 7123, 7138, 7171, 7180, 7277.
STALOFF (Darren), 4778.
STAMM (Konrad), 5285.
STAMOVA (Marijana), 7357.

STANBURY (Sarah), 3482.
STANCIU (Cezar), 7358.
STANĚK (Tomáš), 3785, 6370.
STANFORS (Maria), 6191.
STANOMIR (Ioan), 6523.
STAPLES (Anne), 6241.
STARGARDT (Nicholas), 6371.
STÄRK (Ekkehard), 2110.
STARKEY (Kathryn), 2942.
STARN (Randolph), 301.
STASIAK (Arkadiusz Michał), 4407.
Statius (Publius Papinius), 1842, 2026, 2074, 2115.
STAUBACH (Nikolaus), 2908.
STAUBER (Reinhard), 5479.
Staudinger (Julius Von), 6583.
STEEDMAN (Carolyn), 292.
STEEL (Catherine E. W.), 2111.
STEELE (Stephen), 452.
STEELE (Val), 1115.
STEELMAN (K. L.), 1072.
STEENBAKKERS (Piet), 5355.
STEFAN (Alexandre Simon), 1949.
STEFANOVIC (Djordje), 4525.
STEFANSKI (Valentina Maria), 6372.
STEGER (Florian), 5423.
STEHLÍK (Eduard), 3786.
ŠTEIMANIS (Josifs), 4268.
STEIN (Edith), 5340.
Stein (Lorenz von), 5356.
STEINBACH (Matthias), 5186.
STEINBERG (John W.), 6988.
STEINBERG (Sylvie), 6186.
STEINEGGER (Catherine), 5710.
STEINER (Ann), 1775.
STEINER (Kilian J. L.), 5711.
STEINER (Richard C.), 1329.
STEINER (Zara S.), 7094.
STEINFÜHRER (Henning), 2751.
STEIN-HÖLKESKAMP (Elke), 2048.
STEINICKE (Marion), 566.
STEINIGER (Judith), 1842.
STEININGER (Rolf), 7425.
STEINWEIS (Alan E.), 3934.
STELLA (Angelo), 2997.
STELLADORO (Maria), 2365.
STELLNER (František), 5720.
STELZL-MARX (Barbara), 7145, 7375.
STEMPIN (Arkadiusz), 6700.
Stendhal, 5508, 5526.
STENHOUSE (William), 429.
STENUIT (Bernard), 5175.
STENZEL (Oliver), 5458.
STEPANOVA (Larisa G.), 200.
STERN (B.), 1084.
STERN (Ben), 1115.
STERN (Lewis M.), 7541.
STETTBERGER (Herbert), 2366.

STEVANOVITCH (Colette), 2982.
STEVENSON (John), 624.
STEWART (Andrew), 1631.
STEYMANS (Hans Ulrich), 2367.
STIEGLITZ (Robert R.), 1330.
ŠTIH (Peter), 2611.
Stilicho (Flavius), 1816.
ŠTILLA (Miloš), 4532.
STIRPE (Paola), 1525.
STITES (Richard), 5567.
STITZIEL (Judd), 5922.
STOBART (Jon), 6373.
STÖBER (Rudolf), 5303.
STOCK (Christian), 1835.
STOECKER (Holger), 4949.
STOIANOV (Carmen Antoaneta), 843.
STOIDE (Constantin A.), 5871.
STOJČEVSKI (Gjorgji), 6949.
STOJČEVSKI (Petar), 6949.
STOKES (Doug), 7359.
STOLER (Mark A.), 6694, 7165, 7197.
STOLJAROVA (Ljubov' V.), 2520.
STOLL (Ulrike), 5157.
STOLLBERG-RILINGER (Barbara), 439.
STOLLEIS (Michael), 6481, 6538, 6552.
STOLTE (Stefan), 6603.
STÖLTING (Ulrike), 3473.
STÖLTZNER (Michael), 105.
STOLZ (Michael), 58.
STOLZI (Irene), 6547.
STÖMER (Wilhelm), 121.
STONE (Andrew F.), 2504.
STONE (Anne), 3242.
STONE (David), 3133.
STONE (Glyn), 7232.
Stone (Shepard), 5170.
STOPANI (Antonio), 6604.
STOPFORD (Jennie), 3219.
STORA (Benjamin), 430.
STORA-LAMARRE (Annie), 6496.
STOREY (Ian C.), 1387.
STORY (Joanna), 2612.
Stosch (Albrecht von), 6993.
STOURAITIS (Ioannis), 2505.
STOURZH (Gerald), 3633, 7417.
ŠŤOVÍČEK (Ivan), 7177.
STOYLE (Mark), 4089.
Strabo, 1442.
Strachey, family, 112.
STRAKA (Tomás), 4798.
STRAKER (V.), 1078.
STRANGE (Julie-Marie), 6374.
STRANO (Gioacchino), 2506.
STRASSER (Bruno J.), 818, 5459.
STRASSER (Christian), 4014.
STRATHERN (Marilyn), 772.
STRÄTLING (Susanne), 5568.

STRAUMANN (Lukas), 5873.
STRAUSS CLAY (Jenny), 1632.
STRICKER (Stefanie), 2966.
STRICKHAUSEN (Waltraud), 5081.
STRINATI (Maria Gabriella), 3094.
STRINGER (Martin D.), 949.
STROBEL (Thomas), 5251.
STRÖBELE (Ute), 4930.
STROHSCHNEIDER (Peter), 233.
STRONACH (David), 1367.
STROSETZKI (Christoph), 5495.
STRÖTZ (Jürgen), 4015.
STROUD (Ronald S.), 1404.
STROUMSA (Guy G.), 2208.
STRUEVER (Nancy S.), 482.
STRUPP (Christoph), 1001, 5201.
STRUVE (Kai), 4711.
STRUVE (Lynn A.), 7563.
STRZELCZYK (Jerzy), 3412.
STUARD (Susan Mosher), 2937.
STÜBER (Gabriele), 5.
STUBER (Martin), 5361.
STUCHTEY (Benedikt), 5452.
STUDER (Walter), 2507.
STUSSI (Alfredo), 96, 5544.
STYKA (Jerzy), 2113.
Suárez (Francisco), 6475.
SUCEVEANU (Alexandru), 2020.
SUCH-GUTIÉRREZ (Marcos), 1235.
SUCIU (Ioan Alexandru), 6695.
SUDMANN (Stefan), 3337.
SUDRY (Yves), 6808.
SUGIHARA (Kaoru), 5748.
SUGIYAMA (Saburo), 7887.
SUGIZAKI (Taiichiro), 3398.
Suharto, 4156.
SUHRAWARDY (Shahid), 2716.
SUI (Xiaozuo), 7720.
SUITNER (Franco), 3096.
Suleiman I, Sultan of the Turks, 6881.
Sulla Felix (Lucius Cornelius), 1909, 1913, 2118.
SULLIVAN (Denis F.), 2576.
SULLIVAN (Karen), 3097.
SUMI (Geoffrey S.), 1950.
SUMMERHILL (Thomas), 5966.
Sumner (Charles), 4765.
SUN (Lijuan), 7721.
SUNDBERG (Ulf), 4653.
SÜNDERHAUF (Eshter Sophia), 5171.
SUNDIN (Jan), 819.
ŠUNDRICA (Z.), 2744.
SUNTRUP (Rudolf), 2916.
SUPPAN (Arnold), 7376, 7417.
SURDICH (Luigi), 3098.
SURIÁ I VENTURA (Núria), 2524.
SURIANO (Juan), 698, 3580, 3598, 3608.

SURIKOV (Igor' E.), 1472.
SURMANN (Beate), 2245.
SÜSSMANN (Johannes), 218.
SUTTIE (Andrew), 7095.
SUTTON (Anne F.), 2864.
SUZUKI (Hiroyuki), 7814.
SUZUKI (Kosetsu), 7577.
SUZUKI (Tamon), 4245.
SUZUKI (Tetsuo), 7840.
ŠVÁCHA (Rostislav), 5554.
SVANIDZE (Ada A.), 3529.
SVÅSAND (Lars), 4357.
SVERRIR (Jakobsson), 3099.
SVOBODOVÁ (Jitka), 3754.
ŠVORC (Peter), 6674.
SWAIN (Shurlee), 6189.
SWANN (C. P.), 1347.
SWEET (Julie Anne), 6851.
SWENSON-WRIGHT (John), 7426.
SYED (Yasmin), 2114.
SYKES (Alan), 4090.
SYLOS LABINI (Paolo), 773.
Symeon Magister, 2415.
Synesius Cyrenaicus, episcopus, 2251.
SYRETT (David), 6896.
SZABÓ (Csaba), 7542.
SZABÓ (Péter), 2865.
SZAIVERT (Wolfgang), 2049.
SZALONTAI (Balazs), 7543.
SZÁNTAY (Antal), 4129.
SZÁRAZ (Peter), 3787.
SZAREK (Jarosław), 4406.
SZAYNOK (Bożena), 7510.
SZEGHYOVÁ (Blanka), 582.
SZEPTYCKI (Andrzej), 7544.
SZERWINIACK (Olivier), 3433.
SZIJÁRTÓ (István M.), 4130.
SZIJJ (Jolán), 4115.
SZIRMAI (Julia C.), 2962.
SZKIET (Christine), 50.
SZÖLLÖSI-JANZE (Margit), 5177.
SZRETER (Simon), 5460.
SZUBTARSKA (Beata), 7233.
SZUROMI (Anzelm Sz.), 2772.
SZWAGRZYK (Krzysztof), 4385.

T

TAAVITSAINEN (Irma), 3294.
TABARÉS DOMÍNGUEZ (Marta), 1076.
TABBACH (Vincent), 78.
TABEAU (Ewa), 6052, 6116.
TACCONI (Marica), 3474.
TACHIBANA (Makoto), 7578.
Tacitus (Cornelius), 478, 1843, 1844, 1874, 1918, 1941.
TAGESSON (Göran), 661.
TAGLIABUE (John), 2948.
TAGLIACOZZO (Eric), 5923.
TAGUCHI (Kojiro), 7722.

TAGUIEFF (Pierre-André), 4486.
TAHBERER (Bekircan), 149.
TAISNE (Anne-Maria), 2115.
TAKÁCS (Gábor), 1331.
Takahashi (Keisaku), 5393.
TAKAHASHI (Kiyonori), 2875.
Takano (Chei), 5393.
TAKAYAMA (Hiroshi), 2600, 2601.
TALBOT (Alice-Mary), 2402, 2576.
TALBOTT (John E.), 3567.
TALIERCIO MENSITIERI (Marina), 2181.
TALL (Ahmad Yūsuf), 4252.
TALL (Aminatou), 247.
TALLON (Alain), 4812.
TAMAYAMA (Kazuo), 7288.
TAMBRUN-KRASKER (Brigitte), 2391.
TAMER (Georges), 2692.
TAMM (Ditlev), 2773.
TAN (Elaine S.), 2774.
TAN (Tai Yong.), 6755.
TANAKA (Ryuichi), 4246.
TANCK DE ESTRADA (Dorothy), 248.
TANEJA (Anup), 4151.
TANG (Frank), 2673.
TANG (Leilei), 5990.
TÄNGEBERG (Peter), 3220.
TANGHERONI (Marco), 2835.
TANII (Toshihito), 7723.
TANIUCHI (Yuzuru), 4513.
TANSELLE (George T.), 101.
TANTLEVSKIJ (Igor' R.), 1332.
TANWAR (Shyam Singh), 4152.
TAO (Feiya), 7724.
TAQAVĪ (Muhammad'Alī), 4167.
TARDELLI (Marcello), 254.
TARLING (Nicholas), 7427.
TARQUINI (Stefania), 3511.
TARRAGÓ (Myriam Noemí), 698.
TARUSKIN (Richard), 828.
TARVER (Hollis Micheal), 4799.
Tasca (Angelo), 5154.
TASCHLER (Daniela), 7445.
TASSIGNON (Isabelle), 1750.
Tassilo III von Bayern, 2618.
TATARANNI (Francesca), 1759.
Tatishchev (Vasily N.), 479.
TAURISSON (Dominique), 5052.
TAUSSIG (Michael), 3724.
TAVARES (João Moreira), 4426.
TAVARES DE ALMEIDA (Pedro), 4420.
TAVONI (Maria Gioia), 66.
TAYLOR (Alan M.), 6022.
TAYLOR (Amy Murrell), 6376.
TAYLOR (Andrew), 4091.
TAYLOR (Barry), 3022.
TAYLOR (Claire), 3475.
TAYLOR (David), 6734.

TAYLOR (Gillian), 1045, 1069.
TAYLOR (Jon), 1249.
TAYLOR (Matthew), 6377.
TAYLOR (Nikki M.), 6378.
TAYLOR (Richard C.), 956.
Taylor (Richard), 4767.
TAYLOR (Stephen), 4077.
TAYLOR ALLEN (Ann), 6379.
Taylor Tichenor (Isaac), 5012.
TEBINKA (Jacek), 7545.
TEDESCHI (A.), 1951.
TEDESCHI (Antonella), 1808.
TEDESCHI (John), 454.
TEECE (David J.), 5880.
TEITLER (Hans Carel), 1795.
TEJERA (Eduardo J.), 6852.
TEKA (Zelalem), 1033.
TEKIPPE (Rita), 3157.
TEKLAK (Czeslaw), 433.
TELESKO (Werner), 5569.
TELI DIBRA (Pranvera), 7096.
TELLIER (Thibault), 3896.
TELLIEZ (Romain), 2775.
TELLO DÍAZ (Carlos), 7546.
TEMGOUA (Albert-Pascal), 409.
Tempesta (Antonio), 5605.
TENFELDE (Klaus), 6408, 7097.
TEPPERBERG (Christoph), 4115.
Terentius Afer (Publius), 2087.
TERMIS SOTO (Fernando), 7428.
TERMORSHUIZEN (Gerard), 5264.
TERNAVASIO (Marcela), 3578.
TERPSTRA (Nicholas), 6380.
TERRETTA (Meredith), 3690.
TER-SARKISJANTS (Alla E.), 688.
Tertullianus (Quintus Septimius Florens), 2252, 2348.
TERVOORT (Ad), 5200.
TERWIEL (B. J.), 4670.
TESI (Riccardo), 201.
TESSAROLO (Luigi), 2958.
TESSITORE (Fulvio), 484.
TEYSSIER (Éric), 2050.
THAL (Sarah), 7841.
THALI (Johanna), 3039.
THATCHER (Ian), 4487.
THEIS (Lioba), 2508.
THEISEN (Frank), 2776.
THELIANDER (Claes), 3338.
THÉNAULT (Sylvie.), 6809.
Theodoretus, bishop of Cyrus, 2323.
Theodorus Mopsuestenus, 2271.
Theodorus Scutariotes, 2510.
Theodosius II, Roman emperor, 1985.
Theodosius Princeps, 2453.
THÉRIAULT (Gaétan), 1405.
THÉVANEZ (Clémence), 3422.
THIEL (Andreas), 2368.
THIEMEYER (Guido), 5304.

THING (Morten), 5042.
THOMA (Grid), 5728.
THOMAS (Avril), 225.
THOMAS (Carol G.), 1526.
THOMAS (David), 2693.
THOMAS (Hugh), 3742.
THOMAS (James), 4092.
THOMAS (Ken), 1360.
THOMAS (Markus), 7429.
THOMAS (Martin), 5158, 6732.
THOMAS (Matthew), 4093.
THOMAS (Michael L.), 1775.
THOMAS (Richard), 2866.
THOMAS (Wyndham), 3230.
Thomas Aquinas, 3257, 3263, 3273, 3279, 3301, 3311.
Thomas, 1162.
THOMPSON (Andrew), 6733.
THOMPSON (Augustine), 3477.
THOMPSON (Isobel), 3486.
THOMPSON (John A.), 7166.
THOMPSON (John J.), 31.
Thompson (Marion), 6381.
THOMPSON (Nicholas), 5011.
THOMPSON (Paul), 6381.
THOMPSON (Victoria E.), 6427.
THOMPSON FRIEND (Craig), 4779.
THOMS (Ulrike), 5390.
THOMSON (Alistair), 431.
THOMSON (Elizabeth), 6079.
THOMSON (J. K. J.), 5874.
THORNTON (Martin), 7360.
Thucydides, 480, 1539, 1634.
THUILLIER (Jacques), 848.
THUMFART (Alexander), 722.
THÜMMEL (Hans Georg), 3339.
THÜR (Gerhard), 1479.
THYEN (Hartwig), 2265.
THYS (Walter), 7014.
Thyssen (August), 5813.
Tiberius Claudius Nero, Roman emperor, 1877, 1929, 1942.
Tibullus (Albius), 2055.
TIDL (Georg), 3634.
TIEDEMANN (Joseph S.), 6102.
TIETZ (Werner), 1135.
TIGNÈRES (Serge), 6754.
TIKHONOV (Ju. A.), 5969.
TIKKANEN (Henrikki), 5788.
TILATTI (Andrea), 552, 3478.
TILLEY (Virginia Q.), 3829.
Tillich (Paul), 5717.
TILLIETTE (Jean-Yves), 3264.
TIMM (Stefan), 2226.
TIMMERMAN (Freddy), 3711.
TIMMERMANN (Heiner), 3756, 3775, 7328.
TIMMERMANS (Linda), 5252.
TIMMS (Edward), 5536.
TIMOC (Călin), 2188.
TIMOFEEVA (Natalja P.), 7374.

TIMOSCHENKO (Tatjana), 6382.
TIMUŞ (Mihaela), 912.
TINAZ (Nuri), 950.
Tinbergen (Niko), 5372.
TINNEFELD (F.), 2509.
TINTI (Francesca), 3366.
TÍO BELLÍDO (Ramón), 5570.
ȚÎRĂU (Liviu), 7430.
TIRELLI (Aldo), 1438.
TISCHEL (Alexandra), 5180.
TISCHLER (Matthias M.), 3340.
TISCHNER (Wolfgang), 4835.
TISMĂNEANU (Vladimir), 4456.
Tiso (Jozef), 7313.
TISSA KAIRO (Eusébio), 4327.
Tito (Josip Broz), 4523, 7333, 7389.
TITONE (Fabrizio), 2674.
Titsingh (Isaac), 7832.
Titus, Roman emperor, 1319.
TJUMENTSEV (Igor' O.), 4506.
TJUTJUKIN (Stanislav V.), 4494.
TLOKA (Jutta), 2369.
TOADER (Mihaela), 4436.
TOBLER (Felix), 3622.
TOCCHINI (Gerardo), 5502, 5719.
TOCCI (Raimondo), 2510.
TOCK (Benoît-Michel), 15, 3326.
Tocqueville (Alexis de), 481, 5357.
TODD (Ian A.), 1052.
TODD (Selina), 6383.
TODESCHINI (Giacomo), 2796.
TODEV (Ilija T.), 3686.
TODISCO (Elisabetta), 1981.
TODOROVA (Cvetana), 7008.
Toesca (Pietro), 853, 854.
TOFTEGAARD JENSEN (Jens), 6117.
TOFTGAARD (Anders), 3103.
Togliatti (Palmiro), 5331, 7312, 7333.
TOGNETTI (Sergio), 2867.
Tojo (Hideki), 4245, 7171.
TOKAREVA (I. L.), 32.
TOLOCHKO (Aleksej P.), 479.
TOLOSA I MAITE FRAMIS (Lluïsa), 3154.
Tolsdorff (Theodor), 4010.
TOLSTOGUZOV (Aleksandr A.), 435.
TOMALA (Mieczysław), 7452.
TOMÁŠEK (Dušan), 7098.
TOMASIN (L.), 5544.
TOMBEUR (Paul), 3476.
TOMCZAK (Maria), 7112.
TOMEI (Alessandro), 3155.
TONAGA (Yasushi), 7579.
TONG (Zhiqiang), 7725.
TONIOLO (Federica), 3242.
TONIOLO (Gianni), 6026.
TONO (Haruyuki), 7842.
TOOMASPEG (Kristjan), 2795.

TOOZE (J. Adam), 5875.
TORGAL (Luís Reis), 3561, 4427.
TORP (Cornelius), 5769.
Torquemada (Tomás de), 3442.
TORRANCE (Richard), 5537.
TORRE (Angelo), 4926.
TORRES SÀNCHEZ (Rafael), 6027.
TORRI (Giulia), 1271.
TORSTENDAHL (Rolf), 4507.
TORTORELLA (Stefano), 1751.
TORTZEN (Chr. Gorm), 1431.
TOSTI (Jean), 5855.
Totone di Campione, 2792.
TOURATSOGLOU (Yannis), 1744.
Touré (Ahmed Sékou), 4108.
TOURNÈS (Ludovic), 537.
TOURNOY (Gilbert), 5131.
TOUROVETS (Alexandre), 1368.
TOUSSAINT (Gia), 3206.
Toussaint Louverture (François Dominique), 6921.
TOUZARD (Anne-Marie), 6912.
TOZZI (Carlo), 1024.
TRABA (Robert), 4016.
TRABANT (Jürgen), 199, 5327.
TRACY (James D.), 4339.
TRACY (Sarah W.), 6384.
Traianus Germanicus Dacicus (Marcus Ulpius), Roman emperor, 1949, 2036.
TRAMONTANA (Salvatore), 689.
TRAMPEDACH (Kai), 2370.
TRAMPUS (Antonio), 699, 6474.
TRAN (My-Van), 4803.
TRANFAGLIA (Nicola), 5291, 5302.
TRAŞCĂ (Ottmar), 5970, 7235.
TRAUSCH (Gilbert), 6688.
TRAVERS (Emeric), 538.
TRAVERS (Robert), 6756.
TRAVERSO (Antonella), 1074.
TRAVERSO (Enzo), 3560.
TREFOUSSE (Hans L.), 4780.
TREJO BARAJAS (Dení), 6118.
TREMBATH (Richard), 7431.
TREMBLAY (Florent A.), 3106.
TREMBLAY (Victor J.), 5876.
TRESSERRAS (Jordi Juan), 1055.
TŘEŠTÍK (Jan), 270.
TRICOMI (Antonio), 5538.
TRILLO SAN JOSÉ (Carmen), 3135.
TRINCHESE (Stefano), 5070, 5117.
TRINKUNAS (Harold A.), 4800.
TRISTANO (Caterina), 34.
TRISTRAM (Frédéric), 6028.
TROCHET (Jean-René), 236.
Troeltsch (Ernst), 486.
TROGER (Vincent), 5253.
Trogus (Pompeius), 1819.
TROIANI (Lucio), 346, 1333.
TROILO (Simona), 317.

TRONCARELLI (Fabio), 3312.
TRONCOTĂ (Cristian), 4431.
TROOST (Wout), 4094.
TROPPER (Josef), 1250.
TROW (M. J.), 2728.
TRUBEK (David M.), 486.
Trudeau (Pierre Elliott), 3693.
TRUELOVE (Alison), 2898.
TRUETT (Samuel), 213.
TRUJILLO BOLIO (Mario), 5877.
Trujillo Molina (Rafael Leónidas), 3813, 7380.
Truman (Harry S.), 4747, 4750, 7138, 7386, 7415.
TRUMPP (Thomas), 5.
TRÜPER (Henning), 449.
TRZCIONKA (Silke), 2457.
TSAI (Shih-Shan Henry), 4665.
TSANG (Steven Yui-Sang), 7432.
TSCHOPP (Silvia Serena), 539.
TSCHUBARJAN (Alexandr O.), 7374, 7375.
TSETSKHLADZE (Gocha R.), 1100.
TSHONDA OMASOMBO (Jean), 3728.
TSUGEI (Yukio), 7843.
TSUKADA (Takashi), 7806.
TSYPKIN (G.), 7855.
TUCCI (Pier Luigi), 2189.
TUCCILLO (Fabiana), 1989.
TUCHTENHAGEN (Ralph), 690.
TUCK (Michael W.), 290.
TUCK (Steven L.), 1990.
TUCKER (Jennifer), 5461.
TUCKER (Spencer C.), 635.
TUCZAY (Christa), 3054.
TUDORANCEA (Radu), 5770.
Tuhadelini (Noah Eliaser), 4328.
TULARD (Jean), 6925.
TULLY (John Andrew), 3688.
TŮMA (Oldřich), 3754.
TURAN (Ömer), 6994.
TURCAN (Robert), 1720.
TURCESCU (Lucian), 2233.
TURCHETTI (Mario), 6943.
TURCHI (Roberta), 5502.
TURCHIANO (Maria), 3505.
TURCUŞ (Şerban), 3341.
TURI (Gabriele), 476.
TÜRKDOĞAN (Orhan), 4701.
TURKELTAUB (Daniel), 1633.
TURNBULL (Jean), 6325.
TURNER (George), 5260.
TURNER (Henry Ashby Jr.), 5878.
TURNER (John D.), 6002.
TURPIN (Frédéric), 6757.
TURREL (Denise), 133.
TURVILLE-PETRE (Thorlac), 3036.
TUSAN (Michelle Elizabeth), 5305.
TUSELL (Javier), 4604, 4607, 4621.
TUTEN (Eric Engel), 4364.

Tuthaliya I, king of Hittites, 1254, 1257.
Tuthaliya II, king of Hittites, 1257.
Tuthaliya III, king of Hittites, 1257.
TUTTLE (Gray), 7726.
TVEIT (Odd Karsten), 7548.
Twain (Mark), 984.
TWARDZIK (Stefano), 5173.
TWEED (Thomas A.), 951.
TWOHIG (Peter L.), 5462.
TYBOUT (Rolf A.), 1404.
TYRRELL (Ian), 437.
TYSZKIEWICZ (Jakub), 7510.

U

Uccello (Paolo), 3162.
UDAL (John O.), 4637.
UDOH (Fabian E.), 2051.
UEBERSCHÄR (Gerd R.), 7167.
UECKER (Heiko), 5496.
UEDA (Atsuko), 1003.
UEDA (Tomoaki), 7595.
UEHARA (Mahito), 7833.
UERTZ (Rudolf), 6524.
UGALDE (Luis), 7099.
UGÉ (Karine), 3399.
UGGLA (Andrzej Nils), 6684.
Ugolini (Luigi Maria), 1726.
UĞURLU (Ö. Andaç), 6951.
UHL (Matthias), 286, 7109, 7454.
UHLÍŘ (Jan B.), 7168.
UJVÁRY (Gábor), 3536.
UL'JANOV [Ulyanov] (Mark Ju.), 7597.
UL'JANOVA (Galina N.), 4508.
ULIANOVA (Olga), 7010.
ULLMANN (Hans-Peter), 6029.
ULRICH (Bernd), 6161.
ULUNJAN (Arutjun A.), 4105.
UMBACH (Maiken), 6119.
UNDERHILL (John), 1540.
UNGER (Dagmar), 6605.
UNGER (Helga), 3400.
UNGERN-STERNBERG (Jürgen von), 5254.
UNGUREANU (George), 7236.
UNOWSKY (Daniel L.), 3635.
ÜNSALDI (Levent), 4702.
'Uqbānī, family, 7570.
URBÁNEK (Miroslav), 3788.
URBÁŠEK (Pavel), 3790.
Urbicius, 2416.
URÍAS MARTÍNEZ (Rafael), 2306.
URSINUS (Michael), 4703.
USMANOVA (Diljara M.), 4509.
USUI (Sachiko), 7727.
UTHEMANN (Karl-Heinz), 2371.
UVAROVA (Tat'jana B.), 585.
UZEL (İlter), 149.
UZUN (Ahmet), 5924.

V

VACCA (Giuseppe), 5331.
VACCARO (Luciano), 4934.
VACULÍK (Jaroslav), 6056.
VAGET (Hans Rudolf), 5534.
VAHED (Goolam), 4541.
VAILLANT (Alain), 5539.
VAÏSSE (Maurice), 7436, 7554.
VAJNSHTEJN (Ol'ga B.), 5159.
VAKHTINA (Marina Ju.), 1462.
VALBELLE (Dominique), 1154.
VALCÁRCEL MARTINEZ (Vitalino), 3071.
VALDÉS GUÍA (Miriam), 1721.
VALE (Malcolm), 203.
VALE (Teresa Leonor M.), 5610.
VALENCIA (Marta), 5971.
VALENTI (Marco), 3494.
VALENTINITSCH (Tatjana), *I*.
VALENZUELA LAFOURCADE (Mario), 6952.
Valerius Flaccus Setinus Balbus (Gaius), 1845, 1846.
Valerius Maximus, 1847.
Valla (Lorenzo), 482.
VALLANCE (Edward), 4095.
VALLEJO RUIZ (José María), 2052.
VALLERANI (Massimo), 2777.
Valois, 3132.
VAN BELLE (G.), 857.
VAN BELLE (Gilbert), 2258.
VAN BERKEL (Klaas), 3268.
VAN BRUAENE (Anne-Laure), 2896.
VAN BUEREN (Truus), 2886.
VAN BUREN (Anne), 834.
VAN DAMME (Stéphane), 5160, 5320.
VAN DE MORTEL (Johannes Petrus Maria), 6606.
VAN DEN ABEELE (Baudouin), 2880, 3266.
VAN DEN DUNGEN (Pierre), 5306.
VAN DER GRIJP (Klaus), 4958.
VAN DER HEYDEN (Ulrich), 4949, 6725, 7362.
VAN DER KOOIJ (A.), 940.
VAN DER LINDEN (Marcel), 6462.
VAN DER STEEN (Eveline J.), 1334.
VAN DER STOCK (Jan), 39, 3197.
VAN DER STROOM (Gerrold), 4331.
VAN DER WATT (Jan Gabriël), 2372.
VAN DER WEE (Herman), 667, 5771, 6030.
VAN DEUN (Peter), 2387.
VAN DEURSEN (Arie Theodorus), 4340.
VAN DONZEL (E.), 900.
VAN DYKE (Carolynn), 3109.

VAN EENOO (Romain), *II*.
VAN GELDER (Geert Jan H.), 2717.
VAN HAEPEREN (Françoise), 2148.
VAN HARTESVELDT (Fred R.), 7006.
VAN HENTEN (Jan Willem), 1335.
VAN KALVEEN (C. A.), 3344.
VAN KEULEN (Percy S. F.), 2270.
VAN LAAK (Dirk), 6696.
Van Leerdam (Andrea), 2886.
VAN METER (Robert H.), 7433.
VAN MINGROOT (Erik), 2527.
VAN NIEL (Robert), 6758.
VAN NIEUWENHOVE (Rik), 3311.
VAN ORDEN (Kate), 5712.
VAN POPPEL (Frans), 6385.
VAN RAAY (Rob), 6141.
VAN RAHDEN (Till), 5014.
VAN RIEL (Gerd), 1679.
VAN SCHIE (Patricius Gerardus Cornelis), 4341.
VAN TILBURG (Marja), 6386.
VAN UYTFANGHE (Marc), 2381.
VAN VEEN (Henk), 834.
VAN WIJNENDAELE (Jacques), 3342.
VÁŇA (Daniel), 5720.
VÁŇA (Josef), 7289.
VANAUTGAERDEN (Alexandre), 4809.
VANDENBUSSCHE (Robert), 7308.
VANDERHEYDE (Catherine), 2512.
VANDERJAGT (Arie Johan), 3268.
VANDERLIPPE (John M.), 4704.
VANDIVER NICASSIO (Susan), 4231.
VANĚK (Miroslav), 3790.
VANÍČKOVÁ (Vladimíra), 3789.
VANOLI (André), 6031.
VANOVA (Liljana), 7011.
VANSINA (Jan), 6810.
Vansittart (Robert Gilbert Vansittart, Baron), 5717.
VAQUERO DÍAZ (María Beatriz), 2548.
VARANINI (Gian Maria), 2750, 2796, 2855, 3321.
Varenne (Alexandre), 7143.
Vargas (Getúlio), 7379.
VARGAS (Zaragosa), 6463.
Varnhagen (Francisco Adolfo de, Visconde de Porto Seguro), 6978.
VARSORI (Antonio), 6675.
VARTOLOMEI (Luminița), 832.
Varus (Publius Quinctilius), 1955.
VARVARCEV (Mikola Mikolajovic), 4712.
VAŠČENKO (Elena), 5346.

VASCONCELOS DE SALDANHA (António), 6895.
VASIL'EV (Vladimir I.), 93.
VASILE (Cristian), 4956.
VASILIOU (D.), 5751.
VASILIU (Anca), 3259.
VASINA (Ljudmila), 5346.
VASSALLO (Aude), 5713.
VASSIS (Ioannis), 2417.
VATLIN (Aleksandr Ju.), 4019.
VAUCHEZ (André), 3422.
VAUCHEZ (Antoine), 6497.
VAUGELADE (Daniel), 6897.
VAUGHAN (James R.), 7434.
VAVOURANAKIS (Giorgos), 1398.
VAVRA (Elisabeth), 2941.
VAYSSIÈRE (Bertrand), 7237.
VAYSSIÈRE (Pierre), 3712.
VÁZQUEZ OSUNA (Federico), 4608.
VEDJUSHKIN (Vladimir A.), 3529.
VEENSTRA (Jan R.), 2916.
VEGA SOMBRÍA (Santiago), 4609.
VEGETTI (Matteo), 5334.
VEISSE (Anne-Emanuelle), 1199.
VELASCO GODOY (María de los Angeles), 6853.
VELAZA (Javier), 1831.
Velázquez (Diego), 5592.
VELÁZQUEZ CASTRO (Marcel), 4382.
VELÁZQUEZ MORALES (Catalina), 4321.
VELÁZQUEZ SORIANO (Isabel), 3419.
VELKOV (Asparukh), 51.
VELLA (Nicholas C.), 1336.
Velleius Paterculus, 1848.
VELTEN (Hans Rudolf), 994.
VENBORG PEDERSEN (Mikkel), 3808.
VENCHIERUTTI (Massimo), 5588.
VENNING (Timothy), 2430.
VENTRONE (Angelo), 4232.
Venturi (Franco), 483.
Venturi (Lionello), 854.
VENUGOPAL REDDY (Kanchi), 6464.
VENUH (Neivetso), 6759.
VENUS (Theodor), 3569.
VERA CRUZ (Elizabeth Ceita), 6811.
VERBREYT (Monique), 6030.
VERDERAME (Lorenzo), 1243.
VERDERY (Katherine), 5968.
VERDEYEN (Paul), 3249.
Verdi (Giuseppe), 5658, 5709.
VERDIER (Caroline), 5347.
VERDON (Laure), 2760.
Vergilius Maro (Publius), 1849-1851, 2068, 2097, 2099, 2114.

VERGINE (Lea), 5611.
VERGINELLA (Marta), 438.
VERHAEGEN (Benoît), 3728.
VERHEYDEN (J.), 857.
VERHEYDEN (Jozef), 2258.
VERHULST (Gilliane), 2149.
VERLAGUET (Waltraud), 3313.
VERMA (B. R.), 6741.
Vernadski (Georgy), 338.
VERNASSA (Maurizio), 6995.
VERNER (Lisa), 3110.
VERNEUIL (Yves), 5203.
VERNON (James), 6387.
VERPEAUX (Michel), 6505.
VERPOORTE (Alexander), 1038.
Verwoerd (Hendrik), 4542.
VESELÝ (Zdeněk), 3753, 7238.
Vespasianus (Titus Flavius), Roman emperor, 1861.
VESSEY (Mark), 2289.
VETH (Peter), 1034.
VETTA (Massimo), 823.
VIAENE (Vincent), 6996.
VIALA (Alain), 1004.
VIALES HURTADO (Ronny José), 6323.
VIARRE (Simone), 1834.
VIAZZO (Pier Paolo), 6189.
VICARD (Patrick), 637.
VICKERS (Adrian), 4157.
VIDA (István), 7371.
VIDAL (Frédéric), 6060.
VIDAL (Laurent), 6854.
VIDAL (Mary), 5094.
VIDAL MANZANARES (César), 1005.
VIDAL-NAQUET (Pierre), 6763.
VIDAURRI (Santiago), 4285.
VIDMAR (John C.), 4959.
VIDYASAGAR (K.), 4153.
VIELBERG (Meinolf), 2382.
VIÉVILLE (Dominique), 308.
VIGARELLO (Georges), 564.
VIGEZZI (Brunello), 7364.
Vigilius Thapsensis, episcopus, 2272.
VIGLIARDI (A.), 1019.
VIGNAUD (Laurent-Henri), 5463.
VIJGEN (Ingeborg), 6812.
VILANOVA (Francesc), 4610, 7239.
VILAR (Hermínia Vasconcelos), 3438.
VILAR (Jean), 6824.
VILAR (Pierre), 7135.
VILCHEZ (Roberto), 630.
VILENSKY (Joel A.), 5464.
VILHJALMSSON (Viljamur Örn), 5044.
VILINBAKHOV (G. V.), 32.
VILKUL (Tat'jana L.), 2582.
VILLADS JENSEN (Kurt), 607.

VILLANI (Giovanni), 2583.
VILLANON (Andrew), 2664.
VILLANUEVA (Marie-Christine), 1722.
VILLAR I TORRENT (Joan), 2524.
Villari (Pasquale), 484.
VILLARRUEL (Roberto), 3581.
VILLATOUX (Marie-Catherine), 7436.
VILLATOUX (Paul), 7436.
VILLAUME (Poul), 3809, 7321.
VILLEGAS REVUELTAS (Silvestre), 6997.
VILLERS (Jürgen), 1680.
VILLING (Alexandra), 1453.
VILLORESI (Marco), 3111.
VINCENT-MUNNIA (Nathalie), 5529.
VINCIGUERRA (Lorenzo), 5355.
VINDING (Niels), 2731.
VINGTAIN (Dominique), 3221.
VINOGRADOV (Jurij A.), 1462.
VINOGRADOV (Vladelen N.), 6998.
VINOVSKIS (Maris A.), 5255.
VINYOLES I VIDAL (Teresa-Maria), 2870.
VIRET (Jacques), 3244.
VISCIDO (Lorenzo), 1802.
VISSER (Joseph), 6918.
VISSER (Reidar), 4170.
VISY (Zsolt), 1912.
VITA (Juan Pablo), 1250.
VITALI (Camilla), 1991.
VITCU (Dumitru), 5307.
VITIELLO (Massimiliano), 2515, 2619.
VITOLO (Giovanni), 2793, 3489.
VITOU (Luc), 5855.
VITTU (Jean-Pierre), 821.
VITZ (Evelyn Birge), 3068.
VIVARELLI (Roberto), 3563.
Vivès (Jean Luis), 1854.
VIVES RIERA (Antoni), 5972.
VLADIMIR (Angelo), 4906.
VLASOV (Leonid V.), 7240.
VLASOVA (Marina A.), 7240.
VLČEK (Radomír), 3791.
VLČEK (Vojtěch), 4931.
VOCELKA (Karl), 691.
VODOPIVEC (Peter), 4540.
VODOSEK (Peter), 299.
VOELKE (Jean-Daniel), 5465.
Vöge (Wilhelm), 844.
VOGT (Ernst), 681.
VOGT (Helle), 2773, 2871.
VOGT (Paul), 6503.
VOILLERY (P.), 6999.
VOILLIOT (Christophe), 3898.
VOJNOVIĆ (Branislava), 3736.
VOJVODIC (Mihailo), 7000.
VOLAN (Angela M.), 2516.

VOLANTO (Keith J.), 5973.
VOLOKHINE (Youri), 1200.
Voloshin (Maximilian), 5541.
VOLPE (Giuliano), 3505.
VOLPI (Alessandro), 6388.
VOLPINI (Paola), 6893.
VOLT (Ivo), 2419.
Volta (Alessandro), 5388.
VOLTAIRE (François Marie Arouet de), 485, 5358, 5500.
VONAU (Jean-Laurent), 3900.
VORÁČEK (Emil), 3756.
VORMS (Charlotte), 6121.
VOẞ (Matthias), 7555.
VÖSSING (Konrad), 1372, 1634.
VOSSLER (Frank), 4020.
VOTHKNECHT (Michael), 6607.
VOURI (Sophia), 5256.
VOUT (Caroline), 2190.
VRINTS (Antoon), 3645.
VRTEĽ (Ladislav), 131.
VUILLEUMIER (Marc), 481.
VULPIUS (Ricarda), 4713.
VYBÍRAL (Jan), 793.

W

Wach (Adolf), 6605.
WACHT (Manfred), 1846.
WADA (Haruki), 4510.
WADAUER (Sigrid), 6389.
Wade (Abdoulaye), 4516.
WAECHTER (Hans), 3245.
WAGNER (Christian), 7549.
WAGNER (David), 6390.
WAGNER (Hans-Ulrich), 5664.
WAGNER (Marc-André), 586.
WAGNER (Patrick), 6391.
Wagner (Richard), 5534, 5704.
WAGNER-PACIFICI (Robin), 692.
WAHLE (Stephan), 2332.
WAHNICH (Sophie), 319.
WAISMAN (Sergio Gabriel), 5540.
WALDE MOHENO (Lillian von der), 3101.
WALDREP (Christopher), 4782.
WALEY (Muhammad Isa), 25.
WALGENBACH (Katharina), 5162.
WALKER (Barbara), 5541.
WALKER (Brett L.), 7844.
WALKER (Greg), 5497.
WALKER (Jonathan), 7550.
Walker (Roger M.), 3022.
WALL (Cynthia), 623.
WALL (Richard), 6189.
WALLACE (Rex), 1793.
WALLACE (Robert W.), 1507.
WALLINGA (H. T.), 1473.
WALLIS (Faith), 3290.
WALLIS (John Joseph), 6009, 6525.
WALLSTEN (Scott), 5308.
WALSER (Gerold), 1786.

INDEX OF NAMES

Walsh (Edmund A.), 4894.
WALSH (Kevin), 1039.
WALSH (Margaret), 5163.
WALSH (Richard J.), 2653, 2675.
WALTER (François), 778.
WALTER (Richard J.), 6122.
WALTER (Robert), 6526.
WALTER (Uwe), 465.
WALTON (John K.), 6175.
Wamba, king of the Visigoths, 2577.
WANG (Chunxia), 7728.
WANG (Dong), 6697.
WANG (Fansen), 7729.
Wang (Fuzhi), 5359.
WANG (Jianlang), 7730.
WANG (Jinlong), 7731.
Wang (Pengshen), 7741.
WANG (Q. Edward), 400.
WANG (Qingcheng), 7732.
WANG (Qingjia), 7733.
WANG (Wei), 7768.
WANG (Wenguang), 7734.
WANG (Xiangyuan), 7735.
WANG (Xiaohua), 7736.
WANG (Ying), 7737.
WANG (Youming), 7738.
WANG (Yuquan), 7739.
WANG (Zhongshu), 7740.
WANNENMACHER (Julia Eva), 3314.
WAQUET (Jean-Claude), 6914.
Warburg (Aby), 298.
WARD (Paul), 4096.
WARD (Richard H.), 4140.
WARDEN (P. Gregory), 1775.
WARDLE (David), 1847.
WARD-PERKINS (Bryan), 2209.
WARE (James P.), 2373.
WARNATSCH-GLEICH (Friederike), 3401.
WARNER (Geoffrey), 7373.
WARNKE (Martin), 848.
WARREN (J. Benedict), 6855.
WARREN (Joyce W.), 5481.
WARREN (Louis S.), 5164.
WARREN (Nancy Bradley), 3479.
WARREN (Stephen), 4783.
WARTENBERG (Günther), 4975.
Waschek (Matthias), 489, 848.
WASCHKUHN (Arno), 722.
WASE (Dick), 1794.
Washington (George), 4734.
WASMER (Caterina), 3717.
Wasserstorm (Steven), 925.
WASSON (Haidee), 5714.
WATANABE (Hisashi), 7845.
WATANABE (Miki), 7742.
WATENPAUGH (Keith David), 5165.
WATERMAN (David), 5715.

WATERS (Robert), 7551.
WATSON (Fiona), 5964.
WATSON (Patricia Anne), 2053.
WATSON (Robert P.), 7415.
WATSON (Wilfred G. E.), 1337.
WATT (Mary Alexandra), 3116.
WATZKA (Carlos), 5466.
WAWRYKOW (Joseph), 3311.
WAZANA (Nili), 1338.
WEBB (Jennifer M.), 1389.
WEBBER (Nick), 2647.
WEBER (David J.), 6856.
WEBER (Elka), 3480.
Weber (Friedrich Christian), 6908.
WEBER (Hartmut), 3912, 7374.
WEBER (Jacques), 331.
Weber (Max), 396, 486, 767.
WEBER (Reinhold), 4004.
WEBER (Thomas P.), 5467.
WEBER (Wolfgang), 2748.
WEBSTER (Andrew), 7100.
Webster (Charles), 5440.
WEBSTER (Jane), 2191.
WEBSTER (Wendy), 6735.
WEEKS (Charles A.), 4322, 6857.
WEGENER (Lydia), 3292.
WEGENER FRIIS (Thomas), 7365, 7552.
WEGMANN (Bodo), 4021.
WEGMANN (Milene), 3117.
WEHRLE (Edmund F.), 6465.
Wei (Yuan), 7707.
Wei Zi, 7613.
WEIDENMIER (Marc), 5755.
WEIGEL (Petra), 3402.
WEIGL (Michael), 7101.
WEIL (Patrick), 6715.
WEILHARTNER (Jörg), 1401.
Weill (Kurt), 5698.
WEINACHT (Paul-Ludwig), 5347.
WEINANDY (Thomas G.), 3263.
WEINBERG (Carl R.), 5881.
WEINBERG (Gerhard L.), 7171.
WEINBERGER (Jerry), 5166.
WEINFURTER (Stefan), 566, 2602.
WEINRICH (Lorenz), 3412.
WEIS (Cédric), 3901.
WEIS (Hélène), 300.
WEISMANN (Itzchak), 4697.
WEISNER (Annette), 251.
WEISS (Dieter J.), 4824.
WEISS (Roslyn), 1681.
WEISS (Zeev), 1339.
WEISSENRIEDER (Annette), 2261.
WEISZ (George), 135, 5468.
WEITEKAMP (Margaret A.), 5469.
WEITLAUFF (Manfred), 3365.
WEITZ (Mark A.), 4784, 6608.
WEITZMAN (Steven), 1340.
WELCH (Evelyn), 5925.
WELLINGTON (I. J.), 2150.

WELLMANN (Marc), 5612.
WELLS (Cheryl A.), 4785.
WEN (Shangqing), 7769.
Wendland (Paul), 487.
WENDT (Friederike), 2261.
Weng (Wenhao), 7677.
WENGST (Klaus), 2266.
WENGST (Udo), 7351.
WENZEL (Horst), 2942, 3118.
WENZEL (Knut), 4856.
WENZEL (Siegfried), 3427, 3481.
WERNER (Ingrid), 1776.
WERNER (Yvonne Maria), 5167.
WERUM (Stefan Paul), 6466.
WESS (Timothy), 1069.
WESSON (Cameron B.), 7888.
WEST (Geoffrey), 3022.
WEST (Harry G.), 4327.
WEST (Nigel), 4097, 7244.
WESTAD (Odd Arne), 7366.
WESTERMAN (Jeroen), 3192.
WESTERMANN (Edward B.), 7290.
WESTERN (Jon), 7367.
WESTPHAL (Siegrid), 6194.
WESTPHALEN (Stephan), 2304.
WETHERINGTON (Mark V.), 4786.
WETTERSTEN (John), 5349.
WETTIG (Gerhard), 4511.
WETZEL (David), 7001.
WETZEL (René), 3039.
WEYSSENHOFF BROŻKOWA (Christina), 179.
WHALEN (Logan E.), 3090.
WHEATLEY (Michael), 4188.
WHELAN (Irene), 4825.
WHELPTON (John), 4330.
WHITAKER (Matthew C.), 6609.
WHITBY (Michael), 1894.
WHITE (Colin), 4034.
WHITE (G. Edward), 6507.
WHITE (Hayden), 518.
WHITE (Jeffrey A.), 2961.
WHITE (John F.), 1952.
WHITE (Paul Andrew), 3119.
WHITEHEAD (David), 1478.
WHITEHEAD (John S.), 4787.
WHITEHORN (John), 1635.
WHITELAW (Gavin), 7854.
WHITMARSH (Tim), 1636.
WHITTLE (Alasdair), 1045.
WICH-REIF (Claudia), 2966.
WICKE (Christina), 6610.
WICKHAM (Chris), 540, 2603, 2834.
WICKI (Nikolaus), 3315.
WIDDER (Ellen), 3514.
WIDDER (Roland), 3622.
WIEBE (Donald), 952.
Wied (Hermann von), 5005.
WIEGREFE (Klaus), 7553.
WIELAND (Karin), 3977.

WIEMER (Hans-Ulrich), 1474.
WIENER S. (Guillermo), 3657.
WIERLING (Dorothee), 3932.
WIERMANN (Barbara), 5716.
WIERZBICKI (Leszek Andrzej), 4408.
WIESEHÖFER (Josef), 471.
WIESINGER (Peter), 161.
WIGG (Richard), 7241.
WIGGENHORN (Harald), 6611.
WIKANDER (Stig), 912.
Wilamowitz-Moellendorff (Ulrich von), 487.
WILD (Mark), 6392.
WILDENTHAL (Lora), 6719.
WILDER (Gary), 3565.
WILDI (Tobias), 5470.
WILDT (Michael), 509.
WILENTZ (Sean), 4788.
WILFORD (Hugh), 4781.
WILHELM (Gernot), 1272.
Wilhelm II, Dt. Kaiser u. König von Preußen, 3974, 7012.
WILKES (J. J.), 2192.
WILKINS (Mira), 6032.
WILKINSON (Charles), 4789.
WILKINSON (Sam), 1953.
WILLEBALD (Henry), 5745.
WILLEMEZ (Laurent), 6497.
William III, king of Great Britain, 4094.
WILLIAMS (Abigail), 5501.
WILLIAMS (David Lay), 5350.
WILLIAMS (David), 4238, 5321.
WILLIAMS (Heather Andrea), 5257.
WILLIAMS (John Matthew), 6504.
WILLIAMS (John), 19.
WILLIAMS (Michael E. Sr.), 5012.
WILLIAMSON (Callie), 1992.
WILLIAMSON (Jeffrey G.), 5743.
WILLIFORD (Thomas J.), 3725.
WILLIOT (Jean-Pierre), 5829.
WILLIS (Alan Scot), 5013.
WILLIS (Justin), 6815.
WILLIS (Laura H.), 1045.
WILLMS (Christoph), 3127.
WILLMS (Johannes), 3902.
WILLS (Abigail), 6393.
WILLS (Jocelyn), 5772.
WILLS (John), 799.
WILSON (Daniel J.), 5471.
WILSON (Gordon Anthony), 3250.
Wilson (Harold), 4032.
WILSON (Jon E.), 5974.
WILSON (Penelope), 1201.
WILSON (Trevor), 7085.
Winckelmann (Johann Joachim), 5171.
WIND MEYHOFF (Karsten), 758.
WINKELMANN (Thomas), 251.
WINKLER (Gerhard), 1786.

WINKLER (Henry Ralph), 7102.
WINKLER (Kenneth P.), 957.
WINKS (Robin W.), 2604, 3566, 3567.
WINSTON-ALLEN (Anne), 3403.
WINTER (Barbara), 3618.
WINTER (Engelbert), 1120.
WINTER (Jay), 327.
WINTER (Karin), 6736.
WINTER (Nancy A.), 1777.
WIPF (Matthias), 4661.
WIPPICH (Rolf-Harald), 6666.
WIRNT VON GRAFENBERG, 2977.
WIRSCHING (Andreas), 6115.
WITCHER (Rob), 2193.
Witherspoon (John), 4756.
WITHINGTON (Phil), 4098.
WITTCHOW (Frank), 1954.
WITTEK (Thomas), 5309.
WITTGENSTEIN (Ludwig), 5360.
Wittkower (Rudolf), 853.
WLADIKA (Michael), 3637.
Władysław IV Zygmunt, king of Poland, 4403.
WŁODARKIEWICZ (Wojciech), 7114.
WLOSOK (Antoine), 2288.
WOERTHER (Frédérique), 1637.
WOGAN-BROWNE (Jocelyn), 2904.
WOHL (Robert), 5472.
WOHL (Victoria), 1528.
WOLBOLD (Matthias), 5717.
WOLF (Hubert), 4857, 4898.
WOLF (Manfred), 2565, 2567.
WOLFE (Thomas C.), 5310.
WOLFF (Charlotta), 6394.
WOLFF (Guntram B.), 5727.
WOLFF (Stephan), 6641.
Wölfflin (Heinrich), 848.
WOLFSDORF (David), 1682.
WOLFZETTEL (Friedrich), 240, 3076.
WOLGAST (Eike), 4550.
WOLLER (Hans), 7341.
WOLSZA (Tadeusz), 7439.
WOLTERS (Reinhard), 2049.
WOLVAARDT (Pieter), 7556.
WOLZ (Robert J.), 7415.
WONG (John Chi-Kit), 6395.
WOOD (Betty), 6858.
WOOD (Denis), 266.
WOOD (Elizabeth A.), 6612.
WOOD (J. R. T.), 6816.
WOOD (John H.), 6033.
WOODALL (Ann M.), 6396.
WOODARD (James P.), 3678.
WOODFORD (Chris), 797.
WOODS (Christopher), 1251.
WOODWARD (Peter), 7368.
WOOLF (Greg D.), 2054.
WORK (Clemens P.), 4790.

WORLEY (Matthew), 4067, 4099.
Worringer (Wilhelm), 855.
WORTLEY (John), 2517.
WÓYCICKA (Zofia), 440.
WOYKE (Johannes), 2374.
WRATHALL (Mark A.), 959.
WREDE (Martin), 4024.
WRIEDT (Klaus), 3150.
WRIGHT (Alison), 3223.
WRIGHT (Christopher), 6449.
WRIGHT (Conrad Edick), 5205.
WRIGHT (David Curtis), 7744.
WRIGHT (Donald A.), 441.
WRIGHT (Matthew), 1638.
WRIGHT (Robert E.), 6034.
WRÓBEL (Janusz), 4410.
WU (Tiaopo), 7745.
WU (Yixiong), 7746.
WU (Yongming), 7747.
WU (Yongreng), 5774.
WU (Zengfeng), 7748.
WULFF-KUCKELSBERG (Susanne), 6548.
Wulfstan, saint, 3397.
WUNDERLI (Peter), 3121.
WÜRFEL (Walter), 1955.
WÜST (Wolfgang), 3518.
Wyclif (John), 3481.
WYKES (David L.), 4077.

X

XELLA (Paolo), 885.
Xenophon Ephesius, 1424, 1443.
Xenophon, 1610, 1651.
Xerxes, king of Persia, 1473.
XIA (Yafeng), 7557.
XIAN (Bo), 7749.
XIAO (Zhizhi), 7750.
XIAO (Zili), 7751.
XIE (Liping), 7626.
XIE (Zhonghou), 7752.
XING (Tie), 7753.
XINNHOBLER (Rudolf), 4811.
XIONG (Yaping), 7657.
XU (Chang), 7754.
XU (Guoqi), 7104.
Xu (Heng), 7774.
XU (Xing), 7755.
XU (Zhuoyun), 7756.

Y

YAGIL (Limore), 3903.
YAKUT (Esra), 4705.
Yamagishi (Masako), 4242.
YAMAGISHI (Tuneto), 7814.
YAMAMOTO (Shin), 7758.
YAMANAKA (Yutaka), 7819.
YAMANE (Samuel), 1033.
YAMAZAKI (Kei), 7846.
Yan (Huiqing), 7615.
YAN (Lieshang), 7759.

Yan (Xishan), 7659.
YANG (Hegao), 747.
YANG (Kuisong), 7760.
YANG (Shenqing), 7793.
YANG (Tianhong), 7761.
YANG (Tianshi), 7762.
YANG (Zhenhong), 7763.
YANGWEN (Zheng), 6397.
YANKELEVICH (Pablo), 3581.
YANNOU (Hervé), 4907.
YARDENI (Myriam), 5168.
YAŞAR (Ahmet), 6127.
YASUI (Moyuru), 1956.
YASUR-LANDAU (Assaf), 1341.
YATES (JoAnne), 6398.
YAZIDI (Béchir), 6817.
YEDİYILDIZ (Bahaeddin), 810.
YEE (Carolyn B.), 7448.
YEE (Tet-Lim N.), 2376.
YEILBURSA (Behçet Kemal), 7558.
YI (Huili), 7788.
YIZHAR (Uri), 4195.
YOCUM (John), 3263.
YOFFE (Norman), 723.
YOKKAICHI (Yasuhiro), 7581.
YOKOTE (Shinji), 7002.
YOKOYAMA (Korenori), 7832.
YOKOYAMA (Yuriko), 4247.
Yongzheng, Emperor of China, 6895.
YORMAZ (Abdullah), 541.
YOSHIDA (Nobuyuki), 7803.
Yoshida (Shigeru), 4237.
YOSHIKAWA (Shinji), 7833.
YOSHIMURA (Takehiko), 7833, 7834.
YOSHIZAWA (Seichiro), 7764.
YOSIEF (Desale), 1033.
YOU (Ziyong), 7765.
YOUNG (Christopher), 3062.
YOUNG (Elliott), 213.
YOUNG (Simon), 2944.
YOUNGER (Kenneth Gilmour, sir), 7373.
YOUNG-IOB (Chung), 6035.
YOUNGMAN (Paul A.), 5482.
YPSILANTI (Maria), 2116.
YSÀS (Pere), 6469.
YU (Shicun), 7766.
YU (Wei), 7767.
YU (Xue), 7173.
YU (Yongding), 5776, 7768.
YU (Yuan), 7757.
YU (Zhanghua), 7785.
YUAN (Li), 7769.
Yuan (Shikai), 7783.
YUAN (Yongmin), 7769.
YUASA (Haruhisa), 7847.
Yuen (Francisco L.), 4321.
YUN (Bartolomé), 4630.
YUNIS (Harvey), 1425, 1683.

YURDAKUL (İlhami), 7003.
YUSTE DE PAZ (Miguel Ángel), 7440.

Z

ZABBIA (Marino), 2945.
ZABORSKI (Andrzej), 1342.
ZACCARELLO (M.), 5544.
ŽÁČEK (Pavel), 3759, 3760, 3792, 6534, 6582.
ZACH (Krista), 4859.
Zachaeus, 2287.
ZACHARIADOU (Elizabeth), 2429.
ZACHS (Fruma), 4697.
ZAGDOUN (Mary-Anne), 1684.
ZAGORIN (Perez), 480.
ZAIM (Sabahattin), 4706.
ZAJĄC (Justyna), 7316.
ZAJTSEV (V. N.), 32.
ZAKHARINE (Dmitri), 6399.
ZAMAGNI (Claudio), 2229.
ZAMARAEVA (Natal'ja A.), 4362.
ZAMFIRESCU (Dinu), 4448, 4457.
ZAMIR (Meir), 7174.
ZAMORA (Kurt A.), 1325.
ZAMUNER (Ilaria), 3316.
ZANGARO (Pierantonio), 2584.
ZANINI (Andrea), 5775.
ZANOBONI (M. Paola), 5571.
ZAPICO BARBEITO (María Pilar), 2529.
ZARKA (Yves Charles), 5353.
ZARROW (Peter), 7771.
ZATLOUKAL (Klaus), 2965.
ZAUZICH (Karl-Theodor), 1180.
ZAV'JALOV (Vladimir I.), 493.
ZAVACKÁ (Marína), 7559.
ZAWADZKI (Hubert), 697.
Zbigniew, książę Polski, 2625.
ZDANEK (Maciej), 3151.
ZDRADA (Jerzy), 4411.
ZECCHINI (Giuseppe), 346.
ZEGHAL (Malika), 4325.
ZEIDLER (Manfred), 7175.
ZEIDNER (Robert Farrer), 7105.
ZEITLIN (Judith Francis), 4323.
Zelikow (Philip), 7433.
ZELIN (Madeleine), 5882.
ZELIZER (Viviana A.), 774.
ZELLER (Joachim), 6725.
ZELLER (Thomas), 3964, 5082.
Zeller (Wolfgang), 5700.
ZELNÍČEK (Jakub), 270.
ZELNICK-ABRAMOVITZ (Rachel), 1508.
ZENG (Xiongsheng), 7772.
ZENG (Zhigong), 7773.
ZENTZ (Frank), 6613.
ZERBST (Arne), 855.
ŻERKO (Stanisław), 7176, 7181.
ZHANG (Fan), 7774.

ZHANG (Fuxiang), 7775.
ZHANG (Guogang), 7776.
ZHANG (Haipeng), 7777.
ZHANG (Hong Bo), 7441.
ZHANG (Lipeng), 7631.
ZHANG (Qing), 442.
ZHANG (Qingjun), 7736.
ZHANG (Ruide), 7778.
ZHANG (Shouchang), 7779.
ZHANG (Shude), 7780.
ZHANG (Taiyuan), 7781.
ZHANG (Xianwen), 7782.
ZHANG (Xueji), 7783.
Zhang (Xueliang), 7659, 7760.
ZHANG (Xuexu), 7784.
ZHANG (Yinke), 7785.
ZHANG (Yusheng), 7786.
Zhang (Zhidong), 7748.
ZHAO (Zhihua), 7787.
ZHELEZNOVA (Natal'ja A.), 953.
ZHEN (Zhang), 5718.
Zheng (He), 7610, 7790.
ZHENG (Shiqu), 7789.
ZHENG (Yijun), 7790.
ZHENG (Zelong), 7791.
ZHENPING (Wang), 7792.
ZHIROMSKAJA (V. B.), 6099.
Zhou (Enlai), 7685.
ZHOU (Suiyuan), 7795.
ZHOU (Xin), 7796.
ZHU (Hu), 7797.
ZHU (Lexian), 7614.
Zhu (Shisan), 7634.
ZHU (Ying), 7798.
ZHU (Yingui), 7799.
ZHUMANALIEV (Asanbek), 4264.
ZICHE (Paul), 5208.
ZIEGLER (Dieter), 5883.
ZIEGLER (Karl-Heinz), 6618.
ZIESMANN (Sonja), 150.
ZIFFER (Irit), 1075.
ZILMER (Kristel), 2874.
ZIMINA (O. G.), 32.
ZIMMER (Oliver), 716.
ZIMMER (Petra), 4808.
ZIMMER (Stefan), 191.
ZIMMERMAN (Andrew), 5976.
ZIMMERMAN (Joshua D.), 4217.
ZIMMERMANN (Klaus), 1957.
ZIMMERMANN (Martin), 1150.
ZIMMERMANN (Mosche), 7399.
ZINCONE (Sergio), 954.
ZINELLI (Fabio), 3377.
ZINK (Michel), 2900.
ZINSSER (Judith P.), 5424.
ZIPF (Karin L.), 6467.
ZISCHKE (Birgit), 1001.
Ziya Gökalp, 5520.
ŽIŽEK (Aleksander), 4536.
ZNAMIEROWSKA-RAKK (Elżbieta), 7442.

ZÖBERLEIN (Renate), 4908.
ZOLOTAREV (Vladimir A.), 7004.
ZORODDU (Donatella), 1639.
ZOTZ (Nicola), 3122.
ZOTZ (Thomas L.), 3380.
ZOUHAR (Jan), 5322.
ZUBAR' (Vitalij M.), 1723.

ZUBER (Henri), 3905.
ZUBER (Valentine), 4904.
ZUCCA (Raimondo), 1757.
ZUCCARELLO (Ugo), 4926.
ZUCCHINI (Stefania), 2754.
ZUCKER (Arnaud), 1685.
ZUCKERMAN (Constantine), 2518.

ZUGASTI (Miguel), 5572.
ZUNTZ (Günther), 1686.
ZUO (Shuangwen), 7800.
ŽUPANOV (Ines G.), 4950.
ZUREK (Robert), 4909, 7443.
ZURLI (Loriano), 1797.
ZYWIETZ (Michael), 3231.

GEOGRAPHICAL INDEX

A

Abruzzo, 2532, 3155, 5731.
Acarnania, 2512.
Acemhöyük, 1075.
Actium, 2097.
Adana, 6128.
Addis Ababa, 6048.
Adelaide (South Australia), 1186.
Aden, 7508, 7550.
Adriatique (mer), 2459, 2611.
Adwa, 4201, 6955.
Aegaeum (mare), 1065, 2418.
Aegina, 1436.
Aegyptus, 833, 1151-1201, 1979, 2287, 2465, 2496. – v. Egypt.
Aethiopia, 2349.
Afghanistan, 1455, 7368.
Africa nova, 1865.
Afrique, 1154, 1176, 1791, 1853, 1869, 1911, 229, 2308, 2364, 247, 2786, 3565, 4949, 6712, 6760-6817, 7331, 7368, 7478, 7525, 7718, 7854-7873. – A. centrale, 6760, 7871. – A. du sud-ouest, 6797. – A. occidentale, 5835, 6703. – A. orientale, 245. – A. septentrionale, 6870, 6928. – A. subsaharienne, 769, 974, 7860.
Agadé, 1203.
Agenais, 3475.
Agra, 1721.
Aigle, 3848.
Ain (France), 1785.
Aizu district (Japan), 6203.
Akgérie, 6808.
Aki, 7825.
Akra (Northwest Pakistan), 1360.
Al Madam (Sharjah, United Arab Emirates), 1079.
Alaska, 4787.
Alatri, 3513.
Alava, 2820.
Albania v. Shqipëri.
Alberta, 6280.
Alexandria, 1282, 1327, 1461, 1689.
Alger, 6768.
Algérie, 444, 725, 3573, 3574, 5491, 6585, 6763, 6773, 6793, 6795, 6796, 6809, 7468, 7512.

Alicante (España), 1046.
Alpi (catena montuosa), 2129, 6189.
al-Qāhira, Caire, 4911.
Alsace, 2561, 2613, 3900, 5237.
Altbayern, 121.
Alto Uruguai, 3664.
Amasya province, 1082.
Amathus (Cyprus), 1750.
Ambaca, 6810.
Amérique, 794, 978, 4894, 5699, 6341, 6712, 6720, 6726, 6818-6858, 7874-7888. – A. latine, 209, 224, 265, 364, 644, 979, 3526, 3540, 3554, 4379, 5137, 5150, 5299, 5572, 5753, 6138, 6152, 6196, 6485, 6488, 6531, 6543, 6564, 6693, 6699, 7346, 7368, 7881. – A. septentrionale, 5790, 6829, 7368, 7888. – v. Hispanoamérica. – v. Iberoamérica.
Amiternum, 1792.
Ammaedara (Tunisia), 1780.
Amsterdam, 4962, 6011.
Amud Cave (Israel), 1023.
Anatolie, 1223, 1271, 1348, 1706, 5726, 6037, 7565.
Andalucía, 3082, 3135, 4561, 4562, 4583.
Andean countries, 6831.
Andes (cordillera de los), 581, 3550, 3552, 7874, 7885.
Aneho (Togo), 7857.
Angola, 3575-3577, 6729, 6762, 6800, 6811, 7489.
Anhalt, 5820.
Antalya, 6042, 6188.
Antilles, 128.
Antiochia, 2326.
Appennini, 5898.
Apulia, 2046.
Aquileia, 2363.
Aquitaine, 1856, 3475.
Arab (countries), 7401, 7561.
Arados, 1130.
Aragón, 645, 2660, 4615, 4624.
Araucanía, 3708.
Ardea, 1997.
Ardennes, 7285.
Arezzo, 3414.

Argentina, 373, 630, 698, 3578-3608, 5395, 5483, 5662, 5686, 5860, 6045, 6243, 6595, 6633, 7379, 7412.
Århus, 6117.
Armenija, 688, 1098, 7081.
Asante, 5927, 6410.
Ashkhabad, 4707.
Asie, 762, 4949, 5278, 6712, 6738-6759, 7368, 7400, 7562-7853. – A. centrale, 196, 753, 5046, 7564-7581. – A. centrale-méridionale, 1077. – A. du sud-est, 721, 791, 5923, 7118, 7368, 7420, 7582-7595. – A. méridionale, 415, 1360, 7681, 7427, 7449, 7582-7595. – A. minor, 150, 1119, 1124, 1126, 1128, 1259, 1964, 2330, 6044. – A. occidentale, 7564-7581. – A. orientale, 721, 7563, 7654, 7805, 7806, 7824. – A. septentrionale, 7003. – v. Eurasie. – v. Far East. – v. Middle East. – v. Near East.
Assada, 1278.
Assam, 6412.
Assisi, 4903.
Athenae, 1444, 1445, 1448, 1458, 1467, 1485, 1486, 1494, 1495, 1700, 1704, 1728, 1730, 1744, 1769.
Atlanta, 4743, 4754, 6211.
Atlantic City, 6362.
Atlantique (océan), 3519, 5725, 6217.
Augsburg, 3365, 6344, 679.
Augustenborg, 3808.
Austerlitz, 6925, 6929.
Australia, 3609-3618, 5380, 6504, 6189, 6416, 6417, 6422, 6439, 6449, 6460, 6677, 6861, 6862, 6865, 7431, 7447.
Austrasie, 2869.
Avila, 2580.
Ayia Triada, 1391.

B

Babylon, 1202, 1583.
Baden, 3942, 4884, 6509.

Baghdad (Iraq), 1220.
Baja Caifornia, 5375, 6118.
Baleares (islas), 2433.
Bălgarija, Bulgaria, 572, 3679-3686, 4280, 4281, 5669, 6620, 6664, 6671, 6954, 6991, 6999, 7008, 7009, 7013, 7048, 7060, 7063, 7091, 7236, 7309, 7368, 7442, 7459, 7511, 7538.
Balkans, 2788, 3523, 3541, 3548, 3568, 5307, 6630, 6701, 6998, 7023, 7032, 7034, 7083, 7106, 7107, 7160, 7357.
Baltique (mer, pays), 690, 3319, 3542, 3568, 5704, 6112, 6965.
Bamberg, 3367, 3984.
Bangladesh, 7465, 7583.
Barbados, 3638.
Barcelona, 4564, 4610, 5178, 6119.
Basel, 4883.
Basrah, 4170.
Bayern, 3636, 3971, 4014, 5108, 6356, 6942, 7101, 7133. – v. Altbayern.
Beer-Sheaba, 1318.
Beigua (mons), 1050.
Beijing, Peking, 3716, 5191, 7602, 7615, 7622, 7650, 7687, 7730.
Beirut, 81, 6067, 6077.
Belgique, 312, 667, 790, 2920, 3639-3648, 5306, 5556, 5920, 5980, 5983, 5987, 6030, 6326, 6594, 6783, 6812, 7026, 7063, 7151, 7308, 7368.
Belladere, 6643.
Belluno, 2733.
Bengal, 419, 4134, 4136, 5212, 5615, 5974, 6756, 7594.
Berlin, 207, 299, 3926, 3969, 5079, 5174, 5292, 5368, 6083, 6097, 6150, 6725.
Bhutan, 6619.
Białostocczyzna, 7119.
Bigorre, 2523.
Bihar, 4143.
Bilbao, 2811.
Birkenau, 440.
Birmania, 3649.
Black Sea Region, 567, 810, 1462, 1713, 2598.
Blaj, 5236.
Bolivia, 3650-3657, 3819, 4918, 6004, 6406, 6828, 6977.
Bologna, 3147, 3204, 6189, 6424.
Bonn, 5157.
Bosna-Hercegovina, 903, 2849, 6116, 6954. – v. Jugoslavija.

Boston (England), 2858.
Boston, 4838, 5238, 6166.
Boukfouka, 6857.
Bourgogne, 78, 2759, 2829.
Boyacá, 3719, 3722.
Brandenburg, 3986.
Brasil, 310, 650, 3595, 3658-3678, 4862, 5235, 5277, 5620, 6179, 6224, 6242, 6243, 6595, 6840, 6841, 6849, 6930, 6959, 6978, 7379, 7517.
Bratislava, 2860, 5093.
Braunau am Inn, 3632.
Bremen, 3961, 6162, 6189.
Brescia, 86, 2584.
Bretagne, 6900.
Bristol, 2669.
Britannia, 1043, 1069, 1866, 1896, 1993, 2162, 2191.
British Guiana, 7551.
British Isles, 1031, 3157, 6047.
Brno, 3781, 5677, 7276.
Broadway, 1567.
Bronx (New York city), 6072.
Bruges, 2843.
Bruxelles, 3370, 5175.
Buchenwald, 7303.
Budapest, 4122, 6100.
Buenos Aires, 3583, 3587, 3594, 4892, 5929, 5971, 6045, 6236.
Bulgaria v. Bălgarija.
Burgau, 4900.
Burgenland, 3622.
Burma, 7288, 7368, 7458.
Burundi, 6812.
Buthrotum, 1726.
Buya (Dancalia, Eritrea), 1033.
Byzantium, Constantinopolis, Empire byzantine, 2286, 2340, 2383-2518, 2715, 3222.

C

Cáceres, 4616.
Caere, 1765, 1769.
Caire v. al-Qāhira.
Cajamarca, 6847.
Calabria, 1093.
California, 4937, 5830, 5863, 6184.
Calvari del Molar (Priorat, Tarragona), 1055.
Cambodia, 3687, 3688, 7536.
Cameroon, 409, 3689, 3690.
Campania, 1970, 2207, 3489.
Canada, 75, 441, 3691-3697, 4932, 4995, 5279, 5551, 6074, 6287, 6426, 6432, 6644, 6830, 6835, 7033, 7194, 7195, 7252, 7404, 7461.
Cancho Roano, 1089.

Cannae, 1899.
Canterbury, 3349, 3353.
Canton, 7645, 7698.
Capo Verde, 3698.
Cappadocia, 2479, 2488.
Carambolo Alto (Camas, Sevilla), 1083.
Caria, 1123, 2495.
Caribbean area, 5767, 6095, 6714, 6842.
Carlisle (diocese), 3345.
Carniola, 3498.
Carnthia, 3498.
Carpathian area, 6869.
Carthago, 339, 1288, 1519, 1883, 1957.
Cassiciacum, 2289.
Castilla, 302, 2608, 2660, 2783, 3108, 3169, 6541.
Cataluña, 181, 2870, 4554, 4572, 4581, 4598, 4606, 4608, 5874, 6027.
Catamarca, 3597.
Catania, 2365.
Caucasus (mons), 1100, 1357.
Cavarzere, 2735.
Čechy, 3527, 3771, 3785, 3793, 5111, 5239, 6012, 6311, 6515, 7185, 7271, 7278.
Central Africa, 569, 655.
Centroamérica, 3544.
Cerro de la Encina (Monachil, Granada), 1054.
Československo, Czech Republic, 355, 387, 404, 3753-3793, 4127, 4398, 4834, 4841, 4864, 5099, 5136, 5271, 5286, 5297, 5322, 5455, 5554, 5674, 5720, 5747, 5750, 5807, 5946, 6086, 6202, 6283, 6324, 6331, 6332, 6472, 6484, 6510, 6622, 6641, 6658, 7016, 7028, 7029, 7110, 7139, 7144, 7158, 7177, 7211, 7230, 7238, 7258, 7268, 7269, 7307, 7319, 7342, 7368, 7406, 7511, 7530, 7538, 7540.
Chaco, 7875.
Chad, 3699.
Chalcis, 1415.
Chanbrook (Kent), 5918.
Changsha, 7669.
Charcas, 124, 6198.
Charleston, 6259.
Chascomús, 5929.
Chechnia, 652.
Chelyabinsk Region, 276, 4464.
Chemnitz, 5867.
Chersonesos Taurica, 147, 1723.
Chiapas, 4296, 4311.
Chicago, 1236, 6046, 6071, 6106, 6109, 6114.

Chile, 373, 605, 3539, 3582, 3700-3712, 4791, 5283, 5845, 6122, 6595, 6648, 6952, 6977, 6978, 7010, 7030, 7412, 7495.
China, 362, 400, 603, 708, 803, 2830, 3713-3717, 4939, 5386, 5449, 5748, 5776, 5882, 5948, 6043, 6059, 6397, 6624, 6638, 6697, 6895, 6946, 6982, 7005, 7018, 7047, 7104, 7173, 7193, 7212, 7227, 7265, 7315, 7349, 7368, 7372, 7432, 7450, 7455, 7482, 7490, 7505, 7557, 7571, 7596-7801, 7813.
Chios, 1489.
Chongqing, 7620.
Cieszyn region, 7029.
Cilicia, 149, 1148, 7105.
Cincinnati, 6378.
Città del Vaticano, 3872, 4827, 4830, 6025, 6996, 7067, 7192, 7368, 7421, 7422, 7542.
Claiborne County (Mississippi), 6174.
Cluj, 5970.
Cluny, 3374.
Cnidus, 1146.
Coburg, 3919.
Colli Euganei (Veneto), 1027.
Cologna Veneta, 2750.
Colombia, 234, 3718-3725, 5933, 6846, 7359, 7877.
Commagene, 1948.
Congo, 3726-3728, 3576, 6783.
Constantinopolis v. Byzantium.
Copán, 7876.
Córdoba (Argentina), 3589.
Córdoba (España), 2553, 4556.
Corduba (Hispania), 1907.
Corse, 2661, 3497, 6642, 7295.
Costa Rica, 626, 3729-3735, 6323, 6980.
Côte d'Ivoire, 6769.
Cotswold Hills, 2821.
Coventry, 3510.
Crécy, 2650.
Cremona, 1918.
Creta, 1728.
Crna Gora, Montenegro, 85, 6686, 6938. – v. Jugoslavija.
Croatia v. Hrvatska.
Crotone, 1093.
Cuba, 3742-3749, 5488, 5921, 6319, 6357, 6672, 6985, 7448, 7454, 7460, 7467, 7489, 7546.
Cueva Morín, 1032.
Çukurova, 6128.
Cumberland, 2789.
Cuzco, 4371, 6308.
Cyprus v. Kypros.

Czech Republik v. Česko-slovensko.

D

Dacia, 598, 1790, 1920, 1921, 2143, 2155.
Daghestan, 570, 4480.
Dalécarlie, 2670.
Danmark, 607, 628, 801, 804, 2621, 3794-3809, 5043, 5044, 5078, 5411, 5621, 5791, 5828, 5887, 5957, 6132, 6135, 6621, 6902, 7108, 7121, 7141, 7147, 7149, 7164, 7191, 7219, 7264, 7321, 7327, 7344, 7365, 7368, 7501, 7509, 7533, 7552.
Danube (river), 6695.
Danubian area, 2611, 6869.
Danuvius (flumen), 1787, 1996, 2192, 2474. – v. Danube (river). – v. Danubian area.
Dar es Salaam, 6054.
Darfur, 4635.
Dauphiné, 3503.
Daye, 7618.
Dayr Mār Ya' qūb (Syria), 2304.
Deccan, 756.
Delhi, 6064.
Delos, 1512.
Derry, 351.
Deutschland, 57, 80, 116, 127, 180, 185, 197, 208, 260, 263, 305, 309, 335, 336, 356, 369, 371, 372, 374, 386, 397, 409, 418, 586, 604, 647, 742, 845, 846, 1734, 1888, 2044, 2542, 2633, 2677, 2837, 2894, 2942, 2967, 3003, 3102, 3456, 3524, 3525, 3530, 3797, 3806, 3866, 3908-4024, 4127, 4215, 4390, 4429, 4490, 4550, 4642, 4655, 4710, 4835, 4886, 4891, 4899, 4930, 5014, 5024, 5029, 5075, 5079, 5081, 5082, 5089, 5127, 5162, 5172, 5189, 5207, 5224, 5226, 5227, 5240, 5246, 5250, 5251, 5259, 5271, 5280, 5282, 5297, 5303, 5309, 5370, 5402, 5403, 5443, 5450, 5482, 5514, 5528, 5552, 5584, 5587, 5622, 5627, 5642, 5653, 5664, 5675, 5679, 5694, 5700, 5711, 5717, 5732, 5747, 5764, 5769, 5773, 5794, 5795, 5796, 5802, 5812, 5813, 5827, 5834, 5862, 5866, 5875, 5878, 5885, 5909, 5922, 5943, 5944, 5962, 5986, 6010, 6049, 6079, 6081, 6129, 6130, 6144, 6158, 6206, 6228, 6238, 6246, 6266, 6271, 6273, 6313, 6340, 6341, 6349, 6358, 6367, 6382, 6402,

6405, 6428, 6431, 6434, 6443, 6466, 6495, 6501, 6511, 6512, 6521, 6522, 6524, 6544, 6546, 6556, 6562, 6565, 6572, 6581, 6587, 6588, 6590, 6593, 6613, 6622, 6637, 6641, 6666, 6689, 6691, 6696, 6700, 6719, 6725, 6731, 6737, 6738, 6786, 6797, 6866, 6879, 6894, 6908, 6949, 6976, 6993, 7001, 7008, 7044, 7046, 7054, 7056, 7071, 7110, 7111, 7123, 7131, 7134, 7140, 7142, 7144, 7146, 7148, 7149, 7153, 7156, 7158, 7163, 7170, 7175, 7176, 7178, 7182, 7185, 7191, 7192, 7196, 7200, 7210, 7213, 7215, 7223, 7224, 7234, 7243, 7247, 7250, 7269, 7272, 7279, 7290, 7302, 7317, 7318, 7328, 7331, 7341, 7344, 7348, 7351, 7352, 7361, 7362, 7363, 7365, 7368, 7374, 7387, 7396, 7405, 7406, 7424, 7425, 7438, 7443, 7444, 7445, 7452, 7469, 7492, 7499, 7501, 7514, 7517, 7518, 7552, 7553, 7554, 7555.
Devesa do Rei, 1076.
Dexter Avenue King Memorial Baptist Church (Montgomery, Ala.), 5001.
Dien Bien Phu, 6754.
Disentis/Mustér (Switzerland), 2507.
Dobrogea, 2020.
Donbass, 7222.
Dortmund, 2805.
Dragør, 6080.
Drehem, 1243.
Dronero, 2749.
Dudley, 2866.
Dülük Baba Tepesi, 1120.
Dunkerque, 7287.
Dura-Europos, 1246.
Durham (diocese), 3345.
Durham (North Carolina), 6230.
Dur-Katlimmu, 1131.
Dyrrachium, 1732.

E

East Anglia, 2594.
East European Plain, 1037.
East Timor, 3816.
Easter Island, 7892.
Ebla, 1233.
Ecuador, 3817-3820, 6978.
Edinburgh, 5412.
Edom, 1290.
Egée (Mer), 7060.
Egypt, 22, 2682, 3822-3827, 4817, 6683, 6802, 6956, 6966, 7050, 7275, 7368. – v. Aegyptus.

El Salvador, 3828, 3829.
Elam, 1367.
Elazig, 1209.
Elche, 4588.
Elle, 5267.
Elorus, 1554.
Ely (diocese), 3345.
Emar, 1215, 1255.
Emilia Romagna, 2887.
England, 7, 47, 87, 91, 110, 119, 168, 333, 413, 466, 591, 760, 2531, 2545, 2610, 2632, 2644, 2663, 2669, 2737, 2740, 2746, 2753, 2755, 2781, 2797, 2802, 2808, 2822, 2833, 2846, 2847, 2857, 2861, 2885, 2898, 2919, 2928, 2939, 2978, 2982, 2987, 3015, 3059, 3060, 3085, 3093, 3193, 3219, 3318, 3345, 3366, 3395, 3410, 3428, 3446, 3481, 4039, 4040, 4047, 4054, 4058, 4076, 4089, 4095, 4098, 4842, 4961, 4967, 4982, 5000, 5003, 5008, 5065, 5181, 5449, 5474, 5497, 5576, 5647, 6015, 6146, 6171, 6189, 6214, 6232, 6247, 6325, 6373, 6377, 6383, 6393, 6416, 6542, 6568, 6642, 6698, 6883, 6892, 7027, 7524.
Ephesus, 2359, 2368.
Epirus, 2512.
Erfurt, 3292.
Eritrea, 6300, 6813.
Ešnunna, 1244.
España, 30, 90, 117, 184, 220, 237, 262, 306, 311, 321, 356, 393, 445, 640, 705, 1010, 1085, 2698, 2700, 2716, 2756, 2764, 2892, 2914, 2957, 2980, 2996, 3007, 3022, 3322, 3372, 3419, 3443, 3444, 3488, 3522, 4058, 4548-4630, 4813, 4836, 4897, 4902, 4920, 4927, 4958, 5027, 5060, 5072, 5137, 5155, 5261, 5295, 5447, 5570, 5616, 5618, 5655, 5917, 6027, 6112, 6143, 6250, 6421, 6468, 6469, 6577, 6578, 6645, 6652, 6711, 6721, 6850, 6852, 6856, 6875, 6878, 6882, 6883, 6884, 6889, 6911, 6917, 6940, 7135, 7202, 7216, 7217, 7232, 7239, 7241, 7242, 7274, 7368, 7380, 7387, 7428, 7440, 7521. – v. Hispania.
Esterlla, 2535.
Estland v. Ëstonija.
Estonie v. Ëstonija.
Ëstonija, Estland, Estonie, 3830, 3831.
Ethiopia, 1018, 3832, 3833, 7152, 7368.

Etolia, 2512.
Etruria, 1020, 1764, 1765, 1769, 1771, 1776, 1864.
Euphrates (flumen), 1219, 1231, 1246, 1251.
Eurasie, 6682.
Europe, 111, 137, 166, 176, 195, 223, 241, 252, 319, 693, 702, 706, 708, 716, 718, 726, 728, 738, 762, 771, 793, 804, 814, 821, 825, 967, 1038, 1045, 1050, 1064, 1786, 1889, 2205, 2418, 2595, 2600, 2601, 2603, 2627, 2786, 2907, 3324, 3368, 3398, 3440, 3528, 3530, 3531, 3534, 3535, 3549, 3551, 3553, 3560, 3562, 3566, 3567, 4839, 4934, 5002, 5009, 5035, 5037, 5134, 5147, 5437, 5466, 5681, 5696, 5721, 5754, 5829, 5833, 5846, 6167, 6169, 6183, 6243, 6427, 6488, 6506, 6538, 6669, 6687, 6688, 6716, 6931, 7094, 7126, 7201, 7237, 7238, 7312, 7328, 7336, 7368, 7383, 7385, 7403, 7446, 7480, 7504, 7513, 7535. – E. centrale, 309, 646, 812, 850, 2809, 3538, 3564, 5104, 5992, 6327, 6674, 7055, 7070. – E. du centre-est, 2588, 4119, 4389. – E. du sud-est, 2592, 6668, 6994. – E. méridionale, 6098. – E. occidentale, 32, 696, 2656, 2839, 2895, 5056, 5319, 6277, 6379, 6399, 6407, 7368, 7487. – E. orientale, 328, 646, 737, 850, 988, 2588, 4042, 4986, 5034, 5056, 5304, 6922, 7290, 7337, 7368, 7461. – E. septentrionale, 2727, 2809, 3157, 6309, 6906. – v. East European Plain. – v. Eurasie.
Euskadi, País Vasco, 2591, 6136, 6203.
Évora, 6133.
Extremadura, 2593.

F

Falkland (Islands) v. Malvinas (islas).
Falset, 2743.
Far East, 5022, 7368.
Farasan islands, 6962.
Faravel Plateau (Hautes-Alpes), 1039.
Faubourg Saint-Marcel (Paris), 3847.
Ferrara, 4214.
Ferrol, 6092.
Fezzan, 1019, 1113.

Filador (Margalef de Montsant, Tarragona), 1028.
Filipinas, Philippines, 4383, 4384, 7368.
Finland v. Suomi.
Firenze, 53, 707, 2583, 2799, 2854, 2924, 3080, 3126, 3170, 3178, 3195, 3223, 3439, 3474, 3502, 4871, 5124, 5508, 5604, 5687, 5890, 6164.
Flanders, 2801, 3168, 3399, 6878, 7014, 7024.
Florida, 4936, 6294.
Forum Novum, 1874.
France, 15, 57, 64, 133, 134, 167, 177, 222, 238, 256, 281, 282, 315, 328, 334, 335, 342, 347, 348, 357, 378, 395, 445, 459, 629, 633, 637, 653, 683, 714, 741, 777, 789, 794, 845, 907, 987, 1085, 1371, 2150, 2613, 2633, 2665, 2668, 2775, 2797, 2812, 2815, 2859, 2872, 2875, 2928, 2987, 2991, 2994, 3061, 3063, 3097, 3171, 3218, 3351, 3525, 3565, 3573, 3639, 3835-3905, 4084, 4390, 4815, 4833, 4860, 4866, 4876, 4890, 4915, 4924, 4926, 4977, 5033, 5037, 5074, 5079, 5092, 5113, 5122, 5127, 5135, 5158, 5168, 5203, 5211, 5231, 5245, 5252, 5287, 5335, 5454, 5529, 5586, 5608, 5624, 5633, 5661, 5663, 5689, 5691, 5712, 5765, 5792, 5799, 5897, 5978, 5986, 6028, 6050, 6079, 6112, 6207, 6210, 6226, 6240, 6281, 6287, 6296, 6312, 6318, 6334, 6350, 6363, 6368, 6505, 6522, 6544, 6548, 6554, 6566, 6662, 6703, 6706, 6707, 6711, 6715, 6717, 6724, 6727, 6728, 6732, 6746, 6757, 6787, 6795, 6796, 6819, 6889, 6894, 6897, 6903, 6904, 6906, 6907, 6911, 6912, 6914, 6918, 6920, 6923, 6924, 6926, 6927, 6936, 6965, 6996, 6999, 7001, 7016, 7019, 7021, 7038, 7050, 7051, 7053, 7058, 7064, 7066, 7105, 7132, 7135, 7136, 7137, 7150, 7151, 7154, 7163, 7174, 7177, 7179, 7208, 7216, 7231, 7261, 7289, 7297, 7298, 7302, 7305, 7308, 7345, 7356, 7361, 7368, 7390, 7391, 7393, 7397, 7405, 7411, 7422, 7436, 7438, 7446, 7463, 7479, 7481, 7485, 7512, 7544, 7554, 7632, 7893.
Franken, 121.

Frankfurt am Main, 6890.
Frankfurt am Oder, 3910.
Freiburg, 4023, 4659.
Fujian, 7758.

G

Gabon, 3906, 5858, 6801.
Gaeta, 2615.
Galicia, 1076, 3390, 6092.
Galilaea, 2302.
Galizien, 3624, 4711, 5816.
Gallia Cisalpina, 1959, 2151.
Gallia Narbonensis, 1967.
Gallia Transpadana, 1959.
Gallia, 1888, 1911, 1995, 1999, 2000, 2154, 2489.
Gandìa, 2550.
Gansu, 7706.
Gargano, 1035.
Gaza, 1285, 2325.
Gdańsk, 4396.
Gedaref, 4633.
Genève, 4691, 5010.
Genova, 2813, 4218, 4926, 5775.
Georgia v. Sakartvelo.
Georgia (United States of America), 4786, 5214, 5847, 6851.
Gesher Benot Ya'aqov (Israel), 1275.
Gettysburg, 4732.
Ghana, 673, 4025-4028, 5927, 6410, 7864.
Girona, 2524, 2831.
Girsu, 1216, 1241.
Glasgow, 1389.
Goeteborg, 1738.
Gold Coast, 5927.
Gondelsheimer, 3918.
Gorica (region) 4535.
Gorizia, 7253.
Gortyna, 1414, 1487.
Graecia, 150, 1081, 1114, 1395, 1450, 1493, 1509, 1526, 1564, 1695, 1698, 1710, 1733, 1744, 1924.
Graecia, v. Hellas.
Granada, 2680.
Graupius (mons), 1892.
Great Britain, 57, 158, 335, 615, 625, 666, 675, 694, 2574, 2635, 2665, 2726, 2819, 2944, 3025, 3381, 3521, 3617, 3639, 3742, 3861, 4029-4099, 4171, 4172, 4520, 5305, 5420, 5435, 5443, 5684, 5707, 5757, 5822, 5886, 5894, 5902, 5905, 5986, 6003, 6009, 6033, 6144, 6148, 6149, 6157, 6163, 6173, 6234, 6275, 6374, 6387, 6415, 6417, 6422,
6435, 6439, 6449, 6450, 6451, 6460, 6615, 6619, 6623, 6626, 6627, 6634, 6650, 6677, 6683, 6687, 6704, 6705, 6708, 6710, 6715, 6726, 6728, 6733, 6734, 6735, 6740, 6741, 6745, 6748, 6749, 6756, 6759, 6760, 6765, 6788, 6805, 6851, 6852, 6865, 6870, 6899, 6908, 6910, 6932, 6933, 6939, 6940, 6945, 6947, 6948, 6951, 6953, 6956, 6968, 6973, 6983, 6986, 6992, 6997, 7006, 7009, 7020, 7037, 7049, 7051, 7070, 7074, 7079, 7096, 7102, 7120, 7165, 7174, 7188, 7198, 7199, 7207, 7210, 7212, 7213, 7217, 7224, 7230, 7231, 7241, 7244, 7252, 7257, 7273, 7275, 7279, 7286, 7311, 7337, 7352, 7364, 7368, 7370, 7373, 7396, 7409, 7420, 7427, 7432, 7434, 7439, 7457, 7458, 7466, 7471, 7498, 7508, 7545, 7558, 7595, 7681. – v. Britannia. – v. British Isles.
Great Lakes Basin, 6159.
Greece v. Hellas.
Grenoble, 7296.
Grotta del Romito (Cosenza), 1025.
Grotta Marisa (Lecce, Puglia), 1024.
Grotta Paglicci, 1035.
Grottaferrata, 3387.
Guadalquivir (flumen), 1083.
Guadelupe, 5759.
Guangdong, 7659, 7667, 7746.
Guatemala, 373, 4106, 4107, 6820.
Guerrero, 5724.
Guiana, 7527.
Guinea ecuatorial, 3821, 6770.
Guinea, 671, 4108, 4109, 6777, 6779.
Guizhou, 7699.
Guyana, 7448.

H

Hadhramaut, 1328.
Haft Tappeh, 1345.
Haiti, 4110-4112, 6643, 6672, 6921, 6930, 7448.
Halifax, 6297.
Halle, 718.
Hamburg, 298, 3959, 6119.
Hankou, 7618.
Hannover, 6910, 6899, 6939.
Hansa towns, 6964.
Hanse, 2767.
Harrogate, 2575.
Hatay, 7019.
Haute-Vienne, 3852.
Hawaii, 4787.
Hayy el Sellom (Beirut), 6067.
Heliopolis, 1186.
Hellas, Greece, 141, 348, 406, 557, 602, 680, 748, 803, 823, 1369-1751, 4100-4105, 5630, 5751, 5770, 6189, 6919, 6991, 7000, 7045, 7368, 7459, 7474, 7493. – v. Graecia.
Heracleopolis, 1164.
Herculaneum, 1793, 2178, 2179.
Hertfordshire, 2566.
Hesse, 5364.
Hessen, 6509.
Hessen-Darmstadt, 6610.
Heusden, 6606.
Hidalgo, 4289.
Hierakonpolis, 1165.
Hierapolis di Frigia, 1147.
Hierosolyma, 2427.
Hildesheim, 4997.
Himalaya, 6619.
Hispania, 1856, 1902, 2197.
Hispanoamérica, 663, 5194, 6201, 6250, 6822.
Histria, 1959.
Hollywood, 5715.
Honduras, 4113, 4114, 5965.
Hong Kong, 6638, 6749, 7605, 7698.
Hongrie v. Magyarország.
Honolulu, 5426.
Höyücek, 1015.
Hrvatska, Croatia, 2589, 3736-3741, 5869, 5995, 6596, 6664, 7064. – v. Jugoslavija.
Huanta, 4376.
Huasteca (México), 5900.
Huesca, 4571.
Huizhou, 7637, 7690, 7727.
Hungary v. Magyarország.
Hydisos (Caria), 1127.
Hyrcania, 1091.

I

Iaşi, 4432.
Iasus, 2499.
Iatmul (East Sepik Province, Papua New Guinea), 6462.
Iberia, 1072, 1317.
Iberica (paeninsula), 1914.
Iberoamérica, 6845, 938.
Iceland, 2720, 2721, 3099, 4131, 6005, 7368.
Idaho, 5787.
Igboland, 4351.
Il Cairo, 1166.
Île de Pâques, 886.
Île-de-France, 6227.
Illinois, 5881.

Illyricum, 1786, 2490.
Imola, 2563, 3347.
India, 2, 343, 596, 659, 687, 1017, 1059, 1157, 2707, 4132-4153, 4934, 4950, 5216, 5409, 5625, 5823, 6412, 6464, 6692, 6726, 6741-6743, 6745, 6746, 6750, 6753, 6759, 7227, 7315, 7340, 7450, 7458, 7549, 7592, 7595.
Indiana, 5956.
Indianapolis, 6321.
Indochine, 6751, 6752, 6757, 7368, 7450, 7593.
Indonesia, 4154-4157, 5264, 6739, 6748, 7368, 7584, 7589.
Indus (flumen), 1059.
Ionia, 1740.
Irak, 304, 316, 350, 4168-4170, 4683, 6364, 7196, 7368, 7407, 7576.
Iran, 204, 806, 1068, 1080, 1092, 1343-1368, 4158-4167, 4169, 5589, 6303, 6626, 6912, 6974, 6992, 7368, 7484.
Ireland, 48, 76, 642, 664, 676, 2944, 4171-4188, 4818, 4825, 5934, 6002, 6204, 6677, 7310, 7347. – v. Ulster.
Isoso, 7875.
Israel, 1307, 1332, 2322, 4189-4195, 6625, 7368, 7398, 7399, 7401, 7464, 7477, 7499, 7548, 7561.
İstanbul, 232, 2467, 5413, 5577.
Istria, 2176, 5991. – v. Histria.
Italia, 3, 9, 34, 42, 53, 62, 86, 99, 154, 183, 193, 200, 246, 285, 296, 317, 335, 356, 365, 368, 432, 438, 634, 722, 763, 784, 788, 928, 1022, 1024, 1051, 1752, 1756, 1970, 1994, 2014, 2016, 2024, 2031, 2040, 2045, 2118, 2130, 2131, 2198, 2465, 2476, 2564, 2585, 2607, 2636, 2637, 2655, 2672, 2675, 2784, 2790, 2793, 2814, 2825, 2855, 2928, 2937, 2988, 2990, 3072, 3095, 3098, 3103, 3121, 3174, 3321, 3477, 3485, 3525, 3534, 3568, 4129, 4196-4234, 4535, 4828, 4849, 4879, 4919, 4944, 5063, 5073, 5075, 5083, 5102, 5117, 5125, 5147, 5149, 5200, 5242, 5259, 5270, 5275, 5291, 5293, 5302, 5388, 5479, 5490, 5531, 5556, 5575, 5595, 5623, 5690, 5739, 5740, 5803, 5870, 5872, 5885, 5925, 5936, 5940, 5950, 5981, 5986, 6055, 6205, 6219, 6243, 6271, 6276, 6322, 6358, 6380, 6495, 6547, 6568, 6576, 6654, 6675, 6722, 6776, 6803, 6873, 6882, 6937, 6975, 6981, 7042, 7066, 7080, 7088, 7122, 7129, 7130, 7159, 7160, 7202, 7206, 7220, 7245, 7255, 7314, 7329, 7333, 7341, 7363, 7368, 7369, 7377, 7392, 7421, 7512, 7523, 7526.
Ithaca, 1540.
Iudaea, 1876.
Ixtlahuaca (Valle de), 6853.
İzmir, 4687.

J

Jáchymov, 5808.
Jaffa, 228.
Jamaica, 5945, 6335.
Japan, 217, 243, 435, 528, 682, 735, 4235-4247, 4258, 4803, 5114, 5321, 5393, 5748, 5811, 5825, 5831, 5926, 5953, 6189, 6203, 6359, 6409, 6419, 6471, 6483, 6532, 6666, 6872, 6947, 6970, 6972, 6982, 6984, 6987, 6988, 7002, 7005, 7027, 7039, 7078, 7138, 7173, 7193, 7207, 7218, 7257, 7263, 7288, 7353, 7368, 7426, 7441, 7450, 7465, 7625, 7673, 7702, 7710, 7735, 7740, 7741, 7752, 7757, 7802-7849.
Játiva, 4557.
Java, 6739, 6758, 5186.
Jena, 5208.
Jerusalem, 1276, 1327, 2646, 4251, 4692, 5041.
Jezira, 1134.
Jiangdong, 7671.
Jiangnan, 7797.
Jiangsu, 7798.
Jindřichův Hradec, 6263.
Jodhpur (Princely State), 4152.
Johannesburg, 6222.
Jordan, 1287, 1325, 4248-4252, 6269, 7339, 7368, 7457, 7466.
Judea, 1332.
Jugoslavija, Yugoslavia, 4517, 4520, 4522, 4523, 5076, 7122, 7323, 7333, 7368, 7389, 7435, 7442.
Junan, 7738.
Jutland, 7031.

K

Kabnak, 1345.
Kalavasos-Tenta, 1052.
Kaman-Kalehöyük (central Anatolia), 1111.
Kamisato, 4243.
Kampot (Cambodia), 7588.
Kano, 5961.
Kanto, 7840.
Kargaly archaeological site, 1063.
Karlsbad, 6111.
Karnak, 1184.
Kassala, 4633.
Katanga, 5980.
Katowice, 6372.
Katyn, 7262.
Kaunas, 6070.
Kavkaz, 1090, 4681, 6679.
Kawachi, 7826.
Kazakh, 7574.
Kazakhstan, 6268.
Kempten, 2877.
Kenya, 4253-4257, 6772, 6782.
Khazars, 2685.
Khirbat en-Nahas, 1290.
Kiaochow, 6738.
Kinet Höyük, 1087.
Kish, 1217.
Kleve (Herogtum), 4012.
Klimataria-Manares, 1398.
København, 5042.
Kolberg, 2741.
Köln, 43, 2171, 3501, 4922, 5005, 5177, 5187, 6552, 6874.
Kommos, 1400.
Konya, 5229, 6599.
Kordofan region (Sudan), 6805.
Korea, 4258-4262, 6035, 6971, 6972, 6984, 7039, 7322, 7368, 7388, 7408, 7413, 7427, 7431, 7470, 7543, 7755, 7850-7853.
Košice, 4529.
Kostroma Region, 274.
Kotor, 6938.
Kraichgau, 5040.
Kraków, 3151, 4386.
Krokodeilon Polis, 1330.
Krym, 6944, 6975, 7146.
Kuban Region, 4477.
Kurmainz, 6580.
Kurpfalz, 657.
Kursk Region, 275.
Kuwait, 4263, 7522.
Kypros, Cyprus, 1071, 1088, 1399, 2624, 3750-3752, 5760, 7491, 7500, 7513.
Kyrgyzstan, 4264.
Kyushu (Japan), 1095.

L

L'Arbocer (Font de la Figuera, Valencia), 1060.
La Mancha, 6143.
Lagaba, 1217.
Languedoc, 6156.
Lansing, 5815.
Larsa, 1217.
Latium Vetus, 1020.

Latium, 1997.
Latmos, 1140.
Latvia, Lettland, Lettonie, 4265-4269, 7068.
Lausanne, 2827.
Lebanon, 4270-4272, 7174, 7368.
Lebus, 3363.
Leeds, 5757.
Leeward islands, 6864.
Leipzig, 3923, 3992, 5645, 6611.
Lena Hara Cave (East Timor), 1034.
León (Nicaragua), 4347.
Lesotho, 6799, 7854.
Lettland v. Latvia.
Lettonie v. Latvia.
Lezghins, 570.
Liberia, 7859.
Libya, 4273, 7368.
Liechtenstein, 4274, 6503, 7214, 7215.
Liège, 6904.
Lietuva, Lithuania, 4275-4278, 6887, 7111.
Lima, 4366, 6088, 6330.
Lincolnshire, 5932.
Lindos, 1749.
Lingnan (South China), 1036.
Linköping, 661, 3330.
Lipari (isole), 1409.
Lisboa, 6090.
Lithuania v. Lietuva.
Liverpool, 4069, 6189.
Livonia, 3373.
Ljubljana, 7253.
Łódź, 4410.
Loire, 3848.
Lombardia, 62, 4129.
Łomżyńskie, 7119.
London, 55, 2864, 3173, 4050, 5369, 5410, 5643, 5840, 5912, 6014, 6063, 6105, 6200, 6245, 6396, 6932, 7117.
Longdendale, 2540.
Lorca, 4579.
Los Alamos, 5407.
Los Angeles, 6104, 6110, 6392.
Los Perales (Buenos Aires), 6039.
Louisiana, 5893, 5941, 6357, 6833, 6936.
Lourdes, 4882.
Lucania, 1782.
Lucca, 1, 2562, 2787, 3156, 3448.
Lüdenscheid, 6533.
Lund, 1738.
Luristan (Iran), 1347.
Lusitania, 2052, 2183.
Lutry, 2827.
Luxembourg, 2920, 4279, 7368.
Luzern, 4947.
Lycia, 1144.

Lydia, 1144.
Lyon, 4917, 5160, 7296.

M

Macedonia v. Makedonija.
Madagascar, 595, 4282, 6771.
Madina, 2427.
Madras, 5912.
Madrid, 60, 3787, 4601, 5634, 6121.
Maestrazgo, 4624.
Magdeburg, 3985.
Maghreb, 30, 6098, 6621.
Magnesia apud Maeandrum, 1137.
Magyarország, Hongrie, Hungary, 408, 2589, 2631, 2865, 3621, 3625, 4115-4130, 5487, 6100, 6131, 6288, 6503, 6627, 7040, 7235, 7311, 7320, 7368, 7371, 7391, 7394, 7410, 7496, 7523, 7542.
Maine, 6203.
Makedonija, Macedonia, 170, 1744, 4280, 4281, 5256, 6949, 7000. – v. Jugoslavija.
Malacca, 7610.
Malatya, 1209.
Malaya, 7585.
Malaysia, 5938, 7447, 7537.
Mali, 4283, 6774.
Mallorca, 2473, 2688, 5972.
Malmesbury, 2526.
Malmoe, 1738.
Małopolska, 4406, 7072.
Malta, 1336, 4284.
Malvinas (islas), Falkland (islands), 7486.
Manchester, 6106.
Manchukuo, 4246.
Manchuria, 7743, 7851.
Manhattan (New York), 4746, 5119.
Manipur, 4137, 4141.
Manresa, 5814.
Mantova, 61, 2547.
Manzanares (flumen), 1030.
Mar del Plata, 6178.
Marche, 5898.
Mari, 1211, 1219, 1222, 1224.
Mark (Grafschaft), 4012.
Maroc, 4325, 5558, 6628, 6631, 7368.
Marroquíes Bajos, 1070.
Martinique, 5759, 6823.
Massachusetts, 4751.
Massif Calcaire (Syria), 1118.
Mazagão, 6854.
Méditerranée (mer), 776, 813, 899, 1064, 1108, 1374, 1752, 2603, 5070, 5117, 5165.

Mediterraneum (mare), 1099, 1315, 2465.
Megara, 1518.
Megiddo, 1075, 1334.
Melanesia, 6866.
Memphis, 1201.
Mesopotamia, 1202-1251.
Messenia, 1707.
Mettray, 6571.
México D.F., 4287, 6089, 6094, 6302.
México, 213, 4285-4323, 4383, 4861, 5301, 5693, 5763, 5789, 5877, 5895, 6006, 6023, 6241, 6519, 6820, 6836, 6960, 6997, 7546.
Michoacán de Ocampo, 6855.
Middle East, 74, 618, 1064, 3533, 3537, 6098, 6182, 6625, 7335, 7368, 7416, 7420, 7434, 7472, 7488, 7524.
Milano, 2557, 5173, 5571, 6069, 6937.
Milwaukee, 5218.
Minas Gerais, 3672, 6255.
Minden, 6252.
Minneapolis, 5772.
Mississippi, 4993, 5213, 5639.
Moldova, 2863, 4324, 4872, 4935, 5266.
Mongolia, 7505, 7578.
Monselice, 2742.
Montana, 6280.
Monte d'Accoddi (Sassari), 1074.
Monte, 5929.
Montenegro v. Crna Gora.
Monteroduni-Paradiso (Isernia), 1058.
Monviso (mons), 1050.
Moravia, 802, 2851, 3793, 5045, 6305, 7185.
Mordvins, 574, 575.
Morelos, 4307.
Moskva, 207, 3500, 5841, 6455.
Mozambique, 402, 4326, 4327, 7331, 7555.
Mühlhausen, 2748.
München, 1734, 5157, 6356, 7190.
Munda (Hispania), 1907.
Münster, 4858, 4888, 5220.
Murcia, 4618.
Myus, 1121.

N

Nagasaki, 7832.
Namibia, 4328, 6764, 6786, 6807, 7062.
Nanfeng (Jiangxi), 7773.
Nanjing, Nanking, 7659, 7708, 7760, 7764, 7799, 7800.

Napoli, 2530, 5124, 5270, 5317, 5981, 7262.
Narbonne, 2992.
Natchez, 6857.
Navarra, 2770, 4575.
Naxus, 2478.
Near East, 141, 617, 1114, 1118-1150, 7368.
Neckar (river in Germany), 1940.
Nederland, Pays Bas, 386, 931, 2745, 2842, 2920, 3031, 3191, 3197, 3208, 4129, 4331-4341, 5145, 5200, 5233, 5766, 5804, 5977, 6141, 6385, 6386, 6489, 6550, 6631, 6842, 6884, 6907, 6913, 6942, 7115, 7368.
Negev Highlands, 1323.
Nepal, 662, 4329, 4330, 6619, 7315, 7368.
Neuchâtel, 5298.
New England, 6260, 6361, 6844.
New Haven, 5230.
New Mexico, 3556, 4936, 5274, 6292.
New York, 219, 4877, 5109, 5228, 5966, 6046, 6091, 6107, 6113, 6221, 6315, 6337, 6401, 6452, 6513, 6589.
New Zealand, 651, 686, 4053, 4342-4344, 6279, 6867, 7528.
Nicaragua, 3731, 4345-4350, 5244, 6655, 6980, 7539.
Nicolás Ailío (Chile), 3709.
Nicopolis, 2512.
Niger, 7872.
Nigeria, 4351-4354, 5729.
Nijmegen, 4335.
Nile (river), 4637.
Nilus (flumen), 1153.
Ningxia, 7706.
Nioro du Sahel, 4283.
Nitra, 4531.
Nizhni Novgorod, 4464.
Nogales, 6857.
Norfolk, 2639.
Norge, Norway, 616, 2586, 2719, 2720, 2721, 2762, 2816, 3083, 4355-4357, 5128, 5564, 5899, 6656, 6676, 7292, 7301, 7368, 7548.
Noricum, 1786.
Normandie, 3183, 7157, 7282.
Norrland, 4649.
North America, 2731.
North Carolina, 4761, 6467.
North Saqqara, 1193.
Northampton, 2537.
Norway v. Norge.
Norwich, 5000.
Nouvelle-Calédoine, 331, 5247, 7154, 7890, 7894.

Nova Scotia, 4751.
Novgorod, 2568, 3222, 6880.
Nový Jičín, 3762.
Nubia, 1168, 1170.
Nueva España, 248, 4383, 4940, 6820.
Nueva Granada, 5377.
Nuevo León, 4303, 4320.
Numidia, 2356.
Nürnberg, 4963, 5148, 6123, 6500.
Nush-i Djan, 1368.

O

Oaxaca, 4305.
Oberpfalz, 6142.
Oceania, 6859-6867, 7889-7894.
Odenwald (mountains in Germany), 1940.
Ohio, 6220, 6591.
Olba, 1129.
Olbia (Sardinia), 1753.
Old Rus', 2582, 2596.
Opava, 3762.
Ordos, 7577.
Oruro (Bolivia), 5818.
Orvieto, 1771.
Osaka, 5537.
Osnabrück, 6353.
Österreich, 115, 691, 3569, 3619-3637, 4124, 4129, 4550, 5100, 5110, 5196, 5733, 5810, 5852, 6017, 6079, 6503, 6574, 6659, 6678, 6690, 6736, 6886, 6919, 6969, 7040, 7101, 7203, 7334, 7368, 7375, 7376, 7382, 7390, 7395, 7399, 7414, 7417, 7425, 7496.
Ostrava, 3762, 7268.
Oświęcim, 440.
Ourense, 2548, 4597.
Ovamboland, 6789, 6798.
Owari, 7825.
Oxford, 5183, 5192.

P

Pacific islands, 239.
Pacifique (océan), 4948, 5449, 6386, 6821, 6859, 7248, 7889.
Padova, 3160.
País Vasco v. Euskadi.
Pakistan, 1059, 4358-4362, 7340.
Palaestina, 424, 1960, 2051, 2292, 2370, 4248, 4363, 4364.
Palatinat, 6901.
Palermo, 2674, 2703, 4219, 4926.
Palestine, 267, 421, 3436, 6653, 7120, 7339, 7347, 7409, 7485. – v. Palaestina.
Pamphilia, 1857.
Panjab, 4150, 5952.
Pannonia, 1781.

Papua New Guinea, 6860, 6862.
Paracelsus, 5371.
Paraguay, 214, 4365, 6406, 6819, 6838.
Paris, 207, 2765, 3140, 3143, 3890, 3893, 4810, 4823, 4906, 5109, 5320, 5573, 5637, 5650, 6014, 6057, 6073, 6105, 6112, 6150, 6176, 6272, 6333.
Pashto, 4361.
Patagonia, 3582, 3607, 600.
Pays Bas v. Nederland.
Peking v. Beijing.
Peloponnesos, 1701.
Península Ibérica, 1061.
Pennsylvania (U.S.A.), 1768, 5730, 5806.
Pergamon, 1142.
Perm Region, 4464.
Persia, 1446.
Perú, 631, 3703, 3819, 4366-4382, 4873, 6429, 6578, 6847, 6933, 6978.
Perugia, 2556.
Pessinous (Anatolia), 2128.
Petare, 3720.
Petra, 1287.
Pfalz, 6315.
Pharsalus (Graecia), 1945.
Phellos, 1150.
Philadelphia, 1218, 6034, 6153.
Philippines v. Filipinas.
Piacenza, 1, 4206.
Pianacci dei Fossi (Marche), 1056.
Pienza-Cava Barbieri, 1044.
Pingxiang, 7618.
Pisa, 3176.
Pisidia, 1138, 1857.
Pistoia, 2442.
Pitiusas (islas), 2433.
Platani (fiume), 1755.
Podhale, 7072.
Poggio Colla (Toscana), 1775.
Pohronie, 2826.
Pola, 7253.
Poland v. Polska.
Polska, Poland, 59, 697, 2625, 2631, 3927, 3966, 4127, 4275, 4385-4413, 4506, 4711, 4886, 4901, 4909, 5048, 5246, 5251, 6400, 6567, 6579, 6640, 6641, 6658, 6684, 6689, 6700, 6877, 6887, 6910, 7028, 7029, 7073, 7057, 7075, 7113, 7114, 7134, 7162, 7181, 7188, 7211, 7223, 7233, 7240, 7258, 7271, 7293, 7299, 7306, 7316, 7354, 7368, 7393, 7439, 7443, 7452, 7510, 7518, 7520, 7545.
Pommern, 3966.

Geographical INDEX

Pompeii, 1793, 1794, 1936, 2161, 2174, 2177, 2179, 2181.
Pontic Olbia, 138.
Pontus Euxinus, 170.
Pontus, 1996, 2090.
Portella a Salina (Isole Eolie), 1067.
Porto Alegre, 3658, 3670.
Portugal, 211, 314, 501, 669, 3007, 4414-4427, 5610, 5744, 5838, 6649, 6711, 6713, 6895, 6979, 7178, 7182, 7232, 7310, 7368.
Portus, 2182.
Potosí, 6308.
Potsdam, 3910.
Poznań, 4389, 5193, 7112.
Praha, 3458, 5204, 6093, 6202, 7168, 7267, 7304.
Preußen, 656, 3866, 3930, 3931, 3958, 4016, 5879, 6391, 6423, 6539, 6553, 6592, 6958.
Primorska (region), 7437.
Propontide, 1709.
Provence, 5979.
Puerto Rico, 593, 3810, 5701, 6448, 6818, 6832, 6843.
Puglia, 2795, 3201.
Pula (Istria), 2176.
Punjab, 4132, 6755.
Pusht-i Kuh (Luristan), 1363.
Pyrénées, 6203.

Q

Qasr Ibrim, 1193, 1201.
Qatar, 4428.
Qatna, 1219.
Québéc, 3696, 4881, 5067, 6835.
Queensland, 6420.
Querétaro, 4299.
Qumran, 1208.

R

Rakitovec, 573.
Ranchos, 5929.
Rastatt, 6096.
Ravenna, 2447, 2513, 3199.
Recklinghausen, 6108.
Rekhes Napha, 1323.
Rennes, 3904.
República Dominicana, 3810-3815, 6636, 6643, 7380, 7448.
Retia, 1786.
Réunion, 6778, 6784.
Rheinland, 800, 3975, 5408, 7058, 7059, 7079.
Rhodus, 1397, 1408.
Rhône (vallée du), 3375.
Riga, 6070.
Rijeka, 7253.
Rio de Janeiro, 3673, 5907, 6729.
Rio de la Plata, 4940.

Rio Grande do Sul, 3663, 3671.
Riohacha, 6846.
Ripabianca di Monterado (Ancona), 1051.
Rojas, 3588.
Roma, 67, 141, 423, 620, 643, 729, 1181, 1299, 1327, 1405, 1752-2209, 2308, 2340, 2619, 2752, 3200, 3223, 3385, 3511, 4231, 4850, 4867, 5007, 5054, 5124, 5157, 5272, 5463, 5485, 5547, 5590, 5605, 5609, 5614, 5641, 6101, 6355, 6937, 6976.
România, 4, 384, 546, 562, 805, 808, 832, 840, 843, 4429-4457, 4524, 4956, 5069, 5222, 5268, 5307, 5619, 5656, 5770, 5850, 5891, 5930, 5959, 5968, 6366, 6523, 6624, 6639, 6640, 6673, 6685, 6695, 7036, 7083, 7087, 7093, 7116, 7123, 7200, 7220, 7235, 7236, 7251, 7280, 7281, 7323, 7358, 7368, 7398, 7430, 7462, 7519, 7530.
Rossija, Russia, 93, 144, 206, 217, 274-276, 284, 322, 337, 338, 349, 358, 389, 422, 472, 479, 568, 572, 575-577, 585, 678, 721, 1063, 2599, 2752, 3222, 3553, 3555, 3907, 4243, 4458-4513, 4951-4954, 4969, 5016, 5020, 5022, 5047, 5102, 5115, 5118, 5179, 5276, 5503, 5541, 5557, 5559, 5567, 5568, 5722, 5741, 5837, 5841, 5853, 5884, 5908, 5931, 5939, 5951, 5969, 6099, 6165, 6209, 6285, 6295, 6347, 6400, 6437, 6455, 6602, 6612, 6669, 6686, 6821, 6880, 6898, 6905, 6908, 6920, 6932, 6938, 6958, 6967, 6972, 6975, 6984, 6987-6989, 6998, 7002, 7004, 7018, 7057, 7084, 7090, 7161, 7240, 7260, 7272, 7274, 7354, 7372, 7375, 7385, 7520. – v. SSSR.
Rostov, 4493.
Rouen, 3211.
Roussillon, 5855.
Rovereto, 3405.
Ruhrgebiet, 5854, 5857.
Rumelia, 4703.
Ruscuk, 3684.
Russia v. Rossija.
Rwanda, 6812.
Ryukyu, 7702.

S

Sabah, 5938.
Sabina, 3153.

Sachsen, 3946, 4973, 5182, 6910, 6916.
Sachsen-Anhalt, 3962.
Sagalassos, 1139.
Saguntum (Hispania), 2009.
Saint-Louis (Senegal), 6790.
Saint-Marcel (Normandie), 6050.
Saint-Pierre-de-Préaux, 3391.
Saint-Pol-sur-Ternoise, 2641.
Saint-Victor (abbey, Paris), 2985.
Sais, 1201.
Sakartvelo, Georgia, 3907.
Salvatierra, 2536.
Salzburg, 3631.
San Andrés de Fanio, 20.
San Antonio, 4895.
San Fernando de las Barrancas, 6857.
San Giovanni Rotondo, 4870.
San Isidro (Madrid), 1030.
San Millán de la Cogolla, 19.
San Sebastián, 2811.
Sankt-Peterburg, 568, 6053, 6932, 7259, 7270.
Santa Marta (Colombia), 3723, 6846.
Santander (Colombia), 6834.
Santarém, 4425.
Santiago Atitlán, 4107.
Santiago de Chile, 3706, 6122.
Santo Antão Island (Cape Verde), 3698.
São Paulo, 3665, 3678, 5235, 6001.
Saqqara, 1201.
Sardegna (regno di), 6950.
Sardegna, 2867, 6975.
Sardinia, 1073, 1757, 1758.
Sardis, 1122, 1310.
Saudi Arabia, 4514, 7368.
Savoia, 4926, 4905.
Sayn-Hachenburg, 3994.
Scandinavie, 551, 865, 2871, 3038, 5167, 6922.
Scania, 6203.
Schaffhausen, 4661.
Schlesien, 6391.
Schleswig-Holstein, 3797, 3806, 3936, 4011, 6354, 6447.
Schwaben, 121.
Schwäbischen Alb (mountains in Germany), 1940.
Schwarzenberg, 4005.
Schweiz, Suisse, Svizzera, 164, 481, 649, 3559, 4655-4661, 5459, 5873, 6016, 6287, 6775, 6943, 6979, 7206, 7330, 7336, 7503, 7560.
Scotland, 110, 1892, 2642, 2652, 3079, 4039, 4040, 4054, 4078, 4087, 5181, 5317, 5934, 5964, 6416, 6891.

Sées (diocese), 3346.
Segovia, 3352, 4609.
Selinus, 1406.
Senegal, 4515, 4516.
Sepphoris, 1285, 1339.
Serbia v. Srbija.
Serpis (flumen), 1046.
Ses Païsses (Artà, Mallorca), 1086.
Sevilla, 2853, 5597, 5916.
Shandong, 7738.
Shanghai, 5718, 7020, 7663, 7700, 7850.
Shanxi, 7441.
Shatila, 431.
Shenxi, 7706.
Shijiazhuang, 7657.
Shqipëri, Albania, 1725, 2512, 3570-3572, 5728, 6629, 7042, 7096, 7129, 7357, 7368.
Siam, 7368.
Sibir', 576, 5022, 5837.
Sicilia, 1006, 1706, 2314, 2355, 2795, 3043, 3158, 4868, 5270, 5919.
Side, 1133.
Sidney, 5692.
Siena, 279.
Sikkim, 6619.
Silesia, 4395, 6305, 802.
Sinai, 1062.
Sinaloa, 5375.
Sindh, 4358.
Singapore, 7586.
Sippar, 1202, 1221.
Sippir, 1217.
Siracusa, 1459.
Slovakia v. Slovensko.
Slovenija, 438, 4533-4540, 5084, 6013, 6208, 7133, 7377, 7419. – v. Jugoslavija.
Slovensko, Slovakia, 131, 355, 3508, 4526-4532, 5123, 5864, 5960, 6000, 6658, 6661, 7098, 7559. – v. Československo.
Smakkerup Huse (Northwest Zealand, Denmark), 1029.
Sodor and Man (diocese), 3349.
Somme, 7085.
South Africa, 3539, 4541-4547, 4946, 5243, 5668, 5738, 6187, 6446, 7062, 7362, 7384, 7503, 7525, 7556, 7854, 7866.
Sparta, 1411, 1496, 5318.
Spitzberg, 7252.
Srbija, Serbia, 4517-4525, 6986, 7000, 7084. – v. Jugoslavija.
SSSR, 280, 358, 4236, 4461-4463, 4471, 4474, 4476, 4477, 4479, 4485, 4491, 5115, 5310, 5764, 5937, 6066, 6441, 6620, 6657,
6692, 7010, 7013, 7073, 7091, 7114, 7119, 7123, 7136, 7162, 7188, 7198, 7205, 7218, 7233, 7245, 7247, 7324, 7368, 7371, 7374, 7376, 7381, 7385, 7397, 7404, 7408, 7414, 7430, 7433, 7439, 7454, 7460, 7502, 7520, 7532, 7540, 7543, 7644.
St. Albans, abbey, 29.
St. Gallen, 4657.
St. George Chapel, 3217.
St. Paul, 5772.
Stockholm, 4641, 4653.
Strasbourg, 5193, 5197.
Styria, 3498.
Sudan, 1178, 1197, 4631-4637, 6598, 6815.
Sudeten, 6663, 7065, 7089.
Suez, 6634.
Suhum, 1219.
Suisse v. Schweiz.
Suomi, Finland, 3834, 5704, 5788, 6079, 6375, 6639, 6698, 7143, 7204, 7221, 7226, 7240, 7368, 7429.
Susa, 1203.
Sverdlovsk Region, 4464.
Sverige, Sweden, 41, 563, 779, 819, 1738, 3325, 4355, 4357, 4638-4654, 4889, 4964, 5152, 5188, 5438, 5564, 5648, 5704, 5737, 5788, 6040, 6103, 6155, 6213, 6253, 6262, 6278, 6375, 6394, 6558, 6656, 6676, 6684, 6879, 6880, 6886, 6890, 6898, 6908, 6909, 7156, 7187, 7301, 7343, 7368, 7474, 7480, 7514, 7547.
Svizzera v. Schweiz.
Swabian Jura (Germany), 1038.
Swaziland, 7854.
Sweden v. Sverige.
Syria, 1149, 2304, 4662-4664, 6623, 7368.
Syros, 1416.

T

Taiwan, 4665, 5774, 7612, 7662, 7741.
Tall-e Bakun, 1102.
Tallin, 6070.
Tanger, 7198.
Tanzania, 4666-4668, 5632.
Tappeh Bormi (Iran), 1237.
Tappeh Shizar, 1096.
Țara Bârsei, 5871.
Tarquinii, 1760, 1765.
Tasmania, 622.
Tavium, 1125.
Tehuantepec (istmo de), 4323.
Tel Arad, 1062.
Tel Aviv, 228.
Tel Tanninim, 1330.
Tel-Hazor, 1277.
Tell el-Amarna, 1201.
Tennessee, 4773.
Teotihuacan, 7887.
Tepe Nush-I Jan, 1080.
Tepoztlán, 4312.
Terni, 431.
Terqa, 1145.
Teutoburgiensis saltus, 1867.
Texas, 3556, 4722, 5801, 5963, 5973.
Thailand, 685, 1042, 4669, 4670, 7288.
Thaiti, 6386.
Thebae, 1388, 1393.
Thera, 1520.
Thessalonikē, 783, 2441, 4102.
Thracia, 170, 1412.
Thüringen, 3998, 6915.
Tianjin, 7661, 7755.
Tiberis (flumen), 2173.
Tibet, 6619, 6983, 7571, 7710, 7726.
Tibiscum-Jupa, 2188.
Tirol, 3620, 3636, 4220.
Tiwanaku (Valle de, Bolivia), 7879.
Tlemcen, 7570.
Togo, 648, 5976, 6794.
Torino, 178, 5195, 5579, 6134.
Toronto, 6063, 6286.
Torre de Moncorvo, 2739.
Toscana, 3213, 4230, 5143, 6570, 6604, 6995.
Tours, 5985.
Trabzon, 5889, 7418.
Trans-Appalachian West, 4779.
Transilvania, 14, 2002, 2863, 3341, 3495, 4452, 4955, 5232, 5954, 6172, 6291, 7235.
Transjordan, 7196.
Transvaal, 6792.
Trasano (Matera), 1040.
Traunstein, 4010.
Treblinka, 395.
Treli, 1100.
Trento, 72, 3405.
Trieste, 5947, 6660, 6950, 7253, 7368.
Tübingen, 5206.
Tucumán, 3579, 3601, 5865.
Tunis, 6140.
Tunisie, 4671-4674, 6817, 6995, 7481.
Turkey v. Türkiye.
Türkiye, 152, 559, 920, 955, 961, 993, 2927, 4101, 4102, 4169, 4281, 4675-4706, 5064, 5117, 5217, 5314, 5473, 5519, 5520,

Geographical INDEX 413

5565, 5613, 5723, 5749, 5809,
5924, 5982, 5989, 5993, 6036,
6038, 6041, 6127, 6261, 6299,
6317, 6480, 6601, 6657, 6678,
6877, 6919, 6948, 6951, 6963,
6964, 6994, 6999, 7003, 7004,
7017, 7019, 7022, 7037, 7054,
7067, 7071, 7081, 7105, 7132,
7229, 7368, 7418, 7491, 7493,
7494, 7516, 7558.
Turkmenistan, 594, 4707.
Tyrus, 1286.

U

U.S.A., 213, 257, 338, 437, 577,
606, 636, 755, 761, 768, 1001,
3558, 3611, 4288, 4310, 4715-
4790, 4832, 4845, 4978, 5004,
5012, 5025, 5031, 5059, 5071,
5086, 5098, 5108, 5112, 5151,
5163, 5201, 5225, 5249, 5255,
5257, 5278, 5373, 5381, 5383,
5425, 5430, 5464, 5469, 5561,
5562, 5607, 5617, 5635, 5636,
5640, 5646, 5671, 5678, 5695,
5697, 5707, 5722, 5746, 5782,
5793, 5812, 5819, 5821, 5831,
5851, 5876, 5896, 5906, 5921,
5936, 5958, 5965, 5996, 5999,
6009, 6032, 6033, 6034, 6055,
6075, 6078, 6082, 6102, 6139,
6163, 6183, 6211, 6212, 6231,
6233, 6239, 6248, 6249, 6251,
6254, 6264, 6274, 6282, 6320,
6334, 6338, 6339, 6342, 6345,
6348, 6365, 6369, 6376, 6384,
6390, 6436, 6442, 6453, 6459,
6463, 6465, 6506, 6507, 6514,
6518, 6525, 6608, 6609, 6613,
6625, 6628, 6635, 6636, 6645,
6647, 6655, 6672, 6680, 6681,
6683, 6685, 6693, 6699, 6715,
6749, 6781, 6821, 6825, 6897,
6900, 6918, 6924, 6927, 6928,
6936, 6945, 6953, 6960, 6990,
7036, 7081, 7090, 7099, 7154,
7165, 7166, 7205, 7212, 7213,
7219, 7224, 7231, 7250, 7256,
7265, 7286, 7314, 7322, 7325,
7329, 7335, 7338, 7342, 7346,
7359, 7360, 7367, 7368, 7370,
7372, 7384, 7386, 7388, 7400,
7401, 7403, 7410, 7411, 7415,
7416, 7421, 7424, 7426, 7428,
7430, 7433, 7434, 7444, 7448,
7449, 7454, 7455, 7456, 7457,
7463, 7467, 7470, 7473, 7475,
7483, 7484, 7490, 7495, 7497,
7498, 7500, 7502, 7504, 7515,
7522, 7524, 7527, 7532, 7536,
7539, 7541, 7551, 7553,
7557.
Udine, 7253.
Uganda, 283, 4708, 7868.
Ukraine, 658, 4503, 4709-4713,
5102, 6975, 7113, 7148.
Ulster, 4186, 6635.
Ulug Depe, 1077.
Umbria, 1864, 4913, 5898, 724.
Umma, 1205, 1216.
United Arab Emirates, 4714.
Urals, 1063, 5837.
Urartu, 1207, 1368.
Urmia (lacus), 1358.
Uruguay, 4791-4794, 5492, 5745,
5758, 5780, 6193, 6406, 6458.
Uruk, 1233.
Utrecht, 3344.

V

Valachia, 162, 2863, 5910.
Valais, 1050.
Valcamonica, 1007.
Valencia, 2522, 2832, 2838, 3154,
4573, 4626.
Valle de las Higueras (Huecas,
Toledo), 1057.
Van (lacus), 1105.
Vancouver island, 6826.
Vardar River Valley, 4280.
Varennes-en-Argonne, 3884.
Värmland, 5564.
Vasilikos valley, 1052.
Västergötland, 3338.
Vaud, 2780, 4966.
Veii, 1765.
Venetia, 1959.
Venezia, 2513, 3484, 5296, 5892,
6124, 6301, 6871, 6969.
Venezuela, 4795-4800, 6968,
6985, 7099, 7531, 7877.
Veracruz, 6310.
Verdun, 3872.
Viareggio, 6388.
Vicksburg, 4782.
Vietnam, 4744, 4777, 4801-4803,
6465, 7265, 7322, 7356, 7456,
7497, 7504, 7515, 7528, 7536,
7541.
Villa Hayes (Paraguay), 6819.
Virginia, 4731, 6528.
Vitoria, 2811.
Vizcaya, 2732.
Volci, 1765.

W

Wadi El-Hudi, 1169.
Waiblingen, 3514.
Wales, 1016, 2538, 2571, 2638,
2952, 3028, 3063, 3065, 6180,
6416.
Wallonia, 3202.
Warri (Nigeria), 6814.
Warszawa, 5263.
Weimar, 2751.
Westphalia, 2565, 2567, 3975,
5015, 6203.
Wielkopolska, 7112.
Wien, 5652.
Wiltshire, 2558.
Wittenberg, 5370.
Württemberg, 6509.

X

Xanthos, 1124, 1135, 1405.

Y

Yangzi, 1049, 7754.
Yemen, 4804, 7550.
York, 2669, 2840, 3432.
Yorkshire, 2558.
Yucatán, 4301, 4306.
Yugoslavia v. Jugoslavija.
Yverdon, 2827.

Z

Zacatecas, 4298, 4317.
Zamalek (Cairo, Egypt), 3826.
Zamalou, 7669.
Zambia, 4805.
Zigong, 5882.
Zimbabwe, 4806, 4807, 6816.
Zürich, 5285, 6517.